D0321671

THE EUROPEAN GARDEN FLORA

THE EUROPEAN GARDEN FLORA

A manual for the identification of plants cultivated in Europe, both out-of-doors and under glass

VOLUME II

Monocotyledons (Part II)

edited by

S.M. Walters, A. Brady, C.D. Brickell,
J. Cullen, P.S. Green, J. Lewis, V.A. Matthews,
D.A. Webb, P.F. Yeo and J.C.M. Alexander

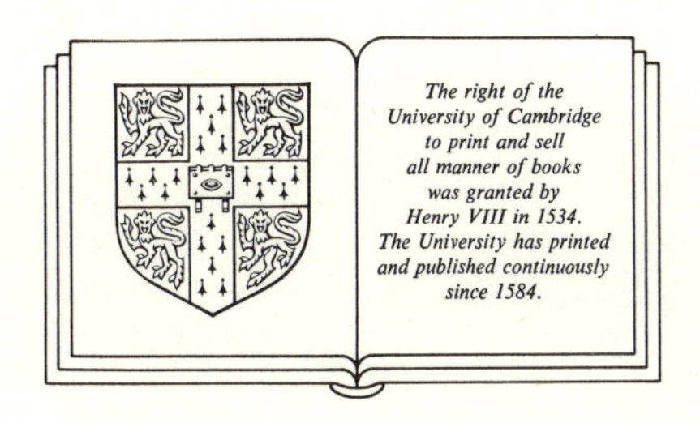

CAMBRIDGE UNIVERSITY PRESS
Cambridge
London New York New Rochelle
Melbourne Sydney

Published by the Press Syndicate of the University of Cambridge
The Pitt Building, Trumpington Street, Cambridge CB2 1FP
32 East 57th Street, New York, NY 10022, USA
296 Beaconsfield Parade, Middle Park, Melbourne 3206, Australia

First published 1984

Printed in Great Britain
at the University Press, Cambridge

Library of Congress catalogue card number: 83–7655

British Library Cataloguing in Publication Data
The European garden flora.
Vol. 2: Monocotyledons (Part 2)
1. Gardening – Dictionaries
2. Plants, Cultivation – Dictionaries
1. Walters, S. M.
635'.03'21 SB450.95

ISBN 0 521 25864 2

CS

CONTENTS

List of maps and figures vii

Organisation and Advisers viii

Editors and Contributors to Volume II ix

Acknowledgements x

Introduction 1

XVII JUNCACEAE 10

XVIII BROMELIACEAE 10

XIX COMMELINACEAE 25

XX GRAMINEAE 31

XXI PALMAE 65

XXII ARACEAE 75

XXIII LEMNACEAE 112

XXIV PANDANACEAE 112

XXV SPARGANIACEAE 113

XXVI TYPHACEAE 113

XXVII CYPERACEAE 114

XXVIII MUSACEAE 117

XXIX STRELITZIACEAE 119

XXX ZINGIBERACEAE 120

XXXI CANNACEAE 129

XXXII MARANTACEAE 130

XXXIII ORCHIDACEAE 137

Glossary 291

Index 301

MAPS AND FIGURES

Map 1. Mean January isotherms for Europe and hardiness codes 4

FIGURES

1 Diagram of some characters used in the identification of Juncaceae and Cyperaceae 11
2 Diagram of some characters used in the identification of Gramineae 32
3 Leaves and scale-leaves of Araceae 76
4 Inflorescences of Araceae 77
5 Flowers of Zingiberaceae 121
6 Leaves of *Calathea* species 132
7 Leaves of various Marantaceae 134
8 Diagram of some characters used in the identification of Orchidaceae (stems and pseudobulbs) 138
9 Diagram of some characters used in the identification of Orchidaceae (leaves) 139
10 Diagram of some characters used in the identification of Orchidaceae (inflorescences and flowers) 140
11 Diagram of some characters used in the identification of Orchidaceae (flowers) 141
12 Diagram of some characters used in the identification of Orchidaceae (columns and pollinia) 142
13 Lips of *Calanthe* species 174
14 Lips of *Encyclia* species 187
15 Lips of *Dendrobium* species 208
16 Upper sepals and petals of *Cirrhopetalum* species 221
17 Lips of *Lycaste* species 228
18 Lips of *Catasetum* species 263
19 Lips of *Odontoglossum* species 275
20 Lips of *Oncidium* species 283
21 Lips of *Oncidium* species 284
22–25 Diagrams illustrating terms defined in the glossary 292–295

ORGANISATION AND ADVISERS

Sponsor: The Royal Horticultural Society

Editorial Committee

A. Brady, National Botanic Gardens, Glasnevin, Dublin
C.D. Brickell, Royal Horticultural Society's Gardens, Wisley
J. Cullen (Secretary), Royal Botanic Garden, Edinburgh
P.S. Green, Royal Botanic Gardens, Kew
J. Lewis, Natural History Museum, London
V.A. Matthews, Royal Botanic Garden, Edinburgh
S.M. Walters (Chairman), University Botanic Garden, Cambridge
D.A. Webb, Trinity College, University of Dublin
P.F. Yeo, University Botanic Garden, Cambridge
J.C.M. Alexander, Research Associate

Advisers

Professor C.D.K. Cook, Zürich, Switzerland
Professor H. Ern, Berlin, Germany
Dr H. Heine, Paris, France
Sr A. Pañella, Barcelona, Spain
Dr D. Wijnands, Wageningen, Netherlands
The late Professor P. Wendelbo, Göteborg, Sweden

EDITORS AND CONTRIBUTORS TO VOLUME II

The various sections of this volume were edited at the following institutions:
Royal Botanic Garden, Edinburgh: Gramineae, Palmae, Pandanaceae, Musaceae, Strelitziaceae, Zingiberaceae, Cannaceae, Marantaceae, Orchidaceae.
Royal Botanic Gardens, Kew: Bromeliaceae, Commelinaceae.
University Botanic Garden, Cambridge: Araceae, Lemnaceae, Cyperaceae.
National Botanic Gardens, Glasnevin, Dublin: Juncaceae, Sparganiaceae, Typhaceae.

Contributors

J.C.M. Alexander (University Botanic Garden, Cambridge/RBG, Edinburgh)
G.C.G. Argent (RBG, Edinburgh)
G.S. Bunting (Jardín Botánico, Maracaibo, Venezuela)
W.D. Clayton (RBG, Kew)
E.J. Cowley (RBG, Kew)
T.B. Croat (Missouri Botanical Garden, USA)
J. Cullen (RBG, Edinburgh)
J. Dransfield (RBG, Kew)
D.R. Hunt (RBG, Kew)
N. Jacobsen (Institute of Systematic Botany, Copenhagen, Denmark)
J.M. Lamond (RBG, Edinburgh)
D.J. Leedy (Los Angeles, USA)
D. McClintock (Platt, Kent)
M.T. Madison (Marie Selby Botanic Garden, Sarasota, USA)
V.A. Matthews (RBG, Edinburgh)
E.C. Nelson (National Botanic Gardens, Dublin, Eire)
D.H. Nicolson (Smithsonian Institution, Washington, USA)
D. Philcox (RBG, Kew)
J.A. Ratter (RBG, Edinburgh)
S. Renvoize (RBG, Kew)
R.M. Smith (RBG, Edinburgh)
D.M. Synnott (National Botanic Gardens, Dublin, Eire)
S.A. Thompson (Carnegie Museum of Natural History, Pittsburgh, USA)
S.M. Walters (University Botanic Garden, Cambridge)
D.A. Webb (Trinity College, Dublin, Eire)
P.J.B. Woods (RBG, Edinburgh)
P.F. Yeo (University Botanic Garden, Cambridge)

ACKNOWLEDGEMENTS

The European Garden Flora project has received substantial support since 1978 from the following:

(*a*) *The institutions to which members of the Editorial Committee belong*: staff time, support and facilities.

(*b*) *The Stanley Smith Horticultural Trust* (Director, Sir George Taylor, FRS): financial support which enabled the project to employ a post-doctoral Research Associate from 1978 to 1982.

(*c*) *The Wolfson Industrial Research Fellowship Scheme*: financial support from October 1982 for the continuing employment of the Research Associate.

(*d*) *The Council of the Royal Horticultural Society*: financial support during 1981 and 1982.

(*e*) *The Cory Fund of Cambridge University Botanic Garden*: financial support for travel by the Research Associate, and other uses.

(*f*) *The Department of Agriculture and Fisheries for Scotland*: financial support for two vacation students working on the preliminary stages of the project.

(*g*) *The Edinburgh Botanic Garden Trust*: financial support for two further vacation students.

The Editorial Committee gratefully acknowledges all this generous support.

Particular thanks are due to J. Bogner (Munich), P. Cribb and S. Mayo (Royal Botanic Gardens, Kew), R.L. Shaw, L. Buchan, R. Kerby, R.U. Cranston and A. Paxton (Royal Botanic Garden, Edinburgh) and Luwasa (Hydroculture) Ltd. (Feltham), for specialised taxonomic and horticultural advice. Many other individual botanists and horticulturists have provided advice and information: they are too numerous to list here and we hope that they will accept this general indication of our gratitude. Special mention should, however, be made of the vacation students who have worked with the project. In its early stages, particularly, they produced the basic lists of species and literature on which the project is founded, and, later, helped with the preparation of the index and other editorial matters. We are especially grateful to them for their enthusiasm and willingness to undertake long, unstimulating but necessary tasks: Bridie Andrews, Elaine Campbell, Christine Couper, S.J. Droop, M.J. Howard, Jane Lees, Margaret McDonald.

We are also grateful to Elaine Campbell, Frances Hibberd, Sally Mackay, Victoria A. Matthews and Rosemary M. Smith for the preparation of the illustrations; to Linda McGhee, who efficiently prepared the whole of the typescript for this volume; and to Hazel Hamlet who read through the whole text.

INTRODUCTION

Amenity horticulture (gardening, landscaping, etc.) touches human life at many points. It is a major leisure activity for a very large number of people, and is a very important means of improving the environment. The industry that has grown up to support this activity (the nursery trade, landscape architecture and management, public parks, etc.) is a large one, employing a considerable number of people. It is clearly important that the basic material of all this activity, i.e. plants, should be readily identifiable, so that both suppliers and users can have confidence that the material they buy and sell is what it purports to be.

The problems of identifying plants in cultivation are many and various, and derive from several sources, which may be summarised as follows:

(*a*) Plants in cultivation have originated in all parts of the world, many of them from areas whose wild flora is not well known. Many have been introduced, lost and then re-introduced under different names.

(*b*) Plants in gardens are growing under conditions to which they are not necessarily well adapted, and may therefore show morphological and physiological differences from the original wild stocks.

(*c*) All plants that become established in cultivation have gone through a process of selection, some of it conscious (selection of the 'best' variants, etc.), some of it unconscious (by methods of cultivation and particularly, propagation), so that, again, the populations of a species in cultivation may differ significantly from the wild populations.

(*d*) Many garden plants have been 'improved' by hybridisation (deliberate or accidental), and so, again, differ from the original stocks.

(*e*) Finally, and perhaps most importantly, the scientific study of plant classification (taxonomy) has concentrated mainly on wild plants, largely ignoring material in gardens.

Nevertheless, the classification of garden plants has a long and distinguished history. Many of the Herbals of pre-Linnaean times (i.e. before 1753) consist partly or largely of descriptions of plants in gardens, and this tradition continued, and perhaps reached its peak in the late eighteenth and early nineteenth centuries – the period following the publication of Linnaeus's major works, when exploration of the world was at its height. This is the period that saw the founding of *Curtis's Botanical Magazine* (1787) and the publication of J.C. Loudon's *Encyclopaedia of plants* (1829 and many subsequent editions).

The further development of plant taxonomy, from about the middle of the nineteenth century to the present, has seen an increasing divergence between garden and scientific taxonomy, leading on the one hand to such works as the *Royal Horticultural Society's dictionary of gardening* (1951 and reprinted, itself based on G. Nicholson's *Illustrated dictionary of gardening*, 1884–1888) and the very numerous popular, usually illustrated works on garden flowers available today, and, on the other hand, to the Floras, Revisions and Monographs of scientific taxonomy.

Despite this divergence, a number of plant taxonomists realised the importance of the classification and identification of cultivated plants, and produced works of considerable scientific value. Foremost among these stands L.H. Bailey, editor of *The standard cyclopedia of horticulture* (1900, with several subsequent reprints and editions), author of *Manual of cultivated plants* (1924, edn 2 1949), and founder of the journals *Gentes Herbarum* and *Baileya*. Other important workers in this field are T. Rumpler (*Vilmorin's Blumengärtnerei*, 1879), L. Dippel (*Handbuch der Laubholzkunde*, 1889–1893), A. Voss and A. Siebert (*Vilmorin's Blumengärtnerei*, edn 3, 1894–1896), C.K. Schneider (*Illustriertes Handbuch der Laubholzkunde*, 1904–1912), A. Rehder (*Manual of cultivated trees and shrubs*, 1927, edn 2 1947), J.W.C. Kirk (*A British garden flora*, 1927), F. Enke (*Parey's Blumengärtnerei*, 1958), B.K. Boom (*Flora Cultuurgewassen*, 1959 and proceeding) and V.A. Avrorin & M.V. Baranova (*Decorativn'ie Travyanist'ie Rasteniya Dlya Otkritogo Grunta SSSR*, 1977).

The present Flora, which, of necessity, is based on original taxonomic studies by many workers, attempts to provide a scientifically accurate and up-to-date means for the identification of plants cultivated for amenity in Europe (i.e. it does not include crops, whether horticultural or agricultural, or garden weeds), and to provide what are currently thought to be their correct names, together with sufficient synonymy to make sense of catalogues and other horticultural works. The needs of the informed amateur gardener have been borne in mind at all stages of the work, and it is hoped that the Flora will meet his needs just as much as it meets the needs of the professional plant taxonomist. The details of the format and use of the Flora are explained in section 2 below (pp. 2–5).

In writing the work, the Editorial Committee has been fully aware of the difficulties involved. Some of these have been outlined above; others derive from that fact that herbarium material of cultivated plants is scanty and usually poorly annotated, so that material of many species is not available for checking the use of names, or for comparative purposes. Because of these factors, attention has been drawn to numerous problems which cannot be solved but can only be adverted to. The solution of such problems requires much more taxonomic work.

The form in which contributions appear is the responsibility of the Editorial Committee. The vocabulary and the technicalities of plant description are therefore not necessarily those endorsed by the contributors.

1. SELECTION OF SPECIES

The problem of determining which species are in cultivation is complex and difficult, and has no complete and final answer. Many species, for instance, are grown in botanic gardens but not elsewhere; others, particularly orchids, succulents and some alpines, are to be found in the collections of specialists but are not available generally. Yet others have been in cultivation in the past but are now lost, or perhaps linger in a few collections, unrecorded and unpropagated. Further problems arise from the fact that the identification of plants in collections is not always as good as it might be, and some less well-known species probably appear in published lists under the names of other, well-known species (and vice versa).

The Flora attempts to cover all those species that are likely to be found in general collections (i.e. excluding botanic gardens and specialist collections) in Europe, whether they are grown out-of-doors or under glass. In order to produce a basic working list of such species, a compilation of all European nursery catalogues available to us was made in 1978 by Margaret McDonald, a vacation student working at the Royal Botanic Garden, Edinburgh. Since then, numerous additions have been made by Hazel Hamlet. This list, (known as the 'Commercial List'), which includes well over 12 000 specific names, forms the basis of the species included here. In addition to the 'Commercial List', several works on the flora of gardens have been consulted, and the species covered by them have been carefully considered for inclusion. These works are: Wehrhahn, H.R., *Die Gartenstauden* (1929–1931); *The Royal Horticultural Society's dictionary of gardening*, edn 2, (1956, supplement revised 1969); Enke, F. (ed.), *Parey's Blumengärtnerei* (1956); Boom, B.K., *Flora Cultuurgewassen* (1959 and proceeding); Bean, W.J., *Trees and shrubs hardy in the British Isles* (edn 8, 1970-1981); Krüssmann, G., *Handbuch der Laubgehölze* (edn 2, 1976–1978); Enke, F., Buchheim, G. & Seybold, S. (eds.), *Zander's Handwörterbuch der Pflanzennamen* (edn 12, 1980). Most of the names included in these works are covered by the present Flora, though some have been rejected as referring to plants no longer in general cultivation.

As well as the works cited above, several works relating to plants in cultivation in North America have been consulted: Rehder, A., *Manual of cultivated trees and shrubs* (edn 2, 1947) and *Bibliography of cultivated trees and shrubs* (1949); Bailey, L.H., *Manual of cultivated plants* (edn 2, 1949); *Hortus Third* (edited by the staff of the L.H. Bailey Hortorium, Cornell University).

The contributors have also drawn on their own experience, as well as that of the family editors, European advisers and other experts, in deciding which species should be included.

Because some species are not very widely grown, but, by the criteria mentioned above, have had to be included, two levels of treatment have been used. Most species have a full entry, being keyed, numbered and described as set out under Section 2c below (p. 3). Less commonly cultivated species are not keyed or numbered individually, but are described briefly under the full-entry species to which they key out in the formal key (this, of course, will not necessarily be the one to which the additional species is thought to be most closely related). The name of any full-entry species which has additional species attached to it is distinguished in the formal keys by an asterisk (*); and the names of additional species are preceded by asterisks in the main descriptive text.

2. USE OF THE FLORA

a. *The taxonomic system followed in the Flora.* Plants are described in this work in a taxonomic order, so that similar genera and species occur close to each other, rendering comparison of descriptions more easy than in a work where the entries are alphabetical. The families (and higher groups) follow the Engler & Prantl system as expressed in H. Melchior's edition (edn 12, 1964) of *Syllabus der Pflanzenfamilien*. The exceptions are minor, apart from the placing of the monocotyledons before the dicotyledons; this has been done purely for convenience, and has no other implications. The assignment of genera to families also usually follows the *Syllabus*.

The order of the species within each genus has been a matter for the individual author's discretion. In general, however, some established revision of the genus has been followed, or, if no such revision exists, the author's own views on similarity and relationships have governed the order used.

b. *Nomenclature.* The arguments for using Latin names for plants in popular as well as scientific works are often stated and widely accepted, particularly for Floras such as this, which cover an area in which several languages are spoken. Latin names have therefore been used at every taxonomic level. A concise outline of the taxonomic hierarchy and how it is used can be found in C. Jeffrey's *An introduction to plant taxonomy* (1968, edn 2, 1982). Because of the difficulties of providing vernacular names in all the necessary languages (not to say dialects), they have not been included. A supplementary volume including them may be possible after the systematic part of the Flora has been completed.

Many horticultural reference works omit the authority which should follow every Latin plant name. Knowledge of this authority prevents confusion between specific names that may have been used more than once within the same genus, and makes it possible to find the original description of the species (using *Index Kewensis*, which lists the original references for all Latin plant names published since 1753). Authorities are therefore given for all names at or below the genus level. These are unabbreviated to avoid the obscure contractions which mystify the lay reader, and, on occasions, the professional botanist. In most cases we have not thought it necessary to include the initials or qualifying words and letters which often accompany author names, e.g. A. Richard, Reichenbach filius, fil. or f. (the exceptions involve a few, very common surnames).

In scientific taxonomic literature, the authority for a plant name sometimes consists of two names joined together by *ex* or *in*. Such formulae have not been used here; the authority has been shortened in accordance with *The international code of botanical nomenclature* (ed. F.A. Stafleu, 1978), e.g. *Capparis lasiantha* R. Brown ex de Candolle becomes *Capparis lasiantha* de Candolle; *Viburnum ternatum* Rehder in Sargent becomes *Viburnum ternatum* Rehder. The abbreviations *hort.* and *auct.*, which sometimes stand in place of the authority after Latin names, have not been used in this work as they are often obscure or misleading. The situations described by them can be clearly and unambiguously covered by the terms *invalid*, *misapplied* or *Anon. Invalid* implies that the name in question has never been validly published in accordance with the Code of Nomenclature, and therefore cannot be accepted. *Misapplied* refers to names which have been applied to the wrong species in gardens or in the literature. *Anon.* is used with validly published names for which there is no apparent author.

Gardeners and horticulturists complain bitterly when long-used and well-loved names are replaced by new and unfamiliar ones. These changes are unavoidable if *The international code of botanical nomenclature* is adhered to. Taxonomic research will doubtless continue to unearth earlier names and will also continue to re-align or split up existing groups, as relationships are further investigated. However, the previously accepted names are not lost; in this work they appear as synonyms, given in brackets after the currently accepted name; they are also included in the index. Dates of publication are not given either for accepted names or for synonyms.

c. *Descriptions and terminology*. Families, genera and species included in the Flora are generally provided with full-length descriptions. Shorter, diagnostic descriptions are, however, used for genera or species which differ from others already fully described in only a few characters, e.g.:

3. P. vulgaris Linnaeus. Like *P. officinalis* but leaves lanceolate and corolla red . . .

This implies that the description of *P. vulgaris* is generally similar to that of *P. officinalis* except in the characters mentioned; it should not be assumed that plants of the two species will necessarily look very like each other. Additional species (see p. 2), subspecies, varieties and cultivars (see p. 5) are described very briefly and diagnostically.

Unqualified measurements always refer to length (though 'long' is sometimes added in cases where confusion might arise); similarly, two measurements separated by a multiplication sign indicate length and breadth respectively.

The terminology used has been simplified as far as is consistent with accuracy. The technical terms which, inevitably, have had to be used are explained in the glossary (p. 291). Technical terms restricted to particular families or genera are defined in the observations following the family or genus description, and are also referred to in the glossary.

d. *Informal keys*. For most genera containing 5 to 20 species (and for most families containing 5 to 20 genera) an informal key is given; this will not necessarily enable the user to identify precisely every species included, but will provide a guide to the occurrence of the more easily recognised characters. A selection of these characters is given, each of which is followed by the entry-numbers of those species which show that character. In some cases, where only a few species of a genus show a particular character, the alternative states are not specified, e.g.:

Leaves. Terete: **18,19**.

This means that only species **18** and **19** in the particular genus have terete leaves; the other species may have leaves of various forms, but they are not terete. No distinction is made between full-entry and additional species. The occurrence of an entry-number merely means that one or more species found under that entry-number will have the particular character, but does not imply that they will all do so.

e. *Formal keys*. For every family containing more than one genus, and for every genus containing more than one species, a dichotomous key is provided. This form of key, in which a series of decisions must be made between pairs of contrasting character-states, should lead the user step by step to an entry-number followed by the name of a full-entry species. If that name is followed by an asterisk it will be necessary to read the descriptions of the additional species under that entry-number, as well as that of the full-entry species, in order to decide which one fits the plant in question. A similar key to all the families of monocotyledons included in this work (those in volume I as well as in the present volume) is also provided (p. 6).

f. *Horticultural information*. Notes on the cultural requirements and methods of propagation are usually included in the observation to each genus; more rarely, such information is given in the observations under the family description. These are necessarily brief and very generalised, and merely provide guidance. Reference to general works on gardening is necessary for more detailed information.

g. *Citation of literature*. References to taxonomic books, articles and registration lists are cited for each family and genus, as appropriate. No abbreviations are used in these citations (though very long titles have been shortened). The citation of a particular book or article

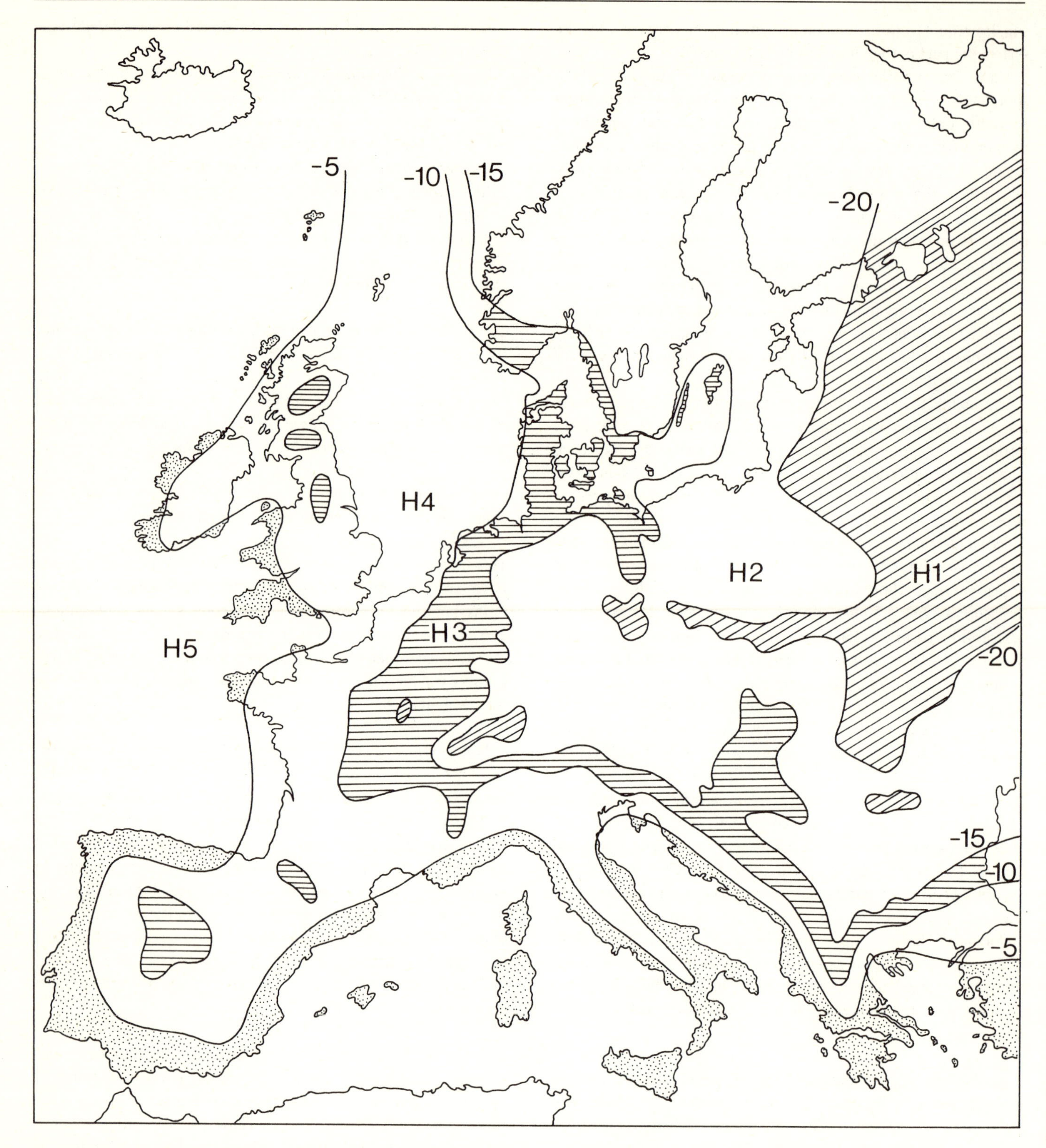

Map 1. Mean January isotherms for Europe (hardiness codes).
(After Krüssmann, *Handbuch der Laubgehölze*, 1960.)

does not necessarily imply that it has been used in the preparation of the account of the particular genus or family in this work.

h. *Citation of illustrations*. References to good illustrations are given for each species (or subspecies or variety); the names under which they were originally published (which may be different from those used here) are not normally given. They may be coloured or black and white, and may be drawings, paintings or photographs. Usually, up to four illustrations per species have been given, and an attempt has been made to choose pictures from widely available, modern works. Where no illustrations are cited, they either do not exist, as far as we know, or those that do are considered to be of doubtful accuracy.

In searching for illustrations, use was made of *Index Londinensis* (1929–1931, supplement 1941) and R.T. Isaacson's *Flowering plant index of illustration and information* (1979). Readers are referred to these works if they wish to find further pictures.

Several pages of illustrations of diagnostic plant parts are included with various groups in the Flora, and should be particularly helpful when plants are being identified by means of the keys. Some of these illustrations are diagrammatic, others have been redrawn from various sources, and others are original.

i. *Geographical distribution*. The wild distribution, as far as it can be ascertained, is given in italics at the end of the description of each species (or subspecies or variety). The choice and spelling of place names in general follows *The Times Atlas*, Comprehensive Edition (1980), except:

(1) Well-established English forms of names have been used in preference to unfamiliar vernacular names, e.g. Crete instead of Kriti, Naples instead of Napoli, Borneo instead of Kalimantan;

(2) New names or spellings will be adopted as soon as they appear in readily available works of reference.

j. *Hardiness* (see map, p. 4). For every species a hardiness code is given. This gives a tentative indication of the lowest temperatures that the particular species can withstand:

G2 – needs a heated glasshouse even in south Europe.

G1 – needs a cool glasshouse even in south Europe.

H5 – hardy in favourable areas; withstands 0–5 °C minimum.

H4 – hardy in mild areas; withstands −5 to −10 °C minimum.

H3 – hardy in cool areas; withstands −10 to −15 °C minimum.

H2 – hardy almost everywhere; withstands −15 to −20 °C minimum.

H1 – hardy everywhere; withstands −20 °C and below.

The map of mean winter minima (p. 4) shows the isotherms corresponding to these codes. It must be understood that H4 includes H5, H3 includes H4 and H5, H2 includes H3–5 and H1 includes H2–5.

k. *Flowering time*. The terms spring, summer, autumn and winter have been used as a guide to flowering times in cultivation in Europe. It is not possible to be more specific when dealing with an area extending from northern Scandinavia to the Mediterranean. In cases where plants do not flower in cultivation, or flower rarely, or whose time of flowering is not recorded, no flowering time is given.

l. *Subspecies, varieties and cultivars*. Subspecies and varieties are described, where appropriate. This is done in various ways, depending on the number of such groups; all these ways are, however, self-explanatory.

No attempt has been made to describe the range of cultivars of a species, either partially or comprehensively. The former is scarcely worth doing, the latter virtually impossible. Reference to individual, commonly grown cultivars is, however, made in various ways:

(1) If a registration list of cultivars exists, it is cited in the 'Literature' paragraph (see section 2g) following the description of the genus.

(2) If a particular cultivar is very widely grown, it may be referred to, either in the description of the species to which it belongs (or which it most resembles), or in the observations to that species.

(3) If, in a particular species, cultivars are numerous and fall into reasonably distinct groups based on variation in some striking character, then these groups may be referred to, together with an example of each, in the observations to the species.

m. *Hybrids*. Many hybrids between species (interspecific hybrids) and some between genera (intergeneric hybrids) are in cultivation, and some of them are widely grown. Commonly cultivated interspecific hybrids are, where possible, included as though they were species. Their names, however, include the '×' sign indicating their hybrid origin; the names of their parents (when known or presumed) are given at the beginning of the paragraph of observations following the description. Other hybrids which are less frequently grown are mentioned in the observations to the individual parent species that they most resemble. In some genera (e.g. *Odontoglossum*), where the number of hybrids is very large, only a small selection of those most commonly grown is mentioned.

MONOCOTYLEDONS
(Part II)

KEY TO FAMILIES

Families included in this volume are provided with page numbers.

1a. Ovary superior or flowers completely without perianth (including all aquatics with totally submerged flowers) 2
b. Ovary inferior or partly so (if aquatic, flowers borne at or above water level) 27

2a. Trees, shrubs or prickly scramblers with large, pleated, usually palmately or pinnately divided leaves; flowers more or less stalkless in fleshy spikes or panicles with large basal bracts (spathes) **XXI. Palmae** (p. 65)
b. Plant without the above combination of characters 3

3a. Small, usually floating aquatic plants not differentiated into stem and leaves **XXIII. Lemnaceae** (p. 112)
b. Plants various, rarely floating aquatics, plant body differentiated into stem and leaves 4

4a. Perianth entirely scarious or reduced to bristles, hairs, narrow scales, or absent 5
b. Perianth well developed, though sometimes small, never entirely scarious 11

5a. Flowers in small, 2-sided or cylindric spikelets provided with overlapping bracts (spikelets sometimes 1-flowered) 6
b. Flowers arranged in heads, superposed spikes, racemes, panicles or cymes, never in spikelets as above 7

6a. Leaves alternate, in 2 ranks, on a stem which is usually hollow and with cylindric internodes; leaf-sheath usually with free margins, at least in the upper part; flowers arranged in 2-sided spikelets (sometimes 1-flowered) each usually subtended at the base by 2 sterile bracts (glumes); each flower usually enclosed by a lower lemma and an upper palea (sometimes absent); perianth of 2–3 concealed scales (lodicules), more rarely 6 or absent; styles generally 2, feathery **XX. Gramineae** (p. 31)
b. Leaves usually spirally arranged on 3 sides of the cylindric or more usually 3-angled stems which usually have solid internodes; young leaf-sheath closed though sometimes splitting later; flowers arranged in 2-sided or cylindric spikelets often with a 2-keeled or 2-lobed glume at the base; each flower subtended only by a glume; perianth of several bristles, hairs or scales, or absent; style 1 with 2 or 3 papillose stigmas **XXVII. Cyperaceae** (p. 114)

7a. Dioecious trees or shrubs with stiffly leathery, sharply toothed leaves, often supported by stilt-roots; fruit a syncarp, often woody **XXIV. Pandanaceae** (p. 112)
b. Plant without the above combination of characters 8

8a. Inflorescence a simple, fleshy spike (spadix) of inconspicuous flowers subtended by or rarely joined to a large bract (spathe); leaves often net-veined or lobed (plant rarely a small, evergreen, floating aquatic) **XXII. Araceae** (p. 75)
b. Plant without the above combination of characters 9

9a. Flowers bisexual; perianth segments 6, scarious; ovary with 3–many ovules **XVII. Juncaceae** (p. 10)
b. Flowers unisexual; perianth segments a few threads or scales; ovary with 1 ovule 10

10a. Flowers in 2 superposed, elongate, brownish or silvery spikes; ovary borne on a stalk with hair-like branches **XXVI. Typhaceae** (p. 113)
b. Flowers in spherical heads; ovary not stalked **XXV. Sparganiaceae** (p. 113)

11a. Carpels free or slightly united at the base 12
b. Carpels united for most of their length though the styles may be free, or carpel solitary 15

12a. Inflorescence a spike, sometimes bifid; perianth segments 1–4 13
b. Inflorescence not a spike; perianth segments 6 14

13a. Stamens 6 or more; carpels 3–6; perianth segments 1–3, petaloid **IV. Aponogetonaceae**
b. Stamens 4; carpels 4; perianth segments 4, not petaloid **V. Potamogetonaceae**

14a. Stamens 9 or more **II. Butomaceae**
b. Stamens 6 **I. Alismataceae**

15a. All perianth segments similar 16
b. Perianth segments of the outer and inner whorls conspicuously different, the former usually sepal-like, the latter usually petal-like 25

16a. Inflorescence subtended by an entire, spathe-like sheath; plants aquatic **XV. Pontederiaceae**
b. Inflorescence not as above; plants terrestrial 17

17a. Perianth persistent, covered by branched hairs; stamens 3; sap usually orange **VIII. Haemodoraceae**
b. Plant without the above combination of characters 18

18a. Plant woody, or not woody and bearing rosettes of long-lived, fleshy or leathery leaves, at or near ground level 19
b. Plant herbaceous, leaves usually not long-lived and in rosettes, if so, then deciduous and not fleshy 23

19a. Leaf-stalk bearing 2 tendrils; leaves net-veined **VI. Liliaceae**
b. Leaf-stalk without tendrils; leaves parallel-veined 20

20a. Leaves very small, scale-like or spiny, their function taken over by flattened stems (cladodes) on which the inflorescences are borne **VI. Liliaceae**
b. Plant with true leaves; cladodes absent 21

21a. Shrubs or woody climbers with scattered stem-leaves; flowers solitary, usually large and hanging; placentation mostly parietal **VI. Liliaceae**
b. Plant without the above combination of characters 22

22a. Leaves leathery and more or less thin, if succulent, then with a spine-like or cylindric tip; flower usually green or whitish, bell- or cup-shaped or with a narrow tube and spreading lobes, often more than 1 to each bract **VII. Agavaceae**
b. Leaves succulent, usually without a spine-like or cylindric tip; flower usually red, yellow or orange, tubular, the lobes not spreading, always 1 to each bract **VI. Liliaceae**

23a. Leaves very small, scale-like or spiny, their function taken over by flattened or needle-like stems (cladodes) on which the inflorescences are borne **VI. Liliaceae**
b. Plant with true leaves; cladodes absent 24

24a. Leaves evergreen, clearly stalked; flowers more than 1 to each bract, with a narrow tube as long as or longer than the spreading lobes **VII. Agavaceae**
b. Leaves deciduous, usually without distinct stalks; flowers of various shapes, rarely as above, always 1 to each bract **VI. Liliaceae**

25a. Flowers solitary or in umbels; leaves broad, opposite or in a single whorl near the top of the stem **VI. Liliaceae**
b. Flowers in spikes, heads, cymes or panicles; leaves not as above 26

26a. Stamens 6 or 5–3 with 1–3 staminodes; anthers basifixed; leaves usually borne on the stems, often with closed sheaths, never grey with scales; bracts neither overlapping nor conspicuously coloured **XIX. Commelinaceae** (p. 25)
b. Stamens 6, staminodes 0; anthers dorsifixed; leaves mostly in basal rosettes, often rigid and spiny-margined, when on the stems usually grey with scales; bracts usually overlapping and conspicuously coloured **XVIII. Bromeliaceae** (p. 10)

27a. Flowers radially symmetric or weakly bilaterally symmetric; stamens 6, 4, 3, or rarely many 28
b. Flowers strongly bilaterally symmetric or asymmetric; stamens usually 5, 2 or 1 (very rarely 6) 40

28a. Unisexual climbers with heart-shaped or very divided leaves; rootstock tuberous or woody **XIV. Dioscoreaceae**
b. Plant without the above combination of characters 29

29a. Perianth persistent, variously hairy; sap usually orange **VIII. Haemodoraceae**
b. Plant without the above combination of characters 30

30a. Rooted or floating aquatics; stamens 2–12; ovules distributed all over the carpel walls (placentation diffuse-parietal) **III. Hydrocharitaceae**
b. Terrestrial or marsh plants, or epiphytes; stamens 3 or 6, rarely many; placentation axile or parietal (ovules restricted to a few rows on the carpel walls) 31

31a. Stamens 3, staminodes absent; leaves often sharply folded, their bases overlapping; style branches often divided **XVI. Iridaceae**
b. Stamens 6, or 3 plus 3 staminodes; leaves not usually as above; style branches not divided 32

32a. Placentation parietal; flowers in an umbel with the inner bracts long, thread-like and hanging **XIII. Taccaceae**
b. Placentation usually axile; inflorescence and bracts not as above 33

33a. Perianth consisting of an outer, calyx-like whorl and an inner, corolla-like whorl; bracts usually overlapping and conspicuously coloured **XVIII. Bromeliaceae** (p. 10)
b. Segments of the perianth not in 2 dissimilar whorls; bracts not as above 34

34a. Ovary half-inferior 35
b. Ovary fully inferior 36

35a. Anthers opening by pores **X. Tecophilaeaceae**
b. Anthers opening by slits **VI. Liliaceae**

36a. Leaves long-persistent, evergreen 37
b. Leaves dying down annually 38

37a. Leaves fleshy or leathery, thick, rigid or flexible, spine-tipped, often with spines or teeth on the margins **VII. Agavaceae**
b. Leaves neither fleshy, leathery nor spine-tipped, without spines or teeth on the margins **XII. Velloziaceae**

38a. Flowers in a spike; leaves fleshy, often spotted with brown, the margins more or less rolled around each other in bud **VII. Agavaceae**
b. Flowers in umbels or solitary; leaves not usually fleshy or spotted with brown but flat, pleated or with the margins folded outwards in bud 39

39a. Leaves all basal, densely hairy, pleated or with prominent veins **XI. Hypoxidaceae**
b. Leaves various, not usually densely hairy, pleated or with prominent veins, basal or not **IX. Amaryllidaceae**

40a. Fertile stamens 6; perianth segments all similar, tube curved and unevenly swollen; stems below ground, fleshy **VII. Agavaceae**
b. Fertile stamens 5, 2 or 1, very rarely 6; staminodes, which may be petal-like, often present; perianth segments usually differing among themselves; fleshy underground stems rare 41

41a. Fertile stamens 2 or 1, united with the style to form a column; pollen usually borne in masses (pollinia); leaf-veins, when visible, all parallel to margins **XXXIII. Orchidaceae** (p. 137)
b. Fertile stamens 5 or 1, rarely 6, not united to the style; pollen granular; leaf with a distinct midrib more or less

parallel to the margins, the secondary veins parallel, running from the midrib to the margins 42

42a. Fertile stamens 5 or rarely 6 43
b. Fertile stamen 1, the remainder transformed into petal-like staminodes 44

43a. Leaves and bracts spirally arranged; flowers unisexual **XXVIII. Musaceae** (p. 117)
b. Leaves and bracts in 2 ranks; flowers bisexual **XXIX. Strelitziaceae** (p. 119)

44a. Fertile stamen with normal structure, not petal-like **XXX. Zingiberaceae** (p. 120)
b. Fertile stamen in part petal-like, and with only 1 pollen-bearing anther-lobe 45

45a. Leaf-stalk with a swollen band (pulvinus) at the junction with the blade; ovary smooth, with 1–3 ovules **XXXII. Marantaceae** (p. 130)
b. Leaf-stalk without a pulvinus at the junction with the blade; ovary usually warty, with numerous ovules **XXXI. Cannaceae** (p. 129)

Many monocotyledonous plants which are woody, or form large or small rosettes of fleshy, leathery, long-lived leaves, flower only irregularly, or at long intervals, or take some years to reach flowering size. These plants will therefore not be easily identifiable with the key above. The following informal key will provide some guidance for the identification of the families to which such plants belong.

Habit. Plants tree-like, with a distinct, woody trunk: **Liliaceae, Agavaceae, Palmae, Pandanaceae**; plants tree-like with a trunk which is not woody, but given rigidity by being made up of overlapping leaf-bases (sheaths): **Musaceae, Strelitziaceae**; plants shrubby, with woody stems but without a distinct, tall trunk: **Liliaceae, Agavaceae, Strelitziaceae**; plants forming large or small rosettes at substrate level, or borne on the ends of the branches of a short, erect or prostrate woody stem: **Liliaceae, Agavaceae, Bromeliaceae, Palmae** (young specimens). Plant with conspicuous stilt-roots: **Pandanaceae.**

Leaf arrangement. Leaves borne on 2 sides of the stem only: **Strelitziaceae**; leaves borne all round the stem, or in a crown at the stem apex, or in a basal rosette: **Liliaceae, Agavaceae, Bromeliaceae, Palmae, Musaceae**; leaves borne all round the stem in a crown at the apex, very regularly arranged, so that their spiral arrangement is clearly visible: **Pandanaceae.**

Leaves. Pleated, usually palmately or pinnately divided or lobed: **Palmae**; not pleated, entire, with a well-developed midrib running parallel with the margins, from which the parallel, lateral veins diverge: **Musaceae, Strelitziaceae**; not pleated, entire, usually leathery, fleshy and persistent, venation not visible, or not as above: **Liliaceae, Agavaceae, Pandanaceae**; neither pleated nor fleshy, but thin, leathery and persistent: **Agavaceae, Bromeliaceae.** With solid, spine-like or cylindric tips: **Agavaceae.** With spiny margins: **Agavaceae, Bromeliaceae, Pandanaceae.** Margins splitting off as threads: **Agavaceae.**

XVII. JUNCACEAE

Annual or perennial herbs with erect or horizontal rhizomes. Leaves basal, long and narrow with sheathing bases, or reduced to scales. Flowers bisexual, crowded into heads. Perianth segments 6, in 2 whorls. Stamens 3 or 6. Ovary superior with 1–3 cells; stigmas 3, ovules 3 or numerous. Fruit a dehiscent capsule. Figure 1, p. 11.

A family of 9 genera distributed throughout the world, but absent from the tropical lowlands. In Europe, *Juncus* and *Luzula* are native, and these are also the only cultivated genera.

1a. Leaves hairless; capsule with numerous seeds **1. Juncus**
b. Leaves hairy; capsule with 3 seeds **2. Luzula**

1. JUNCUS Linnaeus
E.C. Nelson
Hairless perennial herbs with rhizomes. Stamens 6 or fewer, opposite the perianth segments and attached to their bases. Capsule with numerous seeds.

Only 1 species is widely cultivated in Europe, but others occur as weeds, especially in damp gardens and garden ponds in western regions. The genus contains about 225 species and is widely distributed. Propagation is by division of the rhizomes.

1. J. effusus Linnaeus. Illustration: Keble Martin, The concise British flora in colour, 87 (1965).
Stems 50–150 cm. Leaves cylindric, as long as stems, with continuous white pith. Inflorescence dense or loose, with numerous pale brown flowers. Stamens 3, occasionally 6. *Cosmopolitan*. H1. Summer.

This species is represented in gardens only by cultivars, except when it is a weed. 'Spiralis' has spirally twisted leaves and stems; 'Vittatus' has leaves striped longitudinally with yellow. Both cultivars are suitable for pool margins and bog gardens.

2. LUZULA de Candolle
E.C. Nelson
Hairy perennial herbs with rhizomes and stolons. Leaves flat; leaf-sheath closed. Flowers in cymes, sometimes in dense heads. Capsule with 3 seeds.

Only 1 species is commonly cultivated in Europe, but 2 other native European species, *L. nivea* (Linnaeus) de Candolle and *L. luzuloides* (Lamarck) Dany, with decorative, whitish flowers, are sometimes grown. Propagation is by division of the rhizomes.

1. L. sylvatica (Hudson) Gaudin. Illustration: Grounds, Ornamental grasses, 192, 200 (1979).
Perennial, grass-like herb, with stolons. Leaves 5–30 mm wide, with white hairs. Perianth segments brown. Anthers up to 6 times as long as filaments. *Europe, SW Asia*. H1. Summer.

A plant suitable for growing in woodland gardens, where it will form good ground cover. A cultivar with white leaf margins ('Marginata') is available.

XVIII. BROMELIACEAE

Perennial herbs, rarely shrubby; terrestrial, growing on rocks or epiphytic. Roots usually present, but acting as holdfasts in all but terrestrial species. Leaves usually in rosettes, sheathing at the base, simple, entire or variously toothed or spiny, usually bearing peltate scales (modified hairs). Inflorescence terminal or lateral, stalkless or borne on a scape which may bear primary bracts, simple or compound, composed of racemes, spikes, panicles, heads or solitary flowers, usually with brightly coloured floral bracts. Sepals 3, free or united. Petals 3, free or united, sometimes bearing appendages (basal scales) near the base. Stamens 6 in 2 whorls. Ovary superior to inferior. Fruit a capsule or berry.

A large and mostly tropical family of 46 genera and 2100 species. All but one of the species (a *Pitcairnia* found in West Africa) are natives of the New World. Some 17 genera, covering about 80 species, are commercially available in Europe; however, some specialist collections may contain genera other than are included here. For further information, reference should be made to either of the two works cited below.

The inflorescences of the Bromeliaceae are very variable; they are basically spike-like racemes or panicles, but they may be reduced to a solitary flower, or condensed into a head-like structure which may be stalked or stalkless and borne between the leaf-sheaths. Panicles are described as 'bipinnate' when the axis regularly bears branches which are themselves regularly branched.

In cultivation the species flower at irregular intervals, so no flowering times are given in the account below.

Cultural requirements for these plants are of two kinds. On the one hand are the epiphytic species, which differ from most other plants in having specialised root-systems. For most of these, a well-drained medium is required, while the green-leaved species of *Tillandsia* and species of *Vriesea* and *Guzmania* require one which retains a considerable amount of water; Osmunda fibre, for example, is most useful. The terrestrial species, on the other hand, are more robust, and can easily be cultivated in general potting composts. However, species of *Aechmea*, *Billbergia* and *Dyckia* can be kept much drier, in a coarse gravel and sand mixture – as long as the tubular rosettes of those species which have them are regularly filled with water.
Literature: Smith, L.B. & Downs, R.J., Bromeliaceae, *Flora Neotropica* **14**, parts 1–3 (1974–79), from which most of the information presented here has been extracted; Rauh, W., *Bromeliads for home, garden and greenhouse*, English edn (1979).

Figure 1. Diagram of some characters used in the identification of Juncaceae and Cyperaceae. 1, Inflorescence of *Luzula*. 2, Flower of *Luzula*. 3, Flower of *Scirpus* (a, perianth bristles). 4, Inflorescence of *Cyperus*. 5, Spikelet of *Cyperus*. 6, Flower of *Cyperus* (b, glume). 7, Inflorescence of *Carex* (c, male spike; d, female spike). 8,9, Portion of stem and leaf-base of *Carex* (e, leaf-sheath; f, ligule). 10, Male flower of *Carex*. (g, glume). 11, Female flower of *Carex*.

1a. Leaf margin entire, without spines or teeth 2
b. Leaf margin with spines or teeth 6
2a. Petals with their upper parts somewhat irregularly placed so that the corolla is not radially symmetric **3. Pitcairnia**
b. Petals with their upper parts regularly placed so that the corolla is radially symmetric 3
3a. Bases (claws) of the petals united into a tube which is longer than the sepals; flowers in many ranks **7. Guzmania**
b. Bases (claws) of the petals free or united into a tube which is shorter than the sepals; flowers usually in 2 ranks 4
4a. Petals with basal scales **6. Vriesea**
b. Petals without basal scales 5
5a. Inflorescence a loose panicle; flowers pendent, white **2. Fosterella**
b. Inflorescence spike-like or reduced to a solitary flower; flowers not pendent, usually coloured **5. Tillandsia**
6a. Scape absent or short and not immediately visible 7
b. Scape well-developed and evident 10
7a. Leaves rigid with stout marginal spines **12. Fascicularia**
b. Leaves flexible, margins toothed 8
8a. Inflorescence not sunk in the centre of the rosette, spherical, held above the leaf-sheaths **9. Cryptanthus**
b. Inflorescence sunk in the centre of the rosette, between the leaf-sheaths 9
9a. Inflorescence with the flowers stalkless, in flattened bundles covered by large bracts **10. Nidularium**
b. Inflorescence various, not as above, flowers stalked **8. Neoregelia**
10a. Inflorescence cone-like, usually crowned with a tuft of sterile, leaf-like bracts 11
b. Inflorescence not as above 12
11a. Inflorescence terminal, 10–25 cm when in fruit; scape bearing bracts **17. Ananas**
b. Inflorescence apparently lateral, 3–7 cm when in fruit; scape without bracts **11. Acanthostachys**
12a. Inflorescence with an involucre 13
b. Inflorescence without an involucre 14
13a. Petals without basal scales **10. Nidularium**
b. Petals with basal scales **13. Canistrum**
14a. Sepals each with a small apical spine, or at least sharply pointed **14. Aechmea**
b. Sepals without spines, not sharply pointed 15
15a. Sepals to 1 cm; petals to 3 cm 16
b. Sepals more than 1 cm; petals more than 3 cm 17
16a. Ovary completely superior **4. Dyckia**
b. Ovary completely inferior **14. Aechmea**
17a. Ovary completely superior to, at most, half-inferior 18
b. Ovary completely inferior 19
18a. Petal blades twisted together spirally after flowering **1. Puya**
b. Petal blades not twisted together spirally after flowering **3. Pitcairnia**
19a. Petals erect or almost so, corolla radially symmetric **15. Quesnelia**
b. Petals irregularly placed, corolla bilaterally symmetric or petals each recurved in a spiral **16. Billbergia**

1. PUYA Molina

J. Cullen

Perennial, terrestrial herbs, usually with stems. Leaves very leathery, in dense rosettes, with distinct sheaths and spiny-margined blades which are not narrowed towards the base. Inflorescence usually a panicle. Sepals much shorter than the petals. Petals with distinct claw and usually broad blade, the blades coiled together spirally after flowering. Stamens usually somewhat shorter than the petals. Ovary superior or almost entirely so. Fruit a capsule.

A genus of about 170 species from S America. Only a small number is commercially available, and these are rarely seen outside specialist collections. Some of them are particularly prized for their metallic blue or blue-green petals.

1a. Branches of the inflorescence with flowers throughout (rarely flowerless towards the base) 2
b. Branches of the inflorescence flowerless in their upper halves or more, bearing reduced bracts there 5
2a. Branches of the inflorescence more than twice as long as the bracts which subtend them 3
b. Branches of the inflorescence at most one and a half times as long as the bracts which subtend them 4
3a. Petals twice as long as the sepals, with basal scales **2. coerulea**
b. Petals at most one and a half times as long as the sepals, without basal scales **1. spathacea**
4a. Inflorescence branches widely spreading; sepals covered with fine hairs when young, these falling quickly; scape less than 1 m **3. venusta**
b. Inflorescence branches erect; sepals persistently woolly; scape to 6 m **4. gigas**
5a. Leaves of the same colour on both surfaces **5. chilensis**
b. Leaves green above, white beneath **6. alpestris**

1. P. spathacea (Grisebach) Mez. Illustration: Flora Neotropica **14**: f. 20I–L (1974).
Stem erect, short; scape to 1 m. Leaves with spreading or recurved blades, 60–110 × 3–5 cm, grey-green above, pale-scaly beneath, margins with hooked spines to 4 mm. Panicle bipinnate with 10–20 lateral branches, covered with fine, white, stellate hairs; branches 7–30 cm, loosely flowered. Flower-stalks curved, *c.* 1.5 cm. Sepals 1.5–2.2 cm, pinkish red. Petals tongue-shaped, 2.5–3.3 cm, blue to dark green. *North-central Argentina.* H5–G1.

2. P. coerulea Lindley. Illustration: Edwards's Botanical Register **26**: t. 11 (1840); Botanical Magazine, 8194 (1908); Flora Neotropica **14**: f. 21C–G (1974).
Stem well developed, erect; scape to 2 m. Leaves in a rosette at the stem apex, 40–60 × 1–2 cm, sheaths brown, blades ashy white, margins with hooked, reddish brown spines to 5 mm. Inflorescence bipinnate or rarely simple, branches 5–30 cm, sometimes without flowers towards the base. Floral bracts broad, nearly as long as the sepals. Flower-stalks 1–2 cm. Sepals 1.2–2.4 cm. Petals to 5 cm, elliptic, obtuse, dark blue. *C Chile.* H5–G1.

A variable species; var. **violacea** (Brongniart) Smith & Looser (*P. violacea* (Brongniart) Mez), which has the floral bracts much shorter and narrower, is sometimes grown.

3. P. venusta Philippi. Illustration: Flora Neotropica **14**: f. 30A–C (1974).
Stem to 40 cm, scape less than 1 m. Leaves to 30 × 3 cm, greyish on both surfaces, margins with stout, hooked spines to 5–7 mm. Inflorescence with numerous, widely spreading branches, each to 10 cm. Flower-stalks *c.*7 mm. Sepals 1.5–2 cm, covered with fine, stellate hairs when young, soon hairless. Petals to 3.5 cm, narrowly elliptic, obtuse, deep pinkish purple. *Chile.* H5–G1.

4. P. gigas André. Illustration: Revue Horticole **53**: f. 74 (1881); Flora Neotropica **14**: f. 33I (1974).

Scape to 6 m or more. Leaves 1–1.3 × *c.* 4 cm, hairless above, pale-scaly beneath, margin with hooked spines to 1.1 cm. Inflorescence dense, bipinnate, with more or less erect branches which are 13–15 cm, flowerless towards the base. Flower-stalks 1.5–2 cm. Sepals *c.* 2.6 cm, persistently woolly throughout. Petals to 5 cm. *Colombia.* G1.

5. P. chilensis Molina. Illustration: Botanical Magazine, 4715 (1863); Flora Neotropica **14**: f. 59 (1974).
Stem woody, prostrate, simple or branched, to 5 m. Leaves in a dense rosette, *c.* 1 m × 5 cm, erect or stiffly spreading, soon hairless, margin with large (the lower to 1 cm), hooked spines. Inflorescence with many branches, the upper parts of these sterile and covered with reduced bracts. Flowers dense, their stalks *c.* 1.5 cm. Sepals to 3.5 cm, green. Petals narrowly elliptic, *c.* 5 cm, bright yellow or greenish yellow. *C Chile.* G1.

6. P. alpestris (Poeppig) Gay (*P. whytei* Hooker). Illustration: Botanical Magazine, 5732 (1868); Flora Neotropica **14**: f. 62A–D (1974).
Stem prostrate, branched. Leaves rather few in each rosette, to 60 × 2.5 cm, arching and recurved, green above, white beneath from a dense covering of stellate hairs, margins with hooked spines to 4 mm. Inflorescence with rather few branches, each with the upper half to two-thirds sterile and covered with bracts. Flower-stalks *c.* 1 cm. Sepals 2–2.5 cm. Petals *c.* 5 cm, elliptic, olive green. *S Chile.* H5–G1.

2. FOSTERELLA L.B. Smith
D. Philcox
Terrestrial herbs, stemless or nearly so. Leaves usually few in each rosette; sheaths distinct, blades thin, somewhat narrowed towards the base, entire. Scape well developed, slender. Inflorescence a loose panicle, hairless or finely scaly. Floral bracts inconspicuous; flower-stalks distinct, slender. Flowers small, pendent (in ours). Sepals free, shorter than the petals. Petals free, without basal scales, white. Ovary superior.

A genus of 12 or more species from C & western S America.

1. F. penduliflora (Wright) L.B. Smith (*Catopsis penduliflora* Wright; *Lindmania penduliflora* (Wright) Stapf). Illustration: Botanical Magazine, 9029 (1924); Flora Neotropica **14**: f. 73A–E (1974).
Leaves to 35 × 3.5 cm, linear–lanceolate, thin, with a soft apex, hairless above, scurfy beneath. Scape slender, hairless. Inflorescence to 20 cm, without an involucre, many-flowered. Flower-stalks 2–3 mm, slender. Flowers pendent. Sepals *c.* 3.5 mm, lanceolate, obtuse, green with white margins. Petals 8–9 mm, lanceolate–oblong, white. Stamens included in the corolla. *C Peru to NW Argentina.* G2.

3. PITCAIRNIA L'Héritier
J. Cullen
Terrestrial herbs. Leaves in many ranks, forming a dense rosette, mostly narrow, persistent or deciduous along a distinct, transverse line, rarely of 2 kinds. Scape usually present. Inflorescence a raceme or head-like spike (in ours). Flowers stalked or almost stalkless. Sepals free. Petals free, long and narrow, often with their upper parts rather irregularly arranged, with or without basal scales. Stamens about as long as petals. Ovary half-inferior to almost completely superior (in ours). Fruit a capsule.

A genus of 260 species, all but 1 from C & S America. Only 3 are widely grown, but more may be found in specialist collections.

1a. Leaves of 2 kinds, the outer reduced to spiny-margined spines, the inner linear, deciduous before flowering, along a distinct, transverse line; petals usually red **3. heterophylla**
b. Leaves all similar, persistent; petals yellow or orange 2
2a. Leaves to 1 m or more; sepals 1.5–2 cm, orange; petals 4.5–5 cm, pale yellow **1. xanthocalyx**
b. Leaves to 35 cm; sepals *c.* 2.3 cm, green; petals *c.* 6.5 cm, orange with yellow apices **2. andreana**

1. P. xanthocalyx Martius (*P. flavescens* Baker, not K. Koch). Illustration: Botanical Magazine, 6318 (1877); Flora Neotropica **14**: f. 116C–E (1974).
Stemless, scape to 1.3 m. Leaves all similar, persistent, 1 m or more, entire or margins minutely spiny. Inflorescence a raceme. Flower-stalks 1.5–2 cm. Sepals 1.5–2 cm, orange. Petals 4.5.–5 cm, pale yellow. Ovary half-inferior. *Mexico.* G2.

2. P. andreana Linden (*P. lepidota* Regel). Illustration: Illustration Horticole **20**: t. 189 (1873); Botanical Magazine, 6480 (1880); Gartenflora **22**: t. 772 (1873); Flora Neotropica **14**: f. 120I (1974).
Stemless, scape to 20 cm. Leaves all similar, persistent, to 35 × 3 cm, white beneath due to a dense covering of scales, entire. Inflorescence a raceme. Flower-stalks *c.* 1 cm. Sepals *c.* 2.3 cm, green. Petals *c.* 6.5 cm, bright orange with yellow apices. Ovary mostly superior. *Colombia.* G2.

3. P. heterophylla (Lindley) Beer. Illustration: Flora Neotropica **14**: f. 141A–F (1974).
Stemless, scape to 20 cm. Leaves of 2 kinds, the outer reduced to spiny-margined spines, the inner green, linear, to 70 × 1.3 cm, deciduous before flowering, along a distinct, transverse line, entire above this line, spiny-margined below it. Inflorescence a condensed, head-like spike. Flower-stalks at most 3 mm. Sepals *c.* 3 cm. Petals to 5.5 cm, usually red. *C & S America, as far south as Peru.* G2.

A variable species; a white-flowered variant (var. **albiflora** Standley & Smith) is known in the wild, and may be occasionally grown.

4. DYCKIA Schultes
D. Philcox
Small to large herbs, terrestrial or growing on rocks. Leaves in dense rosettes, with fleshy sheaths and blades not narrowed at the base, usually spine-toothed, rigid. Scape conspicuous. Inflorescence simple or compound, without an involucre; primary bracts inconspicuous. Floral bracts variable in length, from shorter than the flower-stalks to longer than the flowers. Flowers relatively small, yellow to red. Sepals usually less than 1 cm, overlapping, usually free, shorter than the petals. Petals usually less than 3 cm, overlapping. Stamens longer or shorter than the corolla. Ovary superior.

A genus of over 100 species from Brazil, Paraguay, Uruguay and N Argentina. They are not widely acceptable as houseplants owing, in the main, to their rigid, inflexible leaves which have vicious spines on the margins.

1a. Flowers in a compound panicle **1. frigida**
b. Flowers in a simple raceme 2
2a. Axes of the inflorescence densely and persistently scaly, scales sometimes felt-like **2. remotiflora**
b. Axes of the inflorescence hairless or sparsely scaly with the scales soon falling 3
3a. Leaf-blades very finely and regularly white-lined beneath, with scales in the deep furrows between the green nerves **3. brevifolia**
b. Leaf-blades not regularly white-lined beneath, the scales more or less

covering the nerves **2. remotiflora**

1. D. frigida Hooker. Illustration: Botanical Magazine, 6294 (1877); Flora Neotropica **14**: f. 184E–I (1974).
Leaves *c.* 1 m, sheaths *c.* 9 cm wide, dark purplish brown; blades *c.* 3.5 cm wide, narrowly triangular, sharply pointed, closely adpressed-scaly beneath, spines 2–5 mm on outer leaves. Scape stout, ascending, scaly when young. Inflorescence paniculate with branches to 30 cm, many-flowered, more or less shortly brown-felted. Flower-stalks to 3 mm. Flowers 1.1–1.8 cm long. Sepals 5–8 mm, ovate, acute. Petals yellow or pale yellow. Stamens much shorter than the petals. *Brazil.* G1.

2. D. remotiflora Otto & Dietrich. Illustration: Flora Neotropica **14**: f. 187I–M (1974).
Leaves 10–25 cm with large, ovate to circular sheaths and blades 8–12 mm wide, arching, narrowly triangular, sharply pointed, flat, dark green and without regular white lines beneath, with slender, curved spines 1–3 mm. Scape stout. Inflorescence 12–20 cm, simple, loose, densely to sparsely but persistently scaly. Flower-stalks very short, stout. Sepals 6–10 mm, ovate. Petals 1.1–2.3 cm, dark orange. Stamens not projecting from corolla. *Brazil, Uruguay & Argentina.* G1.

3. D. brevifolia Baker. Illustration: Flora Neotropica **14**: f. 192A–D (1974).
Leaves 10–20 cm, in a dense rosette with sheaths scarcely broader than the blades; blades 2.5–3.5 cm wide, lanceolate-triangular, acute, very thick, hairless above, minutely pale-scaly between the prominent nerves beneath, producing a finely and regularly white-lined appearance, with distant hooked spines to 2 mm. Scape stout. Inflorescence simple, loose or dense, many-flowered, scaly when young. Flower-stalks 2–4 mm, elongating in fruit. Flowers spreading then erect. Sepals to 8 mm, ovate, acute or obtuse. Petals *c.* 10 mm, bright yellow. Stamens not projecting from corolla. *Brazil.* G1.

5. TILLANDSIA Linnaeus
D. Philcox
Epiphytic herbs with or without stems. Leaves in rosettes, clustered or distributed along a stem, in 2 or many ranks, entire; blades strap-shaped to narrowly triangular or linear. Scape usually distinct. Inflorescence varied; usually a many-flowered spike with the flowers in 2 to many ranks, or the inflorescence reduced to a single flower. Floral bracts conspicuous to minute. Flowers mostly shortly stalked. Sepals usually symmetric, free or united to various degrees. Petals lacking basal scales, free. Ovary superior, hairless.

This is both the largest and most widespread genus of Bromeliaceae. It consists of about 400 species and covers the whole wide range of distribution throughout the Americas. Strangely though, only some 6 or 7 species are commonly found on the general market in Europe, although of recent date a few more are appearing ephemerally, attached to certain *objets d'art* as coffee table conversation pieces. Others not mentioned here are occasionally seen, having originated from the private, specialist collections in Europe. The identification of these should adequately be covered, if the need arises, by reference to Rauh's *Bromeliads*.

1a. Flowers borne all round the inflorescence axis; inflorescence unbranched 2
b. Flowers in 2 ranks on the inflorescence axis or solitary; inflorescence sometimes branched 4
2a. Plant stemless or almost so, leaves spreading in all directions; lower floral bracts gradually tapering into a narrow tail **1. stricta**
b. Plant with a distinct stem, leaves all curved to one side; lower floral bracts rounded, then abruptly narrowed into a tail 3
3a. Petals 2.8–3 cm, white **2. araujei**
b. Petals *c.* 2 cm, blue, pink or white **4. tenuifolia**
4a. Plant with a distinct stem bearing alternating bunches of leaves, forming a lichen-like festoon **3. usneoides**
b. Stem and plant not as above 5
5a. Leaf-sheaths dark, contrasting with the blades **5. tricolor**
b. Leaf-sheaths pale, of the same colour as the blades 6
6a. Floral bracts with prominent nerves; inflorescence-stalk elongate; petals dark blue, each with a small white spot at the base **6. lindenii**
b. Floral bracts without prominent nerves; inflorescence-stalk very short; petals entirely deep violet **7. cyanea**

1. T. stricta Solander. Illustration: Botanical Magazine, 1529 (1813); Edwards's Botanical Register **16**: t. 1338 (1830); Flora Neotropica **14**: f. 259A–D (1977).
Plant with a very short stem. Leaves 6–18 cm, many, in dense rosettes, covered with adpressed grey scales; blades very narrowly triangular, 4–11 mm wide at the base, attenuate. Scape erect to arching, slender; bracts overlapping, the lower leaf-like, the upper elliptic with linear blades, scaly. Inflorescence 2–7 cm long, 1.5–3.5 cm in diameter, simple, dense, with flowers borne all round the axis in several ranks. Lower floral bracts gradually tapering into narrow tails. Sepals 6–13 mm, lanceolate–ovate, hairless, united for up to 4 mm. Petals 1.5–2.5 cm, blue or purple. *Venezuela, Trinidad, Guyana, Surinam, Brazil to northern Argentina.* G2.

2. T. araujei Mez. Illustration: Flora Neotropica **14**: f. 260A–E (1977).
Plant with stem to 30 cm, simple or with few branches. Leaves 3–7 cm, in many ranks along the stem but all curved to one side; sheaths short, white, hairless at the base; blades stiff, attenuate to a sharply pointed apex, covered with fine, pale, brown-centred scales. Scape slender, ascending, hairless. Inflorescence 3–5 cm, simple, more or less loose, with 5–12 flowers borne all round the axis in several ranks. Lower floral bracts rounded, then abruptly narrowed into tails. Sepals 1.2–1.5 cm, lanceolate, acute, united for most of their length. Petals 2.8–3 cm, white, narrow. *Central E Brazil.* G2.

3. T. usneoides (Linnaeus) Linnaeus (*Renealmia usneoides* Linnaeus; *T. filiformis* invalid). Illustration: Botanical Magazine, 6309 (1877); Flora Neotropica **14**: f. 287A–C (1977).
Plant with long stems hanging in festoons up to 8 m; stems usually 1 mm or less in diameter, with the internodes 3–6 cm. Leaves 2.5–5 cm, in 2 ranks, densely grey-scaly, the sheaths to 8 mm, elliptic, inrolled; blades less than 1 mm wide, thread-like. Scape almost absent. Inflorescence reduced to a single flower. Flower almost stalkless. Sepals to 7 mm, narrowly ovate, acute, shortly united. Petals 9–11 mm, pale green or blue. *South-eastern USA to C Argentina & Chile.* G2.

4. T. tenuifolia Linnaeus. Illustration: Flora Neotropica **14**: f. 261B–E (1977).
Plant very variable with stem to 25 cm, often branching. Leaves 5–10 cm, dense and in many ranks along the stem, all curved to one side, minutely adpressed-scaly; sheaths barely distinct from the blades, which are 2–7 mm wide, narrowly triangular, channelled above. Scape erect, slender, largely concealed by the leaves. Inflorescence simple with 4–10 flowers

borne all round the axis. Floral bracts rounded, then abruptly tapering into narrow tails. Sepals *c.* 10 mm, lanceolate, acute, hairless. Petals *c.* 2 cm, blue, white or rose. *West Indies to Bolivia & Argentina.* G2.

5. T. tricolor Schlechtendal & Chamisso (*Vriesea xiphostachys* Hooker). Illustration: Flora Neotropica **14**: f. 301D–F (1977).
Plant stemless. Leaves many, in dense rosettes, equalling or shorter than the inflorescence, the outer curving; sheaths elliptic-oblong, large, dark green; blades *c.* 1 cm wide at the base, linear-triangular, long-tapering, glaucous green. Scape slender, erect. Inflorescence simple or weakly branched, flowers in 2 ranks; spikes erect or diverging, 6–18 × 1.8–2.5 cm, dense, many-flowered, compressed with flat sides. Floral bracts erect, densely overlapping, ovate, acute, *c.* 3 × 1.8 cm, much longer than the sepals, incurved, leathery, smooth, hairless. Flowers stalkless. Sepals *c.* 2 cm, leathery, Petals *c.* 7 cm, erect, violet. Stamens projecting from corolla. *C America.* G2.

6. T. lindenii Regel (*T. lindeniana* Regel). Illustration: Gartenflora **18**: t. 619 (1869); Flora Neotropica **14**: f. 268A–C (1977).
Plant stemless. Leaves to 40 cm, many, arching, curved downwards, somewhat longitudinally striped with red-purple toward the base; sheaths small, elliptic, blades 1.2–1.8 cm wide, linear-triangular, more or less hairless above, minutely scaly beneath. Scape slender, erect, elongate; bracts overlapping. Inflorescence to 20 × 5 cm, simple, lanceolate, acute, with up to 20 flowers in 2 ranks, hairless. Floral bracts overlapping, 4–4.5 cm, slightly longer than the sepals, keeled, more or less leathery, prominently nerved, green to rose. Sepals *c.* 3.5 cm, elliptic, obtuse. Petals to 4 cm, deep blue, each with a small white spot at the base. *NW Peru.* G2.

7. T. cyanea K. Koch (*T. lindenii* Morren not Regel; *T. morreniana* Regel). Illustration: Flora Neotropica **14**: f. 269C–E (1977).
Plant stemless. Leaves to 35 cm, many, more or less erect, somewhat red-striped towards the base, finely scaly; sheaths *c.* 6 cm, distinct; blades 1–1.5 cm wide, linear-triangular. Scape erect or inclined, very short and almost hidden by the leaves; bracts overlapping. Inflorescence to 16 × 7 cm, simple, elliptic, obtuse or broadly acute, with up to 20 flowers. Floral bracts elliptic, keeled, without conspicuous nerves, exceeding the sepals, leathery, rose or red. Sepals *c.* 3.5 cm, free, obtuse or broadly acute. Petals 2–2.5 cm, deep violet. *Ecuador & Peru.*

This species is frequently confused with *T. lindenii*, but is easily distinguished by the apparent absence of a scape, giving the impression that the inflorescence is completely stalkless, while the inflorescence of *T. lindenii* is clearly stalked.

6. VRIESEA Lindley

D. Philcox

Stemless, perennial epiphytes. Leaves in rosettes, many-ranked, entire; blades usually tongue-shaped and sparsely scaly. Scape usually conspicuous. Inflorescence usually of spikes of flowers in 2 ranks, rarely of 1 or more spikes of many-ranked flowers. Floral bracts usually conspicuous. Flowers mostly shortly stalked. Sepals free or almost so, symmetric or almost so. Petals free or united into a tube which is shorter than the sepals, each petal with 2 scales at the base. Ovary nearly or completely superior.

A genus of some 250 species from C & S America and the West Indies of which only 6 are commonly cultivated in Europe. Many other species, however, have been, or are grown in specialist collections, and many hybrids have been produced. The identification of the species in cultivation is therefore often very difficult.

1a. Inflorescence compound 2
b. Inflorescence simple 4
2a. Panicle to 40 × 4.5 cm; leaves to 5 cm wide **1. saundersii**
b. Panicle much larger, to about 1 m; leaves 6–10 cm wide 3
3a. Leaves wholly pale green, checkered; petals 4–5 cm; filaments dilated towards the apex **2. gigantea**
b. Leaves green with irregular, dark green or purple, transverse bands; petals *c.* 3.5 cm; filaments not dilated **3. hieroglyphica**
4a. Flowers all erect and overlapping; bracts keeled and incurved, longer than the sepals **4. splendens**
b. Flowers (at least the lowest) not overlapping, spreading and sometimes with evident space between them; bracts not keeled, not incurved, shorter than the sepals 5
5a. Stamens longer than the petals; petals bright yellow, sometimes green-tipped **5. psittacina**
b. Stamens shorter than the petals; petals greenish white **6. fenestralis**

1. V. saundersii (Carrière) Mez (*Encholirion saundersii* Carrière). Illustration: Flora Neotropica **14**: f. 390E–F (1977).
Leaves 20–30 cm, adpressed-scaly; sheaths broadly elliptic, covered with brown scales; blades 3.5–5 cm wide, broadly acute, grey-scaly, densely and finely brown-spotted beneath, the apex recurved. Scape erect, stout, hairless. Panicle to 14 × 4.5 cm, densely bipinnate, loose, with few flowers. Flowers yellow, stalks short, stout. Sepals *c.* 2 cm, obtuse. Petals *c.* 3.5 cm. Stamens projecting from corolla. *Brazil.* G2.

2. V. gigantea Gaudichaud (*V. tessellata* (Linden) Morren). Illustration: Illustration Horticole **14**: t. 516 (1867); Flora Neotropica **14**: f. 354D (1977).
Leaves 60–100 cm, many, in a large, broadly funnel-like rosette, minutely scaly on both sides; sheaths large, broadly elliptic, blades 6–9 cm wide, broadly acute, strap-shaped, wholly pale green, checkered. Scape erect, stout. Panicle loosely bipinnate, *c.* 100 × 40 cm, almost hairless, with many flowers in 2 ranks. Sepals to 3.7 cm, elliptic. Petals 4–5 cm long, obtuse, greenish yellow. Filaments dilated towards the apex. *S Brazil.* G2.

3. V. hieroglyphica (Carrière) Morren (*Massangea hieroglyphica* Carrière). Illustration: Illustration Horticole **31**: t. 514 (1884), **42**: t.318 (1895); Flora Neotropica **14**: f. 372H–I (1977).
Leaves 50–80 cm, many, in a funnel-like rosette, minutely scaly, the inner erect, the outer recurving; sheaths slightly wider than the blades, dark brown beneath, the blades 7–10 cm wide, strap-shaped, marked with dark green or purple transverse bands. Scape erect, stout. Panicle to 80 cm, bipinnate or further branched, with many flowers. Sepals *c.* 2.5 cm, narrowly elliptic. Petals *c.* 3.5 cm, yellowish. *Brazil.* G2.

4. V. splendens (Brongniart) Lemaire (*Tillandsia splendens* Brongniart). Illustration: Flore des Serres **2**: t. 4 (1846); Flora Neotropica **14**: f. 396A–D (1977).
Leaves 40–80 cm, in dense funnel-like rosettes, arching; sheaths indistinct, brown-scaly; blades 4–6 cm wide, strap-shaped, broadly acute or rounded, with broad, dark, irregular cross-bands, sparsely scaly. Scape erect; bracts clasping the scape, densely overlapping. Raceme to 55 × 6 cm, dense, lanceolate, strongly compressed, with many flowers. Floral bracts densely overlapping, bright red, exceeding the sepals, hairless. Flower-stalks *c.* 4 mm. Sepals *c.* 2.5 cm, wholly yellow, or red at the apex. Petals to

8 cm, yellow. Stamens projecting from corolla. *Venezuela to the Guianas*. G2.

There are several other varieties of this species, but only one is known so far to be cultivated, other than the one described above. This is var. **formosa** Witte (*V. longibracteata* (Baker) Mez) in which the leaf-blades are wholly green. *Venezuela, Guyana, Trinidad & Tobago.*

5. V. psittacina (Hooker) Lindley. Illustration: Edwards's Botanical Register **29**: t. 10 (1843); Flora Neotropica **14**: f. 400B–E (1977).
Leaves 30–50 cm, in a broad funnel-like rosette; sheaths 7–9 cm, elliptic, green or pale brown, scaly, blades 2.2–2.8 cm wide, strap-shaped, somewhat narrowed toward the base, broadly acute, wholly green. Scape erect, slender; bracts overlapping, the lower green, the upper yellow and green. Raceme 17–23 cm, the flowers spreading and not overlapping. Floral bracts slightly shorter than the sepals, red with yellow apices. Flower-stalks to 5 mm, stout. Sepals 3.5–4 cm. Petals *c.* 6 cm, bright yellow or with green apices. Stamens projecting from corolla. *Brazil.* G2.

6. V. fenestralis Linden & André. Illustration: Flora Neotropica **14**: f. 360F–G (1977).
Leaves 35–50 cm, many, in a broad rosette, arching; sheaths large, 10–12 cm wide, marked with small red-brown spots; blades 6.5–8 cm wide, broadly rounded, pale green with narrow dark green lines, hairless above, minutely scaly beneath. Scape erect, green. Raceme to 50 cm, loose, with many spreading, non-overlapping flowers. Floral bracts much shorter than the sepals. Flower-stalks to 1 cm. Sepals *c.* 6.5 cm, green, spotted. Petals *c.* 4.5 cm, longer than the stamens, greenish white. *Brazil.* G2.

7. GUZMANIA Ruiz & Pavon
D. Philcox
Epiphytic herbs without (or rarely with) stems. Leaves entire, in many ranks. Scape usually conspicuous. Inflorescence bipinnate or simple, branches in many ranks, with many flowers. Sepals free to united for most of their length, symmetric or almost so. Petals with their claws united into a tube distinct from the blades, without scales near the base. Ovary superior. Fruit a capsule.

A genus of over 125 species mostly from C and tropical northwest S America.

1a. Sepals longer than the petals; leaves irregularly and transversely marked **3. musaica**
b. Sepals shorter than the petals; leaves not as above 2
2a. Inflorescence branched; leaf-blades gradually tapered **1. zahnii**
b. Inflorescence unbranched; leaf-blades broadly acute or rounded 3
3a. Inflorescence an elongate spike, not surrounded by bracts **2. monostachia**
b. Inflorescence corymbose, surrounded by large, brightly coloured bracts, forming a cup-like structure **4. lingulata**

1. G. zahnii (Hooker) Mez (*Caraguata zahnii* Hooker). Illustration: Flora Neotropica **14**: f. 436F–J (1977); Rauh, Bromeliads, t. 57 (1979).
Leaves to 60 cm, 20–30 in a rosette; sheaths obscurely brown-scaly, yellowish with dark red longitudinal stripes; blades *c.* 2.7 cm wide, purple-red throughout or green near the apex, marked with red-brown longitudinal stripes. Scape erect, hairless, red, with bright scarlet bracts. Panicle to 25 × 10 cm, bipinnate, pyramid-shaped, hairless, bright yellow throughout. Flowers almost stalkless. Sepals *c.* 1.8 cm, united for 2.5 mm. Petals *c.* 3 cm, united for 7 mm, elliptic, obtuse. *Costa Rica & Panama.* G2.

2. G. monostachia (Linnaeus) Mez (*Renealmia monostachia* Linnaeus). Illustration: Flora Neotropica **14**: f. 444C–E (1977); Rauh, Bromeliads, t. 9 (1979).
Leaves to 40 cm, in dense rosettes, scaly, soon hairless; sheaths brownish; blades *c.* 2.5 cm wide, acute, yellow-green. Scape erect, hairless, with pale green bracts. Spike 8–15 × 2–3 cm, many-flowered, sterile toward the apex. Flowers 2.3–3 cm, white, erect. Sepals *c.* 1.8 cm, leathery, united for a quarter of their length. Petals united for most of their length. *USA (Florida), West Indies & Nicaragua to N Brazil & Peru.* G2.

3. G. musaica (Linden & André) Mez (*Tillandsia musaica* Linden & André). Illustration: Flora Neotropica **14**: f. 449G–J (1977); Rauh, Bromeliads, t. 56 (1979).
Leaves to 70 cm, in a spreading rosette; sheaths purplish brown toward the base; blades 4–8 cm wide, broadly acute or rounded, marked with irregular, transverse lines, and with a short spine at the apex. Scape erect; bracts bright pinkish red. Spike more or less spherical, 12–25-flowered, hairless. Flowers more or less without stalks. Sepals 2.5–4.5 cm, united, leathery, yellowish. Petals shorter than the sepals and not projecting at flowering. *Panama & Colombia.* G2.

4. G. lingulata (Linnaeus) Mez (*Tillandsia lingulata* Linnaeus; *Caraguata latifolia* Beer). Illustration: Rauh, Bromeliads, t. 5, 6 (1979).
Leaves 30–45 cm, in a dense rosette, sheaths conspicuous, more or less purplish brown at the base, occasionally marked beneath with fine, purple, longitudinal lines; blades to 4 cm wide, acute. Scape erect, stout; upper bracts usually reddish and forming an involucre. Raceme corymbose, to 7 cm wide, 10–50-flowered. Flowers erect, *c.* 4.5 cm; stalks short and stout. Sepals to 1.7 cm, free, hairless. Petals linear, hooded, white. *Belize, West Indies to Bolivia & Brazil.* G2.

Several varieties are found in cultivation in Europe, and generally are more commonly met with than the wild variety described above. The largest and most brilliant of these, var. **cardinalis** (André) Mez (*G. cardinalis* (André) Mez; *Caraguata cardinalis* (André) André) is very similar to the above but is distinguished by its showy, spreading, bright scarlet bracts. Var. **splendens** (Planchon) Mez, which until recently was commonly known in cultivation as *G. peacockii* (Morren) Mez, is a red-leafed plant in which the lower leaves are bright purple-red and the upper reddish green; the inflorescence terminates in a funnel-shaped, purplish red spike, with the small bracts in the centre yellowish and white-tipped. Var. **minor** (Mez) L.B. Smith (*G. minor* Mez) is a much smaller and more delicate plant, having thin, yellowish leaves *c.* 30 × 1.5 cm; the inflorescence is at the end of a short, stout scape, appearing as a raised cup of yellow or red bracts and whitish yellow flowers.

8. NEOREGELIA L.B. Smith
D. Philcox
Epiphytes with or without stems. Leaves dense, forming flat or tubular rosettes, usually broadly linear, flexible, spiny-margined, sheaths usually large and distinct, scaly or not, often variously spotted or banded, the inner often reddish or purplish, the apex rounded and with a small point. Inflorescence sunk centrally in the rosette, simple to densely umbellate, racemose or corymbose; bracts entire or toothed. Flowers stalked. Sepals united at least at the base. Petals violet, blue or white, rarely red, united at least at the base, without basal scales, blades spreading, acute to acuminate. Stamens

attached to the petals, not projecting. Ovary inferior. Fruit a berry.

A genus, formerly known as *Aregelia* Mez, of more than 70 species largely natives of eastern Brazil, with a few species found in Amazonia and eastern Colombia and Peru. Very similar to *Nidularium* (p. 18); there is some confusion in the naming of cultivated material.

1a. Inner leaves of the rosette bright red at least at the base, unlike the outer 2
b. Inner leaves of the rosette like the outer, not bright red 5
2a. Leaf-blades sparsely and inconspicuously scaly beneath; sepals rounded or acute **2. carolinae**
b. Leaf-blades covered, at least beneath, with adpressed, grey scales; sepals acuminate 3
3a. Floral bracts minutely toothed; leaf-blades 1.5–2 cm wide **1. pineliana**
b. Floral bracts entire; leaf-blades 3–5 cm wide 4
4a. Leaves each with a bright red spot at the tip; petals wholly blue **8. spectabilis**
b. Leaves without red spots at the tips; petals blue above, white below **3. princeps**
5a. Leaf-sheaths forming a cylindric or ellipsoid tank distinctly higher than wide and often more or less constricted at the mouth 6
b. Leaf-sheaths forming a saucer- or funnel-shaped tank about as wide as high, or at least flaring above 7
6a. Leaves with spots or bands darker than the rest of the blade **4. ampullacea**
b. Leaves with spots or bands paler than the rest of the blade **6. chlorosticta**
7a. Sepals 1.2–1.6 cm; leaf-blades 2–3 cm wide **5. tristis***
b. Sepals 1.8–3 cm; leaf-blades 2–10 cm wide 8
8a. Leaves variegated with either light or dark spots **7. marmorata**
b. Leaves of one colour or banded, but not spotted 9
9a. Each leaf with a red spot at the apex **8. spectabilis**
b. Each leaf with a purple spot at the apex **9. concentrica**

1. N. pineliana (Lemaire) L.B. Smith (*Nidularium pinelianum* Lemaire; *Aregelia morreniana* (Antoine) Mez; *Neoregelia morreniana* (Antoine) L.B. Smith; *A. pineliana* (Lemaire) Mez). Illustration: Flora Neotropica **14**: f. 498A–D (1979).
Plant with a short stem or stemless, spreading by short runners. Leaves many, to 50 cm × 15 mm, distributed along the stem; sheaths purplish, densely scaly; blades 1.5–2 cm wide, linear, rounded and with a small point, green with greyish scales on both surfaces, inner leaves red. Inflorescence many-flowered, sunk between the sheaths of the inner leaves; floral bracts green, minutely toothed, hairless. Flowers 4–6 cm. Sepals 1.5–2 cm, acuminate, asymmetric, green, joined for 3–4 mm at the base. Petals more or less erect, to 4 cm, joined to above the middle, blades blue, tube white. *Known only from cultivation.* G2.

A variant in which the inflorescence is replaced by a tuft of enlarged, coloured, sterile bracts, forma **phyllanthidea** (Baker) L.B. Smith, is also known from cultivation.

2. N. carolinae (Beer) L.B. Smith (*Bromelia carolinae* Beer; *Nidularium carolinae* Baker; *Aregelia carolinae* (Beer) Mez).
Plant stemless. Leaves about 20, to 40–60 cm, forming a broad rosette; sheaths large, sparsely covered with minute, dark brown scales; blades 2.5–3.5 cm wide, somewhat densely toothed, glossy green, hairless above, obscurely scaly beneath, the outer leaves wholly green, the inner brilliant red at the base. Inflorescence many-flowered, sunk in the centre of the rosette; bracts entire or a few minutely toothed towards the apex. Flowers to 4 cm. Sepals 1.6–2 cm, asymmetric, each rounded and with a small point, united at the base for *c.* 5 mm. Petals 2.4–2.8 cm, joined to above the middle, dark blue. *Brazil.* G2.

A variant in which the leaf-blades are longitudinally striped white, green and reddish pink is known in cultivation as forma **tricolor** (Foster) L.B. Smith.

3. N. princeps (Baker) L.B. Smith (*Aregelia princeps* Baker; *Karatas meyendorfii* Antoine, in part; *A. marechalii* Mez, in part). Illustration: Flora Neotropica **14**: f. 499A–F (1979).
Stemless plant. Leaves 15–20, to 50 cm, forming a dense, spreading rosette; sheaths large, the outer circular in outline, green, densely scaly; blades 3–4.5 cm wide, with very short spines, the outer blades green, the inner bright red and smaller. Inflorescence sunk in the rosette, many-flowered; floral bracts entire. Flowers to 4 cm. Sepals *c.* 2.4 cm, asymmetric, acuminate, united for about 1.5 mm at the base, hairless, red. Petals *c.* 3.5 cm, linear, acuminate, united for much of their length, dark blue towards the apex, white below. *Brazil.* G2.

4. N. ampullacea (Morren) L.B. Smith (*Nidularium ampullaceum* Morren; *Aregelia ampullacea* (Morren) Mez). Illustration: La Belgique Horticole **35**: t. 14 (1885); Flora Neotropica **14**: f. 500 (1979).
Plant stemless, spreading by runners. Leaves few, 15–20 cm, in a dense rosette; sheaths large, forming an ellipsoid tank constricted at the mouth, densely scaly with brown to purple scales; blades to 1.6 cm wide, linear, rounded, spreading, green, hairless and with dark, narrow bands above, finely toothed. Inflorescence sunk deeply in the rosette, few-flowered. Flowers to 2.5 cm. Sepals *c.* 1.5 cm, green with white margins. Petals *c.* 2 cm, acute, united for three-quarters of their length, blue above, whitish below. *Brazil.* G2.

5. N. tristis (Beer) L.B. Smith (*Bromelia tristis* Beer; *Nidularium cyaneum* Linden & André; *Aregelia tristis* (Beer) Mez).
Plant stemless. Leaves 20–60 cm, in a dense, slender, funnel-like rosette; sheaths large, dark with pale spots, forming a saucer- or funnel-shaped tank; blades 2–3 cm wide, finely spiny with spines to 0.5 mm, green, hairless and with purple-brown spots above, sparsely scaly and broadly dark-banded beneath. Inflorescence sunk deeply in the rosette, many-flowered. Flowers to 3 cm. Sepals asymmetric, to 1.6 cm, united at the base for 3 mm. Petals to 2 cm, united at the base, violet-blue. *Brazil.* G2.

***N. fosteriana** L.B. Smith. Similar, but with coppery leaves tipped with red above, purplish beneath and greyish scaly on both surfaces. *Brazil.* G2.

6. N. chlorosticta (Baker) L.B. Smith (*Karatas chlorosticta* Baker; *Neoregelia sarmentosa* (Regel) L.B. Smith var. *chlorosticta* (Baker) L.B. Smith). Illustration: Rauh, Bromeliads, t. 106 (1979).
Plant stemless, forming a loose rosette. Leaves 20–30 cm, the inner purplish red with large, pale green spots; sheaths almost equalling blades, ovate to elliptic, sparsely brown-scaly, forming a cylindric or ellipsoid tank which is more or less constricted at the mouth; blades *c.* 2.5 cm wide, linear, rounded and with a small point, minutely toothed, scaly especially beneath. Inflorescence many-flowered, sunk in the rosette. Sepals 1.2–1.8 cm, united at the base for up to 7 mm, asymmetric, shortly acuminate. Petals to 1.8 cm, blades elliptic, acute, blue-violet. *Brazil.* G2.

7. N. marmorata (Baker) L.B. Smith (*Karatas marmorata* Baker; *Aregelia marmorata* (Baker) Mez). Illustration: Rauh, Bromeliads, t. 105 (1979).
Plant stemless. Leaves 20–60 cm, about 15 in a dense, broad, funnel-like rosette; sheaths large, broadly elliptic, dark purple with pale green spots, more or less densely covered with dark brown scales, forming a saucer- or funnel-shaped tank; blades broadly rounded, slightly notched and with a small point, to 8 cm wide, toothed, green or spotted with purple, obscurely scaly. Inflorescence sunk deeply in the rosette, usually many-flowered. Sepals to 2.4 cm, green, more or less hairless, united at the extreme base, asymmetric. Petals 2.4–2.7 cm, acute, the lobes spreading with reflexed tips, pale violet to white. *Brazil.* G2.

8. N. spectabilis (Moore) L.B. Smith (*Nidularium spectabile* Moore; *Aregelia spectabilis* (Moore) Mez). Illustration: Botanical Magazine, 6024 (1873); Rauh, Bromeliads, t. 107 (1979).
Plant stemless. Leaves many, 40–45 cm, forming a broad, funnel-shaped rosette, with grey scales and pale-banded beneath; sheaths ovate or broadly elliptic; blades 4–5 cm wide, spreading, loosely toothed to often almost entire, outer leaves green above, lilac-purple with light cross-banding beneath, with a bright red spot at the tip, inner leaves red, or whitish towards the base. Inflorescence many-flowered, deeply sunk in the rosette. Flowers 4–4.5 cm. Sepals 1.8–2.4 cm, very asymmetric, more or less elliptic with a large semi-circular wing and long, linear, hooked blade, shortly united at the base. Petals to 3 cm, acuminate, blades spreading, blue. *Brazil.* G2.

9. N. concentrica (Vellozo) L.B. Smith (*Tillandsia concentrica* Vellozo; *Nidularium acanthocrater* Morren; *Nidularium laurentii* Regel; *Aregelia concentrica* (Vellozo) Mez). Illustration: La Belgique Horticole **34**: t. 9 (1884).
Plant stemless, usually large. Leaves 20–40 cm, up to 30 in a dense, broad, funnel-like rosette; sheaths large, broadly elliptic, covered with pale brown, adpressed scales; blades 5–10 cm wide, broadly rounded and with a small point, wholly green except for a purple spot at the apex, or with dark purple spots above, or the inner leaves at times wholly purple, pale-scaly especially beneath, densely spiny with stout, black spines to 4 mm long. Inflorescence many-flowered, sunk in the centre of the rosette. Flowers 4–5 cm. Sepals 2.2–3 cm, slightly asymmetric, free, hairless, green, hardened at the tip. Petals to 4 cm, narrow, acuminate, united to above the middle, white or pale blue. *Brazil.* G2.

9. CRYPTANTHUS Otto & Dietrich

D. Philcox

Small epiphytes with or without stems. Leaves in rosettes or many-ranked along the stem; sheaths inconspicuous; blades narrowly triangular or somewhat contracted into a more or less stalk-like base, flexible, spiny-margined, often variegated. Inflorescence spherical, either stalkless in the centre of the rosette or at the apex of the stem, but not sunk below the leaf-sheaths, compound. Sepals united. Petals white or yellow, united at the base. Stamens projecting from the corolla. Ovary completely inferior.

A genus of 20 species from Brazil, grown mainly for their attractive rosettes.

1a. Leaf-blades without spots or lines **1. acaulis**
b. Leaf-blades with spots or lines 2
2a. Leaf-markings consisting of irregular, dark, transverse bands **2. zonatus**
b. Leaf-markings consisting of pale longitudinal stripes **3. bivittatus**

1. C. acaulis (Lindley) Beer. Illustration: Flora Neotropica **14**: f. 518A (1979); Rauh, Bromeliads, t. 59 (1979).
Plant rarely stemless. Leaves 10–20 cm, narrowed between the sheath and the blade but not really stalked; sheaths short, blades 2–3 cm wide, narrowly lanceolate, wavy-margined, minutely toothed, covered beneath by a membrane of fused scales, scaly above, green. Inflorescence few-flowered. Flowers to 4 cm. Sepals *c.* 1.5 cm, united for more than half their length. Petals shortly united, spreading, white. *Brazil.* G2

The plants marketed under this name usually belong to 3 distinct varieties. In var. **ruber** Beer the leaves are tinged with red, while in var. **argenteus** Beer, the upper surface of the leaves is green and hairless; var. **acaulis** is described above.

2. C. zonatus (Visiani) Beer. Illustration: Flora Neotropica **14**: f. 518B–E (1979).
Plant stemless. Leaves to 20 cm, in a spreading rosette, narrowed between the sheath and the blade but not stalked; sheaths very short, toothed towards the apex; blades 2–4.5 cm wide, linear-lanceolate, wavy-margined, densely and minutely spiny with spines to 1 mm, dark-banded above, appearing white beneath. Inflorescence few-flowered. Flowers to 3.2 cm. Sepals to 1.9 cm, united for about three-quarters of their length. Petals united for more than half their length, white, slightly exceeding the stamens. *Brazil.* G2.

Two major variants of this species (other than the wild one described above) are known to exist in cultivation, although their availability on the European market is uncertain. They are forma **viridis** Mez, in which the leaves lack scales on the underside and appear green, and forma **fuscus** (Visiani) Mez, in which the leaf-blades are strongly tinged with red.

3. C. bivittatus (Hooker) Regel.
Plant stemless. Leaves 18–25 cm, the outer slightly narrowed between sheath and blade, sheaths short, blades to 4 cm wide, minutely toothed, wavy-margined, dark green with 2 paler, longitudinal stripes above, and with a membrane of pale brown scales beneath. Inflorescence with few flowers. Flowers *c.* 2.6 cm. Sepals united for more than half their length. Petals *c.* 2 cm, white. *Known only from cultivation, almost certainly from Brazil.* G2.

10. NIDULARIUM Lemaire

D. Philcox

Stemless perennial epiphytes. Leaves flexible, equalling to much exceeding the inflorescence; sheaths mostly forming funnel-like tanks; blades toothed or minutely toothed, spreading. Scape shorter than, or exceeding the leaf-sheaths, mostly covered by bracts. Inflorescence compound with flowers in flattened bundles and covered by large bracts, axis usually very short. Floral bracts well developed but not visible. Flowers stalkless. Sepals more or less united at the base, symmetric or strongly asymmetric, with or without hardened points. Petals white to dark blue (in ours), without basal scales, united for half their length or more to form a tubular base, blades spreading or erect, rounded and somewhat hooded. Stamens not projecting from corolla. Ovary inferior.

A genus of more than 20 species, all natives of eastern Brazil. Very similar to *Neoregelia* (p. 16), but unlike that genus, there has been little confusion in the naming of cultivated material.

1a. Petal-blades spreading, acute; inflorescence woolly, reddish brown; sepals 1–1.4 cm, united for 2–3 mm 2
b. Petal-blades erect, obtuse; inflorescence mostly adpressed-scaly

to smooth; sepals generally larger and united for more than 3 mm 3

2a. Floral bracts minutely toothed; petals shortly united at the base; outer bracts of the inflorescence small and not forming an obvious involucre **1. burchellii**

b. Floral bracts entire; petals united for half their length; outer bracts of the inflorescence forming an evident involucre **2. microps**

3a. Scape without bracts below the inflorescence for most of its length **3. seidelii**

b. Scape, when visible, completely covered by bracts 4

4a. Leaf spines to 4 mm **4. fulgens**

b. Leaf spines to *c.* 1 mm **5. innocentii**

1. N. burchellii (Baker) Mez (*Aechmea burchellii* Baker; *Cryptanthus emergens* Lindman; *Aregelia burchellii* (Baker) Mez). Illustration: Flora Neotropica **14**: f. 523A–E (1979).
Stemless plant with long, slender runners. Leaves 20–50 cm, few, in a more or less cylindric rosette, short and broad to long and narrow; sheaths entire or toothed toward the apex, scaly with brown, more or less adpressed scales; blades 2–5 cm wide, somewhat narrowed towards the base, acute with a small point, hairless to obscurely scaly, green, tinged with red beneath, minutely toothed. Scape short, about equalling the leaf-sheaths, reddish brown-scurfy, upper bracts ovate, acute, shorter than the inflorescence, not forming an obvious involucre. Inflorescence spherical, densely reddish brown-woolly. Floral bracts minutely toothed. Sepals *c.* 10 mm, acute, symmetric or almost so, united at the base for 2 mm. Petals *c.* 1.8 cm, united for 1–2 mm, white; blades spreading, flat, acute. *Brazil.* G2.

2. N. microps Mez (*Aregelia microps* (Mez) Mez). Illustration: Flora Neotropica **14**: f. 523F–G (1979).
Like *N. burchellii* but leaves 30–60 cm, more or less erect in a tubular rosette, outer bracts forming a distinct involucre, floral bracts entire, sepals 1–1.4 cm, united at the base for *c.* 3 mm, petals 2–2.5 cm, united for about half their length. *Brazil.* G2.

3. N. seidelii L.B. Smith & Reitz. Illustration: Flora Neotropica **14**: f. 525A–C (1979).
Plant stemless. Leaves 50–80 cm, many, in a broadly funnel-shaped rosette, much longer than the inflorescence; sheaths entire, densely covered with brown-centred, grey scales; blades to 4 cm wide, broadly acute with a fine point, narrowed towards the base, obscurely scaly. Scape erect, finely reddish brown-woolly, but becoming almost hairless, bearing 2 bracts. Inflorescence more or less cylindric, 11–15 × up to 8 cm. Flowers stalkless. Sepals *c.* 1.4 cm, slightly asymmetric, united at the base for 2.5 mm, spathulate. Petals united at the extreme base, blades 10 mm, obtuse, white. *Brazil.* G2.

4. N. fulgens Lemaire (*Karatas fulgens* (Lemaire) Antoine).
Plant stemless. Leaves 30–40 cm, many, in a broad, spreading rosette; sheaths much wider than the blades, entire, densely scaly with brown, adpressed scales; blades slightly narrowed toward the base, broadly acute, acuminate and ending in a short spine, 3–6 cm wide, sparsely spiny with spines to 4 mm, hairless above, obscurely spotted-scaly beneath, pale green. Scape very short. Inflorescence sunk in the centre of the rosette, many-flowered. Flowers stalkless. Sepals 2–2.4 cm, acute, shortly united, red. Petals *c.* 5 cm, united for about half their length or more, the tube white, the blades erect, dark blue with white margins, broadly rounded. *Brazil.* G2.

5. N. innocentii Lemaire (*Karatas innocentii* (Lemaire) Antoine). Illustration: Illustration Horticole **9**: t. 329 (1862); Flora Neotropica **14**: f. 526A–D (1979).
Plant stemless, propagating by runners. Leaves 20–60 cm, many, in a dense, spreading rosette, very variable in colour; sheaths entire or minutely toothed toward the apex, pale green or purplish, with small brown scales; blades 4–5.5 cm wide, acute to broadly rounded, hairless, shiny, with short spines. Inflorescence sunk in the centre of the rosette. Flowers to 6 cm. Sepals slightly asymmetric, 2.2–3 cm, acute, united for 3–9 mm, white or sometimes reddish. Petals *c.* 5 cm, united at the base, white, green at the base. *Brazil.* G2.

Several varieties are known in cultivation, all of which are very striking plants. Var. **innocentii** is clearly distinguished by the leaves being dark red beneath or on both surfaces, while the young bracts in the centre of the rosette are red, with or without a green apex – in all other varieties the leaf-blades are green. In var. **wittmackianum** (Harms) L.B. Smith the young bracts are wholly or mostly reddish purple while var. **striatum** Wittmack is similar to the last but has the leaf-blades with longitudinal white lines. Very close to var. *striatum* is var. **lineatum** (Mez) L.B. Smith which differs in the colour of the young bracts, which are green with red near the apex. This bract colouring also applies to var. **paxianum** (Mez) L.B. Smith, in which the leaves are green, each with a broad, white longitudinal line.

11. ACANTHOSTACHYS Klotzsch
D. Philcox
Herbs, usually epiphytic, initially erect but often eventually pendent, spreading by short, stout, basal runners. Leaves few, to 1 m, not clearly contracted between the small sheath and the blade, covered by a membrane of fused scales, densely so beneath, less densely above; sheaths dark purplish brown, entire; blades linear, to 1.2 cm wide at base, much narrower above, strongly channelled with margins inwardly directed, olive green to reddish brown, sparsely spiny with spines *c.* 1.5 mm. Scape with pale scales, slender, with 2 bracts at the apex. Inflorescence appearing lateral, simple, dense, ovoid or cylindric, cone-like, 3–7 × 2–3.5 cm, with a small tuft of reduced, sterile bracts at the apex. Floral bracts broadly ovate with a spreading, triangular-acuminate apex, toothed or entire, bright orange to red, at first with pale scales, soon hairless. Flowers stalkless, 2.2–2.5 cm. Sepals free, nearly symmetric, 8–11 mm, acute, bright yellow. Petals *c.* 1.6 cm, erect, free, yellow, with 2 scales above the base. Stamens not projecting from the corolla. Ovary with just the upper part superior, the rest inferior.

A genus of 1 species, very popular for use in hanging baskets because of its pendent habit and long-lasting inflorescence.

1. A. strobilacea (Schultes) Klotzsch (*Hohenbergia strobilacea* Schultes). Illustration: Flora Neotropica **14**: f. 560A–G (1979).
Brazil, Paraguay & NE Argentina. G2.

12. FASCICULARIA Mez
D. Philcox
Terrestrial, stemless or shortly stemmed epiphytes. Leaves forming very dense rosettes, but without a central cup; blades linear, rigid, strongly spiny-margined. Inflorescence simple, dense, sunk in the centre of the rosette. Flowers stalkless or shortly stalked. Sepals free, keeled, scaly, especially toward the apex. Petals free, obtuse, fleshy, blue, with 2 small scales near the base. Stamens not projecting from the corolla. Ovary completely inferior.

A genus of 5 species restricted in origin to Chile.

1a. Bracts all shorter than the flowers; flowers 4 cm or more **1. pitcairniifolia**
b. Outer bracts clearly longer than the flowers; flowers 3.3–4 cm **2. bicolor**

1. F. pitcairniifolia (Verlot) Mez (*Hechtia pitcairniifolia* Verlot; *Rhodostachys pitcairniifolia* (Verlot) Baker). Illustration: Botanical Magazine, 8087 (1906); Flora Neotropica **14**: f. 571A–F (1979).
Plant stemless. Leaves to 1 m, spreading; sheaths conspicuous, triangular-ovate, brownish beneath, white above, covered on both sides with a white membrane, minutely toothed; blades to 1.8 cm wide, linear, leathery, glaucous-green, more or less hairless at maturity above, sparsely white-scaly beneath, margin with spines 1.5 mm, those at the base spreading, the others forwardly directed. Inflorescence many-flowered, surrounded by the bright red inner leaves. Floral bracts shorter than the flowers. Sepals 1.7–2.2 cm, rounded, often slightly notched and incurved at the apex, with narrow translucent margins, scaly towards the apex. Petals 2.2–2.4 cm, erect with recurved, spreading apices. *Chile*. H5–G1.

2. F. bicolor (Ruiz & Pavon) Mez (*Bromelia bicolor* Ruiz & Pavon; *Rhodostachys bicolor* (Ruiz & Pavon) Baker). Illustration: Rauh, Bromeliads, t. 92 (1979).
Plant stemless or with a short, stout, erect stem, the rosette spreading to over 1 m in diameter when flowering. Leaves to 50 cm or more, very numerous; sheaths elongate, mostly covered with coarse, reddish brown scales; blades *c.* 1.5 cm wide, linear, attenuate, the apex with a strong spine, hairless above, brown-scaly beneath, glaucous-green becoming red, spiny with the basal spines recurved, the rest spreading or forwardly directed. Inflorescence 20–40-flowered, with the outer bracts forming an involucre, bracts longer than the flowers, ivory white. Sepals to 1.2 cm, obtuse and irregularly twisted at the apex, the margins more or less translucent, generally scaly but scales deciduous towards the base. Petals *c.* 2 cm, with slightly recurved or spreading apices. *Chile*. H5–G1.

13. CANISTRUM Morren
D. Philcox

Perennial, usually epiphytic herbs. Leaves in rosettes, variously spiny-toothed. Scape evident. Inflorescence compound, densely corymbose, bracts of involucre covering all but the petal apices. Flowers shortly stalked or almost stalkless. Sepals free or nearly so. Petals free, with appendages near the base. Stamens not projecting from corolla. Ovary completely inferior.

A small genus of some 7 species from Brazil.

1. C. lindenii (Regel) Mez (*Nidularium lindenii* Regel).
Leaves in a broad funnel-like rosette; sheaths large, entire, covered with adpressed, dark brown scales; blades to 10 cm wide, strap-shaped, broadly acute or rounded, green with spots of darker green, with broad spines to 3 mm. Scape stout, covered with dark brown wool. Inflorescence 7–8 cm in diameter without the bracts, with about 100 flowers, bracts yellowish white to almost white, faintly green at the apex. Flowers 3–3.5 cm, shortly stalked. Sepals 1.2–1.8 cm, free, whitish. Petals free, slightly longer than the sepals, white or pale green toward the apex, with 2 fringed scales at the base. *Brazil*. G2.

A very variable plant. In cultivation, the most common variety appears to be var. **roseum** (Morren) L.B. Smith (*C. roseum* Morren), which differs from the variety above in having a smaller inflorescence of some 50–90 flowers, surrounded by primary and outer bracts of rose to bright red. This variety is illustrated in Rauh, Bromeliads, t. 94 (1979).

14. AECHMEA Ruiz & Pavon
D. Philcox

Epiphytes, usually stemless, of medium size, frequently spreading by short, basal runners. Leaves usually many, either in dense rosettes or grouped together with the sheaths often forming a tank, usually spiny-margined. Scape usually well developed. Inflorescence branched or simple, flowers stalkless or stalked, arranged in 2 opposite rows or in many rows. Sepals free or united, usually asymmetric, each sharply pointed or with an apical spine. Petals free, usually with obvious basal scales. Stamens shorter than the petals. Ovary inferior.

A genus of more than 170 species from C & S America and the West Indies. Many species not marketed have very striking inflorescences and would make ideal plants for cultivation, but their large size at flowering prevents this.

1a. Flowers stalked, stalks *c.* 6 mm **1. lueddemanniana**
b. Flowers stalkless or stalks much less than 6 mm 2
2a. Floral bracts attached to the axis by at least the lower parts of their sides as well as by their bases 3
b. Floral bracts attached to the axis only by their bases, sometimes very small or absent 5
3a. Flowers in many ranks; floral bracts finely toothed **4. fasciata**
b. Flowers in 2 ranks; floral bracts entire 4
4a. Floral bracts narrowly ovate, their margins attached to the axis only towards their bases **6. chantinii**
b. Floral bracts attached to the axis by most of their margins, forming cups **5. distichantha**
5a. Sepals acute but not spine-tipped 6
b. Sepals spine-tipped 8
6a. Inflorescence branched, at least in the lower half 7
b. Inflorescence unbranched **8. bromeliifolia**
7a. Inflorescence branched throughout **2. miniata**
b. Inflorescence branched only in the lower half **3. fulgens**
8a. Sepals free or almost so 9
b. Sepals joined for at least one-third of their length 10
9a. Inflorescence branched **7. purpureo-rosea**
b. Inflorescence completely unbranched **9. nudicaulis**
10a. Scape completely hidden by the leaf-sheaths **10. recurvata**
b. Scape visible above the leaf-sheaths 11
11a. Inflorescence branched, at least at the base 12
b. Inflorescence completely unbranched 14
12a. Petals yellow, drying purple **11. caudata**
b. Petals blue 13
13a. Scape persistently woolly **12. coelestis**
b. Scape hairy at first, soon hairless **13. gracilis**
14a. Petals yellow 15
b. Petals blue or purple 17
15a. Inflorescence loose, the axis visible **14. blumenavii**
b. Inflorescence dense, the axis mostly hidden by flowers and bracts 16
16a. Floral bracts broadly ovate, enfolding the base of the ovary; sepal blade *c.* 2 times as long as spine **15. lindenii**
b. Floral bracts narrowly triangular, not enfolding the base of the ovary; sepal spine and blade more or less equal **16. calyculata**

17a. Inflorescence 5–6 cm, with few (to 25) flowers; flowers 2–2.5 cm, petals 1.4–1.6 cm **13. gracilis**
b. Inflorescence 9–26 cm, with many (to 100) flowers; flowers *c.* 1.5 cm, petals *c.* 9 mm **17. gamosepala**

1. **A. lueddemanniana** (K. Koch) Mez (*Pironneava luddemanniana* K. Koch; *A. caerulea* Morren). Illustration: Flora Neotropica **14**: f. 597D–F (1979); Rauh, Bromeliads, t. 192 (1979).
Plant to 70 cm. Leaves 30–60 cm, forming a cup-shaped rosette, almost straight, densely pale-scaly, especially beneath; sheaths large; blades *c.* 4.5 cm wide, acute or rounded with a short strong point, margin with spines 1–2 mm. Scape slender, white-mealy. Inflorescence paniculate, cylindric, 12–30 × 5–10 cm, white-mealy. Flower-stalks *c.* 6 mm. Sepals *c.* 3.5 mm, each with a broad, lateral wing, mucronate, free. Petals *c.* 9 mm, pink and blue turning dark carmine-red. Fruit white becoming bright purple. *Mexico, Guatemala & Belize.* G2.

2. **A. miniata** (Beer) Baker (*Lamprococcus miniatus* Beer).
Leaves 25–54 cm, exceeding the inflorescence, in a dense funnel-like rosette; sheath scaly with minute brown scales; blades 3–4 cm wide, narrowed toward the base, more or less acute or rounded and shortly pointed, hairless or finely scaly, green, more or less densely spiny with minute spines. Scape slender, hairless. Inflorescence 8–16 cm, branched to the top, ellipsoid or pyramidal, entirely red except for the petals; branches 2–10-flowered, spreading. Floral bracts minute. Flowers to 1.4 cm, stalkless. Sepals free, *c.* 4 mm, asymmetric, without spine-tips. Petals *c.* 10 mm, obtuse, blue. *Brazil.* G2.

A variety more commonly found in cultivation than that described above is var. **discolor** (Beer) Baker which differs in having the leaves tinged with red, and shiny rather than scaly (see Rauh, Bromeliads, t. 73, 1979).

3. **A. fulgens.** Brongniart.
Leaves 30–40 cm, in a dense, funnel-like rosette; sheaths elliptic, minutely brown-scaly; blades 4–6.5 cm wide, broadly acute or rounded and shortly pointed, green, margins with minute spines. Scape red, hairless. Inflorescence 15–20 × 6–7 cm, narrowly pyramidal, branched at the base, but simple from the middle to the top, hairless. Floral bracts absent. Flowers stalkless, spreading, to 2.2 cm. Sepals to 5 mm, nearly free, asymmetric, without spine-tips, dark purple. Petals *c.* 1.2 cm, obtuse, blue or purple at first, soon turning red. *Brazil.* G2.

The commonest plant marketed is var. **discolor** (Morren) Brongniart, which has the undersurface of the leaves reddish purple (see Flora Neotropica **14**: f. 603A–E, 1979; Rauh, Bromeliads, t. 73, 1979).

4. **A. fasciata** (Lindley) Baker (*Billbergia fasciata* Lindley; *B. rhodocyanea* Lemaire; *A. rhodocyanea* invalid). Illustration; Botanical Magazine, 4883 (1855); Flora Neotropica **14**: f. 644C (1979); Rauh, Bromeliads, t. 68 (1979).
Leaves 30–100 cm, forming a cylindric or funnel-like rosette, covered with pale, adpressed scales, pale-banded beneath; sheaths slightly wider than the blades, entire, sometimes purple-tinged; blades 3–8 cm wide, broadly rounded and short-tipped, margins with black spines which are 2–4 mm. Scape slender, white-woolly. Inflorescence 7–8 cm, simple or with a few branches at the base, densely pyramidal, white, shortly woolly. Floral bracts exceeding the sepals, pinkish red, the margins shortly attached to the axis and forming a pouch. Flowers 3–3.5 cm, in several ranks, stalkless. Sepals 1–1.2 cm, united for half their length, asymmetric, acute or with a short point. Petals 2.5–3 cm, blue or purple, red when dry. *Brazil.* G2.

This species has been sold for a number of years by many of the major suppliers under the invalid and incorrect name *A. rhodocyanea*. It is one of the best-known bromeliads, having been cultivated for over 150 years, and almost certainly the most popular.

5. **A. distichantha** Lemaire (*A. excavata* Baker; *A. platyphylla* Hassler). Illustration: Botanical Magazine, 5447 (1864); Rauh, Bromeliads, t. 71 (1979).
Leaves 30–100 cm, forming a very dense rosette, covered with a membrane of fused scales; sheaths usually much wider than the blades, toothed toward the apex, entire elsewhere; blades 2.5–8 cm wide, rounded and short-tipped, margins with stout spines 3–5 mm long. Scape slender, white-woolly. Inflorescence always bipinnate, dense or loose, ovoid to cylindric, pink and white-woolly, except for the petals; spikes erect to spreading, the laterals with 2–12 stalkless flowers in 2 ranks, the terminal with more and many-ranked flowers. Floral bracts entire, attached to the axis by their margins for most of their length, forming a cup. Sepals 5–13 mm, free or shortly united, asymmetric. Petals purple or blue. *Brazil, Paraguay, Uruguay & Argentina.* G2.

6. **A. chantinii** (Carrière) Baker (*Billbergia chantinii* Carrière; *A. amazonica* Ule). Illustration: Rauh, Bromeliads, t. 66 (1979).
Leaves 40–100 cm, densely scaly; sheaths large, chestnut-brown; blades 6–9 cm wide, acute or rounded and shortly spine-tipped, pale olive green with white banding beneath (though in some the leaf colour can range from almost black to red-tinged). Scape slender, white-woolly; bracts lanceolate, bright red. Inflorescence bipinnate; spikes long-stalked, with *c.* 12 flowers borne in 2 ranks. Floral bracts 1–1.3 cm, slightly longer than the ovary, scaly, attached to the axis by the bases of their margins. Flowers to 3.2 cm, stalkless. Sepals 1–1.2 cm, asymmetric, shortly united, without points. Petals *c.* 2 cm, obtuse, orange. *Colombia to Peru & Amazonian Brazil.* G2.

7. **A. purpureo-rosea** (Hooker) Wawra (*Billbergia purpureo-rosea* Hooker; *A. suaveolens* Knowles & Westcott). Illustration: Botanical Magazine, 3304 (1834).
Leaves 50–100 cm, in a dense tubular rosette; sheaths large, entire, covered with minute, adpressed, brown scales; blades of 2 types, the outer narrowly triangular and recurved, the inner strap-shaped, rounded, spine-tipped and erect, 3–6 cm wide, margins with black spines which are 2–5 mm, shiny green above, usually hairless. Scape slender, red, white-woolly. Inflorescence paniculate, many-flowered, 15–30 × 8–12 cm, white-woolly on the axis, elsewhere glandular-hairy; lower bracts shorter than the branches; branches spreading, each with up to 12 stalkless flowers. Floral bracts ovate, ending in long, slender spines. Sepals to 6 mm, asymmetric, ending in short spines, very shortly united, pinkish red. Petals to 1.5 cm, purple or blue. *Brazil.* G2.

8. **A. bromeliifolia** (Rudge) Baker (*Tillandsia bromeliaefolia* Rudge; *A. tinctoria* (Martius) Mez). Illustration: Flora Neotropica **14**: f. 661A–H (1979).
Leaves 60–100 cm, in a tubular rosette, covered with a membrane of white, united scales; sheaths usually much broader than the blades, entire except towards the top; blades 4–9 cm wide, acuminate to rounded with a short spine-tip or notched, green or rarely tinged with red, margins with spines

to 1 cm. Scape stout, densely woolly. Inflorescence to 15 × 3–4 cm, simple, cylindric or ellipsoid, densely white-woolly with only the petals exposed at first. Floral bracts 2-keeled, leathery, shorter than the sepals. Flowers stalkless. Sepals *c.* 7 mm, leathery. Petals *c.* 1.5 cm, greenish yellow soon turning black. *C. America to Argentina.* G2.

9. A. **nudicaulis** (Linnaeus) Grisebach (*Bromelia nudicaulis* Linnaeus; *Billbergia pyramidata* Beer). Illustration: Edwards's Botanical Register **3**: t. 203 (1817); Reichenbach, Flora Exotica **3**: t. 185 (1835); Rauh, Bromeliads, t. 74 (1979).
Leaves 30–100 cm; sheaths large, forming a pitcher-shaped rosette, purple to purplish brown, densely and finely brown-scaly; blades 6–10 cm wide, broadly obtuse and shortly pointed, leathery, pale-scaly beneath, hairless above, margins coarsely spiny. Scape slender, erect to arching, white-woolly. Inflorescence simple, 5–25 cm, cylindric, pale woolly-scaly, soon becoming hairless. Floral bracts short and entire, or absent. Flowers *c.* 2.2 cm, stalkless. Sepals 5–10 mm, asymmetric, spine-tipped, free. Petals *c.* 1.2 cm, yellow. *Mexico & West Indies to northern S America & Brazil.* G2.

Several varieties of this species are known but it is uncertain how many of them have reached the European market. They are not totally dissimilar but var. **aureo-rosea** (Antoine) L.B. Smith is particularly outstanding by reason of its red petals and red-tinged calyx.

10. A. **recurvata** (Klotzsch) L.B. Smith (*Macrochordium recurvatum* Klotzsch; *A. legrelliana* (Baker) Baker). Illustration: Flora Neotropica **14**: f. 632H–K (1979); Rauh, Bromeliads, t. 80 (1979).
Leaves 25–40 cm, forming compact, dense rosettes; sheaths large, sometimes longer than the blades, ovate and much broader than the blades, forming an ellipsoid tank, densely scaly; blades 1–2 cm wide, narrowly triangular, spreading abruptly or recurving just above the sheath, thick, green to reddish green, those in the centre becoming bright red during flowering, toothed. Scape short, completely hidden by leaf-sheaths. Inflorescence simple, few-flowered, ellipsoid or obovoid, scape white-woolly or hairless. Floral bracts equalling or slightly longer than the sepals, red. Flowers stalkless, 3.5–4.5 cm. Sepals 1.1–1.5 cm, excluding the spine which is *c.* 4 mm, slightly asymmetric, united from the base for more than one-third of their length. Petals to 3 cm, purple or pinkish red. *S Brazil, Paraguay, Uruguay & NE Argentina.* G2.

Other than the variety above, 2 others are met with in cultivation and both have their inflorescences within the leaf-sheaths. One of these, var. **ortgiesii** (Baker) Reitz (*A. ortgiesii* Baker) is distinguished by having strongly toothed leaves and bracts, while var. **benrathii** (Mez) Reitz (*A. benrathii* Mez) has leaves and bracts entire or nearly so.

11. A. **caudata** Lindman. Illustration: Flora Neotropica **14**: f. 634A–E (1979); Rauh, Bromeliads, t. 182 (1979).
Leaves 50–100 cm, forming broad, funnel-like rosettes, minutely adpressed-scaly; sheaths ovate or elliptic, brown toward the base, entire; blades to 8 cm wide but often narrower, broadly rounded and sharply pointed, margins with spines to 1 mm. Scape white-woolly. Inflorescence paniculate, 10–25 × *c.* 11 cm, bipinnate to the middle or above, white-woolly. Flowers stalkless. Floral bracts 7–17 mm, red, ovate, each tipped by a long brown spine. Sepals 7–11 mm, including the long spine, united. Petals 1.2–1.5 cm, yellow, turning purplish on drying. *Brazil.* G2.

Var. **variegata** Foster is commonly seen on the market and is easily distinguished by its dark green leaves with longitudinal, creamy stripes and a pink tinge on new growth and near the base of older plants.

12. A. **coelestis** (K. Koch) Morren (*Hoplophytum coeleste* K. Koch). Illustration: Flora Neotropica **14**: f. 636A–B (1979); Rauh, Bromeliads, t. 183 (1979).
Leaves 30–100 cm, more or less erect in a dense rosette, scaly, especially beneath; sheaths large, scarcely wider than the blades, entire; blades 3–5 cm wide, broadly acute or rounded, margin with spines *c.* 0.5 mm, sometimes dark green-banded beneath. Scape slender, densely and persistently white-woolly. Inflorescence to 12 cm, paniculate, bipinnate at the base, simple above, densely white-woolly. Floral bracts ovate, red, each ending in a slender spine. Flowers stalkless, *c.* 2 cm. Sepals 3.5–6 mm excluding the long terminal spine, asymmetric, united for about one-third of their length. Petals *c.* 1.3 cm, blue. *Brazil.* G2.

Var. **albo-marginata** Foster is a very attractive plant which has dark green leaves banded with a regular and clearly defined, white margin.

13. A. **gracilis** Lindman. Illustration: Flora Neotropica **14**: f. 637A–F (1979).
Leaves 50–70 cm, in a more or less cylindric rosette; sheaths elliptic, purple-tinged; blades 3–4 cm wide, broadly rounded and shortly pointed, minutely toothed or almost entire, pale beneath and sometimes banded. Scape slender, woolly at first, the scales soon deciduous. Inflorescence 5–6 cm long, simple or with a few branches at the base, ovoid to pyramidal, hairless, with few flowers. Floral bracts ovate, slender-spined, red. Flowers 2–2.5 cm, stalkless, spreading. Sepals 6–7 mm (excluding the terminal spine which is 3–5 mm), united for 3 mm. Petals 1.4–1.6 cm, pale blue with a darker apex. *Brazil.* G2.

14. A. **blumenavii** Reitz. Illustration: Flora Neotropica **14**: f. 638G–J (1979).
Leaves 40–70 cm, in a funnel-like rosette, curved, spreading; sheaths entire, scaly, tinged with violet above; blades *c.* 4 cm wide, broadly rounded or more or less truncate, shortly pointed, minutely toothed, often white-banded beneath. Scape slender, more or less hairless, dark violet. Inflorescence to 9 × 3.5 cm, simple, cylindric, with the flowers more or less whorled, loose, the axis visible. Floral bracts triangular, acuminate, with the lower quarter enfolding the ovary. Flowers *c.* 1.6 cm, stalkless. Sepals 4 mm (with a spine *c.* 3 mm), asymmetric with a large wing, united for half their length. Petals free, yellow, hooded at the apex. *Brazil.* G2.

15. A. **lindenii** (Morren) Baker (*Hoplophytum lindenii* Morren). Illustration: Botanical Magazine, 6565 (1881); Flora Neotropica **14**: f. 640A–E (1979).
Leaves 1–1.5 m in a rosette; sheaths broadly ovate, minutely brown-scaly, sometimes purplish above; blades to 5.5 cm wide, broadly rounded and shortly spine-tipped, green, covered with a whitish coat of more or less fused scales (especially beneath), margins with spines to 2.5 mm. Scape white-hairy. Inflorescence 7–8 × 4 cm, simple, cylindric or ellipsoid, with many flowers and bracts obscuring the axis. Floral bracts broadly ovate, enfolding the ovary, entire, red. Flowers *c.* 2 cm, stalkless. Sepals *c.* 5 mm (excluding the spine which is *c.* 2 mm), asymmetric, shortly united, red. Petals yellow, turning brown. *Brazil.* G2.

A variety with the leaf-blades striped with yellow, var. **makoyana** Mez, is known from cultivation.

16. A. **calyculata** (Morren) Baker (*Hoplophytum calyculatum* Morren).

Illustration: La Belgique Horticole **15**: 162 (1865); Flora Neotropica **14**: f. 640F–K (1979).
Leaves 30–100 cm, in a funnel-like rosette, minutely scaly; sheaths inconspicuous, slightly darker than the blades; blades 3–4 cm wide, broadly rounded, minutely spine-tipped, minutely toothed. Scape slender, white-woolly. Inflorescence 4–6.5 × 3–5 cm, usually simple, dense, the axis obscured by flowers and bracts, more or less white-woolly, except for the petals. Floral bracts narrowly triangular with a slender-spined tip, entire. Flowers stalkless. Sepals 3–5 mm (terminal spines slender, 3–4 mm), asymmetric, united for one-third to half their length, yellow. Petals *c.* 1.2 cm, yellow. *Brazil & N Argentina.* G2.

17. **A. gamosepala** Wittmack. Illustration: Flora Neotropica **14**: f. 642A–E (1979).
Leaves 25–55 cm, in a dense funnel-like rosette, pale-scaly especially beneath; sheaths narrow; blades 3–5 cm wide, nearly entire or with a few minute spines near the apex, broadly rounded, shortly spine-tipped. Scape slender, sparsely white-woolly, soon hairless. Inflorescence 9–26 cm, simple, cylindric, with numerous flowers. Floral bracts narrowly triangular, spine-tipped, entire, the lower exceeding the flower, the upper shorter than the ovary. Flowers *c.* 1.5 cm, stalkless. Sepals 4–5 mm (including the spine, which is 2.5–4 mm), asymmetric, united for half their length, hairless, red. Petals *c.* 9 mm, purple or blue. *Brazil.* G2.

15. QUESNELIA Gaudichaud

D. Philcox

Epiphytes, stemless, or rarely with a long stem. Leaves few or many, forming open, tubular or funnel-like rosettes; blades broadly linear, minutely toothed. Scape evident, well developed. Inflorescence simple or rarely with a few branches, the flowers in more than 2 ranks. Floral bracts usually showy. Sepals 1–3 cm, asymmetric or symmetric, free or shortly united. Petals free, 2.5 cm or more, usually erect, each with 2 scales near the base. Stamens shorter than the petals and not projecting from the corolla. Ovary completely inferior. Fruit a berry.

A genus of some 14 species from Brazil.

1a. Inflorescence dense **1. seideliana**
b. Inflorescence loose 2
2a. Leaves banded, in 2 ranks; sepals to 10 mm; floral bracts minute **2. marmorata**
b. Leaves not banded, not obviously in 2 ranks; sepals to 2.3 cm; floral bracts evident **3. liboniana**

1. **Q. seideliana** L.B. Smith & Reitz. Illustration: Flora Neotropica **14**: f. 681A–C (1979).
Leaves 35–40 cm, few in a tufted rosette, covered with white, closely adpressed, dark-centred scales; sheaths to 20 cm, tinged with dark purple; blades to 4.5 cm wide, rounded and spine-tipped, green, margins with dark spines to 1 mm. Scape slender, erect, finely white-woolly, becoming hairless; bracts few, remote, thin, enfolding the scape, entire. Inflorescence *c.* 3 cm, dense, simple, ellipsoid. Floral bracts *c.* 2 cm, broadly ovate, acute, about as long as the sepals. Flowers stalkless. Sepals *c.* 1.5 cm, white-scaly, united at the base for 3 mm. Petals *c.* 2.5 cm, bright sky-blue. *Brazil.* G2.

2. **Q. marmorata** (Lemaire) Read (*Billbergia marmorata* Lemaire; *Aechmea marmorata* (Lemaire) Mez). Illustration: Illustration Horticole **2**: t. 48 (1855); Flora Neotropica **14**: f. 682A–E (1979).
Leaves 40–60 cm, in a more or less cylindric rosette, clearly 2-ranked, irregularly banded with green and brown beneath; sheaths nearly as long as the blades, entire; blades 5–7 cm wide, oblong, broadly rounded or truncate, apically spined, distantly toothed, green. Scape short and erect, or longer, slender and curved downwards, hairless, red. Inflorescence 12–21 cm, loose, pyramidal, bipinnate at the base, hairless; bracts all large, some shorter than the branches, some longer, spreading or reflexed. Floral bracts minute, broadly triangular, entire. Flowers to 3 cm, stalkless. Sepals to 10 mm, strongly asymmetric, obtuse, unarmed, shortly united at the base, purple. Petals to 2.5 cm, purple or blue. *Brazil.* G2.

3. **Q. liboniana** (De Jonghe) Mez (*Billbergia liboniana* De Jonghe). Illustration: Botanical Magazine, 5090 (1858); Flora Neotropica **14**: f. 683A–E (1979).
Leaves to 80 cm, in a slender, funnel-like rosette, not clearly in 2 ranks; sheaths elongate, covered with pale, brown-centred scales; blades 3–4.5 cm wide, broadly linear, acute, toothed, pale-scaly, sometimes banded beneath. Scape slender, erect. Inflorescence simple or sometimes with a 1–2-flowered branch at the base. Floral bracts narrowly triangular, the lower exceeding the ovary, the upper acute, much shorter. Flowers stalkless. Sepals to 2.3 cm, shortly united, red, hairless. Petals to 5 cm, dark blue. *Brazil.* G2.

16. BILLBERGIA Thunberg

D. Philcox

Epiphytes, without stems or nearly so. Leaves forming broadly funnel-like to long-tubular rosettes; blades mostly broadly linear with broad apices, often banded or spotted, variously toothed or very rarely entire. Scape erect or curved downwards; bracts large, thin, usually red. Inflorescence compound or simple. Flowers large, showy. Sepals 1 cm or more, free, hairless to densely mealy or woolly. Petals 2.5 cm or more, free, each with 2 scales near the base, blade long and narrow. Stamens projecting.

A genus of over 50 species mostly native to eastern Brazil but with some species found in Central America and south along the Atlantic coast to Argentina.

1a. Petals straight or loosely recurved, contorted as they die off; inflorescence often compound, axis often hairless 2
b. Petals coiled in a tight spiral, remaining so as they die off; inflorescence always simple, densely mealy 9
2a. Inflorescence completely hairless or rarely with a few scales on the sepal apices 3
b. Inflorescence scaly, at least on the bracts or sepals 8
3a. Flowers stalkless 4
b. Flowers stalked 6
4a. Floral bracts all large, equalling or exceeding the sepals **1. iridifolia**
b. Floral bracts much reduced toward the apex of the inflorescence, or all much shorter than the sepals 5
5a. Ovary truncate at the base; petals blue at the apex or wholly green; leaf-blades to 5 cm wide **2. distachia**
b. Ovary attenuate at the base; petals blue on the margins; leaf-blades 6–17 mm wide **3. nutans**
6a. Petals blue on the margins **3. nutans**
b. Petals blue at the apex only 7
7a. Sepals acute; leaves not spotted **4. lietzei**
b. Sepals broadly rounded and minutely spine-tipped; leaves spotted **5. leptopoda**
8a. Flower-stalks 5–10 mm; inflorescence more or less pendent; leaves spotted **6. chlorosticta**
b. Flowers stalkless or stalks less than 5 mm; inflorescence erect; leaves not spotted **7. pyramidalis**

9a. Ovary slightly or not at all furrowed or grooved, wholly white-mealy **8. decora**
b. Ovary furrowed or grooved with the ridges hairless, smooth and dark **9. zebrina**

1. B. iridifolia (Nees & Martius) Lindley (*Bromelia iridifolia* Nees & Martius). Illustration: Edwards's Botanical Register **13**: t. 1068 (1827); Flora Neotropica **14**: f. 688A–C (1979).
Leaves to 60 cm, few, in a short tubular rosette, minutely pale-scaly beneath; sheaths large, entire; blades 2.5–4.5 cm wide, recurved, acute or slenderly acuminate, sparingly and minutely toothed to more or less entire. Scape slender, curved downwards, red, hairless; bracts large, acute, red, the upper overlapping. Inflorescence simple, pendent, loose and few-flowered, hairless. Floral bracts all exceeding the sepals. Flowers 5–6 cm, very shortly stalked. Sepals 2–2.5 cm, rounded, with a fine soft apex, more or less symmetric, usually dark blue toward the apex, red elsewhere. Petals 4–5 cm, blue at the apex, yellow elsewhere. *Brazil*. G2.

In var. **concolor** L.B. Smith, the petals are pale yellow throughout.

2. B. distachia (Vellozo) Mez (*Tillandsia distachia* Vellozo; *Billbergia caespitosa* Lindman). Illustration: Flora Neotropica **14**: f. 688D–E (1979).
Leaves 25–90 cm, few, in a slender, cylindric rosette; sheaths much wider than the blades; blades to 5 cm wide, narrowly triangular and attenuate, sometimes linear and rounded, sparingly and minutely toothed or entire, sparsely scaly beneath, not banded. Scape slender, curved downwards, hairless; bracts erect, overlapping, pinkish red, white-scaly beneath. Inflorescence with short, spreading branches, loose and few-flowered, hairless. Floral bracts much shorter than the ovary. Flowers stalkless. Sepals 1.7–2.5 cm, pale green, blue at the apex. Petals 4–5 cm, linear, obtuse, pale green, blue at the apex. *Brazil*. G2.

3. B. nutans Regel. Illustration: Flora Neotropica **14**: f. 688F–H (1979); Rauh, Bromeliads, t. 88 (1979).
Leaves 30–70 cm, in clusters, rarely in rosettes; sheaths narrowly ovate, densely covered with small, white scales; blades linear or narrowly triangular, 6–17 mm wide, scaly beneath, green, margins with short, slender spines. Scape curved downwards, very slender, hairless; bracts erect, overlapping, pinkish red. Inflorescence with very short branches, usually few-flowered, hairless. Floral bracts much shorter than the ovaries. Flowers stalkless or shortly stalked, in 2 ranks. Sepals 1.5–2 cm, obtuse, pinkish red with dark blue margins. Petals 3.3–4.6 cm, linear, obtuse, pale green with dark blue margins, except for the extreme apex. *Brazil, Paraguay, Uruguay & Argentina*. G1.

Var. **schimperiana** (Baker) Mez has entire leaves and the petals blue at the extreme apex as well as on the margins.

The species has been crossed with *Cryptanthus bueckeri* and the resulting hybrid grown under the name × *Cryptbergia meadii* Wilson & Wilson.

4. B. lietzei Morren. Illustration: La Belgique Horticole **31**: t. 5–7 (1881); Flora Neotropica **14**: f. 689A–D (1979).
Leaves 35–55 cm, in a short, tubular rosette, whitish beneath but not banded; blades 2–3 cm wide, linear, acute and mucronate, margins with spines to 2 mm. Scape slender, red, hairless; bracts overlapping, papery, red, sometimes scaly towards the apex. Inflorescence simple, loose, 6–13-flowered. Floral bracts lanceolate, like the scape-bracts. Flower-stalks 3–5 mm, slender. Flowers 5–6 cm. Sepals 1.5–2 cm, red. Petals *c.* 4 cm, linear, blue toward the apex, the remainder greenish yellow or sometimes wholly green. *Brazil*. G2.

5. B. leptopoda L.B. Smith. Illustration: Flora Neotropica **14**: f. 690A–B (1979); Rauh, Bromeliads, t. 86 (1979).
Leaves 20–30 cm, few, in a short, tubular rosette, pale-scaly beneath, pale-spotted; sheaths large, pale violet; blades 2.5–3.5 cm wide, linear, acute, margins with spines to 2 mm. Scape erect, slender, hairless; bracts 4–5 cm, pinkish red, the upper overlapping. Inflorescence simple, few-flowered, hairless. Floral bracts exceeding the ovaries. Flower-stalks slender, 5–20 mm. Sepals 1.8–2.1 cm. Petals 3.5–4.5 cm, linear, green except for the dark blue apex. *Brazil*. G2.

6. B. chlorosticta Saunders (*B. saundersii* Dombrain). Illustration: Flora Neotropica **14**: f. 690A–B (1979).
Leaves 40–80 cm, usually 5–7, forming a tubular rosette; sheaths large; blades 3.5–4 cm wide, linear, broadly rounded or acute, shortly pointed, margins with spines to 1 mm. Scape slender, more or less curved downwards, white-mealy; bracts large, lanceolate, red. Inflorescence more or less pendent, simple, 7–17-flowered, white-mealy. Lowest floral bracts large, like the scape-bracts, the others minute, acuminate. Flower-stalks slender, 5–10 mm. Flowers 5–6 cm. Sepals *c.* 2 cm, asymmetric, red. Petals to 5.5 cm, linear, greenish yellow, dark blue towards the apex. *Brazil*. G2.

7. B. pyramidalis (Sims) Lindley (*Bromelia pyramidalis* Sims; *Billbergia thyrsoidea* Schultes). Illustration: Botanical Magazine, 1732 (1815); Flora Neotropica **14**: f. 701A–D (1979); Rauh, Bromeliads, t. 89 (1979).
Leaves 40–100 cm, in a tubular rosette, often white-banded beneath; sheaths large, entire, somewhat purple-tinged, with a membrane of fused scales; blades 4–6 cm wide, broadly acute or more or less rounded, spine-tipped, minutely toothed. Scape erect, stout, white-mealy; bracts overlapping, with the upper massed below the inflorescence, rose. Inflorescence erect, to 15 cm, densely pyramidal or cylindric, densely mealy. Floral bracts minute, ovate, acute. Flowers with short stalks. Sepals 1.3–1.8 cm, slightly asymmetric, shortly united, pale red. Petals to 5 cm, red, blue toward the apex. *E Brazil*. G2.

Also recorded from the West Indies, Venezuela, Paraguay and W Brazil, but doubtfully native from these areas. A variety having petals which are red throughout, var. **concolor** L.B. Smith, is known from cultivation.

8. B. decora Poeppig & Endlicher. Illustration: Botanical Magazine, 6937 (1887); Flora Neotropica **14**: f. 708 (1979).
Leaves to 70 cm, in a tubular or funnel-like rosette, white-banded beneath and sometimes yellow-spotted; sheaths large; blades to 6 cm wide, broadly acute or rounded and shortly spine-tipped. Scape curved downwards, white-mealy; bracts large, entire, bright pink, the upper ones massed below the inflorescence. Inflorescence pendent, 10–15 cm, simple, densely white-mealy. Floral bracts minute or lacking. Flowers 9–10 cm, stalkless or almost so. Sepals *c.* 1.2 cm, oblong to ovate. Petals to 7 cm, spirally recurved, green. Stamens green or greenish yellow. Ovary hardly or not at all grooved, without dark, hairless lines. *Amazonian Peru, Bolivia & Brazil*. G2.

9. B. zebrina (Herbert) Lindley (*Bromelia zebrina* Herbert). Illustration: Botanical Magazine, 2686 (1826); Flora Neotropica **14**: f. 710A–G (1979); Rauh, Bromeliads, t. 91 (1979).

Leaves to 1 m, about 6 in a tubular rosette, usually white-banded; sheaths large, entire; blades to 8 cm wide, broadly acute, adpressed-scaly, margins with spines to 4 mm. Scape curved downwards, pale-woolly; bracts large, entire, papery, pink, massed below the inflorescence. Inflorescence 15–40 cm, pendent, simple, many-flowered, white-mealy. Floral bracts minute, hidden by the meal. Flowers 7–8 cm. Sepals *c.* 8 mm, truncate. Petals to 6.5 cm, spirally recurved, linear, green becoming yellow. Ovary with thick, dark, longitudinal, hairless folds near the apex. *Brazil, Paraguay, Uruguay & NE Argentina.* G2.

17. ANANAS Miller

D. Philcox

Terrestrial herbs, not producing runners. Leaves scarcely enlarged at the base, forming a dense rosette. Scape evident, erect. Inflorescence terminal on the scape, densely cone-like, usually crowned with a tuft of sterile, leaf-like bracts. Flowers stalkless. Sepals free, obtuse, slightly asymmetric. Petals free, erect, violet or red, each bearing 2 slender, funnel-like scales toward the base. Stamens not projecting from the corolla. Ovaries combining with each other, the bracts and the axis to form a fleshy, compound fruit (syncarp).

A genus of 8 species from S America, of which only 2 are to be commonly found in cultivation, including the common edible pineapple.

1a. Floral bracts conspicuous, overlapping and covering the ovaries **1. bracteatus**
b. Floral bracts inconspicuous, not covering the ovaries at maturity **2. comosus**

1. A. bracteatus (Lindley) Schultes (*Ananassa bracteata* Lindley; *Ananas sativus* var. *bracteatus* (Lindley) Mez). Illustration: Edwards's Botanical Register **13**: t. 1081 (1827); Botanical Magazine, 5025 (1858); Flora Neotropica **14**: f. 729A–B (1979).

Leaves to 1 m or more × 4 cm, margins coarsely spiny, spines all ascending and up to 10 mm. Scape to 50 cm. Floral bracts conspicuous, to 3.5 cm, coarsely toothed, overlapping and covering the ovaries, coloured at maturity. Flowers to 2.5 cm. Petals violet, becoming reddish. Fruit 15 × 12 cm or more, succulent, orange-red. *Brazil, Paraguay & Argentina.* G2.

A variety (var. **tricolor** (Bertoni) L.B. Smith) with the leaves variegated with longitudinal stripes, is sometimes grown.

2. A. comosus (Linnaeus) Merrill (*Bromelia comosa* Linnaeus; *A. sativus* Schultes). Illustration: Flora Neotropica **14**: f. 730 (1979).

Leaves to 1 m or more, scarcely more than 1.5 cm wide, bright green, channelled, margins with ascending spines to 2 mm. Scape 30–50 cm, stout. Floral bracts relatively inconspicuous, soon exposing the apices of the ovaries, weakly and minutely toothed or entire. Flowers *c.* 1.5 cm. Petals violet, to 1.2 cm. Fruit 20–25 × 12 cm at maturity, succulent, sweet and edible. *Origin almost certainly Brazil.*

This is a very variable species with many cultivars and forms. A variety having attractively variegated leaves in green, white and pink, is commonly cultivated as var. **variegatus** (Lowe) Moldenke. As with *A. bracteatus*, they can easily be grown by planting the top which has been removed from the ripe fruit.

XIX. COMMELINACEAE

Perennial or annual herbs, often fleshy, stems often swollen at the nodes. Leaves simple, entire, with tubular sheathing bases, margins of young leaves inrolled. Flowers radially or bilaterally symmetric, bisexual, or rarely with bisexual and unisexual flowers on the same plant, in terminal or axillary 1 to many-flowered, 1-sided, or coiled cymes (cincinni), often aggregated or fused in pairs. Sepals 3. Petals 3, rarely 1 of them much reduced or absent. Stamens 6 or fewer, free or borne on the petals, all similar or variously differentiated, or some reduced to staminodes. Ovary superior, 3- or rarely 2-celled with several to 2, or rarely 1, ovules per cell. Fruit a capsule, dehiscent or sometimes indehiscent, dry or fleshy; seeds few, with endosperm, position of embryo marked by a thickening (embryotega).

A mainly tropical family of about 35 genera and 600 species, with species of horticultural value in *Tradescantia* and several other genera. The flowers, which open in strict succession along the cymes (cincinni), last a few hours only, the petals then deliquescing into a mass with the stamens and ovary which makes dissection and study well-nigh impracticable. The cymes are also often so condensed and umbel-like as to make their construction difficult to discern. The diagnostic feature of *Tradescantia* and allied genera, however, is the fusion of the cymes in pairs to make a 2-sided unit, recognisable in *Tradescantia* itself by the accompanying paired bracts.

Literature: Clarke, C.B., Commelinaceae, in de Candolle, A., *Monographiae Phanerogamarum* **3** : 115–324 (1881); Rohweder, O., Die Farinosae, in der Vegetation von El Salvador, C. Commelinaceae, *Abhandlungen aus dem Gebiet der Auslandskunde* **61** (ser. C., 18): 98–178 (1956).

Habit. Stemless or nearly so: **5–7.**
Leaves. In stout rosettes: **4,10** (No. **2**), **12** (No. **1**). In 2 ranks: **1,8–11.** Striped or variegated: **1,3,5,6,10,11,12**(Nos. **1–4**).
Cymes. Numerous, aggregated in compound cymes: **3,4,7,10.** Fused in pairs: **10–12.**
Petals. United to form distinct tube: **7,8,12**(No. **2**).

1a. Flowers bilaterally symmetric, the stamens of 2 or more types 2
b. Flowers radially symmetric, stamens all similar 7
2a. Inflorescence more or less enclosed by a folded bract **1. Commelina**
b. Inflorescence-bracts (if present) not folded 3
3a. Cymes condensed and fused in pairs, resembling small umbels; bracts none **11. Tripogandra**
b. Cymes not fused in pairs 4
4a. Petals hairless 5
b. Petals ciliate 6
5a. Cymes 1–2, rarely more in an umbel **2. Tinantia**
b. Cymes numerous, aggregated **3. Dichorisandra**
6a. Large, bromeliad-like epiphyte **4. Cochliostema**
b. Low, stemless or creeping plant with few leaves **5. Geogenanthus**
7a. Plants nearly stemless 8
b. Plants with evident stem, at least when flowering 9
8a. Petals free; leaves elliptic, stalked, clothed with brown hairs **6. Siderasis**
b. Petals united into a long tube; leaves linear, not conspicuously hairy **7. Weldenia**
9a. Cymes simple, stalked or stalkless 10
b. Cymes stalkless and associated or fused in pairs or clusters 11

10a. Cymes solitary, axillary and terminal, usually stalkless **8. Cyanotis**
b. Cymes stalked, borne in pairs or umbels on a terminal inflorescence-stalk **9. Gibasis**
11a. Leaves succulent; bracts inconspicuous, or leaves and bracts 2 cm or less **10. Callisia**
b. Leaves only slightly succulent; bracts conspicuous, paired **12. Tradescantia**

1. COMMELINA Linnaeus
D.R. Hunt
Usually perennial, often tuberous-rooted. Flowers variously coloured, bisexual or some unisexual, bilaterally symmetric, in single or paired cymes more or less enclosed by a folded, keeled, spathe-like bract, a few species also with subterranean, cleistogamous flowers. Sepals 3, free, unequal, the outermost hooded; petals 3, free, usually the upper 2 clawed, the lower reduced, more rarely all equal; fertile stamens 3, together on one side of the flower, often of 2 kinds, filaments naked; staminodes 3–2, with sterile anthers, filaments naked. Ovary 3–2-celled, with 1–2 ovules per cell.

About 200 species, mainly in the tropics, including some used for food, medicine or animal fodder. A few are grown in gardens for their bright, gentian-blue flowers.

1a. Bract 'closed', the back edges fused **1. erecta***
b. Bract 'open', the back edges free 2
2a. Plants prostrate or with very short stems 3
b. Plants erect, 30 cm or more 4
3a. Plant without tubers; lowermost petal reduced **2. communis***
b. Plant tuberous; petals equal **3. tuberosa***
4a. Leaves, at least the lower, almost flat, mostly 1–4 cm broad; bract not or shortly beaked **4. coelestis**
b. Leaves inrolled, recurved and sickle-shaped, 4–10 mm broad; bract often with a beak of 1.5–4 cm **5. dianthifolia**

1. C. erecta Linnaeus (*C. elegans* Kunth). Illustration: Rickett, Wild flowers of the United States **3**: t. 11 (1969).
Stems sprawling or erect, to 70 cm or more from slender, tuberous roots. Leaves lanceolate to ovate-lanceolate, acute or acuminate, rounded at base, 7–10 × 1–3 cm, thin, usually hairless. Inflorescence-stalk usually less than 1 cm; folded bract semi-ovate, shortly acuminate, 2–3.5 cm × 6–12 mm, hairy or hairless, back edges fused. Main cyme with 3–6 flowers. Flowers 2.5 cm in diameter; upper petals pale to deep blue, lower much reduced, white. *America, from southern USA to Argentina.* G1. Summer.

***C. benghalensis** Linnaeus. Leaves ovate, 5–7 × 2–4 cm, hairy. As well as normal flowers, cleistogamous, subterranean flowers are produced. *Old World tropics.* G2. Summer.

'Variegata', with cream-variegated leaves, is grown.

2. C. communis Linnaeus. Illustration: Addisonia **1**: t. 20 (1916).
Stems sprawling, to 70 cm, branching, sometimes rooting at the nodes, tubers absent. Leaves oblong–lanceolate, acute to acuminate, rounded and slightly stalked at base, mostly 6–10 × 1.5–3 cm, thin, nearly hairless. Inflorescence-stalks usually *c.* 2 cm, slightly thickened towards top; folded bract semi-ovate, shortly acute, rounded and open basally, 2–3 cm × 8–13 mm. Main cyme with 3–4 flowers. Flowers 2 cm in diameter; petals blue, the lower reduced. *China, Japan, naturalised in S Europe & USA.* H4. Summer–autumn.

***C. diffusa** Burmann (*C. nudiflora* misapplied). Similar but generally smaller. *Tropics.* G2. Summer–autumn.

3. C. tuberosa Linnaeus. Illustration: Botanical Magazine, 1695 (1814).
Prostrate, branching, roots tuberous. Leaves narrowly lanceolate, almost flat, 6–9 × 1–2 cm, hairy towards the translucent margins, sometimes slightly rough with short, stiff hairs. Inflorescence stalks erect, 3–6 cm, elongating to 10 cm or more in fruit. Folded bract ovate-lanceolate, more or less obtuse or shortly beaked, cordate and open towards the base, 2–3 × 1–1.5 cm, green, usually suffused or streaked with dark purple-blue, sparsely to densely hairy. Main cyme with 6–10 flowers. Flowers 3 cm in diameter, petals equal, *C & S America.* G1. Summer.

This, and *C. elliptica, C. coelestis* and *C. dianthifolia,* are the best-known elements of a large complex of intergrading forms often united as *C. tuberosa*, here treated separately in view of their horticulturally important differences in habit and leaves.

***C. elliptica** Kunth (*C. alpestris* Standley & Steyermark). Almost stemless. Leaves mostly basal, lanceolate, to 11 × 2.5 cm. Inflorescence-stalks scape-like, elongating to 20 cm or more in fruit. Folded bract to 4 × 3 cm. Flowers blue or white. *Mexico to Peru, Bolivia.* G1. Summer.

A readily recognised alpine species within the *C. tuberosa* complex.

4. C. coelestis Willdenow. Illustration: Marshall-Cavendish encyclopedia of gardening, 343 (1968); Everett, Encyclopedia of horticulture, 846 (1981).
Erect, to 1 m or more, roots tuberous. Leaves oblong-lanceolate, the lower rounded to cordate at base, 8–18 × 1–4 cm, rough with fine, stiff hairs. Inflorescences crowded towards top of stem; stalks 1–2 cm, elongating to 5–8 cm. Folded bract semi-ovate, more or less obtuse or abruptly and shortly acuminate, almost cordate and open basally, suffused or streaked with dark purplish blue. Main cyme with 4–10 flowers. Flowers 2–3 cm in diameter, usually blue. *C & S America.* H5. Summer.

'Alba' is a white-flowered cultivar. A robust form has been distinguished as var. **bourgeaui** Clarke.

5. C. dianthifolia Delile. Illustration: Rickett, Wild flowers of the United States **3**: t. 11 (1969).
Erect, to 60 cm or more, branching at base from tuberous roots. Leaves distant, linear, inrolled and recurved,8–12 cm × 4–10 mm, rough with short, stiff hairs. Inflorescence-stalks 2–4 cm, elongated to 10 cm or more. Folded bract 3–6 × 1–1.5 cm, usually with a long beak. Main cyme with 4–8 flowers. Flowers 2–3 cm in diameter; petals equal, blue or white. *Mexico, Southwestern USA.* G1. Late summer.

2. TINANTIA Scheidweiler
D.R. Hunt
Annuals or short-lived perennials. Leaves elliptic to oblanceolate. Flowers bisexual or some unisexual, bilaterally symmetric, in short to elongate terminal cymes borne singly or aggregated in pairs, umbels or rarely whorls. Sepals 3, petals 3. Stamens 6, diverse, all fertile, free or the upper 3 united at base, anther-connective narrow. Ovary usually 3-celled with up to 5 ovules per cell.

An American genus of about 10 species. Besides the following, *T. anomala* (Torrey) Clarke (*Commelinantia anomala* (Torrey) Tharp) and *T. pringlei* (Watson) Rohweder (*C. pringlei* (Watson) Tharp) are occasionally cultivated.

1. T. erecta (Jacquin) Schlechtendal (*T. fugax* Jacquin). Illustration: Botanical Magazine, 1340 (1810).
Erect annual, to 1 m tall, with fleshy stems. Leaves elliptic, acuminate and narrowed to the base, 6–16 × 2–6 cm, thin, sparsely hairy. Inflorescence glandular-hairy; cymes in groups of 1–4, each with 3–20 flowers. Flowers *c.* 3 cm in diameter, stalks 1–2 cm, petals blue or purplish, *c.* 1.5 × 1.5 cm; stamens on the same radii as the sepals, bearded, the uppermost short; those on the same radii as the upper petals with appendages, the lowermost hairless. Capsule oblong, 7–11 × 4–5 mm, seeds 2–3 per cell, *c.* 3–3.5 mm, pale grey-brown, rough. *Tropical America.* G2. Late summer.

Usually treated as a half-hardy annual.

3. DICHORISANDRA Mikan
D.R. Hunt
Stout perennials, erect or rarely climbing, or almost without stems. Leaves spirally arranged or more or less in 2 ranks, relatively broad and thin. Flowers in terminal or rarely axillary compound cymes; sepals free, unequal, the upper hooded, often coloured like the petals; petals free or slightly united with the filament-bases, violet-blue, and often white at base, or white. Stamens 6–5, subequal or the 2 whorls more or less unequal; filaments hairless; anthers opening by apical pores (exceptionally by slits), connective narrow. Ovary 3-celled with several ovules per cell, stigma point-like. Fruit a capsule; seeds with fleshy, orange or red, false aril.

A genus of perhaps 30 species, mostly in tropical S America, one (*D. hexandra* (Aublet) Standley) extending to C America and the West Indies. Numerous species are listed in the horticultural reference books, but only 2 are found in cultivation today.

Dichorisandras require warm, humid greenhouse conditions with some shade in summer. They are usually propagated by cuttings or division. Species of the tropical African genus *Palisota* Endlicher are occasionally grown in botanical collections: they are similar but almost stemless, cymes densely congested; stamens 3, staminodes 3 or 2, bearded. Fruit an orange, pink or red berry.

1. D. thyrsiflora Mikan. Illustration; Botanical Magazine n.s., 590 (1971); Graf, Tropica, 305 (1978).
Erect, to 2.5 m but flowering when much smaller, rootstock shortly rhizomatous. Leaves spirally arranged, elliptic-lanceolate, acuminate, wedge-shaped and almost stalked at base, 20–30 × 5–8 cm, hairless. Compound cyme *c.* 20 cm, dense; bracts subulate, cymes shortly stalked; flowers several, with very short stalks. Sepals 1 cm, violet with darker veins; petals almost circular, *c.* 1.2 × 1.1 cm, violet, white at base. Stamens 6, of two kinds, 4 erect, touching, with filaments 2 mm, anthers 5.5 mm, 2 spreading with filaments 3.5 mm, anthers 4 mm. Capsule 1.2 cm; seed with a red false aril. *SE Brazil.* G2. Autumn.

***D. reginae** (Linden & Rodigas) H.E. Moore. Illustration: Baileya **5**: 120, 123 (1957). Leaves in 2 ranks, often silver-striped and flecked. *Peru.* G2.

4. COCHLIOSTEMA Lemaire
D.R. Hunt
Robust, epiphytic perennials with bromeliad-like habit and large, oblong or lanceolate leaves. Flowers in axillary compound cymes; sepals free, more or less equal; petals free, more or less equal or 1 slightly larger, shortly clawed, violet-blue, long-ciliate. Fertile stamens 3, highly specialised, filaments united at the base, much dilated above, enveloping the anthers of the lateral stamens and produced into a tubular horn, anthers coiled; staminodes 3, the lateral well developed, hairy, the other rudimentary. Ovary 3-celled, hairless; ovules numerous, in 2 series in each cell. Capsule narrowly oblong, 3-valved, many-seeded.

A small genus of 2 species confined to Ecuador and Colombia and remarkable for the very curious, coiled anthers.
Literature: Read, R.W., Cochliostema velutinum (Commelinaceae), a new species from Colombia, *Baileya* **13**: 9–15 (1965).

1. C. odoratissimum Lemaire (*C. jacobianum* K. Koch & Linden). Illustration: Illustration Horticole, t. 217 (1859); Everard & Morley, Wild flowers of the world, t. 187 (1970).
Rosette very large, to 50–80 cm tall and 2 m in diameter. Leaves narrowly oblong-lanceolate, acute, dilated at base, 60–100 × 10–15 cm, leathery, hairless, midrib prominent beneath. Inflorescence-stalk 25–40 cm, flower-bearing part 15–25 cm. Flowers strongly scented, 3–6 cm in diameter; stalks 1.5–2.5 cm. Sepals 1.5–2.5 cm × 5–10 mm. Larger petal 1.5–3 cm × 5–12 mm, the other slightly shorter and broader; hooded stamens 1.2–2 cm overall, anthers 3–5 mm; lateral staminodes 5–10 mm, long-ciliate with pale violet hairs; ovary 2.5–4 mm, style 9–16 mm. *Colombia, Ecuador.* G2. Autumn.

5. GEOGENANTHUS Ule
D.R. Hunt
Stem short or creeping. Leaves few, at the stem-apex, very broad, stalked. Cymes with few flowers, borne near the stem-base. Flowers long-stalked, bisexual. Sepals 3, free; petals 3, free, equal or almost so. Stamens 5, free, unequal, 3 shorter, bearded, 2 longer, hairless; connectives narrow, cells dehiscing by slits. Ovary 3-celled, brown-hairy, with 4–5 ovules per cell. Capsule opening by 3 splits.

A small genus of 2 or 3 species in Amazonian Brazil, Colombia, Ecuador and Peru.

1. G. poeppigii (Miquel) Faden (*G. undatus* (K. Koch & Linden) Mildbraed & Strauss; *Dichorisandra musaica* var. *undata* (K. Koch & Linden) Miller). Illustration: Graf, Tropica, 305, 306 (1978).
Branched at base, to 15–25 cm. Leaves widely spreading, circular, almost heart-shaped, 6–12 × 4–10 cm, thin, transversely wavy, dark green above with several longitudinal silvery stripes, reddish beneath, finely hairy; stalk 1–1.5 cm, channelled. Inflorescence-stalk 8–15 mm, cyme with 2–5 flowers; flower-stalks 1.5–4 cm. Flowers *c.* 2.5 cm in diameter; sepals 8–10 mm, brown-hairy; petals violet-blue, 1.1–1.3 cm × 8–11 mm, ciliate at apex. Short stamens *c.* 6 mm with yellow-tipped hairs; long stamens 8–9 mm, white. Style 3–4 mm. *Brazil, Peru.* G2.

6. SIDERASIS Rafinesque
D.R. Hunt
Densely and softly reddish brown-hairy perennial, nearly stemless, shortly rhizomatous. Leaves in a rosette, elliptic to elliptic-obovate, abruptly acute, 15–25 × 5–9 cm, green with a silvery mid-stripe above, reddish beneath, conspicuously parallel-veined; stalk to 5 cm. Inflorescences axillary, stalks slender, 3–10 cm, cyme with 1–4 flowers. Flowers bisexual, 2.5–4 cm in diameter. Sepals free, 1–1.4 cm; petals violet to purplish pink, 1.2–2 × 1–1.8 cm. Stamens 6, equal, all fertile, filaments hairless, connectives narrow, anthers dehiscent by slits. Ovary 3-celled, with 4–5 ovules per cell. Capsule-cells with 1–4 seeds or 1 cell empty.
Literature: Moore, H.E., Siderasis fuscata, *Baileya* **4**: 27–30 (1956).

A genus of 1 species.

1. S. fuscata (Loddiges) H.E. Moore, (*Pyrrheima fuscata* (Loddiges) Hasskarl). Illustration: Botanical Magazine, 2330 (1822); Graf, Tropica, 306 (1978).
E Brazil. G1. Summer.

7. WELDENIA Schultes
D.R. Hunt
Perennial, nearly stemless, roots tuberous. Leaves tufted or in rosettes, linear-lanceolate or broadly strap-shaped. Cymes stalkless, crowded into a dense head surrounded by the leaves. Flowers regular; calyx tubular, 3–4 cm, tips of lobes free; corolla white or very rarely blue, with a tube 4–6.5 cm, 1–2 mm in diameter, and ovate lobes 1–2 × 1 cm. Anthers projecting, the cells of each touching. Ovary cylindric, 3-celled with several ovules per cell. Capsule with many seeds.

A genus of a single species, distinctive on account of its long-tubed flowers.

Weldenia is a choice subject for the alpine house. It needs a deep pot and well-drained compost and must be kept dry and frost-free in winter. Propagation is by division of the rootstock or seeds.

1. W. candida Schultes (*Lampra volcanica* Bentham). Illustration: Botanical Magazine, 7405 (1895).
Mountains of Mexico & Guatemala. G1. Spring–summer.

8. CYANOTIS D. Don
D.R. Hunt
Usually perennial, often tuberous-rooted. Leaves narrow, often ciliate. Flowers violet-blue or purplish pink, bisexual, radially symmetric or nearly so, generally almost stalkless in dense, single, axillary cymes subtended by a leaf-like or reduced bract, and with conspicuous, sickle-shaped bracteoles. Sepals 3, free or joined below. Petals 3, free or joined with the filament-bases to form a short tube. Stamens 6, equal, fertile, filaments nearly always bearded, connectives narrow. Ovary 3-celled with 2 ovules per cell. Capsule dry, dehiscent.

A genus of about 30 species in the Old World tropics. Besides the 2 species described below, several others have been grown in botanical collections, including *C. barbata* D. Don (*C. hirsuta* Fischer & Meyer), *C. cristata* Schultes, *C. fasciculata* Schultes, *C. moluccana* (Roxburgh) Merrill and *C. speciosa* (Linnaeus) Hasskarl (*C. nodiflora* (Lamarck) Kunth).

1a. Leaves flat, brown-hairy; cymes open; petals free **1. kewensis**
b. Leaves channelled, hairs white; cymes congested; petals united at base **2. somaliensis**

1. C. kewensis (Hasskarl) Clarke (*Belosynapsis kewensis* Hasskarl). Illustration: Botanical Magazine, 6150 (1875); Graf, Exotica, edn 3, 602 (1963).
Prostrate perennial, creeping from its initial rosette, internodes and leaf-sheaths densely brown-hairy. Leaves of rosette lanceolate, acute, rounded at base, to 5 × 2 cm, those of side shoots overlapping in 2 ranks, often smaller, fleshy, dark green above, deep purple beneath, densely velvety. Cymes open, to 3 cm, with up to 8 flowers, the flowers often paired, *c.* 8 mm in diameter, bracteoles elliptic, acute, 5–8 × 1.5–4 mm; flower-stalks 3–4 mm. Sepals *c.* 3.5 × 1.5 mm, densely hairy. Petals free, ovate-elliptic, *c.* 5 × 2.5 mm, purplish pink. Stamens *c.* 6 mm, filaments bearded with violet hairs, anthers yellow with orange pollen. Ovary 1.5 mm, densely hairy; style *c.* 4 mm. *India*. G2. Flowers all year.

2. C. somaliensis Clarke. Illustration: Graf, Exotica, edn 3, 602 (1963); Kakteen und andere Sukkulenten **28**(4): 98 (1977).
Stems of 2 kinds, some with non-flowering basal rosettes (often lacking in cultivated specimens), others stolon-like, creeping shoots, the latter potentially flowering or rooting and forming new rosettes. Leaves oblong-linear, acute and with a small point, stalkless, to 12 × 1.5 cm on rosettes, 1.5–4 cm × 5–10 mm on lateral shoots, these U-shaped in section, and often recurved, succulent, densely hairy below, margins long-ciliate, sheaths inflated, persistent, becoming papery. Bracts 1–1.5 cm; cymes with several flowers. Flowers *c.* 5 mm in diameter, purplish blue. *Somalia*. G2.

Close to *C. foecunda* Hasskarl, but distinct in the basal rosettes and overlapping, persistent leaf-sheaths. According to Faden & Suda in Biological Journal of the Linnaean Society **81**: 311 (1980), the identity of plants cultivated as *C. somaliensis* is uncertain.

9. GIBASIS Rafinesque
D.R. Hunt
Perennial (exceptionally annual), tufted from tuberous roots or creeping and rooting at nodes. Cymes individually shortly stalked, in pairs or umbels, fertile zone dense, bracteoles small, overlapping. Flowers regular; sepals and petals free. Stamens 6, equal, all fertile, bearded. Ovary 3-celled with 2 ovules per cell. Capsule dry, dehiscent.

A genus of about 11 species mostly confined to Mexico, one throughout tropical America.
Literature: Hunt, D.R., Revision of Gibasis, *Kew Bulletin* (in press).

1. G. pellucida (Martens & Galeotti) Hunt (*G. schiedeana* (Kunth) Hunt). Illustration: Botanical Magazine n.s., 636 (1973).
Creeping and soon leggy, rooting at nodes. Leaves in 2 ranks, obliquely lanceolate, acute to acuminate, rounded to cordate at the base, stalkless, 2.5–10 cm × 8–30 mm. Cymes paired. Flowers white, 7–14 mm in diameter. *Mexico*. G1. Flowering all year.

Formerly confused with *G. geniculata* and superficially similar to *Tripogandra multiflora*. More commonly grown in USA than in Europe. It is readily propagated by cuttings.

10. CALLISIA Linnaeus
D.R. Hunt
Perennial herbs with succulent leaves. Flowers usually radially symmetric, bisexual, stalked or stalkless. Cyme-pairs without conspicuous bracts, stalkless in the axils of the upper leaves, these normal or reduced almost to the sheath, or in stalked compound cymes. Sepals 3 (very rarely 2), usually translucent; petals 3 (very rarely 2), white or pink. Stamens 6 or 3–1, filaments typically hairless, rarely bearded, anther-connectives usually broad; anthers dorsifixed; stigma brush-like or papillose; ovary 3–2-celled, ovules 2 per cell. Capsule small, spherical, dehiscent.

As here understood in a broad sense, this is a genus of about 20 species native to tropical America. It includes species with the cymes fused in pairs (in 1 species associated in pairs or threes), regular flowers and the stamens all similar (though sometimes reduced in number), but lacking the conspicuous paired bracts beneath the paired cymes which characterise *Tradescantia*.
Literature: Hunt, D.R., Amplification of Callisia L., *Kew Bulletin* (in press).

1a. Leaves to 30 cm, in a rosette 2
b. Leaves 1–8 cm, spaced along the stem 3
2a. Inflorescence apparently axillary; flowers *c.* 1 cm in diameter; petals purplish pink; stamens 5 mm **1. warscewicziana**
b. Inflorescence terminal; flowers 5–8 mm in diameter; petals white; stamens 8–12 mm, projecting **2. fragrans**

3a. Leaves V-shaped in section; flowers purplish pink **3. navicularis**
b. Leaves flat; flowers white or colourless 4
4a. Fertile stamens 6 **4. elegans**
b. Fertile stamens 5 or fewer 5
5a. Paired cymes stalkless, axillary; flowers odourless; petals narrow **5. repens**
b. Paired cymes stalked, in delicate, many-flowered compound cymes; flowers scented; petals with a broad blade **6. multiflora**

1. C. warscewicziana (Kunth & Bouché) Hunt. Illustration: Botanical Magazine, 5188 (1860); Baileya **10**: 130 (1962).
Stem stout, eventually 1 m or more. Leaves narrowly oblong, acuminate, stalkless, to 30 × 6.5 cm, fleshy, hairless except for the sometimes ciliate, often purple margin. Inflorescence apparently axillary, to 35 cm overall, branched below and sometimes viviparous. Cymes fused at base in pairs or threes, stalkless or almost so, cyme-axes to 2 cm, cymes with many flowers; bracteoles to 7 mm, stalks 1–1.3 cm. Flowers *c.* 1 cm in diameter. Sepals fleshy, persistent, 4–5 mm, hairless; petals purplish pink, 6 × 6 mm. Stamens 6, *c.* 5 mm, usually hairless, connectives broad, yellow. Ovary glandular-hairy towards apex. Capsule 3.5 mm in diameter. *Guatemala.* G2. Spring–autumn.

2. C. fragrans (Lindley) Woodson. Illustration: Edwards's Botanical Register **26**: t. 47 (1840).
Main stem stout, flowering stems to 1.5 m, branching and producing stolons. Leaves elliptic-lanceolate, acute, stalkless, to 30 × 7 cm. Inflorescence terminal, branched, 50–100 cm overall. Paired cymes clustered, stalkless, chaffy, 6 mm. Flowers stalkless, 5–8 mm in diameter, scented. Sepals chaffy, scarcely exceeding the bracteoles. Petals white, lanceolate 5 × 2.5 mm. Stamens 6, 8–12 mm, projecting, filaments hairless, anther-connective broad, thin, white. Ovary 3-celled, hairy on top, with 2 ovules per cell, stigma brush-like. *S Mexico.* G2. Winter–spring.

3. C. navicularis (Ortgies) Hunt (*Tradescantia navicularis* Ortgies). Illustration: Jacobsen, Handbook of succulent plants **2**: 893, 1134 (1960); Graf, Exotica, edn 3, 604 (1963).
Prostrate, with fleshy shoots of 2 intergrading types: short-shoots with tightly overlapping leaves, and 'stolons' with long internodes rooting and producing short-shoots or inflorescences at the nodes. Leaves ovate, acute, stalkless and boat-shaped, 2–3 × 1–2 cm, very fleshy, hairless except on the margins, often longitudinally lined, minutely green-dotted. Cyme-pairs terminal and axillary; bracts resembling the leaves. Bracteoles subulate, 7 mm, green. Flowers 1.5–2 cm in diameter, stalks 1–1.5 cm. Sepals 3, boat-shaped. Petals 3, purplish pink. Stamens 6, filaments bearded. *Mexico.* G1. Summer–autumn.

4. C. elegans Moore (*Setcreasea striata* invalid). Illustration: Baileya **6**: 141 (1958); Graf, Exotica, edn 3, 601, 602 (1963).
Creeping, to 60 cm, branching. Leaves in 2 ranks, oblong-ovate, acute, rounded at base, usually 5–7.5 × 2–3 cm, stiff, fleshy, dark green above with about 6 narrow whitish bands, often purplish beneath, velvety to the touch, and margins minutely ciliate. Inflorescences terminating all main shoots. Cyme-pairs terminal and axillary, stalkless; cymes with 4–5 flowers, bracteoles subulate, 6 mm. Flowers stalkless, 1 cm in diameter. Sepals 3.5–6.5 mm, ciliate on the keel. Petals white, clawed, blade 4.5–6.5 × 5–6.5 mm, claw 2.5–3 × 2–2.5 mm. Stamens 6, subequal, 6–8 mm, those opposite the petals slightly longer, filaments hairless, connective broadly triangular, acute, white or pale yellow. Ovary 3-celled with 2 ovules per cell, hairy at apex; stigma brush-like. *Mexico.* G2. Winter.

5. C. repens Linnaeus. Illustration: Graf, Tropica, 307 (1978).
Slender-stemmed, prostrate, rooting at nodes and forming mats. Leaves ovate, acute, rounded to almost cordate at base, stalkless, usually 1–4 × 1–2 cm, fleshy, hairless except for the minutely ciliate margins. Flowering branches usually ascending, with progressively smaller leaves, the cyme-pairs scarcely projecting from the sheaths. Sepals 3, 2–3 mm. Petals 3, narrow, colourless, slightly shorter than the sepals. Stamens 3, or 6, *c.* 6–8 mm, with 1 or more staminodes projecting; connective broad, rounded, thin. Stigma brush-like. *Tropical America.* G1.

The common clone in cultivation, whose identity has been confirmed by study of its chromosomes, is very compact in habit and seems to be non-flowering, but the small, broad, pointed, fleshy leaves make confusion with other species unlikely.

6. C. multiflora (Martens & Galeotti) Standley. Illustration: Botanical Magazine, 4849 (1855); Baileya **9**: 16 (1961).
Creeping, to 80 cm, branching. Leaves oblong–lanceolate, acute or acuminate, rounded to almost cordate at base, stalkless, 3–9 × 1–2.5 cm, fleshy, rather pale green, velvety to the touch. Inflorescence much-branched. Cyme-pairs stalked; cymes with *c.* 3 flowers. Flowers smelling of violets, 1 cm in diameter; stalks 5 mm, glandular-hairy. Sepals 2–3 mm, green. Petals elliptic, 3–4 mm. Stamens 3, opposite the sepals; filaments hairless, connectives narrow. Ovary 3-celled, with 2 ovules per cell, hairless; style short; stigma papillose, somewhat 3-lobed. Capsule 2 mm. *S Mexico, Guatemala.* G2. Winter.

11. TRIPOGANDRA Rafinesque

D.R. Hunt

Annual or perennial, without tubers. Habit and leaves diverse. Inflorescences terminal and axillary, cyme-pairs stalked, not subtended by leaf-like bracts. Flowers somewhat bilaterally symmetric; sepals and petals equal, free. Stamens 6, dimorphic, 3 with short filaments and 3 with long filaments. Ovary 3-celled with 2 ovules per cell, style short. Capsule dry, dehiscent.

A genus of 21 species in tropical America, from Mexico and the West Indies to Argentina.

Literature: Handlos, W., The taxonomy of Tripogandra (Commelinaceae), *Rhodora* **77**: 213–333 (1975).

1. T. multiflora (Swartz) Rafinesque. Illustration: Baileya **8**: f. 27 (1960).
Creeping perennial, rooting at the nodes, flowering stems ascending, to 80 cm. Leaves ovate, acute, 3–9 × 1–3 cm, rather fleshy, hairless or hairy. Inflorescence-stalks to 3.5 cm; cymes with up to 16 flowers; flower-stalks 2–4 mm. Flowers 5–8 mm in diameter; sepals 2–4 mm; petals 2–4 × 1–2.2 mm, white or pink. Longer stamens 1.5–3 mm, bearded, shorter stamens 0.5–1.3 mm, hairless. Capsule 1.5–2.5 mm in diameter. *Jamaica, Trinidad, Costa Rica, W tropical S America to Argentina.* G2. Flowering all year.

One of several species grown in botanical collections. It is listed in some reference books as being in cultivation and alleged to be widely naturalised in the Azores, but these reports require confirmation. Some illustrations (e.g. in Graf, Exotica, edn 3, 604, 1936) depict *Gibasis pellucida* (p. 28), which is superficially similar but has radially symmetric flowers and the cymes simple, not fused in pairs.

12. TRADESCANTIA Linnaeus

D.R. Hunt

Perennials, habit diverse, roots fibrous or tuberous. Inflorescences mostly terminal, the cymes fused in pairs and subtended by paired, boat-shaped bracts similar to or differentiated from the leaves. Flowers bisexual, radially symmetric. Sepals 3, usually free. Petals 3, usually free sometimes united at base, blue-violet, purple, pink or white. Stamens 6, all fertile, similar and more or less equal; anthers versatile with broad connectives. Ovary 3-celled with 2 ovules (rarely 1) in each cell; stigma knob-like. Capsule dehiscent.

As here understood, *Tradescantia* is a genus of about 65 species native to N and S America. The winter-hardy *T. virginiana* group has long been popular in European gardens, and several non-hardy species (some of them often assigned to small segregate genera) are decorative foliage plants for the house or greenhouse.

Literature: Anderson, E. & Woodson, R.E., The species of Tradescantia indigenous to the United States, *Contributions from the Arnold Arboretum* **9**: 1–132 (1935); Hunt, D.R., Sections and series in Tradescantia (American Commelinaceae IX), *Kew Bulletin* **35**: 437–42 (1980); Hunt, D.R., Campelia, Rhoeo and Zebrina united with Tradescantia, *Kew Bulletin* (in press).

Habit. Bromeliad-like: **1**; creeping, growth non-seasonal: **2,3**; tufted, growth seasonal: **5,6**.
Leaves. White-woolly: **6**; silver-striped above: **2**.
Corolla. Petals united into slender tube below: **2**.

1a. Plant bromeliad-like with short stem and numerous stiff, overlapping, linear-lanceolate leaves; bracts almost enclosing the flowers **1. spathacea**
b. Plant without the above combination of characters 2
2a. Creeping, growth usually non-seasonal; leaves lanceolate to broadly ovate, often longitudinally striped or variegated 3
b. Tufted or shortly creeping and ascending from rhizomatous base, growth seasonal; leaves linear to oblong-elliptic, rarely variegated 4
3a. Petals pink or violet-blue, united at base into a slender tube; leaves typically green, silver-striped above, purplish beneath **2. zebrina**
b. Petals white, free; leaves not silver-striped **3. fluminensis***
4a. Hardy species; leaves linear, green **5. virginiana**
b. Tender species; leaves elliptic or narrowly oblong 5
5a. Leaves densely white-woolly beneath **6. sillamontana**
b. Leaves hairless or hairy but not woolly beneath 6
6a. Flowers distinctly stalked, petals and stamens free; leaves elliptic-oblong to narrowly ovate **4. cerinthoides***
b. Flowers almost stalkless, petals clawed at base and shortly united with filament bases; leaves narrowly oblong **7. pallida**

1. T. spathacea Swartz (*Rhoeo spathacea* (Swartz) Stearn). Illustration: Baileya **5**: f. 57 (1957); Graf, Exotica, edn 3, 601, 604 (1963); Graf, Tropica, 307 (1978).

Erect, with short stem to about 20 cm. Leaves crowded in a rosette, linear-lanceolate, acuminate, scarcely narrowed above the sheath, 20–35 × 3.5–5 cm, fleshy, usually green above and purple beneath, hairless. Inflorescences axillary, stalk short, 2–4.5 cm, bracts deeply boat-shaped, 2–4.5 × 2.5–5 cm. Flowers numerous, scarcely projecting, 1–1.5 cm in diameter, white. Flower-stalks *c.* 1.5 cm, recurved in fruit. Sepals, petals and stamens free, equal. Ovary 3-celled with a single ovule per cell. *S Mexico, Belize, Guatemala; naturalised in West Indies.* G2. Flowering all year.

Widely cultivated; a variant with wholly green leaves is 'Concolor', and another with pale yellow stripes is 'Vittata'.

2. T. zebrina Bosse (*Zebrina pendula* Schnizlein). Illustration: Everard & Morley, Wild flowers of the world, t. 187 (1970); Graf, Exotica, edn 3, 603 (1963); Graf, Tropica, 305, 306 (1978).

Creeping or pendent, rooting at nodes. Leaves ovate-oblong to broadly ovate, acute, rounded at base and stalkless, 4–10 × 1.5–3 cm, slightly fleshy, usually bluish green and often silver-striped above, purplish beneath. Inflorescences terminal with stalks to *c.* 10 cm; bracts unequal, the outer 2–5 cm, the inner 1–2 cm; flower-stalks short. Flowers 1–1.5 cm in diameter, pink or violet-blue; calyx irregularly lobed, 5–8 mm, transparent; corolla-tube to 1 cm, slender, lobes ovate, 5–10 × 3–7 mm. Stamens borne on the petals, anther-connective white. *Mexico; widely naturalised in the tropics.* G2. Flowering all year.

Several clones are widely cultivated, including 'Quadricolor' and 'Purpusii'.

3. T. fluminensis Vellozo (*T. albiflora* Kunth). Illustration: Graf, Exotica, edn 3, 603 (1963); Graf, Tropica, 305, 306 (1978).

Creeping and rooting at the nodes. Leaves ovate to ovate-oblong, acute, asymmetric and sometimes slightly stalked at base, 2–10 × 1–3 cm, thin to fleshy, all green, or purplish beneath, or variegated with several cream or white longitudinal stripes, usually hairless except for the sheath. Paired cymes subtended by bracts similar to or somewhat smaller than the leaves. Flowers 1–1.5 cm in diameter; stalks 5–10 mm, recurved after flowering; sepals to 9 mm, petals to 1.2 cm, white. *SE Brazil to N Argentina.* H5–G1. Flowering all year.

There are several clones in cultivation as houseplants, the variegated, small-leaved variants being preferred. Some clones are hardier than others, and the base will overwinter out of doors in sheltered places as far north as SE England. Although *T. fluminensis* may prove divisible into more than one species, the distinctions between *T. fluminensis* and *T. albiflora* often reported in horticultural literature are false and misleading.

4. T. cerinthoides Kunth (*T. blossfeldiana* Mildbraed). Illustration: Botanical Magazine n.s., 247 (1955); Graf, Exotica, edn 3, 603 (1963).

Creeping or ascending, branching, rooting at nodes. Leaves elliptic-oblong to narrowly ovate, acute, base constricted into the sheath, to 15 × 3.6 cm, glossy dark green and hairless above, purple and densely hairy beneath, ciliate. Inflorescences terminal and axillary from upper nodes. Inflorescence-stalk to 4 cm; bracts unequal, like the leaves but smaller; flower-stalks to 1.5 cm, hairy. Sepals free, *c.* 5.5 × 3 mm, purplish, hairy; petals free, 7–8 × 6 mm, pink above, white towards base. Stamens free, *c.* 4.5 mm, filaments white-bearded, anthers yellow; ovary hairless. Capsule 4 mm; seeds 2 mm. *N Argentina(?).* G1.

A variegated form is also grown. The species was described from cultivated material and is closely allied to the following:

***T. crassula** Link & Otto. Illustration: Botanical Magazine, 2935 (1829); Graf,

Exotica, edn 3, 604 (1963). Leaves green, usually hairless. Flowers white. *N Argentina, Uruguay.* G1. Flowering all year.

5. T. virginiana Linnaeus (*T.* × *andersoniana* Ludwig & Rohweder, invalid). Illustration: Hay & Synge, Dictionary of garden plants, t. 1400 (1969); Perry, Flowers of the world, 77 (1972).
Tufted, stem-base shortly rhizomatous with slightly fleshy roots; flowering stems erect, 30–60 cm, withering after flowering. Leaves linear-lanceolate, 15–35 cm × 5–25 mm, slightly fleshy. Cymes many-flowered, subtended by bracts similar to the leaves but smaller. Flowers 2.5–3.5 cm in diameter with stalks 2–3 cm. Sepals 1–1.5 cm, hairy; petals 1.5–2 × 1–1.5 cm, violet, purplish or rarely white; stamens 1–1.5 cm, filaments very hairy.

The common Tradescantias of gardens are mostly hybrids of *T. virginiana* with related species such as the hairless *T. canaliculata* Rafinesque. The name *T.* × *andersoniana* has been proposed for complex hybrids involving *T. ohiensis* Rafinesque, *T. subaspera* Ker Gawler and *T. virginiana*, but is invalid for lack of description. There are numerous named clones such as 'Isis'.

6. T. sillamontana Matuda (*T. pexata* H.E. Moore; *T. velutina* misapplied; *Cyanotis veldthoutiana* invalid). Illustration: Baileya 8: 98, 101 (1960); Graf, Exotica, edn 3, 605 (1963); Botanical Magazine n.s., 706 (1976).
Stems ascending and spreading from a persistent base and thick non-tuberous roots, to 20–30 cm, withering after flowering. Leaves more or less in 2 ranks, overlapping, elliptic or ovate, acute, rounded or sheathing at base, 4–6 × 1.5–2.5 cm, somewhat fleshy, sparsely hairy above, densely white-woolly beneath. Cyme-pairs terminal, with 6–10 flowers per cyme; bracts similar to the leaves; bracteoles thread-like, 5 mm, hairy; flower-stalks 1–1.5 cm, woolly, recurved after flowering. Flowers 2 cm in diameter, sepals 6–7 mm; petals 9–14 × 9–12 mm, purple-pink; stamens 6–10 mm, filaments hairless. *N Mexico.* G1. Summer.

7. T. pallida (Rose) Hunt (*Setcreasea purpurea* Boom). Illustration: Baileya 5: 150 (1957); Graf, Exotica, edn 3, 602 (1963); Everard & Morley, Wild flowers of the world, t. 187 (1970).
Stem branching and ascending from a persistent base, 20–40 cm. Leaves narrowly oblong, acute, trough-shaped, 8–15 × 2.5–3 cm when flat, slightly fleshy, pale glaucous green to deep purple, hairy or hairless. Cyme-pairs terminal, inflorescence-stalk to 10 cm; bracts 2, rarely 3, leaf-like, the outer longer and up to 4 × 2 cm, the inner smaller, to 1.8 × 2 cm; bracteoles broad, translucent. Flowers several, 1.5–2 cm in diameter; flower-stalks to 6 mm, hairy or hairless. Sepals 5.5 × 2 mm. Petals pink, or pink and white, 1.5–1.8 cm × 8–12 mm, clawed at base and united with the filament bases for *c.* 1 mm; stamens attached to base of petals, filaments bearded. *Mexico.* G1. Summer–autumn.

The widely cultivated form with dark purple stems and leaves ('Purpurea' or 'Purple Heart') was introduced from Mexico but is not known in the wild. It has become popular not only as a pot plant but also for summer bedding.

XX. GRAMINEAE (POACEAE)

Compiled at the Royal Botanic Gardens, Kew

Annual or perennial herbs, rarely woody, often with rhizomes or stolons. Flowering stems (culms) usually cylindric, with hollow internodes and solid nodes, rarely solid throughout, with a growing point at the base of each internode. Leaves alternate, 2-ranked, consisting of sheath, ligule and blade. Sheaths surrounding the stem, with free, overlapping or united margins, sometimes with auricles at the mouth. Ligule situated at the junction of sheath and blade, usually membranous, sometimes a row of hairs or rarely absent. Blade linear to thread-like, rarely lanceolate to ovate, usually with more or less prominent, parallel ribs, generally constricted at the junction with the sheath, rarely shortly stalked. Flowers usually bisexual, consisting of 1–3 (rarely to 6 or more) stamens, and an ovary with 2 (rarely 3) styles; 2 (rarely 0, 3 or 6) small, translucent scales (lodicules) present near base of filaments; each flower enclosed in 2 bracts (a lower lemma and an upper palea), the whole structure forming a floret. Florets 1–many, inserted alternately on 2 sides of a slender, jointed axis (rhachilla) and subtended by usually 2 (rarely 0, 1 or 3) bracts (glumes), the whole forming a spikelet. Glumes sometimes represented by 1 or more bristles forming an involucre. Lemma often with a bristle (awn); palea usually membranous or transparent, 2-keeled, sometimes very small or absent. Lemma with a thickened, sometimes elongate and pointed base (callus). Ovary 1-celled, superior; ovule 1, attached on the inner side of the cell, to a point or line visible in fruit as the hilum. Fruit a grain (caryopsis) or rarely with a free, membranous pericarp. Figure 2, p. 32.

A large cosmopolitan family of about 660 genera and 9000 species. Over 500 species are mentioned in horticultural literature, but these are mainly employed in lawns, sports grounds and landscaping, or offered as novelties of special interest to flower arrangers. Relatively few of them can fairly lay claim to a place in the garden as ornamentals. Many are important cereal and fodder crops.

The bamboos (woody grasses) are placed here at the end of the family (pp. 55–65), with a separate general description and key to the genera.

Literature: The literature on the Gramineae is very extensive and most of it is not especially relevant to grasses cultivated for amenity. Good accounts of most of these (by C.E. Hubbard) can be found in the *Supplement to the Royal Horticultural Society's Dictionary of Gardening* (1969). Hubbard's *Grasses* (1954; edn 2, 1968) is an excellent source of good illustrations and accurate descriptions of the species native to and naturalised in the British Isles.

KEY TO GROUPS

1a. Culms not woody **Group A** (Herbaceous grasses)
b. Culms woody **Group B** (Bamboos, p. 55)

Group A (Herbaceous grasses)

1a. Spikelets unisexual, male spikelets different in appearance from female spikelets 2
b. All or some of the spikelets bisexual, rarely unisexual but then male spikelets similar to female spikelets 4
2a. Inflorescences bisexual, female below, male above, projecting **16. Tripsacum**
b. Inflorescences unisexual, the female enclosed 3
3a. Female inflorescence enclosed by leafy husks **17. Zea**
b. Female inflorescence enclosed by a hard globular jacket **15. Coix**
4a. Spikelets 2-flowered, falling entire at maturity, with the upper floret bisexual and the lower male or sterile; spikelets usually flattened from top to bottom 5

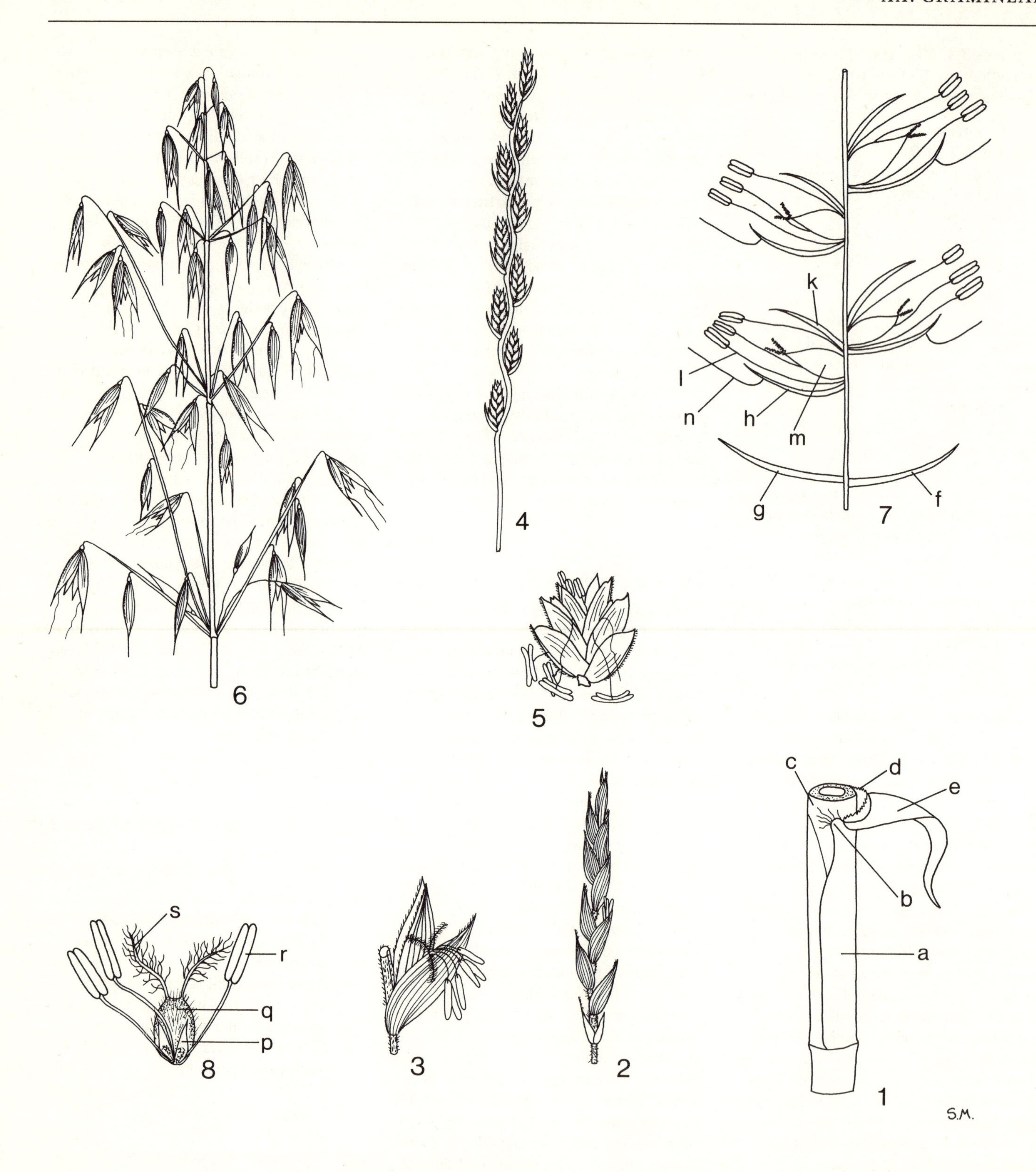

Figure 2. Diagram of some characters used in the identification of Gramineae. 1, Portion of the culm of a bamboo (a, culm-sheath; b, auricle; c, bristles; d, ligule; e, blade of leaf). 2, Spikelet of a bamboo. 3, Flower of a bamboo. 4, A spike of stalkless spikelets. 5, A stalkless spikelet. 6, A panicle of stalked spikelets. 7, Exploded diagram of a spikelet (f, lower glume; g, upper glume; h, lemma; k, palea; l, stamen; m, ovary; n, awn). 8, Flower of *Avena sativa* (p, lodicule; q, ovary; r, anther; s, stigma).

b. Spikelets with 1–many florets, breaking up at maturity above the more or less persistent glumes, or if falling entire then not with 2 florets with the upper floret bisexual and the lower male or sterile; spikelets usually flattened from side to side or terete 24
5a. Upper lemma firmer in texture than the lower glume; spikelets seldom paired 6
b. Upper lemma thinner in texture than the lower glume; spikelets paired 15
6a. Spikelets subtended by an involucre of 1 or more bristles 7
b. Spikelets not subtended by bristles 9
7a. Bristles retained on the branches after the spikelets have fallen **33. Setaria**
b. Bristles falling with the spikelet 8
8a. Involucral bristles free throughout, more or less thread-like **36. Pennisetum**
b. Involucral bristles flattened and united into a little disc below **37. Cenchrus**
9a. Inflorescence a panicle 10
b. Inflorescence composed of racemes 12
10a. Upper lemma leathery to hard and somewhat brittle **32. Panicum**
b. Upper lemma cartilaginous 11
11a. Upper floret flattened from side to side, the stigmas emerging laterally **34. Rhynchelytrum**
b. Upper floret flattened from top to bottom, the stigmas emerging terminally **35. Tricholaena**
12a. Spikelet with a bead-like swelling at its base **29. Eriochloa**
b. Spikelet passing smoothly into its stalk without a bead-like swelling 13
13a. Upper lemma cartilaginous with flat, thin margins covering most of the palea and often overlapping **28. Digitaria**
b. Upper lemma leathery to hard and somewhat brittle with narrow, inrolled margins clasping only the edges of the palea 14
14a. Glumes 2 **31. Echinochloa**
b. Glume 1 **30. Paspalum**
15a. Spikelets in each pair alike 16
b. Spikelets in each pair different 19
16a. Raceme axis tough 17
b. Raceme axis fragile 18
17a. Inflorescence spike-like, cylindric **19. Imperata**
b. Inflorescence a group of elongate racemes **20. Miscanthus**
18a. Lower glume papery, convex, the veins raised **18. Spodiopogon**
b. Lower glume membranous to almost leathery, veins not raised **21. Saccharum**
19a. Internodes and stalk of the raceme with a translucent median line **24. Bothriochloa**
b. Internodes and stalk of the raceme not as above 20
20a. Racemes borne in a panicle 21
b. Racemes paired 22
21a. Spikelets flattened from side to side **23. Chrysopogon**
b. Spikelets flattened from top to bottom **22. Sorghastrum**
22a. Lower glume of stalkless spikelet rounded on back **27. Hyparrhenia**
b. Lower glume of stalkless spikelet laterally 2-keeled 23
23a. Leaf-blades not aromatic **25. Andropogon**
b. Leaf-blades aromatic **26. Cymbopogon**
24a. Inflorescence composed of spikes or racemes 25
b. Inflorescence a panicle 36
25a. Spikes or racemes several in a group 26
b. Spikes or racemes single 28
26a. Spikelets falling entire at maturity **11. Spartina**
b. Spikelets breaking up at maturity 27
27a. Racemes arranged along a central axis **13. Bouteloua**
b. Racemes all arising at the top of the stem **12. Chloris**
28a. Spikelets 1 at each node of the spike 29
b. Spikelets 2–4 at each node of the spike 32
29a. Inflorescence axis tough 30
b. Inflorescence axis fragile 31
30a. Spikelets 1-flowered **61. Mibora**
b. Spikelets 2- or more-flowered **38. Brachypodium**
31a. Spikelets with the side of the lemma towards the axis **39. Aegilops**
b. Spikelets with the back of the lemma towards the axis **52. Gaudinia**
32a. Spikelets with 1 floret **44. Hordeum**
b. Spikelets with 2 or more florets 33
33a. Inflorescence axis fragile **41. Sitanion**
b. Inflorescence axis tough 34
34a. Glumes absent or reduced to 2 short bristles **42. Hystrix**
b. Glumes well developed 35
35a. Spikelets with 2–6 fertile florets **40. Elymus**
b. Spikelets with 1 fertile and 1 sterile floret **43. Taeniatherum**
36a. Spikelets with 1 floret 37
b. Spikelets with 2 or more florets, or of 2 kinds 48
37a. Glumes feathery; panicle spherical to oblong **66. Lagurus**
b. Glumes hairless to shortly hairy 38
38a. Lemma surrounded by fine white hairs from the base, the hairs one-third as long as to much longer than the lemma 39
b. Lemma hairless at base or with only a tuft of very short hairs there 40
39a. Lemma about half the length of the glumes **64. Calamagrostis**
b. Lemmas as long as the glumes **10. Muhlenbergia**
40a. Spikelets falling entire 41
b. Spikelets breaking up at maturity 43
41a. Glumes conspicuously awned **63. Polypogon**
b. Glumes not or shortly awned 42
42a. Lemma awned **59. Alopecurus**
b. Lemma awnless **78. Phaenosperma**
43a. Inflorescence a cylindric, spike-like panicle **60. Phleum**
b. Inflorescence an open or contracted panicle 44
44a. Lemma not strongly hardened and shiny when mature 45
b. Lemma strongly hardened and shiny when mature 46
45a. Lemma awnless or with a sharply bent awn **62. Agrostis**
b. Lemma with a long straight awn **65. Apera**
46a. Lemma awnless **57. Milium**
b. Lemma awned 47
47a. Lemma hairless; awn often falling off **77. Piptatherum**
b. Lemma hairy; awn persistent **76. Stipa**
48a. Spikelets of 2 different kinds, mostly in groups of 3–7 with 1 central bisexual spikelet and 2–6 male or sterile spikelets surrounding it 49
b. Spikelets not obviously of different kinds 51
49a. Male or sterile spikelets persistent **69. Cynosurus**
b. Sterile and fertile spikelets falling in groups 50
50a. Lemma of fertile spikelets awned **68. Lamarckia**
b. Lemma of fertile spikelets awnless **58. Phalaris**
51a. Inflorescence a dense, ovoid to spherical head of stalkless spikelets **67. Echinaria**
b. Inflorescence not a dense head of stalkless spikelets 52
52a. Ligule a row of hairs, sometimes surmounting a short membrane 53
b. Ligule membranous 60
53a. Lowest floret male or sterile 54

b. Lowest floret bisexual 55
54a. Inflorescence feathery **6. Phragmites**
b. Inflorescence not feathery **2. Uniola**
55a. Lemmas, at least of the female florets, hairy 56
b. Lemmas hairless 59
56a. Awn of lemma short, scarcely exceeding the teeth **5. Arundo**
b. Awn of lemma conspicuous 57
57a. Lemma with 7–9 veins **8. Chionochloa**
b. Lemma 3-veined 58
58a. Plant tufted **4. Cortaderia**
b. Plant with creeping rhizome **7. Hakonechloa**
59a. Lemma firm, somewhat cartilaginous **14. Eragrostis**
b. Lemma membranous **9. Molinia**
60a. Lowest floret male or sterile 61
b. Lowest floret bisexual 66
61a. Spikelets with 1 male or sterile floret below the fertile 62
b. Spikelets with 2 male or sterile florets, these often much reduced 63
62a. Spikelets awned **51. Arrhenatherum**
b. Spikelets awnless **3. Chasmanthium**
63a. Inflorescence an open or contracted panicle 64
b. Inflorescence a spike-like panicle 65
64a. Lemma of second floret from the base transversely wrinkled **1. Ehrharta**
b. Lemma of second floret from the base not transversely wrinkled **54. Hierochloe**
65a. Lemma awned **53. Anthoxanthum**
b. Lemma awnless **58. Phalaris**
66a. Lemma circular to broadly ovate, cordate **71. Briza**
b. Lemma usually lanceolate to ovate, not cordate 67
67a. Both glumes, or upper only, as long as or longer than the lowest lemma and usually as long as the whole spikelet and enclosing all the lemmas 68
b. Glumes usually shorter than lowest lemma; the other lemmas distinctly exceeding the glumes (except *Melica*) 74
68a. Lemmas awnless **55. Koeleria**
b. Lemmas awned 69
69a. Spikelets 1.1–3.2 cm; awns 1.2–5.5 cm 70
b. Spikelets 2–11 mm; awns to 1.7 cm 71
70a. Perennials **50. Helictotrichon**
b. Annuals **49. Avena**
71a. Spikelet falling entire at maturity **48. Holcus**
b. Spikelet breaking up at maturity 72
72a. Perennials **47. Deschampsia**
b. Annuals 73
73a. Awn arising on back of lemma **46. Aira**
b. Awn at apex of lemma **56. Rostraria**
74a. Lemmas keeled 75
b. Lemmas rounded on the back 77
75a. Spikelets borne in dense 1-sided clusters on the branches of a panicle **70. Dactylis**
b. Spikelets borne in loose or contracted panicles 76
76a. Lemmas 3-veined **14. Eragrostis**
b. Lemmas with 5–7 veins **72. Poa**
77a. Lemmas awned or with pointed tips 78
b. Lemmas awnless with blunt tips 79
78a. Ovary crowned by a fleshy hairy cap, the styles arising from beneath it **45. Bromus**
b. Ovary hairless or hairy, the styles arising from its tip **73. Festuca**
79a. Spikelets with 1–3 florets **75. Melica**
b. Spikelets with 3–20 florets **74. Glyceria**

Guide to the occurrence of some important characters in the genera of herbaceous grasses

Plants. Dioecious: **4**; monoecious: **15–17**.
Ligule. A line of hairs: **3–8,11–14,20,29, 31,33, 34,36,37**.
Inflorescence. A panicle: **1–10,14,17,18, 21–23, 32,34,35,45–51,54,57,62,64, 65,70–78**; a spike-like panicle: **19,33,36, 37,53,55,56,58–61,63,66,67,69**; composed of racemose spikes or racemes: **11,13,24,29,30,31**; composed of palmately arranged spikes or racemes: **12,16,20,24–28**; of single spikes or racemes: **13,15,30,38–44,52**.
Spikelets. Falling entire: **2,11,15–27, 30–37,39,41,43,44,58,59,63,68,78**. Of 2 different kinds: **11,15–17,22, 23,25–27,44,68,69,76**. With lowest floret male or sterile: **1,3,6,15–37, 51,53,54,58**.
Glumes. The lower suppressed: **28–30,35, 36,42**. Exceeding the lowest lemma: **11,18,20–26,43,46–52,54–65,75–77**. Awned: **15,34,39–41,44,46,63,66**.
Lemmas. With straight or flexuous awn: **4, 7,8,10,12, 31,38**.

1. EHRHARTA Thunberg

Annuals or perennials. Leaves flat or rolled; ligule membranous. Inflorescence a panicle or raceme. Spikelets stalked, flattened from side to side, with 3 florets, the 2 lowest reduced to the lemma, the uppermost complete; glumes persistent, membranous, shorter than the florets; lower lemmas tough, often hairy, awned or awnless; fertile lemma smaller and thinner.

A genus of 25–30 species in South Africa; a few naturalised in N America and Australia. The one cultivated species may be grown in the open in ordinary garden soil, the seed being sown in late spring.

1. E. erecta Lamarck.
Loosely tufted perennial, with ascending, branched stems to 90 cm. Leaf-blades 5–15 cm × 3–9 mm; ligules to 3 mm. Inflorescence to 5–20 cm long, loose, pale green. Spikelets 3–4 mm, awnless; second lemma transversely wrinkled. *South Africa.* H5–G1. Autumn.

2. UNIOLA Linnaeus

Perennials, tufted or rhizomatous, with slender to stout stems. Leaf-blades linear to narrowly lanceolate, flat or rolled. Inflorescence a loose or contracted and dense panicle. Spikelets with few to many florets, strongly compressed from side to side, often large, falling entire. Glumes persistent, keeled, rigid, lemmas compressed and keeled, firm to tough, many-veined, the lower 1–6 sterile.

A genus of 4 species from N & S America which are useful as sand-binders and also very decorative on account of the beautifully formed spikelets. The species below is best sown in a sand-based seed compost.

1. U. paniculata Linnaeus. Illustration: Gould, The grasses of Texas, f. 199 (1975).
Tufted and with extensively creeping rhizomes. Stems stout, erect, 90–240 cm. Leaves smooth, the ligule a dense fringe of hairs; blades to 40 cm × 8 mm, stiff to rigid, tough. Panicles dense, erect or nodding, 16–40 cm long, to 10 cm wide. Spikelets oblong-lanceolate to ovate, much compressed, shortly stalked, with 8–24 florets, 2–3.7 × up to 1.2 cm, straw-coloured; lemmas lanceolate, mucronate, closely overlapping, 8–10 mm, the lower 4–6 sterile. *Eastern USA to the West Indies.* H2. Summer.

3. CHASMANTHIUM Link

Tufted to loosely rhizomatous perennials of diffuse habit. Leaf-blades linear to narrowly lanceolate, flat. Inflorescence an open or contracted panicle. Spikelets with few to many florets, strongly compressed from side to side. Glumes shorter than lemmas, 3–7-veined, acute to acuminate. Lemmas 5–15-

veined, the keels toothed or ciliate, acuminate, entire or bifid, the lower 1–4 sterile.

A genus of five N American species. The panicles are very handsome, for use either in the open or in bouquets or vases. The species below may be propagated by seed or by division; it is suitable for partially shaded borders or open woods on good soil.

1. C. latifolium (Michaux) Yates (*Uniola latifolia* Michaux). Illustration: Gould, The grasses of Texas, f. 16 (1975).
Loosely tufted plant with short, scaly rhizomes. Stems slender to stout, erect or spreading, 45–150 cm. Leaves smooth; ligules extremely short, minutely ciliate; blades 10–25 cm × 5–20 mm, spreading. Panicles loose, drooping, 10–30 cm, of relatively few spikelets. Spikelets oblong-lanceolate to broadly ovate, strongly flattened, with 6–18 florets, long-stalked, 1.5–4.5 × 1–2 cm, green or tawny; lemmas lanceolate, pointed, closely overlapping, 8–15 mm, the lowest sterile. *Eastern USA.* H2. Summer–autumn.

4. CORTADERIA Stapf
Large tufted perennials with separate male and female plants in some species and in others female and bisexual plants. Leaves flat; ligule a line of hairs. Inflorescence a large panicle. Spikelets compressed from side to side with 2–7 (rarely 1) florets. Glumes unequal, membranous, 1-veined. Lemma membranous, with shaggy hairs, 3-veined, with a terminal awn. Rhachilla with shaggy hairs, breaking above the glumes and just above each floret.

A genus of about 15 species in south tropical and temperate S America and New Zealand. Since each plant is usually unisexual, uniformity of habit, flowering period, stature, and size and colour of the inflorescence can only be assured if the plants are propagated by division of the rootstock. Division and replanting is best carried out in April, and the site should be thoroughly prepared. A well-drained loam or deep sandy soil are the most suitable and protection from frost is advisable.
Literature: Stapf, O., The pampas grasses, Cortaderia Stapf, *Flora and Sylva* **3**: 171–6 (1905).

1a. Spikelets apparently all unisexual; male and female inflorescences on different plants, with 3–7 florets; lemma acuminate 2
b. Spikelets either bisexual or female, borne on different plants, with 3–7 florets; lemma bifid, the lobes bristle-like **1. richardii**
2a. Panicles erect or becoming 1-sided, moderately dense; glumes and lemmas 1.2–1.6 cm **2. selloana**
b. Panicles loose and nodding; glumes *c.* 10 mm; lemmas to 1.2 cm **3. jubata**

1. C. richardii (Endlicher) Zotov (*Arundo richardii* Endlicher).
Plants either female or bisexual, densely tufted, with stout, erect, rigid stems 1.2–3 m high. Leaves numerous, 60–120 cm, narrow, curving, leathery. Panicles dense, much-branched, silvery white or yellowish white, 30–60 cm. *New Zealand.* H5. Summer.

2. C. selloana (Schultes & Schultes) Ascherson & Graebner. Illustration: Grounds, Ornamental grasses, t. 5–8 (1979).
Plants dioecious, densely tufted, with stout, erect stems 1.2–3 m high. Leaves numerous, crowded, narrow, 90–270 cm, glaucous, more or less arching, with very rough margins. Panicle 45–120 cm, oblong to pyramidal, silvery white or tinged with red or purple. *S America.* H5. Autumn.

Many cultivars are available, some with variegated leaves, some with coloured panicles, others rather dwarf.

3. C. jubata (Lemoine) Stapf. Illustration: Botanical Magazine, 7607 (1898).
Plants dioecious, densely tufted with stout, erect stems to 3 m high. Leaves numerous, crowded at the base, to 1.2 m × 1.2 cm, drooping, tough, rough with short, stiff hairs. Panicles 30–60 cm, with nodding or drooping branches, the lower to 35 cm, yellowish or tinged with mauve, red or purple. *Western tropical America.* H5. Autumn.

5. ARUNDO Linnaeus
Large rhizomatous perennials. Leaves flat. Inflorescence a loose panicle. Spikelets compressed from side to side, with few, mostly bisexual florets. Glumes almost equal, membranous, equalling the florets, with 3–5 veins, persistent. Lemma with 3–5 veins and with long soft hairs on the lower half of the back, not keeled. Rhachilla hairless.

A genus of about 3 species of reed grasses distributed from the Mediterranean region to China and Japan. They should be grown on damp soils in a warm situation.

1. A. donax Linnaeus. Illustration: Hegi, Flora von Mitteleuropa, edn 2, **1**: f. 202 (1957).
Rootstock knotty, large. Stem 2–6 m, stout. Leaves alternate, regularly arranged, grey-green, almost smooth, arching, to 60 cm or more × 6 cm. Panicle much branched, erect and dense or somewhat drooping, at first reddish then white. Lemmas to 8 mm, bearing white hairs at base. *Mediterranean region.* H5. Autumn.

A variant with variegated leaves is available and is often grown.

6. PHRAGMITES Adanson
Rhizomatous perennials, with erect, stout or robust stems. Leaf-blades linear to narrowly lanceolate, flat, deciduous. Panicle large, dense, profusely branched, silkily hairy. Spikelets with 3–11 florets, the lowest male or sterile, the middle bisexual, the uppermost more or less reduced; rhachilla bearded above with long silky hairs; glumes with 3–5 veins; lemmas narrow, rounded or slightly keeled on the back, hairless; lowest lemma much longer than the glumes, with 3–7 veins, more or less persistent; fertile lemmas acuminate, with 1–3 veins.

A genus of 3 to 4 aquatic species, widely distributed in temperate and tropical regions. *P. australis* is useful for the edges of lakes and large ponds on account of its ornamental habit and large plumose panicles.

1. P. australis (Cavanilles) Steudel (*P. communis* Trinius; *P. vulgaris* (Lamarck) Crépin). Illustration: Hubbard, Grasses, edn 2, 348 (1968); Christiansen, Grasses, sedges and rushes in colour, f. 29 (1979).
Stems robust, 1.2–3 m, closely sheathed. Leaf-blades to 60 cm or more × 1–3 cm, tough, greyish green, long-tapering to a flexuous or curved tip. rough on the margins. Panicles erect or nodding, loose to dense, soft, silkily hairy, purplish or brownish, 15–45 cm. Spikelets 1–1.8 cm, with 3–6 florets; fertile lemmas surrounded by hairs 6–10 mm long. *Widespread in temperate regions and extending into the tropics.* H1. Autumn.

7. HAKONECHLOA Makino
Perennial with long scaly rhizomes, moderately slender stems, linear-lanceolate, finely pointed leaf-blades and minutely ciliate ligules. Inflorescence a loose panicle. Spikelets stalked, slightly compressed from side to side, with 3–5 florets; joints of the axis bearded at the tip; glumes persistent, pointed, membranous, with 3–5 veins, unequal, the lower shorter; lemmas much exceeding the glumes, overlapping, linear-lanceolate in side view, with a short

straight awn from the tip, rounded on the back, membranous, with 3 veins, narrowed at the base into a bearded stalk.

A genus of 1 species of forest grass from the mountains of Japan. It is reasonably hardy, and flourishes in the partial shade of bamboos, in ordinary garden soil, spreading quickly by means of its rhizomes.

1. H. macra (Munro) Makino. Illustration: Grounds, Ornamental grasses, t. 38, 39 (1979).
Stems leafy, ascending or spreading, 30–75 cm. Leaf-blades long and finely pointed, narrowed at the base, 4–25 cm × 5–15 mm, smooth. Panicle lanceolate to ovate, nodding, 9–17 cm; spikelets pale green, narrowly oblong to oblong, 1–2 cm; lemmas 5–9 mm, hairy on the margins, tipped with fine awns 3–4 mm long. *Japan*. H5. Autumn.

8. CHIONOCHLOA Zotov
Coarse perennials, forming tussocks from 20 cm to 2.5 m tall. Leaves mostly deeply grooved. Inflorescence a panicle. Spikelets with several florets. Glumes shorter than spikelets. Lemmas mostly with 7–9 veins, distinctly lobed, the awn a conspicuous prolongation of the middle vein, mostly bent or twisted at the base.

A genus of about 20 species from the mountains of New Zealand and southern Australia. It includes a number of handsome grasses, whose large, graceful panicles are very ornamental, particularly when the grass is planted in rock crevices so that it may be viewed against the sky. *C. conspicua* may be propagated by seed or by division of the rootstock.

1. C. conspicua (Forster) Zotov (*Arundo conspicua* Forster; *Cortaderia conspicua* (Forster) Stapf; *Danthonia cunninghamii* Hooker). Illustration: Buchanan, Indigenous grasses of New Zealand, t. 29 (1880).
Densely tufted, forming large tussocks, with stout stems to 60–150 cm high. Leaves 45–60 cm × 6–8 mm, rigid, tough, flat or concave. Panicle erect or nodding, yellowish, 15–45 cm, loose or compact. Spikelets 8–12 mm, with 3–7 florets; lemmas silky-hairy at the base and on the lower half of the margins; awns straight, 1.8–2.2 cm. *New Zealand*. H4. Summer.

9. MOLINIA Schrank
Tufted perennials. Leaves flat; ligule a row of hairs. Inflorescence a panicle. Spikelets compressed from side to side, with 1–4 florets. Glumes almost equal, membranous, shorter than the florets, the lower with 1 vein or vein absent, the upper with 1–3 veins. Lemma membranous, with 3 veins, rounded on the back. Palea about equalling lemma.

A genus of 2 species in Europe, N and SW Asia. *M. caerulea* is well suited to damp acid soils, does well by the sides of streams and lakes, and is readily propagated by division of the rootstock.

1. M. caerulea (Linnaeus) Moench. Illustration: Hubbard, Grasses, edn 2, 350 (1968); Christiansen, Grasses, sedges and rushes in colour, f. 30 (1979).
Stems 15–120 cm, often forming large tussocks. Leaf-blades 10–45 cm × 3–10 mm, finely pointed, ligules densely ciliate. Panicles very variable, ranging from very dense and spike-like to open and very loose, light to dark purple, brownish or green. Spikelets 4–9 mm; lemmas pointed or blunt, 4–6 mm; stigmas purple. *Europe, N and SW Asia*. H1. Summer–autumn.

10. MUHLENBERGIA Schreber
Annuals or perennials, with mostly slender stems. Leaf-blades thread-like or linear, flat or rolled; ligules membranous. Inflorescence a loose or contracted and dense panicle. Spikelets stalked, small, with 1 floret, awned or awnless. Glumes persistent, thin, with 1–3 veins, shorter than or as long as the lemma. Lemma narrow, membranous to firm, usually with 3 veins, awned or awnless, with the awn straight or flexuous.

A large genus of 120–130 species mainly in warm temperate and tropical America, a few in C and E Asia. *M. mexicana* can be raised from seed sown outdoors in spring in light, well-drained soils in open borders.

1. M. mexicana (Linnaeus) Trinius. Illustration: Britton & Brown, Illustrated flora of the United States and Canada, edn 2, 1: 185 (1913).
Tufted perennial, greyish green, 30–90 cm high, with short, stiff, curved, scaly rhizomes. Stems slender, erect or ascending, branched. Leaf-blades linear, finely pointed, 5–20 cm × 2–6 mm, flat. Panicles linear to narrowly lanceolate, erect or nodding, green or purplish, 4–15 cm, the branches densely covered with spikelets. Spikelets almost stalkless, 2–3 (rarely to 4) mm; glumes linear-lanceolate, finely pointed; lemma about as long as the glumes, shortly bearded at the base. *S Canada & USA*. H2. Summer–autumn.

11. SPARTINA Schreber
Rhizomatous perennials. Leaves flat or rolled. Inflorescence a number of spikes arranged in a raceme. Spikelets strongly compressed from side to side, in 2 rows, closely adpressed to 1 face of the triangular spike-axis, each with 1 or rarely 2 florets. Glumes unequal, papery, the lower 1-veined, the upper with 1–3 (rarely to 6) veins, about as long as lemma. Lemma 1–6-veined, leathery, with a wide membranous margin. Palea slightly shorter than lemma.

A genus of about 14 species of saline or freshwater grasses from the coasts of W & S Europe, NW and S Africa, and inland in America and on the S Atlantic islands. All species of the genus have ornamental inflorescences of decorative value, but only 1 is suitable for cultivation in gardens. They are all propagated by division.
Literature: Saint-Yves, A., Monographia Spartinarum, *Candollea* **5**: 19–100 (1932); Mabberley, D.G., Taxonomy and distribution of the genus Spartina, *Iowa State Journal of Science* **30**: 471–574 (1956).

1. S. pectinata Link. Illustration: Grounds, Ornamental grasses, t. 13 (1979).
Densely tufted, 60–180 cm high, with stout, hard, scaly rhizomes. Stems stout. Leaf-blades 30–120 cm × 4–15 mm, flat, long and finely pointed, rough on the margins. Inflorescence erect, narrow; spikes 5–30, 2.5–10 cm, dense. Glumes toothed on the keels, the upper tapering into an awn 3–8 mm. *Canada & USA*. H2. Autumn.

12. CHLORIS Swartz
Annuals or perennials, tufted or with stolons, with slender stems. Leaves flat or folded; sheaths compressed and keeled; ligule a ciliate rim. Inflorescence of 1–many spikes, usually all arising from the apex of the stem, rarely racemose. Spikelets with 1 complete floret and 1 or more sterile florets above, stalkless, borne in 2 rows on 1 side of the spike-axis, compressed from side to side. Lemmas keeled, 3-veined, awned, usually ciliate on the keel and marginal veins.

A genus of about 50 species. Several species are ornamental and can be grown in the open border during summer, a light well-drained soil suiting them best, the seed being sown outside in May, or earlier in pots in a greenhouse.

1. C. virgata Swartz (*C. elegans* Kunth). Illustration: Chippindal & Cook, Grasses of southern Africa **1**: t. 4 (1973).
Annual, tufted, with ascending or horizontal stems 20–60 cm. Leaf-sheaths somewhat inflated; leaf-blades flat, 2–6 mm wide. Spikes 5–12, all arising from the stem-apex, silky or feathery, 2.5–7.5 cm, green or purplish. Spikelets bearded; lowest lemma *c.* 3 mm, with an awn to 10 mm. *Widespread in the tropics*. H5–G1. Autumn.

13. BOUTELOUA Lagasca
Annuals or perennials, tufted or with stolons, with stiff, slender stems and narrow, flat or rolled leaves. Inflorescence of 1–many, dense, 1-sided spikes or racemes borne on a common axis. Spikelets stalkless or nearly so, in 2 rows on 1 side of the axis, with 1 bisexual floret, and with 1 or more rudimentary florets above it; lemmas 3-veined, the fertile one short-awned or mucronate, the rudimentary usually 3-awned.

A genus of 45–50 species, confined to warm temperate and tropical America, well represented in the USA. They may be propagated by division of the rootstock, or raised from seed sown in the open or in pots in loamy soil.
Literature: Gould, F.W., The genus Bouteloua, *Annals of the Missouri Botanical Garden* **66**: 348–416 (1979).

1a. Racemes persistent **1. gracilis**
b. Racemes deciduous **2. curtipendula**

1. B. gracilis (Kunth) Steudel (*B. oligostachya* Torrey). Illustration: Gould, The grasses of Texas, f. 185 (1975).
Densely tufted perennial, 25–55 cm, with slender, erect stems and linear, flat or inrolled leaves which are 1–2 mm wide. Inflorescence of usually 2, sometimes 1 or 3, dense, 1-sided, spreading or recurved spikes, these 2.5–5 cm long. Spikelets numerous, closely packed, *c.* 5 mm. *C & S USA & Mexico*. H5. Summer.

2. B. curtipendula (Michaux) Torrey. Illustration: Gould, The grasses of Texas, f. 178 (1975).
Perennial with scaly rhizomes; culms erect, tufted, 50–80 cm. Leaf-blades flat or almost rolled, 3–4 mm wide, rough with short, stiff hairs. Inflorescence 15–25 cm, of 35–50 spikes. Spikes each 1–2 cm long, purplish, spreading or pendent, mostly twisted to 1 side of the slender axis. Spikelets 5–8 per spike, adpressed or ascending, 6–10 mm; fertile lemma acute, mucronate; rudimentary lemma with 3 awns and more or less acute intermediate lobes, often reduced and inconspicuous. *Canada to Argentina*. H5. Summer.

14. ERAGROSTIS Wolf
Annuals or perennials. Leaves flat or rolled, narrow; ligule a ring of hairs (rarely membranous). Inflorescence a panicle. Spikelets compressed from side to side, with 2–many florets. Glumes unequal to almost equal, membranous, shorter than the florets, 1-veined. Lemma 3-veined, membranous, usually awnless, falling separately or with the palea and grain. Palea transparent, 2-keeled, often persistent. Stamens 3 or 2.

A genus of about 300 species in the tropics and subtropics. Annual species can be grown in light garden soil in the front of a border or bed, the seeds being sown in pots or in the open in late spring.

1a. Annual, without non-flowering shoots 2
b. Perennial, with non-flowering shoots present 3
2a. Lemmas to 0.8 mm **1. japonica**
b. Lemmas 2–2.5 mm **4. mexicana**
3a. Lemmas 3–3.5 mm; glumes about as long as the lowest lemmas **2. trichodes**
b. Lemmas 2–2.5 mm; glumes shorter than the lowest lemmas **3. curvula**

1. E. japonica (Thunberg) Trinius.
Tufted annual with slender, erect stems, 15–60 cm. Leaf-blades pointed, flat, to 20 cm × 4 mm, hairless. Panicles usually open and loose, lanceolate to ovate-oblong, 7.5–25 cm, mostly over half the height of the plant, rather stiff, with spreading, profusely divided, whorled branches. Spikelets shortly stalked, loosely scattered, linear, with 4–10 florets, pale or purplish, to 4 × 1.3 mm; glumes to 0.8 mm long; lemmas lanceolate, oblong, very blunt, to 0.8 mm long, falling with the paleas at maturity. *Asia & Australia*. H3. Autumn.

2. E. trichodes (Nuttall) Wood. Illustration: Grounds, Ornamental grasses, t.22 (1979).
Tufted perennial, with stiff, erect stems, 45–120 cm. Leaf-blades finely pointed, mostly flat, to 60 cm × 2–6 mm; leaf-sheaths bearded at the top. Panicles loose and open, oblong or elliptic, much-branched, 30–60 cm, about half the height of the plant. Spikelets 4–10 mm, long-stalked, lanceolate to lanceolate-oblong, with 3–9 florets, purple or green; glumes narrowly lanceolate, finely pointed, 3–3.5 mm; lemmas narrowly ovate or ovate–oblong, pointed, 3–3.5 mm. *USA*. H4. Autumn.

3. E. curvula (Schrader) Nees. Illustration: Chippindal & Cook, Grasses of southern Africa **1**: t. 17 (1973); Grounds, Ornamental grasses, t. 21 (1979).
Densely tufted perennial with erect, slender to stout stems, 30–120 cm. Leaf-blades with long, fine points, to 30 cm or more, rolled or opening out and 2–3 mm wide, rough above; sheaths usually hairy at the base. Panicles open and loose, or contracted, 7.5–30 cm, erect or nodding. Spikelets 4–12 × 1.6–2 mm, shortly stalked, narrowly oblong to oblong, dark olive-grey, with 3–18 florets; glumes to 2 mm; lemmas lanceolate-oblong, blunt, 2–2.5 mm. *S Africa*. H5. Autumn.

4. E. mexicana (Hornemann) Link.
Loosely tufted annual, with slender erect or ascending, branched or simple stems, 15–90 cm high. Leaf-blades finely pointed, to 23 cm × 6 mm, flat, rough above. Panicles open and very loose, oblong to ovate, small to large, to 38 × 20 cm. Spikelets 5–8 mm, long-stalked, narrowly ovate to narrowly oblong, with 4–12 florets, purplish or greyish; glumes lanceolate, finely pointed, to 2 mm; lemmas ovate or ovate-oblong, pointed, 2–2.5 mm. *USA & Mexico*. H5. Summer-autumn.

15. COIX Linnaeus
Annuals. Leaves flat. Inflorescence of 3 spikelets, 1 female and 2 sterile, enclosed in a bony jacket, and a terminal spike of male spikelets. Male spikelets in groups of 2 or 3, flattened from top to bottom; glumes papery; lemma and palea membranous; stamens 3. Female spikelets with almost circular, beaked glumes and membranous lemma and palea. Lodicules absent.

A genus of 3 or 4 species from tropical eastern Asia. *C. lacryma-jobi* may be cultivated as a curiosity in an open sunny border, the seed being sown under glass in heat in February or March, or in the open in May. The fruits can be strung and used for necklaces or curtains.

1. C. lacryma-jobi Linnaeus. Illustration: Hegi, Flora von Mitteleuropa, edn 2, **1**: f. 164 (1957).
Annual with stout, erect, much-branched stems, 60–150 cm. Leaves narrowly lanceolate, 10–60 × 2.5–5 cm. Male racemes 1.2–5 cm. Spikelets 8–10 mm, fruit mostly ovoid-spherical, 6–12 mm, bony at maturity, shining, white, bluish white, grey, brown or black. *Widespread in the tropics*. H5–G1. Autumn.

16. TRIPSACUM Linnaeus
Perennial, often broad-leaved and robust, usually rhizomatous. Inflorescences terminal and axillary, of palmately arranged racemes, each raceme female below, with fragile, swollen internodes, male above, with tough, narrow internodes. Female spikelets single without a trace of pairing, deeply sunk in the internodes, the callus transversely truncate with central ridge or peg; lower glume hard and somewhat brittle, smooth, slightly winged at tip; lower floret sterile, without palea. Male spikelets paired, both stalkless or 1 raised on a free stalk; glumes papery; both florets male.

A genus of 13 species mainly in C America. They are propagated by seed or by division, and can tolerate quite severe frosts. They flourish in ordinary garden soil and are grown for their curious decorative inflorescences.

1. T. dactyloides (Linnaeus) Linnaeus. Illustration: Gould, The grasses of Texas, f. 330 (1975).
Forming large clumps, with short thick rhizomes. Stems erect, 1.2–3 m. Leaves green, hairless; blades finely pointed, to 60 × 1–3.5 cm, rough on the margins. Spikes usually 2–3 at the summit of the stem and solitary on the branches, 1.5–2.8 cm, hairless. Female spikelets 8–12 mm, with an ovate, rounded, polished lower glume. Male spikelets oblong, 6–8 mm. *USA to West Indies & tropical S America*. H5. Autumn.

17. ZEA Linnaeus
Annual. Leaves flat. Male inflorescence a terminal panicle of spike-like racemes; female inflorescence axillary, of numerous spikelets arranged in longitudinal rows on a thickened axis, the whole enclosed in leaf-sheaths. Male spikelets in pairs, 1 almost stalkless, the other stalked; florets 2; glumes equal, membranous; lemma and palea transparent; stamens 3, lodicules 2. Female spikelets with 2 florets, the lower sterile; glumes wider than long, fleshy below, transparent above; lemmas short, transparent; lodicules absent; style long, shortly bifid at apex, hairy throughout its length.

A genus of 4 species in C America. *Z. mays* is a staple cereal of all tropical regions and is also extensively grown for forage and silage and as a source of oil, syrup and alcohol. It can be grown from seed outdoors in countries with a warm summer, but without a reasonable sequence of warm, sunny days the grain will not ripen properly. In cooler countries it should be treated as half-hardy and raised under glass before planting out in May.

1. Z. mays Linnaeus. Illustration: Hegi, Flora von Mitteleuropa, edn 2, 1: t. 22 (1957).
Stems to 4 (rarely to 9) m, 2–6 cm in diameter, solid, erect, rooting at the lower nodes. Leaves 3–12 (rarely to 15) cm wide, wavy; ligule 3–5 mm, truncate. Male inflorescence to 20 × 20 cm, erect; spikelets 6–15 mm. Female inflorescence *c.* 20 cm long, stalk with short internodes; styles 15–25 (rarely to 40) cm long, projecting from the apex of the sheaths at and after flowering; grain 5–10 mm, compressed, usually wedge-shaped. *Unknown from the wild; first cultivated in Mexico*. H5–G1. Summer.

Many cultivars, with different ranges of hardiness, are available.

18. SPODIOPOGON Trinius
Perennial or annual, stems slender to stout; ligules membranous. Leaf-blades flat, linear to narrowly lanceolate. Inflorescence a loose or contracted panicle of spike-like racemes. Spikelets in pairs, 1 stalkless, the other stalked, in few-jointed, fragile racemes, all alike, falling entire at maturity, with 2 florets, the lower floret male and the upper bisexual; glumes firm, hairy; lemmas transparent, enclosed by the glumes, the upper awned.

A genus of 8 species in temperate and tropical Asia. Propagated from seed and increased by division. May be grown in ordinary garden soil, in a border, for its decorative, hairy inflorescence.

1. S. sibiricus Trinius.
Perennial, with long, creeping, scaly rhizomes, and erect stems, 60–150 cm; ligules to 1.5 mm. Leaf-blades linear-lanceolate, finely pointed, 15–37.5 cm × 8–18 mm, green or purplish, mostly hairless. Panicles narrowly lanceolate, open or contracted, 10–20 cm; racemes to 1.8 cm. Spikelets 5 mm, narrowly ovate; glumes densely hairy with white hairs; awns 8–12 mm. *E Asia*. H3(?). Summer.

19. IMPERATA Cirillo
Rhizomatous perennials. Inflorescence a narrow, often spike-like panicle with tough branches bearing paired, similar spikelets, each borne upon a slender stalk. Spikelets with 2 florets, terete; glumes as long as spikelet, membranous; lower floret reduced to a transparent lemma; upper floret bisexual with unawned, transparent lemma and shorter palea; stamens 1–2.

A genus of 8 species in warm temperate and tropical regions. Propagated by seed or division of the rootstock.

1. I. cylindrica (Linnaeus) Räuschel. Illustration: Bonnier, Flore Complete **11**: t. 655 (1931).
Stems slender to moderately stout, to 1.2 m. Leaf-sheaths usually hairless; ligule to 2 mm, blades narrowly linear, sharply pointed, tightly rolled or rarely flat, to 60 cm × 3–6 mm, rigid, glaucous. Panicles very dense, spike-like, cylindric, to 20 cm × 25 mm; spikelets 4–5 mm, surrounded by white hairs which are 12–16 mm. *Mediterranean region*. H5–G1. Autumn.

20. MISCANTHUS Andersson
Tufted or rhizomatous, usually with tall, stout, reed-like stems, short membranous ligules, and long, narrow, flat or folded leaf-blades. Inflorescence an oblong-elliptic or fan-shaped panicle of slender, hairy, spike-like racemes, the latter with a continuous axis. Spikelets surrounded with long hairs from the base, falling entire at maturity, paired, those of each pair unequally stalked, with 2 florets but with only the upper floret bisexual; glumes firm, awnless, enclosing the lemmas, the latter delicate, transparent, and the upper usually awned.

A genus of about 12 species in C, E and SE Asia and Polynesia. They are hardy and suitable for most types of soil, propagation being by division of the rootstock.

1a. Spikelets awnless, surrounded by hairs to 12 mm long **1. sacchariflorus**
b. Spikelets awned, surrounded by hairs to 8 mm long **2. sinensis**

1. M. sacchariflorus (Maximowicz) Hackel.
Plant with long, creeping, stout, scaly rhizomes; stems erect, leafy, stout, 1.5–3 m. Leaf-blades linear, very finely pointed, 45–90 × 1–3 cm, flat, spreading, rigid, smooth except for the very rough margins. Panicles ovate or oblong, open and loose, or contracted, 20–40 × 7.5–13 cm, white, reddish or purplish, densely silkily hairy; racemes numerous, 7.5–30 cm, very slender. Spikelets narrowly lanceolate, 4–5 mm, awnless, surrounded by soft, white or purplish hairs which are

6–12 mm. *E Asia* H5–G1. Autumn.

2. M. sinensis Andersson. Illustration: Grounds, Ornamental grasses, t. 12 (1979).
Plant tufted or clump-forming, leafy, with stout rhizomes, and erect, stout stems, 1–3 m. Leaf-blades linear, finely pointed, erect or spreading, 45–75 cm × 8–20 mm, hairy above or hairless, with very rough margins. Panicles dense to somewhat loose, fan-shaped, 25–40 cm long, to 15 cm wide, whitish, brownish or purplish, with a short main axis; racemes numerous, 10–30 cm, slender, erect, or slightly spreading. Spikelets lanceolate, 4–6 mm, awn 6–10 mm, surrounded by fine, white or purplish hairs up to 8 mm long. *China, Japan*. H5. Autumn.

Very variable. Numerous cultivars exist and several may still be available.

21. SACCHARUM Linnaeus
Perennials, often tall. Inflorescence a large, often plumose panicle with persistent branches bearing fragile racemes of paired, similar spikelets. Spikelets with 2 florets, lanceolate, flattened from top to bottom, enveloped in long hairs from the callus; glumes as long as spikelet; lower floret reduced to a transparent lemma; upper floret bisexual, the lemma transparent.

A genus of about 30 species indigenous to tropical and warm temperate regions, several of which are worthy of cultivation as border plants on account of the decorative value of their colourful panicles. They require sunny positions in borders or beds on well-drained soils, where they form dense stately tussocks.

1. S. ravennae (Linnaeus) Murray (*Erianthus ravennae* (Linnaeus) Palisot de Beauvois). Illustration: Coste, Flore de France **3**: 559 (1906).
Stems usually robust, 1–2 m or more. Leaf-blades linear, to 1 m. Panicle 25–60 cm, narrowly ovate, plumose, lobed; racemes 1–2 cm, much shorter than the flexuous supporting branches. Spikelets 3.5–5 mm, the callus hairs about as long as the spikelet; lower glume membranous, hairy on the back or hairless, attenuate to an acute apex; upper lemma 2.5–3 mm, lanceolate, transparent, with a terminal awn 3–4 mm. *Mediterranean region to C Asia, NW India & Somalia*. H5–G1. Summer–autumn.

22. SORGHASTRUM Nash
Annual or perennial; stems tufted, slender to stout, erect; ligules scarious. Leaf-blades narrow, flat or rolled. Inflorescence a loose or dense, narrow panicle, its branches terminating in few-jointed racemes. Spikelets solitary and stalkless at each raceme node, accompanied by the hairy stalk of a suppressed spikelet, falling entire at maturity, with 2 florets, the upper floret alone fertile; glumes leathery, the lower rounded or flattened on the back, the upper boat-shaped; lemmas transparent, enclosed by the glumes, the upper with a bent and twisted awn.

A genus of about 20 species in tropical Africa, warm temperate and tropical N & S America. They are propagated by seed sown under glass, or planted out in the open in the spring. Some of the tropical species can be grown, but they require warm greenhouse treatment. All have beautiful, brownish, yellowish or reddish inflorescences, and may be cut and used for interior decoration, either fresh or dry.

1. S. nutans (Linnaeus) Nash. Illustration: Gould, The grasses of Texas, f. 303 (1975).
Stems 60–120 cm. Leaf-blades finely pointed, to 60 cm long, mostly 5–10 mm wide, rough. Panicle narrow, yellowish or brownish, contracted, rather dense, 7.5–35 cm; raceme-axis and spikelet-stalks greyish-hairy. Spikelets 6–8 mm, lanceolate, hairy; awn bent, 1–1.5 cm. *E & C USA*. H5. Summer–autumn.

23. CHRYSOPOGON Trinius
Perennial, mostly tufted. Leaf-blades linear, often harsh and glaucous. Inflorescence a terminal panicle with whorls of slender, persistent branches bearing terminal racemes, these reduced to a triad of 1 stalkless and 2 stalked spikelets. Stalkless spikelet compressed from side to side. Lower glume rounded on back. Lower floret reduced to a transparent lemma; upper lemma 2-toothed or entire, with a sharply bent awn.

A genus of 24 species from tropical and warm temperate regions of the world. They should be sown in late spring, when clear of frosts, or under glass.

1. C. fulvus (Sprengel) Chiovenda.
Perennial, 30–150 cm. Panicle ovate, 3–15 cm, with fine, delicate branches which are brown-bearded at the tip. Stalkless spikelet narrowly oblong with a shortly pointed callus 1–1.5 mm; upper glume awned; upper lemma with an awn 2–3 cm. Stalked spikelet purplish or pallid, the lower glume awned. *India*. H5. Summer.

24. BOTHRIOCHLOA Kuntze
Annuals or perennials. Inflorescence of 1–many fragile racemes bearing paired, dissimilar spikelets, 1 stalkless, the other stalked; internodes of raceme-axis and spikelet-stalks with a translucent median line. Stalkless spikelet with 2 florets, flattened from top to bottom; lower floret reduced to a transparent lemma; upper floret bisexual with entire lemma passing into a hairless, sharply bent, awn. Stalked spikelet male or sterile, unawned.

A genus of 25–30 species in the tropical and warm temperate regions of both hemispheres. They should be sown in a warm place in spring.

1a. Inflorescence of racemes arising close together near the apex of the stem **1. ischaemum**
b. Inflorescence with a long central axis **2. saccharoides**

1. B. ischaemum (Linnaeus) Keng (*Andropogon ischaemum* Linnaeus). Illustration: Hegi, Flora von Mitteleuropa, edn 2, **1**: t. 22 (1957).
Tufted perennial; stems 15–100 cm, ascending. Inflorescence of 3–15 racemes arising close together near the apex of the stem, each 3–7 cm. Spikelets 3.5–4.6 mm, narrowly elliptic, papery, acute; awn 10–15 mm. *Europe northwards to N France & Poland* H2. Summer–autumn.

2. B. saccharoides (Swartz) Rydberg (*Andropogon saccharoides* Swartz; *A. argenteus* de Candolle; *A. laguroides* de Candolle).
Tufted perennial; stems 60–120 cm. Leaves often glaucous, hairless, 3–5 mm wide. Panicle oblong, dense, silvery white, silkily hairy, to 15 cm; racemes numerous, erect, 1.8–5 cm. Spikelets 4 mm; awn 10–15 mm. *N America to Brazil* H5(?). Summer–autumn.

25. ANDROPOGON Linnaeus
Annuals or perennials. Inflorescence of paired or palmately arranged, fragile racemes bearing paired, dissimilar spikelets, 1 stalked, the other stalkless. Stalkless spikelet with 2 florets, the callus inserted in the cup-like apex of the internode; lower glume with 2 keels; lower floret reduced to a transparent lemma; upper floret bisexual with a 2-toothed lemma passing into a hairless, sharply bent awn. Stalked spikelet male or sterile, awnless or with a very fine hair at the apex.

About 120 species, natives of warm temperate and tropical regions, few of which are of horticultural value, on account of their frequently coarse foliage. They can be grown in a warm border in deep fertile soil.

1. A. gerardii Vitman. Illustration: Gould, The grasses of Texas, f. 304 (1975).
A robust, glaucous perennial, often in large tufts, sometimes with short stolons; stems 1–2 m, usually sparingly branched towards the apex. Lower sheaths and blades sometimes with shaggy hairs, occasionally densely so, the blades flat, mostly 5–10 mm wide, the margins very rough. Racemes on the long-projecting terminal stalk, mostly 3–6, 5–10 cm long, usually purplish, sometimes yellowish. Stalkless spikelet 7–10 mm, the lower glume slightly furrowed, usually rough with short, stiff hairs, the awn sharply bent and tightly twisted below, 1–2 cm. Stalked spikelet not or only slightly reduced, male. *Canada to Mexico*. H5. Summer.

26. CYMBOPOGON Sprengel
Aromatic perennials, with tufted stems. Leaves linear to lanceolate; ligule membranous. Inflorescence usually profusely branched, with spathe-like bracts, each bearing in its axil a pair of slender, spike-like racemes. Spikelets in pairs on the raceme-axis, those of each pair dissimilar, 1 stalkless and bisexual, the other stalked and male, all 2-keeled; stalked spikelets always awnless.

A genus of 45 species in warm regions of the Old World. Many of the species produce aromatic oils used in perfumery. They can be grown in pots in a warm greenhouse and are propagated by division of the rootstock.
Literature: Soenarko, S., The genus Cymbopogon, *Reinwardtia* **9**: 225–375 (1977).

1. C. nardus (Linnaeus) Rendle.
Densely tufted, leafy perennial, with erect, stout, smooth stems to 2.2 m. Leaves to 90 cm or more × 2 cm; ligule blunt, 2–4 mm. Inflorescence much-branched, dense, to 60 cm; spathes elliptic, 1.2–2 cm; racemes 1.2–1.5 cm, spreading, the axis ciliate. Spikelets awnless. *Tropical Asia; cultivated throughout the tropics*. G2.

The commercial source of oil of citronella; grown for its scented leaves.

27. HYPARRHENIA Fournier
Annuals or perennials. Inflorescence of pairs of fragile racemes bearing paired, dissimilar spikelets. Stalked spikelet with 2 florets, the callus oblique, lower glume rounded on the back, without keels; lower floret reduced to a transparent lemma; upper floret bisexual, with a 2-toothed lemma passing into a hairy, sharply bent awn. Stalked spikelets male or sterile, usually longer than the stalkless, unawned or with a fine hair at the apex.

A genus of 60–70 species of tropical and warm temperate countries. They can be raised from seed and put into very large pots in good loamy soil in a tall, warm greenhouse.
Literature: Clayton, W.D., A revision of the genus Hyparrhenia, *Kew Bulletin Additional Series* II (1969).

1. H. cymbaria (Linnaeus) Stapf. Illustration: Die Natürlichen Pflanzenfamilien, edn 2, **14c**: 175 (1940).
Perennial, with slender to stout stems 1.8–6 m. Leaf-blades to 45 cm or more, 6–16 mm wide. Inflorescence often very large and dense, much-branched, to 60 cm or more; bracts subtending the racemes boat-shaped, pointed, horizontally spreading, thin, often red or purple, hairless; racemes 5–8 mm; awns 6–16 mm. *Tropical & S Africa*. G2. Summer–autumn.

28. DIGITARIA Haller
Annuals or perennials. Leaves flat; ligule membranous. Inflorescence of 1-sided racemes arranged palmately or upon a short central axis and bearing the spikelets in adpressed groups of 1–5 or more. Spikelets flattened from top to bottom with 2 florets; lower glume small or absent; upper glume as long as or much shorter than the spikelet; lower floret sterile, represented only by a lemma which is usually as long as the spikelet; upper floret bisexual, the lemma cartilaginous, with its transparent margins enfolding and concealing most of the palea.

A genus of more than 250 species in tropical and warm temperate regions. The seeds of annual species should be sown outside in late spring in a light soil.
Literature: Henrard, J.T., *A monograph of the genus Digitaria* (1950).

1. D. sanguinalis (Linnaeus) Scopoli. Illustration: Hubbard, Grasses, edn 2, 372 (1968).
Green or purplish annual, 10–30 cm, with slender, ascending stems and mostly hairy leaves. Blades narrowly lanceolate, 2.5–10 cm × 3–8 mm, flat. Racemes 4–10, clustered at or near apex of stem, 4–15 cm, very slender, spreading. Spikelets in pairs, 2.5–3 mm, more or less elliptic, lower glume minute. *Temperate regions*. H5. Summer–autumn.

29. ERIOCHLOA Kunth
Annuals or perennials. Ligule a line of hairs. Inflorescence of 1-sided racemes arranged along a central axis. Spikelets flattened from top to bottom, with 2 florets; lower glume vestigial, attached to the swollen, bead-like, lowest rhachilla internode; upper glume almost as long as the spikelet; lower floret male or empty, the lemma resembling the upper glume; upper floret bisexual, the lemma leathery, clasping only the margins of the palea.

A genus of about 30 species in the tropics. The 1 cultivated species needs a warm place in good garden soil.

1. E. villosa (Thunberg) Kunth.
Loosely tufted annual, 10–60 cm. Inflorescence of 3–15 racemes each 1–2 cm; stalks with dense, long hairs. Spikelets 4.5–6 mm, elliptic; lower floret reduced to a lemma; upper lemma mucronate. *Asia*. H5. Summer.

30. PASPALUM Linnaeus
Hairless or sparsely hairy annuals or perennials. Leaves flat or rolled; ligule membranous; sheaths often bearded at the mouth. Inflorescence of 1-sided racemes arising like fingers from the apex of the stem or along a central axis, bearing adpressed spikelets. Spikelets circular to oblong, flattened from top to bottom, usually plano-convex with 2 florets; lower glume usually absent; upper glume as long as the spikelet; lower floret sterile, represented by a lemma as long as the spikelet; upper floret bisexual, the lemma more or less hard and brittle, its narrowly folded margins leaving the palea exposed.

A genus of more than 250 species. It includes several valuable pasture and fodder grasses. Many species have decorative inflorescences but most of these require warm greenhouse cultivation. They may be propagated from seed and increased by division, the half-hardy species being sown under glass and planted out after frosts have ceased.

1. P. ceresia (Kuntze) Chase (*P. membranaceum* Lamarck; *P. elegans* Roemer & Schultes). Illustration: Britton & Brown, Illustrated flora of the United States and Canada **1**: 106 (1896).
Loosely tufted perennial, hairy at the base, with slender ascending stems to 75 cm. Leaves glaucous; ligules very short; blades linear–lanceolate, finely pointed, 4–20 cm × 3–10 mm, flat, hairless or

loosely hairy above. Inflorescence of usually 1–4 dense, ascending or arching racemes, these 2.5–8.5 cm; raceme-axis ribbon-like, purplish or bronze-green, with yellowish brown, translucent margins to 8 mm wide. Spikelets solitary, lanceolate or oblong, 3 mm, hidden by long, silvery-silky hairs; anthers bright yellow; stigmas purple. *Tropical S America*. H5–G1. Summer.

31. ECHINOCHLOA Palisot de Beauvois
Annuals or perennials. Ligule absent or a line of hairs. Inflorescence of 1-sided racemes arranged along a central axis. Spikelets in several rows, convex beneath, flattened above, toothed or awned, with 2 florets. Glumes unequal, the lower shorter than, the upper as long as the spikelet. Lower floret male or sterile, the lemma resembling the upper glume; upper floret bisexual, the lemma hard and rather brittle, smooth, clasping only the margins of the palea.

A genus of 20–25 species in the tropics and warm temperate regions. Seed should be sown under glass and the plant protected from winter frosts.
Literature: Gould, F.W., Ali, M.A. & Fairbrothers, D.E., A revision of Echinochloa in the United States, *American Midland Naturalist* **87**: 36–59 (1972).

1. E. polystachya (Humboldt, Bonpland & Kunth) Hitchcock.
Perennial, with coarse stems 90–180 cm; stem nodes densely bristly with adpressed, yellowish hairs; ligule a line of hairs. Leaf-blades to 2.5 cm wide, rough above. Panicle 10–30 cm, rather dense, with the lower racemes 4–7 cm. Spikelets *c.* 5 mm, awn to 10 mm. *USA (Texas), tropical America from Mexico & the West Indies to Argentina*. G1. Summer.

32. PANICUM Linnaeus
Annuals or perennials. Inflorescence a panicle. Spikelets more or less flattened from top to bottom, unawned, with 2 florets; lower glume shorter than, upper glume as long as the spikelet; lower floret male or sterile, its lemma resembling the upper glume, with or without a palea; upper floret bisexual, the lemma hard and rather brittle, clasping only the margins of the palea.

A genus of more than 400 species. It includes many valuable fodder grasses such as Guinea Grass, *P. maximum* Jacquin, a few cereals such as Common Millet, *P. miliaceum* Linneaus, and several species of ornamental value for gardens. The hardy species and the annuals flourish in ordinary garden soil, the annuals being sown in pots and transplanted to the open border when frosts have ceased, or sown outside in late spring. If required for interior decoration or bouquets, the panicles should be cut when first developed, otherwise their spikelets will be shed.

1a. Leaf-blades lanceolate to narrowly ovate, rounded at the base; spikelets minutely hairy; panicles terminal and axillary, the latter wholly or partly enclosed in the leaf-sheaths **1. clandestinum**
b. Leaf-blades linear to linear–lanceolate; spikelets hairless; axillary panicles, when present, never permanently enclosed in the sheaths 2
2a. Plants with long, creeping stolons; panicle of few, erect, spike-like racemes; spikelets blunt; lower glume almost as long as the spikelet **2. obtusum**
b. Plants tufted or rhizomatous; panicles generally open and loose, not of spike-like racemes; lower glume at most three-quarters of the length of the spikelet 3
3a. Perennial with stout, scaly rhizomes; leaves mostly hairless; lower glume two-thirds to three-quarters of the length of the spikelet **3. virgatum**
b. Annuals; leaves stiffly hairy; lower glume about half the length of the spikelet 4
4a. Panicles erect with widely spreading, much divided, stiff branches; spikelets 2–2.5 mm **4. capillare**
b. Panicles usually nodding; spikelets 5–6 mm **5. miliaceum**

1. P. clandestinum Linnaeus. Illustration: Britton & Brown, Illustrated flora of the United States and Canada, edn 2, **1**: 162 (1913).
Perennial, forming dense clumps. Stems slender to stout, erect or spreading, branched, 60–120 cm. Leaf-sheaths stiffly hairy to nearly hairless; blades lanceolate to narrowly ovate, rounded at the base, finely pointed, 7.5–20 cm × 12–30 mm, green, ciliate at the base. The terminal panicle open, loose, ovate, 7.5–15 × 3.5–10 cm, with smaller panicles wholly or partly enclosed in the leaf-sheaths. Spikelets elliptic, blunt, 3 mm, minutely hairy; lower glume up to one-third of the length of the spikelet; upper glume about equal to the lower lemma. *SE Canada & E USA*. H4. Autumn.

2. P. obtusum Humboldt, Bonpland & Kunth. Illustration: Gould, The grasses of Texas, f. 241 (1975).
Tufted perennials with long, widely creeping stolons which root at the densely hairy nodes. Stems wiry, erect or spreading, 20–75 cm. Leaf-blades linear, finely pointed, 3.5–30 cm × 2–8 mm, flat, hairless or slightly hairy. Panicles narrow, 3.5–12.5 cm long, to 12 mm wide, formed of few, erect, dense, 1-sided racemes. Spikelets shortly stalked, on 1 side of the raceme-axis, elliptic, blunt, 3–4 mm; lower glume slightly shorter than the spikelet, blunt; upper glume and lower lemma about equally long, blunt. *USA & Mexico*. H3. Autumn.

3. P. virgatum Linnaeus. Illustration: Gould, The grasses of Texas, f. 242 (1975).
Tufted perennial, 60–180 cm, purplish, glaucous or green, with stout, creeping, scaly rhizomes. Stems slender to stout. Leaves mostly hairless, blades linear, finely pointed, 15–45 cm × 3–15 mm, flat, erect, firm. Panicles very loose and open, 10–50 × 7.5–25 cm, with widely spreading, stiff, divided branches. Spikelets shortly stalked, elliptic-ovate, acuminate, 3–6 mm; lower glume two-thirds to three-quarters of the length of the spikelet, pointed; upper glume a little longer than the lower lemma, both acuminate. *S Canada & USA to C America*. H3. Autumn.

4. P. capillare Linnaeus. Illustration: Grounds, Ornamental grasses, t. 30 (1979).
Tufted annual, 22–90 cm. Stems erect or spreading, often branched, slender to somewhat stout. Leaf-sheaths usually densely, stiffly rough from minute warts; leaf-blades linear or narrowly lanceolate, finely pointed, 10–30 cm × 5–15 mm, flat, stiffly hairy. Panicles very loose, open, one-third to half or more of the length of the plant, purple or green, with widely spreading, fine, much-divided branches. Spikelets stalked, elliptic, pointed, 2–2.5 mm; lower glume about half the length of the spikelet; upper glume and lower lemma about equal, acuminate. *S Canada & USA*. H3. Summer–autumn.

5. P. miliaceum Linnaeus. Illustration: Hegi, Flora von Mitteleuropa, edn 2, **1**: f. 169 (1957).
Annual, 30–120 cm. Stems solitary or tufted, stout, branched. Leaf-sheaths stiffly hairy, blades linear to linear-lanceolate, rounded at the base, finely pointed, 15–30 × up to 2 cm. Panicles loose or dense, usually nodding, green or purple,

10–30 cm. Spikelets stalked, ovate to elliptic, acuminate, 5–6 mm, prominently nerved; lower glume about half the length of the spikelet, pointed; upper glume and lower lemma equally long, acuminate; upper floret white, yellow to reddish or dark brown. *Warm regions of the world.* H4. Summer–autumn.

33. SETARIA Palisot de Beauvois
Annuals or perennials. Inflorescence a panicle, usually spike-like, the spikelets subtended by bristles which persist on the axis after the spikelets fall. Spikelets oblong to ovate; glumes unequal; lower floret male or sterile, as long as the spikelet; upper floret bisexual, its lemma hard and rather brittle, boat-shaped, often with a wrinkled surface.

A genus of about 100 species, in tropical and warm temperate regions. Several species are grown in warm, humid greenhouses, mainly for their foliage. They require rich loams and a plentiful supply of water. Many annual species have decorative spike-like panicles, with yellow, reddish or purplish bristles; when used for interior decoration the inflorescences should be cut when first developed, otherwise the spikelets are soon shed.

1a. Leaf-blades folded like a fan (pleated) between the veins, especially when young; bristles solitary **1. plicatilis**
b. Leaf-blades flat, not pleated; bristles usually several together 2
2a. Upper glume half to two-thirds of the length of the spikelet; upper lemma transversely wrinkled **2. pumila**
b. Upper glume as long as or nearly as long as the spikelet; upper lemma smooth or almost so 3
3a. Glumes and lower floret more or less persistent, the upper floret deciduous; panicle often lobed or interrupted, erect or nodding, purple or yellow **3. italica**
b. Spikelets falling entire at maturity; spikes very bristly, erect, cylindric or tapering upwards, green or purplish **4. viridis**

1. S. plicatilis (Hochstetter) Hackel.
Loosely tufted perennial, with erect or ascending, slender to moderately stout stems, 45–150 cm. Leaf-sheaths compressed and keeled; ligule a dense fringe of hairs; leaf-blades linear to linear-lanceolate, 15–37.5 × 1–2.5 cm, narrowed to the base, finely pointed, closely pleated, green, rough. Panicles linear to lanceolate-oblong, loose or dense, erect or nodding, to 30 × 7 cm; bristles 6–16 mm. Spikelets narrowly ovate to elliptic-oblong, 3–3.5 mm; upper glumes half to three-quarters of the length of the spikelet; upper lemma smooth or nearly so. *E African highlands.* G2. Summer.

2. S. pumila (Poiret) Roemer & Schultes (*S. glauca* misapplied; *S. pallidefusca* (Schumacher) Stapf & Hubbard).
Annual, with slender, erect or ascending, tufted or solitary stems, 15–45 cm. Leaves green, sheaths compressed and keeled; ligule a dense fringe of hairs; blades linear, finely pointed, 5–30 cm × 2–10 mm, flat or folded, sparsely hairy or hairless. Panicles spike-like, very dense, erect or nodding, very bristly, 1.2–7.5 cm × 4–8 mm (excluding bristles); bristles several, 3–10 mm, yellowish to reddish yellow or orange-brown. Spikelets ovate-elliptic to broadly elliptic, blunt or slightly pointed, 2–3 mm; upper glume half to two-thirds of the length of the spikelet; upper lemma transversely wrinkled. *Warm temperate parts of the Old World.* G1–2. Summer–autumn.

3. S. italica (Linnaeus) Palisot de Beauvois.
Illustration: Christiansen, Grasses, sedges and rushes in colour, f. 91 (1979).
Annual, with stout, erect or ascending stems to 150 cm. Leaf-blades linear to narrowly lanceolate, finely pointed, to 45 cm × 4–20 mm, flat, hairless, rough. Panicles erect or nodding, cylindric or lobed, dense, 2.5–30 cm × 8–30 mm; bristles to 15 mm. Spikelets broadly elliptic, blunt, 2–2.5 mm, persistent; upper glume two-thirds to four-fifths of the length of the spikelet; upper floret falling separately when ripe, smooth, variously coloured, white, cream, yellow, red, brown or black. *Warm temperate Asia.* H5–G1. Summer–autumn.

4. S. viridis (Linnaeus) Palisot de Beauvois.
Illustration: Hubbard, Grasses, edn 2, 366 (1968); Christiansen, Grasses, sedges and rushes in colour, f. 90 (1979).
Loosely tufted annual with slender, erect stems, 10–60 cm. Leaves green; ligule a dense fringe of hairs; blades finely pointed, 3.5–30 cm × 4–10 mm, flat, hairless, minutely rough. Panicles spike-like, very bristly, cylindric or tapering upwards, very dense, erect, 2.5–10 cm × 4–10 mm (excluding bristles), greenish or purplish; bristles 1–3 together, to 10 mm. Spikelets elliptic-oblong, blunt, 2–3 mm; upper glumes as long as the spikelet; upper lemma very finely wrinkled. *Warm temperate Europe & Asia.* H5. Summer–autumn.

34. RHYNCHELYTRUM Nees
Annuals or perennials. Leaf-blades flat or thread-like; ligule a ciliate rim. Inflorescence a panicle with slender branches. Spikelets more or less asymmetric in profile, compressed from side to side, often silky-hairy; lower glume small, sometimes distant from the upper; upper glume as long as the spikelet, firmly membranous to almost leathery, becoming thinner towards the tip, with 5 veins, more or less swollen on one side below the middle and sometimes tapering to a beak above, notched to 2-lobed (rarely entire), often awned from the sinus; lower floret usually male, its lemma resembling the upper glume or somewhat narrower and less swollen, palea usually ciliate; upper lemma smaller than the lower, often falling before the rest of the spikelet, cartilaginous, smooth, compressed from side to side.

A genus of about 15 species in tropical and S Africa, Madagascar, Arabia and India. Only 1 species is in cultivation, but there are many others suitable for greenhouse or open borders on account of their beautiful, colourful inflorescences.

1. R. repens (Willdenow) Hubbard (*R. roseum* (Nees) Stapf & Hubbard).
Illustration: Chippindal & Cook, Grasses of southern Africa 1: t. 18 (1973).
Annual or short-lived perennial, with slender, erect or ascending stems 30–120 cm. Leaf-blades finely pointed, to 30 cm × 3–10 mm, flat. Panicle oblong to ovate, loose or dense, 5–20 × up to 10 cm. Spikelets 2.5–6 mm, densely silky-hairy with adpressed or spreading, purple, reddish or white hairs. *Tropical & S Africa.* H5–G1. Summer–autumn.

35. TRICHOLAENA Schrader
Perennials, rarely annuals. Inflorescence a panicle. Spikelets oblong in outline, awnless; lower glume very small or absent; upper glume as long as the spikelet, thinly membranous, indistinctly 5-veined, slightly notched to acute; lower floret male, its lemma similar to the upper glume; upper floret bisexual, the lemma thinly cartilaginous, clasping only the margins of the palea.

A genus of 4 species from Africa and the Mediterranean to India. They are best grown in light, well-drained loam in a greenhouse, to afford protection from rain and wind which would damage the panicles. They are propagated by seed.

1. T. teneriffae (Linnaeus) Link.
Perennial, with slender, hard, rigid,

ascending stems, 22–60 cm. Leaf-blades 2–15 cm × up to 5 mm, with a hard pointed tip, glaucous, downy above or hairless. Panicles linear, oblong or ovate, dense or loose, 2.5–12 cm. Spikelets ovate to oblong, *c.* 3 mm, pale green or purplish, silkily hairy with hairs exceeding the tip by up to 4 mm. *Mediterranean region & NE Africa.* G1. Summer–autumn.

36. PENNISETUM Richard
Annuals or perennials. Inflorescence spike-like, each spikelet or cluster of spikelets enclosed by an involucre of slender bristles which are free throughout and fall with the spikelets. Spikelets lanceolate to oblong, with 2 florets; lower glume often minute; upper glume very small to as long as the spikelet; lower floret male or sterile, its lemma as long as the spikelet or much reduced; upper floret bisexual, its lemma membranous to thinly leathery.

A genus of about 120 species in tropical and warm temperate regions. Most of the wild species have decorative inflorescences, but few are cultivated. The species may be propagated by seed, or the perennials by division of the rootstock. Those suitable for cultivation in the open need a sunny position and a well-drained soil; the tender kinds require a warm greenhouse.

1a. Bristles, especially the inner, hairy in the lower part 2
b. Bristles hairless, rough with minute, stiff projections 4
2a. Spikelets 9–15 mm; bristles to 7.5 cm; inflorescences broadly cylindric to almost spherical, 2–9 cm **1. villosum**
b. Spikelets 4–6 mm; bristles to 4 cm; inflorescences cylindric, 5–30 cm 3
3a. Inflorescence loose, flexuous; inner bristles densely ciliate in the lower part, to 12 mm **2. orientale**
b. Inflorescences moderately dense, not flexuous; inner bristles loosely hairy in the lower part, to 4 cm **3. setaceum**
4a. Bristles to 3 cm, spreading; leaf-sheaths compressed; inflorescences 2.5–5 cm wide **4. alopecuroides**
b. Bristles to 1.5 cm; leaf-sheaths not compressed; inflorescences 8–20 mm wide **5. macrourum**

1. P. villosum Fresenius. Illustration: Grounds, Ornamental grasses, t. 31 (1979).
Loosely tufted perennial with erect or ascending stems, 15–60 cm, loosely to densely hairy below the inflorescences. Leaf-blades finely pointed, 6.5–15 cm or more long, 2–6 mm wide, flat or folded. Inflorescence broadly cylindric to almost spherical, dense, erect or nodding, feathery, 2–9 × up to 5 cm, tawny or purplish; bristles numerous, spreading, loosely to densely hairy below the middle, the longer to 7.5 cm. Spikelets lanceolate, pointed, 9–15 mm. *Mountains of NE tropical Africa.* H5–G1. Summer–autumn.

2. P. orientale Richard.
Tufted perennial, with short, stout rhizomes. Stems slender, erect or spreading, 15–90 cm, with bearded nodes. Leaf-blades to 10 cm or more × 4 mm, narrowly linear, with very finely pointed, straight or flexuous tips, rough with very small, stiff projections. Inflorescence loose, flexuous, pale or purplish, silky, 5–12.5 × up to 2.5 cm; axis hairy, beset with scattered minute stumps; bristles numerous, flexuous, 4–5 times as long as the spikelets, the inner densely ciliate in the lower half, the outer rough. Spikelets lanceolate, 5–6 mm, in clusters of 2 to 5. *SW Asia to C Asia & NW India.* H5. Summer–autumn.

3. P. setaceum (Forsskahl) Chiovenda.
Densely tufted perennial, with erect, stiff, slender stems 23–90 cm, rough and sometimes hairy beneath the inflorescence. Leaf-blades finely pointed, to 30 cm × up to 3 mm, rolled or flat, erect, rigid, very rough. Inflorescence moderately dense, erect or slightly inclined, feathery, silky, pale or purplish, 10–30 × up to 4 cm; axis beset with minute stumps; bristles numerous, very slender, straight, loosely hairy in the lower part, one stouter than the rest and 1.6–4 cm, the others to 2.5 cm. Spikelets lanceolate, 4–6 mm, solitary or in clusters of 2 to 3. *N & NE tropical Africa, SW Asia, Arabia.* H5–G1. Summer–autumn.

4. P. alopecuroides (Linnaeus) Sprengel.
Densely tufted perennial, with slender, erect stems 30–150 cm. Leaf-sheaths compressed; blades slender, finely pointed, 30–60 cm long, 4–8 mm wide when opened out, hairless or nearly so. Inflorescence narrowly oblong, cylindric, 5–20 × 2.5–5 cm, pale yellow or greenish to dark purple; axis closely studded with minute stumps; bristles numerous, straight or flexuous, very fine, rough, unequal, to 3 cm, spreading. Spikelets lanceolate, pointed, 6–8 mm, solitary or paired. *E Asia to E Australia.* H5–G1. Summer–autumn.

5. P. macrourum Trinius.
Densely tufted perennial, with slender to stout, erect stems to 1.8 m. Leaves mostly basal, blades linear, very finely pointed, to 60 cm × 4–12 mm, flat or rolled, rough. Inflorescence cylindric, dense, erect or inclined, 7.5–30 cm × 8–20 mm, tawny to light brown or purplish; axis closely beset with minute stumps; bristles numerous, rough with very small, stiff hairs, one longer than the rest and to 1.5 cm long, the rest about as long as the spikelets. Spikelets lanceolate, pointed, 4–6 mm, solitary. *S Africa.* H5. Summer–autumn.

37. CENCHRUS Linnaeus.
Annuals or perennials. Inflorescence spike-like, axis angular; each spikelet or cluster of spikelets enclosed by an involucre of more or less flattened bristles or spines which are united below to form a disc or cup which falls with the spikelets. Spikelets lanceolate or ovate, awnless, with 2 florets; lower glume up to half as long as spikelet, sometimes absent; upper glume as long as spikelet; lower floret male or sterile, its lemma as long as spikelet; upper floret bisexual, its lemma firmly membranous to leathery.

A genus of 25–30 species mainly in warm, dry parts of America, Africa and India. The species below can be raised from seed but needs protection from frost. It is an excellent forage grass and is also grown in gardens for its unusual spike-like inflorescence.

1. C. ciliaris Linnaeus (*Pennisetum cenchroides* Richard; *P. ciliare* (Linnaeus) Link). Illustration: Chippindal & Cook, Grasses of southern Africa **1**: t. 1 (1976).
Perennial, with a tough rootstock and slender, rigid stems, 15–90 cm, ascending from a sharply bent base. Leaves hairy or hairless, 7.5–30 cm × 2–4 mm; ligule a dense row of hairs. Inflorescence cylindric, moderately dense, 2.5–15 × 4–16 mm, pale or purplish. Spikelets solitary or in clusters of 2 to 3, surrounded by numerous bristles, the inner bristles densely ciliate, thickened and united at the base. *Africa eastwards to India.* H5–G1. Summer–autumn.

38. BRACHYPODIUM Palisot de Beauvois
Perennials or annuals, rarely woody below, often with extensively branched rhizomes. Inflorescence a raceme of alternate, shortly stalked spikelets inserted in 2 ranks, with the back of the lemmas towards the axis. Spikelets 1 (rarely to 3) at each node, usually with numerous florets. Glumes unequal, shorter than the lowest floret. Glumes and lemmas acuminate, mucronate or with a straight or flexuous, apical awn. Palea equalling or a little shorter than the

lemma, notched or truncate, the keels ciliate or rough with short, stiff hairs. Ovary hairy at the apex. Grain narrowly elliptic to oblanceolate; hilum linear.

A genus of about 16 species, mainly in temperate regions of the northern hemisphere. The *Brachypodium* described below is especially useful for establishing under trees, either by growing from seed or by dividing existing rootstocks.

Species of the genus *Lolium* Linnaeus are frequently grown as lawn, turf or forage grasses. They will key out to *Brachypodium* in the key to genera (p. 31), but can be distinguished by their spikelets, in which the lower glume is present, the upper absent.

1. B. sylvaticum (Hudson) Palisot de Beauvois. Illustration: Hubbard, Grasses, edn 2, 90 (1968); Christiansen, Grasses, sedges and rushes in colour, f. 67 (1979).

A tufted perennial with erect or spreading slender stems to 1 m. Leaves linear-lanceolate, hairy, to 30 × 4–12 mm, yellow-green. Spikelets 18–45 mm, with 8–16 flowers; lemmas tipped with a fine awn which is up to 12 mm. *Most of Europe, temperate Asia & NW Africa*. H1. Summer.

39. AEGILOPS Linnaeus

Annuals. Leaves usually flat. Inflorescence a spike. Spikelets solitary at the nodes of the axis, all bisexual or the upper sterile and the lower vestigial. Florets 2–8, the upper usually male or vestigial. Glumes equal, leathery, truncate, often with 1 or more teeth or awns, usually rounded on the back. Lemma thin below, leathery, and strongly veined towards the toothed or awned apex. Palea with 2 keels.

A genus of about 20 species. The curious bristly inflorescence may be dried for decorative purposes. Seed should be sown in spring out-of-doors.

1. A. ovata Linnaeus. Illustration: Hegi, Flora von Mitteleuropa, edn 2, **1**: t. 39 (1957).

Stems slender, 10–30 cm. Leaves short, mostly hairy. Spikes short, ovoid or ellipsoid, 1–2.5 cm, of 2–4 spikelets, only the lower 2 fertile, the rest rudimentary. Glumes each with 3–5 long, spreading, rough awns. *Mediterranean region*. H5. Summer.

40. ELYMUS Linnaeus

Tufted or rhizomatous perennials. Leaves flat or more or less rolled; ligule short, membranous. Inflorescence a spike, axis usually tough. Spikelets solitary or in groups of 2–3 at each node, almost stalkless, usually overlapping, with 2–11 florets; rhachilla separating above the glumes and beneath each floret or sometimes the spikelets falling entire at maturity. Glumes with 1–11 veins, awnless or with short awns. Lemma lanceolate, with 5 veins. Palea 2-keeled.

A genus of about 100 species in north temperate regions. Propagated either by division of the rootstock or from seed sown outside in spring.

1a. Lemma awnless or with a short awn to 2.5 cm **1. virginicus**
b. Lemma awned, the awn to 4 cm, becoming curved or flexuous and spreading at maturity **2. canadensis**

1. E. virginicus Linnaeus. Illustration: Gould, The grasses of Texas, f. 84 (1975).

Stems tufted, erect, slender, 30–135 cm. Leaf-blades green or glaucous, finely pointed, to 25 cm × 5–15 mm, flat, rough with small, stiff hairs. Spikes erect, stiff, dense, 5–15 cm. Spikelets 1–1.5 cm paired, with 2–4 florets; glumes linear-lanceolate to narrowly lanceolate, 1–1.2 cm, smooth, yellowish, bowed out or arched at the base, strongly veined, rough with small, stiff hairs, firm and hardened above, tapering into straight awns of varying length; lemmas lanceolate, 1–1.2 cm, rough or hairy, tapering into straight awns to 2.5 cm, or awnless. *Canada & USA*. H2. Summer.

2. E. canadensis Linnaeus. Illustration: Gould, The grasses of Texas, f. 83 (1975).

Tufted, green or glaucous plant with erect, slender to stout stems, 75–180 cm. Leaf-blades very finely pointed, to 45 cm × 8–20 mm, flat or rolled, firm, rough or sparsely bristly above. Spikes dense, very bristly, nodding or drooping, 10–25 cm. Spikelets 1.2–1.6 cm (awns excluded) with 2–5 florets, in clusters of 2 to 4; glumes linear-lanceolate, stiff, rough with small, stiff hairs, awned from the tip with an awn to 1.6 cm; lemmas lanceolate, 1–1.6 cm, hairy, rough or hairless, awned from the tip with the awn 2.5–4 cm, becoming curved or flexuous. *Canada & USA*. H2. Summer.

41. SITANION Rafinesque

Tufted perennials with slender stems; ligules membranous. Leaf-blades flat or rolled. Inflorescence a very bristly spike. Spikelets with 2 to few florets, usually in pairs, alternating on opposite sides at each node of the fragile main axis of the spike, the axis breaking at the base of each joint at maturity; glumes bristle-like, with 1–2 veins, extending into 1–3 or more awns; lemmas rounded on the back, 5-veined, firm, slightly 2-toothed at the apex, the central vein extending into a long slender spreading awn, sometimes with 1 or more lateral awns.

A genus of 2 species in western N America. Propagated from seed. Useful in borders for their very ornamental bristly spikes; these must be cut when young if required for interior decoration as they soon break up.

1. S. hystrix (Nuttall) J.G. Smith.

Stems rather stiff, 10–60 cm, erect or spreading. Leaves hairy or hairless; blades flat or rolled, 5–20 cm × 2–6 mm, rather stiff. Spikes erect, 5–20 cm. Glumes very narrow, extending into long awns which are covered with short, stiff hairs; lemmas hairless or short-hairy, smooth or rough, awned from the tip, the awns of glumes and lemmas finally widely spreading, 2.5–10 cm. *N America*. H2. Summer.

42. HYSTRIX Moench

Erect perennials, with flat leaf-blades. Inflorescence a loose, bristly spike. Spikelets with 2–4 florets, 1–4 at each node of a continuous, flattened axis, horizontally spreading or ascending at maturity. Glumes reduced to short or minute awns, the first usually rudimentary, both often absent from upper spikelets. Lemmas convex, rigid, tapering into long awns, with 5 veins, the veins obscure except towards the tip. Palea about as long as the body of the lemma.

A genus of 6 species in N America, N India, China and New Zealand. They may be propagated by division or from seed, and can be grown in an open border for their decorative spikes which may be cut and used fresh or dried.

1. H. patula Moench.

Stems slender, 60–120 cm. Leaf-blades 8–15 mm wide. Spikes nodding, to 15 cm. Spikelets mostly in pairs, *c.* 1.5 cm, eventually spreading horizontally, the lemmas with awns to 4 cm. *N America*. H5. Summer.

43. TAENIATHERUM Nevski

Annuals. Leaves flat or rolled. Inflorescence a dense spike with spikelets in 2 ranks, separating at the nodes below the spikelets at maturity. Spikelets arranged in pairs at the nodes of the axis; florets 2, the lower bisexual, the upper rudimentary and sterile. Glumes joined at the base, narrowly subulate, rigid. Lemma lanceolate, with a flattened callus and a long, flexuous, rough awn.

A genus with probably only 2 species in S Europe and the Middle East. Seed should be sown in an open sunny border in well-drained soil. The spikes are suitable fresh or dried for interior decoration.

1. T. crinitum (Schreber) Nevski.
Stems erect or ascending, 15–45 cm; ligules very short, truncate. Leaf-blades finely pointed, to 12.5 cm × 3 mm, flat or rolled, hairless or hairy. Spikes erect, 7.5–15 cm. Spikelets very long-awned; glumes very fine, rough, erect or spreading, 2–3 cm; lemmas smooth or rough on the back, 1–1.2 cm, narrowed into an erect or spreading, rough awn to 12 cm. *C & E Mediterranean region.* H5. Summer.

44. HORDEUM Linnaeus
Annuals or perennials. Leaves usually flat. Inflorescence a compressed, linear to oblong spike. Spikelets 3 at each node, dispersed together at maturity, the triplets arranged in 2 longitudinal rows, the fertile florets in 2, 4 or 6 longitudinal rows, each triplet with a central bisexual spikelet and 2 lateral bisexual, male or sterile spikelets. Florets 1 per spikelet, rarely 2. Glumes linear-subulate to lanceolate and awned, free to the base. Lemma 5-veined, ovate. Palea narrowly ovate, keeled. Rhachilla usually prolonged in central spikelets.

A genus of about 25 species in temperate regions of the northern hemisphere and S America. They should be sown outdoors in spring in light garden soil. If grown for interior decoration, the spikes should be cut when young and hung upside down to dry, otherwise they will readily break up.

1a. Perennial; awns fine, hair-like, 3.5–9 cm; lateral spikelets mostly reduced to the awns **1. jubatum**
b. Annual; awns bristle-like, 1–2 cm; lateral spikelets with well-developed lemma **2. hystrix**

1. H. jubatum Linnaeus. Illustration: Britton & Brown, Illustrated flora of the United States and Canada, edn 2, **1**: 287 (1913).
Perennial, 23–60 cm; stems slender, erect or spreading, tufted or solitary. Leaf-blades 5–15 cm × 2–3 mm, flat, rough. Spikes very dense, brittle, very bristly, soft, nodding, 7.5–13 × 4–9 cm (including the awns), pale green or purplish. Lateral spikelets shortly stalked, sterile; glumes like those of the middle spikelet; lemma small when present. Middle spikelet complete: glumes bristle-like, very fine, spreading, minutely rough, to 7.5 cm; lemma lanceolate, *c.* 6 mm, with a very fine hair-like awn 3.5–9 cm. *N America, NE Asia.* H1. Summer–autumn.

2. H. hystrix Roth.
Annual, 10–40 cm, with slender, ascending, solitary or tufted stems. Leaf-blades finely pointed, to 7.5 cm × 3.5 mm, flat, mostly softly hairy. Spikes oblong or ovate-oblong, very bristly, 2.5–6 × 1.2–1.8 cm, greyish green or tinged with purple. Lateral spikelets shortly stalked, sterile; glumes bristle-like, rigid, spreading, rough, 8–20 mm; lemma 8–12 mm. Glumes of middle spikelet bristle-like, 1–2.5 cm; lemma lanceolate, 5–10 mm, with a spreading awn 1–2 cm. *Mediterranean region, C Asia.* H4. Spring–summer.

45. BROMUS Linnaeus
Annuals, biennials or perennials. Leaves flat or somewhat rolled; ligule membranous, often jagged. Inflorescence a panicle. Spikelets with 1–many florets; glumes with 1–9 veins, persistent at maturity, the upper usually the larger; lemma with few to many veins, awnless or with an almost terminal awn. Ovary crowned by a fleshy, hairy cap, the styles emerging from beneath it. Grain normally adherent to the palea and tightly enclosed within the lemma; hilum narrow, elongate.

A genus of over 100 species mainly in north temperate regions but also in S America, S Africa and on high mountains in the tropics. A few species are useful fodder grasses, and those described below all have ornamental panicles, which can be dried. They all thrive in ordinary garden soil and should be sown in autumn or spring.
Literature: Scholz, H., Zur Systematik der Gattung Bromus Subgenus Bromus, *Willdenowia* **6**: 139–59 (1970).

1a. Lemma with 3 awns **1. danthoniae**
b. Lemma with 1 awn or almost awnless 2

2a. Lower glume with 1 vein, upper glume with 3 veins; spikelets wedge-shaped, wider at the tip **2. madritensis**
b. Lower glume with 3–7 veins, upper glume with 5–9 veins; spikelets ovate or lanceolate, tapering towards the tip 3

3a. Lemma awnless or with a short awn to 1 mm **3. briziformis**
b. Lemma with an awn to 1.5 cm, flattened at the base 4

4a. Panicle compact, erect and rather dense; spikelet-stalks mostly equalling or shorter than spikelets **4. lanceolatus**
b. Panicle loose, nodding; spikelet-stalks longer than spikelets or panicle racemose 5

5a. Panicle usually 1-sided, racemose, with few spikelets **5. squarrosus**
b. Panicle usually compound with numerous spikelets **6. japonicus**

1. B. danthoniae Trinius.
Annual with solitary or tufted, slender, erect stems to 45 cm. Leaves narrowly linear, to 10 cm, the blades and sheaths softly hairy. Panicle of 1 to few spikelets, narrow, erect. Spikelets lanceolate or oblong, to 5 cm, green or purplish, glistening, hairless or hairy; lemmas 1–1.3 cm long, *c.* 6 mm wide, with 3 awns from just below the tip, the central awn flattened in the lower half and becoming recurved and twisted at maturity, often purple or reddish purple. *SW & C Asia* H4. Spring–summer.

2. B. madritensis Linnaeus. Illustration: Hubbard, Grasses, edn 2, 66 (1968).
Annual, 15–60 cm, with slender, solitary or tufted stems. Leaves hairy, especially on the lower sheaths; ligule to 4 mm. Panicle erect or slightly inclined, contracted and rather dense or somewhat loose, 3.5–15 cm long, to 5.5 cm wide, purple or green. Spikelets oblong, becoming wedge-shaped and gaping, 3–5.5 cm, compressed, with 6–13 florets; lemmas linear-lanceolate in side view, 1.2–2 cm, tipped with a fine, slightly diverging awn 1.2–2 cm. *Mediterranean region.* H4. Spring–summer.

3. B. briziformis Fischer & Meyer. Illustration: Britton & Brown, Illustrated flora of the United States and Canada, edn 2, **1**: 280 (1913).
Annual with solitary or tufted, slender, ascending stems, 30–60 cm. Leaf-sheaths softly hairy; leaves to 15 cm. Panicle loose, pyramidal, drooping, with the branches bearing 1 or rarely 2 spikelets, green or purplish. Spikelets ovate, 1.5–2.5 × 1–1.5 cm, with 10–15 florets; lemmas diamond-shaped or almost square, 8–15 × 4–8 mm, awnless or almost so. *SW & C Asia.* H4. Summer.

4. B. lanceolatus Roth.
Annual with solitary or tufted, slender, erect stems, 20–60 cm. Leaves hairy; ligule short. Panicle erect, narrow, often compact or rather dense, of relatively few spikelets. Spikelets lanceolate to oblong, to 3 cm, with 10–20 florets, green or purplish;

lemmas 1.2–1.5 cm, with a rather stout awn from just below the tip, the awn flattened and often purplish in the lower half, to 1.5 cm, becoming recurved and slightly twisted at maturity. *Mediterranean region.* H5. Summer.

5. B. squarrosus Linnaeus. Illustration: Hegi, Flora von Mitteleuropa, edn 2, **1**: f. 258 (1957).
Annual or biennial, with solitary or clustered, erect or ascending, slender stems, to 45 cm. Leaves hairy, the blades linear, to 6 mm wide. Panicle loose, of few spikelets, nodding, variable in size. Spikelets ovate to broadly oblong-elliptic, 1.8–2.5 cm long, up to 8 mm wide, with up to 20 florets, green or purplish; lemmas *c.* 1 cm broad, with a spreading or recurved awn to 1 cm, from just below the tip. *Mediterranean region.* H5. Summer.

6. B. japonicus Thunberg.
A variable annual or perennial with solitary or tufted, slender, erect or spreading stems to 60 cm. Leaves finely hairy, the blades to 10 cm × 6 mm; ligule to 4 mm. Panicle loose, nodding, of few to many spikelets, 5–15 cm long. Spikelets lanceolate to oblong, 1.8–2.5 cm, green or purplish, with 7–10 florets; lemmas 8–10 mm, awned from near the tip, with the awn 1–1.5 cm, often becoming slightly recurved. *Mediterranean region & temperate Asia.* H4. Summer.

46. AIRA Linnaeus
Annuals. Leaves often rolled. Inflorescence a panicle. Spikelets compressed from side to side, with 2 bisexual florets, the upper stalkless or almost so. Glumes about equal, membranous, equalling or longer than the florets, with 1–3 veins. Lemma with 5 veins in the basal half, bifid at apex, with an awn arising from below the middle on the back; sometimes the lower or both florets awnless. Palea shorter than lemma. Rhachilla not prolonged. Grain more or less fusiform, longitudinally furrowed, hairless.

A genus of 9 species. Some are grown for their decorative value and for bouquets; they grow well sown in spring, in ordinary garden soil, especially the lighter kinds, and will thrive in semi-shade.

1a. Spikelets 2.5–3 mm **1. caryophyllea**
b. Spikelets 2–2.5 mm **2. elegantissima**

1. A. caryophyllea Linnaeus. Illustration: Hubbard, Grasses, edn 2, 258 (1968); Christiansen, Grasses, sedges and rushes in colour, f. 22 (1979).
Tufted annual or biennial, with slender stems, 8–30 cm. Leaf-blades to 5 cm × 1 mm, minutely rough; ligule to 5 mm. Panicles very loose, with widely spreading branches, to 10 cm; spikelet-stalks 1–10 mm. Spikelets 2.5–3 mm, silvery or purplish; glumes pointed; lemmas both awned, the awns projecting from the tips of the spikelets. *Europe, N & W Asia, N Africa, and mountains of tropical Africa.* H1. Spring–summer.

2. A. elegantissima Schur (*A. capillaris* Host; *A. elegans* invalid). Illustration: Hegi, Flora von Mitteleuropa, edn 2, **1**: f. 188 (1957).
Loosely tufted annual, to 30 cm, with very slender stems. Leaf-blades to 5 cm × 1 mm; ligule to 5 mm. Panicles very diffuse, 2.5–10 cm long and wide, with loosely divided, spreading branches; spikelet-stalks to 1 cm or longer, Spikelets 2–2.5 mm, silvery, scattered, 1–2-awned, the awns shortly projecting from the tips of the glumes. *Mediterranean region.* H4. Spring–summer.

47. DESCHAMPSIA Palisot de Beauvois
Perennials, usually tufted. Leaves flat or bristle-like. Inflorescence a panicle. Spikelets compressed from side to side, with 2 (rarely 3) bisexual florets. Glumes more or less equal, membranous, somewhat shorter than or equalling the florets, with 1–3 veins. Lemma obscurely 5-veined, truncate and irregularly toothed at apex, with short hairs at the base; awn arising on the back, straight or sharply bent. Palea as long as the lemma, with 2 rough keels. Rhachilla prolonged.

A genus of about 40 species in temperate and cold regions, also on mountains in the tropics. The species below can be grown in most good soils and can either be grown from seed sown outside in autumn or spring or propagated by division of the rootstock.

1a. Leaves less than 1 mm wide, tightly inrolled; awn sharply bent **1. flexuosa**
b. Leaves more than 1 mm wide, flat or sometimes inrolled; awn straight **2. cespitosa**

1. D. flexuosa (Linnaeus) Trinius. Illustration: Hubbard, Grasses, edn 2, 250 (1968); Christiansen, Grasses, sedges and rushes in colour, f. 22 (1979).
Loosely to densely tufted perennial, 8–90 cm, with smooth, slender, wiry stems. Leaves hairless; ligules blunt, to 3 mm; blades rather stiff, bristle-like, to 20 × 1 mm, rough towards the tips. Panicles open and very loose, 3.5–15 × 7.5 cm. Spikelets 4–6 mm, purplish, brownish or silvery, with awns 4–8 mm. *Eurasia, most of America, mountains in the Old World tropics.* H1. Summer.

2. D. cespitosa (Linnaeus) Palisot de Beauvois. Illustration: Hubbard, Grasses, edn 2, 252 (1968); Christiansen, Grasses, sedges and rushes in colour, f. 23 (1979).
Densely tufted perennial, 30–150 cm, forming large tussocks, with erect, smooth stems. Leaves hairless; ligules to 1.5 cm; blades 10–60 cm × 2–5 mm, flat or inrolled, coarse, rough above on the ribs and margins. Panicles very loose and open, 10–45 cm long and up to 20 cm wide. Spikelets 4–6 mm, green, silvery, golden, purple or variegated, with awns to 4 mm. *Temperate regions, and mountains in tropical Africa & Asia.* H1. Summer.

48. HOLCUS Linnaeus
Perennials or annuals. Leaves flat, Inflorescence a panicle. Spikelets compressed from side to side, with 2–3 florets, the lower bisexual, the upper usually male. Glumes almost equal, membranous, longer than the florets, strongly keeled, the lower with 1 vein, the upper with 3 veins. Lemma obscurely 3–5-veined, leathery, shiny, the upper or both usually with an awn on the back. Palea membranous, slightly shorter than the lemma. Rhachilla shortly prolonged.

A genus of about 8 species in Europe, temperate Asia and N and S Africa. They often invade neglected gardens and are difficult to eradicate.

1. H. mollis Linnaeus. Illustration: Hubbard, Grasses, edn 2, 264 (1968); Christiansen, Grasses, sedges and rushes in colour, f. 19 (1979).
Perennial, 15–45 cm, with tough creeping rhizomes and erect slender stems. Leaf-blades hairy; panicles pale. *Europe.* H1. Summer.

A variant with variegated leaves is widely grown.

49. AVENA Linnaeus
Annuals. Leaves flat; ligule membranous. Inflorescence a panicle. Spikelets large, with 2–5 (rarely 1, or as many as 7) florets. Glumes lanceolate, acuminate, usually about equal, thinly papery, with a shiny scarious margin and several veins. Lemma usually leathery, 2-toothed or with 2 fine hairs (rarely almost entire), generally with

an awn on the back; awn usually with a thick, twisted lower part, and a thinner, straight upper part, usually sharply bent. Palea tough, shorter than the lemma, with translucent margin and 2 ciliate keels.

A genus of about 10–15 species, cultivated in temperate regions or occurring as a weed in crops. Easily grown from seed sown in spring and autumn in ordinary garden soil.

Literature: Baum, B., *Oats: wild and cultivated* (1977).

1a. Spikelets persistent at maturity, the florets not separating from the glumes; lemmas hairless, awnless or with a straight awn **1. sativa**

b. Spikelets breaking up above the persistent glumes at maturity; lemmas hairy, especially at the base, awned, with the awn sharply bent and twisted below the bend 2

2a. Spikelets 3.3–4.3 cm; awns 5–7.5 cm **2. sterilis**

b. Spikelets and awns shorter **3. ludoviciana**

1. A. sativa Linnaeus. Illustration: Christiansen, Grasses, sedges and rushes in colour, f. 85 (1979).

Stems erect, slender or stout, 1–2 m. Leaves to 60 cm; ligules to 5 mm. Panicle loose or somewhat contracted, variable in size and shape. Spikelets 1.8–2.5 cm, with 1–3 florets, the lower 1–2-awned or all awnless. *Temperate regions.* H1. Summer.

2. A. sterilis Linnaeus. Illustration: Bonnier, Flore Complete **12**: t. 665 (1934).

Stems erect or spreading, solitary or tufted, slender to moderately stout, to 1 m. Leaves to 30 × 1.2 cm; ligules to 5 mm. Panicle loose, to 30 × 20 cm. Spikelets 3.3–4.3 cm with 3–5 florets; lower 2–3 lemmas awned, with the awns 5–7.5 cm. *Atlantic islands, Mediterranean region & SW Asia.* H5. Summer.

3. A. ludoviciana Durieu. Illustration: Hubbard, Grasses, edn 2, 240 (1968).

Stems solitary or tufted, erect, stout, to 2 m. Leaves to 60 cm × 6–15 mm, ligules to 8 mm. Panicle nodding, pyramidal, very loose, to 45 cm. Spikelets 1–3 cm with 2–3 florets; lower 2 lemmas awned, the awns 3–5 cm. *Europe & C Asia.* H1. Summer.

50. HELICTOTRICHON Besser

Tufted perennials. Leaves distinctly ribbed above; ligule of stem-leaves usually truncate or toothed; sheaths of basal leaves usually open to the base. Inflorescence a compound panicle. Spikelets usually numerous, erect or spreading, slightly compressed from side to side, with 2–4 fertile and 1 (or rarely 2) sterile florets; all fertile florets usually awned. Glumes lanceolate, unequal, membranous, acute, the lower with 1 vein, the upper with 3 veins. Lemma lanceolate, with 5–7 veins, hairless, slightly 2-toothed; awn sharply bent, arising at about the middle of the lemma, more or less terete in the lower part. Palea more or less bifid at apex, shortly ciliate on keels. Ovary densely hairy.

About 60 species in temperate regions of the northern hemisphere and S Africa, and on high mountains in the Old World tropics. Seeds should be sown outside in the spring in any good garden soil.

1a. Leaf-blades flat or folded, to 12 mm wide, green; lower sheaths minutely rough; spikelets with 4–7 florets, in stiff, erect, rather dense panicles **1. planiculme**

b. Leaf-blades usually rolled, to 1.6 mm in diameter, glaucous; lower sheaths smooth; spikelets with 2–3 florets, in loose, open panicles **2. sempervirens**

1. H. planiculme (Schrader) Pilger (*Avenula planiculmis* (Schrader) Sauer & Chmelitschek). Illustration: Hegi, Flora von Mitteleuropa, edn 2, **1**: f. 196 (1957).

Compactly tufted, with erect, moderately stout stems, 45–90 cm. Lower leaf-sheaths minutely rough; ligules to 1.2 cm, blades abruptly pointed, 7.5–40 cm × 5–12 mm, flat or folded, green, smooth. Panicles contracted, rather dense, 15–33 × 1.2–3.5 cm. Spikelets linear to oblong, 1.2–2.5 cm, with 4–7 florets, tinged with purple; glumes 8–15 mm; lemmas 1–1.5–cm, narrowly lanceolate-oblong, bearded at the base with hairs to 1.6 mm, awns 1.5–2.5 cm. *C & SE Europe & SW Asia.* H4. Summer.

2. H. sempervirens (Villars) Pilger. Illustration: Grounds, Ornamental grasses, t. 23 (1979).

Densely tufted, with slender, erect, rigid stems, 30–120 cm. Leaf-sheaths hairless; ligules obscure, to 1 mm, blades finely pointed, 10–34 cm long, tightly inrolled and 0.6–1.6 mm thick, or opening out and up to 3 mm wide, glaucous, stiff. Panicles loose, open, 5–18 cm. Spikelets oblong, 10–15 mm, with 2–3 florets, yellowish or brownish, variegated with purple; glumes 1–1.2 cm; lemmas lanceolate, 8–12 mm, bearded at the base with hairs 3–4 mm; awns 1.5–2.5 cm. *SW Europe.* H4. Summer.

51. ARRHENATHERUM Palisot de Beauvois

Perennials; basal internodes often swollen and more or less spherical. Leaves flat, rolled when young, not strongly ribbed above. Inflorescence a panicle. Spikelets slightly compressed from side to side, usually with 3 florets, 2 fertile and 1 rudimentary. Lower floret male, with a sharply bent awn arising from the lower third of the back of the lemma, the upper female or bisexual, usually with a short, slender, straight bristle, rarely both florets bisexual and awned. Glumes unequal, translucent, the lower with 1–3 veins, the upper about as long as the florets, with 3 (rarely 5) veins. Lemma with 5–9 veins. Palea shorter than lemma, shortly 2-toothed at apex, ciliate on keels. Lodicules 2. Stamens 3. Florets falling together at maturity. Grain more or less terete, with hilum extending for one-half to two-thirds of its length.

A genus of about 6 species in Europe, N Africa and N & W Asia. Easily propagated by division.

1. A. elatius (Linnaeus) Presl (*A. avenaceum* Palisot de Beauvois). Illustration: Hubbard, Grasses, edn 2, 234 (1968); Christiansen, Grasses, sedges and rushes in colour, f. 26 (1979).

Loosely tufted. Stems 50–150 (rarely to 180) cm, smooth, shining. Leaves 10–40 cm × 10 mm, green, usually hairless or almost so, more or less rough with short, stiff projections; ligule 1–3 mm, obtuse, usually ciliate. Panicle 10–30 cm (rarely less), oblong or lanceolate, loose to rather dense. Spikelets 7–10 mm, oblong; rhachilla between first and second florets not more than 0.6 mm. Glumes acute, the lower 4–6 mm, oblong-lanceolate, the upper 7–10 mm, ovate-lanceolate. Lemma 7–10 mm, oblanceolate, more or less acute, often shortly 2-toothed, hairless or with sparse hairs which are up to 1 mm. Awn of the lower floret 1–2 cm, arising in the lower third of the lemma. Anthers 4–5 mm. Grain 4–5 mm. *Most of Europe, but absent from much of the north east.* H2. Summer.

52. GAUDINIA Palisot de Beauvois

Annuals or biennials. Leaves flat. Inflorescence a spike with spikelets in 2 ranks; axis fragile, breaking up above the insertion of the spikelet. Spikelets compressed from side to side, stalkless, more or less adpressed to the concave axis, with 3–11 florets. Glumes unequal, shorter than or equalling the spikelet, the lower with 3 (rarely 5), the upper with 4–7 (rarely up to 11) strong veins. Lemma obscurely 7–9-veined, leathery, often with an awn on the back. Palea shorter than lemma.

Three species in southern Europe and the Mediterranean region. Seed should be sown thinly in the open in spring in light garden soils, and the spikes cut when young for use fresh or dried in interior decorations.

1. G. **fragilis** (Linnaeus) Palisot de Beauvois (*Avena fragilis* Linnaeus). Illustration; Bonnier, Flore Complete **12**: t. 694 (1934).
Annual; stems 10–30 cm, erect or spreading. Leaf-blades finely pointed, 2.5–15 cm × 2–4 mm, softly hairy. Spikes 7.5–20 cm, green. Spikelets 1–2 cm; lower glume 3–5 mm, the upper 6–8 mm; lemmas 3–6 mm, lanceolate, bearing an awn 6–20 mm. *S Europe and the Mediterranean region, occasionally occurring elsewhere*. H5. Spring–summer.

53. ANTHOXANTHUM Linnaeus
Annuals or perennials, often smelling of coumarin (new-mown hay). Leaves flat. Inflorescence a dense panicle. Spikelets compressed from side to side, of 2 sterile florets and a terminal, bisexual floret. Glumes membranous, very unequal, the lower with 1 vein, the upper with 3 veins. Lemma of sterile florets membranous, toothed or with 2 oblong, obtuse lobes at apex, with 3 veins and an awn on the back. Lemma of fertile floret somewhat hardened, shorter than the sterile ones, with 5–7 veins. Stamens 2. Lodicules absent.

A genus of about 15 species, native of Europe, N Asia, N & S America and the mountains of E Africa. Easily raised from seed sown in the open in ordinary garden soil.
Literature: Valdes, B., Revision de las Especies anuales del Genero Anthoxanthum, *Lagascalia* **3**: 99–141 (1973).

1. A. **gracile** Bivona.
Slender, tufted annual, *c.* 30 cm, with stems ascending. Leaves linear, short, slender-pointed, ciliate, more or less hairy above; ligule short, much torn. Panicle simple, spike-like, ovate, with few florets. Spikelets awned, awn twice as long as glume. *C & E Mediterranean region*. H5. Spring.

54. HIEROCHLOE R. Brown
Perennials, smelling of coumarin (new-mown hay). Leaves flat or more or less rolled, those on the flowering stem with a short blade; ligule truncate, obtuse or acute. Inflorescence a panicle. Spikelets compressed from side to side, with 2 male florets and 1 terminal, bisexual floret. Glumes membranous, almost equal, ovate, about as long as florets, obscurely 3-veined. Lemmas of male florets membranous, obscurely 3–5-veined, obtuse, mucronate or awned. Palea shorter than lemma, membranous, with 2 veins. Lemma of bisexual floret obscurely 5-veined, hard and shiny in fruit, hairy towards apex. Palea shorter than lemma, membranous, with 1 vein. Stamens 3 in male florets, 2 in bisexual floret.

A genus of about 15 species in temperate and arctic regions. Propagated by division of the rootstock or from seed. Suitable for moist or damp places, although they will grow in most garden soils.
Literature: Weimarck, G., Variation and taxonomy of Hierochloe in the northern hemisphere, *Botanisker Notiser* **124**: 129–75 (1971).

1a. Panicles 4–10 cm; lower lemmas awnless; stems 20–45 cm **1. odorata**
b. Panicles 7.5–30 cm; lower lemmas awned; stems 45–135 cm or more **2. redolens**

1. H. **odorata** (Linnaeus) Palisot de Beauvois. Illustration: Hubbard, Grasses, edn 2, 266 (1968); Christiansen, Grasses, sedges and rushes in colour, f. 2 (1979).
Plant with slender creeping rhizomes, forming dense tufts or patches, with slender, erect stems 20–45 cm. Ligules 2–4 mm. Leaf-blades to 30 × 1 cm. Panicles ovate, 4–10 cm. Spikelets stalked, 3.5–5 mm, golden-brown, tinged with green or purple at the base; glumes ovate, blunt; male lemmas 3.5–5 mm, ciliate on the margins, minutely hairy, awnless or nearly awnless like the terminal lemma. *N America, Asia & Europe*. H1. Spring.

2. H. **redolens** (Vahl) Roemer & Schultes.
Plant with widely creeping rhizomes, forming large tufts, with slender to somewhat stout stems, 45–135 cm or more. Ligules 3–6 mm. Leaf-blades 30–75 cm × 5–12 mm, bright green. Panicles loose or dense, nodding, 7.5–30 cm. Spikelets shortly stalked, 6–10 mm, yellowish brown or golden-brown; glumes broadly lanceolate to ovate, pointed; male lemmas rough with very short, stiff hairs, ciliate on the margins and keel, 5–6 mm, with straight awns to 4 mm; terminal lemma awnless. *S America, SE Australia, New Zealand*. H4(?). Summer.

55. KOELERIA Persoon
Tufted perennials. Leaves flat or rolled. Inflorescence a spike-like or loose panicle. Spikelets compressed from side to side, with 2–5 florets (rarely only 1). Glumes equal or unequal, the upper longer and usually equalling the first floret, the lower about two-thirds as long as the first floret, distinctly keeled, with 1–5 veins. Lemma lanceolate to ovate-lanceolate, keeled, mostly longer than the glumes, usually obtuse, with or without an awn. Palea as long as or shorter than the lemma, 2-keeled, bifid at the apex.

About 20–30 species in temperate regions and on mountains in tropical Africa. All the species are easy to grow and are usually propagated by division. Seed can be sown in the open in spring and will do well in most garden soils, especially those of a calcareous nature.
Literature: Domin, K., Monographic der Gattung Koeleria, *Bibliotheca Botanica*, 65 (1907).

1. K. **macrantha** (Ledebour) Schultes (*K. cristata* (Linnaeus) Persoon; *K. gracilis* Persoon). Illustration: Christiansen, Grasses, sedges and rushes in colour, f. 33 (1979).

Compactly tufted perennial, 10–50 cm, with stiff stems. Ligules to 1 mm. Leaf-blades with a fine blunt tip, to 20 cm long, rolled or flat, 1–2.5 mm wide, finely hairy or hairless. Panicle erect, very dense, narrowly oblong or tapering upwards, 2.5–10 cm × 5–20 mm, silvery, green or purplish. Spikelets with 2–3 florets, 4–6 mm, hairless or downy. *Europe, temperate Asia & N America*. H1. Summer.

56. ROSTRARIA Trinius
Annuals. Inflorescence a panicle. Spikelets laterally compressed, with 2–5 (rarely only 1, or as many as 10) florets. Glumes keeled, cartilaginous at the base. Lemma bifid, usually prominently 5-veined, keeled, with a straight or slightly curved awn inserted near the apex. Palea 2-keeled, bifid. Rhachilla and callus hairless or with short hairs.

A genus of about 10 species in temperate Eurasia and N Africa. All are easy to grow, the seed being sown in spring out-of-doors in any good garden soil.

1. R. **cristata** (Linnaeus) Tzvelev (*Koeleria phleoides* (Villars) Persoon; *Lophochloa phleoides* (Villars) Reichenbach).

Illustration: Bonnier, Flore Complete **12**: t. 668 (1934).
Annual, with loosely tufted or solitary stems 5–45 cm. Ligule *c.* 1 mm. Leaf-blades finely pointed, 2.5–15 cm × 1.6–4 mm, flat, mostly hairy. Panicles linear to lanceolate or narrowly oblong, very dense, cylindric or lobed, 1–10 cm × 6–16 mm, pale green. spikelets 3–8 mm, with 3–8 flowers. *Mediterranean region & NE Africa*. H5. Summer.

57. MILIUM Linnaeus
Annuals or perennials. Leaves flat. Inflorescence a panicle. Spikelets slightly flattened from top to bottom, with 1 floret. Glumes equal, membranous, longer than the floret, with 3 veins. Lemma with 5 veins, leathery and shiny in fruit. Palea as long as the lemma, leathery.

A genus of about 6 species in temperate parts of N America, Europe and Asia. Propagation by seed sown outside in spring, or by division of the rootstock.

1. M. effusum Linnaeus. Illustration: Hubbard, Grasses, edn 2, 276 (1968); Christiansen, Grasses, sedges and rushes in colour, f. 4 (1979).
Loosely tufted perennial, 45–180 cm. Leaf-blades 10–30 cm × 5–15 mm, hairless; ligules 3–10 mm. Panicles lanceolate to ovate or oblong, very loose, nodding, 10–30 × up to 20 cm, pale green, rarely purplish, the branches in clusters. Spikelets 3–4 mm. *Europe, Asia & northeastern N America*. H1. Spring–summer.

A variant with yellow leaves ('Aureum') is widely grown.

58. PHALARIS Linnaeus
Annuals or perennials. Leaves flat. Inflorescence an ovoid or cylindric panicle, usually dense and unlobed. Spikelets strongly compressed from side to side, with 2–3 florets, the lower 1 or 2 reduced to lemmas, the uppermost bisexual. Glumes almost equal, papery, longer than the florets, with 3–5 veins, often winged on the keel. Lower lemmas small, linear to lanceolate; upper lemma with 5 veins, leathery. Palea leathery.

A genus of about 15 species, in temperate regions, mainly in the Mediterranean area. All are easily cultivated in ordinary garden soil, and are most useful for their ornamental inflorescences which may be used fresh, or dried for interior decoration. The annuals can be propagated from seed, sown thinly in spring in the open in a border, or in rows for cutting. The perennials may be increased by division of the rootstock, but are sometimes difficult to keep within bounds, the unsightly variegated form of *P. arundinacea* being a depressing feature of many an ill-tended border.
Literature: Anderson, D.E., Taxonomy and distribution of the genus Phalaris, *Iowa State Journal of Science* **36**: 1–96 (1961).

1a. Perennials with creeping rhizomes **1. arundinacea**
b. Annuals 2
2a. Spikelets of 2 kinds, clustered, 1 of each cluster fertile and surrounded by sterile spikelets, the clusters falling entire at maturity; glumes with toothed wings **2. paradoxa**
b. Spikelets all alike, breaking up at maturity and leaving the persistent glumes 3
3a. Fertile lemma 4–6 mm; sterile lemmas about half the length of the fertile **3. canariensis**
b. Fertile lemma *c.* 3 mm; sterile lemmas very dissimilar, the lower minute, the upper about one-third of the length of the fertile **4. minor**

1. P. arundinacea Linnaeus. Illustration: Hubbard, Grasses, edn 2, 274 (1968); Christiansen, Grasses, sedges and rushes in colour, f.3 (1979).
Perennial, 60–200 cm, with extensively creeping rhizomes. Stems stout, erect. Leaves hairless, green; ligules 6–10 mm; blades linear, finely pointed, 10–35 cm × 6–18 mm, firm, rough in the upper part. Panicles lanceolate to oblong, dense, lobed, 5–25 × up to 4 cm. Spikelets oblong, 5–6 mm, greenish, purplish, or whitish green; glumes narrowly lanceolate, pointed, minutely rough, almost wingless; fertile lemma narrowly ovate, 3–4 mm, finely hairy. *Europe, temperate Asia, N America, S Africa*. H1. Summer.

The variegated cultivar ('Variegata') is widely grown.

2. P. paradoxa Linnaeus. Illustration: Bonnier, Flore Complete **11**: t. 646 (1931).
Annual, 10–75 cm, with tufted or solitary, erect or spreading, slender stems. Leaves green, hairless; upper sheaths inflated and often embracing the base of the panicle; ligules 3–10 mm; blades finely pointed, to 30 cm × 2–8 mm, rough. Panicles obovoid, ellipsoid or cylindric, mostly narrowed at the base, pale green or purplish, 2–9 cm × 8–20 mm. Spikelets in clusters of 6–7 falling together at maturity, with 1 in each cluster fertile and 5–8 mm, the rest sterile and smaller; glumes of fertile spikelet lanceolate, sharply pointed, winged on the keels in the upper part, and with a prominent tooth-like projection there, glumes of sterile spikelets smaller or more or less deformed; fertile lemma narrowly ovate, 3–4 mm. *W & C Europe, & Mediterranean region*. H4. Summer.

3. P. canariensis Linnaeus. Illustration: Hubbard, Grasses, edn 2, 272 (1968).
Annual, 20–120 cm, with slender to rather stout, erect, tufted, or solitary stems. Leaves hairless, green; ligules 3–8 mm; blades linear to linear-lanceolate, finely pointed, 5–25 cm × 4–12 mm, flat, rough. Panicles ovoid to ovoid-cylindric, very dense, erect, 1.5–6.5 × 1.2–2 cm wide, stiff, whitish except for the green nerves, rarely purplish. Spikelets obovate, 6–10 × up to 6 mm; glumes oblanceolate, abruptly pointed, broadly winged on the upper part of the keels, slightly hairy, minutely rough; fertile lemma narrowly ovate, 4–6 mm, hairy, becoming glossy and bright yellow. *W Mediterranean region*. H5. Summer.

4. P. minor Retzius. Illustration: Bonnier, Flore Complete **11**: t. 646 (1931).
Annual, 10–90 cm, with slender, erect or ascending, tufted or solitary stems. Leaves hairless, green; ligules to 6 mm; blades linear, finely pointed, 5–15 cm × 3–6 mm, smooth. Panicles ovoid to cylindric, dense, 1.2–6.5 cm × 8–16 mm, rather soft, whitish green. Spikelets elliptic, 5–6 × 2.5–3 mm; glumes narrowly oblong, pointed, narrowly winged on the keels in the upper part, with the wings slightly toothed; fertile lemma *c.* 3 mm, narrowly ovate, greyish at maturity, finely hairy; sterile lemma to 1.5 mm. *Mediterranean region*. H5. Summer.

59. ALOPECURUS Linnaeus
Usually almost hairless annuals or perennials. Inflorescence a spike-like panicle. Spikelets compressed from side to side, with 1 floret. Glumes almost equal, often united below, with 3 veins. Lemma translucent, with 3 veins, usually awned from the back; margins often united below. Palea usually absent, or if present, small. Ovary hairless; styles usually united below. Rhachilla breaking up below the glumes.

A genus of about 25 species, natives of temperate regions of the northern hemisphere. The variegated forms are useful in bedding, are hardy and not particular as to soil.

1. A. pratensis Linnaeus. Illustration:

Hubbard, Grasses, edn 2, 334 (1968); Christiansen, Grasses, sedges and rushes in colour, f. 9 (1979).
A loosely to compactly tufted perennial with smooth erect stems, 30–120 cm. Leaves flat, rough to touch; sheath smooth; ligule membranous, to 2 mm, obtuse. Panicle 2.5–10 cm, dense, soft, pale green or purple. *Europe, N Asia.* H1. Spring–summer.

60. PHLEUM Linnaeus
Annuals or perennials. Leaves flat. Inflorescence a dense spike-like, ovoid or cylindric panicle. Spikelets strongly compressed from side to side with 1 floret. Glumes membranous, almost equal, longer than the lemma, keeled, shortly awned, with 3 veins, with the margins overlapping along most of their length. Lemma membranous, truncate or obtuse, with 1–7 veins, unawned, hairless to densely ciliate. Palea equalling or almost equalling the lemma, ovate-lanceolate, obtuse, with 2 veins. Stamens 2–3.

A genus of about 15 species from temperate regions, mainly in Europe and the Mediterranean region. All species are hardy in NW Europe and are easy to grow in most garden soils. The annuals should be sown thinly in the open in the spring, the perennials may be propagated from seed or by division. The spike-like inflorescences, either fresh or dried, are useful for mixed bouquets and for vases.

1. **P. pratense** Linnaeus. Illustration: Hubbard, Grasses, edn 2, 322 (1968); Christiansen, Grasses, sedges and rushes in colour, f. 6 (1979).
Loosely to densely tufted perennial, with relatively stout, erect or ascending stems, 40–150 cm, usually swollen or bulbous at the base. Leaves hairless, green; ligules to 6 mm; blades finely pointed, to 45 cm × 3–8 mm, flat, usually minutely rough. Panicles dense, cylindric, 5–35 cm × 6–10 mm, greyish green or purplish, bristly. Spikelets oblong, 3–4 mm; glumes narrowly oblong, truncate, each bearing a rigid awn 1–2 mm, the keels fringed with stiff spreading hairs. *N, W & C Europe.* H3. Summer.

61. MIBORA Adanson
Annuals. Leaves flat. Inflorescence spike-like, 1-sided. Spikelets compressed from side to side, with 1 floret. Glumes more or less equal, membranous, longer than the floret, with 1 vein, persistent. Lemma with 5 veins, truncate, densely hairy, thinner than the glumes. Palea as long as lemma, hairy.

A genus of 1 species. Seed should be sown in late summer or autumn in a light sandy soil.

1. **M. minima** (Linnaeus) Desveaux. Illustration: Hubbard, Grasses, edn 2, 338 (1968).
Compactly tufted, 2.5–15 cm, with numerous hair-like stems. Leaves hairless, blades to 2.5 cm long, flat or rolled and up to 0.5 mm wide. Racemes erect, 5–20 mm long, to 1 mm wide, reddish, purple, or green. Spikelets almost stalkless, adpressed to the axis, oblong, blunt, 2–3 mm; lemma densely hairy. *Mediterranean region, Britain.* H4. Winter–spring.

62. AGROSTIS Linnaeus
Perennials or annuals. Leaves flat or bristle-like. Inflorescence a panicle. Spikelets compressed from side to side, with 1 floret. Glumes more or less equal to somewhat unequal, membranous, longer than the floret, usually with 1 vein. Lemma membranous or scarious, truncate, with 3–5 veins, often with the lateral veins slightly projecting from the margin, sometimes with a sharply bent awn on the back. Palea shorter than the lemma, often very small. Rhachilla prolonged or not. Callus usually hairy.

A genus of about 120 species. Some of the annual species with delicate open panicles are attractive and their inflorescences may be dried for ornamental purposes. Seeds of these should be sown thinly in the open in the spring; they may also be grown in pots for decorative purposes.

1. **A. nebulosa** Boissier & Reuter.
Tufted annual to 35 cm, with flat, narrow leaf-blades; ligule blunt, to 5 mm. Panicles very diffuse and loose, to 15 cm long, with 10 or more basal branches. Spikelets *c.* 1.5 mm, awnless. *Spain, Portugal & Morocco.* H5. Summer.

63. POLYPOGON Desfontaines
Annuals or perennials. Leaves flat. Inflorescence a dense panicle. Spikelets somewhat compressed from side to side, with 1 floret. Glumes more or less equal, with 1 vein, papery, rough with small, stiff hairs, longer than floret, usually with an apical awn. Lemma with 5 veins, transparent, truncate. Palea transparent. Rhachilla not prolonged.

About 10 species, in warm temperate regions. The soft silky inflorescences are very attractive, and the plants are useful for the front of a border or bed. The panicles may also be used when young for interior decoration, but if left too long the spikelets are rapidly shed. Propagated from seed, sown thinly in the open in the spring in light loams.

1. **P. monspeliensis** (Linnaeus) Desfontaines. Illustration: Hubbard, Grasses, edn 2, 310 (1968).
Annual, 7.5–60 cm. Stems tufted or solitary, slender. Leaves hairless; ligules 3–15 mm, blades pointed, 5–15 cm × 2–8 mm. Panicles very dense, narrowly ovate to oblong, cylindric or lobed, 1.8–15 × 1–3.5 cm, pale green or yellowish. Spikelets narrowly oblong, 2–3 mm; glumes blunt and slightly notched at the tip, minutely hairy on the margins, rough below; awns 4–8 mm. *Mediterranean region, NE & S Africa.* H4. Summer.

64. CALAMAGROSTIS Adanson
Perennials. Leaves flat or rolled; ligule membranous. Spikelets narrow, with 1 floret, in dense to rather loose panicles. Lemma membranous, with 3–5 veins, shorter than the glumes, surrounded usually by long hairs from the base, with an awn arising from the back or apex. Rhachilla separating above the glumes.

A genus of over 250 species in temperate regions throughout the world. Both species described below are grasses of open places in damp woods and fens and are therefore suitable for planting in such habitats in parks and wild gardens. Their inflorescences are also very decorative both in the living state and when dried. The plants are best propagated by division of the rootstock.
Literature: Wasiljew, W.N., Das System der Gattung Calamagrostis Roth, *Feddes Repertorium* **63**: 229–51 (1960).

1a. Leaf-blades hairy on the upper surface; panicles rather loose; ligules 2–5 mm **1. canescens**
b. Leaf-blades entirely hairless; panicles dense; ligules 4–12 mm **2. epigejos**

1. **C. canescens** (Weber) Roth (*C. lanceolata* Roth). Illustration: Hubbard, Grasses, edn 2, 282 (1968); Christiansen, Grasses, sedges and rushes in colour, f. 16 (1979).
Stem slender, erect, forming loose tufts or patches, 60–120 cm high; leaves to 45 cm × 3–6 mm, shortly hairy above, rough; ligules 2–5 mm. Panicles lanceolate to oblong, moderately loose, wavy or nodding, 10–25 × 2.5–6 cm, purplish, greenish or yellowish. Spikelets lanceolate,

5–6 mm. *W, N & C Europe*. H3. Summer.

2. C. epigejos (Linnaeus) Roth. Illustration: Hubbard, Grasses, edn 2, 284 (1968); Christiansen, Grasses, sedges and rushes in colour, f. 15 (1979).
Stems erect, forming tufts or tussocks, 60–180 cm high; leaves to 60 cm × 4–10 mm, rough; ligules 4–12 mm. Panicles erect, lanceolate to oblong, dense, 15–30 × 3–5.5 cm, purplish, brownish or green. Spikelets 5–6 mm. *Throughout Europe & temperate Asia*. H1. Summer.

65. APERA Adanson
Annuals. Leaves flat or rolled. Inflorescence a panicle. Spikelets compressed from side to side, with 1 floret (rarely more). Glumes unequal, membranous, the lower with 1 vein, the upper with 3 veins and about as long as the lemma. Lemma obscurely 5-veined, papery, rounded on the back, with a long awn near the apex. Palea about equalling lemma. Rhachilla shortly prolonged.

A genus of 3 species. They are easily raised from seed which may be sown in the open in spring. The inflorescences may be dried for decorative purposes.

1. A. spica-venti (Linnaeus) Palisot de Beauvois (*Agrostis spica-venti* Linnaeus). Illustration: Hubbard, Grasses, edn 2, 290 (1968); Christiansen, Grasses, sedges and rushes in colour, f. 11 (1979).
Annual, 25–30 cm; culms slender, mostly erect. Leaves hairless, to 10 mm wide; ligule oblong, 3–10 mm. Panicle ovate to oblong, 10–25 cm. Spikelets 2.5–3 mm, green or purple, with fine awns 5–10 mm. *Europe & N Asia*. H1. Summer.

66. LAGURUS Linnaeus
Annual. Leaves flat. Inflorescence a dense, ovoid or cylindric to almost spherical panicle. Spikelets compressed from side to side, with 1 floret. Glumes almost equal, linear, membranous, with 1 vein and a long, densely ciliate, bristle-like apex. Lemma with 5 veins, membranous, hairy at the base, with 2 long apical bristles and a sharply bent awn on the back. Palea somewhat shorter than lemma. Rhachilla prolonged, ciliate.

A genus with a single species. Easily grown from seed in any good garden soil, but a light sandy one is best. Seeds sown in autumn should be protected from frost.

1. L. ovatus Linnaeus. Illustration: Hubbard, Grasses, edn 2, 314 (1968); Grounds, Ornamental grasses, t. 24 & 25 (1979).
Softly hairy annual, 15–60 cm. Leaf-blades linear to narrowly lanceolate 1.2–20 cm × 1–12 mm, velvety. Panicles ovoid to oblong-cylindric or spherical, 5–60 × 5–20 mm, bristly, softly hairy, pale, rarely purplish. *Mediterranean region*. H4. Summer.

67. ECHINARIA Desfontaines
Annual. Leaves usually flat. Inflorescence a dense, ovoid or spherical, prickly head. Spikelets almost stalkless, somewhat compressed from side to side, with 3–4 (rarely 1) florets. Glumes more or less equal, membranous, the lower with 2 (rarely 5) strong veins which run out at the margins, the upper with a midrib which runs out at the apex. Lemma leathery, with 5 (rarely 7) very strong veins prolonged as flattened awns, which are deflexed at maturity. Palea as long as the lemma, the 2 (rarely 5) veins prolonged as flattened awns.

A genus of 1 species. Seed should be sown, in the spring, in ordinary garden soil.

1. E. capitata (Linnaeus) Desfontaines. Illustration: Bonnier, Flore Complete **11**: t. 650 (1931).
Stems 5–20 cm, rigid, erect or spreading, leafy towards the base. Leaf-blades to 5 cm × 2 mm, green, downy. Flower-heads green, becoming hard and spiny, to 1.5 cm wide and long. *Mediterranean region*. H5. Summer.

68. LAMARCKIA Moench
Annual. Leaves flat. Inflorescence a rather condensed 1-sided panicle. Spikelets of 2 kinds, all somewhat compressed from side to side, falling in groups of 3–5; fertile spikelets with 1 rudimentary and 1 fertile floret, surrounded by 2–4 sterile spikelets with many florets. Glumes thin, almost equal, acute. Lemma of fertile spikelets papery, awned, of the sterile spikelets membranous, unawned, Palea 2-keeled.

A genus with a single species. Seed should be sown in late spring in a light sandy soil; it can be sown in autumn, but the plants must then be overwintered under glass before planting out in spring.

1. L. aurea (Linnaeus) Moench. Illustration: Bonnier, Flore Complete **12**: t. 677 (1934).
Stems 10–30 cm. Leaf-blades finely pointed, 5–13 cm × 4–8 mm. Panicles oblong, 2.5–7.5 × 1.2–2.5 cm, straw-coloured, golden-yellow or sometimes purplish. Fertile spikelets *c.* 4 mm; sterile spikelets 4–8 mm, hiding the fertile. *Mediterranean region*. H5. Summer.

69. CYNOSURUS Linnaeus
Perennials or annuals. Leaves flat. Inflorescence a more or less 1-sided, dense panicle. Spikelets of 2 different kinds, the fertile compressed from side to side, with 1–5 florets, the sterile below the fertile and consisting of narrow, rigid glumes and lemmas. Glumes of fertile spikelets thin, almost equal, acute. Lemma papery, with an apical awn. Palea 2-keeled, shortly bifid at apex, about as long as lemma.

A genus of 5 species mainly in the Mediterranean region, 1 widespread in Europe. Seed should be sown thinly in the spring, in light loamy soils.

1. C. echinatus Linnaeus. Illustration: Hubbard, Grasses, edn 2, 218 (1968).
Annual, with solitary or tufted, erect or ascending stems, 7–90 cm. Leaves to 20 cm × 3–10 mm; ligule blunt, to 10 mm. Panicles spike-like, dense, erect, 1-sided, bristly, 1–7.5 cm, green or purplish, shining. Spikelets in dense clusters, the fertile wedge-shaped, 8–12 mm, with 1–5 flowers; the lemmas bearing a fine straight awn, 6–15 mm, from near the tip; the sterile broadly obovate, 6–12 mm, consisting of up to 18 awned bracts. *Mediterranean region*. H5. Summer.

70. DACTYLIS Linnaeus
Perennials. Leaves flat or rolled. Inflorescence a panicle with spikelets in dense clusters on the spreading or erect branches. Spikelets compressed from side to side with 2–5 florets. Glumes almost equal, keeled, somewhat curved, the lower with 1 vein, the upper with 3 veins. Lemma with 5 veins, papery, mucronate or shortly awned at apex. Palea about as long as lemma.

A genus of 4 or 5 species in Europe, the Mediterranean region and temperate Asia. The 1 cultivated species thrives on most good soils and is easily raised from seed sown outside in spring.

1. D. glomerata Linnaeus. Illustration: Hubbard, Grasses edn 2, 216 (1968); Christiansen, Grasses, sedges and rushes in colour, f. 36 (1979).
Densely tufted, 15–120 cm, with slender to stout, erect or spreading stems and compressed vegetative shoots. Leaves green or greyish green, sharply pointed, 10–45 cm × 2–12 mm, rough; ligule 2–12 mm. Panicles 1-sided, erect, oblong

or ovate, very variable, 3–30 cm, of few branches, green, purplish or yellowish. Spikelets 5–10 mm, with 2–5 florets; lemmas stiffly ciliate or rough on the keels, tipped with an awn up to 1.5 mm long. *Temperate regions of the Old World*. H1. Summer–autumn.

71. BRIZA Linnaeus

Almost hairless annuals or perennials. Inflorescence usually much-branched. Spikelets ovoid or broadly triangular, compressed from side to side; florets 4–20. Glumes ovate-circular, heart-shaped at the base, with 3–9 veins and with wide scarious margins. Lemma almost circular, heart-shaped at the base, with 7–9 veins, closely overlapping. Palea ovate to obovate, obtuse, shorter than to almost equalling the lemma, with 2 narrowly winged keels. Ovary hairless; styles terminal. Grain flattened on one side.

A genus of about 12 species in temperate regions mainly in the northern hemisphere. Most of the species are ornamentals growing well in any good humus-rich soil in the sun, the seeds being sown in autumn or spring. The panicles, once dried, are useful for decoration and should be gathered and dried as soon as fully grown.

1a. Perennial **1. media**
b. Annual 2
2a. Spikelets 1.2–2.5 cm, fewer than 12 **2. maxima**
b. Spikelets 3–5 mm, numerous **3. minor**

1. B. media Linnaeus. Illustration: Hubbard, Grasses, edn 2, 214 (1968); Christiansen, Grasses, sedges and rushes in colour, f.37 (1979); Grounds, Ornamental grasses, t. 35 (1979).

Perennial, tufted with shortly creeping rhizome and erect, slender, smooth stems 15–100 cm. Leaves to 15 cm × 2–7 mm, rough on the margins; ligule short, truncate. Panicle 4–18 cm, loose, erect at first. Spikelets broadly elliptic to broadly ovate, 4–8 mm long and wide, with 4–12 florets, usually purplish. *Europe, N & SW Asia*. H1. Summer.

2. B. maxima Linnaeus. Illustration; Hubbard, Grasses, edn 2, 212 (1968); Grounds, Ornamental grasses, t. 36 (1979). Annual with slender, smooth, solitary or tufted stems 10–60 cm. Leaves to 20 cm × 8 mm, rough on the margins; ligule 2–5 mm, acute. Panicles 2–10 cm, loose, nodding, with up to 12 large spikelets, these ovate or oblong, 1.2–2.5 cm × 8–15 mm, variously coloured, silvery, reddish brown, purplish or greenish, with 7–20 florets, hairy or hairless. *Mediterranean region*. H5. Spring–summer.

3. B. minor Linnaeus. Illustration: Hubbard, Grasses, edn 2, 210 (1968). Annual with solitary or tufted, slender, smooth stems 10–60 cm. Leaves narrowly lanceolate, to 1.5 cm × 3–8 mm, rough; ligule lanceolate, 3–6 mm. Panicle 5–20 cm, broadly pyramidal, loose. Spikelets circular or triangular-ovate, 3–5 mm long and wide, greenish white or purplish, shining, with 4–8 florets. *W Europe and the Mediterranean region*. H4. Summer.

72. POA Linnaeus

Annuals or perennials. Inflorescence a panicle. Spikelets compressed from side to side, with 2–10 florets (rarely only 1). Glumes keeled, membranous, usually with 3 veins or the lower with 1 vein. Lemma with 5 veins, keeled, membranous, awnless or rarely with a short terminal awn. Palea 2-keeled, the keels with very small teeth or ciliate. Grain ellipsoid; hilum basal.

A genus of more than 250 species in temperate and cold regions, a few on mountains in the tropics. Many species are valuable pasture or hay grasses; others are important for lawns although they are little used for the purpose in Europe; a few are ornamental either on account of their foliage or their graceful panicles. They may be propagated from seed or by division, and generally are tolerant of a wide range of soils.

1a. Leaf-blades bright green, 5–10 mm wide; panicle loose **1. chaixii**
b. Leaf-blades bluish grey, 2–4 mm wide; panicle stiffly erect, loose or dense **2. glauca**

1. P. chaixii Villars. Illustration: Hubbard, Grasses, edn 2, 184 (1968). Compactly tufted perennial, 60–120 cm, with moderately stout, erect stems. Leaves bright green; sheaths strongly compressed and keeled, rough; ligules to 1.5 mm; blades abruptly sharply pointed, to 45 cm long, folded or flat, 5–10 mm wide, rough. Panicles ovate to ovate-oblong, loose, 10–25 × 5–11 cm, green; branches in clusters of up to 7, spreading or the lower drooping. Spikelets ovate to oblong, 5–6 mm, with 3–4 florets; lemma lanceolate-oblong, pointed, 3.5–4 mm, minutely rough, hairless. *Europe & SW Asia*. H1. Spring–summer.

2. P. glauca Vahl. Illustration: Hubbard, Grasses, edn 2, 180 (1968). Tufted, bluish grey perennial, 10–40 cm, with erect, slender, rather stiff stems, 2–3-noded below the middle. Leaves hairless; ligules 1–2.5 mm; blades abruptly pointed, 1.8–7.5 cm × 2–4 mm, stiff, smooth. Panicles stiffly erect, lanceolate to ovate, loose or dense, 1.8–10 × up to 3.8 cm, usually variegated with purple; branches mostly 2–3 together. Spikelets ovate to oblong, 4–6 mm, with 3–6 florets; glumes with 3 veins; lemma oblong, rather blunt, 3–4 mm, silkily hairy on the keels and veins up to middle. *N Eurasia, northern N America, and mountains further south*. H1. Summer.

73. FESTUCA Linnaeus

Tufted or rhizomatous perennials. Leaves folded, rolled or flat; sheaths open or closed. Inflorescence a panicle. Spikelets stalked, with 4 or more florets (rarely only 2), compressed from side to side. Glumes 2, the lower usually with 1 vein, the upper wider, usually with 3 veins. Lemma with rounded back, not keeled, with or without a terminal or almost terminal awn. Palea 2-keeled, scarious.

A large genus of about 300 species in temperate and subtropical regions throughout the world and on mountain tops in the tropics. The species are all hardy and can be sown outdoors in the spring in ordinary garden soil.

Literature: Hackel, E., *Monographia Festucarum europeum* (1882); Tzvelev, N.N., K sistematike i filogenii ovsyanits (Festuca L.) flory SSSR: I., *Botanicheskii Zhurnal* **56**: 1252–62 (1971).

1a. Leaf-blades blunt 2
b. Leaf-blades sharply pointed, very stiff and rigid 4
2a. Leaf-blades compressed from side to side and longitudinally grooved, especially when dry **1. valesiaca**
b. Leaf-blades cylindric 3
3a. Leaves green or greyish green **2. glacialis**
b. Leaves and usually the whole plant, bluish white, covered with a whitish wax **3. glauca**
4a. Ligule very short with 2 auricles; lemmas narrowly scarious **4. punctoria**
b. Ligule longer, without auricles; lemmas with broadly scarious tips **5. eskia**

1. F. valesiaca Gaudin. Illustration: Hegi, Flora von Mitteleuropa, edn 2, **1**: f. 235 (1957).
Densely tufted, with slender stems, 20–50 cm. Leaf-blades blunt, hair-like, to 0.6 mm thick, flaccid or somewhat stiff, rough or very rough, glaucous, with 5 veins, grooved when dry. Panicles erect, oblong or ovate-oblong, rather dense, 5–10 cm. Spikelets with 3–8 florets, ovate or ovate-oblong, *c.* 6 mm, pale green or tinged with purple, more or less glaucous; lemma *c.* 4 mm, mostly hairless and smooth, tipped with an awn to 1.5 mm. *C & S Europe, Asia.* H3. Spring–summer.

2. F. glacialis (Hackel) Richter.
Densely tufted, with slender stems to 15 cm, coated at the base with the remains of old sheaths; blades thread-like, 0.5–0.6 mm in diameter, very blunt, slightly glaucous, smooth, with 5 veins. Panicles narrow, to 2.5 cm, the branches bearing 1, rarely 2 to 3, spikelets. Spikelets with 3–5 clustered florets, *c.* 6 mm; lemma 4 mm, pointed, awnless or very shortly awned. *Europe (Pyrenees and Alps).* H3. Summer.

3. F. glauca Lamarck. Illustration: Bonnier, Flore Complete **12**: t. 681 (1934).
Densely tufted, bluish green, with numerous vegetative shoots and erect slender stems, 7.5–37.5 cm. Leaves smooth, the blades almost cylindric, thread-like, blunt, *c.* 0.8 mm in diameter, stiff, straight or curved, usually with 9 veins, ligules obscure. Panicles rather dense to very dense, ovate-oblong, erect, bluish green, 3–10 cm, with a smooth axis and short branches. Spikelets 5–8 mm, elliptic or oblong, with 4–7 flowers; lemma 4–5 mm, mucronate or shortly awned, hairless. *SE & C Europe.* H3. Spring–summer.

4. F. punctoria Sibthorp & Smith.
Densely tufted, with rigid stems, to 30 cm. Leaf-blades 5–7.5 cm, tightly inrolled, cylindric, 1–1.2 mm thick, sharply pointed, curved, very hard and rigid, smooth outside, bloomed, with 7–9 veins. Panicles narrowly oblong, dense 4–5 cm. Spikelets with 4–7 florets, oblong-ovate, 8–10 mm; lemma *c.* 5 mm, tipped with an awn to 2 mm, bloomed. *Turkey.* H5. Summer.

5. F. eskia de Candolle. Illustration: Bonnier, Flore Complete **12**: t. 680 (1934).
Mat-forming, compact, with a creeping rootstock, and moderately slender stems 23–45 cm. Leaf-blades to 20 cm or more, *c.* 1 mm thick, very rigid, often curved, cylindric, very acute and sharp-pointed at the tip, smooth; ligules 3–8 mm. Panicles loose, nodding, ovate, 4–10 cm. spikelets ovate-lanceolate, *c.* 8 mm, more or less variegated with green, gold and purple, with 6–8 florets; lemma pointed, *c.* 6 mm, smooth or rough, with scarious apex, mucronate or with a very short awn. *Europe (Pyrenees).* H3. Summer.

74. GLYCERIA R. Brown

Hairless perennials. Leaves flat; sheaths often entire; ligule membranous. Inflorescence a panicle with a 3-sided main axis. Spikelets ovate to linear in outline, with 3–many florets. Glumes unequal, transparent, with 1 vein, shorter than the lowest lemma. Lemma awnless, ovate to oblong-lanceolate, rounded on the back, with 5–11 veins. Palea about as long as lemma.

A genus of about 20 species from wet places in the cooler parts of the northern hemisphere, a few in Australia, New Zealand and S America. Propagated by division of the rootstock or from seed. They are useful for planting on the banks of streams or lakes and may be used as border plants, except on dry soils.

1. G. maxima (Hartman) Holmboe (*G. aquatica* (Linnaeus) Wahlenberg; *Poa aquatica* Linnaeus). Illustration: Hubbard, Grasses, edn 2, 124 (1968); Christiansen, Grasses, sedges and rushes in colour, f. 52 (1979).
A stout leafy perennial with erect stems 1–2.5 m. Leaf-blades abruptly pointed, 30–60 cm × 8–20 mm, hairless. Panicle loose to rather dense, broadly ovate to oblong, 15–45 cm, much-branched, green to purplish. Spikelets oblong, 5–12 × 2–3 mm, with 4–10 florets; lemmas 3–4 mm. *Temperate Europe & Asia.* H1. Summer.

75. MELICA Linnaeus

Perennials. Leaves flat or rolled; sheaths entire. Inflorescence a usually rather loose, often simple, panicle. Spikelets slightly compressed from side to side, with 1–several bisexual florets, and 2–3 sterile lemmas forming a terminal, club-shaped structure. Glumes more or less unequal, firmly membranous, with 3–5 veins. Fertile lemma rounded on the back, leathery in fruit, the veins 5–9 (rarely to 13) but sometimes obscure. Palea membranous, 2-keeled, 2-toothed.
Literature: Hempel, W., Vorarbeiten zu einer Revision der Gattung Melica L., *Feddes Repertorium* **81**: 657–86 (1971), **84**: 533–68 (1973).

A genus of 40–50 species in Europe, temperate Asia, temperate America and N and S Africa. They are propagated by division or by seed sown outdoors in a light to medium loamy soil especially of a calcareous nature.

1a. Lemmas densely hairy towards and along the margins **1. ciliata**
b. Lemmas hairless 2
2a. Panicles dense, somewhat cylindric or 1-sided, 10–25 cm, with the branches bearing several spikelets 1–1.2 cm long **2. altissima**
b. Panicles loose, 1-sided, 4–20 cm, bearing few spikelets 4–8 mm long 3
3a. Ligules with a bristle-like outgrowth; panicles with spreading branches: spikelets with 1 fertile floret **3. uniflora**
b. Ligules truncate; inflorescence a raceme; spikelets with 2–3 fertile florets **4. nutans**

1. M. ciliata Linnaeus. Illustration: Christiansen, Grasses, sedges and rushes in colour, f. 40 (1979).
Densely tufted, greyish green; stems slender, erect, 20–45 cm. Ligules to 2 mm. Leaf-blades finely pointed, thread-like, 7.5–18 cm × 0.8–1 mm, or opening out and up to 3 mm wide, smooth beneath, closely ribbed above. Panicles spike-like, cylindric, dense, 5–15 × 1–1.2 cm, whitish or purplish. Spikelets shortly stalked, elliptic-oblong, 6–8 mm, falling entire at maturity; glumes ovate, pointed, 6–8 mm; lemmas pointed, *c.* 6 mm, with long white hairs along the margins. *Europe, N Africa & SW Asia.* H1. Summer.

2. M. altissima Linnaeus.
Stems tufted, leafy, 60–150 cm, with creeping rhizomes. Ligules 4–6 mm. Leaf-blades finely pointed, 10–23 cm × 5–12 mm, green, rough. Panicles erect, 1-sided, interrupted below, dense above 10–25 cm × 8–12 mm. Spikelets shortly stalked, oblong, 1–1.2 cm, with 2 fertile florets; glumes blunt, papery; fertile lemmas pointed, minutely rough. *C & E Europe, temperate Asia.* H2. Summer.

3. M. uniflora Retzius. Illustration: Hubbard, Grasses, edn 2, 226 (1968); Christiansen, Grasses, sedges and rushes in colour, f. 38 (1979).
Plant with slender creeping rhizomes, forming loose patches; stems 20–60 cm. Leaf-sheaths loosely hairy or hairless; ligule produced into a short bristle; blades finely

pointed, 5–20 cm × 3–8 mm, bright green, shortly hairy above. Panicles very loose, sparingly branched, erect or nodding, 2.5–20 × up to 10 cm, with spreading branches. Spikelets 4–8 mm, elliptic-oblong, with 1 fertile floret, on fine stalks which are 2–5 mm, purplish or brownish; glumes equal or slightly unequal; fertile lemmas blunt, smooth. *Europe, SW Asia.* H1. Spring–summer.

4. M. nutans Linnaeus. Illustration: Hubbard, Grasses, edn 2, 224 (1968); Christiansen, Grasses, sedges and rushes in colour, f. 39 (1979).
Plants with slender rhizomes; stems loosely clustered or solitary, 20–60 cm, slender, angular. Ligules very short. Leaf-blades 5–20 cm × 2–6 mm, bright green, shortly haired above. Racemes loose, nodding, 1-sided, 4–15 cm. Spikelets 6–8 mm, elliptic-oblong, blunt, purplish or reddish purple, with 2 or 3 fertile florets on curved hair-like stalks which are 3–15 mm; glumes similar, papery-membranous; fertile lemmas blunt, minutely rough. *Europe, N & SW Asia.* H1. Spring–summer.

76. STIPA Linnaeus
Tufted perennials, rarely annuals. Leaves usually pleated or rolled, at least when dry, with prominent ribs on the upper surface. Inflorescence a panicle. Spikelets somewhat compressed from side to side, with 1 floret. Glumes usually more or less equal, translucent or membranous, much longer than the lemma, with 1–3 veins; lemma usually leathery, rolled, terete, entire or shortly bifid at apex, hairy, awned from apex or sinus; hairs usually confined to the veins and margins, those of the overlapping margins forming a single line. Callus bearded, usually long and pointed; awn usually with a thick twisted lower part (column), and a thinner straight upper part (limb), hairless or hairy, usually twice bent. Palea translucent, 2-veined, usually enclosed by the lemma. Rhachilla not prolonged.

A genus of some 300 tropical and temperate species, mostly showing adaptations to dry conditions. It includes a number of useful ornamentals with attractive feathery awns, which are suitable for a dry, sunny border.

1a. Glumes 1–5 cm (rarely shorter); awn at least 1.5 cm — 2
b. Glumes not more than 9 mm; awn to 1 cm — 4
2a. Awn rough, with fine, stiff projections throughout its length — **1. gigantea**
b. Awn with hairy limb, the hairs 2–7 mm — 3
3a. Leaf with fine, bristle-like tip; ligule scarcely visible — **2. tirsa**
b. Leaf with pointed tip; ligule *c.* 4 mm — **3. pennata**
4a. Ligule minute; glumes almost equal, the lower 8–9 mm (rarely shorter) — **4. calamagrostis**
b. Ligule *c.* 6 mm; upper glume shorter than the lower, the lower 5–6 mm — **5. splendens**

1. S. gigantea Link. Illustration: Grounds, Ornamental grasses, t. 14 (1979).
Stems to 2.5 m. Ligule very short. Panicle 20–50 cm, very loose. Glumes 2.5–3.5 cm, long-attenuate. Lemma 1.4–1.8 cm, membranous, with soft hairs 1–5 mm, 2-toothed at the tip; awn 7–12 cm, rough with minute projections throughout its length. *SW Europe.* H5. Summer.

2. S. tirsa Steven.
Very robust. Leaves rolled, *c.* 0.5 mm in diameter, with a long, bristle-like tip, those at the base to 100 cm; ligule scarcely visible. Panicle dense; glumes long-attenuate; lemma 1.6–2.1 cm; awn *c.* 40 cm, the column hairless, the limb with hairs 5–6 mm long. *C, S & E Europe.* H2. Summer.

3. S. pennata Linnaeus. Illustration: Christiansen, Grasses, sedges and rushes in colour, f. 5 (1979).
Stems 25–40 cm. Leaves with a pointed tip; ligule *c.* 4 mm. Panicle dense *c.* 10 cm; glumes *c.* 5 cm; lemma 1.3–2 cm; awn 16–30 cm, usually 20–28 cm, the column pale, the limb hairy. *S & C Europe eastwards to Himalaya.* H4. Summer.

4. S. calamagrostis (Linnaeus) Wahlenberg.
Robust, tufted perennial, stems 60–120 cm. Leaves 2–3 mm wide; ligule minute. Panicle to 80 cm, rather loose. Spikelets 8–9 mm (rarely shorter), often purplish; glumes lanceolate, acute; lemma 3–4 mm, leathery, with hairs up to 4 mm long on the back; awn *c.* 10 mm, straight or curved. *S & SC Europe.* H3. Summer.

5. S. splendens Trinius.
Robust, tufted perennial, stems to 2.5 m. Ligule *c.* 6 mm. Panicle to 50 cm. Spikelets 5–6 mm; upper glume three-quarters to four-fifths as long as the lower; lemma 5–6 mm. *Siberia & C Asia.* H3. Summer.

77. PIPTATHERUM Palisot de Beauvois
Perennials. Leaves flat or rolled. Inflorescence a loose panicle. Spikelets somewhat compressed from side to side, with 1 floret. Glumes equal, the lower with 5 veins, the upper with 3–5 veins. Lemma obscurely 3–5-veined, leathery, with a terminal awn that is eventually shed. Palea about as long as lemma, leathery. Callus very small. Rhachilla not prolonged.

A genus of about 25 species in the Old World subtropics. Propagation is by seed or by division of the rootstock, and the plants prefer a light, well-drained soil.
Literature: Freitag, H., The genus Piptatherum in SW Asia, *Notes from the Royal Botanic Garden, Edinburgh* **33**: 341–408 (1975).

1a. Spikelets 3–3.5 mm; awns 4–8 mm; panicle branches in clusters of 5–8 — **1. miliaceum**
b. Spikelets 6–8 mm; awns 1–2 cm; panicle branches solitary or in clusters of 2–4 — **2. paradoxum**

1. P. miliaceum (Linnaeus) Cosson. Illustration: Bonnier, Flore Complete **12**: t. 662 (1934).
Loosely tufted with smooth, green, rigid stems, 45–150 cm. Leaf-blades finely pointed, to 30 cm × 3–10 mm; ligules very short. Panicles linear to oblong, loose, more or less nodding, 15–30 cm, the branches in clusters of 5–8. Spikelets stalked, very numerous, 3–3.5 mm, greenish or tinged with purple; lemmas smooth; awn 4–8 mm. *Mediterranean region.* H5. Summer–autumn.

2. P. paradoxum (Linnaeus) Palisot de Beauvois. Illustration: Bonnier, Flore Complete **12**: t. 662 (1934).
Densely tufted perennial, with slender stems 45–120 cm. Leaf-blades broadly linear, pointed, to 30 × 1 cm, flat, smooth on both sides; ligules very short. Panicles loose, 15–30 cm; branches solitary or in clusters of 2–4. Spikelets elliptic-lanceolate, 6–8 mm, shortly stalked, green or tinged with purple; glumes lanceolate, acuminate; lemma narrowly elliptic, turning dark brown; awn 1–2 cm. *SW Europe, NW Africa, Atlantic Islands.* H5. Summer.

78. PHAENOSPERMA Bentham
Tufted perennial with slender to stout stems, scarious ligules and flat, narrowly lanceolate leaf-blades which have numerous cross-veins. Inflorescence a loose panicle with clustered branches. Spikelets falling at maturity, elliptic-oblong, at length widely gaping, awnless, with 1 flower; glumes thinly membranous, the lower about one-half the length of the upper, which is as long as the spikelet;

lemma as long as the spikelet, rounded on the back, becoming firmly papery, with 3–5 veins, hairless; grain spherical or almost so, with longitudinal grooves, and a free pericarp.

A genus containing a single species. Propagated from seed or by division. Hardy in sheltered places in the south and west of Europe, but succeeds well in cool greenhouses, in partial shade in ordinary garden soil. A remarkable grass on account of its unusual grains, and of ornamental value because of its graceful foliage and inflorescences.

1. **P. globosa** Bentham. Illustration: Hooker's Icones Plantarum **20**: t. 1991 (1891).
Stems 45–150 cm, erect. Ligules 5–20 mm. Leaf-blades finely pointed, gradually narrowed to the base, 30–60 × 1.2–3.6 cm, rough, green above, hairless beneath. Panicle lanceolate to narrowly ovate, 20–25 cm with spreading branches. Spikelets *c.* 4 mm, green; grain 2–3 mm long and broad. *NE India, Burma, China, Taiwan, Korea & Japan.* H5–G1. Summer.

Group B (Bamboos)
Woody, evergreen, perennial grasses, Rhizomes sometimes thick and short, sometimes thin and far-running. Culms emerging in spring, summer or autumn, as thick, pointed, asparagus-like shoots, growing rapidly; mature culms 20 cm to 46 m, 3 mm to 65 cm in circumference, usually hollow except at the nodes, which are marked by a more or less prominent line; below this there is usually another line which is the raised scar left by a sheathing leaf (culm sheath); these are either pushed off by the developing branches or are persistent; the culm sheaths are alternate, on 2 sides of the culm, smooth and shining inside, and usually have a stalkless, short-lived blade which often bears auricles and bristles at its junction with the sheath; a ligule is also present. Crowns sometimes pendent with the weight of the leaves and branches. Branches single at each node (except for *Chusquea*) but the branch itself is branched very near its base, so that many branches appear to arise at each node. Foliage leaves largest on new culms, with basal persistent sheaths, frequently with auricles and bristles at the sheath apex and always with a ligule; blades stalked on the sheaths, deciduous at a precise line or joint, with at least one margin with fine, forwardly pointing teeth, the veins parallel but joined by many transverse veinlets, the venation often showing as a distinct mosaic of tessellation. Inflorescence of panicles, spikes or racemes, sporadically produced in Europe. Spikelets various, with 0–4 (usually 2) glumes, and a varying number of florets (sometimes partly abortive in cultivation) which may be in part unisexual or otherwise imperfect. Lodicules usually 3. Stamens 3–6 in cultivated species. Style 1, stigmas 2–3.

Because bamboos flower and fruit uncertainly in Europe (see the paper by McClintock, cited below) the keys and descriptions given here are based almost entirely on vegetative characters. Identification using these is not entirely easy, and there is much confusion in the naming of the cultivated species. In order to use the account properly a hand lens (× 10 or × 15) is necessary, and the plants should be examined whole, if possible, at the most suitable time of the year, that is, soon after the culms have grown to their full height but while they retain their sheaths, which should be examined at the middle of the culm rather than above or below on mature plants. The bristles which occur on the culm and leaf-sheaths are important for identification: they may be absent, or present, when their colour and surface are important. The surface may be entirely smooth, or roughened with small projections which can be seen with the hand lens. In a few cases the bristles are rough near the base and smooth above. The presence or absence of vein tessellation in the leaves may be seen by holding them up to the light and examining them through the lens.

The size and development of many species varies considerably, depending on where they are being grown; a species which may do well to attain 5 m in northern Europe may well exceed 12 m further south; and similar differences in the degree of branching occur.

Though flowering is sporadic in Europe, and the flowers may be imperfect, whether or not and when and how a plant has flowered can be a useful pointer in identification. Despite constantly repeated statements to the contrary, flowering, in Europe at any rate, of any one species is by no means always synchronous, nor does the plant normally die as a result. The effect of flowering is to weaken or kill the culms that bear the flowers, which usually lose their leaves at the same time, but the rootstock can be expected to survive.

Bamboos form most of the subfamily Bambusoideae of the Gramineae; this subfamily is native over wide areas of E Asia, C and S Africa and America from Maryland south to Argentina. Because of uncertainties in classification, it is not possible to state the number of genera and species of bamboos, and there is much still to be learned about their classification.

Bamboos thrive best in sheltered environments which are not too dry; some species will form wind breaks. The safest time for transplanting is as growth is resumed in the spring and early summer, as the ground warms up. Establishment can take a considerable time, and demands watchful care, particularly with watering. Thereafter most will need no more than the old canes removing, but will gain from feeding and mulching. Propagation is usually by division. Pests and diseases are few and rare.

Literature: Munro, W., A monograph of the Bambusaceae, *Transactions of the Linnaean Society* **26**: 1–157 (1868); Rivière, A. & Rivière, C., *Les Bambous* (1878); Gamble, J.S., The Bambusaceae of British India, *Annals of the Royal Botanic Garden Calcutta* **7**: 1–133 (1894–96); Freeman-Mitford, A.B., *The bamboo garden* (1896); Satow, E., The cultivation of bamboos in Japan, *Transactions of the Asiatic Society of Japan* **27**: 1–127 (1899); Houzeau de Lehaie, J., *Les Bambous*, Nos. 1–10 (1906–08); Camus, E.G., *Les Bambusees* (1913); Lawson, A.H., *Bamboos* (1966); McClure, F.A., *The bamboos* (1966); Lin, Wei-chih, Studies on morphology of bamboo flowers, *Bulletin of the Taiwan Forestry Institute* **68** (1974); Suzuki, S., *Index to Japanese Bambusaceae* (1978); McClintock, D., Bamboos, some facts and thoughts on their flowering, *The Plantsman* **1**: 31–50 (1979); Martin, F. & Demoly, J.-P., Bambous, *Bulletin de l'Association des Parcs Botaniques de France* **1**: 7–17 (1979), **2**: 13–25 (1979), **3**: 13–22 (1980), **4**: 25–31 (1981); Crouzet, Y., *Les Bambous* (1981). There is an American society for bamboo enthusiasts, which produces a journal, *The Journal of the American Bamboo Society* (1980 and continuing).

1a. Culms always solid when mature 2
b. Culms sooner or later hollow (except at the nodes) 4
2a. Culms D-shaped in section **80. Shibataea**
b. Culms terete in section 3
3a. Culms to 40 cm; branches 1–3 at each node **87. Pleioblastus**
b. Culms to 6 or rarely to 10 m; branches very numerous at each node **94. Chusquea**
4a. Culms not terete, at least above 5

b. Culms terete throughout 8
5a. Culms obtusely 4-cornered in section **88. Chimonobambusa**
b. Culms at least in part with an obvious groove 6
6a. Culms grooved throughout, or at least between those nodes where branches are borne **79. Phyllostachys**
b. Culms mainly terete below, if grooved then only above 7
7a. Culms with a large hollow, held stiffly erect; leaves 1.6 cm or more broad **81. Semiarundinaria**
b. Culms ultimately bending with the weight of the crown and with a small hollow; leaves at most 6 mm broad **82. Otatea**
8a. Branches always 1 at each node 9
b. Branches mostly more than 1 at each node 13
9a. Culms ascending or flopping; sheath bristles, if present, rough throughout 10
b. Culms erect; sheath bristles absent, or, if present, entirely smooth or rough only at the base 11
10a. Culms persistently ascending; leaves 4–11 on each branch, at most 40 cm **89. Sasa**
b. Culms ultimately flopping; leaves 1–2 on each branch, to 60 cm **90. Indocalamus**
11a. Sheath bristles rough near their bases; culms to 1.5 cm in circumference **91. Sasaella**
b. Sheath bristles smooth or absent; culms more than 3 cm in circumference 12
12a. Culms to 3 m, without white, waxy bloom below the nodes **92. Sasamorpha**
b. Culms to 6 m, with white, waxy bloom below the nodes **93. Pseudosasa**
13a. Leaves without visible tessellation 14
b. Leaves with visible tessellation 17
14a. Culms 15 m or more in height, 20 cm or more in circumference **84. Dendrocalamus**
b. Culms to 15 m in height but usually much less, at most to 10 cm in circumference 15
15a. Leaves and sheaths silvery grey-hairy **83. Bambusa**
b. Leaves and sheaths hairless or hairy but not as above 16
16a. New shoots appearing in late summer or autumn **88. Chimonobambusa**
b. New shoots appearing in spring **85. Arundinaria**
17a. Culm sheaths soon falling, though often hanging from a small attachment at the base for a time; branches forming on the culms from the base upward **81. Semiarundinaria**
b. Culm sheaths persistent, not hanging as above; branches forming on the culms from the top downwards 18
18a. Bristles rough **85. Arundinaria**
b. Bristles smooth 19
19a. Bristles and ligule dark in colour **86. Sinarundinaria**
b. Bristles and ligule whitish in colour 20
20a. Culms finally with a small hollow, emerging in late summer; culm sheaths marbled, blades minute **88. Chimonobambusa**
b. Culms with a wide hollow, emerging in spring; culm sheaths not marbled, blade obvious, not minute **87. Pleioblastus**

79. PHYLLOSTACHYS Siebold & Zuccarini
D. McClintock

Rhizomes short or elongate. Culms close or spaced, with hollow internodes which are grooved on alternate sides above each node (except sometimes for those below the lowest branches); nodes prominent. Branches developing in sequence from below upwards, 3 at each node, the outer unequal, the central small and depauperate, sometimes absent; branchlets numerous. Sheaths soon deciduous (those at the base sometimes more persistent), bristles rough or absent. Leaves 6–20 cm, visibly tessellate, hairless above, paler and sometimes downy near the base beneath; bristles rough but soon deciduous. Inflorescences spicate or racemose. Spikelets with usually 2 glumes and 5–13 florets. Lodicules 3, ciliate. Stamens 3. Stigmas 3.

A genus of perhaps 60 or more species, all originating in China, but very widely cultivated elsewhere.

Literature: McClure, F.A., Bamboos of the genus Phyllostachys under cultivation in the US, *USDA Handbook* No. 114 (1957); Haubrich, R., Handbook of bamboos cultivated in the United States, Part 1: the genus Phyllostachys, *Journal of the American Bamboo Society* **1**: 48–96 (1980); Wang Cheng-Ping and others, A taxonomical study of Phyllostachys, China, *Acta Phytotaxonomica Sinica* **18**: 15–19, 168–93 (1980).

Culms. Well over 10 m (at least in warmer areas); **3,6,9**. Downy: **7**; hairy: **6**; slightly roughened: **2,6,7,9**. Blackish: **5,7**; blotched: **1,3,5,7,8**; bright yellow, often with green stripes: **6,9**; only the groove yellowish: **2**. Lowest internodes very close and asymmetric: **1,6**. Curved at the base: **6,8**. Without white, waxy bloom below the nodes: **1,3**.
Culm sheaths. Hairless: **1–5,9**; hairy: **6-8**. Ciliate: **1,2,6,7**. Without auricles or bristles: **1,3–5,7,9**. Persistent near the base of the culm: **1,6**.
Ligule. Downy: **1,2,4,5,7**; hairless: **3,6,8,9**.

1a. Lowest nodes asymmetric and very close together 2
b. Lowest nodes symmetric and usually distant 3
2a. Most nodes with deep swollen bands immediately beneath **1. aurea**
b. Nodes all without swollen bands beneath **6. heterocycla**
3a. Culms with the groove of a different colour from the rest 4
b. Grooves the same colour as the rest of the culm 5
4a. Culms greyish green, grooves striped with yellow **2. aureosulcata**
b. Culms yellow, grooves green **3. bambusoides**
5a. Culms yellow with varying amounts of green striping 6
b. Culms green or greyish green, not yellow 8
6a. Nodes without white, waxy bloom beneath; green stripes few **3. bambusoides**
b. White, waxy bloom present below the nodes; green stripes many 7
7a. Culms smooth; sheaths hairy **6. heterocycla**
b. Culms initially slightly rough; sheaths hairless **9. viridis**
8a. Culms somewhat zig-zag; leaf-sheaths usually without auricles or bristles **5. flexuosa**
b. Culms straight; leaf-sheaths usually with auricles and bristles, at least when young 9
9a. Nodal ridges not prominent, single at lower, branchless nodes 10
b. Nodal ridges prominent, always 2 11
10a. Culms grey; sheaths hairy and with bristles; small auricles sometimes present; leaves 4–10 cm **6. heterocycla**
b. Culms green; sheaths hairless and without auricles or bristles; leaves 7–12 cm **9. viridis**
11a. Culm sheaths usually without spots; leaves often less than 5 cm 12
b. Culm sheaths spotted or streaked; leaves mostly more than 7 cm 13

12a. Culms grey-green, sheaths hairless, usually without auricles or bristles **4. congesta**
b. Culms black, blotched and spotted, or green; sheaths downy and usually with conspicuous auricles and bristles **7. nigra**
13a. Culms emerging in early summer, erect, without white, waxy bloom below the nodes; sheaths usually without auricles; leaves to 4.5 cm wide; sheaths with conspicuous bristles to 1.2 cm **3. bambusoides**
b. Culms emerging in spring, often curved at the base, with white, waxy bloom below the nodes; sheaths usually with auricles; leaves to 3.8 cm wide; sheaths with smaller deciduous bristles or bristles absent **8. viridi-glaucescens***

1. P. **aurea** (Carrière) Rivière (*P. formosana* Hayata). Illustration: Lawson, Bamboos, 117 (1966); Suzuki, Index to Japanese Bambusaceae, 11, 14, 72–73 (1978).
Culms 4–8 m, 9–12 cm in circumference, erect, smooth, hairless, greenish, greyish or brownish yellow, rarely blotched; each node mostly with a deep swollen band beneath; lowest internodes on many culms very short, and variously bulging and asymmetric. Sheaths (the lowermost often persisting on stouter culms) hairless except for a very narrow fringe of white hairs at the base, lightly spotted and streaked, without auricles or bristles, ciliate above. Leaves 6–15 × 1–2 cm, somewhat yellowish green, downy near the midrib beneath; sheaths usually with auricles and bristles which are often long and erect; ligule *c.* 1 mm, downy. *SE China.* H2.

This species, which is widely grown, especially in continental Europe, recently flowered in 1950 and 1964. A variant with leaves variegated with white ('Albo-variegata') is sometimes found; 'Holochrysa', with orange-yellow culms is more frequent.

2. P. **aureosulcata** McClure (*P. nevinii* misapplied). Illustration: USDA Handbook No 114: 19, 20 (1957); Crouzet, Les Bambous, 68 (1981).
Culms 3–6 m, 3–10 cm in circumference, sometimes zig-zag, greyish green, slightly rough and with white, waxy bloom below the nodes when young, grooves with green and yellow stripes which soon pale. Sheaths hairless, striped with pale yellow and olive green, those at mid-culm with auricles and bristles, blade small. Branches almost erect. Leaves to 15 × 2 cm, but usually much smaller, sheaths sometimes without auricles or bristles; ligule *c.* 2 mm, downy. *NE China.* H2.

3. P. **bambusoides** Siebold & Zuccarini (*P. quiloi* (Carrière) Rivière; *P. mazelii* Rivière; *P. reticulata* misapplied). Illustration: Polunin & Everard, Trees and bushes of Europe, t. 182d (1976) – as P. nigra; Suzuki, Index to Japanese Bambusaceae, 11, 13, 74–5 (1978); Kitamura & Murata, Coloured illustrations of woody plants of Japan, t. 132 (1979); Crouzet, Les Bambous, 36 (1981).
Culms 6–20 m, 2–30 cm in circumference, hairless, green, stiffly erect. Sheaths almost hairless, obtuse above, with small auricles and kinked bristles, heavily marked and spotted; ligule short, blunt, blades mostly short. Nodes almost without waxy bloom. Leaves 6–20 × 1–4.5 cm, stalk to 8 mm; sheaths with auricles and prominent, persistent bristles to 1.2 cm; ligule to 2 mm, hairless. *China.* H2.

This species, which is quite commonly grown, has flowered synchronously since 1962. Shoots usually emerge later than those of the other species. It differs from the earlier emerging *P. viridis* 'Robert Young' in its smooth culms. For more detail see Soderstrom & Calderon, Pacific Horticulture 37(3): 7–14 (1976). It is variable and a number of clones have been given names. 'Castillonis' (*Bambusa castillonis* Carrière; *P. castillonis* (Carrière) Mitford) has culms which are yellow with a green groove; it flowered synchronously in 1965–66 and 1979–80 and may have died out. 'Allgold' ('Holochrysa'; 'Sulphurea' misapplied) is smaller with sparse green stripes on its yellow culms.

4. P. **congesta** Rendle. Illustration: USDA Handbook No. 114: 28 (1957); Crouzet, Les Bambous, 73 (1981).
Culms 3–8 m, 6–16 cm in circumference, erect, tapered, grey-green; nodes with white, waxy bloom. Sheaths hairless, dark green suffused with pale purple, usually without auricles or bristles, blades short, somewhat wavy. Leaves to 15 × 1.5 cm but often much smaller, markedly downy near the base beneath; sheaths without auricles and with few or no bristles; ligule to 2 mm, minutely downy. *China.* H2.

Not commonly cultivated.

5. P. **flexuosa** (Carrière) Rivière. Illustration: Rivière, Les Bambous, 269, 275 (1879); USDA Handbook No. 114: 35, 36 (1957).
Culms 3–7 m, 6–12 cm in circumference, erect or spreading, slightly ribbed, liable to darken with age to almost black, with white, waxy bloom below the nodes. Sheaths tapered above, hairless, smooth, with some dark markings, without auricles or bristles but with a tuft of hairs from the ligule. Leaves to 15 × 1.5 cm, usually smaller, often somewhat downy near the midrib beneath, sheaths without bristles or auricles (except in the case of fresh, young shoots which also have purple ligules at first); ligule *c.* 2 mm, downy. *China.* H2.

6. P. **heterocycla** (Carrière) Mitford (*P. pubescens* Houzeau de Lehaie; *P. edulis* invalid; *P. mitis* misapplied). Illustration: Suzuki, Index of Japanese Bambusaceae, 10, 12, 70–1 (1978); Crouzet, Les Bambous, 57 (1981).
Culms 5–20 m, 5–30 cm in circumference, tapered, distant, grey ultimately green or almost orange, curved near the base where the nodes are close together, terete and branchless in the lower part where the nodes have only 1 ridge; internodes velvety when young, soon becoming hairless, white-waxy. Sheaths to 1 m, thick, dark with many markings and covered with erect, brown hairs, the basal sometimes overwintering, auricles none or very small, bristles prominent; blades small, hairless. Leaves usually 4–10 cm × 4–10 mm; sheaths with no auricles or auricles poorly developed, and bristles; ligule *c.* 1 mm, hairless. *E China.* H3.

This species flowered in 1937. The variant described above, which is forma **pubescens** (Houzeau de Lehaie) McClintock, is commonly grown in warmer areas, but is difficult to propagate; the wild variant, often found in the literature as *P. pubescens* 'Heterocycla' or 'Kikkochiku' (Illustration: National Geographic Magazine, October 1980: 516, 520; Crouzet, Les Bambous, 36, 1981), has the lower internodes short, bulging on one side and grossly asymmetric on some culms (see Oshima, J., Journal of the American Bamboo Society 3: 2–28, 1982). Forma **nabeshimana** (Muroi) Muroi (*P. pubescens* 'Bicolor') has the culms striped with yellow to varying degrees.

7. P. **nigra** (Loddiges) Munro. Illustration: Botanical Magazine, 7994 (1905); Lawson, Bamboos, 129 (1966); Suzuki, Index of Japanese Bambusaceae, 11, 15, 78–9 (1978).
Culms 3–5 m, 6–8 cm in circumference, slightly ribbed, olive green at first with white, waxy bloom, later usually shining black. Sheaths with rapidly deciduous hairs, unspotted except sometimes near the apex, usually with prominent, spreading

auricles and bristles but sometimes these are absent; blades broad. Leaves 6–13 cm × 8–18 mm, with auricles and bristles absent or poorly developed or bristles erect; ligule to 1.5 mm, downy. *E China*. H2.

A commonly cultivated species, striking when in mature colour, which last flowered between 1932 and 1935. Several variants are grown. 'Punctata' (*P. nigra* var. *punctata* Bean) has culms finally heavily speckled with purplish brown. Var. **henonis** (Mitford) Stapf (*P. henonis* Mitford; 'Henonis'; *P. puberula* Makino). Illustration: Hooker's Icones Plantarum **28**: t. 2614 (1901). Culms to 10 m, unblotched and erect, downy and rough when young, broader, leafier and arching above. Widely grown. 'Boryana' (*P. boryana* Mitford) has culms finally marked with dark streaks.

8. P. viridi-glaucescens (Carrière) Rivière (*Bambusa viridi-glaucescens* Carrière). Illustration: USDA Handbook No. 114: 61, 62 (1957); Polunin & Everard, Trees and bushes of Europe, t. 182e (1976). Culms 4–7 m, 9–13 cm in circumference, often curved at the base, streaked, with white, waxy bloom below the nodes; sheaths downy, but the hairs soon falling, variable and usually rather lightly marked, tapered above, usually with auricles and bristles, ligule normally long. Leaves to 2.8 cm broad, bristles short and soon falling. *E China*. H2.

Flowered in 1979.

***P. nidularia** Munro. Illustration: Camus, Les Bambusees, t. 36A (1913); USDA Handbook No. 114: 43, 44 (1957). Culm-sheaths with large blades and prominent nodal ridges which often bear thick, brownish hairs, sheaths with auricles and few or no bristles. *NE China*.

Has flowered occasionally since 1968.

9. P. viridis (Young) McClure (*P. mitis* misapplied; *P. edulis* misapplied). Illustration: USDA Handbook No. 114: 63 (1957); Crouzet, Les Bambous, 40, 58 (1981).
Similar to *P. heterocycla* but smaller, culms hairless, olive green, minutely pitted, often curved at the base and also above; sheaths ciliate, without other hairs, auricles or bristles; leaves fewer, 7–12 × 1.5–4 cm. *E China*. H3.

Variable. Somewhat similar to a large *P. heterocycla* and widely grown in warmer areas. 'Robert Young' ('Sulphurea' misapplied) has smaller culms which are yellowish green at first (later golden yellow), variably striped with green and with a dark band below each node. It differs from *P. bambusoides* 'Allgold' in its somewhat rough culms and plentiful green stripes.

80. SHIBATAEA Nakai
D. McClintock
Culms D-shaped in section, usually tufted (though rhizomes occasionally running), slender, at most 6 mm in circumference, solid, with prominent nodes. Sheaths fairly persistent, hairless, without auricles or bristles. Branches 2–5 at each node, short, all more or less equal. Leaves visibly tessellate, broadly lanceolate, without bristles. Inflorescence a panicle. Spikelets with 2–3 glumes and 1–2 florets. Lodicules 3. Stamens 3. Stigmas 3.

A genus of 5 species, 1 from Japan, the rest from China. They are attractive, distinct and easy to grow as long as the site is not too dry.

1. S. kumasasa (Zollinger) Nakai (*Bambusa ruscifolia* Munro; *Phyllostachys kumasasa* (Zollinger) Munro; *P. ruscifolia* (Munro) Satow). Illustration: Suzuki, Index of Japanese Bambusaceae, 100–1 (1978); The Plantsman **1**: 46 (1979); Kitamura & Murata, Coloured illustrations of woody plants of Japan, t. 132 (1979).
Culms to 1.5 m, to 6 mm in circumference, somewhat zig-zag. Sheaths olive green with purplish ribs, hairless except for cilia, blade minute. Leaves 5–11 × 1.5–2.5 cm, often dead at apex, hairless above, hairy beneath when young, the hairs soon deciduous. *Japan*. H2.

Fairly frequently grown. Flowers from time to time.

81. SEMIARUNDINARIA Nakai
D. McClintock
Culms usually tightly clumped (but capable of spreading), hollow, terete or with the upper internodes grooved, held stiffly erect. Sheaths soon falling, except on late shoots, but often hanging by their bases for some time, the inner surface highly polished, usually purplish red, without bristles. Branches erect, 3–8 at each node, developing from the base of the culm upwards, the lower nodes without branches. Leaves to 20 cm, visibly tessellate and with rough bristles. Inflorescence a number of spikes. Spikelets with glumes absent or rarely 1 present, with 3–6 flowers of which the upper 1–3 are male only. Lodicules 3, ciliate. Stamens 3. Stigmas 3.

A genus of perhaps 20 species from E Asia. They are easy to grow and excellent for tall hedging.

1. S. fastuosa (Mitford) Makino (*Bambusa fastuosa* Mitford; *Arundinaria narihita* Makino; *Phyllostachys fastuosa* (Mitford) Nicholson; *A. fastuosa* (Mitford) Makino; *A. narihira* Makino). Illustration: Suzuki, Index of Japanese Bambusaceae, 16, 84–7 (1978); Horticulture **7**: 27 (1979); Kitamura & Murata, Coloured illustrations of woody plants of Japan, t. 133 (1979); Crouzet, Les Bambous, 38 (1981).
Culms 3–6 m, 10–25 cm in circumference, held stiffly erect, and with a large hollow. Sheaths thick, hairless or slightly downy, auricles and bristles very small or absent; blades short, narrow. Leaves mostly 12–15 × 1.6–2 cm, broader on young shoots, thick, hairless, without auricles or bristles, often purple. *Japan*. H2.

Var. **yashadake** (Makino) Makino (*S. yashadake* (Makino) Makino). Illustration: Suzuki, Index of Japanese Bambusaceae, 16, 88–9 (1978); Kitamura & Murata, Coloured illustrations of woody plants of Japan, t. 132 (1979). Has leaves to 4 cm broad, leaves and sheaths with plentiful brown hairs at the base.

This species has flowered since 1957.

82. OTATEA (McClure & E.W. Smith) Calderon & Soderstrom
D. McClintock
Rhizome short or elongate. Young plants tufted, becoming more open with age. Culms slender, solid at first, later hollow, the upper internodes each with a shallow groove on one side, the whole culm ultimately arching with the weight of the crown. Sheaths soon deciduous, without bristles. Branches 3 at each node initially, later more numerous. Leaves without visible tessellation, 3–6 mm broad, the sheaths mostly without auricles or bristles. Inflorescence a panicle. Spikelets with 2 glumes and 3–7 florets. Lodicules 3. Stamens 3. Stigmas 2.

A genus of 2 species from C America (Mexico to Honduras), suitable only for very warm areas.
Literature: McClure, F.A., Genera of bamboos native to the New World, *Smithsonian Contributions to Botany* **9**: 116–19 (1973); Calderon, C.E. & Soderstrom, T.R., The genera of Bambusoideae (Poaceae) of the American continent, *Smithsonian Contributions to Botany* **44**: 21 (1980); Haubrich, R., The American bamboos, Part 1, *Journal of the American Bamboo Society* **1**: 29–30 (1980); Dunmire, J.R., Weeping Mexican bamboos, *Pacific Horticulture* **42**: 18 (1981).

1. O. acuminata (Munro) Calderon & Soderstrom (*Arundinaria longifolia* Fournier: *Arthrostylidium longifolium* (Fournier) Camus; *Bambusa longifolia* (Fournier) McClure; *Yushania aztecorum* McClure & E.W. Smith; *Otatea aztecorum* (McClure & E.W. Smith) Calderon & Soderstrom). Illustration: Smithsonian Contributions to Botany **9**: 117 (1973).
Culms flexuous, ultimately bending with the weight of the leaves, 2–6 m, 2–12 cm in circumference. Sheaths more or less hairy, sometimes the hairs deciduous, without auricles and bristles, or these poorly developed; blade elongate–triangular. Leaves 9–18 cm × 3–6 mm, numerous on non-flowering culms, very long-acuminate, often hairy towards the base beneath, hanging; sheaths downy; ligule 1–2 mm, downy. *Mexico*. H5.

83. BAMBUSA Schrader
D. McClintock
Rhizomes much-branched, short. Culms clump-forming to 15 m, but usually much less, to 10 cm in circumference, hollow, terete; new shoots emerging in late summer. Sheaths thick, deciduous, with rough bristles or bristles absent. Branches numerous at each node, of varying sizes. Leaves without visible tessellation, their sheaths with auricles and rough bristles. Inflorescence a number of spikes. Spikelets without glumes, or rarely 1 glume present, with 5–7 florets. Lodicules 3, ciliate. Stamens 3. Stigmas 3.
A genus of *c.* 100 or more species in subtropical Asia, Africa and America. At one time or another, most bamboos have been placed in this genus.
Literature: Holttum, R.E., On the identification of the common hedge bamboo of SE Asia, *Kew Bulletin* **11**: 207–11 (1956); Haubrich, R., Handbook of bamboos cultivated in the United States, Part II, *Journal of the American Bamboo Society* **2**: 2–20 (1981).

1. B. glaucescens (Willdenow) Holttum (*B. multiplex* (Loureiro) Steudel; *B. nana* Roxburgh; *Leleba multiplex* (Loureiro) Nakai). Illustration: Annals of the Royal Botanic Garden Calcutta **7**: t. 38 (1896); Suzuki, Index to Japanese Bambusaceae, 102–5 (1978); Kitamura & Murata, Coloured illustrations of woody plants of Japan, t. 134 (1979).
Culms 3–5 m, 6–9 cm in circumference, ascending, thin-walled, rough when young, somewhat arching above when mature, branched from the base. Sheaths soon hairless, unmarked, usually with conspicuous auricles and smooth bristles; ligule very short; blade triangular, decurrent on the sheath. Nodes rather prominent with white, waxy bloom beneath. Branches very numerous at each node, slender. Leaves variable in number and size, but generally *c.* 11 × 1.2 cm, occasionally to 16 × 2 cm, hairless above, silvery grey-hairy beneath; sheaths smooth, usually without auricles and with smooth bristles; ligule *c.* 0.5 mm. *China, widely cultivated elsewhere*. H5–G1.
A variable species with numerous cultivars. 'Alphonse Carr' has the mature culms yellow with green stripes, and 'Fernleaf' ('Gracillima') has very numerous, small leaves, markedly arranged in 2 ranks on each branch.

84. DENDROCALAMUS Nees
D. McClintock
Very similar to *Bambusa*, from which it is differentiated mainly by floral characters and its great size. Culms 15 m or more tall, 20 cm or more in circumference. Spikelets forming rounded structures in long spikes, with 2–3 glumes and up to 6 florets. Lodicules absent. Stamens 6. Stigma 1.
A genus of about 30 species, native to Asia.
Literature: Holttum, R.E., The bamboo of the Malay Peninsula, *Gardens' Bulletin Singapore* **16**: 89–104 (1958); Haubrich, R., Handbook of bamboos cultivated in the United States, Part II, *Journal of the American Bamboo Society* **2**: 2–20 (1981).

1. D. asper (Schultes) Heyne. Illustration: Annals of the Royal Botanic Garden Calcutta **7**: t. 76 (1896); Ochse & Bakhuizen van den Brink, Vegetables of the Dutch E Indies, 308, 309 (1931); Sunset Magazine (July 1980); Crouzet, Les Bambous, 37 (1981).
Culms 15–20 m, rarely to 30 m, 20–30 cm (rarely to 75 cm) in circumference, tufted, hairless, thin-walled, glaucous at first, covered at first with brown, closely adpressed hairs which persist near the base. Sheaths large, soon falling, pale green at first, thick and hard, ciliate and sparsely hairy, sometimes with small auricles and curved bristles; blade triangular, acute, adpressed-hairy on both sides. Nodes rather prominent, often with the base of the sheath still attached, hairy below and with white, waxy bloom; lower nodes with prominent root buds. Leaves 4–30 cm (rarely to 55 cm) × 1.5–4 cm (rarely to 11 cm), rough above, hairy towards the base or ultimately hairless beneath. Sheaths almost completely hairless, ciliate; auricles small or absent, bristles absent; ligule to 2.5 mm, downy. *Burma, China, much cultivated elsewhere*. H5–G1.

85. ARUNDINARIA Michaux
D. McClintock
Rhizomes slender and elongate, or compact and much branched. Culms erect, hollow, terete; new shoots emerging in spring. Sheaths persistent with well-developed blades and dark, rough bristles. Branches several at each node, foliage developing from the top downwards. Leaves variably tessellate or not; sheaths bearing dark, tough bristles (rarely bristles absent); ligule often dark-coloured. Inflorescence racemose or paniculate. Spikelets with 2 or rarely 3 glumes and several florets of which the upper may be unisexual. Lodicules 3 or rarely 6. Stamens usually 3 or 6. Stigmas 3.
A genus of an uncertain number of species, mainly from the Himalaya, but also in America, China and South Africa. Species 4–7 are generally included in the genus *Thamnocalamus* Munro. This differs from *Arundinaria* essentially only in floral characters, so is not recognised here, though synonyms under it are cited.

1a. Leaves without visible tessellation **5. falconeri**
b. Leaves with visible tessellation 2
2a. Rhizomes far-running, culms spaced **3. jaunsarensis**
b. Rhizomes short, culms tufted 3
3a. Culms rough to touch **2. maling**
b. Culms smooth to touch 4
4a. Culm sheaths whitish, fairly persistent **7. tessellata**
b. Culm sheaths not whitish, deciduous 5
5a. Branches usually 1–3 at each node **1. amabilis**
b. Branches numerous at each node 6
6a. Culm sheaths with auricles and bristles **4. aristata***
b. Culm sheaths without auricles or bristles **6. spathacea**

1. A. amabilis McClure. Illustration: McClure, The bamboos, f. 67, 68 (1966).
Culms 3–12 m, 10–20 cm in circumference. Sheaths hairy. Nodes without white, waxy bloom below. Branches usually 1–3 at each node. Leaves *c.* 18 × 2 cm, occasionally to 36 × 4 cm, pale beneath, downy towards the base, with bristles and auricles; ligule *c.* 1 mm. *China, where it is much cultivated*. H2.

The species, which is seen occasionally in Europe, flowered in 1979–80.

2. A. maling Gamble (*A. racemosa* misapplied). Illustration: Annals of the Royal Botanic Garden Calcutta **7**: t. 8 (1896).
Culms to 5 m, *c.* 10 cm in circumference, glaucous, rough to the touch. Sheaths persistent, tapered and truncate, bristly hairy, downy inside, blades long and narrow; ligule fringed. Nodes with white, waxy bloom below. Leaves to 18 × 2 cm, sometimes downy towards the base beneath; sheaths ciliate above, with auricles and a few short bristles; ligule 1.5–2 mm, minutely downy. *NE Himalaya.* H2.

3. A. jaunsarensis Gamble (*A. anceps* Mitford; *Yushania anceps* (Mitford) Lin). Illustration: Annals of the Royal Botanic Garden Calcutta **7**: t. 22 (1896); Bulletin of the Taiwan Forestry Institute **68**: t. 3 (1974); The Plantsman **1**: 37 (1979).
Rhizomes extensively creeping from the central tuft. Culms 3–6 m, arching when mature, 2–5 cm in circumference, with longitudinal lines, rough and with white, waxy bloom below the nodes. Sheaths shorter than the internodes, pale or purplish brown, not overwintering except sometimes on late shoots, ciliate and with auricles and bristles. Leaves usually *c.* 10 × 1 cm, sometimes to 16 × 2 cm, hairless, tessellate; sheaths hairless, usually with auricles and bristles; ligule *c.* 1 mm, downy. *C & NW Himalaya.* H2.

Flowering more or less freely, but by no means everywhere, since 1957. It is very invasive, and liable to suffer in hard winters. 'Pitt White' (*A. niitikayamensis* misapplied) has taller, stouter culms to 10 m, 12 cm in circumference; it has flowered since 1981.

4. A. aristata Gamble (*Thamnocalamus aristatus* (Gamble) Camus). Illustration: Annals of the Royal Botanic Garden Calcutta **7**: t. 17 (1896); The Plantsman **1**: 38 (1979).
Culms 5–8 m, rarely to 10 m, 4–6 cm in circumference, yellowish, soon smooth, speckled when old. Sheaths deciduous, tapering above, sparsely hairy, bristly towards the base, with auricles and bristles at the apex, blade short. Nodes with white, waxy bloom below. Branches reddish. Leaves *c.* 10 × 1 cm, downy beneath, with minute glands on the stalk, often twisted at the apex, visibly tessellate; sheaths dull purple, hairless but with a hairy callus at the apex, bearing bristles. Ligule to 3 mm, downy. *W Himalaya.* H3.

This species flowered in 1950–52, but is now uncommon in Europe.

***A. spathiflora** Trinius (*Thamnocalamus spathiflorus* (Trinius) Munro). Illustration: Annals of the Royal Botanic Garden Calcutta **7**: t. 16 (1896). Similar, but culms grey, white-bloomed at first, later pinkish in the sun, with dark bands below the nodes, somewhat zig-zag, branches 2–3, sheaths ciliate, with longer blades and without an apical callus, leaves greyer. *NW Himalaya.* H2.

5. A. falconeri (Munro) Rivière (*Thamnocalamus falconeri* Munro; *A. nobilis* Mitford; *A. pantlingii* misapplied). Illustration: Annals of the Royal Botanic Garden Calcutta **7**: t. 18 (1896); Botanical Magazine, 7947 (1904).
Culms 7–10 m, rarely to 20 m, 6–10 cm in circumference, the outer often spreading, greenish yellow, purple below the nodes, with no white waxy bloom. Sheaths deciduous, minutely downy, not ciliate, purplish, much longer than the internodes, without auricles or bristles; ligule broad, very shortly ciliate. Leaves to 16 × 1.6 cm but usually much smaller, hairless, pale green without visible tessellation; sheaths hairless, shortly ciliate, without bristles; ligule to 1.5 mm, downy. *NE Himalaya.* H2–3.

Very decorative, but suffering badly in harsh winters. The species flowered synchronously in 1929–31 and 1964–69, all the plants dying afterwards. It may be best placed in a separate genus together with some species currently in *Chimonobambusa.*

6. A. spathacea (Franchet) McClintock (*Fargesia spathacea* Franchet; *A. murielae* Gamble; *Sinarundinaria murielae* (Gamble) Nakai; *Thamnocalamus spathaceus* (Franchet) Soderstrom). Illustration: Hay & Synge, Dictionary of garden plants, t. 1901 (1969); The Garden **4**: 22, 24, 26 (1979).
Culms to *c.* 4 m, *c.* 4 cm in circumference, greenish with white, waxy bloom below the nodes, branching in the first year, arching when mature. Sheaths deciduous, rounded above and without auricles or bristles, downy and greenish purple when young, soon hairless and pale brown, noticeable for a short time while standing away from the culm before falling. Leaves to 15 × 2 cm, mostly much less, visibly tessellate, stalks purplish in the sun; sheaths often purplish, hairless, often without bristles; ligule 1–2 mm, downy. *China.* H2.

Much confused with *Sinarundinaria nitida.* Dwarf variants are grown in Europe. Some of these flowered in 1978–80 and then died.

7. A. tessellata (Nees) Munro (*Nastus tessellatus* Nees; *Thamnocalamus tessellatus* (Nees) Soderstrom & Ellis). Illustration: Hooker's Icones Plantarum **30**: t. 2930 (1911).
Culms tufted, 1.5–7 m, 6–8 cm in circumference, thick-walled, erect, smooth, with a dark purple ring below each node, unbranched and leafless during the first year. Sheaths fairly persistent, mostly longer than the internodes, thick, whitish, slightly downy, with auricles and bristles; blades long and narrow. Branches many, short, erect, purplish. Leaves 5–14 × 1–2 cm, visibly tessellate, long-acuminate, dark green; sheaths ciliate, with auricles and bristles; ligule 1.5 mm, fringed. *South Africa.* H2.

86. SINARUNDINARIA Nakai
D. McClintock
Like *Arundinaria,* but with a much-branched rhizome, habit tufted, branches almost equal, arising horizontally, leaf bristles smooth and white. Inflorescence a small panicle. Spikelets with 2 glumes and 2–4 florets. Lodicules 3. Stamens 3. Stigmas 2–3.

1. S. nitida (Mitford) Nakai (*Arundinaria nitida* Mitford). Illustration: Lawson, Bamboos, 101 & t. 14 (1966); Bean, Trees and shrubs hardy in the British Isles, edn 8, t. 16 (1970); Bartels, Gartengehölze, 416 (1981).
Culms to 6 m, *c.* 5 cm in circumference but often much less, variable in colour from dark purple to grey, arching above when mature. Sheaths soon shorter than the internodes, persistent for a year or longer in most cases, somewhat narrowed above, without auricles or bristles, usually purple, variably hairy or hairless. Nodes with white, waxy bloom beneath. Shoots unbranched till second year. Leaves usually *c.* 5 cm × 5 mm, sometimes to 10 × 1.3 cm, hairless, or hairy towards the base beneath. Sheaths hairless, ciliate, usually purple, sometimes with bristles; ligule 0.5–1 mm, often purple. *N China.* H2.

Various clones are cultivated; they are hardy if protected from wind. The species may be confused with *Arundinaria spathacea,* from which it may be distinguished by its smaller leaves, more tapered, less prominent culm sheaths and its general purple and grey coloration.

87. PLEIOBLASTUS Nakai
D. McClintock
Rhizomes mostly elongate, slender. Culms erect, almost always hollow but thick-walled, terete. New shoots arising in spring, leafing from the top downwards. Sheaths persistent, bearing smooth, whitish bristles. Branches usually 3–7 at each node or sometimes branches only 1–2, all in the lower part of the culm. Leaves with visible tessellation and with smooth, whitish bristles which are occasionally absent. Inflorescence a spike or raceme. Spikelets with 2 glumes and 5–13 florets. Lodicules 3, ciliate. Stamens 3. Stigmas 3.

A genus of about 20 species from China and Japan, formerly considered part of *Arundinaria*, mostly easy to grow, but some can form thickets.

Rhizomes. Markedly running: **1,2,4**.
Culms. More than 2 m: **1,2,5**; 1–2 m: **3,6,7**; less than 1 m: **4**.
Leaf-sheaths. Purple: **6,7**.
Branches. 1–2, rarely 3: **3,4,6,7**; 3–7: **1,2,5**.
Leaves. Variegated: **1,3,6,7**. 12–20 times longer than broad: **2**. At least some with markedly oblique apices: **2,5**. Ligule more than 3 mm: **2**.

1a. Leaves 12–20 times longer than broad; ligule 3 mm or more **2. gramineus***
b. Leaves 6–8 times longer than broad; ligule to 2 mm 2
2a. Upper margin of leaf-sheath oblique **5. simonii**
b. Upper margin of leaf-sheath horizontal 3
3a. Culms 3–4 m; branches numerous **1. chino**
b. Culms to 2 m; branches often fewer than 3 4
4a. Culms to 1 m 5
b. Culms 1–2 m 6
5a. Culms 60 cm or more; leaves variegated **6. variegatus**
b. Culms 50 cm or less; leaves not variegated **4. pygmaeus**
6a. Leaves entirely green **3. humilis**
b. Leaves variegated 7
7a. Leaves soon hairless **1. chino**
b. Leaves persistently downy **7. viridi-striatus**

1. P. chino (Franchet & Savatier) Makino (*Arundinaria chino* (Franchet & Savatier) Makino; *P. maximowiczii* (Rivière) Nakai). Illustration: Suzuki, Index to Japanese Bambusaceae, 304–9 (1978); Kitamura & Murata, Coloured illustrations of woody plants of Japan, t. 135 (1979).
Rhizome far-running. Culms 3–4 m, *c.* 4 cm in circumference, with white, waxy bloom below the nodes, purple when young. Sheaths very slightly downy when young, with white bristles. Branches from upper nodes spreading, rather short. Leaves 15–25 × 1.5–2.2 cm, green on both sides, slightly downy towards the base beneath; sheaths ciliate; ligule very short. *Japan*. H2.

A variable species fairly commonly grown; many variants have been described; they are difficult to distinguish and require further study. The most important are: forma **angustifolius** (Mitford) Muroi (*Bambusa angustifolia* Mitford; *Arundinaria angustifolia* (Mitford) Houzeau de Lehaie; *P. angustifolius* (Mitford) Nakai & Okamura), which is smaller (at most to 2 m), with narrower leaves which are usually striped with white (it has flowered since 1978); forma **humilis** (Makino) Suzuki, which is even smaller and confused with *P. humilis* (see below); forma **vaginatus** (Hackel) Muroi & Okamura (*Arundinaria vaginatus* Hackel; *P. vaginatus* (Hackel) Nakai; *P. chino* var. *vaginatus* (Hackel) Suzuki; *P. gracilis* (Makino) Nakai), which has culms which are somewhat purplish, mature leaves to 15 × 1.5 cm and even larger in young shoots (this is probably the most frequently seen variant, and is probably the one which has flowered sporadically since 1974). In addition to these there are 2 cultivars (probably clones) which are known only in cultivation: 'Argenteo-striatus' (*P. argenteo-striatus* (Regel) Nakai; *Arundinaria chino* var. *argenteo-striata* (Regel) Makino), which is shorter (to 1 m), its leaves striped with cream, and is grown occasionally; 'Laydekeri' (*Arundinaria laydekeri* Bean; *P. chino* var. *laydekeri* (Bean) Nakai), which is also small and has leaves dully and irregularly mottled.

2. P. gramineus (Bean) Nakai (*Arundinaria graminea* (Bean) Makino). Illustration: Suzuki, Index to Japanese Bambusaceae, 23, 292–3 (1978); Kitamura & Murata, Coloured illustrations of woody plants of Japan, t. 134 (1979).
Rhizomes extensive. Culms 3–5 m, 3–6 cm in circumference, with white, waxy bloom below the nodes, hairless. Branches numerous, hanging. Leaves to 28 × 1.2 cm (rarely broader), long-tapered, somewhat drooping and twisted towards the apex; most apices oblique. Sheaths hairless, apex oblique; bristles, if present, long; ligule 3–4 mm, blunt, downy, *Japan*. H2.

This species is occasionally grown and has flowered frequently since 1948.

***P. hindsii** (Munro) Nakai (*Arundinaria hindsii* Munro; *Bambusa erecta* invalid; *B. gracilis* misapplied). Illustration: Suzuki, Index to Japanese Bambusaceae, 23, 288–9 (1978); Kitamura & Murata, Coloured illustrations of woody plants of Japan, t. 134 (1979). Similar, but rhizome shorter, and plant not so copiously leafy. Branches erect, leaves erect, darker, thicker, to 1.5 cm broad, ligule 3–4 mm, acute. *Hong Kong*.

This species flowered between 1905 and 1911, and is relatively uncommon.

3. P. humilis (Mitford) Nakai (*Arundinaria humilis* Mitford; *Sasa humilis* (Mitford) Camus). Illustration: Suzuki, Index to Japanese Bambusaceae, 322–3 (1978); The Plantsman **1**: 41 (1979).
Rhizome running. Culms to 1.5 m, *c.* 3 cm in circumference. Sheaths purple when young. Nodes usually hairless, with white, waxy bloom below. Branches 1–3, long, found low on the culm. Leaves to 20 × 1.8 cm, sometimes downy beneath towards the base; sheaths slightly downy, ciliate; ligule minute, downy. *Japan*. H2.

A variable species; several variants have been named, of which the following are the most important: forma **humilis**, with short, bushy branches well up the culm; a further variant, with no appropriate name in this species, but found as *Arundinaria pumila* Mitford, *A. gauntletti* invalid and *P. pumilus* (Mitford) Nakai, is more robust and compact and has conspicuously bearded bases to the culm sheaths, and fresh green leaves.

4. P. pygmaeus (Miquel) Nakai (*Arundinaria pygmaea* (Miquel) Mitford; *Sasa pygmaea* (Mitford) Camus). Illustration: Suzuki, Index to Japanese Bambusaceae, 310–13 (1978).
Rhizome running. Culms to 40 cm, *c.* 9 mm in circumference, usually solid and flattened above, hairless, with white, waxy bloom. Sheaths hairy. Nodes conspicuous, downy or bristly at least at first. Branches 1–2 at the lower nodes. Leaves 3–8 cm × 4–8 mm, slightly downy, like the sheaths. *Japan*. H2.

This is the smallest species of the genus and various clones are grown, as is var. **distichus** (Mitford) Nakai (*Bambusa disticha* Mitford; *Arundinaria disticha* (Mitford) Pfitzer; *Sasa disticha* (Mitford) Camus; *P. distichus* (Mitford) Muroi & Okamura) in which the culms are always hollow, and up to 1 m tall, with usually hairless nodes, the leaves and their sheaths hairless or not, larger and more numerous, arranged in 2 rows on the culm. This variety flowered synchronously in 1967–70; it grades into the variant described above.

5. P. simonii (Carrière) Nakai (*Arundinaria simonii* (Carrière) Rivière; *Nipponocalamus simonii* (Carrière) Nakai). Illustration: Lawson, Bamboos, 109 & t. 16 (1966); Suzuki, Index to Japanese Bambusaceae, 294–5 (1978).
Rhizomes shortly running. Culms 4–8 m, *c.* 6 cm in circumference, hollow, the outer often arching. Sheaths conspicuous, long-persistent, bearded at the base when young. Nodes with white, waxy bloom below. Leaves to 25 × 3.5 cm, often smaller, minutely downy beneath, sometimes half green and half glaucous; sheath apex usually oblique; ligule *c.* 1 mm, downy. *Japan.* H2.

A commonly cultivated species, whose young leaves may be somewhat variegated at first, which flowered more or less generally in 1965–68 and 1981. 'Variegatus' (*Arundinaria simonii* var. *variegata* Hooker; *P. simonii* forma *variegata* (Hooker) Muroi; *A. simonii* var. *albo-striata* Bean; *P. simonii* var. *heterophyllus* (Makino) Nakai) is less tall and has its mature leaves variously broad or narrow, green or variegated. It is occasionally grown and has flowered since 1980 – see McClintock, Journal of the American Bamboo Society (in press).

6. P. variegatus (Miquel) Makino (*Arundinaria variegata* (Miquel) Makino; *Sasa variegata* (Miquel) Camus; *P. fortunei* invalid; *A. fortunei* invalid). Illustration: Flore des Serres **15**: t. 1535 (1863); Reader's Digest encyclopaedia of garden plants and flowers **2**: 60 (1971); The Plantsman **1**: 45 (1979); Crouzet, Les Bambous, 34 (1981).
Rhizome shortly running. Culms to 75 cm, *c.* 2 cm in circumference. Sheaths minutely downy, ciliate. Nodes hairless, with white, waxy bloom below. Branches usually 1–2, mostly from low down on the culm, erect. Leaves *c.* 14 × 1.3 cm (rarely to 20 × 1.8 cm), tapered below, slightly downy on both sides, dark green with cream stripes; sheaths purplish, minutely downy, ciliate. *Japan.* H2.

This species flowered in about 1970 and has done so again since 1977. It is variable, and several named variants are grown, including dwarfs (less than half the height given above), which may be 'Pygmaeus' ('Nana'). Var. **viridis** (Makino) McClintock (*Arundinaria variegata* var. *viridis* Makino; *P. shibuyanus* Nakai) is the wild variant with leaves entirely green and the nodes sometimes downy. Seedlings from this variant are grown; their leaves are often broader and usually downy beneath; though some with hairless leaves are grown as forma **glaber** (Makino) McClintock.

7. P. viridi-striatus (André) Makino (*Arundinaria auricoma* Mitford; *A. viridi-striata* (André) Makino; *A. fortunei* var. *aurea* invalid; *Bambusa argentea* invalid). Illustration: Hooker's Icones Plantarum **27**: t. 2613 (1901); Lawson, Bamboos, t. 9 (1966); Suzuki, Index to Japanese Bambusaceae, 24 (1978).
Rhizome shortly running. Culms to 1.5 m, *c.* 2 cm in circumference, purple. Sheaths markedly downy when young. Nodes hairy, with white, waxy bloom below. Branches 1–2, erect, from low down on the culm. Leaves *c.* 18 × 2.8 cm (rarely to 20 × 4 cm), rounded below, softly downy on both sides, especially beneath when young, brilliant yellow with green stripes of various breadths; sheaths densely downy, ciliate, purple. *Japan.* H2.

A widely grown species, which has a few spikelets in most years on mature culms. *P. kongosanensis* Makino is a name applied to the wild form, which has leaves entirely green.

88. CHIMONOBAMBUSA Makino

D. McClintock

Rhizomes slender, mostly shortly running. Culms hollow, terete or 4-cornered; new shoots appearing in late summer or autumn. Sheaths without bristles, variably persistent. Branches 3–many, developing from the top downwards during the following year. Leaves visibly tessellate or not, with smooth bristles or bristles absent. Inflorescence a panicle. Spikelets with 0–2 glumes and several florets. Lodicules 3. Stamens 3. Stigmas 2.

A genus of several species from the Himalaya, China and Japan. Their late-emerging culms may not ripen sufficiently to overwinter, which renders them less reliable in cultivation than many others. Some of the species included here, together with *Arundinaria falconeri* (p. 60), should perhaps be transferred to a new genus.

1a. Culms obtusely 4-cornered **4. quadrangularis**
b. Culms terete 2
2a. Leaves visibly tessellate; branches usually 3 at each node; culm sheaths not ciliate **3. marmorea**
b. Leaves not visibly tessellate; branches many at each node; culm sheaths ciliate 3
3a. Culms 5–8 cm in circumference; sheaths soon deciduous; leaves usually less than 16 cm **1. falcata**
b. Culms 9 cm or more in circumference; culm sheaths fairly persistent; leaves 15–30 cm **2. hookeriana**

1. C. falcata (Nees) Nakai (*Arundinaria falcata* Nees; *Thamnocalamus falcatus* (Nees) Camus; *A. gracilis* invalid). Illustration: Annals of the Royal Botanic Garden Calcutta **7**: t. 11 (1896).
Culms 3–6 m, 5–8 cm in circumference, closely clumped, with a large hollow, terete, somewhat lined, glaucous at first, the inner erect, the outer spreading. Sheaths long-tapered, ciliate, not long-persistent, but emerging branches likely to push through them, as long as or longer than the internodes and dark red when young, later straw-coloured; ligule elongate, to 12 mm, toothed, blade narrow. Nodes hairy at first, sometimes with white, waxy bloom below. Branches numerous. Leaves usually *c.* 11 × 1.2 cm, rarely to 16 cm, not visibly tessellate, soon hairless beneath except for the midrib, the apex somewhat twisted; sheaths hairless but ciliate, usually without bristles; ligule 2.5–6.5 mm, downy, pointed. *NW Himalaya.* H3.

This species has flowered since 1978, mostly with specimens cultivated under glass.

2. C. hookeriana (Munro) Nakai (*Arundinaria hookeriana* Munro). Illustration: Annals of the Royal Botanic Garden Calcutta **7**: t. 15 (1896); Camus, Les Bambusees, t. 24b (1913).
Culms 5–10 m or more, 9 cm or more in circumference, with a large hollow, glaucous, usually striped with olive green and pink when young, sometimes with white, waxy bloom below the nodes. Sheaths almost hairless, often longer than internodes, fairly persistent, the broad base frequently remaining; ligule toothed. Nodes hairless, with numerous branches. Leaves 15–30 × 1–4 cm, minutely downy beneath, occasionally with thin, yellow stripes, not visibly tessellate; sheaths hairless, without bristles; ligule 2–3 mm, downy. *E Himalaya.* H3.

Flowered under glass in 1963.

3. C. marmorea (Mitford) Makino (*Bambusa marmorea* Mitford; *Arundinaria marmorea* (Mitford) Makino). Illustration: Suzuki,

Index to Japanese Bambusaceae, 24, 334–5 (1978); Crouzet, Les Bambous, 54 (1981).
Culms 2–3 m, *c.* 2 cm in circumference, very thick-walled, terete, with hairless, purplish, rather short internodes. Sheaths thin, marbled purplish or pinkish brown at first, soon fading, almost hairless above, bushily hairy at the base when fresh, fairly persistent; blade minute. Nodes prominent. Branches usually 3, rather short and erect. Leaves to 16 × 1.2 cm, but often less (about half this size), visibly tesellate, minutely downy beneath; sheaths with bristles, downy; ligule minute. *Japan.* H2.

Flowers can be found in most years on some mature plants. 'Variegata' has white-striped leaves.

4. C. quadrangularis (Fenzi) Makino (*Bambusa quadrangularis* Fenzi; *Arundinaria quadrangularis* (Fenzi) Makino; *B. angulata* invalid). Illustration: Lawson, Bamboos, 31 (1966); Suzuki, Index to Japanese Bambusaceae, 17, 98–9 (1978); Crouzet, Les Bambous, 64 (1981).
Rhizome far-running. Culms 3–7 m, rarely to 10 m, 6–12 cm in circumference, obtusely 4-cornered, slightly rough below the nodes. Sheaths hairless, thin, open, soon falling. Nodes very prominent, with a purple band below and with the bases of incipient aerial roots, which decrease in size up the culm. Branches 3–7 at each node, rather short, brittle. Leaves 14–25 × 1.4–2 cm, visibly tessellate, dark green; sheaths with bristles. *China.* H3.

The flowers of this species are unknown.

89. SASA Makino & Shibata
D. McClintock
Rhizomes slender. Culms distant, ascending, hollow, terete, 2–3 m tall, 3–6 cm in circumference, unbranched below. Sheaths persistent, mostly shorter than the internodes, bristles spreading and rough or rarely absent. Nodes rather prominent, with white waxy bloom below. Branches 1 at each node. Leaves 4–11 per branch, thick and hairless, visibly tessellate; bristles rough, whitish, rigid, borne at right angles to the branch, sometimes absent; ligule downy. Inflorescence a panicle. Spikelets with 2 glumes and 4–10 flowers. Lodicules 3. Stamens 6. Stigmas 3.

A genus of probably 20–30 species, but many more have been described, from Japan and Korea. They grow well in partial shade, but are not easy to control.
Literature: Haubrich, R., The Sasas, *Journal of the American Bamboo Society* **2**: 24–38 (1982).

1a. Rhizomes not far-running; culms with many branches above — 2
b. Rhizomes far-running; culms with few branches, all near the base — 3
2a. Culms to 3 m; nodes hairless; bristles usually absent — **1. kurilensis**
b. Culms to 1.6 m; nodes hairy below; bristles usually present on young leaf-sheaths — **4. tsuboiana**
3a. Leaves to 25 cm, soon with a broad, white edge — **3. veitchii**
b. Leaves to 40 cm, green with a yellow midrib — **2. palmata***

1. S. kurilensis (Ruprecht) Makino & Shibata. Illustration: Suzuki, Index to Japanese Bambusaceae, 19, 118–19 (1978); Kitamura & Murata, Coloured illustrations of woody plants of Japan, t. 137 (1979).
Rhizome shortly running. Culms 50 cm–3 m, 5–30 mm in circumference, with the central hollow about half the diameter, hairless, richly branched above. Sheaths over half the length of the internodes, almost hairless, ciliate, disintegrating irregularly. Nodes hairless. Leaves variable, 15–22 × 2.2–4.8 cm, shining above, sometimes toothed on one side only, veins prominent; sheaths minutely downy, ciliate; auricles and bristles (if any) very soon falling; ligule to 2 mm. *Japan, Korea.* H2.

Relatively uncommon in cultivation; it has flowered from 1978.

2. S. palmata (Burbidge) Camus (*Arundinaria palmata* (Burbidge) Bean; *Bambusa palmata* Burbidge; *S. senanensis* misapplied; *S. cernua* misapplied). Illustration; Suzuki, Index to Japanese Bambusaceae, 20, 170–3 (1978); Crouzet, Les Bambous, 61 (1981).
Rhizome extensive. Culms 2 m (rarely to 4 m), *c.* 3 cm in circumference, often leaning, often hairless, usually soon purple-streaked (forma **nebulosa** (Makino) Suzuki). Sheaths minutely downy towards the apex and the base; auricles and bristles frequently absent; ligule *c.* 1 mm, very broad. Leaves to 40 × 9.5 cm, acuminate, bright, shining green, paler beneath, the midrib yellow, the margin sometimes decaying in hard winters; stalks greenish yellow; sheaths usually without bristles; ligule *c.* 2 mm, downy. *Japan.* H2.

This species began flowering in 1962, but this phase of flowering is now dying out. There is a variant with the leaves striped longitudinally with yellow.

***S. senanesis** (Franchet & Savatier) Rehder. Illustration: Suzuki, Index to Japanese Bambusaceae, 174–7 (1978).
Leaves narrower and densely downy beneath. *Japan.* H2.

3. S. veitchii (Carrière) Rehder (*Arundinaria veitchii* (Carrière) N.E. Brown; *S. albo-marginata* (Miquel) Makino). Illustration: Lawson, Bamboos, 146 (1966); Suzuki, Index to Japanese Bambusaceae, 190–2 (1978); Horticulture 7: 26 (1979).
Rhizome extensive. Culms to 1.3 or rarely to 2 m, *c.* 2 cm in circumference, finally purplish, glaucous, hairy at first, hairs soon deciduous, with white, waxy bloom below the nodes. Lowest internodes short. Sheaths white-hairy at first, with auricles and bristles, sometimes decaying irregularly. Leaves to 25 × 6 cm, abruptly acuminate, rather dull green above, glaucous beneath, soon with a broad, white margin; stalks often purplish. Sheaths hairless, bristles soon falling; ligule to 1.6 mm, dark in colour. *Japan.* H2.

This species has not, so far as is known, flowered outside Japan. A small variant, with culms to 27 cm, leaves 3–14 cm × 8–32 mm, is grown; it is sometimes called forma *nana*, or *S. albo-marginata* forma *nana* Makino.

4. S. tsuboiana Makino. Illustration: Suzuki, Index to Japanese Bambusaceae, 20, 163 (1978).
Rhizomes very shortly running. Culms to 1.6 m, *c.* 1 cm in circumference, erect, with a very small central hollow. Sheaths hairy below at least when young, ciliate, not or hardly disintegrating untidily, less than half the length of the mature internodes. Leaves to 24 × 5 cm, usually with bristles, at least on the young shoots. *Japan.* H2.

90. INDOCALAMUS Nakai
D. McClintock
Differs from *Sasa* essentially in floral characters. The species described below has less prominent nodes without white, waxy bloom, short internodes which are exceeded by the sheaths and very large leaves (the largest of any bamboo hardy in Europe), which are borne 1–2 to a branch; bristles, when present, dark. Stamens 3. Styles 2.

A genus of about 20 species from Malaysia and China. The 1 cultivated species is easy to grow and will tolerate dense shade.

1. I. tessellatus (Munro) Keng (*Bambusa tessellata* Munro; *Sasa tessellata* (Munro) Makino & Shibata; *Sasamorpha tessellata* (Munro) Nakai; *B. ragamowskii* Nicholson; *Arundinaria ragamowskii* (Nicholson)

Pfitzer). Illustration: Bailey, Standard cyclopedia of horticulture **1**: 445 (1942); Arnoldia **6**: t. 3 (1946).
Culms to 2 m, *c.* 4 cm in circumference, thick-walled, soon curved down with the weight of the leaves. Sheaths slightly downy but the hairs soon falling, rather loose. Internodes minutely downy. Leaves to 60 × 8.6 cm or more, sometimes with a line of hairs down one side of the midrib beneath; sheaths often purple towards the apex; ligule to 2 mm, dark in colour. *China.* H2.

This species is occasionally grown, and flowered in the USA between 1979 and 1982.

91. SASAELLA Makino
D. McClintock
Like *Sasa*, but culms erect, to 1.5 m tall and to 1.5 cm in circumference, bristles not spreading, rough below only, smooth above. Leaves 5–8 per branch. Inflorescence a panicle. Spikelets with 2 glumes and 5–10 flowers. Lodicules 3. Stamens 6. Stigmas 3.

A genus of about 12 species from Japan. Literature: McClintock, D., On the nomenclature and the flowering in Europe of the bamboo Sasaella ramosa (Arundinaria vagans), *Kew Bulletin* **38**: 191–5 (1983).

1. S. ramosa (Makino) Makino (*Bambusa ramosa* Makino; *Sasa vagans* misapplied; *Sasa ramosa* (Makino) Makino & Shibata; *Arundinaria vagans* Gamble; *A. ramosa* (Makino) Makino; *Pleioblastus viridi-striatus* forma *vagans* (Gamble) Muroi; *P. kongosanensis* Makino 'Vagans'). Illustration: Makino, Illustrated flora of Japan, t. 2633 (1948); Suzuki, Index to Japanese Bambusaceae, 22, 240–3 (1978); Kitamura & Murata, Coloured illustrations of woody plants of Japan, t. 136 (1979).
Rhizome extensive. Culms to 1.5 m, 1.5 cm in circumference, hairless. Nodes almost globular, hairless. Leaves to 20 × 3 cm, usually hairless above, distinctly downy beneath, the margins sometimes withering in harsh winters; sheath bristles few. *Japan.* H2.

This species first flowered outside Japan in 1981 (see McClintock, Kew Bulletin, **38**: 191–5, 1983). There are dwarf variants, and another with variegated leaves.

92. SASAMORPHA Nakai
D. McClintock
Like *Sasa* but culms erect to 3 m tall and *c.* 3 cm in circumference, sheaths without any bristles, longer than the internodes. Leaves 2–5 per branch, without bristles. Inflorescence a panicle. Spikelets with 2 glumes and 4–8 florets. Lodicules 3. Stamens 6. Stigmas 3.

A genus of 4 species from eastern Asia. Literature: Haubrich, R., The Sasas, *Journal of the American Bamboo Society* **2**: 24–38 (1982).

1. S. borealis (Hackel) Nakai (*Arundinaria borealis* (Hackel) Makino; *Sasa borealis* (Hackel) Makino; *Sasamorpha purpurascens* (Hackel) Nakai; *Sasamorpha purpurascens* var. *borealis* (Hackel) Nakai). Illustration: Suzuki, Index to Japanese Bambusaceae, 23, 270–1 (1978); Kitamura & Murata, Coloured illustrations of woody plants of Japan, t. 137 (1979).
Culms to 3 m, often less, *c.* 6 cm in circumference, minutely downy, well branched above. Sheaths with long and short persistent hairs, tight, disintegrating irregularly. Leaves 2–3 per branch, 22–27 × 2.5–4 cm, long-acuminate, glaucous, slightly paler beneath, usually purplish towards the base and on the stalk and sheath; ligule *c.* 2.5 mm. *Japan, Korea.* H2.

Confused with *Pseudosasa japonica.* Has flowered since 1978.

93. PSEUDOSASA Nakai
D. McClintock
Rhizomes slender, usually clump-forming, occasionally running. Culms erect, to 6 m. Sheaths persistent, longer than the internodes, bristles usually absent, if present then smooth and white. Branches 1 at each node, the lower nodes without branches. Leaves 4–7 per branch, without auricles or bristles. Inflorescence a panicle. Spikelets with 2 glumes and 3–8 florets. Lodicules 3. Stamens 3 or rarely 4. Stigmas 3.

A genus of 6 species from E Asia.

1. P. japonica (Steudel) Makino (*Arundinaria japonica* Steudel; *Sasa japonica* (Steudel) Makino; *Yadakea japonica* (Steudel) Makino; *Bambusa metake* Miquel). Illustration: Lawson, Bamboos, t. 1–2 (1966); Green Fingers **8**: 58 (1976) & **10**: 32 (1978) – both as Phyllostachys nigra var. henonis; The Plantsman **1**: 42 (1979).
Culms to 6 m, *c.* 6 cm in circumference, thin-walled, richly branched above and arching when mature, unbranched in the first year. Sheaths densely and roughly hairy at first, closely persistent, blades rather narrow, entire. Nodes often oblique, with some white, waxy bloom below. Leaves up to 7 on each branch, to 36 × 3.5 cm, though usually less, rather thick, long-tapered, one-third green and two-thirds glaucous beneath; ligule 1.5–3 mm, minutely downy. *Japan, Korea.* H2.

Commonly cultivated, and probably in flower somewhere every year. Like *Sasamorpha borealis*, but larger, with white, waxy bloom below the nodes and the undersides of the leaves bicoloured.

94. CHUSQUEA Kunth
D. McClintock
Rhizomes elongate, slender. Culms solid, terete, more than 1 cm in circumference. Sheaths persistent, without bristles. Branches numerous at each node. Leaves narrow, without bristles. Inflorescence usually a panicle. Spikelets with usually 4 glumes and with 1 (rarely 2) floret(s). Lodicules 3. Stamens 3. Stigmas usually 2, occasionally with 1 branched again.
A genus of over 90 species from Mexico to S Argentina. They are not easy to propagate vegetatively, and take a considerable time to establish. Once established, they can tolerate drier conditions than most bamboos.
Literature: Parodi, L.R., Sinopsis de las Graminas chilenas del Genero Chusquea, *Revista Universitaria, Universidad Catolica de Chile* **30**(1): 61–71 (1945).

1. C. culeou Desvaux (*C. andina* Philippi; *C. breviglumis* Philippi; *C. quila* misapplied). Illustration: Lawson, Bamboos, 149 (1966); Crouzet, Les Bambous, 55 (1981); Everett, Encyclopedia of horticulture **3**: 760 (1981).
Plant loosely tufted with very short rhizomes. Culms 4–6 m, rarely to 10 m, *c.* 11–13 cm in circumference, yellowish, erect, tapering. Sheaths downy above, white, much longer than the internodes; blades narrow. Nodes hairy, quite prominent, with white, waxy bloom below. Branchlets developing in the second year and bursting through the sheaths, to 20 cm or more, bushy, for the most part unbranched. Leaves 5–10 × *c.* 1 cm, hairless, visibly tessellate, horny-edged and with the mid-vein prominent beneath; sheaths hairless, without bristles; ligule 0.5–1 mm, hairless. *Most of Andean South America.* H2.

There are 2 variants at least in cultivation: 1 as described above, the other (var. **tenuis** McClintock) with slenderer, shorter culms which are spreading.

***C. cumingii** Nees. Leaves 2–6 cm × 3–5 mm, midrib hardly apparent. *Chile.* H2.

This may be the same species as *C. culeou*; if the 2 are combined, the correct name would be *C. cumingii*.

XXI. PALMAE (ARECACEAE)

Compiled at the Royal Botanic Gardens, Kew

Evergreen shrub- or tree-like plants or woody climbers, sometimes spiny. Stems solitary or clustered, or colonial from long rhizomes, rarely branched aerially, erect or prostrate, sometimes apparently stemless, smooth, spiny or covered in leaf-sheaths. Leaves alternate; sheaths tubular at first, sometimes forming a column (crownshaft) above the stem or splitting, often toothed or spiny; stalk usually present. Leaf-blade pinnate, bipinnate, palmate, costapalmate (i.e. palmate but with a short extension of the stalk between each pair of leaflets), or undivided with palmately or pinnately arranged folds; sometimes a protrusion (hastula) present on upper surface of leaf-stalk where the leaflets are attached. Leaflets (segments) with 1 to several ribs, acute, bifid, truncate, or irregularly toothed. Inflorescences among or below the leaves, or aggregated into a terminal mass (the stem then dying after fruiting), 1 (rarely more) per node, paniculate to spicate, the stalk bearing a basal bract with 2 keels, and upper bracts few to many (rarely none), all persistent or deciduous; inflorescence branches (where present) subtended by bracts, branching occurring at up to 4 orders. Flowers unisexual (when plant monoecious or dioecious) or bisexual, stalkless or stalked, or sunk into pits, borne singly, in clusters or in 3s (triads) of 2 lateral males and a central female, or in pairs of bisexual and male, or female and neuter, or of male flowers only. Sepals and petals usually 3 in each whorl, free or united, or perianth of one undifferentiated series. Stamens usually 6, sometimes from 3 to almost 1000. Ovary superior with carpels free or united, usually 3 sometimes 1, 2 or more than 3; ovule 1 only in each carpel, variously attached; carpels sometimes infertile and fused into a pistillode. Fruit fleshy or dry, sometimes covered in scales, hairs or spines, sometimes with bony endocarp; seeds 1–3 or more, sometimes with a fleshy seedcoat (sarcotesta); endosperm smooth or ruminate, i.e. furrowed, the furrows penetrated by folds of the seedcoat.

A large, almost entirely tropical and subtropical family of 212 genera and about 2700 species.

It is difficult to ascertain which palms are grown in Europe, particularly in the south. The selection of species presented here is based on what might be expected to grow well in Europe, and is compiled with reference to the articles of Chabaud, Carvalho-e-Vasconcellos & do Amaral-Franco (see below) and on the basis of observations made in north Italy. The account does not deal with palms found only in botanic gardens or in enthusiasts' collections, or with juveniles of many species which enter the pot-plant trade as ephemeral houseplants with little chance of survival. The real palm enthusiast is recommended to read articles on palms in the journals *Gentes Herbarum* and *Principes* (Journal of the Palm Society) and the entries by H.E. Moore in Hortus Third; several reference books such as McCurrach, 'Palms of the world' and its supplement by Langlois are of value, but contain many incorrectly named photographs.

Literature: Chabaud, B., *Les Palmiers de la Côte d'Azur* (1915); Carvalho-e-Vasconcellos, J. de & do Amaral-Franco, E.J., As Palmeiras de Lisboa & Arredores, *Portugaliae Acta Biologica* (Serie B) 2(4): 289–425 (1948); McCurrach, J.C., *Palms of the world* (1960); Moore, H.E., The major groups of palms and their distribution, *Gentes Herbarum* 11(2): 27–140 (1973); Langlois, A.C., *Supplement to palms of the world* (1976).

1a. Leaves bipinnate with fishtail segments **17. Caryota**
b. Leaves pinnate or palmate 2
2a. Leaves palmate or palmately ribbed or divided 3
b. Leaves pinnate or pinnately divided 16
3a. Leaf without a hastula **12. Nannorrhops**
b. Leaf with a hastula 4
4a. Leaf undivided except for very shallow marginal lobing **7. Licuala**
b. Leaf divided into segments 5
5a. Leaf divided into segments along the upper ribs or between the ribs 6
b. Leaf divided into segments along the lower ribs 7
6a. Leaf divided to base along the upper ribs **7. Licuala**
b. Leaf divided not quite to base between ribs **5. Rhapis**
7a. Stem bearing spine-like roots; blade with a major split along the mid-line **4. Cryosophila**
b. Stem without spine-like roots; blade without such a major split 8
8a. Leaf-sheath disintegrating to form erect, needle-like spines **1. Rhapidophyllum**
b. Leaf-sheath without such spines 9
9a. Stem covered with a skirt of dead leaves; inflorescence bracts woody **11. Washingtonia**
b. Stem with or without persistent leaf-bases but not with a skirt; bracts thin or leathery 10
10a. Carpels free 11
b. Carpels joined together at tips or base or throughout 12
11a. Dwarf clustering palm; leaf-stalk with conspicuous spines **3. Chamaerops**
b. Usually erect solitary palms; leaf-stalk with inconspicuous spines **2. Trachycarpus**
12a. Carpels joined throughout **13. Sabal**
b. Carpels joined only in stylar region 13
13a. Seed with ruminate endosperm **10. Copernicia**
b. Seed with smooth endosperm 14
14a. Apparently stemless, clustering palm **8. Serenoa**
b. Erect, solitary (rarely clustering) palms 15
15a. Sepals free, overlapping **9. Brahea**
b. Sepals joined **6. Livistona**
16a. Leaflets V-shaped in cross-section at the base (induplicate) 17
b. Leaflets Λ-shaped in cross-section at the base (reduplicate) 18
17a. Leaflets acute, the lowermost modified as spines **14. Phoenix**
b. Leaflets with irregularly toothed apices **16. Arenga**
18a. Slender, climbing palm with barbed whips **15. Calamus**
b. Barbed whips absent 19
19a. Palms with crownshafts 20
b. Palms without crownshafts 25
20a. Leaflets with irregularly toothed apices **25. Ptychosperma**
b. Leaflets acute or bifid but not irregularly toothed 21
21a. Clustering palms 22
b. Solitary palms 23
22a. Crownshaft green or grey **19. Chrysalidocarpus**
b. Crownshaft orange to red **23. Cyrtostachys**
23a. Flowers in triads only near base of inflorescence (hence fruit only near the base); female flowers otherwise absent **20. Archontophoenix**
b. Flowers in triads (hence fruit) more or less throughout inflorescence 24
24a. Stamens 6 **22. Rhopalostylis**
b. Stamens 9–12 **21. Hedyscepe**

25a. Leaflets dark green on upper surface, chalky-white beneath **30. Microcoelum**
b. Leaflets more or less the same colour above and beneath 26
26a. Leaflets regularly arranged in one plane 27
b. Leaflets grouped and held in several planes 31
27a. Palm with very slender, reed-like stem less than 5 cm in diameter **18. Chamaedorea**
b. Trunk at least 5 cm in diameter 28
28a. Trunk not exceeding 10 cm in diameter; inflorescence spicate **24. Howea**
b. Trunk at least 15 cm in diameter; inflorescence branched 29
29a. Trunk very massive, at least 90 cm in diameter **28. Jubaea**
b. Trunk not exceeding 50 cm in diameter 30
30a. Leaf-stalks with marginal teeth **27. Butia**
b. Leaf-stalks not toothed **26. Cocos**
31a. Stem more or less absent; leaflets often splitting at the tips **31. Allagoptera**
b. Stem well developed, forming a trunk; leaflets not splitting at the tips 32
32a. Trunk smooth; leaf-stalk not toothed **29. Arecastrum**
b. Trunk covered in persistent leaf-bases; leaf-stalk toothed **32. Elaeis**

1. RHAPIDOPHYLLUM Wendland & Drude
Clustering, more or less stemless palm producing large clumps; plants mostly unisexual, or with a few flowers of the opposite sex. Leaf-sheaths disintegrating to form long, sharp, erect, needle-like spines. Leaf palmate, to *c.* 1 m, the stalk without teeth or spines, the blade divided into 5–12 or more spreading, stiff, 3-ribbed segments to 4 cm wide. Inflorescences short, buried in the leaf-sheaths and needles at the base of the leaf-stalks. Male and female flowers similar, wine-red, with 3 sepals, 3 petals, 6 stamens (fertile or sterile) and 3 separate carpels with short styles, or pistillodes. Fruit ovoid, *c.* 2 × 1.5 cm, covered in woolly hairs. Seed with smooth endosperm.

One species native to the southern USA. Apparently the hardiest N American palm. It is slow growing and not widely cultivated. Propagation by seed.
Literature: Shuey, A.G. & Wunderlin, R.P., The needle palm: Rhapidophyllum hystrix, *Principes* **21**: 47–59 (1977).

1. R. hystrix (Pursh) Wendland & Drude (*Chamaerops hystrix* Pursh). Illustration; Principes **21**: 48, 51–3 (1977).
Southern USA. H4–5.

2. TRACHYCARPUS Wendland
Small to moderate, solitary or clustering, dioecious palms, rarely monoecious or with unisexual and bisexual flowers on the same plant; trunks often covered with persistent, fibrous leaf-sheaths. Leaves palmate, divided into 1-ribbed segments. Inflorescences among the leaves, with several thin bracts on the inflorescence-stalk, branching to 3 orders. Flowers in clusters of 2–4, with 3 basally overlapping sepals and 3 overlapping petals. Male flowers with 6 stamens and 3 small pistillodes. Female flowers with 6 staminodes and 3 carpels. Bisexual flowers with 6 stamens and 3 carpels. Fruit spherical to kidney-shaped or oblong-ovoid with remains of stigma at apex. Seed with smooth endosperm and a lateral embryo.

About 6 species native to subtropical Asia, widely cultivated, and 1 species (*T. fortunei*) amongst the hardiest of palms. However, taxonomically the genus is still very poorly known; the differences between some of the species may be very slight, and little is known of their variation outside cultivation.
Literature: Kimnach, M., The species of Trachycarpus, *Principes* **21**: 155 (1977).

1a. Leaf-blade divided by splits of the same length, leaf-bases soon falling; fruit oblong-ovoid; seed with a longitudinal groove **1. martianus**
b. Leaf-blade divided by splits of varying lengths; leaf-bases persistent; fruit spherical to kidney-shaped; seed without a longitudinal groove 2
2a. Fibres on trunk loose and ruffled; leaf-base appendages ribbon-like, recurved, leaf-blades divided to more than half their length **2. fortunei***
b. Fibres on trunk closely adpressed; leaf-base appendages triangular, erect; leaf-blades divided to about half their length **3. takil***

1. T. martianus (Wallich) Wendland (*Chamaerops martiana* Wallich; *T. khasyanus* (Griffith) Wendland). Illustration: Botanical Magazine, 7128 (1890); Principes **21**: 156 (1977).
Trunk to 16 m, *c.* 10 cm in diameter, bare, ringed with leaf-scars. Leaves circular, very regularly divided to the middle into many segments, dark green above, usually glaucous beneath. Fruit glossy blue, oblong-ovoid, *c.* 12 × 10 mm. Seed with a longitudinal groove. *C & E Himalaya, Burma, N Thailand, probably also China.* H5–G1.

2. T. fortunei (W.J. Hooker) Wendland (*Chamaerops fortunei* W.J. Hooker; *C. excelsa* invalid; *T. excelsa* invalid). Illustration: Bean, Trees and shrubs hardy in the British Isles **4**: t. 86 (1980); McCurrach, Palms of the world, 250 (1960).
Trunk solitary, to 20 m, *c.* 10 cm in diameter, conspicuously covered with black, hair-like fibres from the old leaf-sheaths. Leaves circular, to 1.25 m across, dull green, sometimes paler beneath, variously divided to the middle or almost to the base into rather stiff, horizontal or drooping segments. Leaf-stalk to 1 m, finely toothed along its margins. Inflorescences not exceeding the leaves, with masses of golden yellow flowers. Fruit spherical to kidney-shaped, bluish, *c.* 1.2 cm across. Seed depressed in the centre. *Origin unknown, probably N Burma or C & E China; widely cultivated and becoming naturalised in China and Japan.* H4–5.

***T. wagnerianus** Roster. Illustration: Principes **21**: 160 (1977). Said to differ from *T. fortunei* by having smaller, more rigid leaves. It is known only in cultivation and probably represents only a form of *T. fortunei*.

***T. fortunei** var. **surculosa** Henry. This name was given to a clustering plant which still exists at Leonardslee, Sussex, England. It now consists of a few old stems and appears to have lost its suckering potential. In leaf form, this variety is similar to *T. wagnerianus*.

***T. caespitosus** Roster. Illustration: Principes **21**: 155 (1977). Named from plants cultivated in California. This may be the same as *T. fortunei* var. *surculosa*. Clustering *Trachycarpus* has been observed at Abbotsbury, Dorset, England, but it is not clear whether this is true clustering, or growth of seedlings at the base of the trunk.

3. T. takil Beccari. Illustration: Principes **21**: 159 (1977).
Very similar to *T. fortunei* but differing in the tendency of young plants to grow obliquely, the less divided leaf-blades, and the trunk fibres being closely adpressed to the trunk. *W Himalaya*. H5.

Named from plants cultivated in Italy from seed sent from Kumaon (India); the original plant still exists but is now very tall.

***T. nanus** Beccari (probable synonym *T. dracocephalus* Ching & Hsu), from Yunnan,

China, is a very short-stemmed or apparently stemless *Trachycarpus*; it is not known in cultivation, but would obviously be well worth introducing into gardens.

3. CHAMAEROPS Linnaeus
Clustering, shrubby palms, mostly dioecious or rarely with a few flowers of the opposite sex in each inflorescence. Stems suckering and forming low clumps but occasionally producing an erect trunk to 5 m, covered in more or less persistent, fibrous leaf-sheaths. Leaves palmate; stalks with short spines pointing towards the apex; blade held stiffly and deeply divided into 1-ribbed segments. Inflorescences among the leaves, with a single large bract at the base and smaller bracts subtending the few branches. Flowers borne singly, but crowded together, golden yellow. Calyx a low, 3-toothed cup. Petals 3, overlapping. Stamens or staminodes 6. Carpels 3, free or reduced to pistillodes. Fruit rounded, reddish brown, fleshy, with a rancid smell. Seed with a slightly ruminate endosperm.

One species from S Europe and N Africa. One of the hardiest palms, widely cultivated throughout southern Europe. Propagation by seed or offsets.

1. C. humilis Linnaeus. Illustration: McCurrach, Palms of the world, 44 (1960). *S Europe, N Africa.* H4–5.

4. CRYOSOPHILA Blume
Small to medium, solitary, bisexual palms with trunks covered, at least at the base, with root-spines. Leaves palmate, divided more or less to the base along the mid-line and then into 1 to several, ribbed, acute segments; stalk with smooth margins. Inflorescence among the leaves, arched, with densely hairy, loosely sheathing bracts on the stalk and many crowded branches. Flowers solitary, creamy white to purplish. Sepals 3, shortly joined at the base. Petals 3, overlapping. Stamens 6, with filaments joined in their basal halves. Carpels 3, free, with narrow styles. Fruit yellowish or white. Seed spherical with smooth endosperm.

About 8 species in C & northern S America, some of considerable horticultural potential. One species is widely cultivated under glass. Propagation by seed.

1. C. warscewiczii (Wendland) Bartlett (*Acanthorrhiza warscewiczii* Wendland). Illustration: McCurrach, Palms of the world, 71, 72 (1960).
Trunk to 6 m or more, grey, with long root-spines near the base, becoming smooth near the crown. Leaves green on the upper surface, silvery beneath, divided into 50–60 segments which are up to 90 × 3.5 cm. Inflorescence to 60 cm or more. Fruit spherical to pear-shaped, white, to *c.* 2.5 cm. *Costa Rica to Panama.* G2.

5. RHAPIS Linnaeus
Clustering, slender, dioecious palms without teeth or spines, with reed-like or bamboo-like stems. Leaves palmate, divided nearly to the base into 2 to several, ribbed segments, shallowly or sharply toothed at the apex; sheaths fibrous, persistent; stalks well developed. Inflorescences among leaves; stalk with several empty bracts, then branching, the branches several, bearing yellow flowers. Male flowers stalkless with 3-toothed calyx, and 3-lobed corolla. Stamens 6. Female flowers similar but with calyx stalked; stamens absent. Carpels 3, separate. Fruit a small 1-seeded berry. Seed 1, spherical, with smooth endosperm.

A genus of about 9 species from S China to Thailand, mostly very poorly known; 2 species have long been in cultivation. Several plants of as yet unknown identity have recently been introduced into cultivation in Thailand, whence they have spread to specialist palm collections. Seed of at least one has recently been offered for sale on a large scale. Propagation by seed, or by offsets removed with care. Established plants indoors are generally tolerant of neglect.
Literature: Bailey, L.H., Species of Rhapis in cultivation, *Gentes Herbarum* **4**: 199–208 (1939).

1a. Sheaths coarsely fibrous; at least some leaves with only 3–7 segments, others with up to 10, uniformly curving or drooping, 4–8 cm wide **1. excelsa**
b. Sheaths finely fibrous; leaves mostly with 9 or more segments, held rather stiffly, less than 3 cm wide **2. humilis**

1. R. excelsa (Thunberg) Henry (*R. flabelliformis* Aiton; *Chamaerops excelsa* Thunberg; *Trachycarpus excelsus* (Thunberg) Wendland). Illustration: Gentes Herbarum 4: 200, 201, 204 (1939).
Stems numerous, to 3 m or more tall, covered in coarse fibres. Some leaves with 3–7 segments, other with up to 10 segments, uniformly curving or drooping, broad at apex, 4–8 cm wide. *?S China.* G1.

2. R. humilis Blume.
Stems numerous, to 3 m, usually more slender than in *R. excelsa*, covered in very fine fibres. Leaves with 9 or more segments usually held rather stiffly, less than 3 cm wide, very narrow at the apex. *S China.* G1.

6. LIVISTONA R. Brown
Solitary, mostly very tall, bisexual palms. Leaves shortly but conspicuously costapalmate, the stalks usually with marginal spines, the blades deeply divided sometimes almost (but not quite) to the base, with 1-ribbed, bifid segments. Inflorescences among the leaves, with tubular, sheathing bracts subtending branches, branching to up to 4 orders. Flowers solitary or in clusters with the sepals joined into a low, 3-lobed cup, petals 3, free and edge-to-edge above, shortly joined at the base; stamens 6, joined into a ring; carpels 3, free except in the stylar region. Fruit spherical to ellipsoid. Seed with smooth endosperm and a lateral embryo.

About 29 species native to the tropics and subtropics of Asia, and Australasia. Widely cultivated in the tropics, they also make very decorative pot plants in the juvenile state. Propagation by seed.
Literature: Beccari, O., Asiatic Palms – Corypheae, *Annals of the Royal Botanic Gardens Calcutta* **13** (1935).

1a. Leaf-segments stiff, only slightly hanging; ripe fruit orange-red **1. rotundifolia**
b. Upper quarter or third of each leaf-segment conspicuously hanging; ripe fruit bluish green to black 2
2a. Trunk smooth, grey; leaf-segments very neatly and regularly hanging **2. chinensis**
b. Trunk brown, ridged with leaf-scars; leaf-segments more untidily hanging 3
3a. Leaf-stalk with strong spines throughout its length **3. australis**
b. Leaf-stalk with spines in the basal half only **4. decipiens**

1. L. rotundifolia (Lamarck) Martius. Illustration: McCurrach, Palms of the world, 121 (1960).
Trunk to 30 m, prominently ringed. Leaves more or less circular with large, entire central portion and many short, stiff, shortly 2-lobed segments. Inflorescence dividing at the very base into 3 equal branches, these branching to 3 orders. Flowers borne singly. Fruit orange-red, turning black on rotting. *Philippines, Sabah (offshore Islands only), Celebes & Moluccas.* G2.

This species makes a beautiful pot plant.

2. L. chinensis (Jacquin) Martius. Illustration: McCurrach, Palms of the world, 120 (1960).
Trunk to 10 m, becoming smooth and grey. Leaf-stalk toothed near base; blade with large, entire central portion and numerous, deeply cleft, very neatly and regularly hanging segments. Inflorescence with main axis bearing several branches, each branching to 2–3 orders. Fruits bluish green, ellipsoid to almost spherical. *Japan (Ryukyu Islands)*. G1, ?H5.

3. L. australis (R. Brown) Martius. Illustration: McCurrach, Palms of the world, 118 (1960).
Trunk to 20 m, dull brown, conspicuously ridged. Leaf more or less circular in outline, to 1.75 m wide, divided to the middle into *c.* 70 soft, deeply 2-cleft segments with hanging tips; stalk spiny throughout its length. Inflorescence with hairy bracts and hairless, yellow branches, bearing flowers in 2s and 3s. Flowers hairless. Fruit purplish black, to 2.2 × 2 cm. *E Australia*. H5.

4. L. decipiens Beccari. Illustration: McCurrach, Palms of the world, 120 (1960).
Trunk smaller than *L. australis*. Leaves to *c.* 1.5 m, very deeply divided into *c.* 80 very narrow segments with hanging tips, stalk spiny near the base only. Inflorescence with more or less hairless bracts, and hairless branchlets bearing flowers in 2s and 3s. Fruit blackish, *c.* 1.3 cm in diameter. *Originally described from the French Riviera, but thought to have come from Australia.*

7. LICUALA Thunberg
Usually small, single-stemmed or clustered, bisexual palms. Leaves palmate or costapalmate; stalk usually with marginal spines and the blade sometimes undivided, or more often deeply divided into 1 to many, broad, wedge-shaped, ribbed segments with shallow apical lobing. Inflorescence among leaves, the stalk with tight, tubular bracts and ending in a spike; or with bracts subtending branches, themselves spicate or branched to 2 orders. Flowers solitary or in clusters of a few. Calyx cup-shaped with 3 apical lobes. Petals 3, joined in the basal half. Stamens 6, inserted at the mouth of the corolla tube, free, or joined into a ring. Carpels 3, free except for the joined styles, each with 1 ovule, usually only 1 developing. Fruit usually orange or red, with remains of stigma at apex. Seed with smooth endosperm and a lateral embryo.

About 108 species, from SE Asia extending through Malaysia to Australia and the New Hebrides. Many species are extremely decorative but they are often slow growing and most appear to require relatively high temperatures.

1a. Leaf-blade undivided **1. grandis**
b. Leaf-blade divided to the base into many segments **2. spinosa**

1. L. grandis Wendland. Illustration: Botanical Magazine, 6704 (1883).
Stem solitary to *c.* 2 m. Leaves more or less circular, to *c.* 90 cm wide, undivided except for very shallow marginal lobing. Inflorescence to *c.* 1.5 m, branching to 2 orders. Flowers arranged singly. Staminal ring 3-lobed with 3 anthers at the tips of the lobes, 3 between. Fruit orange, *c.* 1.5 cm in diameter. *New Hebrides*. G2.

2. L. spinosa Thunberg. Illustration: McCurrach, Palms of the world, 117 (1960).
Stems clustered, to *c.* 5 m. Leaves more or less circular in outline, to *c.* 1.5 m across, divided into 18–19, several-ribbed segments. Inflorescence to 2 m, with about 10 primary branches, each of these divided in 3–10 hairy, flower-bearing branches. Flowers arranged in clusters or singly, finely hairy. Staminal ring with 6 equal teeth. Fruit *c.* 1 cm in diameter, bright orange-red. *SE Asia*. G2.

8. SERENOA J.D. Hooker
Small, colonial, apparently stemless or very shortly stemmed, bisexual palms, without teeth or spines except for the teeth on the leaf-stalk margins. Leaves palmate, to 75 cm across, divided for more than half their length into *c.* 20, stiff, 1-ribbed, 2-lobed segments, bluish green in colour. Inflorescences to *c.* 60 cm, with tubular bracts on the stalk subtending primary branches, these branching 1–2 times. Flowers creamy white, fragrant, arranged in pairs or singly. Sepals joined into a 3-lobed cup. Petals 3, joined in the basal half, edge-to-edge above. Stamens 6. Carpels 3, free except for the joined styles. Fruit black, 1-seeded, ellipsoid to almost spherical. Seed with smooth endosperm and a lateral embryo.

A genus of 1 species, much cultivated in the USA. Its status in Europe is not known, but it was once cultivated on the French Riviera. Propagation by seed.

1. S. repens (Bartram) Small (*S. serrulata* (Michaux) Nicholson). Illustration: McCurrach, Palms of the world, 234 (1960). *Southeast USA*. H5–G1.

9. BRAHEA Martius
Dwarf to medium, solitary or rarely clustered, bisexual palms. Leaves shortly costapalmate; stalks smooth or with short spines; blade divided into 1-ribbed, bifid segments. Inflorescence elongate, among leaves, usually exceeding them in length, with tubular bracts subtending primary branches, to 2–3 orders. Flowers solitary or in clusters of 3. Sepals 3, free. Petals 3, free or shortly united. Stamens 6. Carpels 3, free except in the stylar region. Fruit ellipsoid to almost spherical with remains of stigma near apex and basal abortive carpels. Seed with smooth endosperm.

About 12 species confined to limestone in Mexico and C America. They are very decorative, and are often called *Erythea* Watson. Propagation by seed; they require full sun and a limey soil.
Literature: Bailey, L.H., Brahea and an Erythea, *Gentes Herbarum* **6**: 177–97 (1943).

1a. Leaves long-persistent, pale bluish green; inflorescences to 5 m, much longer than the leaves **1. armata**
b. Leaves not long-persistent, green or subglaucous; inflorescences to 2 m, not or only slightly longer than the leaves 2
2a. Leaf-blade paler beneath; inflorescence slightly longer than leaves **2. dulcis**
b. Leaf-blade green on both surfaces; inflorescence shorter than leaves **3. edulis**

1. B. armata Watson (*Erythea armata* (Watson) Watson). Illustration: Gentes Herbarum **4**: 106, 107 (1937).
Trunk solitary, to 14 m, robust, covered in old leaves unless cut or burned. Leaves waxy blue, deeply divided into *c.* 50 segments, the stalk with stout, curved teeth. Inflorescence to 5 m, arching out of the crown, with hanging, creamy branches. Flowers in clusters of 3 on softly hairy branches. Fruit spherical, *c.* 2 cm in diameter, yellow, fleshy. *Mexico (Baja California)*. H5.

Slow growing but extremely decorative.

2. B. dulcis (Humboldt, Bonpland & Kunth) Martius (*B. calcarea* Liebmann). Illustration: Gentes Herbarum **4**: 120 (1937).
Trunk to 7 m or more, solitary and erect, prostrate at the base, occasionally suckering. Leaves not long-persistent; stalk with short hooked spines; blade stiff, to 1.7 m long, green above, paler beneath, with *c.* 60 segments. Inflorescences to 2 m,

slightly longer than leaves. Flowers solitary, partly immersed in yellowish hairs. Fruit yellow, succulent, *c.* 1.5 cm in diameter. *W & C Mexico, Guatemala.* H5.

3. **B. edulis** Watson (*Erythea edulis* (Watson) Watson). Illustration: Gentes Herbarum **4**: 84, 86 (1937).
Trunk solitary to 10 m and 40 cm in diameter, becoming smooth. Leaves not long-persistent; stalks to 1.25 m × 6 cm, with short spines. Blades green on both surfaces, to 2 m wide, soft, with *c.* 80 segments. Inflorescence not exceeding leaves, bearing flowers in clusters of 3 on softly hairy branches. Fruit black, *c.* 2.5 cm in diameter. *Mexico (Guadalupe Islands).* H5.

Slow growing but very decorative.

10. COPERNICIA Martius
Moderately short to tall, solitary, bisexual palms with trunks frequently clothed with persistent leaf-bases, rarely with persistent leaves. Leaves palmate, with very short to long stalks which usually have marginal teeth; hastula conspicuous; blades of various shapes, shallowly divided into 1-ribbed segments. Inflorescences among, and scarcely to greatly exceeding the leaves, with tubular bracts on the stalk which subtend primary branches, these unbranched or up to 2 times branched. Flowers solitary or in clusters of 2–4, with sepals joined into 3-lobed cup, petals joined in the basal half, 6 stamens borne on the petals, and 3 carpels, which are more or less free except in the stylar region. Fruit ovoid to spherical, with abortive carpels and remains of stigma at apex. Seed 1, with deeply ruminate endosperm and a basal embryo.

About 29 species native to the West Indies and S America, with an astonishing range of species in Cuba.

Many species are handsome ornamentals, but are generally slow growing. *C. prunifera* (Miller) H.E. Moore is the source of Carnauba wax. One species is said to be in cultivation in Europe. Propagation by seed.
Literature: Dahlgren, B.E. & Glassman, S.F., A revision of the genus Copernicia, *Gentes Herbarum* **9**: 1–232 (1961–63).

1. **C. alba** Morong. Illustration: Gentes Herbarum **9**: 18–19 (1961).
Stems to 30 m, covered (in young trees) with persistent leaf-bases. Leaf-stalk to 60 cm or more, toothed near the base, blade circular, *c.* 60 cm long, covered in white wax. Inflorescence with paired flowers. Fruit ovoid, *c.* 2 cm long. *Brazil to Argentina.* H5–G1.

11. WASHINGTONIA Wendland
Robust, solitary, bisexual palms, with trunks (in nature) clothed in persistent dead leaves. Leaves costapalmate with 1-ribbed segments; stalks with spines along margins. Inflorescences borne among leaves, exceeding them in length and bearing several empty bracts on the stalk, and woody, open, sword-shaped bracts each subtending a hanging inflorescence branch which bears slender, flower-bearing branches. Flowers cream, with tubular calyx; petals 3, deciduous; stamens 6; carpels 3, free, except at the very tip, where united into a common style. Fruit small, 1-seeded. Seed with smooth endosperm.

Two species native to southwest USA and NW Mexico. Widely cultivated in the drier parts of the subtropics, and areas with a Mediterranean climate. The 2 species are difficult to distinguish in the juvenile state. Propagation by seed.
Literature: Bailey, L.H., Washingtonia, *Gentes Herbarum* **4**: 53–82 (1936).

1a. Leaves grey-green, lacking a tawny patch around the hastula; mature leaves with hanging segments and filaments between them **1. filifera**
b. Leaves brilliant green with a conspicuous tawny patch around the hastula; mature leaves with stiff segments and usually without filaments between them **2. robusta**

1. **W. filifera** (Linden) Wendland. Illustration: Gentes Herbarum **4**: 55 (1936).
Robust palm to 25 m or more, with trunk not tapered, in nature covered in a dense, even skirt of old leaves. Crown open and broad. Leaves grey-green without a tawny patch around the hastula on the underside, with conspicuous hanging filaments and gracefully hanging segments. *Southwest USA (California and Arizona).* H5.

2. **W. robusta** Wendland. Illustration: Gentes Herbarum **4**: 478 (1936).
Robust palm to 30 m or more, with trunk tapering from a swollen base, in nature covered in an untidy skirt of old leaves. Leaves brilliant green with a conspicuous tawny patch around the hastula on the underside, without conspicuous filaments and tending to have stiff segments. (Juvenile plants sometimes have leaf filaments.) *NW Mexico.* H5.

12. NANNORRHOPS Wendland
Small, bushy, clustering and stoloniferous, unarmed, bisexual palms, without teeth or spines; stems to 3 m or more, branching by equal forking (dichotomy). Leaves glaucous, shortly costapalmate, lacking a hastula; stalk well developed, smooth, segments 8–20 or more, to 1 m long, deeply 2-cleft to the middle, often becoming pendent. Inflorescence compound, ending the growth of the stem, with many lateral inflorescences borne in the axils of reduced leaves or bracts, branching to 3 orders. Flowers borne in clusters on the branches, white with dark purple anthers and pink carpels. Fruit spherical, orange, *c.* 2 cm in diameter, 1-seeded. Endosperm smooth.

One species in the semi-arid areas of Pakistan, Afghanistan, Iran and Arabia, rarely cultivated in S Europe. This relatively hardy palm deserves to be much more widely grown. Propagation by seed.
Literature: Moore, H.E., *Flora Iranica* **146**: 1–4 (1980).

1. **N. ritchiana** (Griffith) Aitchison (*Chamaerops ritchiana* Griffith). Illustration: Flora Iranica **146**: t. 1–4 (1980).
Pakistan, Afghanistan, Iran & Arabia. H5.

13. SABAL Adanson
Bisexual palms, apparently stemless or with well-developed trunks, without teeth or spines. Leaves costapalmate, blades either regularly divided for half to three-quarters of their length into 1-ribbed segments, or deeply and irregularly divided into less deeply divided pairs of segments; base of leaf-stalk split centrally, often long-persistent. Inflorescences among leaves, with tight tubular bracts on the stalk subtending primary branches, these 1–3 times branched. Flowers creamy white, borne singly. Sepals united into a 3-lobed cup. Petals 3, overlapping, joined basally. Stamens 6, borne on the corolla. Carpels 3, joined. Fruit spherical to pear-shaped with remains of stigma at base. Seed more or less compressed, with smooth endosperm and a lateral embryo.

About 14 species in the tropics and subtropics of C & N America, and in northern S America. Several species are widely cultivated in the tropics and subtropics as ornamentals, but apparently only 2 species are cultivated in Europe, though there is no recent information available. Furthermore, *S. blackburniana* Glazebrook, described from cultivated material, may also be cultivated in S Europe but little is known of this species.

Propagation by seed; germination rapid, but subsequent growth slow.
Literature: Bailey, L.H., Revision of the American Palmettoes, *Gentes Herbarum* **6**: 367–459 (1944).

1a. Dwarf palm with subterranean stem **1. minor**
b. Palms with well-developed trunks **2. palmetto**

1. S. minor (Jacquin) Persoon (*S. adansonii* Guersent). Illustration: McCurrach, Palms of the world, 219 (1960).
Stems subterranean or rarely erect, then very short. Leaves to 1 m, tinged with green, or bluish, stiff and nearly flat, shortly costapalmate, regularly divided to two-thirds or more of their length into 16–40 entire or very shortly lobed, 1-ribbed segments. Inflorescence erect, usually exceeding the leaves. Flowers whitish. Fruit glossy black, to 12 mm in diameter. *SE USA*. H5.

The hardiest species of *Sabal*.

2. S. palmetto (Walter) Loddiges. Illustration: McCurrach, Palms of the world, 223 (1960).
Trunk to 30 m, rough and bare at maturity. Leaves to 2 m long, strongly costapalmate, regularly divided for about two-thirds of their length into 1-ribbed, deeply lobed segments, with filaments in the sinuses between them. Inflorescence usually exceeding the leaves. Flowers whitish. Fruit black, to 1.2 cm in diameter. *SE USA*. H5.

Besides these 2 species, several others are likely to be cultivated in the Mediterranean region; the interested gardener is referred to illustrated accounts of the genus in McCurrach, Palms of the world, and Stevenson, G.B., Palms of south Florida (1974).

14. PHOENIX Linnaeus
Solitary or clustering dwarf to robust, dioecious palms. Leaves pinnate with acute segments which are V-shaped in section, the lowermost modified as spines. Inflorescence among the leaves, subtended by a single, usually deciduous bract, the stalk compressed and bearing single or clustered branches at the tip. Flowers creamy yellow, borne singly. Male flowers with sepals joined into a 3-lobed cup and 3 petals which are edge-to-edge, stamens usually 6. Female flowers with 3 sepals joined into a cup, and 3 overlapping petals; carpels 3, free. Fruit usually developing from only 1 carpel, cylindric to spherical with dry to fleshy mesocarp. Seed single, longitudinally grooved, with smooth endosperm and a basal or lateral embryo.

About 17 species native to drier parts of the Old World tropics and subtropics. Several species are cultivated as ornamentals. One species, *Phoenix dactylifera*, is the date palm, cultivated extensively as a staple and export crop in the Middle East and elsewhere. Despite the economic and horticultural potential of the genus, several of the species of *Phoenix* are poorly known; in the herbarium they are very difficult to distinguish, despite having distinctive features of habit.
Literature: Beccari, O., Rivista monografica delle specie del genere Phoenix Linn., *Malesia* **3**: 345–416 (1890).

1a. Dwarf or apparently stemless palms rarely exceeding 2 m 2
b. Tall trees 3
2a. Leaflets delicate, rather limp, arranged very regularly in 1 plane **5. roebelenii**
b. Leaflets stiff, arranged regularly or in groups, in more than 1 plane **6. loureirii**
3a. Trunk *c.* 80–90 cm in diameter **2. canariensis**
b. Trunk not exceeding 50 cm in diameter 4
4a. Leaves glaucous, bluish green, with white powdery wax **1. dactylifera***
b. Leaves bright green or greyish green 5
5a. Stems clustered **4. reclinata**
b. Stem solitary **3. sylvestris**

1. P. dactylifera Linnaeus. Illustration: McCurrach, Palms of the world, 162–4 (1960).
Trunk to 30 m, rarely exceeding 30 cm in diameter, clustered at base, but suckers frequently removed. Leaf-stalk scars as high as or higher than wide. Leaves bluish green, covered in powdery wax, with leaflets to 45 cm, narrow, stiff, borne in pairs, clusters or more or less regularly arranged, diverging from the axis at an acute angle. Inflorescence elongate; petals of male flowers not acuminate; female flowers with calyx half as long as petals. Fruit cylindric or ellipsoid, 2.5–10 cm, with sweet edible flesh. Seed more or less pointed. *One of the oldest crops, probably originating in W Asia and N Africa*. H5.

***P. theophrastii** Greuter. Differs in having dry fruit. *Crete*.

This recently named species may be relatively hardy.

2. P. canariensis Chabaud. Illustration: Principes **18**: 85 (1974).
Solitary, trunk to 20 m, *c.* 80–90 cm in diameter, in age with leaf-stalk scars broader than high. Leaves very numerous, to 6 m long, sometimes twisted at an angle from the horizontal, with very many crowded, more or less regularly arranged, or paired dark green leaflets in 1 plane. Inflorescence elongate. Petals of male flowers not acuminate. Female flowers with calyx nearly as long as petals. Fruit to 2×1.5 cm, cylindric to ellipsoid, yellow to reddish. *Canary Islands, widely cultivated elsewhere*. H5.

3. P. sylvestris (Linnaeus) Roxburgh. Illustration: McCurrach, Palms of the world, 169 (1960).
Trunk solitary to 16 m, *c.* 30 cm in diameter, at length with leaf-stalk scars higher than wide. Leaves numerous, appearing ragged, to 5 m long, greyish green but lacking white powdery wax; stalk short. Leaflets numerous, arranged in groups in 2–4 planes, to 45×2.5 cm. Inflorescence to 90 cm long. Male flowers with obtuse petals. Female flowers with calyx half as long as petals. Fruit to 3 cm, cylindric, orange-yellow. *India*. H5.

4. P. reclinata Jacquin (*P. spinosa* Schumacher). Illustration: McCurrach, Palms of the world, 165 (1960).
Stems clustered, to 10 m or more, *c.* 15 cm in diameter. Leaves numerous, to 3 m long, graceful, arched, sometimes twisted. Leaflets *c.* 80 on each side, bright green, to 40×2.5 cm, arranged irregularly in several planes, rarely in 1 plane. Inflorescence to 1.2 m. Male flowers with acute or acuminate petals. Female flowers with calyx about half as long as petals. Fruit cylindric-ellipsoid, to 2×1 cm, orange-red to black. *Africa*. G2.

Very variable.

5. P. roebelenii O'Brien. Illustration: McCurrach, Palms of the world, 168 (1960).
Stem solitary (? rarely clustered) to 2 m, 5–10 cm in diameter, rough with remains of leaf-stalks. Leaves to 1.1 m, with *c.* 50 leaflets on each side, regularly arranged, to 20 cm × 5–7 mm, soft, bright green, borne in a single plane. Inflorescence to 45 cm. Male flowers with acuminate petals. Female flowers with calyx less than half as long as petals. Fruit blackish, *c.* 12×4 mm. *Laos*. G1.

A very elegant, dwarf date palm.

6. P. loureirii Kunth (*P. hanceana* Naudin; *P. humilis* Royle).

Variable, dwarf, apparently stemless, or with stem to 4 m, frequently clustering. Leaves more or less glaucous with leaflets in groups. Inflorescence-stalk to 90 cm in fruit. Male flowers with blunt petals. Female flowers with calyx half as long as corolla. Fruit red to black, 1.2–2 cm × 6–12 mm. *India to China*. G1.

A poorly understood species which has been confused with *P. roebelenii*.

15. CALAMUS Linnaeus
Spiny, solitary or clustered, usually climbing, rarely apparently stemless or shrub-like, dioecious palms. Leaves pinnate, sometimes ending in a barbed whip, or with a similar whip borne on the leaf-sheath; leaflets always 1-ribbed, almost always acute, the whole leaf and sheath variously spiny. Inflorescence borne on the leaf-sheath, with tubular, spiny, persistent bracts, rarely the bracts splitting and expanding. Male flowers borne singly, with 3 joined sepals, 3 petals joined together at the base, and 6 stamens. Female flower borne in a pair with a sterile male flower, the ovary 3-celled. Fruit covered in neat vertical rows of reflexed scales. Seed usually 1 only, covered in a fleshy seedcoat, and with smooth or ruminate endosperm.

About 370 species, from W Africa, India and S China through SE Asia to Fiji and NE Australia. Of considerable economic importance as one of the genera producing 'rattan' canes used in furniture making. A few species may be cultivated as pot plants.

1. C. ciliaris Blume. Illustration: McCurrach, Palms of the world, 33 (1960), as Calamus species.
Slender, densely clustering; stems to 10 m, *c.* 1 cm in diameter, including sheaths. Leaves to 30 cm with numerous, very fine, regularly arranged leaflets. Sheaths, stalks and leaflets all densely covered with stiff, fine, greenish bristles. Fruit greenish white, rounded, *c.* 1 cm in diameter. *Indonesia (W Java, Sumatra)*. G2.

16. ARENGA Labillardière
Dwarf to very robust, solitary or clustered, monoecious or dioecious palms. Leaves pinnate or rarely undivided, the leaflets V-shaped in section, with jagged upper margins, often pale on the undersurface. Flowering usually beginning at the top of the stem and then downwards, followed by death of the stem. Inflorescences usually many-branched, rarely unbranched, 1–several at each node, with large scaly bracts and branches bearing triads (2 males, 1 female); inflorescence becoming unisexual by abortion. Female flowers with 3 overlapping sepals, 3 edge-to-edge petals, and ovary 3-celled. Fruit 1–3-seeded. Seeds with smooth endosperm.

About 17 species native to tropical and subtropical Asia. Several species are of economic importance, and, for this reason, may be cultivated in Europe in specialist collections; other species are cultivated as ornamentals in the tropics. Only 1 species is widely cultivated in Europe. Propagation by seed.

1. A. engleri Beccari. Illustration: McCurrach, Palms of the world, 16 (1960).
Stems clustered, often very short. Leaves to 1.75 m, with stalk to 80 cm. Leaflets linear, 28–32 on each side, held in 1 plane, to 3 cm wide, toothed only at the very tip. Inflorescences solitary. Male flowers maroon, very fragrant, with golden pollen. Female flowers green at first. Fruit spherical, red, *c.* 2 cm in diameter. Seeds 3. *Taiwan, Japan (Ryukyu Islands)*. H5–G1.

17. CARYOTA Linnaeus
Solitary or clustered, moderate to robust, monoecious palms with stems flowering from the top downwards and then dying. Leaf twice pinnate, with fishtail-like leaflets, with jagged teeth on the oblique apical margins. Inflorescences branching once, or very rarely unbranched, bearing triads of flowers throughout (2 male, 1 female). Male flowers with 3 overlapping sepals, 3 edge-to-edge petals, and stamens 6–100 or more. Female flowers with 3 overlapping sepals, 3 edge-to-edge petals joined at the base, and ovary with 3 cells. Fruit 1–2-seeded with irritant flesh. Seed with ruminate endosperm.

About 12 species native to Asia, poorly known, but of considerable local economic importance, and very decorative. Two species are widely cultivated. Propagation by seed.

1a. Stems clustering, rarely exceeding 10 m **1. mitis**
b. Stems solitary, to 15 m **2. urens**

1. C. mitis Loureiro. Illustration: McCurrach, Palms of the world, 36 (1960).
Stems clustering, to *c.* 10 m tall and to *c.* 10 cm in diameter. Male flowers with *c.* 12–16 stamens. Fruit reddish, 1-seeded, *c.* 1.5 cm in diameter. *SE Asia*. G2.

2. C. urens Linnaeus. Illustration: McCurrach, Palms of the world, 37 (1960).
Stems solitary to *c.* 15 m and to 30 cm or more in diameter. Male flowers with 40–45 stamens. Fruit reddish, 1–2-seeded, to *c.* 2 cm in diameter. *India, Sri Lanka*. G2.

Apparently more cold-resistant than *C. mitis*.

18. CHAMAEDOREA Willdenow
Solitary or clustering, slender, dioecious palms, without teeth or spines. Leaves pinnate, or bifid with pinnate venation, the sheaths tubular but not forming a crownshaft; leaflets linear or asymmetric at base and apex. Inflorescences among or below the leaves, usually solitary but sometimes several at each node, branched or unbranched, with 3 or more bracts. Flowers solitary or rarely (in male plants) clustered or arranged in vertical rows, in the female rarely also in vertical rows. Branches of female inflorescence becoming orange or red in fruit. Male flowers with cup-like calyx. Petals variously free or united. Stamens 6. Pistillode prominent. Female flower with sepals and petals variously free or united. Staminodes 0–6. Ovary 3-celled. Fruit usually 1-seeded, red or black, with remains of stigma at base. Seed with smooth endosperm and a lateral embryo.

About 133 species, native to C & S America, whose classification is confused. Many species are highly ornamental and are cultivated on a wide scale in the USA and in specialist collections in Europe. Only 1 species is widely grown in Europe, but many more could be tried as pot plants.

1. C. elegans Martius (*Collinia elegans* (Martius) Liebmann; *Neanthe bella* invalid). Illustration: McCurrach, Palms of the world, 57 (1960).
Solitary to 2 m tall, but beginning to flower when still more or less stemless. Leaves dark green with 11–20 leaflets on each side, linear to lanceolate, to 20 × 2 cm. Inflorescence erect, with a long stalk, with 4–9 tubular bracts, and branching to 1 or rarely 2 orders, the male with more flexuous branches than the female. Male flowers yellow, globular; calyx lobed, petals joined except for a triangular opening, stamens minute. Female flower similar but with a flattened spherical ovary. Fruit spherical, black, *c.* 6 mm in diameter. *Mexico, Guatemala*. G1.

This is the commonest pot palm found for sale in chain stores throughout Europe; it is very tolerant of neglect, poor light conditions, and the dry atmosphere of rooms, but responds well to good cultivation. Fruit may be produced after hand pollination.

19. CHRYSALIDOCARPUS Wendland
Solitary or clustered, monoecious shrubs or trees, without teeth or spines. Stems usually conspicuously ringed. Leaf-sheaths open or tubular, sometimes forming a crownshaft. Leaves pinnate, usually with many slender, acute, 1-ribbed leaflets. Inflorescence among leaves, or below them, with 2 unequal, deciduous or more or less persistent bracts, and branching to 2 orders. Flowers borne in triads (2 males and 1 female) near the bases of the branches, with solitary or paired male flowers above. Male flowers with 3 overlapping sepals, 3 edge-to-edge petals, 6 stamens and a prominent pistillode. Female flowers with 3 overlapping sepals, 3 overlapping petals and an ovary with 3 cells and 1 ovule. Fruit with stigma scar at base. Endosperm smooth. Embryo lateral.

About 20 species in Madagascar, the Comoro Islands and Pemba Island. Several species are cultivated in the tropics, but only 1 species is cultivated in Europe as a pot plant. Propagation by seed or offsets.

1. **C. lutescens** Wendland (*Areca lutescens* misapplied). Illustration: McCurrach, Palms of the world, 46 (1960).
Stems many, densely clustering, sometimes aerially branching, to 10 m or more. Crownshaft present, tinged yellowish, covered in wax and black scales when young. Leaf-stalks and axis bright yellow or orange. Leaf ascending or curved. Leaflets narrow, regularly arranged, shining green, 40–60 on each side. Flowers golden yellow. Fruit cylindric-ellipsoid, violet-black, *c.* 1.5 cm long. *Madagascar*. G2.

Sucker shoots are freely produced, even under low light and dry conditions, but mature aerial stems are unlikely to be formed except under hot and humid conditions in specialist collections. Suckering juvenile plants are frequently sold as pot plants. *Areca lutescens* Bory is a synonym of *Hyophorbe indica* Gaertner, which is not generally cultivated.

20. ARCHONTOPHOENIX Wendland & Drude
Solitary, monoecious palms, without teeth or spines, with a smooth trunk ringed with leaf-scars and with a conspicuous crownshaft. Leaves pinnate, with acute, regularly arranged leaflets. Inflorescence below the leaves, subtended by 2 large papery, deciduous bracts, and with a short stalk, and numerous branches, the lowermost further branched. Flowers in triads (2 male and 1 female) near the base, and solitary or paired (male) near the tips. Male flowers lopsided; sepals 3, small, acute and overlapping; petals 3, edge-to-edge, angled; stamens more than 6; pistillode as long as stamens. Female flowers symmetric with overlapping sepals and petals, and 1-celled ovary. Fruit with stigma scar at apex. Seed with ruminate endosperm and a basal embryo.

A genus of 2 or more species from Australia, both widely grown as ornamentals. Propagation by seed; rapidly germinating and fast growing.
Literature: Bailey, L.H., The king palms of Australia – Archontophoenix, *Gentes Herbarum* 3: 391–409 (1935).

1a. Trunk enlarged at base; leaflets greyish white beneath **1. alexandrae**
b. Trunk scarcely swollen at base; leaflets green beneath **2. cunninghamiana**

1. **A. alexandrae** (Mueller) Wendland & Drude (*Ptychosperma alexandrae* Mueller). Illustration: Gentes Herbarum 3: 398 (1935).
Stem to 25 m tall, enlarged towards the base. Leaflets *c.* 5 cm wide, greyish white on the underside. Flowers white or cream-coloured. Fruit red, *c.* 1.25 cm. *Australia (N Queensland)*. G1.

Var. **beatricae** (Mueller) Bailey. Differs in having the trunk deeply ringed, and thus more or less stepped at the base.

2. **A. cunninghamiana** (Wendland) Wendland & Drude (*Ptychosperma cunninghamiana* Wendland; *Seaforthia elegans* misapplied). Illustration: Gentes Herbarum 3: 393 (1935).
Stem scarcely swollen at the base. Leaflets *c.* 10 cm wide, green on the underside. Flowers lilac or purplish. Fruit red, more than 1.25 cm. *Australia (New South Wales, Queensland)*. H5–G1.

21. HEDYSCEPE Wendland & Drude
Solitary, monoecious palms without teeth or spines, with a smooth, green, conspicuously ringed trunk to 10 m tall, and conspicuous silvery blue crownshaft to 40 cm. Leaves pinnate with curved axis bearing regularly arranged leaflets, *c.* 40 on each side, stiff and ascending, to 55 × 2.5 cm. Inflorescence borne below the leaves, to 45 cm with a short stalk bearing 2 deciduous bracts, and branching to 2 orders, the branches hairless and bearing triads (2 male flowers, 1 female) more or less throughout. Male flowers asymmetric with 3 narrow, acute sepals, 3 angled, edge-to-edge petals, 9–12 stamens, and a pistillode as long as the stamens. Female flower with 3 overlapping sepals, 3 overlapping petals which are edge-to-edge at the very tip, 3 staminodes and a 1-celled ovary. Fruit broadly ellipsoid, to *c.* 4 × 2.5 cm, dull deep red, with remains of stigma near apex. Seed with smooth endosperm and a basal embryo.

A genus of 1 species; propagation by seed.

1. **H. canterburyana** (Moore & Mueller) Wendland & Drude (*Kentia canterburyana* Moore & Mueller). Illustration: Principes 10: 16–18 (1966).
Lord Howe Island, off Australia. H5–G1.

A strikingly beautiful palm, but rather slow growing.

22. RHOPALOSTYLIS Wendland & Drude
Solitary, monoecious palms, without teeth or spines, with a smooth trunk and very conspicuous swollen crownshaft. Leaves pinnate, with leaflets acute or acuminate, regularly arranged. Inflorescence below the leaves with 2 deciduous bracts, the upper enclosed within the lower, and a short stalk, the axis branching to 2 orders. Flowers borne in triads (2 male and 1 female). Male flowers asymmetric, with 3 narrow, acute sepals, 3 edge-to-edge petals, 6 stamens, and a pistillode equalling the stamens. Female flowers symmetric with 3 overlapping sepals, 3 overlapping petals with edge-to-edge tips, and a 1-celled ovary. Fruit red with remains of stigma at apex. Seed with smooth endosperm and a basal embryo.

Three species, one each in New Zealand, Norfolk Island and Raoul Island. Propagation by seed. They are relatively hardy and very attractive palms best suited to the oceanic fringe of Europe.
Literature: Bailey, L.H., Rhopalostylis, *Gentes Herbarum* 3: 429–35 (1935).

1a. Fruit spherical; fruiting perianth open and spreading **3. cheesemanii**
b. Fruit ovoid to ellipsoid; fruiting perianth cup-like 2
2a. Inflorescence to 60 cm; seed with adhering fibres, light in colour **1. sapida**
b. Inflorescence to *c.* 90 cm; seed dull brown **2. baueri**

1. **R. sapida** Wendland & Drude. Illustration: Gentes Herbarum 3: 429, 430, 431 (1935).
Stem to 8 m tall, *c.* 23 cm in diameter, strongly and closely ringed. Leaves

ascending, to 2.75 m long, with leaflets to 90 × 5 cm. Inflorescence to 60 cm long. Fruit ellipsoid, to 15 × 7 mm. Seed ellipsoid with tightly adhering fibres. *New Zealand.* H5.

2. R. baueri Wendland & Drude. Illustration: Botanical Magazine, 5735 (1868).
Stems to 12 m tall. Inflorescence to *c.* 90 cm long. Fruit ovoid to ellipsoid, to *c.* 17 × 13 mm. Seed spherical, dull brown. *Norfolk Island.* G1.

3. R. cheesemanii Beccari. Illustration: Cockayne, Vegetation der Erde, **14**, edn 2: f. 94 (1928).
Stems to 12 m tall. Leaves to 3.4 m long, with leaflets to 100 × 5 cm. Inflorescence to 60 cm. Fruit spherical, *c.* 15 mm in diameter. Seed spherical, light brown. *Raoul Island.* G1.

Rare in cultivation.

23. CYRTOSTACHYS Blume
Clustering, monoecious palms, without teeth or spines. Leaves pinnate with sheaths forming a crownshaft; leaflets acute. Inflorescence below the leaves, with short stalk and 2 deciduous bracts, the axis branching to 2 orders. Flowers in triads (2 male and 1 female) in deep, open pits. Male flowers with 3 free, rounded, overlapping sepals, 3 edge-to-edge petals, and 6–15 stamens. Female flowers with 3 overlapping sepals, 3 overlapping petals and ovary 1-celled. Fruit ovoid to ellipsoid with remains of stigma at apex. Seed with smooth endosperm and basal embryo.

About 12 species from New Guinea and offshore islands; 1 species in Sumatra, Malaysia and Borneo. Propagation by seed.

1. C. renda Blume (*C. lakka* Beccari). Illustration: McCurrach, Palms of the world, 73 (1960).
Trunk to 10 m tall, *c.* 15 cm in diameter. Crownshaft brilliant orange to red. Leaves to 2 m long with axis and stalk bright orange, and with about 50 greyish green leaflets regularly arranged on each side, to 45 × 4 cm. Fruit blackish, borne on a dull crimson axis. *Sumatra, Malaysia, Borneo.* G2.

Extremely decorative but requires high temperatures; in Europe it is occasionally listed in nurserymen's catalogues.

24. HOWEA Beccari
Solitary, monoecious palms without teeth or spines, without crownshaft. Leaves pinnate; leaflets 1-ribbed, acute. Inflorescence with 1 or several spikes with long stalks subtended by a common bract in a leaf axil, each spike in bud enclosed by a thin bract at the apex of the stalk. Flowers in triads (2 male, 1 female), sunk in pits. Male flowers greenish brown with overlapping sepals, edge-to-edge petals, numerous stamens and a minute pistillode. Female flowers greenish brown with sepals and petals overlapping, ovary 1-celled. Fruit ellipsoid, *c.* 4 cm long, with remains of stigma at apex. Seed 1, with smooth endosperm and a basal embryo.

A genus of 2 species restricted to Lord Howe Island (off Australia), both widely cultivated as cool greenhouse subjects or houseplants. Propagation by seed; *H. forsteriana* is one of the commonest pot palms.
Literature: Bailey, L.H., Howea in cultivation, *Gentes Herbarum* **4**: 188–98 (1939).

1a. Tree to 20 m; leaves eventually more or less horizontal with well-developed stalks; leaflets horizontal or hanging; inflorescence with 3–8 spikes **1. forsteriana**
b. Tree to 7 m; leaves strongly arched with very short stalks; leaflets crowded, directed upwards; inflorescence with 1 spike **2. belmoreana**

1. H. forsteriana (Moore & Mueller) Beccari (*Kentia forsteriana* Moore & Mueller). Illustration: Gentes Herbarum **4**: 190–1 (1939).
Tree to 20 m. Leaves to 3 m with long stalks, eventually held more or less horizontally and bearing rather distant, horizontal or hanging leaflets. Inflorescences to 1 m, with 3–8 spikes. Fruit narrowly ellipsoid. *Lord Howe Island.* G1.

2. H. belmoreana (Moore & Mueller) Beccari (*Kentia belmoreana* Moore & Mueller). Illustration: Gentes Herbarum **4**: 188 (1939).
Tree to 7 m. Leaves to 2 m, with very short stalks, strongly arched, and bearing crowded leaflets directed upwards from the axis. Inflorescence to 1 m, with 1 spike. Fruit broadly ellipsoid. *Lord Howe Island.* G1.

25. PTYCHOSPERMA Labillardière
Solitary or clustering, monoecious palms, without spines or teeth, with a smooth stem and conspicuous crownshaft. Leaves pinnate, leaflets slender to very broadly wedge-shaped, with conspicuously toothed apices. Inflorescence below the leaves, with a short stalk, and 2 large deciduous bracts, the inner enclosed within the outer; the axis branching to 2 orders. Flowers in triads (2 male flowers, 1 female) almost throughout the inflorescence, with single or paired male flowers confined to the tips. Male flower with 3 overlapping sepals, 3 edge-to-edge petals, numerous stamens and a pistillode usually as long as stamens. Female flower with 3 overlapping sepals, 3 overlapping petals, 1–6 tooth-like staminodes, and ovary 1-celled. Fruit with apical stigma and thin endocarp. Seed with 3–5 weak to very strong longitudinal grooves, smooth or ruminate endosperm, and a basal embryo.

About 30 species native to Australia, New Guinea, Solomon Islands and Micronesia. Propagation by seed.
Literature: Essig, F.G., A revision of the genus Ptychosperma Labill. (Arecaceae), *Allertonia* **1**(7): 415–78 (1978).

1a. Stem solitary; leaflets to *c.* 80 × 8 cm; seed with ruminate endosperm **1. elegans**
b. Stems clustered; leaflets to *c.* 50 × 5 cm; seed with smooth endosperm **2. macarthurii**

1. P. elegans (R. Brown) Blume (*Seaforthia elegans* R. Brown). Illustration: McCurrach, Palms of the world, 193 (1960).
Stem solitary, to 8 m or more. Leaves to 2.5 m, with leaflets regularly arranged, *c.* 28 on each side, to 80 × 8 cm, with many long, twisted scales on the undersurface. Inflorescence green. Male flowers fragrant, green with white stamens. Fruit red, ovoid, *c.* 2 cm long. Seed with ruminate endosperm. *Australia.* G2–H5.

It is possible some records of this palm in outdoor cultivation may refer to *Archontophoenix cunninghamiana* (p. 72).

2. P. macarthurii (Wendland) Nicholson (*Actinophloeus macarthurii* (Wendland) Beccari). Illustration: McCurrach, Palms of the world, 194 (1960).
Stems clustered, to 6 m or more. Leaves to *c.* 1.5 m, with *c.* 28 leaflets regularly arranged on each side, to 50 × 5 cm. Inflorescence green. Male flowers green with white stamens. Fruit red, *c.* 1.5 cm long. Seed with smooth endosperm. *New Guinea.* G2.

26. COCOS Linnaeus
Robust, solitary, monoecious tree, without spines or teeth. Leaves pinnate. Sheaths fibrous, open. Leaf-stalk deeply channelled

on upper surface. Leaflets numerous, 1-ribbed, regularly arranged, acute. Inflorescence among leaves, with 1 order of branching and with 2 bracts, the lower inconspicuous, the upper woody, striped but not furrowed. Flowers cream, borne in triads (2 male, 1 female) at the bases of the branches and solitary or paired above. Male flowers with 3 overlapping, acute sepals, 3 edge-to-edge petals and 6 stamens. Female flower much larger than male, with 3 overlapping sepals and 3 overlapping petals. Ovary 3-celled. Fruit usually 1-seeded, with a thick fibrous husk and a thick bony endocarp with 3 pores near the base. Seed with smooth hollow endosperm partially filled with fluid.

Now considered a genus of 1 species. Origin uncertain, though there is evidence for an origin in the Eastern Indian Ocean or Western Pacific. The coconut is one of the most important tropical crops. Coconuts will not reach maturity in Europe outside specialist collections, but young seedlings still attached to the nut are frequently sold as pot plants. Many other palms were first included in *Cocos* but are now put in different genera. Some of them are still referred to by their old names in the horticultural trade.

1. **C. nucifera** Linnaeus. Illustration: McCurrach, Palms of the world, 53–5, 1960.
Origin uncertain; widely grown in the tropics. G2.

27. BUTIA (Beccari) Beccari
Solitary, monoecious palms. Leaves pinnate, leaflets acute, stalk and open sheath with stout, sharp teeth. Inflorescence among the leaves, with 2 bracts, the lower concealed among the leaves, the upper beaked, tubular in bud, splitting down one side, smooth or woolly but not grooved. Inflorescence axis branching once, with flowers in triads (2 male and 1 female) near the base, and solitary or paired male flowers above. Male flowers with 3 small, acute sepals joined at the base, 3 edge-to-edge petals and 6 stamens. Female flower with 3 free, overlapping sepals, 3 free, overlapping petals with short edge-to-edge tips, staminodes forming a short cup, and 3-celled ovary. Fruit 1-seeded with fleshy mesocarp, and bony endocarp pointed at both ends, slightly ridged and with pores near the base. Seed with smooth endosperm and basal embryo.

About 12 species native to tropical and subtropical S America, from Brazil to Argentina; generally very decorative. Propagation by seed.
Literature: Bailey, L.H., The genus Butia, *Gentes Herbarum* **4**: 16–50 (1936).

1a. Upper bract of inflorescence densely woolly **3. eriospatha**
b. Upper bract of inflorescence smooth 2
2a. Inflorescence branches straight; perianth short in fruit, not more than one-quarter of the length of the fruit **1. capitata**
b. Inflorescence branches flexuous; perianth longer in fruit, at least one-third of the length of the fruit **2. yatay**

1. **B. capitata** (Martius) Beccari (*Cocos capitata* Martius; *C. australis* misapplied). Illustration: Gentes Herbarum **4**: 27–34 (1936).
Trunk to 6 m × 50 cm, covered in persistent leaf-bases. Leaves to 2.5 m, curving, with leaflets regularly arranged (rarely clustered) ascending from the axis, usually whitish beneath, or glaucous, to 75 × 2 cm. Upper bract to 90 cm, smooth. Inflorescence branches straight. Flowers yellow to reddish-tinged. Fruit ovoid, to 3 × 2.75 cm, bright yellow-orange; perianth in fruit not more than one-quarter of the fruit length. *Brazil.* H5.

2. **B. yatay** (Martius) Beccari (*Cocos yatay* Martius). Illustration: Gentes Herbarum **4**: 40 (1936).
Trunk to 10 m × 50 cm, eventually becoming bare in old age except below the crown. Leaflets more or less regularly arranged, glaucous beneath. Upper bract of inflorescence smooth, to 90 cm. Inflorescence branches flexuous. Flowers creamy yellow. Fruit dark yellow to orange or reddish-tinged, ovoid, to 3 × 2.75 cm; perianth in fruit at least one-third of the length of the fruit. *Argentina.* H5.

3. **B. eriospatha** (Drude) Beccari (*Cocos eriospatha* Drude). Illustration: Gentes Herbarum **4**: 43 (1936).
Trunk to 6 m, becoming bare. Leaves ascending then curving outwards, with hairy stalks; leaflets more or less regularly arranged, green on upper surface, glaucous beneath, usually distinctly bifid at tips. Upper bract of inflorescence densely brown woolly-hairy. Fruit yellow, more or less flattened, to 20 × 22 mm, the perianth more or less obscured beneath the fruit. *Brazil.* H5.

28. JUBAEA Humboldt, Bonpland & Kunth
Solitary, monoecious tree palm, without teeth or spines, stem to 10 × 1 m. Leaves numerous, pinnate. Sheath open, fibrous; leaf-stalk short. Leaflets *c.* 120 on each side of axis, regularly arranged, or in groups of 2–5, stiff, green or silvery green, each with 2 apical lobes or a deep cleft. Inflorescence to 1 m or more, among the leaves, with long stalk and 2 major bracts, the lower concealed by the leaf-bases, the upper not deeply grooved externally, the axis branching once. Flowers reddish brown, borne in triads (2 males, 1 female) near the base and with solitary or paired male flowers above. Male flower maroon with yellow centre and many yellow stamens. Female flowers globular with overlapping sepals and petals, which are more or less of the same length, staminodes united into a ring and ovary with 3 cells, each with 1 ovule. Fruit ovoid, *c.* 4 × 2.5 cm, yellow, with thick endocarp and bearing 3 basal pores. Seed 1, with smooth, hollow endosperm.

One species in coastal C Chile, widely cultivated in areas with a Mediterranean climate. Propagation by seed. In Chile the species is used as a source of sap for the production of palm honey and is now very rare in the wild because of overexploitation.

1. **J. chilensis** (Molina) Baillon (*J. spectabilis* Humboldt, Bonpland & Kunth). Illustration: Principes **18**: 85 (1974).
C Chile. H5.

29. ARECASTRUM (Drude) Beccari
Solitary, monoecious tree palm, without teeth or spines, trunk smooth, grey-brown, to 15 m tall. Sheaths fibrous. Leaves pinnate, many borne in the crown, arching or drooping, to 5 m or more; stalk well developed. Leaflets to 100 × 3 cm, numerous, borne in groups of 1–5, and in several planes, giving the whole leaf a plumose effect. Inflorescence among the leaves, to 1 m long, with 2 bracts, the lower concealed, the upper woody and splitting down one side to become boat-shaped, deeply grooved on outside. Branches with creamy yellow flowers in triads (2 male, 1 female) almost throughout, with solitary or paired males near the tip. Male flowers with calyx 3-lobed, petals 3, edge-to-edge, 6 stamens and a minute pistillode. Female flowers with 3 overlapping sepals, and 3 petals which are overlapping except at the tips where they are edge-to-edge, staminodes in a low ring and a 3-celled ovary. Fruit yellow, ovoid, 2.5–4 cm long, with fleshy

mesocarp and thick bony endocarp with 3 basal pores and a prominent intrusion into the single seed on one side. Endosperm smooth.

A genus of 1 species from subtropical S America. Propagation by seed; relatively short-lived.

1. A. **romanzoffianum** (Chamisso) Beccari (*Cocos romanzoffianum* Chamisso; *C. plumosa* W.J. Hooker). Illustration: McCurrach, Palms of the world, 13 (1960).
Brazil to Argentina. H1.

Var. **australe** (Martius) Beccari, which has smaller fruit with the endocarp not narrowed at both ends, is also grown.

30. MICROCOELUM Burret & Potztal
Slender, solitary, monoecious palms, without teeth or spines. Leaves pinnate; stalks without teeth; leaflets 1-ribbed, obliquely 2-cleft or acuminate. Inflorescence among leaves, branched once, with a long stalk and 2 bracts, the lower inconspicuous, the upper woody and deeply furrowed. Flowers borne in triads (2 male and 1 female) near the base of the branches; upper part with solitary or paired male flowers. Male flower asymmetric, with 3 small sepals, 3 edge-to-edge petals and 6 stamens. Female flower with 3 broad, overlapping sepals and 3 broad, overlapping petals with short edge-to-edge tips; staminodes in a low ring; ovary with 3 cells, 1 ovule in each. Fruit 1-seeded, epicarp and mesocarp splitting into 3 parts exposing the smooth papery endocarp with basal pores. Seed with smooth hollow endosperm.

A genus of 2 species native to Brazil, 1 widely cultivated as a pot plant. Propagation by seed.

1. M. **weddellianum** (Wendland) H.E. Moore (*Cocos weddelliana* Wendland; *Syagrus weddelliana* (Wendland) Beccari; *Microcoelum martianum* (Drude) Burret). Illustration: Langlois, Supplement to palms of the world, 171 (1976).
Eventually to *c.* 3 m. Leaves to 3.2 m with purplish scales along the stalk margin and the axis. Leaflets 50–60 on each side, regularly arranged, very narrow, dark green on upper surface, greyish beneath. Inflorescence to 1 m. Fruit *c.* 2.5 cm long. *Brazil*. G1.

31. ALLAGOPTERA Nees
Dwarf, apparently stemless, mostly clustering, monoecious palms, without spines or teeth. Leaves pinnate, with leaflets acute, but often splitting at the tip. Inflorescence among the leaves, erect, spike-like, with 2 bracts, the lower inconspicuous, the upper woody and grooved. Flowers fragrant, cream-coloured, in triads, 2 male and 1 female, very densely arranged. Male flowers lopsided, with 3 slender, elongate sepals joined at the base, 3 edge-to-edge petals, and 6 or more stamens. Female flower with 3 overlapping sepals, 3 overlapping petals, staminodes joined into a low ring and ovary 3-celled. Fruit 1-seeded with remains of stigma at apex, and endocarp with basal pores. Seed with smooth endosperm.

Five species native to Brazil and Paraguay. Propagation by seed.

1. A. **arenaria** (Gomes) Kuntze (*Diplothemium maritimum* Martius). Illustration: Principes **8**: 56, 57 (1964).
Leaves to 1.25 m with long stalks, leaflets *c.* 50 on each side, arranged in groups of 2–5, and borne in several planes, pale on undersurface. Inflorescence with stalk to 90 cm and spike to 15 cm. Male flowers with 10–16 stamens. Fruit covered in brown woolly scales. *Coastal Brazil*. G2 (possibly hardy in warmest parts of Mediterranean coast).

A coastal species, resistant to salt spray.

32. ELAEIS Jacquin
Robust, solitary, monoecious palms. Leaves pinnate, falling untidily leaving persistent bases. Leaf-stalk with fibrous, spine-like processes at the base, the midribs of the short lowermost leaflets also spine-like. Leaflets acute, 1-ribbed. Inflorescences unisexual among the leaves, with very short stalks and very congested branches with spine-like tips. Male flowers surrounded by united bracteoles, and with 3 free sepals, 3 edge-to-edge petals and 6 stamens. Female flower with 3 overlapping sepals, 3 overlapping petals and a 3-celled ovary. Fruit orange-red at base, usually blackish at the tip, with very oily mesocarp, and bony endocarp with 3 apical pores. Seed with smooth, hollow endosperm.

A genus of 2 species; 1 in S America, the other (the oil palm) native to tropical Africa, now very widespread as a tropical crop and becoming naturalised in some areas. Sometimes grown as an ornamental. Propagation by seed.

1. E. **guineensis** Jacquin. Illustration: McCurrach, Palms of the world, 83, 84 (1960).
Stem erect to *c.* 25 m, covered in persistent leaf-bases. Leaves to 5 m, with leaflets numerous, more or less grouped and held in several planes. Inflorescence branches to 15 cm, with sharp tips. Fruit ovoid, *c.* 3 cm long. *Tropical Africa*. G2.

The oil palm is the most important tropical source of vegetable oil. Seedlings grow rapidly and are sometimes sold in Europe as pot plants.

XXII. ARACEAE

Plants perennial, herbaceous or slightly woody, usually hairless, sometimes with a distinct juvenile phase. Leaves alternate or basal, usually stalked and with a broad blade. Scale-leaves present at least at some stage. Leaf-stalk usually with a definite sheath; sheath usually embracing the principal stem but if not, the leaves are interspersed with scale-leaves. Leaf-blade simple, lobed or compound, often with netted veins; primary lateral veins sometimes alternating with smaller straight or sinuous veins (the *interprimary* veins) which originate from the midrib or are formed by amalgamation of branches from the primary laterals; 1 or more marginal veins often present near the edge. Plants bisexual or, rarely, unisexual. Inflorescence consisting of a stalk and a spadix with its spathe. Spadix sometimes more or less united to spathe, sometimes with a stalk (*stipe*) between spathe attachment and lowest flowers, sometimes with a sterile terminal appendix. Fertile florets bisexual or unisexual. Unisexual florets arranged in zones on spadix, the female usually below the male, sometimes few in number, the female occasionally solitary. Sterile florets sometimes present, usually either as filaments or flattish scales, occurring between male and female zones or above male zone. Perianth segments 0, 4, 6 or occasionally more. Stamens 0, 1, 4 or 6 (rarely more), free or (in male florets) sometimes joined; filaments usually short or none. Staminodes sometimes present in female florets. Ovary with 1–4 cells, sometimes with incomplete partitions. Style usually short or absent. Ovules 1–many. Fruit usually a berry. (Figures 3 & 4, pp. 76 & 77).

A family of about 115 genera and 2000 species, mainly tropical but with some subtropical or temperate aquatic and some temperate tuberous representatives. Many tropical Araceae are woody climbers with aerial roots. A few members of the family are grown for the beauty of the inflorescence (e.g. some *Anthurium*, *Arisaema* and *Zantedeschia* species and hybrids) and some are grown for their

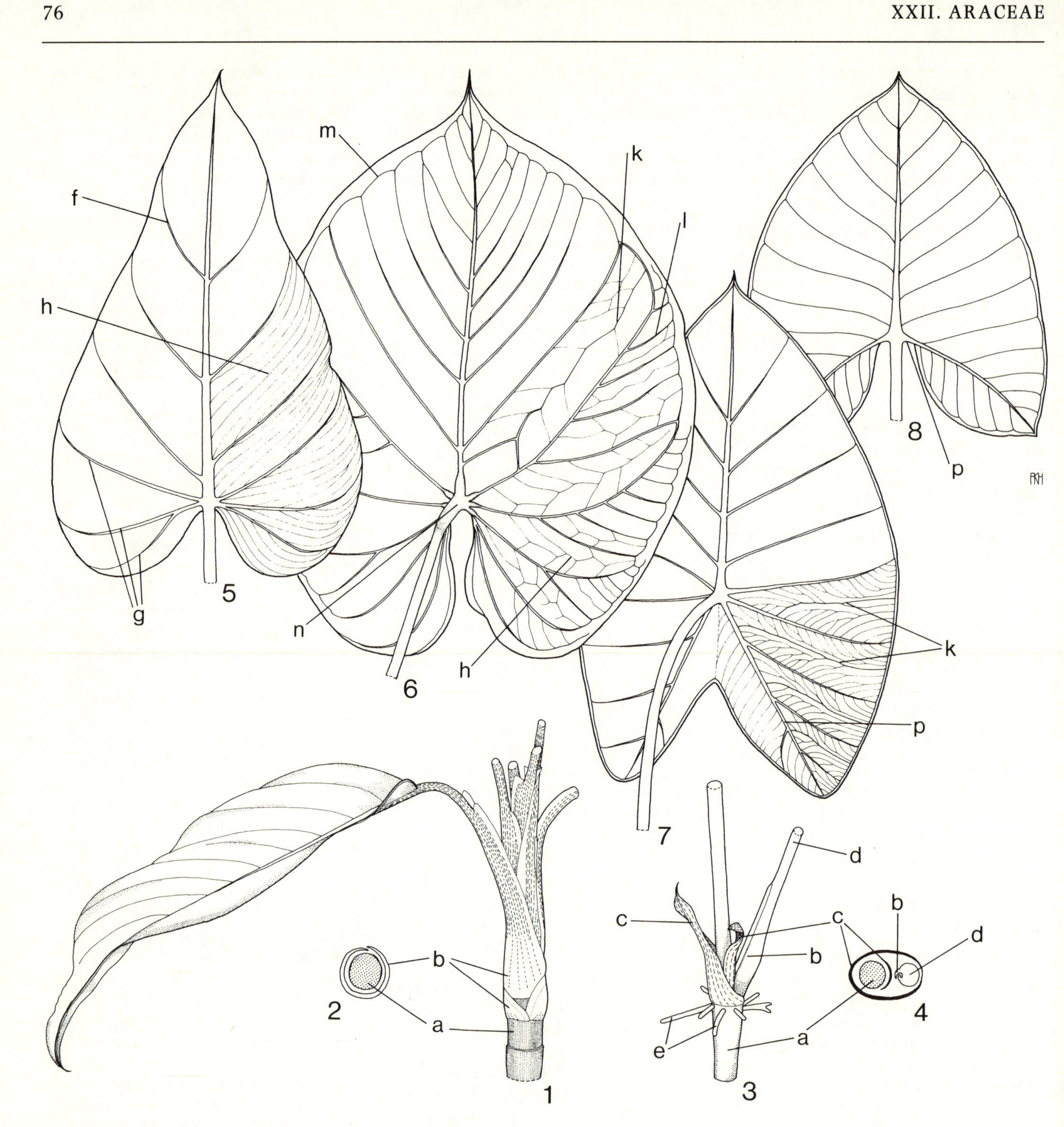

Figure 3. Leaves and scale-leaves of Araceae. 1, *Dieffenbachia*, leaf-sheaths embracing stem. 2, Cross-section of *Dieffenbachia* stem. 3, *Anthurium*, leaf-sheath not embracing stem. 4, Cross-section of *Anthurium* stem. 5, *Philodendron*, leaf, minor veins almost parallel to primary lateral veins. 6, *Anthurium*, leaf with pulvinus at apex of stalk, minor veins netted. 7, *Colocasia*, peltate leaf with midribs in basal lobes. 8, *Xanthosoma*, leaf with midrib in basal lobe exposed in the sinus. a, stem; b, leaf-sheath; c, scale-leaves; d, leaf-stalk; e, aerial roots; f, primary lateral vein; g, basal veins; h, minor veins; k, interprimary veins; l, secondary vein; m, marginal vein; n, pulvinus; p, midrib in basal lobe. (8 after *Das Pflanzenreich* 71.)

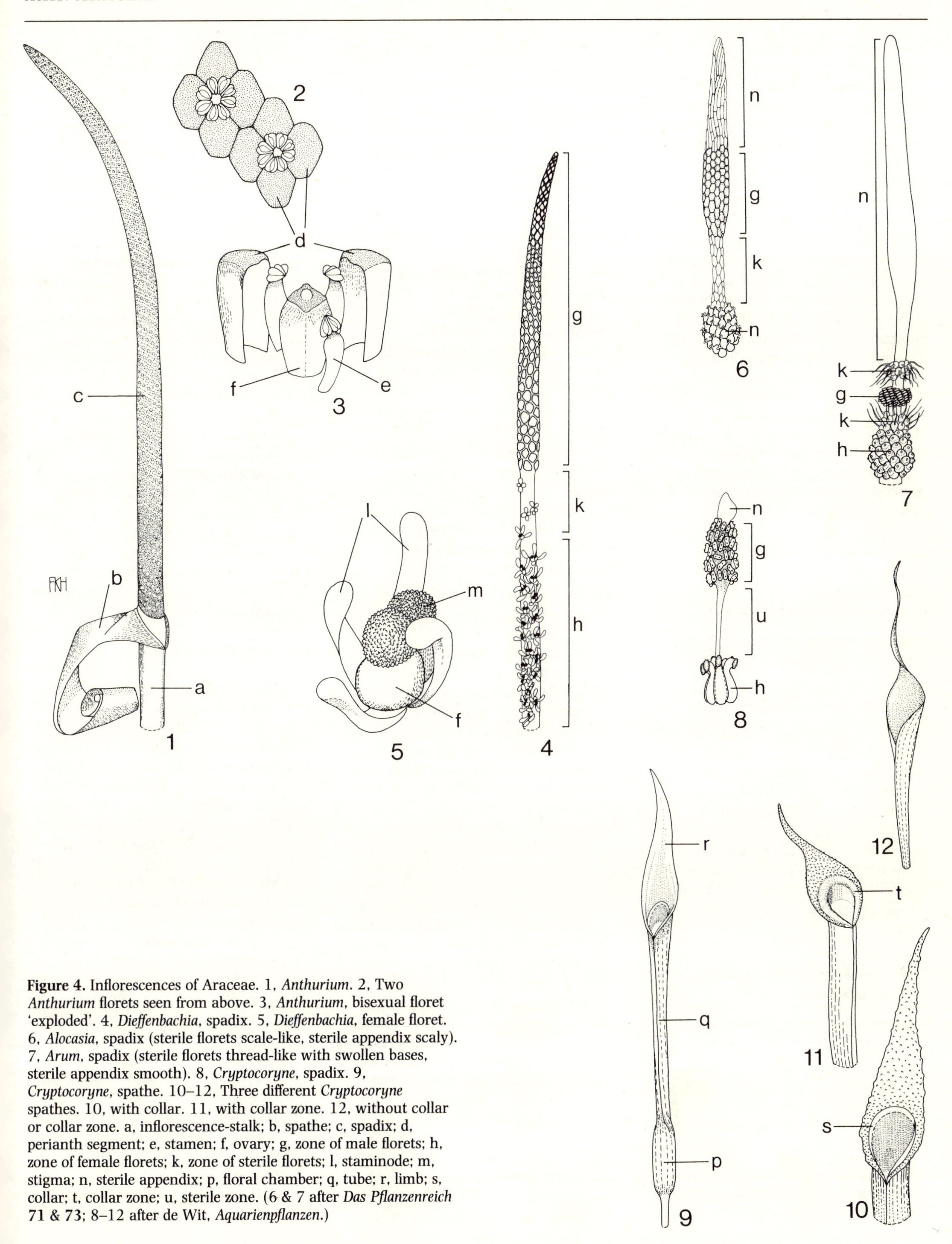

Figure 4. Inflorescences of Araceae. 1, *Anthurium*. 2, Two *Anthurium* florets seen from above. 3, *Anthurium*, bisexual floret 'exploded'. 4, *Dieffenbachia*, spadix. 5, *Dieffenbachia*, female floret. 6, *Alocasia*, spadix (sterile florets scale-like, sterile appendix scaly). 7, *Arum*, spadix (sterile florets thread-like with swollen bases, sterile appendix smooth). 8, *Cryptocoryne*, spadix. 9, *Cryptocoryne*, spathe. 10–12, Three different *Cryptocoryne* spathes. 10, with collar. 11, with collar zone. 12, without collar or collar zone. a, inflorescence-stalk; b, spathe; c, spadix; d, perianth segment; e, stamen; f, ovary; g, zone of male florets; h, zone of female florets; k, zone of sterile florets; l, staminode; m, stigma; n, sterile appendix; p, floral chamber; q, tube; r, limb; s, collar; t, collar zone; u, sterile zone. (6 & 7 after *Das Pflanzenreich* 71 & 73; 8–12 after de Wit, *Aquarienpflanzen*.)

bizarre inflorescences and/or spotted and banded vegetative parts (these are mostly tuberous genera). However, most of the tropical species in cultivation are grown for their foliage, which is often variegated. *Cryptocoryne* and *Lagenandra*, both tropical, are used as aquarium plants.

Literature: Engler, A. (editor), *Das Pflanzenreich*, **11,21,37,48,55,60,64,71,73** (1905–20); Birdsey, M.R., *The cultivated aroids* (1951); Matuda, E., Las Araceas Mexicanas, *Anales del Instituto de Biología, Mexico* **25**: 97–218 (1954); Bunting, G.S., Generic delimitation in Araceae, subfamily Monsteroideae, *Baileya* **10**: 21–31 (1962); Bunting, G.S., Commentary on Mexican Araceae, *Gentes Herbarum* **9**: 290–382 (1965); Bunting, G.S., Addenda to 'Commentary on Mexican Araceae', *Gentes Herbarum* **10**: 115–16 (1967); Madison, M., Notes on Caladium (Araceae) and its allies, *Selbyana* **5**: 342–77 (1981).

The following keys make provision for the identification of juvenile phases when these are prevalent in cultivation but it should not be assumed that all juvenile phases will be identifiable.

Key to genera (not requiring flowering material)

Specimens must first be assigned to one of the five slightly overlapping groups.

Group 1
Temperate aquatic and marsh plants.

Smell of leaves. Fragrant: **2**; fetid when young: **10,11**.
Leaf-stalks. None: **2**. Without sheath: **10,11**; with well-formed sheath: **23**. Widely channelled: **10**; narrowly channelled: **11**.
Leaf-blades. Submerged or floating: **12**. Linear: **2**; narrowly elliptic: **12,23**; broadly ovate or oblong: **10–12**; with sagittate base: **24**; with deeply cordate base: **23**.

Group 2
Tropical and subtropical aquatic and amphibious plants.

Habit. Floating rosettes of approximately truncate, hairy leaves: **42**; amphibious plants with pointed hairless leaves: **40,41**.

Group 4
Temperate tuberous plants grown in the open or under glass.

Plan of leaf-blade. Simple: **23,33,36,37**; palmate: **38,39**; pedate: **34,35,38,39**.
Shape of leaf-blade when simple. Cordate or sagittate: **23,33,37**; elliptic-lanceolate to linear: **23,36**.
Leaf-blade venation. Mainly parallel: **23**.

Group 4
Tropical and subtropical tuberous plants, usually with a dormant phase (young plants of Group 5, which may be stemless, are not included).

Plan of leaf-blade. With 3 or 5 pinnate or pinnately lobed divisions: **14**; pinnatisect: **27**; simple: **25–28,30,31**.
Shape of leaf-blade when simple. Peltate: **25–28**; not peltate but distinctly cordate or sagittate: **25,26,28,31**; base weakly cordate or truncate: **25,30**.
Posture of leaf-blade or its principal lobe. Drooping: **25–28**; not drooping: **28,30,31**.
Leaf-stalk. Striped: **25,30**; mottled: **14**; banded: **27,28**; green: **25–28,31**; violet or purple: **?25,26–28**. Shorter than blade: **30**.

Group 5
Erect, trailing or climbing tropical plants without tubers (young plants of which may lack obvious above-ground stems) or tropical rosette plants.

1a. Sheath of leaf-stalk not embracing principal stem or the sheath next above (figure 3(3 & 4), p. 76); scale-leaves present among foliage leaves; sometimes rosette plants — 2
b. Sheath of leaf-stalk embracing principal stem and often the sheath next above (figure 3(1 & 2), p. 76); scale-leaves not usually present among foliage leaves except at base of plant or among inflorescences; not rosette plants — 3
2a. Leaf-stalk with a pulvinus at the apex (figure 3(6), p. 76) (sometimes not obvious; several leaves should be examined) — **1. Anthurium**
b. Leaf-stalk without a pulvinus — **19. Philodendron**
3a. Leaf-blade deeply lobed or compound, sometimes perforated — 4
b. Leaf-blade simple, sometimes shallowly pinnately lobed — 10
4a. Leaf-blade with 3 leaflets, or pedate — **29. Syngonium**
b. Leaf-blade pinnatisect or pinnatifid — 5
5a. Leaf-stalk with a pulvinus at apex (figure 3(6), p. 76) (sometimes not obvious; several leaves should be examined) — 6
b. Leaf-stalk without a pulvinus — 9
6a. Leaves pinnate with segments tapered to a point — **3. Rhaphidophora**
b. Leaves pinnate or pinnatifid with obliquely truncate segments (sometimes also perforated) or perforated or entire — 7
7a. Leaves with pinprick-like holes near midrib — **4. Epipremnum**
b. Leaves with large perforations or none — 8
8a. Leaves entire — **3. Rhaphidophora; 4. Epipremnum**
b. Leaves perforated or pinnatifid — **3. Rhaphidophora; 8. Monstera**
9a. Plant erect, without aerial roots — **28. Alocasia**
b. Plant eventually climbing, with aerial roots — **16. Rhektophyllum**
10a. Plant climbing or trailing on the ground — 11
b. Plant erect — 18
11a. Leaf-stalk with a pulvinus at the apex (figure 3(6), p. 76) (sometimes not obvious; several leaves should be examined) — 12
b. Leaf-stalk without a pulvinus — 15
12a. Leaf-blade heart-shaped, strongly asymmetric — **5. Scindapsus**
b. Leaf-blade tapered or rounded at the base — 13
13a. Leaf-sheath not reaching nearly to pulvinus — **20. Anubias**
b. Leaf-sheath reaching nearly to pulvinus — 14
14a. Stem shortly creeping — **9. Spathiphyllum**
b. Plant climbing — **7. Rhodospatha**
15a. Leaf-blade sagittate — **15. Nephthytis; 29. Syngonium** (juvenile)
b. Leaf-blade not sagittate — 16
16a. Base of leaf-stalk surrounded by 1 or 2 scale-leaves; leaf-blade net-veined and wrinkled — **32. Callopsis**
b. Base of leaf-stalk not surrounded by scale-leaves; leaf-blade neither net-veined nor wrinkled — 17
17a. Stem climbing; leaves shortly stalked, in 2 rows closely applied to the substrate — **3. Rhaphidophora** (juvenile)
b. Stem trailing; leaves fairly long-stalked, clustered near upturned tips of shoots — **21. Aglaonema**
18a. Leaf-stalk with a pulvinus at the apex (figure 3(6), p. 76) (sometimes not obvious; several leaves should be examined) — **6. Stenospermation**
b. Leaf-stalk without a pulvinus — 19
19a. Leaf-blade with the subsidiary lateral veins mostly arising from midrib and running parallel to the main laterals (figure 3(5), p. 76) — **17. Homalomena; 18. Schismatoglottis; 21. Aglaonema; 22. Dieffenbachia**

b. Leaf-blade with the subsidiary lateral veins mostly arising from the primary laterals and forming a network, often joined to an interprimary vein (figure 3(6 & 7), p. 76) 20
20a. Leaf-blade less than 20 cm **29. Syngonium** (juvenile)
b. Leaf-blade normally exceeding 20 cm 21
21a. Leaf-blade peltate (figure 3(7), p. 76) **28. Alocasia**
b. Leaf-blade not peltate 22
22a. Main vein of basal lobe of leaf-blade exposed in the sinus for part of its length (figure 3(8), p. 76) **16. Rhektophyllum** (juvenile); **26. Xanthosoma**
b. Main vein of basal lobe of leaf-blade not exposed in the sinus **26. Xanthosoma; 28. Alocasia**

Key to genera (requiring flowering material)

(*Scindapsus* (5) & *Rhektophyllum* (16) do not appear in this key as they do not flower in cultivation.)

1a. Floating aquatic plant with rosettes of more or less stalkless downy leaves broadest at the apex **42. Pistia**
b. Plant otherwise 2
2a. Waterside plant with iris-like leaves, fragrant when crushed; spadix protruding sideways from an apparent leaf **2. Acorus**
b. Plant otherwise 3
3a. Stipe several times longer than spadix, part of it wrapped in the sheathing part of the spathe 4
b. Stipe not longer than spadix 5
4a. Spathe reaching fertile part of spadix, its limb well developed, boat-shaped; marsh plant with large leaves **10. Lysichiton**
b. Spathe not reaching fertile part of spadix, its limb like a small bract; aquatic plant with submerged, floating and emerging leaves **12. Orontium**
5a. Spadix united throughout its length to the mainly flat spathe **31. Spathicarpa**
b. Spadix free, or united to spathe for less than half its length 6
6a. Spathe in the form of an upwardly open funnel or chalice, white, yellow, pink or, occasionally, green; spadix without appendix **23. Zantedeschia**
b. Spathe not shaped as above; if upwardly open, then dark red, or spadix with an appendix, or both conditions apply 7
7a. Whole spadix visible at flowering time when viewed from the side, no portion of it concealed by the spathe 8
b. At least the base of the spadix concealed or encircled by the spathe at flowering time 20
8a. Spadix uniform, florets all or nearly all bisexual (figure 4(1–3), p. 77) 9
b. Spadix with zones bearing female, male and sometimes sterile florets exclusively, sometimes with 1 or more non-flowering zones (figure 4(4–8), p. 77) 17
9a. Florets with a perianth (figure 4(2 & 3), p. 77) 10
b. Florets without perianth 11
10a. Sheath of leaf-stalk not embracing principal stem or the sheath next above (figure 3(3 & 4), p. 76) **1. Anthurium**
b. Sheath of leaf-stalk embracing principal stem and often the sheath next above (figure 3(1 & 2), p. 76) **9. Spathiphyllum**
11a. Plant trailing or climbing, often by means of adhesive roots on the stem; leaf-blade simple in the juvenile phase, usually pinnately lobed, pinnatisect, pinnate or perforated in the adult phase 12
b. Plant not trailing or climbing; leaf-blade simple 16
12a. Leaf-blade, except in very young plants, with large perforations **8. Monstera**
b. Leaf-blade with pinprick-like perforations or not perforated 13
13a. Leaf-blade always variegated in cultivation 14
b. Leaf-blade not variegated 15
14a. Leaf-blade broadly ovate-cordate or broadly ovate-oblong, irregularly and extensively variegated with light green and yellow **4. Epipremnum**
b. Leaf-blade elliptic or elliptic-ovate, not cordate, variegated with scattered patches of light mottling between the veins **7. Rhodospatha**
15a. Leaf-blade in the adult phase with a row of pinprick-like holes on each side of the midrib; leaves of the juvenile phase with stalks nearly as long as blades, not arranged in 2 rows **4. Epipremnum**
b. Leaf-blade in the adult phase without pinprick-like holes; leaves of the juvenile phase with stalks much shorter than blades, arranged in 2 rows closely applied to the substrate **3. Rhaphidophora**
16a. Leaf-stalk without a pulvinus; leaf-blade about as wide as long **13. Calla**
b. Leaf-stalk with a pulvinus at the apex (figure 3(6), p. 76) (not always obvious; several leaves should be examined) **6. Stenospermation**
17a. Spathe spreading, flat but folded at base, white **32. Callopsis**
b. Spathe erect, more or less boat-shaped or hooded, green or yellowish, at least outside 18
18a. Spadix with zone of female florets scarcely shorter than zone of male florets **20. Anubias**
b. Spadix with zone of female florets much shorter than zone of male florets 19
19a. Leaf-blade with the base tapered, rounded or slightly cordate; leaf-stalk with a well-developed sheath **21. Aglaonema**
b. Leaf-blade sagittate; leaf-stalk with a very small sheath **15. Nephthytis**
20a. Lower part of spathe in the form of a closed tube, formed by fusion of margins 21
b. Lower part of spathe with free but overlapping margins 25
21a. Aquatic or amphibious plants with more or less vertical rhizomes and long stolons 22
b. Terrestrial and tuberous or shortly rhizomatous plants 23
22a. Each half of leaf-blade separately inrolled when young; female florets 20–50, spirally arranged **40. Lagenandra**
b. One half of leaf-blade rolled round the other when young; female florets 4–7, in a single whorl **41. Cryptocoryne**
23a. Leaves pedate; spadix appendix *c.* 30 cm **35. Sauromatum**
b. Leaves simple; spadix appendix much less than 30 cm 24
24a. Spadix with the zones of male and female florets adjacent; leaf-blade heart-shaped or sagittate **37. Arisarum**
b. Spadix with a gap which sometimes bears sterile florets between the zones of male and female florets; leaf-blade tapered or truncate at base **36. Biarum**
25a. Florets all bisexual; spadix nearly as broad as long; temperate marsh plants flowering early in spring, before their leaves appear **11. Symplocarpus**
b. Florets mostly unisexual; spadix not nearly as broad as long; plants otherwise 26

26a. Spadix with a sterile terminal appendix, sometimes only a few millimetres long (it may be patterned as if with male florets, figure 4(6–8), p. 77) 27
b. Spadix without a sterile terminal appendix 35
27a. Spathe-limb falling off soon after flowering and while still fresh, the persistent remnant truncate **18. Schismatoglottis**
b. Spathe-limb not shed in this manner 28
28a. Female part of spadix completely joined to spathe **39. Pinellia**
b. Female part of spadix free from spathe 29
29a. Leaves compound or deeply lobed 30
b. Leaves simple 32
30a. Spadix unisexual or with male florets not touching or (usually) showing both these conditions **38. Arisaema**
b. Spadix bisexual; male florets more or less touching 31
31a. Leaves palmate with 3 or 5 pinnate or pinnately lobed divisions; only 1 leaf present at a time **14. Amorphophallus**
b. Leaves pedate, several **34. Dracunculus**
32a. Sterile florets between male and female zones of spadix each in the form of a thread or prong, often with a swollen base (figure 4(7), p. 77) **33. Arum**
b. Sterile florets each in the form of a flat plate, occasionally with a rudimentary ovary forming a boss in the middle, or sterile florets absent (figure 4(6), p. 77) 33
33a. Spadix without a constriction between male and female zones; sterile florets few and each with a central boss, or none; female florets each with a cup round the ovary (formed from fused staminodes) **24. Peltandra**
b. Spadix usually constricted in the zone of sterile florets (figure 4(6), p. 77); sterile florets without a central boss; female florets without staminodes 34
34a. Placentation parietal; ovules and seeds several or many **27. Colocasia**
b. Placentation basal; ovules and seeds 1 or few **28. Alocasia**
35a. Male florets consisting of groups of united stamens, the filaments of which form a stalk 6–7 mm long **30. Synandrospadix**
b. Anthers not borne on a long common stalk 36
36a. Leaf-sheath not embracing the principal stem or the sheath next above (figure 3(3 & 4), p. 76), sometimes inconspicuous **19. Philodendron**
b. Leaf-sheath embracing the principal stem and often the sheath next above (figure 3(1 & 2), p. 76) 37
37a. Leaves compound (sometimes simple in juvenile state) 38
b. Leaves simple 39
38a. Stems loosely scrambling or closely applied to the substrate **29. Syngonium**
b. Plants essentially stemless **26. Xanthosoma**
39a. Leaf-blade with the areas between main lateral veins occupied mainly by veins diverging widely from the latter (figure 3(6 & 7), p. 76) 40
b. Leaf-blade with the areas between main lateral veins occupied mainly by veins running nearly parallel to the latter (figure 3(5), p. 76) 41
40a. Stigma undivided, resting on a disc formed by expansion of the style and which coheres with those of neighbouring ovaries; ovary with 20–100 ovules **26. Xanthosoma**
b. Stigma divided by 3 or 4 grooves and placed directly on the ovary (style absent); ovary with 2–20 ovules **25. Caladium**
41a. Leaf-blade sagittate **24. Peltandra**
b. Leaf-blade not sagittate 42
42a. Leaf-stalk with a pulvinus at the apex (figure 3(6), p. 76); plant creeping; female florets numerous; spadix with female zone longer than male **20. Anubias**
b. Leaf-stalk without a pulvinus; plant erect or, if creeping, spadix with not more than 10 female florets; spadix with female zone equal to or shorter than male 43
43a. Female florets with club-shaped staminodes (figure 4(5), p. 77); male florets with anthers united under a common flat-topped connective (figure 4(4), p. 77), the pollen being extruded from pores on the sides **22. Dieffenbachia**
b. Female florets without staminodes; male florets with anthers free, arranged randomly or in groups of 2–4, the pollen being extruded from pores on their tops 44
44a. Ovary with 1 ovule; berry with 1 seed **21. Aglaonema**
b. Ovary with several to many ovules; berry with several to many seeds **17. Homalomena**

1. ANTHURIUM Schott
T.B. Croat

Plants usually epiphytic, erect, climbing or pendent, evergreen; stems non-tuberous. Leaf-stalks with a pulvinus at the apex (figure 3(6), p. 76); leaf-blades entire or palmately lobed, the major veins often joining to form a marginal vein; minor veins usually netted (figure 3(6), p. 76). Spathe flat or hooded, its edges not overlapping at base, erect, reflexed or spreading, often colourful. Spadix nearly cylindric (figure 4(1), p. 77), more or less tapered towards apex. Florets closely packed, bisexual (figure 4(2 & 3), p. 77); perianth segments 4; stamens 4, usually emerging just above the perianth segments. Ovary with 2 cells, usually with 1 ovule per cell, sometimes with 2 or more. Berries usually ovoid to spherical; colour various; flesh gelatinous. Seeds with sticky appendages at base or apex or both.

There are 700–900 species in the New World tropics. Approximately 28 species are widely available. In species having deeply divided leaves, only those of the mature plant are described here; those of younger plants may be less divided. The basal veins are those which arise at the top of the leaf-stalk in leaf-blades which are heart-shaped or sagittate.

Most Anthuriums grow best under greenhouse conditions with high humidity and partial shade, but most of the species listed here adapt themselves well to other growing conditions and can survive considerable abuse. A loose, well-drained potting soil is essential and plants generally respond well to fertiliser.

Literature: Engler, A., Araceae, Pothoideae, *Das Pflanzenreich* **21**: 53–295 (1905); Madison, M., The species of Anthurium with palmately divided leaves, *Selbyana* **2**: 239–81 (1978); Croat, T.B. & Baker, R., Anthurium (Araceae) of Costa Rica, *Brenesia* **16**: 1–174 (1979); Bunting, G.S., Sinopsis de las Araceae de Venezuela, *Revista de la Facultad de Agronomía (Maracay)* **10**: 139–290 (1979); Mayo, S.J., Anthurium acaule (Jacq.) Schott & the Bird's Nest Anthuriums of the West Indies, *Kew Bulletin* **36**: 691–719 (1982); Croat, T.B., A revision of Anthurium for Mexico and Central America, Part I, Mexico and Middle America, *Annals of the Missouri Botanical Garden* (in press), Part II, Panama, *Annals of the Missouri Botanical Garden* (in press).

Habit. Forming a rosette: **21,22,24**; climbing: **4,15,25**.

Leaf-blade. Palmately lobed or divided: **1–7**; entire or lobed at base: **8–20**; elongate, acute at base: **21–27**. Glandular-dotted: **21,25–28**. With velvety surface: **9–13**.
Inflorescence. Longer than leaf: **3,4,9,10,12,13,14** (rarely), **17,18,24,27,28**; shorter than leaf: **1,2,5–8,11,14–16,19,21–23,25–26**; equal to leaf: **4, 8,20,21,24,26,27**.
Spadix. White, creamy white, yellow: **9,11,14,17,18,20,26**; green: **3,4,8,12,13,15,16,22,24,25**; pink, red, red-violet: **19,20,23,27**; purplish: **1–3,5,7,10,16,19–22,25**.

1a. Leaf-blade palmately lobed or divided 2
b. Leaf-blade simple, sometimes lobed 9
2a. Segments of leaf-blade all free to the base 3
b. Segments of leaf-blade united at the base 7
3a. Margins of leaf segments sinuous or deeply lobed **1. clavigerum**
b. Margins of leaf segments entire 4
4a. Leaf-blade with 11–15 segments, up to 1.5 cm wide **4. polyschistum**
b. Leaf-blades with 5–11 segments, these 4–10 cm wide 5
5a. Spathe deciduous after flowering 6
b. Spathe persisting even on the fruiting spadix **2. pentaphyllum** var. **pentaphyllum**
6a. Spadix 5–10 cm in flower, its stalk 2–7 cm **2. pentaphyllum** var. **bombacifolium**
b. Spadix 10–15 cm in flower, its stalk 5–12 cm **5. digitatum**
7a. Segments of leaf-blade commonly pinnatisect with 5–12 narrow divisions; inflorescence-stalk longer than the leaf-stalk **3. podophyllum**
b. Segments of leaf-blade not so deeply divided, sometimes wavy; inflorescence-stalk shorter than the leaf-stalk 8
8a. Central segment of leaf-blade wavy, twice as long as lateral segments; plant climbing; scale-leaves deciduous; berries deep purple **7. pedatum**
b. Central segment of leaf-blade not markedly wavy, not quite as long as lateral segments; plant with short stems; scale-leaves weathering and persisting; berries red **6. palmatum**
9a. Leaf-blade prominently or weakly lobed (cordate or sagittate) at base 10
b. Leaf-blade not lobed at base 22
10a. Leaf-blade either appearing conspicuously blistered or velvety with pale major veins 11
b. Leaf-blade neither blistered nor velvety 16
11a. Leaf-blade appearing conspicuously blistered; major veins usually not pale, all veins including minor veins (figure 3(6), p. 76) conspicuously sunken; marginal vein (figure 3(6), p. 76) arising from near base of blade **8. corrugatum**
b. Leaf-blade velvety, not blistered; major veins usually pale; the smaller veins not at all sunken; marginal vein not arising from base of blade 12
12a. Leaf-blade peltate **9. forgetii**
b. Leaf-blade not peltate (leaf-stalk borne at base of blade) 13
13a. Leaf-stalk with 4 wings or ridges **10. magnificum**
b. Leaf-stalk terete or grooved along upper side but not with 4 ridges or wings 14
14a. Leaf-blade 2 or more times as long as broad; berries red **11. warocqueanum**
b. Leaf-blade 1–1.5 times as long as broad; berries orange or white and purplish 15
15a. Leaf-blade to 25 cm, moderately thick, broadly ovate; major veins silver-grey; berries orange, spherical **12. clarinervium**
b. Leaf-blade to 40 cm, moderately thin, ovate; major veins usually pale green; berries ovoid, white at base, purplish above **13. crystallinum**
16a. Leaf-blade 2 or more times as long as broad; primary lateral veins reflexed towards the leaf-base near the midrib, then prominently arched as they approach the margin **14. veitchii**
b. Leaf-blade less than twice as long as broad; venation not as above 17
17a. Plant creeping, with branching stems to about 1 cm in diameter; leaf-blade less than 20 cm **15. radicans**
b. Plant erect, stems not branching, usually thicker; leaf-blade usually 25 cm or more 18
18a. Basal lobes of leaf-blade usually much longer than broad **16. watermaliense**
b. Basal lobes of leaf-blade about as broad as long, usually rounded 19
19a. Spathe narrowly lanceolate, thin, often twisted, green; leaf-blade dark green above with a satin-like sheen, the veins pale green; scale-leaves weathering to fibres **17. regale**
b. Spathe oblong-elliptic, oblong-lanceolate or wider, moderately thick, never twisted, usually white or reddish, sometimes tinged with green; leaf-blades medium green, the veins not markedly paler than the surface; scale-leaves persisting intact, reddish brown 20
20a. Spathe broadly ovate to almost rounded, bright red and puckered between the veins (though cultivars may vary in colour and are sometimes smooth); spadix white, tipped with yellow (cultivars various) **18. andraeanum***
b. Spathe oblong-elliptic, oblong-lanceolate, or ovate to narrowly obovate, rarely more or less rounded, white or greenish, sometimes tinged with red, rarely pale red 21
21a. Inflorescence shorter than leaf; spathe whitish green; stems to 14 cm; berries orange **19. huixtlense**
b. Inflorescence as long as leaf; spathe white tinged with green to reddish; stems to 100 cm; berries reddish violet **20. nymphaeifolium**
22a. Plant forming a rosette; leaf-blade usually broadest above the middle, usually more than 60 × 20 cm; primary lateral veins mostly extending to the margin; the roots so dense as to be contiguous 23
b. Plant not forming a rosette; leaf-blade usually not broadest above the middle, usually less than 50 × 10 cm; primary lateral veins joining the marginal vein (figure 3(6), p. 76) which arises at or near the base; stems long or short but the roots never contiguous 26
23a. Leaves black-dotted on both sides **21. hookeri**
b. Leaves not black-dotted 24
24a. Spathe moderately thin, often withering and deciduous, green; leaf-blade with the minor veins (figure 3(6), p. 76) prominently raised; leaf-stalk usually rounded on the back, sometimes prominently 3-ribbed 25
b. Spathe thick, usually persisting in fruit, usually tinged with purple in age; usually prominently reflexed; leaf-blade with minor veins (figure 3(6), p. 76) not prominently raised; leaf-stalk usually sharply 4-angled **22. schlechtendalii**
25a. Inflorescence usually as long as leaf, its stalk slender, usually spreading or pendent; spadix commonly 40 times as long as thick **24. crenatum**

b. Inflorescence usually only about half as long as leaf, its stalk stout, usually stiffly erect; spadix commonly 10–30 times (rarely to 35 times) as long as thick **23. crassinervium**
26a. Leaf peltate; surface velvety; major veins pale **9. forgetii**
b. Leaf not peltate; surface not at all velvety; major veins not pale 27
27a. Plant climbing; stem less than 6 mm in diameter; leaf-blade more or less elliptic, less than 16 cm **25. scandens**
b. Plant free-standing with short stems which are often 1 cm or more in diameter; leaf-blade oblong-elliptic to oblanceolate, never elliptic, usually more than 20 cm 28
28a. Spathe yellow-green; leaf-blade with marginal veins much more conspicuously sunken than the primary lateral veins; berries bright red **26. bakeri**
b. Spathe red, orange-red or violet-purple; leaf-blade with marginal veins no more conspicuously sunken than the primary lateral veins; berries white, red or orange 29
29a. Spathe red or orange-red, usually more than 6 cm **27. scherzerianum**
b. Spathe pale violet-purple, less than 4 cm **28. amnicola**

1. A. **clavigerum** Poeppig & Endlicher (*A. panduratum* Martius; *A. holtonianum* Schott; *A. kalbreyeri* N.E. Brown). Illustration: Graf, Exotica, edn 8, 132, 135 (1976); Graf, Tropica, 94, 95 (1978); Brenesia **16**: 120–1 (1979); Everett, Encyclopedia of horticulture **1**: 188 (1980).
Epiphytic creeper. Stem to 2 m, 3–5 cm in diameter. Leaf-blade 80 cm–2 m wide, palmatisect; segments 7–13 with margins sinuous to deeply lobed. Inflorescence pendent, its stalk to 90 cm, shorter than the leaf-stalk. Spathe 30–65 × 3–11 cm, violet-purple, lanceolate. Spadix 20–75 cm, pale purple. Berries obovoid, purple. *Nicaragua to the Guianas, Brazil and Bolivia*. G2.

2. A. **pentaphyllum** (Aublet) G. Don (*A. enneaphyllum* (Vellozo) Stellfeld; *A. undatum* Kunth). Illustration: Selbyana **2**: 255 (1978) – var. pentaphyllum.
Epiphytic creeper with short internodes. Leaf-stalk 19–44 cm, terete. Leaf-blade palmatisect; segments 5–9, 13–30 cm, oblanceolate to narrowly elliptic or ovate, the lateral unequal at base. Inflorescence erect, short, its stalk 1–13 cm. Spathe to 15 × 2.5 cm, pale green to purple, lanceolate to ovate, usually persisting in fruit or (var. **bombacifolium** (Schott) Madison (*A. aemulum* Schott)) falling soon after flowering. Spadix usually less than 10 cm long, 8–20 mm in diameter at base, pale violet-purple. Berries red to dark red-violet, more or less ellipsoid. *Trinidad and Atlantic coast of S America to Brazil* (var. *pentaphyllum*); *Mexico to Panama* (var. *bombacifolium*). G2.

3. A. **podophyllum** (Chamisso & Schlechtendal) Kunth. Illustration: Graf, Exotica, edn 8, 135, 137 (1976); Selbyana **2**: 253 (1978); Annals of the Missouri Botanical Garden **69**: f. 144, 156 (1982).
Terrestrial or on rocky cliffs. Stem short, 4–6 cm in diameter; internodes short. Leaf-stalk 40–80 cm, grooved above. Leaf-blade to 90 cm wide, palmatisect or palmate, thick; segments 7–11, stalkless or with stalk up to 2 cm long, each pinnatisect with 5–12 divisions or rarely with the margins merely wavy. Inflorescence erect-spreading, its stalk 50–100 cm, longer than the leaf-stalk. Spathe 5–8 cm, green, lanceolate. Spadix 6–15 cm, green or purplish, tapered to apex. Berries red. *Mexico (C Veracruz)*. G2.

4. A. **polyschistum** Schultes & Idrobo (*A. putumayo* invalid; *A. pictamayo* invalid). Illustration: Graf, Exotica, edn 8, 1505 (1976); Aroideana **1**: 3 (1978); Selbyana **2**: 254 (1978); Everett, Encyclopedia of horticulture **1**: 189 (1980).
Epiphytic climber or terrestrial creeper. Stem with internodes 6–10 cm × 5–7 mm. Leaf-stalk 6–22 cm, nearly terete but weakly grooved along upper side. Leaf-blade palmate; segments 11–15, 8–15 × 0.5–1.5 cm, linear-lanceolate, thin, with the margins often crisped. Inflorescence-stalk 8–28 cm. Spathe 6–10 cm, green, narrowly lanceolate, reflexed. Spadix 6–14 cm long, 7 mm in diameter at base, pale green, erect. Berries obovoid, violet-purple to red-violet. *Colombia, Ecuador & northern Peru in the Amazon Basin*. G2.

5. A. **digitatum** (Jacquin) G. Don (*A. pentaphyllum* var. *digitatum* (Jacquin) G. Don). Illustration: Graf, Exotica, edn 8, 137 (1976); Everett, Encyclopedia of horticulture **1**: 188 (1980).
Epiphytic creeper. Stem to 2 cm in diameter; internodes short near the apex. Leaf-stalk 20–50 cm. Leaf-blade palmate; leaflets 7–13, 10–22 × 4–10 cm, nearly equal, stalkless or with stalk to 2 cm, thick, dark green. Inflorescence erect, much shorter than the leaf-stalk, its stalk 6–18 cm. Spathe 3–7 cm × 7–30 mm, green or lightly tinged purple, lanceolate to ovate, falling after flowering. Spadix 10–15 cm long, 8–20 mm wide at base, stalkless tapered towards apex, medium purple. Berries spherical, red to purple. *Venezuela*. G2.

Sometimes included in *A. pentaphyllum* (Aublet) G. Don.

6. A. **palmatum** (Linnaeus) G. Don (*A. fissum* Regel). Illustration: Graf, Exotica, edn 8, 1505 (1976); Selbyana **2**: 252 (1978).
Climbing epiphyte to 5 m. Stem with internodes 2–6 × 1.5–3 cm. Leaf-stalk 40–100 cm. Leaf-blade palmatisect; segments 5–11, 25–40 × 4–6 cm, oblanceolate. Inflorescence erect to spreading, its stalk 20–70 cm. Spathe 6–18 cm × 8–20 mm, lanceolate, green. Spadix 8–16 cm × 5–8 mm, tapered towards apex. Berries red, ovoid, rounded at apex. *West Indies (Lesser Antilles: St Lucia to Guadeloupe)*. G2.

7. A. **pedatum** Humboldt, Bonpland & Kunth (*A. elegans* Engler; *A. fortunatum* Bunting). Illustration: Graf, Exotica, edn 8, 137, 140 (1976); Selbyana **2**: 253 (1978); Aroideana **4**: 19 (1981).
Terrestrial. Stem short, 2–6 cm in diameter; internodes short. Leaf-stalk 30–75 cm, terete. Leaf-blade palmatisect, leathery; segments 9–13, with wavy margins, unequal, the central one 20–45 × 5–8 cm, the lateral up to two-thirds as long as the central. Inflorescence-stalk half to five-sixths as long as the leaf-stalk. Spathe 6–10 × 2–4 cm, lanceolate, yellowish green, held horizontally. Spadix 7–12 cm × 5–10 mm, green to purplish, tapered, pendent. Berries deep purple. *Colombia (Western Andes)*. G2.

8. A. **corrugatum** Sodiro (*A. papilionense* invalid; *A. papillosum* Markgraf). Illustration: Graf, Exotica, edn 8, 135 (1976); Graf, Tropica, 95 (1978).
Terrestrial. Stem somewhat climbing; internodes 6–12 × 2–3 cm. Leaf-stalk usually longer than blade, flattened and grooved along upper side. Leaf-blade 35–55 × 15–40 cm, ovate to more or less elliptic, sagittate, the surface appearing closely blistered or coarsely wrinkled with numerous deeply sunken primary lateral veins; the smaller veins sunken above, raised beneath; marginal vein (figure 3(6), p. 76) arising from base of blade. Inflorescence erect, its stalk shorter than or equalling the leaf-stalk. Spathe to 20 × 3 cm, green, reflexed. Spadix about as

long as spathe, 8–15 mm in diameter, stalkless, pale green, slightly tapered to the apex. Berries not known. *Colombia & Ecuador*. G2.

9. A. forgetii N.E. Brown. Illustration: Aroideana **1**: 18 (1978); Graf, Exotica, edn 8, 133 (1976).
Small epiphyte. Stem short with short internodes. Leaf-stalk 15–25 cm. Leaf-blade 20–30 × 15–22 cm, ovate, peltate, acuminate at apex, rounded or weakly notched at base, olive green and velvety on upper surface, with the major veins paler. Inflorescence 40–60 cm, erect, held about twice as high as the leaves. Spathe 8–15 cm, linear-lanceolate, green. Spadix about 15 cm, yellowish, gradually tapered. Berries narrowly ovoid, white, tinged with violet-purple. *Colombia*. G2.

The species readily hybridises with *A. crystallinum* and probably many of the plants sold as *A. forgetii* are such hybrids.

10. A. magnificum Linden. Illustration: Parey's Blumengärtnerei **1**: 169 (1958); Graf, Tropica, 93 (1978); Aroideana **1**: 18 (1978).
Probably terrestrial. Stem to 3 cm in diameter; internodes short. Leaf-stalk 25–40 cm, 4-sided and acutely ribbed. Leaf-blade 30–41 × 20–26 cm, ovate, deeply cordate, thick, olive green with greenish white major veins; upper surface matt, velvety. Inflorescence erect, its stalk 50–61 cm. Spathe 15–20 cm, lanceolate, recurved, green or sometimes tinged with red. Spadix 20–25 cm, tapered to apex, purplish. Berries not known. *Colombia*. G2.

11. A. warocqueanum J. Moore. Illustration: Das Pflanzenreich **21**: 197 (1905); Graf, Exotica, edn 8, 133 (1976); Graf, Tropica, 92, 95 (1978); Everett, Encyclopedia of horticulture **1**: 187 (1980).
Epiphytic creeper. Stem to about 1 m long and 1 cm in diameter; internodes short. Leaf-stalk 15–60 cm, terete, nearly erect. Leaf-blade mostly 50–90 × 11–25 cm, vertically pendent, narrowly ovate to oblong-ovate, somewhat leathery, dark green and velvety above; primary lateral veins pale green. Inflorescence nearly erect in flower, its stalk 6–30 cm. Spathe 8–10 × 1.5–2 cm, narrowly ovate-lanceolate, green. Spadix 6.5–30 cm × 5–12 mm, gradually tapered to apex, yellowish green. Berries red. *Colombia*. G2.

12. A. clarinervium Matuda. Illustration: Graf, Exotica, edn 8, 133 (1976); Aroideana **1**: 19, f. 1 (1978) – as A. leuconeurum; Graf, Tropica, 92 (1978).
Terrestrial, usually on limestone outcrops. Stem short, 1–2 cm in diameter. Leaf-stalk 7–16 cm, terete. Leaf-blade ovate, deeply cordate, dark green, velvety on upper surface, pale green along the major veins; lower surface pale, matt. Inflorescence erect, held well above the leaves, its stalk 16–38 cm. Spathe 3.7–6.5 × 1–1.5 cm, lanceolate, pale green. Spadix green, tinged with violet-purple. Berries bright orange, more or less spherical. *Mexico (C Chiapas)*. G2.

13. A. crystallinum Linden & André. Illustration: Illustration Horticole **20**: 87 (1873); Graf, Exotica, edn 8, 133 (1976); Aroideana **1**: 18 (1978); Everett, Encyclopedia of horticulture **1**: 186 – also as A. clarinervium (1980).
Epiphytic. Stem to 25 cm long, 2–2.5 cm in diameter. Leaf-stalk terete. Leaf-blade 25–39 × 15–22 cm, reflexed more or less parallel to leaf-stalk, ovate to narrowly ovate, deeply cordate; upper surface dark green, velvety, with all major veins outlined in pale green. Inflorescence erect-spreading, its stalk 20–28 cm. Spathe 7–9 × 1.5–2 cm, oblong-lanceolate, tinged with red-violet. Spadix gradually tapered towards apex, green, turning yellow. Berries narrowly ovoid, white, tinged with violet-purple, beaked at apex. *Panama to Peru*. G2.

14. A. veitchii Masters. Illustration: Das Pflanzenreich **21**: 203 (1905); Parey's Blumengärtnerei **1**: 169 (1958); Graf, Exotica, edn 8, 130, 1505 (1976); Graf, Tropica, 92, 95 (1978).
Epiphytic. Stem 2–3 cm in diameter; internodes short. Leaf-stalk 20–70 cm, terete, nearly erect. Leaf-blade 25–100 × 12–30 cm, somewhat leathery, vertically pendent, oblong-ovate, cordate at the base with the sinus usually narrow or closed; primary lateral veins numerous, deeply sunken, each forming a broad arch and uniting with the marginal vein which arises from near the base. Inflorescence erect, shorter or rarely longer than the leaf, its stalk 13–50 cm. Spathe 7–13 cm, ovate to ovate-lanceolate, green, somewhat leathery, spreading. Spadix 5–9 cm long, to 1.6 cm in diameter, cream-coloured, cylindric, tapered slightly at both ends. Berries red. *Colombia*. G2.

15. A. radicans C. Koch. Illustration: Das Pflanzenreich **21**: 200 (1905); Graf, Exotica, edn 8, 130 (1976); Aroideana **1**: 37 (1978).
Decumbent creeper. Stem elongate with short internodes, to about 1 cm in diameter. Leaf-stalk 3–10 cm, grooved along upper side. Leaf-blade 10–16 × 6–9 cm, ovate, cordate, thick, often prominently arched along the midrib, appearing blistered above with the primary lateral and secondary veins deeply sunken. Inflorescence short, often hidden by the leaves, its stalk 2–4 cm, shorter than the leaf-stalk. Spathe 3.5–5 × 2–2.8 cm, thick, broadly ovate, cordate. Spadix 2–4 cm long and up to 1.5 cm in diameter, nearly cylindric but slightly tapered to apex, reddish green to brown. Berries ovoid, rounded at apex with an acute style. *Brazil (Bahia)*. G2.

16. A. watermaliense Bailey. Illustration: Graf, Exotica, edn 8, 131, 134 – also as A. 'Negrito', 135 – as A. longilinguum (1976); Brenesia **16**: 173 (1979); Annals of the Missouri Botanical Garden **69**: f. 224–5 (1982).
Terrestrial. Stem to 25 cm long and 1–2.5 cm in diameter; internodes short. Leaf-stalk 12–88 cm, erect-spreading, weakly flattened or grooved along upper side. Leaf-blade 30–60 × 19–36 cm, moderately thick, ovate-triangular, sagittate; basal lobes usually much longer than broad with a deep sinus between them; upper surface dark green. Inflorescence nearly erect, its stalk 12–66 cm, usually much shorter than the leaf. Spathe 15–21 × 2.3–8.5 cm, lanceolate-triangular, green to dark violet-purple or blackish. Spadix 7–10 cm, whitish to greenish, yellow or purple, prominently tapered to apex. Stamens unusually prominent, projecting well above perianth segments. Berries yellow to orange, ovoid to obovoid, beaked. *Costa Rica to Colombia*. G2.

17. A. regale Linden. Illustration: Graf, Exotica, edn 8, 130 (1976).
Probably epiphytic. Stem moderately short, to 2.5 cm in diameter; internodes short. Leaf-stalk 25–40 cm, grooved along upper side. Leaf-blade 30–57 × 19–36 cm, somewhat leathery, ovate, cordate, dark green on upper surface with a satin-like sheen and pale green veins; lower surface paler green. Inflorescence nearly erect, held above the leaves, its stalk 40–70 cm. Spathe 7–19 cm, green, lanceolate, reflexed. Spadix 10–25 cm, greenish white, prominently tapered. Berries not seen. *Peru (along Rio Huallaga)*. G2.

18. A. andraeanum André. Illustration: Bailey, Manual of cultivated plants, edn 2,

178 (1949); Parey's Blumengärtnerei **1**: 32 (1956); Graf, Exotica, edn 8, 128, 1755 (1976); Graf, Tropica, 92 (1978).
Epiphytic. Stem usually less than 30 cm long, 1–2 cm in diameter; internodes short. Leaf-stalk 20–60 cm, nearly erect, terete. Leaf-blade 17–50 × 11–22 cm, leathery, narrowly ovate, sagittate, with prominent, sometimes overlapping, basal lobes; marginal vein (figure 3(6), p. 76) arising from the first basal vein. Inflorescence more or less erect, usually held above the leaves. Spathe 6–15 × 5–12.5 cm, bright red, broadly ovate to almost rounded, cordate, puckered, often with prominently raised veins. Spadix 4–9 cm × 6–8 mm, usually white, becoming yellowish at least towards the apex, slightly tapered to apex, usually curved slightly and directed downwards. Berries ovoid, reddish. *SW Colombia & NW Ecuador*. G2.

The above description refers to material of *A. andraeanum* collected in the wild. The species has been used in breeding programmes since it was introduced by André in 1876 and because few records were kept it is now difficult to say what the cultivated material bearing the name *A. andraeanum* really is.

No fewer than 26 hybrid names are listed in the Royal Horticultural Society's Dictionary of gardening. Almost half of these involve crosses with *A. nymphaeifolium* (here understood to include *A. ornatum* and *A. lindenianum*). Birdsey has proposed the name *A.* × *cultorum* for at least part of this material.

Independently of hybridisation, *A. andraeanum* has undergone a great deal of spontaneous mutation, resulting in a wide array of flower forms (see 'Anthurium breeding in Hawaii' by H. Kamemoto, Aroideana **4**: 77–86, 1981).

Among the hybrids in cultivation which have *A. andraeanum* as one of the parents are:

*A. × **ferrierense** Masters & T. Moore (*A.* × *carneum* Regel; *A.* × *roseum* Closon). Illustration: Illustration Horticole **32**: 175 (1885) & **33**: 31 (1886); Revue de l'Horticulture Belge **23**: 193 (1897) & **27**: 97 (1901); Graf, Exotica, edn 8, 128, 129 (1976); Graf, Tropica, 92, 94 – 'Rhodochlorum' (1978).

The hybrid between *A. andraeanum* and *A. nymphaeifolium*, usually with longer leaf-stalks and broader leaf-blades than *A. andraeanum*, and with spathe less puckered, or smooth. Inflorescence variable in colour; spathe white, yellowish, pink or red; spadix white, pink or yellowish, or white tipped with pink or yellow.

*A. × **chelsiense** N.E. Brown. Illustration: Graf, Exotica, edn 8, 128, 129, 131 (1976).

This is most similar to *A. veitchii* but differs in having the spathe 13 × 9 cm, broadly heart-shaped, smooth, dark red, the spadix 8 cm long, and the leaf-blade with fewer, less arching primary lateral veins. It is the hybrid *A. andraeanum* × *A. veitchii*, with the former as the male parent.

19. A. huixtlense Matuda (*A. fraternum* misapplied). Illustration: Annals of the Missouri Botanical Garden **69**: f. 85, 89 (1982).
Terrestrial or epiphytic. Stem to 14 cm long and 3 cm in diameter. Leaf-stalk 15–70 cm long, terete, nearly erect. Leaf-blade 17–70 × 10.5–36 cm, more or less leathery, ovate to broadly ovate, sagittate, broadest just below point of attachment to leaf-stalk; sinus triangular to horseshoe-shaped; basal veins (figure 3(5), p. 76) joined at the base for 2–2.5 cm; marginal vein (figure 3(6), p. 76) arising from one of the lower primary lateral veins. Inflorescence erect-spreading, shorter than the leaf, its stalk 21–39 cm. Spathe 4.5–15 × 2.7–4.5 cm, pale whitish green, oblong-lanceolate. Spadix 4.2–14 cm, dark pink to pale purple. Berries orange, spherical. *Mexico to Nicaragua*. G2.

Anthurium fraternum Schott, long confused with this species, is a distinct species from northern Colombia.

20. A. nymphaeifolium C. Koch & Bouché (*A. lindenianum* C. Koch & Augustin; *A. ornatum* Schott). Illustration: Botanical Magazine, 5848 (1870); Das Pflanzenreich **21**: 209 (1905); Graf, Exotica, edn 8, 131, not 141 (1976).
Terrestrial or rarely epiphytic. Stem 30–100 cm long and 2.5 cm in diameter; internodes short. Leaf-stalk 50–90 cm, terete, erect-spreading. Leaf-blade 31–70 × 19–40 cm, broadly ovate, deeply cordate, moderately thick; sinus open or closed; basal veins (figure 3(5), p. 76) united or free to the base; marginal vein usually arising from one of the primary lateral veins, sometimes from the uppermost basal vein (figure 3(6), p. 76). Inflorescence usually erect, equalling the leaves, its stalk 35–95 cm. Spathe 9–21 × 2–7.5 cm, oblong-elliptic to ovate or narrowly obovate, rarely more or less rounded and lobed at the base, white tinged with green to reddish throughout. Spadix 8–14 × 1–1.5 cm, white to pink, purplish or red, slightly tapered to the apex. Berries reddish. *Venezuela; ?Colombia (Cordillera Oriental)*. G2.

Bunting's (1979) broad view of this species, including much variation, is followed here.

21. A. hookeri Kunth (*A. huegelii* Schott). Illustration: Graf, Exotica, edn 8, 138 (1976); Graf, Tropica, 95 (1978).
Epiphytic or on rocks, forming a rosette. Stem short, densely rooted. Leaf-stalk 2–14 × 1.5–2 cm, D-shaped to bluntly triangular in section. Leaf-blade 35–140 × 10–45 cm, usually obovate, rarely elliptic to ovate, black-dotted on both surfaces; primary lateral veins (figure 3(5), p. 76) mostly free to the margin, not linked by secondary lateral veins. Inflorescence more or less erect or spreading, its stalk 45–90 cm. Spathe 9–23 × 1.5–2.7 cm, oblong, green. Spadix 10–29 cm × 5–9 mm, gradually tapered, pale violet-purple. Berries obovate to cylindric, translucent, white, reddish to purplish at tip. *Lesser Antilles from St Kitts southward, Trinidad & Surinam*. G2.

This species has been confused with *A. crassinervium* but its leaves have glandular dots and their secondary veins run almost parallel between the primary laterals.

22. A. schlechtendalii Kunth (*A. crassinervium* misapplied; *A. tetragonum* (Hooker) Schott). Illustration: Graf, Exotica, edn 8, 138, 141 – as A. tetragonum (1976); Annals of the Missouri Botanical Garden **69**: f. 172–3 (1982).
Epiphytic or on rocks. Stem short, 2.5–5.5 cm in diameter, densely rooted. Leaf-stalk 12–23 cm × 5–20 mm, trapezoid in cross-section, the lower side sometimes ribbed, sometimes rounded. Leaf-blade 30–112 × 10–58 cm, obovate-elliptic, leathery; most primary lateral veins (figure 3(5), p. 76) extending to the margin. Inflorescence spreading, shorter than the leaf, its stalk 33–34 cm. Spathe 15–28 × 1.5–5 cm, thick, green to violet-purple, linear-lanceolate to lanceolate, strongly reflexed. Spadix 3–34 × 1.7–2 cm, green, often tinged with violet-purple, conspicuously tapered to apex. Berries 1–1.5 cm long, red, lanceolate-ellipsoid. *Mexico to Nicaragua*. G2.

23. A. crassinervium (Jacquin) Schott (*A. ellipticum* C. Koch & Bouché; *A. rugosum* Schott). Illustration; Graf, Exotica, edn 8, 138, 141 (1976).
Epiphytic, forming a rosette. Stem stout, moderately short, very densely rooted; internodes short. Leaf-stalk much shorter

than blade, usually 5–20 cm, nearly quadrangular, broadly grooved along upper side, 3-ribbed beneath. Leaf-blade 45–110 × 9–35 cm, oblanceolate to obovate; primary lateral veins (figure 3(5), p. 76) mostly extending to the margin; tertiary veins forming a network, prominently raised. Inflorescence more or less erect, its stalk to *c.* 60 cm, stout. Spathe mostly 10–15 × 1–2.5 cm at flowering time, oblong-lanceolate, moderately thin, spreading, green. Spadix 14–20 cm, dull red-violet at flowering time, gradually tapered to the apex. Berries red. *Venezuela (coastal region).* G2.

This species has long been confused with *A. schlechtendalii* and also with *A. hookeri.*

24. A. crenatum Kunth (*A. acaule* misapplied). Illustration: Graf, Exotica, edn 8, 138 (1976).
Epiphytic or on steep slopes or cliffs, with a bird's nest habit. Stem short, thick, densely rooted. Leaf-stalk 5–25 cm, broadly grooved along upper side. Leaf-blade 40–100 × 6–30 cm, oblong-lanceolate to oblanceolate, leathery; primary lateral veins (figure 3(5), p. 76) mostly extending to the margin. Inflorescence mostly spreading, its stalk to *c.* 1 m. Spathe 6–15 cm, linear-lanceolate, soon withering. Spadix 10–35 cm, greenish brown, long-tapered. Berries red, obovoid. *Hispaniola, Puerto Rico & Virgin Islands.* G2.

The species was long confused with the apparently rare *A. acaule* (Jacquin) Schott.

25. A. scandens (Aublet) Engler. Illustration: Graf, Exotica, edn 8, 139 only (1976); Aroideana **3**: 88 (1980); Annals of the Missouri Botanical Garden **69**: f. 169 (1982).
Climbing epiphyte. Stem elongate, slender, with internodes 5–40 × 4–6 mm. Leaf-stalk 1–6 cm, usually grooved along upper side, usually much shorter but occasionally longer than blade. Leaf-blade 4–16 × 1.5–7 cm, elliptic to ovate-lanceolate, densely black-dotted on lower surface; marginal vein (figure 3(6), p. 76) arising from the base. Inflorescence much shorter than the leaf, its stalk 2–8 cm, *c.* 1 mm in diameter. Spathe lanceolate to oblong-lanceolate, green, tightly reflexed after opening. Spadix 7–30 × 2–3 mm, green, sometimes becoming pale purple. Berries depressed-spherical, usually pale violet to almost white, with 2–10 or more seeds. *West Indies & S Mexico to S Brazil.* G2.

26. A. bakeri J.D. Hooker. Illustration: Graf, Exotica, edn 8, 115, 136, 139 (1976); Graf, Tropica, 92 (1978); Brenesia **16**: 112 (1979); Everett, Encyclopedia of horticuture **1**: 187 (1980).
Epiphytic. Stem less than 1.5 cm in diameter. Leaf-stalk to 17 cm, terete to sharply grooved above. Leaf-blade 19–55 × 2.8–9 cm, narrowly lanceolate-elliptic to narrowly oblanceolate, much paler and dotted with reddish brown on lower surface; marginal vein (figure 3(6), p. 76) arising from base of blade, conspicuously sunken and more prominent than the primary lateral veins. Inflorescence erect-spreading, its stalk 5.5–30 cm. Spathe pale yellow-green, oblong-lanceolate. Spadix 2–11 cm long, 5–15 mm wide at base, creamy white. Berries bright red, obovoid, pointed at apex. *Guatemala to Colombia.* G2.

27. A. scherzerianum Schott. Illustration: Graf, Exotica, edn 8, 129 (1976); Graf, Tropica, 92, 93 (1978); Brenesia **16**: 157 (1979); Everett, Encyclopedia of horticulture **1**: 186 (1980).
Epiphytic or terrestrial. Stem short, less than 1.5 cm in diameter. Leaf-stalk 4–20 cm, terete. Leaf-blade 5–26 × 1.5–6.5 cm, linear to elliptic or lanceolate, leathery, densely dotted on both surfaces. Inflorescence as long as or longer than the leaf; stalk 14–52 cm. Spathe 3.7–12 × 2.4–6 cm (often larger in hybrids), elliptic to ovate, bright red or orange-red. Spadix 2–8 cm, orange to red, tapered, usually coiled. Berries orange to red. *C Costa Rica (Cordillera Central & Cordillera Talamanca).* G2.

The group name *A.* × *hortulanum* Birdsey has been proposed for cultivars of *A. scherzerianum.* Two cultivars are worthy of special mention, 'Atrosanguineum', which differs from wild collections in having a much larger, dark red spathe, and 'Rothschildianum' (*A.* × *rothschildianum* Bergman & Veitch), with a large red spathe, heavily spotted with white, and a yellow spadix (Illustration: Graf, Tropica, 92, 1978).

28. A. amnicola Dressler (*A. lilacinum* Dressler not Bunting). Illustration: Selbyana **2**: 301 (1978); Aroideana **1**: 87 (1978).
Less than 20 cm tall. Leaf-stalk 3.5–9 cm, nearly terete, weakly grooved above. Leaf-blade lanceolate-elliptic to oblong-elliptic, sparsely and obscurely gland-dotted on upper surface; marginal vein arising from one of the basal veins (figure 3(6) p. 76). Inflorescence erect, its stalk 2–4 times as long as the leaf-stalk. Spathe pale violet-purple, narrowly to broadly ovate. Spadix 8–20 mm, dark violet-purple. Berries spherical, white. Seeds 1–2. *C Panama.* G2.

The species grows clinging to rocks in rapidly flowing streams.

2. ACORUS Linnaeus

D.A. Webb

Rhizomes branched bearing linear leaves and stems which are leafless except for the spathe. Leaves equitant, without a distinct stalk. Spathe similar to the leaves (except for the basal part, which is narrow and stem-like), continuing the line of the inflorescence-stalk, so that the spadix appears lateral. Spadix stalkless, without appendix. Florets very numerous and densely packed, bisexual, with 6 obovate perianth-segments, 6 stamens, and a broadly ovoid, 2- or 3-celled ovary. Fruits not formed in Europe.

A genus of 2 species from S and E Asia and southern N America. They are easily grown in shallow water, marshy soil or well-watered pots. The tissues of both species (especially the rhizome) give off a spicy fragrance when bruised.

1a. Spadix at least 6 mm wide; leaves rather stiff, with a fairly prominent midrib **1. calamus**
b. Spadix not more than 5 mm wide; leaves soft and flexible, without a distinct midrib **2. gramineus**

1. A. calamus Linnaeus. Illustration: Boswell & Syme, English Botany **9**: 11 (1883); Hegi, Illustrierte Flora von Mitteleuropa, edn 3, **2**(1); 321 (1967–80); Graf, Tropica, 90 (1978).
Leaves to 125 × 2.5 cm, often transversely wrinkled in places, rather stiff, with a distinct and usually prominent midrib. Inflorescence-stalk 2- or 3-angled, narrower than the leaves and somewhat shorter. Spathe to 80 cm. Spadix 6–9 cm × 7–12 mm, pale yellow. *S & E Asia & southeastern USA; very widely naturalised elsewhere, so that native limits are uncertain.* H1. Summer.

A variant with leaves striped with cream and yellow ('Variegatus') is sometimes grown.

2. A. gramineus Aiton. Illustration: Makino, Illustrated flora of Japan, enlarged edn, 779 (1961); Graf, Exotica, edn 8, 143 (1976) – 'Pusillus' and 'Variegatus'; Graf, Tropica, 88 (1978) – 'Variegatus'
Like *A. calamus* but less robust, and smaller in all its parts, and with soft, flexible, grass-

like leaves without a distinct midrib. Leaves to 50 cm × 8 mm. Spathe 7–15 cm. Spadix 5–10 cm × 3–4 mm. *NE India to Taiwan & C Japan*. H3. Summer.

Very variable in size. The smallest variants ('Pusillus') have leaves not more than 8 cm long and do not flower. 'Variegatus' has white-striped foliage.

3. RHAPHIDOPHORA Hasskarl
P.F. Yeo
Creeping or climbing evergreen plants with adhesive roots on the stems, sometimes with distinct juvenile and adult phases. Leaf-stalk with a pulvinus at the apex (figure 3(6), p. 76); sheath soon withering. Leaf-blade entire or pinnately lobed or divided, sometimes perforated; minor veins in leaf of adult parallel to the main lateral veins. Spathe boat-shaped, unfurling throughout its length at flowering time, deciduous. Florets without perianths, bisexual or a few female; stamens 4; ovary sometimes expanded at the top, 1-celled or incompletely divided into 2 or 3 cells, with several to many ovules in each cell. Seeds with a soft coat, straight.

A genus of about 60 species in tropical E Asia and the Pacific islands, a very few of which are cultivated. They need a well-drained but moist, rich soil and a support upon which to grow. Propagation is by dividing root-bearing parts of the stem. The genus name is often spelled *Raphidophora* (references to this issue are found in Flora of Java **3**: 106 (1968) by Backer & Bakhuizen van den Brink).
Literature: Engler, A. & Krause, K., Monsteroideae & Calloideae, *Das Pflanzenreich* **37**: 17–53 (1908).

1a. Leaf-blade of adult phase irregularly pinnatifid or pinnatisect, the lobes obliquely truncate at the apex **1. celatocaulis**
b. Leaf-blade of adult phase regularly pinnatisect, the lobes gradually tapered to an acute point **2. decursiva**

1. R. celatocaulis (N.E. Brown) Knoll (*Pothos celatocaulis* N.E. Brown; *Monstera latevaginata* misapplied; *Marcgravia paradoxa* misapplied). Illustration: Birdsey, The cultivated aroids, 103 (1951); Graf, Exotica, edn 8, 226, 227 (1976).
Climber. Leaves of juvenile phase shortly stalked, arranged in 2 rows pressed to the substrate, with entire, elliptic-ovate blades 8.5–10 × 5.5–6.5 cm, overlapping their neighbours and concealing the stem. Leaves of adult phase with the stalks 10–15 cm, the blades 17–40 × 11–31 cm, entire, pinnatifid or pinnatisect, and sometimes perforated, with tips of lobes obliquely truncate. Inflorescence-stalk about as long as the leaf-stalk. Spathe 15 × 6 cm, very fleshy, yellow. Spadix 12 × 2.5 cm, pale yellow. *Borneo*. G2.

Usually seen in the juvenile phase.

2. R. decursiva (Wallich) Schott. Illustration: Botanical Magazine, 7282 (1893); Birdsey, The cultivated aroids, 105 (1951); Graf, Tropica, 110 (1978).
Tall climber. Leaf-stalk 40–50 cm. Leaf-blade to 90 × 70 cm, regularly pinnatisect; segments 7–21 on each side, lanceolate with constricted base and tapered apex. Inflorescence-stalk much shorter than the leaf-stalk. Spathe *c.* 17.5 × 10 cm, fleshy, yellowish. Spadix *c.* 15 × 2–3 cm, yellowish, becoming grey-green. *India, Sri Lanka, N Burma*. G2.

Not usually seen in the juvenile state.

4. EPIPREMNUM Schott
M.T. Madison
Like *Rhaphidophora*, differing in the 1-celled ovary with usually 2 or 4 ovules, and in the hard-coated, kidney-shaped seeds.

A genus of about 12 species native to tropical SE Asia and the Pacific islands. The plants seldom flower in cultivation outside the tropics, as they require 5–10 m of host tree on which to climb before mature foliage is produced. In horticulture they are grown principally in hanging baskets or on poles for their attractive foliage which is tolerant of low light and low humidity. A moist, well-drained soil rich in humus is appropriate. The plants are propagated by cuttings. Plants growing out of doors in warmer regions will be damaged, but not killed, by a light frost.
Literature: Engler, A. & Krause, K., Monsteroideae & Calloideae, *Das Pflanzenreich* **37**: 54–67 (1908).

1a. Leaf-blade regularly pinnatisect and with tiny holes along the mid-rib, not variegated **1. pinnatum**
b. Leaf-blade entire or irregularly pinnatisect, variegated **2. aureum**

1. E. pinnatum (Linnaeus) Engler. Illustration: Das Pflanzenreich **37**: 61 (1908).
Climbing perennial, to 20 m tall, the stems green or brown, 1–4 cm thick. Leaf-blade dull green, pinnatisect with 8–20 divisions and with a row of pinprick-sized holes along each side of midrib. *Malaysia, Indonesia & New Guinea*. G1.

2. E. aureum (Linden & André) Bunting (*Pothos aureus* Linden & André; *Scindapsus aureus* (Linden & André) Engler; *Rhaphidophora aurea* (Linden & André) Birdsey). Illustration: Illustration Horticole **27**: 69 (1880); Birdsey, The cultivated aroids, 113 (1951); Baileya **10**: 156, 158 (1962); Graf, Exotica, edn 8, 228 (1976).
Climbing or trailing herb with broadly ovate, shiny leaves, more or less cordate at the base, variegated with golden yellow and green. In cultivation usually only the juvenile form with leaf-blades to 15 cm is found. Adult leaf-blade to 80 cm and irregularly pinnatisect. *Solomon Islands*. H5.

'Marble Queen' has cream and green variegation (Illustration: Graf, Exotica, edn 8, 138A, 1976).

5. SCINDAPSUS Schott
M.T. Madison
Plant evergreen, with tough stems climbing by means of adhesive, adventitious roots. Leaves entire, with parallel lateral veins. Spathe boat-shaped, deciduous. Spadix shorter than the spathe. Florets without perianths, each with 4 stamens and a 1-celled ovary with a solitary basal ovule. Seed more or less spherical, with a hard coat.

A genus of about 24 species from tropical Asia and the Pacific; only 1 species, *S. pictus*, is frequent in cultivation. It is slow growing and best grown in a hanging basket or clambering over rocks near a waterfall or pool. In pot-culture it requires a moist, well-drained soil rich in humus, and low to moderate light. Propagation is by cuttings.
Literature: Engler, A. & Krause, K., Monsteroideae & Calloideae, *Das Pflanzenreich* **37**: 67–80 (1908).

1. S. pictus Hasskarl var. **argyraeus** Engler. Illustration: Gartenflora **82**: 245 (1933); Graf, Exotica, edn 8, 228 (1976).
Slender, creeping climber. Stem 2–6 mm thick. Leaves 7–10 × 5–8 cm, ovate, cordate, silvery deep green marked with bright silver spots. Evidently a fixed juvenile form, of which flowers are unknown. *Java to Borneo*. G2.

6. STENOSPERMATION Schott
P.F. Yeo
Plants with horizontal rhizomes and erect, leafy stems. Leaf-stalk with a long sheath and a pulvinus at the apex (figure 3(6), p. 76); leaf-blade simple, with uniform, fine lateral veins. Inflorescence-stalk recurved at the top at flowering time.

Spathe boat-shaped, deciduous. Spadix shorter than spathe, uniform. Florets bisexual, without perianths; stamens 4; ovary with an overhanging rim at the top, 2-celled, each cell with 4 or more ovules.

A genus of about 20 species from central and tropical S America. They are plants of slender habit with small but conspicuous drooping white spathes, and a few are occasionally grown in hothouse collections. Literature: Engler, A. & Krause, K., Monsteroideae & Calloideae, *Das Pflanzenriech* **37**: 81–90 (1908).

1. **S. popayanense** Schott (*S. wallisii* Masters). Illustration: Botanical Magazine, 6334 (1877); Graf, Exotica, edn 8, 229 (1976).
Stem to 1 m. Leaf-stalk 15–17.5 cm. Leaf-blade 22–30 × 5–7 cm, mostly oblong-elliptic or oblong-lanceolate, rather dark green above with a narrow translucent margin; lateral veins uniform, very numerous. Spathe 8–12 cm long and nearly as wide, deeply hollowed, white, facing outwards from the shoot on its drooping stalk. Spadix 2.5–6 cm × 5–10 mm, yellowish white. *Colombia, Ecuador*. G2.

7. RHODOSPATHA Poeppig

P.F. Yeo

Evergreen climbing plants. Stem with leaves attached in 2 rows. Leaf-stalk with a pulvinus at the apex (figure 3(6), p. 76). Leaf-blade simple, with fine veins parallel to the main laterals and joined by small cross-veins. Inflorescence-stalk shorter than the leaf-stalk. Spathe boat-shaped, whitish or pink inside, soon shed. Spadix with a stipe, uniform. Florets without perianths, bisexual, or a few at the base female. Stamens 4, becoming twice as long as ovary. Ovary 2-celled, quadrangular in section, flat-topped. Ovules and seeds numerous.

About 12 species in the American tropics. They are suitable only for large hothouses, and should be treated like *Philodendron* (p. 91). Several have been in cultivation from time to time.
Literature: Engler, A. & Krause, K., Monsteroideae & Calloideae, *Das Pflanzenriech* **37**: 90–6 (1908).

1. **R. picta** Nicholson. Illustration: Graf, Exotica, edn 8, 226 (1976).
Leaf-stalk to 30 cm; sheaths reaching nearly to pulvinus, each to 1.2 cm wide at base; pulvinus *c.* 4 cm. Leaf-blade to 45 × 25 cm, elliptic or elliptic-ovate; upper surface deep green, with golden green mottling here and there between main lateral veins. Spathe 20–22 × 6.5–9 cm, pinkish. Spadix reaching tip of spathe, 1.6–1.8 cm thick, with a stipe *c.* 1.5 cm. *S America (origin not precisely known)*. G2.

8. MONSTERA Adanson

M.T. Madison

Climbing or sprawling, evergreen plants with tough fibrous stems and large perforated leaves. Leaf-stalk with pulvinus at apex (figure 3(6), p. 76). Spathe boat-shaped, deciduous. Spadix shorter than the spathe. Florets without perianths, each with 4 stamens and a 2-celled ovary with 2 ovules in each cell. Seeds soft.

A genus of about 25 species native to the New World tropics. Monsteras are much prized for their striking, perforated leaves and sweet, edible fruits. They are grown in pots, usually climbing on a pole, or rambling over stones near a pool or waterfall in a glasshouse. A moist, well-drained soil rich in humus is preferable. Though tolerant of low light and low atmospheric humidity, Monsteras will grow better in moderate light and with the moist air of a glasshouse. Propagation is by cuttings or seed.
Literature: Madison, M.T., A revision of Monstera (Araceae), *Contributions from the Gray Herbarium* **207**: 1–100 (1977).

1a. Leaf-blade thick, both perforated and pinnatifid **1. deliciosa**
b. Leaf-blade thin, perforated but not pinnatifid **2. adansonii**

1. **M. deliciosa** Liebmann (*Philodendron pertusum* Kunth & Bouché).
Illustration: Revue de l'Horticulture Belge **30**: 125 (1904); Graf, Exotica, edn 8, 181 (1976).
Climber to 20 m. Leaf-blade 25–90 × 25–75 cm, leathery, pinnatifid or pinnatisect, frequently perforated as well. Spadix 10–18 × 2.5–3 cm, swelling in fruit and becoming cream-coloured, the lower parts of the berries edible after the irritant upper portion has fallen away. *Mexico to Panama*. G1.

2. **M. adansonii** Schott var. **laniata** (Schott) Madison (*M. pertusa* (Linnaeus) de Vriese). Illustration: Graf, Exotica, edn 8, 177 (1976) – as M. friedrichsthalii.
Climber to 8 m. Leaf-blade 22–55 × 15–40 cm, ovate, thin, with numerous perforations. Spadix 8–13 × 1.5–2 cm, pale yellow. *Costa Rica to Brazil*. G1.

9. SPATHIPHYLLUM Schott

P.F. Yeo

Plants with rhizomes in or above the soil; with numerous leaves clustered near tips of rhizomes or on short, erect stems. Leaf-stalk nearly as long as blade or longer, with a long sheath embracing the stem at the base and with a pulvinus at the apex (figure 3(6), p. 76). Leaf-blade lanceolate or oblanceolate to oblong-ovate, dark green, often with the main lateral veins numerous and impressed. Inflorescence-stalk longer than the leaf-stalk. Spathe flat or slightly concave, white or greenish, its proportions similar to those of the leaf-blade. Spadix shorter than spathe, with a stipe which is often largely united with the spathe. Florets bisexual with 4–6 free or united, green or white perianth segments, 4–6 stamens with white or yellowish anthers, and a more or less conically tipped, white, incompletely 3-celled ovary; ovules 1–5 per cell.

One species in the Philippines and Indonesia and about 35 in tropical America. About 10 are grown as houseplants or in conservatories for their handsome foliage and inflorescences, which are often fragrant. They may also be grown for cut-flowers. They require a moist soil, rich in organic matter, and shade. Flowering is infrequent in the relatively dry atmosphere of occupied buildings. Dimensions of the plant are much affected by age and environment; those given below apply to well-grown plants. Many hybrids have been and continue to be raised in cultivation.
Literature: Bunting, G.S., The cultivated species of Spathiphyllum, *Baileya* **9**: 109–17 (1960); Bunting, G.S., A revision of Spathiphyllum (Araceae), *Memoirs of the New York Botanical Garden* **10**(3): 1–54 (1960).

Leaf-stalk. With mottled sheath: **4**.
Leaf-blade. Less than 25 cm: **3**. Broadest above the middle: **1**. With 9 or fewer pairs of main veins: **3**.
Stipe. Long and largely joined to spathe: **4–6**; short and free from spathe: **1,3**.
Spathe. Spreading and recurved, not concave: **1–3**; erect or nearly so, concave: **4–6**. Mainly white on both sides: **2,3,5**; white with green back: **1**; yellow-green: **4,6**.

1a. Tops of ovaries scarcely conical so that spadix is nearly smooth; spathe flat, spreading 2
b. Tops of ovaries conical, making the spadix rough; spathe concave, more or less erect 4

2a. Leaf-blade 13–20 cm; perianth segments green, so that spadix is patterned with green and white **3. floribundum***
b. Leaf-blade 25–50 cm; perianth segments white; spathe white or cream-coloured 3
3a. Leaf-blade about 3 times as long as broad; angle of divergence of lateral veins from midrib not more than *c.* 45° **1. cannifolium**
b. Leaf-blade usually about twice as long as broad; angle of divergence of lateral veins from midrib more than 45° **2. commutatum**
4a. Spathe white, becoming green in age, rounded at base **5. wallisii**
b. Spathe yellow-green, often acute at base 5
5a. Spathe 18–33 cm, leathery; ovules 6–16 **4. cochlearispathum**
b. Spathe 10–26 cm, papery; ovules 3–6 **6. phryniifolium**

1. S. cannifolium (Dryander) Schott. Illustration: Botanical Magazine, 603 (1802); Graf, Exotica, edn 8, 232 (1976); Everett, Encyclopedia of horticulture **9**: 3195 (1982).
Leaf-blade 25–45 × 8–16 cm, narrowly to broadly oblanceolate or narrowly elliptic, with tapered base; veins numerous, diverging from midrib at 45° or less and curving towards apex of blade, moderately impressed. Spathe 10–22 × 2.5–6.5 cm, oblong or elliptic, thick, spreading and recurved, its base clasping the stalk, upper surface white, lower green. Flowers fragrant. Spadix 4.5–12 cm × 5–20 mm, nearly smooth, cream-coloured; stipe 5–20 mm, free from spathe. Perianth segments united, white. Ovules usually 6–18. *Northern S America east of the Andes, Trinidad.* G2.

2. S. commutatum Schott. Illustration: Gartenflora **19**: 1 (1870); Baileya **9**: 115 (1961); Graf, Exotica, edn 8, 232 (1976).
Very similar to *S. cannifolium*, but with a relatively broader, elliptic, leaf-blade, 35–50 × 11–24 cm, its veins mostly diverging from midrib at more than 45°, spathe relatively broader, 14–24 × 5.5–8 cm, sometimes with the edges at the base continued down the stalk for up to 2 cm, creamy white with green edge; spadix slightly narrower. *Indonesia (N Celebes and some neighbouring islands), S & E Philippines.* G2.

3. S. floribundum (Linden & André) N.E. Brown. Illustration: Illustration Horticole **21**: 24 (1874); Birdsey, The cultivated aroids, 116 (1951); Baileya **9**: 116 (1961); Everett, Encyclopedia of horticulture **9**: 3194 (1982).
Leaf-blade 13–20 × 5.5–9 cm, elliptic to oblong or oblanceolate, with about 9 pairs of scarcely impressed veins diverging widely from the midrib; upper surface with a velvety lustre. Spathe 4–8 × 1.2–3 cm, lanceolate to oblong-elliptic, white, spreading and recurved, its base clasping the stalk. Spadix 2.5–5 cm × 7–8 mm, nearly smooth, white and green; stipe 3–8 mm. Perianth segments free or adherent in age, green. Ovules 4–6. *Colombia.* H2.

Valued for conservatory use but less suitable as a house-plant.

***S. patinii** (Hogg) N.E. Brown (*S. candidum* (Bull) N.E. Brown). Illustration: Illustration Horticole **27**: 135 (1880); Graf, Exotica, edn 8, 231 (1976). Similar to *S. floribundum* but with much narrower leaf-blades (12–21 × 2.8–5 cm), with 4–6 pairs of main veins diverging from midrib at less than 45°, glossy, without velvety lustre on upper surface, with narrower spathe and smaller spadix. Back of spathe creamy white with green midrib. *Colombia.* H2.

4. S. cochlearispathum (Liebmann) Engler. Illustration: Baileya **9**: 110 (1961); Graf, Exotica, edn 8, 231 (1976).
Leaf-blade 45–65 × 15–23 cm, oblong, oblong-lanceolate or narrowly elliptic, with very numerous, widely diverging, impressed lateral veins; margin wavy; sheath of leaf-stalk and scale-leaves mottled with white and usually crisped. Spathe 18–33 × 6–10 cm, oblanceolate or elliptic, erect, concave, leathery, yellow-green. Spadix 5.5–11.5 cm, very rough from the long ovaries, white; stipe 4.5–8.5 cm, entirely united with spathe or the top 1 cm free. Perianth segments free. Ovules 6–16. *Mexico, near the Caribbean coast.* H2.

The tallest species, easily reaching 1.2 m.

5. S. wallisii Regel. Illustration: Gartenflora **26**: 323 (1877).
Leaf-blade 24–36 × 5–10 cm, lanceolate-elliptic to oblong-elliptic, with *c.* 8–10 pairs of main veins arising from midrib at an angle of 45–50°; margin wavy. Spathe 7–17 × 2.5–7.5 cm, ovate to oblong-elliptic, concave, white, becoming green in age. Flowers fragrant. Spadix 1.5–8 cm, very rough from the long ovaries, white, with midrib green on the back; stipe 2–5 cm, united with spathe except in upper 5–10 mm. Perianth segments free, white. Ovules 6–12. *Columbia?* H2.

Rare in cultivation. The variant most frequently grown (often as *S. wallisii*) is 'Clevelandii' (*S. clevelandii* invalid; *S. kochii* misapplied; *S. candidum* misapplied). Illustration: Birdsey, The cultivated aroids, 118 (1951); Graf, Exotica, edn 8, 233 (1976); Everett, Encyclopedia of horticulture **9**: 3195 (1982). Leaf-blade 36–42 × 9–12 cm, relatively narrower than in *S. wallisii*, thin and drooping, glossy, with *c.* 12–15 pairs of impressed veins. Spathe *c.* 18 × 8.5 cm, erect or slightly spreading. Spadix 7–8.5 × 1.5–2 cm. 'Mauna Loa' (Illustration: Graf, Exotica, edn 8, 233, 1976; Huxley & Gilbert, Success with house-plants, 368, 1979) is probably descended in part from 'Clevelandii', from which it differs in its broader, less drooping leaf-blades and the abundance of erect, broadly elliptic spathes with the spadix centrally placed.

6. S. phryniifolium Schott (*S. friedrichsthalii* misapplied). Illustration: Baileya **9**: 112 (1961); Graf, Exotica, edn 8, 138A, 231, 233 (1976).
Leaf-blade 25–50 × 8–16 cm, lanceolate or sometimes oblong-elliptic, rounded at base, with numerous pairs of impressed veins arising from midrib at an angle of *c.* 60°. Spathe 10–26 × 3–9 cm, lanceolate to oblong-elliptic, papery, pale yellow-green, the edges at the base continued a little way down the stalk. Spadix 2.5–8 cm, rough from the long ovaries, white; stipe 2.5–6 cm, united with spathe except in upper 5–15 mm. Perianth segments free, white. Ovules 3–6. *Panama, Costa Rica.* G2.

10. LYSICHITON Schott
D.A. Webb
Robust, rhizomatous, stemless herbs, forming large clumps. Leaves ovate-oblong, truncate at the base, shortly stalked, without a sheath, appearing about the same time as the flowers. Spathe arising from ground level, withering before the fruit is ripe, the lower part consisting of a narrow sheath enclosing the very long stipe of the spadix, but with free margins, the upper part erect, elliptic, strongly concave. Spadix cylindric, on a long stipe, without appendix. Flowers crowded, bisexual; perianth-segments 4, often unequal; stamens 4; ovary 2-celled. Berries partly embedded in the spadix; seeds 2.

A genus of 2 closely related species from the North Pacific area. They may be cultivated on the banks of streams or the margins of ponds.

1. L. americanus Hultén & St John (*L. camtschatcensis* misapplied). Illustration: Botanical Magazine, 7937 (1904); Hay & Beckett, Reader's Digest encyclopaedia of garden plants and flowers, edn 2, 424 (1978).
Rhizome short, thick, vertical. Leaf-blade 40–100 × 25–70 cm, rather fleshy; leaf-stalk very short, thick. Stipe lengthening during flowering, green above. Lower (sheathing) part of spathe white, wrinkled; upper part 15–35 cm, bright yellow. Spadix 5–21 cm, stout, greenish or yellowish. Berries green. *Western N America, from Alaska to C California*. H2. Spring.

**L. camtschatcensis* (Linnaeus) Schott. Illustration: Makino, Illustrated flora of Japan, 734 (1924); Kitamura et al., Coloured illustrations of herbaceous plants of Japan (Monocotyledoneae), t. 49, f. 317 (1964); Hay & Beckett, Reader's Digest encyclopaedia of garden plants and flowers, edn 2, 424 (1978). Very similar to *L. americanus*, but slightly smaller, with the spathe white instead of yellow and flowering rather later. *NE Asia, from S Kamchatka to C Japan*. H. Late spring.

11. SYMPLOCARPUS Salisbury
D.A. Webb
Fetid, stemless herbs with stout, vertical rhizome. Leaves numerous, forming large clumps; stalk channelled, without a sheath; blade ovate, cordate, acute, entire. Spathes borne at ground level, very strongly concave, somewhat suggestive of a shell. Spadix with a short stipe enclosed within the spathe, shortly cylindric or almost spherical. Florets bisexual; perianth-segments 4; stamens 4; ovary 1-celled. Fruits very deeply embedded in the spadix so as to form a compound fruit, succulent on the surface, dry and spongy within.

A genus of probably 2 species, from temperature E Asia and eastern N America. The species described below should be grown in damp, humus-rich soil in shade.
Literature: Krause, K., Monsteroideae & Calloideae, *Das Pflanzenreich* **37**: 150–2 (1908).

1. S. foetidus (Linnaeus) Nuttall. Illustration: Botanical Magazine, 3224 (1833); Parey's Blumengärtnerei **1**: 176 (1958); Everett, Encyclopedia of horticulture **10**: 3274, 3275 (1982).
Leaves to 80 × 40 cm, appearing after the flowers. Spathe yellowish green, dark brownish purple or spotted and striped with both colours. Spadix 2–3 cm across in flower; greatly enlarged in fruit. *Cool-temperate regions of E Asia; eastern N America, from Nova Scotia to Florida*. H2. Later winter or early spring.

12. ORONTIUM Linnaeus
D.A. Webb
Plant aquatic, rhizomatous. Leaves all basal, with long stalks, clustered, mixed with scale-leaves; blade ovate, entire, submerged, floating or emerging. Inflorescence-stalk a stipe, slender, about as long as leaves, white above, terminating in a slender yellow spadix. Spathe much reduced, consisting of a sheath surrounding the lower part of the very long stipe and bearing a small, bract-like green blade which withers early. Florets numerous, crowded, bisexual, with 4–6 perianth-segments, 4–6 stamens and a 1-celled ovary. Fruit green, deeply embedded in the spadix, containing a single seed.

A single species, cultivated in shallow water overlying a fair depth of loamy soil.
Literature: Krause, K., Monsteroideae & Calloideae, *Das Pflanzenreich* **37**: 152–3 (1908).

1. O. aquaticum Linnaeus. Illustration: Gleason, Illustrated flora of the northeastern United States and adjacent Canada **1**: 369 (1952); Mühlberg, Das grosse Buch der Wasserpflanzen, 368 (1980); Everett, Encyclopedia of horticulture **7**: 2444 (1981).
E United States. H3. Late spring.

13. CALLA Linnaeus
D.A. Webb
Rhizomatous aquatic plants. Leaves all basal, mixed with scale-leaves, long-stalked, broadly ovate to kidney-shaped, shortly pointed, cordate; sheathing bases of the leaf-stalks persisting on the rhizome. Spathe similar to the leaf-blades in shape but smaller, erect, concave but scarcely enfolding the spadix even at its base, greenish outside, white on inner surface during flowering period, persistent in fruit. Two or 3 spathes per spadix are sometimes developed. Spadix green, shortly cylindric, with a short stipe and without appendix. Florets crowded, without perianths, mostly bisexual, but the uppermost usually male; stamens 6 or more; ovary 1-celled. Berry red, with 4–10 seeds.

A single species, easily cultivated in marshy ground or shallow water around the edge of a pond.
Literature: Krause, K., Monsteroideae & Calloideae, *Das Pflanzenreich* **37**: 154–5 (1908).

1. C. palustris Linnaeus. Illustration: Botanical Magazine, 1831 (1816); Hegi, Illustrierte Flora von Mitteleuropa, edn 2, **2**: 165, 172 (1939) and edn 3, **2**(1): 321 (1967–80); Polunin, Flowers of Europe, t. 183 (1969).
Plants 15–30 cm high. Leaves 5–10 cm wide; inflorescence-stalk rising up to 30 cm above water; spathe 3–6 cm wide; spadix 2–3 cm. *Cool temperate and subarctic regions of North America and Eurasia*. H1. Summer.

14. AMORPHOPHALLUS Blume
P.F. Yeo
Plants each with an underground corm, usually hollowed on top, from which arise several scale-leaves, one large deciduous leaf with erect stalk, and one large inflorescence. Leaf-blade with 3 divisions (or by dichotomy 5 or more) which are once or twice pinnately lobed or cut. Spathe funnel-shaped or bell-shaped, often shorter than spadix. Spadix with a sterile appendix. Florets without perianths, unisexual; zones of male and female florets adjacent. Male florets with 1–6 stamens. Ovary with 1–4 cells, each with 1 ovule.

About 90 species in the Old World tropics. Grown mainly for the striking but evil-smelling inflorescence which emerges from the dry corm in spring, without need of soil or water. After flowering the corms are planted in a humus-rich soil and given supplementary fertiliser. The plants must be dried off in autumn and the corms stored at a temperature above 10°C. The species treated here may be placed out of doors in summer and can be increased from cormlets.
Literature: Engler, A., Lasioideae, *Das Pflanzenreich* **48**: 61–109 (1911).

1a. Leaves not bearing cormlets; leaf-divisions beyond the dichotomies profusely bipinnatisect; spadix appendix narrowly conical **1. rivieri**
b. Leaves bearing cormlets; leaf-divisions beyond the dichotomies sparsely pinnate or bipinnatisect; spadix appendix ovoid **2. bulbifer**

1. A. rivieri Durieu. Illustration: Flore des Serres **19**: 43, 45 (1873); Birdsey, The cultivated aroids, 29 (1951); Graf, Exotica, edn 8, 143 (1976).
Corm to 25 cm wide. Leaf-stalk to 100 cm, mottled. Leaf-blade to 1.3 m wide, its divisions once or twice dichotomous and then bipinnatisect. Inflorescence-stalk 50–70 cm; spathe 20–40 cm, dark brownish red inside, edges wavy; spadix nearly twice as long as spathe; appendix

20–55 cm, narrowly conical, dark reddish brown. *SE Asia*. G1. Spring.

The most commonly cultivated variant is 'Konjac', with the largest inflorescences and edible tubers, for which it is widely cultivated in the West Pacific region.

2. A. bulbifer (Curtis) Blume. Illustration: Botanical Magazine, 2072 (1819), 2508 (1824); Parey's Blumengärtnerei **1**: 177 (1958).
Corm to 7 cm wide. Leaves bearing cormlets. Leaf-stalk to 100 cm, mottled; leaf-blade to 60 cm wide; main divisions cut to the base into a few segments, rarely again divided. Inflorescence-stalk to 20 cm; spathe 15–20 cm, whitish with pale green or pink mottling outside and whitish, tinged with pink, inside; spadix about as long as spathe; appendix to 7 cm, ovoid, pinkish. *NE India, Burma; ?SW China*. G1. Spring.

15. NEPHTHYTIS Schott
P.F. Yeo
Plants evergreen. Stem a horizontal rhizome bearing leaves at short intervals. Leaf-stalk terete; sheath very small. Leaf-blade sagittate or hastate, the basal lobes about as long as the principal lobe; primary lateral veins joining the marginal vein (figure 3(6), p. 76), with netted veins between them. Inflorescence-stalk with several scale-leaves at base. Spathe open; hooded, flat, or reflexed. Florets without perianths, unisexual. Spadix with zones of male and female florets adjacent, the female zone much the shorter. Male florets of 1–4 stamens. Ovary containing 1 or 2 ovules.

Six species of West tropical Africa, sometimes cultivated in tropical collections. They require a moist soil high in organic matter, a hot humid atmosphere and some shade. The genus name is sometimes applied informally to the cultivated species of *Syngonium* (p. 103).
Literature: Engler, A., Lasioideae, *Das Pflanzenreich* **48**: 110–14 (1911); Bogner, J., On two Nephthytis species from Gabon and Ghana, *Aroideana* **3**: 75–85 (1980).

1. N. afzelii Schott. Illustration: Das Pflanzenreich **48**: 111 (1911); Graf, Exotica, edn 8, 224 (1976).
Leaf-stalk to 50 cm. Leaf-blade to 35 × 25 cm, sagittate; principal lobe triangular, slightly shorter than the ovate laterals; all lobes sharply pointed. Inflorescence-stalk slightly shorter than the leaf-stalk. Spathe 6–7 cm, hooded, green, unspotted. Spadix slightly shorter than spathe. Male florets with 4 stamens. Stigma disc-like. Berries few, to 9 mm, ovoid, orange. *West tropical Africa*. G2.

16. RHEKTOPHYLLUM N.E. Brown
P.F. Yeo
Evergreen, climbing plants with distinct juvenile and adult phases. Juvenile phase with thin stems and simple leaf-blades. Leaf-blades in juvenile phase hastate or sagittate, in adult phase ovate or oblong-ovate and with perforations and marginal incisions between wedge-shaped segments; main veins of basal lobes exposed in the sinus (figure 3(8), p. 76); minor veins netted. Inflorescence-stalk shorter than leaf-stalk. Spathe 9–10 cm, green outside and reddish purple inside. Florets without perianths, unisexual. Spadix with male zone 3 times as long as female. Male florets with 3 or 4 stamens. Ovary with 1 cell and 1 ovule.

A genus of 1 species from West Africa. Cultivated in the juvenile phase for its beautifully patterned leaves; it requires the same conditions as *Philodendron* (p. 91).
Literature: Engler, A., Lasioideae, *Das Pflanzenreich* **48**: 119–21 (1911).

1. R. mirabile N.E. Brown. Illustration: Journal of Botany **20**: 136, t. 230 (1882); Das Pflanzenreich **48**: 120 (1911); Birdsey, The cultivated aroids, 107 (1951); Graf, Exotica, edn 8, 224 (1976).
Leaf-blade of juvenile phase to 30 cm, hastate or sagittate with coarsely wavy margins; basal lobes broad with main veins exposed at the sinus for part of their length; principal lobe with 2 or 3 pairs of main lateral veins curving towards its apex; upper surface dark green, variegated with whitish between the veins in a pattern suggestive of fern fronds growing inwards from the margin. *Nigeria to Congo*. G2.

17. HOMALOMENA Schott
P.F. Yeo & D.J. Leedy
Plants evergreen, with a short stem above or below ground. Leaf-blade variable in shape; primary and subsidiary lateral veins parallel and joining a marginal vein (figure 3(6), p. 76), only the finest ones forming a network. Inflorescences several from one axil, sometimes accompanied by scale-leaves. Spathe margins overlapping, unrolling slightly above at flowering time. Spadix shorter or longer than spathe. Florets without perianths, unisexual. Zone of female florets shorter than the adjacent zone of male florets. Male florets usually with 2–4 stamens; lowest florets sometimes sterile. Female florets with or without 1 or more staminodes. Ovary incompletely divided into 2–4 cells, each with several to many ovules. Fruit a berry.

About 140 species, mainly in S Asia and the SW Pacific region but a few in tropical S America. They are used and cultivated like *Dieffenbachia* (p. 96), but *H. wallisii* does not respond well to pot-culture.
Literature: Engler, A., Homalomeninae & Schismatoglottidinae, *Das Pflanzenriech* **55**: 25–81 (1912); Bogner, J., Homalomena lindenii (Araceae), *Baileya* **20**: 6–10 (1976).

1a. Leaf-stalk shorter than blade; leaf-blade elliptic-oblong, with blotchy variegation **1. wallisii**
b. Leaf-stalk longer than blade; leaf-blade triangular-ovate, with the main veins white to yellow **2. lindenii**

1. H. wallisii Regel. Illustration: Gardeners' Chronicle **7**: 108 (1877); Birdsey, The cultivated aroids, 68 (1951); Graf, Exotica, edn 8, 176 (1976).
Leaves crowded on a short, erect stem. Leaf-blade 13–20 × 6–8 cm, much longer than the stalk, more or less elliptic-oblong, rounded or cordate at base; upper surface smooth, dark green with a narrow white margin and bold irregular yellow blotches; lower surface rough, glaucous, tinged with red. Inflorescence longer than leaf-stalk. Spathe *c.* 8 cm, pale reddish purple, glossy within, constricted at the middle. Spadix as long as spathe. Female florets without staminodes. *Colombia*. G2.

Differs from *Aglaonema pictum* (p. 95) in the yellow rather than light green or silvery variegation, the white margin of the leaf-blade and the elongated spathe.

2. H. lindenii (Rodigas) Lindley (*Alocasia lindenii* Rodigas). Illustration: Illustration Horticole **33**: 111 (1886); Graf, Exotica, edn 8, 120, 125 (1976); Baileya **20**: 6, 8 (1976).
Stem to 25 cm, erect. Leaf-stalk whitish. Leaf-blade 20–25 × 12–18 cm, shorter than the stalk, triangular-ovate with a deeply cordate base, bright green and slightly glossy above; midrib and main veins impressed, white to yellow except at extremities. Inflorescence much shorter than leaf-stalk. Spathe 5–7 cm, pale green, scarcely constricted at the middle. Spadix as long as spathe. Female florets without staminodes. *New Guinea*. G2.

When damaged the plant smells of aniseed or chervil. The supposed intergeneric hybrid × *Homalocasia miamiensis* is thought by Bogner to be a form of *H. lindenii*.

18. SCHISMATOGLOTTIS Zollinger & Moritzi
P.F. Yeo
Plants evergreen. Stem rhizomatous or above ground and erect. Leaf-blade with minor veins nearly parallel to primary lateral veins (figure 3(5), p. 76); 1 or more marginal veins present. Inflorescence-stalk shorter than the leaf-stalk. Spathe margins overlapping, the upper part open at flowering time, dropping off while both parts are still fresh. Spadix sometimes with a sterile zone between male and female zones, bearing sterile florets sparsely or densely; usually with a club-shaped sterile appendix patterned with rudimentary florets (figure 4(6), p. 77). Florets without perianths, unisexual. Male florets with 2 or 3 stamens, filaments flattened. Female florets (in our species) with staminodes. Ovary 1-celled, ovules many.

About 100 species, mainly in tropical Asia, but a few in tropical America. Several have been brought into cultivation and are grown mainly for their variegated leaves. They grow best in a loose, humus-rich soil. The rhizomatous species are propagated by division and the erect-stemmed species by cuttings.
Literature: Engler, A., Homalomeninae & Schismatoglottidinae, *Das Pflanzenreich* **55**: 1–134 (1912).

1a. Stem erect, above ground; spadix with a sterile zone between male and female zones **1. concinna**
b. Stem an underground rhizome; spadix without a sterile intermediate zone 2
2a. Leaf-blade green, usually variegated irregularly with whitish yellow blotches **2. neoguineensis**
b. Leaf-blade with a bluish white band on either side, about half-way between midrib and margin **3. picta**

1. S. concinna Schott. Illustration: Illustration Horticole **28**: 71 (1881) – as S. lavallei, **29**: 173 (1882); Das Pflanzenreich **55**: 96 (1912); Graf, Exotica, edn 8, 230 (1976).
Stem erect, to *c.* 20 cm. Stem and leaf-stalk often reddish or pink. Leaf-stalk to *c.* 15 cm, not much longer than blade, slightly channelled towards apex. Leaf-blade to *c.* 15 × 4 cm, ovate-lanceolate with rounded or slightly cordate base; upper surface green, often with irregular grey-green blotches; lower surface green or purple. Spathe 6.5 cm; tube *c.* 2 cm. Spadix with female zone to 1.5 cm, intermediate sterile zone *c.* 1 cm, with scattered sterile florets, male zone *c.*1 cm and sterile appendix to 3 cm; male zone and appendix yellowish. *Borneo, Java, Sumatra.* G2.

'Immaculata' has the leaf-blade uniformly green above and purple beneath; 'Purpurea' has the leaf-blade variegated above and purple beneath.

2. S. neoguineensis André. Illustration: Illustration Horticole **27**: 68 (1880); Birdsey, The cultivated aroids, 111 (1951); Graf, Exotica, edn 8, 230 (1976).
Stem an underground rhizome. Leaf-stalk flattened on the upper side towards apex, up to twice as long as blade, green or purplish. Leaf-blade 12–25 × 7–16 cm, ovate, cordate at base, thin; upper surface bright green with irregular blotches of whitish yellow, lower light green. Spathe *c.* 7 cm; tube 2.5–3 cm, green; limb white. Spadix with female zone *c.* 1.5 cm and male zone *c.* 1 cm, adjacent to female; appendix *c.* 1 cm, orange. *New Guinea.* G2.

3. S. picta Schott. Illustration: Graf, Exotica, edn 8, 229, 230 (1976).
Stem an underground rhizome. Leaf-stalk 20–30 cm, or sometimes to 60 cm, flattened on upper side towards apex, green. Leaf-blade 15–20 × 7.5–13, or sometimes to 35 × 26 cm, narrowly ovate or oblong-ovate, cordate at base, thin; upper surface bright green and usually marked with a ragged, pale, glaucous band running along each side about half-way between midrib and margin. Spathe 5–6 cm; tube green; limb greenish yellow. Spadix with female zone 2 cm and male zone 1.5 cm, adjacent to female, pale yellow; appendix 5–7.5 mm. *Borneo, Celebes, Sumatra & Java.* G2.

19. PHILODENDRON Schott
G.S. Bunting
Plants usually climbing, evergreen, with stout stems bearing adventitious roots. Leaf-stalk commonly stout and about as long as the blade, with a well-developed sheath which does not embrace the stem (figure 3(3 & 4), p. 76). Leaf-blade often large, entire or variously lobed to pinnatifid or pedate, with pinnate venation (figure 3(5), p. 76). Inflorescence-stalks 1 or more in leaf-axils. Spathe fleshy, rolled around the spadix, unfurling at flowering time in the upper half or more, sometimes nearly to the base, then hooded or boat-shaped, green or white or variously marked with red or purple, persistent until the fruits ripen. Spadix usually white or whitish. Florets without perianths, unisexual; zones of male and female florets touching; male florets of 2–5 or more stamens; lowest florets sterile; female florets without staminodes. Ovary with 2–many cells, each with 1–many ovules. Berries white to orange or red when ripe.

Perhaps 500 species of the American wet tropical forests, many of which are widely cultivated for their attractive foliage or spathes and durability in interior decorations and also for landscaping in warm climates. Several striking hybrids are in cultivation, as well as some white-variegated forms.

Small plants in containers often have only juvenile foliage which may be distinct from the adult leaf form. Positive identification can be made only with an adult leaf, whose form is noted in the descriptions below.

Philodendrons are of easy culture in a continually humid, rich loam with good drainage, but some tolerate dry soils. Many species thrive in the low atmospheric humidity of heated dwellings, etc. and most tolerate remarkably low light, but some, such as *P. bipinnatifidum (P. selloum)*, require high light intensity, thriving in full sun. Climbing species may be planted with a support to which their aerial roots can cling; if the support is moistened regularly and other conditions are adequate, the erect-growing stem will finally produce leaves of the adult form, while if left to grow pendent, the leaves (except in the case of *P. scandens*) will become very reduced and the plant of limited ornamental value. Propagation is by cuttings or layers. The so-called arborescent species such as *P. bipinnatifidum* are usually propagated by seeds and are often of mixed parentage.
Literature: Krause, K., Philodendrinae, *Das Pflanzenreich* **60**: 1–136 (1913); Bunting, G.S., Vegetative anatomy and taxonomy and the Philodendron scandens complex, *Gentes Herbarum* **10**: 136–68 (1968).

Habit. Prostrate: **5,10,12,13**; erect or decumbent: **2,5**; climbing: **1,3,4,6–9.**
Leaf-blade. Deeply lobed or pinnatisect: **1–4**; sinuate (shallowly lobed): **1**; entire with base acute to truncate or notched and erect on leaf-stalk: **5**; entire with base cordate or sagittate and reflexed on leaf-stalk: **6–13**. Green on upper surface: **1–7,9,10,13**; brown or black-green on upper surface: **8,9** (juvenile); grey-splotched: **11,12**; strongly variegated with white and grey-green: **10**.
Leaf-stalk. More or less terete: **1,3,4,7–9**; prominently flattened or channelled on

upper surface: **2,5,9,10**; narrowly winged near apex: **12**. Surface more or less rough or covered with fleshy scales: **3,6,8,11**.

Spathe. Tube green outside, limb green or white (or pink) on both surfaces: **1–6,8,9,11,13**; entire spathe green outside, limb red inside: **10**; tube purple outside, limb green or white on both surfaces: **1**; entire spathe brown-red, red, or purplish red outside, limb red or white inside: **2,7,12**. Covered with fleshy scales: **6**.

1a. Leaf-blade lobed or pinnatisect 2
b. Leaf-blade entire or margin only shallowly sinuate 5
2a. Each lateral segment of blade (discounting basal lobes) with a single principal vein arising from midrib; blade regularly pinnatifid or pinnatisect 3
b. Each lateral segment of blade (discounting basal lobes) with 2 or more principal veins arising from midrib; blade lobed or irregularly pinnatifid 4
3a. Leaf-blade once pinnatifid or pinnatisect; leaf-stalk more or less terete **1. angustisectum***
b. Leaf-blade twice pinnately cut or lobed; leaf-stalk flat on upper side with angular-ridged borders **2. bipinnatifidum**
4a. Leaf-blade with a prominent elliptic or obovate apical lobe, plus 2–5 or more deeply separated, oblong segments on each side of midrib, in addition to the pinnatifid basal lobes **3. pedatum***
b. Leaf-blade with a prominent elliptic or obovate apical lobe, the remainder shallowly 2-lobed on each side of midrib **4. bipennifolium**
5a. Leaf-blade narrowly obovate-oblong, erect on a much shorter, stout leaf-stalk whose upper surface is flat with angular-ridged borders **5. wendlandii***
b. Leaf-blade ovate or triangular in outline with prominent basal sinus, reflexed; leaf-stalk not as above, about as long as blade or longer 6
6a. Mature leaf-blade brown or black-green or suffused with red-purple on one or both surfaces 7
b. Mature leaf-blade more or less pure green on both surfaces, at least never brown or black-green or suffused with red-purple 10
7a. Leaf-stalks, scale-leaves, inflorescence-stalks and spathes covered with fleshy scales **6. verrucosum**
b. Leaf-stalks, scale-leaves, inflorescence-stalks and spathes never covered with fleshy scales 8
8a. New stems purple or red-purple **7. erubescens**
b. New stems green or brownish 9
9a. Stem marked with longitudinal white lines; leaf-blade narrowly ovate with sagittate base, usually black-green on upper surface; juvenile blade ovate and shallowly cordate with leaf-stalk winged in lower half **8. melanochrysum**
b. Stem not marked with white lines; leaf-blade ovate with cordate base, shiny and brownish to velvety dark green on upper surface; leaf-stalk never winged **9. scandens***
10a. Stem without persistent fibrous remains of old scale-leaves 11
b. Stem with persistent, fibrous remains of old scale-leaves 12
11a. Leaf-blade to *c.* 30 cm with only 3–4 principal veins arising from each side of midrib (discounting basal ribs) **9. scandens***
b. Leaf-blade to *c.* 60 cm with about 5–6 principal veins arising from each side of midrib (discounting basal ribs) **10. domesticum***
12a. Leaf-blade dark green with white midrib and principal veins **13. gloriosum**
b. Leaf-blade with veins not white 13
13a. Plant climbing; leaf-stalk rough **11. ornatum**
b. Plant prostrate; leaf-stalk crinkly-winged near apex **12. mamei**

1. P. angustisectum Engler (*P. elegans* Krause). Illustration: Das Pflanzenreich **60**: 116, 117 (1913); Birdsey, The cultivated aroids, 77 (1951); Graf, Exotica, edn 8, 212 (1976); Graf, Tropica, 113 (1978).
Climber of moderate size. Leaf-blade *c.* 60 × 45 cm, more or less reflexed, ovate in outline, pinnatisect with up to 16 linear, acute segments on each side; segments *c.* 2.5 cm wide, each with one vein; basal sinus triangular. Leaf-stalk terete, nearly as long as blade. Inflorescence-stalk short. Spathe *c.* 15 cm, green outside with pink margins, limb yellowish inside. *W Colombia.* G2.

***P. lacerum** (Jacquin) Schott. Illustration: Birdsey, The cultivated aroids, 85 (1951); Graf, Exotica, edn 8, 193, 221 (1976); Graf, Tropica, 115, 119 (1978). Plant tall, with long internodes. Leaf-blade to *c.* 75 cm, ovate-circular in outline, cordate at base, pinnately lobed less than halfway to midrib; lobes more or less wedge-shaped and obtuse; midrib and principal veins very prominent. Leaf-stalk to 90 cm, terete. Inflorescence-stalk to twice as long as spathe. Spathe *c.* 12 cm, with tube outside dull purple and limb greenish cream. *Cuba, Jamaica, Hispaniola.* G2.

***P. × corsinianum** Senoner. Illustration: Botanical Magazine, 8172 (1908); Birdsey, The cultivated aroids, 75 (1951); Parey's Blumengärtnerei **1**: 181 (1958); Graf, Tropica, 112 (1978). Plant tall with leaf-blade to 75 × 60 cm, ovate, cordate-sagittate at base, shallowly pinnately lobed; marked beneath when new with metallic brownish purple between green veins. Leaf-stalk about as long as blade. Inflorescence-stalk short, with white lines. Spathe *c.* 19 cm, with tube purple outside and limb green. *Garden origin.* G2.

The first recorded hybrid of *Philodendron*, possibly involving *P. verrucosum*. Recommended for cultivation only in a warm, humid environment.

2. P. bipinnatifidum Endlicher (*P. selloum* C. Koch). Illustration: Gartenflora **29**: 353, t. 1029 (1880); Graf, Exotica, edn 8, 183, 185 – as P. × barryi, 187 (1976); Graf, Tropica, 112, 120, 121 (1978).
Huge, tree-like plant, erect to *c.* 2 m, decumbent in age, with very stout aerial roots. Stem to 10 cm in diameter; internodes very short; leaf-scars prominent. Leaf-blade 1 m, more or less reflexed, glossy, ovate in outline, sagittate, regularly pinnatisect into many blunt-tipped segments, each often again pinnatifid or lobed, flat or wavy. Leaf-stalk about as long as blade, upper surface more or less flat between the 2 longitudinal ridges. Inflorescence-stalk *c.* 10 cm, solitary. Spathe *c.* 30 × 7 cm, green to dark purplish red on outside and red-edged, cream inside. *SE Brazil.* G2.

A handsome species whose adult phase is suitable only for a large area with bright light.

3. P. pedatum (W.J. Hooker) Kunth (*P. laciniatum* (Vellozo) Engler). Illustration: Das Pflanzenreich **60**: 112 (1913); Parey's Blumengärtnerei **1**: 182 (1958); Graf, Exotica, edn 8, 202, 212 (1976).
Tall climber of moderate size. Leaf-blade reflexed, glossy, ovate; the part along midrib *c.* 45 × 30 cm, irregularly pinnatifid into up to 5 oblong, obtuse segments on each side, each with 2 or more principal veins, in addition to the elliptic, obovate or rhombic terminal segment; the flaring basal lobes 27 × 17 cm. Leaf-stalk longer than

blade, more or less terete. Spathe 14 cm, green outside with red edge. *S Venezuela to Surinam & SE Brazil.* G2.

The juvenile phase has 5-lobed leaves. This variable species develops long internodes and small leaves in adverse indoor conditions. Based on leaf variations, several varieties and cultivars have been named (e.g. *P.* 'Florida' and *P.* 'Florida Compacta'), and there are some hybrids involving this species.

***P. squamiferum** Poeppig. Illustration: Illustration Horticole **33**: 45 (1886); Birdsey, The cultivated aroids, 97 (1951); Graf, Exotica, edn 8, 202, 212 (1976). Climber of moderate size. Leaf-blade to 60 × 45 cm, reflexed, ovate in outline, deeply lobed or pinnatisect into 5 parts: a prominent, elliptic to rhombic terminal lobe, a pair of oblong-triangular and more or less sickle-shaped lateral segments, and prominent, elliptic basal lobes. Leaf-stalk terete, red near apex, covered with red or green fleshy scales. Inflorescence-stalk red, covered with fleshy scales. Spathe 10–15 cm, with tube red outside and blade cream. *Surinam to NE Brazil.* G2.

4. P. bipennifolium Schott (*P. panduraeforme* misapplied). Illustration: Martius, Flora Brasiliensis 3(2): t. 34 (1878); Graf, Exotica, edn 8, 203, 221 (1976); Graf, Tropica, 118 (1978).
Tall climber of moderate size. Leaf-blade *c.* 45 × 15–25 cm, reflexed, glossy dark green, with a long, terminal, obovate lobe, the remainder much wider and formed by the sinuate lateral margins into a pair of blunt-angled lobes on each side of the midrib; basal sinus open. Leaf-stalk terete, somewhat shorter than blade. Inflorescence-stalk short. Spathe *c.* 11 cm, greenish cream outside. *SE Brazil.* G2.

An excellent species for indoor culture.

5. P. wendlandii Schott. Illustration: Das Pflanzenreich **60**: 27 (1913); Graf, Exotica, edn 8, 213 (1976); Graf, Tropica, 118 (1978).
Plant forming a rosette, erect to prostrate. Leaf-blade to 75 × 20 cm, erect on its stalk, narrowly obovate, approximately truncate or with small auricles at base; midrib 2.5 cm wide. Leaf-stalk *c.* 30 cm, stout, flat along upper side with angular borders. Spathe *c.* 18 cm, with tube light green suffused with purple outside, dark red inside, and limb cream. *S Nicaragua to Panama.* G2.

Several hybrids have been produced from this species.

***P. cannifolium** Kunth, not *P. cannifolium* (Rudge) Engler; (*P. martianum* Engler). Illustration: Martius, Flora Brasiliensis **3**(2): t. 31 (1878); Peyritsch, Aroideae Maximilianae, t. 38, 39 (1879); Graf, Exotica, edn 8, 219 – also as P. warmingii (1976); Graf, Tropica, 115, 117 (1978) – as P. warmingii. Stem prostrate; internodes short. Leaf-blade 45 × 15–20 cm, glossy, leathery, ovate, obtuse to truncate at base, erect on a stout, spongy leaf-stalk somewhat shorter than blade and wide-channelled above with angular margins. Inflorescence-stalks 11–17 cm. Spathe *c.* 17 cm and nearly cylindric, with tube dark red at base inside, otherwise green becoming cream above. *SE Brazil.* G2.

6. P. verrucosum Schott. Illustration: Illustration Horticole **18**: 192 (1871) & **20**: 12 (1873) – as P. daguense; Gartenflora **42**: 257 (1893); Birdsey, The cultivated aroids, 99 (1951); Graf, Tropica, 123 (1978).
Tall climber of moderate size. Leaf-blade to 60 × 40 cm, reflexed, ovate with sagittate-cordate base, very dark shimmering green on upper surface with light green zones along the furrowed midrib and principal veins, red-violet between veins beneath; margins wavy and shallowly sinuate. Leaf-stalk about as long as blade, more or less flattened on upper side and covered with red, green or white fleshy scales as are also the inflorescence-stalk, scale-leaf and spathe. Inflorescence-stalk *c.* 17 cm, marked with elevated, white, interrupted lines below. Spathe *c.* 21 cm, outside red-brown, becoming lime green above, inside pinkish red, becoming white above, or with white edge. *Costa Rica to Ecuador.* G2.

Thrives only under warm, humid conditions.

7. P. erubescens C. Koch & Augustin. Illustration: Botanical Magazine, 5071 (1858); Flore des Serres **14**: 39 (1861); Birdsey, The cultivated aroids, 79 (1951); Graf, Tropica, 114, 116 (1978).
Tall climber of moderate size. New stems purple or red-purple, scale-leaves deep pink or pinkish brown. Leaf-blade to 40 cm, reflexed, slightly leathery, ovate-triangular, with shortly sagittate-cordate base, more or less glossy dark green on upper surface, suffused with purple beneath, especially on principal veins. Leaf-stalk about as long as blade, flattened above towards the blade, suffused with red-purple. Inflorescence-stalk and spathe purplish red outside, somewhat lighter coloured inside. *Colombia.* G2.

'Red Emerald' (Illustration: Graf, Tropica, 118, 1978) is a selection from *P. erubescens*, which is more robust, with red-purple stems and leaf-stalks, and the principal veins red-purple on the undersurface of the leaf-blade.

8. P. melanochrysum Linden & André (*P. andreanum* Devansaye). Illustration: Illustration Horticole **20**: 198 (1873); Revue Horticole **58**: 36 (1886); Birdsey, The cultivated aroids, 73 (1951); Graf, Tropica, 115 (1978).
Tall climber of moderate size. Leaf-blade to 80 × 30 cm, reflexed-pendent, narrowly ovate with sagittate base; upper surface black-green with a velvety appearance and pale green ribs and principal veins, pinkish orange when new. Leaf-stalk to 50 cm, somewhat rough. Spathe *c.* 20 cm, with tube more or less green and limb white, slender-tipped. *Colombia.* G2.

A beautiful species when well grown, but demanding in cultural requirements. Recommended only for a warm, moist environment.

9. P. scandens C. Koch & Sello (*P. cordatum* misapplied). Illustration: Das Pflanzenreich **60**: 57 (1913) – in part as P. oxycardium; Birdsey, The cultivated aroids, 89 (1951) – as P. oxycardium; Graf, Exotica, edn 8, 203, 207 (1976) – as P. oxycardium; Graf, Tropica, 115, 121 (1978).
Tall climber of small size with terminal part of stem finally pendent and nearly reaching ground, where flowering occurs. Internodes elongate. Leaf-blade to 30 × 23 cm, reflexed, more or less glossy, heart-shaped with about 3 principal veins from each side of midrib. Leaf-stalk shorter than blade, slender, nearly terete, channelled on upper side. Inflorescence-stalk short. Spathe to 19 cm, green with more or less white limb. *Mexico & West Indies to Surinam, Peru & SE Brazil.* G2.

Perhaps the commonest philodendron in cultivation, thriving under adverse conditions in soil, water, or even when submerged in aquaria. Several variants occur, especially distinguishable in the juvenile growth phase. Subsp. **scandens** has juvenile leaves with a velvety or metallic appearance on upper surface; its forma **scandens** has leaves which are green above and green or red-purple beneath, while forma **micans** (C. Koch) Bunting (*P. micans* C. Koch) has leaves metallic brown (bronze) above and reddish to brown-red beneath, with more ample basal lobes that slightly overlap. Subsp. **oxycardium** (Schott) Bunting (*P. oxycardium* Schott) has its glossy juvenile leaves tinted brown when immature and green when mature.

***P. imbe** Endlicher. Illustration: Martius, Flora Brasiliensis **3**(2): t. 33 (1878); Peyritsch, Aroideae Maximilianae, t. 34 (1879); Graf, Tropica, 120 (1978).
Climber. Leaf-blade *c.* 33 × 18 cm, reflexed, glossy, narrowly ovate, with only about 3 nearly horizontally spreading principal veins from each side of midrib; base sagittate with narrow sinus *c.* 7 cm long. Leaf-stalk terete, about as long as blade. Spathe *c.* 9 cm, the tube green outside and red inside, the limb cream. *SE Brazil.* G2.

10. P. domesticum Bunting (*P. hastatum* misapplied). Illustration: Baileya **14**: 90 (1966); Graf, Exotica, edn 8, 203, 204 (1976).
Climber of moderate size. Leaf-blade to 60 × 30 cm, reflexed, glossy, long-triangular, with sagittate base and 5–6 principal veins from each side of midrib. Leaf-stalk as long as blade, flattened on upper surface near the blade and with a longitudinal central ridge. Inflorescence-stalk nearly as long as spathe. Spathe *c.* 18 cm, green outside, deep pinkish red inside with green border. *Known only in cultivation.* G2.

P. 'Tuxla' is similar in appearance, but it is not certain from which species or hybrid it was developed.

***P. cordatum** (Vellozo) Kunth. Illustration: Vellozo, Flora Fluminensis **9**: t. 111 (1835). Tall climber. Leaf-blade to 45 × 25 cm or larger, ovate-triangular with sagittate base and more or less oblong, rounded-angular basal lobes separated by a narrow sinus to 15 cm long, or slightly overlapping. Spathe *c.* 15 cm, greenish. *SE Brazil.* G2.

Material of *P. scandens* has been offered for sale under this name.

***P. ilsemannii** Sander (*P. sagittifolium* misapplied). Illustration: Möllers Deutsche Gärtnerei-Zeitung **23**: 257 (1908); Parey's Blumengärtnerei **1**: 180 (1958); Graf, Exotica, edn 8, 201 (1976); Graf, Tropica, 114 (1978). Climber with narrowly ovate leaf-blade that is shallowly cordate at base, often with large white zones and few to numerous smaller white and grey-green irregular spots. Leaf-stalk winged in the lower part. *Origin unknown.* G2.

The juvenile phase of an undetermined species, strongly suggesting the variegated form of *P. cordatum.*

***P. devansayeanum** Linden. Illustration: Illustration Horticole **42**: 376 (1895). Stem more or less prostrate with short internodes and erect, purplish red leaf-stalks to nearly twice as long as blade. Leaf-blade ovate, glossy; principal veins and midrib purplish red on under surface; new leaf orange-brown. *Peru (Andes).* G2.

11. P. ornatum Schott (*P. asperatum* C. Koch; *P. sodiroi* invalid). Illustration: Peyritsch, Aroideae Maximilianae, t. 40–2 (1879) – as P. imperiale; Birdsey, The cultivated aroids, 95 (1951); Graf, Exotica, edn 8, 208, 221, 222, 223 (1976); Graf, Tropica, 118, 123 (1978).
Climber of moderate to large size; upper part of stem hidden by persistent, fibrous remains of scale-leaves. Leaf-blade to 60 cm or more, reflexed, ovate with cordate base, shiny on upper surface; veins on undersurface early suffused with red. Leaf-stalk about as long as blade, flat above in the upper part, more or less rough, purplish when young. Inflorescence-stalk short, reddish. Spathe to 14 cm, ending in a thread-like structure to 2 cm long and having a green tube and white limb. *S Venezuela to Peru & SE Brazil.* G2.

Generally seen in the juvenile phase with leaf-blade blotched with silvery grey between the principal veins.

12. P. mamei André. Illustration: Illustration Horticole **43**: 293 (1896); Birdsey, The cultivated aroids, 87 (1951); Graf, Exotica, edn 8, 217, 223 – centre row only (1976); Graf, Tropica, 114 (1978).
Stem prostrate; internodes very short. Scale leaves reddish, drying and persistent. Leaf-blade to 60 × 45 cm, reflexed, ovate, sagittate at base with more or less angular basal lobes, irregularly spotted with grey-green between the principal veins, these and lesser veins all furrowed on upper surface. Leaf-stalk nearly as long as blade, reddish at apex and base, flat on upperside and narrowly crinkly-winged near apex. Spathe *c.* 15 cm, brownish red outside. *Ecuador.* G2.

13. P. gloriosum André. Illustration: Illustration Horticole **23**: 194 (1876); Birdsey, The cultivated aroids, 81 (1951); Graf, Exotica, edn 8, 205 (1976).
Stem prostrate; internodes short. Dry scale-leaves persistent. Leaf-blade *c.* 40 × 33 cm, reflexed, broadly ovate with cordate-sagittate base; upper surface dark green with a velvety appearance and a white midrib and principal veins. Leaf-stalk much longer than blade, flattened above towards the top. Inflorescence-stalk *c.* 16 cm, reddish towards apex, marked with white lines. Spathe as long as inflorescence-stalk, with the tube green, flushed with pink, and the limb pink. *Colombia.* G2.

20. ANUBIAS Schott
P.F. Yeo

Plants evergreen. Stem a horizontal rhizome. Leaf-stalk long, with a pulvinus at the apex. Leaf-blade entire, elongate, variously shaped at base, with a marginal vein and with minor veins parallel to primary lateral veins. Inflorescence-stalk nearly as long as the leaf-stalk. Spathe open throughout or with margins overlapping in the lower half. Florets without perianths, unisexual. Zones of male and female florets adjacent. Male florets with 3–5 stamens united into a short column, the lowest sometimes sterile. Ovary containing numerous ovules.

About 15 species from tropical Africa. One species is sometimes grown at the borders of aquaria or in pots. It requires a wet soil, a humid atmosphere and some shade.

Literature: Engler, A., Anubiadeae etc., *Das Pflanzenreich* **64**: 2–9 (1915).

1. A. afzelii Schott (*A. lanceolata* misapplied). Illustration: Graf, Exotica, edn 8, 143 (1976); Mühlberg, Das grosse Buch der Wasserpflanzen, f. 170, 171 (1980).
Leaf-stalk to 40 cm, with a sheath more than half its length. Leaf-blade erect, to 55 × 17 cm, tapered at apex and base. Spathe 6–6.5 cm, green, closed at the base, contracted into a short point. Spadix longer than spathe, with a stipe; female zone 3 cm; male zone 3 cm, its basal half sterile. *West tropical Africa.* G2.

21. AGLAONEMA Schott
D.H. Nicolson

Plants evergreen, with erect or sometimes creeping stems. Leaf-stalk with well-developed sheath; leaf-blade entire, slightly asymmetric, sometimes variegated, the primary lateral veins with numerous fine minor veins parallel to them (figure 3(5), p. 76). Spathe with margins usually overlapping at base; upper part open and flat or inrolled. Spadix free from spathe or partially attached at base. Florets without perianths, unisexual. Zone of female florets adjacent to the much longer male zone, in which the stamens are not grouped into recognisable florets. Ovary with 1 cell, with 1 ovule. Differs from *Dieffenbachia* (p. 96) in having stamens free from each other and female florets which lack staminodes.

About 20 species in tropical Asia. The species are commonly grown for interior decoration, mostly for their variegated leaves. A minimum of 15 °C, plenty of water, and shade are required for best growth. They are easily propagated from short sections of stem half-buried in rooting medium or from fresh seeds. The species are variable in nature and the variants brought into cultivation have given rise, through hybridisation and selection, to many cultivars.

Literature: Nicolson, D.H., A revision of the genus Aglaonema (Araceae), *Smithsonian Contributions to Botany* **1**: 1–69 (1969).

Habit. Creeping: **1**,**2**; erect: **3–7**.
Sheath-margin. Scarious: **6**,**7**; membranous: **1–5**.
Spathe/spadix. Spathe shorter than spadix: **1**,**2**,**4**; spathe as long as or longer than spadix: **3**,**5–7**.
Spadix. Broadest at end: **4**; broadest in middle: **1**,**2**; elongate: **3**,**5–7**.
Stipe. Attached to spathe: **3**; free from spathe: **1**,**2**,**4–7**.

1a. Plants creeping; sheath of leaf-stalk not more than 1 cm — 2
b. Plants erect; sheath much more than 1 cm — 3
2a. Scale-leaves subtending foliage leaves persistent on drying; leaf-stalk equalling or longer than blade; leaf-base obtuse, more or less rounded — **1. brevispathum**
b. Scale-leaves subtending foliage leaves immediately deliquescent; leaf-stalk equalling or shorter than blade; leaf-base slightly cordate to rounded — **2. costatum**
3a. Stipe and major part of female zone of spadix united with spathe — **3. modestum**
b. Stipe and female zone of spadix free from spathe — 4
4a. Spadix club-shaped; spathe spherical — **4. pictum**
b. Spadix cylindric; spathe oblong — 5
5a. Sheath of leaf-stalk membranous (remaining green) — **5. commutatum**
b. Sheath of leaf-stalk scarious (drying and cracking) — 6
6a. Leaves elliptic, obtuse at base, with a broad silvery stripe running down each side of the upper surface — **6. crispum**
b. Leaves narrowly elliptic, acute at base, with pale bars or blotches along lateral veins, or unvariegated — **7. nitidum**

1. A. brevispathum (Engler) Engler. Illustration: Graf, Exotica, edn 8, 117 (1976) – as A. hospitum; Everett, Encyclopedia of horticulture **1**: 88 (1980).
Stem creeping. Scale-leaves persistent, like brown paper, clasping the leaf-stalks. Leaf-stalks, as long as or longer than blade. Leaf-blade to 20 cm long, 2.5 times as long as broad or more, usually obtuse at base; surface with scattered blotches on both sides (forma **hospitum** (Williams) Nicolson), unvariegated (forma **brevispathum**), or with midrib only white (unnamed). Spathe to 3.5 cm, usually with a small point; stipe to 1 cm. Spadix as long as or longer than spathe. Female florets fewer than 10. *Continental SE Asia*. G2.

This species is very close to *A. costatum* and, upon further study, its distinguishing characters (persistence of scale-leaves, longer leaf-stalks, obtuse leaf-base) may prove insufficient to distinguish it as a species.

2. A. costatum N.E. Brown. Illustration: Birdsey, The cultivated aroids, 15 (1951); Graf, Exotica, edn 8, 114, 118 (1976) – as A. costatum and A. costatum foxii; Everett, Encyclopedia of horticulture **1**: 88 (1980).
Stem creeping. Scale-leaves quickly deliquescent. Leaf-stalk half as long to as long as the blade. Leaf-blade to 20 cm long, up to 2.5 times as long as broad, usually slightly cordate at base; with white midrib and with scattered spots (forma **costatum**), or without spots (forma **immaculatum** (Ridley) Nicolson); or with green midrib and scattered spots (forma **virescens** (Engler) Nicolson), or without spots (forma **concolor** Nicolson). Spathe to 4 cm, abruptly acuminate; stipe to 1 cm. Spadix as long as or longer than spathe. Female florets fewer than 10. *Continental SE Asia*. G2.

Forma *immaculatum* and forma *virescens* are illustrated in Graf, Exotica, edn 8, 118 (1976).

3. A. modestum Engler (*A. acutispathum* N.E. Brown). Illustration: Addisonia **19**: 5 (1935); Birdsey, The cultivated aroids, 17 (1951); Graf, Exotica, edn 8, 119 (1976); Everett, Encyclopedia of horticulture **1**: 86 (1980).
Stem erect. Leaf-sheaths broad and membranous. Leaf-stalk half as long to as long as the blade. Leaf-blade 14–22 × 8–11 cm, not variegated (except in the rare 'Variegatum' and 'Shiningi'), ovate, the apex long-acuminate. Spathe exceeding spadix by 1–3 cm. Stipe and female zone of spadix joined to spathe. Male florets sterile, not shedding pollen (plant an apomictic triploid). Berry becoming orange. *SE Asia and S China*. G2.

One of the 2 most frequently grown species of the genus. It is considered to be a good-luck plant by the Chinese.

4. A. pictum (Roxburgh) Kunth. Illustration: Graf, Exotica, edn 8, 119 (1976); Everett, Encyclopedia of horticulture **1**: 87 (1980).
Stem erect. Leaf-sheaths broad, translucent to scarious. Leaf-stalk much shorter than blade. Leaf-blade 10–16 × 3–6 cm, dull above, variegated, elliptic to ovate, apex often with a small point. Spathe spherical, to 3 cm long and 7 cm wide when unrolled. Stipe *c.* 1 cm. Spadix club-shaped, projecting from the spathe. *Sumatra*. G2.

The dull leaves of this species are distinctive. The variegation is quite irregular, usually light or silvery green on a dark green background.

5. A. commutatum Schott. Illustration (of most common aspect, A. commutatum var. maculatum): Gartenflora **14**: 130 (1865); Botanical Magazine, 5500 (1865); Addisonia **9**: 27 (1924); Graf, Exotica, edn 8, 119 (1976).
Stem erect. Leaf-sheaths usually membranous. Leaf-stalk usually the same length as the blade. Leaf-blade usually narrowly oblong-elliptic, 13–30 × 4–10 cm, the ashy variegation in irregular bars following the primary lateral veins. Spathe usually 1 cm longer than spadix. Stipe 5–10 mm. Male florets sterile, not shedding pollen (plants apomictic tetraploids and hexaploids). *Philippines*. G2.

Many cultivars, such as: 'Pseudobracteatum' (probably called 'Marmoratum' in France) with leaf-stalk white, blade heavily variegated in 3 shades (including yellow) in a somewhat marbled composition; 'Silver Queen' with leaf mostly silvery green, edged and streaked dark green ('Parrot Jungle' is probably the same); 'Treubii', similar but with very narrow leaves. These and others are illustrated by Graf (Exotica, edn 8, 114–19, 1976). 'Silver King' is patterned in 3 shades of green, and is blotchy rather than streaky. A recent cultivar, 'Manila', should probably be attributed to this species but, unlike others, produces pollen.

6. A. crispum (Pitcher & Manda) Nicolson (*A. roebelinii* Pitcher & Manda). Illustration: Birdsey, The cultivated aroids, 21 (1951); Graf, Exotica, edn 8, 117 (1976).
Stem erect. Leaf-sheaths with scarious margins. Leaf-stalk about half as long as blade. Leaf-blade 14–32 × 7–12 cm, elliptic, the silvery variegation covering most of each half of the leaf, leaving the midrib and

marginal areas free. Spathe 1–2 cm longer than spadix. Stipe 3–7 mm. *Philippines (S Luzon)*. G2.

Probably an apomictic triploid. 'Pewter' is the commonest cultivar, but one with a white leaf-stalk has also been raised.

7. A. nitidum (Jack) Kunth. Illustration: Birdsey, The cultivated aroids, 19 (1951); Graf, Exotica, edn 8, 117 (1976) – both as A. oblongifolium curtisii.
Stem erect. Leaf-sheaths with scarious margins. Leaf-stalk half as long to as long as the blade. Leaf-blade 20–45 × 7–16 cm, narrowly elliptic, not variegated or with bars or blotches along lateral veins. Spathe turning white with age, equalling spadix. Stipe 2–9 mm. Female florets 16–37, scattered. *Malaysia, Sumatra, Borneo*. G2.

This species is common in conservatories but rarely seen in houses.

22. DIEFFENBACHIA Schott
P.F. Yeo
Plants erect, evergreen, with stems becoming slightly woody. Leaf-stalk with a well-developed sheath; leaf-blade entire, slightly asymmetric, the primary lateral veins with numerous fine minor veins parallel to them (figure 3(5), p. 76). Spathe with margins overlapping at base; upper part open and hooded or flat, sometimes recurved at apex. Spadix nearly as long as spathe. Florets without perianths, unisexual. Zones of male and female florets about equal, adjacent or slightly separated, sometimes with a few sterile florets between them. Male florets with 4 or 5 stamens united into a short column. Female florets loosely spaced, each with 4 or 5 staminodes (figure 4(4 & 5), p. 77). Ovary with 2 or 3 cells, each with 1 ovule; stigmatic lobes 2 or 3. Differs from *Aglaonema* (p. 94) in the longer zone of female florets, which have well-developed staminodes.

25–30 species in tropical America. The genus is commonly grown for interior decoration and in glasshouses on account of its wide range of leaf variegations. A minimum temperature of 15 °C, plenty of water and some shade are required. Easily propagated from short sections of stem half-buried in a rooting medium. All parts of the plant are poisonous.

The leaf colour of some species is variable in nature and the variants brought into cultivation have also given rise to new cultivars by selection and hybridisation. Most cultivars belong to *D. seguine*. The leaf shape and colouring are somewhat influenced by the stage of growth of the plant and the degree of shading; the colouring is also influenced by the age of the individual leaf.
Literature: Engler, A., Anubiadeae, Aglaonemateae, Dieffenbachiae etc., *Das Pflanzenreich* **64**: 36–78 (1915).

Habit (inapplicable to young specimens). Branched, and with leaves all the way up the stem: **3,5**; simple, with leaves only at the top: **1,2,4,6**.
Leaf-blade variegation. None and midrib green: **2,3,4**; none and midrib white: **1,2**; in 2 colours or shades: **4,5**; in 3 colours; **4,6**.
Leaf-blade shape. Ovate-lanceolate to lanceolate: **3,4**; oblong, ovate, or broadly ovate: **1,2,4,5,6**.

1a. Leaf-blade green, except sometimes the midrib and its immediate borders 2
b. Leaf-blade variegated or spotted 6
2a. Midrib white or yellowish white 3
b. Midrib green 4
3a. Upper surface of leaf-blade with a sheen, leaf-stalk mottled **1. leopoldii**
b. Upper surface of leaf-blade without a sheen, matt; leaf-stalk not mottled **2. oerstedii** 'Variegata'
4a. Leaf-blade narrowly ovate to oblong-lanceolate, mostly more than 3 times as long as broad **3. humilis**
b. Leaf-blade broadly ovate or oblong-ovate, mostly not more than 3 times as long as broad 5
5a. Plant small, not more than about 1 m tall; leaf-blade to 25 cm **2. oerstedii**
b. Plant large, to 3 m tall; leaf-blade often to 40 cm or more **4. seguine***
6a. Leaf-blade variegated with irregular white or greenish white bands along the veins **4. seguine***
b. Leaf-blade not with pale bands along the veins 7
7a. Stem branched; leaf-blade to 75 cm, thin and drooping **5. bowmannii**
b. Stem unbranched; leaf-blade to 45 cm, firm 8
8a. Leaf-blade light yellowish green with the midrib dark green, and with dark green blotches and smaller white blotches **6. × bausei**
b. Leaf-blade differently variegated **4. seguine***

1. D. leopoldii Bull. Illustration: Graf, Exotica, edn 8, 172 (1976); Rochford & Gorer, The Rochford book of house plants, f. 12 (1963) – as D. oerstedii.
Leaf-stalk green or brownish with white mottling, mostly sheathed to the middle or above. Leaf-blade to 35 cm, 2–2.5 times as long as broad, ovate, rather shortly tapered at apex, rounded or weakly cordate at base; upper surface uniformly dark green with a satiny sheen, midrib white. Spathe to 17 cm, white; female florets with the staminodes 5 mm, weakly club-shaped; stigmas 2-lobed. *Costa Rica*. G2.

2. D. oerstedii Schott. Illustration: Graf, Exotica, edn 8, 168, 173 (1976).
Plant not more than 1 m. Leaf-stalk not mottled, sheathed to the middle or above. Leaf-blade to 25 cm, 2–2.5 times as long as broad, gradually tapered at apex, truncate or cordate at base; upper surface matt, rather dark green, sometimes with white midrib ('Variegata'). Spathe to 15 cm, whitish, becoming orange in fruit. *Mexico to Costa Rica*. G2.

3. D. humilis Poeppig. Illustration: Baileya **14**: 102 (1966).
Plant not more than 45 cm. Stem freely branched. Leaves crowded. Leaf-stalk sheathed nearly to the apex. Leaf-blade to 30 cm, mostly more than 3 times as long as broad, narrowly ovate to oblong-lanceolate, slightly glossy above. Spathe about 9 cm, increasing with age to 16 cm, green. Female florets with club-shaped staminodes. *Brazil, Peru*. G2.

4. D. seguine (Jacquin) Schott (*D. maculata* (Loddiges) G. Don; *D. picta* Schott). Illustration: see citations below under cultivars.
Plants usually attaining 1 m and sometimes 3 m. Leaf-stalk sheathed beyond the middle. Leaf-blade to about 40 cm, 2–2.5 times as long as broad, shortly tapered at apex, cordate at base. Spathe 15–18 cm. Female florets with club-shaped staminodes not exceeding the ovary. *Tropical America*. G2.

There are many cultivars, of which the commonest or most distinctive are described below; most do not exceed 1 m in height.
Leaf-blades unvariegated
'Viridis'. Illustration: Graf, Exotica, edn 8, 170 (1976).
Leaf-blades irregularly white-variegated
'Maculata'. Illustration: Rochford & Gorer, The Rochford book of house plants, f. 18 (1963); Graf, Exotica, edn 8, 168, 170 (1976). Upper surface freely spotted with white or yellowish white, the spots somewhat coalescent, not extending near the margin nor on to the midrib; leaf-stalk mottled.

'Superba' ('Roehrs Superb'; 'Roehrs Superba'). Illustration: Graf, Tropica, 106, 107 (1978). White areas much more extensive, though blotched with green and not affecting the midrib.

'Exotica' (*D. perfecta* invalid; 'Exotica Perfection'?). Illustration: Hay et al., Dictionary of indoor plants in colour, t. 190 (1974); Graf, Exotica, edn 8, 168, 172 (1976); Graf, Tropica, 108 (1978). Similar, but midrib white or variegated.

'Baraquiniana'. Illustration: Birdsey, The cultivated aroids, 65 (1951). White spots very sparse, midrib white, leaf-stalk white and apparently longer than usual.

'Magnifica'. Illustration: Graf, Exotica, edn 8, 174 (1976). Spotted similarly to 'Baraquiniana' but leaf-stalks and midribs green.

'Lancifolia'. Illustration: Graf, Exotica, edn 8, 169 (1976). Leaf spotting like that of 'Maculata' but leaf-blade 5–6 times as long as broad, its veins forming a smaller angle with the midrib.

Leaf-blades with yellowish white bands along the lateral veins

'Amoena' (D. amoena misapplied). Illustration: Graf, Exotica, edn 8, 173 (1976). Plant to 2 m or more.

'Tropic Snow'. Illustration: Graf, Exotica, edn 8, 169 (1976); Graf, Tropica, 106, 107 (1978). Similar to 'Amoena' but less tall and with the bands of variegation coalescent and the leaf-blade wrinkled near the midrib.

'Jenmannii'. Illustration: Graf, Exotica, edn 8, 168–70 (1976). To 70 cm with leaf-blades to 25 cm and 2.5–4 times as long as broad, drooping, on slender, shortly sheathed stalks.

Midrib pale, with a yellowish white, fringed band along either side

'Liturata' (*D. amoena* 'Liturata'). Plant to 3 m.

Leaf-blades variegated in three colours or shades

'Rudolph Roehrs' ('Roehrsii'). Illustration: Hay et al., Dictionary of indoor plants in colour, t. 191 (1974); Graf, Exotica, edn 8, 168, 169 (1976). White with green midrib and irregular green border, the white area becoming pale green by the darkening of the finest veins which appear as striations at close quarters.

'Memoria Corsii' ('Memoria'). Illustration: Birdsey, the cultivated aroids, 59 (1951); Graf, Exotica, edn 8, 170 (1976). Mainly grey-green, developing a darker colour with age by darkening of the finest veins and heavily blotched with dark green and sparsely spotted with white.

Leaf-blades variegated with pale (not yellowish) green

'Shuttleworthii' ('Shuttleworthiana'). Illustration: Graf, Exotica, edn 8, 174 (1976). With pale, scattered spots and a pale fringed border on either side of the midrib.

'Nobilis'. Illustration: Graf, Exotica, edn 8, 170 (1976). Plant to 3 m; leaf-blades with pale blotches.

***D. × splendens** Bull. Illustration: Birdsey, The cultivated aroids, 67 (1951); Graf, Exotica, edn 8, 172 (1976). Resembles *D. seguine* 'Baraquiniana' but has a shorter and broader leaf-blade and leaf-stalks mottled in light and dark green (*D. leopoldii* × *D. seguine* 'Maculata').

5. D. bowmannii Carrière (*D. reginae* invalid). Illustration: Graf, Exotica, edn 8, 173 (1976); Graf, Tropica, 105, 106 (1978).

Plant more than 1 m tall, with branched stem. Leaves persistent; leaf-blade to 75 cm, twice as long as broad or slightly more, shortly tapered at apex, rounded at base, thin and drooping, deep green with extensive paler green variegation and a few white spots, bluish green beneath. Spathe 15–18 cm, pale green. Female florets with staminodes 2 mm long. *Colombia, Brazil.* G2.

Descriptions of this species are inconsistent but the name is widely in use. The branched habit and large thin leaves are characteristic.

6. D. × bausei Masters & Moore. Illustration: Birdsey, The cultivated aroids, 55 (1951); Graf, Exotica, edn 8, 138A, 168, 171 (1976).

Plant to 1 m. Leaf-stalk mottled, fading to white at base where it contrasts with the dark green stem, sheathed to about the middle. Leaf-blade to about 30 cm, mostly 2.5–3 times as long as broad, ovate, tapering gradually from below the middle and ending in a long fine point; surface largely yellowish green, but with a dark green border and midrib and with sparse spots and blotches of both dark green and white. Spathe 12 cm, pale green. Female florets with club-shaped staminodes; stigmas with 2 or 3 lobes. *Garden origin.* G2.

The hybrid between *D. seguine* 'Maculata' and *D. weirii* Berkeley.

23. ZANTEDESCHIA Sprengel

P.F. Yeo

Plants with oblique, tuberous, underground rhizomes, which bear apical clusters of leaves or leaves and inflorescences. Scale-leaves absent. Leaf-stalk spongy. Leaf-blade wavy, lobed at the base or entire; main lateral veins joining a marginal vein, minor ones forming a network. Inflorescence as tall as the leaves or taller, fragrant; stalk spongy. Spathe normally white or brightly coloured, obliquely funnel-shaped, withering but persistent in fruit. Spadix much shorter than spathe, yellowish. Florets without perianths, unisexual. Zones of male and female florets adjacent. Male florets not individually recognisable. Female florets with or without staminodes. Ovary usually with 3 cells, each with 1–8 ovules. Style short; stigma entire. Fruit a berry. Seed leathery.

Six species from southern Africa. They are valued for their showy, fragrant inflorescences and are extensively grown as cut-flowers. The range of variation has been increased by artificial hybridisation. All species require a rich, moist soil, and can be grown under glass with only protection from frost, but winter heating permits continued growth of *Z. aethiopica* and *Z. elliottiana*. *Z. aethiopica* is grown out-of-doors in W Europe as a waterside plant, with the tubers deeply planted to protect them from frost; some of its cultivars are hardy out-of-doors in southern England without this protection. Propagation is by separation of the lateral branches of the rhizomes.

Literature: Traub, H.P., The genus Zantedeschia, *Plant Life* **4**: 8–32 (1949); Shibuya, R., *Intercrossing among pink Calla, white-spotted Calla and yellow Calla* (1956); Letty, C., The genus Zantedeschia, *Bothalia* 11: 5–26 (1973).

1a. Leaf-blade lanceolate, tapered at base **4. rehmannii**

b. Leaf-blade wider, cordate, sagittate or hastate 2

2a. Spathe bright yellow **3. elliottiana***

b. Spathe dull yellow, cream, white or occasionally pinkish or mainly green 3

3a. Staminodes present in female part of spadix; leaves usually unspotted; spathe not purple at base within **1. aethiopica**

b. Staminodes not present in female part of spadix; leaves usually white-spotted; spathe purple at base within **2. albo-maculata**

1. Z. aethiopica (Linnaeus) Sprengel (*Calla aethiopica* Linnaeus; *Richardia africana* Kunth). Illustration: Botanical Magazine, 832 (1805); Birdsey, The cultivated aroids,

130 (1951); Flowering Plants of Africa **30**: t. 1190 (1955); Bothalia **11**: 10, 11 (1973).

Plant to *c.* 1.75 m, evergreen in frost-free conditions. Leaf-blade to 45 × 25 cm, normally unspotted, broadly ovate, cordate with broad obtuse basal lobes which are bent upwards and sometimes slightly turned outwards. Spathe to 25 cm, green at base outside, otherwise white, widely open with a large limb, horizontal towards, and recurved at apex. Female part of spadix with conspicuous yellow staminodes. *South Africa (Cape Province, Natal, Transvaal, Lesotho). Widely naturalised in frost-free regions.* H3–G1. Spring–summer.

'Crowborough' is distinguished by being relatively hardy (H3) as are 'Green Goddess', which also has a mainly green spathe, and the miniature cultivar 'Little Gem'.

2. Z. albo-maculata (W.J. Hooker) Baillon (*Z. oculata* (Lindley) Engler). Illustration: Botanical Magazine, 5140 (1859), 5765 (1869); Birdsey, The cultivated aroids, 132 (1951); Letty, Wild flowers of the Transvaal, 10, 11 (1962); Bothalia **11**: 18–25 (1973).

Plant to *c.* 1.25 m, deciduous in winter. Leaf-blade to 45 × 23 cm, usually with numerous translucent white spots, hastate with main lobe oblong-triangular, or ovate-cordate (subsp. **valida**); basal lobes bent upwards. Spathe to 11.5 cm (or to 15 cm in subsp. **valida**), usually dark purple inside at base, elsewhere white, cream, straw-coloured, greenish yellow or, rarely, pinkish, moderately widely open (more so in subsp. **valida**), not recurved at apex. Female part of spadix without staminodes. *South Africa & E tropical Africa.* H5. Spring–summer.

Letty treats this as a variable species comprising 3 subspecies; unspotted leaves occur rarely in subsp. *albo-maculata*, sometimes in subsp. *macrocarpa* (Engler) Letty, and always in subsp. *valida* Letty.

3. Z. elliottiana (Watson) Engler (*Richardia elliottiana* Watson). Illustration: Botanical Magazine, 7577 (1898); Birdsey, The cultivated aroids, 134 (1951); Graf, Exotica, edn 8, 138A, 240 (1976).

Plant to *c.* 1.5 m, deciduous in winter. Leaf-blade to 28 × 25 cm, very broadly ovate, cordate with basal lobes bent upwards, with numerous translucent white spots. Spathe to 15 cm, bright golden yellow, duller outside, widely open above with narow spreading edge and recurved tip. Female part of spadix without staminodes. *?South Africa (unknown in nature).* H5. Spring–summer.

Letty suggests a hybrid origin for this plant. However, Shibuya found only trivial variation on selfing. R.E. Harrison, on the other hand (Journal of the Royal Horticultural Society **103**: 78, 1978), obtained plants with dark-centred spathes on crossing *Z. elliottiana* with *Z. rehmannii* and suggested that this indicated *Z. pentlandii* in the parentage of *Z. elliottiana*.

***Z. jucunda** Letty. Illustration: Letty, Wild flowers of the Transvaal, 15 (1962); Bothalia **11**: 14 & t. 3 (1973). Leaf-blade triangular-hastate, spotted. Spathe very like that of *Z. elliottiana* but with a dark purple area inside at base; inner surface slightly wrinkled. *South Africa (Transvaal).* H5. Summer.

***Z. pentlandii** (Watson) Wittmack (*Richardia pentlandii* Watson). Illustration: Botanical Magazine, 7397 (1895); Gartenflora **47**: 593 (1898); Letty, Wild flowers of the Transvaal, 14 (1962); Bothalia **11**: 16 & t. 4 (1973). Leaf-blade oblong-elliptic to oblong-lanceolate, sagittate or cordate, rarely spotted. Spathe similar to that of *Z. elliottiana* but with a dark purple area inside at base; apex not usually recurved. *South Africa (Transvaal).* H5. Summer.

The spathe is described and depicted in Botanical Magazine as wrinkled within but is shown as smooth in the plate by Letty.

4. Z. rehmannii Engler. Illustration: Birdsey, The cultivated aroids, 136 (1951); Letty, Wild flowers of the Transvaal, 7 (1962); Bothalia **11**: 12, 13 & t. 2 (1973); Graf, Exotica, edn 8, 240 (1976).

Plant usually to 60 cm, deciduous in winter. Leaf-blade to 40 × 7 cm, asymmetrically lanceolate, gradually tapered at both ends, rarely spotted. Spathe 7.5–12 cm, white to pink or dark purple, rather narrow; apex spreading or recurved. Female part of spadix without staminodes. *South Africa (Transvaal, Natal, Swaziland).* H5. Spring–summer.

Some cultivars are listed by Traub.

24. PELTANDRA Rafinesque

P.F. Yeo

Plants with horizontal rhizomes bearing rosettes of scale-leaves and long-stalked foliage leaves. Leaf-blade sagittate or hastate, with 2 or 3 marginal veins (figure 3(6), p. 76); lateral veins mostly parallel but the smallest joined by cross-veins; each basal lobe with a distinct midrib. Inflorescence-stalk as long as or slightly longer than leaf-stalks. Spathe with margins overlapping in the zone of female florets, open above; upper part decaying in fruit. Florets without perianths, unisexual. Male florets of 4 or 5 stamens united to form a flat plate with anthers around the edge, opening by pores at the top. Female florets with a cup-like structure round the ovary, derived from staminodes. Ovary 1-celled; ovules 1 or few; style present.

Four species of marsh-plants in E North America. They are grown out of doors as foliage plants in shallow water. Propagation is by division of the rhizomes, preferably in spring.

Literature: Engler, A., Philodendroideae, Anubiadeae-Peltandrae, *Das Pflanzenreich* **64**: 72–5 (1915).

1a. Female florets covering nearly half the length of the spadix; spathe-limb whitish, not wavy **1. sagittifolia**

b. Female florets covering one-third to one-fifth of the length of spadix; spathe-limb green with yellowish or whitish wavy edges **2. virginica**

1. P. sagittifolia (Michaux) Morong. Illustration: Das Pflanzenreich **64**: 73 (1915).

Leaf-blade to 15 cm, sagittate or slightly hastate. Spathe 7–10 cm, the tube green and the limb white, opening widely, with flat edges. Spadix covered with female florets for about half its length; female zone separated from male zone by scale-like sterile florets, each with a boss in the middle. Inflorescence-stalk erect in fruit. Berries red. *Southeastern USA.* H3. Early summer.

2. P. virginica (Linnaeus) Kunth. Illustration: Gleason, Illustrated flora of the northeastern United States and adjacent Canada 1: 368 (1952); Rickett, Wild flowers of the United States **1**(1): 87 (1965); Graf, Tropica, 110 (1978); Everett, Encyclopedia of horticulture **8**: 2537 (1981).

Leaf-blade to 30 cm (larger after flowering time), more or less hastate. Spathe 10–20 cm, mainly green; limb with wavy yellowish or whitish margins, gaping only slightly. Spadix covered with female florets for one-fifth to one-third of its length, and with few or no sterile florets; sometimes with a small sterile appendix. Inflorescence-stalk recurved in fruit. *E & Southeastern USA.* H2. Late spring–summer.

25. CALADIUM Ventenat

M.T. Madison & P.F. Yeo

Plants with starchy, cylindric or spherical, subterranean tubers which are dormant in winter. Leaf-blade with base hastate to truncate or rounded, usually variegated with pink, red, white, and green patterns; main lateral veins with netted veinlets between them, joined to a marginal vein (figure 3(6), p. 76). Spathe with the margins overlapping below and the upper part open. Spadix shorter than the spathe, with female, sterile and male florets in adjacent zones. Florets without perianths. Stamens 2–5, joined into a column. Ovary with 2 or 3 cells, incompletely separated, each with several ovules; style none; stigma grooved. Berry whitish.

A genus of 7 or 8 species native to northern S America and naturalised throughout the tropics. Caladiums are cultivated in shady garden borders or as pot plants for their showy, variegated foliage. They grow well in moderate shade with a moist, well-drained soil rich in humus. Tubers should be lifted after the leaves die down in the autumn and stored in a cool dark spot until the following spring. Propagation is by division of tubers, or less often by seed.

Literature: Madison, M.T., Notes on Caladium (Araceae) and its allies, *Selbyana* 5: 342–77 (1981).

1a. Leaf peltate 2
b. Leaf not peltate 3
2a. Leaf-blade not more than 10 cm **1. humboldtii**
b. Leaf-blade more than 15 cm **2. bicolor**
3a. Leaf-blade hastate **3. lindenii**
b. Leaf-blade with base weakly cordate, rounded or truncate **4. schomburgkii**

1. C. humboldtii Schott (*C. argyrites* Lemaire; *C. lilliputiense* Rodigas). Illustration: Illustration Horticole **5**: t. 185, no. 3 (1858); Graf, Tropica, 96, 98 (1976); Everett, Encyclopedia of horticulture **2**: 545 (1981); Selbyana **5**: 370 (1981).
To 25 cm. Tuber 1–2 cm in diameter, yellow within, suckering freely. Leaf-stalk 10–24 cm × 1–1.5 mm. Leaf-blade 5–10 × 2–4.5 cm, ovate, peltate, with sagittate base, dull deep green with white blotches and spots. Flowers unknown. *Brazil & Venezuela*. H5.

2. C. bicolor (Aiton) Ventenat (*C.* × *hortulanum* Birdsey; *C. marmoratum* Mathieu; *C. picturatum* C. Koch; *C. poecile* Schott). Illustration: Botanical Magazine, 2543 (1825); Graf, Exotica, edn 8, 149 (1976).
To 90 cm. Tuber 2.5–6 cm broad, flattened-spherical. Leaf-blade 18–46 × 12–25 cm, peltate, ovate to elliptic, with cordate or sagittate base, dark to medium green with or without white, pink or red spots, bands or blotches. Spathe 6–11 cm. Spadix with male part 2.5–6 cm, white. *Northern S America*. H5. Late spring–autumn.

Has produced an immense number of cultivars, including many in which the leaf-blade is largely white (see Graf, Exotica, edn 8, 147–62, 1976).

3. C. lindenii (André) Madison (*Xanthosoma lindenii* (André) Engler). Illustration: Illustration Horticole **19**: 3 (1872) – as Phyllotaenium lindenii; Birdsey, The cultivated aroids, 127 (1951); Graf, Tropica, 130 (1978); Everett, Encyclopedia of horticulture **10**: 3569 (1982).
To 80 cm. Tuber suckering freely. Leaf-stalk to 40 cm, covered with sparse, short, brown hairs, green with purple stripes. Leaf-blade 16–45 cm long, 10–20 cm wide across principal lobe, hastate, not peltate, hairy on the midrib; basal lobes with a pinnately branched midrib (figure 3(8), p. 76); upper surface green with midrib and main veins white or yellowish, the surface often paler green near midrib. Spathe 7.5–14 cm; tube whitish green; limb 7–9 cm, white. Spadix with male part 7–8 cm × 6–8 mm, white. *Colombia*. G1.

4. C. schomburgkii Schott. Illustration: Illustration Horticole **8**: t. 297, nos. 2 & 3 (1861).
To 45 cm. Tuber spherical. Leaf-stalk 16–23 cm. Leaf-blade about same length as stalk, triangular, ovate, broadly lanceolate or almost rhombic, with weakly cordate, truncate or rounded base, not peltate, usually variegated with pink or white, the margins often ruffled. Inflorescence-stalk 18–26 cm. Spathe *c.* 7 cm. *Guianas*. H5.

Represented by many cultivars (see Graf, Exotica, edn 8, 162–4, 1976).

26. XANTHOSOMA Schott

S.A. Thompson

Plants with starchy, cylindric or spherical, underground corms or with above-ground stems. Leaf-stalk with well-developed sheath; leaf-blade entire, sagittate, hastate, or pedately divided into 3–18 segments; main venation pinnate; minor veins netted. Spathe with margins overlapping at base and upper part open. Spadix shorter than spathe. Florets without perianths, unisexual. Zone of male florets 2 or more times as long as female zone, separated from it by sterile florets. Stamens 4–6, united into a column. Ovary of 2–4 cells with several to many ovules.

A genus of approximately 50 species native to the New World tropics. Several species of *Xanthosoma* are cultivated throughout the tropics for their starchy corms and edible leaves, as well as for ornament.

Xanthosoma species will succeed in most soils, but prefer rich, well-drained soil with abundant water. Propagation is by vegetative means; either the stem is cut into pieces as cuttings, or cormlets are planted directly. Flowering under cultivation is sporadic.

Literature: Engler, A., Colocasioideae, *Das Pflanzenreich* **71**: 41–62 (1920).

Habit. With stem: **1,2**; stemless: **3–7**.
Leaves. Entire and sagittate: **1–4**; entire and hastate: **5**; divided: **6,7**.

1a. Plant with above-ground stem 2
b. Plant without stem 3
2a. Main veins of basal lobes of leaf-blade exposed in the sinus for part of their length (figure 3(8), p. 76) **1. undipes**
b. Main veins of basal lobes of leaf-blade not exposed in the sinus **2. sagittifolium**
3a. Leaves entire 4
b. Leaves divided 6
4a. Plant hairy **3. pilosum**
b. Plant hairless 5
5a. Leaf-blade sagittate **4. violaceum**
b. Leaf-blade hastate **5. brasiliense**
6a. Segments of leaf-blade 5–18, narrow, to 9 cm wide **6. helleborifolium**
b. Segments of leaf-blade 5–9, broad, to 53 cm wide **7. hoffmannii**

1. X. undipes (C. Koch) C. Koch (*X. jacquinii* Schott). Figure 3(8), p. 76. Illustration: Das Pflanzenreich **71**: 46 (1920).
Above-ground stem to 2 m tall and 20 cm thick. Leaf-stalk to 1 m; sheath wavy, green; leaf-blade 50–200 cm long and somewhat less wide at the base, broadly cordate-sagittate, main veins of basal lobes exposed in the sinus for 1–5 cm (figure 3(8), p. 76). Spathe-tube to 7 cm, ovoid, green outside, whitish yellow inside; spathe-limb 15–25 cm, pale yellowish green outside, whitish yellow inside. Female florets orange; sterile florets dull white; male florets yellow. *Tropical America from Mexico south to Peru; cultivated as a crop in the West Indies*. G2.

2. X. sagittifolium (Linnaeus) Schott (*Arum xanthorrhizon* Jacquin; *X. atrovirens* C. Koch). Illustration: Botanical Magazine, 4989 (1857).
Mature plant with above-ground stems 1 m long or more. Leaf-stalk to 1 m, bloomed; leaf-blade to 70 cm long and a little less wide, broadly sagittate-ovate, bloomed; main veins of basal lobes not exposed at the sinus. Spathe-tube to 7 cm, oblong-ovoid, green; spathe-limb to 15 cm, acuminate, greenish white. Spadix much shorter than spathe. *New World tropics, exact native distribution uncertain; widely cultivated in the tropics.* G2.

3. X. pilosum C. Koch & Augustin (*X. holtonianum* Schott; *Caladium puberulum* Engler). Illustration: Croat, Flora of Barro Colorado Island, 226 (1978).
Plant stemless, most parts covered with short fine hairs. Leaf-stalk to 40 cm, sheathed halfway or more; leaf-blade to 35 × 20 cm, sagittate-ovate to cordate-ovate, green, sometimes mottled with white spots; main veins of basal lobes shortly exposed in the sinus (figure 3(8), p. 76). Spathe-tube to 7 cm, oblong, green; spathe-limb to 10 cm, oblong-lanceolate, white. Sterile florets violet. *Costa Rica to Colombia.* G2.

4. X. violaceum Schott. Illustration: Martius, Flora Brasiliensis 3(2): t. 43 (1878); Birdsey, The cultivated aroids, 129 (1951); Graf, Exotica, edn 8, 238 (1976).
Plant stemless, robust, with large tuberous corm. Leaf-stalk 30–85 cm, dark brownish purple, bloomed; leaf-blade to 70 × 45 cm, sagittate-ovate, shortly acuminate at apex, bloomed at least when young, with midrib and main veins purple beneath and margin purple; main veins of basal lobes exposed for a short distance in the sinus in mature leaves (figure 3(8), p. 76). Spathe-tube to 10 cm, greenish violet outside, whitish yellow inside; spathe-limb to 20 cm, whitish yellow. Sterile florets violet to reddish brown. *New World tropics, exact native distribution uncertain; cultivated and naturalised throughout the tropics.* G2.

5. X. brasiliense (Desfontaines) Engler (*X. hastifolium* C. Koch). Illustration: Proceedings of the Florida State Horticultural Society **85**: 90 (1972).
Plant stemless. Leaf-stalk 20–40 cm, narrowly sheathed to less than one-third of its length; leaf-blade of young plant ovate-sagittate, leaf-blade of adult plant to 40 × 15 cm, hastate, with outspread oblong basal lobes; main veins of basal lobes exposed for part of their length in the sinus (figure 3(8), p. 76). Spathe-tube to 5 cm, oblong-ovoid, green; spathe-limb to 12 cm, oblong-lanceolate, yellowish green. Female florets with yellow stigmas; sterile florets pink; male florets dull white. *New World tropics, exact native distribution uncertain; widely cultivated in the tropics, especially in the West Indies.* G2.

6. X. helleborifolium (Jacquin) Schott. Illustration: Croat, Flora of Barro Colorado Island, 226 (1978).
Plant stemless. Leaf-stalk to 85 cm, broadly sheathed in the lower part, mottled with purple and green; leaf-blade kidney-shaped in outline, deeply dissected into 5–18 oblong or lanceolate segments, the central to 36 × 9 cm. Inflorescence-stalks mottled like the leaf-stalks. Spathe-tube 3–6 cm, elliptic-ovoid, green; spathe-limb 6–10 cm, oblong-ovate, yellowish green to white. Spadix with a broad short stipe. *El Salvador to the Guianas & Peru; West Indies.* G2.

7. X. hoffmannii Schott (*X. barilletii* Carrière). Illustration: Revue Horticole **54**: 260 (1882) – as X. barilletii.
Plant essentially stemless. Leaf-stalk to 1 m, sometimes purple at base, broadly sheathed; leaf-blade kidney-shaped in outline, deeply dissected into 5–9 broadly ovate, acuminate segments, to 50 × 53 cm. Spathe-tube to 7 cm, oblong, green; spathe-limb to 12 cm, lanceolate, acuminate, white. Spadix on a short stipe, which is *c.* 5 mm long. *Mexico to Costa Rica; reported from Surinam.* G2.

The plant introduced to Europe in 1882 as *X. barilletii* has been tentatively assigned to *X. hoffmannii.*

27. COLOCASIA Schott

P.F. Yeo

Plants evergreen, in our species without an above-ground stem, sometimes with stolons, rhizomes or tubers. Leaf-blade peltate (figure 3(7), p. 76) and either ovate-cordate or sagittate, usually with 2 or 3 marginal veins (figure 3(6), p. 76); branches of primary lateral veins bent towards margin and joined by cross-veins to form rectangular compartments. Inflorescences 1 or several in each axil, much shorter than the leaf-stalk. Spathe constricted between the short overlapped part and the much longer, hooded or flattened limb, which is finally shed. Spadix shorter than spathe, with a short or long sterile appendix. Florets without perianths, unisexual. Zone of male florets much longer than zone of female florets, separated from it by a constriction bearing scale-like sterile florets (figure 4(6), p. 77). Male florets with 3–6 stamens joined to form a short column. Ovary 1-celled; placentation parietal; ovules several to many. Stigma stalkless or nearly so. Fruit a berry.

Seven species in the Old World tropics. A few have variants with strikingly coloured foliage which are grown for ornament in hothouses or for planting out in summer in warm situations. Rich damp soil and warmth are necessary for vigorous growth. Propagation is by division of the rootstock or by seed.

Literature: Krause, K., Colocasioideae, *Das Pflanzenreich* **71**: 62–71 (1920); Holttum, R.E., Alocasia marshallii, *Gardeners' Chronicle* **150**: 217 (1961).

1a. Leaf-blade to *c.* 16 cm, broadly ovate, cordate **1. affinis**
b. Leaf-blade to *c.* 50 cm, ovate or elliptic, sagittate **2. esculenta**

1. C. affinis Schott (*Alocasia jenningsii* Veitch; *A. marshallii* Bull). Illustration: Flore des Serres **17**: 183 (1868); Illustration Horticole **16**: t. 585 (1869); Gardeners' Chronicle 1872: 801 (1872); Graf, Exotica, edn 8, 126 (1976).
Plant with a round tuber and many small rhizomes. Leaf-stalk 20–30 cm, transversely banded with purple. Leaf-blade 10–16 × 7–10 cm, broadly ovate, cordate, shortly acuminate at apex, with about 3 pairs of main lateral veins; main vein of each basal lobe without major lateral veins on the side towards the sinus; upper surface with large purplish blotches between the main veins, and a small to large central silvery patch. Spathe 6–9 cm, pale green. Appendix of spadix 3–4 cm, longer and much narrower than the remainder. *Tropical eastern Himalaya.* G2.

2. C. esculenta (Linnaeus) Schott (*C. antiquorum* Schott). Illustration: Botanical Magazine, 7364 (1894); Baileya **7**: 67 (1959).
Rootstock variable. Leaf-stalk to 1 m. Leaf-blade to 50 cm, ovate or elliptic, with rounded or angular basal lobes, acute apex and 4–8 pairs of main lateral veins; midrib of one or both basal lobes with a main lateral vein on the side towards the sinus (figure 3(7), p. 76). Spathe 15–35 cm; tube greenish; limb pale yellow. Appendix of spadix very variable in length, only slightly narrower than the remainder unless rudimentary. *Tropical E Asia.* H5.

Widely cultivated in warm-temperate to tropical climates for its edible tubers, 'taro' and 'dasheen'. These occur as numerous cultivars. For ornamental cultivation the 2 following are best known: 'Fontanesii' (Illustration: Botanical Magazine, 7732, 1900; Birdsey, The cultivated aroids, 45, 1951) with the leaf-stalk, inflorescence-stalk and spathe-tube blackish violet and the upper surface of the leaf-blade dark green, with dark margin and veins; and 'Illustris' (Illustration: Birdsey, The cultivated aroids, 45, 1951) with the leaf-stalk brownish purple at the base, greener above, and the blade with a blackish violet blotch between each pair of main veins.

Distinctions from *Alocasia odora* are given under that species (p. 102).

28. ALOCASIA Necker

D.J. Leedy, T.B. Croat & P.F. Yeo

Plants normally evergreen, rhizomatous and tuberous, with erect or ascending, usually short, stems. Leaf-blade entire to pinnatisect, peltate, at least in the juvenile stage, and cordate or sagittate, with netted subsidiary veins, and with interprimary veins and a marginal vein (figure 3(6), p. 76). Leaf-stalk long, cylindric, with a long sheath. Inflorescence-stalk usually much shorter than the leaf-stalk. Spathe with margins overlapping below to form a cylindric or ovoid tube; limb usually boat-shaped, becoming reflexed, falling after flowering. Spadix shorter than spathe, with a sterile appendix. Florets unisexual, without perianths. Zone of female florets enclosed in spathe-tube, usually separated from male zone by sterile florets. Male florets of 3–8 stamens united into a short column. Sterile florets flattened, scale-like (figure 4(6), p. 77). Style present. Stigma with 3–5 lobes. Ovary with a single cell, sometimes with partitions in the upper part. Placentation basal. Berries with 1 or few seeds, ripening to orange or red. Differs from *Colocasia* in the basal, not parietal, placentation.

About 70 species have been named, though some are not now accepted. Species of *Alocasia* in their young stages may be grown as houseplants on account of their strikingly marked foliage. Mature stages of the larger species are suitable for glasshouse collections or conservatories.

Most species thrive in warm and humid conditions; some (*A. odora, A. cucullata, A. macrorrhiza*, etc.) can survive 2–3 °C in winter and many become dormant during the cooler weather. Plants should be under-potted (in small pots) and given a loose, well-drained potting medium. Sufficient crocking in the bottom of the container and up to 50% wood chips (bark or fern wood) in the mix provides the needed drainage, but necessitates frequent watering. Some growers recommend small additions of dolomite lime. The plants have moderate light requirements and thrive in dense shade. Propagation may be achieved from seeds, cuttings, or the corms which most species produce from the rhizomatous stem. The stems topple in old plants of some species.

Many interspecific hybrids of *Alocasia* have arisen in cultivation, only some of which are mentioned below. Their existence should be borne in mind if difficulties in identification are encountered.

Literature: Engler, A. & Krause, K., Colocasioideae, *Das Pflanzenreich* **71**: 71–115 (1920); Bunting, G.S., The genus Schizocasia, *Baileya* **10**: 113–20 (1962); Bunting, G.S. & Nicolson, D.H., The Alocasia plumbea confusion, *Baileya* **10**: 142–6 (1963); Dotort, F. & Thompson, T., Alocasias, *Aroideana* **2**: 35–51 (1979).

Plant size. 2 m or more: **2–5**; 1 to 2 m: **6–9**; less than 1 m: **1,10,11**.
Leaf-blade colour. All green (both surfaces): **1,2,4,7**; purple beneath: **3,6,8,9,11**; lighter coloured midrib and lateral veins: **5,6,8–10**.
Leaf-blade shape. Lobed at the sides: **5,6**.
Leaf-blade texture. Velvety: **10**.

1a. Leaf-blade with margins lobed or pinnatisect 2
b. Leaf-blade entire apart from the basal sinus, sometimes wavy 3
2a. Upper surface of leaf-blade with main veins and margin boldly marked in white; margin sinuous **6. sanderiana**
b. Upper surface of leaf-blade with the veins not or only slightly paler than background; margin deeply sinuous (in young plants) to pinnatifid **5. portei**
3a. Leaf-blade on both sides and leaf-stalk green, without variegation, or the blade with very irregular, more or less segmental variegation and the stalk striped with white 4
b. Leaf-blade on one or both sides purple, or with variegation following the veins; leaf-stalk sometimes variegated or marbled 6
4a. Leaf-blade to 12 cm, heart-shaped or triangular, hollow above attachment of leaf-stalk **1. cucullata**
b. Leaf-blade normally very much larger, ovate–sagittate, without a hollow 5
5a. Leaf of mature plant not peltate; main lobe of blade with 9–12 pairs of veins **2. macrorrhiza**
b. Leaf of mature plant peltate, the basal lobes being shortly united; main lobe of blade with 6–10 pairs of veins **4. odora**
6a. Leaf-blade with the veins curved towards base of blade as they leave midrib, arching towards its apex as they approach the margin, deeply sunken **11. cuprea**
b. Leaf-blade with the veins not curved towards base of blade near midrib 7
7a. Leaf-blade with at least the main veins pale, forming a conspicuous pattern 8
b. Leaf-blade without a conspicuous pale pattern following the veins 10
8a. Leaf not or scarcely peltate, lower surface green **10. micholitziana**
b. Leaf distinctly peltate, basal lobes joined for one-seventh of their length or more, lower surface purple 9
9a. Leaf-blade with the minor veins not pale **8. longiloba***
b. Leaf-blade with the minor as well as the major veins pale **9. veitchii***
10a. Leaf-blade to 35 cm, green on both surfaces **7. zebrina**
b. Leaf-blade to 75 cm, green or purple above, purple beneath **3. plumbea**

1. A. cucullata (Loureiro) G. Don. Illustration: Graf, Exotica, edn 8, 127 (1976); Aroideana **2**: 42 (1979); Everett, Encyclopedia of horticulture **1**: 117 (1980).

Plant less than 1 m. Stem to 40 cm or more with many suckers, branched and inclined. Leaf-blade 6–12 × 4–7 cm, heart-shaped or triangular, cordate, peltate with basal lobes united for less than half their length, somewhat cupped above attachment of stalk, glossy green; veins prominent, strongly curved towards apex of blade, crowded at the base. Spathe fleshy; tube 4–8 cm; limb 5–10 cm. *Sri Lanka to Burma*. H5 (can endure light frost with only leaf damage).

2. A. macrorrhiza (Linnaeus) G. Don (*A. indica* misapplied). Illustration: Edwards's Botanical Register **8**: 641 (1822); Botanical Magazine, 3935 (1842); Graf, Exotica, edn 8, 121 – and as A. odora, 122 – juvenile (1976).

Plant to 3 m or more; vegetative parts bright green. Stem stout, to 1 m or more. Leaf-blade to 125 × 75 cm, erect or spreading, ovate, sagittate, not peltate in the mature plant; main lobe with 9–12 pairs of veins; upper surface smooth, shiny; uniformly green on both surfaces. Leaf-stalk to 1 m or more. Spathe-tube 6–8 cm, limb 12–16 cm. *Sri Lanka, widely naturalised elsewhere in warm regions*. H5.

'Variegata'. Illustration: Birdsey, The cultivated aroids, 25 (1951); Graf, Exotica, edn 8, 122 (1976). Leaf-blade with large or small irregular blotches of creamy white, grey-green and darker green. Leaf-stalk longitudinally striped with green and white.

This species, much cultivated in the tropics and subtropics for its edible (when cooked) rhizomes and shoots, is often confused with *A. odora*, which has peltate leaves in the mature state.

3. A. plumbea (C. Koch) Van Houtte (*A. indica* var. *metallica* misapplied; *A. macrorrhiza* var. *rubra* (Hasskarl) Furtado). Illustration: Flore des Serres **21**: 93 (1875); Graf, Exotica, edn 8, 123 (1976); Aroideana **2**: 40 (1979).
Similar to *A. macrorrhiza*, but purplish in the colour of the vegetative parts. Leaf-blade 75 × 50 cm or more, with 6–8 pairs of veins in the main lobe, sometimes wavy; upper surface glossy purple or green, sometimes with purple veins or blotches; lower surface deep purple. Leaf-stalk purple or variously spotted or marbled. *Java*. G1.

Usually becoming dormant in winter, unless the temperature is maintained at or above 16°C.

4. A. odora (Loddiges) Spach. Illustration: Graf, Exotica, edn 8, 125 (1976); Aroideana **2**: 43 (1979).
Very similar to (and confused with) *A. macrorrhiza*, differing mainly in the peltate leaf-blade of the mature plant, which also has only 6–10 pairs of veins in the main lobe. Spathe-tube 3–4 cm; limb 10–14 cm. *Tropical Asia*. H5.

Distinguishable from *Colocasia esculenta* (p. 100) by its erect or spreading, not drooping leaf-blades, with pinnately veined basal lobes.

5. A. portei Schott (*Schizocasia portei* (Schott) Engler; *S. regnieri* Linden & Rodigas). Illustration: Illustration Horticole **34**: 17 (1887) – leaves of semi-mature plant; Baileya **10**: 114–17 (1962); Graf, Exotica, edn 8, 125 – juvenile, 230 (1976).
Plant to 2 m or more, with stem to 24 cm tall and 10 cm in diameter. Leaf-blade ovate-sagittate; that of juvenile plant with deeply sinuous margins; that of mature plant to *c*. 2 m, pinnatisect, its segments linear-lanceolate, obtuse, strongly wavy; surface dark metallic green with the main veins sometimes slightly lighter in colour. Leaf-stalk to 2 m or more, marbled with red-purple on green. Spathe tube 2–2.5 cm; limb 5–6 cm. *Philippines*. G2.

6. A. sanderiana Bull (*Schizocasia sanderiana* (Bull) Engler). Illustration: Birdsey, The cultivated aroids, 27 (1951); Graf, Exotica, edn 8, 127 (1976).
Plant to 2 m. Leaf-blade 30–40 × 12–20 cm, narrowly sagittate, thinly leathery, with deeply sinuous margin; upper surface deep glossy green, with major lateral veins and margin boldly marked in white; lower surface green or (in 'Gandavensis') purple. Leaf-stalk 25–40 cm or more, brownish green. Inflorescence-stalk nearly as long as the leaf-stalk. Spathe 11 cm, with green tube and cream-coloured limb. *Philippines*. G2.

Plants with shallowly sinuous leaf-blades are likely to be hybrids of this species, for example, with *A. lowii* var. *grandis* (*A.* 'Amazonica'. Illustration: Aroideana **2**: 48, 1979) or with *A. plumbea* (*A.* × *chantrieri* André. Illustration: Illustration Horticole **35**: 79, 1888; Graf, Exotica, edn 8, 125, 1976).

7. A. zebrina C. Koch & Veitch. Illustration: Flore des Serres **15**: 81 (1862); Graf, Exotica, edn 8, 124 (1976).
Plant to 1 m or more. Leaf-blade to 35 × 20 cm, triangular, sagittate, scarcely peltate, green on both surfaces. Leaf-stalk to 1 m, transversely variegated with black and green. Inflorescence-stalk variegated. Spathe-tube 2.5–4 cm; limb 12–14 cm, whitish. Spadix appendix pink. *Philippines*. G2.

8. A. longiloba Miquel. Illustration: Graf, Exotica, edn 8, 124 (1976); Aroideana **2**: 46 (1979).
Stem 30–75 cm. Leaf-blade 30–50 × 15 cm (or to *c*. 70 × 27 cm in 'Magnifica'), narrowly triangular-sagittate; basal lobes more than half as long as principal lobe, joined for one-seventh to one-fifth of their length, acute; main veins 6–7 pairs; midribs of basal lobes with no main lateral veins towards the sinus. Upper surface of blade rich deep glossy green with margin and main veins light greyish; lower surface glossy purple. Leaf-stalk 25–32 cm, slender, not patterned. Spathe to 10 cm; limb much longer than tube. *Malaysia, Borneo, Java*. G2.

This plant was described from the Malay peninsula; whether specimens from Borneo and Java are the same is uncertain. The group of species dealt with here under Nos. 8 and 9 is represented in nature by populations isolated by the sea and by their preference for limestone outcrops.

***A. lowii** W.J. Hooker. Illustration: Botanical Magazine, 5376 (1863). Leaf-blade to *c*. 50 × 18 cm; basal lobes one-third to half as long as principal lobe, joined for about one-quarter of their length, tapered to an obtuse tip. Main veins up to 13 pairs. Spathe *c*. 11 cm, greenish white; limb *c*. 3 times as long as tube. *Borneo*. G2.

Hybrids of this species with *A. cuprea* (*A.* × *sedenii* Veitch. Illustration: Illustration Horticole **24**: 154, 1877; Graf, Exotica, edn 8, 125, 1976) have the veins less distinctly coloured and show the impressed, curved veins and nearly united basal lobes of *A. cuprea*. For the plant known as *A. lowii* (var.) *grandis*, see under No. 9.

9. A. veitchii (Lindley) Schott (*A. lowii* W.J. Hooker var. *veitchii* (Lindley) Engler). Illustration: Botanical Magazine, 5497 (1865) – as A. lowii var. picta; Das Pflanzenreich **71**: 107 (1920).
Stem 30–75 cm. Leaf-blade to *c*. 45 × 20 cm, narrowly triangular-sagittate; basal lobes about half as long as principal lobe, joined for one-sixth to one-fifth of their length; main veins about 9 pairs; midribs of basal lobes with no main lateral veins towards the sinus. Interprimary veins slightly sinuous. Upper surface of blade deep green with margin, main veins (together with the surface adjacent to them) and minor veins light greyish or grey-green; lower surface purple. Leaf-stalk *c*. 45 cm, greenish to purplish with darker transverse bands. Spathe *c*. 11 cm; tube greenish; limb about 3 times as long as tube, pale yellow, tinged with red at tip. *Borneo*. G2.

See remarks under *A. longiloba*. The following all have the minor veins pale:

***A. watsoniana** Masters. Illustration: Gardeners' Chronicle **13**: 569 (1893); Graf, Exotica edn 8, 123, 125, 127 (1976); Aroideana **2**: 41 (1979). Leaf-blade to 90 × 45 cm, elliptic-ovate, sagittate; basal lobes more than half as long as main lobe, joined for one-third to half of their length; main veins about 7–10 pairs; midrib of one of the basal lobes sometimes with 1 or 2 main lateral veins towards the sinus; main veins of principal lobe often with a conspicuous branch on the basal side, running out to the margin. Interprimary veins sharply zig-zag (figure 3(6), p. 76). Upper surface bluish green with margins, main veins and minor veins silvery. Leaf-stalk to 60 cm. Spathe with limb *c*. 4 times as long as tube. *Sumatra*. G2.

***A. putzeysii** N.E. Brown. Illustration: Illustration Horticole **29**: 11 (1882); Gardeners' Chronicle **19**: 501 (1883). Leaves smaller than in *A. watsoniana*, their blades more elliptic and with only *c.* 6 pairs of main veins; basal lobes joined for one-fifth to one-third of their length, their main veins without branches towards the sinus. Main veins of principal lobe without conspicuous branches; interprimary veins gently sinuous. *Java*. G2.

Plants presently cultivated as *A. putzeysii* may be incorrectly named. Another variant of this alliance in cultivation has the upper leaf surface with a greyish metallic sheen, the basal lobes elliptic and the sinus correspondingly abruptly narrowed towards the base; the midribs of the basal lobes have rather indistinct branches towards the sinus, and the blade is up to 80 cm long and has about 8 pairs of main veins; it is grown as *A. korthalsii* Schott and is perhaps illustrated by the figure labelled *A. lowii* var. *veitchii* in Graf, Exotica, edn 8, 123 (1976). (*A. thibautiana* Masters (Illustration: Illustration Horticole **18**: 72, 1881) was stated by Engler and Krause to be synonymous with *A. korthalsii* but it is not the same as the plant just described.)

The plant known as *A. lowii* (var.) *grandis* (Illustration: Graf, Exotica, edn 8, 123, 1976) belongs here; it has broadly ovate-cordate leaf-blades with a narrow sinus, and rounded basal lobes joined for about one-third of their length. The midribs of the basal lobes give off 3 or 4 main veins towards the sinus. The subsidiary venation is very regular and the interprimary veins are nearly straight (figure 3(7), p. 76).

10. A. micholitziana Sander. Illustration: Botanical Magazine, 8522 (1913); Aroideana **2**: 37 (1979); Everett, Encyclopedia of horticulture **1**: 117 (1980). Plant less than 1 m. Stem 30–35 cm. Leaf-blade to 25 cm or more, narrowly sagittate, sometimes peltate, with the basal lobes shortly united, wavy, rich green with white to cream midrib and lateral veins; upper surface with velvety texture. Leaf-stalk to 35 cm or more, paler green, with brown to purple marbling. Spathe-tube 2.5–3 cm, green, limb 9–10 cm, paler green than tube outside, whitish or yellowish inside. Spadix almost as long as spathe. *Philippines*. G2.

'Maxkowskii' has the leaf-blade less wavy and more cordate, darker green with whiter veining. Leaf-stalk clear green without marbling; a very popular cultivar known in commerce as "African Mask".

11. A. cuprea (C. Koch & Bouché) C. Koch. Illustration: Botanical Magazine, 5190 (1860); Birdsey, The cultivated aroids, 23 (1951); Graf, Exotica, edn 8, 124 (1976); Aroideana **2**: 39 (1979).
Plant less than 1 m. Leaf-blade usually 20–30 × 10–18 cm, oblong-ovate or elliptic, cordate, peltate, with basal lobes united for most of their length, shortly and abruptly acuminate at apex; upper surface greenish with purplish and pinkish orange metallic sheen (coppery) and darker main veins; lower surface intensely purple; major veins markedly sunken, curved towards the base of the blade as they leave the midrib and then towards its apex as they approach the margin. Leaf-stalk and inflorescence-stalk purple. Spathe 5–7 cm, purple; limb about same length as tube, greenish. *Malaysia, Borneo*. G2.

There is also a variant in which the leaf-blade is green on both surfaces. The hybrid with *A. lowii* is mentioned under that species (p. 102).

29. SYNGONIUM Schott
T.B. Croat
Plants evergreen, epiphytic or rooted in soil, the stems loose and climbing when young, adpressed to trees, and with short internodes when adult. Sap milky. Juvenile plants with simple, usually ovate to sagittate leaf-blades; pre-adult leaf-blade larger, commonly sagittate. Leaf-blade of adult plant (in cultivated species) with 3 or sometimes 5 leaflets, the lateral often unequal, sometimes cleft or divided (blade pedate). Marginal veins usually several. Inflorescence-stalk erect in flower, pendent in fruit. Lower portion of spathe with overlapping edges; limb of spathe usually boat-shaped, opening broadly, usually thin, falling after a few days. Spadix shorter than spathe. Florets unisexual, without perianths. Zone of female florets enclosed within spathe-tube; male and female zones separated by a few sterile florets. Male florets with 4 united stamens. Fruit compound, pulpy.

A genus of 33 species throughout the New World tropics. Relatively few species are commercially available and plants are usually sold in juvenile or pre-adult stages. Variation in leaf-markings has led to the naming of cultivars in species Nos. 2 and 3.

The plants are difficult to pollinate and the fruits are never seen in cultivation. They grow best under conditions of high humidity and partial shade but species currently in cultivation do remarkably well in average home or office conditions. The potting medium should be loose and well drained but plants should be watered frequently.
Literature: Croat, T.B., A revision of Syngonium (Araceae), *Annals of the Missouri Botanical Garden* **68**: 565–651 (1981).

1a. Each lateral division of the leaf-blade entire or with a subsidiary lobe or leaflet; inflorescences 3–5 per axil; spathe-tube nearly cylindric, frequently more than 5 cm; spathe-limb usually twice as long as visible part of spadix **1. auritum**
b. Each lateral division of the leaf-blade with 2–5 connected lobes or, more usually, leaflets; inflorescences 4–11 per axil; spathe-tube ellipsoid, less than 5 cm; spathe-limb only slightly longer than visible part of spadix 2
2a. Stem sometimes with numerous rough projections; leaf-stalk usually with a sharp rib along the upper side; middle segments of leaf-blade nearly equal in size and shape; male florets deeply hollowed at apex, showing no indication of fusion between anthers **2. angustatum**
b. Stem smooth, lacking any projections; leaf-stalk with at most an obtuse rib along the upper side; middle segments of leaf-blade usually somewhat unequal; male florets truncate at apex with an indication of the line of fusion between anthers **3. podophyllum**

1. S. auritum (Linnaeus) Schott (*Philodendron auritum* invalid, not Lindley; *P. trifoliatum* invalid). Illustration: Birdsey, The cultivated aroids, 123 (1951); Graf, Exotica, edn 8, 234, 237 (1976); Annals of the Missouri Botanical Garden **68**: 611 (1981)
Stem of adult plant slightly glaucous. Leaf-blade of juvenile plant ovate, sagittate to hastate at base. Leaf-blade of pre-adult sagittate to hastate. Leaf-stalk of adult 15–48 cm, sheathed usually for four-fifths of its length. Leaf-blade of adult with 3 or sometimes 5 leaflets, the laterals sometimes with an external subsidiary lobe or leaflet; central segment 20–30 × 6–20 cm, broadly elliptic. Inflorescences up to 5 per axil; stalks 7–13 cm. Spathe with tube 4–8.5 cm, nearly cylindric, dark green, tinged reddish within at base; limb 12–18 cm, mucronate at apex, cream outside, faintly tinged with violet-purple within. Spadix extending to half the height of the spathe-limb; female part 3.5–5 cm; male part 3.5–9 cm, creamy white. *Jamaica & Hispaniola*. G2.

2. S. angustatum Schott (*S. oerstedianum* Schott; *S. podophyllum* var. *oerstedianum* (Schott) Engler; *S. albolineatum* invalid; *Nephthytis triphylla* Bailey). Illustration: Birdsey, The cultivated aroids, 121 (1951); Annals of the Missouri Botanical Garden **68**: 607, 614, 619 (1981).
Stem not glaucous, usually sparsely warty in juvenile plant. Leaf-blade of juvenile plant heart-shaped, dark green, marked on upper surface along principal veins with grey-green. Leaf-blade of pre-adult sagittate or hastate. Leaf-stalk of adult 15–40 cm, sometimes bloomed, sheathed for half to four-fifths of its length. Leaf-blade of adult pedate; central segment 11–31 × 4–10 cm, elliptic to oblong-elliptic or oblanceolate; lateral divisions with 2–5 segments. Inflorescences about 7 per axil; stalk to 17 cm. Spathe 7–14 cm; tube 2–5 cm, ellipsoid, green and usually glaucous; limb cream. Spadix with female part 1.3–2.6 cm × 5–10 mm, greenish, male part 2.5–6.5 cm, pale yellow. *Mexico to Costa Rica*. G2.

3. S. podophyllum Schott (*S. gracile* (Miquel) Schott; *S. riedelianum* Schott; *S. vellozianum* Schott). Illustration: Birdsey, The cultivated aroids, 125 (1951); Graf, Exotica, edn 8, 234, 237 (1976); Aroideana **1**: 81 (1978) – as S. auritum; Annals of the Missouri Botanical Garden **68**: 632, 638, 641 (1981).
Stem sometimes glaucous. Leaves sometimes with grey-green markings. Leaf-blade of juvenile plant 7–14 cm, heart-shaped, of pre-adult sagittate or hastate. Leaf-stalk of adult 15–60 cm, sheathed for two-thirds of its length. Leaf-blade of adult pedate; central segment 16–38 × 6–17 cm, obovate to elliptic; lateral divisions with 2–5 united or free segments, the outermost variously auricled. Inflorescences 4–11 per axil; stalk usually less than 9 cm when in flower, sometimes glaucous. Spathe 9–11 cm; tube 3–4 cm, ovoid to ellipsoid, green, sometimes glaucous; limb greenish white to creamy white or rarely yellowish. Spadix with female part 1–2 cm × 6–9 mm, greenish white, male part 4–7 cm. *Mexico to the Guianas, Brazil & Bolivia*. G2.

30. SYNANDROSPADIX Engler
P.F. Yeo
Plant tuberous, with a resting period. Leaves few. Leaf-stalk to 30 cm. Leaf-blade 20–45 × 15–20 cm, ovate with cordate base, or sagittate. Inflorescences sometimes emerging before the leaves; stalk shorter than leaf-stalk, pale green with darker green stripes. Spathe to *c*. 15 cm, obliquely funnel-shaped, tailed, curved forwards so that open side is inclined downwards; margins recurved except at base where they overlap (some descriptions indicate that spathe unfurls completely); outer surface pale green with darker green stripes, inner brownish red with darker stripes of same colour. Spadix *c*. 12 cm, with a stipe; lower part united to spathe. Florets without perianths, mostly unisexual. Male florets with 4–5 united stamens; common filament 6–7 mm. Female florets with 3–5 staminodes in the form of ovate scales. Ovary with 3–5 cells, each with 1 ovule; style well developed.

One species from the eastern side of the Andes. It may be grown in a cool greenhouse for its striking inflorescences which are quite freely produced, but emit a powerful smell of decaying flesh.

1. S. vermitoxicus (Grisebach) Engler. Illustration: Botanical Magazine, 7242 (1892); Everett, Encyclopedia of horticulture **10**: 3276 (1982).
Argentina (Tucuman Province). G1.

31. SPATHICARPA W.J. Hooker
P.F. Yeo
Small evergreen plants, producing leaves and inflorescences from a short tuberous rhizome. Leaf-stalk with a long sheath. Leaf-blade with netted veins. Inflorescence-stalk longer than the leaf-stalk or the leaf, slender. Spathe opening to expose all the florets, green. Spadix half-cylindric, united to spathe. Florets without perianths, unisexual. Male florets of 3–4 united stamens. Female florets each with 3 small staminodes. Ovary 1-celled; ovule 1.

About 6 species from tropical South America, one of which is occasionally cultivated in tropical collections for its curious inflorescences which are produced almost continuously. It needs damp, shady conditions in summer and lighter and drier conditions in winter. Propagation is by division or from seed, which, however, rapidly loses viability.
Literature: Engler, A., Aroideae & Pistioideae, *Das Pflanzenreich* **73**: 53–7 (1920).

1. S. sagittifolia Schott. Illustration: Das Pflanzenreich **73**: 55 (1920); Parey's Blumengärtnerei, **1**: 193 (1958).
Leaf-stalk about twice as long as blade. Leaf-blade to 12 × 4 cm, sagittate or hastate, conspicuously net-veined, rich green. Inflorescence-stalk inclined. Spathe *c*. 5 cm, lanceolate, inclined so that inner surface faces upwards, recurved at apex, green. Spadix with florets in 4 or 5 rows, the 2 outer female, the others male. Male florets peltate. *Brazil, Paraguay, Argentina*. G2.

32. CALLOPSIS Engler
P.F. Yeo
Plants evergreen. Stem a horizontal rhizome with crowded, erect, lateral shoots, each consisting of 1–2 sheathing scale-leaves and a leaf or inflorescence-stalk. Leaf-stalk to 13 cm. Leaf-blade to 13 cm, entire, spreading, ovate with cordate base, wrinkled; main veins joined by cross-veins and joined to the marginal vein. Inflorescence borne above the leaves. Spathe 3–4 cm, white, flat but folded at the base. Spadix 1.5 cm, yellow, cylindric, joined to the spathe at the base. Florets without perianths, unisexual; zones of male and female florets adjacent. Female florets only 3–12, with prominent style and stigma turned to one side. Ovule 1.

One species. Sometimes cultivated in tropical collections. Requires a soil high in organic matter, shade from strong sun and a humid atmosphere.
Literature: Engler, A., Aroideae & Pistioideae, *Das Pflanzenreich* **73**: 58–9 (1920).

1. C. volkensii Engler. Illustration: Botanical Magazine, 8071 (1906); Das Pflanzenreich **73**: 59 (1920); Graf, Exotica, edn 8, 166 (1976).
Tanzania. G2. Winter.

33. ARUM Linnaeus
D.A. Webb & P.F. Yeo
Tuberous herbs. Leaves all basal; stalk shortly sheathing at base, usually about twice as long as blade; blade hastate to sagittate, entire. Inflorescence-stalks shorter than leaf-stalks or nearly equalling them, usually visible above the ground. Lower part of spathe narrow, with overlapping margins, enclosing the floral region of the spadix and usually the lower part of its appendix; limb usually lanceolate, concave, longer than the spadix. Florets without perianths, unisexual. Spadix with zones of male and female florets usually separated by a zone of sterile florets; sterile florets usually present also above the male (figure 4(7), p. 77). Male florets with 3 or 4 stamens; ovary 1-celled. Berry red; seeds 1–6.

A genus of 12 species, ranging from the Canary Islands to Iran, and northwards to Sweden. *A. orientale, A. maculatum* and *A.*

italicum are readily naturalised in shade or semi-shade and occur as native plants in gardens in some parts of Europe. The other species should be planted in brighter and warmer situations, though they may need some shade in the Mediterranean region. The floral odour is usually fetid, *A. dioscoridis* being particularly offensively scented, but in *A. creticum* the scent is sweet. Propagation is by division of the tubers.

Literature: Engler, A., Aroideae & Pistioideae, *Das Pflanzenreich* **73**: 67–99 (1920); Prime, C.T., *Lords & ladies* (1960).

Season of growth. Leaves appearing in autumn: **1–3,8**; leaves appearing in spring: **4–7**. Flowers appearing in autumn: **2**.
Tuber. Horizontal: **7,8**; vertical: **1–6**.
Spathe. Mainly white, yellow or green, though sometimes with purple spots on margin: **1,3,5–8**; mainly purple inside: **2–4,6**.
Appendix of spadix. With a stalk-like basal part much narrower than the upper part: **2,4,6–8**; base more or less uniform in diameter, not stalk-like: **1,3,5,6**.

1a. Flowering in autumn **2. pictum**
b. Flowering in spring or early summer 2
2a. Female zone of spadix much shorter than male **1. creticum**
b. Female zone of spadix about as long as male or longer 3
3a. Spathe-limb uniformly dark purple inside and pale greenish outside; female zone of spadix about same length as male **4. palaestinum**
b. Spathe-limb with inner and outer surfaces not strongly contrasted or, if they are, part of the inner surface is boldly spotted; female zone of spadix much longer than male 4
4a. Spathe-limb marked in part with bold coalescent spots, usually of dark purple on a greenish or yellowish background but sometimes the reverse **3. dioscoridis**
b. Spathe-limb with scattered spots or none 5
5a. Leaves appearing in autumn **8. italicum**
b. Leaves appearing in spring 6
6a. Spadix with little or no space between male zone and the upper and lower zones of sterile florets; spathe 5–10 × 2–3 cm **5. hygrophilum**
b. Spadix with a clear space between male zone and the upper and lower zones of sterile florets; spathe usually more than 10 × 3 cm 7
7a. Tuber horizontal; leaves often with blackish spots; fruiting spike 3–5 cm **7. maculatum**
b. Tuber vertical; leaves unspotted; fruiting spike 6–8 cm **6. orientale**

1. A. creticum Boissier & Heldreich. Illustration: Botanical Magazine n.s., 101 (1950); Huxley & Taylor, Flowers of Greece and the Aegean, f. 418 (1977); Grey-Wilson & Mathew, Bulbs, pl. 35 (1981).
Tuber vertical. Leaves appearing in autumn, hastate-sagittate, the blade 10–20 × 7–14 cm. Inflorescence-stalk usually exceeding the leaf-stalk. Spathe 15–25 cm, bright yellow, greenish yellow, white or rarely pale green. Zone of female florets much shorter than that of male; sterile florets few and small, sometimes absent. Appendix of spadix orange-yellow or purple, almost equalling the spathe. *Crete & Karpathos.* H4. Late spring.

2. A. pictum Linnaeus filius. Illustration: Coste, Flore de la France **3**: 433 (1906); Grey-Wilson & Mathew, Bulbs, pl. 35 (1981).
Tuber vertical. Leaves appearing in autumn, hastate; basal lobes short and rounded; stalk only slightly longer than blade. Inflorescence-stalk mainly below ground. Spathe 15–25 cm, dark reddish purple. Zone of female florets longer than that of male; sterile florets above the male but absent between male and female. Appendix of spadix dark purple, rather shorter than the spathe, the basal half forming a narrow stalk. *Islands of W Mediterranean.* H4. Autumn.

3. A. dioscoridis Sibthorp & Smith. Illustration: Das Pflanzenreich **73**: 70 (1920); Polunin & Huxley, Flowers of the Mediterranean, f. 224 (1965); Huxley & Taylor, Flowers of Greece and the Aegean, f. 419 (1977); Rix & Phillips. The bulb book, 157 (1981).
Tuber vertical. Leaves appearing in autumn, hastate-sagittate, acute; basal lobes triangular, acute. Inflorescence-stalk mainly below ground. Spathe 15–35 cm, usually for the most part dark purple, but sometimes pale green only slightly tinged with purple, always to some extent spotted with brown, yellow, black or dark purple, some of the spots coalescent. Zone of female florets twice as long as that of male. Sterile florets numerous, short. Appendix of spadix dark purple, not much shorter than the spathe. *From Turkey & Iraq to Israel.* H5. Spring.

4. A. palaestinum Boissier. Illustration: Botanical Magazine, 5509 (1865); Birdsey, The cultivated aroids, 41 (1951); Graf, Exotica, edn 8, 144 (1976); Rix & Phillips, The bulb book, 156 (1981).
Tuber vertical. Leaves appearing in spring, hastate; lobes broad, obtuse or acute. Spathe 15–20 cm, yellowish green at the base, the blade spreading, greenish outside, uniformly dark purple inside. Zones of male and female florets of about equal length. Appendix of spadix slender, nearly as long as the spathe, the lower part narrowed into a stalk. *Israel.* H4. Spring.

5. A. hygrophilum Boissier. Illustration: Das Pflanzenreich **73**: 71 (1920); Everett, Encyclopedia of horticulture **1**: 261 (1980).
Tuber vertical. Leaves appearing in spring, hastate, with obtuse basal lobes. Spathe 5–12.5 cm, mostly green, but usually with purple margin and sometimes inner surface suffused with purple; limb not much longer than tube. Zone of female florets twice as long as that of male; upper and lower zones of sterile florets adjacent or nearly so to male zone. Appendix of spadix dark purple, moderately or considerably shorter than spathe, of uniform diameter at base. *Cyprus to Syria & Israel; ?Morocco.* H4. Spring.

6. A. orientale Bieberstein. Illustration: Huxley & Taylor, Flowers of Greece and the Aegean, f. 418 (1977).
Tuber vertical. Leaves appearing in spring, sagittate. Spathe 7–15 cm, green, variably tinged with purple and sometimes almost entirely purple, but not spotted. Zone of female florets much longer than that of male; upper and lower zones of sterile florets clearly separated from the male zone. Appendix of spadix dark purple, with or without a stalk-like base, somewhat shorter than the spathe. Fruiting spike 6–8 cm. *C & E Europe, northwards to Sweden; SW Asia, eastwards to Iran.* H3. Early summer.

7. A. maculatum Linnaeus. Illustration: Prime, Lords & ladies, t. 1, 57 (1960); Proctor & Yeo, The pollination of flowers, 290 (1970); Ross-Craig, Drawings of British plants, part 30; 39 (1973); Grey-Wilson & Mathew, Bulbs, pl. 35 (1981).
Very like *A. orientale*, and only distinguished with certainty by the horizontal tuber. It often also differs, however, in having blackish spots on the leaves and spathe, which is otherwise greenish, in having the appendix to the spadix occasionally yellow, and always with a stalk-like base, and in the longer

spathe (10–20 cm) and shorter fruiting spike (3–5 cm). *Europe*. H2. Late spring.

8. A. italicum Miller. Illustration: Botanical Magazine, 2432 (1823); Polunin & Huxley, Flowers of the Mediterranean, f. 220 (1965); Graf, Tropica, 95 (1978); Grey-Wilson & Mathew, Bulbs, pl. 35 (1981).
Tuber horizontal. Leaves appearing in autumn or early winter, hastate or sagittate, sometimes veined with white, rarely with dark spots. Spathe 15–40 cm, greenish yellow, rarely white, sometimes edged with purple. Zone of female florets much longer than that of male. Appendix of spadix yellow, with distinct, stalk-like base, much shorter than spathe. *From the Canary Islands to Cyprus, & northwards to England*. H3. Late spring.

34. DRACUNCULUS Schott

D.A. Webb & P.F. Yeo

Tuberous plants, *c.* 70–180 cm, with several large leaves. Leaves basal, their stalks with long, overlapping sheaths which together resemble a stem. Leaf-blade pedately divided or deeply pedately lobed. Lower part of spathe with overlapping margins enclosing lower part of spadix, limb large, widely open. Florets without perianths, unisexual. Spadix with zones of male and female florets adjacent or separated; appendix long. Male florets with 2–4 stamens. Ovary 1-celled. Berry green; seeds 6.

A genus of 3 species, all described below. They should be cultivated in fairly moist, humus-rich soil in a sheltered position in sun or semi-shade.

Literature: Engler, A., Aroideae & Pistioideae, *Das Pflanzenreich* **73**: 99–104 (1920) – as *Dracunculus* and *Helicodiceros*; Henderson, A., More on dragon plants, *The Garden* **107**: 498–9 (1982).

1a. Appendix of spadix and inner surface of spathe-limb covered with hairs or filaments **3. muscivorus**
b. Appendix of spadix and inner surface of spathe-limb without hairs or filaments 2
2a. Spathe within, and appendix of spadix dark purplish red **1. vulgaris**
b. Spathe white within; appendix of spadix yellow **2. canariensis**

1. D. vulgaris Schott (*Arum dracunculus* Linnaeus). Illustration: Fiori & Paoletti, Iconographia Florae Italicae, edn 3, 67 (1933); Polunin & Huxley, Flowers of the Mediterranean, f. 223 (1965); Huxley & Taylor, Flowers of Greece and the Aegean, f. 242 (1977); Rix & Phillips, The bulb book, 157 (1981).
Leaf-stalk marbled or spotted with dark purple; blade often spotted or striped with white, with 11–15 lanceolate segments, the largest to 20 cm long. Inflorescence-stalk longer than leaf-sheaths. Spathe 25–40 cm; tube 6 cm; limb 10–15 cm wide, tilted backwards, with wavy margins, uniformly dark reddish purple inside. Spadix with zones of male and female florets adjacent; sterile florets present above the male zone; appendix almost equalling the spathe, stout, blackish red. *Mediterranean region, from Corsica to S Turkey*. H3. Early summer.

2. D. canariensis Kunth. Illustration: Bramwell & Bramwell, Wild flowers of the Canary Islands, 108 (1974); The Garden **107**: 499 (1982) – lower figure.
Segments of leaf-blade 7–11, slightly smaller than in *D. vulgaris*, unspotted. Inflorescence-stalk purple-spotted. Spathe 25–35 cm; limb 5–6 cm wide, erect, with slightly inrolled and wavy margins, white inside. Spadix without sterile florets and with a slender yellow appendix, its male florets not enclosed in base of spathe. *Madeira & the Canary Islands*. H5. Early summer.

3. D. muscivorus (Linnaeus filius) Parlatore (*D. crinitus* Schott; *Helicodiceros muscivorus* (Linnaeus filius) Engler). Illustration: Edwards's Botanical Register **10**: 831 (1824); Flore des Serres **5**: t. 445, 446 (1849); Fiori & Paoletti, Iconographia Florae Italicae, edn 3, 67 (1933); Graf, Exotica, edn 8, 145 (1976).
Leaf-stalk, inflorescence-stalk and outside of spathe-tube pale green with short darker green stripes. Leaf-blade deeply pedately cut into narrow lobes, the 3 central ones hastately arranged, the others disposed at various angles owing to twisting of each lateral axis. Spathe situated below the leaves; tube 15 cm; limb *c.* 25 × 20 cm, bent backwards horizontally, purplish brown with darker or lighter mottling, covered with coarse reddish hairs. Spadix with sterile florets above and below the male zone; appendix dark green or yellow, about half as long as spathe-limb, mainly horizontal, covered with bristle-like filaments. *Corsica, Sardinia & Balearic Islands*. H5. Early summer.

35. SAUROMATUM Schott

P.F. Yeo

Tuberous herbs, bearing one leaf at a time. Leaf-blade pedately cut or divided. Inflorescence appearing before the leaves; stalk short, so that the spathe may rest on the ground. Spathe with margins fused below to form a tube; limb lanceolate; both parts withering after flowering. Florets without perianths, unisexual. Spadix with zones of male and female florets about equal, with a long sterile zone between them which bears club-shaped sterile florets below, and a long curved appendix. Male florets indistinct. Ovary 1-celled, with 2–4 ovules.

A genus of 2 species from E and W Africa and S Asia, both cultivated. That described below is grown for its strikingly spotted leaf-stalks and spathes, which give it a sinister appearance. The tubers are sold as a curiosity for flowering indoors without soil or water, though the floral odour is sometimes offensive. Plants may be grown in the open or in greenhouses in a humus-rich well-watered soil. They are dormant in winter, when the tubers may be lifted if desired. Propagation is by offsets from the tubers.

Literature: Engler, A., Aroideae & Pistioideae, *Das Pflanzenreich* **73**: 122–7 (1920).

1. S. venosum (Aiton) Kunth (*S. guttatum* (Wallich) Schott; *Arum venosum* Aiton; *A. guttatum* Wallich; *S. nubicum* Schott). Illustration: Edwards's Botanical Register **12**: t. 1017 (1826); Botanical Magazine, 4465 (1849); Flore des Serres **13**: 79 (1858); Everett, Encyclopedia of horticulture **9**: 3063 (1982).
Tuber to *c.* 15 cm in diameter. Leaf-stalk to 50 cm, green, often with darker, grey-green spots. Leaf-blade pedately cut or divided, with 7–15 lobes or leaflets. Inflorescence-stalk *c.* 5 cm. Spathe-tube 5–10 × 2–2.5 cm; limb 30–70 × 8–10 cm, channelled, becoming twisted, often with wavy margins; outer surface of limb dull purplish, inner surface yellowish or greenish, variously spotted with brownish purple. Spadix appendix to *c.* 30 cm, green, flushed with purple. *E & W Africa, Himalaya from Kashmir to N Burma, peninsular India*. H3–4. Spring–early summer.

Very variable in colour and extent of spotting; probably varying in hardiness according to origin.

36. BIARUM Schott

P.F. Yeo

Plants with small flattened-spherical tubers producing roots above. Scale-leaves conspicuous around bases of foliage leaves and inflorescences. Leaf-blade entire, not lobed. Inflorescence-stalk concealed below ground or slightly exposed in fruit. Spathe with margins of lower part united to form a tube, withering before fruit is ripe. Florets unisexual, without perianths. Spadix with a wide space between zones of male and female florets on which are usually borne some slender sterile florets; appendix as long as remainder of spadix or longer. Male florets with 1 or 2 stamens. Ovary 1-celled, with 1 ovule. Fruiting spadix approximately spherical. Berry white or pale green, sometimes striped with purple.

A genus of about 12 species in the Mediterranean region and W Asia, of which a few are occasionally cultivated. The inflorescences have been described as looking like 'black holes in the ground with a tail protruding'. They are hardy and need hot conditions when dormant in summer to promote flowering. Identification in this genus is difficult, partly because leaves and flowers are not usually present at the same time.

Literature: Engler, A., Aroideae & Pistioideae, *Das Pflanzenreich* **73**: 133–43 (1920); Talavera, S., Revision de las especies españolas del genero Biarum Schott, *Lagascalia* **6**: 275–96 (1976).

1a. Spadix with sterile florets above and below male zone; spadix appendix to 4 mm in diameter **1. tenuifolium**
b. Spadix with sterile florets only below male zone; spadix appendix 6–8 mm in diameter **2. eximium**

1. B. tenuifolium (Linnaeus) Schott. Leaf usually 5–20 cm × 3–15 mm, narrowly lanceolate, often without a distinct stalk. Inflorescence appearing before the leaves. Spathe-tube to 6 cm; limb to 30 cm, purple inside and green outside. Spadix with zone of female florets 3 mm and zone of male florets 6–8 mm; sterile florets present above and below male zone; appendix to 4 mm in diameter, exceeding the spathe, much longer than remainder of spadix. *Mediterranean region from Greece westwards to Portugal.*

Very variable; some populations have been separated as distinct species or treated as varieties.

Var. **tenuifolium**. Illustration: Botanical Magazine, 2282 (1821); Huxley & Taylor, Flowers of Greece, f. 422 (1977). Plant relatively large. Spathe-limb many times longer than wide, erect, usually twisted. Spadix appendix 20–30 cm, erect. *C & E Mediterranean region.* H3. Autumn.

Var. **abbreviatum** (Schott) Engler. Illustration: Mathew, Dwarf bulbs, 50 (1973); Grey-Wilson & Mathew, Bulbs, pl. 35 (1981). Plant smaller. Spathe-limb less than 10 cm, about half as wide as long, hooded. Spadix appendix *c.* 10 cm, emerging almost horizontally from spathe. *Italy, Yugoslavia, N Greece.* H3. Late summer.

2. B. eximium (Schott & Kotschy) Engler. Illustration: Das Pflanzenreich **73**: 138 (1920); Mathew, Dwarf bulbs, 50 (1973). Leaf-stalk to 10 cm; leaf-blade to 9 × 4 cm. Inflorescence appearing before the leaves. Spathe *c.* 12 cm; limb 9 × 4 cm, nearly flat, velvety and blackish purple within, pale outside with dark brownish purple spots. Spadix about as long as spathe; female zone 5 mm, sterile zone and male zone each 1–1.5 cm; sterile florets present only below male zone; appendix 6–8 mm in diameter, not exceeding the spathe, about as long as remainder of spadix. *Turkey.* H3. Autumn.

37. ARISARUM Targioni-Tozzetti

D.A. Webb

Rootstock a tuber or rhizome. Leaves long-stalked, sagittate, entire. Spathe withering before fruit is ripe, the lower part with margins united so as to form a closed tube, the upper part open in front but curved forwards at the tip so as largely to conceal the spadix. Spadix shorter than or equalling the spathe, with a short floral zone at the base and a long appendix. Florets without perianths, unisexual, the female few, the male situated immediately above the female. Stamen 1; ovary 1-celled. Berry greenish, with about 6 seeds.

Two species from the Mediterranean region and the Atlantic islands. They are best cultivated in semi-shade in the rock garden or in a cool house in soil fairly rich in leaf-mould.

1a. Spathe acute to mucronate; appendix of spadix greenish **1. vulgare**
b. Spathe terminating in a long, thread-like tip; appendix of spadix whitish **2. proboscideum**

1. A. vulgare Targioni-Tozzetti (*A. simorrhinum* Durieu). Illustration: Botanical Magazine, 6023 (1873); Fiori & Paoletti, Iconografia della Flora Italiana, 62 (1895); Rix & Phillips, The bulb book, 156 (1981), Grey-Wilson & Mathew, Bulbs, pl. 35 (1981).
Rootstock an irregular tuber, from which arise an inflorescence and a single leaf. Leaves 5–12 × 3–10 cm; stalk to 30 cm, purple-spotted. Inflorescence-stalk 7–30 cm, slender. Spathe 4–6 × 1–1.5 cm, greenish, variably tinged or striped with dark purple. Appendix of spadix greenish. *Mediterranean region, Canaries & Azores.* H4–5. Late autumn–spring.

2. A. proboscideum (Linnaeus) Savi. Illustration: Botanical Magazine, 6634 (1882); Fiori and Paoletti, Iconografia della Flora Italiana, 62 (1895); Rix & Phillips. The bulb book, 159 (1971); Grey-Wilson & Mathew, Bulbs, pl. 35 (1981). Rootstock a slender rhizome. Leaves 6–15 × 4–10 cm; stalk to 25 cm, not spotted. Inflorescence-stalk 8–15 cm, often inclined or prostrate. Spathe dark brown, paler towards the base, the hooded part and tube 2–4 cm, the thread-like tip 5–15 cm. Appendix of spadix whitish, concealed within the spathe. *C & S Italy, S Spain.* H4. Spring.

Cultivated mainly as a curiosity, the upper part of the spathe (which is usually all that protrudes from beneath the leaves) being strongly suggestive of the tail and hindquarters of a mouse.

38. ARISAEMA Martius

D.A. Webb

Plants with tubers (often with offsets or stolons), or rarely rhizomes. Leaves solitary or few, compound, long-stalked, all basal, though their sheathing base sometimes conceals the lower part of the inflorescence-stalk so as to make the leaves appear to be borne on a stem. Two or more whitish, papery, narrow-oblong sheathing scale-leaves are usually present, enfolding the bases of the leaf-stalks and the inflorescence-stalk. Lower part of spathe narrow, with overlapping margins, surrounding the floral part of the spadix and sometimes much of its appendix as well; upper part concave or nearly flat, erect or bent forwards at the tip, but seldom completely concealing the spadix. Spadix with a long appendix. Florets without perianths, unisexual, the male usually not touching; borne, in monoecious plants, immediately above the female. Stamens 1–5, united when more than 1. Ovary 1-celled; seeds few.

Most species are normally dioecious, including all those described below except *A. dracontium*. The condition is, however, variable in some species. Dioecious species occasionally bear some abortive male florets or filamentous sterile florets above the female.

A genus of 120–150 species, mainly from E Asia and the Himalayan region, but a few

from N America and E Africa. They should be cultivated in semi-shade in damp, but not sodden, peaty soil.
Literature: Engler, A., Aroideae & Pistioideae, *Das Pflanzenreich* **73**: 149–220 (1920); Hara, H., A revision of the eastern Himalayan species of the genus Arisaema (Araceae), *Flora of Eastern Himalaya, Second Report (University Museum, Tokyo, Bulletin No. 2)*: 320–54 (1971); Ohashi, H. & Murata, J., Taxonomy of the Japanese Arisaema (Araceae), *Journal of the Faculty of Science of the University of Tokyo*, Section III, **12**: 281–336 (1980); Mayo, S.J., A survey of cultivated species of Arisaema, *The Plantsman* **3**: 193–209 (1982).

Tip of spathe. Very long and thread-like: **9**.
Tip of spadix. Very long and thread-like: **1,2,8**.
Leaves. Pedate: **7,8**; palmate with 9 or more leaflets: **9**; with 3, rarely 5, leaflets: **1–6**.
Sexuality of plant. Usually monoecious: **8**.

1a. Spathe terminating in a thread-like tip nearly as long as the rest of the spathe **9. consanguineum***
b. Spathe acute, acuminate, or with a slender tip much shorter than the rest of the spathe 2
2a. Spadix prolonged into a long, thread-like tip 3
b. Spadix obtuse or acute, but without a thread-like tip 5
3a. Leaves 2; limb of spathe broader than long, deeply 2-lobed, with a short filament arising from the sinus **1. griffithii**
b. Leaf solitary; limb of spathe longer than broad, not 2-lobed 4
4a. Leaf-blade pedate, with 7–15 leaflets; appendix of spadix whitish with green tip **8. dracontium**
b. Leaf-blade with 3 leaflets; appendix of spadix yellowish pink, with a dark purple tip **2. speciosum**
5a. Leaf-blade pedate, with 5–11 leaflets **7. serratum**
b. Leaf-blade with 3 (rarely 5) leaflets 6
6a. Upper part of spathe white, cream or pale pink **3. candidissimum**
b. Upper part of spathe mostly green or dark purple, though sometimes striped with white 7
7a. Spathe with conspicuous rounded auricles at base of blade **5. ringens**
b. Spathe without auricles 8
8a. Appendix of spadix white, much thicker at the tip than at the base **6. sikokianum**
b. Appendix of spadix brownish, more or less cylindric **4. triphyllum**

1. A. griffithii Schott. Illustration: Botanical Magazine, 6491 (1880); Rix & Phillips, The bulb book, 158 (1981); The Plantsman **3**: 201 (1982).
Leaves 2; stalks stout, 12–20 cm, not spotted; leaflets 3, to 25 × 20 cm, broadly diamond-shaped to nearly circular. Inflorescence-stalk shorter than the leaf-stalks. Lower part of spathe *c.* 10 × 2.5 cm, dark purple with longitudinal white stripes; limb to 15 × 20 cm, dark purple with a network of green veins, deeply divided into 2 rounded lobes by a sinus 8–10 cm deep, from which protrudes a slender filament 5–10 cm long. Spadix with a stout floral zone surmounted by a dark purple appendix 5 cm long (excluding the thread-like tip), expanded into a wide disc near its base; thread-like tip 30 cm or more. *E Himalaya*. H4. Early summer.

2. A. speciosum (Wallich) Martius. Illustration: Botanical Magazine, 5964 (1872); Graf, Exotica, edn 8, 145 (1976); The Plantsman **3**: 207 (1982).
Leaf solitary; stalk 25–40 cm, stout, boldly spotted with purplish brown; leaflets 3, 20 × 12 cm, shortly stalked, triangular-ovate to lanceolate, acuminate, edged with reddish purple. Inflorescence-stalk much shorter than the leaf-stalk. Spathe 12–20 cm, the tubular base pale green, striped with dull red; limb ovate-lanceolate, long-acuminate, deep purple inside, striped with pale green, greenish outside. Appendix of spadix 5–7 cm, yellowish pink, terminating in a dark purple, thread-like tip 30–50 cm. *C & E Himalaya*. H5. Spring.

3. A. candidissimum Wright Smith. Illustration: Botanical Magazine, 9549 (1939); Bulletin of the Alpine Garden Society **38**: 409 (1970); Rix & Phillips, The bulb book, 158 (1981).
Leaf solitary, appearing with the flower or just after; stalk 25–35 cm; leaflets 3, 10–20 × 7–18 cm, the middle one the largest. Inflorescence-stalk 12–20 cm. Spathe 8–15 cm, the lower part green, striped with white; limb ovate, acuminate, white, cream or pale pink. Male florets touching. Appendix of spadix 5–6 cm, tapered towards the tip, green. *S China (Yunnan and S Sichuan)*. H4. Early summer.

4. A. triphyllum (Linnaeus) Schott (*A. atrorubens* (Aiton) Blume). Illustration: Botanical Magazine, 950 (1806); Britton & Brown, Illustrated flora of the northern States and Canada **1**: 361 (1896); Rix & Phillips, The bulb book, 154 (1981).
Leaves usually 2, rarely 1 or 3; stalk 30–60 cm, slender, spotted; leaflets 3, 8–15 × 3–7 cm, ovate, acute. Inflorescence-stalk slightly shorter than leaves. Spathe 10–15 cm, green, dark purple, or green with purple stripes, the tip arching over the spadix but not concealing it. Appendix of spadix about 7 cm, brownish. *Eastern N America, from S Canada to Louisiana, and westwards to Kansas*. H2. Summer.

5. A. ringens (Thunberg) Schott (*A. praecox* Koch). Illustration: Botanical Magazine, 5267 (1861); Rix & Phillips, The bulb book, 155 (1981).
Similar to *A. triphyllum*, and formerly confused with it, differing chiefly in the long-acuminate or tailed leaflets and the spathe, which has 2 rounded auricles at the base of the blade, and whose upper part is curved sharply downwards so as to conceal the spadix. *S Japan, S Korea & E China*. H4. Spring.

6. A. sikokianum Franchet & Savatier (*A. sazensoo* misapplied). Illustration: Botanical Magazine, 9589 (1940); Graf, Tropica, 97 (1978); Rix & Phillips, The bulb book, 159 (1981); The Plantsman **3**: 205 (1982).
Leaves 2; stalk long, but colourless and sheathing in its lower part so that the leaves appear relatively short-stalked and borne on a stem; leaflets 3–5, to 15 × 10 cm, oblong-elliptic to broadly ovate, shortly acuminate. Inflorescence-stalk 10–30 cm. Spathe 15–20 cm, dark purple with green and white stripes; limb narrowly ovate, long-acuminate. Appendix of spadix 5 cm, slender below, thickened above and terminating in a spherical knob closing the throat of the spathe. *S Japan*. H4. Early summer.

7. A. serratum (Thunberg) Schott (*A. japonicum* Blume). Illustration: Botanical Magazine, 7910 (1903); Kitamura et al., Coloured illustrations of herbaceous plants of Japan (Monocotyledoneae), t. 53, f. 338 (1970).
Leaves 2; stalks 40–70 cm, sheathing for more than half their length; blade pedate with 5–11 lanceolate, more or less stalkless leaflets. Spathe 8–12 cm, the basal, tube-like portion constituting more than half the total length, shortly acuminate, green or purple with white stripes. Appendix of spadix 4–5 cm, slender, yellow. *C & S Japan*. H4. Spring.

8. A. dracontium (Linnaeus) Schott. Illustration: Edwards's Botanical Register **8**: 668 (1822); Gleason, Illustrated flora of the northeastern United States and adjacent

Canada 1: 367 (1952).
Leaf solitary; stalk to 80 cm, sheathing at the base; blade pedate, with 7–15 oblong-elliptic, shortly stalked segments. Spathe 5–7 cm, narrow, without any clear distinction between basal tubular part and blade, greenish. Spadix often with male and female florets; appendix white at the base, prolonged into a slender, green, thread-like tip about twice as long as the spathe.. *Eastern USA & SE Canada*. H2. Late spring.

9. A. consanguineum Schott. Illustration: Gardeners' Chronicle **62**: 127 (1917); Das Pflanzenreich **73**: 175 (1920); The Plantsman **3**: 197 (1982).
Leaf solitary; stalk to 80 cm, sheathing in its lower two-thirds, often spotted with dark red; blade palmate, with 11–21 linear-lanceolate, long-acuminate leaflets to 40 cm long. Inflorescence-stalk shorter than leaf-stalk. Spathe *c.* 20 cm, including the long, thread-like tip, which is nearly as long as the rest of the spathe, dark purple with fine white stripes. Spadix with a few sterile florets above the fertile; appendix 3 cm, slender, mostly concealed within the tube of the spathe. *From C China (Hubei) to NE India & NW Thailand*. H4. Early summer.

***A. concinnum** Schott. Illustration: Hara, Flora of eastern Himalaya, Second Report (University Museum, Tokyo, Bulletin No. 2), t. 19 (1971); Rix & Phillips, The bulb book, 158 (1981). Has the leaf-blade with 7–11 leaflets, which are broader (obovate-lanceolate) and thinner than in *A. consanguineum*, with wavy margins; spadix appendix thickened and roughened at the tip. *Himalaya & N Burma*. H5.

39. PINELLIA Tenore
D.A. Webb & P.F. Yeo
Slender plants growing from tubers. Leaves basal, their blades simple, lobed or compound. Inflorescence arising from tuber separately from the leaves. Spathe narrow, with margins overlapping below and with a flat or channelled limb above. Florets without perianths, unisexual. Spadix with zone of female florets entirely united with the spathe; male zone free; appendix slender. Male florets with 1 or 2 stamens. Ovary 1-celled. Berry with a single seed.

A genus of about 6 species from E Asia. They may be cultivated in well-drained, sandy soil, but *P. ternata* is apt to become a weed on account of the bulbils on the leaf-stalks.
Literature: Engler, A., Aroideae & Pistioideae, *Das Pflanzenreich* **73**: 220–5 (1920).

1a. Leaflets 3, not more than 12 cm; spadix appendix far exceeding spathe **1. ternata**
b. Leaflets 7–11, 15–18 cm; spadix appendix shorter than spathe **2. pedatisecta**

1. P. ternata (Thunberg) Breitenbach. Illustration: Das Pflanzenreich **73**: 223 (1920); Gleason, Illustrated flora of the northeastern United States and adjacent Canada **1**: 367 (1952); Makino, Illustrated flora of Japan, enlarged edn, 776 (1961); Kitamura et al., Coloured illustrations of herbaceous plants of Japan (Monocotyledoneae), t. 50, f. 323, 195 (1964).
Tuber 1–2 cm wide. Leaf often with a bulbil on the stalk or at base of blade. Leaf-blade simple in young plants, with 3 leaflets in adult plants. Leaflets 3–6 (rarely to 12) cm, ovate-elliptic to lanceolate. Inflorescence-stalk longer than leaf-stalk. Spathe 5–7 cm, green; limb two-thirds to twice as long as tube, downy inside, its apex clasping appendix of spadix. Spadix appendix 1–2 mm thick, much longer than the spathe, purple below, green above. *China, Korea & Japan*. H3. Summer.

2. P. pedatisecta Schott. Illustration: Das Pflanzenreich **73**: 223 (1920).
Tuber to 4 cm wide. Leaflets 7–11, the central ones 15–18 cm, ovate-lanceolate to lanceolate. Inflorescence-stalk shorter than leaf-stalk. Spathe 12–19 cm; limb 2–4 times as long as tube. Spadix appendix 1–2 mm thick, shorter than spathe, yellowish green. *N & W China*. H2.

40. LAGENANDRA Dalzell
N. Jacobsen
Plants evergreen, with creeping or erect rhizomes. Scale-leaves always present. Leaf-stalks shortly sheathed at the base. Leaf-blade with inrolled margins when young, narrowly elliptic to ovate, 10–50 cm long, with main veins running to the tip of the blade. Inflorescence-stalk much shorter than the leaf-stalk. Spathe of 2 parts, a basal floral compartment with united margins, and a limb with a narrow or broad opening, and usually with a tail-like appendage. Florets without perianths, unisexual. Spadix with zones of male and female florets separated, the sterile part slender. Female florets spirally arranged, each with a few ovules. Male florets each of 1 stamen which extrudes pollen through 2 horns. Fruiting spadix spherical; fruits somewhat fleshy, shed individually. Seeds large, ellipsoid, distinctly longitudinally ridged. The fruit takes about 6–9 months to mature, and the seeds germinate within a few days of being shed, without drying out.

About 10 species from Sri Lanka to southern India and tropical northeastern India, where they grow in and along shaded rivers. Several species are imported from Sri Lanka, but only 3 are commonly available. Generally, flowering specimens are necessary for a correct determination. Some species are poisonous.

In cultivation, Lagenandras require humid conditions and a rich soil, preferably without calcium. They are fine foliage plants for the aquarium, and the 3 numbered species can withstand being submerged for long periods. The rhizome should be placed at the surface of the soil. Flowering usually occurs from November to April. Propagated by cutting up the (frequently sprouting) rhizome.
Literature: De Wit, H.C.D., *Aquarienpflanzen*, Stuttgart (1971); De Wit, H.C.D., Revisie van het Genus Lagenandra Dalzell (Araceae), *Mededelingen Landbouwhogeschool Wageningen* **13**: 1–48 (1978); De Wit, H.C.D., *Aquariumplanten* (1983).

1a. Spathe smooth or finely wrinkled outside **2. thwaitesii***
b. Spathe rough to warty outside 2
2a. Spathe more than 12 cm **1. praetermissa***
b. Spathe less than 10 cm **3. lancifolia**

1. L. praetermissa De Wit (*L. ovata* misapplied).
Leaf-blade 20–50 × 4–12 cm, narrowly ovate, smooth, fleshy, and shining green. Spathe rough, brownish to greenish outside, more or less red inside. *Sri Lanka*. G2.

***L. ovata** (Linnaeus) Thwaites (*Arum ovatum* Linnaeus; *Cryptocoryne ovata* (Linnaeus) Schott; *L. insignis* Trimen). Illustration: De Wit, Aquarienpflanzen, 233 (1971). Differs only in the large, inflated, roughly warted, black-purple spathe. *Sri Lanka to S India*. G2.

2. L. thwaitesii Engler. Illustration: De Wit, Aquarienpflanzen, 235 (1971).
Leaf-blade 10–20 × 1–4 cm, narrowly ovate, rough, dull dark green, mostly with a silvery margin. Leaf-stalk usually short. Spathe finely wrinkled. *Sri Lanka*. G2.

***L. koenigii** (Schott) Thwaites (*Cryptocoryne koenigii* Schott). Leaves 20–50 × 1–2 cm. Spathe completely smooth, often curved. *Sri Lanka*. G2.

3. L. lancifolia (Schott) Thwaites (*Cryptocoryne lancifolia* Schott). Illustration: De Wit, Aquarienpflanzen, 229 (1971).

Leaf-blade narrowly ovate, 6–15 × 2–8 cm, sometimes with leaf-stalk twice as long as blade or more. Spathe usually with a series of warts. *Sri Lanka*. G2.

41. CRYPTOCORYNE Wydler
N. Jacobsen
Plants evergreen, with short, stout, more or less vertical rhizomes and long stolons. Scale-leaves present in flowering specimens, sometimes also in sterile specimens. Leaves basal, in rosettes, with distinct stalks. Leaf-stalk shortly sheathed at the base. Leaf-blade with one half rolled round the other when young; main veins running to tip of blade; subsidiary veins connecting the main ones. Inflorescence-stalk much shorter than the leaf-stalk. Spathe of 3 parts, a basal floral compartment with fused margins, a tube with overlapping margins, and a limb (figure 4(9), p. 77). Floral compartment (containing spadix) separated from tube by a flap; mouth of tube usually marked by a collar or collar-zone (figure 4(10 & 11), p. 77). Florets without perianths, unisexual. Spadix with zones of male and female florets separated (figure 4(8), p. 77), the sterile part slender, with scent-producing bodies above the female zone. Female florets only 4–7, in a single whorl, united, each with many ovules. Male florets each of 1 stamen. Fruit compound, spherical, star-like when open. Seeds large, ovoid to kidney-shaped, germinating a few days after being shed, without drying out.

About 50 species in tropical areas from India and Sri Lanka to New Guinea. Aquatic to amphibious herbs growing in fast- or slow-running rivers in the rain forest. Many species are imported from Colombo, Bangkok, Singapore and Jakarta, but generally only a dozen are available at any one time. Flowering specimens are, in most cases, necessary for correct determination. Despite great variability, the leaves are characteristic for each species, but they are difficult to describe.

Cryptocorynes are cultivated in aquaria for their spectacular foliage, and they are among the more prized aquarium plants, even though many species are difficult to cultivate. Generally, they are not given adequate conditions (resulting in small, starved growth); they should have a rather rich, preferably calcium-free, mineral soil (but some species, as noted below, tolerate substantial amounts of calcium). Other species prefer more acid conditions and can be cultivated in almost pure sphagnum-peat. Water temperature should be 24–28 °C but the hardier species can tolerate 20 °C (e.g. *C. ciliata, C. undulata, C. wendtii, C. beckettii, C. lutea, C. parva, C.* × *willisii, C. pontederiifolia*, forms of *C. cordata, C. affinis, C. usteriana*, and forms of *C. crispatula*).

Propagation is by the usually plentiful long runners. Seed propagation is unimportant for commercial use. All species can be grown emerging from the water, which encourages flowering. Flowering occurs all the year but is most marked during winter.

Literature: De Wit, H.C.D., *Aquarienpflanzen* (1971); Rataj, K., Revision of the genus Cryptocoryne Fischer, *Studie Československé Akademie Véd, Praha*, 3 (1975); Jacobsen, N., Notes on Cryptocoryne of Sri Lanka (Ceylon), *Botaniska Notiser* **128**: 179–90 (1976); Rataj, K. & Horemann, T.J., *Aquarium plants, tropical fish hobby* (1977); Jacobsen, N., Chromosome numbers and taxonomy in Cryptocoryne (Araceae), *Botaniska Notiser* **130**: 71–87 (1977); Jacobsen, N., *Akvarieplanter i farver*, (1977), translated into: English (1979), Dutch (1979), German (1979) and Swedish (1980); Jacobsen, N., *Cryptocoryner* (1979), translated into: German (1982) and Dutch (1982); Jacobsen, N., The Cryptocoryne albida group of mainland Asia (Araceae), *Mededelingen Landbouwhogeschool Wageningen* **19**: 183–204 (1980); Arends, J.C., Bastmeijer, J. & Jacobsen, N., Chromosome numbers and taxonomy in Cryptocoryne (Araceae) II, *Nordic Journal of Botany* **2**: 453–63 (1982); De Wit, H.C.D., *Aquariumplanten (1983)*.

Leaves. Green (sometimes brownish) without any purple: **1,6–8,13,14,16**; with a few purple markings: **2–5, 8–12,15**; with more or less dense purple markings: **2–4,9–12,15**.
Scale-leaves. Usually present in non-flowering specimens: **8,13,15**.
Spathe. Tube not more than 1.5 cm: **7–9,15**; tube at least 2 cm: **1–6,9–14,16**.
Limb of spathe. With a collar: **1–9,12,13**; with a collar-zone: **5,8,10,11,13–15**; without a collar-zone: **15,16**. With a rough surface: **1,6,7,9,10,13**.

1a. Scale-leaves usually present 2
b. Scale-leaves usually absent, except in flowering specimens 3
2a. Leaf-blade ovate (see also No. 15. *thwaitesii*) **8. pontederiifolia***
b. Leaf-blade narrowly ovate to linear **13. usteriana**
3a. Leaf-blade narrowly ovate to linear, more than 5 times as long as broad; limb of spathe more or less spirally twisted 4
b. Leaf-blade elliptic to ovate, usually not more than 4 times as long as broad; limb of spathe not spirally twisted 5
4a. Leaf narrowly ovate, with a distinct blade, green to dark green above, green with reddish veins or dark red beneath (see also No. 2. *undulata*); limb of spathe black-purple **12. affinis**
b. Leaf narrowly elliptic to linear, often without a distinct blade, green to rust-brown, smooth, wavy or blistered; limb of spathe whitish with purple lines **16. crispatula**
5a. Leaf-blade evenly green, without any purple 6
b. Leaf-blade purple, purple-marbled, purple-tinged, or with purple veins on the surface 9
6a. Plants usually more than 15 cm; leaf-blade narrowly ovate to linear, base cordate to truncate; limb of spathe ciliate **1. ciliata**
b. Plants usually smaller than 15 cm; leaf-blade narrowly ovate to obovate; limb of spathe not ciliate 7
7a. Leaf-blade spongy, obovate, 2–2.5 cm wide; limb of spathe with a long tail; collar absent **14. lingua**
b. Leaf-blade not spongy, narrowly ovate to ovate, 5–10 mm wide; limb of spathe short; collar present 8
8a. Plant 7–15 cm high; leaf-blade narrowly ovate to ovate; spathe 4–6 cm, with the limb (1–2 cm) upright **6. × willisii**
b. Plant 2–10 cm high; leaf-blade narrowly ovate to narrowly elliptic; spathe 2–3 cm, with the limb (3–8 mm) obliquely twisted **7. parva**
9a. Limb of spathe without a collar, but sometimes with a collar-zone 10
b. Limb of spathe with a pronounced collar 12
10a. Leaf-blade elliptic to ovate, with margin finely scalloped, appearing finely toothed; limb of spathe mostly with a long tail; collar-zone usually absent **15. thwaitesii**
b. Leaf-blade ovate to heart-shaped, margin entire; limb of spathe without a prolonged tail; collar-zone conspicuous 11
11a. Limb of spathe yellow, sometimes brown-tinged, surface more or less smooth **11. cordata**
b. Limb of spathe red, surface more or less rough **10. purpurea**

12a. Leaf-blade narrowly ovate to ovate; limb of spathe red-brown to purple, surface with small warts **9. minima***
b. Leaf-blade narrowly ovate, sometimes ovate; limb of spathe brownish to yellowish; surface more or less smooth 13
13a. Limb of spathe yellowish; collar of the same colour 14
b. Limb of spathe yellowish to dark brown; collar dark brown to black-purple 15
14a. Limb of spathe flat or recurved, slightly twisted; leaf-blade narrowly ovate to ovate, more or less stiff **5. walkeri**
b. Limb of spathe more or less obliquely twisted; leaf-blade narrowly ovate, soft **2. undulata**
15a. Limb of spathe flat or twisted once, not obliquely twisted; leaf-blade mostly brown to greenish, more or less stiff **4. beckettii**
b. Limb of spathe obliquely twisted; leaf-blade very variable, narrowly ovate to ovate, green to brown or red-brown, more or less flexible **3. wendtii**

1. C. ciliata (Roxburgh) Schott (*Ambrosina ciliata* Roxburgh; *C. elata* Griffith; *C. drymorrhiza* Zippelius). Illustration: De Wit, Aquarienpflanzen, 138–40 (1971).
To 1 m. Leaf-blade green, smooth, narrowly ovate to linear, with inconspicuous venation. Spathe with a long tube; limb with a characteristic ciliate margin. *India to New Guinea*. G2.

Rather variable in leaf form and size. A form with narrow leaves and long rooting runners (var. **ciliata**) is diploid, while a broader leaved form with short, fragile runners (var. **latifolia** Rataj) is triploid, but several other forms exist. It is a plant from the inner mangrove swamps and it grows well in a rich soil.

2. C. undulata Wendt (*C. willisii* Baum; *C. axelrodii* Rataj). Illustration: De Wit, Aquarienpflanzen, 205 (1971); Botaniska Notiser **129**: 186 (1976).
Submerged leaves 5–15 cm, narrowly ovate, green to brownish, with a distinct reddish venation and a truncate to cordate base. Emerged leaves shorter and broader, green with reddish veins, especially on the lower surface. (A triploid form has longer and broader leaves.) Limb of spathe twisted, dull yellowish to light brownish; collar of the same colour. *Sri Lanka*. G2.

Submerged, starved specimens often develop long internodes, 'stilt-roots', and narrow leaves, while plants under optimal conditions have larger and much more colourful leaves with a wavy margin.

3. C. wendtii De Wit. Illustration: De Wit, Aquarienpflanzen, 203 (1971); Botaniska Notiser **129**: 186 (1976).
Leaf-blade 5–25 × 1–3 cm, narrowly ovate to ovate, very variable in size and colour, green to brown or red-brown; base more or less cordate. Submerged leaves larger and more flexible. Limb of spathe more or less obliquely twisted, yellowish to brownish or reddish; collar black-purple. *Sri Lanka*. G2.

A most variable species with several different diploid and triploid forms. In some forms the leaves can change size and colour when transferred to new conditions of cultivation. One of the most undemanding species.

4. C. beckettii Trimen (*C. petchii* Alston). Illustration: De Wit, Aquarienpflanzen, 129 (1971); Botaniska Notiser **129**: 186 (1976).
Leaf-blade 5–10 cm, narrowly ovate to ovate, greenish to light or dark brown, with a truncate to slightly cordate base. Limb of spathe yellowish to brownish, more or less twisted; collar black-purple. *Sri Lanka*. G2.

Several leaf forms are found within both diploids and triploids.

5. C. walkeri Schott (*C. lutea* Alston). Illustration: De Wit, Aquarienpflanzen, 163, 169, 201 (1971); Botaniska Notiser **129**: 187 (1976).
Leaf-blade 5–10 cm, more or less ovate, rather stiff, more or less dark green; lower surface more brownish, often with reddish venation; base more or less cordate. Limb of spathe yellow (to greenish), slightly twisted or recurved; collar yellow. *Sri Lanka*. G2.

Submerged and emerged leaves rather similar, and stiffer than in the 3 preceding species.

6. C. × willisii Reitz (*C. lucens* De Wit; *C. nevillii* misapplied). Illustration: De Wit, Aquarienpflanzen, 167 (1971); Botaniska Notiser **129**: 186 (1976).
Leaf-blade 1–3 cm, narrowly ovate to ovate, green, inconspicuously veined; base truncate. Leaf-stalk sometimes with a reddish tinge. Limb of spathe somewhat twisted or recurved, more or less wrinkled, purple; collar yellowish to dark purple. *Sri Lanka*. G2.

Hybridisation experiments have shown that *C. × willisii* and *C. lucens* belong to a hybrid complex derived from *C. parva* on one side and at least *C. beckettii* and *C. walkeri* on the other. The hybrids are sterile.

7. C. parva De Wit. Illustration: De Wit, Aquarienpflanzen, 175 (1971); Botaniska Notiser **129**: 186 (1976).
Leaf-blade 5–15 mm, narrowly ovate to ovate, green, inconspicuously veined; base truncate. Limb of spathe more or less wrinkled, purple; collar dark purple, obliquely twisted. *Sri Lanka*. G2.

8. C. pontederiifolia Schott (*C. sulphurea* De Wit). Illustration: De Wit, Aquarienpflanzen, 179 (1971).
Leaf-blade 5–10 cm, ovate, cordate, green, sometimes slightly purple-marbled. Scale-leaves almost always present, even in non-flowering specimens. Limb of spathe upright, wrinkled, yellow; collar prominent and large; tube short. *Sumatra*. G2.

***C. longicauda** Engler (*C. caudata* N.E. Brown; *C. johorensis* Ridley). Illustration: De Wit, Aquarienpflanzen, 165 (1971). Leaf-blade 5–10 cm, ovate, cordate, green (to dull purple), sometimes strongly blistered. Limb of spathe red to dark purple, wrinkled, developed into a very long tail; collar present. *Borneo & Malaysia (Johore)*. G2.

Requires acid conditions.

9. C. minima Ridley. Illustration: De Wit, Aquarienpflanzen, 169 (1971).
Leaf-blade 2–4 cm, ovate, more or less cordate, green, sometimes with reddish marbling; lower surface reddish, often with reddish veins. Limb of spathe recurved, less than 1 cm broad, wrinkled, mostly reddish; collar present; tube short. *Malay peninsula*. G2.

***C. griffithii** Schott. Illustration: De Wit, Aquarienpflanzen, 159 (1971). Leaf-blade 4–8 cm, ovate; upper surface dark green to purplish; lower red. Limb of spathe recurved, more than 1 cm broad, strongly wrinkled, red to dark purple; collar present. *Malay peninsula*. G2.

10. C. purpurea Ridley (*C. hejnyi* Rataj; *C. griffithii* misapplied). Illustration: De Wit, Aquarienpflanzen, 181, 189 (1971).
Leaf-blade 4–8 cm, ovate, more or less cordate, green, sometimes reddish-marbled; lower surface reddish, often with reddish veins. Limb of spathe somewhat wrinkled, red; collar-zone prominent, sometimes yellowish. *Malay peninsula*. G2.

11. C. cordata Griffith (*C. siamensis* Gagnepain; *C. kerrii* Gagnepain; *C. blassii* De Wit; *C. evae* Rataj; *C. stonei* Rataj; *C. siamensis* Gagnepain var. *schneideri* Schöpfel). Illustration: De Wit, Aquarienpflanzen, 145, 189 (1971).
Leaf-blade 3–15 cm, narrowly ovate to ovate, more or less cordate, green to

brown-purple or marbled, often more reddish beneath than above, sometimes somewhat blistered; texture varying from fragile to tough. Limb of spathe yellow, sometimes more or less brown-tinged, especially towards the margin; collar-zone conspicuous. *Malay peninsula*. G2.

Occurs in the diploid, tetraploid and hexaploid state. Leaves very polymorphic. The forms with a tougher leaf texture are generally the best in cultivation; the more fragile-leaved forms require acid conditions. Recently a form with variegated, whitish veins has come into cultivation, under the cultivar name 'Rosaefolia' (*C. siamensis* var. *schneideri*).

12. C. affinis N.E. Brown (*C. haerteliana* Milkuhn). Illustration: De Wit, Aquarienpflanzen, 125 (1971).
Leaf-blade 5–20 cm, narrowly ovate, more or less cordate, sometimes strongly blistered; upper surface green or tinged with red, with the veins often distinctly lighter coloured; lower surface red or green and tinged with red. Limb of spathe black-purple, spirally twisted; collar present. *Malay peninsula*. G2.

A fast-growing species that prefers somewhat calcareous conditions.

13. C. usteriana Engler (*C. aponogetifolia* Merrill). Illustration: De Wit, Aquarienpflanzen, 197 (1971).
Leaf-blade 20–100 × 1–4 cm, narrowly ovate to linear, green, blistered, with 2–6 strong veins. Scale-leaves always present. Limb of spathe wrinkled, dull violet, somewhat twisted; collar present. *Philippines*. G2.

A species that grows well under calcareous conditions.

14. C. lingua Engler (*C. spathulata* Engler). Illustration: De Wit, Aquarienpflanzen, 159 (1971).
Leaf-blade 1–3 cm wide, obovate, green and spongy. Limb of spathe yellowish with red spots and with a long tail; collar absent. *Borneo, introduced into Singapore*. G2.

15. C. thwaitesii Schott. Illustration: De Wit, Aquarienpflanzen, 193 (1971); Botaniska Notiser **129**: 187 (1976).
Leaf-blade elliptic to ovate, more or less cordate, green to purplish; margin often finely scalloped. Limb of spathe white, sometimes more or less spotted with red, developed into a shorter or longer tail that often bends forward; collar usually absent. *Sri Lanka*. G2.

16. C. crispatula Engler (*C. sinensis* Merrill; *C. balansae* Gagnepain; *C. tonkinensis* Gagnepain; *C. longispatha* Merrill; *C. bertelihansenii* Rataj; *C. yunnanensis* Li; *C. kwangsiensis* Li). Illustration: De Wit, Aquarienpflanzen, 129, 185 (1971).
Leaf-blade 10–70 cm × 2–40 mm, more or less linear, green to rust-brown; smooth to blistered; margin sometimes with fine, forward-pointing teeth. Limb of spathe spirally twisted, whitish with short or long purple lines; collar absent. *E India to Thailand, Vietnam & S China*. G2.

A very variable species, of which some of the narrow-leaved, smooth forms and some of the broad-leaved, blistered forms are well suited for cultivation, the latter preferring not too acid conditions. *C. crispatula* has sometimes been included in *C. retrospiralis* (Roxburgh) Engler.

42. PISTIA Linnaeus
P.F. Yeo
Evergreen floating aquatics with a rosette of leaves and feathery roots below. Scale-leaves delicate, present on stolons which produce new rosettes. Leaves 20 × 7 cm, stalkless or nearly so, broadly wedge-shaped with truncate, sometimes notched, apex and 7–15 nearly parallel veins which are prominent beneath, but with no distinct midrib; both surfaces densely covered with minute hairs; upper surface light green. Spathe 1.5 cm, almost stalkless, hairy outside; margins overlapping below; upper half opening, acute. Spadix not evident; florets attached to back of spathe, without perianth, unisexual; two dissimilar scales present between male florets and female. Male florets few, combined into a single whorl on a short stalk, borne in the open part of the spathe. Female floret solitary. Ovary large, asymmetric, occupying almost the whole of the closed part of the spathe, to which it is extensively attached. Style present. Ovules numerous.

One rather variable species distributed throughout the tropics and extending into the subtropics, and characterised by the rosettes of distinctively shaped leaves. It is grown as an aquarium plant and requires a temperature of 21–26 °C. For overwintering, plants must be placed in shallow water and allowed to root into the soil. Alternatively, seed may be collected in summer and sown next spring in a fine wet soil at 30 °C.
Literature: Hooker, W.J. & Smith, J., Pistia stratiotes, *Botanical Magazine*, 4564 (1851); Engler, A., Aroideae & Pistioideae, *Das Pflanzenreich* 73: Araceae, 250–62 (1920).

1. P. stratiotes Linnaeus. Illustration: Botanical Magazine, 4564 (1851); Das Pflanzenreich **73**: 251 (1920); Graf, Tropica, 129 (1978).
Tropics and subtropics. G2. Summer.

XXIII. LEMNACEAE

S.M. Walters
Small aquatic herbs, usually floating on the surface of still water, with a green plant body undifferentiated into stem and leaf. Inflorescence minute, in a pocket on the margin of the plant body, and consisting of flowers all without perianths, 1 or more male flowers, each with 1 or 2 stamens, and a single female flower with a solitary ovary. Flowers and seeds are rarely produced in northern Europe.

The family is related to Araceae, in which family the floating aquatic genus *Pistia* gives some indication of possible affinities. Species of *Lemna* are familiar to all gardeners and aquarists, but they are rarely cultivated and even more rarely distinguished from each other. The genera can be distinguished as follows:

1a. Plant with 1 or more rootlets on its lower side — 2
b. Plant rootless — **Wolffia (& Wolffiella)**
2a. Plant with a single rootlet — **Lemna**
b. Plant with 2 or more rootlets — **Spirodela**

Literature: Landolt, E., Key to the determination of taxa within the family of Lemnaceae, *Veröffentlichungen des Geobotanischen Institutes der ETH, Stiftung Rübel, in Zürich* **70**(1): 13–21 (1980).

XXIV. PANDANACEAE

Plants woody, trees, shrubs or climbers, often with apparently dichotomous branching and often with numerous stilt-roots. Leaves stiff, leathery, evergreen, long and narrow, usually with spiny margins. Plants dioecious, the individual flowers closely massed together and difficult to distinguish. Perianth absent. Stamens numerous, apparently in spikes or panicles. Ovary superior with 1–several cells, ovules 1–many per cell, placentation basal or parietal. Fruit compound (a syncarp) made up of woody or fleshy units.

A family of 3 genera from the Old World tropics and Hawaii. Only a few species of *Pandanus* are grown, and these are more often seen in botanic gardens and specialist collections than in general cultivation.

1. PANDANUS Parkinson
J. Cullen
Plants woody, dwarf or forming trees, usually dichotomously branched and with stilt-roots. Trunk with prominent, ring-like leaf-scars. Leaves basically arranged in 3 ranks which are, however, twisted so that the leaves appear very conspicuously spirally arranged, in rosettes at the ends of the branches; margins and the underside of the midrib spiny. Stamens in elongate spikes. Female flowers aggregated into large, spherical or oblong heads. Fruit a syncarp of drupes, hard, spherical, conical or obconical.

A genus of 250 or more species from the Old World tropics. They are valued for their striking appearance, but most of them are too large for any but the largest greenhouses. The dwarfer species are occasionally grown as pot plants, but these tend to remain in a juvenile phase, without flowers or fruits, and their identification, and relationships to the wild species, are difficult and imperfectly understood. The striking, corkscrew-like, spiral arrangement of the leaves distinguishes them easily from other woody monocotyledons.

They are relatively easy to cultivate, but require high temperatures and an abundance of water, particularly when actively growing. A rich compost is required, and pot plants should be repotted each year before active growth begins. Good ventilation is essential, and light shade is required during very bright days.
Literature: There are numerous papers by H. St John, under the general title 'Revision of the genus Pandanus', covering species from various areas; most of these were published in the journal *Pacific Science* between 1960 and 1973.

1a. Leaves to 1.5 cm broad **4. pygmaeus**
b. Leaves at least 5 cm broad 2
2a. Plant tree-like; leaves uniformly glaucous green **1. utilis**
b. Plant dwarf; leaves variegated with white, silver-white or yellow 3
3a. Leaves with white or silver-white margins **2. veitchii**
b. Leaves with several longitudinal yellow stripes **3. sanderi**

1. P. utilis Bory. Illustration: Bailey, Standard cyclopedia of horticulture (popular edn) **3**: f. 2743–5 (1935); Graf, Exotica, edn 8, 1372, 1373 (1976).
Tree to 20 m. Leaves 50–200 × 6–10 cm, glaucous, with reddish spines on the margins and midrib beneath. Male flowers fragrant. Female inflorescences long-stalked, solitary, pendent. Syncarp more or less spherical, 15–20 cm in diameter. *Madagascar or possibly Mauritius but widely cultivated elsewhere in the tropics*. G2.

2. P. veitchii Masters & Moore. Illustration: Step, Favourite flowers of garden and greenhouse **4**: t. 293 (1897); Graf, Exotica, edn 8, 1373 (1976).
Plant to 60 cm high, but forming a large clump if given space. Leaves 50–100 × *c*. 7.5 cm, with spiny margins, dark green with white or silver-white bands along the margins. *?Polynesia*. G2.

A species of uncertain origin and status. It is usually grown as a pot plant and rarely reaches its mature development. A very dwarf variant ('Compactus') is available.

3. P. sanderi Masters. Illustration: Gardeners' Chronicle **23**: 256 (1898); Graf, Exotica, edn 8, 1373 (1976).
Like *P. veitchii* but generally somewhat smaller, leaf margins minutely spiny, leaves dark green variegated with several longitudinal yellow stripes. *Malaysia, ?Indonesia*. G2.

Another species of uncertain origin and status. Several cultivars, differing in the degree and nature of the variegation, are available.

4. P. pygmaeus Thouars. Illustration: Botanical Magazine, 4736 (1853); Graf, Exotica, edn 8, 1373 (1976).
Shrub to 60 cm. Leaves 30–60 × at most 1.5 cm, green, minutely spiny on the margins and midrib beneath. *Madagascar*. G2.

XXV. SPARGANIACEAE

Aquatic or semi-terrestrial, monoecious, perennial herbs. Stems floating or erect, from a creeping rhizome. Leaves linear, in 2 ranks, hairless, sheathing the stem at the base. Flowers in globular heads (capitula), of which the upper are male and the lower female. Perianth scales brown, 3–4 in female flowers, 1–6 in male flowers. Stamens 1–8. Ovary superior with 1 cell. Fruit indehiscent, with 1 seed.

A family of only 1 genus.

1. SPARGANIUM Linnaeus
D.M. Synnott
A genus of about 20 species from north temperate regions. They are easily grown in shallow water, e.g. at lake margins, where they provide useful cover for wild fowl. Propagation by seed.
Literature: Cook, C.D.K., Sparganium in Britain, *Watsonia* **5**: 1–10 (1961).

1a. Inflorescence branched, stems erect **2. erectum**
b. Inflorescence not branched, stems erect or floating 2
2a. Leaves triangular in cross-section; stems erect or floating **1. emersum**
b. Leaves not triangular in cross-section; stems floating **3. minimum**

1. S. emersum Rehmann (*S. simplex* Hudson). Illustration: Ross-Craig, Drawings of British plants, part 30: t. 36 (1973); Keble Martin, The concise British flora in colour, 88 (1965).
Stems erect or floating, 20–60 cm. Leaves both erect and floating, triangular in cross-section. Inflorescence not branched, exceeded by the lowest leafy bract. Female capitula 3–6, the lower usually stalked. Male capitula 3–10, separated and stalkless on main axis. Fruit ellipsoid, 4–6 × 2–3 mm, stalked, with long, persistent style. *Europe, W & C Asia, N America*. H1. Summer.

2. S. erectum Linnaeus (*S. ramosum* Hudson). Illustration: Ross-Craig, Drawings of British plants, part 30: t. 35 (1973); Keble Martin, The concise British flora in colour, 88 (1965).
Stems erect, 50–100 cm. Leaves erect, triangular in cross-section, keeled below. Inflorescence branched. Fruits ellipsoid to pyramidal, 6–9 × 2–6 mm. *Europe, W & C Asia*. H1. Summer.

Sometimes divided into subspecies on the basis of the shape of the fruits.

3. S. minimum Wallroth. Illustration: Ross-Craig, Drawings of British plants, part 30: t. 38 (1973); Keble Martin, The concise British flora in colour, 88 (1965).
Stems floating, 10–30 cm. Leaves floating or submerged, thin, flat, usually translucent. Inflorescence not branched; lowest leafy bract scarcely longer than inflorescence. Female capitula 2–3, stalkless and widely separated. Male capitulum 1. Fruit obovoid, stalkless, with short beak-like, persistent style. *N temperate regions*. H1. Summer.

XXVI. TYPHACEAE

Perennial, monoecious herbs, aquatic or semi-terrestrial. Flowers in cylindric spikes (partial inflorescences), male above female. Male flowers with stamens in clusters of 1–3, rarely as many as 8; filaments irregularly fused. Female flowers on stalks which bear hair-like branches; ovary superior with 1 cell; style persistent. Fruit dry, usually dehiscent, with 1 seed.

A family of 1 genus, distributed throughout the world in suitable freshwater habitats, but generally absent from the tropics.

1. TYPHA Linnaeus
E.C. Nelson
Hairless perennial herbs, with stout, creeping rhizomes. Stems erect. Leaves mostly basal, in 2 ranks, erect, with sheath closely enclosing stem.

A widely distributed genus of about 15 species. They are suitable for the margins of lakes and ponds, and may be propagated by division of the rhizomes or from seed.

1a. Leaves not more than 3 mm wide; female spike to 5 cm, oblong to cylindric **3. minima**
b. Leaves more than 3 mm wide; female spike more than 5 cm, not oblong 2
2a. Leaf-sheath usually with auricles; male and female spikes separated by 3–8 cm **1. angustifolia**
b. Leaf-sheath tapering into blade; male and female spikes touching or separated by less than 3 cm 3
3a. Female spike brown and mottled at maturity **2. latifolia**
b. Female spike silvery grey at maturity **4. shuttleworthii**

1. T. angustifolia Linnaeus. Illustration: Keble Martin, The concise British flora in colour, 88 (1965).
Perennial herb to 1.5 m. Leaves 3–6 mm wide. Male and female spikes separated by 3–8 cm; female spike 8–20 × 1.3–2.5 cm, dark or reddish brown at maturity. *Cosmopolitan*. H1. Summer.

More slender and less invasive than *T. latifolia*; suitable for small ponds.

2. T. latifolia Linnaeus. Illustration: Keble Martin, The concise British flora in colour, 88 (1965).
Robust perennial herb to 2 m or more. Leaves 8–20 mm wide. Male and female spikes touching or separated by not more than 2.5 cm; female spike 8–15 × 1.8–3 cm, dark brown but becoming mottled with age. *Almost worldwide but absent from the southern hemisphere S of 30° S.*

3. T. minima Funck. Illustration: Grounds, Ornamental grasses, 204 (1979).
Perennial herb 25–75 cm. Leaves not more than 3 mm wide. Male and female spikes touching or separated; female spike 1.5–5 cm long, oblong to cylindric, dark brown, sometimes subtended by a leaf-like bract. *C & S Europe, Asia eastwards to N China*. H1. Summer.

Suitable for small ponds.

4. T. shuttleworthii Koch & Sonder. Illustration: Coste, Flore de France **3**: 437 (1906).
Similar to *T. latifolia* but not as tall or robust. Female spike becoming silvery grey with age. *S Europe*. H1. Summer.

XXVII. CYPERACEAE

Usually perennial herbs, with solid stems which are often triangular in section, and linear leaves sometimes reduced to sheaths. Florets inconspicuous, wind-pollinated, bisexual or unisexual, subtended by a small bract (glume) and arranged in spikes. Perianth absent or represented by bristles persisting round the base of the 1-seeded indehiscent fruit (nut).

A large cosmopolitan family with about 100 genera and 4000 species, mostly growing in wet places, and of little horticultural importance. In addition to the 4 genera treated here, species of *Eriophorum* are sometimes grown for their decorative, cottony fruiting heads; they would perhaps be more widely cultivated if they did not require wet, peaty soils.

1a. Florets all unisexual; female florets enclosed in a sac (utricle) which persists round the fruit **4. Carex**
b. Florets mostly bisexual; utricle absent 2
2a. Inflorescence a simple terminal spike 3
b. Inflorescence of several (often many) spikes 4
3a. Fruit with persistent style-base; spike without a basal bract **2. Eleocharis**
b. Fruit without persistent style-base; spike with a more or less leafy basal bract **1. Scirpus**
4a. Spikes more or less cylindric; glumes and florets not in 2 ranks **1. Scirpus**
b. Spikes compressed; glumes and florets in 2 ranks **3. Cyperus**

1. SCIRPUS Linnaeus
S.M. Walters
Rhizomatous or tufted, mostly perennial herbs of varied habit, sometimes more or less leafless. Inflorescence usually compound, with bisexual florets arranged spirally in spikes. Perianth represented by bristles at base of ovary, sometimes absent. Stamens and stigmas 3 or 2. Fruit a nut.

Treated here in its wide sense, a large cosmopolitan genus. A few species are grown in gardens, especially at the water's edge or in damp, shady places. Propagation is by division, except for *S. cernuus*, which is by seed.

1a. Delicate tufted plant with thread-like stems; inflorescence usually a single spike, sometimes 2 or 3 **4. cernuus**
b. Robust, tufted or rhizomatous plants with inflorescence of several spikes 2
2a. Stems leafy, 3-angled **1. maritimus**
b. Stems leafless or nearly so, cylindric 3
3a. Spikes globular; inflorescence an umbel **2. holoschoenus**
b. Spikes ovoid; inflorescence not an umbel **3. lacustris**

1. S. maritimus Linnaeus. Illustration: Roles, Illustrations to Clapham et al., Flora of the British Isles **4**: 53 (1956).
Robust, shortly rhizomatous, with 3-angled, leafy stems to 1 m. Leaves to 10 mm wide, keeled. Inflorescence dense, corymbose, with rather few, reddish brown, ovoid, stalkess spikes to 2 cm. Nut *c.* 3 mm, obovate, flattened. *Cosmopolitan, in saline habitats*. H2. Summer.

Grown by lakes and ponds, more often as 'Variegatus', with striped foliage.

2. S. holoschoenus Linnaeus (*Holoschoenus vulgaris* Link). Illustration: Roles, Illustrations to Clapham et al., Flora of the British Isles **4**: 54 (1956).
Tufted, with smooth, cylindric stems to 1 m. Only the upper leaf-sheaths with blades, which are rigid and half-cylindric. Inflorescence apparently lateral, consisting of an umbel of several globular spikes up to 10 mm, subtended and exceeded by a long, pointed bract. Nut 1 mm, more or less 3-angled. *From England & the Canary Islands, through S Europe & N Africa to C Asia*. H3. Summer.

A very distinctive sedge, easy to cultivate and tolerant of relatively dry soils, sometimes grown as 'Variegatus' with alternate rings of green and yellow tissue on leaf and stem.

3. S. lacustris Linnaeus (*Schoenoplectus lacustris* (Linnaeus) Palla; *Scirpus tabernaemontani* Gmelin). Illustration: Everett, Encyclopedia of horticulture **9**: 3090 (1982).
Rhizomatous, with smooth, cylindric, leafless stems to 2.5 m. Leaves, if present, long, slender, only produced under water.

Inflorescence dense, terminal, with several ovoid, reddish brown spikes up to 8 mm. Nut *c.* 2 mm, broadly elliptic, flattened. *Cosmopolitan, though uncommon in the tropics.* H2. Summer.

Subsp. **lacustris**, with green stems and a smooth glume subtending the 3-angled nut, is occasionally planted (and often found) by lakes and in slow-moving streams. The commonly planted cultivars belong to subsp. **tabernaemontani** (Gmelin) Palla, a smaller plant with glaucous stems, papillose glumes and a biconvex nut, occurring mainly in coastal habitats in the wild; 'Albescens' has creamy white stems with narrow, green longitudinal stripes, whilst 'Zebrinus' has stems with alternate transverse bands of green and white.

4. S. cernuus Vahl (*S. gracilis* Koch; *S. savii* Sebastiani & Mauri; *Isolepis cernua* (Vahl) Roemer & Schultes). Illustration: Graf, Tropica, 390 (1978); Everett, Encyclopedia of horticulture **9**: 3091 (1982).
Delicate, tufted, often annual plant with very slender, thread-like, almost leafless stems to 30 cm. Inflorescence usually a solitary (sometimes 2 or 3) ovoid, greenish spike to 5 mm, apparently lateral with a short, pointed bract continuing the stem. Nut less than 1 mm, almost spherical, reddish brown. Perianth-bristles absent. *Widespread, mainly in temperate regions.* H4. Summer.

Commonly grown in conservatories as a decorative pot plant or for edging.

2. ELEOCHARIS R. Brown

S.M. Walters

Tufted or rhizomatous, leafless herbs, with cylindric or angled stems sheathed at base and bearing solitary, bractless spikes. Florets as in *Scirpus*. Fruit a nut with persistent style-base.

A large cosmopolitan genus of aquatic and marsh plants, of very little horticulture value.

1. E. acicularis (Linnaeus) Roemer & Schultes. Illustration: De Wit, Aquarium plants, English edn, 72–3 (1964).
Delicate, rhizomatous perennial with tufts of 4-angled, thread-like stems, to 30 cm if submerged, as is usual in cultivation. Spikelets (produced only when plant is out of water) 3–4 mm; nut *c.* 1 mm, with clearly separated style-base, and fine, deciduous bristles. *Subarctic and temperate regions of the N Hemisphere.* H3.

Commonly grown rooted in sand in aquaria, and tolerant of water temperatures up to 30 °C. Propagation is by division.

3. CYPERUS Linnaeus

S.M. Walters

Tufted or rhizomatous herbs with terminal, umbellate or capitate inflorescences. Spikes many-flowered, flattened; florets bisexual, in 2 ranks. Perianth absent; stamens usually 3. Fruit a nut.

A large, mainly tropical or subtropical genus of about 500 species with few temperate representatives. Propagation is usually by division, but seed can also be used.
Literature: Kükenthal, G., Cyperus, *Das Pflanzenreich* **101** (1935–36).

Stems. Leafless: **1,5**. More than 1.5 m: **1**.
Leaves. With 3 prominent, whitish nerves: **6**.
Stolons. With terminal tubers: **3**.
Inflorescence. A compound umbel with many slender, drooping branches: **1**.
Spikelets. In dense, nearly spherical heads: **4**.

1a. Stems leafless or nearly so, with basal sheaths only — 2
b. Stems obviously leafy, at least at base — 3
2a. Inflorescence-branches very slender and drooping, much exceeding the few bracts — **1. papyrus**
b. Inflorescence-branches rather stout, more or less equalling the numerous bracts — **5. involucratus**
3a. Plant bearing slender underground stolons terminating in brown tubers — **3. esculentus**
b. Plant without tubers — 4
4a. Spikes in dense heads — **4. eragrostis**
b. Spikes solitary or loosely arranged — 5
5a. Leaves usually more than 10 mm wide; spikes narrowly lanceolate, often solitary — **6. albostriatus**
b. Leaves usually less than 10 mm wide; spikes linear, in a loose raceme — **2. longus**

1. C. papyrus Linnaeus. Illustration: Gardeners' Chronicle **3**: 81 (1875) – habit of Sicilian plant; Everett, Encyclopedia of horticulture **3**: 983 (1981).
Giant rhizomatous, perennial 'reed' with bluntly 3-angled, leafless stems often exceeding 2 m. Inflorescence a compound umbel with very many long, slender, drooping primary branches much exceeding the few bracts. Spikes slender, brownish, 20–30 together in larger spikes grouped at the ends of the branches. *S & tropical Africa, extending northwards to Egypt; naturalised in Sicily.* H5–G1.

Often cultivated as an ornamental aquatic in greenhouses; the classical source of Egyptian papyrus. *C. papyrus* 'Nanus' (*C. isocladus* Knuth; '*C. haspan viviparus*' misapplied) is a dwarf variant, sometimes treated as a distinct species, with stems to 80 cm; it is native to southern Africa, northwards to Zanzibar, and is grown as a pot plant in conservatories.

2. C. longus Linnaeus (*C. maximus* Anon.). Illustration: Roles, Illustrations to Clapham et al., Flora of the British Isles **4**: 56 (1965).
Rhizomatous, with smooth, 3-angled, leafy stems to 1 m. Inflorescence a compound umbel, with several leafy bracts often exceeding the inflorescence, and unequal primary branches to 10 cm. Spikes *c.* 10 mm, more or less linear, reddish-brown, in loose racemes at the ends of the secondary branches. *Europe, SW & C Asia, N Africa.* H3. Summer.

The hardiest species, commonly grown in ponds, lakes and water-gardens.

3. C. esculentus Linnaeus. Illustration: Coste, Flore de la France **3**: 463 (1906).
Rhizomatous, with slender stolons bearing terminal brown tubers, and leafy stems to 50 cm. Inflorescence a few-branched umbel with distant, grouped, yellowish brown spikes. *Native in the Mediterranean region and perhaps elsewhere; now widely naturalised in subtropical and warm temperate regions.* H5.

The cultivated plant, var. **sativus** Boeckeler, which yields the commercial edible tuber ('Chufa' or 'Tiger Nut'), has numerous tubers crowded on short stolons, and rarely flowers. The wild plant is not cultivated in gardens, though it may occur as a weed.

4. C. eragrostis Lamarck (*C. vegetus* Willdenow). Illustration: Coste, Flore de la France **3**: 463 (1906).
Shortly rhizomatous, with leafy stems to 60 (rarely to 90) cm. Leaves 4–8 mm wide, often equalling stems. Inflorescence a simple or compound umbel with 8–10 branches and 5–8 (rarely to 11) bracts exceeding the branches; spikes 8–15 mm, in dense, nearly spherical heads 1–2 cm in diameter, pale brown or straw-coloured. *Western USA and warm-temperate S America; naturalised in SW Europe.* H4. Summer.

5. C. involucratus Rottboell (*C. flabelliformis* Rottboell; *C. alternifolius* misapplied). Illustration: Graf, Tropica, 390 (1978); Mühlberg, Das grosse Buch der Wasserpflanzen, t. 183 (1980).
Tufted, with short, erect, ridged, leafless stems to 1.5 m, clothed below with brown sheaths. Inflorescence a terminal umbel with an involucre of many long, spreading,

leaf-like bracts; spikes 6–8 mm, mostly bunched at the ends of the branches, pale brown. *Africa; widely naturalised in the tropics and subtropics*. H5.

Commonly grown as a pot plant standing in shallow water; in frost-free regions also in water-gardens. A dwarf variant, 'Gracilis', is usually attributed to this species; the stems are *c.* 40 cm and the bracts narrower than the type, and flowers are rarely produced. 'Variegatus' has stems and bracts variously striped and mottled with white.

6. C. albostriatus Schrader (*C. diffusus* misapplied; *C. elegans* misapplied). Illustration: Everett, Encyclopedia of horticulture **3**: 984 (1981).
Rhizomatous, with slender, leafy stems to 60 cm. Leaves and involucral bracts 8–16 mm wide, with 3 prominent, whitish nerves. Inflorescence a loose terminal umbel with thread-like branches and 6–9 bracts; spikes 6–13 mm, narrowly lanceolate, single or in groups of 2–4, pale brown. *South Africa*. G1.

Grown as a decorative pot plant; 'Variegatus' has longitudinally striped leaves and bracts.

4. CAREX Linnaeus
S.M. Walters
Rhizomatous or tufted perennials, with grass-like leaves in 3 rows, and stems often triangular in section. Inflorescence compound, terminal, containing (in all cultivated species) both male and female florets grouped in spikes. Perianth absent; male florets consisting of 2 or 3 stamens, female consisting of an ovary with 2 or 3 stigmas and enclosed in a sac or utricle. Fruit a small nut enclosed in the persistent utricle. In the key and descriptions, 'fruit' denotes the utricle and the enclosed nut.

A very large genus with more than 1500 species, principally in the North temperate region, and mostly characteristic of wet habitats. The few species grown in gardens are mainly used on the edges of lakes, ponds and streams. Since several other species not included in this account are widespread and common native European plants, they may be present naturally in woodland, marsh, stream and lakeside habitats, and tolerated in such places within gardens. Propagation is by division of rootstock or rhizome.
Literature: Kükenthal, G., *Das Pflanzenreich* **38** (1909).

Habit. More or less rhizomatous: **1,2,11**.
Leaves. Very narrow (less than 1.5 mm wide): **8,9**; evergreen, stiff: **1,7**.
Inflorescence. A terminal spike with no subtending bract: **1**; whitish: **1,2**; with pendent female spikes: **5,10**.
Fruit. Berry-like, red or purplish when ripe: **3**; flattened (ovary with 2 stigmas): **4,8**.

1a. Fruit berry-like, red or purplish when ripe **3. baccans**
b. Fruit not berry-like, brownish, greenish or whitish 2
2a. Inflorescence a dense, whitish head 3
b. Inflorescence not capitate, usually greenish or brownish 4
3a. Inflorescence with 2–5 leaf-like bracts **2. baldensis**
b. Inflorescence without bracts **1. fraseri**
4a. Leaves not more than 1.5 mm wide, half-circular in section 5
b. Leaves more than 2 mm wide, flat 6
5a. Stems and leaves brownish; fruit flattened (ovary with 2 stigmas) **8. buchananii***
b. Stems and leaves pale green; fruit not flattened (ovary with 3 stigmas) **9. comans**
6a. Leaves long, stiff, evergreen **7. morrowii**
b. Leaves not evergreen 7
7a. Dwarf 'alpine' with short, stiff leaves and stems to 20 cm **6. firma**
b. Tall plant with long leaves and stems much exceeding 20 cm 8
8a. Fruit flattened; ovary with 2 stigmas **4. elata**
b. Fruit ovoid or 3-angled, not flattened; ovary with 3 stigmas 9
9a. Female spikes distant, pendent, to 25 cm **5. pendula**
b. Female spikes not all pendent, not more than 9 cm 10
10a. Male spike solitary **10. pseudocyperus**
b. Male spikes 2 or more **11. riparia***

1. C. fraseri Andrews (*Cymophyllus fraseri* (Andrews) Mackenzie). Illustration: Botanical Magazine, 1391 (1811) – as C. fraseriana; Kükenthal, Das Pflanzenreich **38**: 95 (1909).
Shortly rhizomatous, with leafless flowering stems to 40 cm, and stiff, thick, evergreen leaves to 5 cm wide, appearing after flowering. Inflorescence a single terminal, whitish, ovoid-spherical or oblong spike without any subtending bract. Fruit *c.* 6 mm, inflated. *Eastern USA, from Pennsylvania to South Carolina*. H4. Summer.

Suitable for damp, peaty, shady places; an unusually attractive species.

2. C. baldensis Linnaeus. Illustration: Kükenthal, Das Pflanzenreich **38**: 185 (1909).
Rhizomatous, with stems to 30 cm, and leaves 3–4 mm wide. Inflorescence a small, dense, whitish head of 3–9 spikes, subtended by 2–5 pale, spreading, linear bracts. Fruit 4–5 mm, pale. *E Alps*. H1. Summer.

3. C. baccans Nees. Illustration: Botanical Magazine, 7288 (1893).
Tufted, with stems to 1 m, and long, curving leaves to 1.5 cm wide. Inflorescence a bracteate panicle of clustered spikes, bearing shiny, berry-like fruits at first pale red, later purplish. *From the Himalaya & Taiwan to Sri Lanka & Java*. H5–G1.

Cultivated as a decorative pot plant in the cool greenhouse; can be grown outside in relatively frost-free areas.

4. C. elata Allioni (*C. stricta* Goodenough; *C. hudsonii* Bennett). Illustration: Jermy et al., Sedges of the British Isles, edn 2, 205 (1982).
Robust, tufted, capable of forming large tussocks, with stems to 1 m, and long, sharply keeled leaves to 6 mm wide. Inflorescence of 1 or 2 (rarely 3) terminal male spikes, and 2–4 rather distant female spikes below, the latter each with a small, very narrow bract. Fruit *c.* 3 mm, flattened, with short, entire beak. *Europe, extending to N Africa & the Caucasus*. H5. Summer.

Usually seen in gardens as 'Aurea', with yellowish foliage.

5. C. pendula Hudson. Illustration: Jermy et al., Sedges of the British Isles, edn 2, 127 (1982).
Robust, tufted, often forming large tussocks, with slender, graceful flowering stems to 1.5 m. Leaves shorter than stems, to 2 cm wide. Inflorescence with 1 terminal male spike and 4 or 5 distant, pendent, cylindric female spikes to 25 cm. Fruit *c.* 3 mm, ellipsoid, with short beak. *Europe from Denmark southwards, extending to N Africa & SW Asia*. H3. Summer.

Much the most familiar species in cultivation in W Europe, often planted in wet shady places, but surprisingly tolerant of drier ground.

6. C. firma Host. Illustration: Coste, Flore de la France **3**: 514 (1906).
Densely tufted or mat-forming plant, with stems to 20 cm and short, stiff, pointed leaves to 4 mm wide in rosettes. Inflorescence with 1 terminal male spike and 1–3 lateral, ovoid female spikes. Fruit

c. 4 mm, oblong-ellipsoid. *Mountains of Europe, from the Pyrenees to the Carpathians; only on limestone.* H1. Summer.

The usual plant has blue-green leaves; in 'Variegata' these are longitudinally striped with white. Both are grown as 'alpines'.

7. C. morrowii Boott (*C. foliosissima* misapplied; *C. japonica* misapplied). Illustration: Graf, Tropica, 390 (1978).
Tufted, with stems to 30 cm, and long, stiff, long-pointed evergreen leaves to 8 mm wide. Inflorescence with 1 terminal male spike and 2–4 slender, long-stalked, cylindric female spikes to 2.5 cm. Fruit *c.* 3 mm, with long beak. *C & S Japan.* H5. Summer.

Grown as a pot plant for the conservatory, usually as 'Variegata' or 'Albo-mediana' with longitudinal white stripes on the leaves. It can be used as a border plant in frost-free regions.

8. C. buchananii Berggren.
Tufted, with stems to 60 cm, and very narrow, hard, reddish brown leaves half-circular in section. Inflorescence pale brownish, with a terminal male spike and 5–8 lateral female spikes. Fruit 3 mm, flattened. *New Zealand.* H4. Summer.

Used as a coloured edging plant by water.

***C. flagellaris** Colenso. A coarser plant with stems to more than 1 m, but otherwise rather similar. *New Zealand.*

9. C. comans Berggren (*C. vilmorinii* Mottet).
Tufted, with slender stems to 40 cm and very narrow, drooping, pale green leaves half-circular in section. Inflorescence brownish, with a terminal male spike and about 5 lateral female spikes. Fruit *c.* 2 mm, narrowly ovoid. *New Zealand.* H4. Summer.

Very common in New Zealand, in both reddish brown and green variants; only the latter seems to be in cultivation in Europe.

10. C. pseudocyperus Linnaeus.
Illustration: Jermy et al., Sedges of the British Isles, edn 2, 117 (1982).
Tufted, with stout, sharp-angled stems to 80 cm; leaves often exceeding stems and to 12 mm wide. Inflorescence with a single terminal male spike, and 3–5 broad, cylindric female spikes to 5 cm, the upper clustered, the lower nodding, all subtended by long, leafy bracts. Fruit 4–6 mm, ovoid, shining, with long beak. *Widespread in the temperate regions of both hemispheres; also in tropical America.* H1. Summer.

An attractive species sometimes grown by or in water; it prefers rather acid, peaty soils.

11. C. riparia Curtis. Illustration: Jermy et al., Sedges of the British Isles, edn 2, 115 (1982).
Robust, somewhat tufted plant with short rhizomes, rough, sharply angled stems to 1.5 m, and long, keeled, greyish green leaves to 2 cm wide. Inflorescence with 2–6 large male spikes at top, and 1–5 more distant, lateral female spikes to 9 cm, almost erect or lower nodding, subtended by leafy bracts. Fruit *c.* 6 mm, ovoid with short beak. *Most of Europe, extending to N Africa & W Asia.* H1. Summer.

One of the commonest reed-swamp species by lakes and slow-flowing rivers in much of Europe; often present in suitable places in large gardens; if specially planted, usually as 'Aurea' ('Bowles' Yellow') or 'Variegata', with golden or longitudinally striped leaves respectively.

***C. acutiformis** Ehrhart. Very similar, but can be distinguished by its narrower leaves (not more than 1 cm wide), smaller fruits (*c.* 4 mm) and blunt (not pointed) male glumes. *Eurasia to S Africa.* H1.

It is as likely as *C. riparia* to occur in reed-swamps in large gardens and parks.

XXVIII. MUSACEAE

Large, evergreen perennial herbs. False stems made up of overlapping leaf-sheaths. Leaves spirally arranged, lanceolate or oblong, pinnately veined, entire, without stipules or ligules, but divided into sheath, stalk and blade. Inflorescences terminal with flowers arranged in 1 or 2 rows in the axils of large, spirally arranged, often coloured bracts. Female or bisexual flowers under the basal bracts, male flowers under the more terminal. The unopened male part of the inflorescence is termed the 'male bud'. Flowers bilaterally symmetric with a variously fused, 3–5-pointed perianth often forming a 3-sided hood (the upper lip) opposite a single free segment, which may be 3-lobed. Stamens 5 (rarely 6) in male or bisexual flowers. Ovary inferior, 3-celled, ovules numerous, placentation axile. Fruit a berry or fleshy capsule.

A small tropical family of 2 genera and about 60 species, commercially important because it contains the cultivated bananas.

1a. Plant solitary, not normally suckering, false stems distinctly swollen at the base; basal bracts persisting until the fruit is ripe; seeds usually more than 10 mm in diameter **1. Ensete**

b. Plant suckering freely; false stems cylindric, scarcely swollen at the base; basal bracts usually quickly deciduous; seeds rarely more than 6 mm in diameter **2. Musa**

1. ENSETE Horaninow
G. Argent
Solitary perennial herbs, flowering only once and then dying, not suckering from the base (unless damaged). Leaf-sheaths often spreading, rarely completely encircling the false stem. Inflorescence hanging, bracts usually persistent. Flowers many to each bract, in 2 rows. Upper lip often deeply lobed, the free segment 3-pointed or entire. Seeds usually more than 10 mm in diameter.

A genus of about 7 species from tropical Africa and Asia; they are rarely cultivated in Europe, but can make attractive architectural features.
Literature: Cheesman, E.E., Classification of the bananas I: the genus Ensete Horaninow, *Kew Bulletin* **2**: 97–106 (1947); Baker, R.E.D. & Simmonds, N.W., The genus Ensete in Africa, *Kew Bulletin* **8**: 405–16 (1953).

1a. False stem rarely more than 1 m; leaves entirely green; upper lip 3-lobed almost to the base, the free segment with toothed and pointed lateral lobes; seeds 1–1.2 cm in diameter, rarely less **2. superbum**

b. False stem to 5 m; leaves usually with reddish midribs; upper lip 3-lobed to less than half way, the free segment with rounded lateral lobes; seeds 1.2–2.3 cm in diameter **1. ventricosum**

1. E. ventricosum (Welwitsch) Cheesman (*E. edule* Horaninow; *Musa ventricosa* Welwitsch; *M. ensete* Gmelin; *M. arnoldiana* De Wildeman). Illustration: Botanical Magazine, 5223, 5224 (1861).
Swollen false stem to 5 m, only suckering if damaged. Leaf-blade to 6 × 1 m from a stalk which is up to 40 cm. Inflorescence more than 1 m, emerging horizontally, eventually drooping. Male bracts broadly lanceolate, tapering towards the apex for about two-thirds of their length. Basal flowers bisexual, the apical male. Upper lip 3-lobed to less than half way, the free segment with entire margins, rounded lateral lobes and a long, subulate central point. Stamens 6, 1 much smaller than the other 5. Fruit oblong, 6–10 cm. Seeds black, glossy, 1.2–2.3 cm in diameter. *Ethiopia to Angola.* H5–G1. Flowering irregular.

2. E. superbum (Roxburgh) Cheesman (*Musa superba* Roxburgh). Illustration: Botanical Magazine, 3849, 3850 (1841). False stem very swollen, to 1 m. Leaf-blade *c.* 150 × 50 cm, stalk *c.* 50 cm. Inflorescence arching, later hanging vertically, rather short. Male bracts ovate, concave, tapering from about half way to the broad apex which may have a small acute point. Basal flowers female with staminodes, the apical male. Upper lip 3-lobed almost to the base, the free segment with toothed margins, pointed lateral lobes and a long, subulate central point. Stamens 5. Fruit with numerous seeds which are mostly 1–1.2 cm in diameter (rarely less). *S India*. H5–G1. Flowering irregular.

2. MUSA Linnaeus
G. Argent
Plants suckering and often forming large clumps. False stem cylindric, made up of completely encircling leaf-sheaths which are hardly swollen at the base. Bracts mostly deciduous. Flowers few to many per bract, in 1 or 2 rows. Upper lip with minute lobes, the free segment with a single point. Seeds rarely more than 5 mm in diameter.

A genus of about 40 species from SE Asia, commonly cultivated for fruit and ornament.

Literature: Moore, H.E., Musa and Ensete, the cultivated bananas, *Baileya* **5**: 167–94 (1957) & **6**: 69–70 (1958). Simmonds, N.W., *The evolution of the bananas* (1962). Simmonds, N.W., *Bananas*, edn 2 (1966).

Height. Less than 2 m: **1,3,5**; 3 m or more: **1,2,4**.
Inflorescence. Hanging: **1,2**; erect: **3–5**.
Bracts. Yellow: **1,2**; dark purple: **1**; scarlet: **5**; pink or purplish pink: **3,4**.
Fruit. Hairy: **3**.

1a. False stem less than 2 m — 2
b. False stem more than 2 m — 4
2a. Inflorescence hanging; flowers in 2 rows per bract — **1. acuminata**
b. Inflorescence erect; flowers in 1 row per bract — 3
3a. Bracts pale purple; inflorescence-stalk and fruit hairy; leaves *c.* 2.5 times as long as broad with midrib purplish-flushed — **3. velutina**
b. Bracts red; inflorescence-stalk and fruits hairless; leaves *c.* 3.5 times as long as broad, completely green — **5. coccinea**
4a. Inflorescence erect; bracts purplish pink — **4. ornata**
b. Inflorescence hanging; bracts not purplish pink — 5
5a. Leaves and false stems without a glaucous waxy deposit; bracts dull yellow to brownish; leaf-stalks with broad, spreading, green, non-waxy margins — **2. basjoo**
b. Leaves and false stems with a glaucous, waxy deposit; bracts variable, usually dark purple; wings of leaf-stalk usually erect, if spreading then waxy and with a dead, brown margin — **1. acuminata**

1. M. acuminata Colla (*M. cavendishii* Paxton; *M. zebrina* Planchon). Illustration: Simmonds, Bananas, 20–1, 81 (1966); Graf, Tropica, 672 (1978).
Usually 4–6 m, mostly glaucous green. Upper edge of leaf-sheath with a brown, papery margin *c.* 1 cm wide. Inflorescence pendent, flowers in 2 rows per bract. Male bud pear-shaped, acutely pointed. Bracts variable in colour, usually dull purple outside, rolling back on opening to expose the flowers. Flowers white, cream or yellow, the free segment with a marked subapical wrinkle and the point turned inwards. Fruit variable, usually ripening yellow, with pale yellowish pulp. *SE Asia, from Thailand to N Australia*. H5–G2. Flowering irregular.

A very variable species which has given rise to most of the sweeter dessert bananas of commerce. The wild species is fully fertile with seeded fruits, but is unlikely to be found in cultivation. Various cultivars and hybrids with seedless fruits which are formed without fertilisation may be found in greenhouses, and are difficult to name precisely. 'Dwarf Cavendish' (*M. cavendishii* Paxton) is one of the more distinct and common; it has a false stem *c.* 2 m high and leaf-blades about 2 times longer than wide and is sometimes known as the Canary or Chinese banana; it is cultivated on a moderate scale in Sicily and Crete, and is often planted in greenhouses because of its small size, resistance to cool conditions and sweet fruit.

M. × paradisiaca Linnaeus with orange-yellow flowers and long, narrow, starchy, edible fruit is a commonly used name, which strictly refers to the cultivar 'French Plantain', which is of hybrid origin (*M. acuminata* × *M. balbisiana* Colla).

M. × sapientum Linnaeus (*M. rosacea* Jacquin; *M. violacea* Baker) with white or pinkish flowers and short, stumpy, starchy, edible fruit is another commonly grown variant; the name strictly applies to the cultivar 'Silk Fig', again thought to be of the same hybrid origin as *M.* × *paradisiaca*. A variant of *M.* × *sapientum* whose stem, leaves and fruit are tinged with purplish blue is often sold as *M. violacea* Baker (an invalid name). The name *M. rosacea* correctly applies to this hybrid, but is commonly and incorrectly applied to *M. ornata*.

2. M. basjoo Siebold (*M. japonica* Thibaut & Keteleer). Illustration: Botanical Magazine, 7182 (1891); Noailles & Lancaster, Mediterranean plants and gardens, 101 (bottom right, without name), 1977.
Plants 3.5–5 m, green, waxless. Leaf-sheath with a spreading, green upper edge. Inflorescence pendent, flowers in 2 rows per bract. Male bud broadly ovoid, bluntly pointed, bracts dull yellowish to somewhat brownish, rolling back on opening. Flowers cream or pale yellow, the free segment reflexed backwards from a subapical wrinkle. Fruit green, seeded, inedible. *S Japan (Ryukyu Islands)*. H5–G1. Summer.

3. M. velutina Wendland & Drude. Illustration: Clay & Hubbard, The Hawaii garden, tropical exotics, 149 (1977); Graf, Tropica, 671, 673 (1978).
Plant 1–1.5 m, purplish green, sparsely waxy. Leaves *c.* 2.5 times as long as broad with a flat, green or brownish edge to the leaf-sheath. Inflorescence erect, flowers in a single row per bract. Male bud slender, pointed, the bracts pale purple, margins slightly rolled under. Fruit pink, hairy, *c.* 9 × 3 cm, splitting when ripe to reveal white flesh in which are embedded numerous black seeds. *NE India*. G2. Flowering irregular.

4. M. ornata Roxburgh (*M. rosacea* misapplied; *M. rosea* misapplied). Illustration: Edwards's Botanical Register **9**: t. 706a, b (1823).
False stems slender, 2–3 m, rather glaucous, waxy green. Leaf-blade 5–6 times as long as broad, to 2 m × 35 cm, upper edge of the sheath smooth, green and adpressed. Inflorescence erect, flowers in 1 row per bract. Male bud slender, acute, with pale purplish pink bracts which sometimes roll slightly before falling. Flowers deep orange-yellow with a paler free segment which is without a subapical wrinkle. Fruit 6–8 × 1.5–2 cm, greenish yellow when ripe, with white pulp and numerous black seeds. *Bangladesh (Chittagong hills)*. G2. Flowering irregular.

5. M. coccinea Andrews (*M. uranoscopus* Loureiro). Illustration: Botanical Magazine, 1559 (1813); Graf, Tropica, 666, 673 (1978).

Plant *c.* 1 m, green, waxless. Leaves *c.* 3.5 times as long as broad and up to 1 m long, upper margin of the sheath with a flat green edge. Inflorescence erect, flowers in a single row per bract. Male bud rather cone-like, due to the persistence of the open, scarlet bracts which do not curl back. Fruit hairless, 4–5 × 2–2.5 cm, orange-yellow when ripe, pulp white with numerous black seeds. *S China, Vietnam, Laos, Kampuchea.* G2. Flowering irregular.

Names of doubtful application

M. mannii Hooker. Illustration: Botanical Magazine, 7311 (1893). An imperfectly understood small species up to 1.3 m high with purplish red bracts which do not curl back. *NE India.*

M. sanguinea Hooker. Illustration: Botanical Magazine, 5975 (1872). Similar to both *M. coccinea* and *M. mannii* but differing from both in having strongly rolled bracts and a hairy inflorescence-stalk. *NE India.*

XXIX. STRELITZIACEAE

Evergreen trees or perennial herbs. Leaves 2-ranked, simple, pinnately veined, entire, without stipules or ligules. Inflorescences arising within large, 2-ranked, green or coloured spathes. Flowers bisexual, bilaterally symmetric, with 3 free calyx lobes and 3 fused or free corolla lobes. Stamens 6, or 5 functional often with an additional staminode. Ovary inferior, 3-celled, placentation axile; ovules few to numerous. Fruit a capsule or berry, seeds often surrounded by filamentous or fleshy, coloured arils.

A small, mostly tropical family of 4 or 5 genera and just over 100 species, usually conspicuous because of the 2-ranked arrangement of the leaves and bracts.

1a. Inflorescence of 1–5 spathes; stamens 5; aril reddish brown **1. Strelitzia**
b. Inflorescence of more than 5 spathes; stamens 6; aril ciliate, bluish **2. Ravenala**

1. STRELITZIA Aiton

G. Argent

Palm-like trees forming trunks, often suckering from the base, or herbs with short rhizomes. Leaves with open, partly sheathing bases, grooved stalks and mostly with oblong or lanceolate, broad blades. Inflorescence axillary, usually more or less horizontal, simple or compound. Flowers bilaterally symmetric, the calyx with 3 narrow segments, the corolla of 2, often arrow-shaped, paired parts (the 'tongue'), which enclose the stamens and style, and a shorter upper part. Stamens 5. Fruit a 3-celled capsule with numerous seeds surrounded by reddish brown, filamentous arils.

A genus of 4 species from southern Africa.

Literature: Moore, H.E. & Hyypio, P.A., Some comments on Strelitzia, *Baileya* **17**: 65–74 (1970).

1a. Herbaceous plants, without woody trunks; calyx orange, yellow or reddish **1. reginae**
b. Plant with a woody trunk; calyx white or purplish blue 2
2a. Inflorescence compound with more than 1 spathe; the unpaired, inner corolla segment to 2 cm **2. nicolai**
b. Inflorescence simple with only 1 sheathing spathe, although often with additional sterile bracts; the unpaired, inner corolla segment more than 2.5 cm 3
3a. Stem solitary; tongue forming a narrow, boat-shaped structure up to 1.25 cm broad, with smooth margins **3. alba**
b. Stem suckering; tongue forming a boat-shaped structure 2–3 cm wide, with wavy margins **4. caudata**

1. S. reginae Aiton (*S. parvifolia* Aiton; *S. juncea* Link). Illustration: Botanical Magazine, 119 (1797); Graf, Tropica, 669, 670 (1978).
Almost stemless and more or less herbaceous, 1–1.5 m. Leaf-blade oblong-lanceolate, 25–50 × 10–25 cm, acute to rounded or even notched at the apex, rounded or tapering and often wrinkled at the base, stalk oval in cross-section, not channelled, 25–100 cm. Inflorescence with flowers from a solitary green or purplish-flushed spathe *c.* 12 cm. Flowers opening in succession, usually with orange calyx and blue corolla, *c.* 10 cm. *South Africa.* H5–G1. Spring.

A variable species in flower colour, stature and size of the leaves; var. **parvifolia** (Aiton) Anon. has small, lanceolate leaf-blades, and var. **juncea** (Link) Ker Gawler has no blades at all.

2. S. nicolai Regel & Körnicke. Illustration: Botanical Magazine, 7038 (1889); Dyer, The flowering plants of Africa **25**: t. 996 (1946); Graf, Tropica, 664 (1978).
Stems woody, in clumps, to 10 m. Leaf-blade oblong or ovate-oblong, *c.* 1.5 m × 60 cm, rounded or cordate at the base, stalk to 2 m. Inflorescences axillary with 3–5 purplish red spathes which are 40–45 cm. Flowers *c.* 20 cm, the calyx white, the corolla white or light purplish blue, tongue boat-shaped, arrowhead-shaped at the base with the margins turned inwards and meeting in the middle. *South Africa (coastal regions of S Natal & NE Cape Province).* G1. Flowering intermittent.

3. S. alba (Linnaeus) Skeels (*S. augusta* Wright). Illustration: Botanical Magazine, 4167, 4168 (1845); Dyer, The flowering plants of Africa **25**: t. 995 (1946); Graf, Tropica, 665, 669 (1978).
Stems woody, in clumps, to 10 m. Leaf-blade oblong-lanceolate, *c.* 2 m × 45–60 cm. Leaf-stalk *c.* 1 m. Inflorescence with 1 purplish red, glaucous spathe 25–30 cm, and some sterile bracts. Flowers *c.* 20 cm, calyx and corolla white, tongue not arrow-shaped at the base but margins there turned inwards and overlapping. *South Africa (south coast of Cape Province).* G1. Flowering intermittent.

4. S. caudata Dyer. Illustration: Dyer, The flowering plants of Africa **25**: t. 997 (1946).
Stems woody, suckering, to 6 m. Leaf-blade ovate to oblong, 1.5–1.75 m × 80–85 cm. Leaf-stalk 1.5–1.75 m. Inflorescence bracteate with a single, dark purplish red, somewhat glaucous spathe *c.* 30 cm. Flowers *c.* 20 cm, calyx white, corolla white or sometimes tinged with bluish purple at the base. Tongue forming a boat-shaped structure with obtuse basal lobes and a wavy margin. *South Africa (N Transvaal), Swaziland.* G2. Flowering intermittent.

2. RAVENALA Adanson

G. Argent

Palm-like tree with solitary, unbranched trunk to 25 m. Leaves like those of a banana, stalked, borne in a 2-ranked fan, *c.* 3 m. Inflorescence lateral, compound, composed of 6–12, 2-ranked spathes in which are compact, cymose inflorescences. Flowers white with 3 calyx and 3 corolla segments, 2 of the corolla segments larger than the third. Stamens 6. Fruit a capsule with numerous seeds with bright blue arils.

A genus of 1 species from Madagascar.

1. R. madagascariensis Sonnerat. Illustration: Graf, Tropica, 672 (1978). *Madagascar.* G2.

Plants of uncertain identity from SE Asia are often grown under this name; they are similar, except in producing occasional suckers and a terminal inflorescence. They may be a species of *Ravenalopsis* Nakai.

Species of the genus **Heliconia** Linnaeus are sometimes recorded as being in cultivation, but it is extremely doubtful if they are seen outside botanic gardens or specialist collections. The genus consists of herbaceous plants with false stems which are made up of concentric circles of overlapping leaf-bases (a feature not seen in *Strelitzia* and *Ravenala*). Their classification is difficult, and it is not certain which species may be cultivated.

XXX. ZINGIBERACEAE

Perennial herbs, often highly aromatic; rootstock usually fleshy. Leaves arranged in 2 ranks or in a spiral, leaf-sheaths encircling the stem, each commonly with a membranous outgrowth (ligule) at the junction of sheath and leaf. Inflorescence terminal on a tall or short leafy shoot, or from a separate, leafless shoot. Flowers usually arising in the axils of bracts, each flower often subtended by a bracteole. Calyx and corolla tubular, petals 3. Functional stamen 1, sometimes with a crest at apex, the remainder of the staminal whorl modified into 1 usually showy staminode (lip), and, in most cases, 2 variously formed lateral staminodes. Ovary inferior, commonly 3-celled with axile placentation. Fruit a dehiscent capsule or fleshy berry.

A family with over 40 genera and about 1000 species, found mainly in the tropics of the Old World but with some representatives in the New World and in subtropical Asia. It is remarkably uniform in vegetative structure, so this aspect is described very briefly in the generic descriptions. An important spice plant family, yielding ginger, turmeric and cardamom. *Costus* and *Tapeinochilos* are placed in a separate family, *Costaceae*, by some authors.

The Zingiberaceae are plants of the forest floor, requiring shade; they are very sensitive to high light intensity, particularly if grown in pots where the root-run is restricted. Damage resulting from this appears as yellowing of the foliage and browning of the leaf tips. Pest attack under glass is problematic with pot culture, particularly from red spider mite. Spraying has a limited effect, and biological control may be more suitable. A moist atmosphere, obtained by overhead spraying of the foliage and frequent damping down will also be helpful.

Literature: Schumann, K., *Das Pflanzenreich* **20** (1904); Holttum, R.E., The Zingiberaceae of the Malay Peninsula, *Gardens Bulletin of Singapore* **13**: 1–249 (1950); Smith, R.M., Synoptic keys to the genera of the tribes Zingibereae, Globbeae, Hedychieae and Alpineae in part, *Royal Botanic Garden Edinburgh, Departmental Publication Series No. 2* (1981).

1a. Leafy shoots well developed, leaf-blades spirally arranged; plants not aromatic 2
b. Leafy shoots well developed, leaf-blades arranged in 2 ranks, or leafy shoots not well developed, leaves appearing tufted; plants aromatic 3
2a. Flowers much larger than bracts; lip showy; lateral staminodes absent; plant rarely branched **1. Costus**
b. Flowers barely exceeding the bracts; lip inconspicuous; lateral staminodes present as small, tooth-like appendages; plants often branched **2. Tapeinochilos**
3a. Inflorescence borne separately from the leaves 4
b. Inflorescence terminal on the leafy shoot or from the centre of a tuft of leaves 10
4a. Inflorescence long, slender, prostrate; bracts well spaced **15. Elettaria**
b. Inflorescence congested; bracts often overlapping 5
5a. Anther crest wrapped around the projecting style **4. Zingiber**
b. Anther crest, if present, not as above 6
6a. Lateral staminodes large, petal-like 7
b. Lateral staminodes absent or reduced to small appendages 8
7a. Bracts joined to each other laterally in the lower part, forming pouches; anther versatile **11. Curcuma**
b. Bracts free to the base; anther not versatile **5. Kaempferia**
8a. Inflorescence held well above the ground, surrounded by an involucre of sterile bracts which are showy **14. Nicolaia**
b. Inflorescence often held at ground level, showy sterile bracts absent 9
9a. Lip large, almost circular, to *c.* 6 cm in diameter; fruit flask-shaped **13. Aframomum**
b. Lip elliptic, *c.* 1 cm wide; fruit spherical **12. Amomum**
10a. Filament arched; anther with lateral appendages; flowers often replaced by bulbils **3. Globba**
b. Filament not arched; anther without lateral appendages; bulbils rare 11
11a. Bracts joined to each other laterally in the lower part, forming pouches **11. Curcuma**
b. Bracts free to the base 12
12a. Lateral staminodes showy, petal-like 13
b. Lateral staminodes absent or small and tooth-like 18
13a. Anther versatile, almost always with conspicuous basal spurs 14
b. Anther never versatile, lacking conspicuous basal spurs 15
14a. Fruit elongate, slow to split; flowers usually purple, rarely yellow **9. Roscoea**
b. Fruit short, splitting readily; flowers yellow **10. Cautleya**
15a. Short-stemmed plants, often appearing tufted 16
b. Tall plants with distinct leafy stems 17
16a. Lip deeply 2-lobed; anther crest prominent **5. Kaempferia**
b. Lip entire or slightly lobed; anther crest not prominent **6. Boesenbergia**
17a. Lip showy; filament elongate **7. Hedychium**
b. Lip much reduced, at most 3–4 mm or entirely absent; filament not elongate **8. Brachychilum**
18a. Petals much larger than the lip; fruit elongate; flowers orange **16. Burbidgea**
b. Petals smaller than the lip; fruit round or oval; flowers rarely orange **17. Alpinia**

1. COSTUS Linnaeus

R.M. Smith

Plants to 3 m. Leaves spirally arranged. Inflorescence terminal or on a separate leafless shoot, forming a dense head of flowers. Bracts often with a leaf-like extension at tip, each subtending 1 or 2 flowers. Bracteoles tubular or open to the base. Flowers showy, lip more or less trumpet-shaped. Lateral staminodes absent, stamen petal-like, the anther usually attached in the middle.

A genus of about 90 species, the majority from tropical America and Africa, a few occurring in tropical Asia and Australia. The genus requires hot and humid conditions where a night minimum temperature of 18 °C can be maintained. Direct planting in a border, where the plants can enjoy a free root-run, is preferable to pot culture. Regular

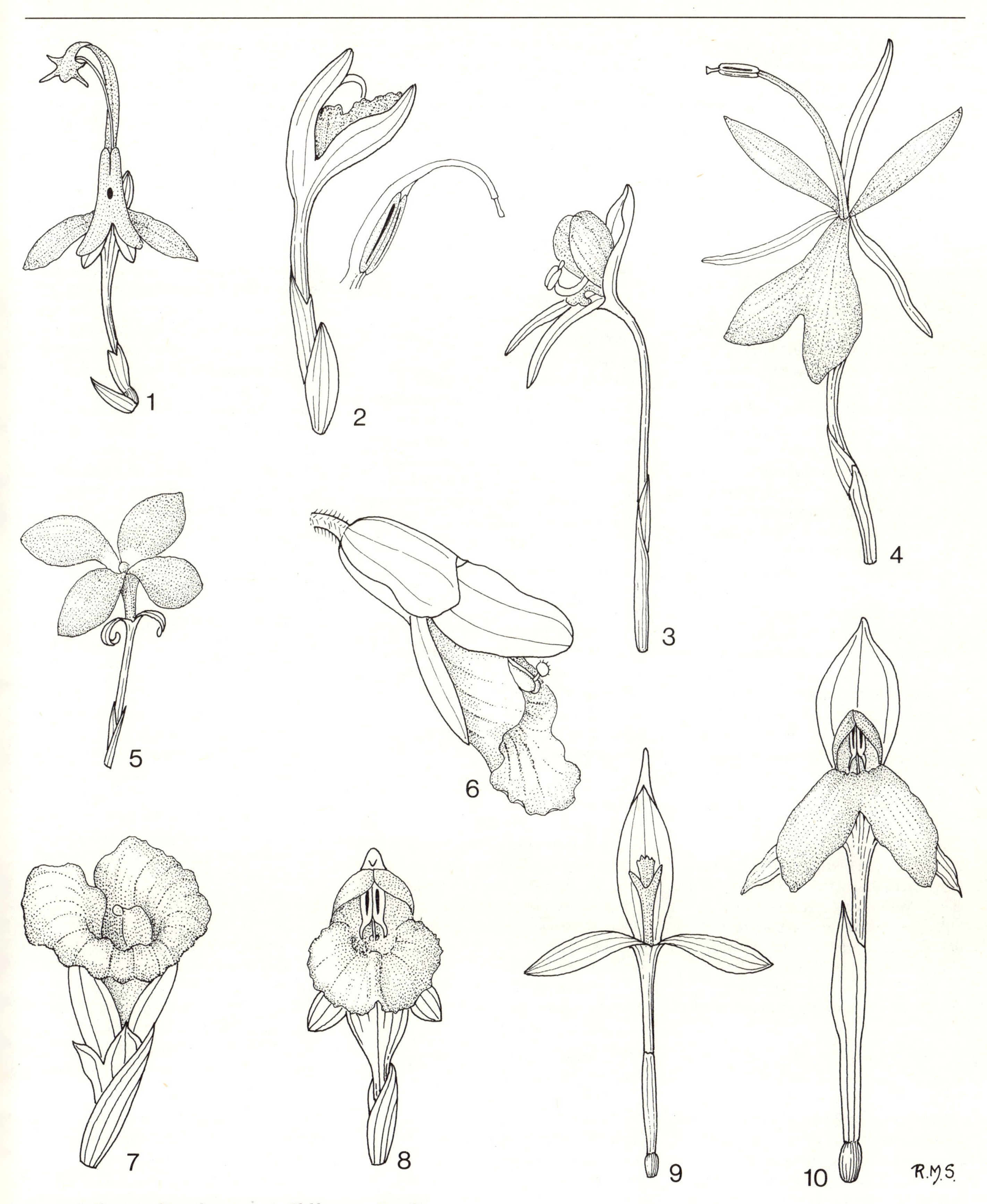

Figure 5. Flowers of Zingiberaceae. 1, *Globba marantina*. 2, *Zingiber zerumbet* (with stamen enlarged). 3, *Brachychilum horsfieldii*. 4, *Hedychium coccineum*. 5, *Kaempferia elegans*. 6, *Alpinia zerumbet*. 7, *Costus afer*. 8, *Curcuma zedoaria*. 9, *Burbidgea schizocheila*. 10, *Roscoea purpurea*. (Not to scale.)

mulching with peat and rich feeding are essential for healthy growth. Plants in pots suffer from root restriction and usually suffer heavy infestations of red spider mite. Propagation by division.

Leaves. With light and dark green bands: **2**. Sheaths with a ring of bristles near the top: **4**.
Bracts. Subtending 1 flower: **1–3**; subtending 2 flowers: **4–5**. Tips leaf-like: **1**. Tips spiny: **3**.
Flowers. Plain orange: **1**.

1a. Plant to 50 cm; flowers plain orange **1. cuspidatus**
b. Plant to 1 m or more; flowers variously coloured, never plain orange 2
2a. Flowers golden yellow with orange-red venation; leaves densely downy, dark green with lighter bands between the veins above **2. malortieanus**
b. Flowers white, variously marked; leaves downy or not, not banded 3
3a. Bracts spine-tipped; flowers 1 per bract; lip white to pinkish white **3. speciosus**
b. Bracts never spine-tipped; flowers 2 per bract; lip white tinged with yellow, sometimes with purple stripes 4
4a. Leaf-sheaths with a row of bristles just below the top; calyx longer than the bracteole **4. lucanusianus**
b. Leaf-sheaths without bristles; calyx shorter than the bracteole **5. afer**

1. C. cuspidatus (Nees & Martius) Maas (*C. igneus* R. Brown). Illustration: Botanical Magazine, 6821 (1885).
Plant to 50 cm, stems usually purple-red. Leaves very short-stalked, 10–20 × 4–8 cm, elliptic, with a long, narrow tip. Inflorescence terminal, bracts with leaf-like tips. Flowers 1 per bract, orange, each with a tubular bracteole. Lip more or less circular, 5–7 cm across when spread out. *Brazil.* G2. Flowering throughout the year.

2. C. malortieanus Wendland. Illustration: Botanical Magazine, 5894 (1871).
Plant to 1 m. Leaves short-stalked, to *c.* 20 × 18 cm, broadly elliptic, upper surface densely downy, banded light and dark green. Inflorescence terminal, bracts rounded at their tips. Flowers 1 per bract, yellow, each with a tubular bracteole. Lip broadly ovate, to 5 cm across when spread out, striped with red. *C America.* G2. Flowering throughout the year.

3. C. speciosus (Koenig) J.E. Smith.
Plant to 3 m. Leaves short-stalked, 12–25 × 3–6 cm, narrowly elliptic with pointed tips, softly hairy beneath. Inflorescence terminal, bracts with spine-like tips. Flowers 1 per bract, white or pinkish white, each with a non-tubular bracteole. Lip broadly obovate, to 10 cm across when spread out, yellow in the centre. *SE Asia to New Guinea; much cultivated in the tropics.* G2. Flowering throughout the year.

4. C. lucanusianus Braun & Schumann.
Plants 1–2 m. Leaf-sheaths with a conspicuous rim of bristles just below the top. Leaves 12–25 × 5–7 cm, lanceolate-oblong, downy beneath. Inflorescence terminal, bracts rounded at their tips. Flowers 2 per bract, white, each with a non-tubular bracteole, calyx tube longer than the bracteoles. Lip broadly obovate, 5 cm across when spread out, yellowish in the centre, striped with purple. *Tropical W Africa.* G2. Flowering throughout the year.

5. C. afer Ker Gawler. Figure 5(7), p. 121. Illustration: Botanical Magazine, 4979 (1857).
Like *C. lucanusianus* but without the bristles on the leaf-sheaths, the calyx tube shorter than the bracteoles. Flowers with or without purple markings. May produce terminal and separate inflorescences on the same plant. *Tropical W Africa.* Flowering throughout the year.

2. TAPEINOCHILOS Miquel
R.M. Smith

Plants sometimes branched. Leaves spirally arranged. Inflorescence terminal on a leafy shoot or borne separately, forming a dense head. Bracts overlapping, each subtending a single flower. Bracteoles absent. Flowers not, or scarcely exceeding the bracts. Lip inconspicuous. Lateral staminodes reduced to small appendages.

A genus of about 15 species from Indonesia, New Guinea and Australia. Cultivation as for *Costus* (p. 120).

1. T. ananassae Hasskarl (*T. pungens* Teysmann & Binnendijk). Illustration: Journal of the Royal Horticultural Society **83**: 71 (1958).
Plant to 2 m. Inflorescence resembling a red pineapple, borne separately from the leaves on a stalk 4–125 cm, the spike ovoid, to 12 cm. Bracts scarlet, strongly recurved, hard. Flowers yellow. *Indonesia.* G2.

3. GLOBBA Linnaeus
R.M. Smith

Plants slender, usually less than 1 m. Inflorescence terminal on the leafy shoot. Bracts often brightly coloured, each subtending 2–many flowers. Bracteoles open to the base. Flowers to 3 cm, the tube slender, much longer than the petals. Lip 2-lobed at apex. Lateral staminodes about the same size as the petals. Stamen with a long, curved filament, anther (in the cultivated species) with 2 small triangular appendages on either side. Flowers sometimes replaced by bulbils.

A genus of about 70 species, all from Asia. They require hot and humid conditions with a night minimum temperature of 18 °C. The plants must be dried out in the winter, allowing them to go dormant. This is most easily done with pot cultivation, but they may be planted out in a border or, preferably, in a raised bed which can be kept dry during the winter, and whose soil can be renewed periodically.

1a. Bracts green; bulbils always copiously produced **1. marantina**
b. Bracts red-purple or red; bulbils very rarely produced 2
2a. Leaves with stalks to 10 cm; bracts red-purple **2. winitii**
b. Leaves without stalks; bracts red **3. atrosanguinea**

1. G. marantina Linnaeus (*G. bulbifera* Roxburgh; *G. schomburgkii* Hooker). Figure 5(1), p. 121. Illustration: Botanical Magazine, 6298 (1876).
Plant to 75 cm. Leaves more or less stalkless, elliptic, to 20 cm. Inflorescence to 10 cm, eventually hanging, bracts green, to 1.5 cm. Flowers yellow, often with orange markings at the base of the lip, sometimes entirely replaced by spherical, white bulbils. *SE Asia.* G2. Autumn.

2. G. winitii Wright. Illustration: Botanical Magazine, 9314 (1926). Plant to 1 m. Leaf stalks to 10 cm, blade to 20 cm, lanceolate, cordate at the base. Inflorescence to 15 cm, hanging. Bracts red-purple, to 3.5 cm, reflexed, all usually fertile, each with a spike of 5–7 yellow flowers. *Thailand.* G2. Autumn–winter.

3. G. atrosanguinea Teysmann & Binnendijk. Illustration: Botanical Magazine, 6626 (1892).
Plant to *c.* 75 cm, stems purplish. Leaves more or less stalkless, to 20 cm, elliptic-lanceolate. Inflorescence erect, to 7 cm. Bracts bright red, all fertile, each subtending 6–7 flowers. *Borneo.* G2.

4. ZINGIBER Boehmer
R.M. Smith
Inflorescence cone-like, elliptic or cylindric, usually borne separately from the tall, leafy shoots. Flowers produced singly in the axils of broad, overlapping bracts. Bracteoles open to the base. Lip 3-lobed, the lateral lobes representing the lateral staminodes. Anther with an elongate crest which enfolds the upper part of the style. Fruit fleshy, held within the persistent bracts.

A genus of about 100 species, widely distributed in tropical Asia. *Z. officinale* is of economic importance as the source of commercial ginger. Cultivation as for *Globba* (p. 122).

Leaves. With 2-lobed ligule: **1,2,5**; with entire ligule: **3,4**. Less than 1.5 cm wide: **2**.
Inflorescence. Stalkless: **1**. Cylindric: **3**.
Lip. Plain yellow: **1,4,5**; dark purple with yellow spots: **2,3**.

1a. Inflorescence stalkless, arising at ground level **1. mioga**
b. Inflorescence held above ground on a distinct, erect stalk 2
2a. Lip dark purple with pale yellow spots; bracts not closely overlapping 3
b. Lip plain yellow; bracts closely overlapping 4
3a. Inflorescence robust, cylindric, 10–30 cm, the upper edges of the bracts stiffly incurved, forming pouches **2. spectabile**
b. Inflorescence ovoid, to 5 cm, margins of the bracts only slightly incurved **3. officinale**
4a. Ligule entire; leaves 5–8 cm wide **4. zerumbet**
b. Ligule 2-lobed; leaves 2–4 cm wide **5. purpureum**

1. Z. mioga (Thunberg) Roscoe. Illustration: Botanical Magazine, 8570 (1914).
Leafy shoots to 1 m. Leaves stalkless, to 30 cm, linear-lanceolate, ligule 2-lobed. Inflorescence more or less stalkless, ellipsoid, to *c.* 10 cm. Bracts ovate-oblong, green with reddish spots. Flowers yellow. *Japan.* G2.

2. Z. spectabile Griffith. Illustration: Botanical Magazine, 7967 (1904).
Leafy shoots to at least 2 m. Leaves stalkless, to 50 cm, oblong-lanceolate to elliptic, ligule entire. Inflorescence cylindric, to 30 cm, held on a robust stalk which is 30–100 cm. Bracts yellowish, usually turning red with age, their tips incurved. Flowers pale yellow, the lip dark purple with pale yellow spots. *Malaysia.* G2.

3. Z. officinale Roscoe.
Leafy shoots to 1.5 m. Leaves stalkless, to 15 × 1.5 cm, linear-lanceolate, ligule slightly 2-lobed. Inflorescence to 5 cm, held on an erect stalk which is 15–30 cm. Bracts green, broadly ovate. Flowers greenish yellow, lip dark purple with pale yellow spots. *Asia, precise area of origin unknown.* G2.

The rhizome is the source of commercial ginger and the plant is widely cultivated throughout the tropics.

4. Z. zerumbet (Linnaeus) J.E. Smith. Figure 5(2), p. 121. Illustration: Botanical Magazine, 2000 (1818).
Leafy shoots to 2 m. Leaves more or less stalkless, to 35 cm, lanceolate, ligule entire, prominent. Inflorescence cone-like, to 13 cm, held on an erect stalk which is 20–45 cm. Bracts ovate, green becoming red. Flowers white to pale yellow. *India.* G2.

Widely cultivated as a medicinal plant. 'Darceyi' with variegated yellow-green leaves is occasionally seen.

5. Z. purpureum Roscoe (*Z. cassumunar* Roxburgh). Illustration: Botanical Magazine, 1426 (1811).
Leafy shoots to 1.5 m. Leaves stalkless, to 35 cm, linear-lanceolate, ligule very small, 2-lobed. Inflorescence cone-like, to 12 cm, held on an erect stalk which is 15–25 cm. Bracts ovate, slightly hairy, brownish purple. Flowers pale yellow. *India.* G2.

5. KAEMPFERIA Linnaeus
R.M. Smith
More or less stemless. Leaves often appearing tufted, sometimes variegated. Inflorescence terminal and appearing with the leaves, or borne on a separate leafless shoot, when the flowers usually appear before the leaves. Flowers single in the bract axils. Bracteoles open to the base, very thin, deeply split. Petals narrow. Lateral staminodes large and petaloid, often similar to the halves of the deeply 2-lobed, showy lip. Anther with a prominent, reflexed crest.

A genus of about 40 species widely distributed in tropical Asia. The African species formerly included in the genus have recently been transferred to *Siphonochilus* Wood & Franks, which is not in general cultivation.

Leaves. Lanceolate-oblong, held erect: **1,2,4,5**; more or less circular, held horizontally: **3,6,7**. Red or purple beneath: **1,6**; green beneath: **2–4,7**; variegated green and creamy white on both surfaces: **5**.
Flowers. Borne on a leafless shoot: **1,2**; terminal on a leafy shoot: **3–7**. Lilac: **3,4**; white: **6**; white with some lilac on the lip: **1,2,5,7**.

1a. Flowers borne on a separate leafless shoot, usually before the leaves 2
b. Flowers terminal on the short leafy shoot 3
2a. Leaves variegated light and dark green above, purple beneath; lip *c.* 5 cm **1. rotunda**
b. Leaves plain green on both surfaces; lip *c.* 2.5 cm **2. parishii**
3a. Flowers lilac 4
b. Flowers white, usually with some lilac on the lip 5
4a. Leaves held horizontally; stalk of inflorescence enclosed by the leaf-sheaths **3. pulchra**
b. Leaves more or less erect; stalk of inflorescence projecting beyond the leaf-sheaths **4. elegans**
5a. Leaves more or less erect, oblong-lanceolate, variegated creamy white and green **5. gilbertii**
b. Leaves held horizontally, more or less circular; if variegated then not as above 6
6a. Leaves green with paler green variegation above, reddish beneath; flowers pure white **6. roscoeana**
b. Leaves green and sometimes reddish on the margin above, paler green beneath; lip with lilac bands **7. galanga**

1. K. rotunda Linnaeus. Illustration: Botanical Magazine, 6054 (1873).
Leaves to 45 cm, erect, short-stalked, lanceolate, variegated green above, purple beneath. Inflorescence appearing separately from the leaves. Flowers white, lip lilac, deeply 2-lobed, lateral staminodes white. *SE Asia; widely cultivated, precise origin unknown.* G2. Summer.

2. K. parishii Hooker (*K. ovalifolia* Roscoe; *K. ovalifolia* Roxburgh misapplied). Illustration: Botanical Magazine, 5763 (1869).
Leaves erect, short-stalked, 15–20 cm, oblong-lanceolate. Inflorescence appearing separately from the leaves. Flowers white, lip bright purple, yellow in the throat, deeply 2-lobed. Lateral staminodes white. *Burma.* G2.

3. K. pulchra Ridley.
Leaves 2 or 3, held horizontally, to 14 cm, elliptic, dark green with paler variegation

above. Inflorescence central to the leaf tuft and enclosed by the leaf-sheaths. Flowers lilac, lip deeply 2-lobed. *Thailand, Malay peninsula.* G2.

4. K. elegans (Wallich) Baker. Figure 5(5), p. 121.
Leaves 1 or 2, more or less erect, to 15 cm, elliptic, sometimes with greyish spots above. Inflorescence central to the leaf tuft and shortly projecting from the leaf-sheaths. Flowers lilac. *Burma to Malay peninsula.* G2.

Very close to, and perhaps not distinct from, *K. pulchra.*

5. K. gilbertii Bull. Illustration: Gartenflora **32**: 215 (1883).
Leaves more or less erect, short-stalked, to 10 cm, oblong-lanceolate. Inflorescence central to the leaf tuft. Flowers white, the lip striped with violet, 2.5 cm, deeply 2-lobed. Lateral staminodes white. *India.* G2.

6. K. roscoeana Wallich. Illustration: Botanical Magazine, 1212 (1829).
Leaves usually 2, held horizontally, 9–11 × 8–9 cm, almost circular, deep green with pale green variegation above, reddish beneath. Inflorescence central to the leaf tuft. Flowers pure white. Lip deeply lobed. *Burma.* G2. Autumn.

7. K. galanga Linnaeus. Illustration: Botanical Magazine, 850 (1805).
Leaves 2 or 3, held horizontally, to 15 × 10 cm, almost circular, green, sometimes with a red margin above, paler green beneath. Flowers white, lip with violet bands towards the base, deeply 2-lobed. *India.* G2. Summer-autumn.

6. BOESENBERGIA Kuntze
R.M. Smith
Short-stemmed plants with 1–several, usually long-stalked leaves. Inflorescence terminal. Flowers single in the axils of usually thin, lanceolate bracts. Bracteoles thin, non-tubular. Lateral staminodes petaloid. Lip often rather basin-shaped, entire or slightly lobed. Anther with or without a small crest.

A genus of about 30 species, all Asiatic, formerly known as *Gastrochilus* Wallich. Difficult in cultivation, both in pots and in borders. The plants are rarely vigorous and are particularly vulnerable to excessive watering during the winter. An open compost with a high proportion of peat appears to suit them best. Propagation is by division, but this is very slow.

1a. Leaves plain green; flowers pink **1. rotunda**
b. Leaves variegated; flowers white and yellow or yellow-orange 2
2a. Leaves dark green with a grey stripe along the mid-vein above, greyish green beneath; flowers white and yellow **2. vittata**
b. Leaves green with an irregular silver-white band along the mid-vein above, purple-brown beneath, flowers yellow and orange **3. ornata**

1. B. rotunda (Linnaeus) Mansfield (*B. pandurata* (Roxburgh) Schlechter; *Kaempferia pandurata* Roxburgh; *Gastrochilus panduratus* (Roxburgh) Ridley).
Leaves *c.* 4 per shoot, stalks 5–12 cm, blades to 28 × 10 cm, ovate-oblong, green. Flowers pink, lip *c.* 2.5 cm, slightly reflexed at the shortly 2-lobed tip. *Indonesia.* G2.

2. B. vittata (N.E. Brown) Loesener (*Kaempferia vittata* N.E. Brown; *Gastrochilus vittatus* (N.E. Brown) Valeton).
Leaves *c.* 4–6 per shoot, stalks 8–13 cm, blades 8–12 × 2–3 cm, elliptic, dark green and with a velvety sheen above, the mid-vein with a grey stripe along its length, grey-green beneath. Flowers white, lip *c.* 1.5 cm, basin-shaped, yellow in the centre. *Indonesia (Sumatra).* G2.

3. B. ornata (N.E. Brown) R.M. Smith (*Kaempferia ornata* N.E. Brown; *Gastrochilus ornatus* (N.E. Brown) Valeton). Illustration: Illustration Horticole **31**: t. 537 (1884).
Leaves several per shoot, stalks to 14 cm, blades 10–20 × 2.5–3 cm, lanceolate, upper surface with an irregular silver-white band along the mid-vein, purple-brown beneath. Flowers yellow, lip *c.* 1.8 cm, spotted with orange. *Borneo.* G2.

7. HEDYCHIUM Koenig
R.M. Smith
Leafy shoots to at least 2 m, with many stalkless or shortly stalked leaves. Inflorescence terminal, showy, with many flowers. Flowers 1–6, in the axils of overlapping or well-spaced bracts. Bracteoles usually tubular. Corolla tube slender, petals long and narrow, reflexed. Lip notched or deeply 2-lobed, narrowed at the base. Lateral staminodes petaloid. Filament usually long and slender. Anther without a crest.

A genus of about 40 species, all from Asia. Cultivation essentially as for *Costus* (p. 120). *H. gardnerianum* may be grown outside in favoured localities: a sheltered border out of direct sunlight is ideal, with rich dressings of humus in the form of compost, peat or leaf mould incorporated into the soil.
Literature: Schilling, T., A survey of cultivated Himalayan and Sino-Himalayan Hedychium species, *The Plantsman* **4**(3): 129–49 (1982).

Bracts. Broad and overlapping: **1–4**; well spaced: **5–10**.
Flowers. Deep orange or red: **3,5,6,8–10**; white, often with a flush of another colour: **1,2,7**. One per bract: **5,6**; two to six per bract: **1–4,7–10**.
Calyx. Shorter than bract: **1,2,4,5,7,9**; as long as, or longer than bract: **3,6,8,10**.
Corolla tube. Shorter than to as long as bract: **4,5**; longer than bract: **1–3,6–10**.
Lip. Entire: **8**. About 5 cm across: **1**.
Stamen. Much exceeding lip: **4,5,8–10**; shorter than to scarcely longer than lip: **1–3,5–7**. White: **1**.
Staminodes. Linear: **8**.

1a. Bracts broad and overlapping, hiding the main axis 2
b. Bracts narrow, well spaced, main axis clearly visible 5
2a. Flowers red, corolla tube hidden within the bract **4. greenei**
b. Flowers white, yellow or white flushed with yellow, orange or pink; corolla tube much longer than the bract 3
3a. Lip and staminodes yellow; calyx hairy; stamen distinctly longer than the lip **3. flavescens**
b. Lip and staminodes white, sometimes flushed with greenish yellow, yellow-orange or pink towards the base; calyx more or less hairless; stamen not, or scarcely longer than the lip 4
4a. Lip to 3.5 cm across, flushed with orange-yellow or pink **2. chrysoleucum**
b. Lip *c.* 5 cm across, yellow at base or the whole pure white **1. coronarium**
5a. Stamen shorter than or scarcely longer than lip 6
b. Stamen conspicuously longer than the lip 8
6a. Inflorescence narrowly cylindric, dense, at least 3 times as long as wide; corolla tube shorter than bract; flowers orange-red **5. densiflorum**
b. Inflorescence usually no more than 2 times as long as wide; corolla tube much longer than bract; flowers white or yellow-pink and white 7
7a. Flowers yellow, pink and white; bracts to 2.5 cm **6. spicatum**
b. Flowers white; bracts 4–5 cm **7. forrestii**

8a. Lip entire **8. speciosum**
b. Lip deeply 2-lobed 9
9a. Flowers white and yellow; leaves usually over 10 cm wide **9. gardnerianum**
b. Flowers red or varying shades of orange, red or purple; leaves to 4 cm wide **10. coccineum**

1. H. coronarium Koenig. Illustration: Botanical Magazine, 708 (1803).
Plant to 3 m. Leaves to 60 × 11 cm, lanceolate-acuminate. Inflorescence to 20 × 11 cm, bracts 4–6 cm, overlapping each other, each subtending 2–6 flowers. Flowers white. Calyx hidden by the bract. Corolla tube to 7 cm. Staminodes 3–5 cm, oblong-lanceolate. Lip about 5 cm across, usually yellow-green in the centre, deeply 2-lobed. Stamen white, shorter than the lip. Flowers very fragrant. *India to Indonesia.* G2. Spring.

Widely cultivated in the tropics. Var. **maximum** (Roscoe) Baker (Illustration: Edwards's Botanical Register **12**: t. 1022, 1826) is also cultivated; it has broader leaves, bracts with conspicuously hairy margins, larger flowers and the filament tinged with pink. It originated in India.

2. H. chrysoleucum Hooker (*H. coronarium* Koenig var. *chrysoleucum* (Hooker) Baker). Illustration: Botanical Magazine, 4516 (1850).
Stems to 1.5 m. Leaves to 40 × 8 cm, oblong-lanceolate, acuminate. Inflorescence to 20 cm. Bracts *c.* 3 cm, ovate to lanceolate, overlapping each other, each subtending 2–6 flowers. Flowers white, calyx hidden by the bract. Corolla tube to 8 cm. Staminodes to 4 cm, flushed with orange-yellow or pink. Lip to 3.5 cm across, coloured as the staminodes, 2-lobed. Stamen as long as or scarcely longer than the lip, orange-pink. *India.* G2. Summer.

3. H. flavescens Roscoe (*H. flavum* Roxburgh misapplied; *H. coronarium* Koenig var. *flavescens* (Roscoe) Baker).
Stems to 3 m. Leaves to 60 × 8 cm, lanceolate, acuminate. Inflorescence to 20 × 8 cm. Bracts *c.* 3.5 cm, ovate, overlapping each other, each subtending at least 4 flowers. Flowers yellow, calyx about the same length as the bract. Corolla tube 7–8 cm. Staminodes lanceolate, 2.5–3 cm. Lip c. 3 cm across, 2-lobed. Stamen longer than the lip, yellow-orange. *Himalaya.* G2. Spring.

4. H. greenei W.W. Smith.
Stems to 2 m. Leaves 20–25 × *c.* 5 cm, oblong, acuminate. Inflorescence to *c.* 12 cm. Bracts 5–7 cm, ovate, overlapping each other, each subtending 2–3 flowers. Flowers red, calyx and corolla tube shorter than the bracts. Staminodes 1.5–2 cm, spathulate. Lip 3–4 cm across, darker than the rest, shortly 2-lobed. Stamen shorter than the lip, red. *Himalaya.* G2.

This species is said to form bulbils.

5. H. densiflorum Wallich. Illustration: Botanical Magazine n.s., 325 (1958).
Stems to 5 m. Leaves 30–40 × 5–6 cm, oblong-lanceolate. Inflorescence to 14 × 4 cm, dense, cylindric. Bracts *c.* 1.5 cm, oblong, not overlapping, each usually subtending a single flower. Flowers orange-red, calyx and corolla tube shorter than the bract. Staminodes elliptic, *c.* 1 cm. Lip to 7 mm across. Stamen red, just longer than the lip. *Himalaya.* G2.

6. H. spicatum J.E. Smith. Illustration: Botanical Magazine, 2300 (1822).
Stems to 1 m. Leaves 10–40 × 3–10 cm, oblong-lanceolate, acuminate. Inflorescence to 20 cm, loose. Bracts to 2.5 cm, lanceolate, well spaced, each subtending a single flower. Flowers yellow, pink and white, calyx longer than the bract. Corolla tube to 8 cm. Staminodes *c.* 1 cm, white and pink, spathulate. Lip 1.5–2 cm across, yellow, 2-lobed. Stamen shorter than the lip, red-orange. *Himalaya.* G2. Autumn.

7. H. forrestii Diels.
Stems to 1.25 m. Leaves 30–50 cm, narrowly lanceolate. Inflorescence to *c.* 25 cm, with many flowers. Bracts 4–5 cm, spreading, each subtending 2–3 flowers. Flowers white, sometimes flushed with pink, calyx shorter than the bract. Corolla tube 3–4.5 cm. Staminodes elliptic. Lip *c.* 3 cm across, 2-lobed. Stamen not exceeding the lip. *SW China (Yunnan).* G2.

8. H. speciosum Wallich. Illustration: Wallich, Plantae Asiaticae Rariores **3**: t. 285 (1832).
Stems to 2.5 m. Leaves to 30 × 9 cm, oblong-lanceolate, acuminate. Inflorescence to 30 cm, moderately dense. Bracts 3–4 cm, oblong-ovate, well spaced, each subtending 2 flowers. Flowers yellow, calyx about as long as the bract. Corolla tube 4–5 cm. Staminodes linear, shorter than the petals. Lip *c.* 2.5 × 1.5 cm, oblong, entire. Stamen red, longer than the lip. *Himalaya.* G2. Summer.

9. H. gardnerianum Ker Gawler. Illustration: Edwards's Botanical Register **9**: t. 774 (1824).
Stems to 2 m. Leaves 25–40 × 10–15 cm, lanceolate, acuminate. Inflorescence 25–35 cm, with many flowers. Bracts 3–5 cm, oblong, each subtending 1–2 flowers. Flowers white and yellow, calyx shorter than the bract. Corolla tube 5–6 cm. Staminodes 2.5–3 cm, lanceolate. Lip 1–2 cm across, 2-lobed. Stamen bright red, longer than the lip. *Himalaya.* H5–G1. Summer–autumn.

10. H. coccineum J.E. Smith. Figure 5(4), p. 121. Illustration: Edwards's Botanical Register **14**: t. 1209 (1829).
Stems to 3 m. Leaves 30–50 × *c.* 3.4 cm, linear-lanceolate, acuminate. Inflorescence at least 20 cm, with many flowers. Bracts 1.5–2.5 cm, oblong, well spaced, each subtending 2–4 flowers. Flowers orange-red to purple, calyx and corolla tube about the same length as the bract. Staminodes *c.* 2 cm, oblong. Lip 2 cm across, deeply 2-lobed. Stamen much longer than lip. *Himalaya.* G2. Autumn.

Very variable in flower colour and leaf size. Var. **angustifolium** Roxburgh has very narrow leaves; var. **carneum** Roscoe has leaves to 6 cm wide, white to pink flowers and a spotted lip. The orange-flowered *H. aurantiacum* Roscoe is perhaps merely a variety of *H. coccineum.*

8. BRACHYCHILUM Petersen
R.M. Smith

Stems to 1 m. Leaves short-stalked, lanceolate to linear-lanceolate. Inflorescence terminal on the leafy shoot, rather loose. Bracts ovate, each subtending at least 2 flowers. Bracteoles tubular. Corolla tube long and slender. Petals hanging, twisted. Lip minute. Lateral staminodes elliptic. Stamen shorter than the petals.

A genus of only 1 species closely related to *Hedychium.* Cultivation as for *Costus* (p. 120).

1. B. horsfieldii (R. Brown) Petersen. Figure 5(3), p. 121.
Leaves 18–30 cm. Flowers yellow, lip 3–4 mm, occasionally entirely suppressed. Stamen orange. Capsule orange, red within. *Indonesia (Java).* G2.

9. ROSCOEA J.E. Smith
E.J. Cowley

Inflorescence terminal on the leafy stem, sometimes long-stalked. Bracts overlapping, each subtending a single flower. Bracteoles absent. Upper petal hooded. Lip entire or 2-lobed, sometimes each lobe notched. Staminodes petaloid. Anther versatile, spurred at the base. Fruit elongate, not splitting readily.

A genus of 17 species from China and the Himalaya. Plants are hardy in western Europe, and require a cool situation out of direct sunlight. They respond to rich feeding and compost, peat or leaf mould should be incorporated into the soil when planting. Propagation by division in the spring, or by seed.

Literature: Cowley, E.J., A revision of the genus Roscoea, *Kew Bulletin* **36**: 747–77 (1982).

Leaves. Conspicuously auricled at base: **2**.
Inflorescence. Usually stalked: **3,6**.
Bracts. The lowest tubular: **3,4**. Hairy at the top: **3,5**.
Calyx lobes. Broad, rounded: **5**.
Corolla tube. Projecting from calyx: **5,6**.
Petals. Pink: **3,6**; yellow: **4**; purple and/or white: **1,2,5,6**. The uppermost circular: **6**.
Anther spurs. With rounded ends: **4**.

1a. Bracts acute at apex 2
b. Bracts obtuse to truncate at apex 5
2a. Staminodes at least 2.5 cm, narrow with a long claw **1. purpurea**
b. Staminodes to 2 cm, relatively broad, with a short claw 3
3a. Leaves conspicuously auricled at the base **2. auriculata**
b. Leaves not auricled at the base 4
4a. Bracts equal to, or longer than the calyx, hairy at the top **3. scillifolia**
b. Bracts shorter than the calyx, hairless **4. cautleoides**
5a. Upper petal broadest above the middle, 2.5 cm or more, mucronate at the apex **5. humeana**
b. Upper petal circular or elliptic, less than 2.2 cm, not mucronate at the apex 6
6a. Upper petal circular; first bract much shorter than the calyx **6. alpina**
b. Upper petal elliptic; first bract equal to or much longer than the calyx **3. scillifolia**

1. R. purpurea J.E. Smith (*R. procera* Wallich; *R. purpurea* var. *procera* (Wallich) Baker). Figure 5(10), p. 121. Illustration: Edwards's Botanical Register **26**: t. 61 (1840); Hooker, Exotic flora, t. 144 (1825).
Stems to 55 cm, sometimes flushed purple. Leaves 4–8, lanceolate to oblong-ovate, 7–25 cm. Inflorescence enclosed in the upper leaf-sheaths. Flowers pale purple or white with dark purple markings, 1 open at a time; upper petal narrow, 3.2–6 × 1–2.8 cm; lip 4.5–6.5 cm including the claw which is 8–12 mm, entire or 2-lobed. Staminodes 2.5–4 cm, narrow. *Himalaya.* H5. Summer–autumn.

2. R. auriculata Schumann (*R. sikkimensis* invalid; *R. purpurea* misapplied). Illustration: The Garden **78**: 158 (1914).
Stems to 55 cm, sometimes flushed with purple. Leaves 3–10, linear to broadly lanceolate, auriculate at the base, 5–27 cm. Inflorescence enclosed in the upper leaf-sheaths. Flowers bright purple, 1 open at a time; upper petal broad, 2.8–3.7 × 1.5–2.8 cm. Lip 3.3–4.8 cm including the claw which is 7–12 mm, deflexed. Staminodes white, 1.5–2 cm. *E Nepal, NE India (Sikkim).* H5. Summer–autumn.

R. 'Beesiana', mentioned in nurserymen's catalogues, is a possible hybrid between *R. auriculata* and *R. cautleoides*, having stems up to 43 cm, with conspicuously auriculate leaf-bases and yellow flowers, sometimes with purple streaks. There is no known wild species which matches this description.

3. R. scillifolia (Gagnepain) Cowley (*R. capitata* var. *scillifolia* misapplied; *R. capitata* misapplied). Illustration: The New Flora and Silva **11**: opposite p. 22 (1938).
Stems to 37 cm. Leaves 1–5, sometimes not fully developed at flowering, linear to linear-lanceolate, 6–12 cm. Inflorescence usually exposed on a stalk. First bract tubular, hairy at the top. Flowers pink, 1 open at a time. Upper petal elliptic, 1.4–2 cm × 6–10 mm. Lip narrow, entire or 2-lobed, 1.3–2 cm. *SW China (Yunnan).* H5. Summer.

4. R. cautleoides Gagnepain (*R. chamael* Gagnepain; *R. yunnanensis* Loesener; *R. sinopurpurea* Stapf invalid). Illustration: Botanical Magazine, 9084 (1926).
Stems to 55 cm. Leaves 1–4, sometimes not fully developed at flowering, linear to lanceolate, 7.5–42 cm. Inflorescence exposed on a stalk. First bract tubular. Flowers yellow, usually more than 1 open at a time; upper petal wider above the middle, mucronate, 2–3.8 × 1–2.7 cm. Lip deflexed, 2.6–4 cm (including the claw). Staminodes 1.2–1.8 cm. *SW China (Yunnan, Sichuan).* H5. Spring.

5. R. humeana Balfour & Smith. Illustration: Botanical Magazine, 9075 (1925).
Stems to 34 cm. Leaves 1–2 (rarely 3), usually not fully developed at flowering, oblong to ovate, 2–22 cm. Inflorescence enclosed in the upper leaf-sheaths. Bracts hairy at the top, much shorter than the broadly lobed calyx. Flowers purple, many open at the same time. Corolla tube clearly projecting from the calyx. Upper petal wider above the middle, mucronate, 2.5–4.5 × 1.9–3 cm. Lip 2-lobed, deflexed, 2.2–4.5 × 2.8–3.5 cm including the claw. Staminodes small, shortly clawed. *SW China (Yunnan, Sichuan).* H5. Spring.

There is also a yellow-flowered form of this species.

6. R. alpina Royle (*R. longifolia* Baker; *R. intermedia* Gagnepain). Illustration: Royle, Illustrations of the botany of the Himalayan Mountains, t. 89 (1839).
Stems to 30 cm. Leaves 1–2, usually not fully developed at flowering, up to 4 after flowering. Inflorescence very short, not exposed. Flowers pink to purple, 1 open at a time. Corolla tube clearly projecting from the calyx. Upper petal circular, 1.3–2.2 × 1.1–2.4 cm. Lip narrow, 1.4–2.6 × 1–2 cm. Staminodes circular to elliptic, 6–12 × 4–8 mm. *Himalaya.* H5. Summer.

10. CAUTLEYA Hooker

R.M. Smith

Stems to 1 m. Inflorescence terminal on the leafy stem. Bracts each subtending a single yellow flower. Bracteoles absent. Lip deeply 2-lobed. Staminodes petaloid. Anther versatile, spurred at the base. Fruit splitting readily.

A genus of 2 or 3 species from the Himalaya. Cultivation as for *Roscoea*, but plants are less hardy and can be grown only in mild localities in sheltered borders.

1a. Bracts green, much shorter than the calyx; inflorescence with up to 12 flowers; plant very slender **1. gracilis**
b. Bracts green tinged with red, or entirely red, equalling or longer than the calyx; inflorescence with many flowers; plant robust **2. spicata**

1. C. gracilis (J.E. Smith) Dandy (*C. lutea* (Royle) Hooker). Illustration: Botanical Magazine, 6991 (1888).
Slender plant to 40 cm. Leaves 4–6, stalkless, up to 20 cm, ovate to oblong-lanceolate, usually purple beneath. Inflorescence axis dark red. *Himalaya.* H5–G1. Summer.

2. C. spicata (J.E. Smith) Baker.
Stems to 1 m. Leaves more or less stalkless or with stalks to *c.* 3 cm, the whole to 30 cm, lanceolate to oblong-lanceolate. Inflorescence robust, with many flowers,

axis reddish green or dark red. *Himalaya.* H5–G1. Summer.

The plant often grown as *C. robusta* (a misapplied name) is probably no more than a larger form of *C. spicata.*

11. CURCUMA Linnaeus
R.M. Smith
Plants to 1 m. Inflorescence terminal on a usually short-stemmed leafy shoot or on a separate leafless shoot and sometimes appearing before the leaves. Bracts always of 2 kinds: flower-bearing (fertile), which are joined to each other for about a third to a half of their length forming pouches which contain 2–7 flowers, and sterile (coma), which arise at the top of the inflorescence and are usually of a different colour. Bracteoles open to the base. Lip with a thickened central part. Lateral staminodes petaloid, folded under the hooded upper petal. Stamen with a versatile, usually spurred, anther.

A genus of perhaps 40 species from SE Asia, including the spice, turmeric (*C. longa*). This and *C. zedoaria* have been cultivated in SE Asia for a very long time, and are especially difficult for the botanist to classify. The modern Flora of Java regards both as 'collective species', and a similar treatment has been followed here. More recent work in Indonesia suggests that many of these plants are sexually sterile triploids or pentaploids. Cultivation essentially as for *Globba* (p. 122).

Stem. Less than 45 cm: **2,6**; larger: **1,3–5,7,8.**
Leaves. Rounded or cordate at base: **4.** Downy beneath: **7.**
Bracts. All orange-red: **1.** Sterile bracts absent: **2**; sterile bracts whitish green: **3**; sterile bracts tinged with purple-pink: **4–8.**
Anther spurs. Absent: **1.**

1a. Bracts all orange-red; anther spurless **1. roscoeana**
b. Bracts variously coloured, never orange; anther spurred 2
2a. Inflorescence without sterile bracts at the top; plant to 30 cm **2. albiflora**
b. Inflorescence with sterile bracts at the top; plant larger 3
3a. Inflorescence terminal on the leafy shoot 4
b. Inflorescence borne on a leafless shoot 6
4a. Sterile bracts white or greenish white **3. longa**
b. Sterile bracts with at least some purple-pink markings 5
5a. Leaves rounded to cordate at the base **4. petiolata**
b. Leaves narrowed towards the base **5. amada**
6a. Plant slender, to 45 cm; rhizome small **6. angustifolia**
b. Plant much larger; rhizome robust 7
7a. Leaves downy beneath, often with light green variegation **7. aromatica**
b. Leaves hairless beneath, usually with a purple-brown 'cloud' down the centre, occasionally plain green **8. zedoaria**

1. C. roscoeana Wallich. Illustration: Botanical Magazine, 4667 (1852).
Stems to *c.* 90 cm. Leaves long-stalked, blades to 30 cm. Inflorescence terminal, fertile and sterile bracts orange. Flowers yellow. Anther lacking basal spurs but with a prominent crest. *Burma.* G2. Summer–autumn.

2. C. albiflora Thwaites. Illustration: Botanical Magazine, 5909 (1871).
Stems to *c.* 30 cm. Leaves shortly stalked, blades to 15 cm. Inflorescence either terminal or lateral to the developing leaves. Sterile bracts absent, fertile bracts green. Flowers white, lip tinged with yellow. *Sri Lanka.* G2.

3. C. longa Linnaeus (*C. domestica* Valeton; *C. purpurascens* Blume).
Stems *c.* 1 m. Leaves long-stalked, blades to 50 cm. Inflorescence terminal. Sterile bracts white or greenish white, fertile bracts pale green. Flowers white and yellow. *Origin unknown, widely cultivated.* G2.

The turmeric of commerce.

4. C. petiolata Roxburgh. Illustration: Botanical Magazine, 5821 (1870).
Stems to 60 cm. Leaves long-stalked, blades to 25 cm, rounded or cordate at the base. Inflorescence terminal, sterile bracts purple, fertile bracts green. Flowers yellow and white. Anther spurs rather small, but distinct. *Burma.* G2.

5. C. amada Roxburgh. Illustration: Roscoe, Monandrian plants, t. 99 (1826).
Stems to 60 cm. Leaves long-stalked, blades to 40 cm. Inflorescence terminal. Sterile bracts pink-tinged, fertile bracts green. Flowers yellow and white. *India.* G2.

6. C. angustifolia Roxburgh. Illustration: Asiatic Researches **11**: t. 3 (1810).
Slender plant, stems to *c.* 45 cm. Leaves short-stalked, blades to 40 cm. Inflorescence borne separately from and before the leaves. Sterile bracts purple-pink, fertile bracts green. Flowers yellow. *Himalaya.* G2.

7. C. aromatica Salisbury. Illustration: Botanical Magazine, 1564 (1813).
Stems to 1 m. Leaves very long-stalked, blades to 60 cm, usually variegated light and dark green. Inflorescence borne separately from and before the leaves. Sterile bracts pink-tinged, fertile bracts pale greenish white. Flowers yellow-white, petals tinged with pink. *?India; widely naturalised.* G2.

8. C. zedoaria (Christmann) Roscoe (*C. heyneana* Valeton; *C. mangga* Valeton & van Zijp; *C. xanthorrhiza* Roxburgh). Figure 5(8), p. 121. Illustration: Roscoe, Monandrian plants, t. 109 (1807).
Stems to 1.5 m. Leaves long-stalked, blades to 80 cm, usually with a purple-brown coloration ('cloud') running down either side of the mid-vein. Inflorescence borne separately from and usually appearing before the leaves. Sterile bracts whitish, stained with pink or red, fertile bracts pale green with reddish margins. Flowers white and pale yellow usually tinged with pink. *SE Asia; precise origin unknown, widely cultivated and naturalised.* G2.

12. AMOMUM Roxburgh
R.M. Smith
Leafy shoots well developed. Inflorescence borne separately on a leafless stalk, usually cone-like. Bracts each subtending a single flower. Bracteoles usually tubular, rarely absent. Lip showy. Lateral staminodes reduced to very small appendages, or absent. Anther often crested.

A genus of about 90 species from Asia and Australasia. Cultivation as for *Costus* (p. 120).

1. A. compactum Maton (*A. cardamomum* Linnaeus misapplied; *A. kepulaga* Sprague & Burkill).
Stems to 3 m. Leaves stalkless, to 25 cm, narrowly lanceolate. Inflorescence congested, held on a stalk which is 5–10 cm. Flowers yellow. Lip *c.* 1 cm, elliptic. Anther crest 3-lobed. *Indonesia.* G2.

The round cardamom of Java.

13. AFRAMOMUM Schumann
R.M. Smith
Leafy shoots well developed. Inflorescence borne separately on a leafless stalk, congested. Bracts each subtending a single flower. Bracteoles open to the base. Lip (in the cultivated species) very showy, almost circular. Lateral staminodes reduced to small appendages. Anther crested.

A genus of perhaps 50 species from tropical Africa, the Mascarenes and Madagascar. Cultivation as for *Costus* (p. 120).

1. A. melegueta Schumann.
Stems to 2 m. Leaves more or less stalkless, to 20 cm, linear. Inflorescence fusiform, with few flowers. Flowers pale violet, the lip yellow in the centre. Fruit flask-shaped, bright red. *W Africa*. G2.

14. NICOLAIA Horaninow
R.M. Smith
Leafy shoots to 4 m. Inflorescence borne on a separate, erect, leafless stalk, surrounded by a conspicuous involucre of sterile bracts. Fertile bracts each with a single, tubular bracteole and a single flower. Lip short, erect, joined at the base to the lower part of the stamen and forming a distinct tube above the petals. Lateral staminodes absent. Anther without a crest.

A genus of 7 or 8 species from Malaysia, Indonesia and New Guinea, formerly called *Phaeomeria* Lindley. Cultivation as for *Costus* (p. 120).

1. N. elatior (Jack) Horaninow (*Phaeomeria speciosa* (Blume) Merrill; *P. magnifica* (Roscoe) Schumann; *N. imperialis* Horaninow; *N. speciosa* (Blume) Horaninow). Illustration: Botanical Magazine, 3192 (1832).
Inflorescence stalk to 1 m, head pyramidal, to 12 cm. Sterile bracts brownish red to pink, spreading. Flowers pink, lip deep brownish red with a white or yellow margin. Fruit spherical. *Malaysia*. G2.

15. ELETTARIA Maton
R.M. Smith
Leafy shoots to *c.* 4 m. Inflorescence borne separately from the leaves on a long, prostrate axis (rarely erect), often subterranean. Bracts well spaced, each subtending several flowers with tubular bracteoles. Flowers not showy, lateral staminodes reduced to small appendages. Anther with a small crest. Fruit spherical or oblong.

A genus of 3 or 4 species from tropical Asia. Cultivation as for *Boesenbergia* (p. 124).

1. E. cardamomum (Linnaeus) Maton.
Leaves to 1 m. Flowers white, lip streaked with violet. *India*. G2.

The fruits are the cardamoms of commerce.

16. BURBIDGEA Hooker
R.M. Smith
Leafy stems to 1 m. Inflorescence terminal on the leafy shoot, bracts absent or falling very quickly. Flowers borne singly (rarely in pairs), orange or yellow-orange, sometimes tinged with pink. Petals more conspicuous than the narrow, erect lip. Lateral staminodes absent. Anther with an elongate crest. Fruit elongate.

A genus of 5 or 6 species from Borneo. Cultivation as for *Costus* (p. 120).
Literature: Smith, R.M., The genus Burbidgea, *Notes from the Royal Botanic Garden Edinburgh* **31**: 297–306 (1972).

1. B. schizocheila Hackett (*B. nitida* Hooker, misapplied). Figure 5(9), p. 121. Illustration: Botanical Magazine, 8009 (1905).
Stems to 50 cm. Leaves to 15 cm, elliptic, purplish beneath. Flowers orange. *Borneo*. G2.

17. ALPINIA Roxburgh
R.M. Smith
Leafy shoot well developed, simple or branched. Inflorescence terminal on the leafy shoot. Flowers arising singly, in pairs, or in spikes with few to many flowers; often in the axils of bracts, though these are sometimes absent. Bracteoles tubular, open to the base, or absent. Lip often showy. Lateral staminodes reduced to small appendages, occasionally absent. Fruit usually spherical. Some species were formerly placed in *Catimbium* Jussieu, or *Languas* Small.

A very polymorphic genus of about 200 species from Asia and Australasia, Cultivation essentially as for *Costus* (p. 120).

Leaves. Hairy all over the lower surface: **3,5,6**; hairless on lower surface: **1,2,4,5,7–9**. With small bristles on the margins: **4**.
Inflorescence. Erect: **1,6–9**; drooping or horizontal: **2–5**.
Flowers. Borne singly: **6,8**; borne 2 or more together: **1–5,7,9**.
Bract. Absent: **4–7**; small, falling quickly: **4,8,9**; conspicuous and persistent: **1–3**. Bright red: **1**. Green and pink: **2**.
Bracteoles. Tubular: **1–3**; open to the base (often quickly deciduous): **4–9**.

1a. Bracts bright red, persistent, inflorescence erect; flowers white **1. purpurata**
b. Bracts, if present, never bright red, if persistent then inflorescence arching over; flowers never pure white 2
2a. Leaves variegated; bracts persistent and conspicuous, green tinged pink **2. vittata**
b. Leaves plain green; bracts, if present, not conspicuous and persistent 3
3a. Inflorescence tightly congested; flowers orange **3. rafflesiana**
b. Inflorescence loose; flowers variously coloured, never orange 4
4a. Lip 2.5–6 cm, yellow heavily patterned with red 5
b. Lip to 2 cm, white lined with pink or red 8
5a. Leaves up to 2.5 cm wide, margins with well-spaced bristles **4. calcarata**
b. Leaves 5–20 cm, margins sometimes hairy but not with bristles 6
6a. Inflorescence drooping **5. zerumbet**
b. Inflorescence erect 7
7a. Leaves softly hairy beneath; lip to 6 cm **6. malaccensis**
b. Leaves more or less hairless, except for the margin; lip to 3.5 cm **7. mutica**
8a. Inflorescence simple; flowers produced singly **8. officinarum**
b. Inflorescence often branched; flowers 3–5 **9. galanga**

1. A. purpurata (Vieillard) Schumann (*Guillainia purpurata* Vieillard).
Stems 3–4 m. Leaves stalked, to 80 cm, lanceolate. Inflorescence to 50 cm, erect, cylindric. Bracts bright red, each subtending 4–5 white flowers. Bracteoles tubular. Corolla tube slender, lip petaloid, usually 3-lobed. Lateral staminodes absent. Stamen with a small, almost circular crest. Capsule spherical. *Islands of SE Pacific*. G2.

Occasionally short leafy stems may replace the flowers in the axils of the lowermost bracts; these shoots may terminate in small inflorescences. In cultivation, completely sterile inflorescences can occur.

2. A. vittata Bull (*A. sanderae* Sander; *A. tricolor* Sander). Illustration: Graf, Exotica, edn. 8, t. 268A, 1490 (1976).
Stems to *c.* 1 m. Leaves to *c.* 15 cm, narrowly lanceolate, green with white or cream variegation. Inflorescence nodding, to *c.* 15 cm. Bracts well spaced and persistent, green tinged with pink, each with 2–5 greenish flowers. Bracteoles tubular. Lip rather fleshy, not petaloid. *Islands of the SE Pacific*. G2.

Grown mainly for its attractive foliage.

3. A. rafflesiana Baker. Illustration: Hooker's Icones Plantarum, t. 1963 (1891).
Stems to *c.* 1.5 m. Leaves with stalks to 4 cm, blades to 60 cm, lanceolate, hairy

on both surfaces. Inflorescence to 8 cm, tightly congested, held horizontally. Bracts 5–10 mm, each subtending 1–3 orange flowers. Bracteoles tubular. Lip 3-lobed. Lateral staminodes reduced to triangular appendages. Stamen with a minute tooth on either side at the top of the anther. Capsule spherical, green. *Malaysia*. G2.

4. A. calcarata Roscoe (*A. cernua* Sims). Illustration: Botanical Magazine, 1900 (1817).
Stems to 1 m, of rather slender habit. Leaves stalkless, to 30 cm, narrowly lanceolate, the margins with well-spaced, minute bristles. Inflorescence to 10 cm, held horizontally. Bracts absent, or very small and quickly deciduous. Bracteoles open to the base, quickly deciduous. Flowers usually arising in pairs. Lip to 3 cm, yellow, strongly patterned with red-purple lines. Lateral staminodes reduced to small appendages. Anther without a crest. *?China*. G2.

5. A. zerumbet (Persoon) Burtt & Smith (*A. speciosa* (Wendland) Schumann; *A. nutans* misapplied). Figure 5(6), p. 121. Illustration: Botanical Magazine, 1817 (1816).
Stems to 3 m. Leaves more or less stalkless, to 60 cm, lanceolate, margin always hairy, the lower surface sometimes so. Inflorescence to 40 cm, drooping, main axis very hairy. Bracts absent. Flowers usually in pairs, each with a pink-tinged bracteole which enfolds the bud. Corolla white. Lip to 4 cm, yellow heavily marked with brownish red. Lateral staminodes reduced to small appendages, exceptionally up to 1 cm. Anther without a crest. Capsule spherical, red. *Tropical and subtropical Asia*. G2. Spring–summer.

6. A. malaccensis (Burmann) Roscoe, Illustration: Edwards's Botanical Register **4**: t. 328 (1818).
Very similar to *A. zerumbet* but leaves always softly hairy beneath, inflorescence erect and flowers borne singly. *India*. G2.

7. A. mutica Roxburgh. Illustration: Botanical Magazine, 8621 (1915).
Similar to *A. zerumbet* and *A. malaccensis* but less robust. Leaves to 50 cm, margins hairy. Inflorescence erect. Flowers arising in 2s or 3s. Bracteoles falling very quickly. Lip 3–3.5 cm. Lateral staminodes absent. *India, Malaysia*. G2.

8. A. officinarum Hance. Illustration: Botanical Magazine, 6995 (1888).
Stems to 1.5 m. Leaves more or less stalkless, to 30 cm, linear-lanceolate. Inflorescence erect. Flowers arising singly. Bracts minute, falling quickly. Bracteoles absent. Lip to 2 cm, white with red veins. Lateral staminodes reduced to small appendages. *China*. G2.

9. A. galanga (Linnaeus) Willdenow (*A. zingiberina* Hooker).
Stems to *c*. 2 m. Leaves very shortly stalked, to 50 cm, lanceolate. Inflorescence erect, usually branched: bracts small and falling quickly, each with 3–5 flowers. Bracteoles inconspicuous, open to the base. Corolla greenish white. Lip *c*. 2.5 cm, white with pinkish lines. Lateral staminodes reduced to small appendages. Capsule spherical, red. *Tropical and subtropical Asia*. G2.

XXXI. CANNACEAE

Perennial rhizomatous herbs with generally unbranched, erect, leafy shoots. Leaves large, alternate, pinnately veined, with sheathing bases. Inflorescence a spike, raceme or panicle, commonly with 2 flowers in the axil of each bract. Flowers bisexual, asymmetric. Calyx of 3 usually free sepals. Corolla of 3 petals, united below into a tube of varying length. Stamens forming the showy part of the flower; in the cultivated species represented by 4 petaloid staminodes (one of which, the lip, is usually smaller and reflexed) and a fertile, petaloid stamen bearing a 1-celled anther on its margin, all united below to form a tube or a short united region. Ovary inferior, 3-celled, with many ovules, papillose or warty; style single, flat or club-shaped, conspicuous in the centre of the flower. Fruit a 3-valved capsule bearing the persistent sepals at its apex.

A family consisting of a single genus with about 50 species, native in the tropics, subtropics and warm-temperate regions of the New World.

1. CANNA Linnaeus

J.A. Ratter

Description as for family.
Literature: Kränzlin, F., *Das Pflanzenreich* **56** (1912); Winkler, H., in Engler & Prantl, *Die Natürlichen Pflanzenfamilien, edn 2*, **15**a: 650–4 (1930).

Several species of this genus are grown as ornamentals, and numerous hybrids have been produced. The spectacular flowers and foliage of the hybrid Cannas make them important as summer bedding plants in the warmer parts of Europe. Most of them are complex hybrids, involving in their origin the species included here as well as *C. coccinea* Aiton and various others, and are extremely variable in height, foliage colour and flower colour and form. They have been grouped as *C.* × *generalis* Bailey and *C.* × *orchioides* Bailey, the latter comprising the orchid-flowered Cannas derived from hybrids of *C. flaccida*, and showing the distinctive broad, wavy-margined staminodes derived from that species. Good colour photographs of a number of named varieties of hybrid Cannas are given in Graf, Tropica, 288 & 289 (1978).

1a. Flowers more or less hanging, 8–11 cm, long-tubed, waxy, reddish pink **4. iridiflora**
b. Flowers without the above combination of characters 2
2a. Petals reflexing; tube long, prominent; flower pale yellow, flaccid, reminiscent of that of a yellow flag-iris **1. flaccida**
b. Petals remaining erect or ascending; tube very short or absent; flowers yellow or not 3
3a. Plant glaucous; staminodes yellow **2. glauca**
b. Plant not glaucous; staminodes red or orange **3. indica**

1. C. flaccida Salisbury. Illustration: Rosoe, Monandrian plants, t. 18 (1828); Das Pflanzenreich **56**: 51 (1912).
To 2 m. Leaves green, ovate-lanceolate, 20–50 × 10–13 cm. Racemes erect, loose and few-flowered. Flowers *c*. 12 cm, borne singly at each node of the inflorescence, yellow, soft or flaccid in texture, reminiscent of those of a yellow flag-iris. Sepals *c*. 3 cm, reflexed. Petals forming a tube of *c*. 4 cm, and with strongly reflexed lobes of about the same length. Staminodes forming a tube considerably exceeding that of the petals, the 3 upper very broad, rounded and showy, with scalloped margins. *USA (South Carolina to Florida)*. G1–2. Summer.

2. C. glauca Linnaeus. Illustration: Roscoe, Monandrian plants, t. 19 (1828); Botanical Magazine, 3437 (1835).
Plant glaucous, to 2 m. Leaves lanceolate, long-pointed, to 50 × 15 cm, hairless. Raceme usually simple. Flowers *c*. 9 cm, pale lemon yellow to darker yellow, more rarely with a tint or mottling of red or orange. Petals *c*. 8 cm, spoon-shaped but with pointed tips, united at the base into a tube *c*. 1.5 cm. Upper staminodes *c*. 8 cm, obtuse. *Tropical & subtropical America, from the West Indies to Bolivia & Argentina*. G1. Summer–autumn.

3. C. indica Linnaeus. Illustration: Edwards's Botanical Register **9**: t. 776 (1823) – colour atypical; Hooker, Exotic flora **1**: t. 53 (1823) – as var. maculata, with upper staminodes very speckled; Die Natürlichen Pflanzenfamilien, edn 2, **15a**: 649 (1930).
Shoots slender, usually to 1.7 m. Leaves to 50 × 25 cm, ovate-lanceolate to oblong with an uneven, pointed tip, green, hairless. Inflorescence a raceme or somewhat branched panicle. Flowers to 6.5 cm. Sepals green. Petals *c.* 4 cm, spoon-shaped, pointed, united at the base into a tube *c.* 4 mm. Staminodes (including the lip) free almost to the base, the 3 upper straight, spreading, bright red or orange (rarely mottled or streaked), lip curled, yellow or orange with darker mottling or streaks; fertile stamen patterned like the lip. *Tropical and subtropical America.* G1. Autumn.

4. C. iridiflora Ruiz & Pavon. Illustration: Botanical Magazine, 1968 (1818); Edwards's Botanical Register **8**: t. 609 (1822); Graf, Tropica, 288 (1978) but not Graf, Exotica, edn 8, 586 (1976).
Robust, shoots to 3 m. Leaves to 100 × 40 cm, broadly elliptic with a long, uneven, pointed tip, green, hairless. Inflorescence a drooping panicle. Flowers to 11 cm, waxy, reddish pink, more or less hanging. Petals *c.* 7 cm, united into a tube for about half their length, lobes narrow, pointed. Staminodes to 12 cm, forming a tube for about half their length, all reflexed above. *Peru.* G1. Autumn.

XXXII. MARANTACEAE

Perennial, rhizomatous herbs. Leaves generally in 2 ranks, winged sheath usually well developed; blade pinnately veined, often noticeably asymmetric, with a pulvinus at its junction with the petiole. (The leaf-stalk in Marantaceae consists of 3 parts: the winged sheath, the unwinged petiole and the pulvinus. The relative proportions of these vary from species to species; in the descriptions the terms 'pulvinus', 'petiole' and 'sheath' are used to prevent confusion.) Inflorescence a spike, raceme or head made up of 2-ranked or spiral bracts with paired flowers, or groups of paired flowers in their axils; rarely a more diffuse panicle. Flowers bisexual, asymmetric, in mirror-image pairs. Calyx of 3 nearly free sepals. Corolla of 3 petals united below into a tube of variable length. Stamens borne on the corolla, basically in 2 whorls, the outer whorl represented by 1 or 2 petaloid staminodes (rarely absent), the inner by a fertile stamen with a 1-celled anther, often bearing a petaloid structure in addition, and 1 or 2 petaloid staminodes of specialised structure. Ovary inferior, 3-celled or 1-celled by abortion, with a single ovule in each cell. Fruit a capsule, berry or nut.

A family of 32 genera and about 350 species occurring in the tropics and subtropics of both the New and Old Worlds, but with its principal development in the former. The species are mainly characteristic of forest floor habitats. Many are grown under glass, mainly for the decorative, coloured and patterned foliage, and some flower only rarely or never. They require shade, as exposure to bright sunlight will give rise to yellowing or blotching of the leaves, particularly if the plants are in pots.

Information on the wild habitat of many of the species in cultivation is scanty or altogether lacking; their geographical ranges can therefore be described only in very general terms.

Literature: Schumann, K., *Das Pflanzenreich* **11**: 1–184 (1902); Loesener, T., in Engler & Prantl, *Die Natürlichen Pflanzenfamilien*, edn 2, **15a**: 654–93 (1930).

1a. Ovary 3-celled, with 3 ovules; flowers in compact heads or dense spikes (usually with spiral bracts); inflorescence often basal (see also p. 136) **1. Calathea**
b. Ovary 1-celled, with 1 ovule; inflorescences various, but never basal, if a raceme or spike then bracts 2-ranked, never spiral 2
2a. Outer staminode 1; tall, erect, semi-aquatic plants (see also p. 137) **5. Thalia**
b. Outer staminodes 2, rarely absent; not erect, semi-aquatic plants 3
3a. Corolla tube elongate; inflorescence a raceme or spike with 2–4, 2-ranked bracts which are sometimes much exceeded by the branched flower axes **3. Maranta**
b. Corolla tube very short, wide; inflorescence a tight spike or raceme with many bracts, or a scarlet panicle 4
4a. Inflorescence a dense spike with persistent green, purple-tinged membranous bracts; flowers borne on 1 side; outer staminodes purple **2. Ctenanthe**
b. Inflorescence a panicle, raceme or spike with red, deciduous bracts; outer staminodes small or absent **4. Stromanthe**

1. CALATHEA Meyer

J.A. Ratter

Herbs with or without stems, with green or attractively marked and/or coloured leaves, often forming low clumps. Inflorescence usually a head or short spike with spiral bracts, more rarely a narrower spike with 2-ranked bracts. Outer staminode 1. Ovary 3-celled, each cell with 1 ovule. Capsule usually with 3 seeds.

A genus of about 300 species from tropical America. More than 90 specific names have been recorded in cultivation, but the accuracy of many of these is doubtful (as is the persistence of the plants). The 23 species included here cover those most likely to be found in general collections.

The plants require hot and humid conditions, and enjoy an open, loose compost which allows root movement and free passage of air. Any peat in the mixture should be left in a rough state to prevent stagnation. They are easily propagated by division into single crowns. Shading is necessary for quick establishment, which will take place very quickly in a case or growth chamber.

1a. Leaves not variegated 2
b. Leaves variegated 6
2a. Leaves linear-lanceolate, reddish-pilose along petiole and midrib **17. rufibarba**
b. Leaves without the above combination of characters 3
3a. Leaves red-purple beneath 4
b. Leaves greenish beneath 5
4a. Plant to 30 cm; leaves with pulvinus borne almost directly on the winged sheath, blade 10–14 cm; bracts and corolla orange **2. crocata**
b. Plant to 2 m; leaves with petioles *c.* 1 m, blade 20–60 cm; bracts yellowish; corolla white and violet **11. ornata**
5a. Plant to 2 m; leaf-blade to 80 cm, elliptic or oblong **3. cylindrica**
b. Plant to 30 cm; leaf-blade to 30 cm, nearly circular **16. rotundifolia**
6a. Lower surface of leaves purple, red or variously spotted or patterned in these colours 7
b. Lower surface of leaves green or green-variegated 22
7a. Upper surface of leaves with a distinct row of darker blotches or stripes on either side of the midrib 8
b. Leaves not as above 14

8a. Leaves with a row of distinct, usually elliptic blotches (often large and small alternating with each other) along each side of the midrib 9
b. Leaves not as above 11
9a. Leaves broadly ovate **8. makoyana**
b. Leaves linear-lanceolate to ovate-oblong 10
10a. Leaves linear-lanceolate, blade usually more than 20 cm, erect, conspicuously wavy **4. lancifolia**
b. Leaves ovate-oblong to oblong-lanceolate, blade 10–20 cm **22. wiotii**
11a. Leaf-blades less than 30 cm 12
b. Leaf-blades 40–60 cm 13
12a. Upper surface of leaf with dark and pale green bands alternating to give a pinnate pattern; pulvinus *c.* 2 cm, minutely downy; inflorescence and its subtending foliage leaf borne on a long scape; bracts in 2 rows **6. lietzii**
b. Upper surface of leaf with a jagged pale pattern along the midrib and a sinuous, feathered, often interrupted, pale line running parallel to it along each side of the blade; pulvinus *c.* 8 cm, hairless; spike with bracts spirally arranged and a crown of sterile bracts above **20. veitchiana**
13a. Upper surface of leaf with a pinnate pattern of pale stripes running from midrib to margin; pulvinus hairless **23. zebrina**
b. Upper surface of leaf with a row of triangles of pale colour running out from the midrib but usually not reaching the margin; pulvinus downy **21. warscewiczii**
14a. Upper surface of leaf with a pinnate pattern of thin, pink or whitish stripes between the veins, often in groups; petiole to 1 m **11. ornata**
b. Leaf not as above 15
15a. Upper surface of the leaf with a dark green, feathered stripe along the midrib and margins and a pale yellowish green zone in between **14. princeps**
b. Upper surface of leaf with a pale or pinkish, often feathered, median stripe 16
16a. Upper surface of leaf with a distinct, narrow, often sinuous and feathered, pale whitish, silvery or pinkish arc on each side between the median stripe and margin 17
b. Such pale areas absent or somewhat indistinct 19
17a. Median stripe along midrib narrow, often rose-coloured; plant to 20 cm **15. roseo-picta**
b. Median stripe wider, whitish, silvery or pale green, feathered; plant usually taller 18
18a. Pulvinus *c.* 1.5 cm, minutely downy; median marking feathered but usually little lobed **13. picturata**
b. Pulvinus up to 8 cm, hairless; median marking feathered and deeply lobed **20. vietchiana**
19a. Midrib with deep yellow-green triangles on each side which extend more than half way to the margin **21. warscewiczii**
b. Midrib without yellow-green triangles on each side 20
20a. Median stripe broad, often occupying most of the leaf surface **13. picturata**
b. Median stripe narrower 21
21a. Upper surface of leaf dark green with a jagged median band which is whitish; blade *c.* 10 cm with petiole a little shorter; plant to 20 cm **18. undulata**
b. Upper surface of leaf with midrib feathered pale green and an indistinct band of the same colour close to the margin; blade to 40 cm with petiole about the same length; plant taller than 30 cm **7. lindeniana**
22a. Upper surface of leaf deep green with a white median band 23
b. Variegation patterns various but not as above 24
23a. Plant to 20 cm; leaf-blades to *c.* 13 × 4 cm; spike ellipsoid, *c.* 1.5 cm on a slender stalk to 12 cm; flowers white or pale violet **10. micans**
b. Plant to 60 cm; leaf-blades to 20 × 9 cm; spike almost spherical or ovoid, *c.* 5 cm on a stalk of the same length or longer; flowers violet **9. medio-picta**
24a. Leaf-blade *c.* 35 cm, oblong, acute, glossy dark green with a median band of black-green above; petiole very long **19. variegata**
b. Without the above combination of characters 25
25a. Leaf almost circular, with a pinnate pattern of alternating silvery white and green stripes **16. rotundifolia**
b. Leaves narrower, with patterns of dark lateral blotches or bands, but not as above 26
26a. Paired blotches in a row more or less midway between midrib and margin on each side; spike few-flowered, enclosed by the lowermost bract and borne on a long stalk **12. pavonii**
b. Blotches touching or almost touching the midrib, not in pairs on each side; spike usually on a stalk *c.* 3 cm 27
27a. Dwarf plant; leaf-blades 10–20 cm, pulvinus less than 1 cm **22. wiotii**
b. Plant taller; leaf-blades more than 20 cm, pulvinus less than 1 cm or not 28
28a. Background of upper surface of leaf bright pale green; pulvinus *c.* 5 mm **5. leopardina**
b. Background of upper surface of leaf grey-green; pulvinus *c.* 2 cm **1. bachemiana**

1. C. bachemiana Morren (*C. bella* (Bull) Regel; *Maranta bachemiana* invalid; *M. kegeliana* invalid). Figure 6(2), p. 132. Illustration: Pflanzenfamilien, edn 2, **15a**: 679 (1930); Graf, Exotic plant manual, edn 4, 177 (1974); Graf, Exotica, edn 8, 1140, 1143 (1976); Graf, Tropica, 632 – as C. kegeliana, 634–as C. bella (1978).
Stemless, to 50 cm. Leaves with petiole *c.* 15 cm and pulvinus downy, *c.* 2 cm. Blade leathery, to *c.* 35 × 12 cm, ovate-lanceolate, narrowly acuminate, greenish silver-white above with dark green bands and often spots and sometimes a dark margin; green, rarely slightly purplish, beneath. Spike emerging directly from the soil, almost cylindric, *c.* 6 cm on a stalk *c.* 2.5 cm (sometimes to 10 cm). Flowers *c.* 5 cm (large for the genus). *Brazil.* G2.

2. C. crocata Morren & Jorisenne. Illustration: Botanical Magazine, 7820 (1902); Graf, Tropica, 631 (1978).
To 20 cm or taller. Leaves forming a clump, blade oblong or ovate-elliptic, *c.* 12 cm, green above, sometimes with a grey feather pattern, red-purple beneath, pulvinus borne almost directly on the top of the winged sheath. Inflorescence an ovoid or spherical head *c.* 4 cm, borne above the leaves on a scape. Bracts and corollas bright orange. *Brazil.* G2.

3. C. cylindrica (Roscoe) Schumann (*Phrynium cylindricum* Roscoe; *C. grandifolia* Lindley). Illustration: Edwards's Botanical Register **14**: t. 1210 (1827); Graf, Tropica, 634 (1978) – as C. grandiflora.
Robust plant with a stem to 2 m. Leaves with long petioles, pulvinus 2.5–5 cm, hairless, blades to 80 × 33 cm, elliptic or oblong, shortly acuminate, rounded at the base, green on both sides. Inflorescence a cylindric spike *c.* 15 cm on a stalk to 30 cm, emerging from the central leaf-sheaths. Bracts green, spirally arranged. Flowers yellow or white. *S Brazil.* G2.

This species is sometimes misnamed *C. grandiflora* (Roscoe) Schumann, which is, in fact, a much smaller plant with a stalkless

Figure 6. Leaves of *Calathea* species. 1, *C. zebrina*. 2, *C. bachemiana*. 3, *C. ornata*. 4, *C. makoyana*. 5, *C. pavonii*. 6, *C. warscewiczii*. 7, *C. lancifolia*.

basal inflorescence bearing yellow flowers which are *c.* 6 cm. The confusion has no doubt arisen because of the similarity of the names 'grandiflora' and Lindley's synonym for the species, 'grandifolia'.

4. C. lancifolia Boom (*C. insignis* invalid, not *C. insignis* Petersen). Figure 6(7), p. 132. Illustration: Graf, Exotic plant manual, edn 4, 175 (1974); Graf, Exotica, edn 8, 1140 (1976); Graf, Tropica, 631, 632, 634 (1978).
Stemless, forming a compact clump to 60 cm. Petiole 15–40 cm; leaf-blade about the same length, linear-lanceolate, wavy, erect, upper surface with a row of alternating large and small, dark green, elliptic spots on either side of the midrib, background yellow-green, becoming darker at the margins, red-purple beneath. *Brazil.* G2.

5. C. leopardina (Bull) Regel. Illustration: Gartenflora **26**: t. 893 (1877); Graf, Exotic plant manual, edn 4, 177 (1974); Graf, Exotica, edn 8, 1143 (1976); Graf, Tropica, 633 (1978).
To 40 cm. Petiole about as long as the leaf-blade with pulvinus *c.* 5 mm, downy; blade to *c.* 20 × 9 cm, oblong or ovate-lanceolate, acuminate, bright pale green above with a row of dark green blotches along the midrib, pale green with deeper green marginal zones beneath, sometimes tinted bronze-purple. Inflorescence a more or less spherical, shortly stalked spike *c.* 5 cm, borne amongst the leaf-bases. Bracts transparent and lined green. Flowers yellowish. *Brazil.* G2.

6. C. lietzii Morren. Figure 7(3), p. 134. Illustration: La Belgique Horticole **25**: t. 15–17 (1875); Gartenflora **27**: t. 935 (1878); Graf, Exotic plant manual, edn 4, 177 (1974); Graf, Exotica, edn 8, 1141 (1976); Graf, Tropica, 632 (1978).
Clump-forming plant to 60 cm with conspicuous, long, erect stalks bearing the inflorescence and subtending leaf, later growing on to produce stalked plantlets. Petioles short, the minutely downy pulvinus which is *c.* 2 cm long often borne directly on the sheath; leaf-blades oblong-lanceolate, wavy, pointed at the apex, rounded to cordate at base, with somewhat ragged, dark and light green stripes alternating along the upper surface to give a pinnate pattern, red-purple beneath. Spikes *c.* 6 cm, narrow, subtended by a leaf which is usually solitary at flowering; stalk to 40 cm; bracts green, few, 2-ranked. Flowers white. *Brazil.* G2.

7. C. lindeniana Wallis. Illustration: Gartenflora **18**: t. 601 (1869); Graf, Exotic plant manual, edn 4, 176 (1974); Graf, Exotica, edn 8, 1141 (1976); Graf, Tropica, 632 (1978).
Clump-forming plant to 1 m. Petiole to 40 cm including the hairless pulvinus which is 5 cm. Leaf-blade to about 40 cm long, elliptic, somewhat pointed, upper surface dark green with a paler green, feathered median stripe and rather indistinct pale green zones running just inside the margins on each side, undersurface with red-purple zones corresponding to the dark green ones above. Spike ellipsoid, to 10 cm, almost stalkless, emerging directly from the soil. Bracts red-purple, spirally arranged. Flowers pale yellow. *NW Brazil, Peru.* G2.

8. C. makoyana Morren (*Maranta makoyana* Morren). Figure 6(4), p. 132. Illustration: La Belgique Horticole **22**: t. 24, 25 (1872); Parey's Blumengärtnerei **1**: 408 (1958); Graf, Exotic plant manual, edn 4, 175 (1974); Graf, Exotica, edn 8, 1143, 1186A (1976); Graf, Tropica, 632, 633, 635, 639 (1978).
Stemless clump-forming plant to 60 cm. Petiole shorter than or about as long as the leaf-blade; blade to 33 × 16 cm but usually much smaller, broadly ovate, pale green or cream above with a row of alternating large and small, dark green, more or less elliptic blotches on either side of the midrib, underside showing the same pattern in dark red-purple. Inflorescence emerging directly from the soil; spike ovoid, 2 cm, stalk *c.* 4 cm. Bracts green, spirally arranged. Flowers *c.* 2 cm, white, petals tinged with pink at the tips. *E Brazil.* G2.

9. C. medio-picta (Morren) Regel. Illustration: Gartenflora **28**: t. 934 (1878); Graf, Exotica, edn 8, 1136, 1139 (1976).
To 60 cm. Leaves with long petioles, blades to *c.* 20 × 9 cm, oblong to ovate-lanceolate with a short point, dark green above with the midrib feathered white, pale green beneath. Spike borne within the leaf-sheaths, more or less spherical or ovoid, *c.* 5 cm, stalk 5–10 cm. Bracts whitish or yellowish, spirally arranged. Flowers violet. *Brazil.* G2.

10. C. micans (Mathieu) Koernicke (*C. albicans* Schumann). Figure 7(5), p 134. Illustration: Martius, Flora Brasiliensis **3**(3): t. 20 (1890); Graf, Exotic plant manual, edn 4, 176 (1974); Graf, Exotica, edn 8, 1141 (1976); Graf, Tropica, 632 (1978).
Small plant to 20 cm, but usually lower, with leaves close to the ground. Petiole to 5 cm; leaf-blade to *c.* 13 × 4 cm, oblong-lanceolate to elliptic, dark green above with midrib feathered silvery white, grey-green beneath. Spike borne within the leaf-sheaths, ellipsoid, *c.* 1.5 cm, on a slender stalk to 12 cm. Flowers white or pale violet. *Mexico to Brazil.* G2.

11. C. ornata (Planchon & Linden) Koernicke (*C. sanderiana* invalid). Figure 6(3), p. 132. Illustration: Flore des Serres **4**: t. 413, 414 (1848); Graf, Exotic plant manual, edn 4, 177 (1974); Graf, Exotica, edn 8, 1136 (1976); Graf, Tropica, 636 (1978).
Robust stemless plant to 2 m. Leaves emerging from soil separately but usually forming a clump; petioles over 1 m, pulvinus 5–10 cm; blade often erect, to 60 cm, ovate-lanceolate to broadly ovate, leathery, green above; usually with a pinnate pattern of thin, white or pinkish stripes, often in groups, starting well out from the midrib; red-purple beneath. Spike ovoid, *c.* 8 cm, borne on a stalk *c.* 30 cm. Bracts spirally arranged, yellowish. Flowers white and violet. *Tropical S America from Ecuador to Guyana.* G2.

Several variants of this species are recognised, e.g. 'Roseo-lineata' and 'Sanderiana'.

12. C. pavonii Koernicke (*C. tubispatha* Hooker). Figure 6(5), p. 132. Illustration: Botanical Magazine, 5542 (1865); Graf, Exotica, edn 8, 1148 (1976).
Plant 30–50 cm. Petiole very short, the pulvinus being borne almost directly on the winged sheath; leaf-blade to *c.* 25 × 14 cm, oblong-acute to ovate-elliptic, thin, pale green above with a row of dark brown paired blotches on each side between the midrib and margin, greenish beneath. Spike borne within the leaf-sheaths, cylindric, to 3 cm, inclined on a long, slender stalk. Bracts 2 or 3, green, the lowest forming an involucre around the others. Flowers yellow, few. *Peru.* G2.

13. C. picturata (Linden) Koch & Linden (*C. vandenheckei* (Lemaire) Regel). Figure 7(2), p. 134. Illustration: Graf, Exotic plant manual, edn 4, 177 (1974); Graf, Exotica, edn 8, 1136, 1141 (1976); Graf, Tropica, 636 (1978).
Plant to 35 cm. Petiole to 30 cm with a pulvinus *c.* 1.5 cm; leaf-blade to *c.* 25 × 10 cm, elliptic, dark green with feathered whitish lines along the midrib and as arcs somewhat inside the margin (in 'Argentea' a single median whitish feathered area occupies most of the leaf surface), red-purple (sometimes very pale) or rarely green beneath. Spike narrowly

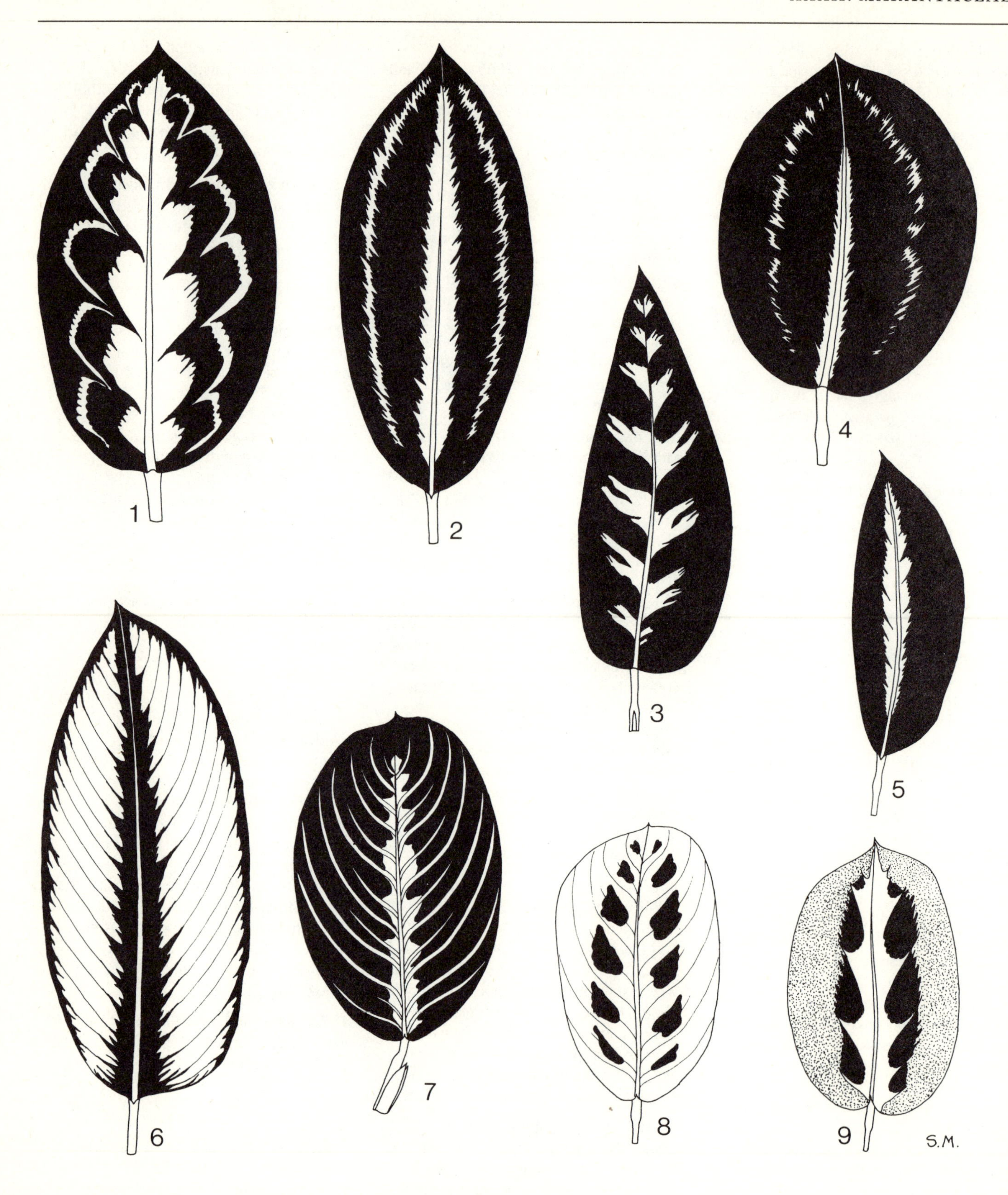

Figure 7. Leaves of species of Marantaceae. 1, *Calathea veitchiana*. 2, *C. picturata*. 3, *C. lietzei*. 4, *C. roseo-picta*. 5, *C. micans*. 6, *C. princeps*. 7, *Maranta leuconeura* var. *massangeana*. 8, *M. leuconeura* var. *kerchoviana*. 9, *M. bicolor*.

cylindric, up to 10 cm, shorter than its stalk. Flowers white. *NW Brazil.* G2.

C. vandenheckei is a small, juvenile form of this species.

14. C. princeps (Linden) Regel. Figure 7(6), p. 134. Illustration: Graf, Exotic plant manual, edn 4, 175 (1974); Graf, Exotica, edn 8, 1136, 1140 (1976); Graf, Tropica, 631, 633, 636 (1978).
Robust plant up to 2 m tall, but often smaller. Leaves often strikingly 2-ranked, reminiscent of the crown of *Ravenala* (p. 119). Petioles to *c.* 40 cm, pulvinus *c.* 5 cm; leaf-blades about as long as petioles, oblong-lanceolate, rounded at the base, upper surface dark green at the midrib and margins with a pale yellowish green zone between, red-purple beneath. *NW Brazil.* G2.

Graf (Tropica, 960, 1978) suggests that this species is a juvenile form of *C. altissima* (Poeppig & Endlicher) Koernicke.

15. C. roseo-picta (Linden) Regel. Figure 7(4) p. 134. Illustration: Gartenflora **18**: t. 610 (1869); Graf, Exotic plant manual, edn 4, 177 (1974); Graf, Exotica, edn 8, 1141, 1186A (1976); Graf, Tropica, 631, 636 (1978).
Plant scarcely reaching 20 cm. Petiole usually short but sometimes up to 12 cm, deep red; leaf-blade to 20 × 15 cm, broadly oval, obtuse, dark green above with a narrow pink or whitish stripe along the midrib and a thin, feathered, somewhat wavy line of the same colour (or whiter) running in an arc inside the margin, lower surface red-purple. Spikes *c.* 9 cm, more or less cylindric, stalk *c.* 9 cm or more. Bracts green, spirally arranged. Flowers white and violet. *NW Brazil.* G2.

16. C. rotundifolia (Koch) Koernicke (*C. fasciata* (Linden) Koernicke). Illustration: Gartenflora **13**: t. 452 (1864); Graf, Exotica, edn 8, 1143, 1146, 1186A (1976); Graf, Tropica, 633, 636, 640 (1978).
Low, stemless plant to 30 cm, forming clumps of leaves. Petiole up to 9 cm, leaf-blade to 30 × 27 cm, almost circular, often somewhat cordate at the base, leathery, green on both surfaces. Spike almost stalkless, nearly spherical, emerging directly from the soil. Bracts brown, spirally arranged. Flowers transparent white. *E Brazil.* G2.

The leaf of var. **fasciata** (Linden) Petersen has a pattern of silvery white, pinnately arranged bands and the undersurface sometimes tinged purple.

17. C. rufibarba Fenzl. Illustration: Botanical Magazine, 7560 (1897); Graf, Exotica, edn 8, 1142, 1146 (1976).
Stemless, forming large dense clumps to 50 cm. Petiole and leaf-midrib (sometimes almost the whole plant) densely covered in reddish hairs. Petiole up to 20 cm; leaf-blade to 25 cm, linear-lanceolate to lanceolate, wavy, glossy green above, tinged with purple beneath. Spike basal, to 7 cm, borne on a stalk *c.* 5 cm, bracts spirally arranged, violet with reddish hairs. Flowers yellow, projecting. *E Brazil.* G2.

18. C. undulata Linden & André. Illustration: Gartenflora **25**: t. 852 (1876); Graf, Exotic plant manual, edn 4, 176 (1974); Graf, Exotica, edn 8, 1136, 1144 (1976); Graf, Tropica, 636 (1978).
Small clumped plant to *c.* 20 cm. Petiole very short with pulvinus borne almost on the sheath; leaf-blade to *c.* 10 × 5 cm, oblong-elliptic, undulate, dark green above with a jagged whitish band along the midrib, purplish beneath. Spike *c.* 2 cm borne on a stalk which is *c.* 7.5 cm. Bracts spirally arranged, green edged and spotted with white. *Peru, Brazil.* G2.

19. C. variegata (Koch) Koernicke.
Robust plant. Petiole very long, with hairless pulvinus *c.* 6 cm; leaf-blade to 35 × 17 cm, oblong, acute, glossy deep green with a median band of black-green above, paler green with deeper bands beneath. Spike ellipsoid, *c.* 5 cm, borne on a stalk to 12 cm. Bracts spirally arranged. Flowers pale yellow. *Origin unknown.* G2.

20. C. veitchiana Hooker. Figure 7(1) p. 134. Illustration: Botanical Magazine, 5535 (1865); Graf, Exotic plant manual, edn 4, 175 (1974); Graf, Exotica, edn 8, 1139 (1976); Graf, Tropica, 636, 639 (1978).
Robust plant to 1 m. Petiole to 12 cm with hairless pulvinus which is up to 8 cm. Leaf-blade to 30 × 15 cm, ovate-elliptic or oblong, obtuse, glossy green above with a median deeply lobed or jagged, pale stripe along the midrib and a pale sinuous stripe lying parallel to it on each side, pattern repeated in red-purple beneath. Spike top-shaped with a crown of sterile bracts above, *c.* 8 cm, on a stalk to 20 cm. Bracts spirally arranged, green. Flowers white and violet. *Peru.* G2.

21. C. warscewiczii (Mathieu) Koernicke. Figure 6(6), p 132. Illustration: Gartenflora **15**: t. 515 (1866); Graf, Exotic plant manual, edn 4, 175 (1974); Graf, Exotica, edn 8, 1139 (1976); Graf, Tropica, 635 (1978); University of California Publications in Botany **71**: t. 20 (1978).
Robust plant to 1 m. Petiole 10 cm or more, with the pulvinus downy, *c.* 3 cm. Leaf-blade to *c.* 60 cm, oblong-lanceolate, dark velvety green above, feathered along the midrib with yellow-green triangles (like a row of shark-fins on either side), red-purple beneath. Spikes ellipsoid, *c.* 6 cm, borne on long scapes bearing 2 leaves near their apices, the upper leaf subtending the spike. Bracts broad, spirally arranged, pale yellow outside. Flowers yellowish. *Costa Rica.* G2.

22. C. wiotti (Morren) Regel (*C. wiotiana* Morren). Illustration: La Belgique Horticole **25**: t. 15–17 & f. 7 (1875); Pflanzenfamilien, edn 2, **15a**: f. 298a (1930); Graf, Exotic plant manual, edn 4, 177 (1974); Graf, Exotica, edn 8, 1140 (1976); Graf, Tropica, 632 (1978).
Dwarf plant forming patches from a branching and spreading rhizome. Petiole *c.* 8 cm with a hairless pulvinus which is *c.* 8 mm. Leaf-blade 10–20 cm, ovate-oblong to oblong-lanceolate, rounded to almost cordate at the base, wavy, light green above with a row of darker green blotches (often alternating larger and smaller) along each side of the midrib, dull greyish green beneath with patches of yellow-green, or sometimes purple. *E Brazil.* G2.

23. C. zebrina (Sims) Lindley. Figure 6(1), p. 132. Illustration: Botanical Magazine, 1926 (1817); Graf, Exotic plant manual, edn 4, 175 (1974); Graf, Exotica, edn 8, 1146, 1186A (1976); Graf, Tropica, 635, 636, 639, 640 (1978).
Robust plant up to 80 cm. Leaves with a hairless pulvinus, *c.* 4 cm, borne directly on the broadly winged sheath; leaf-blade to 65 × 25 cm, oblong-lanceolate, velvety dark green above with yellow stripes from the midrib to margins giving a pinnate effect, and shorter stripes bifurcating the ends of the green areas, red-purple beneath. Spike rather massive, *c.* 10 cm, almost spherical or ovoid, dense, with many flowers, borne on a stalk to 35 cm. Bracts broad, green tinted with violet. Flowers *c.* 4.5 cm, dark purple and white. *E. Brazil.* G2.

'Binotii' (var. *binotii* Bailey) is somewhat larger; it has narrower leaves to 90 cm, with the dark green areas much reduced, and the under surface greyish green.

Two other species, which are doubtfully in cultivation, will key out to *Calathea* in the generic key (p. 130). Both have 2 outer staminodes, whereas *Calathea* has only 1. They are:

Ataenidia conferta (Bentham) Milne-Redhead (*Phrynium confertum* (Bentham) Schumann), which has a usually simple stem up to 1 m, ovate-lanceolate leaves to 45 × 20 cm which have long petioles and are green on both sides, and the inflorescence consisting of stalkless or shortly stalked heads in the centre of the apical groups of leaves. The flowers are pale pink or purple. *W tropical Africa from Cameroun to Angola.*

Stachyphrynium jagorianum (Koch) Schumann, a much smaller plant, up to 30 cm, with linear-oblong to oblong-lanceolate leaves which have a pinnate pattern of blackish stripes from the midrib on the upper surface and a pale green lower surface. The flowers are white on a nearly stalkless, cylindric spike. *Malaysia & Indonesia.*

2. CTENANTHE Eichler

J.A. Ratter

Perennial herbs, often with basal leaves with long petioles, and leaves on the flowering shoots with short petioles. Inflorescence of dense spikes or racemes with closely overlapping, persistent bracts; flowers usually borne on 1 side only. Corolla tube very short. Outer staminodes 2, petaloid. Ovary 1-celled, with 1 ovule.

A genus of about 20 species, 1 in Costa Rica, the rest in Brazil. Cultivation as for *Calathea* (p. 130).

1a. Leaves uniformly green or rarely with cream-coloured strips along the margins **3. setosa***
b. Leaves otherwise variegated 2
2a. Leaves deep red beneath **2. oppenheimiana***
b. Leaves pale yellow-green beneath **1. lubbersiana**

1. C. lubbersiana (Morren) Eichler. Illustration: Graf, Exotic plant manual, edn 4, 176 (1974); Graf, Exotica, edn 8, 1136, 1145 (1976); Graf, Tropica, 637 (1978).

Herb to *c.* 80 cm with widely branched shoots leaning under their own weight. Basal leaves with petioles to 10 cm. Lower internodes of shoots long; stem-leaves clustered above, with the green pulvinus borne almost directly on the broad, membranous-margined sheath. Leaf-blades 30 × 14 cm, oblong to oblong-lanceolate, abruptly contracted to a pointed apex, upper surface with pale yellow-green and dark green in an irregular speckling, paler beneath. Flowers white. *Brazil.* G2.

2. C. oppenheimiana (Morren) Schumann (*Stromanthe porteana* var. *oppenheimiana* invalid). Illustration: Graf, Exotic plant manual, edn 4, 175 (1974); Graf, Exotica, edn 8, 1140, 1186A (1976); Graf, Tropica, 637, 638, 639 (1978).

Vigorous herb to 1 m. Petiole hairless or downy, to *c.* 16 cm, pulvinus *c.* 3.5 cm. Leaf-blade leathery, to 40 × 12 cm, ovate-lanceolate with a long, slender point, pinnately patterned with dark and pale green bands above, sometimes rather faint (and in 'Variegata', 'Tricolor', etc. with large cream areas reminiscent of variegated ivy), deep blood-red beneath. Spikes solitary or in pairs, to *c.* 5 cm. Bracts ovate, acute, hairless, membranous, pale green with pale purple along the margins. Flowers white. *E Brazil.* G2.

***C. kummeriana** (Morren) Eichler. Illustration: Gartenflora **79**: 378 (1930); Martius, Flora Brasiliensis **3**(3): t. 44 (1890). Similar but petioles of the basal leaves conspicuously hairy. *SE Brazil.*

3. C. setosa (Roscoe) Eichler. Illustration: Das Pflanzenreich **11**: 154 (1902).

Robust rhizomatous herb, to 1.5 m. Basal leaves with conspicuously hairy sheaths and usually long petioles up to 16 cm; stem leaves and some basal leaves with petioles as short as 2–3 cm; blades to *c.* 50 × 15–23 cm, oblong-lanceolate to oblong with a well-developed, crooked point, green on both surfaces (or rarely with cream-coloured strips along the margins). Inflorescence generally a cluster of spikes, each up to 7 cm, looking like a group of plump centipedes or woodlice. Bracts green, bristly, with long points. Flowers white and violet, borne on 1 side of the inflorescence only. *SE Brazil.* G2.

***C. compressa** (Dietrich) Eichler. Spikes solitary or paired and bracts ovate, acute *Brazil.*

3. MARANTA Linnaeus

J.A. Ratter

Erect or prostrate herbs, often with slender and much-branched stems (sometimes apparently dichotomous). Leaves often patterned. Inflorescence a spike, raceme or panicle with bracts in 2 rows, often with few nodes. Outer petaloid staminodes 2. Ovary 1-celled due to abortion of the other 2 cells, with 1 ovule. Fruit indehiscent, nut-like.

A genus of 30 species from tropical America. Cultivation as for *Calathea* (p. 130).

1a. Robust, erect plant up to 1 m or more; leaves ovate-lanceolate, acute at the apex, the basal with long petioles above the winged sheaths **1. arundinacea**
b. Small, low, often prostrate plants; leaves oblong, very obtuse at the apex; pulvinus borne almost directly on top of the sheath in all leaves (petiole at most 2 cm) 2
2a. Leaf with a broad median pale green stripe divided into a deep 'fishtail' pattern, and a dark green zone outside, sometimes divided by the tips of the fishtail into separate blotches, but these never clearly demarcated; bracts 4 or more **2. bicolor**
b. Leaf with a row of clearly demarcated brownish blotches on either side of the midrib and/or a herring-bone pattern picked out in white or red along lateral veins; bracts 2 **3. leuconeura**

1. M. arundinacea Linnaeus. Illustration: Botanical Magazine, 2307 (1822); Das Pflanzenreich **11**: 127 (1902); Graf, Exotic plant manual, edn 4, 176 (1974); Graf, Exotica, edn 8, 1145 (1976); Graf, Tropica, 637, 638, 640 (1978).

Erect, much-branched herb up to 1 m or more, with thick tuberous, starch-filled rhizomes. Basal leaves with long petioles, stem-leaves with pulvinus borne more or less on the leaf-sheath, blade up to *c.* 30 × 8 cm, ovate-lanceolate, green on both sides or sometimes variegated with large creamy areas. Inflorescence often branched. Flowers white. *Tropical America; widely cultivated in the tropics.* G2.

The source of West Indian arrowroot, which comes from the rhizomes. 'Variegata' has leaf variegations in dark and light greens and greenish yellow.

2. M. bicolor Ker Gawler. Figure 7(9), p. 134. Illustration: Edwards's Botanical Register **10**: t. 786 (1824); Graf, Exotica, edn 8, 1147 (1976); Graf, Tropica, 637 (1978).

Low plant to *c.* 20 cm with prostrate shoots. Leaves with pulvinus borne almost directly on the winged sheath, or more rarely with a short petiole up to 2 cm. Leaf-blade more or less horizontal, to *c.* 16 × 12 cm, oblong to elliptic-oblong, upper surface with broad median, pale green, fishtail patterned stripe and a dark green zone outside, the latter sometimes partially divided by the fishtails into blotches but these never with a clear demarcation; red-purple beneath. Spike solitary, bracts usually 4, sometimes more. Flower stalks not or little exceeding the bracts. Flowers white and violet. *Brazil & Guyana.* G2.

3. M. leuconeura Morren. Illustration: La Belgique Horticole **25**: 172, t. 9 (1875); Graf, Exotic plant manual, edn 4, 176 (1974); Graf, Exotica, edn 8, 1147, 1186A (1976); Graf, Tropica, 637, 638, 639, 640 (1978).
Low plant, often much branched at the base, with well-developed prostrate or sometimes semi-erect shoots. Leaves similar to *M. bicolor* but with a variety of colour patterns (see below) and usually rather smaller. Spike solitary, very slender, with only 2 rather distinct bracts which are often considerably exceeded by the flower stalks. Flowers white or violet. *Brazil.* G2.

Several cultivated variants, based on differences in leaf pattern, are recognised:

Var. **kerchoviana** Morren. Figure 7(8), p. 134. Illustration: La Belgique Horticole **29**: t. 5 (1879); Graf, Exotic plant manual, edn 4, 176 (1974); Graf, Exotica, edn 8, 1147 (1976); Graf, Tropica, 637 (1978). Leaves vivid to pale greyish green above with a row of clearly defined dark brown-green blotches along each side between midrib and margin, tinged with purple beneath, at least on the undersides of the blotches.

Var. **massangeana** Morren. Figure 7(7), p. 134. Illustration: La Belgique Horticole **25**: t. 10 (1875); Graf, Exotic plant manual, edn 4, 176 (1974); Graf, Exotica, edn 8, 1147, 1186A (1976); Graf, Tropica, 637 (1978). Leaves usually with a pale green or whitish jagged stripe along the midrib and the lateral veins picked out to give a white herring-bone pattern on a dark, velvety green background which often becomes paler towards the margins; small leaves often showing a row of blotches as in var. *kerchoviana*; underside red-purple, at least beneath the dark areas on the upper surface. In the frequently cultivated variant known as 'Erythroneura' or 'Erythrophylla' the veins are picked out in red and the midrib is pale yellow-green.

4. STROMANTHE Sonder

J.A. Ratter

Herbs with stems, basal leaves with long petioles. Stem-leaves in clumps on the flowering shoots. Inflorescence a raceme or fairly loose panicle with coloured deciduous bracts. Outer staminodes small, 2 or absent. Ovary 1-celled due to abortion of the other cells, with 1 ovule. Fruit a capsule.

A genus of 13 species in tropical S America. Cultivation as for *Calathea* (p. 130).

1a. Inflorescence a raceme with sometimes a single side branch at the base; petals violet-purple **1. porteana**
b. Inflorescence a branching panicle; petals white **2. sanguinea**

1. S. porteana Grisebach Illustration: Martius, Flora Brasiliensis 3(3): t. 42 (1890).
Leaf-blades ovate-oblong or lanceolate, green above, red-purple beneath. Inflorescence a raceme, sometimes with a single side branch at the base. Bracts red. Petals violet-purple. *E Brazil.* G2.

2. S. sanguinea Sonder. Illustration: Botanical Magazine, 4646 (1852); Graf, Exotica, edn 8, 1147, 1148 (1976); Graf, Tropica, 637 (includes one photograph labelled '*Ctenanthe oppenheimiana* inflorescence'), 638 (1978).
Robust plants to 1.5 m with basal, 2-ranked clumps of leaves producing long, rigid scapes on which are borne terminal groups of stem-leaves and inflorescences. Sheaths broad, membranous, pinkish or deep red, often hairy along the margins; petioles of some basal leaves to 25 cm, almost absent in others and in the stem-leaves (which have the pulvinus borne almost directly on the sheath), pulvinus to *c.* 5 cm; blade to *c.* 55 × 12 cm, lanceolate to linear-lanceolate, dark green with slightly paler midrib above, red-purple beneath. Inflorescence paniculate. Bracts and sepals scarlet. Petals white. *SE Brazil.* G2.

5. THALIA Linnaeus

J.A. Ratter

Tall marsh plants with leaves basal, with long petioles. Inflorescence a rather loose panicle with many flowers, borne on a long scape. Bracts deciduous. Corolla tube short. Outer staminode 1, petaloid. Ovary 1-celled due to abortion of the other cells, with 1 ovule. Stigma 2-lipped. Fruit indehiscent.

A genus of about 12 species from tropical and warm-temperate America, 1 species extending into tropical Africa. The species will grow in a cool glasshouse, and will survive outside in warmer areas, provided they are planted *c.* 60 cm below the surface of the water. Compost containing raw humus or peat should be avoided, as these encourage anaerobic conditions under water. Frequent feeding is, however, necessary, and fertiliser mixed with puddled loam can easily be dropped at the base of the plants.

1a. Inflorescence, and sometimes the whole plant, usually covered in white meal; bracts bluish green, crowded, forming dense spikelets; flowers deep purple **1. dealbata**
b. plant not mealy; panicle very diffuse with rather distant green bracts, not or hardly forming spikelets; flowers blue-violet **2. geniculata**

1. T. dealbata Fraser. Illustration: Botanical Magazine, 1690 (1815); Das Pflanzenreich 11: t. 22 (1902).
Erect marsh plant to 2 m, usually covered in white meal. Leaves with hairless, green sheaths; petiole long, pulvinus *c.* 3 cm; blades to *c.* 50 × 25 cm, ovate-lanceolate with pointed apex, green on both surfaces. Inflorescence a panicle, up to 20 cm, subtended by a bladeless sheath. Bracts mealy, pale bluish green, crowded and overlapping to form spikelets. Flowers deep purple. *Southeastern USA.* H5–G1. Summer.

2. T. geniculata Linnaeus. Illustration: Roscoe, Monandrian plants, t. 49 (1828). Differs from *T. dealbata* in the absence of meal, in the inflorescence, which is very diffuse (to *c.* 40 cm), the green bracts which are up to 1 cm apart so that they do not form crowded spikelets, and in the blue-violet flowers. *Subtropical & tropical America from Florida to SE Brazil & N Argentina; also W Tropical Africa.* G2. Summer.

Ischnosiphon bambusaceus (Poeppig & Endlicher) Koernicke is said to be found sometimes in cultivation. It will key out to *Thalia* in the generic key, but is a tall climbing herb reaching 10 m or more, with a bamboo-like habit; the leaves are small, ovate-lanceolate, green above and bluish beneath. It is native to Peru.

XXXIII. ORCHIDACEAE

Plants usually with rhizomes, sometimes with root-tubers, growing in soil, on rocks or as epiphytes, some (not ours) without chlorophyll. Aerial roots frequent. Stems usually present, often fleshy and swollen, when referred to as pseudobulbs. Leaves deciduous or persistent, borne on the stems (more rarely directly on the rhizomes) in 2 ranks or spirally arranged, rolled or folded when young, often hard and leathery. Flowers bilaterally symmetric or rarely asymmetric, solitary or in racemes or panicles borne terminally or laterally, usually on scapes which bear bracts; usually the flowers are turned upside down by a twist of 180° in the flower-stalk and

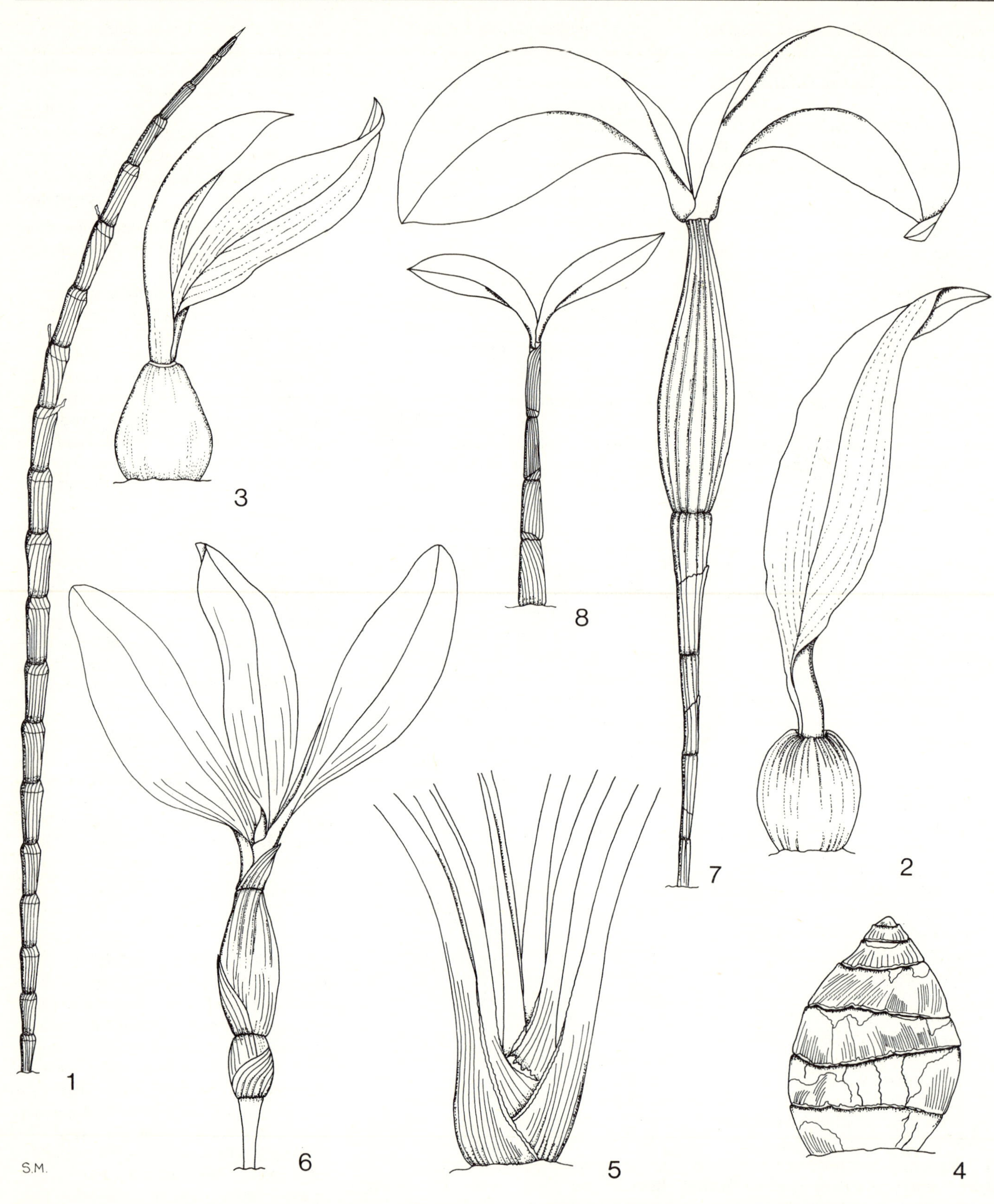

Figure 8. Diagram of some characters used in the identification of Orchidaceae. 1, A cane-like stem. 2, A simple pseudobulb bearing a single leaf. 3, A simple pseudobulb bearing 2 leaves. 4, A compound pseudobulb. 5, A compound pseudobulb enclosed in leaf-bases. 6, A fusiform compound pseudobulb. 7, A club-shaped compound pseudobulb. 8, A compound pseudobulb swollen at the base. (6, 7 & 8 are treated as pseudobulbs in the keys to the genera and species.)

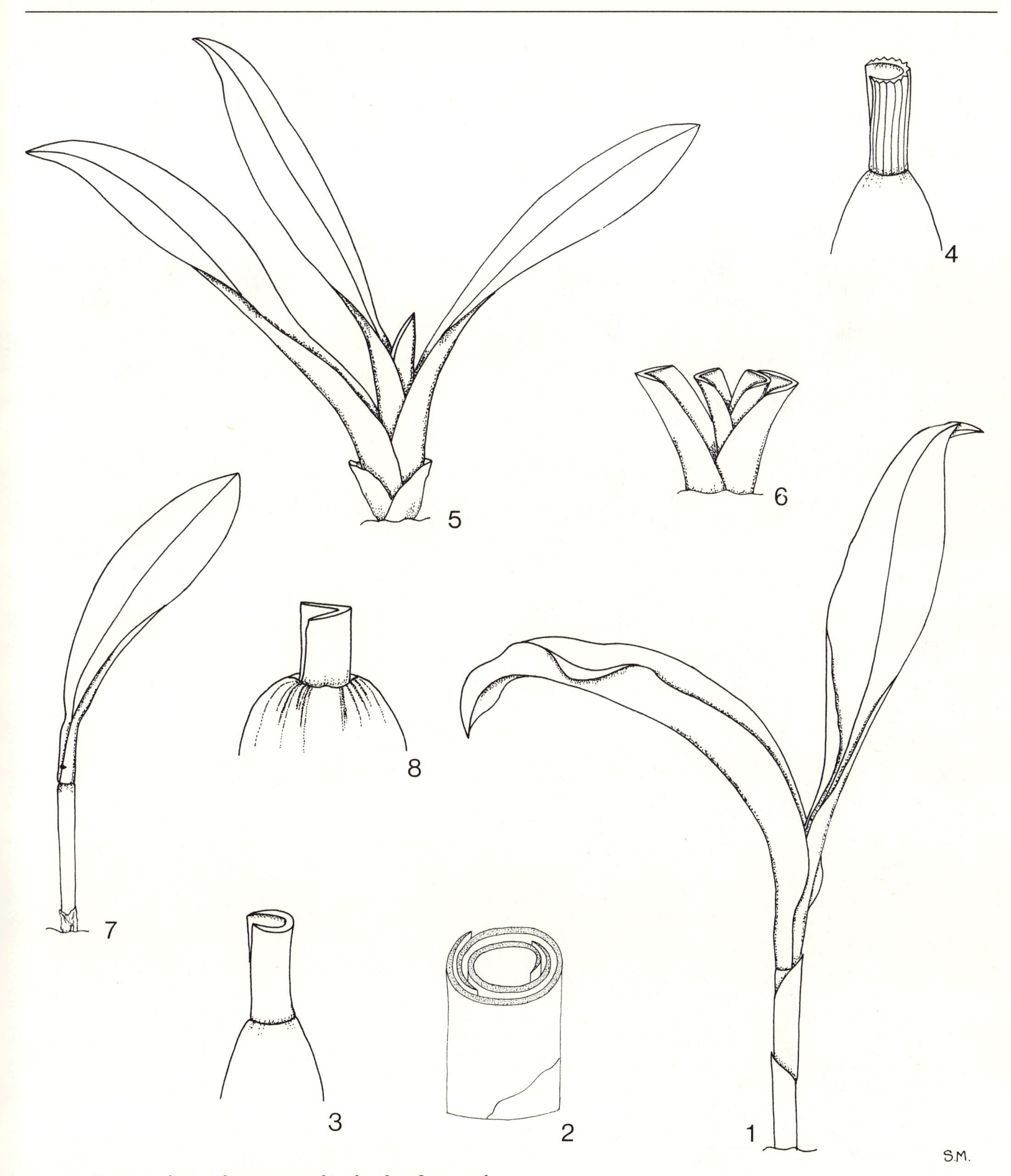

Figure 9. Diagram of some characters used in the identification of Orchidaceae. 1, A stem with rolled leaves. 2, Cross-section of the bases of 2 rolled leaves. 3, A pseudobulb with the base of a rolled leaf. 4, A pseudobulb with the base of a rolled and pleated leaf. 5, A stem with folded leaves. 6, Cross-section of the bases of folded leaves. 7, A stem of 1 internode with a single folded leaf. 8, A pseudobulb with the base of a folded leaf.

Figure 10. Diagram of some characters used in the identification of Orchidaceae. 1, An inflorescence terminal on a stem. 2, An inflorescence terminal on a pseudobulb. 3,4, Inflorescences lateral in the leaf axils on stems. 5, An inflorescence lateral to a leaf-scar on a compound pseudobulb. 6, An inflorescence basal to a pseudobulb. 7, Inflorescence erect, flowers with lip lowermost (resupinate). 8, Inflorescence erect, flowers with lip uppermost (non-resupinate). 9, Inflorescence hanging, flowers with lip lowermost (non-resupinate). 10, Inflorescence hanging, flowers with lip uppermost (resupinate). 11, Inflorescence arching, flowers with lip lowermost (a, resupinate; b, non-resupinate).

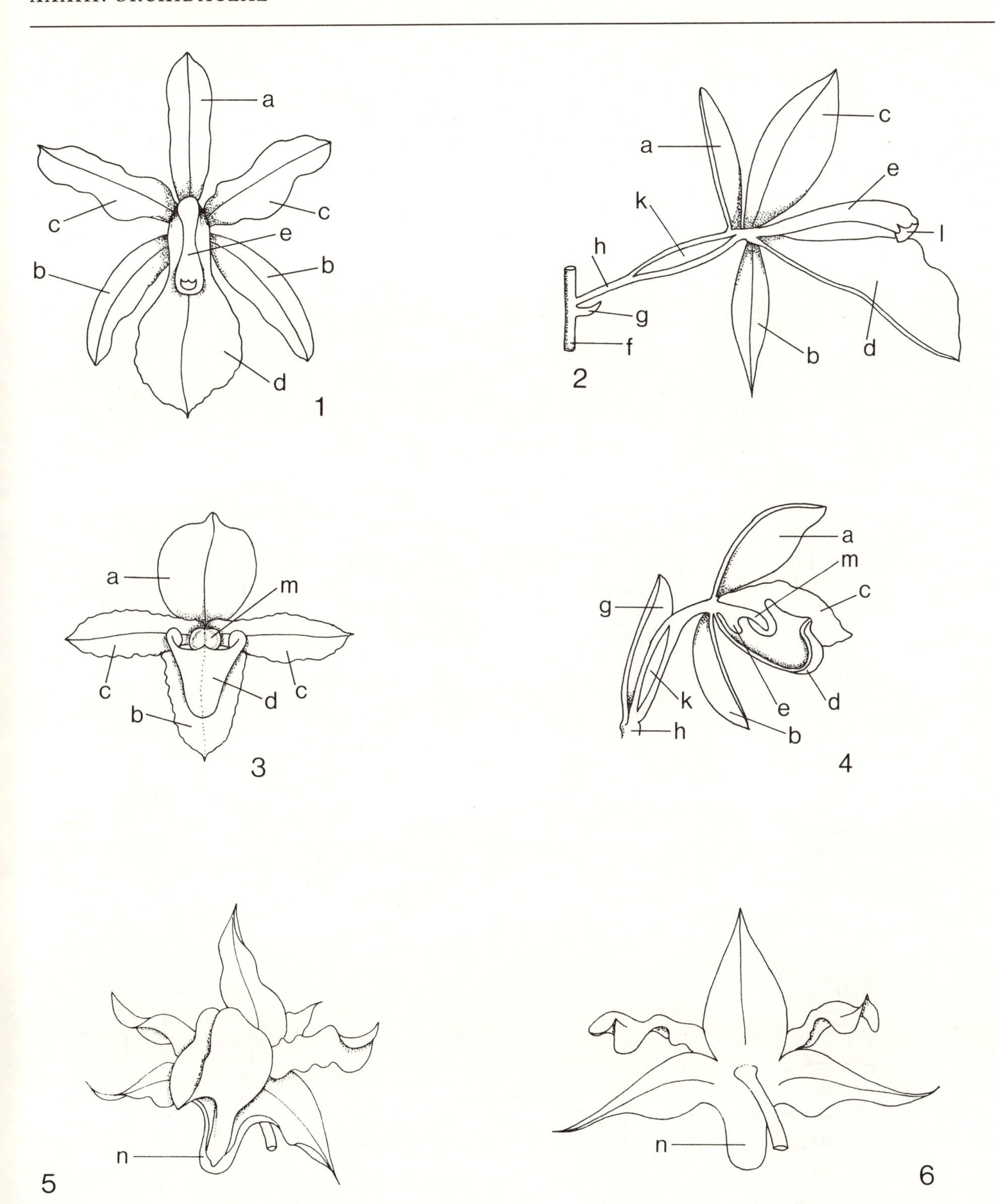

Figure 11. Diagram of some characters used in the identification of Orchidaceae. 1, A generalised orchid flower (a, upper sepal; b, lateral sepal; c, petal; d, lip; e, column). 2, Section of a generalised orchid flower (a, upper sepal; b, lateral sepal; c, petal; d, lip; e, column; f, inflorescence-stalk; g, bract; h, flower-stalk; k, ovary; l, anther). 3, Flower of *Paphiopedilum* (a, upper sepal; b, lower sepal; c, petal; d, lip; m, staminode). 4, Section of flower of *Paphiopedilum* (a, upper sepal; b, lower sepal; c, petal; d, lip; e, column; g, bract; h, flower-stalk; k, ovary; m, staminode). 5, Flower of a *Dendrobium* from the front (n, mentum). 6, Flower of a *Dendrobium* from the back (n, mentum).

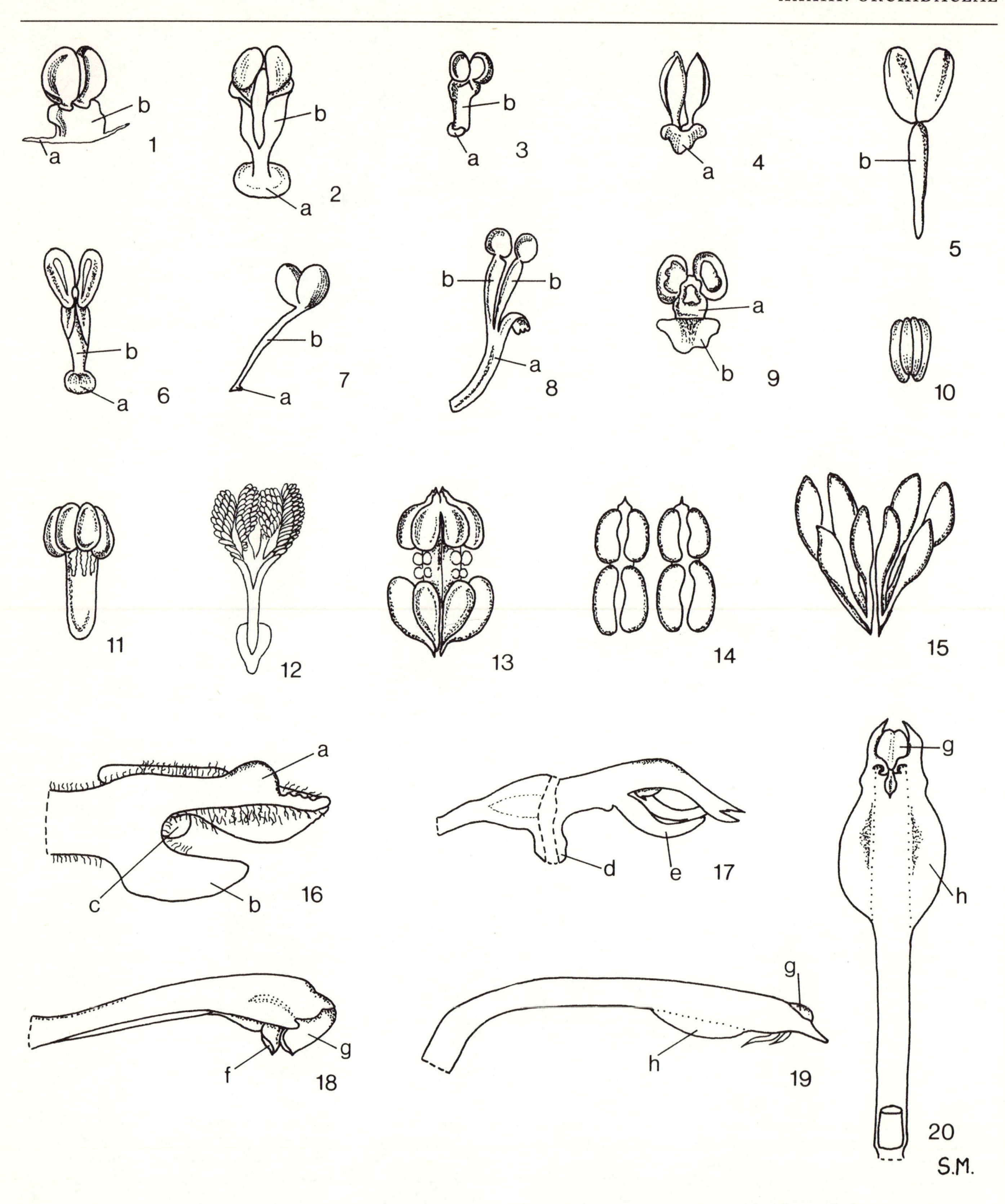

Figure 12. Diagram of some characters used in the identification of Orchidaceae. 1–15, Pollinia of various types. 1–9, Pollinia borne in groups of 2 (a, viscidium; b, stipe). 10,11, Pollinia borne in groups of 4. 12, Pollinia borne in a group of 4, each made up of easily separable masses. 13–15, Pollinia in groups of 8. 16–20, Columns of various types. 16, Column and staminode of *Paphiopedilum* in section (a, staminode; b, column, c, pollinium). 17, Column with foot and lobes (d, foot; e, lobes). 18, Column with a beaked rostellum (f, rostellum; g, anther). 19, Side view of a winged column (g, anther; h, wing). 20, Lower view of a winged column (g, anther; h, wing).

ovary (flowers resupinate). Sepals 3, free or united. Petals 3, 2 of them usually similar to each other and to the sepals, the other (the lowermost in resupinate flowers) usually different in shape, size and colour, and referred to as the lip. Stamen(s), style and stigmas united into a solid structure known as the column. Anthers 2 or more frequently 1, borne at or near the apex of the column. Pollen aggregated into powdery or waxy masses known as pollinia, of which there may be 2, 4, 6, 8 or more. Pollinia usually borne on a sticky pad (the viscidium), or more rarely on 2 viscidia, to which they may be attached by pollinial tissue, or by a structure of different hardness and texture known as the stipe. Stigmas 3, the lateral 2 receptive and usually joined, forming a hollow on the inner surface of the column below the anther, or rarely borne on stalks; the median stigma not receptive, variously modified, borne just above the fertile 2 and beneath the anther, forming a structure known as the rostellum which is sometimes conspicuous, and contributes material to the viscidium. Ovary inferior, usually 1-celled, more rarely 3-celled, placentation parietal. Ovules numerous. Fruit a capsule. Seeds numerous, very small and dust-like.

The Orchidaceae, with about 20 000 species, is probably the largest flowering plant family. It is extremely variable both vegetatively and florally, and the flowers show numerous adaptations to insect pollination. They vary from the extremely beautiful to the bizarre, and the plants hold a rather special place in gardening, with many enthusiasts growing them (and little else) on a large scale. The literature on the family is extensive, including many illustrated works, taxonomic accounts and horticultural guides, and there are several journals devoted solely to the family (see below).

Cultivation. The cultivation of orchids is a subject which has generated much discussion, controversy and publication. It is not possible to go into detail here, and the reader is referred to the books and journals cited below for further information. Most of the species included here are of tropical origin, and require greenhouse protection throughout the whole of Europe; the hardy species (of which there is a relatively small number, mainly in genera Nos. 1–28) can be grown outside.

Two different composts are used for the tropical species. The first is for species which are terrestrial in their native habitats (terrestrial compost), the second for those which are epiphytes (epiphytic compost).

Terrestrial compost is made up of 2 parts (by bulk) of fibrous loam, 2 parts of finely crushed bark or medium-grade peat, and 1 part of grit or sand; 50 grams of a slow-release general fertiliser should be added to each cubic metre of the mixture. Species with particularly vigorous root systems will benefit from an increase in the amount of bark or peat in the mixture.

Epiphytic compost consists of 3 parts of pine bark mixed with 1 part of rough peat or sphagnum moss; 50–70 grams of slow-release general fertiliser are added to each cubic metre of this mixture. Some epiphytic species grow best when attached to a piece of cork-oak bark, using the smallest possible amount of compost.

Control of pests and diseases in orchids is a rather complex and specialised matter, and the relevant literature should be consulted for precise information. Good hygiene and cultivation are, of course, essential.

Literature. As mentioned above, the literature on orchids is very extensive; only relatively recent publications are listed here, in 3 categories:

General taxonomic accounts: Schlechter, R., *Die Orchideen* (1914); edn 2, edited by E. Miethe (1927); edn 3, edited by F.G. Brieger, R. Maatsch & K. Senghas, appearing in parts and continuing (1972–present). Holttum, R.E., *A revised flora of Malaya*, I, *Orchids of Malaya* (1953). Hawkes, A.D., *Encyclopaedia of cultivated orchids* (1965). Sheehan, T. & Sheehan, M., *Orchid genera illustrated* (1979). Dietrich, H., *Bibliographia Orchidacearum* (1980). Bechtel, H., Cribb, P. & Launert, E., *Manual of cultivated orchid species* (German edn, 1980, entitled *Orchideenatlas*; English edn, 1981). Dressler, R.L., *The orchids: natural history and classification* (1981).

Horticultural guides: Northen, R., *Home orchid growing* (1962). Richter, W., *Orchideen* (1969). Hartmann, W.I., *Introduction to the cultivation of orchids* (1971). Sander, D., *Orchids and their cultivation*, edn 9 (1979). Thomson, P.A., *Orchids from seed* (1979). Black, P.M., *The complete book of orchid growing* (1980). Williams, B.A. and others, *Orchids for everyone* (1980). Williams, B.S., *The orchid grower's manual*, edn 7 (1897, reprinted 1982).

Hybrids: *Sander's list of orchid hybrids* – first published in 1905, supplements are published as necessary by the Royal Horticultural Society (the registration authority for the names of hybrid orchids).

Several journals are devoted entirely to the family, containing articles of both taxonomic and horticultural interest; the most important of these are: *The Orchid Review*, 1893, and continuing (British); *Orchis*, 1906–20 (German); *The American Orchid Society Bulletin*, 1932 and continuing (American); *Orchideen*, 1934 and continuing (Dutch); *Australian Orchid Review*, 1936 and continuing (Australian); *Orchid Digest*, 1937 and continuing (American); *Die Orchidee*, 1949 and continuing (German); *Orchid Journal*, 1952–55 (American); *The Orchadian*, 1963 and continuing (Australian); *Orchidata*, 1965 and continuing (American); *South African Orchid Journal*, 1970 and continuing (S African); *Orquidea*, 1971 and continuing (Mexican).

The identification of orchids can be difficult, particularly at the genus level, and because the plants are so variable and so different from those of most other plant families, a specialised terminology has been developed. The important terms used in this account are briefly defined below, and some indication is given of how the various characters are to be observed.

Growth habit. Orchids grow either monopodially (i.e. with a growing point that continues vegetative growth from year to year, the inflorescences being borne laterally), or sympodially (with a growing point that ceases growth after some time, generally by flowering, further growth being continued by a lateral bud formed on the older growth). This feature is often quite difficult to appreciate in individual specimens, and is not used, as such, in the key to genera (pp. 144–149). It is important, however, in that monopodial orchids do not produce pseudobulbs, whereas sympodial orchids may do so.

Pseudobulbs (see figure 8, p. 138). A pseudobulb is a swollen fleshy stem; whether or not a particular plant has pseudobulbs or more normal stems is a very important feature in its generic identification. In a few species and genera there is no sharp dividing line between a 'normal' stem and a pseudobulb; the keys to the genera have been constructed, as far as possible, to allow for these to be identified whether they have been considered as having a stem or a pseudobulb. The pseudobulbs, if present, are borne on the rhizomes, and may consist of several internodes, when they will bear leaves or leaf-scars along their lengths; such pseudobulbs are described here as 'compound'. Alternatively, they may consist of a single internode, when they bear leaves or leaf-scars only at the apex; such pseudobulbs are described here as 'simple'. The pseudobulbs are borne on the rhizomes and may be close together or distant, or, in a few genera, they form chains built up of several pseudobulbs borne one on top of the other.

Leaves (see figure 9, p. 139). The manner in which the leaves are packed when young is an important character in identification of the genus. The young leaves may be rolled, so that one margin overlaps the other and the back of the young leaf is rounded, or they may be folded once longitudinally. This character is generally easy to see in most plants when young leaves are present, and it is usually quite easy to see in mature leaves, as the rolling or folding persists at the base. In some genera which bear few leaves it may be difficult to decide on this character when the leaves are mature; the following features help in making a decision. Folded leaves generally retain a longitudinal line or small fold down the middle over most of their length, and usually have a definite keel on the outside towards the base; very hard, leathery leaves are almost all folded. Rolled leaves are often pleated or have prominent veins, do not usually have a single line or fold down the middle, and are rounded on their backs towards the base. In some genera and species the leaves are terete or cylindric and with a groove or grooves; such leaves cannot be said to be either rolled or folded when young, but they are keyed here with genera with folded leaves.

Inflorescences (see figure 10, p. 140). These may be terminal on the stem or pseudobulb, or lateral; in the latter case the inflorescence can arise in a leaf axil or a leaf-scar axil on the pseudobulb or stem, or from the rhizome at the bases of the pseudobulbs. In a few genera in which the stems bear only 1 leaf (e.g. *Pleurothallis*), the racemes are considered here to be terminal, even though they might be interpreted as axillary to the leaf.

Flowers (see figures 10 & 11, pp. 140 & 141). In all the keys and descriptions the flowers are considered to be resupinate (i.e. with their stalks and ovaries twisted through 180° so that the lip petal is lowermost with the column above it) unless the contrary is stated. In species with arching or hanging inflorescences the degree of resupination of the individual flowers may vary, those flowers towards the hanging apex being non-resupinate (they are effectively turned upside down by the hanging of the inflorescence), while those at the more erect base may be properly resupinate. However, in a few genera with hanging inflorescences (e.g. *Paphinia*) the flowers are all strictly resupinate.

In many genera the base of the column is prolonged at an angle to the rest, forming a column foot (figure 12, p. 142). In all genera with a foot the lip is attached at or near the end of it. In many of these genera the lateral sepals are also borne on the sides of the foot, which results in a humped, angular, chin-like projection being visible on the outside of the back of the flower. This projection is known as a mentum (figure 11, p. 141), and is of some importance in identification.

Pollinia (see figure 12, p. 142). Much of the classification of the family is based on the structure of the pollinia and their associated organs. In this account we have attempted to use these characters as little as possible, but the pollinia cannot be completely ignored, particularly their number. They can be extracted from the flower by sliding a needle or a match slowly upwards along the inner side of the column, when they will attach themselves to the needle, and can be examined through a hand-lens (a magnification of $\times 10$ or $\times 15$ is usually sufficient). The number of pollinia can usually easily be seen, though care must be taken, as the pollinia are occasionally not all extracted at once.

A few species have 2 deeply bilobed pollinia, which can look like 4; again, other species may have extra, infertile pollinia which are shrunken and colourless. In most of the tropical species in cultivation the pollinia are waxy and of well-defined shape. In some hardy species they are granular or powdery and in a few others they are each made up of several to many individual packets (known technically as massulae) which separate very easily from each other.

The column (see figures 11 & 12, pp. 141 & 142). These are very variable in shape, size and details of structure; a selection is shown in the illustration.

The key to the genera which follows is divided into 5 group keys, with a key to the groups at the beginning. It makes use, as far as possible, of easily observed characters. Genera which are variable in some of the characters used in the key to the groups (e.g. whether or not they have pseudobulbs) have been deliberately keyed out in several of the groups. Many hybrids are cultivated in the Orchidaceae, including intergeneric hybrids to which 2 or more genera have contributed genetic materal. It is not, of course, possible to deal with these hybrids in this account. Reference should be made to *Sander's list of orchid hybrids* and its supplements for further information.

KEY TO GROUPS

1a. Annual aerial growth dying off completely each autumn or winter, leaving no part above ground except, occasionally, a rosette of small leaves or a solitary leaf, formed after the current year's growth has died down **Group A**
b. Annual growth persisting throughout the year 2
2a. Leaves rolled when young (see figure 9, p. 139) 3
b. Leaves folded once along the mid-line when young, or leaves terete or cylindric with a narrow groove (see figure 9, p. 139) 4
3a. Plant with distinct pseudobulbs or stems swollen above or below **Group B** (p. 145)
b. Plant without pseudobulbs or swollen stems **Group C** (p. 146)
4a. Plant with distinct pseudobulbs or stems swollen above or below **Group D** (p. 147)
b. Plant without pseudobulbs or swollen stems **Group E** (p. 148)

Group A

1a. Leaves absent at flowering; lip with 3 ridges at the base which become fleshy plates on the central lobe **80. Aplectrum**
b. Leaves present at flowering; lip not ridged as above 2
2a. Lip slipper-like 3
b. Lip not slipper-like 4

3a. Stem with 2 or more separated nodes and 2 or more leaves; lateral sepals pointing downwards, usually fused **1. Cypripedium**
b. Stem with 1 node and 1 leaf; lateral sepals pointing upwards, free **79. Calypso**
4a. A spur, or rarely 2 spurs, present, arising from the base of the lip or the base of the central sepal 5
b. Spur(s) absent 19
5a. Spurs 2 6
b. Spur 1 7
6a. Plant with 1 leaf which is held flat against the ground; flower 1 or rarely 2 **7. Corybas**
b. Leaves several, borne on erect stems; flowers numerous **28. Satyrium**
7a. Spur arising from the base of the concave or hooded central sepal **26. Disa**
b. Spur arising from the base of the lip 8
8a. Flower with the lip, which is similar to the petals, uppermost (i.e. flower not resupinate); spur 2 mm or less **20. Nigritella**
b. Flower with the lip, which is different from the petals, lowermost (i.e. flower resupinate); spur more than 2 mm 9
9a. Spur 2.5 cm or more 10
b. Spur less than 2.5 cm 12
10a. Lip strap-shaped, unlobed **23. Platanthera**
b. Lip not strap-shaped, variously lobed and dissected 11
11a. Uppermost leaf tightly sheathing the stem; column with 1 stigmatic area **18. Orchis**
b. Uppermost leaf not tightly sheathing the stem; column with 2 stalked stigmatic areas **24. Habenaria**
12a. Sepals united at their bases to the column and lip **25. Stenoglottis**
b. Sepals free from column and lip 13
13a. Spur 10 mm or more, slender 14
b. Spur less than 10 mm, slender or not 16
14a. Spur cylindric, obtuse; flowers usually yellow **19. Dactylorhiza**
b. Spur thread-like, acute; flowers purple, pink or white 15
15a. Raceme conical; flowers with a foxy smell; lip deeply 3-lobed with 2 basal ridges **17. Anacamptis**
b. Raceme cylindric; flowers fragrant; lip shallowly 3-lobed without basal ridges **21. Gymnadenia**
16a. Lip spirally twisted or its lateral lobes with wavy margins 17
b. Lip not spirally twisted, margins not wavy 18
17a. Central lobe of lip 2 or more times longer than the lateral lobes, strap-shaped, spirally twisted; bracts shorter than the flowers **15. Himantoglossum**
b. Central lobe of lip less than 2 times longer than the lateral lobes, oblong, not spirally twisted; bracts longer than the flowers **16. Barlia**
18a. Lip strap-shaped, terminating in 3 short teeth; flowers greenish **22. Coeloglossum**
b. Lip not strap-shaped, variously lobed; flowers not greenish 19
19a. Root-tubers ovoid; leaves mostly in a basal rosette, the uppermost usually completely sheathing the stem; all bracts membranous, shorter than the flowers **18. Orchis**
b. Root-tubers lobed; leaves not in a basal rosette; upper leaves sheathing the stem at its base only, often transitional to the leaf-like bracts; bracts usually as long as, or longer than the flowers **19. Dactylorhiza**
20a. Stem with 2 more or less opposite leaves; lip strap-shaped, 2-lobed at the apex **6. Listera**
b. Stem with more than 2 leaves which are distributed along it or are in a basal rosette; lip not as above 21
21a. Flowers arranged in a tight spiral **8. Spiranthes**
b. Flowers not arranged in an obvious spiral 22
22a. Lip divided into 2 parts by a constriction at the middle 23
b. Lip not divided into 2 parts 25
23a. Roots tuberous; leaves not pleated; bracts large, coloured like the sepals, sheathing the basal parts of the flowers **13. Serapias**
b. Roots fibrous, tubers absent; leaves pleated; bracts green and leaf-like, not sheathing the basal parts of the flowers 24
24a. Flowers shortly stalked, borne mostly on 1 side of the axis and horizontal or downwardly pointing; sepals and petals spreading **4. Epipactis**
b. Flowers not stalked, borne all round the axis, pointing upwards; sepals and petals not widely spreading **5. Cephalanthera**
25a. Lip small, usually sensitive to touch, hidden within a hood formed by the sepals and petals **27. Pterostylis**
b. Lip as large as or larger than, and not concealed by the sepals and petals, not sensitive to touch 26
26a. Lip velvety or covered with short hairs or with a geometric pattern or with a shining area or any combination of these, if lobed the lobes not long and slender **12. Ophrys**
b. Lip without any of the above characteristics, the lateral and central lobes slender, like the arms and legs of a man **14. Aceras**

Group B

1a. Pollinia 2 2
b. Pollinia 4 or 8 20
2a. Pseudobulbs compound 3
b. Pseudobulbs simple 9
3a. Flowers asymmetric, due to the twisting of the column to 1 side **137. Mormodes**
b. Flowers symmetric about the vertical plane 4
4a. Lip uppermost (i.e. flower not resupinate); flowers usually unisexual 5
b. Lip lowermost (i.e. flower resupinate); flowers bisexual 6
5a. Column thin but expanded conspicuously towards the apex, arching, without 2 backwardly directed, bristle-like outgrowths **139. Cycnoches**
b. Column equally thick throughout, straight, usually with 2 backwardly projecting, bristle-like outgrowths **138. Catasetum**
6a. Raceme terminal; lip rolled around and partly concealing the column **128. Galeandra**
b. Raceme lateral; lip not rolled around and concealing the column 7
7a. Lip with a distinct, though short, spur **129. Eulophia**
b. Lip without a spur 8
8a. Lip with a distinct and conspicuous claw **132. Cyrtopodium**
b. Lip without a claw **131. Euophiella**
9a. Sepals and petals directed forwards, incurved, the flower bell-shaped 10
b. Sepals and petals spreading or reflexed, the flower not bell-shaped 13
10a. Lip with a distinct claw which bears the lateral lobes near its middle **141. Acineta**
b. Lip with a short claw or none, the lateral lobes borne at the extreme base 11
11a. Anterior part of the lip freely movable on the rest **142. Peristeria**
b. Anterior part of the lip not movable on the rest 12

12a. Pseudobulbs with 1–2 leaves; central lobe of lip not downy **143. Lacaena**
b. Pseudobulbs with 3–5 leaves; central lobe of lip downy **144. Lueddemannia**
13a. Column with a reflexed apex; lip very complex, bearing 2 nectar-secreting, peg-like glands near the base, which drip nectar into the bucket-like anterior part **151. Coryanthes**
b. Column without a reflexed apex; lip variously complex, but not as above 14
14a. Flowers in hanging racemes, the column (as viewed in the growing position) below the lip (i.e. flowers resupinate) 15
b. Flowers in hanging, arching or erect racemes, the lip (as viewed in the growing position) below the column (i.e. flowers not resupinate) 17
15a. Apex of the lip bearing white, club-shaped, hair-like processes **147. Paphinia**
b. Apex of the lip without such processes 16
16a. Upper sepal fused to the column for about half of the length of the latter; anther at the apex of the column **149. Gongora**
b. Upper sepal free from column; anther apparently on the back of the column **150. Cirrhaea**
17a. Basal part of the lip, or the whole lip, hollowed out and saccate **148. Stanhopea**
b. Lip not at all hollowed out 18
18a. Pseudobulbs with 2–4 leaves **95. Eriopsis**
b. Pseudobulbs with 1 leaf 19
19a. Lateral lobes of lip narrow, curving backwards and upwards and then forwards; axis of raceme without hairs **145. Houlletia**
b. Lateral lobes of lip oblong-triangular, curving upwards; axis of raceme downy **146. Polycycnis**
20a. Pollinia 8 21
b. Pollinia 4 29
21a. Racemes apparently terminal 22
b. Racemes obviously lateral 23
22a. Column with a distinct foot to which the lateral sepals are attached, forming a mentum **61. Eria**
b. Column without a foot, mentum absent **30. Bletilla**
23a. Column with a conspicuous foot to which the lateral sepals are attached, forming a mentum 24
b. Column without a foot, mentum absent 25
24a. Pseudobulbs simple **31. Bothriochilus**
b. Pseudobulbs compound **36. Chysis**
25a. Lip with a projecting spur 26
b. Lip without a spur though a pouch is sometimes present which does not project from the flower 27
26a. Leaves absent at flowering; axis of raceme, flower-stalks and backs of flowers hairy **32. Calanthe**
b. Leaves present at flowering; axis of raceme, flower-stalks and backs of flowers not hairy **34. Phaius**
27a. Central lobe of lip with a long, narrow claw **35. Spathoglottis**
b. Central lobe of lip without a long, narrow claw 28
28a. Inflorescence bracts broad, almost as long as the flower-stalks, often falling early **34. Phaius**
b. Inflorescence bracts narrow, much longer than the flower-stalks, persistent **33. Bletia**
29a. Column with a distinct foot 30
b. Column without a foot 39
30a. Flower solitary 31
b. Flowers few to many, in racemes 32
31a. Perianth segments all incurved, flower almost spherical **85. Anguloa**
b. Perianth segments all spreading, flower widely open **84. Lycaste**
32a. Lip freely movable on the column; flowers many, 2-ranked in slender racemes **44. Dendrochilum**
b. Lip not movable on the column; flowers various, not as above 33
33a. Callus of lip divided into finger-like projections **88. Acacallis**
b. Callus of lip not divided as above 34
34a. Callus of lip transverse, forming a curved frill across the base of the lip **87. Zygopetalum**
b. Callus of lip elongate or formed of a few longitudinal ridges or plates 35
35a. Pollinia 4 in 2 pairs, each pair borne on a separate stipe; column winged above 36
b. Pollinia 4, all borne on a common stipe (which may be very small); column not winged 37
36a. Pseudobulbs with 1 leaf **83. Bifrenaria**
b. Pseudobulbs with 2–3 leaves **96. Mendoncella**
37a. Petals more or less equal in size to the upper sepal **95. Eriopsis**
b. Petals smaller than the upper sepal 38
38a. Leaves in 2 ranks; column somewhat swollen above **81. Warrea**
b. Leaves not in 2 ranks; column cylindric **82. Xylobium**
39a. Lip uppermost (i.e. flower not resupinate), with auricles which extend downwards below the column **77. Malaxis**
b. Lip lowermost (i.e. flowers resupinate), without auricles 40
40a. Lip pouched or sac-like towards the base; racemes dense, with many flowers **45. Pholidota**
b. Lip not pouched or sac-like towards the base; racemes with 1–4 flowers, or if more, then loose 41
41a. Lip bent sharply backwards at or about the middle; flowers at most 2.5 cm in diameter **78. Liparis**
b. Lip not bent sharply backwards; flowers more than 2.5 cm in diameter 42
42a. Leaves membranous, deciduous; lip unlobed or very shallowly 3-lobed, base rolled around and concealing the column **43. Pleione**
b. Leaves not membranous; lip clearly 3-lobed, lateral lobes erect or spreading, not concealing the column 43
43a. Central lobe of lip hairy, lateral lobes spreading **86. Pabstia**
b. Central lobe of lip hairless, lateral lobes erect **42. Coelogyne**

Group C

1a. Leaves with veins differently coloured from the rest of the leaf; pollinia made up of easily separable masses 2
b. Leaves with veins the same colour as the rest; pollinia granular or waxy, not made up of easily separable masses 4
2a. Lip uppermost (i.e. flower not resupinate); flowers asymmetric, due to twisting of the lip and column in opposite directions **10. Macodes**
b. Lip lowermost (i.e. flower resupinate); flower symmetric about the vertical plane 3
3a. Lip not widening towards the apex, claw not fringed or toothed **9. Goodyera**
b. Lip abruptly widened to a transverse blade, claw fringed or toothed **11. Anoectochilus**
4a. Pollinia 2, borne on a triangular stipe **132. Cyrtopodium**
b. Pollinia 4 or 8, not borne on a stipe 5
5a. Axis of raceme, flower-stalks and the backs of the flowers hairy **32. Calanthe**
b. Plants not hairy as above 6
6a. Lip with an obtuse spur 7

b. Lip without a spur 8
7a. Inflorescence axillary, erect **34. Phaius**
b. Inflorescence terminal, hanging **37. Thunia**
8a. Leaves many in 2 ranks; column without a foot, mentum absent **39. Sobralia**
b. Leaves 2 or 4 in opposite pairs; column with a distinct foot to which the lateral sepals are attached, forming a mentum **61. Eria**

Group D

1a. Column with a distinct foot, mentum often conspicuous 2
b. Column without a foot, mentum absent 18
2a. Pseudobulbs compound 3
b. Pseudobulbs simple (all nodes at, or very near the apex), sometimes in chains 6
3a. Panicle large, terminal **133. Ansellia**
b. Racemes lateral (rarely apparently terminal) 4
4a. Petals less than one-third of the length of the upper sepal; raceme arising from the rhizome near the base of the pseudobulb **135. Grammangis**
b. Petals at least two-thirds of the length of the upper sepal; racemes usually borne on the pseudobulbs 5
5a. Pollinia 4 **73. Dendrobium**
b. Pollinia 8 **61. Eria**
6a. Pollinia 2 or 4 7
b. Pollinia 8 16
7a. Pseudobulbs borne one on top of another, forming erect chains 8
b. Pseudobulbs not as above 9
8a. Flowers white to purplish pink **59. Scaphyglottis**
b. Flowers bright reddish orange **46. Hexisea**
9a. Flowers or racemes terminal 10
b. Flowers or racemes lateral or basal 12
10a. Axis of raceme downy; flower with lip uppermost (i.e. not resupinate) **127. Polystachya**
b. Axis of raceme hairless; flower with lip uppermost (i.e. resupinate) 11
11a. Flowers more than 4.5 cm in diameter; lip 3-lobed, not pouched at apex **74. Epigeneium**
b. Flowers less than 2 cm in diameter; lip entire, pouched at apex **60. Nageliella**
12a. Lateral sepals much larger than upper sepal **76. Cirrhopetalum**
b. All sepals more or less equal 13
13a. Bracts on the flowering scape clasping, overlapping or almost so, rarely with their apices spreading **98. Maxillaria**
b. Bracts on the flowering scape not clasping, distant, spreading 14
14a. Lip not 3-lobed, sometimes fringed **75. Bulbophyllum**
b. Lip distinctly 3-lobed 15
15a. Leaves terete **97. Scuticaria**
b. Leaves flat **89. Promenaea**
16a. Flower solitary **63. Meiracyllium**
b. Flowers in dense racemes 17
17a. Sepals fused at the base into a spherical or urn-shaped cup which conceals the petals **62. Cryptochilus**
b. Sepals almost free from each other but fused to the column foot; petals exposed **61. Eria**
18a. Flower with a spur, which may be formed by the bases of the sepals or by the base of the lip 19
b. Flower without a spur of any kind 24
19a. Spur formed by the base of the lip 20
b. Spur formed by the bases of the sepals 21
20a. Flowers 1 or 2, 2.5 cm or more in diameter **152. Trichocentrum**
b. Flowers 5 or more, up to 2.5 cm in diameter **130. Oeceoclades**
21a. Spur attached to the ovary **49. Broughtonia**
b. Spur free from the ovary 22
22a. Lip without appendages which project back into the sepal spur **153. Ionopsis**
b. Lip with 1 or 2 appendages which project back into the sepal spur 23
23a. Lip with 1 appendage **154. Rodriguezia**
b. Lip with 2 appendages **155. Comparettia**
24a. Sides of the lip attached to the column, at least at the base 25
b. Sides of lip entirely free from the column (though the point of lip attachment may be a little way up the column) 31
25a. Pollinia 2 26
b. Pollinia 4 or 8 28
26a. Pseudobulbs compound **134. Cymbidium**
b. Pseudobulbs simple 27
27a. Stigmatic surface divided into 2; sepals and petals free from the column **157. Cochlioda**
b. Stigmatic surface not divided; upper sepal and petals fused to the back of the column **161. Aspasia**
28a. Pollinia 4 29
b. Pollinia 8 30
29a. Sides of the lip attached to the column for almost the whole length of the latter **47. Epidendrum**
b. Sides of the lip attached to the column for at least the lower half of the latter **48. Encyclia**
30a. Leaves ovate, elliptic or oblong, leathery **56. Sophronitis**
b. Leaves linear, fleshy **57. Sophronitella**
31a. Pollinia 2 32
b. Pollinia 4 or more (sometimes some sterile) 42
32a. Pseudobulbs compound 33
b. Pseudobulbs simple 34
33a. Stipe of pollinia divided; plants usually very large (raceme to 2 m) **136. Grammatophyllum**
b. Stipe of pollinia not divided; plant much smaller **134. Cymbidium**
34a. Sides of the lip rolled over the column **156. Trichopilia**
b. Sides of the lip not rolled over the column 35
35a. Sepals and petals directed forwards, incurved **162. Ada**
b. Sepals and petals spreading or reflexed 36
36a. Sepals and petals at least 6 times as long as broad **163. Brassia**
b. Sepals and petals relatively much broader 37
37a. Base of lip ascending more or less parallel with the column 38
b. Base of lip directed away from the column from its base, at an angle of *c.* 90° 40
38a. Lip with 2 basal auricles which are fused to the base of the column **159. Helcia**
b. Lip without auricles 39
39a. Ridges on the base of the lip reaching up to and slightly clasping the column; upper sepal and petals directed upwards, lateral sepals directed downwards; flowers basically yellow **158. Gomesa**
b. Ridges on the base of the lip not reaching the column; sepals and petals usually evenly spreading; flowers usually not basically yellow **160. Odontoglossum**
40a. Lip with a distinct, parallel-sided claw **166. Sigmatostalix**
b. Lip scarcely clawed, or if with a claw then the claw not parallel-sided and conspicuous 41
41a. Lip usually simple, with 2 (rarely more) longitudinal ridges or lines of hairs near the base; flowers more or

less flattened, not basically yellow; column not thickened towards the base **164. Miltonia**
b. Lip usually lobed and bearing a complex callus of tubercles or ridges and tubercles; flowers not flattened, basically yellow; column thickened, with a fleshy plate below the stigma **165. Oncidium**
42a. Fertile pollinia 8 (sometimes some extra and sterile) 43
b. Fertile pollinia 4 (sometimes some extra and sterile) 45
43a. Column with narrow, acute teeth projecting beyond the anther but not adpressed to it; stigma irregularly lobed **55. Brassavola**
b. Column with small, blunt, apical teeth which are adpressed to the anther; stigma triangular or semicircular 44
44a. Column slender with 3 blunt apical teeth; sepals and petals rarely wavy **53. Laelia**
b. Column stout with 1 blunt apical tooth; sepals and petals usually wavy **54. Schomburgkia**
45a. Sepals united into a tube at the base **100. Trigonidium**
b. Sepals completely free from each other 46
46a. Lip freely movable on its attachment to the base of the column **99. Mormolyca**
b. Lip not freely movable on its attachment to the column 47
47a. Flowers to 2.5 cm in diameter; basal part of the lip not rolled around the column **78. Liparis**
b. Flowers 3 cm or more in diameter; basal part of the lip rolled around the column 48
48a. Leaves sharply toothed **50. Laeliopsis**
b. Leaves entire **52. Cattleya**

Group E

1a. A broad, flat staminode present on the column and extending beyond it; fertile anthers 2 2
b. Staminode absent; fertile anther 1 3
2a. Ovary 3-celled; sepals edge-to-edge in bud **2. Phragmipedium**
b. Ovary 1-celled; sepals overlapping in bud **3. Paphiopedilum**
3a. Inflorescence lateral, usually axillary 4
b. Inflorescence terminal, or borne directly on the rhizome 45
4a. Lip consisting mainly of a vertical pouch or sac, its blade relatively small, consisting of 3 lobes borne on the margins of the pouch 5
b. Lip with a conspicuous blade which is lobed or not, spurred or not, but not as above 10
5a. Column without a foot; pollinia 2 6
b. Column with a foot; pollinia 4, often in 2 pairs 7
6a. Lateral lobes of lip fused to the column, central lobe hairy or fringed **118. Gastrochilus**
b. Lateral lobes of lip free from the column, central lobe neither hairy nor fringed **117. Ascocentrum**
7a. Plant leafless when flowering (though with small, green bracts on the scape) **104. Chiloschista**
b. Plant leafy when flowering 8
8a. Spur or pouch of lip without any calluses inside **105. Pteroceras**
b. Spur or pouch of lip with calluses inside 9
9a. Leaves in 2 ranks; central lobe of the lip smaller than the lateral lobes **103. Sarcochilus**
b. Leaves borne all around the stem, or 1-sidedly; central lobe of lip larger than the lateral lobes **120. Cleisostoma**
10a. Flower with a distinct spur projecting from the back 11
b. Flower without a spur, though the base of the lip sometimes slightly hollowed 24
11a. Lip distinctly and obviously 3-lobed 12
b. Lip entire or very slightly and inconspicuously 3-lobed 18
12a. Column without a foot 13
b. Column with a foot 16
13a. Sepals and petals 2–5 mm; rostellum linear, bifid, very conspicuous **119. Schoenorchis**
b. Sepals and petals at least 10 mm, often much larger; rostellum not as above 14
14a. Pollinia 4; flowers in panicles **116. Renanthera**
b. Pollinia 2; flowers in racemes 15
15a. Flower at most 3 cm in diameter, entirely white **123. Neofinetia**
b. Flower more than 4 cm in diameter, coloured, at least in part **112. Vanda**
16a. Lip freely movable on its attachment to the column foot; spur pointing forwards **107. Sedirea**
b. Lip not movable on its attachment to the column foot; spur pointing backwards or downwards 17
17a. Pollinia borne on a short, broad stipe; flower 8–10 cm in diameter **112. Vanda**
b. Pollinia borne on a long, slender stipe; flowers to 5 cm in diameter **106. Aerides**
18a. Column with a long foot **106. Aerides**
b. Column without a foot, or if with a short foot then lateral sepals not attached to form a mentum 19
19a. Rostellum small, rather inconspicuous, bifid **121. Angraecum**
b. Rostellum large, beaked, conspicuous, not bifid 20
20a. Sepals conspicuously broader than the petals; flower usually white, spotted or blotched with red or pinkish purple **108. Rhynchostylis**
b. Sepals and petals equally broad; flower white or greenish white, sometimes with the lip yellowish at the base but not spotted as above 21
21a. Pollinia 2, each borne on a separate stipe 22
b. Pollinia 2, both borne on a common stipe 23
22a. Lip about the same size as the sepals and petals, without a tooth-like callus in front of the mouth of the spur **124. Cyrtorchis**
b. Lip much broader than the sepals and petals, with a tooth-like callus in front of the mouth of the spur **125. Diaphananthe**
23a. Spur with a distinct, narrow mouth; lip equalling the sepals and petals in breadth or very slightly broader **122. Aerangis**
b. Spur with a wide, indistinct mouth; lip considerably broader than the sepals and petals **126. Eurychone**
24a. Column without a foot 25
b. Column with a distinct foot 33
25a. Pollinia 2 26
b. Pollinia 4 29
26a. Sepals and petals 7–15 mm; leaves strongly keeled, with overlapping bases **140. Lockhartia**
b. Sepals and petals much larger; leaves various, not strongly keeled, sheaths usually not overlapping 27
27a. Claw of lip fused to column; leaves very fleshy or leathery; plant a climber **29. Vanilla**
b. Claw of lip free from column; leaves usually not very fleshy or leathery; plant not a climber 28
28a. Leaves conspicuously veined, the apex acute or 2-lobed; lateral lobes of the lip erect and partially encircling the column **134. Cymbidium**
b. Leaves not conspicuously veined, the

apex irregularly toothed; lateral lobes of lip not as above **112. Vanda**
29a. Flowers of 2 kinds in each raceme; raceme axis and the backs of the flowers woolly **115. Dimorphorchis**
b. Flowers all similar in each raceme; axis of raceme and the backs of the flowers not woolly 30
30a. Lip above the column (i.e. flower not resupinate); rostellum long-beaked, anther with a long, terminal appendage **102. Ornithocephalus**
b. Column above the lip (i.e. flowers resupinate); rostellum not long-beaked, anther without an appendage 31
31a. Flower solitary; lip usually without a callus or ridges, not hollowed out at the base **101. Dichaea**
b. Flowers in racemes; lip with a callus or longitudinal ridges, hollowed out near the base 32
32a. Lip freely movable on the column, with erect, oblong lateral lobes which are free from each other **114. Esmeralda**
b. Lip not movable on the column, the lateral lobes small, joined together by a fleshy plate above the ridge of the central lobe **113. Vandopsis**
33a. Pollinia 2 34
b. Pollinia 4 or 8 36
34a. Leaves flat **110. Phalaenopsis**
b. Leaves terete, or cylindric with a narrow, longitudinal groove 35
35a. Sepals and petals greenish yellow, brownish yellow or whitish, sometimes flushed with pink or violet **111. Paraphalaenopsis**
b. Sepals and petals yellow or greenish yellow, blotched with reddish brown **97. Scuticaria**
36a. Pollinia 8 **61. Eria**
b. Pollinia 4 37
37a. Racemes with few to many flowers 38
b. Flower solitary 39
38a. Lip 3-lobed, with 2 linear appendages on the claw; leaves clearly in 2 ranks **109. Doritis**
b. Lip 3-lobed, without appendages on the claw; leaves usually not obviously in 2 ranks **73. Dendrobium**
39a. Lip at most 1 cm **98. Maxillaria**
b. Lip at least 2 cm 40
40a. Callus of lip fringed **94. Huntleya**
b. Callus of lip not fringed 41
41a. Petals yellow with fringed or toothed margins **90. Chondrorhyncha**
b. Petals basically white, cream, greenish, pink or purple, sometimes spotted with yellow at the base, margins entire 42
42a. Lip distinctly 3-lobed 43
b. Lip not 3-lobed 44
43a. Lateral lobes of the lip rolled over the column and overlapping; lip with little or no claw **91. Cochleanthes**
b. Lateral lobes of the lip spreading; lip with a conspicuous claw **92. Pescatoria**
44a. Column hooded, protruding over the lip callus **93. Bollea**
b. Column neither hooded nor protruding over the lip callus **91. Cochleanthes**
45a. Pollinia 2, 4 or 6, sometimes with some extra and sterile 46
b. Fertile pollinia 8 59
46a. Lateral sepals conspicuously fused at the base (the upper sepal may be fused to the laterals or free from them) 47
b. Lateral sepals free or very shortly and inconspicuously fused at the base 55
47a. Pollinia 4 48
b. Pollinia 2 49
48a. Stem with several leaves which are up to 3 mm broad; upper sepal 8–9 mm **40. Isochilus**
b. Stems each with 1 leaf which is at least 2.5 cm wide; upper sepal at least 2 cm **70. Restrepia**
49a. Upper sepal free, or almost so, from the fused lateral sepals **69. Pleurothallis**
b. Upper sepal clearly fused at the base to the fused lateral sepals 50
50a. Sepals fused at base and towards apex, leaving 3 'windows' to the interior of the flower **71. Cryptophoranthus**
b. Sepals fused only at the base 51
51a. Sepal tube forming a 3-angled cup slightly constricted at the mouth; column without a foot **64. Physosiphon**
b. Sepal tube not as above; column with a foot (sometimes small) 52
52a. Lip with a distinct claw 53
b. Lip without a claw 54
53a. Leaves papillose on the upper surface; lip sensitive to touch; flowering stem densely hairy **67. Porroglossum**
b. Leaves not papillose; lip not sensitive; stem not densely hairy **66. Dracula**
54a. Fused lateral sepals forming an oblong, acute blade bearing 2 lateral tails **68. Triaristella**
b. Fused lateral sepals not as above, with a sinus or notch between the lateral lobes, which may be tailed or not **65. Masdevallia**
55a. Lip fused at least to the lower half of the column 56
b. Lip entirely free from the column 57
56a. Lip fused to the column for most of the length of the latter **47. Epidendrum**
b. Upper half of the column free from the lip **51. Barkeria**
57a. Pollinia 2; leaves, and racemes or panicles, borne directly on the rhizome (stem absent) **165. Oncidium**
b. Pollinia 4 or 6; stems present 58
58a. Pollinia 6, 4 large and 2 small; stems at most 4 cm **58. Leptotes**
b. Pollinia 4, all similar; stems more than 4 cm (often much more) **52. Cattleya**
59a. Sepals to 2 cm; lip not rolled around column 60
b. Sepals 2.5 cm or more; lip rolled around, and often concealing the column 61
60a. Lip deeply concave or pouched; flowers pinkish purple, at least in part **41. Arpophyllum**
b. Lip flat; flowers white, yellow or greenish yellow **72. Octomeria**
61a. Lip conspicuously 3-lobed, without a claw, its lateral lobes rolled around the column **53. Laelia**
b. Lip entire or shallowly 3-lobed, sometimes with a conspicuous claw, its base rolled round the column 62
62a. Lateral sepals spreading **55. Brassavola**
b. Lateral sepals lying close together under the lip **38. Arundina**

1. CYPRIPEDIUM Linnaeus

J.C.M. Alexander

Terrestrial, aerial parts dying away in winter. Leaves deciduous, pleated (except No. 9), rolled when young, spirally arranged or in a single, almost opposite pair. Flowers persistent, solitary or up to 12 per raceme; bracts leaf-like. Sepals edge-to-edge in bud. Petals and upper sepal free, lateral sepals fused into a single lower sepal below the lip (free in No. 6). Lip extended into a sac-like pouch with an inturned rim (except No. 13), the opening partially blocked by a staminode. Stigmatic surface, 2 fertile anthers and staminode borne on the column. Rostellum absent. Pollen in exposed granular masses. Seeds fusiform in a single-celled ovary.

A genus of about 35 species, widely distributed throughout the northern

hemisphere, mostly north of 35° N, though extending further south in Asia and Mexico. They can be cultivated out-of-doors on well-drained, humus-rich soils, performing best in partial shade. Though mostly hardy, they may need protection from severe frost, if not covered by snow. The name *Cypripedium* is still sometimes applied to species of *Paphiopedilum* and *Phragmipedium*, though the latter have long been considered distinct.
Literature: Franchet, A., Les Cypripedium de l'Asie Centrale et de l'Asie Orientale, *Journal de Botanique* 8: 225–33, 249–56, 265–71 (1894); Luer, C.A., *Native orchids of the United States and Canada* (1975).

Leaves. Paired: **8–13**; spirally arranged: **1–7**. Basal: **12**. Fan-shaped: **8**. 3-veined: **9**. Spotted: **10**. Membranous: **9**.
Lower stem. Hairless: **9,10**; sparsely to moderately hairy: **1–7,12,13**; shaggy: **8,11**.
Floral bract. Linear: **9**. Absent: **10**.
Flowers. Solitary: **2,3,6,8–13**; 1–3 per raceme: **1,7,11**; more than 4 per raceme: **4,5,11**. Clustered: **11**.
Lateral sepals. Free: **6**; fused into a rounded lower sepal: **7,9**; fused into a pointed or bifid lower sepal: **1–5,8,10–13**.
Petals. Longer than lip: **1–3,6,8–11**; equal to or shorter than lip: **4,5,7,12,13**. Linear or narrowly lanceolate: **1,2,6**; spathulate: **13**; acute: **1–4,6,8,10–12**; rounded: **5,7,9,13**. Twisted or curled: **1**.
Lip. Furrowed in front: **8–12**. With upstanding or turned-out rim: **13**. Spurred: **6**.

1a. Leaves 3 or more, spirally arranged — 2
b. Leaves 2, almost opposite, basal or higher on the stem — 8
2a. Lateral sepals free; lip with a blunt, conical spur — **6. arietinum**
b. Lateral sepals fused; lip without a spur — 3
3a. Petals linear to narrowly lanceolate, more than 2 cm, more than 3 times as long as broad — 4
b. Petals oblong, elliptic or ovate, (if lanceolate then less than 2 cm), less than 3 times as long as broad — 5
4a. Petals twisted or curled; lower sepal entire, or bifid to less than 5 mm — **1. calceolus***
b. Petals not twisted or curled; lower sepal usually bifid to more than 5 mm — **2. cordigerum**
5a. Petals acute — 6
b. Petals rounded — 7
6a. Leaves 3 or 4; flowers usually solitary; lip red, purple, blackish or greenish brown — **3. macranthon***
b. Leaves 5–16; flowers 1–8; lip yellow — **4. irapeanum**
7a. Flowers 1–3 (rarely 4); petals more than 2 cm, upper sepal more than 2.5 cm — **7. reginae***
b. Flowers 5–10; petals less than 2 cm; upper sepal less than 2.5 cm — **5. californicum**
8a. Petals and sepals equal to or shorter than lip — 9
b. Petals and sepals longer than lip — 10
9a. Petals rounded, less than 2.5 cm; flower blotched; lip with out-turned rim — **13. guttatum**
b. Petals acute, more than 3.5 cm; flower neither spotted nor blotched; lip with in-turned rim — **12. acaule**
10a. Petals less than 3.5 cm; upper sepal less than 3 cm — 11
b. Petals more than 3.5 cm; upper sepal more than 4 cm — 12
11a. Lower stem hairless; leaves heart-shaped, membranous; petals rounded — **9. debile**
b. Lower stem shaggy; leaves elliptic, pleated; petals acute — **11. fasciculatum**
12a. Leaves spotted, broadly elliptic, obtuse; stem hairless below; floral bract absent — **10. margaritaceum**
b. Leaves not spotted, fan-shaped; stem shaggy below; floral bract present — **8. japonicum**

1. C. calceolus Linnaeus.
To 80 cm, sparsely to moderately hairy. Leaves 3–5, 5–20 × 2–10 cm, ovate to elliptic, acute, spirally arranged. Flowers 1 or 2. Upper and lower sepals ovate, acuminate. Petals 3–7 cm, linear to narrowly lanceolate, longer than the lip, twisted or curled, acute. Sepals and petals greenish yellow with purple markings or evenly purplish or reddish brown. Lip dull to bright yellow, spotted inside with red or purple.

Var. **calceolus** (*C. microsaccus* Kränzlin). Illustration: Sundermann, Europäische und mediterrane Orchideen, 216 (1975); Williams et al., Orchids of Britain and Europe, 25 (1978). To 30 cm, sparsely hairy. Leaves 10–18 × 5–6 cm. Sepals and petals purplish brown. *From N & C Europe across Siberia to Korea.* H2. Spring–summer.

Var. **pubescens** (Willdenow) Correll (*C. pubescens* Willdenow; *C. veganum* Cockerell Barker; *C. luteum* Aiton). Illustration: Botanical Magazine, 911 (1806); American Orchid Society Bulletin **38**: 905 (1969); Luer, Native orchids of the United States and Canada, 45, 47 (1975). Taller and hairier than var. *calceolus* and larger in its floral parts. Sepals and petals greenish yellow with purple markings. *N America.* H2. Spring–summer.

Var. **parviflorum** (Salisbury) Fernald (*C. parviflorum* Salisbury). Illustration: Botanical Magazine, 3024 (1830); American Orchid Society Bulletin **38**: 904 (1969); Luer, Native orchids of the United States and Canada, 51 (1975). Very similar to var. *calceolus* though smaller in its floral parts, sparsely downy. Petals reddish brown, very twisted. *Northeast N America.* H2. Summer.

***C. candidum** Willdenow. Illustration: Botanical Magazine, 5855 (1870); American Orchid Society Bulletin **38**: 906 (1969); Luer, Native orchids of the United States and Canada, 53 (1975). Flowers similar in size to var. *parviflorum*. Petals and sepals green marked with purple, or almost completely purple; lip white, sometimes veined with purple, with red or purple spots inside, and with up to 5 leaves held very upright on each stem. *E & C USA, adjacent Canada.* H2. Spring–summer.

***C. montanum** Lindley. Illustration: Botanical Magazine, 7319 (1893); American Orchid Society Bulletin **38**: 904 (1969); Luer, Native orchids of the United States and Canada, 55 (1975). Up to 6 leaves on each stem; flowers similar in size to var. *pubescens*. Petals and sepals uniformly brownish purple; lip compressed, white flushed with purple near the base. *Western N America east to Wyoming.* H2. Summer.

2. C. cordigerum Don. Illustration: Botanical Magazine, 9364 (1934); Pradhan, Indian Orchids I: 35 (1976); Senghas (editor), Proceedings of the 8th world orchid conference, t. 4 (1976).
Differs from *C. calceolus* and its allies in its shorter, broader petals; lower sepal bifid to 1.8 cm; sepals and petals pale yellow, white or pale green; lip white with a few purple spots inside. *Himalaya.* H2. Spring.

3. C. macranthon Swartz (*C. speciosum* Rolfe; *C. franchetii* Rolfe). Illustration: Botanical Magazine, 2938 (1829); American Orchid Society Bulletin **44**: 769 (1975); Williams et al., Orchids of Britain and Europe, 27 (1978).
To 50 cm, sparsely to moderately hairy. Leaves 3 or 4, 8–16 × 4–7 cm, ovate to elliptic, acute, spirally arranged. Flowers usually solitary. Upper and lower sepals broadly ovate, acuminate, violet to purplish pink or greenish brown, with darker veins.

Petals 4–6.5 × 1.5–3 cm, broadly ovate, acute, a little longer than lip. Lip similar in colour, paler round the rim. *USSR South of c. 60° N, Mongolia, N & W China, Korea, Japan.* H2. Spring–summer.

*C. **tibeticum** Hemsley. Illustration: Botanical Magazine, 8070 (1906); Pradhan, Indian Orchids I: 35 (1976); Senghas (editor), Proceedings of the 8th world orchid conference, t. 8 (1976). Similar but distinguished by its larger flowers; petals and sepals pinkish yellow marked with a reddish purple network. Lip uniformly dark (sometimes almost black). *SW China.* H2.

*C. **himalaicum** Hemsley. Illustration: Botanical Magazine, 8965 (1922); Morley & Everard, Wild flowers of the world, t. 109 (1970). Distinguished by its slightly smaller flowers and narrower sepals and petals which are greenish yellow, strongly marked with dark red lines; lip purple, paler behind and around the mouth. *Himalaya & W China.* H2.

*C. **fasciolatum** Franchet (*C. wilsonii* Rolfe). Described as having an almost spherical lip spotted with violet below and broadly striped above, as are the sepals and petals. *W. China.* H2.

C. × ventricosum Swartz (Illustration: Williams et al., Orchids of Britain and Europe, 26, 1978) described as the hybrid between *C. macranthon* and *C. calceolus* is reported from the wild and in cultivation; it may be merely a colour variant of *C. macranthon.*

4. C. irapeanum Llave & Lexarza. Illustration: Luer, Native orchids of the United States and Canada, 59 (1975); Senghas (editor), Proceedings of the 8th world orchid conference, t. 4 (1976); Orchid Digest **41**: 209 (1977).
To 1.5 m, moderately hairy, especially below. Leaves 5–16, 5–15 × 3–9 cm, ovate to elliptic, acute, spirally arranged. Flowers 1–8. Upper sepal lanceolate, acuminate; lower sepal ovate, acuminate, entire. Petals 3–6.5 cm, oblong, acuminate, equal to or shorter than lip. Sepals and petals bright yellow. Lip bright yellow, flushed with red inside. *C & S Mexico, Guatemala.* G1. Summer.

5. C. californicum Gray. Illustration: Botanical Magazine, 7188 (1891); Luer, Native orchids of the United States and Canada, 63 (1975); Orchid Digest **43**: 112 (1979).
To 1.2 m, sparsely hairy. Leaves 5–10, 5–15 × 2–6 cm, elliptic to lanceolate, acute, spirally arranged. Flowers 5–10. Upper sepal oblong to elliptic, acuminate; lower sepal elliptic, shortly bifid. Petals 1.4–2 cm, oblong-lanceolate, blunt, equal to or shorter than lip. Sepals and petals yellowish green. Lip white. *USA (S Oregon, N California).* H2. Spring–summer.

6. C. arietinum Brown (*C. plectrochilum* Franchet). Illustration: Botanical Magazine, 1569 (1813); Luer, Native orchids of the United States and Canada, 43 (1975); American Orchid Society Bulletin **44**: 384 (1977).
To 30 cm, sparsely hairy. Leaves 3–4, 1.5–3.5 cm, elliptic to lanceolate, acute, spirally arranged. Flower solitary. Upper sepal ovate-lanceolate, acuminate, green with purple veins; lateral sepals free, linear-lanceolate, greenish purple. Petals 1.3–2.2 cm, linear, acute, greenish purple, a little longer than lip. Lip with a blunt conical spur, white above with purplish veins coalescing below. *East & Central N America, from 42° to 52° N; C & SW China, N Burma.* H2. Spring–summer.

7. C. reginae Walter (*C. album* Aiton; *C. spectabile* Salisbury; *C. canadensis* Michaux). Illustration: Botanical magazine, 216 (1793); Shuttleworth et al., Orchids, 17 (1970); Luer, Native orchids of the United States and Canada, 57 (1975).
To 1 m, densely hairy. Leaves 3–7, 10–25 × 6–16 cm, ovate to lanceolate, spirally arranged. Flowers 1–3. Upper sepal broadly elliptic to circular, acuminate; lower sepal similar, blunt. Petals 2.5–5 cm, oblong-elliptic, rounded. Sepals and petals white; petals shorter than lip. Lip pink streaked with white, or pure white. *Eastern N America.* H2. Spring–summer.

The name *C. reginae* has apparently been misused for the hybrid *Paphiopedilum leeanum × fairrieanum.*

*C. **flavum** Hunt & Summerhayes (*C. luteum* Franchet, not Aiton, not Rafinesque). Illustration: Gardeners' Chronicle **57**: 257 (1915). Similar but slightly smaller (to 60 cm), with less dense, brownish hair, and bright yellow flowers which are faintly spotted or striped with purplish brown, particularly on lip. *W China.* H2.

8. C. japonicum Thunberg (*C. formosanum* Hayata). Illustration: Botanical Magazine, 9520 (1938); Shuttleworth et al., Orchids, 19 (1970); American Orchid Society Bulletin **44**: 768 (1975).
To 50 cm, shaggy below, sparsely hairy above. Leaves 2, in the upper half of the stem, 13–20 × 13–20 cm, broadly fan-shaped. Flower solitary. Sepals and petals green with purple spots at base; the acute lower sepal sometimes slightly bifid. Petals pointed, 4–6 cm, just longer than lip. Lip pale pink with purple spots, deeply furrowed in front. *Japan, Taiwan, S & C China.* H2.

9. C. debile Reichenbach. Illustration: Botanical Magazine, 8183 (1908); Maekawa, The wild orchids of Japan, 85 (1971); Kitamura et al., Herbaceous plants of Japan (Monocotyledons), t. 1 (1978).
To 25 cm, hairless. Leaves 2, about halfway up stem, 3–6 cm, heart-shaped, membranous, with 3–7 prominent veins. Bracts linear. Flower solitary, pendent. Sepals and petals pale green. Petals 1.3–2 cm, lanceolate, rounded, just longer than lip. Lip spherical, white marked with purple around the mouth. *Japan, W China.* H2.

10. C. margaritaceum Franchet (*C. ebracteatum* Rolfe). Illustration: Gardeners' Chronicle **46**: 419 (1909).
To 15 cm, hairless below. Leaves 2, about halfway up stem, 10–18 × 10–18 cm, broadly elliptic, spotted with purple. Bract absent. Flower solitary. Sepals and petals broadly ovate, acute, yellowish green spotted with purple. Petals 4–5 cm, longer than lip. Lip fleshy, angular, glandular-hairy, yellow with purple spots. *W China (Yunnan).* H2. Summer.

11. C. fasciculatum Watson (*C. knightae* Nelson). Illustration: Botanical Magazine, 7275 (1893); American Orchid Society Bulletin **38**: 906 (1969); Luer, Native orchids of the United States and Canada, 65 (1975).
To 30 cm, shaggy below. Leaves 2, in middle or upper half of stem, 5–12 × 3–8 cm, elliptic, obtuse. Flowers in clusters of up to 4, purplish brown to green. Sepals and petals acute, about twice as long as lip; petals 1.5–2.5 cm. *Western N America.* H2. Spring–summer.

C. elegans Reichenbach from the Himalaya is similar and should perhaps be regarded as the same species.

12. C. acaule Aiton. Illustration: Botanical Magazine, 192 (1792); Shuttleworth et al., Orchids, 19 (1970); Luer, Native orchids of the United States and Canada, 41 (1975).
To 40 cm, sparsely hairy. Leaves 2, 10–28 × 5–15 cm, basal, elliptic, obtuse. Flower solitary. Sepals and petals shorter than lip, lanceolate, acute, yellowish green to purple; petals 4–6 cm. Lip pink or white,

deeply furrowed in front. *Northeastern N America*. H2. Spring–summer.

13. C. guttatum Swartz (*C. yatabeanum* Makino). Illustration: Botanical Magazine, 7746 (1900); Luer, Native orchids of the United States and Canada, 67 (1975); Williams et al., Orchids of Britain and Europe, 29 (1978).
To 30 cm, sparsely hairy. Leaves 2, 5–12 × 2.5–7 cm, bluish green drying black, elliptic, acute. Flowers solitary, white blotched with purple or brown. Upper sepal elliptic, acuminate; lower sepal lanceolate, shortly bifid. Petals 1.3–2 cm, spathulate or with oblong, rounded tips, shorter than lip. Lip a deep open pouch with upstanding or out-turned rim. *From European Russia across temperate Asia to Alaska & NW Canada*. H2. Summer.

2. PHRAGMIPEDIUM (Pfitzer) Rolfe
J.C.M. Alexander
Terrestrial, rarely epiphytic or on rocks. Leaves persistent, strap-shaped, leathery, 2-ranked, folded when young. Flowers in terminal racemes or panicles, soon falling. Sepals edge-to-edge in bud. Petals and upper sepal free, lateral sepals fused into a single lower sepal. Lip a sac-like pouch (rarely petal-like), sometimes with 2 triangular appendages on the rim; opening of lip partially blocked by a staminode. Stigmatic surface, 2 fertile anthers and staminode borne on the column. Rostellum absent. Pollen in exposed granular masses. Seeds fusiform in a 3-celled ovary.

A genus of 15–20 species from C & S America, most of which were originally placed in *Cypripedium* or *Selenipedium*; the latter name is frequently used for them in the horticultural trade. No true *Selenipedium* species are cultivated. *Phragmipedium* species require cultivation conditions similar to those for the unmottled species of *Paphiopedilum* (p. 153). Literature: Garay, L.A., The genus Phragmipedium, *Orchid Digest* **43**: 133–48 (1979).

Leaves. 30 cm or less: **1**; more than 30 cm: **2–6**. Less than 2 cm wide: **5**; 2–5 cm wide: **1,2,4,6**; more than 5 cm wide: **3**. With yellow margins: **2**.
Flowers. 2–4 per inflorescence: **2,3,4**; 6 or more per inflorescence: **1,2,5,6**. 8 cm wide or less: **1–5**; more than 8 cm wide: **2–6**.
Petals. 20 cm or more, more than 3 times longer than sepals: **3,4**; 15 cm or less, less than 3 times longer than sepals: **1,2,5,6**. Similarly shaped to sepals: **1**. Wavy: **2–6**; crisped: **6**; not wavy or crisped: **1**.
Staminode. Elliptic to circular: **1**; diamond-shaped to 3-lobed: **2–4**; triangular to semicircular: **5,6**.

1a. Lip not pouched, similar to petals **4. lindenii**
b. Lip pouched 2
2a. Petals many times longer than sepals **3. caudatum**
b. Petals less than 3 times longer than sepals 3
3a. Petals similar to sepals in shape, but larger **1. schlimii**
b. Petals longer and narrower than sepals 4
4a. Opening of lip without lobes 5
b. Opening of lip with 2 vertical, triangular lobes **6. longifolium***
5a. Petals not tapering; leaves more than 3 cm wide, with yellow margins **2. lindleyanum***
b. Petals narrow and tapering; leaves less than 2 cm wide, without yellow margins **5. caricinum**

1. P. schlimii (Linden & Reichenbach) Rolfe. Illustration: Botanical Magazine, 5614 (1866); Orchid Digest **43**: 134 (1979); Bechtel et al., Manual of cultivated orchid species, 262 (1981).
Leaves 4–8, 20–30 × 2.5–3 cm, strap-shaped, acute. Racemes or panicles to 50 cm, bearing 6–10 flowers; stems hairy. Flowers 5–6 cm wide. Sepals *c*. 2 cm, ovate, obtuse, white flushed with pink. Petals similar but larger, speckled with pink at base. Lip an ellipsoid pouch, mottled dark on pale pink. Staminode elliptic to circular, bright yellow. *Colombia*. G2. Summer.

2. P. lindleyanum (Lindley) Rolfe. Illustration: American Orchid Society Bulletin **44**: 235 (1975); Orchid Digest **43**: 134 (1979).
Leaves 4–7, 30–50 × 4–6.5 cm, lanceolate, acute. Racemes or panicles to 80 cm, bearing 3–7 flowers; stems reddish green, downy. Flowers 7–8 cm wide. Sepals elliptic, slightly hooded, yellowish green; upper sepal 3 × 2 cm; lower sepal a little broader. Petals *c*. 5 cm, oblong, rounded, with wavy margins, yellowish green at the base fading to greenish white at the tips; tips strongly purple-veined. Lip a broad shallow pouch with incurved margins, yellow with brown veins. Staminode diamond-shaped to weakly 3-lobed, yellow. *Venezuela & Guyana*. G2.

***P. sargentianum** (Rolfe) Rolfe. Illustration: Botanical Magazine, 7446 (1895); Orchid Digest **43**: 134 (1979). Differs from *P. lindleyanum* in its narrower lip and shorter raceme (*c*. 40 cm), bearing 2–4 larger flowers (to 10 cm wide). *Brazil*. G2.

3. P. caudatum (Lindley) Rolfe (*P. warscewiczianum* (Reichenbach) Schlechter). Illustration: Orchid Digest **43**: 138 (1979); Bechtel et al., Manual of cultivated orchid species, 262 (1981); Shuttleworth et al., Orchids, 20 (1970).
Epiphytic or on rocks. Leaves 5–9, 40–60 × 6 cm, strap-shaped, light green; apex irregularly toothed. Racemes to 1 m, bearing 2–4 flowers; stems downy. Sepals lanceolate, tapering, off-white to yellowish green with darker green or orange veins; upper sepal 10–15 × 2–3 cm, curled forward over flower; lower sepal 7–10 × 3–4 cm. Petals to 60 × 1 cm, trailing, dark reddish to greenish brown, paler at the base. Lip a deep pinkish white pouch with pink or brown net-venation; rim turned out, yellow. Staminode diamond-shaped, off-white with dark purple corners. *Mexico through C America to Peru, Ecuador, Colombia & Venezuela*. G2. Spring–summer.

4. P. lindenii (Lindley) Dressler & Williams. Illustration: Orchid Digest **43**: 139 (1979); Bechtel et al., Manual of cultivated orchid species, 262 (1981).
Differs from *P. caudatum* in having a simple unpouched lip similar in shape and colour to the petals, though a little broader. *Colombia, Ecuador & Peru*. G2. Summer.

5. P. caricinum (Lindley) Rolfe. Illustration: Botanical Magazine, 5466 (1864); American Orchid Society Bulletin **44**: 236 (1955); Orchid Digest **43**: 143 (1979).
Leaves 3–7, 35–50 cm × 5–15 mm, linear, stiff and sedge-like. Raceme or panicle 30–40 cm, bearing several flowers in succession; stems downy. Flowers 5–8 cm long. Sepals 2–3 × 1–1.5 cm, ovate to lanceolate with wavy margins, yellowish or purplish green with a green base. Petals 5–10 cm × 5 mm, tapering, twisted, similar to sepals in colour. Lip a rounded greenish yellow pouch with brown veins. Staminode triangular, green. *Peru, Bolivia & Brazil*. G2. Spring–autumn.

6. P. longifolium (Warscewicz & Reichenbach) Rolfe. Illustration: Botanical Magazine, 5970 (1872); American Orchid Society Bulletin **44**: 225 (1975); Orchid Digest **43**: 146 (1979).
Leaves 5–8, 40–60 × 2–3 cm, strap-shaped,

tapering to an acute tip. Raceme to 60 cm, bearing several flowers in succession; stems reddish purple, sparsely downy. Flowers to 18 cm wide. Sepals green or greenish yellow with dark green or greenish red veins; upper sepal ovate-lanceolate, 5 × 1.5 cm; lower sepal broadly ovate. Petals to 12 × 1 cm, tapering and twisted with wavy margins, greenish yellow with purple margins, solid purple at tips. Lip a deep pouch with 2 vertical triangular lobes on the rim, greenish yellow flushed with purple. Staminode broadly triangular to semicircular, greenish yellow; upper edge deep purple. *Costa Rica to Colombia*. G2. Winter.

***P. boissierianum** (Reichenbach) Rolfe. Illustration: Orchid Digest **43**: 146 (1979). Differs from *P. longifolium* in its shorter leaves and drooping petals with crisped margins. *Peru*. G2. Summer–autumn.

3. PAPHIOPEDILUM Pfitzer

J.C.M. Alexander

Epiphytic, terrestrial or on rocks. Stems very short, enclosed in leaf-bases. Leaves 4–several, ovate to strap-shaped, persistent, leathery, often mottled, in 2 opposite ranks, folded when young. Flowers solitary, or in terminal racemes bearing 2–8 flowers all open together, or many flowers in succession over several months. Sepals overlapping in bud. Petals and upper sepal free; lateral sepals united into a single lower sepal under the lip. Lip extended into a sac-like pouch usually with side lobes on the rim; opening of pouch partially blocked by a broad staminode borne on the column. Rostellum absent. Pollen in exposed granular masses. Seeds fusiform; ovary 1-celled. (Figure 11(3 & 4), p. 141.)

A genus of about 60 species from the Himalaya and S India through SE Asia to the Philippines, New Guinea and the Solomon Islands. Most of the species were originally placed in *Cypripedium* which now contains only temperate species with rolled leaves. Many artificial hybrids have been produced which exploit the showiness of the upper sepal seen in the wild species. The petals are often strikingly marked with spots or warts and frequently bear marginal hairs. The flowers persist for several weeks if not pollinated.

Plants can be grown on bark or in well-drained pots in epiphytic compost. Lacking pseudobulbs they must not be allowed to dry out, though less water should be given in winter. They need shading from direct summer sunlight. Species with mottled leaves do best in a warm, moist house; those with plain leaves generally prefer cooler conditions. Propagation is usually effected by division of large plants when repotting in early spring; some species have been successfully grown from seed, e.g. *P. delenatii*.

Literature: Asher, J.R., A checklist for the genus Paphiopedilum, *Orchid Digest* **44**: 175–84, 213–28 (1980) & **45**: 15–26, 57–65 (1981); Graham, R. & Roy, R., *Slipper orchids* (1982).

1a. Lip with 2 side lobes on rim 2
b. Lip without lobes on rim 32
2a. Flowers 2 or more, arising from different points on flowering stem 3
b. Flower solitary, rarely 2 or 3 arising from the same point on the flowering stem 5
3a. Petals spathulate, flat or slightly twisted **36. lowii***
b. Petals linear to oblong, twisted and wavy 4
4a. Flowers open 1 at a time; upper sepal almost circular; petals horizontal, hairy on margins **34. victoria-regina**
b. Flowers open all together; upper sepal elliptic; petals drooping, not hairy on margins **35. parishii**
5a. Leaves mottled on upper surface 6
b. Leaves not mottled on upper surface 25
6a. Lip with prominent tooth on outer edge of rim **26. acmodontum**
b. Lip without tooth 7
7a. Petals without spots or warts **16. dayanum***
b. Petals with spots or warts 8
8a. Each petal with 50 or more spots or warts, evenly distributed over at least the basal half of blade 9
b. Each petal with 40 or fewer spots or warts, usually concentrated near margins, mid-line or upper basal corner of blade; if more than 40 then never evenly distributed 14
9a. Spots or warts extending well into apical half of petal, though often absent from extreme tip 10
b. Spots or warts mostly in basal half of petal 13
10a. Petals distinctly broader in apical half **27. javanicum**
b. Petals oblong, elliptic or tapering from the base 11
11a. Petals acute, gradually tapering, almost flat, not curled, twisted or curved downwards **33. sukhakulii***
b. Petals obtuse or acute, curled, twisted or curved downwards 12
12a. Petals evenly marked with many small spots or warts **31. superbiens***
b. Petals irregularly marked with large spots, some with pale centres **32. argus**
13a. Petals elliptic, mostly purplish brown with dark spots or warts; upper sepal with *c.* 12 distinct purple and green veins **29. purpuratum**
b. Petals spathulate, mostly pinkish purple with faint pinkish purple spots; upper sepal with 15–20 indistinct veins **22. hookerae**
14a. Petal spots or warts almost all on or near margin 15
b. Some petal spots or warts away from margin, often on mid-line or in upper basal corner 20
15a. Upper sepal green to greenish yellow, often brown at base, indistinctly veined; lip with bright green rim **20. bullenianum**
b. Upper sepal white, sometimes flushed with pink and green, distinctly veined with green or red and green; lip without green rim 16
16a. Upper sepal ovate, gradually tapering, white or off-white with green veins 17
b. Upper sepal almost circular, usually with a short abrupt tip, white usually flushed with pink and green, veined with purple or green 18
17a. Petals 4–5 cm, spathulate **25. venustum**
b. Petals to 10 cm, narrowly elliptic **16. dayanum***
18a. Upper and lower petal margins with 6–10 warts **19. lawrenceanum**
b. Upper petal margin with 4–8 warts (sometimes 1 or 2 on blade or lower margin) 19
19a. Petals almost horizontal, with dark purple veins; warts *c.* 1 mm in diameter **17. barbatum**
b. Petals curved downwards, with indistinct green veins; warts *c.* 3 mm in diameter **18. callosum***
20a. Petals with large, black, often coalescing spots; lip dull yellow to reddish brown, strongly veined with greenish brown **25. venustum**
b. Not the above combination of characters 21
21a. Petals green and bright pink or pinkish purple 22
b. Petals pale green, brown or brownish purple 24
22a. Spots or warts all in basal half of petal **21. appletonianum**
b. Spots or warts extending into apical half of petal 23

23a. Petals descending, broadly spathulate, with a prominent white patch **23. volonteanum**
b. Petals almost horizontal, narrowly spathulate, lacking a white patch **28. virens**
24a. Petals pale green with *c.* 10 large black spots on upper margin and midline; upper sepal with broad white margin **24. tonsum**
b. Petals reddish brown and dull green with many small spots in basal half, often concentrated in upper corner **30. mastersianum**
25a. Petal tips strongly recurved; upper sepal with bold interconnecting veins **15. fairrieanum**
b. Not the above combination of characters 26
26a. Upper sepal unspotted, finely or indistinctly speckled or with several indistinct stripes 27
b. Upper sepal with large dark spots or 1 distinct central stripe 29
27a. Petals oblong, 4 cm; upper sepal *c.* 6 cm wide **9. charlesworthii**
b. Petals spathulate, 6 cm or more; upper sepal less than 4 cm wide 28
28a. Bract much shorter than ovary; petals with black or purple specks at base **10. hirsutissimum**
b. Bract almost equal to ovary; petals without specks **11. villosum**
29a. Upper sepal with dark central stripe **12. spicerianum***
b. Upper sepal with large, dark spots 30
30a. Flower-stalk and ovary with shaggy hairs **11. villosum**
b. Flower-stalk and ovary at most downy or felted 31
31a. Flowers 8 cm wide or less **13. exul**
b. Flowers 10 cm wide or more **14. insigne***
32a. Petals similar to sepals in shape and size; flowers solitary, rarely 2 or 3 arising from the same point on the flowering stem 33
b. Petals longer and thinner than sepals; flowers 2 or more, arising from different points on the flowering stem 37
33a. Lip cup-shaped to spherical; petals usually 4 cm or less; upper sepal unspotted 34
b. Lip deeply conical; petals more than 4 cm; upper sepal spotted or blotched 35
34a. Lip the same colour as petals and sepals; staminode white and yellow **4. niveum**
b. Lip pink; petals and sepals pink or white; staminode pink and yellow **5. delenatii**
35a. Flowers pale to deep yellow, finely flecked **1. concolor**
b. Flowers white or pale pink with large spots or blotches 36
36a. Flowering stem 3–10 cm; flower marked with coalescing blotches **3. godefroyae**
b. Flowering stem very short; flower marked with spots which coalesce only at bases of sepals and petals **2. bellatulum**
37a. Upper sepal unstriped or with 1–5 bold stripes **8. stonei**
b. Upper sepal with 10–20 stripes 38
38a. Basal part of petal evenly spotted with purple **6. rothschildianum**
b. Basal part of petal unspotted or with a few uneven spots **7. philippinense***

1. P. concolor (Lindley) Pfitzer. Illustration: Botanical Magazine, 5513 (1865); Orchid Digest **42**: 72, 74 & 75 (1978); Bechtel et al., Manual of cultivated orchid species, 253 (1981).
Leaves 4–5, 10–15 × *c.* 6 cm, elliptic to oblong, obtuse, mottled greyish green above, reddish purple below. Flower-stalk to 50 cm, purple, hairy, shorter than leaves, 1- or 2-flowered. Flower *c.* 5 cm wide, pale to deep yellow, finely flecked with reddish brown. Upper sepal circular, hooded; lower sepal ovate, acute. Petals elliptic to oblong, slightly curved downwards. Lip blunt, conical, equal to or shorter than the petals, without side lobes. *Burma, Thailand, Cambodia, S Laos & S China.* G2. Winter–spring.

2. P. bellatulum (Reichenbach) Stein. Illustration: Orchid Digest **41**: 36 (1977); Graf, Tropica, 749 (1978); Bechtel et al., Manual of cultivated orchid species, 252 (1981).
Leaves 4–6, *c.* 25 × 8 cm, elliptic to oblong, mottled pale and dark green above, reddish below. Flower-stalk very short, 1- or 2-flowered. Flower to 8 cm wide, white or very pale yellow with bold reddish brown spots which coalesce only at the bases of the petals and sepals. Upper sepal almost circular, hooded. Petals almost circular, sometimes shallowly notched. Lip blunt, conical, without side lobes. *Burma (Maymo plateau & Shan Highlands); adjacent Thailand.* G2. Spring.

3. P. godefroyae (Godefroy-Lebeuf) Stein (*P. leuchochilum* (Rolfe) Fowlie). Illustration: Botanical Magazine, 6876 (1886); Orchid Digest **41**: 33 (1977) & **44**: 178 (1980).
Similar to *P. bellatulum*. Leaves *c.* 4 cm wide, oblong. Flower-stalk 3–10 cm. Flower irregularly marked with coalescing spots and blotches. *Burma, Thailand (Kra isthmus & Birdsnest Islands) & S Vietnam.* G2. Spring.

Plants with unspotted, pure white lips have been called *P. leucochilum*: Illustration: Rittershausen & Rittershausen, Orchids in colour, 94 (1979).

The name *P. ang-thong* Fowlie has been given to plants with fewer and finer spots on the flowers; these may be hybrids between *P. godefroyae* and *P. niveum*.

4. P. niveum (Reichenbach) Stein. Illustration: Botanical Magazine, 5922 (1871); Orchid Digest **41**: 33 (1977); Bechtel et al., Manual of cultivated orchid species, 255 (1981).
Leaves 6–8, 10–15 cm, elliptic to oblong, mottled grey and green above, deep purple below. Flower-stalk 12–15 cm, 1- or 2-flowered, deep purple with short hairs. Flower 6–8 cm wide, white with a fine stippling of reddish brown near the base of the petals. Upper sepal circular, hooded; lower sepal smaller, ovate, acute, bifid. Sepals streaked with purple near base. Petals broadly ovate to circular, scarcely pointed, slightly curled at the margins. Lip cup-shaped, with an incurved margin, without side lobes. Staminode yellow and white. *Thailand, Malay peninsula & offshore islands.* G2. Spring–summer.

5. P. delenatii Guillaumin. Illustration: Botanical Magazine, 89 (1950); Orchid Digest **41**: 35 (1977); Bechtel et al., Manual of cultivated orchid species, 253 (1981).
Similar to *P. niveum*. Petals and sepals pure white; lip pink; staminode pink and yellow. *C Vietnam.* G2. Spring–summer.

6. P. rothschildianum (Reichenbach) Stein (*P. elliotianum* misapplied, not *P. elliotianum* (O'Brien) Pfitzer). Illustration: Botanical Magazine, 7102 (1890); Orchid Digest **40**: 180 (1980); Bechtel et al., Manual of cultivated orchid species, 256 (1981).
Leaves to 50 × 7 cm, elliptic to oblong, unmottled, shiny. Flowering stem longer than leaves, reddish, downy, with 2–6 flowers. Flowers *c.* 13 cm from top to bottom. Upper sepal *c.* 8 cm, ovate, acute, cream-coloured with *c.* 15 black stripes, margins white. Petals to 15 cm, narrow, tapering, drooping, wavy with fine blunt tips and marginal hairs, cream with purple spots at base, yellowish green apically, veined with purple. Lip pointed, without

side lobes, purplish brown, yellow around rim; flattened sideways. *N Borneo (false records from New Guinea)*. G2. Summer.

7. P. philippinense (Reichenbach) Stein (*Cypripedium laevigatum* Bateman). Illustration: Botanical Magazine, 5508 (1865); Orchid Digest **40**: 183 (1980); Bechtel et al., Manual of cultivated orchid species, 255 (1981).
Leaves to 30 × 4 cm, strap-shaped, unmottled, shiny. Flowering stem to 50 cm, hairy, bearing 2–6 flowers. Flowers *c.* 8 cm from top to bottom. Upper sepal *c.* 3.5 cm, ovate, acute, pale yellow with *c.* 12 brownish purple stripes. Petals to 15 cm, narrow, tapering, drooping, twisted and wavy, with marginal hairs, yellow at base with a few uneven spots, reddish purple to green at tips. Lip rounded, without side lobes, yellow with green veins. *Philippines*. G2. Summer.

***P. praestans** (Reichenbach) Pfitzer (*P. glanduliferum* (Blume) Pfitzer). Illustration: Gardeners' Chronicle **26**: 776 (1886); Graf, Tropica, 746 (1978); Orchid Digest **40**: 182 (1980). Upper sepal 5 cm; petals to 12 cm, less drooping than in *P. philippinense*. *New Guinea*. G2. Summer.

8. P. stonei (Hooker) Stein. Illustration: Botanical Magazine, 5349 (1862); Orchid Digest **44**: 184 (1980); Bechtel et al., Manual of cultivated orchid species, 256 (1981).
Leaves 30–40 × 4 cm, strap-shaped, rounded, unmottled. Flowering stem to 60 cm, greenish purple, bearing 3–5 flowers. Flowers 14 cm from top to bottom. Upper sepal 6 cm, ovate, acuminate, white, unstriped or with 1–5 thick dark stripes. Petals 12–15 cm, narrow, tapering, drooping, not twisted, dull yellow to green, with purple blotches, lacking hairs or warts. Lip rounded, without side lobes, dull pink, yellowish green below, yellow around rim. *NW Borneo*. G2. Summer–autumn.

9. P. charlesworthii (Masters) Pfitzer. Illustration: Botanical Magazine, 7416 (1895); Orchid Digest **40**: 152 (1976); Bechtel et al., Manual of cultivated orchid species, 252 (1981).
Leaves 12.5–20 cm, linear to oblong, acute, hairless not mottled above, spotted with brownish purple beneath. Flower-stalk to 20 cm, shortly hairy, red-spotted. Flower 8–9 cm wide. Upper sepal *c.* 6 cm wide, almost circular, pink with a fine network of darker veins, white below, slightly reflexed above. Lower sepal much smaller, yellowish green. Petals *c.* 4 cm, oblong, obtuse-tipped, yellowish green with a fine network of brown veins. Lip greenish yellow to reddish brown. *Burma & NE India*. G2. Autumn.

10. P. hirsutissimum (Hooker) Stein. Illustration: Botanical Magazine, 4990 (1857); Orchid Digest **44**: 218 (1980); Bechtel et al., Manual of cultivated orchid species, 254 (1981).
Leaves 20–35 cm, oblong, not mottled above. Flower-stalk 15–30 cm, green with purple hairs. Bract shorter than ovary. Flower 10–12 cm wide, all parts shaggy on back. Upper sepal *c.* 4 × 3 cm, circular to ovate, wavy, green or greenish yellow with a central purplish brown patch, indistinctly speckled and streaked. Lower sepal smaller, pale green. Petals 6–7 cm, spathulate with hairy margins, base green to brown, finely spotted with black or purple, margins wavy, tip bright purple. Lip green to greenish brown. Staminode green with 2 basal warts. *NE India, Indo-Burmese border, Thailand & S China*. G1–2. Spring.

11. P. villosum (Lindley) Stein (*P. boxallii* (Reichenbach) Pfitzer). Illustration: Orchid Review **77**: 273 (1969); Orchid Digest **36**: 148 (1972); Bechtel et al., Manual of cultivated orchid species, 257 (1981).
Leaves 25–40 × 2–3 cm, oblong, uniformly dull green, purple-spotted underneath at the base. Flower-stalk 20–35 cm, green spotted with brown, shaggy. Bract almost equal to ovary. Flower glossy, 12–15 cm from top to bottom. Upper sepal to 5 cm, circular to ovate, reflexed at the sides, yellow or greenish yellow with a brownish purple base diffusing into indistinct stripes above. Petals 6–7 cm, spathulate, obtuse, sometimes notched, with a dark brownish purple mid-line separating an upper yellowish brown half from a lower greenish yellow half. Lip conical, yellow to green, flushed with brown. Staminode obovate, yellowish brown. *NE India, Burma, N Thailand & Laos*. G1–2. Winter–spring.

Variants from Burma with boldly mottled upper sepals and pink net-venation on the paler lip have been called *P. boxallii* (Illustration: Orchid Digest **36**: 148, 1972).

12. P. spicerianum (Masters & Moore) Pfitzer. Illustration: Botanical Magazine, 6490 (1880); Orchid Digest **36**: 28 (1972); Bechtel et al., Manual of cultivated orchid species, 256 (1981).
Leaves 12.5–25 × 3 cm, oblong, uniformly dark green above, mottled beneath at the base. Flower-stalk to 30 cm, purple, slightly hairy. Bract much shorter than ovary. Flower 7–8 cm wide. Upper sepal broadly ovate, to 5 cm across, reflexed at the sides, bent forward, keeled above, white with a green base and a central dark purple stripe. Lower sepal narrower, greenish yellow with a purple stripe. Petals to 5 cm, oblong, obtuse, wavy, yellowish green with a dark purple stripe, flecked with purple at the base; margins wavy. Lip greyish brown. Staminode purple and white, circular with recurved margins. *NE India*. G2. Winter–spring.

***P. druryi** (Beddome) Stein. Illustration: Orchid Digest **38**: 35 (1974); Botanical Magazine n.s., 764 (1978); Bechtel et al., Manual of cultivated orchid species, 253 (1981). Leaves light green with dark veins. Upper sepal yellow or yellowish green with a broad, purplish brown stripe. Petals elliptic to oblong, not wavy, hairy, yellow with dark central stripe. Lip pale yellow. *S India; possibly extinct in the wild*. G2. Spring–summer.

13. P. exul (Ridley) Kerchove. Illustration: Botanical Magazine, 7510 (1896); Orchid Digest **40**: 149 (1976); Bechtel et al. Manual of cultivated orchid species, 253 (1981).
Leaves 20–30 × 2–2.5 cm, oblong, obtuse, uniformly bright green. Flower-stalk 15–30 cm, with sparse red hairs. Bract nearly as long as ovary. Flowers to 8 cm wide. Upper sepal 3–4 × 2–3 cm, broadly ovate, greenish yellow, heavily spotted with brownish purple, with a broad white wavy margin. Petals *c.* 5 × 2 cm, oblong to elliptic with hairy margins, yellow with a faint central band of purplish brown. Lip yellow, suffused with brown. Staminode yellow, spotted with brown. *S Thailand*. G2. Spring–summer.

Very similar to *P. insigne* and perhaps a subspecies of it.

14. P. insigne (Lindley) Pfitzer. Illustration: Botanical Magazine, 3412 (1835); American Orchid Society Bulletin **42**: 809 (1973); Bechtel et al., Manual of cultivated orchid species, 254 (1981).
Leaves 20–30 cm, linear to lanceolate, uniformly yellowish green. Flower-stalk to 30 cm, purple, felted. Flower *c.* 10 cm wide. Upper sepal *c.* 5 × 4 cm, ovate, obtuse, wavy, mostly green or greenish yellow with brown spots; tip and some of margin white, unspotted. Lower sepal smaller, green with brown streaks. Petals 6–7 cm, wavy, oblong to narrowly spathulate, greenish yellow with a network of brown veins. Lip greenish or reddish brown, with a yellow rim. Staminode heart-shaped. *E. Himalaya*. G1–2. Autumn–spring.

Over 100 varieties have been described of which the following are among the best known: var. **sanderae** Reichenbach. Illustration: Graf, Tropica, 749 (1978); Bechtel et al., Manual of cultivated orchid species, 254 (1981). Upper sepal primrose yellow with reddish brown spots; petals and lip yellow. Var. **sanderianum** Sander is similar to var. **sanderae** but lacks spots.

P. × leeanum (Reichenbach) Kerchove. Illustration: Cogniaux & Goossens, Dictionnaire Iconographique des Orchidées, Cypripedium Hybrides, t. 3 (1903). The hybrid between *P. insigne* and *P. spicerianum*. Upper sepal white with lines of purple spots; petals slightly drooping; lip greenish brown. *Himalaya & garden origin.* G2.

15. P. fairrieanum (Lindley) Stein. Illustration: Botanical Magazine, 5024 (1857); Orchid Digest **40**: 155 (1976); Bechtel et al., Manual of cultivated orchid species, 253 (1981).
Leaves 9–15 cm, oblong, unmottled. Flower-stalk 9–15 cm, pale purple, sparsely hairy. Flower 6–8 cm from top to bottom. Upper sepal circular with a broad acuminate tip, pale green with *c.* 15 finely branched dark purple veins interconnecting near the wavy margin. Lower sepal smaller with unconnected veins. Petals 4–5 cm, oblong to lanceolate, curved downwards and strongly recurved at the tips, hairy and wavy on margins, green and white with green and purple veins. Lip inflated, greenish yellow, with a network of purplish brown veins. *Sikkim, Bhutan & NE India.* G2. Autumn.

Lindley in his original description misspelt the name 'fairieanum'; the man after whom he named this species was a Mr Fairrie.

16. P. dayanum (Lindley) Stein (*Cypripedium 'spectabile'* Reichenbach var. *dayanum* Lindley). Illustration: Botanical Magazine n.s. 594 (1971); Williams et al., Orchids for everyone, 152 (1980); Orchid Digest **39**: 157 (1975) & **45**: 24 (1981).
Leaves 12–20 × 5 cm, oblong to lanceolate, strongly mottled pale and bluish green above. Flower-stalk 20–30 cm, downy, purple. Flower 10–12 cm wide. Upper sepal *c.* 6 × 3 cm, narrowly ovate, acuminate, white with *c.* 20 narrow green veins. Lower sepal similar but smaller and more pointed. Petals to 10 cm, narrowly elliptic to spathulate, with long marginal hairs, greenish brown at the base to pink at the tip, veined with green and purple, sometimes with small marginal warts. Lip purplish brown, conical. *N Borneo.* G2. Summer (may also flower in winter).

P. violascens Schlechter (*P. bougainvilleanum* Schoser). Illustration: Orchid Digest **42**: 100 (1978); Millar, Orchids of Papua New Guinea, 86 (1978); Bechtel et al., Manual of cultivated orchid species, 257 (1981). Upper sepal *c.* 3 × 3 cm, petals 4–4.5 cm, sickle-shaped, purple. Lip greenish brown. *Indonesia (West Irian).* G2. Autumn.

17. P. barbatum (Lindley) Stein. Illustration: Botanical Magazine, 4234 (1846); Orchid Digest **41**: 61 (1977); Bechtel et al., Manual of cultivated orchid species, 251 (1981).
Leaves 10–15 × 2–3 cm, oblong to lanceolate, mottled. Flower-stalk 20–30 cm. Bract ovate-lanceolate, a quarter of the ovary length; flower 8–10 cm wide. Upper sepal almost circular with a short abrupt tip, green at the base, white or pale pink above with *c.* 20 dark purple veins. Lower sepal shorter and narrower with greenish purple veins. Petals 5 cm, oblong, slightly wider at the tip, hairy on margins, held just below horizontal, reddish brown at the base, shading to purplish red at the tips, dark-veined, with 4–8 dark hairy warts about 1 mm in diameter on the upper margin. *S Thailand & Malay peninsula.* G2. Spring–summer.

18. P. callosum (Reichenbach) Stein. Illustration; Botanical Magazine, 9671 (1946); Orchid Digest **36**: 9 (1972); Bechtel et al., Manual of cultivated orchid species, 252 (1981).
Leaves 10–25 × 4–6 cm, oblong to elliptic, greyish green with dark mottling. Flower-stalk to 30 cm, brownish purple, densely downy; bract 2–3 cm, ovate, acute. Flower 7–9 cm wide. Upper sepal 6 cm, almost circular, with a short tip, white with *c.* 25 dark purple veins which are green below. Lower sepal shorter and narrower. Petals 5–7 cm, strongly curved downwards, indistinctly green-veined, pinkish yellow at the tips, with 5–8 black hairy warts *c.* 3 mm in diameter on the upper margin (occasionally 1 or 2 on the blade). Lip reddish brown tinged with green. *E Thailand, Cambodia & S Vietnam.* G2. Spring.

Var. **sanderae** Anon. lacks red or yellow pigment and is white with brilliant green veins.

P. × maudiae Anon. Illustration: American Orchid Society Bulletin **46**: 1089 (1977); Graf, Tropica, 746 (1978). Petals less deflexed than in *P. callosum*, with 1 or 2 warts on the lower margin as well as those on the upper. This cross was originally made between 'albino' varieties of *P. callosum* and *P. lawrenceanum* (lacking red or yellow pigment) resulting in an 'albino' hybrid. The name, however, strictly applies to all hybrids between these parents. *Garden origin.* G2.

P. callosum and *P. barbatum* probably represent the extreme forms of a single species which runs from pure *P. barbatum* at the southern end of the Malay peninsula to *P. callosum* in N Thailand. Intermediates from the Kra isthmus have been referred to as *P. × siamense* Rolfe (*P. sublaeve* (Reichenbach) Fowlie).

19. P. lawrenceanum (Reichenbach) Pfitzer. Illustration: Botanical Magazine, 6432 (1879); Orchid Digest **41**: 61 (1977); Bechtel et al., Manual of cultivated orchid species, 254 (1981).
Leaves 15–30 × 5–7 cm, oblong to ovate with long sheathing bases, mottled dark and yellowish green. Flower-stalk 30–40 cm, purple, hairy. Bract ovate-oblong, much shorter than ovary. Flower 10–12 cm wide. Upper sepal 5 cm, circular, white flushed with pink; veins *c.* 20, purple above, green below. Petals 5–6 × 1.5 cm, horizontal, oblong to narrowly elliptic, obtuse, dull green with purplish veins; tips tinged with reddish purple. Each margin with 6–10 blackish purple hairy warts. Lip brownish purple. *N Borneo.* G2. Summer.

Var. **hyeanum** (Reichenbach) Pfitzer lacks red pigment and is green, white and yellow.

20. P. bullenianum (Reichenbach) Pfitzer (*P. linii* Schoser). Illustration: Orchid Digest **42**: 29 (1978); Bechtel et al., Manual of cultivated orchid species, 252 (1981).
Leaves 10–17.5 cm, faintly mottled with bluish green, often red beneath. Flower-stalk to 30 cm, hairy. Flower 7–8 cm wide. Upper sepal 3 cm, acute, reflexed above, green to greenish yellow, often brown at the base with *c.* 20 indistinct darker green veins. Petals spathulate, obtuse, more or less horizontal, apically pinkish purple with green margins, shading through brown to green at the base where there are a few dark shiny warts on marginal undulations. Lip slightly inflated, buff below, darker brown above with a green rim. Staminode with a narrow parallel-sided sinus. *Tioman Isles off Malay peninsula, Borneo & Indonesia (Sulawesi & Ambon).* Summer–autumn.

21. P. appletonianum (Gower) Rolfe (*P.

wolterianum (Kraenzlin) Pfitzer). Illustration: Rittershausen & Rittershausen, Orchids in colour, 92 (1979); Orchid Digest **44**: 227 (1980); Bechtel et al., Manual of cultivated orchid species, 251 (1981).
Leaves *c.* 20 × 3–5 cm, elliptic, mottled. Flower-stalk to 50 cm, hairy. Flower *c.* 8 cm from top to bottom. Upper sepal *c.* 3.5 × 2.5 cm, ovate, green, paler above with darker green veins. Lower sepal smaller and narrower, acute. Petals 5–6 × 1–2 cm, elliptic to spathulate, obtuse to rounded, slightly wavy, drooping and half twisted; basal half bright green with brown spots concentrated near margins and mid-line; apical half pinkish purple. Lip 4–5 cm, pale greenish brown below, dark brown near the rim. Staminode sinus broad, diverging. *Laos, Thailand & Cambodia*. G2. Winter–spring.

Taller plants with broader, more drooping petals have often been treated separately as *P. wolterianum*. However, there are no consistent differences between the 2 species.

22. P. hookerae (Reichenbach) Stein. Illustration: Botanical Magazine, 5362 (1863); Orchid Digest **40**: 155 (1976) & **45**: 17 (1981).
Leaves 10–15 × 3–4 cm, ovate to oblong, mottled greyish on dark green. Flower-stalk to 20 cm, greyish purple, hairy. Flower to 10 cm wide. Upper sepal ovate, acute, yellowish with a green centre, indistinctly veined with darker green. Petals spathulate, acute, 4–5 cm, held below horizontal, bright pinkish purple at the tips shading to dull greenish yellow with many small pinkish spots in basal half. Lip yellowish green, tinged with reddish brown above. Sinus on staminode closed by overlapping lobes. *N Borneo*. G2. Spring–summer.

23. P. volonteanum (Masters) Pfitzer (*P. hookerae* (Reichenbach) Stein var. *volonteanum* (Masters) Hallier). Illustration: Orchid Digest **39**: 166 (1975) & **45**: 18 (1981).
Similar to *P. hookerae*. Petal spots purplish black, extending into outer half of petal but concentrated near margins and mid-line. Green and purple areas of petal separated by a patch of white or very pale green. *N Borneo*. G2.

24. P. tonsum (Reichenbach) Stein. Illustration: Orchid Digest **40**: 85 (1976); Bechtel et al., Manual of cultivated orchid species, 257 (1981); Botanical Magazine n.s., 838 (1982).
Leaves 12–20 cm, oblong, ovate, mottled dark and light green. Flower-stalk 20–45 cm, purplish green. Flower 12 cm wide. Upper sepal 5–6 cm, broadly ovate, pale green with 25–30 green and purple veins. Petals 6–7 cm, slightly below horizontal, narrowly spathulate, similar in colour and venation to the upper sepal but with a few large dark spots on the upper margin and in the mid-line; margins hairy near the tip. Lip greenish yellow flushed with pink. *Sumatra*. G2. Summer-autumn.

25. P. venustum (Sims) Pfitzer. Illustration: Botanical Magazine, 2129 (1820); Orchid Digest **36**: 177 (1972); Bechtel et al., Manual of cultivated orchid species, 257 (1981).
Leaves 15–25 cm, ovate-lanceolate, mottled dark on light green above, streaked and spotted with red beneath. Flower-stalk 20–30 cm, purple. Flower 8–10 cm wide. Upper sepal ovate, acute, white or off-white with *c.* 20 green veins. Petals 4–5 cm, horizontal or slightly descending, spathulate, hairy on margins, yellowish green or green at the base, reddish brown at the tip, with several irregular dark hairy spots on blade and margins. Lip dull yellow to reddish brown, strongly veined with greenish brown. *Nepal, Bhutan, Bangladesh & NE India*. G1–2. Winter-spring.

26. P. acmodontum Wood. Illustration: Orchid Review **84**: 350 (1976); Orchid Digest **41**: 60 (1977); Bechtel et al., Manual of cultivated orchid species, 251 (1981).
Leaves 15–20 × 4–5 cm, oblong to elliptic, mottled. Flower-stalk to 25 cm, sparsely hairy. Flowers 7–8 cm wide. Upper sepal *c.* 4 × 3 cm, broadly ovate, acuminate, off-white with *c.* 15 purple veins, flushed with pink at base. Petals 4–4.5 × 1.5–2 cm, obovate to spathulate, acute to acuminate, greenish yellow at base, reddish brown in outer half. Inner half irregularly spotted with purplish black. Lip *c.* 4 cm, yellowish green below, brown towards rim, with a prominent acute tooth on outer edge of rim. *Philippines*. G2. Spring.

27. P. javanicum (Lindley & Paxton) Pfitzer (*P. virens* misapplied; *P. purpurascens* Fowlie). Illustration: Orchid Review **85**: 159 (1977); Orchid Digest **45**: 24 (1981); Bechtel et al., Manual of cultivated orchid species, 251 (1981).
Leaves to 20 × 5 cm, oblong to elliptic, mottled. Flower-stalk to 25 cm. Flower *c.* 10 cm from top to bottom. Upper sepal *c.* 4.5 × 2.5 cm, ovate, acuminate, greenish yellow with 20–25 dark green veins; lower sepal similar but smaller. Petals *c.* 5 × 1.5 cm, narrowly obovate, broader in apical half, drooping, dull green, pink at tips, most of blade covered in small purplish black warts. Lip *c.* 4.5 cm, dull green to brownish green with darker veins. *Java & Borneo*. G2. Flowers intermittently.

28. P. virens (Reichenbach) Pfitzer (*P. javanicum* (Lindley & Paxton) Pfitzer var. *virens* (Reichenbach) Stein). Illustration: Reichenbach, Xenia Orchidacea, t. 162 (1870); Orchid Digest **40**: 237 (1976) & **45**: 22 (1981).
Very similar to *P. javanicum* and often considered a subspecies of it. Petals narrowly spathulate, horizontal; basal two-thirds bright green, apical third pinkish purple; warts concentrated on margins and mid-line. Lip pale brown, darker near rim. *N Borneo*. G2.

Frequently confused with *P. javanicum* and thus mistakenly said to be native to Indonesia.

29. P. purpuratum (Lindley) Stein. Illustration: Botanical Magazine, 4901 (1856); Orchid Review **82**: 39 (1974); Bechtel et al., Manual of cultivated orchid species, 255 (1981).
Leaves 8–12 cm, oblong-elliptic, mottled dark and pale green. Flower-stalk 20–25 cm, deep purple, hairy. Flower 8–10 cm wide. Upper sepal circular, acuminate, white with about 12 alternating purple and green veins. Petals 4–5 cm, elliptic, hairy on margins, purplish brown with many small dark warts in basal half, paler at the tip. Lip purplish brown. *Hong Kong & adjacent China*. G2. Summer–winter.

30. P. mastersianum (Reichenbach) Stein. Illustration: Botanical magazine, 7629 (1898); Orchid Digest **45**: 19 & 20 (1981); Bechtel et al., Manual of cultivated orchid species, 255 (1981).
Leaves 20–24 × 3–4 cm, oblong-elliptic, mottled dark and yellowish green. Flower-stalk 30–45 cm, deep purple, hairy. Flower to 10 cm wide, fleshy. Upper sepal circular, 5 cm broad, greenish yellow with about 20 dark green veins and a broad white or off-white margin. Petals 5–6 × 2 cm, horizontal, oblong to slightly spathulate, round-tipped, reddish brown in apical half, remainder dull green with many fine purplish warts above the mid-line. Lip dull reddish purple, yellow round the rim. *Indonesia (Borneo & Moluccas)*. G2. Spring–summer.

31. P. superbiens (Reichenbach) Stein (*P. curtisii* (Reichenbach) Stein). Illustration: Orchid Review **83**: 394 (1975); Orchid Digest **42**: 177 (1978) & **45**: 60 (1981); Williams et al., Orchids for everyone, 152 (1980).
Leaves 15–20 cm, oblong-lanceolate, strongly mottled. Flower-stalk to 30 cm, purple, downy; bract ovate, a third of the ovary length; flower to 8 cm wide. Upper sepal broadly ovate, acuminate, pale green, purple at the base with about 25 dark green to purple veins. Lower sepal small, ovate-acuminate. Petals oblong, drooping, curled, hairy on margins, purple at the base shading to pale green, evenly marked with numerous small purplish black warts. Lip brownish purple. *Sumatra*. G2. Summer.

***P. ciliolare** (Reichenbach) Stein (*P. superbiens* (Reichenbach) Stein subsp. *ciliolare* (Reichenbach) Wood). Illustration: Orchid Digest **41**: 47 (1977); Rittershausen & Rittershausen, Orchids in colour, 90 (1979); Bechtel et al., Manual of cultivated orchid species, 252 (1981). Petals not curled, with dense marginal hairs; warts absent from outer third. *Philippines*. G2. Spring–summer.

32. P. argus (Reichenbach) Stein. Illustration: Botanical Magazine, 6175 (1875); Orchid Digest **41**: 61 (1977) & **45**: 57 (1981); Rittershausen & Rittershausen, Orchids in colour, 90 (1979).
Leaves 12–20 cm, oblong-lanceolate, pale green with darker mottling. Flower-stalk to 30 cm, purplish, glandular hairy. Bract pale green, half the ovary length. Flower 10 cm wide. Upper sepal broad, acuminate, off-white with about 20 dark green or purplish veins. Petals 7–8 cm, oblong, slightly tapering, hairy on margin, descending, off-white at the base, shading to pink at the tip, boldly marked all over with dark purple spots, some having pale centres. Lip yellow with dark green net-venation, red above, flushed with brown. *Philippines*. G2. Spring.

33. P. sukhakulii Schoser & Senghas. Illustration: Orchid Review **77**: 145 (1969); Orchid Digest **39**: 208 (1975); Bechtel et al., Manual of cultivated orchid species, 256 (1981).
Leaves 12–15 × *c*. 4.5 cm, narrowly elliptic, mottled. Flower-stalk to 25 cm. Flower *c*. 10 cm wide. Upper sepal 3.5–4 × 3 cm, broadly ovate, acuminate, yellowish green with 15–20 green veins. Petals 5–6 × 1.3–1.7 cm, oblong to narrowly elliptic, tapering to an acute point, almost horizontal, yellowish green, closely spotted with purplish black except at extreme tip. Lip *c*. 4.5 cm, conical, greenish brown. *Thailand*. G2. Autumn.

***P. wardii** Summerhayes. Illustration: Botanical Magazine, 9481 (1937); Orchid Digest **36**: 85 (1972) & **45**: 15 (1981); Pradhan, Indian orchids, t. 19 (1976). Flowers usually browner than in *P. sukhakulii*; petals descending. *N Indo-Burmese border*. G2. Winter.

Very similar to *P. sukhakulii*. They should probably be considered as one species under *P. wardii*, which is the older name.

34. P. victoria-regina (Sander) Wood (*P. victoria-mariae* invalid). Illustration: Botanical Magazine, 7573 (1898); Orchid Digest **44**: 217 (1980).
Leaves oblong, to 40 × 6 cm, often faintly mottled. Flowering stem 30–60 cm or more, reddish brown, bearing several flowers in succession over several months. Flowers 7–8 cm wide. Upper sepal 3–4 cm wide, circular to broadly ovate, greenish yellow with 5–8 purple veins and a white to creamy yellow border. Petals 5 cm, narrowly oblong, horizontal, hairy on margins, twisted and wavy, greenish yellow, veined with reddish purple and with an unbroken border of reddish purple. Lip slightly compressed, purple, unspotted, with a green and yellow rim. *Sumatra*. G2. Intermittent flowering.

The subspecies described above is endemic to a small area in central Sumatra and is uncommon in cultivation. The following subspecies are more commonly grown:

Subsp. **chamberlainianum** (Sander) Wood (*P. chamberlainianum* (Sander) Stein). Illustration: Botanical Magazine, 7578 (1898); Orchid Review **84**: 140 (1976); Bechtel et al., Manual of cultivated orchid species, 252 (1981). Leaves to 10 cm broad; upper sepal yellow and purple; lip inflated, evenly spotted with purple. *Sumatra*. G2. Intermittent flowering.

Subsp. **glaucophyllum** (J.J. Smith) Wood (*P. glaucophyllum* J.J. Smith). Illustration: Orchid Review **84**: 140 (1976); Orchid Digest **44**: 216 (1980); Bechtel et al., Manual of cultivated orchid species, 253 (1981). Leaves glaucous; petals slightly deflexed, marked with discontinuous purple spots on a white background. *E & C Java*. G2. Intermittent flowering.

35. P. parishii (Reichenbach) Stein. Illustration: Botanical Magazine, 5791 (1869); Orchid Digest **40**: 213 (1980); Bechtel et al., Manual of cultivated orchid species, 255 (1981).
Epiphytic. Leaves to 30 cm, strap-shaped, unmottled, bright green. Flowering stem to 60 cm, pale green, bearing 4–7 flowers. Flowers 8 cm wide. Upper sepal elliptic, pale yellowish green with faint veins. Petals to 15 cm, linear, gently tapering, drooping and twisted, green at base with marginal purple spots, purple at tips. Lip green to greenish purple. *Burma, NW Thailand & S China*. G2. Spring–summer.

36. P. lowii (Lindley) Stein. Illustration: Gardeners' Chronicle, 765 (1847); Orchid Digest **44**: 214 (1980); Williams et al., Orchids for everyone, 154 (1980).
Epiphytic or on cliffs. Leaves oblong, to 40 × 4 cm, unmottled, light green. Flowering stem to 1 m, arched, brownish purple, bearing 2–6 flowers. Flowers 15 cm across. Upper sepal 5 cm, broadly elliptic, short-pointed, concave above, with reflexed margins below, yellowish green flushed with purple, streaked with brown at base. Petals 8 cm, spathulate, spreading, slightly drooping at tips, slightly twisted, yellowish green with brown spots at base; tips bright purple. Lip blunt, oblong, pale brownish green with brown markings, flushed with purple near rim. *Malaysia & Indonesia*. G2. Spring–summer.

***P. haynaldianum** (Reichenbach) Stein. Illustration: Botanical Magazine, 6296 (1877); Orchid Digest **44**: 214 (1980); Bechtel et al., Manual of cultivated orchid species, 254 (1981). Upper sepal pale, marked with large irregular brown blotches. *Philippines*. G2. Spring.

Possibly better considered as a subspecies of *P. lowii*.

4. EPIPACTIS Zinn

P.J.B. Woods

Perennial herbs with creeping or vertical rhizomes and numerous fleshy but not tuberous roots. Aerial growth dying off in winter. Leaves spaced along the stem, spirally arranged or in 2 opposite rows, strongly nerved, pleated. Flowers held horizontally or hanging, often only on 1 side of the axis. Bracts green and leaf-like, not sheathing the basal parts of the flowers. Sepals and petals spreading or sometimes scarcely opening. Lip in 2 distinct parts, the basal cup-like, the apical flat and variously shaped. Spur absent. Pollinia rapidly breaking up.

A genus of about 20 species from north temperate regions. Cultivation as for *Dactylorhiza* (p. 167); some species, e.g. *E. gigantea* and *E. palustris*, prefer moister conditions.

Rhizome. Creeping: **4,5**; not creeping: **1–3**.

Leaves. Spirally arranged: **1,4,5**; in opposite rows: **2,3.**
Lip. Lateral lobes present: **4,5**; lateral lobes absent: **1–3.**

1a. Rhizome short, not creeping; lip without lateral lobes 2
b. Rhizome creeping, elongate; lip with lateral lobes 4
2a. Leaves spirally arranged **1. helleborine**
b. Leaves arranged in 2 opposite rows 3
3a. Axis, flower-stalks and ovary downy; flowers purple **2. atrorubens**
b. Axis, flower-stalks and ovary not downy; flowers yellowish green sometimes tinged with red or reddish violet **3. phyllanthes**
4a. Basal part of lip narrower than the ovate apical part **4. palustris**
b. Basal part of lip broader than the elongate apical part **5. gigantea**

1. E. helleborine (Linneaus) Crantz (*E. latifolia* (Linnaeus) Allioni). Illustration: Danesch & Danesch, Orchideen Europas, Mitteleuropa, edn 3, 190, 191 (1972); Sundermann, Europäische und mediterrane Orchideen, 206 (1975); Williams et al., Orchids of Britain and Europe, 15 (1978); Mossberg & Nilsson, Orchids of northern Europe, 35 (1979).
Rhizome short. Stems 1–3, arising close together, 35–90 cm, sparsely hairy. Leaves 3–10, 5–17 × 2.5–10 cm, spirally arranged, ovate or broadly elliptic, the upper lanceolate, acute or acuminate. Raceme 10–40 cm with 15–50 flowers, borne on 1 side of the axis, lowest bracts longer than flowers, the upper shorter. Sepals 1–1.3 cm, ovate, greenish; petals slightly smaller, pinkish or purplish. Apical part of the lip ovate, acute, recurved, with 2 basal bosses. *Europe, N Africa, SW Asia to Himalaya.* H2. Summer.

2. E. atrorubens (Hoffmann) Besser (*E. atropurpurea* Rafinesque). Illustration: Danesch & Danesch, Orchideen Europas, Mitteleuropa, edn 3, 186, 187, 194 (1972); Sundermann, Europäische und mediterrane Orchideen, 212 (1975); Williams et al., Orchids of Britain and Europe, 158 (1978); Mossberg & Nilsson, Orchids of northern Europe, 33 (1979).
Rhizome short. Stems 15–60 cm, downy, purplish-tinged. Leaves 5–10, 4–10 × 1.5–4.5 cm, arranged in opposite ranks along the stem, ovate to ovate-lanceolate, acute. Raceme to 25 cm with 8–18 flowers, lower bracts as long as flowers. Flowers wine-red, faintly fragrant. Sepals 6–7 mm, ovate, acuminate, petals slightly narrower. Lip 5.5–6.5 mm, basal part cup-shaped, apical part wider than long, with 3 rough basal bosses, apex acute or blunt, recurved. *Europe to C Asia.* H2. Summer.

3. E. phyllanthes G.E. Smith. Illustration: Landwehr, Wilde Orchideeën van Europa **2**: 493, 495 (1977); Williams et al., Orchids of Britain and Europe, 155 (1978); Mossberg & Nilsson, Orchids of northern Europe, 41 (1979).
Rhizome short. Stems 1 or occasionally 3, 20–45 cm, hairless or slightly hairy. Leaves 3–16, arranged in opposite rows on the stem, 3.5–7 × 3–5 cm, ovate or lanceolate, acute. Raceme to 15 cm with 15–35 flowers, bracts longer than flowers. Flowers greenish, hanging, opening only a little. Sepals 8–10 mm, petals sometimes slightly purplish-tinged. Basal and apical parts of the lip not always clearly distinct. *NW & C Europe.* H2. Summer.

4. E. palustris (Miller) Crantz. Illustration: Danesch & Danesch, Orchideen Europas, Mitteleuropa, edn 3, 183–5 (1972); Sundermann, Europäische und mediterrane Orchideen, 214 (1975); Williams et al., Orchids of Britain and Europe, 149 (1978); Mossberg & Nilsson, Orchids of northern Europe, 43 (1979).
Rhizome creeping. Stems to 50 cm, downy. Leaves 4–8, 5–15 × 2–4 cm, spirally arranged and often crowded, oblong to lanceolate, acuminate. Raceme to 15 cm, with up to 15 flowers, lower bracts equalling the flowers. Sepals 8–12 mm, ovate-lanceolate, greenish lined with red. Petals whitish flushed with pink. Lip 1–1.2 cm, basal part concave with erect lateral lobes, pinkish white, purple-lined and orange-dotted, narrower than the apical part which is ovate, as broad as long, wavy-margined, white with a yellow bar across the base. *Eurasia, N Africa.* H2. Summer.

5. E. gigantea Hooker (*E. royleana* Lindley). Illustration: Jahresberichte des Naturwissenschaftlichen Vereins in Wuppertal **23**: t. 5 (1970); American Orchid Society Bulletin **71**: 240 (1971); Luer, Native orchids of the United States and Canada, 79 (1975); Flora Iranica **126**: t. 15, 16 (1978).
Rhizome creeping. Stems to 1 m, hairless. Leaves 4–12, 5–20 × 2–7 cm, spirally arranged, ovate to lanceolate, acuminate. Raceme to 30 cm with up to 15 flowers, bracts longer than flowers. Sepals 1.5–2 cm, ovate-lanceolate, greenish yellow, purple-veined. Petals ovate, 1.3–1.5 cm, flushed with pink. Lip 1.4–1.5 cm, basal part concave with red warts, lateral lobes bluntly triangular, yellowish with reddish nerves, separated by a rigid fold from the apical part which is oblong-lanceolate with 2 basal ridges, orange or yellowish, the apex suffused with pink. *NW America, Himalaya.* H2. Summer.

5. CEPHALANTHERA Richard

P.J.B. Woods

Perennial herbs with erect or short, creeping rhizomes. Roots fibrous. Aerial growth dying down in winter. Leaves evenly spaced along stem, pleated. Flowers stalkless or very shortly stalked in a loose terminal spike, pointing upwards; bracts green and leaf-like, not sheathing the basal parts of the flowers. Sepals and petals hooded, scarcely opening. Lip constricted at the middle into 2 distinct parts, the base concave and clasping the base of the column, the apex with several ridges. Spur absent. Pollinia 2, club-shaped, each longitudinally divided, powdery.

A genus of about 14 species from north temperate regions. Cultivation as for *Dactylorhiza* (p. 167).

1a. Flowers pinkish red; ovary hairy **3. rubra**
b. Flowers white; ovary not obviously hairy 2
2a. Bracts as long as or much longer than ovary; lip with 3–5 ridges **1. damasonium**
b. Bracts (except the lowermost) distinctly shorter than ovary; lip with 4–6 ridges **2. longifolia**

1. C. damasonium (Miller) Druce. Illustration: Landwehr, Wilde Orchideeën van Europa **2**: 513 (1977); Williams et al., Orchids of Britain and Europe, 145 (1978); Mossberg & Nilsson, Orchids of northern Europe, 31 (1979).
Stems 15–70 cm. Leaves 3–6, 5–8 × 2–3 cm, ovate-lanceolate to oblong-ovate, the upper narrowest. Spike with 3–5 or more flowers. Flowers *c.* 2 cm, loosely arranged, creamy white. Bracts as long as the ovary, lowermost much longer, ovate-lanceolate to lanceolate. Sepals and petals oblong to lanceolate, incurved, petals slightly shorter than sepals. Lip shorter than sepals and petals, yellowish orange at base; apical half heart-shaped, recurved, with 3–5 yellowish orange ridges. *Europe, N Africa, SW Asia.* H2. Spring–summer.

2. C. longifolia (Linnaeus) Fritsch. Illustration: Landwehr, Wilde Orchideeën van Europa **2**: 515 (1977); Williams et al., Orchids of Britain and Europe, 145 (1978); Mossberg & Nilsson, Orchids of northern Europe, 29 (1979).
Stems 10–50 cm. Leaves 4–12, 4–18 cm × 8–40 mm, linear-lanceolate to lanceolate. Spike with 10–20 or more flowers; flowers *c.* 1.8 cm, white, sepals and petals incurved but more open than in *C. damasonium*. Bracts, except for the lowermost, much shorter than ovary, mostly linear. Sepals lanceolate, petals broader and shorter. Lip shorter than sepals and petals, yellowish orange at base; apical half heart-shaped, recurved, with 4–6 yellowish orange ridges. *Europe, N Africa, to the Middle East & temperate Asia*. H2. Spring.

3. C. rubra (Linnaeus) Richard. Illustration: Danesch & Danesch, Orchideen Europas, Mitteleuropa, edn 3, 195 (1972); Landwehr, Wilde Orchideeën van Europa **2**: 517 (1978); Mossberg & Nilsson, Orchids of northern Europe, 27 (1979).
Stems 10–60 cm, upper part hairy. Leaves 2–8, 3–14 × 1–3 cm, oblong to lanceolate. Spike with 2–15 flowers, hairy; flowers 2–2.5 cm, opening fairly widely, pinkish red. Bracts mostly longer than ovary, but the upper much shorter, linear. Sepals and petals to 2.5 cm, ovate to lanceolate. Lip as long as sepals, white, with several brownish ridges at base; apical half lanceolate, recurved, red-violet with *c.* 10 yellowish brown ridges. *Europe, N Africa, SW Asia east to Iran*. H2. Spring–early summer.

6. LISTERA R. Brown
P.J.B. Woods
Herbaceous perennials with short rhizomes. Aerial growth dying down in winter. Leaves 2 (rarely 3–4), ovate, almost opposite, arising at about the middle of the stem or below it. Flowers in a spike-like raceme. Sepals and petals more or less equal in size. Lip longer than the sepals and petals, strap-shaped, its apex 2-lobed. Spur absent. Pollinia 2, club-shaped, each longitudinally divided and made up of easily separable masses.

A genus of about 10 species distributed throughout the temperate regions. Cultivation as for *Dactylorhiza* (p. 167).

1a. Plants to 20 cm, leaves to 2.5 cm; raceme with 4–12 flowers **1. cordata**
b. Plants to 60 cm, leaves 5–20 cm; raceme with many flowers **2. ovata**

1. L. cordata (Linnaeus) R. Brown. Illustration: Danesch & Danesch, Orchideen Europas, Mitteleuropa, edn 3, 135, 137 (1972); Landwehr, Wilde Orchideeën van Europa **2**: 533 (1977); Williams et al., Orchids of Britain and Europe, 141 (1978); Mossberg & Nilsson, Orchids of northern Europe, 49 (1979).
Rhizome slender, creeping. Stems slender, sometimes copper-coloured, 4.5–20 cm, upper part minutely downy. Leaves 2 (occasionally 4), ovate, shortly pointed, 1–2.5 cm. Raceme 1–6 cm with 4–12 flowers. Bracts minute. Sepals greenish, oblong-elliptic, 2–2.5 mm, obtuse. Petals reddish. Lip purplish, linear, 3.5–4.5 mm, lateral lobes small and arising at the base, central lobe divided to about the middle, its segments diverging. *Europe, N America*. H2. Summer.

2. L. ovata (Linnaeus) R. Brown. Illustration: Danesch & Danesch, Orchideen Europas, Mitteleuropa, edn 3, 134, 137 (1972); Landwehr, Wilde Orchideeën van Europa **2**: 531 (1977); Williams et al., Orchids of Britain and Europe, 141 (1978); Mossberg & Nilsson, Orchids of northern Europe, 47 (1979).
Rhizome moderately thick. Stems 20–60 cm, hairless below, downy above. Leaves 2 (occasionally 3–4), ovate to broadly elliptic, distinctly 3–5-veined, obtuse, 5–20 cm. Raceme 7–25 cm, with numerous greenish or brownish green flowers. Bracts minute. Sepals incurved, ovate, obtuse, 4–5 mm. Petals about the same length as sepals but narrower. Lip yellowish green, strap-shaped, 7–15 mm, the apex divided for one-third to half its length, lateral lobes vestigial or absent. *Eurasia*. H1. Summer.

7. CORYBAS Salisbury
P.J.B. Woods
Small, 1-leaved, terrestrial perennials. Aerial growth dying off in winter. Root tuberous, small and round. Leaves ovate to heart-shaped, held flat against the ground. Flower 1 (rarely 2 in a few species), large compared to the leaf size, shortly stalked (rarely not stalked), the stalk elongating in fruit. Upper sepal conspicuous, broad, concave or hooded, lateral sepals and petals thread-like, sometimes very small. Lip conspicuous, the basal part erect under the upper sepal, inrolled and tube-like around the column, the apical part expanded into a broad, often abruptly bent blade, the margin entire or fringed. Spurs 2, sometimes inconspicuous, placed on either side of the column. Pollinia 2, granular.

A genus of about 60 species from the Himalaya and E Asia, extending to Australasia. Cultivation as for *Pleione* (p. 182), or they can be grown in peaty compost, with high humidity when growing. When dormant, they should be kept cool and dry.

1. C. dilatatus (Rupp & Nicholls) Rupp. Illustration: American Orchid Society Bulletin **44**: 993 (1975), **46**: 989 (1977), **47**: 1120, 1121 (1978); Botanical Magazine n.s., 836 (1981).
Plant to 3 cm. Leaf to 3 cm wide, ovate to heart-shaped. Flower solitary, reddish purple. Upper sepal to 2.5 cm. Lip with the apical part white or white tinged with crimson, papillose towards the apex; margins crimson, translucent, coarsely toothed, blotched and purple-veined. *SE Australia, including Tasmania*. G1. Summer.

8. SPIRANTHES Richard
J. Lamond & P.J.B. Woods
Perennial herbs with slender or sturdy, fusiform tubers. Aerial growth dying off in winter. Leaves all basal in a rosette or evenly distributed on the stem, ovate-elliptic or lanceolate, rolled when young. Inflorescence elongate, twisted, with many flowers which are close, spirally arranged and often scented. Sepals and petals almost equal, incurved, the tips free. Lip entire, its apical margins variously frilled or not. Spur absent. Pollinia 2, each made up of easily separable masses.

A genus of about 30 species mainly in temperate zones, the majority occurring in N America. They should be grown in pots containing a well-drained compost and protected in a cold-frame or greenhouse.

1a. Basal leaves ovate-elliptic, forming a flat rosette; inflorescence lateral **1. spiralis***
b. Basal leaves linear-lanceolate, not forming a rosette; inflorescence terminal **2. cernua***

1. S. spiralis (Linnaeus) Chevallier (*S. autumnalis* Richard). Illustration: Ross-Craig, Drawings of British plants, part 28: 47 (1971); Landwehr, Wilde Orchideeën van Europa **2**: 534, 535 (1977); Grey-Wilson & Mathew, Bulbs, t. 47 (1981).
Tubers stout, 5–13 mm thick. Leaves 3–7, basal, 2–4 cm × 5–15 mm, ovate-elliptic, shortly and broadly stalked; upper leaves scale-like, sheathing the stem. Inflorescence to 20 cm, glandular-hairy above, produced laterally from below the leaves. Flowers

7–20, white, scented, in 1 spiral row. Bracts about as long as ovary. Sepals and petals 4–7 mm, the upper 3 and the lip forming a tube around the column. Lip apex truncate and scalloped. *Europe & N Africa to N Iran*. H2. Late summer–autumn.

***S. lacera** (Rafinesque) Rafinesque (*S. gracilis* (Bigelow) Beck). Illustration: Luer, Native orchids of the United States and Canada, 111 (1975). Tubers slender, 3–6 mm thick; inflorescence to 50 cm, bearing up to 40 flowers. *Eastern N America*. H1. Late summer.

2. S. cernua (Linnaeus) Richard (*S. odorata* (Nuttall) Lindley). Illustration: Luer, Native orchids of the United States and Canada, 119, 121 (1975).
Tubers slender, elongate. Basal leaves 2–6, 5–24 cm × 2–20 mm, linear-oblanceolate, the upper leaves small, sheath-like. Inflorescence to 50 cm, with up to 60 flowers; upper parts downy, hairs sometimes glandular. Flowers white with yellowish centres, arranged in 3–4 rows. Bract shorter or longer than ovary. Sepals and petals 1.1–1.2 cm, sepals forming a tube with the lip. Lip apex rounded, scalloped, recurved. *Eastern N America*. H4. Autumn.

***S. aestivalis** (Poiret) Richard. Illustration: Ross-Craig, Drawings of British plants, part 28: 48 (1971); Landwehr, Wilde Orchideeën van Europa **2**: 537 (1977); Williams et al., Orchids of Britain and Europe, 139 (1978). Flowers 6–7 mm, arranged in 1 spiral row. *C & S Europe*. H2. Summer.

***S. sinensis** (Persoon) Ames (*S. australis* Lindley). Illustration: Annals of the Royal Botanic Garden Calcutta **8**: t. 369 (1898); Dansk Botanisk Arkiv **32**: 105 (1978); Williams et al., Orchids of Britain and Europe, 139 (1978). Flowers white or pink. Lip with 2 basal calluses. *Asia to Australasia*. G2. Summer.

9. GOODYERA R. Brown

P.J.B. Woods

Terrestrial herbs, rhizomes creeping, sometimes above ground, aerial growth persistent through the winter. Leaves membranous or fleshy, often asymmetric, mostly basal in rosettes, more rarely scattered on erect stems, often with differently coloured veins, rolled when young. Inflorescence erect, flowers few to many, arranged spirally or on 1 side, or uniformly in a cylindric spike or raceme. Sepals directed forwards or spreading. Petals directed forwards over the column. Lip saccate, the sac usually with hairs inside, entire, the tip usually pointed and reflexed. Column short, blunt or with 2 long or short teeth on the rostellum. Pollinia 2, club-shaped, each longitudinally divided and made up of easily separable masses.

A genus of about 80 species from the north temperate regions, and SE Asia and Australasia. As most of them are of creeping habit, they are best grown in wide pans, using terrestrial compost which should be kept moist throughout the year. Most of the species require shade. A periodic dressing of pine-needles is beneficial (see Christian, P., Quarterly Bulletin of the Alpine Garden Society **43**: 322–4, 1975).

1a. Flowers in a cylindric spike, not spirally arranged — 2
b. Flowers in a 1-sided spike, or spirally arranged — 3
2a. Sepals with few hairs, reddish brown, tips white; leaves very dark, main veins red — **1. colorata**
b. Sepals densely hairy, white with green mid-vein; leaves green with white veins — **2. pubescens**
3a. Flowers 5 mm or more; leaves 3–11 cm, green, usually with a broad, white main vein, the rest net-veined — **3. oblongifolia**
b. Flowers less than 5 mm; leaves 1–4 cm, usually green or dark, with or without net-venation — **4. repens***

1. G. colorata (Blume) Blume. Illustration: Blume, Flore Javae **4**: t. 9b, f. 2 (1858); La Belgique Horticole **12**: 1, f. 4 (1862); J.J. Smith, Orchids of Java **2**: f. 93 (1909). Stems *c.* 15 cm. Leaves *c.* 6, scattered on the lower part of the stem, *c.* 6 × 2.5 cm, blackish green with a red main vein, narrowly ovate, pointed, sheathing at base. Inflorescence *c.* 8 cm, hairy, with many flowers. Flowers *c.* 6 mm, sepals not spreading. Base of lip sac-like, hairy inside, the tip pointed and deflexed. *Indonesia (Java & Sumatra)*. G2.

2. G. pubescens (Willdenow) R. Brown. Illustration: Botanical Magazine, 2540 (1825); Flore des Serres **15**: t. 1555 (1862–65); Luer, Native orchids of the United States and Canada, 141 (1975); American Orchid Society Bulletin **47**: 400 (1978), **48**: 1115 (1979).
Rhizomes creeping, sometimes above ground. Stems to 50 cm, densely hairy. Leaves in rosettes, to 9 × 4 cm, oblong-elliptic, blunt, net-veined, veins white. Inflorescence cylindric, with up to 80 flowers, dense. Sepals *c.* 5 × 3 mm, ovate, bluntly pointed. Petals narrower, spathulate. Lip sac-like, the outer surface minutely warty, the tip pointed and recurved. Column with blunt rostellum. *Eastern USA*. H1. Summer.

3. G. oblongifolia Rafinesque (*G. decipiens* (Hooker) Hubbard; *G. menziesii* Lindley). Illustration: Luer, Native orchids of the United States and Canada, 143 (1975).
Rhizomes creeping, stoloniferous. Stems to 48 cm, densely hairy. Leaves to 11 × 4 cm, in rosettes, oblong-elliptic, blunt, mid-vein usually marked by a broad white line, the other veins forming a white, net-like pattern. Inflorescence 1-sided, flowers loose or dense, scented. Sepals 6–10 × 3–4 mm, oblong to narrowly ovate. Petals spathulate. Lip sac-like, its tip oblong, blunt, recurved. Column with pointed rostellum. *Western N America*. H2. Summer.

4. G. repens (Linnaeus) R. Brown. Illustration: Danesch & Danesch, Orchideen Europas, Mitteleuropa, edn 3, 201, 202 (1972); Landwehr, Wilde Orchideeën van Europa **2**: 541 (1977); Williams et al., Orchids of Britain and Europe, 137 (1978); Mossberg & Nilsson, Orchids of northern Europe, 52, 53 (1979).
Rhizome creeping, stoloniferous. Stems to 25 cm, glandular-hairy. Leaves 1.5–4 × 1–2 cm, ovate, blunt, forming a rosette, green or with a white, net-like pattern (var. **ophioides** Fernald). Inflorescence spirally arranged, or 1-sided, dense. Flowers glandular-hairy, cream-white, fragrant. Sepals 3–4.5 × 3 mm, ovate. Petals oblong-spathulate. Lip sac-like, its tip pointed, recurved. Column with a blunt rostellum. *Eurasia, N America*. H1. Summer.

***G. hispida** Lindley. Illustration: Annals of the Royal Botanic Garden Calcutta **8**: t. 375 (1898); Bechtel et al., Manual of cultivated orchid species 215 (1981). Leaves reddish, elliptic-lanceolate, with a pink or white net-like pattern, rostellum pointed. *NE India, Malaysia, Thailand*. G2. Summer.

10. MACODES Lindley

P.J.B. Woods

Terrestrial. Stems short, fleshy, rooting at basal nodes. Leaves few, mostly basal, somewhat fleshy, attractively coloured and veined, rolled when young. Inflorescence erect, flowers few to about 40, asymmetric due to the twisting in opposite directions of the lip and column. Lip uppermost, 3-lobed, its base concave and containing 2 glands.

Column with a single stigmatic area and with 2, thin, parallel wings on the front. Pollinia 4, club-like, made up of small, easily separable masses of pollen.

A genus of about 10 species, superficially similar to *Anoectochilus*, and distributed from Malaysia to New Guinea.

Cultivation as for tropical terrestrial species with the substitution of beech or oak leaf-mould for peat and bark and the addition of perlite to improve drainage. High humidity and protection from slugs are essential.

1. M. petola (Blume) Lindley. Illustration: Blume, Flore Javae **4**: t. 31 (1858); Orchid Digest **38**: 29 (1974); Bechtel et al., Manual of cultivated orchid species, 224 (1981).
Leaves *c.* 6 × 4 cm, ovate, velvety, dark green, the main veins and cross-veins beautifully marked with bright golden yellow. Inflorescence *c.* 20 cm, downy, with 10–40 flowers. Flowers reddish brown, lip white. Sepals, petals and lip *c.* 5 mm. *Malaysia, Philippines to Indonesia (Sumatra).* G2. Autumn–spring.

11. ANOECTOCHILUS Blume
P.J.B. Woods
Terrestrial. Stems short, basal part rooting at nodes, upper part erect, fleshy. Leaves few, mostly basal, thin, fleshy, with distinctively coloured veins, rolled when young. Inflorescence erect, flowers few, symmetric. Lip lowermost, the apex widening abruptly to a transverse blade, the claw fringed, the base with a spur containing 2 glands. Column with 2 wings on the front and 2 distinct stigmatic areas. Pollinia 2, club-like, made up of small, easily separable masses of pollen on a common stipe.

A genus of perhaps 40 species distributed from India to NE Australia. Cultivation as for *Macodes* (p. 161).

1. A. roxburghii (Wallich) Lindley. Illustration: Annals of the Royal Botanic Garden Calcutta **8**: t. 390 (1898); Botanical Magazine, 9529 (1938); Orchid Digest **39**: 132 (1975); Dansk Botanisk Arkiv **32**: 55 (1978).
Stems to 30 cm. Leaves 7 × 5.5 cm, ovate, velvety, brownish or pinkish green; veins marked in an attractive pink or yellowish net-like pattern, lower surface of leaf paler than upper. Inflorescence to 15-flowered, glandular-hairy. Sepals *c.* 1 cm, reddish or greenish red, reflexed; petals white or pinkish, forming a hood with the middle sepal, tips sickle-shaped. Lip *c.* 2 cm, inclined upwards. Spur bent abruptly away from ovary. *NE India to Vietnam.* G2. Autumn–spring.

***A. setaceus** (Blume) Lindley (*A. regalis* Blume). Illustration: Edwards's Botanical Register **23**: t. 2010 (1837); Botanical Magazine, 4123 (1844); Flore de Serres, t. 6 (1846). Very similar, but lip downward pointing and spur parallel to ovary. *Sri Lanka, Java.* G2. Spring.

12. OPHRYS Linnaeus
J. Lamond & P.J.B. Woods
Perennial herbs. Tubers 2 (occasionally 3), spherical, fleshy. Leaves lanceolate, oblong or ovate, rolled when young, mostly in a basal rosette, sometimes spaced along the stem, when usually sheathing, usually appearing in autumn after the current season's growth has died back. Flowers few to *c.* 15 in a short or long, loose or rarely dense spike. Bracts leaf-like, often inrolled. Sepals spreading, usually greenish or pinkish, oblong or ovate, obtuse, all equal in length, the upper concave, erect or curved over the column. Petals usually smaller and narrower than the sepals, often hairy or velvety. Lip spurless, hairy, complex in structure, often like a bee, wasp or other insect, oblong, square, rounded or diamond-shaped, usually marked with a coloured, hairless, often complex, patterned area (the speculum), lobed or not, with or without humps or horn-like basal protuberances; apex often with a short, tooth-like, deflexed or upcurved appendage. Column erect or curved forwards; anther-connective beak-like, its apex obtuse or pointed. Pollinia 2, club-like, made up of small, easily separable masses.

A genus of between 40 and 50 species, many of which are superficially similar, thus making identification difficult. They are distributed from NW Europe throughout the Mediterranean region to the Middle East. There is some degree of specificity between pollinators (mostly bees and wasps) and the species of *Ophrys*. The male insects are attracted and sexually stimulated by the scent, shape, colour and the tactile stimuli of the hairs on the lip of the orchid. Fuller descriptions of these complex flowers are given in Danesch & Danesch (cited below) and Wood, Orchid Review **89**: 298 (1981). Literature: Nelson, E., *Gestaltwandel und Abbildung erörtert am Beispiel der Orchidaceen Europas und der Mittelmeerländer mit einer Monographie und Ikonographie der Gattung Ophrys* (1962) – all the species included in the following account are illustrated in this work, to which no further references are made; Danesch, E. & Danesch, O., *Orchideen Europas, Ophrys-Hybriden* (1972).

Sepals. Pinkish or whitish: **6,7**.
Petals. Blackish or brownish purple: **1,2**; greenish or yellowish: **3–7**; brownish: **4,5**; pink or purplish: **6,7**. Hairy or velvety: **1,2,6,7**; hairless or papillose: **3–5**. Rounded: **2**; triangular: **6,7**; oblong to lanceolate: **2–7**; linear: **1,7**.
Lip. 3-lobed: **1–4,6,7**; entire: **5,6**. Mostly yellow: **3**. Densely fringed: **2**. Speculum bright, shiny blue: **2,4**; speculum H-shaped or of parallel lines: **5,7**. With an apical appendage: **5–7**; without an apical appendage: **1–5**.
Anther-connective. Obtuse: **1–4,6**; acute: **5–7**.

1a. Lip with a tooth-like appendage at the apex 2
b. Lip without a tooth-like appendage at the apex 4
2a. Speculum of 2 more or less parallel lines, or if H-shaped then petals hairless **5. sphegodes**
b. Speculum various, not as above 3
3a. Lip distinctly 3-lobed **7. apifera***
b. Lip not 3-lobed, or lobes very indistinct **6. holoserica***
4a. Lip 3-lobed, central lobe notched or not 5
b. Lip not lobed **5. sphegodes**
5a. Lip, at least at the margin, bright yellow and usually hairless **3. lutea**
b. Lip blue, violet, purple, reddish, brownish or black, lateral lobes occasionally yellow but then margins densely fringed 6
6a. Speculum bright, shining blue, covering almost the whole lip; margins of lip densely fringed with long hairs **2. vernixia**
b. Speculum blue or not, restricted to the basal half of the lip; margins of lip not fringed 7
7a. Petals slender, antenna-like, blackish purple; lip sharply cleft at the apex **1. insectifera**
b. Petals oblong, greenish; lip notched at the apex **4. fusca**

1. O. insectifera Linnaeus (*O. muscifera* Hudson). Illustration: Danesch & Danesch, Orchideen Europas, Südeuropa, 128 (1969); Danesch & Danesch, Orchideen Europas, Mitteleuropa, edn 3, 222, 223 (1972); Ross-Craig, Drawings of British plants, part 28: t. 9 (1971); Mossberg & Nilsson, Orchids of northern Europe, 130, 131 (1979).

Plant to 60 cm. Leaves linear to oblong. Spike with up to 10 flowers. Sepals 6–8 mm, oblong-ovate, green. Petals 4–6 mm, linear, antenna-like, blackish purple, velvety. Lip 9–10 × 6–7 mm, oblong, 3-lobed, central lobe cleft at apex, blackish purple or dark violet, papillose, lateral lobes oblong. Speculum square or kidney-shaped, pale bluish violet. Anther-connective obtuse. *Most of Europe, uncommon in the extreme south and north and absent in the southeast.* H2. Spring–summer.

2. O. **vernixia** Brotero (*O. speculum* Link). Illustration: Danesch & Danesch, Orchideen Europas, Südeuropa, 126 (1969); Sundermann, Europäische und mediterrane Orchideen, 82 (1975); Landwehr, Wilde Orchideeën van Europa **2**: 377, 379 (1977); Grey-Wilson & Mathew, Bulbs, t. 37 (1981).
Plant 7–30 cm. Leaves oblong to lanceolate. Spike with up to 15 flowers. Sepals 6–8 mm, oblong-ovate, greenish or yellowish green, sometimes lined with brown. Petals one-third to half the length of the sepals, lanceolate, ovate-lanceolate or rounded, dark purple, hairy. Lip 1.3–1.5 cm, 3-lobed, margins densely fringed, hairs brownish purple to blackish purple, yellowish or brownish red, central lobe ovate, usually notched; lateral lobes oblong, yellowish. Speculum large, shining blue, with a yellow margin. Anther-connective obtuse. *Portugal, Mediterranean region.* H3. Spring.

3. O. **lutea** (Gouan) Cavanilles. Illustration: Danesch & Danesch, Orchideen Europas, Südeuropa, 123, 124 (1969); Landwehr, Wilde Orchideeën van Europa **2**: 391 (1977); Williams et al., Orchids of Britain and Europe, 57 (1978); Grey-Wilson & Mathew, Bulbs, t. 37 (1981).
Plant 7–30 cm. Basal leaves ovate. Spike with up to 7 flowers. Sepals *c.* 1 cm, green. Petals one-third to half the length of the sepals, oblong, yellowish green, hairless or margins papillose. Lip 9–18 mm, oblong or rounded, 3-lobed, lateral lobes rounded, central lobe kidney-shaped, marginal zone broad, bright yellow, centre dark brown or blackish purple. Speculum oblong, 2-lobed, greyish or bluish grey. Anther-connective obtuse. *Portugal, Mediterranean region.* H3. Spring.

4. O. **fusca** Link. Illustration: Danesch & Danesch, Orchideen Europas, Südeuropa, 118 (1969); Landwehr, Wilde Orchideeën van Europa **2**: 383, 385, 387 (1977); Williams et al., Orchids of Britain and Europe, 55 (1978); Grey-Wilson & Mathew, Bulbs, t. 37 (1981).
Plant 10–40 cm. Leaves oblong or lanceolate. Spike with up to 8 flowers. Sepals 9–11 mm, greenish or yellowish green. Petals 6–8 mm, oblong, yellowish green or brownish green, indistinctly papillose. Lip 1.3–1.5 cm × 9–12 mm, oblong or obovate, purplish, purplish red or reddish brown, velvety, 3-lobed, lateral lobes not spreading, oblong, rounded, central lobe rounded, notched. Speculum oblong, usually divided down the middle, bluish, often with a paler or yellowish margin. Anther-connective obtuse. *Portugal, Mediterranean region, SW Romania.* H3. Spring.

Subsp. **iricolor** (Desfontaines) Swartz (*O. iricolor* Desfontaines). Illustration: Danesch & Danesch, Orchideen Europas, Südeuropa, 119 (1969); Sundermann, Europäische und mediterrane Orchideen, 84 (1975); Landwehr, Wilde Orchideeën van Europa **2**: 383 (1977); Williams et al., Orchids of Britain and Europe, 55 (1978). Spike with up to 4 flowers, petals often papillose, lip 2.5–3.1 cm, speculum bright blue. *C & E Mediterranean region.*

5. O. **sphegodes** Miller (*O. aranifera* Hudson). Illustration: Danesch & Danesch, Orchideen Europas, Südeuropa, 100–2 (1969); Ross-Craig, Drawings of British plants, part 28: t. 8 (1971); Landwehr, Wilde Orchideeën van Europa **2**: 397 (1977); Mossberg & Nilsson, Orchids of northern Europe, 132, 133 (1979); Orchid Review **89**: 295 (1981).
Plant 10–45 cm. Basal leaves ovate-lanceolate. Spike with up to 10 flowers. Sepals 6–12 mm, greenish. Petals 4–8 mm, hairless, oblong-lanceolate, greenish or occasionally brownish, margins sometimes wavy. Lip 1–1.2 cm × 8–12 mm, entire, rounded, pale or dark brown or blackish brown, velvety, with or without basal protuberances. Speculum usually H-shaped, the cross-line sometimes absent, greyish or purplish blue. Anther-connective acute. *Europe.* H3. Spring–summer.

A variable species (see Wood, J.J., The subspecies of O. sphegodes in Cyprus and the eastern Mediterranean, Orchid Review **89**: 292–9, 1981). Subsp. **litigiosa** (Camus) Becherer (*O. litigiosa* Camus) has the lip 5–10 × 5–10 mm, pale or yellowish brown, usually without basal protuberances. Subsp. **mammosa** (Desfontaines) Soó (*O. mammosa* Desfontaines) has the lip 8–17 × 8–17 mm, blackish brown to blackish purple, usually with large basal protuberances. Both of these subspecies are from southern Europe. The specific name is sometimes misspelled 'sphecodes'.

6. O. **holoserica** (Burmann) Greuter (*O. fuciflora* (Schmidt) Moench). Illustration: Danesch & Danesch, Orchideen Europas, Südeuropa, 79a, b, c (1969); Danesch & Danesch, Orchideen Europas. Mitteleuropa, edn 3, 224–7 (1972); Ross-Craig, Drawings of British plants, part 28: t. 7 (1971): Landwehr, Wilde Orchideeën van Europa **2**: 449, 461, 469 (1977); Mossberg & Nilsson, Orchids of northern Europe, 134, 135 (1979); Grey-Wilson & Mathew, Bulbs, t. 39 (1981).
Plant 15–55 cm. Basal leaves ovate-oblong. Spike with 2–6 flowers (occasionally more). Sepals 9–13 mm, pinkish, whitish or greenish. Petals about one-quarter to one-third of the length of the sepals, triangular to oblong-lanceolate, pink or purplish, hairy. Lip 9–13 mm, ovate or almost square, entire or rarely indistinctly 3-lobed, brownish or brownish purple, velvety; central part papillose, basal protuberances distinct, apical appendage upcurved, 3-toothed. Speculum purplish or bluish grey bordered by a bold pattern of vertical and horizontal bands of pale green or yellow. Anther-connective acute. *W, SW & C Europe.* H3. Spring–summer.

*O. **tenthredinifera** Willdenow. Illustration: Danesch & Danesch, Orchideen Europas, Südeuropa, 92, 93 (1969); Sundermann, Europäische und mediterrane Orchideen, 98 (1975); Landwehr, Wilde Orchideeën van Europa **2**: 469, 471 (1977); Williams et al., Orchids of Britain and Europe, 43 (1978); Grey-Wilson & Mathew, Bulbs, t. 39 (1981). Similar, but petals broadly triangular, lip almost square with a notched apex, dark brown, marginal area usually yellowish; speculum violet or bluish with a pale greenish border, pattern usually simple; anther-connective obtuse. *Portugal, Mediterranean region.* H3. Spring.

7. O. **apifera** Hudson. Illustration: Danesch & Danesch, Orchideen Europas, Südeuropa, 94 (1969); Danesch & Danesch, Orchideen Europas, Mitteleuropa, edn 3, 228–33 (1972); Ross-Craig, Drawings of British plants, part 28: t. 6 (1971); Quarterly Bulletin of the Alpine Garden Society **44**: 59 (1976); Mossberg & Nilsson, Orchids of northern Europe, 136, 137 (1979).
Plants 15–50 cm. Basal leaves lanceolate to ovate. Spike with up to 10 flowers. Sepals 1–1.5 cm, pink or purplish, rarely whitish. Petals less than half as long as sepals, triangular to linear-lanceolate, greenish or

rarely purplish, hairy. Lip 1–1.3 cm, ovate, 3-lobed, brownish or brownish purple; lateral lobes triangular, deflexed; basal protuberances *c.* 3 mm, apical appendage longer than wide, yellow, deflexed. Speculum H-shaped or shield- or W-shaped, violet-purple with a pale green margin, sometimes indistinct. Anther-connective pointed. *W, S & C Europe.* H2. Spring–summer.

*__O. scolopax__ Cavanilles. Illustration: Danesch & Danesch, Orchideen Europas, Südeuropa, 72–4 (1969); Landwehr, Wilde Orchideeën van Europa **2**: 437, 438 (1977); Williams et al., Orchids of Britain and Europe, 35 (1978); Grey-Wilson & Mathew, Bulbs, t. 39 (1981). Sepals 8–10 mm, the upper erect. Lip 8–12 mm, longer than wide, oblong-elliptic, with a narrow, hairless or papillose marginal zone, basal protuberances short, obtuse, apical appendage short, wider than long. Speculum often X-shaped. *Portugal, W & C Mediterranean region.* H3. Spring.

A very variable species; several of its subspecies are occasionally grown in specialist collections, e.g. subsp. **cornuta** (Steven) Camus (*O. oestrifera* Bieberstein), which has a lip with a wide, hairless marginal zone and horn-like basal protuberances to 10 mm, and is found in the eastern Mediterranean region, the Caucasus and Iran; subsp. **heldreichii** (Schlechter) Nelson (*O. oestrifera* subsp. *heldreichii* (Schlechter) Soó) with lip 1.3–1.5 cm, also from the eastern Mediterranean region; and subsp. **orientalis** (Renz) Nelson (*O. umbilicata* Desfontaines), with sepals pale greenish, whitish or rarely reddish pink, the central hooded over the column, from Cyprus, Turkey, Syria and the Lebanon.

13. SERAPIAS Linnaeus

J. Lamond & P.J.B. Woods

Terrestrial herbs. Roots tuberous, tubers 2–4, occasionally 5, stalkless and in some species also produced at the end of root-like stolons, fleshy, ovoid. Stems erect. Leaves lanceolate, mostly basal or spaced along the stem, rolled when young, later often folded. Aerial growth dying off during the current year. Spike loose- or dense-flowered; bracts very conspicuous, boat-shaped, usually purplish or glaucous. Sepals and petals forming a pointed hood, usually coloured like the bracts, veins distinct; sepals lanceolate; petals ovate at base, tapering to a long point. Lip 3-lobed, composed of a basal and an apical part, usually hairy on the upper surface, lateral lobes upturned, central lobe usually abruptly bent downwards, tongue-like, the basal part with a swollen area or a double ridge at the base. Pollinia 2, club-shaped, consisting of easily separable masses. Column apex beak-like.

A genus of about 6 species between which the differences are not always clear, distributed from the Azores through the Mediterranean region to the Caucasus. Alpine house or cold-frame protection is necessary for plants grown in northern Europe. Pot culture and a compost incorporating coarse sand and limestone chips is recommended. New growth is produced during the autumn when water must be given sparingly and increased only when the plants are growing vigorously. As the growth dies down and the plants become dormant they require a warm, dry rest.

Literature: Nelson, E., *Monographie und Ikonographie der Orchidaceen Gattungen Serapias, Aceras, Loroglossum, Barlia* 3–45 (1968); Christian, P., The genus Serapias in the wild and in cultivation, *Quarterly Bulletin of the Alpine Garden Society* **43**: 188–98 (1975).

1a. Lip with a single, black, hump-like swelling at base **1. lingua**
b. Lip with 2 dark or pale ridges at base 2
2a. Apical part of lip about as broad as the basal part when flattened **2. cordigera***
b. Apical part of lip narrower than the basal part when flattened **3. vomeracea***

1. S. lingua Linnaeus. Illustration: Danesch & Danesch, Orchideen Europas, Südeuropa, 196, 197 (1969); Sundermann, Europäische und mediterrane Orchideen, 116 (1975); Landwehr, Wilde Orchideeën van Europa **2**: 355 (1977); Williams et al., Orchids of Britain and Europe, 67 (1978); Grey-Wilson & Mathew, Bulbs, t. 40 (1981).

Plant to *c.* 30 cm. Tubers 2–5, 1 stalkless. Leaves 3–6, to *c.* 15 cm × 5–12 mm, linear to lanceolate, sheathing bases usually unspotted. Spike with 2–5 (sometimes as many as 9) flowers, loose. Bracts *c.* 3–4 cm, about as long as the flowers. Hood *c.* 2 cm. Lateral lobes of lip purple; apical part to *c.* 1.9 cm × 7 mm, broadly lanceolate to narrowly ovate, hairless or shortly hairy, reddish, pinkish, violet or yellowish. *Portugal, N Africa, Mediterranean regions of S Europe east to Greece.* H5. Spring.

2. S. cordigera Linnaeus. Illustration: Danesch & Danesch, Orchideen Europas, Südeuropa, 192, 193 (1969); Landwehr, Wilde Orchideeën van Europa **2**: 369 (1977); Williams et al., Orchids of Britain and Europe, 65 (1978); Grey-Wilson & Mathew, Bulbs, t. 40 (1981).

Plant to 40 cm. Tubers 2–3, 1 stalkless. Leaves 3–8, to *c.* 15 × 1–1.5 cm, lanceolate, sheathing bases usually red-spotted. Spike 2–10-flowered, dense. Bracts *c.* 3.4–4.5 cm, shorter or scarcely longer than flowers. Hood *c.* 2.5 cm. Basal ridges of lip diverging, lateral lobes dark purple or reddish; apical part 2–3 × 1.6–2 cm, heart-shaped, very hairy, purplish to maroon. *Spain, Portugal, Mediterranean region to W Turkey.* H5. Spring.

*__S. neglecta__ de Notaris. Illustration: Danesch & Danesch, Orchideen Europas, Südeuropa, 189–91 (1969); Landwehr, Wilde Orchideeën van Europa **2**: 363 (1977); Williams et al., Orchids of Britain and Europe, 63 (1978); Grey-Wilson & Mathew, Bulbs, t. 40 (1981). Plant usually 15–25 cm, sturdy; sheathing leaf-bases unspotted; spike dense; bracts greenish. Basal ridges of lip parallel; apical part heart-shaped, pale yellowish or orange. *C Mediterranean regions of Europe.* H5. Spring.

3. S. vomeracea (Burman) Briquet (*S. longipetala* (Tenore) Pollini; *S. pseudocordigera* Moricand). Illustration: Danesch & Danesch, Orchideen Europas, Südeuropa, 185 (1969); Sundermann, Europäische und mediterrane Orchideen, 113 (1975); Landwehr, Wilde Orchideeën van Europa **2**: 357, 359 (1977); Grey-Wilson & Mathew, Bulbs, t. 40 (1981).

Plant 10–60 cm. Tubers 2, stalkless. Leaves 6–8, *c.* 6–11 cm × 5–12 mm, linear to lanceolate, sheathing bases green, unspotted. Spike to 10-flowered, loose. Bracts to 5 cm, longer than the flowers. Lip reddish or brownish, distinctly hairy, basal ridges parallel, lateral lobes often blackish; apical part *c.* 1.5–2 cm × 6–10 mm, lanceolate. *S Europe.* H5. Spring.

Subsp. **orientalis** Greuter (*S. orientalis* invalid). Illustration: Landwehr, Wilde Orchideeën van Europa **2**: 367 (1977); Williams et al., Orchids of Britain and Europe, 61 (1978). Plant 15–35 cm. Spike 5–7-flowered, dense. Bracts scarcely as long as or slightly longer than flowers, purplish. Apical part of lip broadly lanceolate.

Although usually treated separately, we find it almost impossible to distinguish between this subspecies and *S. neglecta* but have followed recent accounts in placing it with *S. vomeracea*.

***S. olbia** Verguin (*S. gregaria* Godfery). Illustration: Danesch & Danesch, Orchideen Europas, Südeuropa, 195 (1969); Sundermann, Europäische und mediterrane Orchideen, 116 (1975); Landwehr, Wilde Orchideeën van Europa **2**: 365, t. 3, 373 (1977); Williams et al., Orchids of Britain and Europe, 67 (1978). Plant slender. Tubers 3–4, 1 stalkless. Spike few flowered. *Southeast France*. H5. Spring.

Thought to be a natural hybrid between *S. lingua* and *S. parviflora*.

14. ACERAS R. Brown
P.J.B. Woods
Herbaceous perennial. Aerial growth usually dying down in winter or plant sometimes overwintering as a basal rosette. Stems to *c*. 50 cm arising from 2 ovoid tubers. Leaves unspotted, the lower crowded, spreading, 5–12 × *c*. 2.5 cm, lanceolate, obtuse, the upper smaller, erect, sheathing the stem. Spike to 20 cm, bearing 50 or more flowers. Sepals obovate-lanceolate, 6–7 mm, petals linear, slightly shorter, all yellowish or greenish yellow, hooded over the column. Lip *c*. 1.2 cm, shaped like a man, the lateral lobes being the arms and the divided central lobe the legs. Spur absent. Pollinia 2, club-shaped, composed of easily separable masses.

A genus of a single species rarely seen in cultivation. It is similar to some species of *Orchis*, differing in the lack of a spur. In nature it occurs on chalk or limestone. Cultivation as for *Orchis*.

1. A. anthropophorum (Linnaeus) Aiton. Illustration: Danesch & Danesch, Orchideen Europas, Mitteleuropa, edn 3, 131 (1972); Danesch & Danesch, Orchideen Europas, Südeuropa, 240 (1967); Williams et al., Orchids of Britain and Europe, 67 (1979); Mossberg & Nilsson, Wilde Orchideeën van Europa **2**: t. 153 (1977).
W Europe, Mediterranean region. H5. Summer.

15. HIMANTOGLOSSUM Koch
P.J.B. Woods
Herbaceous perennials with 2 ovoid tubers. Aerial growth dying down in winter. Leaves in a basal rosette and arranged along the stem. Spike elongate, bracts equalling the flowers. Sepals and petals free, forming a hood over the column. Lip long, strap-shaped, 3-lobed, the central lobe bifid and spirally twisted, 2 or more times longer than the lateral lobes. A single spur present, arising from the base of the lip. Pollinia 2, club-shaped, composed of easily separable masses, attached to a single viscidium.

A genus of 3 to 5 species from Europe and SW Asia not often found in cultivation. Cultivation as for *Orchis*.

1. H. hircinum (Linnaeus) Sprengel. Illustration: Danesch & Danesch, Orchideen Europas, Mitteleuropa, edn 3, 130 (1972); Danesch & Danesch, Orchideen Europas, Südeuropa, 241 (1967); Williams et al., Orchids of Britain and Europe, 69 (1978); Mossberg & Nilsson, Orchids of northern Europe, 125 (1979).
Stem to 90 cm, purplish-blotched. Lower leaves 4–6, 5–51 × 3–5 cm, elliptic-oblong, obtuse, the upper leaves smaller, acute, their bases clasping the stem. Spike 10–50 cm, cylindric, with 15–18 loosely arranged flowers which emit a goat-like smell. Bracts shorter than the flowers. Sepals and petals pale green, sometimes purplish-lined, hooded, ovate, obtuse, the petals linear, 7–10 mm. Lip 3–5 cm × 2 mm, lateral lobes strap-shaped, central lobe ribbon-like, coiled in bud, spirally twisted when open, apex bifid, pale green suffused with purple, purple-spotted at base. Spur 2–2.5 mm, conical, downwardly pointing. *W, C & S Europe*. H4. Summer.

16. BARLIA Parlatore
P.J.B. Woods
Very similar to *Himantoglossum* but bracts as long as or longer than the flowers, the sepals erect, not hooded, and the lip much shorter, its central lobe not spirally twisted.

A genus of a single species, requiring cultural conditions similar to those for *Orchis*.

1. B. robertiana (Loiseleur) Greuter (*Himantolossum longibracteatum* (Bernardi) Schlechter). Illustration: Danesch & Danesch, Orchideen Europas, Südeuropa, 242, 243 (1967); Sundermann, Europäische und mediterrane Orchideen, 122 (1975); Landwehr, Wilde Orchideeën van Europa **2**: 345 (1977); Williams et al., Orchids of Britain and Europe, 97 (1978).
Mediterranean region, Canary Islands. H5. Summer.

17. ANACAMPTIS Richard
P.J.B. Woods
Perennial herbs. Tubers spherical or ovoid. Stems to 75 cm with up to 8 leaves. Leaves 15–25 × *c*. 2 cm, mostly basal, unspotted, lanceolate, upper leaves shorter, rolled when young. Aerial growth dying down in winter. Spike to 8 cm, conical, becoming shortly cylindric, dense. Bracts linear, slightly longer than ovary. Flowers pink, red or purple, occasionally white, emitting a faint fox-like smell. Sepals and petals 6–8 × *c*. 3 cm, upper sepal and petals incurved, lateral sepals spreading. Lip deeply 3-lobed, 6–9 × 8–10 mm, as broad as long or broader than long, lobes oblong, *c*. 4 mm, apex obtuse or truncate; base of lip with 2 erect ridges leading towards the mouth of the spur. Spur 1–1.4 cm, slender, downward pointing. Pollinia 2, club-like, made up of easily separated masses.

A genus of a single species, in nature occurring in open grassland and calcareous dunes. Although occurring naturally in northern Europe and therefore probably hardy enough for normal outdoor culture, the species is perhaps better suited for pot-culture as recommended for *Orchis*.

1. A. pyramidalis Richard. Illustration: Danesch & Danesch, Orchideen Europas, Mitteleuropa, edn 3, 146, 147 (1972); Landwehr, Wilde Orchideeën van Europa **2**: 341, 343 (1977); Mossberg & Nilsson, Orchids of northern Europe, 123 (1979); Williams et al., Orchids of Britain and Europe, 81 (1978).
Europe, Mediterranean region, SW Asia. H1. Summer.

18. ORCHIS Linnaeus
P.J.B. Woods
Perennial herbs. Aerial growth dying down in winter. Tubers 2–3, spherical or ovoid. Leaves mostly in a basal rosette, spotted or unspotted, stem-leaves usually sheath-like. Flowers in a short or long, dense- or loose-flowered spike. Bracts membranous. Sepals and petals almost equal, free, all curved forward over the column, or the lateral sepals spreading and the upper sepal and petals incurved and hooded over the column. Lip mostly 3-lobed, sometimes unlobed. Spur 8–25 mm, arising from the base of the lip. Pollinia 2, club-like, powdery.

A genus of about 35 species from Europe to Asia. *Dactylorhiza* (p. 167) is often included under *Orchis* and most nurserymen still offer plants under the latter name.

Most species, apart from those with northerly distributions, are best suited for growing in pots in a cold-frame or unheated greenhouse. The compost should be extremely well drained and kept moist only while the plants are in active growth. The tubers should be repotted annually and kept almost dry while dormant.

Leaves. Spotted: **1–2**; unspotted: **1–7**.
Bracts. Usually equal to ovary: **1–3**; shorter: **2,6,7**; longer: **1,2,4,5**.
Flower colour. Deep pink, purple, red or brownish: **1–7**; yellow: **1**; greenish: **3,5**; white: **1–3,6,7**.
Spur. Length 2 mm: **6**; 5–12 mm: **1–5,7**; 15–25 mm: **2,3**. Erect: **1–3**; horizontal: **1–3**; downwardly pointing: **2,4–7**.
Lateral sepals. Spreading: **1,2**; hooded: **3–7**.
Lip. Unlobed: **1–4**; lobed: **1–3,5**.

1a. Lateral sepals spreading, only the petals and sometimes the upper sepal hooded over column 2
b. Sepals and petals all hooded or forwardly directed over column 3
2a. Spur to 1.2 cm, cylindric, not narrowed towards apex; or flowers yellow **1. mascula***
b. Spur 1.5 cm or more, slender, narrowed towards apex; flowers not yellow **2. anatolica***
3a. Lip unlobed, ovate to fan-shaped **4. papilionacea**
b. Lip lobed 4
4a. Spur pointing upwards, or occasionally horizontal **3. morio***
b. Spur pointing downwards 5
5a. Central lobe of lip entire, at most indistinctly spotted **5. coriophora**
b. Central lobe of lip divided or notched, distinctly spotted 6
6a. Sepals and petals much darker than lip; central lobe of lip divided into smaller lobes which are about as long as broad **6. purpurea***
b. Sepals and petals not darker than lip; central lobe of lip divided into smaller lobes which are twice or more longer than broad **7. militaris***

1. O. mascula Linnaeus. Illustration: Sundermann, Europäische und mediterrane Orchideen, 146 (1975); Landwehr, Wilde Orchideeën van Europa **1**: 261 (1977); Williams et al., Orchids of Britain and Europe, 101 (1978).
Stem to 60 cm. Leaves 3–11, 5–20 × 1.5–4.5 cm, lanceolate or oblanceolate to narrowly obovate, purplish-spotted or unspotted, mostly basal, upper leaves sheathing stem. Spikes to 32 cm, few to many, loosely or densely flowered; flowers purple, lilac, or sometimes white. Bract about as long as the ovary. Upper sepal to 10 mm, oblong, curved forwards; lateral sepals to 1.3 cm, obliquely oblong or ovate, erect or reflexed. Petals to 9 × 7 mm, obliquely ovate, hooded (with the upper sepal) over the column. Lip to 1.5 cm, 3-lobed or sometimes scarcely lobed, edges usually folded back, centre finely papillose, with purple lines or spots. Spur to 1.2 cm, about equal to ovary, cylindric, horizontal or erect. *Europe to W Iran.* H1. Spring.

***O. provincialis** Balbis. Illustration: Danesch & Danesch, Orchideen Europas, Südeuropa, 210, 211 (1967); Landwehr, Wilde Orchideeën van Europa **1**: 279 (1977); Williams et al., Orchids of Britain and Europe, 103 (1978). Flowers bright or pale yellow. *S Europe, N Africa to E Mediterranean region.* H5. Spring.

***O. quadripunctata** Tenore. Illustration: Danesch & Danesch, Orchideen Europas, Südeuropa, 206–8 (1967); Sundermann, Europäische und mediterrane Orchideen, 144 (1975); Landwehr, Wilde Orchideeën van Europa **2**: 297 (1977); Williams et al., Orchids of Britain and Europe, 99 (1978). Spur to 1.2 cm, curved downwards, uniformly narrow; lip base with 2–4 small purple spots. *E Mediterranean region.* H5. Spring.

2. O. anatolica Boissier. Illustration: Sundermann, Europäische und mediterrane Orchideen, 146 (1975); Landwehr, Wilde Orchideeën van Europa **1**: 283 (1977); Williams et al., Orchids of Britain and Europe, 99 (1978).
Stem to 40 cm. Leaves 3–8, to 13 × 2.5 cm, lanceolate to broadly lanceolate, unspotted, or blotched or spotted with purple, basal; stem with 1–2 sheath-like leaves. Spike to 15 cm, loose, with few to many flowers, flowers purplish to pale pink, sometimes white. Bract as long as or shorter than ovary. Sepals *c.* 10 × 4 mm, oblong to narrowly ovate; upper sepal slightly broader, hooded (with the slightly smaller petals) over the column. Lip to 1.4 cm, 3-lobed, edges often folded back, centre white with purple spots or lines. Spur 1.5–2.5 cm, longer than ovary, slender, wider at mouth, narrow at apex, upwardly curved or horizontal. *E Mediterranean region to W Iran.* H5. Summer.

***O. joo-iokiana** (Makino) Maekawa (*Ponerorchis joo-iokiana* (Makino) Maekawa). Illustration: Kitamura et al., Coloured illustrations of herbaceous plants of Japan **17**: t. 2 (1964); Maekawa, The wild orchids of Japan in colour, t. 19 (1971). Leaves 1–4, not in a basal rosette, flowers mostly on one side of the inflorescence; bracts longer than ovary. *Japan.* H5. Summer.

3. O. morio Linnaeus. Illustration: Danesch & Danesch, Orchideen Europas, Südeuropa, 222 (1967); Landwehr, Wilde Orchideeën van Europa **1**: 245, 247 (1977); Williams et al., Orchids of Britain and Europe, 93 (1978).
Stem to 40 cm. Leaves to 10, basal, to 16 × *c.* 1.3 cm, lanceolate to broadly oblong; 1–2 sheath-like leaves on stem. Spike 3–13 cm, loose or dense, with few to many flowers; flowers reddish purple, mauve, or sometimes greenish or white. Bract about as long as ovary. Sepals 7–12 × 4–6 mm, oblong, apex rounded, often greenish- or purplish-veined, lateral sepals a little longer. Petals to 7 × 3 mm, oblong. Lip *c.* 10 × 10 mm, often broader than long, edges often folded back. Spur *c.* 8 mm, shorter than or as long as lip, cylindric, horizontal or upwardly curved. *Europe to W Iran.* H1. Spring.

***O. longicornu** Poiret. Illustration: Danesch & Danesch, Orchideen Europas, Südeuropa, 225 (1967); Landwehr, Wilde Orchideeën van Europa **1**: 251 (1977); Williams et al., Orchids of Britain and Europe, 95 (1978). Spur to 1.6 cm, much longer than lip. *W Mediterranean region.* H5. Spring.

4. O. papilionacea Linnaeus. Illustration: Danesch & Danesch, Orchideen Europas, Südeuropa, 220 (1967); Sundermann, Europäische und mediterrane Orchideen, 138 (1975); Landwehr, Wilde Orchideeën van Europa **1**: 233 (1977).
Stem to 40 cm. Leaves to 10, to *c.* 10 × 1 cm, lanceolate, mostly basal. Spike ovoid with up to 15 flowers: flowers purplish, red or brownish. Bract longer than ovary. Sepals and petals 1–1.8 cm, narrowly ovate, strongly veined, upper sepal and petals forming a hood. Lip 1.2–2.5 × 1.2–2.5 cm, entire, fan-shaped or ovate, distinctly lined, margin shallowly toothed. Spur *c.* 1.2 cm, shorter than ovary, cylindric, downwardly curved. *S Europe to SW Asia.* H5. Spring.

5. O. coriophora Linnaeus. Illustration: Danesch & Danesch, Orchideen Europas, Südeuropa, 226 (1967); Landwehr, Wilde Orchideeën van Europa **2**: 301, 303 (1977); Williams et al., Orchids of Britain and Europe, 83 (1978).
Stem to *c.* 50 cm. Leaves to 7, *c.* 15 × 2 cm, linear to lanceolate, erect, on basal part of stem, often turning brown before completion of flowering. Spike 3–12 cm,

ovoid to cylindric; flowers loose or dense, red, brownish or purplish, sometimes greenish. Bract a little longer than ovary. Sepals and petals hooded, tips free, sepals to 10 × 4 mm, ovate to lanceolate, the laterals wider than the upper; petals to 6 × 1–2 mm. Lip 5–11 mm, longer than broad, 3-lobed. Spur 5–9 mm, shorter than or equalling ovary, conical or cylindric, downwardly curved. *S, C & E Europe*. H2. Spring.

6. O. purpurea Hudson. Illustration: Danesch & Danesch, Orchideen Europas, Mitteleuropa, edn 3, 114–17 (1972); Landwehr, Wilde Orchideeën van Europa **2**: 325 (1977); Mossberg & Nilsson, Orchids of northern Europe, 115 (1979).
Stem to 80 cm or more, robust. Leaves to 6 or more, to 23 × 6 cm, broadly lanceolate to elliptic, shining, unspotted, forming a basal rosette, 1–2 leaves sheathing the lower part of the stem. Spike to 17 × 6 cm, broadly cylindric, many-flowered; flowers brownish or reddish purple, rarely white, lip white or pinkish, its edges darker, with purple spots. Bract to 5 mm, much shorter than ovary. Sepals to 1.4 cm, ovate, acute. Petals shorter, linear-lanceolate. Lip to 1.5 cm, longer than broad, 4-lobed; lateral lobes oblong, central lobe divided into 2 smaller, oblong lobes which are about as long as broad and between which is a small tooth. Spur *c*. 5 mm, less than half the length of ovary, cylindric, downwardly curved. *Europe to N Africa*. H1. Spring–summer.

***O. ustulata** Linnaeus. Illustration: Danesch & Danesch, Orchideen Europas, Südeuropa, 230 (1967); Landwehr, Wilde Orchideeën van Europa **2**: 311 (1977); Mossberg & Nilsson, Orchids of northern Europe, 113 (1979). Stem to 25 cm. Sepals to 3.5 mm; lip with no tooth between the lobes of middle lobe; spur 2 mm. *Europe to the Caucasus*. H1. Spring–summer.

7. O. militaris Linnaeus. Illustration: Danesch & Danesch, Orchideen Europas, Mitteleuropa, edn. 3, 119–24 (1972); Landwehr, Wilde Orchideeën van Europa **2**: 319 (1977); Williams et al., Orchids of Britain and Europe, 89 (1978).
Stem to 45 cm, robust. Leaves 3–5, to 13 × 4 cm, broadly elliptic to oblanceolate, obtuse, unspotted, basal; 1–2 leaves sheathing lower part of stem. Spike to 10 × 4.5 cm, broadly cylindric, many-flowered; flowers whitish pink, rarely white, veins darker, lip purplish, white towards base. Bracts *c*. 3 mm, much shorter than ovary. Sepals to 1.5 cm, ovate-lanceolate, acute. Petals shorter, linear. Lip to 1.5 cm, longer than broad, 4-lobed; lateral lobes linear, central lobe divided into 2, almost transverse, oblong lobes which are about twice as long as broad and between which is a small tooth. Spur to 7 mm, about half the length of the ovary, cylindric, downwardly curved. *C Europe to the Caucasus*. H1. Spring–summer.

***O. italica** Poiret. Illustration: Sundermann, Europäische und mediterrane Orchideen, 134 (1975); Landwehr, Wilde Orchideeën van Europe **2**: 321 (1977); Williams et al., Orchids of Britain and Europe, 91 (1978). Leaves 5–8, margins wavy. Lobes of central lobe of lip linear, 4 or more times longer than broad. *Portugal & Mediterranean region*. H5. Spring.

19. DACTYLORHIZA Nevski

J.C.M. Alexander

Terrestrial. Root-tubers palmately lobed or divided into finger-like sections. Aerial parts dying down in winter. Leaves 3–15, spirally arranged on flowering stem, not forming a basal rosette at flowering time, sometimes spotted or blotched with brownish purple; upper leaves often small and bract-like. Flowers several to many, in compact spherical, conical or cylindric racemes. Bracts leaf-like, green or purplish green. Sepals and petals free; lateral sepals spreading or erect; upper sepal and petals curved forwards forming a hood over the column. Lip simple or 3-lobed, spurred, usually marked with lines, dots or streaks. Column short; rostellum 3-lobed.

A temperate genus of 20–30 species from Europe, N Africa, Asia and N America. They are generally hardy and can be cultivated in full sun out-of-doors in a peat-bed. Cultivation conditions for the more southerly species are similar to those for *Serapias* and the less hardy species of *Ophrys* (see p. 162). The species comprising *Dactylorhiza* were long considered to be part of *Orchis*, and were separated as the genus *Dactylorchis* by Vermeulen in 1947 who was presumably unaware of Nevski's earlier name.

Hybridisation is rife in *Dactylorhiza* and the resulting crosses are often as fertile and plentiful in the wild as their parents. In addition, many frequently used characters are very variable. As a result the delimitation of species and other categories is not well defined and the classification is confused. The absence or presence of spots on the leaves is not a reliable character when attempting to identify wild populations but should suffice for the species in cultivation.
Literature: Nelson, E., *Monographie und Ikonographie der Orchidaceen-Gattung Dactylorhiza* (1976).

Height. Over 70 cm: **1,2**.
Stem. Hollow: **1–3**; solid: **4**.
Leaves. Yellowish green: **1,2**; mid to dark green: **1,3,4**. Spotted: **3,4**; unspotted: **1,2**. Hooded at tip: **2,4**; not hooded at tip: **1,3,4**. Not bract-like near top of stem: **2**. Overtopping inflorescence: **2**.
Bracts. Longer than flowers: **1–3**; equal to or shorter than flowers: **1,4**.
Inflorescence. More than 15 cm long: **1**.
Flowers. White: **4**; pink: **1,2,4**; purplish pink: **1,3**; crimson, brownish pink or brick-red: **2**; yellow: **2**.
Lip. More than 1.2 cm wide: **1**.
Spur. 1.2–1.5 cm: **1,2**. Conical: **1–3**; cylindric: **2–4**.

1a. Leaves mid or dark green, spotted or blotched with brownish purple 2
b. Leaves bright yellowish green, unspotted 3
2a. Stem hollow; spur conical to cylindric **3. majalis**
b. Stem solid; spur cylindric **4. maculata***
3a. Lip more than 11 mm wide; leaves 8–14 **1. foliosa***
b. Lip less than 10 mm wide; leaves 4–7 **2. incarnata***

1. D. foliosa (Solander) Soó (*Orchis maderensis* Summerhayes). Illustration: Botanical Magazine, 5074 (1858); Landwehr, Wilde Orchideeën van Europa **2**: 219 (1977); Williams et al., Orchids of Britain and Europe, 119 (1978).
Stem to 70 cm, hollow, Leaves 8–14, to 20 cm, narrowly lanceolate to triangular, keeled, acute, stiff, ascending or arched, yellowish green, unspotted; upper leaves bract-like. Bracts equal to or shorter than the flowers and scarcely protruding from the inflorescence. Inflorescence 5–13 cm, cylindric, densely many-flowered. Flowers bright purplish pink. Sepals and petals 1–1.3 cm, lanceolate to ovate, rounded; lateral sepals a little broader. Lip 1–1.1 × 1.2–1.3 cm, always wider than long, transversely oblong or elliptic, shallowly 3-lobed, marked with lines and streaks; central lobe triangular. Spur 6–8 mm, narrowly conical, almost straight. *Madeira*. H5–G1. Spring–summer.

***D. elata** (Poiret) Soó. Illustration: Landwehr, Wilde Orchideeën van Europa **2**: 205–13 (1977); Williams et al., Orchids of Britain and Europe, 112 (1978); Grey-

Wilson & Mathew, Bulbs, f. 44 (1981). To 110 cm. Bracts longer than flowers and protruding from inflorescence. Inflorescence 15–25 cm. Flowers pink to purple. Spur 1.2–1.5 cm. *SW Europe*. H5–G1. Spring–summer.

2. D. incarnata (Linnaeus) Soó. Illustration: Landwehr, Wilde Orchideeën van Europa **2**: 127–47 (1977); Williams et al., Orchids of Britain and Europe, 111 (1978); Grey-Wilson & Mathew, Bulbs, f. 44 (1981). Stem to 80 cm, hollow. Leaves 4–7, 10–20 cm, narrowly triangular to lanceolate, keeled, stiff, hooded at tip, ascending or arched, bright yellowish green, unspotted, often overtopping the inflorescence. Upper 1 or 2 leaves bract-like. Lower bracts longer than flowers and usually protruding from inflorescence. Inflorescence 4–10 cm, conical to cylindric, densely many-flowered. Flowers pale to bright pink or brick-red. Sepals and petals 5–8 mm, lanceolate to ovate, acute or obtuse, usually erect or ascending. Lip 5–7 × 5–9 mm, diamond-shaped, sometimes 3-lobed, marked with hieroglyphic-like lines of darker pink. Spur 5–8 mm, broadly conical, slightly curved. *Europe & W Asia, uncommon in the south*. H2. Summer.

***D. sambucina** (Linnaeus) Soó. Illustration: Landwehr, Wilde Orchideeën van Europa **2**: 59–61 (1977); Williams et al., Orchids of Britain and Europe, 109 (1978); Grey-Wilson & Mathew, Bulbs, f. 43 (1981). To 30 cm. Leaves in lower half of the stem only, all much larger than bracts, never overtopping inflorescence. Flowers yellow, crimson or reddish brown, in small, spherical to ovoid heads. Spur 1.2–1.5 cm, equal to or longer than ovary, cylindric, obtuse. *Europe, N Africa & Turkey*. H5–G1. Summer.

3. D. majalis (Reichenbach) Hunt & Summerhayes (*D. latifolia* (Linnaeus) Soó – confused name). Illustration: Landwehr, Wilde Orchideeën van Europa **2**: 173–81 (1977); Williams et al., Orchids of Britain and Europe, 113 (1978); Grey-Wilson & Mathew, Bulbs, f. 44 (1981).
Stem to 60 cm, hollow. Leaves 4–8, to 15 × 2.5 cm, broadly lanceolate to elliptic or oblong, acute, not hooded at tip, mid to dark green with irregular brownish purple spots and blotches; sometimes 1 or 2 upper leaves bract-like. Lower bracts usually longer than flowers. Inflorescence 6–12 cm, cylindric, dense, many-flowered; flowers dark purplish pink. Sepals and petals 6–12 mm, lanceolate to oblong, obtuse, ascending or erect. Lip 6–9 × 9–12 mm, circular, diamond-shaped or transversely elliptic, simple or 3-lobed, with broad lateral lobes and a short, triangular central lobe, marked with lines and dots or dots alone. Spur 7–10 mm, conical to cylindric, equal to or shorter than ovary. *Europe (except extreme south), N Russia, Turkey & the Baltic*. H2. Summer.

4. D. maculata (Linnaeus) Soó (*Orchis ericetorum* (Linden) Marshall; *O. elodes* Grisebach; *D. maculata* (Linnaeus) Soó subsp. *ericetorum* (Linden) Hunt & Summerhayes). Illustration: Landwehr, Wilde Orchideeën van Europa **2**: 73–5, 95–7 (1977); Williams et al., Orchids of Britain and Europe, 121 (1978); Grey-Wilson & Mathew, Bulbs, f. 44 (1981). Stem to 40 cm, solid. Leaves 5–12, to 20 × 3 cm, linear-lanceolate to oblong, acute to rounded, sometimes slightly hooded at tip, mid to dark green with irregular spots and blotches of brownish purple; some small upper leaves bract-like. Bracts about equal to flowers and not protruding from inflorescence. Inflorescence 5–10 cm, conical, ovoid or cylindric, dense, many-flowered; flowers white to pink. Sepals and petals 7–11 mm, narrowly lanceolate, obtuse, usually horizontal. Lip 7–8 × 10–11 mm, diamond-shaped to transversely elliptic, usually marked with darker pink dots and short lines, shallowly 3-lobed; central lobe shorter than lateral lobes. Spur 4–8 mm, cylindric, equal to or shorter than ovary. *Europe, N Africa & W Asia*. H2. Summer.

***D. fuchsii** (Druce) Soó (*D. maculata* subsp. *fuchsii* (Druce) Hylander). Illustration: Landwehr, Wilde Orchideeën van Europa **2**: 103–5, 107–9 (1977); Williams et al., Orchids of Britain and Europe, 121 (1978); Grey-Wilson & Mathew, Bulbs, f. 44 (1981). Very similar to *D. maculata* and often regarded as a subspecies of it. Lip more deeply 3-lobed, marked with hieroglyphic-like lines; central lobe narrow and usually longer than lateral lobes. *Europe (except extreme south) & W Asia*. H2. Summer.

20. NIGRITELLA Richard

P.J.B. Woods

Herbaceous, terrestrial perennial. Tubers 2, palmately lobed. Stems to 28 cm. Leaves 6–11, crowded, to 11 cm × 1–9 mm, linear to narrowly lanceolate, channelled, mostly basal, unspotted. Aerial growth dying down in winter. Spike 1–3 cm × 7–25 mm, conical becoming ovoid or elongate, dense. Bracts slender, as long as or longer than flowers. Flowers *c*. 5 mm, blackish crimson, red, yellowish or rarely white, vanilla-scented; sepals and petals linear to lanceolate; tips acute. Lip uppermost, triangular to lanceolate-ovate, entire. Spur *c*. 2 mm, conical, shorter than ovary. Pollinia 2, granular.

A genus of 1 species, cultivated as *Dactylorhiza* (p. 167).

1. N. nigra (Linnaeus) Reichenbach. Illustration: Sundermann, Europäische und mediterrane Orchideen, 178 (1975); Landwehr, Wilde Orchideeën van Europa **1**: 229, 231 (1977); Williams et al., Orchids of Britain and Europe, 123 (1978). *Scandinavia to N Spain & Greece*. H1. Summer.

21. GYMNADENIA R. Brown

P.J.B. Woods

Herbaceous, terrestrial perennials. Tubers palmately lobed. Leaves linear to lanceolate, arranged along lower part of stem, unspotted. Aerial growth dying down in winter. Flowers in a cylindric spike, sweetly scented. Upper sepal and petals incurved to form a hood over the column; lateral sepals spreading almost horizontally. Lip shallowly 3-lobed. Spur slender, shorter or longer than ovary. Pollinia 2, club-like, powdery.

A genus of perhaps 10 species from Europe and temperate Asia. Cultivation as for *Orchis* (p. 165).

1a. Spur much longer than ovary; leaves oblong to lanceolate **1. conopsea**
b. Spur not longer than ovary; leaves linear **2. odoratissima**

1. G. conopsea (Linnaeus) R. Brown. Illustration: Landwehr, Wilde Orchideeën van Europa **1**: 221, 223 (1977); Williams et al., Orchids of Britain and Europe, 125 (1978); Mossberg & Nilsson, Orchids of northern Europe, 81 (1979).
Stems to 50 cm. Leaves to 7, arising along the lower third of the stem, to 25 cm × 6–35 mm, oblong to lanceolate. Spike to 25 cm, flowers crowded, rose or purple, sometimes white. Bracts lanceolate, as long as or longer than ovary. Sepals and petals 4–7 mm, upper sepal and petals elliptic to ovate, lateral sepals oblong, obtuse, spreading, margins slightly reflexed. Lip 5–7 mm, 3-lobed, lobes almost equal. Spur to 2 cm, about twice as long as ovary, downwardly curved. *Eurasia*. H1. Spring–summer.

2. G. odoratissima (Linnaeus) Richard.

Illustration: Danesch & Danesch, Orchideen Europas, Mitteleuropa, edn 3, 150, 151 (1972); Landwehr, Wilde Orchideeën van Europa 1: 223, 225 (1977); Mossberg & Nilsson, Orchids of northern Europe, 83 (1979).
Stems to 40 cm. Leaves to 7, mostly basal, to 18 × 1.6 cm, linear, slightly infolded. Spike to *c.* 10 cm, flowers crowded, pale pink, purple, sometimes white. Bracts lanceolate, longer than ovary. Sepals and petals to 3 mm, upper sepal and petals elliptic to ovate, lateral sepals oblong, obtuse, spreading. Lip *c.* 3 mm, 3-lobed, lateral lobes shorter than central lobe. Spur to 5 mm, as long as or shorter than ovary, horizontal or downwardly curved. *Sweden to N Spain, N Italy & W Russia.* H1. Spring–summer.

22. COELOGLOSSUM Hartmann
P.J.B. Woods
Herbaceous, terrestrial perennials with 2 ovoid tubers. Leaves unspotted, the lower ovate or oblong, obtuse, the upper smaller, lanceolate and pointed. Aerial growth dying down in winter. Flowers slightly scented, 5–25 in a loose cylindric spike. Sepals and petals hooded over the column, greenish or yellow-green often tinged reddish, sepals ovate, petals relatively narrower, almost linear. Lip strap-shaped, the apex shortly 2-lobed with or without a short middle tooth, with a short, rounded spur. Pollinia 2, club-shaped, attached to separate viscidia.

A north temperate genus of 2 species. Cultivation as for *Dactylorhiza* (p. 167); in nature the 1 cultivated species occurs in calcareous grassland and at woodland edges.

1. C. viride (Linnaeus) Hartmann.
Illustration: Danesch & Danesch, Orchideen Europas, Mitteleuropa, edn 3, 138 (1972); Landwehr, Wilde Orchideeën van Europa 1: 227 (1977); Williams et al., Orchids of Britain and Europe, 127 (1978); Mossberg & Nilsson, Orchids of northern Europe, 74 (1979).
Stems 10–35 cm. Leaves 2–6, 3–10 cm. Spike 2–10 cm. Sepals and petals 4.5–6 mm. Lip 3.5–6 mm. *Eurasia, N America.* H1. Summer.

23. PLATANTHERA Richard
P.J.B. Woods
Herbaceous, terrestrial perennials. Tubers 2 or more, spindle-shaped, elongate. Stems erect, unbranched. Leaves basal or on the stem. Aerial growth dying down in winter. Raceme usually with many small or medium-sized, white or greenish flowers. Sepals and petals free, upper sepal and petals usually incurved over the column, lateral sepals spreading or recurved. Lip entire with the base extended into a spur. Anthers distinctly separated. Pollinia club-shaped, granular.

A genus of between 80 and 100 species mostly from temperate areas of Europe, Asia and N & S America. Cultivation as for *Dactylorhiza* (p. 167).

1a. Anther cells distinctly parallel, *c.* 2 mm apart at the base; spur 1.5–2.5 cm **1. bifolia**
b. Anther cells distinctly divergent towards the base, where they are *c.* 4 mm apart; spur 1.5–3.5 cm **2. chlorantha**

1. P. bifolia (Linnaeus) Richard.
Illustration: Sundermann, Europäische und mediterrane Orchideen, 182 (1975); Williams et al., Orchids of Britain and Europe, 129 (1978); Mossberg & Nilsson, Orchids of northern Europe, 68–70 (1979).
Stems to 50 cm. Tubers 2, ovoid, the tips narrow and elongate. Leaves 2 (rarely 3), basal, almost opposite, 7–25 × 2.5–8 cm, circular or elliptic or oblong-lanceolate, narrowing into a stalk; stem-leaves small and bract-like. Raceme 2.5–3.8 cm wide, cylindric, loose or dense, with many flowers; bracts lanceolate, as long as or slightly longer than the ovaries. Flowers 1.2–2 cm wide, white, scented. Lip 6–15 mm, strap-shaped, directed downwards, white, greenish towards the tip. Spur 1.5–2.5 cm, narrowly cylindric, slightly inflated, greenish towards the apex, often straight. Anther cells parallel, *c.* 2 mm apart. *Europe, N Africa, SW Asia.* H1. Summer.

2. P. chlorantha (Custer) Reichenbach.
Illustration: Sundermann, Europäische und mediterrane Orchideen, 182 (1975); Landwehr, Wilde Orchideeën van Europa 1: 31–3 (1977); Mossberg & Nilsson, Orchids of northern Europe, 70, 71 (1979); Williams et al., Orchids of Britain and Europe, 129 (1978).
Stems to 60 cm. Tubers 2, ovoid, the tips narrow, elongate. Leaves 2 (rarely 3), basal, almost opposite, 6–20 × 2.5–8 cm, oblong to oblong-lanceolate or obovate, narrowing into a stalk; stem-leaves small and bract-like. Raceme 3–4.5 cm wide, cylindric, loose or dense, with many flowers; bracts lanceolate, the lower somewhat longer than the ovaries. Flowers to 2.5 cm wide, greenish white, scented. Lip 8–18 mm, strap-shaped, directed downwards, greenish, the tip darker. Spur 1.5–3.5 cm, thickened towards the tip, greenish, darker at the tip. Anther cells divergent towards the base and *c.* 4 mm apart there. *Europe, SW Asia.* H1. Summer.

24. HABENARIA Willdenow
P.J.B. Woods
Terrestrial herbs. Roots tuberous, fleshy, elongate or ovoid. Stems erect. Leaves spaced along stem or sometimes basal, linear to lanceolate, sheathing at base, rolled when young. Aerial growth dying down in winter. Raceme few- to many-flowered. Upper sepal often forming a hood with the petals; lateral sepals spreading (in *H. bonatea* the basal parts of the sepals, petal lobes, column and lip are united). Lip variously lobed or fringed, base spurred, spur slender, long. Column short or long, slender or thick. Pollinia 2, club-like, breaking up into easily separable masses. Stigmas 2, borne on club-like arms, or stalkless.

The genus contains between 600 and 800 species distributed throughout temperate N America and east Asia and the subtropics and tropics of the Old and New World. Generic limits within this group are somewhat confused; Luer, for instance, place all the North American species under *Platanthera.* For the purpose of this Flora, *Bonatea* Willdenow and *Pecteilis* Rafinesque have been included in *Habenaria.* Most species require greenhouse or alpine-house protection and are best grown in terrestrial compost or in one containing peat, leaf-mould, loam and sand in equal parts. Adequate water is necessary while the plants are actively growing, with cool, almost dry conditions when dormant.
Literature: Luer, C., *Native orchids of the United States and Canada,* 175–242 (1975); Seidenfaden, G., Orchid genera in Thailand V, *Dansk Botanisk Arkiv* **31**: 22–6, 65–137 (1977).

1a. Lobes of lip entire **2. carnea***
b. Lobes of lip fringed, laciniate or thread-like 2
2a. Petals 2-partite, the lobes linear; lateral lobes of lip linear to thread-like **1. bonatea**
b. Petals entire, oblong or ovate, at most finely toothed at apex; lateral lobes of lip fringed or laciniate **3. radiata***

1. H. bonatea Reichenbach (*Bonatea speciosa* (Linnaeus) Willdenow). Illustration: Botanical Magazine, 2926 (1829); Bolus,

Orchids of South Africa **2**: t. 44, 45 (1911); Schelpe, An introduction to the South African orchids, 75 (1966); South African Orchid Journal: cover (June 1971); American Orchid Society Bulletin **47**: 994 (1978).
Plant 30–80 cm. Leaves many, often crowded along stem, to *c.* 15 × 5 cm, ovate-oblong or oblong, sometimes with purplish marks, obtuse. Raceme dense, bracts *c.* 3 cm, ovate-acuminate. Flowers to 15 or more, scented, greenish, the lip and stigmatic processes white tipped with green. Sepals ovate, 1.7–2.2 cm, lateral sepals longer than the upper. Petals 2-partite, the upper segment as long as the upper sepal, sickle-shaped; the lower segment as long as or longer than the lateral sepal, linear or filiform. Lip 3–3.5 cm, 3-lobed, the lobes thread-like, curved or bent; spur 3–3.5 cm, toothed at the mouth, thickened towards tip, greenish. *South Africa.* H5–G1. Spring–summer.

2. H. carnea R. Brown. Illustration: Gardeners' Chronicle **2**: 729 (1891); The Garden **47**: facing 182 (1895); Holttum, Flora of Malaya, edn 3, **1**: 86 (1964); Dansk Botanisk Arkiv **31**: 136 (1977).
Plant 15–30 cm. Leaves 4–6, to 10 × 4 cm, lanceolate, mostly basal, olive green with pale spots. Raceme 4–15-flowered; flowers pink or reddish. Sepals and petals *c.* 1 cm, lateral sepals broader and longer than the upper; petals broad, ovate, veins several. Lip 3-lobed, *c.* 3 × 2.5 cm, sometimes as wide as long, lobes oblong or ovate, central lobe notched, surface and margins finely papillose-ciliate. Spur to 6 cm. *Malay Peninsula (Langkawi Islands), Thailand.* G2. Summer–autumn.

*__H. rhodocheila__ Hance (*H. militaris* Reichenbach). Illustration: Botanical Magazine, 7571 (1897); American Orchid Society Bulletin **48**: 366, 367 (1979); Dansk Botanisk Arkiv **31**: 135 (1977); Bechtel et al., Manual of cultivated orchid species, 216 (1981). Sepals and petals green. Petals slender, oblong-spathulate, 1-veined. Lip yellowish orange or bright red, margins not papillose-ciliate. *China, Thailand, Malay Peninsula, Laos, Vietnam, Philippines.* G2. Summer–autumn.

3. H. radiata Thunberg (*Pecteilis radiata* (Thunberg) Rafinesque). Illustration: Australian Orchid Review **33**: 5 (1968); American Orchid Society Bulletin **44**: 809 (1975); Maekawa, The wild orchids of Japan in colour, 90 (1971); Bechtel et al., Manual of cultivated orchid species, 257 (1981).
Plant *c.* 30 cm. Leaves 3–7, to 10 cm × *c.* 6 mm, linear, acute, the lowermost and uppermost shortest. Inflorescence 1–2-flowered, Sepals and petals *c.* 1 cm, ovate, margins of petals finely and irregularly toothed. Lip *c.* 1.5 × 2.5 cm or larger, 3-lobed; lateral lobes obovate, their outer margins laciniate; central lobe linear or oblong, shorter than lateral lobes, margins entire. *Korea, Japan.* H5–G1. Summer.

*__H. ciliaris__ (Linnaeus) R. Brown (*Blephariglottis ciliaris* (Linnaeus) Linnaeus; *Platanthera ciliaris* (Linnaeus) Lindley). Illustration: Botanical Magazine, 1668 (1814); Shuttleworth et al., Orchids, 25 (1970); Luer, Native orchids of the United States and Canada, 181 (1975); Die Orchidee **27**: 155 (1976), **32**: 103 (1981). Leaves lanceolate to oblong. Flowers orange; lip oblong, unlobed. *Eastern USA.* H1. Summer.

*__H. psycodes__ (Linnaeus) Sprengel (*Platanthera psycodes* (Linnaeus) Lindley). Illustration: American Orchid Society Bulletin **39**: 789 (1970); Luer, Native orchids of the United States and Canada, 196, 199 (1975); Die Orchidee **27**: 154, 155 (1976). Leaves oblong to obovate. Flowers purplish, occasionally white; lip 3-lobed, lobes obovate or fan-shaped. *SE Canada, Northeastern USA.* H1. Summer.

25. STENOGLOTTIS Lindley

P.J.B. Woods

Herbaceous perennials, terrestrial, growing on rocks or epiphytic. Roots several, tuberous, oblong, fleshy. Leaves numerous, mostly basal and forming a rosette, membranous, rolled when young, usually dying back after flowering. Inflorescence erect, many-flowered, bearing several sheaths, upper ones smaller, not clasping stem. Basal part of sepals slightly joined to the lip and column. Petals often slightly broader than sepals, hooded over the column. Lip joined to the column, longer than sepals and petals, oblong, widening towards apex which is 3- to 5-, occasionally 7-lobed or toothed. Spur absent (in ours). Column short. Pollinia 2, club-like, attached to a short stalk, easily separable into distinct masses.

A genus of 3 species confined to east and southern Africa. Cultivation in terrestrial compost. Plants should be given moist, shady conditions while growing vigorously.

1a. Flowers to *c.* 20, arranged mostly to 1 side of inflorescence; lip 3-, occasionally 4-toothed **1. fimbriata**
b. Flowers to *c.* 100, evenly arranged in a cylindric inflorescence; lip 5-, occasionally 7-toothed **2. longifolia**

1. S. fimbriata Lindley. Illustration: Botanical Magazine, 5872 (1870); Flowering Plants of South Africa **15**: t. 585 (1935); American Orchid Society Bulletin **48**: 159 (1979); Stewart & Hennessy, Orchids of Africa, 51 (1981).
Leaves 5–12, to 12 × *c.* 2 cm, lanceolate or oblanceolate, arranged in a basal rosette, purplish-spotted or not, margins often wavy. Raceme loose, 1-sided, to 38 cm. Flowers few to *c.* 20, pale pink to mauve, lip paler with darker spots. Sepals 3–5 mm. Petals slightly broader than sepals. Lip 5–9 mm, its apical third 3- or occasionally 4-lobed or toothed; lobes almost equal, 2–4 mm. The ovary, sepals and petals are covered with small, crystal-like papillae. *Tropical E Africa to S Africa.* G1. Autumn.

2. S. longifolia Hooker. Illustration: Botanical Magazine, 7186 (1891); Flowering Plants of South Africa **24**: t. 933 (1944); Orchid Review **87**: 62 (1979); Bechtel et al., Manual of cultivated orchid species, 275 (1981).
Leaves many, to 18 × 2.5 cm, lanceolate, gradually narrowed at either end, tips slightly recurved, occasionally blackish-spotted, margins often wavy. Flowers to *c.* 100, crowded in a cylindric spike to 70 cm, pale pink, occasionally white, with scattered, dark marks or dots. Sepals *c.* 8 mm. Petals smaller than sepals. Lip to *c.* 1.2 cm, its apical part 5-, occasionally 7-toothed, teeth unequal, to *c.* 6 mm. *S Africa.* G1. Autumn.

26. DISA Bergius

P.J.B. Woods

Herbaceous, terrestrial perennials, occasionally evergreen, with tuberous roots, sometimes stoloniferous. Sterile shoots usually rosette-like. Leaves basal or on the stem, ovate to linear, mostly lanceolate. Inflorescence a dense or loose raceme or corymb. Flowers usually with lip lowermost (not obvious in single-flowered or corymbose inflorescences), less commonly with lip uppermost. Median sepal concave and usually hood-like, and with a basal spur, lateral sepals spreading. Petals smaller and narrower than sepals, often partially attached to the column. Lip small and narrow. Column short; pollinia 2, granular.

A genus of over 100 species from tropical and southern Africa and Madagascar. Cultivation can be difficult, and full details are given in the papers by Vogelpoel

and Stoutamire cited below.
Literature: Linder, H.P., A revision of Disa excluding Sect. Micranthae, *Contributions from the Bolus Herbarium* **9**: 1–370 (1981); Vogelpoel, L., Disa uniflora – its propagation and cultivation, *American Orchid Society Bulletin* **49**: 961–72 (1980); Vogelpoel, L., Disa species and their hybrids, *American Orchid Society Bulletin* **49**: 1084–92 (1980); Stoutamire, W., Cultivated Disas in Ohio, *American Orchid Society Bulletin* **50**: 1195–200 (1981).

1. D. uniflora Bergius (*D. grandiflora* Linnaeus). Illustration: Botanical Magazine 4073 (1844); Orchid Review **78**: 95 (1970), **85**: 229 (1977), **86**: 127 (1978); Bechtel et al., Manual of cultivated orchid species, 204 (1981).
Stems to 60 cm. Leaves of sterile shoots in a loose basal rosette; leaves on flowering stems to 8, 15 × 1.25 cm, lanceolate, acuminate, the upper 3–4 smaller, sheathing. Inflorescence with 1–3 (rarely to 10) flowers, bracts longer than ovaries. Flowers 8–12 cm wide, lip lowermost, faintly scented, scarlet or carmine, orange to yellowish inside (pure white or yellow forms are occasionally recorded). Upper sepal *c.* 5 × 3 cm, ovate, the hooded basal part forming a narrow conical, laterally flattened spur, 1–1.5 cm long. Lateral sepals broadly elliptic, acuminate. Petals *c.* 2 cm × 8 mm, narrowly obovate, erect. *South Africa (Cape Province).* G1. Winter.

27. PTEROSTYLIS R. Brown
J. Lamond & P.J.B. Woods
Terrestrial herbs. Roots thread-like and fleshy with ovoid or spherical tubers. Stems erect. Leaves deciduous, mostly basal and forming a flat rosette, with 1–2 bract-like stem-leaves, or spaced along stem and linear to lanceolate; rolled when young. Flowers 1 or few, erect or pendent. Upper sepal and petals forming a hood, erect or curved forwards; lateral sepals joined along their basal inner margins, the tips tail-like or filament-like, diverging widely, erect or deflexed. Lip linear or oblong, apart from the tip which is mostly hidden inside hood; base with a small, erect appendage variously divided at its tip. The lip is sensitive to touch so that a small insect alighting at the tip is projected upwards against the column, where it is trapped between column and lip; in squeezing its way out it detaches the pollinia. The lip gradually returns to its original position in 0.5–2 hours. Column slender, variously winged towards apex and extended into a short foot at base. Pollinia 4, powdery.

A genus of about 95 species distributed from Australia to New Zealand and extending to New Guinea and New Caledonia. A compost of loam, sphagnum moss, peat and sand in equal parts is adequate for all species; it should be kept dry and frost-free while the plants are dormant.

1a. Leaves linear or lanceolate, more than 5 times longer than broad, evenly spaced on the stem **1. banksii**
b. Leaves oblong to elliptic, less than 5 times longer than broad, mostly in a basal rosette **2. curta***

1. P. banksii Hooker. Illustration: Botanical Magazine, 3172 (1832); Orchadian **1**: 129 (1965); Salmon, New Zealand flowers and plants in colour, edn 2, 44 (1967); Cooper, A field guide to New Zealand native orchids, 60 (1981).
Plant robust, to 35 cm. Leaves to 6, to 25 × 1.5 cm, pointed, closely spaced along stem and extending above flower. Flower solitary, to 7 cm long, erect, pale green with darker stripes; tips of sepals and petals pale orange to pink. Tips of lateral sepals filament-like, bent around hood and diverging with age; tips of upper sepal and petals curving forwards and downwards. Lip with greenish margins and reddish central part. *New Zealand.* H5–G1. Summer–winter.

2. P. curta R. Brown. Illustration: Rupp, Orchids of New South Wales, edn 2, 81 (1969); Nicholls, Orchids of Australia, t. 301 (1969); American Orchid Society Bulletin **41**: 804, 805, (1972); Pocock, Ground orchids of Australia, t. 113 (1972).
Plant robust, to 30 cm. Leaves usually 4, blade to 4 × 2 cm, ovate to elliptic, margins often wavy; stem-leaves 1–2, bract-like. Flowers usually solitary, *c.* 3 cm, erect, green flushed with brown. Tips of lateral sepals tailed, not extending beyond hood. Lip *c.* 2 cm, oblong-linear, characteristically twisted above the middle, reddish brown. *Australia, New Caledonia.* H5–G1. Summer–autumn.

***P. pedunculata** R. Brown. Illustration: Australian Orchid Review **34**: 142 (1969); Nicholls, Orchids of Australia, t. 309 (1969): Pocock, Ground orchids of Australia, t. 125 (1972). Flower to 2 cm, erect, green flushed with reddish brown or dark brown. Lateral sepals filament-like, extending beyond the hood. Lip *c.* 5 mm, ovate, not twisted, dark reddish brown. *Australia.* H5–G1. Winter–spring.

***P. nutans** R. Brown. Illustration: Botanical Magazine, 3085 (1831); Australian Orchid Review **34**: 142 (1969); Nicholls, Orchids of Australia, t. 292 (1969); Pocock, Ground orchids of Australia, t. 123 (1972); American Orchid Society Bulletin **44**: 992 (1975). Flower to 3 cm, conspicuously nodding, green, tips of sepals and petals pale brownish. *Australia & North Island of New Zealand.* H5–G1. Summer–autumn.

28. SATYRIUM Swartz
P.J.B. Woods
Herbaceous, terrestrial perennials. Roots tuberous, ovoid. Stems unbranched, leafy. Aerial growth dying down in winter. Flowers several to many, in racemes or spikes, not resupinate, variously coloured. Lip uppermost, forming a hood. Spurs 2, attached to base of lip. Pollinia 2, granular.

A genus of perhaps 100 species, mainly from Africa, though 2 occur in Asia. Crossland (Quarterly Bulletin of the Alpine Garden Society **46**: 301, 1978) recommends 'a compost providing humus and drainage . . . , that being used being one part each of loam, leaf mould, peat and sand. A position in semi-shade where the plant can remain cool and moist throughout growth, and dry but not arid when dormant is most suitable'. Annual potting should be done before growth commences. Apparently plants may remain dormant over several years.

1. S. nepalensis Don. Illustration: Annals of the Royal Botanic Garden Calcutta **8**: t. 443, 444 (1898); Orchid Review **85**: 149 (1977); Norwegian Journal of Botany **26**: 286 (1979).
Tubers ovoid, produced singly during the flowering season at the ends of vigorous roots. Stem 10–75 cm, bearing 2–3 leaves. Leaves 7–25 × 3–9 cm, sheathing, oblong-ovate to lanceolate, obtuse, the upper shorter. Spike cylindric, 5–15 cm, flowers numerous, pink or white, scented, bracts longer than the flowers, deflexed. Sepals and petals to 8 mm, linear-oblong, pointed, slightly recurved, petals narrower than sepals (which have hairy margins in var. **ciliata** (Lindley) King & Pantling). Lip to 8 mm, as broad as or broader than long, concave, keeled outside, the base with 2 spurs which are 8–12 mm. *N India to Sri Lanka & Burma.* H5–G1. Summer–autumn.

Plants offered under this name in the trade have occasionally proved to be *Herminium angustifolium, Habenaria arietina* or *Platanthera clavigera.*

29. VANILLA Miller
J.C.M. Alexander
Perennial branched climbers with long green stems, initially terrestrial, often becoming epiphytic, bearing a leaf and 1 or 2 roots at each node. Leaves alternate, fleshy or leathery, not sheathing, sometimes reduced to scales, folded when young. Flowers in axillary spikes or racemes. Sepals and petals similar, fleshy, spreading; petals with distinct midrib. Lip with a narrow claw fused to the lower part of the column and an entire or obscurely lobed blade enclosing the upper part of the column. Pollinia 2, powdery-granular. Fruit a fleshy pod.

Between 70 and 100 species distributed throughout the tropics. *V. planifolia* is the source of natural vanilla essence and is widely cultivated in the range of the genus. The long climbing stems of *Vanilla* species must be firmly supported, and can be attached to bark at various points along their length. Much of their moisture is obtained from the atmosphere so the pots, containing epiphytic compost, can be quite small. The plants should be frequently syringed, even in winter. Propagation is effected by cutting up the stem into small sections which can be planted.

1. V. planifolia Andrews (*Myrobroma fragrans* Salisbury; *V. fragrans* Ames). Illustration: Botanical Magazine, 7167 (1891); Shuttleworth et al., Orchids, 30 (1970); Orchid Digest **41**: 181 (1977); Luer, Native orchids of Florida, 75 (1971).
Stem fleshy, terete. Leaves 15–20 × 3–5 cm, fleshy, ovate-elliptic to oblong, acute. Flowers 10–15 per spike, pale green or greenish yellow. Bracts 7 mm, triangular to ovate. Sepals and petals to 6 × 1 cm, linear-oblong to narrowly obovate. Lip to 4.5 × 2 cm; blade triangular with longitudinal papillose veins in the mid-line and a blunt, reflexed apical lobe. Column 3.5 cm, slender, downy. Ovary 4 cm, curved, maturing into a black fleshy cylindrical pod 15–25 cm. *From C America and the West Indies south to Bolivia, Paraguay & N Argentina; naturalised in Madagascar, the Seychelles and elsewhere.* G2. Intermittent flowering.

***V. pompona** Schiede. Illustration: Bechtel et al., Manual of cultivated orchid species, 280 (1981). Similar to *V. planifolia* but generally more robust. Sepals and petals 8 cm, greenish yellow. Lip yellow or orange. Pod 3-sided. *From C America and the Lesser Antilles south through Peru & W Brazil to Bolivia.* G2.

***V. aphylla** Blume. Stem somewhat flattened. Leaves reduced to small green triangular scales. Flowers 3 or 4 together; sepals and petals pale green; lip white with pink hairs on the central lobe. *Burma (Tenasserim), N Malay peninsula & Java.* G2.

30. BLETILLA Reichenbach
J.C.M. Alexander
Terrestrial, with corm-like pseudobulbs. Stems erect, produced laterally from the bases of old pseudobulbs. Leaves alternate, sheathing, thin and pleated, rolled in bud. Flowers several, in loose terminal racemes. Petals and sepals similar, free, pointing forward around lip. Lip 3-lobed; lateral lobes curled around column; central lobe with warts or ridges. Column free, long and narrow. Pollinia 8.

A genus of about 10 species from E Asia, named for its apparent similarity to the genus *Bletia* from C & S America with which it is often confused. Species of *Bletilla* have sometimes been classified under *Limodorum, Arethusa* and *Cymbidium*. Pseudobulbs should be planted in terrestrial compost and kept in a cold-frame. In mild areas, plants will flourish in the open. Very little water should be given in winter.

1. B. striata (Thunberg) Reichenbach (*B. hyacinthina* (Smith) Pfitzer). Illustration: Botanical Magazine, 1492 (1812); Hay & Synge, Dictionary of garden plants, 126 (1969); Bechtel et al., Manual of cultivated orchid species, 171 (1981).
Plant to 60 cm. Pseudobulbs horizontally flattened. Leaves 3–6, deciduous, 30–50 × 4–7 cm, elliptic, acute. Flowers 3–10 in a loose raceme, magenta. Sepals 2–3 cm × 5–10 mm, elliptic, acute; petals similar but broader. Lip 2.5–3 cm; central lobe longitudinally ridged, shallowly notched at apex. Column 2.5–3 cm, narrowly winged, purple at tip. *China & Japan.* H5–G1. Spring–summer.

***B. sinensis** (Rolfe) Schlechter. Differs in its smaller, rose-red flowers with darker tips to the sepals, petals and lip. *W China (Yunnan).* H5–G1. Early summer.

31. BOTHRIOCHILUS Lemaire
J.C.M. Alexander
Epiphytic or terrestrial, often on rocks; rhizomes creeping. Pseudobulbs spherical to ovoid, smooth, simple, bearing several leaves. Leaves linear to lanceolate, acute, pleated, rolled when young. Flowers few to many, in robust lateral racemes or spikes, on sheathed stems. Upper sepal free; lateral sepals fused to column foot, forming a mentum. Petals free, lip sharply bent or pouched at junction of blade and claw; blade shallowly 3-lobed; central lobe triangular, acute. Column long and slender; column foot equalling it in length. Pollinia 8, waxy.

A genus of 4 species from C America frequently considered part of *Coelia* from which it is distinguished by the length of the column foot, the presence of a distinct mentum and the complicated lip. Cultivation requirements are similar to those for *Lycaste* (p. 227), though *B. machrostachyus*, having fairly hard pseudobulbs, benefits from a longer dry period.
Literature: Williams, L.O., The orchid genera Coelia & Bothriochilus, *Botanical Museum Leaflets, Harvard University* **8**: 145–8 (1940).

1a. Pseudobulbs 5 cm or less; flowers few, about 5 cm long **1. bellus**
b. Pseudobulbs 6 cm or more; flowers many, 1–1.5 cm long **2. macrostachyus**

1. B. bellus Lemaire (*Coelia bella* (Lemaire) Reichenbach). Illustration: Botanical Magazine, 6628 (1882); Orchid Review **82**: 314 (1974).
Terrestrial, to 1 m. Pseudobulbs 3–5 × 1.5–2.5 cm, spherical to ovoid, surmounted by a short stalk, bearing 3–5 leaves. Leaves 15–60 × 1–2 cm, linear-lanceolate, acuminate, with 3–5 veins, pale green. Flowering stems to 20 cm, heavily sheathed, bearing 2–6 fragrant flowers; flowers 5 × 3.5–5 cm, funnel-shaped. Sepals elliptic to oblong, obtuse, off-white with purple tips; upper sepal 3.5 × 1 cm, lateral sepals 5 × 1 cm, mentum 2 cm. Petals 3.5 × 1 cm, oblong to spathulate, obtuse or rounded, off-white with a central patch of purple. Lip pouched; blade 4.5 × 1.5 cm with small lateral lobes; central lobe narrowly triangular, inner face an orange, granular callus. Column 1.5 cm. *Mexico, Guatemala & Honduras.* G2. Summer.

2. B. macrostachyus (Lindley) Williams (*Coelia macrostachya* Lindley). Illustration: Botanical Magazine, 4712 (1853); Shuttleworth et al., Orchids, 71 (1970); Ospina & Dressler, Orquideas de las Americas, t. 40 (1974).
Epiphytic, to 1 m. Pseudobulbs 6–10 cm, spherical to ovoid, surmounted by a short stalk, bearing 3–6 leaves. Leaves to 100 × 2.5–4 cm, linear-lanceolate, acuminate, fleshy. Flowering stems to

60 cm, sheathed. Flowers many in a dense cylindric spike, 10–15 × 5–7 mm, pale to dark pink or purple. Sepals fleshy, warty on outer face; upper sepal 8–10 mm, ovate-lanceolate; outer part of lateral sepals 1.5 cm, semi-ovate to triangular, inner part forming a mentum. Petals 9 mm, oblong-lanceolate, asymmetric, membranous. Lip pouched; blade 12 mm, shallowly 3-lobed; central lobe narrowly triangular, acute. Column 1.5 cm. *Mexico to Panama*. G2. Summer.

32. CALANTHE R. Brown
J. Cullen
Usually terrestrial. Pseudobulbs usually inconspicuous, more rarely large and grooved. Leaves large, thin, pleated, rolled when young, mostly evergreen, deciduous in some species with large pseudobulbs. Racemes arising from the leaf axils, each with a long axis which is usually hairy. Bracts conspicuous, persistent or soon falling, usually hairy. Sepals and petals similar, spreading, usually directed upwards, often hairy outside. Lip simple or 3-lobed, usually spurred, with warts or ridges at the base. Column united to the lip for most of its length (except in No. 13). Pollinia 8 in 2 groups of 4, attached to a common base.

A genus of about 50 species with a distribution extending from Madagascar and the Mascarene Islands to the Himalaya, SE Asia generally, Japan and Australia. Many hybrids are cultivated, and Calanthes were among the first orchids to be deliberately hybridised.

Two types of plant occur, deciduous and evergreen. Both should be grown in terrestrial compost. Deciduous species should be given some shade and abundant moisture when growing; after the fall of the leaves they should be kept cool and dry. The evergreen species require more shade, and should not be dried out.
Literature: Seidenfaden, G., Orchid genera in Thailand I: Calanthe, *Dansk Botanisk Arkiv* **29**(2), (1975).

Pseudobulbs. Conspicuous: **1–3**; inconspicuous: **4–13**.
Leaves. Fallen at flowering: **1–3**; present at flowering: **4–13**.
Raceme. Arising from pseudobulb: **1–3**; arising from a rosette of leaves: **4–12**; arising from the lower part of an elongate leafy stem: **13**. Corymbose; **8**.
Lip. Simple: **1**; 3-lobed: **2–13**. Without spur: **11–13**; spur 2–12 mm: **9,10**; spur 1.5–2.9 cm: **1–3,6–8**; spur 3–5 cm: **4**. With central lobe itself 2-lobed or notched: **2–6,8–10,12,13**; with central lobe entire: **7–11**. Free from column: **13**.

1a. Bracts falling as soon as the flower opens; lip free from the column; plant with conspicuous leafy stem **13. gracilis**
b. Bracts persistent; lip attached to column; leafy stem inconspicuous, leaves in a rosette or borne on pseudobulbs 2
2a. Plants with large pseudobulbs, leafless when flowering 3
b. Plants with inconspicuous pseudobulbs, leafy when flowering 5
3a. Lip undivided, without any lateral lobes **1. rosea**
b. Lip divided, with 2 distinct lateral lobes 4
4a. Sepals, petals and usually the lip white; sepals and petals 2 cm or more **2. vestita**
b. Sepals, petals and lip pink; sepals and petals to 1.5 cm **3. rubens**
5a. Spur absent 6
b. Spur present, long or short 7
6a. Sepals and petals spreading; upper surface of lip with 3 conspicuous ridges **11. tricarinata**
b. Sepals and petals reflexed; upper surface of lip without ridges **12. reflexa**
7a. Spur 2–12 mm 8
b. Spur 1.5–5 cm 9
8a. Spur to 3 mm, blunt, oblong, straight **10. brevicornu**
b. Spur 5–12 mm, pointed, tapering, curved **9. discolor**
9a. Lip with the central lobe itself unlobed **7. cardioglossa**
b. Lip with the central lobe itself lobed or notched 10
10a. Flowers all white, except for yellow warts on the base of the lip; raceme corymbose **8. triplicata**
b. Flowers with at least the sepals and petals green, yellow or violet; raceme not corymbose 11
11a. Lip white, the central lobe itself deeply lobed, its lobes oblong, diverging at a wide angle; sepals and petals greenish or yellow **6. herbacea***
b. Lip violet or greenish tinged with purple, the central lobe notched, its lobes rounded, scarcely diverging; sepals and petals usually violet, rarely greenish 12
12a. Spur 2–2.7 cm, broadening towards its apex **5. sylvatica**
b. Spur 3–5 cm, not broadening towards its apex **4. masuca**

1. C. rosea (Lindley) Bentham & Hooker. Figure 13(1), p. 174. Illustration: Dansk Botanisk Arkiv **29**(2): 41 (1975).
Pseudobulbs 10–15 cm, conspicuous, ovoid to oblong, grooved, narrowed just below the apex. Leaves lanceolate, fallen by flowering time. Raceme to 45 cm with 7–12 flowers. Bracts persistent, 1.5–2 cm. Flowers 4–5 cm in diameter, red-pink. Sepals and petals lanceolate, acute. Lip simple, larger than the petals, its base rolled around the column. Spur *c.* 1.5 cm, straight, oblong. *Thailand*. G1–2. Winter.

2. C. vestita Lindley. Figure 13(2), p. 174. Illustration: Botanical Magazine, 4671 (1852); Dansk Botanisk Arkiv **29**(2): 29 & t. 4 (1975); Bechtel et al., Manual of cultivated orchid species, 175 (1981).
Pseudobulbs to 8 cm, ovoid, grooved. Leaves lanceolate, fallen by flowering time. Raceme to 70 cm with 6–12 flowers. Bracts 2–3 cm, persistent. Flowers 6–6.5 cm in diameter, white or cream. Lip white blotched with yellow or rarely entirely pink, 3-lobed, the lateral lobes about the same size as the divergently notched central lobe. Spur 1.8–2.5 cm, tapering, curved. *SE Asia, from Thailand to Sulawesi*. G2. Winter–spring.

Variable in the precise coloration of the lip, a character formerly used in the recognition of varieties (cultivars). 'Regnieri' (*C. regnieri* Reichenbach), which has a pink lip, is frequently grown.

3. C. rubens Ridley. Figure 13(3), p. 174. Illustration: Dansk Botanisk Arkiv **29**(2): 33 & t. 5 (1975); Bechtel et al., Manual of cultivated orchid species, 175 (1981).
Similar to *C. vestita* but flowers *c.* 3 cm in diameter, pink. *Thailand, NE Malaysia*. G2.

4. C. masuca (Don) Lindley. Figure 13(4), p. 174. Illustration: Botanical Magazine, 4541 (1850); Dansk Botanisk Arkiv **29**(2): t. 2 (1975).
Pseudobulbs inconspicuous. Leaves 20–30 × 7–11 cm, narrowly elliptic. Raceme arising from the leaf rosette, to 70 cm, with many flowers. Bracts persistent, 1.3–2 cm. Flowers to 6 cm in diameter, pale reddish purple. Lip as long as the petals, 3-lobed, with oblong lateral lobes and a somewhat notched central lobe, deep reddish violet with yellow warts towards the base. Spur 3–5 cm, curved. *Himalaya, SW China, Thailand & Malaysia*. G1–2. Summer.

This species is now usually considered part of *C. sylvatica*. However, the

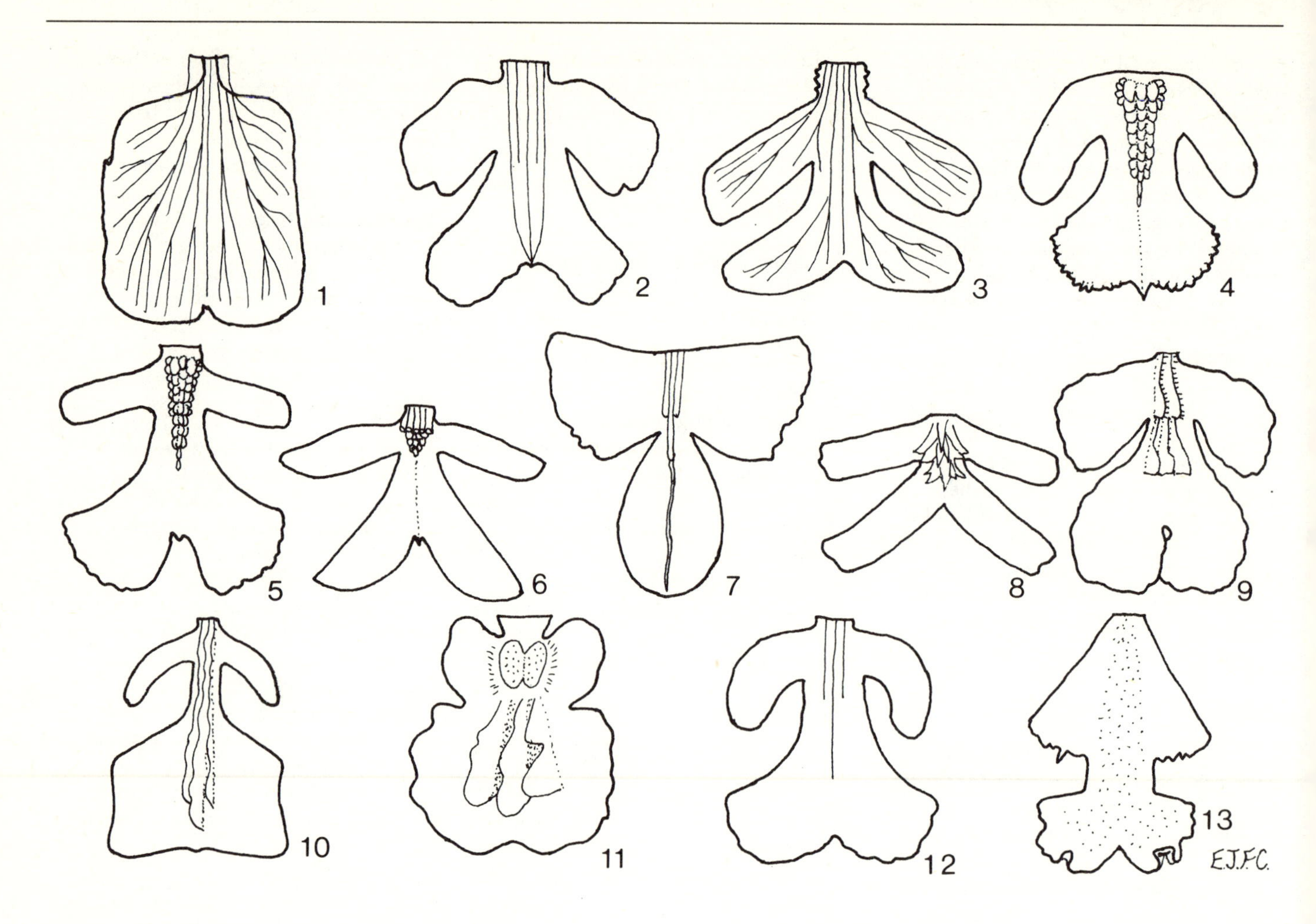

Figure 13. Lips of *Calanthe* species. 1, *C. rosea*. 2, *C. vestita*. 3, *C. rubens*. 4, *C. masuca*. 5, *C. sylvatica*. 6, *C. herbacea*. 7, *C. cardioglossa*. 8, *C. triplicata*. 9, *C. discolor*. 10, *C. brevicornu*. 11, *C. tricarinata*. 12, *C. reflexa*. 13, *C. gracilis*. (Compiled from various sources; not to scale.)

distinctions between them, though imprecise, have some horticultural importance, so the 2 are maintained here separately.

5. C. sylvatica (Thouars) Lindley. Figure 13(5). Illustration: Bechtel et al., Manual of cultivated orchid species, 175 (1981).
Pseudobulbs inconspicuous. Leaves 15–25 × 5–7 cm, narrowly elliptic. Raceme arising from the leaf rosette, 20–60 cm, with numerous flowers. Bracts persistent, 1.5–3 cm. Flowers 3–4 cm in diameter, usually reddish purple, more rarely greenish with the lip tinged with purple. Lip as long as the petals, 3-lobed, the lateral lobes small, oblong, the central reversed-heart-shaped, notched at the apex. Spur 2–2.7 cm, curved, broadening towards the apex. *N. Madagascar, Mascarene Islands*. G2. Spring–summer.

6. C. herbacea Lindley. Figure 13(6). Illustration: Annals of the Royal Botanic Garden Calcutta **5**: t. 44 (1895).
Pseudobulbs inconspicuous. Leaves 20–25 × 5–8 cm, elliptic. Raceme arising from the leaf rosette, 30–40 cm, with numerous flowers. Bracts persistent, 1.8–2.5 cm. Sepals and petals greenish or yellowish. Lip longer than the petals, white with yellow warts at the base, 3-lobed, the lateral lobes oblong, the central deeply and divergently 2-lobed. Spur straight or S-shaped, 2–3 cm. *N India*. G1–2.

***C. izu-insularis** (Satomi) Ohwi. Illustration: Maekawa, Wild orchids of Japan in colour, t. 142 (1979). Similar but with pink sepals and petals, the lip white flushed with violet and the spur 1.6–2 cm. *Japan*.

7. C. cardioglossa Schlechter. Figure 13(7). Illustration: Dansk Arkiv **29**(2): 37 & t. 6 (1975).
Pseudobulbs inconspicuous. Leaves to 30 × 11 cm, lanceolate to elliptic. Raceme arising from the leaf rosette, to 70 cm, with 10–20 flowers. Bracts persistent, 1.5–2 cm. Flowers to 2 cm in diameter, pale reddish purple. Lip about as long as the petals, 3-lobed, with reversed-triangular, upcurving lateral lobes which are pale violet spotted darker, and an entire, diamond-shaped or elliptic central lobe. *Vietnam, Laos, Kampuchea & Thailand*. G2. Winter.

8. C. triplicata (Willemet) Ames (*C. veratrifolia* (Willdenow) R. Brown; *C. furcata* Lindley). Figure 13(8). Illustration: Dansk Botanisk Arkiv **29**(2): 16 & t. 1 (1975); Maekawa, Wild orchids of Japan in colour, t. 150 (1979); Bechtel et al., Manual of cultivated orchid species, 175 (1981).
Pseudobulbs inconspicuous. Leaves elliptic, to 40 × 13 cm. Raceme arising from the

leaf rosette, to 1 m, the flowers corymbosely arranged at the top. Bracts persistent, 8–15 mm. Flowers 3–5 cm in diameter, white except for the yellow warts on the lip. Lip longer than the petals, 3-lobed, the lateral lobes oblong, the central deeply and divergently lobed. Spur 1.5–2.5 cm, curved, tapering. *SE Asia, Japan & E Australia*. G1–2. Spring.

9. C. discolor Lindley (*C. lurida* Decaisne; *C. amamiana* Fukuyama). Figure 13(9), p. 174. Illustration: Botanical Magazine, 7026 (1888); Lin, Native orchids of Taiwan, t. 46 (1975); Maekawa, Wild orchids of Japan in colour, t. 141, 143 (1975); Bechtel et al., Manual of cultivated orchid species, 174 (1981).
Pseudobulbs inconspicuous. Leaves narrowly elliptic, 17–23 × 4–7 cm. Raceme arising from the leaf rosette, to 40 cm, with many flowers. Bracts persistent, 4–8 mm. Flowers 3–5 cm in diameter, reddish purple or yellow. Lip about as long as the petals, 3-lobed, the lateral lobes oblong or broadening towards their apices, the central reversed-heart-shaped, deeply notched, bearing 3 prominent ridges towards the base. Spur 5–12 mm, curved, tapering. *Japan, Korea & Taiwan*. G1–2. Spring.

The yellow-flowered variant is called var. *flava* Yatabe in recent Japanese Floras. It has also been called *C. striata* (Banks) R. Brown, *C. bicolor* Lindley, *C. sieboldii* Decaisne and various combinations of these names at the level of variety or forma.

10. C. brevicornu Lindley. Figure 13(10), p. 174. Illustration: Annals of the Royal Botanic Garden Calcutta **8**: t. 227 (1898).
Pseudobulbs inconspicuous. Leaves 20–30 × 5–6 cm, narrowly elliptic. Raceme arising from the leaf rosette, to 50 cm, with many flowers. Bracts persistent, *c.* 1.5 cm. Flowers 2.5–4 cm in diameter, sepals and petals brown. Lip a little shorter than the petals, 3-lobed, the lateral lobes rounded, the central reversed-heart-shaped, deeply notched, purplish red with white margins, and with 3 yellow ridges near the base. Spur to 3 mm, oblong, blunt, straight. *Nepal, NE India (Sikkim)*. G1. Summer.

11. C. tricarinata Lindley. Figure 13(11), p. 174. Illustration: Annals of the Royal Botanic Garden Calcutta **8**: t. 223 (1898); Botanical Magazine, 8803 (1919); Lin, Native orchids of Taiwan, t. 49, 50 (1975); Maekawa, Wild orchids of Japan in colour, t. 138 (1979).
Pseudobulbs inconspicuous. Leaves 18–30 × 5–9 cm, elliptic. Raceme arising from the leaf rosette, 30–40 cm, with several flowers. Bracts 8–20 mm, persistent. Flowers 2.5–4 cm in diameter, sepals and petals greenish yellow. Lip about as long as the petals, red-brown, 3-lobed, the lateral lobes small, rounded, the central reversed-heart-shaped, notched, margins wavy, bearing 3 very pronounced ridges. Spur absent. *Himalaya, China, Japan & Taiwan*. G1–2.

12. C. reflexa Maximowicz. Figure 13(12), p. 174. Illustration: Botanical Magazine, 9648 (1943); Lin, Native orchids of Taiwan, t. 44, 45 (1975); Maekawa, Wild orchids of Japan in colour, t. 137 (1979).
Like *C. tricarinata* but generally smaller; sepals and petals reflexed, each narrowed into a bristle-like point, lip pale purple or white with oblong lateral lobes and an obovate, scarcely wavy, irregularly toothed but not notched central lobe which bears no ridges. *Japan, Korea (Quelpaert Islands) & Taiwan*. G1–2. Summer.

13. C. gracilis (Lindley) Lindley. Figure 13(13), p. 174. Illustration: Botanical Magazine, 4714 (1853); Annals of the Royal Botanic Garden Calcutta **8**: t. 222 (1898).
Pseudobulbs inconspicuous. Leafy stems to 40 cm. Leaves 15–20 × 3–4 cm, narrowly elliptic. Racemes arising laterally near the bases of the leafy stems and shorter than them, with 10–15 flowers. Flowers 2–3 cm in diameter, greenish yellow. Sepals and petals spreading. Lip free from the column, white with a yellow band, 3-lobed, the lateral lobes obliquely triangular, the central finely irregularly toothed and deeply notched. Spur absent. *Himalaya*. G1–2. Autumn.

Var. **venusta** (Schlechter) Maekawa (*C. venusta* Schlechter). Illustration: Maekawa, Wild orchids of Japan in colour, t. 136 (1979). Larger, with stems to 1 m, leaves 20–40 × 4–7 cm, sepals longer than the petals, lateral lobes of the lip ovate. *Japan*.

33. BLETIA Ruiz & Pavon

J.C.M. Alexander

Terrestrial, rarely epiphytic. Pseudobulbs almost spherical, slightly flattened, corm-like, bearing 1–several leaves. Leaves linear to lanceolate, pleated, rolled when young, deciduous. Flowers many, in loose racemes or panicles; flowering stems produced from lower nodes of pseudobulbs. Sepals and petals similar, free, spreading; lateral sepals closely pressed together at base but not fused. Lip free, simple or 3-lobed; lateral lobes erect or curved around column, central lobe broad, spreading, notched or more deeply 2-lobed, with 3–7 longitudinal toothed or scalloped keels. Column arched, apically winged. Pollinia 8, hard and compressed.

Between 25 and 50 species ranging from Mexico, Florida and the West Indies south to Argentina. Several species now placed in other genera were originally described in *Bletia* and the name is still sometimes used for plants quite unlike the genus as currently recognised. *Bletia* species should be grown in terrestrial compost and repotted when new growth first appears; the pseudobulbs should be partially covered. Little water should be given once growth has ceased. Plants can be divided after flowering.
Literature: Dressler, R.L., Notes on Bletia, *Brittonia* **20**: 182–90 (1968).

1a. Lip 3.5 × 2–2.5 cm, central lobe deeply notched or 2-lobed **1. catenulata**
b. Lip 1.5–2 × 1.2–1.5 cm, central lobe almost square **2. purpurea**

1. B. catenulata Ruiz & Pavon (*B. sanguinea* Poeppig & Endlicher; *B. sherrattiana* Lindley; *B. watsonii* Hooker). Illustration: Botanical Magazine, 5646 (1867); Pabst & Dungs, Orchidaceae Brasilienses **1**: 229 (1975); Bechtel et al., Manual of cultivated orchid species, 170 (1981).
To 2 m. Pseudobulbs 2–6 cm, bearing 1–6 leaves in 2 opposite ranks. Leaves 20–100 × 1–10 cm, variable in shape, linear to lanceolate or elliptic, acuminate. Flowering stem to 2 m, bearing 3–15 flowers; flowers 5–6.5 cm wide. Sepals 3–3.5 cm × 7–10 mm, oblong, acute, red to pinkish purple. Petals 3–3.5 × 1.7–2 cm, ovate, rounded, darker than sepals. Lip 3.5 × 2–2.5 cm, circular to oblong, deeply 3-lobed; lateral lobes semicircular to oblong, erect to spreading, purple with darker veins; central lobe deeply notched to 2-lobed, rounded, pink or pinkish purple with 3–5 pale to bright yellow ridges. Column club-shaped. *Colombia, Ecuador, W Brazil & Bolivia*. G2. Summer.

2. B. purpurea (Lamarck) de Candolle (*B. verecunda* (Salisbury) R. Brown; *B. shepherdii* Hooker). Illustration: Botanical Magazine, 3319 (1834); Dunsterville & Garay, Venezuelan orchids illustrated **2**: 41 (1961); Luer, Native orchids of Florida, 227 (1972).
Terrestrial, rarely epiphytic, to 1.5 m. Pseudobulbs 2–4 cm, bearing 3–5 leaves.

Leaves to 100 × 5 cm, linear-lanceolate, tapering, pale green. Flowering stem to 1.5 m, sometimes branched, bearing 3–80 flowers; flowers 2–2.5 cm wide, opening in succession. Sepals and petals 1.5–2.5 cm × 6–8 mm, pink to pinkish purple; upper sepal oblong to lanceolate; lateral sepals and petals ovate-lanceolate, acuminate, asymmetric; lateral sepals often with a central yellow stripe. Lip 1.5–2 × 1.2–1.5 cm, 3-lobed; lateral lobes triangular to diamond-shaped, scalloped, curved over column, purple; central lobe almost square, scalloped, purple, with 5–7 yellow scalloped ridges. Column white, 1–1.3 cm. *USA (Florida), Bahamas & Mexico, south through C America and the West Indies to Colombia, Venezuela & Guyana*. G2. Flowers throughout the year.

34. PHAIUS Loureiro
J.C.M. Alexander
Terrestrial, rarely epiphytic, robust. Pseudobulbs spherical to ovoid, (sometimes stem-like), bearing 3–10 leaves. Leaves large, lanceolate to elliptic, alternate, stalked, pleated, persistent, rolled when young. Flowering stems lateral, from pseudobulb bases or lower axils. Flowers few to many in loose or dense racemes. Sepals and petals similar, fleshy, free. Lip simple or 3-lobed with lateral lobes erect or curved around column, spurred or pouched below (except No. 3), shortly fused to column at base. Column short. Pollinia 8, in 2 groups of 4.

A widespread genus of 30–50 species ranging from tropical Africa eastwards to the Himalaya, N India, China & Japan, and south through SE Asia to N Australia and the Pacific islands. The spurless species, formerly treated as the separate genus *Gastrorchis* Schlechter, are now usually considered part of *Phaius*. The terrestrial species should be grown in large, well-drained pots and kept moist and partially shaded during the growing season, though overhead spraying should be avoided. During the resting period small amounts of moisture can still be given and plants can be exposed to stronger light. The epiphytic species require more shading and are easily harmed by draughts.

1a. Pseudobulbs stem-like, narrowly fusiform to cylindric **1. mishmensis**
b. Pseudobulbs spherical, conical or ovoid 2
2a. Flowers 10 cm wide or more; sepals and petals lanceolate, more than 3 times as long as broad **2. tankervilleae**
b. Flowers 8 cm wide or less; sepals and petals obovate, elliptic or oblong, less than 3 times as long as broad **3. humblotii***

1. P. mishmensis (Lindley) Reichenbach. Illustration: Botanical Magazine, 7497 (1896); Bose & Bhattacharjee, Orchids of India, 431 (1980); Bechtel et al., Manual of cultivated orchid species, 258 (1981).
Terrestrial. Stems to 140 cm, narrowly fusiform to cylindric, fleshy, bearing 6–8 leaves. Leaves 15–30 × 8–14 cm, lanceolate to ovate or oblong, acuminate. Flowering stems 1 or more from lower axils of pseudobulbs, to 60 cm, bearing 7–10 flowers; flowers 5–6 cm wide. Sepals and petals similar, pale pink to darker reddish brown; sepals to 3.5 × 1 cm, oblong, acuminate; petals somewhat smaller. Lip 3 × 2.5 cm, 3-lobed; lateral lobes oblong, rounded, pink speckled with red or purple, curved around column; central lobe square to oblong, shallowly lobed, with a central hairy ridge, pink speckled with red or purple, margins yellow. Spur 1.5–1.7 cm, tapering, curved. Column 1.2 cm, straight. *NE India, Burma, Thailand, Taiwan & Philippines*. G2. Summer–autumn.

2. P. tankervilleae (Banks) Blume (*P. bicolor* Lindley; *P. grandifolius* Loureiro; *P. blumei* Lindley; *P. wallichii* Hooker). Illustration: Botanical Magazine, 4078 (1844), 7023 (1888); Sheehan & Sheehan, Orchid genera illustrated, 133 (1979); Bechtel et al., Manual of cultivated orchid species, 258 (1981).
Terrestrial. Pseudobulbs 2.5–6 cm, conical to ovoid, few-leaved. Leaves 30–120 × 9–20 cm, elliptic to lanceolate, acuminate, membranous. Flowering stem lateral, from base of pseudobulb, to 2 m, bearing 10–20 flowers; flowers 10–12.5 cm wide. Sepals and petals similar, 4.5–6.5 × 1–1.3 cm, lanceolate, acuminate, reddish or yellowish brown, margin paler, greenish red on back. Lip 5 × 4.5 cm, 3-lobed; lateral lobes oblong, rounded, curved around column, pink to purplish red, yellow below; central lobe broadly ovate, acute or truncate, pink and yellow; throat yellow and purplish red; callus oblong; spur 1 cm, cylindric, curved. Column 2 cm, club-shaped. *C China, N India, Sri Lanka, through SE Asia to N Australia*. G2. Spring–summer.

3. P. humblotii Reichenbach (*Gastrorchis humblotii* (Reichenbach) Schlechter). Illustration: Cogniaux & Goossens, Dictionnaire Iconographique des Orchidées, Phaius, t. 1 (1898); Veitch, Manual of orchidaceous plants, part 6: 12 (1890).
Terrestrial. Pseudobulbs 4–6 × 3–4 cm, spherical to conical, ridged, few-leaved. Leaves 30–60 × 6–10 cm, broadly lanceolate, acuminate, membranous. Flowering stem robust, 30–80 cm, from base of pseudobulb, bearing 7–12 flowers; flowers 5–6 cm wide. Sepals and petals similar, 2.5–3 × 1.5–2 cm, obovate to elliptic, acuminate, pale pink with darker streaks of purple. Lip 3-lobed, lacking a spur, strongly crisped; lateral lobes oblong, erect to spreading, pinkish purple to reddish brown; central lobe semicircular to oblong, pink with white streaks; callus yellow, divided into 2 tooth-like ridges. Column club-shaped. *Madagascar*. G2. Summer.

***P. tuberculosus** (Thouars) Blume (*Gastrorchis tuberculosa* (Thouars) Schlechter). Illustration: Botanical Magazine, 7307 (1893); Proceedings of 8th world orchid conference, t. 7 (1976); Bechtel et al., Manual of cultivated orchid species, 258 (1981). Pseudobulbs *c.* 2.5 cm × 5 mm. Sepals and petals white; lip with 3 yellow keels; lateral lobes yellow with red spots. *Madagascar*. G1.

***P. flavus** (Blume) Lindley (*P. maculatus* Lindley). Illustration: Botanical Magazine, 3960 (1842); Cogniaux & Goossens, Dictionnaire Iconographique des Orchidées, Phaius, t. 3 (1904). Leaves with white spots. Sepals and petals yellow to greenish yellow, oblong, obtuse. Lip spurred, greenish yellow with reddish brown streaks in the heavily crisped margin. *SE Asia*. G2. Spring

35. SPATHOGLOTTIS Blume
J.C.M. Alexander
Terrestrial. Pseudobulbs conical to ovoid, sheathed, bearing 1–several leaves. Leaves narrowly lanceolate, tapering gradually to the stalk, pleated, prominently veined, rolled when young. Flowering stems robust, borne laterally from basal leaf axils, unbranched, basally sheathed, many-flowered. Sepals spreading, free; petals similar, often wider. Lip deeply 3-lobed; lateral lobes erect or spreading; central lobe lanceolate to spathulate, with a long narrow claw and a prominent basal callus and small basal teeth. Column slender, arched. Pollinia club-shaped, 8, in 2 groups of 4.

A genus of about 40 species distributed from N India through SE Asia and C China to Australia and the Pacific islands.

Spathoglottis species perform best if repotted every spring in well-drained terrestrial compost. Plenty of water can be supplied through the summer but this should be decreased when the plants are dormant. Growing temperatures depend on the country of origin. Propagation is effected by dividing sections of rhizome with pseudobulbs.

1a. Flowers yellow **1. aurea***
b. Flowers purple **2. plicata***

1. S. aurea Lindley. Illustration: Holttum, Flora of Malaya I: Orchids, 162 (1964); Shuttleworth et al., Orchids, 86 (1970). Pseudobulbs to 20 cm, ovoid, with 2–4 leaves. Leaves to 100 × 5 cm. Flowering stem to 60 cm or more, bearing 4–10 flowers; flowers 6–7 cm wide, deep yellow; bracts 1–2 cm, spathulate, concave, spreading. Sepals and petals 2.5–3 × 1.2–1.5 cm, elliptic to ovate, obtuse, slightly concave. Lip 2.5–3 × 2.5–3 cm, deeply divided into 3 main lobes; lateral lobes oblong, rounded, broadest at base, erect; central lobe 3–4 mm wide, lanceolate, rounded, with 2 small triangular lobes or teeth at base; callus of 2 erect triangular teeth. Lateral lobes of lip, callus and base of central lobe spotted with crimson. Column 1–5 cm, narrowly club-shaped, arched. *Malay peninsula*. G2. Summer.

***S. kimballiana** Hooker. Illustration: Botanical Magazine, 7443 (1895); Gardeners' Chronicle **4**: 93 (1888). Flowers to 7.5 cm wide, pale yellow. Lateral lobes of lip broadly spathulate. Sepals flecked with purple beneath. *Borneo*. G2. Spring.

***S. fortunei** Lindley. Illustration: Edwards's Botanical Register **31**: t. 19 (1845); Schlechter, Die Orchideen, edn 2, 308 (1927). Leaves 15–20 × 1.5–2 cm. Bracts on flowering stem narrowly ovate, attenuate, not spreading. Flowers 3–4 cm wide. Lateral lobes of lip oblong, rounded, strongly sickle-shaped. *Hong Kong*. G2. Winter.

2. S. plicata Blume. Illustration: Holttum, Flora of Malaya I: Orchids, 6 (1964); Shuttleworth et al., Orchids, 86 (1970); Bechtel et al., Manual of cultivated orchid species, 274 (1981).
Pseudobulbs to 7 × 5 cm, with 4–7 leaves. Leaves 30–120 × 2–7 cm. Flowering stem to 150 cm, bearing 5–25 flowers; bracts lanceolate, acuminate; flowers 3.5–4.5 cm wide, purple to pinkish purple. Sepals 2.3 × 1.2 cm, elliptic, acute; lateral sepals asymmetric; petals 2 × 14 cm, elliptic to ovate, obtuse. Lip 1.7 × 1.8 cm, 3-lobed; lateral lobes ovate to narrowly spathulate, rounded to truncate; central lobe spathulate, long-clawed, with 2 small teeth at base; callus triangular to heart-shaped, yellow. Column 1.2 cm, club-shaped, arched. *SE Asia from Sumatra to the Philippines (?India); naturalised in Florida, Hawaii and parts of tropical Africa*. G2. Flowers continuously.

***S. vieillardii** Reichenbach. Illustration: Botanical Magazine, 7013 (1888); Graf, Tropica, 754 (1978). Larger in all parts than *S. plicata*; flowers pale purple, to 6 cm wide. Lateral lobes of lip pinkish or brownish purple. *Malaysian & Indonesian islands*. G2. Autumn.

36. CHYSIS Lindley

J.C.M. Alexander

Epiphytic or on rocks. Pseudobulbs cylindric to fusiform or narrowly club-shaped, compound, fleshy, bearing several leaves in upper half, sheathed below. Leaves in 2 opposite rows, pleated, rolled when young, prominently veined, deciduous. Flowering stems lateral from lower leaf axils; flowers several in loose racemes. Sepals and petals free, spreading; lateral sepals arising obliquely from column foot forming a mentum. Lip 3-lobed, basally fused to column foot; lateral lobes broad, erect; central lobe spreading or reflexed, entire or notched; base of lip with keels or thickened veins. Column short and thick, 2-winged, with a prominent foot. Pollinia 8, often fused to column causing self-fertilisation.

A genus of about 6 species from Mexico, south through Central America to Venezuela and Peru. They should be grown in epiphytic compost in pots or baskets. A warm shady humid environment gives best results when the plants are actively growing though lower temperatures and much less water should be applied once growth ceases and the leaves fall.

1. C. aurea Lindley. Illustration: Botanical Magazine, 3617 (1837); Skelsey, Orchids, 101 (1979); Shuttleworth et al., Orchids, 87 (1979); Bechtel et al., Manual of cultivated orchid species, 183 (1981). Epiphytic. Pseudobulbs to 50 × 4 cm, cylindric to fusiform, pendent, bearing 4–15 leaves. Leaves to 45 × 7 cm, lanceolate, acuminate (when young) to attenuate. Flowering stem to 45 cm, bearing 3–12 flowers; bracts *c.* 2 cm, nearly equal to ovary; flowers to 6 cm wide, yellow with reddish lines and spots on petals and lip. Upper sepal 3–3.5 × 1.5 cm, oblong, bluntly tapering; lateral sepals 2.5 × 1.7 cm, triangular, asymmetric, obtuse. Petals 2.5–3 × 1–1.5 cm, spathulate to narrowly diamond-shaped, obtuse or rounded. Lip 2–2.5 × 2.5–3 cm, 3-lobed, very fleshy and crisped, lateral lobes oblong to ovate, sickle-shaped, erect; central lobe ovate to circular. Callus of 5 fleshy, hairy ridges. Column white, with a prominent foot. *Mexico to Venezuela, Colombia & Peru*. G2. Spring.

Var. **bractescens** (Lindley) Allen (*C. bractescens* Lindley). Illustration: Botanical Magazine, 5186 (1860); Williams et al., Orchids for everyone, 184 (1980); Bechtel et al., Manual of cultivated orchid species, 183 (1981). Bracts on flowering stem 2.5 cm or more, leaf-like. Flowers 6–8 cm wide; sepals and petals white with yellow streaks. *Mexico, Guatemala & Belize*. G2. Winter–spring.

***C. limminghii** Linden & Reichenbach. Illustration: Botanical Magazine, 5265 (1861); Cogniaux & Goossens, Dictionnaire Iconographique des Orchidées, Chysis, t. 3 (1901). Sepals and petals white with purple or purplish brown streaks; lip margin not crisped. *C Mexico*. G2. Summer.

***C. laevis** Lindley. Illustration: Cogniaux & Goossens, Dictionnaire Iconographique des Orchidées, Chysis, t. 2 (1901); Ospina & Dressler, Orquideas de las Americas, t. 42 (1974). Bracts about half the ovary length. Flowers yellow to orange with streaks of reddish purple. Lip pale yellow with pinkish purple spots and blotches; callus ridges hairless. *Mexico & Costa Rica*. G2. Summer.

37. THUNIA Reichenbach

J.C.M. Alexander

Terrestrial. Stems tall, tufted, cane-like, sheathed below, densely leaved above. Leaves long, acuminate, slightly bluish, spreading, in 2 opposite ranks, rolled when young. Flowers densely clustered in pendent, terminal racemes; bracts large, boat-shaped, white to pale green. Sepals and petals free, similar. Lip unlobed, rolled around column, with 5–7 narrow fringed keels and a short obtuse spur. Column 2-winged at apex. Pollinia 8.

A genus of about 6 species from India, Burma and SE Asia, which could all be regarded as colour variants of a single species. They perform best if repotted annually on the appearance of new growth, and should then be kept shaded and slightly moist. Once the new growth is well established more water can be given. After leaf-fall plants should be kept cool and dry. Propagation by division of cuttings.

1. **T. marshalliana** Reichenbach.
Illustration: Shuttleworth et al., Orchids, 43 (1970); Skelsey, Orchids, 140 (1979); Bechtel et al., Manual of cultivated orchid species, 276 (1981).
To 150 cm. Leaves 15–20 cm, lanceolate, acuminate, pale green. Flowers 5–12, to 12.5 cm wide, fragrant; bracts to 5 cm, white, boat-shaped. Sepals and petals 6.5–7.5 × 1.3–1.5 cm, lanceolate to ovate, acuminate, white. Lip 5–6 × 4.5 cm, broadly elliptic to oblong, yellow with a white border, margin irregular and fringed; blade with orange, branching, hairy ridges; spur *c.* 1 cm, shallowly 2-lobed. Column 2–2.5 cm. *Burma, Thailand & S China*. G2. Summer.

*T. **bracteata** (Roxburgh) Schlechter (*T. alba* Reichenbach: *Phaius albus* Wallich). Illustration: Botanical Magazine, 3991 (1843); Graf, Tropica, 754 (1978). Robust, to 60 cm. Flowers 4–10, to 7.5 cm wide; lip less than 4 cm wide, oblong to ovate; ridges orange or purple. *N. India, Burma & Thailand*. G2. Summer.

*T. **bensoniae** Hooker. Illustration: Botanical Magazine, 5694 (1868). Sepals and petals pale purple with darker tips. Lip white at base; blade dark purple with bright yellow, hairy ridges. *Burma*. G2. Summer.

38. ARUNDINA Blume

J.C.M. Alexander

Terrestrial, to 3 m. Stems erect, slender and rigid, tufted, unbranched, many-leaved. Leaves 12–30 × 1.5–4 cm, linear to lanceolate, grass-like, in 2 opposite ranks, folded when young; sheaths overlapping at base. Flowers many, 6–7 cm wide, in simple or branched terminal spikes, fragrant; bracts prominent. Sepals and petals *c.* 4 cm, white to pale pinkish purple, acute; sepals *c.* 1 cm wide, lanceolate to elliptic; lateral sepals lying close together under lip; petals *c.* 2 cm wide, ovate to circular, wavy. Lip 4 × 3.5 cm, shallowly 3-lobed, oblong, rounded, base rolled around column, crisped, deeply notched at apex, 3-keeled; throat pale, usually with a yellow patch; blade white to bright pink. Column slender, without foot. Pollinia 8, flattened.

Formerly considered to contain several species, *Arundina* is now usually regarded as a single widespread and highly variable species. Distinctive forms from the extremes of the range appear to be interconnected by intermediates. Cultivation conditions are similar to those for *Sobralia* though a slightly higher minimum temperature is required.

1. **A. graminifolia** (Don) Hochreutiner (*Bletia graminifolia* Don; *A. speciosa* Blume; *A. chinensis* Blume; *A. bambusifolia* Lindley; *A. revoluta* Hooker). Illustration: Botanical Magazine, 7284 (1893); Shuttleworth et al., Orchids, 43 (1970); Graf, Tropica, 709 (1978).
Himalaya & S China, Sri Lanka, SE Asia & Pacific islands. G1. Flowers continuously.

39. SOBRALIA Ruiz & Pavon

J.C.M. Alexander

Terrestrial, rarely epiphytic. Stems erect, slender and unbranched, many-leaved. Leaves leathery, pleated, rolled when young, long-sheathed, in 2 opposite ranks. Flowers few or solitary, terminal or in the upper leaf axils. Bracts usually well developed, sometimes several to each flower. Sepals fused at base; petals free, broader than sepals. Lip slightly fused to base of column, simple or 2-lobed, rolled around column; blade expanded, wavy; small calluses and ridges often present. Column slender, without foot, 3-lobed at apex. Pollinia 8, in 2 groups of 4, granular.

A genus of 35–50 species from C America and tropical S America. They perform best if repotted very infrequently and should be allowed to become well established in large pots or beds. When actively growing, plants should be kept well shaded and watered. Once growth ceases watering can be considerably reduced.

1a. Flowers less than 12 cm wide **1. sessilis***
b. Flowers more than 12 cm wide 2
2a. Bracts 2.5–5 cm; column *c.* 2.5 cm; lip 8 cm or less **2. leucoxantha**
b. bracts 10 cm or more; column *c.* 3.5 cm; lip 8 cm or more 3
3a. Flowers stalked, purple or pinkish purple **3. macrantha**
b. Flowers almost stalkless, yellow **4. xantholeuca**

1. **S. sessilis** Lindley. Illustration: Dunsterville & Garay, Venezuelan orchids illustrated **1**: 397 (1959); American Orchid Society Bulletin **42**: 395 (1973).
To 90 cm. Leaves to 22 × 8 cm, stiff, narrowly elliptic to ovate, acute or obtuse. Flowers solitary, terminal, tapering smoothly into ovary and stalk; bracts several, dark brown. Sepals 4.5–5 × 1.5–2 cm, lanceolate, acute; petals 4–5 × 1–2 cm, lanceolate to ovate, acute or obtuse, strongly recurved; petals and sepals white. Lip 4.2–5 × 2.8–3.5 cm, ovate to circular, shallowly notched, slightly scalloped, rolled around column, white with yellow stripe in throat and patch on blade; callus of 2 small basal projections. Column 2.5 cm. *Brazil, Guyana & Venezuela*. G2. Summer.

*S. **decora** Bateman. Illustration: Hamer, Las Orquideas de El Salvador, t. 34 (1974); Graf, Tropica, 759 (1978). Occasionally epiphytic. Leaves with warts and short stiff hairs beneath and on sheaths. Flowers to 10 cm wide, pale to dark purple. *Mexico to Honduras*. G2. Spring–summer.

*S. **cattleya** Reichenbach. Illustration: American Orchid Society Bulletin **44**: 195 (1975). Flowers in several axillary panicles, each with 5–12 flowers, pinkish brown inside, white or cream outside. Lip maroon with yellow ridges. *Venezuela & Colombia*. G2.

2. **S. leucoxantha** Reichenbach.
Illustration: Botanical Magazine, 7058 (1889); Graf, Tropica, 759 (1978); Bechtel et al., Manual of cultivated orchid species, 273 (1981).
Terrestrial. Stems to 150 cm, clustered. Leaves 20–30 × 3–4 cm, elliptic to lanceolate, acuminate, scurfy beneath; sheaths warty. Flowers solitary, terminal, 15–18 cm wide; bracts 2.5–5 cm. Sepals to 7 × 2.5 cm, linear to lanceolate; petals to 7 × 3–3.5 cm, lanceolate to oblong; petals and sepals white. Lip 6–8 × 5–6 cm, ovate, notched, wavy, rolled around column, white with a yellow to reddish orange throat. Column 2.5 cm. *Costa Rica & Panama*. G2. Summer–autumn.

3. **S. macrantha** Lindley. Illustration: Botanical Magazine, 4446 (1849); Skelsey, Orchids, 137 (1979); Bechtel et al., Manual of cultivated orchid species, 273 (1981).
Terrestrial or on rocks. Stems to 2 m, erect, clustered. Leaves 13–30 × 2–7.5 cm, lanceolate, acuminate, dark green above, paler beneath. Flowers solitary, 15–18 cm wide, fragrant; bracts to 11 cm, green. Sepals to 9 × 2.5 cm, lanceolate to elliptic, obtuse or acute, with curled tips; petals to 9 × 3–3.5 cm, elliptic, asymmetric, obtuse or rounded; sepals and petals pinkish purple. Lip to 10 × 7 cm, ovate, rolled around column, deeply scalloped and notched, slightly darker than petals, paler in throat, with 7 darker ridges; callus of 2 white, fleshy teeth near lip base. Column 3–4 cm, club-shaped. *Mexico to Costa Rica*. G2. Summer.

4. **S. xantholeuca** Reichenbach.

Illustration: Botanical Magazine, 7332 (1894); Hamer, Las Orquideas de El Salvador, t. 34 (1974); Graf, Tropica, 759 (1978).
Epiphytic or on rocks. Stems to 180 cm, thick. Leaves 15–28 × 3–7 cm, lanceolate to oblong, acuminate; sheaths speckled with reddish brown. Flowers solitary, terminal, *c.* 15 cm wide, almost stalkless, bright yellow. Sepals 8–11 × 1.5–2 cm, lanceolate, acuminate; petals 8–10 × 2–2.5 cm, lanceolate to elliptic, acute or acuminate. Lip 8–11 × 6–7 cm, ovate to oblong, rolled around column, deeply notched, scalloped; throat streaked with deeper yellow. Column 2.5 cm. *Guatemala & El Salvador*. G2. Spring–summer.

40. ISOCHILUS R. Brown
J.C.M. Alexander
Epiphytic, terrestrial or on rocks. Stems slender, erect and clustered or creeping. Leaves many, in 2 opposite ranks, linear to oblong, folded when young, with sheathing bases, often notched. Flowers terminal in 1- or 2-ranked racemes, sometimes solitary, white to pinkish purple. Upper sepal free; lateral sepals almost always at least half fused, inflated below, keeled or occasionally winged. Petals free, shorter and broader than sepals, short-clawed. Lip fused to base of column or column foot, short-clawed, narrow, often S-shaped. Column erect, sometimes with a small foot. Pollinia 4, waxy.

Now regarded as 2 species from tropical America with rather variable floral characters interconnecting them. This variation has resulted in the use of many different names at species level. The degree of sepal fusion may vary considerably, even on a single plant. Cultivation conditions are generally similar to those for *Cattleya* (p. 191) though water should be given throughout the year.

1. I. linearis (Jacquin) R. Brown.
Illustration: Dunsterville & Garay, Venezuelan orchids illustrated **1**: 178 (1959); Shuttleworth et al., Orchids, 71 (1970); Skelsey, Orchids, 115 (1979).
Epiphytic. Stems to 50 cm, erect, covered in grey leaf-sheaths. Leaves to 6 cm × 3 mm, linear, slightly tapering, asymmetrically notched, with prominent midrib. Flowers 1–several, in dense 1- or 2-sided racemes, *c.* 1 cm long, pale pink to dark pinkish purple (rarely orange). Sepals 8–9 × 3 mm, triangular to narrowly ovate; lateral sepals fused to about half their length, inflated and forming a small mentum with column foot. Petals 6–8 × 3–4 mm, lanceolate, irregular, short-clawed. Lip 8 × 2 mm, linear to lanceolate, constricted in the middle, short-clawed. Column 3–4 mm, with a short foot. *Mexico to Peru & Brazil, West Indies*. G2. Summer.

41. ARPOPHYLLUM Llave & Lexarza
J.C.M. Alexander
Epiphytic or on rocks, with creeping, simple or branched rhizomes. Stems cane-like, sheathed, 1-leaved. Leaves fleshy or leathery, folded when young. Flowers many, small in dense terminal racemes; flowering stem with large basal sheath. Sepals spreading; lateral sepals inflated below and fused to column; petals smaller than sepals. Lip uppermost (flower non-resupinate), usually longer than sepals and petals, inflated and pouched below. Column slightly arched, with small foot. Pollinia 8, waxy, pear-shaped.

A small genus of 2 to 5 species from tropical America. Floral differences between the species are slight and specific distinctions are made largely on vegetative characters. Cultivation conditions are generally similar to those for *Cattleya* (p. 191), though a simpler epiphytic compost of equal parts of sphagnum and tree-fern fibre is adequate. Good ventilation is important.
Literature: Garay, L., Synopsis of the genus Arpophyllum, *Orquidea (Mexico)* **4**: 16–19 (1974).

1. A. spicatum Llave & Lexarza.
Illustration: Botanical Magazine, 6022 (1873); Shuttleworth et al., Orchids, 70 (1970); Graf, Tropica, 710 (1978).
To 75 cm. Stem flattened, almost concealed in sheaths. Leaf to 50 × 4 cm, oblong, slightly sickle-shaped, obtuse, strongly folded along midrib, fleshy. Inflorescence to 30 × 3 cm, cylindric; stalks and ovaries with black warts; basal sheath to 15 × 2 cm. Flowers 8–10 mm wide, purplish pink. Upper sepal 4.5–5.5 × 1.5–3 mm, oblong to ovate, obtuse or with a small point; lateral sepals 5–6 × 2–3 mm, elliptic to oblong, acute to rounded, 3-veined, asymmetric, inflated at base. Petals 5–5.5 × 1.5–2 mm, linear to narrowly elliptic, obtuse or rounded, 1-veined. Lip 5.5–6 × 3.5 mm, constricted above pouch; blade ovate, curved over column, 7-veined. Column 3.5–4 mm. *Mexico*. G2. Spring.

***A. giganteum** Lindley. Illustration: Dunsterville & Garay, Venezuelan orchids illustrated **6**: 50 (1976); Bechtel et al., Manual of cultivated orchid species, 168 (1981). To 1 m. Leaf leathery, flattened, folded only in basal part, straight-sided. Inflorescence to 18 cm long. *Mexico to Costa Rica; Jamaica & Colombia*. G2. Spring.

42. COELOGYNE Lindley
J.C.M. Alexander & P.J.B. Woods
Epiphytic. Pseudobulbs cylindric, ovoid or fusiform, clustered or distant, with 1 or 2 leaves. Leaves linear to elliptic, leathery and pleated, rolled when young. Flowering stems erect to pendent, variously sheathed, terminal on pseudobulbs or basal and sometimes later developing into pseudobulbs. Flowers 1–many in loose racemes, often fragrant. Sepals and petals free; sepals often keeled; petals usually narrower, sometimes similar. Lip 3-lobed; lateral lobes erect; disc with entire, warty, toothed or fringed keels which may extend onto central lobe. Column arched, exposed. Pollinia. 4.

A genus of 100–50 species from India, China, SE Asia and the Pacific islands, many of which have showy and long-lasting flowers.

Himalayan species such as *C. cristata* and *C. barbata* perform best in cool conditions never exceeding 25 °C. They have a very definite resting period when no water should be given; this may last for several months. If the pseudobulbs shrivel during this time, light syringing will supply all the moisture required until growth begins again. Tropical species, e.g. *C. speciosa* and *C. dayana*, grow throughout the year and should be kept constantly moist with a minimum temperature of 15 °C. Good drainage is essential. The spreading habit of many of these species is best accommodated on bark or in hanging baskets. Well-established plants flower heavily and repotting should be avoided unless absolutely necessary.
Literature: Seidenfaden, G., Orchid genera in Thailand 3, Coelogyne, *Dansk Botanisk Arkiv* **29**(4), (1975).

Pseudobulbs. Clustered: **1,2,5–7,9,10,12,15**; distant: **3,4,8,11,13,14,16**. One-leaved: **1**.
Leaves. 3 cm wide or less: **2,3,15**.
Flowering stem. Basal: **1,5–16**; on mature pseudobulbs: **2–4,11**. Erect or arched (majority of flowers held at or above level of pseudobulbs): **1–4,11–16**; pendent (majority of flowers held below level of pseudobulbs): **5–10,16**. 5 cm or less: **13**; more than 50 cm: **4,5,7,9,11**; more than 100 cm: **7**. Densely hairy: **6**; with a few scattered hairs: **7,9**. Sheathed at base

only: **1,3,5,6,8–13,15,16**; with a cluster of scales just below flowers: **4**; sheathed from base up to lowest flowers: **2,7,14**.
Flowers. Solitary: **1–3**; 10 or fewer per stem: **1,3–6,8,9,11–13,15,16**; more than 25 per stem: **5,7,9,14**. About 2 cm wide: **2,5**; 3–4.5 cm wide: **2,3,6,8,10,12, 15,16**; more than 5 cm wide: **1–4,7, 9–16**.
Bracts. Persistent: **5–7,9,11,13,14**; deciduous: **1–4,8,12,15,16**.
Petals. 2 cm or less: **3,5,6,12,13**; 5 cm or more: **11,12,13**. Narrower than sepals: **1–4,8,10,12,14–16**; about as wide as sepals: **5–7,9,11–13**.
Lip. 4.5 cm or more: **1,2,11**. Margin entire: **2,5–10,12,14,15**; margin toothed: **1,4**; margin fringed: **3,4,13**; margin scalloped: **11**. With a distinct waist: **11**. With ringed patches: **12**. With an even number of keels: **2–4,7,14**; with an odd number of keels: **1–6,8–10,12,13,15**. Keels entire: **3,5,12,14**; keels wavy: **2,8,15,16**; keels papillose: **1,6,9,11**; keels toothed: **7,13**; keels fringed: **1,2,4,10,13**.

1a. Flowers solitary, if more, then not all open at the same time 2
b. Flowers more than 1, all open at the same time 5
2a. Lip margin not fringed 3
b. Lip margin fringed 4
3a. Pseudobulbs 1-leaved **1. speciosa**
b. Pseudobulbs 2-leaved **2. lawrenceana***
4a. Flowering stem short, with scales at base only **3. ovalis***
b. Flowering stem long, with a cluster of overlapping scales just below the flowers **4. barbata***
5a. Flowering stem pendent; majority of flowers held below level of pseudobulbs 6
b. Flowering stem erect or arched; majority of flowers held at or above level of pseudobulbs 11
6a. Flowers 2 cm wide or less; keels on lip not reaching base of central lobe **5. veitchii**
b. Flowers 3 cm wide or more; keels on lip reaching base of central lobe or further 7
7a. Inflorescence-stalk densely hairy **6. tomentosa**
b. Inflorescence-stalk hairless or with scattered hairs 8
8a. Main keels on lip 2; raceme to 100 cm with up to 100 flowers **7. dayana**
b. Main keels on lip 3; raceme 20–60 cm with 9–30 flowers 9
9a. Flowers 10 or fewer, 4 cm wide or less **8. flaccida**
b. Flowers more than 10, 5 cm wide or more 10
10a. Lip *c.* 3 times wider than central lobe; sepals and petals pale yellowish **9. massangeana**
b. Lip *c.* 2 times wider than central lobe; sepals and petals whitish **10. swaniana**
11a. Flowers predominantly green **11. pandurata***
b. Flowers white, cream, yellow or brown 12
12a. Lip with ringed spots **12. nitida***
b. Lip without ringed spots 13
13a. Petals and lateral sepals more or less equal in width **13. cristata**
b. Petals distinctly narrower than lateral sepals 14
14a. Flowers 6–7 cm wide; flowering stem to 30 cm, with 10–18 flowers **14. asperata**
b. Flowers 5 cm wide or less; flowering stem 18 cm or less, with 10 flowers or fewer 15
15a. Leaves less than 2 cm wide **15. viscosa**
b. Leaves 3 cm wide or more **16. trinervis***

1. C. speciosa (Blume) Lindley. Illustration: Botanical Magazine, 4889 (1855); Die Orchidee **29**: 18 (1978); Bechtel et al., Manual of cultivated orchid species, 187 (1981).
Pseudobulbs 2–8 cm, ovoid, with 4 slightly concave sides, clustered, 1-leaved. Leaf-blade to 35 × 9 cm, narrowly elliptic, acuminate; leaf-stalk *c.* 5 cm. Flowering stem to 20 cm, basal, erect or arched, sheathed at base, bearing 1–3 flowers; bracts deciduous: flowers *c.* 6 cm wide, drooping. Sepals 5–7 × 1–1.8 cm, narrowly ovate, acuminate; petals 5–6 cm × 2–3 mm, linear, petals and sepals yellowish green. Lip 5 × 3–3.5 cm, 3-lobed, dark reddish brown with a white or yellow tip; keels 3, 2 long and 1 short, papillose, fringed; central lobe with a short broad claw. Column 3.5 cm, arched. *Java & Sumatra* (Malaysian records refer to *C. xyrekes* and those from Vietnam to *C. lawrenceana*). G2. Flowers continuously.

2. C. lawrenceana Rolfe. Illustration: Botanical Magazine, 8164 (1907); Gardeners' Chronicle **47**: 335 (1910); Bechtel et al., Manual of cultivated orchid species, 186 (1981).
Pseudobulbs 5–8 × 2.5–3 cm, ovoid to cylindric, clustered, 2-leaved. Leaves 20–30 × 3 cm, elliptic, acuminate, tapering into stalk. Flowering stem 15–20 cm, erect, on mature pseudobulb, sheathed throughout, 1-flowered; flower 5–7 cm wide. Sepals 5 × 1.5 cm, triangular to ovate, obtuse; petals 4.5 cm × 3–4 mm, linear, drooping; petals and sepals yellow. Lip 5 × 2.5–3 cm, 3-lobed, with 3 long, fringed keels and 2 shorter basal ones; disc and lateral lobes purplish red; central lobe white with a yellow stripe. Column 3–4 cm, club-shaped. *Vietnam*. G2. Spring.

***C. holochila** Hunt & Summerhayes (*C. elata* misapplied; *C. stricta* misapplied). Illustration: Botanical Magazine, 5001 (1857). Flowers 6–10 per raceme, *c.* 4.5 cm wide; sepals and petals white. Lip obscurely 3-lobed, yellow or orange with 2 sinuous, fringed keels. *NE India & Burma*. G2. Spring–summer.

***C. miniata** (Blume) Lindley (*Hologyne miniata* (Blume) Pfitzer). Illustration: Cogniaux & Goossens, Dictionnaire Iconographique des Orchidées, Coelogyne, t. 7 (1907); Van Steenis, Mountain flora of Java, t. 35 (1972). Pseudobulbs oblong to fusiform, distant, enclosed in scaly sheaths. Flowers several, *c.* 2 cm wide, bright red. Column with small foot. *Indonesia (Java & Bali; ?Sumatra)*. G2. Summer.

3. C. ovalis Lindley. Illustration: Botanical Magazine, 9255 (1931); Orchid Review **80**: 98 (1976); Bechtel et al., Manual of cultivated orchid species, 187 (1981).
Pseudobulbs 3–9 × 2 cm, fusiform to ovoid, sheathed below, distant, 2-leaved. Leaves 7–17 × 1.5–4.5 cm, lanceolate, acute; leaf-stalk short. Flowering stem to 12 cm, erect, on mature pseudobulb, 1–3 flowered, sheathed below; flowers to 6 cm wide; bracts deciduous. Sepals 3–3.5 × 1.2 cm, ovate, acute; petals 3 cm × 1 mm, linear, acute; sepals and petals pinkish brown. Lip 3–3.5 × 2–2.5 cm, oblong, 3-lobed, finely fringed, yellow to pinkish brown with purplish brown patches; keels 3, wavy, 2 long and 1 short. Column 1.5–1.8 cm, apically winged. *W Himalaya to Burma, China & Thailand*. G2. Summer.

***C. fimbriata** Lindley. Illustration: Edwards's Botanical Register **11**: t. 868 (1825); Dansk Botanisk Arkiv **29**(4): 18 (1975); Bechtel et al., Manual of cultivated orchid species, 186 (1981). Flowering stem to 5 cm; flowers 3–4 cm wide, pale yellow. Upper sepal 1.5–2.5 mm. Lip marked with purplish brown, with 2 extra, short keels on the central lobe. *NE India to China & Malaysia*. G2. Summer–autumn.

***C. fuliginosa** Lindley. Illustration:

Botanical Magazine, 4440 (1849); Dansk Botanisk Arkiv **29**(4): 21 (1975); Bechtel et al., Manual of cultivated orchid species, 186 (1981). Flowers 5–6 cm wide; sepals and petals yellowish orange; lip dark brown; keels 4, slightly wavy, 2 long and 2 short. *Burma; ?Java*. G2. Winter.

4. C. barbata Griffith. Illustration: Veitch, Manual of orchidaceous plants: Eriae, 32 (1890); Lindenia **16**: 735 (1901); Orchid World **6**: 274 (1916).
Pseudobulbs to 9 × 2.5 cm, conical to ovoid, sheathed below when young, distant, 2-leaved. Leaves to 40 × 5 cm, oblong to lanceolate, acute or acuminate; leaf-stalk *c.* 5 cm. Flowering stem to 60 cm, erect, on mature pseudobulb, up to 10-flowered, with a cluster of overlapping scales just below the flowers; bracts deciduous. Sepals 3.5–4 × 1.5 cm, ovate to oblong, acute; petals 3.5–4 cm × 2–3 mm, linear, acute; sepals and petals white. Lip 3.5 × 3 cm, 3-lobed; central lobe laciniate; disc and central lobe dark brown to black, lateral lobes white; keels 3, fringed, middle one shorter. Column 2.5–3 cm. *Bhutan & NE India*. G2. Winter.

*__C. stricta__ (D. Don) Schlechter (*C. elata* Lindley). Illustration: Orchid Review **80**: 210 (1972). Lip less than 3 cm, keels 2, margin finely and shortly toothed. *Himalaya*. G2. Spring–summer.

The illustration under this name in Botanical Magazine, 5001 (1857) is not of this species.

5. C. veitchii Rolfe. Illustration: Botanical Magazine, 7764 (1901); Orchid Review **33**: 219 (1925).
Pseudobulbs 7.5–10 × 2–4 cm, fusiform, becoming grooved, at first almost enclosed in sheathing scales, clustered, 2-leaved. Leaf-blade to 30 × 8 cm, broadly elliptic, acute, distinctly 5-veined beneath; leaf-stalk *c.* 10 cm. Flowering stem 15–60 cm, basal, pendent, sheathed at base, bearing up to 30 flowers or more; bracts persistent; flowers *c.* 2 cm wide, white. Sepals 12–14 × 5 mm, oblong-lanceolate, acute, incurved and concave; petals 10–13 × 3 mm, lanceolate, acute, incurved and concave. Lip 1.5 × 1.3 cm, shallowly 3-lobed, with 3 short basal keels. Column 7 mm, widening apically with a winged slightly wavy margin. *New Guinea*. G2. Summer–autumn.

6. C. tomentosa Lindley. Illustration: Dansk Botanisk Arkiv **29**(4): 61 (1975).
Pseudobulbs to 6 × 3 cm, ovoid, clustered, 2-leaved. Leaf-blade to 38 × 10 cm, elliptic; leaf-stalk 8 cm. Flowering stem to 40 cm, basal, pendent, densely covered in short black hairs, base loosely sheathed, bearing up to 20 flowers or more; bracts persistent, hairy; flowers 4–4.5 cm wide, pale orange or salmon-coloured. Sepals 2.5–3 cm × 7–9 mm, lanceolate; petals slightly shorter and narrower. Lip 2.5 × 1.8 cm, 3-lobed, yellowish; lateral lobes brown with paler veins, keels 5, 3 long and 2 short, papillose at base, warty at apex. Column *c.* 6 mm. *Malaysia, Thailand; ?Borneo*. G2. Spring.

7. C. dayana Reichenbach. Illustration: Schlechter, Die Orchideen, edn 2, 135 (1927); Botanical Magazine n.s., 309 (1958); Graf, Tropica, 715 (1978); Bechtel et al., Manual of cultivated orchid species, 186 (1981).
Pseudobulbs 10–22 × 2–5 cm, fusiform to cylindric, clustered, 2-leaved. Leaf-blade 25–75 × 5–10 cm, elliptic, acute, with 7–9 prominent veins beneath; leaf-stalk *c.* 9 cm. Flowering stem to 100 cm, basal, pendent, with short scattered hairs, sheathed at base and above, bearing up to 50 flowers; bracts moderately persistent; flowers *c.* 7 cm wide, pale yellowish, brownish yellow or whitish. Sepals 3–3.5 cm × 5–7 mm, lanceolate, acute; petals a little smaller. Lip *c.* 3.5 × 2.5–3 cm, 3-lobed; lateral lobes brown with white veins; central lobe brown with a narrow white margin, keels 6, 2 long and 4 short. *Malay peninsula, Sumatra, Borneo; ?Java*. G2. Winter–spring.

8. C. flaccida Lindley. Illustration: Botanical Magazine, 3318 (1834); Schlechter, Die Orchideen, edn 2, 136 (1927); Orchid Review **80**: 190 (1972); Bechtel et al., Manual of cultivated orchid species, 186 (1981).
Pseudobulbs to 12 × 2.5 cm, cylindric to fusiform, grooved, sheathed in shining purple scales below, 2-leaved. Leaf-blade to 16 × 3.6 cm or more, lanceolate, narrowing gradually to base, apex acuminate. Flowering stem 20–25 cm, basal, pendent, sheathed at base, bearing *c.* 8 flowers; bracts deciduous; flowers to 4 cm wide, white. Sepals 3–4 cm × 9 mm, lanceolate. Petals 2.7–3.5 cm × 3 mm, linear. Lip 1.7 × 1.2 cm, 3-lobed, white; base of central lobe yellow; lateral lobes veined with reddish brown; disc with 3 wavy keels. Said to have an unpleasant smell. *E Himalaya*. G2. Winter–spring.

9. C. massangeana Reichenbach. Illustration: Botanical Magazine, 6979 (1888); Dansk Botanisk Arkiv **29**(4): 62 (1975); Williams et al., Orchids for everyone, 75 (1980); Bechtel et al., Manual of cultivated orchid species, 187 (1981).
Pseudobulbs 5–10 × 2.5 cm, fusiform to ovoid, clustered, 2-leaved. Leaf-blade to 50 × 12 cm, elliptic; leaf-stalk 5–10 cm. Flowering stem to 60 cm, basal, pendent, with scattered short black hairs, sheathed at base, bearing up to 30 flowers; bracts persistent; flowers to 7 cm wide, pale yellowish. Sepals to 3.5 × 1 cm, lanceolate, petals slightly smaller. Lip 2–2.8 × 1.9 cm, 3-lobed; lateral lobes brown with white or yellowish veins inside, whitish outside; central lobe brown and yellow; keels 5, 3 long and 2 short, warted except for crested basal area. *Thailand, Malay peninsula, Java; ?Sumatra*. G2. Spring–summer.

10. C. swaniana Rolfe. Illustration: Botanical Magazine, 7602 (1898); Orchid Review **83**: 150 (1975).
Pseudobulbs to 10 × 3.5 cm, fusiform, 4- to 6-angled, clustered, 2-leaved. Leaf-blade to 25 × 6 cm, elliptic; leaf-stalk 5–7.5 cm. Flowering stem to 40 cm, basal, pendent, with minute purple spots, loosely sheathed at base, bearing *c.* 20 flowers; bracts apparently deciduous; flowers *c.* 5 cm wide. Sepals 2.5 cm × 7–10 mm, lanceolate, white, or white tinged with pale brown; petals narrower, white. Lip 2.5 × 1.6 cm, 3-lobed, pale brown, darker at the margins and lobe tips; lateral lobes white-veined; keels 5, 3 long and 2 short. *Malay peninsula, Sumatra, Borneo; ?Philippines*. G2. Spring–summer.

11. C. pandurata Lindley. Illustration: Botanical Magazine, 5084 (1858); Shuttleworth et al., Orchids, 41 (1970); Skelsey, Orchids, 103 (1979).
Pseudobulbs 7–12 × 2–3 cm, cylindric to fusiform or ovoid, compressed, 2-leaved, distant. Leaf-blade 20–50 × 7 cm, lanceolate to elliptic, acute to acuminate; leaf-stalk *c.* 6 cm. Flowering stem to 50 cm, basal, erect or arched, sheathed below, with up to 15 flowers, bracts persistent; flowers *c.* 10 cm wide, fragrant, pale green. Sepals and petals 3.5–5.5 × 1.2–1.5 cm; sepals lanceolate to ovate, acute; petals lanceolate to slightly spathulate, acute. Lip 4.5 × 1.5–2.5 cm, 3-lobed with a long waist, mottled with purplish brown or black, 2-keeled at base; disc with warty ridges. *Malaysia, Sumatra & Borneo*. G2. Summer.

*__C. parishii__ Hooker. Illustration: Botanical Magazine, 5323 (1862); Graf, Tropica, 715 (1978); Bechtel et al., Manual of cultivated orchid species, 187 (1981).

Flowering stem on mature pseudobulb, with 3–5 flowers; flowers 6–8 cm wide. Lip to 2.5 cm. *Burma*. G2. Spring–summer.

12. C. nitida (Wallich) Lindley (*C. ochracea* Lindley; not *C. nitida* misapplied, which is probably *C. punctulata* Lindley). Illustration: Botanical Magazine, 4661 (1852); Hay & Synge, Dictionary of garden plants in colour, 60 (1969); Bechtel et al., Manual of cultivated orchid species, 187 (1981). Pseudobulbs 2–8 × 1–1.5 cm, cylindric to fusiform or narrowly ovoid, clustered, 2-leaved. Leaves 8–30 × 2–6 cm, lanceolate to elliptic, acute to acuminate; stalk short or absent. Flowering stem to 20 cm, erect or arched, basal, with 2–8 flowers, sheathed below, bracts deciduous; flowers 3–4 cm wide, fragrant. Sepals 2–3 cm × 7–9 mm, lanceolate, acuminate, upper sepal almost spathulate; petals 2–2.5 cm × 5 mm, lanceolate, acute; sepals and petals white. Lip 2 × 1–1.5 cm, 3-lobed, ovate, white with large yellow or orange patches ringed with reddish orange; keels 3, 2 long and 1 short. Column c. 1 cm. *W Himalaya to Burma, China, Thailand & Laos*. G2. Spring–summer.

*C. **corymbosa** Lindley. Illustration: Botanical Magazine, 6955 (1887); Northen, Miniature orchids, 57 (1980); Bechtel et al., Manual of cultivated orchid species, 185 (1981). Pseudobulbs 2.5–4 × 1.5–2.5 cm, ovoid to fusiform. Flowers to 7 cm wide; lip 3 cm. *E Himalaya & NE India*. G2. Summer.

*C. **corrugata** Wight (*C. nervosa* Richard). Illustration: Wight, Icones Plantarum Indiae Orientalis, t. 1639 (1851); Botanical Magazine, 5601 (1866). Pseudobulbs wrinkled. Sepals and petals more than 1 cm wide, elliptic. Lip with 3 equal wavy keels; central lobe longer than broad, mostly yellow with orange streaks. *S India*. G2. Autumn.

13. C. cristata Lindley. Illustration: Botanical Magazine, 8477 (1913); Williams et al., Orchids for everyone, 185 (1980); Bechtel et al., Manual of cultivated orchid species, 185 (1981).
Pseudobulbs 3.5–7.5 × 2–4 cm, cylindric to fusiform or ovoid, irregularly ridged, distant, 2-leaved. Leaves 12–30 × 2–3 cm, lanceolate, acute; stalk short or absent. Flowering stem 15–30 cm, erect or arched, basal, with 3–10 flowers, sheathed at base, bracts persistent; flowers 7–9 cm wide, white, very fragrant. Sepals and petals similar, 3.5–5.5 × 1.5–2 cm, oblong to lanceolate, acute or obtuse. Lip 3–4 × 3–3.5 cm, 3-lobed; keels 3, yellow and fringed, 2 long and 1 short. Column 2.5 cm. *E Himalaya*. G2. Winter–spring.

14. C. asperata Lindley. Illustration: Orchid Review **84**: 338 (1976); Millar, Orchids of Papua New Guinea, 74 (1978).
Differs from *C. pandurata* in its pale cream-coloured flowers 6–7 cm wide; lip *c.* 3 cm, lined and speckled with brownish orange. *Malaysia, Sumatra to New Guinea*. G2. Spring–summer.

The illustration under this name in Sheehan & Sheehan, Orchid genera illustrated, 67 (1979) is probably of *C. pandurata*.

15. C. viscosa Reichenbach (*C. graminifolia* Parish & Reichenbach). Illustration: Botanical Magazine, 7006 (1885); Dansk Botanisk Arkiv **29**(4): 39 (1975).
Pseudobulbs 2.5–8 cm, ovoid, ribbed when mature, clustered, 2-leaved. Leaves 20–45 × 1–3 cm, linear. Flowering stem 10–15 cm, basal, erect, sheathed at base, bearing 2–7 flowers; bracts deciduous; flowers *c.* 5 cm wide, white, apparently fragrant. Sepals 2–3.5 cm × 6–9 mm, lanceolate; petals narrower. Lip 2 × 1–1.5 cm, 3-lobed, white; lateral lobes brownish-veined; central lobe dark yellow; keels 3, finely wavy, the middle one less distinct, purplish apically. *SW China, N India, Burma, Vietnam & Malaysia*. G2. Winter.

16. C. trinervis Lindley (*C. cinnamomea* Teijsmann & Binnendijk). Illustration: Smith, Orchideen von Java **2**: t. 109 (1905); Dansk Botanisk Arkiv **29**(4): 47 (1975).
Pseudobulbs to 6 × 3 cm, ovoid to narrowly ovoid, *c.* 2 cm apart, 2-leaved. Leaf-blade to 40 × 4.5 cm, lanceolate, acute, distinctly 3-veined beneath, leaf-stalk *c.* 7 cm. Flowering stem to 18 cm, basal, erect, sheathed at base, bearing 6–8 flowers; bracts deciduous; flowers *c.* 3 cm wide, cream or pale brown. Sepals 2.2 cm × 6 mm, lanceolate; petals 2.2 cm × 3 mm, linear. Lip to 1.8 × 1.4 cm, 3-lobed; keels 7, 3 long, 4 short. Column 1.4 cm. *Burma, Cambodia, Vietnam, Thailand, Malaysia & Java*. G2. Winter.

*C. **fuscescens** Lindley. Illustration: Annals of the Royal Botanical Garden Calcutta **8**: t. 181 (1898); Orchid Review **80**: 188 (1972); Dansk Botanisk Arkiv **29**(4): 27 (1975). Flowers to 5 cm wide, pale brown. Lip *c.* 4 × 1.5 cm; lateral lobes forward pointing, less than 5 mm long. *NE India*. G2. Winter.

Var. **brunnea** (Lindley) Lindley. Illustration: Botanical Magazine, 5494 (1865); Dansk Botanisk Arkiv **29**(4): 29 (1975). Forward pointing lateral lobes more than 5 mm. *Burma, Thailand, Laos & Vietnam*. Possibly not distinct.

*C. **lactea** Reichenbach (*C. huettneriana* misapplied). Illustration: Cogniaux & Goossens, Dictionnaire Iconographique des Orchidées, Coelogyne, t. 5 (1902); Dansk Botanisk Arkiv **29**(4): 42 (1975). Leaves to 30 × 3–5 cm. Flowers to 4 cm wide, white, sweetly scented. Lip 2 × 1.4–1.7 cm. *Burma, Thailand, Laos & Vietnam*. G2. Spring.

Further study may show this to be merely a variety of *C. flaccida* (No. 8).

43. PLEIONE D. Don

J.C.M. Alexander

Epiphytic or terrestrial, often on rocks. Pseudobulbs clustered, spherical, ovoid or barrel-shaped, often sheathed, formed from bases of previous season's flowering stems. Leaves membranous, pleated, rolled when young, deciduous, 1 or 2 per flowering stem and produced with or soon after the flowers, or, in autumn-flowering species, several months later on mature pseudobulbs (Nos. 1 and 2). Flowering stems usually 1-flowered (occasionally 2- or 3-flowered in No. 5), sheathed below. Sepals and petals free, similar. Lip entire or very shallowly 3-lobed; lateral lobes or basal portion rolled or upturned around column, often concealing it; central lobe irregularly toothed, lacerate or fringed, with 2–9 longitudinal keels. Column slender, hooded, often winged at the apex. Pollinia 4.

A genus of about 12 Asian alpines, growing mostly between 1000 and 3500 m, most of which were originally described in *Coelogyne*. They should be grown in epiphytic compost in well-drained, shallow pots or wide pans in cool conditions with temperatures never exceeding 20–25°C. In winter they should be protected from frost. Plants perform best if repotted annually; this should be done just after flowering has finished when new growth has just appeared. They should then be kept in shady, humid conditions, though no water should be given until the new roots are actively growing. Once growth ceases and the leaves drop, temperatures should be lowered and water withheld.

Literature: Hunt, P.F. & Vosa, C.G., The cytology and taxonomy of the genus Pleione, *Kew Bulletin* **25**: 423–32 (1971); Hunt, P.F. & Butterfield, I., The genus Pleione, *The Plantsman* **1**(2): 112–23 (1979).

Pseudobulbs. Barrel-shaped: **1**,**2**; spherical, slightly flattened: **6**; ovoid or pear-shaped: **3–7**. Lacking sheaths: **3**.
Leaves. Two on pseudobulbs: **1**,**2**; one on flowering stems: **3–7**. Fully developed at flowering: **3**; fully developed soon after flowering: **4–7**; fully developed several months later: **1**,**2**.
Flowering stem. 4 cm or less: **4**; over 15 cm: **5**,**6**.
Flower. 5–7 cm wide: **1**,**3–7**; 8 cm wide or more: **2**,**4**,**5**. Yellow: **7**.
Lip keels. Papillose: **1**,**2**,**6**; finely toothed; **6**; sinuous or toothed: **5**; finely barbed: **3**,**4**; irregularly and bluntly toothed: **7**.
Lip margin. Irregular: **1**; irregularly toothed: **2**; finely fringed: **3**,**4**,**6**,**7**; finely toothed: **5**.

1a. Pseudobulbs barrel-shaped, narrowing abruptly above; leaves 2, on mature pseudobulbs 2
b. Pseudobulbs spherical, ovoid or pear-shaped; leaves 1, on flowering stems 3
2a. Flowers to 6 cm wide; keels on lip running almost to apex; flowering stem sheaths not warty **1. maculata**
b. Flowers to 8 cm wide; keels on lip two-thirds of lip length; flowering stem sheaths warty **2. praecox**
3a. Keels on lip barbed with fine, backwardly curving teeth 4
b. Keels on lip sinuous, irregularly cut or with upright teeth 5
4a. Leaves produced with flowers; flowering stems more than 3 times longer than pseudobulbs **3. hookeriana**
b. Leaves produced after flowers; flowering stem about 2 times longer than pseudobulbs **4. humilis**
5a. Leaves produced with flowers **5. bulbocodioides**
b. Leaves produced after flowers 6
6a. Flowers red, pink or purple **6. yunnanensis**
b. Flowers yellow to orange **7. × confusa**

1. P. maculata (Lindley) Lindley. Illustration: Botanical Magazine, 4691 (1853); The Plantsman **1**(2): 113 (1979); Bechtel et al., Manual of cultivated orchid species, 263 (1981).
On rocks. Pseudobulbs 3–4 cm, barrel-shaped, narrowing abruptly above, warty, 2-leaved. Leaves 15–30 × 2.5–3 cm, lanceolate to elliptic, acute. Flowering stems 5–10 cm, sheaths not warty, with a leafy bract above. Flowers 5–6 cm wide. Sepals and petals 3–3.5 cm × 5–7 mm, lanceolate, acute, white or pale cream; petals sometimes streaked with pink. Lip 2–3.5 × 2.5–3.5 cm, shallowly 3-lobed; lateral lobes triangular, curled around column; central lobe semicircular to oblong, wavy, margin irregular, slightly notched at tip, white with purple blotches and a central yellow patch. Keels 5–7, papillose, running almost to tip. *NE India, Bhutan, Burma & Thailand.* H5–G1. Winter.

2. P. praecox (J.E. Smith) D. Don (*Coelogyne wallichiana* Lindley; *P. lagenaria* Lindley; *P. reichenbachiana* (Moore & Veitch) Williams). Illustration: Botanical Magazine, 5370 (1863); The Plantsman **1**(2): 121 (1979); Bechtel et al., Manual of cultivated orchid species, 263 (1981).
Epiphytic or on rocks. Pseudobulbs 2–2.5 cm, barrel-shaped, narrowing abruptly above, purple with small greenish warts, 2-leaved. Leaves to 20 × 3.5 cm, lanceolate to narrowly elliptic, acute. Flowering stems to 15 cm, basal sheaths warty. Flowers *c.* 8 cm wide. Sepals and petals 4–6 cm × 5–15 mm, lanceolate, acute, pink to bluish purple. Lip 5–6 × 3 cm, shallowly 3-lobed; lateral lobes oblong to triangular, rolled around column; central lobe oblong, irregularly toothed, strongly notched, pinkish white with large marginal purple blotches and a central patch of yellow. Keels 5, papillose; lateral keels running about two-thirds of lip length. *Nepal, NE India, China & Burma.* H5–G1. Autumn.

3. P. hookeriana (Lindley) Williams. Illustration: Botanical Magazine, 6388 (1878); The Plantsman **1**(2): 114 (1979); Bechtel et al., Manual of cultivated orchid species, 263 (1981).
Epiphytic or on rocks. Pseudobulbs 2–3.5 cm, conical to ovoid, not sheathed. Leaves 1 per flowering stem, 3–15 × 3–4 cm, lanceolate to elliptic, acuminate. Flowering stems to 15 cm, bearing closely adpressed, smooth bracts below the leaf. Flowers 5–6.5 cm wide. Sepals and petals 2.5–3.5 cm, pink to pale purple, sepals broader than petals, lateral sepals sickle-shaped. Lip 2–2.5 × 3–4 cm, shallowly 3-lobed; lateral lobes bluntly triangular, not concealing column, white to pale pink; central lobe semicircular to oblong, notched, finely fringed, pink with brown blotches and a central yellow patch. Keels 5–7, finely barbed. Column winged. *NE India, Nepal, Bhutan, Thailand & Laos.* H5–G1. Spring.

4. P. humilis (J.E. Smith) D. Don. Illustration: Botanical Magazine, 5674 (1867); The Plantsman **1**(2): 117 (1979).
Epiphytic or on rocks. Pseudobulbs 2.5–5 cm, pear-shaped, sheathed. Leaves 1 per flowering stem, 20–30 × 4–5 cm, produced after flowering, lanceolate. Flowering stem *c.* 4 cm, sheathed at base. Flowers 7–9 cm wide. Sepals and petals 3.5–4.5 cm, lanceolate to narrowly oblong, pale pinkish white, sepals broader than petals. Lip 3–4 × 2.5–3 cm, oblong, more or less unlobed, base rolled around column, apical portion finely fringed, pale pink strongly marked with crimson or orange. Keels 5–7, barbed. *NE India, Nepal & Burma.* H5–G1. Winter–spring.

5. P. bulbocodioides (Franchet) Rolfe (*Coelogyne pogonioides* Rolfe; *P. delavayi* (Rolfe) Rolfe; *P. formosana* Hayata; *P. pricei* Rolfe; *P. henryi* (Rolfe) Schlechter; *P. limprichtii* Schlechter). Illustration: Botanical Magazine, 8729 (1917) & n.s., 397 (1962), 421 (1963).
Epiphytic or on rocks. Pseudobulbs 2–4.5 cm, conical, ovoid or pear-shaped, pale or dark green to purple, sheathed. Leaves 1 per flowering stem, 10–50 × 3–5 cm, lanceolate to elliptic, acute, reaching full size after flowering. Flowering stems 10–20 cm, sheathed, with 1 (rarely 2 or 3) flowers. Flowers 5–12 cm wide. Sepals and petals 2.5–4 cm × 5–15 mm, lanceolate to asymmetrically elliptic, acute, pink to pale purple. Lip 2.5–4.5 × 3–4 cm, shallowly 3-lobed to bluntly diamond-shaped or almost circular, base rolled around column, apical portion finely toothed to lacerate, shallowly notched, white to pink, heavily spotted with pale brown or purplish pink or both. Keels 2–7, wavy or toothed. *W China, Burma & Taiwan.* H5–G1. Spring.

Variable. The synonyms quoted are frequently used in the horticultural trade to describe well-known forms which were formerly regarded as separate species; recent research suggests that some of them should be resurrected as valid species.

6. P. yunnanensis (Rolfe) Rolfe. Illustration: Botanical Magazine, 8106 (1906); Orchid Review **83**: 9 (1975).
On rocks. Pseudobulbs 2–3 cm, spherical, flattened, shiny, sheathed. Leaves 1 per flowering stem, 20–30 × 2–3 cm, lanceolate to elliptic, produced after flowering. Flowering stem 10–20 cm, sheathed at base. Flowers 5–7 cm wide. Sepals and petals 3–4 cm, oblong to lanceolate, obtuse, pinkish purple. Sepals

broader than petals. Lip oblong to circular; base rolled around column; apical portion finely fringed, purplish pink with purple blotches. Keels 5, finely toothed or papillose. *China (Yunnan)*. H5–G1. Spring.

7. P. × confusa Cribb & Tang (*P. forrestii* misapplied). Illustration: Botanical Magazine n.s., 501 (1967); Orchid Review **83**: 10 (1975).
On rocks. Pseudobulbs 2.5 cm, pear-shaped, sheathed. Leaves 1 per flowering stem, to 15 × 4 cm, lanceolate to elliptic, produced after flowering. Flowering stem 2–5 cm, sheathed at base. Flowers 5–6 cm wide. Sepals and petals *c.* 4 cm, lanceolate to elliptic, yellow or orange. Lip almost circular, obscurely 3-lobed, its base rolled around column; blade finely fringed, yellow with purple blotches; keels 5 or 6, irregularly and bluntly toothed. *Burma & China (Yunnan)*. H5–G1. Spring.

Recent cytological and morphological studies have shown that this plant, generally known as *P. forrestii*, is a natural hybrid between *P. forrestii* Schlechter and *P. albiflora* Cribb & Tang. True *P. forrestii* has only recently been re-introduced to cultivation.

44. DENDROCHILUM Blume
J.C.M. Alexander
Epiphytic, rarely on rocks. Pseudobulbs fusiform, conical or ovoid, clustered or distant, 1- or 2-leaved. Leaves flat, narrow, leathery, stalked, tapered at both ends, rolled in bud. Flowering stems lateral, long and slender, arched, bearing many flowers in dense, 2-ranked racemes or spikes. Flowers small, fragrant. Sepals and petals spreading, similar; upper sepal often keeled; lateral sepals slightly fused to column foot. Lip simple or 3-lobed, fleshy below, loosely hinged to column foot, often with 2 or 3 keels. Column short, with narrow side lobes and apical wings. Pollinia 4, pear-shaped.

A genus of 120–150 species from SE Asia. Most of the commonly cultivated species are in section *Platyclinis* which has sometimes been recognised as a distinct genus. Cultivation conditions are similar to those for the tropical species of *Coelogyne* (p. 179).

1a. Flowers less than 1 cm wide **1. filiforme**
b. Flowers more than 1 cm wide 2
2a. Lip distinctly shorter than petals and sepals; lateral lobes about half as long as central lobe; central lobe obtuse or rounded **2. glumaceum***
b. Lip about equal to petals and sepals; lateral lobes very small or absent; central lobe wedge- or fan-shaped **3. cobbianum**

1. D. filiforme Lindley (*Platyclinis filiformis* (Lindley) Bentham). Illustration: Orchid Review 77: 154 (1969); Skelsey, Orchids, 109 (1979); Sander, Orchids and their cultivation, t. 39 (1979).
Epiphytic. Pseudobulbs 2.5 × 1 cm, ovoid, densely clustered, 1- or 2-leaved. Leaves 12–20 × 1–2 cm, linear-lanceolate, acute. Flowering stem 20–45 cm, thread-like, arched, bearing up to 100 flowers or more. Flowers *c.* 6 mm wide in 2 ranks, fragrant, pale yellow or greenish yellow. Sepals and petals 2–3 × 1–1.5 mm, lanceolate to ovate or elliptic. Lip 2 × 2 mm, heart-shaped, shallowly notched, 2-ridged. Column *c.* 1 mm. *Philippines*. G2. Summer.

2. D. glumaceum Lindley (*Platyclinis glumacea* (Lindley) Bentham). Illustration: Botanical Magazine, 4853 (1855); Graf, Tropica, 724 (1978); Bechtel et al., Manual of cultivated orchid species, 203 (1981).
Epiphytic. Pseudobulbs 2–5 × 1–2 cm, ovoid, densely clustered, 1-leaved, fusiform, covered in red sheaths when young. Leaves 25–30 × 2–4 cm, narrowly lanceolate, acute, with a short stalk. Flowering stem to 40 cm, thread-like, arched, bearing many flowers; bracts spreading, papery, acute. Flowers 1.2–1.8 cm wide, 2-ranked, fragrant, off-white to pale yellow. Sepals 7–10 × 1–2 mm, linear-lanceolate, acute; petals 4–7 × 1 mm, linear-lanceolate. Lip 3–4 × 2 mm, 3-lobed, pale green, yellow or white, with 2 thick basal ridges; lateral lobes broadly triangular, rounded or obtuse, central lobe circular to broadly ovate, curved downwards. Column erect. *Philippines*. G2. Spring.

***D. latifolium** Lindley (*Platyclinis latifolia* (Lindley) Hemsley). Illustration: Orchid Review 77: 155 (1969). Flowers greenish yellow. Lip striped with brown; lateral lobes linear to narrowly triangular, acute, deeply and finely toothed; central lobe obovate, obtuse. *Philippines*. G2. Spring.

3. D. cobbianum Reichenbach (*Platyclinis cobbiana* (Reichenbach) Hemsley). Illustration: Sheehan & Sheehan, Orchid genera illustrated, 83 (1979); Williams et al., Orchids for everyone, 187 (1980); Bechtel et al., Manual of cultivated orchid species, 203 (1981).
Epiphytic. Pseudobulbs 2.5–8 × 1–2 cm, narrowly conical, tapering above, 1–leaved. Leaves 6–35 × 2.5–6 cm, lanceolate to oblong, acuminate, softly leathery. Flowering stem 35–50 cm, slender, arched; bracts 4–7 mm, papery. Flowers 1.2–1.8 cm wide, in 2 twisted ranks, slightly spaced out, fragrant. Sepals and petals 7–10 × 2–3 mm, oblong to elliptic, obtuse, white. Lip 6–8 × 3 mm, wedge- or fan-shaped, yellow to orange; callus oblong. *Philippines*. G2. Autumn.

45. PHOLIDOTA Hooker
J.C.M. Alexander
Epiphytic with creeping rhizomes. Pseudobulbs clustered, distant, or in branching chains, 1- or 2-leaved; leaves rolled when young. Flowering stems terminal, slender, erect, arched or pendent, several- to many-flowered; axis often zig-zag; bracts conspicuous, sometimes falling early. Flowers small, alternate in 2 ranks. Sepals and petals spreading or pointing forwards; sepals concave, lateral sepals often keeled; petals similar to sepals or narrower. Lip fused to column base, pouched below, with a small deflexed blade, often 2- or 3-lobed. Column short, winged around anther. Pollinia 4.

About 40 species from India, China and SE Asia. Cultivation conditions are generally similar to those for the tropical species of *Coelogyne* (p. 179) though less water should be given from when the pseudobulbs are fully formed until new growth appears.

1a. Pseudobulbs developed on top of older pseudobulbs, forming long, pendent chains **1. articulata**
b. Pseudobulbs developed side by side **2. pallida***

1. P. articulata Lindley. Illustration: Smith, Orchideen von Java **2**: t. 115 (1909); Orchid Review **85**: 300 (1977); Banerjee & Thapa, Orchids of Nepal, 122 (1978).
Pseudobulbs to 10 × 1.5 cm, cylindric, 2-leaved, developed on top of older pseudobulbs, forming long pendent, sometimes branched chains. Leaves 8–16 × 3–5 cm, elliptic, acute; leaf-stalk *c.* 1 cm. Flowering stem arched, *c.* 15 cm; bracts *c.* 1 cm, falling as flowers open. Flowers 10–20, *c.* 1.2 cm wide, fragrant, pink to dull brown. Sepals and petals similar, 7–8 mm, spreading, ovate to elliptic, acute or obtuse. Lip *c.* 5 mm; pouch with 5 yellow ridges, tapering to junction with blade; blade 3 × 5 mm, kidney-shaped, 2-lobed, orange at base. Column 3–4 mm. *Himalaya through SE Asia to Java*. G2. Spring–summer.

2. P. pallida Lindley (*P. imbricata* Lindley 1827, not 1824). Illustration: Shuttleworth et al., Orchids, 42 (1970); Sheehan & Sheehan, Orchid genera illustrated, 141 (1979); Bechtel et al., Manual of cultivated orchid species, 261 (1981).
Pseudobulbs 3–6 cm conical, 1-leaved. Leaves 30 × 6 cm, lanceolate to elliptic, acute; leaf-stalk to 5 cm. Flowering stem to 30 cm, pendent, many-flowered; bracts 8–10 mm, persistent. Flowers 6–7 mm wide, off-white to pale pink. Upper sepal 5 × 3 mm; lateral sepals to 6 mm, fused below, keeled. Petals 4 × 1.5 mm, linear to oblong. Upper sepal and petals curved over column. Column almost circular, 3–4 mm. *Burma & S China to Australia*. G2. Spring–summer.

***P. ventricosa** (Blume) Reichenbach. Illustration: Smith, Orchideen von Java 2: t. 114 (1909). Pseudobulbs 2-leaved. Leaf-stalks to 15 cm. Flowering stem erect; flowers 1–1.2 cm wide, greenish white. *Sumatra, Java & Borneo*. G2.

***P. chinensis** Lindley. Illustration: Sander, Orchids and their cultivation, t. 35 (1979); Williams et al., Orchids for everyone, 192 (1980). Pseudobulbs spherical to oblong, 2-leaved. Flowering stem pendent; flowers 1.2–2 cm. Sepals and petals pinkish brown, petals linear; lip white. *China & Hong Kong*. G2. Spring–summer.

46. HEXISEA Lindley

J.C.M. Alexander

Epiphytic. Stems sheathed, sometimes branched, composed of chains of successive pseudobulbs. Leaves linear to strap-shaped, folded when young, in opposite pairs at the apex and upper nodes of the stems. Flowers few, in short terminal racemes on short scale-bearing stalks. Sepals and petals similar, spreading; lateral sepals fused to column foot, forming a small mentum. Lip entire or shallowly lobed, divided into a narrow basal claw and a sharply deflexed blade; claw pressed against or fused to column forming a nectary. Column club-shaped, with a broad hollow foot, trifid. Pollinia 4 (in ours).

A genus of 5 species from tropical America, similar to *Scaphyglottis* but lacking a movable lip. Cultural requirements are similar to those for *Cattleya* (p. 191), though they can also be grown on bark slabs, and generally require shadier conditions throughout the year. Propagation is best effected by separating the stems when repotting.
Literature: Dressler, R.L., The genus Hexisea, *Orquidea (Mexico)* 4: 197–200 (1974).

1. H. bidentata Lindley. Illustration: Botanical Magazine, 7031 (1888); Sheehan & Sheehan, Orchid genera illustrated, 105 (1979); Bechtel et al., Manual of cultivated orchid species, 216 (1981).
To 50 cm. Pseudobulbs to 10 × 1 cm, brownish purple, grooved. Leaves to 15 × 1.5 cm, strap-shaped, dark green, bifid or trifid at apex. Flowers 2.5–3 cm wide, bright reddish orange. Sepals 1.5–2 cm × 5 mm, lanceolate, acute or obtuse; petals similar but smaller. Lip to 1.5 cm × 5 mm, oblong, obtuse with a narrow claw fused to the column. Column 3–4 mm. *C America to Venezuela, Colombia & Peru*. G2. Flowering intermittent.

47. EPIDENDRUM Linnaeus

J. Cullen

Terrestrial or epiphytic. Stems usually elongate and leafy, more rarely pseudobulb-like with 1–2 leaves, erect or hanging. Leaves acute or notched at the apex, folded when young. Flowers in racemes or panicles which are rarely umbel-like, bracts small or conspicuous. Flowers resupinate or not, variable in size. Sepals and petals similar, spreading. Lip joined to the column for most of the length of the latter, variable in shape, entire or 3-lobed. Pollinia 4, laterally flattened, parallel.

A genus of about 750 species from the New World. Its relationships with other genera are controversial: here, *Epidendrum* is restricted to plants with the column united for all or most of its length to the lip. Species with the lip and column mostly free from each other are included in *Encyclia* (p. 186) and *Barkeria* (p. 191).

Plants should be potted in early spring in an epiphytic compost, using the smallest size of pot possible. Propagation is by division or by means of the young plantlets produced by some species after the flowering spikes have been cut back.
Literature: Ames, O., Hubbard, F.T. & Schweinfurth, C., *The genus Epidendrum in the United States and Middle America* (1936).

Main axis. Stem-like: **2–4,7**; pseudobulb-like: **1,5,6**.
Inflorescence. Panicle with many flowers: **1**; spherical raceme with many flowers: **2,3**; elongate raceme with up to 5 flowers: **4–7**.
Sepals and petals. To 3 cm: **1–3**; more than 3 cm: **4–7**.
Lip. Not lobed: **7**; lobes more or less equal, the lip forming a cross-shaped structure at the top of the column: **2,3**; lobes unequal, not as above: **1,4–6**.

1a. Lip not lobed, though sometimes fringed or notched at the apex **7. medusae***
b. Lip distinctly 3-lobed 2
2a. Central lobe of lip narrowly triangular, narrowly oblanceolate or linear, projecting between and beyond the lateral lobes 3
b. Central lobe of lip various but not as above 5
3a. Stem elongate, bearing several leaves along its length **4. nocturnum**
b. Stem short, pseudobulb-like, bearing 1–2 leaves at its apex 4
4a. Lateral lobes of the lip deeply fringed along their outer margins **5. ciliare**
b. Lateral lobes of the lip entire or at most toothed along their outer margins **6. parkinsonianum***
5a. Inflorescence an almost spherical raceme borne on a long stalk covered with bladeless sheaths, which terminates the leafy shoot; lobes of the lip approximately equal in size, forming a cross at the apex of the column, often all fringed 6
b. Inflorescence usually a panicle, rarely an elongate raceme, borne directly on the leafy shoot or on an entirely leafless shoot; lobes of the lip various, usually unequal, rarely fringed, and not forming a cross at the top of the column **1. stamfordianum***
6a. Stem arching or scrambling, rooting from many of the nodes **2. ibaguense**
b. Stem erect, not rooting at all from the nodes **3. arachnoglossum***

1. E. stamfordianum Bateman. Illustration: Fieldiana 26: 381 (1953); Bechtel et al., Manual of cultivated orchid species, 210 (1981).
Stems to 25 cm, erect, increasing in thickness upwards, then tapering, bearing 4–6 leaves near the apex. Leaves 13–24 × 3–7 cm, linear-oblong to linear-elliptic, slightly notched at the apex. Panicle (rarely raceme) arising on a leafless stalk from the base of the leafy stem or, more rarely, terminating a leafy shoot. Flowers greenish yellow spotted with red. Sepals and petals similar, 1.4–2.5 cm. Lip deeply 3-lobed, the lateral lobes oblong, recurved, the central deeply notched with divergent lobes, margins fringed. *Guatemala*. G2. Spring.

Two other species which are in

cultivation will key out here:

*E. **paniculatum** Ruiz & Pavon (*E. floribundum* Humboldt, Bonpland & Kunth). Illustration: Botanical Magazine, 3637 (1838). With tall, cane-like stems terminated by a panicle of flowers with greenish brown, reflexed sepals and very narrow, reflexed petals and a white lip with very divergently lobed central lobe. *C America, northern S America, West Indies*. G2.

*E. **myrianthum** Lindley (*E. verrucosum* Swartz var. *myrianthum* (Lindley) Ames & Correll). Illustration: Bateman, Second century of orchidaceous plants, t. 163 (1867). Like *E. paniculatum* but with warty leaf-sheaths and lilac-purple sepals, petals and lip. *Guatemala*. G2.

2. E. ibaguense Humboldt, Bonpland & Kunth (*E. radicans* Ruiz & Pavon). Illustration: Hamer, Orquideas de El Salvador **1**: 211 (1974); Bechtel et al., Manual of cultivated orchid species, 209 (1981).
Stems to 1 m, scrambling or arching, producing roots from many of the nodes. Leaves ovate-oblong to oblong, notched at the apex. Raceme more or less spherical, borne on a long stalk covered with bladeless sheaths, which terminates the leafy shoot. Flowers variable in colour, orange, scarlet red, purplish red or yellow. Sepals and petals erect, 1–2.2 cm. Lip forming a cross at the top of the column, 3-lobed, the lobes all similar or the central larger, deeply fringed, the central 2-lobed. *C America, northern S America, West Indies*. G2. Flowers most of the year.

E. × o'brienianum Rolfe (*E. ibaguense* × *E. evectum* Hooker) is similar but has a longer inflorescence, the sepals and petals orange and the column tipped with yellow.

3. E. arachnoglossum Reichenbach. Illustration: Revue Horticole 1882: 554. Very similar to *E. ibaguense* but stems erect, not rooting, and flowers purplish pink throughout. *Colombia*. G2. Summer.

*E. **xanthinum** Lindley. Illustration: Botanical Magazine, 7586 (1898). Similar, but flowers yellow and the apex of the column bright red. *Brazil*.

*E. **cnemidophorum** Lindley. Illustration: Botanical Magazine, 5656 (1867). Also similar but with a more elongate inflorescence and the lip more obscurely 3-lobed, with entire lobes. *C America*.

4. E. nocturnum Jacquin. Illustration: Fieldiana **26**: 347 (1953); Bechtel et al., Manual of cultivated orchid species, 210 (1981).
Stem to 1 m, erect, leafy. Leaves oblong-lanceolate to elliptic, unequally notched at the apex. Raceme terminal with up to 5 flowers, borne directly on the leafy stems, with conspicuous, boat-shaped bracts. Sepals greenish white, petals white, all 3.5–9 cm, spreading. Lip white, 3-lobed, the lateral lobes reversed-triangular with finely toothed outer margins, the central lobe 2–2.5 cm, very narrowly triangular, projecting between and beyond the lateral lobes. *USA (Florida), C America, northern S America, West Indies*. G2. Summer–autumn.

5. E. ciliare Linnaeus. Illustration: Botanical Magazine, 463 (1799); Cogniaux & Goossens, Dictionnaire Iconographique des Orchidées, Epidendrum, t. 6 (1898); Fieldiana **26**: 319 (1953); Bechtel et al., Manual of cultivated orchid species, 209 (1981).
Stems short, pseudobulb-like, bearing 1–2 leaves at the apex. Leaves oblong-elliptic, somewhat unequally notched at the apex. Raceme terminal, with few flowers and conspicuous, boat-shaped bracts. Sepals and petals 4–9 cm, erect, pale green or yellowish. Lip white, 3-lobed, the lateral lobes narrowly reversed-triangular with deeply fringed outer margins, the central lobe 2.5–6 cm, linear, projecting between and beyond the lateral lobes. *C America, northern S America, West Indies*. G2. Winter.

6. E. parkinsonianum Hooker (*E. falcatum* Lindley). Illustration: Botanical Magazine, 3778 (1840).
Stems short, pseudobulb-like, usually hanging, bearing 1–2 or rarely more leaves. Leaves linear-lanceolate to lanceolate, acute. Raceme terminal with 1–3 flowers and conspicuous, boat-shaped bracts. Sepals and petals 5.5–8.5 cm, spreading, white, yellowish or yellowish green, becoming pale orange if not fertilised. Lip white or yellowish, deeply 3-lobed, the lateral lobes reversed-triangular with finely toothed outer margins, the central lobe 3.5–5.5 cm, projecting between and beyond the lateral lobes. *C America*. G2. Autumn.

*E. **oerstedtii** Reichenbach (*E. costaricense* Reichenbach). Illustration: Cogniaux & Goossens, Dictionnaire Iconographique des Orchidées, Epidendrum, t. 3 (1897); Botanical Magazine, n.s., 375, (1962). Similar but with elliptic-oblong leaves and the central lobe of the lip narrowly oblanceolate. *Costa Rica*.

7. E. medusae (Reichenbach) Veitch (*Nanodes medusae* Reichenbach). Illustration: Botanical Magazine, 5723 (1868); Bechtel et al., Manual of cultivated orchid species, 210 (1981).
Stems to 20 cm, usually leafy, hanging. Leaves greyish green, lanceolate, tapering to an acute apex. Raceme with 1–3 large flowers. Sepals and petals 3–4 cm, spreading, yellowish green flushed with red. Lip broadly kidney-shaped, broader than long (to 6 cm broad), purplish brown, deeply fringed all round. *Ecuador*. G2. Summer.

There are several other rarely cultivated species, which, though unlike each other and *E. medusae*, will key out here because of their simple lips:

*E. **diffusum** Swartz. Illustration: Botanical Magazine, 3565 (1837). Stems erect, leaves ovate, panicle wide, of small flowers (sepals and petals *c.* 8 mm) which are entirely yellowish green. *C America*.

*E. **eximium** Williams. Illustration: Orquidea **2**: 245 (1972). Stems erect, cane-like, raceme of several flowers with yellow sepals and petals (*c.* 2.5 cm) and an oblong, entire, yellow, red-lined lip. *Mexico, El Salvador*.

*E. **latilabrum** Lindley. Illustration: Martius, Flora Brasiliensis **3**(5): t. 31 (1898). Stems erect, zig-zag, cane-like, raceme few-flowered, umbel-like, sepals and petals 1.8–3.3 cm and the lip entire, broader than long, deeply notched. *Brazil*.

*E. **loefgrenii** Cogniaux. Illustration: Martius, Flora Brasiliensis **3**(5): t. 49 (1898). Stems hanging with broad, overlapping leaves and a raceme of few, very small (*c.* 1 cm in diameter) greenish flowers with a broadly obovate, unfringed, notched lip. *Brazil*.

48. ENCYCLIA Hooker
J.C.M. Alexander
Epiphytic, with creeping rhizomes. Pseudobulbs erect, swollen, basally enclosed in chaffy sheaths. Leaves 1–4, folded when young, borne on top of the pseudobulb. Flowers in terminal racemes or panicles. Sepals and petals free, often similar. Lip often 3-lobed, the lobes clasping of the column. Claw of lip joined to the column for up to half the length of the latter, column usually terete, often apically winged. Pollinia 4; rostellum at right angles to column.

A genus of about 150 species from C & S America, most of which were originally in *Epidendrum*. Cultivation as for *Epidendrum* (p. 185).
Literature: Dressler, R.L. & Pollard, G.E., *The genus Encyclia in Mexico* (1974).

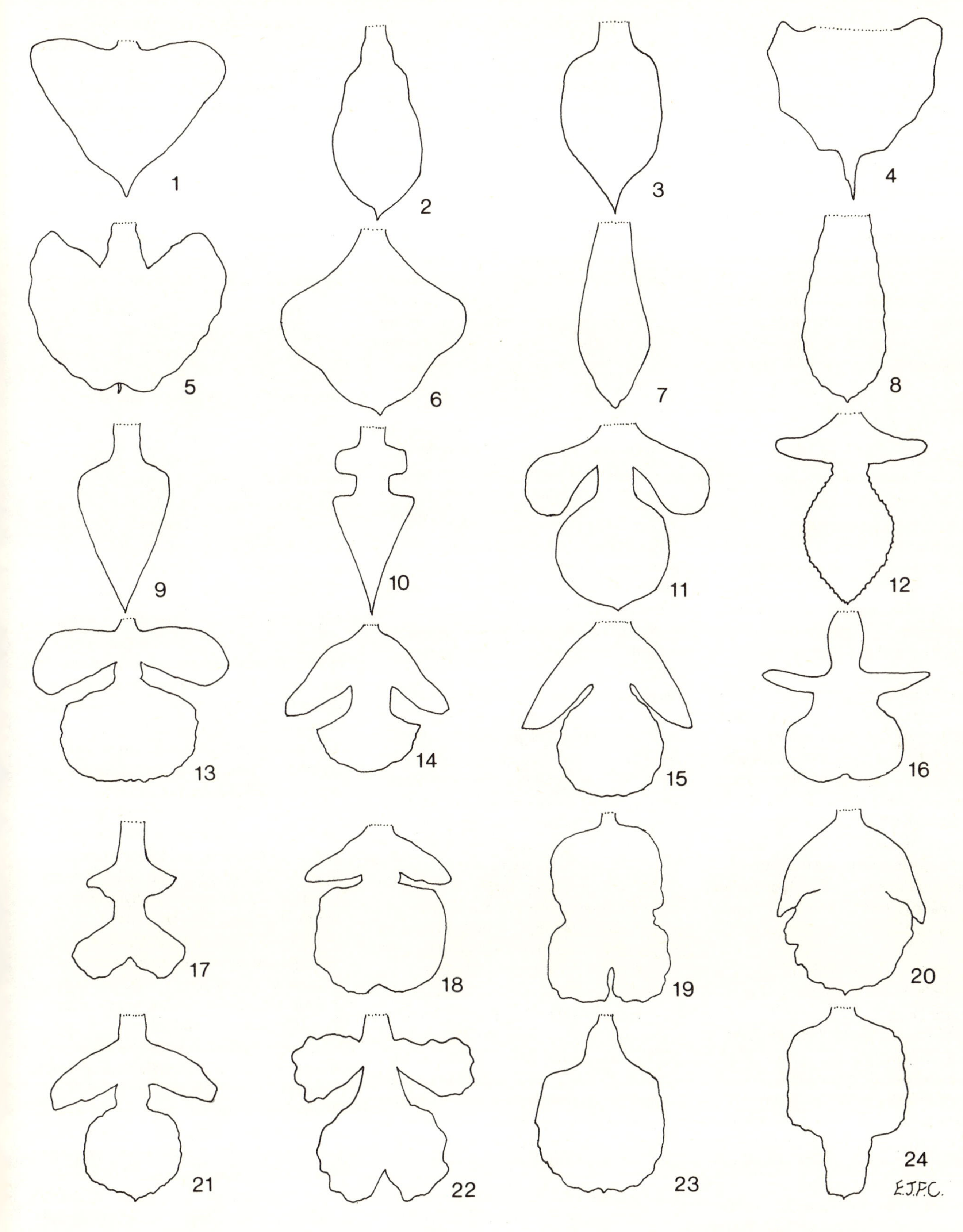

Figure 14. Lips of *Encyclia* species. 1, *E. cochleata*. 2, *E. glumacea*. 3, *E. fragrans*. 4, *E. baculus*. 5, *E. radiata*. 6, *E. vespa*. 7, *E. vitellina*. 8, *E. boothiana*. 9, *E. brassavolae*. 10, *E. prismatocarpa*. 11, *E. patens*. 12, *E. adenocaula*. 13, *E. alata*. 14, *E tampensis*. 15, *E. bractescens*. 16, *E. michuacana*. 17, *E. varicosa*. 18, *E. cordigera*. 19, *E. mariae*. 20, *E. aromatica*. 21, *E. selligera*. 22, *E. advena*. 23, *E. polybulbon*. 24, *E. citrina*. (Compiled from various sources; not to scale.)

Pseudobulbs. Clustered: **1,5–14**; separated: **1–4,6,7,10,15**. Bearing a single leaf: **2,10**.
Flower stalks. Warty: **8**.
Flowers. In racemes: **1–7,10–14**; in panicles: **5,8–11,14**. Solitary: **13,15**; Two or three per spray: **2,3,5,6,10,11–13**; four or more per spray: **1,2,4–12,14**.
Lip. At the top of flower: **1–4**; at bottom of flower: **5–15**. Unlobed: **1–6,12,15**; lobed: **7–14**. More than 3 cm: **6,8,11–14**; less than 3 cm: **1–5,7,9,10,14,15**. With blade or central lobe tapering to an acute or obtuse point: **1–8**; with blade or central lobe not tapering to an acute or obtuse point: **4,9–15**. Papillose or warty: **10**.
Petals. More than 4 times as long as broad: **1–3,6–12,15**; less than 4 times as long as broad: **4,5,7,10,11,13,14**.

1a. Flowers with lip at top (not resupinate) 2
b. Flowers with lip at bottom (resupinate) 5
2a. Pseudobulbs bearing a single leaf **2. fragrans**
b. Pseudobulbs bearing 2 or more leaves 3
3a. Sepals twisted or drooping, at least 5 times as long as broad 4
b. Sepals firm and erect, at most 3 times as long as broad **4. radiata***
4a. Petals at most 6 times as long as broad; flowers 2 or 3 per raceme; lip cream-coloured with radiating purple lines **3. baculus**
b. Petals at least 8 times as long as broad; flowers 4 or more per raceme; lip purplish brown, flushed with yellowish green **1. cochleata***
5a. Lip blade or central lobe more or less isodiametric, not tapering to a point 6
b. Lip blade or central lobe usually longer than broad, tapering to an obtuse or acute point 12
6a. Lip (including claw) 3 cm or more 7
b. Lip (including claw) 2.5 cm or less 9
7a. Flowers erect or spreading, 3–15 per raceme; lip 4 cm or less, deeply 3-lobed, pink **11. cordigera**
b. Flowers pendent, 1–4 per raceme; lip 4.5 cm or more, unlobed or shallowly 3-lobed, white or yellow 8
8a. Flowers green and white; lip notched **12. mariae**
b. Flowers yellow; lip truncate **13. citrina**
9a. Lip without lateral lobes on blade or claw **15. polybulbon**
b. Lip with lateral lobes on blade or claw 10
10a. Petals hooded at tip **14. selligera***
b. Petals not hooded at tip 11
11a. Lateral lobes of lip 12 × 6 mm or more, strongly constricted at base **9. alata**
b. Lateral lobes of lip 10 × 4 mm or less, triangular, lanceolate or narrowly spathulate **10. tampensis***
12a. Lip without lateral lobes on claw 13
b. Lip with lateral lobes on claw 14
13a. Flower-stalks and ovaries covered in warts **8. adenocaula**
b. Flower-stalks and ovaries not covered in warts **7. prismatocarpa***
14a. Lip (including claw) 3 cm or more; petals and sepals linear, 3 cm or more, yellow or yellowish green **6. brassavolae**
b. Lip (including claw) 2 cm or less; petals and sepals elliptic or ovate, 2.5 cm or less, scarlet or orange **5. vitellina***

1. E. cochleata (Linnaeus) Lemee (*Hormidium cochleatum* (Linnaeus) Brieger). Figure 14(1), p. 187. Illustration: Botanical Magazine, 572 (1801); Luer, Native orchids of Florida, t. 60 (1972); Dressler & Pollard, The genus Encyclia in Mexico, t. 7 (1974).
Pseudobulbs 5–26 cm, ellipsoid to pear-shaped, sometimes stalked, loosely clustered. Leaves 2–4 per pseudobulb, 20–33 × 3–4 cm, elliptic to lanceolate, acute. Raceme to 50 cm, many-flowered. Sepals and petals 3–7.5 cm × 3–7 mm, narrowly lanceolate, acute, twisted and drooping, pale green. Lip uppermost, 1–2.1 × 1.3–2.6 cm, triangular to heart-shaped, concave, deep purple flushed with yellowish green; base white with deep purple veins. Column 7–9 mm, green with purple spots; base fused to claw of lip. *USA (Florida); S & W Mexico to Colombia & Venezuela.* G2. Flowers continuously.

*E. **glumacea** (Lindley) Pabst (*Hormidium glumaceum* (Lindley) Brieger). Figure 14(2), p. 187. Illustration: Edwards's Botanical Register **26**: t. 6 (1840). Differs in its white flowers streaked with pink and its narrower acuminate lip, 1.8–2.5 cm × 8–12 mm. Base of raceme enclosed in long chaffy scales. *E Brazil & Ecuador.* G2.

2. E. fragrans (Swartz) Lemee (*Hormidium fragrans* (Swartz) Brieger). Figure 14(3), p. 187. Illustration: Botanical Magazine, 1669 (1814); Dunsterville & Garay, Venezuelan orchids illustrated **4**: 87 (1966); Dressler & Pollard, The genus Encyclia in Mexico, t. 4 (1974).
Pseudobulbs 4.5–11 cm, narrowly ovoid to ellipsoid, 1–4 cm apart. Leaves 9–35 cm, solitary, elliptic to strap-shaped, obtuse or acute. Raceme 5–20 cm, with 2–6 white, cream or greenish, very fragrant flowers. Sepals 2–3.5 cm × 4–9 mm, lanceolate to elliptic, acuminate; petals 2–3 cm × 7–12 mm, elliptic to diamond-shaped, acuminate. Lip uppermost, 1.7–2.4 cm × 9–17 mm, ovate-acuminate with purple radial lines. Column 6–8 mm, basal half fused to lip. *Greater Antilles, S. Mexico to Brazil.* G2. Spring–summer.

3. E. baculus (Reichenbach) Dressler & Pollard (*Epidendrum pentotis* Reichenbach; *Hormidium baculus* (Reichenbach) Brieger). Figure 14(4), p. 187. Illustration: Botanical Magazine, 702 (1975); Dressler & Pollard, The genus Encyclia in Mexico, t. 1 (1974); Graf, Tropica, 729 (1978).
Similar to *E. fragrans* but differing in its larger, gradually tapering pseudobulbs 15–35 cm, bearing 2 or 3 leaves and in its short 2- or 3-flowered racemes. Petals and sepals more than 3.5 cm. Lip blade with prominent basal auricles. *S Mexico to Brazil.* G2. Spring–summer.

4. E. radiata (Lindley) Dressler (*Hormidium radiatum* (Lindley) Brieger). Figure 14(5), p. 187. Illustration: Botanical Magazine n.s., 662 (1974); Dressler & Pollard, The genus Encyclia in Mexico, t. 8 (1974); Graf, Tropica, 729 (1978).
Pseudobulbs 7–11 cm, ellipsoid to ovoid, grooved, sometimes stalked, about 2.5 cm apart. Leaves 2–4 per pseudobulb, 14–35 × 1.5–3 cm, lanceolate to elliptic, obtuse or acute. Raceme 7–20 cm, with 4–12 cream or greenish white flowers. Sepals 1.5–2 cm × 5–7 mm, elliptic; petals 1.5–2 cm × 8–11 mm, broadly elliptic to diamond-shaped. Lip uppermost, 1–1.3 × 1.3–2 cm, semicircular to kidney-shaped, notched, with purple radial lines. Column 8–10 mm, basal half fused to lip. *C & S Mexico, Guatemala, Honduras & Costa Rica.* G2. Autumn–winter.

*E. **vespa** (Vellozo) Dressler (*Epidendrum variegatum* Hooker; *Hormidium variegatum* (Hooker) Brieger). Figure 14(6), p. 187. Illustration: Botanical Magazine, 3151 (1832); Pabst & Dungs, Orchidaceae Brasilienses **1**: 187 (1975); Graf, Tropica, 729 (1978). Differs in its small pale yellowish green lip scarcely longer than the column and its yellowish green petals and sepals densely spotted with blackish purple. Leaves faintly mottled. *West Indies, Venezuela, Guyana, Surinam, Ecuador, Peru & Brazil.* G2. Autumn–summer.

5. E. vitellina (Lindley) Dressler. Figure 14(7), p. 187. Illustration: Botanical Magazine, 4107 (1844); Die Orchidee **22**: 45 (1971); Dressler & Pollard, The genus Encyclia in Mexico, t. 31 (1974).
Pseudobulbs 2.5–5 cm, conical to ovoid, clustered. Leaves 1–3, 7–22 × 1–4 cm, lanceolate to elliptic, obtuse or acute. Raceme or panicle 12–30 cm, with 4–12 flowers. Sepals and petals lanceolate to elliptic, acute, orange to bright scarlet; sepals 1.5–2.5 cm × 3–8 mm; petals 1.5–2.5 cm × 6–11 mm. Lip 1.1–1.6 cm, yellow or orange with a red tip; blade elliptic to oblong, acute with deflexed sides. Column 6–7 mm; base fused to lip. *S Mexico & Guatemala*. G2. Spring–summer.

Var. **majus** Veitch has shorter, denser racemes of larger, more brightly coloured flowers.

***E. boothiana** (Lindley) Dressler (*Diacrium bidentatum* (Lindley) Hemsley; *Hormidium boothianum* (Lindley) Brieger). Figure 14(8), p. 187. Illustration: Luer, Native orchids of Florida, t. 61 (1972); Dressler & Pollard, The genus Encyclia in Mexico, t. 29 (1974). Flowers smaller and fleshier. Sepals and petals 8–14 × 2–5 mm, rounded; yellowish green heavily spotted with purplish brown. Lip 7–10 mm, diamond-shaped or weakly 3-lobed, with down-curled margins, white to greenish yellow. *USA (Florida), S Mexico, Belize, Guatemala & West Indies*. G2. Winter.

6. E. brassavolae (Reichenbach) Dressler (*Hormidium brassavolae* (Reichenbach) Brieger). Figure 14(9), p. 187. Illustration: Botanical Magazine, 5664 (1867); Dressler & Pollard, The genus Encyclia in Mexico, t. 32 (1974); Rittershausen & Rittershausen, Orchids in colour, 177 (1979).
Pseudobulbs 9–18 cm, conical, ovoid or fusiform, clustered or up to 4.5 cm apart. Leaves 2 or 3 per pseudobulb, 15–30 × 3.5–5 cm, oblong to lanceolate, obtuse. Raceme 15–100 cm with 3–15 flowers. Sepals and petals linear-lanceolate, acuminate, pale yellowish green to brown. Sepals 3.5–5.5 cm × 3–6 mm; petals 3.2–4.7 cm × 2–4 mm. Lip 3–4.5 cm, cream-coloured with a purple tip; blade ovate to triangular, acuminate. Column 1.2–1.5 cm; basal half fused to lip. *S Mexico to W Panama*. G2. Summer–autumn.

7. E prismatocarpa (Reichenbach) Dressler (*Hormidium prismatocarpum* (Reichenbach) Brieger). Figure 14(10), p. 187. Illustration: Botanical Magazine, 5336 (1862); Shuttleworth et al., Orchids, 48 (1970); Graf, Tropica, 729 (1978).
Pseudobulbs to 30 cm, ovoid with a long 'neck'. Leaves 2–4 per pseudobulb, 25–40 × 3–6 cm, strap-shaped, obtuse. Raceme to 40 cm, many-flowered. Sepals and petals similar, 2–2.5 cm × 3–6 mm, linear-lanceolate to oblong, acute, pale greenish yellow with large purple blotches. Lip 2–2.5 cm, 3-lobed; lateral lobes *c.* 5 mm, obtuse and rounded, yellow or greenish yellow; central lobe narrowly triangular, acute, pale yellow or white with a large central patch of purple and a yellow tip. Column with basal third fused to lip, lobed and fringed at tip. *Costa Rica, Panama*. G2. Flowers intermittently.

***E. patens** Hooker (*Epidendrum odoratissimum* Lindley; not *Epidendrum patens* Swartz). Figure 14(11), p. 187. Illustration: Botanical Magazine, 3013 (1830); Pabst & Dungs, Orchidaceae Brasilienses **1**: 197 & 298 (1975). Differs in its unspotted, greenish brown, spathulate petals and shorter (1–1.5 cm) cream-coloured lip with oblong to obovate lateral lobes and an obovate central lobe. *SW Brazil*. G2. Summer.

The Botanical Magazine plate shows the raceme growing from beside the pseudobulb, which would exclude this species from *Encyclia*. Subsequent authors make no reference to this; presumably the illustration is inaccurate.

8. E. adenocaula (Llave & Lexarza) Schlechter (*Epidendrum nemorale* Lindley; *Epidendrum verrucosum* Lindley not Swartz). Figure 14(12), p. 187. Illustration: Botanical Magazine, 4606 (1851); Orchid Digest **37**: 45 (1973); Rittershausen & Rittershausen, Orchids in colour, 177 (1979).
Pseudobulbs 5–8 cm, conical to ovoid, clustered. Leaves 2 or 3 per pseudobulb, 11–35 cm × 5–30 mm, strap-shaped, acute or obtuse. Panicles 30–100 cm, many-flowered; stem, branches and ovaries covered in warts. Sepals and petals linear to lanceolate, acute, pale purplish pink; sepals 2.5–5.5 cm × 5–7 mm; petals 3–5 cm × 4–8 mm. Lip 3.5–4.5 cm, 3-lobed; lateral lobes lanceolate to oblong, *c.* 10 × 5 mm; central lobe circular to ovate, regularly scalloped, mucronate, with 1 or more streaks of dark red. Column 1.3–1.5 cm, club-shaped, apically winged; basal quarter fused to lip. *W Mexico*. G2. Spring–summer.

9. E. alata (Bateman) Schlechter. Figure 14(13), p. 187. Illustration: Botanical Magazine, 3898 (1842); Dressler & Pollard, The genus Encyclia in Mexico, t. 53(1974); Hamer, Las Orquideas de El Salvador **1**: t. 14 (1974).
Pseudobulbs 6–12 cm, conical to ovoid, clustered. Leaves 2 or 3 per pseudobulb, 20–60 × 1.5–3.5 cm, strap-shaped to narrowly elliptic, acute. Panicles 30–200 cm, many-flowered. Sepals and petals 2.2–3 cm × 5–8 mm, greenish yellow with a central patch of reddish brown near the tip; margins curled back. Sepals lanceolate to elliptic, obtuse; petals elliptic to spathulate, obtuse or mucronate. Lip 1.8–2.3 cm, 3-lobed, pale greenish yellow, lateral lobes 1.2–1.6 cm, oblong to obovate, rounded, flared; central lobe circular to transversely diamond-shaped, fleshy, veined with dark red. Column 1–1.2 cm, apically winged; basal quarter fused to lip. *S Mexico to Costa Rica*. G2. Summer.

10. E. tampensis (Lindley) Small. Figure 14(14), p. 187. Illustration: Shuttleworth et al., Orchids, 46 (1970); Luer, Native orchids of Florida, t. 57 & 58 (1972); American Orchid Society Bulletin **43**: 808 (1974).
Pseudobulbs 1–7 cm, ovoid, slightly separated. Leaves 1–3 per pseudobulb, 30–40 × 1.5–2 cm, linear to narrowly lanceolate. Panicles to 1 m, few- or many-flowered. Sepals and petals 1.2–2.2 cm × 4–6.5 mm, lanceolate to obovate, greenish yellow to brown suffused with purple. Lip 1.5–2 cm, white, 3-lobed; lateral lobes triangular, upright or slightly flared; central lobe semicircular with a central patch of pink or pink streaks. Column 10 mm, apically winged; base fused to claw of lip. *USA (Florida)*. G2. Flowers intermittently.

***E. aromatica** (Bateman) Schlechter. Figure 14(20), p. 187. Illustration: Dressler & Pollard, The genus Encyclia in Mexico, t. 63 (1974); Hamer, Las Orquideas de El Salvador **1**: t. 15 (1974). Differs in its smaller (1–1.5 cm) cream or yellow lip with red or brown veins and an oblong to diamond-shaped, wavy-edged central lobe. *SE Mexico*. G2. Spring–summer.

***E. bractescens** (Lindley) Hoehne. Figure 14(15), p. 187. Illustration: Botanical Magazine, 4572 (1851); Dressler & Pollard, The genus Encyclia in Mexico, t. 46 (1974); Northen, Miniature orchids, C-31 (1980). Pseudobulbs 3 cm or less; leaves to 25 × 1 cm. Sepals and petals narrowly lanceolate, purplish brown with greenish yellow tips. Lip with long narrow lateral lobes and circular central lobe. *S Mexico*. G2. Spring–summer.

***E. michuacana** (Llave & Lexarza) Schlechter (*Epidendrum virgatum* Lindley). Figure 14(16), p. 187. Illustration: Dressler & Pollard, The genus Encyclia in Mexico, t. 20 (1974); Hamer, Las Orquideas de El Salvador **1**: t. 23 (1974). Lip 10–11 mm, with narrow lateral lobes arising from base of blade rather than from claw. Sepals and petals dark brown. Lip cream or pale yellow often spotted with purple. *S Mexico & Guatemala*. G2. Flowers intermittently.

***E. varicosa** (Lindley) Schlechter. Figure 14(17), p. 187. Illustration: Dressler & Pollard, The genus Encyclia in Mexico, t. 22a (1974); Hamer, Las Orquideas de El Salvador **1**: t. 23 (1974). Differs from *E. tampensis* in its papillose or warty notched lip with triangular lateral lobes arising from claw, and blade of 2 oblong truncate lobes. Sepals and petals brown. Pseudobulbs with long slender necks. *S & W Mexico to Panama*. G2. Mainly winter.

11. E. cordigera (Humboldt, Bonpland & Kunth) Dressler (*Epidendrum atropurpureum* misapplied: *Epidendrum macrochilum* Hooker). Figure 14(18), p. 187. Illustration: Botanical Magazine, 3534 (1836); Dressler & Pollard, The genus Encyclia in Mexico, t. 57 (1974); Sheehan & Sheehan, Orchid genera illustrated, 89 (1979).
Pseudobulbs 3–11 cm, conical to ovoid, clustered. Leaves 2 or 3 per pseudobulb, 12.5–45 × 1.5–4.5 cm, narrowly elliptic. Racemes or panicles 15–75 cm with 3–15 flowers. Sepals and petals fleshy, 2.5–3.5 cm × 5–11 mm, green streaked with brown. Sepals obovate to elliptic, obtuse or mucronate; petals elliptic to spathulate. Lip 3.3–4 cm, cream or pink, centrally streaked with magenta, 3-lobed, lateral lobes oblong to lanceolate, central lobe broadly obovate. Column 1.5–1.7 cm; base fused to lip, white. *S Mexico to Colombia & Venezuela*. G2. Spring–summer.

12. E. mariae (Ames) Hoehne (*Hormidium mariae* (Ames) Brieger). Figure 14(19), p. 187. Illustration: American Orchid Society Bulletin **41**: 7 (1972); Dressler & Pollard, The genus Encyclia in Mexico, t. 36 (1974); Graf, Tropica, 729 (1978).
Flowers pendent, 2–4 per inflorescence. Petals and sepals green, spreading; lip white with green or yellowish green throat, unlobed or shallowly 3-lobed, apex notched. *E Mexico*. G2. Summer.

13. E. citrina (Llave & Lexarza) Dressler (*Cattleya citrina* (Llave & Lexarza) Lindley; *Epidendrum citrinum* (Llave & Lexarza) Reichenbach). Figure 14(24), p. 187. Illustration: Botanical Magazine, 3742 (1839); Dressler & Pollard, The genus Encyclia in Mexico, t. 37 (1974); Bechtel et al., Manual of cultivated orchid species, 207 (1981).
Flowers bright yellow, pendent, 1 or 2 per inflorescence. Lip shallowly 3-lobed, central lobe oblong, truncate, *Mexico*. G1. Spring.

14. E. selligera (Lindley) Schlechter (*E. oncidioides* misapplied). Figure 14(21), p. 187. Illustration: Reichenbach, Xenia Orchidacearum **3**: t. 233 (1892); Hamer, Las Orquideas de El Salvador **1**: t. 17 (1974); Dressler & Pollard, The genus Encyclia in Mexico, t. 59 (1974).
Pseudobulbs 4.5–8 cm, conical to ovoid, clustered. Leaves 2 per pseudobulb, 15–25 × 1.5–4 cm, elliptic to strap-shaped, obtuse. Panicles 25–80 cm, many-flowered. Sepals and petals pale green, strongly flushed with purplish brown. Sepals 1.8–2.4 cm × 5–9 mm, lanceolate to spathulate, obtuse. Petals 1.8–2.1 cm × 7–9 mm, spathulate, hooded, obtuse or mucronate. Lip 1.7–2.2 cm, 3-lobed, cream, pink or lavender; lateral lobes to 10 × 5 mm, broad-tipped, constricted at base; central lobe ovate to circular. Callus elliptic, running into 2 ridges on central lobe of lip. Column 1.1–1.3 cm; base fused to claw of lip. *S Mexico & Guatemala*. G2. Winter–spring.

Frequently confused with *E. oncidioides* (Lindley) Schlechter, which has smaller narrowly ellipsoid pseudobulbs and smaller yellow flowers flushed with brown. Dressler considers the latter to be very rare in cultivation. The following 2 species are also very similar.

***E. hanburii** (Lindley) Schlechter. Illustration: Orchid Digest **34**: 307 (1970); Dressler & Pollard, The genus Encyclia in Mexico, t. 60 (1974). Differs in its reddish purple sepals and petals and its pink, heavily veined lip. Callus ovate to rhombic, running into a single ridge on the central lobe of the lip. *S Mexico*. G2. Spring–summer.

***E. advena** (Reichenbach) Dressler (*Epidendrum capartianum* Lindley; *Epidendrum osmanthum* Barbosa-Rodrigues). Figure 14(22), p. 187. Illustration: Botanical Magazine, 7792 (1901). Larger in all its parts than *E. selligera*. Pseudobulbs 7.5–13 cm. Petals spathulate to circular, about 1.5 cm wide, yellow streaked with pink. Lip more than 3 cm, wavy and scalloped, white with pink streaks. *Brazil*. G2. Autumn–winter.

15. E. polybulbon (Swartz) Dressler (*Dinema polybulbon* (Swartz) Lindley. Figure 14(23), p. 187. Illustration: Botanical Magazine, 4067 (1844); Die Orchidee **29**: lxxxvii (1978); Dressler & Pollard, The genus Encyclia in Mexico, t. 38 (1974).
Pseudobulbs 1–1.5 cm, ovoid, 5–40 mm apart. Leaves 2 or 3 per pseudobulb, 1–5 cm × 5–10 mm, elliptic to ovate, notched. Flowers solitary. Sepals 9–17 × 1.5–3 mm, linear-lanceolate, acute, brownish yellow. Petals similar, narrower with slightly hooked tips. Lip 1.5 cm, white, with a wavy, more or less circular, sometimes mucronate blade. Column *c.* 4 mm, fused to lip, winged, with two 3 mm apical processes. *S Mexico, Guatemala, Honduras, Cuba & Jamaica*. G2. Flowers intermittently.

49. BROUGHTONIA R. Brown
J.C.M. Alexander
Epiphytic, to 50 cm. Pseudobulbs 3–6 cm, spherical to ovoid, slightly flattened, with 1 or 2 apical leaves. Leaves 7.5–15 × 2–4 cm, oblong, stiff and leathery, folded when young. Flowering stems to 50 cm, borne on top of pseudobulbs, with a small scale at each node, bearing 5–12 purplish red flowers; flowers 2–3 cm wide. Sepals 1.5–2.5 cm × 5 mm, lanceolate, acute, extending into a long narrow spur which is fused to the ovary. Petals 1.5–2.5 × 1 cm, elliptic to ovate, rounded. Lip 2.5 × 2.5 cm, more or less circular, notched and irregularly toothed. Column free, exposed. Pollinia 4.

A single species from the West Indies which grows best on bark or rafts suspended near the roof. General conditions are similar to those for *Epidendrum* (p. 185); it should not be heavily shaded, nor should too much water be given in winter.

1. B. sanguinea (Swartz) R. Brown (*B. coccinea* Lindley). Illustration: Botanical Magazine, 3536 (1836); Skelsey, Orchids, 96 (1979); Bechtel et al., Manual of cultivated orchid species, 173 (1981). *Jamaica & Cuba*. G2. Intermittent flowering.

50. LAELIOPSIS Lindley
J.C.M. Alexander
Pseudobulbs ovoid to fusiform. Leaves 1–3, apical, rigid, sharply toothed. Flowers in terminal spikes; spur absent. Lip rolled around column. Column without foot, concealed. Pollinia 4.

Two species from the West Indies. Cultivation as for *Broughtonia*.

1. L. domingensis Lindley (*Broughtonia domingensis* (Lindley) Rolfe; *B. lilacina*

Henfrey). Illustration: Shuttleworth et al., Orchids, 52 (1970); Orchid Digest **40**: 212 (1976).
Pseudobulbs 5–8 cm. Flowers 3–8, 3.5 cm wide, pale pinkish purple; lip yellow-throated, finely toothed. *Dominican Republic*. G2. Spring–summer.

51. BARKERIA Knowles & Westcott
J. Cullen

Epiphytes. Stems erect, elongate, cane-like, sometimes somewhat swollen at the base, leafy. Leaves leathery, folded when young. Raceme (rarely panicle) with few to many flowers. Flowers resupinate. Sepals and petals similar or the petals somewhat broader. Lip united to the column for up to half the length of the latter, entire. Column winged. Pollinia 4, laterally flattened, parallel.

A genus of about 10 species from C America, closely related to and often included in *Epidendrum*, but differing in that at least half of the column is free from the lip; differing from *Encyclia* in the lack of true pseudobulbs.

Cultivation as for *Epidendrum* (p. 185), but the plants are better grown on rafts or bark, as their elongate rhizomes are restricted in pots.

Literature: Halbinger, F., The genus Barkeria, *American Orchid Society Bulletin* **42**: 620–6 (1973).

1a. Lip joined to the column for almost half the length of the latter, its blade 1.3–1.8 cm, bearing 3 yellow ridges **3. skinneri**
b. Lip joined to the column for at most one-third of the length of the latter, its blade 2–3.5 cm, without ridges or the ridges not yellow 2
2a. Lip without ridges, though the main vein sometimes thickened **1. elegans**
b. Lip with 3–5 ridges **2. spectabilis**

1. B. elegans Knowles & Westcott (*Epidendrum elegans* (Knowles & Westcott) Reichenbach). Illustration: Botanical Magazine, 4784 (1854); American Orchid Society Bulletin **42**: 621 (1973).
Stems to 25 cm, erect, thick, leafy. Leaves linear-lanceolate to elliptic-lanceolate, acute. Raceme loose, with few flowers. Sepals and petals similar or the petals broader, all narrowly elliptic, acute, 2–3.5 cm. Lip 2–3.5 cm, obovate-elliptic to oblong, broadly rounded but with a small point, white with a large purple blotch and with a callus but without ridges. Column joined to the lip for up to one-third of its length. *Mexico, Guatemala*. G2. Spring.

2. B. spectabilis Lindley. Illustration: Botanical Magazine, 6098 (1874); Fieldiana **26**: 342 (1953); American Orchid Society Bulletin **42**: 619 (1973); Bechtel et al., Manual of cultivated orchid species, 170 (1981).
Very similar to *B. elegans* but often larger, raceme often with many flowers, flowers white, pink or deep purple, lip often spotted, bearing 3–5 ridges which are the same colour as the rest. *C America*. G2. Winter.

Variable in flower colour and the precise shape of the lip. Halbinger recognises *B. lindleyana*, which is usually sunk in *B. spectabilis*, as a distinct species on the basis of lip shape and column length (15–18 mm in *B. spectabilis*, 6–14 mm in *B. lindleyana*).

3. B. skinneri (Lindley) Richard & Galeotti (*Epidendrum skinneri* Lindley). Illustration: Botanical Magazine, 3951 (1843); American Orchid Society Bulletin **42**: 624 (1973); Bechtel et al., Manual of cultivated orchid species, 170 (1981).
Very similar to *B. elegans* but smaller, inflorescence sometimes paniculate, flowers red to purple, lip 1.3–1.8 cm with 3 yellow ridges. Column joined to the lip for almost half its length. *Guatemala*. G2. Winter.

52. CATTLEYA Lindley
J.C.M. Alexander

Epiphytic or terrestrial, often on rocks. Rhizomes creeping, bearing erect, swollen or cane-like stems with several nodes. Leaves 1 or 2 (rarely 3 or 4) borne on top of the stems, folded when young. Flowers solitary or in terminal racemes (except No. 10). Sepals and petals free. Lip 3-lobed or entire; lateral lobes or basal part rolled around the column, often concealing it. Column free from lip, slightly flattened, often winged. Pollinia 4 (except No. 4), cohering in pairs.

A genus of about 40 species from C & S America; often used in bi- and trigeneric hybrids. They require maximum light throughout the year though new growth should be protected from strong sunlight. In the growing season water should be applied freely and the atmosphere kept moist with some air movement. Watering should be reduced during the winter but stems should not be allowed to shrink. Clusters of stems can be divided when repotting.

Literature: Withner, C., The genus Cattleya, *American Orchid Society Bulletin* **17**: 296–306, 363–9 (1948); Fowlie, J.A., *Brazilian bifoliate Cattleyas and their color varieties* (1977).

Stems. Cane-like: **6–9**; swollen: **1–5,7,8,10–12**; swollen above; **4,5,7,8,11,12**; swollen at base: **2**. Less than 15 cm; **4,6,10**; more than 15 cm; **1–9,11,12**.
Leaves. 1 per stem: **10–12**; 2 or 3 per stem: **1–10**. Spotted: **6,7,11**.
Flowering stems. Lateral: **10**.
Flowers. 1 or 2 per stem: **4,6,7,9**; 3–9 per stem: **2–5,7–12**; 10 or more per stem: **2,3,7,8**. Less than 5 cm wide: **1**; 6–7.5 cm wide: **2,3,6–8**; 8–10 cm wide: **3–11**; 11–13 cm wide: **4,5,10,12**; 14 cm wide or more: **12**.
Petals or sepals. Spotted: **5–8**; unspotted: **1–5,7,9–12**.
Lip. Unlobed: **1–3,9**; shallowly or obscurely lobed: **2–4,11,12**; 3-lobed: **5–8,10**; with very small basal lateral lobes: **9**. Broadest across lateral lobes: **1–5,7,8,11,12**; broadest across central lobe: **6,9,10**.
Lip lateral lobes. Semicircular: **6**. Overlapping central lobe: **4,5**.
Lip central lobe. Fan-shaped: **5,6,9**; kidney-shaped: **6,8,10**; fishtail-shaped: **5,7,8**.
Lip sinus. Broad and rounded: **10**; narrow and rounded: **5**; short and narrow: **4**; triangular: **6–8**.
Column. Fully exposed: **6,9,10**; partially or fully concealed: **1–5,7,8,11,12**.

1a. Flowers lateral on slender leafless shoots **10. walkeriana***
b. Flowers terminal on leafy stems 2
2a. Stems bearing 2 or more leaves 3
b. Stems bearing a single leaf 12
3a. Lip simple, shallowly lobed, or with very small basal lateral lobes, never with deep sinuses 4
b. Lip strongly 3- or 4-lobed 7
4a. Flowers red to orange, to 4.5 cm across; lip less than 2.5 × 1.5 cm, obtuse or acute, not pale-throated **1. aurantiaca**
b. Flowers magenta to purple, 4.5 cm across or more; lip more than 2.5 × 1.5 cm, rounded or notched, pale-throated 5
5a. Lip clearly divided into claw and fan-shaped blade **9. bicolor**
b. Lip irregularly oblong to ovate 6
6a. Flowers 6–8 cm across; stems 1 cm or less in diameter, swollen at base **2. bowringiana**
b. Flowers 9–12 cm across; stems 2–3 cm in diameter, not swollen at base **3. skinneri**
7a. Claw of central lobe of lip at least one-quarter of lobe length 11
b. Central lobe of lip shortly clawed or unclawed, separated from lateral lobes

by narrow sinuses 8
8a. Lateral lobes of lip strongly angled **7. amethystoglossa***
b. Lateral lobes of lip rounded 9
9a. Lip widest across central lobe; sinus broadly triangular or oblong **6. aclandiae***
b. Lip widest across lateral lobes; sinus narrow or concealed by overlapping lobes 10
10a. Petals narrower than sepals, usually 1 cm wide or less **4. forbesii***
b. Petals equal to or broader than sepals, 1.5 cm wide or more **5. intermedia***
11a. Column totally exposed; lateral lobes of lip absent or reduced to small basal projections **9. bicolor**
b. Column partially or totally concealed; lateral lobes of lip well developed, triangular to oblong, strongly angled **8. guttata***
12a. Flowers 10 cm wide or less **11. lawrenceana***
b. Flowers more than 10 cm wide **12. labiata***

1. C. **aurantiaca** (Lindley) P.N. Don. Illustration: Shuttleworth et al., Orchids, 58 (1970); Williams et al., Orchids for everyone, 105 (1980); Bechtel et al., Manual of cultivated orchid species, 178 (1981).
Epiphytic. Stems 20–35 cm, club-shaped, 2-leaved. Leaves 5–20 × 2.5–5.5 cm, ovate to elliptic, notched. Flowers several per raceme, 3–4 cm wide, orange to scarlet. Sepals 1.5–2.5 cm × 5 mm, lanceolate, acute. Petals 1.5–2.5 cm × 5 mm, ovate to lanceolate, acute. Lip 1.5–2.5 × 1 cm, ovate to elliptic; base rolled around column. Column 1 cm. *Mexico, Guatemala, Honduras & El Salvador*. G2. Summer.

2. C. **bowringiana** O'Brien. Illustration: Botanical Magazine n.s., 451 (1964); Shuttleworth et al., Orchids, 60 (1970); Williams et al., Orchids for everyone, 105 (1980).
Terrestrial, on rock or sand. Stems 20–35 cm × 5–10 mm, 4- or 5-noded, swollen at base, bearing 2 (rarely 3) leaves. Leaves 13–20 × 5–6 cm, oblong to elliptic, leathery, dark green. Flowers 5–20 (rarely to 45) per raceme, 6–7.5 cm wide, purple to magenta; lip darker, pale-throated. Sepals 3.8–5 × 1–1.4 cm, oblong to elliptic. Petals 4–5 × 2.5–3 cm, oblong to ovate. Lip 3.5–4 × 2.8–3.5 cm, oblong to ovate, base rolled around column. Column 1–1.2 cm, winged. *Belize & Guatemala*. G2. Autumn–winter.

3. C. **skinneri** Bateman. Illustration: Botanical Magazine, 4270 (1846); Williams et al., Orchids for everyone, 106 (1980); Bechtel et al., Manual of cultivated orchid species, 181 (1981).
Epiphytic or terrestrial. Similar to *C. bowringiana* but stems pseudobulbous (2–3 cm in diameter) and flowers larger (7.5–10 cm wide). Flowers 4–12 per raceme. *Mexico to Costa Rica; ?Panama*. G2. Winter–spring.

4. C. **forbesii** Lindley (*C. pauper* (Vellozo) Stellfeld). Illustration: Botanical Magazine, 3265 (1833); Williams et al., Orchids for everyone, 105 (1980); Bechtel et al., Manual of cultivated orchid species, 179 (1981).
Epiphytic. Stems 10–20 × 1–1.5 cm, 3- or 4-noded, slightly swollen above, 2-leaved. Leaves 9–15 × 3–7 cm, oblong, leathery. Flowers 1–5, 9–11 cm wide. Sepals and petals 5–6 cm, pinkish yellow to pale brown; sepals to 1.5 cm wide; lateral sepals sickle-shaped; petals to 1 cm wide. Lip 4.5–5.5 × 3–3.5 cm, 3-lobed, off-white with pinkish brown veins and a central yellow stripe; lateral lobes semicircular to oblong, rolled around column, separated from the small central lobe by a short narrow sinus; central lobe 1.4–1.6 cm wide, white, almost circular. Column 4 cm. *S Brazil*. G2. Spring–summer.

The name *C. pauper*, first published in *Epidendrum* by Vellozo and transferred into *Cattleya* by Stellfeld, has chronological priority and may be the correct name for this species.

*C. **dormaniana** Reichenbach. Illustration: American Orchid Society Bulletin **45**: 978 (1976); Fowlie, Brazilian bifoliate Cattleyas, 17 (1977); Bechtel et al., Manual of cultivated orchid species, 178 (1981). Central lobe of lip larger (2.3–2.8 cm wide), dark purple, overlapped by lateral lobes when flattened out. Sometimes with 4 extra sterile pollinia. *S Brazil*. G2. Summer–autumn.

5. C. **intermedia** Graham. Illustration: Botanical Magazine, 2851 (1821); Fowlie, Brazilian bifoliate Cattleyas, 110 (1977); Bechtel et al., Manual of cultivated orchid species, 179 (1981).
Epiphytic or on rocks. Stems 15–40 × 1.5 cm, club-shaped, 2-leaved. Leaves 7–15 × 3.5–7 cm, elliptic to oblong, obtuse, fleshy. Flowers 3–10, 10–12.5 cm wide. Upper sepal 6–7 × 1.5 cm; lateral sepals shorter and broader. Petals 5.5–7 × 1.5–1.8 cm. Petals and sepals off-white to pale pink. Lip 5–6 × 4–4.6 cm, 3-lobed. Lateral lobes asymmetrically ovate, rolled around column, off-white to pale pink with irregularly toothed tips, separated from the central lobe by a narrow rounded sinus. Central lobe fishtail-shaped, purplish pink, minutely toothed. Column 2.5–3.5 cm, winged. *S Brazil*. G2. Autumn.

*C. **loddigesi** Lindley. Illustration: Fowlie, Brazilian bifoliate Cattleyas, 123 (1977); Williams et al., Orchids for everyone, 106 (1980); Bechtel et al., Manual of cultivated orchid species, 180 (1981). Sepals and petals 2 cm wide or more, sparsely and finely spotted. Lateral lobes of lip broadly triangular, overlapping the semicircular central lobe. *S Brazil*. G2. Summer–winter.

*C. **harrisoniana** Lindley (*C. loddigesii* var. *harrisoniana* Rolfe; *C. harrisoniae* Bateman). Illustration; Botanical Magazine, 4085 (1844); Fowlie, Brazilian bifoliate Cattleyas, 126 (1977). Stems slightly thicker above. Flowers deep pink, unspotted. Petals strongly sickle-shaped. Lateral lobes of lip rounded, pink with yellow veins; central lobe pale, wavy, slightly notched. *S Brazil*. G2. Spring & autumn.

6. C. **aclandiae** Lindley. Illustration: Botanical Magazine, 5039 (1858); Fowlie, Brazilian bifoliate Cattleyas, 32 (1977); Bechtel et al., Manual of cultivated orchid species, 178 (1981).
Epiphytic. Stems to 12 cm, 1- or 2-noded, slightly thicker above, 2-leaved. Leaves 4–10 × 2–3.5 cm, elliptic, fleshy, sparsely spotted with purple. Flowers 1 or 2 per raceme, 7–10 cm wide. Sepals and petals 4.5 × 1.5–2 cm, elliptic, yellowish green to brown, densely spotted with purplish brown; sepals acuminate; petals obtuse or rounded, wavy. Lip 4.5–5 × 3–3.5 cm, 3-lobed; lateral lobes semicircular, white to pale pink, not concealing column, separated from central lobe by a broad triangular sinus; central lobe kidney-shaped with a deep pointed notch, pink, minutely toothed, wavy. Column 2.5 cm, winged, deep pinkish purple. *E Brazil*. G2. Spring.

*C. **velutina** Reichenbach. Illustration: Shuttleworth et al., Orchids, 62 (1970); Fowlie, Brazilian bifoliate Cattleyas, 20 (1977); Bechtel et al., Manual of cultivated orchid species, 182 (1981). Stems 20–30 cm, cane-like. Sepals and petals yellow with fine purple spots, wavy. Central lobe of lip fan-shaped, pale pink with darker veins and a broad yellow margin. *S Brazil*. G2. Summer.

7. C. **amethystoglossa** Linden & Reichenbach. Illustration: Botanical

Magazine, 5683 (1868); Williams et al., Orchids for everyone, 105 (1980); Bechtel et al., Manual of cultivated orchid species, 178 (1981).
On rocks. Stems to 1 m, cane-like, slightly swollen above, with 4–6 nodes, 2-leaved. Leaves 16–23 × 5–7.5 cm, oblong to narrowly elliptic, obtuse, leathery, dark green. Flowers 6–12 per raceme, 7–10 cm wide, very pale pinkish purple spotted with dark purple; lip with darker lobes. Upper sepal 3.8–4.5 × 1.4–1.7 cm, ovate to oblong; lateral sepals shorter and broader, curved downwards. Petals 3.5–4.5 × 2–2.5 cm, lanceolate to obovate. Lip 3–4 × 2.7–3.5 cm, 3-lobed; lateral lobes ovate to triangular, with warty, slightly spathulate tips, rolled around column, separated from the central lobe by a narrow triangular sinus. Central lobe fishtail-shaped with warty ridges. Column 2 cm. *E Brazil.* G2. Spring–summer.

*C. **schilleriana** Reichenbach. Illustration: Shuttleworth et al., Orchids, 61 (1970); Fowlie, Brazilian bifoliate Cattleyas, 36 (1977); Bechtel et al., Manual of cultivated orchid species, 181 (1981). Stems 15–20 cm, swollen. Leaves 6–10 cm, elliptic, spotted. Flowers 1 or 2 per raceme, to 9 cm wide. Sepals and petals yellow to brown, spotted. *E Brazil.* G2. Winter.

*C. **violacea** (Humboldt, Bonpland & Kunth) Rolfe (*C. superba* Lindley). Illustration: Botanical Magazine, 4083 (1944); Fowlie, Brazilian bifoliate Cattleyas, 48 (1977); Bechtel et al., Manual of cultivated orchid species, 182 (1981). Flowers 1–5 per raceme. Sepals and petals pink, unspotted. *Venezuela, Guyana, Peru & Brazil.* G2. Spring–summer.

8. C. guttata Lindley (*C. elatior* Lindley). Illustration: Botanical Magazine, 3693 (1839); American Orchid Society Bulletin **45**: 979 (1976); Fowlie, Brazilian bifoliate Cattleyas, 102 (1977).
Epiphytic or on rocks. Stems 20–80 (rarely to 150) × 1–3 cm, slightly swollen above, with 3–8 nodes, 2- or 3-leaved. Leaves 15–25 × 5–7 cm, elliptic, leathery. Flowers 3–20 per raceme, 7.5–10 cm wide. Sepals and petals 3.5–4.5 cm × 8–10 mm, yellowish green to brown, boldly spotted with purplish red; lateral sepals shorter, sickle-shaped; petals slightly wavy. Lip 2.6–3.3 × 2.5–3 cm, 3-lobed; lateral lobes diamond-shaped, rounded at base, rolled around column, pale pink; central lobe fishtail-shaped, tapering gradually into a long claw, ridged, reddish purple. Column 2–2.5 cm. *S Brazil.* G2. Winter.

*C. **granulosa** Lindley. Illustration: Botanical Magazine, 5048 (1858); Fowlie, Brazilian bifoliate Cattleyas, 71 (1977); Bechtel et al., Manual of cultivated orchid species, 179 (1981). Sepals and petals green, yellow or orange, finely spotted with red. Leaves 7–12 × 3.5–5 cm. Lip off-white to pale yellow; lateral lobes more or less triangular, not fully concealing column; central lobe kidney-shaped, tapering abruptly into a long narrow claw, pale pink, densely and finely spotted with purple. *E Brazil.* G2. Spring.

*C. **leopoldii** Lemaire (*C. guttata* var. *leopoldii* (Lemaire) Rolfe). Illustration: American Orchid Society Bulletin **45**: 978 (1976); Fowlie, Brazilian bifoliate Cattleyas, 98 (1977); Bechtel et al., Manual of cultivated orchid species, 180 (1981). Flowering stem 10–25 cm. Central lobe of lip broad, fishtail-shaped, almost touching lateral lobes when flattened, abruptly contracted into claw, purple, ridged. *S Brazil.* G2. Summer.

The name *C. tigrina* Richard has chronological priority and may be the correct name for this species.

*C. **elongata** Barbosa Rodrigues. Illustration: Botanical Magazine, 7543 (1897); Fowlie, Brazilian bifoliate Cattleyas, 87 (1977). Very similar to *C. leopoldii.* Differs in its longer flowering stem (35–60 cm) and its unspotted, copper-coloured petals and sepals. *E Brazil.* G2. Autumn–winter.

9. C. bicolor Lindley. Illustration: Botanical Magazine, 4909 (1856); Fowlie, Brazilian bifoliate Cattleyas, 23 (1977); Bechtel et al., Manual of cultivated orchid species, 178 (1981).
Epiphytic. Stems 30–80 cm, slender, slightly thicker above, with 4–6 nodes, 2-leaved. Leaves 12–20 × 2–2.5 cm, oblong, leathery. Flowers 2–9 per raceme, 8–9 cm wide. Sepals 4.5–6 × 1–1.5 cm; lateral sepals broader, sickle-shaped. Petals 4.5–5 × 1.5–2 cm, wavy. Sepals and petals coppery to greenish brown. Lip 3.5 × 1.5 cm, without lobes, or with small basal flaps, long-clawed; blade fan-shaped, shallowly notched, wavy, minutely toothed, pink. Column 3 × 1.2 cm, pale pink, winged. *S Brazil.* G2. Summer–autumn.

10. C. walkeriana Gardner (*C. bulbosa* Lindley). Illustration: Shuttleworth et al., Orchids, 62 (1970); Fowlie, Brazilian bifoliate Cattleyas, 56 (1977); Bechtel et al., Manual of cultivated orchid species, 182 (1981).
Epiphytic or on rocks. Pseudobulbs 3–9 × 1–2 cm, 2- or 3-noded, bearing a single leaf (rarely 2-leaved). Leaves 5–10 × 3–5 cm, elliptic to ovate, rounded or notched. Flowers 1–3 per raceme, 8–9 cm wide, borne on slender, leafless, lateral shoots. Sepals and petals 4.5–5.5 cm, pinkish purple; sepals 1.2–1.5 cm wide, lanceolate, acuminate; petals 2.5–3.5 cm wide, diamond-shaped to obovate, rounded. Lip 3.5–4.5 × 2.5–3.5 cm, 3-lobed, broadest across central lobe. Lateral lobes triangular to obovate, pink, asymmetrically notched; sinus broad and rounded. Central lobe almost circular, notched, dark purplish red with a narrow central patch of yellow. Column 2.5–3 cm, pink, winged, exposed. *E Brazil.* G2. Spring.

*C. **nobilior** Reichenbach. Illustration: Orchid Digest **36**: 33 (1972); Fowlie, Brazilian bifoliate Cattleyas, 68 (1977); Pabst & Dungs, Orchidaceae Brasilienses **1**: 192 (1975). Leaves 2 or 3. Flowers 10–12 cm wide. Petals and sepals pink; central lobe of lip pink with a large central yellow patch; lateral lobes of lip almost concealing column. *C Brazil.* G2. Summer.

11. C. lawrenceana Reichenbach. Illustration: Botanical Magazine, 7133 (1890); Bechtel et al., Manual of cultivated orchid species, 180 (1981).
Epiphytic. Stems to 25 cm, swollen above, 2- or 3-noded, bearing a single leaf. Leaves 8–20 × 4–5 cm, oblong, hard and leathery, sometimes purple-spotted. Flowers 3–8 per raceme, to 10 cm wide, pale to dark reddish purple, sometimes almost white; lip with a pale throat and a central reddish brown patch. Sepals 6.5–7 × 1.5–2 cm, elliptic to oblong; petals 6.5 × 2.5–4 cm, oblong to broadly elliptic, obtuse, wavy. Lip 5.5–6.3 × 4 cm, oblong to ovate, rolled around column. Column 2 cm. *Venezuela & Guyana.* G2. Spring–summer.

*C. **luteola** Lindley. Illustration: Botanical Magazine, 5032 (1858); Bechtel et al., Manual of cultivated orchid species, 180 (1981). Flowers yellow, sepals and petals 5 cm or less. Lip to 3 cm, obscurely 3-lobed. *Ecuador, Peru, Brazil & Bolivia.* G2. Autumn.

12. C. labiata Lindley var. **labiata**. Illustration: Botanical Magazine, 3998 (1843); Bechtel et al., Manual of cultivated orchid species, 180 (1981).
Epiphytic. Stems to 30 cm, club-shaped, bearing a single leaf. Leaves 15–30 × 4–7 cm, oblong, obtuse, leathery. Flowers 3–5 per raceme, *c.* 15 cm wide. Sepals and petals 7–8 cm, pale pink; sepals

2–2.5 cm wide, lanceolate, acute; petals 5–7 cm wide, elliptic, wavy. Lip 7–9 × 4.5–7 cm, ovate, shallowly 3-lobed, rolled around column, bright reddish purple with a pink margin; throat white and yellow. Column 3–3.5 cm. *E Brazil.* G2. Autumn.

Many different colour forms from other parts of C & S America are cultivated; they are often treated as separate species. As they are morphologically very similar to *C. labiata* var. *labiata* it seems best to regard them as varieties. The following are among the most common:

Var. **dowiana** (Bateman & Reichenbach) Veitch. Illustration: Botanical Magazine, 5618 (1867); Bechtel et al., Manual of cultivated orchid species, 178 (1981). Sepals and petals bright yellow; lip crimson to purple. *Costa Rica & Colombia.* G2. Autumn.

Var. **gaskelliana** (Reichenbach) Veitch. Illustration: Bechtel et al., Manual of cultivated orchid species, 179 (1981). Stems fusiform. Flowers 15–18 cm wide. Sepals and petals white to pale pink; lip white to pale pink with a yellow throat and a small indistinct central patch of purplish pink separated from the yellow by an area of white or pale pink. *Venezuela.* G2. Summer.

Var. **mendelii** (Backhouse) Reichenbach. Illustration: Bechtel et al., Manual of cultivated orchid species, 181 (1981). Differs from var. *gaskelliana* in its red-veined throat and larger central patch of purplish pink on the lip. *Colombia.* G2. Spring–summer.

Var. **mossiae** (Hooker) Lindley. Illustration: Bechtel et al., Manual of cultivated orchid species, 181 (1981). Very similar to vars. *gaskelliana* and *mendelii*. Sepals and petals pink. Lip white; throat yellow with purple veins which merge with a large central patch of purplish pink. *Venezuela.* G2. Spring–summer.

Var. **schroederae** (Reichenbach) Anon. Illustration: Bechtel et al., Manual of cultivated orchid species, 181 (1981). Very fragrant. Sepals and petals white or very pale pink. Lip pale pink with a purple tip. *Colombia.* G2. Winter–spring.

Var. **trianae** Linden & Reichenbach. Illustration: Bechtel et al., Manual of cultivated orchid species, 182 (1981). Differs from var. *schroederae* in its less fragrant flowers with a purple lip. *Colombia.* G2. Winter–spring.

Var. **warneri** (Moore) Veitch. Illustration: Bechtel et al., Manual of cultivated orchid species, 182 (1981). Flowers purplish pink; lip darker with a white and yellow throat. *S Brazil.* G2. Summer.

Var. **warscewiczii** (Reichenbach) Reichenbach. Illustration: Bechtel et al., Manual of cultivated orchid species, 182 (1981). Flowers 17–20 cm wide. Lip mostly deep purplish crimson; throat greenish yellow, veined with crimson. *Colombia.* G2.

Var. **percivalliana** Reichenbach. Illustration: Bechtel et al., Manual of cultivated orchid species, 181 (1981). Sepals and petals pinkish purple. Lip maroon, with a yellow throat, margin pink. *Venezuela.* G2. Winter–spring.

***C. rex** O'Brien. Illustration: Cogniaux & Goossens, Dictionnaire Iconographique des Orchidées, Cattleya, t. 22 (1899); Botanical Magazine, 8377 (1911). Sepals and petals white to very pale yellow. Lip purple-veined, yellow in basal half, apically pink with a white border. *Peru & Colombia.* G2. Summer.

***C. maxima** Lindley. Illustration: Bechtel et al., Manual of cultivated orchid species, 180 (1981). Petals about twice the width of the sepals. Flowers pale to deep pink. Lip strongly veined with purple, with a central yellow stripe. *Ecuador, Colombia & Peru.* G2.

53. LAELIA Lindley

J.C.M. Alexander

Epiphytic or terrestrial, often on rocks. Rhizomes creeping, bearing erect pseudobulbous or cane-like stems of several nodes (except No. 11). Leaves 1–3 per stem, folded when young. Flowers in terminal racemes. Sepals and petals free, rarely wavy. Lip 3-lobed or simple, internally ridged or keeled (except Nos. 1–3); lateral lobes or basal part curved around column, often concealing it. Column free from lip; apical teeth lacking or adpressed to anther cap. Pollinia 8.

A genus of about 50 species extending from C America and the West Indies south to Peru and Brazil, often used in bi- and trigeneric hybrids with *Cattleya* and *Brassavola*. Many of the species were originally described in *Bletia* or *Cattleya* and can only reliably be distinguished from the latter on the number of pollinia. Species Nos. 4, 5 & 6 have been described under the new genus *Hoffmannseggella* Jones but seem most conveniently kept in *Laelia*. Cultural treatment should be as for *Cattleya* (p. 191), though the Mexican species need a longer and cooler resting period.

Literature: Jones, H.G., Review of sectional division in the genus Laelia of the Orchidaceae, *Botanischer Jahrbücher* **97**: 309–16 (1976).

Stem. Club-shaped: **1–3**; swollen at base: **4,5**; conical, ovoid or fusiform: **6–11**; very slender throughout: **4**. Ribbed: **7,9**. Of 1 internode: **11**. 7 cm or less: **5–11**; 7 cm or more: **1–8**. 1-leaved: **1–8,10,11**; 2-leaved: **4,7–9,11**; 3-leaved: **7**.

Leaves. 20 cm or more: **1–4,7,8**; less than 10 cm: **6**. Less than 5 mm wide: **11**; 1.5–2.5 cm wide: **5,9,11**; 2.5–4.5 cm wide: **1,2,4–11**; more than 4.5 cm wide: **3**.

Flowering stem. 10 cm or less: **1,10,11**; 10–25 cm: **1–3,5,6,9–11**; 25–50 cm: **2–5,7–9**; more than 50 cm: **7,8**. Unsheathed: **10**; with large basal sheaths: **1–6,11**; with several small sheathing scales: **7–9**.

Flowers. 1 per stem: **7,10,11**; 2–7 per stem: **1–11**; 8–15 per stem: **4–7,9**. 6 cm wide or less: **1,4–6,8,9**; 7–12 cm wide: **1,2,7,8,11**; 13 cm wide or more: **1–3,10**.

Ovary. Glandular: **7**.

Sepals and petals. Pink to purple: **1,3,6–11**; white: **1,3,6,9,11**; yellow, orange or red: **1,2,4–6**. 4.5 cm or less: **1,4–6,9,11**; 5–7 cm: **1,2,7,8,11**; more than 7.5 cm: **2,3,10**.

Petals. Wider than sepals: **1–3,7–11**; as wide as sepals: **4–6,11**. Less than 1 cm wide: **4–6,9,11**; more than 4 cm wide: **3,10**. With recurved margins: **1**. Crisped: **1**.

Lip. Clearly 3-lobed: **1,2,4–11**; unlobed or obscurely 3-lobed: **3,11**. Crisped: **6**. With reflexed central lobe: **7,9**. 3 × 2 cm or less: **4–6,9**; 4 × 2.5 cm or more: **1–3,7,8,10,11**. Lacking keels or ridges: **1–3**; 1-ridged: **8,10**; 2-ridged: **6,10**; 3-ridged: **4,5,7,9–11**; 4 or 5 ridged: **11**; 7-ridged: **11**.

Column. Concealed: **3,6,8,9,11**; exposed or partly exposed: **1,2,4,5,7,10**.

1a. Lip without internal keels or ridges 2
b. Lip with internal keels or ridges 4
2a. Column about half as long as lip or more **1. perrinii***
b. Column distinctly less than half as long as lip 3
3a. Petal length more than twice petal width **2. grandis***
b. Petal length twice petal width or less **3. purpurata***
4a. Flowering stem sheathed at base or with a small sheathing scale at each node 5
b. Flowering stem without sheaths or scales 11
5a. Sepals and petals similar in shape 6
b. Sepals distinctly narrower than petals 8

6a. Flowers 5 cm wide or more; sepals and petals orange to red **4. cinnabarina***
b. Flowers less than 5 cm wide; sepals and petals white, yellow, pink or purple 7
7a. Lip more than 2 × 1.5 cm; sepals and petals yellow **5. flava**
b. Lip 1.6 × 1.2 cm or less; sepals and petals white, pink or purple **6. crispata***
8a. Flowers 7 cm wide or more; lip 4 × 2.5 cm or more 9
b. Flowers 6 cm wide or less; lip 2.5 × 1.6 cm or less **9. rubescens***
9a. Lip clearly 3-lobed 10
b. Lip unlobed or obscurely 3-lobed **11. pumila***
10a. Lip with 3 longitudinal ridges; central lobe acute or acuminate **7. autumnalis**
b. Lip with 1 longitudinal ridge; central lobe blunt **8. anceps**
11a. Lateral lobes of lip flared out exposing column **10. speciosa**
b. Lateral lobes of lip not flared; column concealed or very slightly exposed **11. pumila***

1. L. perrinii Bateman. Illustration: Botanical Magazine, 3711 (1840); Orchid Digest **41**: 149 (1977); Bechtel et al., Manual of cultivated orchid species, 219 (1981).
Epiphytic. Stem 7–30 cm, ovoid to club-shaped, 1-leaved. Leaves 20–30 × 3–4.5 cm, oblong to strap-shaped, obtuse or slightly notched, leathery, often spotted underneath. Flowering stem 10–25 cm, robust, 2- or 3-flowered, basally sheathed; flowers 10–13 cm wide. Upper sepal 6.5 × 1.5 cm, lanceolate; lateral sepals smaller, sickle-shaped; petals 6.5 × 2 cm, ovate; sepals and petals pink. Lip 5 × 3.5 cm, 3-lobed; lateral lobes obtuse, white to pale pink, sometimes edged with purple; central lobe oblong to circular, crisped, with a broad purple margin and a pale yellow throat, lacking keels or ridges. Column 3–4 cm, white to pale pink, partially concealed. *SE Brazil*. G2. Autumn.

***L. xanthina** Lindley. Illustration: Botanical Magazine, 5144 (1859); Pabst & Dungs, Orchidaceae Brasilienses **1**: 210 (1975); Bechtel et al., Manual of cultivated orchid species, 220 (1981). Flowers 4 or 5 per flowering stem, less than 8 cm wide; sepals and petals yellow; petal margins recurved. Lip white and yellow, striped with pink. *E Brazil*. G2. Spring–summer.

***L. crispa** (Lindley) Reichenbach. Illustration: Botanical Magazine, 3910 (1842); Orchid Digest **41**: 144 (1977); Bechtel et al., Manual of cultivated orchid species, 218 (1981). Sepals and petals white or very pale purple, tinged with pink at base; petals crisped and wavy. Lip strongly crisped, lateral lobes white, central lobe purple with darker veins and a white margin, throat yellow. *S Brazil*. G2. Summer.

2. L. grandis Lindley & Paxton. Illustration: Botanical Magazine, 5553 (1866); Graf, Tropica, 733 (1978); Bechtel et al., Manual of cultivated orchid species, 219 (1981).
Epiphytic. Stem 20–30 cm, club-shaped, 1-leaved. Leaves 20–35 × 3–4.5 cm, strap-shaped, obtuse. Flowering stem to 30 cm, basally sheathed, bearing 2–5 flowers; flowers 10–13 cm wide. Sepals 5–7 × 1–1.5 cm, lanceolate; petals 5–5.5 × 1.8–2 cm, diamond-shaped to elliptic, wavy. Petals and sepals greenish yellow to orange. Lip 5.5–6 × 3.5 cm, 3-lobed, white with purple veins; lateral lobes obtuse; central lobe oblong, obtuse, crisped and wavy, lacking keels or ridges. Column 2 cm, partially concealed. *E Brazil*. G2. Summer.

***L. tenebrosa** Rolfe (*L. grandis* var. *tenebrosa* Gower). Illustration: Ospina & Dressler, Orquideas de las Americas, t. 64 (1974); Orchid Digest **41**: 144 (1977); Bechtel et al., Manual of cultivated orchid species, 220 (1981). Flowers 14–19 cm wide. Sepals 8–10 × 1.6–2 cm; petals 7–10 × 2.5–3.5 cm; petals and sepals greenish purple. Lip 6.5–8.5 × 4–6 cm, pale pinkish purple, darker inside with a deep purple throat. *E Brazil*. G2. Summer.

3. L. purpurata Lindley. Illustration: Graf, Tropica, 737 (1978); Williams et al., Orchids for everyone, 107 (1980); Bechtel et al., Manual of cultivated orchid species, 220 (1981).
Epiphytic. Stem to 30 × 3 cm, fusiform to club-shaped, 1-leaved. Leaves 20–30 × 5 cm, oblong, rounded, leathery. Flowering stem to 30 cm, basally sheathed, bearing 2–7 flowers; flowers 15–20 cm wide. Sepals 7.5–11 × 2 cm, lanceolate to strap-shaped; petals 8–10 × 4–6 cm, oblong to ovate; petals and sepals white to very pale pink. Lip 7–10 × 6.5–8 cm, obscurely 3-lobed, wavy; lateral lobes curled around column; central lobe more or less circular; throat yellow, lateral lobes deep purple, margins paler, all veined with darker purple, lacking keels or ridges. Column 2.5–3 cm, pale green, concealed. *S Brazil*. G2. Early summer.

***L. lobata** (Lindley) Veitch (*L. boothiana* Reichenbach). Illustration: Pabst & Dungs, Orchidaceae Brasilienses **1**: 208 (1975); Orchid Digest **41**: 144 (1977). Flowers 13–15 cm wide, pink. Lip 6 × 5 cm with dark veins, throat deep pink. *Brazil*. G2. Spring.

4. L. cinnabarina Lindley. Illustration: Botanical Magazine, 4302 (1847); Williams et al., Orchids for everyone, 191 (1980); Bechtel et al., Manual of cultivated orchid species, 218 (1981).
On rocks. Stems 10–25 cm, swollen below, 1- or 2-leaved. Leaves 10–25 × 3–3.5 cm, linear to oblong, acute, often purplish green. Flowering stems 35–50 cm, basally sheathed, bearing 5–15 flowers; flowers 5–6 cm wide, orange to red. Sepals and petals 3–3.5 cm × 5 mm, linear to lanceolate, acute. Lip 2–2.5 × 1 cm, 3-lobed, lateral lobes acute, curled around column; central lobe narrowly diamond-shaped to ovate, acute, with 3 ridges, crisped and reflexed. Column club-shaped, mostly concealed. *SE Brazil*. G2. Spring.

***L. harpophylla** Reichenbach. Illustration: Ospina & Dressler, Orquideas de las Americas, t. 63 (1974); Graf, Tropica, 733 (1978); Bechtel et al., Manual of cultivated orchid species, 219 (1981). Epiphytic; stems very slender. Flowering stem shorter than leaves, bearing 4–7 flowers. *SE Brazil*. G2. Winter–spring.

5. L. flava Lindley. Illustration: Orchid Digest **38**: 69 (1974); Graf, Tropica, 733 (1978); Bechtel et al., Manual of cultivated orchid species, 218 (1981).
On rocks. Stems 5–20 cm, clustered, swollen below, purplish green, 1-leaved. Leaves 8–15 × 2.5–3 cm, lanceolate to oblong, obtuse, leathery, purple beneath. Flowering stem 20–45 cm, basally sheathed, bearing 5–10 flowers; flowers 3–4.5 cm wide, yellow. Sepals and petals 2–2.5 cm × 6 mm, linear to narrowly lanceolate, acute. Lip 2.5–3 × 2 cm, 3-lobed; lateral lobes blunt, curled around column; central lobe oblong, obtuse, crisped, wavy and slightly reflexed, 3-ridged. Column 9 mm, mostly concealed. *SE Brazil*. G2. Winter, spring & early summer.

6. L. crispata (Thunberg) Garay (*L. rupestris* Lindley). Illustration: American Orchid Society Bulletin **43**: 992 (1974); Pabst & Dungs, Orchidaceae Brasilienses **1**: 212 (1975); Northen, Miniature orchids, C-37 (1980).
On rocks. Stems 4–10 cm, not swollen,

1-leaved. Leaves 10–15 × 3–3.5 cm, strap-shaped to oblong, obtuse, leathery. Flowering stem 15–25 cm, with a single basal sheath, bearing 2–10 flowers; flowers 4–5 cm wide. Sepals and petals 2–2.5 cm × 8–10 mm, oblong to elliptic, obtuse to rounded, pink with pale centres. Lip 1.5–2 × 1.2 cm, 3-lobed; lateral lobes oblong, curled around column, pale pink; central lobe ovate to oblong, crisped and wavy, yellow with a pink margin, 2-ridged. Column concealed. *SE Brazil*. G2. Winter.

*L. **crispilabia** Richard. Illustration: American Orchid Society Bulletin **43**: 993 (1974); Orchid Digest **38**: 69 (1974). Stems pear-shaped. Leaves 6–8 cm. Sepals and petals pinkish purple with darker veins. Lip colour similar with dark margins. *SE Brazil*. G2. Summer.

*L. **longipes** Reichenbach. Illustration: Botanical Magazine, 7541 (1897); American Orchid Society Bulletin **43**: 993 (1974); Bechtel et al., Manual of cultivated orchid species, 218 (1981). Stems conical to ovoid. Sepals and petals white, very pale pink or pale yellow. Lip pale yellow with a deep yellow throat. *SE Brazil*. G2. Summer.

7. L. **autumnalis** Lindley (*L. gouldiana* Reichenbach). Illustration: Botanical Magazine, 3817 (1841); Orchid Digest **42**: 20 (1978); Bechtel et al., Manual of cultivated orchid species, 218 (1981).
Epiphytic or on rocks. Stems 10–15 cm, conical to pear-shaped, 2- or 3-leaved, ribbed. Leaves 12–20 × 3–4 cm, oblong to lanceolate, acute or obtuse, leathery. Flowering stems 30–100 cm, with several small sheathing scales, bearing 4–10 flowers; flowers 7.5–12 cm wide. Sepals and petals 5–6.5 cm, acuminate, pinkish purple with darker streaks; sepals 1.3 cm wide, lanceolate; petals 1.8 cm wide, ovate. Lip 4–5 × 2.5–3 cm, 3-lobed; lateral lobes rounded, partially enclosing column, white; central lobe oblong, acuminate, reflexed, purple with a white and yellow base, 3-ridged. Column 2.5–3 cm. *S Mexico*. G2. Winter.

Var. **furfuracea** (Lindley) Rolfe (*L. furfuracea* Lindley). Illustration: Botanical Magazine, 3810 (1841); Northen, Miniature orchids, C-40 (1980). Stems 4–6 cm, ovoid, deeply grooved, 1-leaved. Flowers 1–3 per flowering stem; ovary mealy-glandular. *S Mexico*. G2. Winter.

8. L. **anceps** Lindley. Illustration: Botanical Magazine, 3804 (1841); Williams et al., Orchids for everyone, 107 (1980); Bechtel et al., Manual of cultivated orchid species, 218 (1981).
Similar to *L. autumnalis*. Stems 7–12 cm, flattened, 1- or 2-leaved. Flowering stem bearing 2–5 flowers. Lip lateral lobes pinkish purple outside, yellow with purple veins and border inside; central lobe blunt, dark reddish purple, 1-ridged. *C Mexico*. G2. Winter.

9. L. **rubescens** Lindley (*L. acuminata* Lindley; *L. peduncularis* Lindley). Illustration: Botanical Magazine, 4099 (1844); Orchid Digest **42**: 20 (1978); Graf, Tropica, 733 (1978).
Epiphytic or on rocks. Stems 2.5–4.5 × 1.5–2.5 cm, clustered, ovoid, flattened, smooth, becoming furrowed, 1-leaved. Leaves 10–14 × 3–3.5 cm, oblong to lanceolate, obtuse, channelled, leathery. Flowering stem 25–50 cm, with several small sheathing scales, bearing 2–10 flowers; flowers 5–6 cm wide. Sepals 3.5 cm × 5 mm, lanceolate, acute; petals 2.5–3 cm × 8–10 mm, ovate to oblong, acute, wavy; sepals and petals white to pinkish purple. Lip 2.5–3 × 1.5 cm, 3-lobed; lateral lobes oblong, tips flared; central lobe oblong, acute, reflexed, wavy, white to pinkish purple; throat with dark purple patch surrounded by yellowish green. Column 8 mm, concealed. *S Mexico, Guatemala, Nicaragua & Costa Rica*. G2. Winter.

*L. **albida** Lindley. Illustration: Botanical Magazine, 3957 (1842); Graf, Tropica, 733 (1978); Bechtel et al., Manual of cultivated orchid species, 217 (1981). Leaves 2 per stem, less than 1.5 cm wide, linear to lanceolate. Flowers white, faintly flushed with pink; lip 3-ridged. *Mexico*. G2. Winter.

10. L. **speciosa** (Humboldt, Bonpland & Kunth) Schlechter (*L. grandiflora* Lindley; *L. majalis* Lindley). Illustration: Botanical Magazine, 5667 (1867); Die Orchidee **22**: 239 (1974); Orchid Digest **42**: 21 (1978).
Epiphytic. Stems 3–5 cm, ovoid to fusiform, clustered, 1-leaved. Leaves 10–15 cm, oblong to lanceolate, acute or obtuse, stiff. Flowering stems 10–20 cm, lacking sheathing scales, bearing 1 or 2 flowers; flowers 15–20 cm wide. Sepals 10–12 × 1.5–2.5 cm, lanceolate, acute; Petals 8–10 × 4.5–6 cm, ovate to diamond-shaped, obtuse, wavy; sepals and petals pale pinkish purple. Lip 7.5–9 × 4.5–5.5 cm, deeply 3-lobed; lateral lobes ovate to oblong, rounded, flared away from column, white to pale pinkish purple, edged with darker pink; central lobe oblong to circular, with 1–3 yellow keels, notched, white with pink streaks and a broad pink or pinkish purple border. Column exposed. *C Mexico*. G2. Spring–summer.

11. L. **pumila** (Hooker) Reichenbach. Illustration: Botanical Magazine, 3656 (1839); Northen, Miniature orchids, C-40 (1980); Bechtel et al., Manual of cultivated orchid species, 220 (1981).
Epiphytic. Stems 2–3 cm, ovoid to fusiform, 1-leaved. Leaves 10–13 × 2–2.5 cm, linear to oblong, obtuse. Flowering stem 4–10 cm, 1(rarely 2)-flowered; flowers 8–12 cm wide. Sepals 4.5 × 1–1.5 cm, oblong to lanceolate, acute; petals 3.5–4.5 × 2–3 cm, elliptic, acute or obtuse, wavy; petals and sepals pinkish purple. Lip 4–4.5 × 3 cm, shallowly 3-lobed or unlobed, basal part pinkish purple, curled around column; central lobe oblong to circular, with 3–5 ridges, wavy, sometimes notched, dark purple with a white or pale yellowish green throat. Column 2–2.5 cm, concealed. *SE Brazil*. G2. Autumn.

*L. **jongheana** Reichenbach. Illustration; Botanical Magazine, 6038 (1873); Northen, Miniature orchids, C-41 (1980); Bechtel et al., Manual of cultivated orchid species, 285 (1981). Leaves 3.5–4 cm wide. Sepals and petals 6–7 cm. Lip 5.5 × 3 cm; central lobe white, throat white and yellow with 7 wavy keels. *SE Brazil*. G2. Spring.

*L. **lundii** Reichenbach & Warming (*L. regnellii* Rodrigues). Illustration: Northen, Miniature orchids, C-42 (1980); Skelsey, Orchids, 117 (1980); Bechtel et al., Manual of cultivated orchid species, 219 (1981).

Leaves 2 per stem, 8–10 cm × 5 mm, linear, channelled and pointed. Sepals and petals 2–2.5 cm × 3–4 mm, white. Lip clearly 3-lobed, white with pink veins, wavy. *SE Brazil*. G2.

54. SCHOMBURGKIA Lindley
J.C.M. Alexander

Epiphytic or on rocks. Rhizomes branched, creeping, bearing erect, cylindric to fusiform, few-noded pseudobulbs. Leaves 2 or 3 (rarely 4) per pseudobulb, folded when young, narrowly ovate to oblong. Flowers several, in tall terminal racemes or panicles. Flowering stems with a sheathing scale at each node. Flower-stalks with long or short bracts. Sepals and petals similar, free, spreading, wavy. Lip 3-lobed; lateral lobes loosely enclosing column, becoming erect or spreading; central lobe rounded or notched, often curled downwards, sometimes with keels or ridges. Column stout, with a single, adpressed, apical tooth. Pollinia 8.

A genus of 12–15 species from the American tropics, many of which were originally described in *Cattleya* or *Laelia*. Cultural requirements are similar to those for *Cattleya* (p. 191) though they perform best if repotted less frequently.

1a. Bracts more than half flower-stalk length 2
b. Bracts less than half flower-stalk length **4. tibicinis**
2a. Flowers more than 9 cm wide **1. superbiens**
b. Flowers less than 9 cm wide 3
3a. Flowers less than 6 cm wide; lip whitish, sometimes spotted **2. gloriosa***
b. Flowers 6 cm wide or more; lip pinkish purple **3. undulata**

1. S. superbiens (Lindley) Rolfe. Illustration: Botanical Magazine, 4090 (1844); Orchid Digest **42**: 21 (1978); Bechtel et al., Manual of cultivated orchid species, 272 (1981).
Pseudobulbs to 1 m or more, fusiform, flattened, 1- or 2-leaved. Leaves to 35 × 6 cm, oblong, acute, leathery and stiff. Flowering stem to 1 m or more, robust, bearing 12–20 flowers; flowers 10–12 cm wide, fragrant; ovary plus flower-stalk 6–8 cm; bracts 5–7 cm. Sepals and petals 5–7 × 1.3–1.8 cm, oblong to narrowly lanceolate, acute, pink to pinkish purple, paler at base; petals slightly wavy. Lip 4–5.5 × 3–3.5 cm; lateral lobes rounded, crisped, greenish yellow with dark purple veins and a pink margin; central lobe ovate, notched, wavy, with 5 yellow ridges and a pink blade. Column 2.5–3 cm, club-shaped, curved. *Mexico, Guatemala & Honduras*. G2. Winter.

2. S. gloriosa Reichenbach (*S. crispa* Lindley 1844 not 1838; *S. crispa* misapplied, not *Laelia crispa* (Lindley) Reichenbach). Illustration: Pabst & Dungs, Orchidaceae Brasilienses **1**: 218 (1975); Dunsterville & Garay, Venezuelan orchids illustrated **6**: 398 (1976); Bechtel et al., Manual of cultivated orchid species, 383 (1981).
Epiphytic. Pseudobulbs to 30 × 5 cm, club-shaped with long tapering bases, ridged, 2- or 3-leaved. Leaves to 35 × 6 cm, oblong to lanceolate, acute or obtuse. Flowering stem to 75 cm, slender, bearing 8–15 flowers; flowers 7.5–8.5 cm wide; ovary plus flower-stalk to 9 cm; bracts to 6 cm. Sepals and petals 3.5–4 × 1 cm, oblong, obtuse (sepals each with a small point), wavy, yellowish brown with darker veins. Lip 2 × 2 cm, white to pale pink with yellowish brown margins and central lobe; lateral lobes kidney-shaped to ovate; central lobe broadly triangular, rounded, crisped, curled downwards, 3-ridged. Column 2 cm. *Venezuela, Guyana & Surinam*. G2. Winter.

***S. lyonsii** Lindley. Illustration: Botanical Magazine, 5172 (1860); Gardeners' Chronicle **26**: 203 (1899). Flowers *c.* 5 cm wide, cream-coloured with purple spots. *Jamaica*. G2. Summer–autumn.

3. S. undulata Lindley. Illustration: Dunsterville & Garay, Venezuelan orchids illustrated **2**: 256 (1961); Skelsey, Orchids, 38 (1979); Bechtel et al., Manual of cultivated orchid species, 272 (1981).
Epiphytic. Pseudobulbs to 25 cm, club-shaped with long tapering bases, ridged, 2-leaved. Leaves to 30 × 5 cm, lanceolate to oblong, obtuse, leathery. Flowering stem to 50 cm, slender, bearing 5–20 flowers; flowers 5–6 cm wide; ovary plus flower-stalk 5–7 cm, purplish green; bracts *c.* 5 cm. Sepals and petals 3–3.5 × 1.5 cm, oblong, acute, crisped and very wavy, purplish brown. Lip 3 × 1.5–2 cm, purple; lateral lobes broad and shallow, erect; central lobe oblong, shallowly notched, slightly curled downwards, wavy, 5-ridged. Column 1.5 cm. *Venezuela & Trinidad*. G2. Summer.

4. S. tibicinis Lindley. Illustration: Botanical Magazine, 4476 (1849); Shuttleworth et al., Orchids, 66 (1970); Skelsey, Orchids, 137 (1979).
Epiphytic, very large and robust. Pseudobulbs to 60 × 8 cm or more, fusiform, hollow, 2- or 3-leaved. Leaves to 45 × 8 cm, elliptic to oblong, rounded or obtuse. Flowering stems to 5 m, branched, bearing 10–20 flowers; flowers 6–8 cm wide, opening in succession; ovary plus flower-stalk 3–5 cm; bracts 1 cm. Sepals and petals 4–5 × 1.5–2 cm, oblong to spathulate, obtuse or rounded, purplish blue flushed with brown, paler beneath. Lip 4–5 × 3–4 cm, 3-lobed, yellow, white and purple; lateral lobes semicircular to oblong; central lobe diamond-shaped, curled downwards. *Mexico to Panama*. G2. Spring.

55. BRASSAVOLA R. Brown

J.C.M. Alexander

Epiphytic or on rocks. Rhizomes creeping, bearing erect, narrow stems or pseudobulbs. Stems or pseudobulbs sheathed, 1-leaved. Leaves fleshy, mostly cylindric or linear, sometimes flattened and leathery, folded when young. Flowers terminal (except No. 6), solitary or in racemes. Sepals and petals free and spreading, similar, linear to narrowly lanceolate. Lip unlobed, or obscurely 3-lobed, base or claw rolled around and often concealing column, blade or central lobe oblong, circular or heart-shaped. Column free from lip, with acute projecting teeth at tip. Pollinia 8 (often a few extra abortive).

A genus of 15–20 species extending from C America and the West Indies south to Brazil and Argentina, often used as a parent in bi- and trigeneric hybrids with *Cattleya* and *Laelia*. Species Nos. 1 and 2 are sometimes treated as the separate genus *Rhyncholaelia* Schlechter, as they are distinct in stem and leaf shape. Cultural treatment should be as for *Epidendrum* (p. 185) though the *Brassavola* growth habit is more easily accommodated on blocks of tree fern or cork bark. They should have a definite winter rest.
Literature: Jones, H.G., Nomenclatural revision of the genus Brassavola of the Orchidaceae, *Annalen des Naturhistorischen Museums in Wien* **79**: 9–22 (1975).

Plant. Erect: **1–5**; pendent: **6**.
Stems. Pseudobulbous: **1,2**; cane-like: **3–6**.
Leaves. Flat: **1,2**; cylindric or linear: **3–6**. Glaucous: **1,2**. Less than 12 cm long: **1,4**; 12–40 cm long: **1–5**; more than 40 cm: **4,6**. Less than 1 cm wide: **3–5**; 1–2 cm wide: **6**; 2–4 cm wide: **1–4**; 5 cm wide or more: **2**.
Flowers. Lateral: **6**; terminal: **1–6**. 1–3 per raceme: **1–3,5,6**; 4 or more per raceme: **4,5**. Less than 7 cm wide: **4,5**; more than 7.5 cm wide: **1–3,5,6**.
Petals and sepals. Spotted or warty: **4,5**. Less than 5 cm long: **4**; 5 cm long or more: **1–3,5,6**. Less than 1 cm wide: **3–6**; more than 1 cm wide: **1–3**.
Lip. Unclawed or with very short claw: **1–4**; with claw about one-third of total lip length: **5,6**. Lacerate: **2,3**. Long and gradually tapering: **3**. 3–4 cm long: **4**; 3.5–5.5 cm long: **1,5**; 6 cm long or more: **2,3,6**. Less than 4.5 cm wide: **1,3–5**; more than 4.5 cm wide: **2,6**.

1a. Leaves flattened; stems pseudobulbous 2
b. Leaves cylindric to linear; stems cane-like 3
2a. Lip margin entire; column less than 2 cm **1. glauca**
b. Lip margin deeply lacerate; column more than 2 cm **2. digbyana**
3a. Lip 6 cm or more, long and gradually tapering **3. cucullata**
b. Lip 5 cm or less, acute or obtuse, not gradually tapering 4
4a. Lip unclawed or with a very short claw **4. perrinii***

b. Lip claw about one-third of total lip length 5
5a. Plant erect; flowering stems terminal **5. nodosa***
b. Plant pendent; flowering stems lateral **6. acaulis**

1. B. glauca Lindley (*Rhyncholaelia glauca* (Lindley) Schlechter; *Laelia glauca* (Lindley) Bentham). Illustration: Botanical Magazine, 4033 (1843); Shuttleworth et al., Orchids, 68 (1970); Bechtel et al., Manual of cultivated orchid species, 172 (1981).
Epiphytic or terrestrial. Pseudobulbs 2–10 cm, fusiform to club-shaped, 1-leaved. Leaves 5–12.5 × 2.5–3.5 cm, oblong to elliptic, obtuse, leathery and glaucous. Flowering stems to 10 cm, terminal, 1-flowered; flower 8–10 cm wide, very fragrant. Sepals and petals similar, 5–6.5 cm, linear to lanceolate, obtuse, white to pale green or purple; sepals 1–1.5 cm wide, margins reflexed; petals 1.5–2 cm wide, margins wavy. Lip 5–5.5 × 4 cm, obscurely 3-lobed, wavy; lateral lobes rounded, bases rolled around column; central lobe oblong, abruptly pointed, throat with a narrow purple patch. Column 5 mm, partly exposed. *Mexico to Panama*. G2. Winter–spring.

2. B. digbyana Lindley (*Rhyncholaelia digbyana* (Lindley) Schlechter; *Laelia digbyana* (Lindley) Bentham). Illustration: Botanical Magazine, 4474 (1849); Williams et al., Orchids for everyone, 107 (1980); Bechtel et al., Manual of cultivated orchid species, 171 (1981).
Epiphytic. Pseudobulbs 12–20 cm, club-shaped, 3- or 4-noded, 1-leaved. Leaves 12–20 × 5–6 cm, elliptic, obtuse, leathery and slightly glaucous. Flowering stem to 15 cm, terminal, 1-flowered; flower 10–12 cm wide, fragrant. Sepals 8–9.5 × 2–2.5 cm, lanceolate, obtuse; petals somewhat wider; sepals and petals very pale green to greenish yellow, margins wavy. Lip 7–8 × 7.5 cm, obscurely 3-lobed, almost circular, margins deeply and finely lacerate; lateral lobes semicircular, bases rolled around but not concealing column; central lobe rectangular to oblong. Column 3–3.5 cm. *Mexico, Belize & Honduras*. G2. Summer.

3. B. cucullata (Linnaeus) R. Brown (*B. cuspidata* Hooker). Illustration: Botanical Magazine, 3722 (1839); Dunsterville & Garay, Venezuelan orchids illustrated **2**: 43 (1961); Bechtel et al., Manual of cultivated orchid species, 171 (1981).
Epiphytic. Stems to 25 cm, cylindric, thin, several-noded, 1-leaved. Leaves 20–40 cm × 5–10 mm, cylindric to linear, fleshy. Flowering stems short, bearing 1–3 flowers; flowers 8–10 cm wide, fragrant, white and greenish yellow tinged with pink or brown. Ovary plus flower-stalk 15–25 cm. Sepals and petals similar, 7–14 cm × 5–10 mm, lanceolate, tapering. Lip 6.5–10 × 1.5–2.5 cm, base circular to heart-shaped with lacerate margin; tip long and tapering. Column 1.5–2 cm, partly exposed. *Mexico & West Indies south to Ecuador*. G2. Autumn–winter.

4. B. perrinii Lindley. Illustration: Botanical Magazine, 3761 (1839); Cogniaux & Goossens, Dictionnaire Iconographique des Orchidées, Brassavola, t. 2 (1904).
Epiphytic or on rocks. Stems to 20 cm × 5 mm, cylindric or slightly angular, branched, 3- or 4-noded, 1-leaved. Leaves 10–25 cm × 5–80 mm, circular to triangular in cross-section, acute, deeply grooved, erect. Flowering stem 2–8 cm, terminal, 1- or 2-flowered; flowers 5–6 cm wide. Ovary plus flower-stalk 5–10 cm. Sepals and petals 3.5–4.5 cm × 4–6 mm, narrowly lanceolate, acute, greenish yellow. Lip 3–4 × 2–2.5 cm, ovate, acute, base rolled around column, white with a central patch of greenish yellow and green veins. Column 1 cm, partly exposed. *SE Brazil & Paraguay*. G2. Spring–summer.

***B. tuberculata** Hooker (*B. fragrans* Lemaire). Illustration: Botanical Magazine, 2878 (1829); Pabst & Dungs, Orchidaceae Brasilienses **1**: 219 (1975); Bechtel et al., Manual of cultivated orchid species, 172 (1981). Flowers 1–12; sepals cream-coloured, marked with reddish purple spots and warts. Lip ovate, blunt, white with a central patch of yellow. *Brazil*. G2. Summer–autumn.

***B. flagellaris** Rodrigues. Illustration: Bechtel et al., Manual of cultivated orchid species, 171 (1981). Flowers 5–15. Leaves 35–50 cm × 5 mm. Petals 2–3 mm wide. *E Brazil*. G2. Spring.

5. B. nodosa (Linnaeus) Lindley. Illustration: Botanical Magazine, 3229 (1833); Williams et al., Orchids for everyone, 182 (1980); Bechtel et al., Manual of cultivated orchid species, 172 (1981).
Epiphytic or on rocks. Stems 8–16 cm, cylindric, slightly thicker above, 4- or 5-noded, 1-leaved. Leaves 20–35 × 2.5–3.5 cm, linear, acute, fleshy and grooved. Flowering stem 8–22 cm, terminal, bearing 1–several flowers; flowers 7.5–8.5 cm wide, fragrant. Sepals and petals 6–9 cm × 5 mm, linear, acute, pale green to greenish yellow. Lip 3–5.5 × 2–4 cm, white; claw about one-third of total lip length, rolled around column; blade circular to ovate, acuminate. Column *c*. 1 cm, concealed. *Mexico & West Indies south to Colombia, Venezuela & Surinam*. G2. Winter.

***B. subulifolia** Lindley (*B. cordata* Lindley). Illustration: Botanical Magazine, 3782 (1840); Shuttleworth et al., Orchids, 68 (1970); Skelsey, Orchids, 45 (1979). Stems 2–8 cm, distinctly thicker above. Leaves to 20 × 1.5 cm. Flowers 3–5, 5–7 cm wide, spotted with red. Column about 5 mm. *Jamaica*. G2. Autumn.

6. B. acaulis Lindley & Paxton (*B. lineata* Hooker). Illustration: Botanical Magazine, 4734 (1853); Shuttleworth et al., Orchids, 67 (1970); Graf, Tropica, 708 (1978). Similar to *B. nodosa*. Whole plant pendent. Leaves cylindric, grooved. Flowering stems lateral, 1- or 2-flowered. Flowers pale yellow to pale green. *Guatemala & Belize to Panama*. G2. Autumn.

56. SOPHRONITIS Lindley

J.C.M. Alexander

Epiphytic or on rocks. Rhizomes creeping, branched, bearing clusters of 1-leaved pseudobulbs. Leaves ovate, elliptic or oblong, folded when young, leathery. Flowers terminal, solitary or in few-flowered racemes. Sepals and petals free; petals broader than sepals. Lip small, unlobed or 3-lobed, shortly fused to column at base and partly enclosing it. Column 2-winged. Pollinia 8.

A genus of about 8 species from Paraguay and E Brazil; often used to give bright red colours in intergeneric hybrids with *Cattleya, Laelia, Brassavola, Epidendrum*, etc. Cultivation as for *Cattleya* (p. 191), using tree-fern blocks or rafts in cool, shady conditions.

1a. Leaves less than 3 cm wide; flowers 2–5, 2–3 cm wide; petals less than 1 cm wide **1. cernua**
b. Leaves 3 cm wide or more; flowers solitary, 6–7 cm wide; petals 2 cm wide or more **2. coccinea**

1. S. cernua Lindley. Illustration: Botanical Magazine, 3677 (1839); Shuttleworth et al., Orchids, 69 (1970); Bechtel et al., Manual of cultivated orchid species, 274 (1981).
Pseudobulbs 1–2 cm, ovoid to cylindric, compressed. Leaves 2–2.8 × 1.5–2 cm, elliptic to oblong, slightly cordate at base, obtuse or abruptly pointed. Flowering stem

to 6 cm, bearing 2–5 flowers; flowers 2–3 cm wide, reddish orange. Sepals and petals 1–1.5 cm × 5 mm; sepals ovate to lanceolate, acute; petals ovate to diamond-shaped, acute. Lip 1 cm × 8 mm, triangular to lanceolate, unlobed or shallowly 3-lobed, acute. Column 5 mm, club-shaped with 2 small, purple-tipped wings. *E Brazil*. G2. Winter.

2. S. coccinea (Lindley) Reichenbach (*S. grandiflora* Lindley). Illustration: Botanical Magazine, 3709 (1839); Williams et al., Orchids for everyone, 196 (1980); Bechtel et al., Manual of cultivated orchid species, 274 (1981).
Pseudobulbs 1.5–3.5 cm, ovoid to fusiform. Leaves 3–6 × 1–1.5 cm, ovate to elliptic, fleshy, sometimes glaucous. Flowering stem to 8 cm, 1-flowered; flowers 6–7 cm wide, yellow to red or pinkish red. Sepals 1.5–2.5 × 1 cm, elliptic to oblong, acute; petals 2.5–3.5 × 2–3 cm, circular to broadly ovate, rounded. Lip 1.5–2 × 1.5–2 cm, 3-lobed; lateral lobes triangular, rolled around column; central lobe narrowly triangular. Column 5–10 mm, club-shaped, partly concealed. *E Brazil*. G2. Winter.

57. SOPHRONITELLA Schlechter
J.C.M. Alexander
Epiphytic with creeping rhizomes. Pseudobulbs 1.5–3 cm × 3–10 mm, ovoid to fusiform, clustered, ridged, 1-leaved. Leaf 4–8 × 1 cm, linear, acute, channelled, folded when young, fleshy. Flowering stem to 5 cm, 1- or 2-flowered; flowers 2.5–4 cm wide, purple. Sepals and petals 1.7–2 cm × 3–5 mm, oblong to lanceolate, acute, free. Lip 1.7–2 cm × 5 mm, unlobed, lanceolate to ovate or diamond-shaped, abruptly acute, fused to column at base. Column 5 mm, club-shaped. Pollinia 8.

A single species from Brazil, originally described in *Sophronitis*. Cultivation as for *Sophronitis*.

1. S. violacea (Lindley) Schlechter (*Sophronitis violacea* Lindley). Illustration: Botanical Magazine, 6880 (1886); Ospina & Dressler, Orquideas de las Americas, t. 71 (1974); Bechtel et al., Manual of cultivated orchid species, 274 (1981).
E Brazil. G2. Winter.

58. LEPTOTES Lindley
J.C.M. Alexander
Dwarf epiphytes with creeping rhizomes. Stems short and cane-like, sheathed, 1-leaved, sometimes branched. Leaves cylindric to linear, acute, fleshy and channelled, folded when young. Flowers few or solitary, in terminal racemes. Sepals and petals free, similar, or petals broader than sepals. Lip free, 3-lobed; lateral lobes small, sometimes enfolding but not concealing the short fleshy column. Pollinia 6 (4 large and 2 small), waxy.

Three species from Brazil, Paraguay and Argentina. Cultivation requirements are similar to those for *Cattleya* (p. 191) though their dwarf habit is best accommodated on hanging rafts or baskets; the roots should be covered with moss.

1a. Flowers erect, 4–5 cm wide; sepals similar to petals; lip with a dark purple patch at base **1. bicolor**
b. Flower pendent, 3–4 cm wide; sepals wider than petals; lip uniformly pink **2. unicolor**

1. L. bicolor Lindley (*Tetramicra bicolor* (Lindley) Bentham). Illustration: Botanical Magazine, 3734 (1840); Shuttleworth et al., Orchids, 44 (1970); Pabst & Dungs, Orchidaceae Brasilienses 1: 222 (1975); Skelsey, Orchids, 118 (1979).
Stem 1–4 cm, cylindric, fleshy. Leaves 5–15 cm × 5–10 mm. Flowering stem to 6 cm, bearing 2–4 flowers; flower-stalk plus ovary 4–5 cm; flower 4–5 cm wide, erect, white to very pale pink, yellow or green. Sepals and petals 2–2.5 cm × 3–5 mm, linear to narrowly lanceolate, acute with curled tips. Lip 1.5–2 cm × 5–8 mm, 3-lobed; lateral lobes triangular to circular, white or pale green, curved around column; central lobe oblong to narrowly ovate, acute, pale to deep purple with darker veins. Column 5 mm, greenish brown, exposed. *S Brazil & Paraguay*. G2. Winter–spring, autumn.

2. L. unicolor Rodrigues. Illustration: Pabst & Dungs, Orchidaceae Brasilienses **1**: 223 (1975); Northen, Miniature orchids, 100 (1980); Bechtel et al., Manual of cultivated orchid species, 221 (1981).
Stem 1–3 cm, cylindric. Leaves 2.5–6 cm × 4–8 mm, arched. Flowering stem 1–3 cm, 2-flowered; flowers 3–4 cm wide, pendent, white to pale pink or purple. Sepals 2–2.5 cm × 5 mm, lanceolate, acute; petals 1.5–2 cm × 3 mm, strap-shaped, acute. Lip 1.5–2 × 1–1.3 cm, 3-lobed; lateral lobes triangular, obtuse, flat, not enclosing column; central lobe diamond-shaped, acute. Column 5 mm, brown, exposed. *S Brazil & Argentina*. G2. Winter.

59. SCAPHYGLOTTIS Poeppig & Endlicher
J.C.M. Alexander
Epiphytic or on rocks. Pseudobulbs in superposed chains; leaves all apical, folded when young. Flowers terminal or at stem nodes, in racemes or clusters, white or pale shades of blue, green, yellow or purple, never bright orange or red. Column base enlarged into a foot to which the lateral sepals are basally fused to form a mentum. Pollinia 4 (in ours), waxy.

A tropical American genus of about 40 species. Cultivation as for *Cattleya* (p. 191), though its straggling habit makes it more easily grown on rafts or bark suspended near the roof.

1. S. amethystina (Reichenbach) Schlechter. Illustration: Bechtel et al., Manual of cultivated orchid species, 272 (1981).
Leaves 1 or 2, to 15 × 1.5 cm, apical, linear, notched at apex. Flowers 1–1.5 cm wide, in terminal clusters, white to pale purplish pink. Lip 3-lobed. Pollinia 4. *Guatemala to Panama*. G2. Spring.

60. NAGELIELLA Williams
J.C.M. Alexander
Epiphytic or terrestrial. Rhizomes creeping, bearing simple, 1-leaved, club-shaped stems. Leaves fleshy, mottled with purple, folded when young. Flowers pinkish purple, terminal, in tall, slender, branched racemes or clusters. Sepals pointing forwards; lateral sepals fused basally to column foot forming a mentum. Lip unlobed, bent downwards, basally fused to column foot, with a small apical pouch. Column slender, slightly arched over upper part of lip. Pollinia 4, waxy.

A small genus of 2 species from Mexico and C America; long known as *Hartwegia* Lindley though this name had already been used for a genus in the Liliaceae. Cultivation requirements are similar to those for *Cattleya* (p. 191). Propagation is easily effected by division, though this should not happen frequently as the plants perform better when the pseudobulbs are densely clustered.

1. N. purpurea (Lindley) Williams (*Hartwegia purpurea* Lindley). Illustration: Hamer, Orquideas de El Salvador **2**: 139 (1974); Orquidea (Mexico) **4**: 115 (1974); Bechtel et al., Manual of cultivated orchid species, 234 (1981).
Stems 2–8 cm, narrowly fusiform to club-shaped, shallowly ridged. Leaves 3–12 × 1–2.5 cm, elliptic to ovate with sheathing bases. Flowering stems

10–50 cm with small, chaffy bracts, gradually lengthening over a few months. Flowers several, 1–1.5 cm wide, opening in succession. Sepals 7–10 × 3 mm, lanceolate to ovate, concave, acute or obtuse. Petals shorter and narrower than sepals, linear to lanceolate. Lip 10 × 4 mm, basally fused to column foot; middle portion linear, S-shaped; apex a small triangular pouch. Column *c.* 7 mm. *Mexico to Honduras*. G2. Summer.

61. ERIA Lindley

J.C.M. Alexander

Epiphytic or occasionally terrestrial, often densely hairy. Pseudobulbs ovoid, fusiform or cylindric, sometimes slender and cane-like, often densely sheathed. Leaves 1–several, folded when young (rolled in Nos. 1 & 2). Flowers in terminal or axillary racemes. Upper sepal free; lateral sepals united at base with column foot forming a pronounced mentum. Petals similar to sepals. Lip often 3-lobed, borne on column foot. Pollinia 8, waxy, pear-shaped or ovoid.

A very diverse genus of about 500 species from India, SE Asia and the Pacific islands. It is generally similar to *Dendrobium* but the flowers have 8 pollinia and are never spurred. Cultivation details are similar to those for the tropical species of *Dendrobium* (p. 207); the choice of pot, raft or bark will be governed by the stature and habit of the individual species. The resting period will depend on the fleshiness of the pseudobulb.

Literature: Seidenfaden, G., Orchid genera in Thailand X, *Opera Botanica* **62**: 1–157 (1982).

1a. Leaves rolled when young 2
b. Leaves folded when young 3
2a. Stems short, concealed; flowers 10–30 per raceme; hairy outside **1. javanica**
b. Stems long, exposed; flowers 2–6 per raceme, not hairy **2. coronaria**
3a. Flowers in dense racemes; flower-stalks hairy **4. spicata***
b. Flowers in loose racemes; flower-stalks not hairy **3. bractescens**

1. E. javanica (Swartz) Blume (*E. fragrans* Reichenbach; *E. stellata* Lindley). Illustration: Botanical Magazine, 3605 (1837); Williams et al., Orchids for everyone, 188 (1980); Bechtel et al., Manual of cultivated orchid species, 212 (1981).
Epiphytic. Pseudobulbs 5–8 × 1–2 cm, narrowly ovoid, of 1 internode, sheathed, bearing 1 or 2 leaves. Leaves 20–50 × 3.5–6 cm, lanceolate, acute, rolled when young. Flowering stem 30–60 cm, erect to arched, terminal, bearing up to 30 flowers. Flowers *c.* 4 cm wide, fragrant, white or cream-coloured. Flower-stalks and outside of sepals hairy. Sepals 2–2.5 cm, lanceolate to narrowly triangular, acute; petals smaller, slightly sickle-shaped. Lip 1–1.5 cm, oblong to lanceolate, 3- or 5-ridged, shallowly lobed, central lobe narrowly triangular, obtuse or acute; lateral lobes semi-ovate, erect. Column *c.* 8 mm; foot *c.* 5 mm. *Himalaya, Thailand, Malaysia, Indonesia & Philippines*. G2. Spring–summer.

2. E. coronaria (Lindley) Reichenbach (*Trichosma suavis* Lindley). Illustration: Orchid Review **82**: 283 (1974); Bechtel et al., Manual of cultivated orchid species, 212 (1981); Opera Botanica **62**: 41 & 151 (1982).
Terrestrial or epiphytic. Stems 7.5–25 × 5 mm, sheathed at base only, bearing 2 leaves near the apex. Leaves 10–25 × 3–5 cm, narrowly elliptic to ovate, leathery, acute, rolled when young. Flowering stem to *c.* 15 cm, erect, robust, terminal, bearing 2–6 flowers. Flowers 2.5–4 cm wide, white or pale yellow with crimson markings on the lip, fragrant. Sepals 1.5–2.2 cm × 5–8 mm, elliptic to ovate; petals *c.* 2 × 1 cm, oblong to elliptic, obtuse. Lip *c.* 2 × 1 cm, oblong to ovate, 3-lobed; lateral lobes oblong, rounded, erect; central lobe short, almost circular, with 5–7 wavy keels. Column longer than column foot. *E Himalaya to Thailand & Malay peninsula*. G2. Autumn–winter.

3. E. bractescens Lindley. Illustration: Botanical Magazine, 4163 (1845); Holttum, Flora of Malaya I: 386 (1953); Millar, Orchids of Papua New Guinea, 62 (1978).
Pseudobulbs 6–15 cm, cylindric to fusiform, sheathed below, bearing 2 or 3 leaves. Leaves 12–20 cm, oblong to narrowly elliptic, obtuse. Flowering stems usually 2 per pseudobulb, 15–20 cm, erect to horizontal, many-flowered; bracts conspicuous, to 2.5 cm, oblong, rounded, cream-coloured to pale green. Flowers *c.* 2.5 cm wide and long, white to pale green, sometimes flushed with pink. Sepals narrowly triangular, acute; mentum pronounced. Petals lying close under upper sepal. Lip oblong, rounded, shallowly 3-lobed; lateral lobes oblong, rounded, pale purple; central lobe almost circular, with 3 or 5 pink, wavy keels. Column shorter than column foot. *E Himalaya, Andaman Islands, SE Asia, Indonesia & Philippines*. G2. Summer.

4. E. spicata (D. Don) Handel-Mazzetti (*E. convallarioides* Lindley). Illustration: Shuttleworth et al., Orchids, 72 (1970); Graf, Tropica, 732 (1978); Bechtel et al., Manual of cultivated orchid species, 213 (1981).
Pseudobulbs 5–15 × 1.5–3 cm, densely sheathed, bearing up to 4 leaves. Leaves narrowly elliptic, folded when young. Flowering stems shorter than leaves, densely many-flowered, erect. Flowers *c.* 1 cm wide, almost spherical, white to pale yellow. *E Himalaya & Burma*. G2. Spring.

***E. floribunda** Lindley. Illustration: Edwards's Botanical Register **30**: t. 20 (1844); Orchid Review **76**(9): cover (1968). Stems to 30 × 1 cm, cylindric, slender, bearing 3–7 leaves in upper half. Flowering stems several, *c.* 15 cm, axillary, horizontal to pendent, densely many-flowered. Flowers *c.* 1 cm wide, white to pinkish red. Column dark purple. *Himalaya to Vietnam, Malay peninsula, Sumatra & Borneo*. G2. Summer–autumn.

62. CRYPTOCHILUS Wallich

J.C.M. Alexander

Epiphytic. Pseudobulbs cylindric to spherical, simple, crowded, developing from flowering stems. Leaves 1 or 2 per pseudobulb, leathery, usually shortly stalked, folded when young, appearing after flowering. Flowering stems sheathed below, with many flowers in dense 2-rowed spikes; flowers equal to or shorter than the persistent bracts. Sepals fused into a urn-shaped to spherical tube with free, spreading, triangular points. Petals linear to oblong, concealed in sepal-tube. Lip fused to column foot. Column short, straight, smooth at apex. Pollinia 8.

A genus of 2 species confined to the Himalaya. The affinities of *Cryptochilus* are uncertain and it was previously considered a member of the tribe *Vandae*; it is now thought to be allied to *Eria*. The flowers show adaptations to bird pollination. Cultivation conditions are similar to those for Himalayan species of *Dendrobium* (p. 207), allowing plants to become quite dry between waterings during the winter.

1. C. sanguinea Wallich. Illustration: Orchid Review **79**: 227 (1971); American Orchid Society Bulletin **46**: 244 (1977).
Pseudobulbs to *c.* 7 cm, ovoid, slightly compressed. Leaves 9–20 × 2–4 cm, oblong to lanceolate, acute. Flowers *c.* 2 cm long; bracts about equal to flowers in length.

Sepals united into a scarlet, urn-shaped tube, swollen at base; tips of lobes acute. Petals small, oblong to obovate, obtuse, yellow, flushed with brown. Lip oblong, broader at tip, channelled, yellow. Column about half as long as lip, straight. Pollinia greenish blue. *Nepal & NE India (Sikkim, Meghalaya).* G1. Summer.

*C. **lutea** Lindley. Illustration: Annals of the Royal Botanic Garden Calcutta **8**: t. 221 (1898). Pseudobulbs cylindric. Leaves 10–15 × 1.5–2.5 cm, narrowly elliptic, acute. Sepal-tube *c.* 1 cm, spherical to cylindric, not swollen at base, yellow. Bracts longer than flowers. Petals diamond-shaped. Pollinia yellow. *NE India (Sikkim, Naga hills), Bhutan.* G1. Summer.

63. MEIRACYLLIUM Reichenbach

J.C.M. Alexander

Creeping rhizomatous epiphytes. Stems short, scaly, 1-leaved, slightly swollen, simple. Leaves stalkless, thick and leathery, folded when young. Flowers solitary or several, borne at junction of leaf and stem. Sepals similar, free except at the base where they are fused to the column foot, forming a short mentum. Petals narrower than sepals, spreading. Lip simple, deeply concave or pouched. Pollinia 8, in 2 groups of 4, narrowly club-shaped, waxy, unequal. Rostellum prominent.

A genus of 2 species from Mexico, Guatemala and El Salvador. Cultivation conditions are similar to those for *Cattleya* (p. 191), though night temperatures should be higher in winter.

Literature: Dressler, R.L., The relationships of Meiracyllium, *Brittonia* **12**: 222–5 (1960).

1. M. wendlandii Reichenbach (*M. gemma* Reichenbach). Illustration: Orchid Digest **41**: 69 (1977); Die Orchidee **30**: 76 (1979); Northen, Miniature orchids, C-57 (1980).

Stems to 1 cm, ascending. Leaves to 5 × 2.3 cm, oblong, obtuse to rounded. Flowering stems to *c.* 4 cm, 1-flowered. Flowers pinkish purple. Upper sepal 1–1.7 cm × 5 mm, ovate, acute or slightly acuminate; lateral sepals similar but narrower and slightly asymmetric. Petals 8–15 × 1–2.5 mm, oblong to lanceolate, acute, finely toothed. Lip 1–1.3 cm × 7 mm, ovate, concave; apex acuminate and curved downwards. Column *c.* 6 mm, acute, slightly narrowed at base. *Mexico, Guatemala & El Salvador.* G2. Summer.

***M. trinasutum** Reichenbach. Illustration: Orchid Digest **41**: 69 (1977); Northen, Miniature orchids, C-56 (1980); Bechtel et al., Manual of cultivated orchid species, 231 (1981). Leaves 2.8–5 × 1.5–3.5 cm, elliptic to circular. Flowers several, pale pinkish purple, almost white below. Sepals 8–11 × 3.5–5 mm, oblong to elliptic, acuminate. Petals 7–10 × 3 mm, elliptic to spathulate, acute. Lip 7–9 × 4–5 mm, deeply pouched, acuminate. Column broad at base, round-tipped. *Mexico & Guatemala.* G2. Summer.

64. PHYSOSIPHON Lindley

J.C.M. Alexander

Tufted epiphytes. Stems slender, erect or ascending, sheathed, 1-leaved at apex. Leaves tapering to the base, fleshy or leathery, folded when young. Flowers many, in long, narrow, erect or arching racemes, borne at the junction of leaf and stem. Sepals equally fused for more than half their length, forming a 3-angled, inflated tube, slightly constricted at the mouth; free portions spreading. Petals and lip very small, concealed in sepal-tube. Lip 3-lobed or simple, fleshy, channelled. Column small, 3-lobed. Pollinia 2, ovoid, waxy.

A genus of about 6 species from the New World tropics. Cultivation conditions are similar to those for *Masdevallia.*

1. P. tubatus (Loddiges) Reichenbach (*Stelis tubatus* Loddiges; *P. loddigesii* Lindley: *P. guatemalensis* Rolfe). Illustration: Botanical Magazine, 4869 (1855); Orchid Review **79**: 220 (1971); Bechtel et al., Manual of cultivated orchid species, 262 (1981). Stems 1.5–12 cm. Leaves 4–17 × 1.5–3 cm, elliptic to narrowly obovate. Flowering stem 8–45 cm; flowers 20–100, greenish yellow to brick-red. Sepals 6–22 mm; free portion oblong to lanceolate or elliptic. Petals 1.5–2.5 × 0.6–1 mm, obovate to spathulate, sometimes obscurely 3-lobed. Lip 2–3.2 × 1–2.5 mm, 3-lobed; lobes rounded. Column 2–3.5 mm. *Mexico & Guatemala.* G2. Summer.

Very variable in size and colour of flowers.

65. MASDEVALLIA Ruiz & Pavon

J.C.M. Alexander

Epiphytic or on rocks. Stems short and slender, erect, tufted, concealed in papery sheaths. Leaves 1, apical, linear to lanceolate, somewhat fleshy, leathery, minutely 3-toothed at tip, folded when young. Flowers 1–several, in terminal racemes borne at the leaf-base. Sepals fused below into a narrow or cup-shaped tube; lateral sepals usually further fused into a broad oblong blade; free portions of sepals spreading and usually long-tailed, separated by a sinus or notch. Petals and lip small, often concealed in the sepal-tube. Lip oblong, sometimes 3-lobed or fiddle-shaped. Column toothed at apex; column foot forming a short mentum with base of lateral sepals. Pollinia 2.

About 300 species from Mexico, C and S America, mostly from the Andean cloud-forests, frequently grown for their large and brightly coloured sepals, whose tails often continue to grow after the flower has opened. They perform best if grown in epiphytic compost or on tree-fern bark in cool, moist conditions with plenty of ventilation. Having no pseudobulbs, plants must never be allowed to dry out; they should be shaded from direct sunlight. Well-established plants can be divided in early spring.

Literature: Woolward, F.H., *The genus Masdevallia* (1890–6); Kraenzlin, F.W.L., Masdevallia, *Feddes Repertorium Beihefte* **34** (1925); Braas, L., The genus Masdevallia, *Orchid Review* **86**: 30–43 (1978).

1a. Flowering stems winged or sharply angled — 2
b. Flowering stems not winged or sharply angled — 3
2a. Lateral sepals fused into a blade 2 cm or more wide — **1. tovarensis***
b. Lateral sepals fused into a blade less than 2 cm wide — **2. infracta**
3a. Two or more flowers open together on the same stem — 4
b. Flowers open 1 at a time — 5
4a. Flowering stems 20 cm or more; lateral sepals fused above sepal-tube; petals not toothed — **8. schlimii***
b. Flowering stems less than 20 cm; sepals all equally fused; petals toothed — **7. caloptera***
5a. Sepal-tube narrow and slightly curved — 6
b. Sepal-tube broad and cup-shaped — 9
6a. Lateral sepals not drawn out into fine tails — **9. coccinea***
b. Lateral sepals drawn out into fine tails — 7
7a. Flowering stems about equal to leaves; lip fringed — **10. rosea**
b. Flowering stems longer than leaves; lip not fringed — 8
8a. Free part of upper sepal (excluding tail) 2 cm or more — **11. veitchiana**
b. Free part of upper sepal (excluding tail) 1 cm or less — **12. amabilis***
9a. Free part of upper sepal (including

tail) more than 7 cm **3. macrura***
b. Free part of upper sepal (including tail) less than 6 cm 10
10a. Apex of lip without papillae **4. rolfeana***
b. Apex of lip with papillae 11
11a. Upper sepal abruptly contracted into a thick fleshy tail; flowering stem equal to or slightly shorter than leaves **5. coriacea**
b. Upper sepal gradually tapered into a slender tail; flowering stem distinctly shorter than leaves **6. peristeria***

1. M. tovarensis Reichenbach. Illustration: Botanical Magazine, 5505 (1865); Skelsey, Orchids, 120 (1979); Bechtel et al., Manual of cultivated orchid species, 227 (1981).
Epiphytic. Leaves to 15 × 2 cm, lanceolate to obovate, obtuse or rounded. Flowering stem to 20 cm, angled, bearing up to 4 white flowers, all open together. Sepal-tube *c.* 6 mm; upper sepal 4–5 cm, finely tapered. Lateral sepals 3.5–4.5 cm, triangular to ovate, slender-tailed, fused for 2–2.5 cm. Petals *c.* 6 × 1.5 mm, linear, acute, asymmetric. Lip *c.* 5 × 2 mm, oblong to spathulate, rounded or acute. Column 4.5 mm, finely toothed. *Venezuela*. G2. Winter.

***M. ephippium** Reichenbach. Illustration: Botanical Magazine, 6208 (1876); Dunsterville & Garay, Venezuelan orchids illustrated **1**: 213 (1959); American Orchid Society Bulletin **45**: 104 (1976). Flowering stem 20–30 cm, 3-angled, bearing several flowers in succession. Flowers purplish brown with yellow tails. Upper sepal to 14 cm, blade small, circular, abruptly narrowed to a long tail. Lateral sepals fused into a deep cup; tails 7–10 cm. *Colombia & Ecuador*. G2. Spring–summer.

***M. maculata** Klotzsch & Karsten. Illustration: Orchid Digest **36**: 24 (1972); Bechtel et al., Manual of cultivated orchid species, 227 (1981). Flowering stem to 25 cm, bearing several flowers in succession. Sepals orange to purplish brown with yellow tails; sepal-tube *c.* 1.5 cm. Upper sepal 6–8 cm, narrowly triangular. Lateral sepals *c.* 8 cm, narrowly ovate, fused for *c.* 3 cm. Petals *c.* 6 mm, oblong, asymmetric, each with a small point. Lip *c.* 6 mm, spathulate to fiddle-shaped, papillose. Column *c.* 6 mm. *Venezuela, Colombia & Peru*. G2. Summer–autumn.

2. M. infracta Lindley. Illustration: Pabst & Dungs, Orchidaceae Brasilienses **1**: 225 (1975); Graf, Tropica, 737 (1978); Bechtel et al., Manual of cultivated orchid species, 225 (1981).
Epiphytic. Leaves 8–14 × 1.5–2.5 cm, lanceolate to oblong, obtuse. Flowering stem to 25 cm, angled, bearing several flowers in succession. Sepals pinkish purple with greenish yellow tails; sepal-tube 1–1.5 cm. Upper sepal 4–5 cm, triangular to ovate, slender-tailed. Lateral sepals 4–5 cm, oblong to ovate, fused for about 2 cm, slender-tailed. Petals 6–7 mm, linear, each with a small point, pink. Lip *c.* 6 mm, oblong, with a small point, pink and orange. Column *c.* 6 mm. *Peru & Brazil*. G2. Summer.

3. M. macrura Reichenbach. Illustration: Botanical Magazine, 7164 (1891); Die Orchidee **30**: 105 (1979); Bechtel et al., Manual of cultivated orchid species, 226 (1981).
On rocks. Leaves 20–30 × 5–7 cm, oblong to elliptic, obtuse, often notched. Flowering stem 20–30 cm, 1-flowered. Sepals red to brownish yellow with prominent veins and dark warts; tails long and slender, yellow; sepal-tube 1–1.5 cm, cup-shaped. Upper sepal 10–15 cm, lanceolate to narrowly triangular, tapered; lateral sepals 9–13 cm, ovate to oblong, fused for 2–5 cm, tapered. Petals 8–10 mm, lanceolate to rhombic, rounded, yellow with dark stippling. Lip 6–8 mm, oblong, rounded, papillose, yellow with thick red margins. Column 6–8 mm, yellow. *Colombia & Ecuador*. G2. Winter–spring.

***M. velifera** Reichenbach. Illustration: Gardeners' Chronicle **1**: 745 (1887); Journal of the Royal Horticultural Society **33**: 393 (1908). Leaves 15–20 × 2.5 cm, lanceolate, obtuse. Flowering stem 6–8 cm. Sepal-tube *c.* 2 cm. Upper sepal 7–8 cm, ovate, slender-tailed, brownish yellow spotted with brown; lateral sepals 7–8 cm, oblong, fused for *c.* 5 cm, shiny reddish brown with yellow tails. Petals *c.* 1.2 cm, oblong, greenish yellow. Lip *c.* 1.2 cm, oblong, dark purple, fringed. *Colombia*. G2. Winter.

4. M. rolfeana Kraenzlin. Illustration: Woolward, The genus Masdevallia, 55 (1893); Orchid Digest **32**: 188 (1968).
Epiphytic. Leaves 11–14 × 2 cm, obovate to elliptic, obtuse, keeled. Flowering stem *c.* 8 cm, 1-flowered. Sepals dark purple, tapered into slender green and yellow tails. Sepal-tube to 1 cm, cup-shaped. Upper sepal *c.* 4 cm, triangular to ovate; lateral sepals *c.* 6 cm, oblong to ovate, fused for about 2 cm. Petals *c.* 6 mm, oblong to elliptic, obtuse, narrowed at base, purple. Lip 8–10 mm, oblong, apiculate, pink with darker spots. Column *c.* 8 mm, finely toothed at apex. *Costa Rica*. G2. Spring–summer.

***M. estradae** Reichenbach. Illustration: Botanical Magazine, 6171 (1875); Journal of the Royal Horticultural Society **26**: 146 (1901). Leaves 5–8 × 2 cm. Flowering stem *c.* 10 cm. Sepals equally fused into a cup-shaped tube *c.* 6 mm long. Upper sepal purple with a yellow margin, abruptly contracted into a slender, yellow tail. Lateral sepals oblong, purple below, pink above, tapered to a slender yellow tail. *Colombia*. G2. Spring.

***M. erinacea** Reichenbach (*M. horrida* Teuscher & Garay). Illustration: American Orchid Society Bulletin **29**: 23 (1960) & **37**: 520 (1968); Die Orchidee **30**: CXXV (1979); Northen, Miniature orchids, C-51 (1980). Dwarf epiphyte. Leaves 3–6 cm × 1–3 mm, linear, acute. Flowering stems 2.5–6 cm; bract *c.* 5 mm. Sepals *c.* 1.5 cm, equally fused into a shallow cup *c.* 8 mm long, densely hairy on the veins; tails club-shaped, pendent. Upper sepal white to greenish yellow; lateral sepals densely spotted with maroon. *Costa Rica & Ecuador*. G2. Summer.

5. M. coriacea Lindley. Illustration: Gardeners' Chronicle **21**: 95 (1897); Orquideologia **9**: 34 (1974); Bechtel et al., Manual of cultivated orchid species, 225 (1981).
Epiphytic or on rocks. Leaves 12.5–20 × 1–2 cm, linear to lanceolate, obtuse, keeled. Flowering stem 17–20 cm, 1(rarely 2)-flowered; flowering stem, bract and ovary flecked with purple. sepal-tube 1.5 cm, cup-shaped; sepals greenish white with lines of crimson spots. Upper sepal 3–4 cm, ovate, abruptly contracted into a broad, green, purple-flecked tail. Lateral sepals *c.* 4 cm, narrowly triangular, fused for 2–2.5 cm, without distinct tails. Petals 1.2–1.5 cm, oblong to spathulate, obtuse, white with a central purple stripe. Lip *c.* 1.2 cm, oblong, obtuse, papillose at tip, purple and green. Column *c.* 1 cm, finely toothed at tip, pale green. *Colombia*. G2. Spring–summer.

6. M. peristeria Reichenbach. Illustration: Botanical Magazine, 6159 (1875); Orquideologia **10**: 159 (1975); Die Orchidee **27**(6): 20 (1976).
Epiphytic. Leaves 10–15 × 2.5 cm, linear to lanceolate or oblong, obtuse. Flowering stem 5–8 cm, 1-flowered. Sepal-tube 1.5–2 cm, cup-shaped; sepals 4.5–6 cm, triangular to narrowly ovate, finely

tapered, greenish yellow spotted with crimson; lateral sepals fused for *c.* 1.5 cm. Petals 1–1.2 cm, lanceolate to spathulate, obtuse, green. Lip 1.2–1.5 cm, oblong, narrowed at base, rounded to truncate, papillose, purple. Column *c.* 1.2 cm, finely toothed, greenish white. *Colombia.* G2. Spring–summer.

*M. **civilis** Reichenbach & Warscewicz (*M. leontoglossa* misapplied). Illustration: Botanical Magazine, 5476 (1864); not Dunsterville & Garay, Venezuelan orchids illustrated **4**: 128 (1966). Leaves 15–25 × 1–1.5 cm, linear, acute. Flowering stem 2.5–7 cm; sepals 3.5–4 cm, ovate, contracted into slender tails; outside of flower brownish purple below, green above, with crimson spots. Petals white with a central purple stripe. Lip oblong with a rounded, papillose tip, purple. *Peru.* G2. Spring–summer.

The name *M. leontoglossa* Reichenbach is frequently found in horticultural literature and catalogues. It is usually treated as a synonym of *M. civilis* though the leaf shape is quite distinct in Reichenbach's original description. Illustrations in the Botanical Magazine (7245) and Woolward (The genus Masdevallia, 20) show strongly pendent flowering stems. This is not referred to by Reichenbach.

*M. **corniculata** Reichenbach. Illustration: Botanical Magazine, 7476 (1896); Die Orchidee **30**: 105 (1979). Leaves 20–25 × *c.* 4 cm, lanceolate. Flowering stem 7–10 cm, enclosed in an inflated sheath which also conceals the leaf-base. Flowers yellow, spotted and blotched with orange. Sepal-tube *c.* 2 cm, cup-shaped. Lateral sepals fused for *c.* 3 cm. *Colombia.* G2. Summer–autumn.

7. M. caloptera Reichenbach (*M. biflora* Regel). Illustration: Gartenflora **40**: 1341 (1891); Bechtel et al., Manual of cultivated orchid species, 224 (1981).
Epiphytic. Leaves 6–8 × 1.4–2 cm, oblong to lanceolate, obtuse to rounded, older leaves tinged with red. Flowering stem 8–12.5 cm, bearing 2 or more flowers, 2 or more open together. Sepals *c.* 2 cm, equally fused for *c.* 6 mm, white with purple streaks, margins finely toothed; tails slender, yellow. Sepal-tube sharply indented above mentum. Upper sepal almost circular; lateral sepals triangular to oblong. Petals 4–5 mm, toothed, white with a central purple stripe. Lip *c.* 5 mm, oblong to shallowly 3-lobed, yellow and crimson. Column *c.* 4 mm, green and purple. *Colombia, Ecuador & Peru.* G2. Summer.

*M. **abbreviata** Reichenbach (*M. polysticta* Reichenbach; *M. melanopus* misapplied). Illustration: Botanical Magazine, 6258 (1876) & 6368 (1878). Leaves 12.5–15 cm. Flowering stem 15–20 cm, bearing many flowers, several open together. Sepals *c.* 3 cm. Lip yellow. *Peru.* G2. Autumn–winter.

Very similar to *M. caloptera*. The name *M. melanopus*, which was given to a similar species with a single row of flowers all pointing in the same direction, has frequently been misapplied to *M. abbreviata*.

8. M. schlimii Lindley. Illustration: Botanical Magazine, 6740 (1884); Dunsterville & Garay, Venezuelan orchids illustrated **3**: 171 (1965); Bechtel et al., Manual of cultivated orchid species, 227 (1981).
Epiphytic. Leaves to 20 × 5 cm, obovate to elliptic, rounded or obtuse. Flowering stem to 35 cm, bearing up to 8 flowers, several open together. Sepals 5–6 cm, with long yellow tails; sepal-tube *c.* 6 mm, cup-shaped. Upper sepal triangular to ovate, finely tapered, greenish yellow, spotted with maroon below. Lateral sepals fused for 1.5–2 cm, oblong. Petals *c.* 6 mm, linear, pale yellow. Lip *c.* 6 mm, oblong to fiddle-shaped, acute, pink and yellow with purple spots and blotches. Column 6–8 mm. *Colombia & Venezuela.* G2. Summer.

*M. **racemosa** Lindley. Illustration: Die Orchidee **23**: 141 (1972) & **30**: 105 (1979); Orchid Review **86**: 35 (1978); Bechtel et al., Manual of cultivated orchid species, 227 (1981). Creeping terrestrial, rarely epiphytic. Leaves 8–12.5 × 2 cm, oblong to lanceolate, obtuse. Sepal-tube *c.* 1.5 cm, straight and narrow; sepals bright scarlet. Upper sepal *c.* 1 cm, triangular, acuminate. Lateral sepals 3–3.5 cm, elliptic to circular, shortly apiculate, fused for *c.* 2.5 cm. *Colombia.* G2. Spring–autumn.

9. M. coccinea Lindley (*M. lindenii* André; *M. harryana* Reichenbach). Illustration: Botanical Magazine, 5990 (1872); Shuttleworth et al., Orchids, 37 (1970); Bechtel et al., Manual of cultivated orchid species, 225 (1981).
Terrestrial; stems *c.* 4 cm. Leaves 15–20 × 2–3 cm, oblong to lanceolate, obtuse to rounded. Flowering stem to 40 cm, 1-flowered, streaked with purple, flowers held well above leaves. Sepal-tube *c.* 2 cm, narrow and curved; upper sepal to 5 cm, narrowly triangular, finely tapered; tail erect or curved back. Lateral sepals *c.* 6 × 2 cm, ovate to oblong, acuminate, fused for 2–3 cm. Sepals purple, crimson, scarlet, orange, yellow or white; base of tube white, greenish yellow or yellow. Petals *c.* 1 cm, linear, notched, off-white. Lip *c.* 8 mm, oblong, shallowly fiddle-shaped, pinkish. Column *c.* 6 mm. *Colombia & Peru.* G2. Spring.

*M. **militaris** Reichenbach (*M. ignea* Reichenbach). Illustration: Botanical Magazine, 5962 (1872). On rocks. Stems 5–8 cm. Flowering stem to 30 cm. Flowers held just above leaves. Very similar to *M. coccinea* but upper sepal tail drooping; sepals usually reddish brown to orange. Petals acuminate. *Colombia & Venezuela.* G2. Spring–summer.

10. M. rosea Lindley. Illustration: Graf, Tropica, 736 (1978); Northen, Miniature orchids, C-52 (1980).
Epiphytic. Leaves 12–20 × 2–3 cm, elliptic to obovate, acute. Flowering stem 12–20 cm, 1-flowered, arched. Sepal-tube 2.5–3.5 cm, narrow, slightly curved. Upper sepal 2.5–3.5 cm, tapering from the base into a long narrow tail. Lateral sepals 4–5 cm, oblong to narrowly triangular, fused for *c.* 2 cm. Sepals reddish purple and red. Petals 4–5 mm, oblong, off-white. Lip *c.* 5 mm, oblong, shallowly fiddle-shaped, pink and yellow at base; tip dark purplish brown, densely fringed. *Colombia & Ecuador.* G2. Spring–summer.

11. M. veitchiana Reichenbach. Illustration: Botanical Magazine, 5739 (1868); Graf, Tropica, 736 (1978); Bechtel et al., Manual of cultivated orchid species, 228 (1981).
On rocks. Leaves 15–25 × 2.5 cm, oblong to narrowly obovate, obtuse. Flowering stem 30–45 cm, erect, 1-flowered. Sepal-tube *c.* 3 cm, cylindric. Upper sepal 5–7 × 2 cm, ovate to triangular, finely tapered. Lateral sepals 6–7.5 × 1.5–2.5 cm, triangular to ovate-oblong, fused for *c.* 3 cm, acuminate. Sepals bright orange inside, with short purplish hairs; yellowish with dark veins outside. Petals 1–1.5 cm, oblong, each with a small point, off-white. Lip 1–1.5 cm, oblong to shallowly 3-lobed, purple and white. *Peru.* G2. Spring–summer.

12. M. amabilis Reichenbach. Illustration: Veitch, Manual of orchidaceous plants **5**: 24 (1889); American Orchid Society Bulletin **42**: 240 (1973).
Leaves 12–18 × 1.5–2 cm, oblong to lanceolate, obtuse to rounded. Flowering stem 25–30 cm, erect, 1-flowered, tinged with pink. Sepal-tube 2–2.5 cm, narrow, slightly curved. Upper sepal *c.* 4 cm,

triangular to ovate, acuminate, long-tailed, reddish orange with darker veins. Lateral sepals 4–4.5 cm, ovate to oblong, acuminate, fused for *c.* 1.5 cm, long-tailed, crimson, Petals *c.* 6 mm, oblong, each with a small point, yellow. Lip *c.* 6 mm, spathulate; base oblong. *Peru.* G2. Winter.

***M. barleana** Reichenbach. Illustration: American Orchid Society Bulletin **42**: 240 (1973); Bechtel et al., Manual of cultivated orchid species, 224 (1981). Leaves 8–12 × 2–2.5 cm. Flowering stem 15–25 cm. Sepal-tube 1.5–2 cm. Lateral sepals abruptly contracted into long tail. Sepals pink with darker veins. Petals with 3 almost equal teeth at apex. *Peru.* G2. Spring–summer.

66. DRACULA Luer
J.C.M. Alexander
Epiphytic, terrestrial or on rocks. Stems short, slender, erect, concealed in papery sheaths. Leaves 1, apical, linear to lanceolate, thinly leathery, sharply keeled, minutely 3-toothed at tip, folded when young. Flowers 1 or a few in succession on erect or arched racemes borne at leaf-base. Sepals fused below into a wide, shallow cup. Lateral sepals usually further fused into a broad blade; free parts of sepals ovate, long-tailed. Petals small, spathulate, split at tip into 2 bivalve-like flaps, often papillose. Lip hinged to column foot, clearly divided into a short claw and cup-shaped or pouched blade, 3-keeled below; marginally toothed. Column short; column foot slender, fused to base of lateral sepals forming a mentum. Pollinia 2.

About 60 species, mostly from Colombia with a few representatives in C America, Ecuador and Peru. The species in this genus were originally described in *Masdevallia* section *Saccilabiatae* Reichenbach, and more recently in section *Chimeroideae* of the same genus. They are, however, quite distinct in floral morphology and appearance. They should be grown in conditions similar to those described for *Masdevallia* (p. 201) though, being native to slightly lower altitudes, they prefer higher temperatures.
Literature: Hawley, R.M., Masdevallia chimaera and the marvellous monsters, *American Orchid Society Bulletin* **46**: 600–9 (1977); Luer, C.A., Dracula, a new genus in the Pleurothallininae (Orchidaceae), *Selbyana* **2**: 190–8 (1978); Kennedy, G., The genus Dracula, *Orchid Digest* **43**: 31–8 (1979).

1a. Lip blade broader than long, kidney-shaped **1. bella**
b. Lip blade isodiametric or longer than broad **2. chimaera***

1. D. bella (Reichenbach) Luer (*Masdevallia bella* Reichenbach). Illustration: Gardeners' Chronicle **16**: 237 (1881); Orchid Review **77**: 207 (1969); American Orchid Society Bulletin **46**: 601 (1977).
Epiphytic or terrestrial. Leaves 12–20 × 2.5 cm, oblong to lanceolate, obtuse. Flowering stems to 18 cm, pendent, purple to purplish green, 1-flowered. Flowers to *c.* 25 cm long. Sepals hairless; sepal-tube 1–1.5 cm. Upper sepal 10–12 × 2–2.5 cm, triangular to ovate, finely tapered, greenish yellow, densely spotted with purplish brown; tail long and slender. Lateral sepals 8–10 × 1.5–2. cm, fused for *c.* 2.5 cm into an almost square blade; tails on outer corners, often crossed. Petals 6–8 mm, divided into a fan-shaped blade and a claw; blade unequally 2-flapped, yellow with brown markings, with papillae and teeth on the flaps. Lip-blade 1.2 × 2 cm, kidney-shaped, concave, white with radiating lines; claw *c.* 4 mm, pink and fleshy. Column *c.* 3 mm, stout, purplish brown. *Colombia.* G2. Winter–summer.

2. D. chimaera (Reichenbach) Luer (*Masdevallia chimaera* Reichenbach; *M. backhousiana* Reichenbach). Illustration: Shuttleworth et al., Orchids, 37 (1970); American Orchid Society Bulletin **46**: 600 (1977); Graf, Tropica, 738 (1978). Terrestrial or on rocks, rarely epiphytic. Leaves 15–25 × 3–5 cm, elliptic to obovate, obtuse. Flowering stems to 60 cm, erect or ascending, green to purplish green, bearing 3–8 flowers. Sepal-tube 1–1.5 cm; lateral sepals fused for *c.* 2.5 cm; sepal blades 5–6 cm, triangular to ovate, tapered, yellow or greenish yellow, densely spotted and blotched with brownish purple or black, hairy and warty; tails 8–20 cm slender, purple. Petals *c.* 8 mm, spathulate, white with purple markings. Lip *c.* 2 cm, oblong to square or circular in surface view, deeply pouched, white to yellow or pinkish brown; claw pink and orange. Column *c.* 1 cm, minutely toothed at apex; column foot pink. *Colombia.* G2. Winter.

***D. radiosa** (Reichenbach) Luer (*Masdevallia radiosa* Reichenbach). Illustration: Orchid Review **38**: 247 (1930); Orquideologia **9**: 44 (1974) & **10**: 81 (1975); American Orchid Society Bulletin **46**: 604 (1977). Flowering stems erect or pendent, bearing 2–4 flowers. Sepals equally fused for *c.* 1.5 cm, brownish yellow with dark purplish brown papillae inside, pinkish or brownish yellow outside; free parts broadly ovate to rectangular, abruptly tapered; tails 7–10 cm, purplish brown. Petals 4–5 mm, yellow with purple spots; flaps dissimilar, papillose. Lip *c.* 6 mm in diameter, pinkish white. *Colombia.* G2. Spring–summer.

***D. erythrochaete** (Reichenbach) Luer (*Masdevallia erythrochaete* Reichenbach; *M. astuta* Reichenbach). Illustration; Orquideologia **9**: 43 (1974); Northen, Miniature orchids, C-30 (1980); Bechtel et al., Manual of cultivated orchid species, 225 (1981). Epiphytic. Leaves 15–25 × 1–1.5 cm. Flowering stem erect to pendent, bearing 1–3 flowers. Sepal-tube *c.* 6 mm; sepals *c.* 5 cm, ovate, abruptly tapered, pale yellow, spotted and veined with pinkish purple; tails purple. Petals *c.* 3 mm, pinkish. Lip 1–1.5 cm, pale pinkish yellow. *Costa Rica to Colombia.* G2. Autumn–winter.

***D. vampira** (Luer) Luer (*Masdevallia vampira* Luer). Illustration: Orchid Digest **43**: 34 (1979). Epiphytic. Flowering stem 20–40 cm or more, horizontal to pendent, bearing 5–7 flowers. Sepals hairless, very pale green with dense purplish black veins and tails. Petals *c.* 6 mm, white marked with purple. Lip 1.5–2.5 cm; blade concave, white with pink or yellow veins. *Ecuador.* G2. Summer.

67. PORROGLOSSUM Schlechter
J.C.M. Alexander
Similar to *Masdevallia*. Leaves obovate, lanceolate or elliptic, long-stalked. Flowers solitary or a few, open together or in succession, on a slender erect or arched stem. Sepals equally fused, forming a cup-shaped tube, free parts spreading. Petals small, meeting around column. Lip S-shaped in side profile, with a distinct claw, attached below apex of column foot, mobile and sensitive. Rostellum upright. Anther erect, on column apex. Pollinia 2, hard.

A genus of 8 or 9 montane species from Venezuela to Peru, originally in *Masdevallia* but distinct in lip and column structure and in the sensitivity of the lip which folds up suddenly against the column if stimulated by touch or shaking. Cultivation should be in cool, moist, shady conditions similar to those for *Masdevallia* (p. 201).
Literature: Sweet, H.R., The genus Porroglossum, *American Orchid Society Bulletin* **41**: 513–24 (1972).

1. P. echidnum (Reichenbach) Garay (*Masdevallia echidna* Reichenbach; *M. muscosa* Reichenbach; *P. muscosum*

(Reichenbach) Schlechter). Illustration: Botanical Magazine, 7664 (1899); Northen, Miniature orchids, 149 (1980); Bechtel et al., Manual of cultivated orchid species, 267 (1981).
Leaves to 10 cm, elliptic, obtuse; upper surface papillose. Flowering stem to 15 cm, densely hairy, bearing 1 or 2 flowers. Sepal-tube 3–4 mm, free parts 2.5–3 cm × 6 mm, triangular to ovate; lateral sepals slightly asymmetric. Sepals yellow or yellowish brown. Petals 5–8 × 1 mm, pinkish. Lip 5–8 × 4 mm, triangular to diamond-shaped, long-clawed, front margin hairy. *Venezuela, Colombia & Ecuador*. G2. Flowering intermittent.

68. TRIARISTELLA (Reichenbach) Luer
J.C.M. Alexander
Similar to *Masdevallia*. Distinguished by its small, almost free, upper sepal and large, almost totally fused lateral sepals with tails attached to side of blade. Lip arrow-shaped.
Cultivation as for *Masdevallia* (p. 201).

1. T. reichenbachii Brieger (*Masdevallia triaristella* Reichenbach). Illustration: Botanical Magazine, 6268 (1876); Bechtel et al., Manual of cultivated orchid species, 228 (1981).
Leaves 2.5–5 cm × 5–8 mm, cylindric, channelled, rounded, tinged with purple. Flowering stems 7.5–10 cm, erect, bearing 1–3 flowers in succession. Upper sepal *c.* 2 cm, triangular to ovate, long-tailed; lateral sepal blade 15 × 2–4 mm, tails 1–1.5 cm. Sepals yellow and pinkish purple. Petals *c.* 4 mm, linear, each with a small point. Lip *c.* 6 mm, pinkish brown. *Costa Rica, Panama & Colombia*. G2. Summer–autumn.

69. PLEUROTHALLIS R. Brown
J.C.M. Alexander
Epiphytic, rarely on rocks. Stems clustered, slender, with a few overlapping sheaths, 1-leaved. Leaves leathery, stalkless or shortly stalked, folded when young. Flowering stems 1–several, apical, erect or arched, bearing 1–many flowers in racemes. Flowers small, usually 2-ranked, not widely open, often pendent. Upper sepal free or very shortly fused to lateral sepals; lateral sepals partially to completely fused, concave, forming a small mentum with the column foot. Petals usually much smaller than sepals. Lip small, loosely hinged to column foot. Pollinia 2 (in our species).
A large genus of about 1000 species, widespread throughout the mountainous regions of the New World tropics and subtropics. Cultivation conditions are similar to those for *Masdevallia* (p. 201).

Stems. 7 cm or less: **1,2,6**; 8 cm or more: **1,3–6**. Pendent: **1**. Sheaths brown: **1,3,5,6**; sheaths grey: **4**; sheaths white: **1,2**.
Leaf. 7 cm or less: **1,2,6**; 8 cm or more: **1,3–6**. Less than 2 cm wide: **1,2,6**. Cone-shaped: **1**. Deeply grooved on midrib above: **1**.
Flowering stems. Solitary: **1–6**; 2 or more: **6**. 6 cm or less: **1,2,6**; shorter than leaf: **1,6**. Zig-zag: **1,2,6**.
Flowers. 3 or fewer: **1**; 25 or more: **4,6**. Yellow or greenish yellow: **1,2,4,6**; greenish brown: **5**; purple or purplish brown: **1,3**; pink: **1**. Translucent: **2,4,6**. Hairy: **6**. With dangling marginal appendages: **6**.
Lateral sepals. Less than 1.5 cm: **1,2,4,6**; more than 2 cm: **3,5**. Almost totally fused: **2–4,6**; fused for two-thirds or more: **1–4,6**; fused for half or less; **5,6**.
Petals. 5 mm or less: **1,2,6**; 6–10 mm: **4,5**; *c.* 1.5 cm: **3**. Almost equal to sepals: **4**; distinctly shorter than sepals: **1–3,5,6**.
Lip. Warty: **1,3**. With erect basal margin: **4**.
Ovary. Spiny: **1**.

1a. Lateral sepals fused for two-thirds of their length or more — 2
b. Lateral sepals fused for half of their length or less — 5
2a. Flowering stem shorter than leaves, if longer then lower part enclosed in a groove running about half way up leaf-blade — **1. immersa***
b. Flowering stem longer than leaves, lower part not enclosed in a groove running on leaf-blade — 3
3a. Stems 1 cm or less; leaf *c.* 1 cm wide or less — **2. grobyi**
b. Stems 2 cm or more; leaf 2 cm wide or more — 4
4a. Flowering stem arched; flowers purple; sepals 2.5–3 cm, twice as long as petals — **3. macrophylla**
b. Flowering stem erect; flowers yellowish green; sepals 5–15 mm, almost equal to petals — **4. quadrifida**
5a. Stems 30 cm or more, flowering stem 1; sepals *c.* 3 cm — **5. grandis**
b. Stems usually less than 30 cm; flowering stems 1 or more; sepals 1 cm or less — **6. gelida***

1. P. immersa Linden & Reichenbach. Illustration: Botanical Magazine, 7189 (1891); Dunsterville & Garay, Venezuelan orchids illustrated 3: 254 (1965); Ospina, Proceedings of 7th world orchid conference, 73 (1974).
Stems 2–7 cm, with brown sheaths. Leaf 7–20 × 2–4.5 cm, lanceolate to narrowly obovate, rounded to acute, with a central winged groove enclosing the lower part of flowering stem. Flowering stem 17–40 cm, erect, zig-zag, sometimes arched above, bearing 12–20 bell-shaped flowers. Flowers purplish brown or bright greenish yellow. Upper sepal 8–14 × 3–4.5 mm, lanceolate, obtuse; lateral sepals fused for *c.* two-thirds of their length into an elliptic blade, 8–13 × 5–7 mm. Sepals stiff, finely hairy. Petals 3–4 × 2–3 mm, spathulate, fleshy with thin margins. Lip *c.* 4 × 2 mm, triangular, clawed, 2-ridged, warty at tip. *Mexico to Colombia & Venezuela*. G2. Winter.
***P. pectinata** Lindley. Illustration: Schlechter, Die Orchideen, edn 2, 176 (1927); Pabst & Dungs, Orchidaceae Brasilienses 1: 226 & 345 (1975). Stems 10–15 cm, pendent. Leaf 7–10 cm, broadly elliptic, cone-shaped by fusion of lower margins, enclosing the base of the flowering stem. Flowering stem to 6 cm; flowers *c.* 15 in 2 dense ranks, yellow to green. *S Brazil*. G2. Spring–summer.
***P. tribuloides** Lindley. Illustration: Reichenbach, Xenia Orchidacearum, t. 275 (1892). Stem less than 1 cm, with white sheaths. Leaf to 7 × 1.5 cm, obovate. Flowering stem to 1 cm, bearing 1–3 pink to brownish red or purple flowers. Lateral sepals fused from one-third of their length to completely so. *C America, West Indies*. G2.

2. P. grobyi Lindley. Illustration: Botanical Magazine, 3682 (1839); Shuttleworth et al., Orchids, 38 (1970); Skelsey, Orchids, 133 (1979).
Stems 1 cm or less, with white sheaths. Leaf to 7 × 1.1 cm, lanceolate to narrowly obovate, obtuse or rounded. Flowering stem to 15 cm, slender, erect, zig-zag, bearing 4–12 flowers in 2 opposite ranks; flowers translucent, pale yellow to green with scattered purplish veins. Upper sepal 4–10 × 1.5–3 mm, ovate, acuminate, rounded. Lateral sepals fused into an oblong to elliptic, slightly bifid blade, 5–14 × 4–7 mm. Petals 1.5–3 × 0.5–1 mm, lanceolate to spathulate. Lip 2–4 × 1 mm, tongue-shaped, arched, rounded. *Mexico & West Indies to Peru & Brazil*. G2. Summer.

3. P. macrophylla Humboldt, Bonpland & Kunth (*P. roezlii* Reichenbach). Illustration: Cogniaux & Goossens, Dictionnaire Iconographique des Orchidées, Pleurothallis, t. 1 (1898); Journal of the

Royal Horticultural Society **26**: 147 (1901); Gartenflora **50**: 272 (1901).
Stems 8–15 cm, with brown sheaths. Leaf 12–20 × 2.5–4 cm, oblong to lanceolate, acute, sometimes notched. Flowering stem to 30 cm, arched to pendent, bearing 5–10 flowers; bracts 1–1.5 cm. Flowers pendent, dark purple. Upper sepal 2.5–3 × 1–1.5 cm, elliptic, acute; lateral sepals fused into an elliptic, obtuse blade, 2.5–3 × 1.4–2 cm. Petals *c.* 1.5 cm, lanceolate, acute, fleshy, prominently 3-veined. Lip *c.* 1.5 cm, spathulate, fleshy, 2-ridged, papillose above. *Colombia.* G2. Winter.

4. P. quadrifida (Llave & Lexarza) Lindley (*P. ghiesbrechtiana* Richard & Galeotti). Illustration: Dunsterville & Garay, Venezuelan orchids illustrated **2**: 299 (1961); Graf, Tropica, 752 (1978); Bechtel et al., Manual of cultivated orchid species, 265 (1981).
Stems to 15 cm, with grey sheaths. Leaf to 17 × 3 cm, oblong to narrowly elliptic, rounded, often notched, sometimes grooved along the back of the midrib. Flowering stem to 45 cm, usually longer than leaf, erect to spreading, bearing 10–30 flowers; flowers pendent, translucent, pale yellow to yellowish green. Upper sepal 7–10 × 1.5–2.5 mm, ovate, acuminate, concave, 3-veined; lateral sepals fused into an elliptic, rounded, minutely notched, 6-veined blade, 5–12 × 4–6 mm. Petals 6–10 × 2–4 mm, lanceolate, acute, 1-veined. Lip 4–6 × 1.5–3 mm, oblong, waisted; margin erect at base. *Mexico & West Indies south to Venezuela & Colombia.* G2. Winter.

5. P. grandis Rolfe. Illustration: Botanical Magazine, 8853 (1920); Orchid Review **30**: 99 (1922).
Stems 30–45 cm, with pale brown sheaths. Leaf 18–35 × 9–18 cm, ovate to broadly elliptic, acute to obtuse. Flowering stem to 50 cm, erect to spreading, loosely many-flowered. Flowers spreading to pendent, greenish brown, flushed and striped with brownish red. Upper sepal 2.5–3 cm, linear to lanceolate, acute; lateral sepals *c.* 3 cm, lanceolate, obtuse, fused for about half their length. Petals *c.* 1 cm, oblong, obtuse. Lip *c.* 2 cm, oblong to spathulate, with 2 fleshy incurved lobes at the tip. *Costa Rica.* G2. Summer–autumn.

6. P. gelida Lindley. Illustration: Rickett, Wild flowers of the United States **2**: 125 (1967); Luer, Native orchids of Florida, 185 (1972); American Orchid Society Bulletin **45**: 194 (1976).
Stems 5–35 cm, with brown sheaths. Leaf to 25 × 6 cm, oblong to elliptic, rounded, shortly stalked. Flowering stems 1–several, 5–30 cm, erect, bearing 5–25 flowers; flowers pale yellowish green, translucent, softly hairy. Upper sepal *c.* 8 × 3 mm, lanceolate to ovate, obtuse, concave. Lateral sepals *c.* 8 × 3 mm, lanceolate, acute, fused for about half their length, pouched below. Petals *c.* 4 × 2 mm, oblong to spathulate, irregularly toothed at tip. Lip *c.* 2.5 mm, oblong, rounded or truncate. *S Mexico & Florida, south through C America & West Indies to Venezuela, Colombia, Ecuador & Peru.* G2. Flowers intermittently.

***P. saurocephala** Loddiges. Illustration: Botanical Magazine, 3030 (1830); Pabst & Dungs, Orchidaceae Brasilienses **1**: 351 (1975). Stems to 15 cm, grooved. Leaf to 15 cm, elliptic. Flowering stems usually 2, equal to or slightly shorter than leaf; inflorescence dense, with fibrous chaffy bracts at the base. Sepals *c.* 1 cm, green outside, yellow spotted with brown inside. Petals minute. *S Brazil.* G2. Spring–autumn.

***P. schiedei** Reichenbach (*P. ornata* Reichenbach). Illustration: Hamer, Las Orquideas de El Salvador **2**: 237 & t. 27 (1974); Northen, Miniature orchids, C-77 (1980). Stems to 1.5 cm. Leaf to 4 cm × 8 mm, lanceolate, spotted on back. Flowering stems 1–several, longer than leaf. Inflorescence loose, bearing 5–12 flowers. Sepals *c.* 7.5 mm, broadly elliptic to spathulate, obtuse to rounded, greenish yellow with purple warts and dangling white marginal appendages. *Mexico.* G2. Spring–summer.

P. tribuloides (see under No. 1) may also key out here as there is some variation in the fusion of its lateral sepals.

70. RESTREPIA Humboldt, Bonpland & Kunth
J.C.M. Alexander
Dwarf, tufted epiphytes, rarely on rocks, occasionally creeping. Stems slender, 1-leaved at apex, concealed in flattened, papery, overlapping sheaths. Leaves stiff and leathery, folded when young. Flowers solitary on slender stalks, borne at the junction of leaf and stem. Upper sepal and petals lanceolate or narrowly triangular, drawn out into long slender tails with club-shaped tips. Lateral sepals almost totally fused into a broad, often bifid lower sepal. Lip oblong, often with fine lateral lobes near base; tip rounded, truncate or notched, papillose. Column slender, arched, club-shaped, with small column foot. Pollinia 4.

A genus of about 40 species from Mexico through C America to Argentina. Cultivation conditions are similar to those for *Masdevallia* (p. 201).

1a. Lower sepal marked with stripes or lines of spots separated by clear areas **3. antennifera**
b. Lower sepal marked with scattered spots without clear areas 2
2a. Leaves at least twice as long as wide **1. guttulata**
b. Leaves less than twice as long as wide **2. elegans**

1. R. guttulata Lindley (*R. maculata* Lindley). Illustration: Dunsterville & Garay, Venezuelan orchids illustrated **4**: 263 (1966); American Orchid Society Bulletin **47**(7): cover (1978); Bechtel et al., Manual of cultivated orchid species, 269 (1981).
Stems 1–15 cm. Leaves 5–8 × 2.5–3 cm, narrowly elliptic, obtuse, slightly keeled. Flowering stems to 7 cm, appearing behind the leaf through twisting of leaf-base. Upper sepal to 2.5 cm × 3 mm, lanceolate, pinkish purple with pale margins, finely tapered into a slender tail with a club-shaped tip. Lower sepal 2.5 × 1 cm, oblong to elliptic, slightly bifid, brownish green spotted with purple. Petals 13 × 1 mm, narrowly triangular, pale pink with a darker central stripe. Lip 10 × 2 mm, oblong with 2 narrow basal lobes, similar in colour to the lower sepal, papillose. Column *c.* 6 mm, club-shaped, pale with purple spots at base. *Venezuela, Colombia & Ecuador.* G2. Winter.

2. R. elegans Karsten. Illustration: Botanical Magazine, 5966 (1872); Dunsterville & Garay, Venezuelan orchids illustrated **1**: 375 (1959); Bechtel et al., Manual of cultivated orchid species, 269 (1981).
Stems 4–6 cm, erect or ascending. Leaves 4–6 × 3–4 cm, elliptic to circular, obtuse to rounded. Flowering stem to *c.* 4 cm, slender. Upper sepal 2.5 × 2–3 mm, lanceolate, attenuate, white with purple veins and tip. Lower sepal 2.4 cm × 9 mm, oblong, obtuse, yellow to brown, spotted with purple. Petals 14 × 1 mm, narrowly triangular, attenuate, pale with purple veins and tip. Lip 11 × 5 mm, 3-lobed; lateral lobes narrow and sickle-shaped, acute; central lobe fiddle-shaped, narrowing rapidly above the lateral lobes, then broadening above into an oblong notched blade. Column *c.* 6 mm, pale green. *Venezuela.* G2. Winter–spring.

3. R. antennifera Humboldt, Bonpland & Kunth. Illustration: Dunsterville & Garay,

Venezuelan orchids illustrated **4**: 259 (1966); Shuttleworth et al., Orchids, 39 (1970); American Orchid Society Bulletin **43**: 621 (1974); not Botanical Magazine, 6288 (1877).
Stems to *c.* 10 cm. Leaves to 5.5 × 3 cm, elliptic to ovate, obtuse or rounded. Flowering stems *c.* 2.5 cm, slender. Upper sepal 2.3 cm × 3 mm, lanceolate, attenuate, white with purple veins; tail green with pink tip. Lower sepal 2.3 × 1.2 cm, oblong to ovate, acute, bifid, pale yellowish brown with purple lines or dense lines of spots. Petals 1.5 cm × 1 mm, lanceolate to narrowly triangular, white with a purple mid-vein. Lip 10 × 3 mm, oblong to fiddle-shaped with very narrow, acute lateral lobes and a short, cylindric claw; tip truncate and papillose, pale brown, spotted with purple. Column *c.* 5 mm, club-shaped, pale brown. *Venezuela, Colombia & Ecuador.* G2. Winter–spring.

71. CRYPTOPHORANTHUS Barbosa Rodrigues
J.C.M. Alexander
Tufted epiphytes, rarely terrestrial. Stems erect or ascending, sheathed, 1-leaved at apex. Leaves erect, usually hard and leathery, folded when young. Flowers 1 to many, clustered at junction of leaf and stem. Sepals fused at base and tip, leaving 2 lateral openings or windows. Lateral sepals forming a short mentum with column foot. Petals very small. Lip very small, free, shallowly 3-lobed. Column erect, club-shaped. Pollinia 2, waxy.

About 25 species from the New World tropics. Cultivation conditions are similar to those described for *Masdevallia* (p. 201). If grown in pots, plants should be well shaded and frequently repotted to avoid the development of stale, soggy compost.

1. C. atropurpureus (Lindley) Rolfe (*Pleurothallis atropurpurea* Lindley; *Masdevallia fenestrata* Hooker; not *C. fenestratus* Barbosa Rodrigues). Illustration: Botanical Magazine, 4164 (1845).
Stem 3–5 cm. Leaves 3–9 × 1.5–3 cm, oblong to elliptic, acute to obtuse, shortly stalked. Flowers solitary, purplish crimson. Sepals *c.* 1.5 cm × 5 mm, acute, deflexed. Petals *c.* 4 × 2 mm, oblong, obtuse to truncate. Lip *c.* 4.5 × 1 mm, spear-shaped. Column *c.* 2 mm. *Cuba & Jamaica.* G2. Autumn.

*__C. maculatus__ Rolfe. Leaves 4–5.5 × 1.8–3.2 cm, spotted with purple above. Flowers many, *c.* 5 mm long, yellow spotted with purple. *Brazil.* G2. Summer.

*__C. dayanus__ Rolfe. Illustration: Botanical Magazine, 8740 (1917); Ospina & Dressler, Orquideas de las Americas, f. 76 (1974). Stem 10–15 cm. Leaves 6–9 × 3.5–5.5 cm, elliptic to circular. Flowers solitary or few, 3.5–4 cm long, white to pale yellow, spotted with pink, purple or crimson. *Colombia.* G2. Autumn.

72. OCTOMERIA R. Brown
J.C.M. Alexander
Creeping epiphytes, rarely on rocks. Stems slender, often densely tufted, erect, sheathed below, bearing a single leaf; sheaths sometimes frayed into chaffy strands. Leaves stalked or stalkless, folded when young, fleshy or leathery, flat to cylindric. Flowers 1–many, small, bell-shaped, in terminal clusters at the leaf-base. Sepals and petals free or slightly fused, lanceolate to narrowly ovate. Lip very small, simple or 3-lobed, sometimes shortly clawed, hinged to the base of the column. Column very short, with a small column foot. Pollinia 8.

A genus of 50–100 species mostly from Brazil with representatives in the West Indies and other parts of C and S America. Cultivation conditions are generally similar to those for *Masdevallia* (p. 201). The plants should never be allowed to dry out, though less water should be given in winter.

1a. Leaves cylindric or semi-cylindric **3. juncifolia***
b. Leaves flat or slightly concave 2
2a. Stems densely tufted, 6–20 cm high **1. grandiflora***
b. Stems 1–4 cm apart, 3–7 cm high **2. graminifolia**

1. O. grandiflora Lindley. Illustration: Dunsterville & Garay, Venezuelan orchids illustrated **1**: 258 (1959) & **6**: 286 (1976); Pabst & Dungs, Orchidaceae Brasilienses **1**: 228 & 365 (1975); Bechtel et al., Manual of cultivated orchid species, 235 (1981).
Densely tufted epiphyte. Stems 6–20 cm, flattened above. Leaves 10–20 cm × 7–17 mm, linear to narrowly lanceolate, acute, stiff and leathery. Flowers several, yellow and greenish yellow or white. Sepals 1–1.2 cm × 3.5–5 mm, elliptic to ovate, obtuse. Petals 9–11 × 2.5–3.5 mm, lanceolate to ovate, acute or obtuse. Lip 6–8 × 4–6 mm, almost oblong, truncate or notched at tip, with small lateral lobes near base, white to pale yellow with 2 purple ridges and spots near base. Column 3–4 mm, curved, often with a pair of reddish spots near tip. *Venezuela, Trinidad, Surinam, Brazil, Bolivia & Paraguay.* G2. Autumn.

*__O. crassifolia__ Lindley. Illustration: Martius, Flora Brasiliensis **3**(4): t. 128 (1896); Pabst & Dungs, Orchidaceae Brasilienses **2**: 363 (1977). Stems 15–20 cm. Leaves 10–18 × 1.2–2 cm. Flowers densely clustered. Sepals and petals *c.* 6 × 2 mm. Lip 3.5–4 × 2.5–3 mm, oblong to elliptic, shallowly 3-lobed, acute to rounded at apex. Column 1.5–2 mm. *Brazil, Uruguay & Paraguay.* G2. Winter–summer.

2. O. graminifolia R. Brown. Illustration: Botanical Magazine, 2764 (1827).
Creeping epiphyte. Stems 3–7 cm, 1–4 cm apart. Leaves 6–11 cm × 5–10 mm, linear to narrowly lanceolate, acute. Flowers 1 or 2 on short individual stalks, pale yellow or greenish yellow, fragrant. Sepals 6–8 × 2.5–3 mm, ovate to lanceolate, acute. Petals 6–7.5 × 2–2.5 mm, lanceolate to elliptic, acute. Lip 4–6 × 2–3 mm, shallowly 3-lobed, pale yellow with purple ridges; central lobe oblong to circular. Column 2–2.5 mm, yellow. *West Indies & Brazil.* G2. Spring.

3. O. juncifolia Barbosa Rodrigues. Illustration: Martius, Flora Brasiliensis **3**(4): t. 132 (1896); Pabst & Dungs, Orchidaceae Brasilienses **1**: 228 & 372 (1975).
Tufted epiphyte. Stems 20–35 cm, very slender. Leaves 30–50 cm × 3–5 mm, very slender, cylindric with a shallow groove, arched. Flowers few (rarely numerous), yellow. Sepals and petals 8–10 × 2.5–3 mm, lanceolate to oblong, obtuse. Lip 4.5–5 × 3–3.5 mm, ovate to elliptic, centrally spotted with red. Column 2–2.5 mm. *S Brazil.* G2. Summer–autumn.

*__O. gracilis__ Lindley (not *O. gracilis* Barbosa Rodrigues). Illustration: Pabst & Dungs, Orchidaceae Brasilienses **1**: 372 (1975). Stems 3–9 cm. Leaves 6–12 × 1.5–2.5 mm, semi-cylindric, grooved. Flowers 1–8, pale yellow. Sepals and petals 3.5–4.5 × 1–1.5 mm, linear to oblong, obtuse. Lip 2.5–3 × 1.5 mm, circular to oblong, shallowly 3-lobed. Column *c.* 1.5 mm. *S Brazil.* G2. Spring.

73. DENDROBIUM Swartz
P.J.B. Woods & J. Cullen
Sympodial epiphytes. Stems thin or more frequently fleshy and pseudobulbous, bearing several leaves or leaf-scars. Leaves variable, with or without a sheath enclosing the stem at the base, folded when young. Flowers 1–many in racemes borne on leafy or leafless stems. Upper sepal free, the laterals attached to the elongate

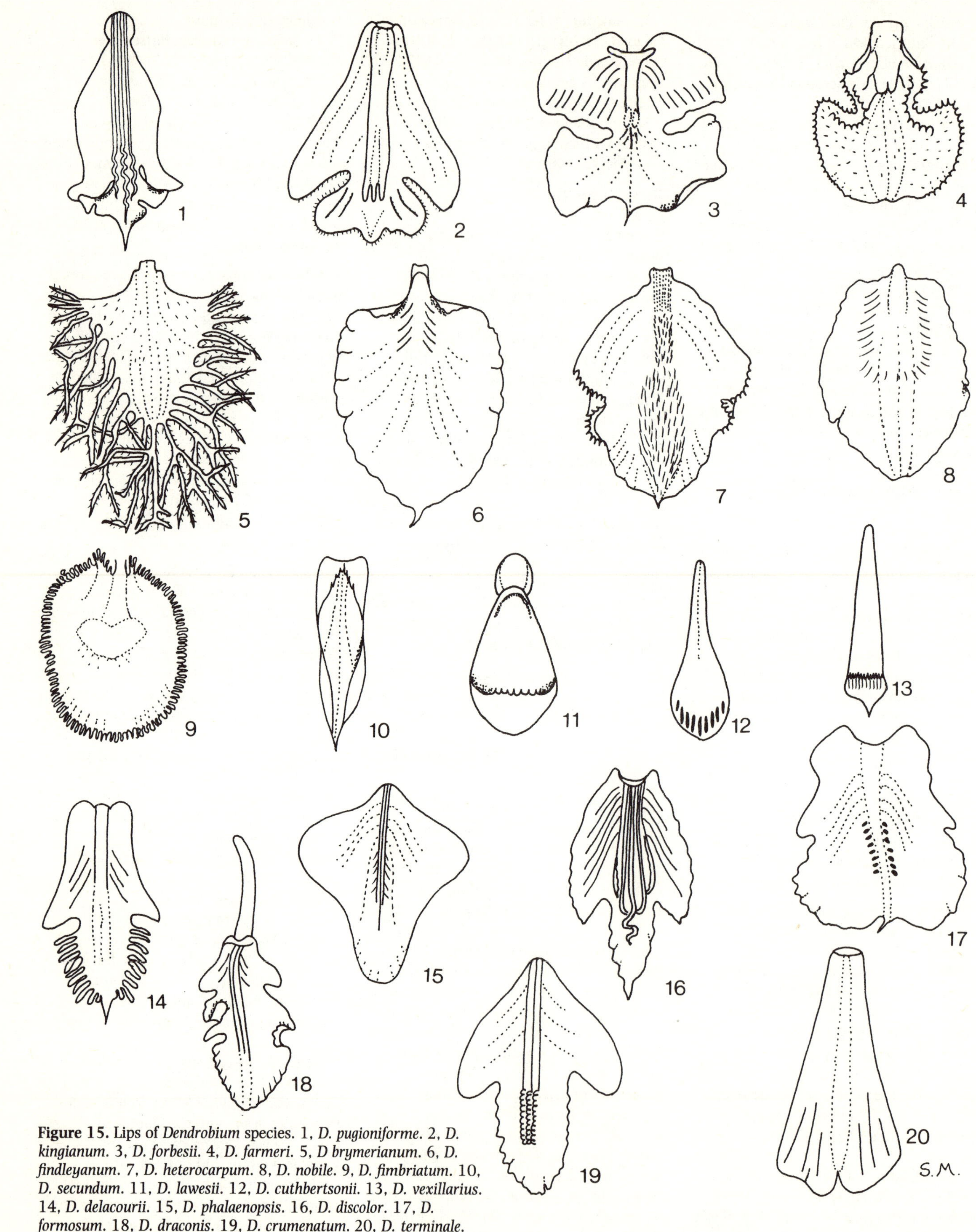

Figure 15. Lips of *Dendrobium* species. 1, *D. pugioniforme*. 2, *D. kingianum*. 3, *D. forbesii*. 4, *D. farmeri*. 5, *D brymerianum*. 6, *D. findleyanum*. 7, *D. heterocarpum*. 8, *D. nobile*. 9, *D. fimbriatum*. 10, *D. secundum*. 11, *D. lawesii*. 12, *D. cuthbertsonii*. 13, *D. vexillarius*. 14, *D. delacourii*. 15, *D. phalaenopsis*. 16, *D. discolor*. 17, *D. formosum*. 18, *D. draconis*. 19, *D. crumenatum*. 20, *D. terminale*. (Compiled from various sources; not to scale.)

column foot, forming a conspicuous, variously shaped mentum which is sometimes spur-like. Petals larger or smaller than the sepals. Lip entire or more frequently 3-lobed, attached to the apex and sometimes to the sides of the column foot, very variable in shape, usually with conspicuous longitudinal ridges. Column short, though with an extended foot, frequently bearing 2 lateral horns. Pollinia 4 in 2 pairs.

A genus of about 900 species, widely distributed in Asia, in particular the Himalaya, Burma, Thailand, Malaysia, Indonesia and New Guinea, and extending to Australia. Its supraspecific classification is rather uncertain, both taxonomically and nomenclaturally. The account below follows the sections (and their order) proposed by Schlechter (Feddes Repertorium Beihefte 1: 440–52, 1912); the sections themselves, however, are referred to as 'groups' because of uncertainties with regard to their correct nomenclature. The sectional names used by Schlechter and Kränzlin are cited as appropriate. In Schlechter (edited by Brieger, Maatsch & Senghas), Die Orchideen, edn 3, **1**(11 & 12): 636–752 (1981), Brieger divides the genus, as recognised here, into numerous segregate genera, formed largely by the raising of the Schlechter sections to generic level. This produces many new names, which are not cited here as synonyms. There is no doubt that the older, inclusive concept of the genus is appropriate here; whether the new genera will be widely accepted remains in doubt. The distinction between stem and pseudobulb is very blurred in this genus. We have used the latter term only those cases in which the stem is conspicuously swollen.

About 80 species are reputedly in cultivation, together with numerous hybrids. Because of the large size of the genus, the methods of cultivation differ somewhat, depending on the origins of the species. The Indian and the Burmese species require heat and moisture when growing, exposure to air and light during autumn, and a cooler, drier, rest period during the winter. The species from Australia and New Guinea require similar treatment, but need more frequent watering and a slightly higher winter temperature. All the species benefit from appropriate shading, especially when the new growth appears. Epiphytic compost should be used, and the containers should be as small as possible. Many species do well on bark, slabs or tree fern. Propagation is mainly by division of old plants, by small pseudobulbs, and by plantlets produced on the stems.

Literature: Kränzlin, F., *Das Pflanzenreich* **45**: 25–313 (1910).

1a. Leaves without sheaths 2
b. Leaves with sheaths 17
2a. Stems wiry, creeping or hanging, branched, each branch terminating in a single, thick, hard-pointed leaf **1. pugioniforme**
b. Stems and leaves various, not as above 3
3a. Lip hairy on the upper surface, often fringed 4
b. Lip hairless on the upper surface, never fringed, sometimes wavy 9
4a. Stems creeping, bearing scattered pseudobulbs, each with 1 leaf **11. jenkinsii**
b. Pseudobulbs clustered, each with several leaves 5
5a. Leaves and pseudobulbs white-hairy **12. senile**
b. Leaves and pseudobulbs hairless 6
6a. Flowers entirely yellow 7
b. Lip yellow, sepals and petals white sometimes flushed with pink 8
7a. Pseudobulbs 4-angled; lip not fringed **14. densiflorum**
b. Pseudobulbs many-angled; lip fringed **13. chrysotoxum**
8a. Pseudobulbs 4-angled; sepals and petals white flushed with pink **16. farmeri**
b. Pseudobulbs not angled or with several slight ridges; sepals and petals white **15. thyrsiflorum**
9a. Basal part of the lip erect, ridges usually conspicuous, close to or contacting the column 10
b. Basal part of the lip diverging from the column, ridges usually shallow and not contacting the column 14
10a. Ovary and flower-stalk hairless 11
b. Ovary and flower-stalk roughly hairy 13
11a. Central lobe of lip obtuse and deeply notched; racemes hanging **10. woodsii***
b. Central lobe of lip acute, not notched; racemes erect 12
12a. Lateral lobes of lip large, spreading to form a cup-like structure around the column; flowers 7–9 cm in diameter **8. spectabile**
b. Lateral lobes of lip small, not as above; flowers *c.* 4 cm in diameter **9. atroviolaceum***
13a. Petals cream with purple spots and lines; lip with purple lines on all lobes; leaves to 30 × 10 cm **6. macrophyllum**
b. Petals cream, without spots or lines; lip cream with purple lines on the lateral lobes only; leaves 10–18 × 5–7 cm **7. forbesii**
14a. Stems drooping, very slender, wire-like in the lower half then dilated and 4-angled above; sepals very acuminate **5. tetragonum**
b. Stems erect, not as above; sepals not acuminate 15
15a. Stems tapered evenly from base to apex; lateral lobes of the lip acute (figure 15(2), p. 208) **4. kingianum**
b. Stems uniformly thick or sometimes narrowed at the base; lateral lobes of the lip obtuse 16
16a. Central lobe of lip acute, upcurved; racemes 8–16 cm with 4–20 flowers; leaves thin **3. falcorostrum**
b. Central lobe of lip rounded and apiculate; racemes 10–60 cm with many flowers; leaves thick and leathery **2. speciosum***
17a. Stems and leaves strongly flattened, leaf-bases folded and overlapping **71. terminale**
b. Stems and leaves not as above 18
18a. Stems thickened only at the base, the rest thin and normally stem-like **70. crumenatum**
b. Stems not at all thickened, or thickened above the base 19
19a. Lip simple, unlobed, its margins attached to the column foot for some distance 20
b. Lip 3-lobed or simple, its margins not attached to the column foot 27
20a. Plants crystalline-papillose on leaves, sheaths, ovaries, flower-stalks and sometimes also on the flowers **49. cuthbertsonii**
b. Plants not at all crystalline-papillose, sometimes warty 21
21a. Lip hollowed, its apical part fringed, upcurved 22
b. Lip flat, its apical part neither fringed nor upcurved 23
22a. Leaf-sheaths and stem covered with fine warts; flowers orange or orange with yellow tips **48. subclausum***
b. Warts absent, except occasionally at the tops of the leaf-sheaths; flowers scarlet or purplish **47. lawesii**
23a. Leaves clustered at stem apex; ovary angled or winged; racemes with 2–4 flowers 24
b. Leaves evenly distributed along the length of the stem; ovary not angled

or winged; racemes with many flowers 25

24a. Leaves linear, terete; ovary with 5 obscure wings (3 close together along the top of the ovary) **51. hellwigianum***

b. Leaves oblong to oblong-lanceolate, flat; ovary with 3 wings **50. vexillarius***

25a. Sepals and petals white (rarely bluish), lip apex pale purple; sepals 2–3 cm **44. amethystoglossum***

b. Flowers differently coloured; sepals less than 2 cm 26

26a. Flowers pink or reddish, rarely white; raceme conspicuously 1-sided **45. secundum**

b. Flowers orange; inflorescence not 1-sided **46. bullenianum**

27a. Mentum double, consisting of 2 swellings at right angles to each other 28

b. Mentum single, consisting of 1 swelling 29

28a. Flowers 3–7 cm in diameter; petals broadly rounded, occasionally with a small point; calluses on lip numerous and dense **53. bigibbum**

b. Flowers 6–10 cm in diameter; petals acuminate; calluses on lip small and obscure **54. phalaenopsis**

29a. Leaf-sheaths hairy 30

b. Leaf-sheaths not hairy 36

30a. Stems no more than 8 cm; mentum broadly conical, blunt **69. bellatulum**

b. Stems taller; mentum spur-like 31

31a. Leaves hairy; petals as broad as or a little broader than sepals 32

b. Leaves hairless; petals much broader than sepals 33

32a. Lateral lobes of lip borne at its base; central lobe not fringed (figure 15(18), p. 208) **67. draconis**

b. Lateral lobes of lip borne approximately half way from base to apex; central lobe fringed **68. williamsonii**

33a. Ovary 3-keeled; base of lip green or purple-streaked 34

b. Ovary not keeled; base of lip red, yellow or orange 35

34a. Base of lip greenish **65. dearei**

b. Base of lip streaked with purple **66. sanderae**

35a. Flower *c.* 10 cm in diameter; lip scarcely 3-lobed, notched **63. formosum**

b. Flower up to 8 cm in diameter; lip clearly 3-lobed, irregularly fringed at the apex **64. infundibulum**

36a. Flowers not resupinate; lip orange with purple or brown markings **17. seidenfadenii***

b. Flowers resupinate; lip various but not as above 37

37a. Lip hairy on the upper surface; mentum relatively short for the size of the flower 38

b. Lip hairless on the upper surface; mentum relatively long for the size of the flower 64

38a. Raceme with 5 or more flowers and an axis longer than the individual flower-stalks 39

b. Flowers borne singly or in racemes of 2, 3 or rarely 4, axis absent or shorter than the individual flower-stalks 42

39a. Petals *c.* 3.5 cm broad, almost 2 times broader than the sepals; leaf-sheaths red-striped **35. pulchellum**

b. Petals at most 2–2.2 cm broad, less than 2 times broader than the sepals; leaf-sheaths not purple-striped 40

40a. Lip with the apical part and the sides curved upwards, forming a pouch, margin entire **18. moschatum**

b. Lip not as above, margin fringed 41

41a. Stems hanging; flowering stems usually leafy; fringing of the lip up to 4 mm deep **19. hookerianum**

b. Stems erect; flowering stems usually leafless; fringing of the lip at most to 2 mm deep **20. fimbriatum**

42a. Lip fringed to the naked eye 43

b. Lip not fringed to the naked eye, entire or finely toothed under a lens 48

43a. Fringe of lip greater in extent than the solid part; petals not broader than sepals (figure 15(5), p. 208) **21. brymerianum**

b. Fringe of lip narrow; petals broader than sepals 44

44a. Sepals and petals yellow 45

b. Sepals and petals white, white tipped with pinkish lilac or entirely reddish purple or red 46

45a. Lip densely hairy beneath **23. ochreatum**

b. Lip more or less hairless beneath **22. chrysanthum**

46a. Sepals and petals 5–6 cm; lip mauve with 2 purple blotches but without any yellow **26. anosmum**

b. Sepals and petals up to 3 cm; lip variously coloured, always with some yellow 47

47a. Stem hanging or arching, 60–100 cm; upper surface of lip minutely hairy **37. devonianum**

b. Stem erect, to 15 cm; upper surface of lip densely velvety **27. loddigesii**

48a. Stem with the internodes thickening upwards conspicuously, thus appearing beaded 49

b. Stem with the internodes more or less equally thick throughout, not appearing beaded 52

49a. Lip without any yellow; racemes usually borne on leafy stems **28. moniliforme**

b. Lip with some yellow; racemes usually borne on leafless stems 50

50a. Leaves small, grass-like, 5–8 cm × 3–4 mm; centre of lip purple, outlined with pale yellow **38. falconeri**

b. Leaves not grass-like, much larger; centre of lip yellow 51

51a. Sepals 2–2.5 cm × 6–8 mm **29. findleyanum**

b. Sepals 3–4 cm × 9–12 mm **39. pendulum***

52a. Flowers mostly yellow, without any mauve or purple except occasionally on the lip, usually without any white or pink 53

b. Flowers basically white or purple or pink, without any yellow except occasionally on the lip 54

53a. Racemes borne on leafy stems **24. luteolum**

b. Racemes borne on leafless stems **25. heterocarpum**

54a. Sepals and petals either entirely white, white flushed with pink or purple, or purple, without tips of contrasting colour 55

b. Sepals and petals white with contrasting mauve, pink or purple tips 61

55a. Flowers without any yellow on the lip 56

b. Flowers with the lip mostly yellow (sometimes very pale) 60

56a. Flowers 5 cm or more in diameter 57

b. Flowers less than 5 cm in diameter 58

57a. Lip trumpet-shaped, its basal part more or less tubular, directed downwards, the apical part bell-shaped, directed forwards **30. lituiflorum**

b. Lip not trumpet-shaped, entirely directed obliquely forwards **36. nobile**

58a. Free part of lip very small, much smaller than mentum **43. aduncum**

b. Free part of lip much larger than the mentum 59

59a. Sepals white; lip white with 2 red blotches **31. transparens**

b. Sepals purple; lip mauve with 2

deeper purple blotches **32. parishii**
60a. Petals broader than sepals; lip with a line of small, papilla-like hairs down the middle **33. aphyllum**
b. Sepals and petals equally broad; lip with most of the upper surface velvety **34. primulinum**
61a. Flowers more than 5 cm in diameter; lip with no yellow blotch **36. nobile**
b. Flowers less than 5 cm in diameter; lip with a yellow blotch 62
62a. Sepals and petals 1.2–1.3 cm; lip not hairy beneath **40. crepidatum**
b. Sepals and petals 1.7–3.5 cm; lip hairy beneath 63
63a. Lip densely velvety above and beneath; petals entire **41. gratiosissimum**
b. Lip densely velvety above, sparsely hairy beneath; petals finely toothed **42. amoenum**
64a. Margin of lip fringed (figure 15(14), p. 208) **52. delacourii**
b. Margin of lip not fringed 65
65a. Petals 1.5–2 times longer than lateral sepals 66
b. Petals as long as or a little longer than lateral sepals 69
66a. Raceme with 5 or more flowers; petals conspicuously twisted 67
b. Raceme with 2–3 flowers; petals not conspicuously twisted 68
67a. Flowers spotted with dark red or brown; lip mauve at base **55. helix**
b. Flowers with dark red venation; lip white at base **56. tangerinum**
68a. Central lobe of lip broader than long **57. minax**
b. Central lobe of lip longer than broad **58. stratiotes**
69a. Sepals greenish white, petals red **59. taurinum**
b. Sepals and petals pale yellow or dull yellow-brown 70
70a. Central lobe of the lip conspicuously undulate-crisped; sepals and petals cream or yellowish **60. crispilinguum**
b. Central lobe of the lip not undulate-crisped; sepals and petals dull yellow-brown or brown 71
71a. Sepals and petals very wavy; racemes with many flowers **61. discolor**
b. Sepals and petals not wavy; racemes with 4–12 flowers **62. mirbelianum**

GROUP A (Section *Rhizobium* Schlechter). Leaves without sheaths; stems hanging, wiry, branched, each branch terminating in a single, hard-pointed leaf.

1. D. pugioniforme Cunningham. Figure 15(1), p. 208. Illustration: Dockrill, Australian indigenous orchids, 369 (1969); Bechtel et al., Manual of cultivated orchid species, 201 (1981).
Stems thin, wiry, forming large hanging masses up to 2 m, each branch terminating in a single leaf. Leaves 1–7 cm × 5–20 mm, flat, ovate-acuminate, very sharply pointed. Raceme with 1–3 flowers. Sepals 8–12 mm, petals slightly shorter, all spreading, pale green. Lip 3-lobed with oblong lateral lobes and a triangular, acute, deflexed central lobe which is notched or wavy on either side of the apex, very pale green with purple or red markings on the lateral lobes and the base of the central. *Australia (New South Wales, Queensland).* G2.

GROUP B (Subgenus *Dendrocoryne* Lindley: Section *Dendrocoryne* (Lindley) Schlechter). Leaves without sheaths; lip shorter than sepals and petals, its basal part diverging from the column, hairless on the upper surface.

2. D. speciosum J.E. Smith. Illustration: Cogniaux & Goossens, Dictionnaire Iconographique des Orchidées, Dendrobium, t. 24 (1900): Dockrill, Australian indigenous orchids, 385 (1969); Bechtel et al., Manual of cultivated orchid species, 202 (1981).
Stems 8–100 cm. Leaves 4–25 × 2–8 cm, usually ovate or oblong, thick and leathery. Racemes 10–60 cm with many flowers. Flowers variable, sepals and petals widely spreading or not. Sepals 2–4 cm, petals slightly shorter and much narrower, all white, cream or yellow. Lip shorter than the petals, white, cream or yellow with red or purple spots, lobes obtuse, the central lobe *c.* 2 times broader than long. *Australia (New South Wales, Queensland, Victoria).* G1. Spring.

A very variable species, divided by Dockrill into 5 varieties; it is uncertain which of these are in cultivation.

***D. ruppianum** Hawkes (*D. speciosum* var. *fusiforme* Bailey). Illustration: Botanical Magazine n.s., 749 (1978). Similar, but with stems swollen at the base and then tapering and then broadening, central lobe of the lip 3 times broader than long. *Australia (Queensland), New Guinea.* G1.

3. D. falcorostrum Fitzgerald. Illustration: Dockrill, Australian indigenous orchids, 407 & t. 11 (1969).
Stems 12–50 cm, erect, aggregated. Leaves 2–4, 6–14 × 1.5–3 cm, thin, narrowly obovate or rarely narrowly ovate. Racemes 8–16 cm with 4–20 flowers. Flowers 2.5–4 cm in diameter. Sepals 3–4 cm, oblong, petals slightly shorter and narrower, all white. Lip shorter than the petals, white with yellow and purple markings, the central lobe acute, upcurved, the lateral lobes obtuse. *Australia (New South Wales, Queensland).* G2. Autumn.

4. D. kingianum Bidwill. Figure 15(2) p. 208. Illustration: Edwards's Botanical Register **31**: t. 61 (1845); Botanical Magazine, 4527 (1850); Dockrill, Australian indigenous orchids, 415 (1969); Bechtel et al., Manual of cultivated orchid species, 198 (1981).
Stem 8–30 cm, broadest at the base, tapering upwards. Leaves 3–10 × 1–2 cm, narrowly ovate to narrowly obovate. Racemes with 2–10 flowers which are 1–2.5 cm in diameter. Sepals 9–16 mm, triangular, the laterals erect and sickle-shaped; petals much narrower, all in various shades of pink or lilac, rarely white. Lip shorter than the petals to almost as long, pink with deep red stripes or blotches, the lateral lobes acute, the central transversely oblong, blunt or minutely apiculate. *Australia (New South Wales, Queensland).* G1. Spring.

5. D. tetragonum Cunningham. Illustration: Botanical Magazine, 5956 (1872); Dockrill, Australian indigenous orchids, 421 (1969); Bechtel et al., Manual of cultivated orchid species, 202 (1981).
Stems hanging, aggregated, thin and wiry in the lower part, then dilated and 4-angled above. Leaves 2–5 near the stem apex, 3–8 × 1.5–2.5 cm, ovate, acuminate. Racemes short with few flowers; flowers 4–9 cm in diameter, very variable, very fragrant. Sepals 2–5 cm, very narrowly lanceolate, very acuminate; petals erect, shorter and narrower than the sepals, all greenish yellow with irregular brown or purple blotches. Lip much shorter than the petals, 3-lobed, very variable in shape, yellow with dark red markings. *Australia (New South Wales, Queensland).* G1. Winter.

GROUP C (Section *Latouria* (Blume) Schlechter).
Leaves without sheaths, mostly borne at the stem apex; lip hairless, as long as the petals, the basal part erect, close to the column, basal callus conspicuous.
Literature: Cribb, P.J., Key to the species of Dendrobium section Latouria with setose ovaries, *Botanical Magazine* **183** (1): 28 (1980); Cribb, P.J., A preliminary key to Dendrobium section Latouria, *The Orchadian* **6**: 277–83 (1981).

6. D. macrophyllum Richard. Illustration: Botanical Magazine, 5649 (1867); Illustration Horticole **35**: t. 57 (1888); American Orchid Society Bulletin **38**: 589 (1969).
Stems to 1 m × 2 cm, club-shaped. Leaves 2–3, to 30 × 10 cm, elliptic to obovate. Racemes to 30 cm with 8–10 (rarely more) flowers, erect. Sepals *c.* 3–3.5 cm, triangular, pale yellowish green, crimson-spotted, bristly outside. Petals shorter than the sepals, spathulate. Lip as long as petals, lateral lobes truncate, white, purple-veined inside, purple-spotted outside, the central lobe transversely elliptic, pale green with dark crimson spots outside. Basal callus 1.5 cm, white, with 3 ridges. *Java to New Guinea & Fiji*. G2. Summer–autumn.

7. D. forbesii Ridley. Figure 15(3), p. 208. Illustration: Botanical Magazine, 8141 (1907); Millar, Orchids of Papua New Guinea, 34 (1978).
Stems 15–45 × 2.5 cm, club-shaped. Leaves 2–3, 10–18 × 5–7 cm, elliptic. Racemes to 17 cm, with 5 or more flowers, erect. Sepals to 3.5 cm, triangular pale greenish yellow or cream, unspotted, bristly outside on the veins. Petals white, longer than sepals. Lip pale greenish cream, unspotted, bases of the lateral lobes purple-veined, the central lobe transversely elliptic. Callus obscurely 3-ridged. *New Guinea*. G2. Summer–spring.

8. D. spectabile (Blume) Miquel. Illustration: Botanical Magazine, 7747 (1900); Cogniaux & Goossens, Dictionnaire Iconographique des Orchidées, Dendrobium, t. 22 (1900); Orchid Review **74**: f. 59 (1966); Millar, Orchids of Papua New Guinea, (1978).
Stems to 60 cm, club-shaped. Leaves 2–5 near the apex of the stem, to 15 × 5 cm, oblong. Raceme to 30 cm, erect, with 5–25 slightly fragrant flowers. Sepals and petals *c.* 3.5 cm, narrowly triangular, pale yellow and marked with purple, margins wavy. Lip *c.* 5 cm, white, finely veined with purple, the lateral lobes large, forming a cup-shaped structure around the column, central lobe elongate, margins wavy. *New Guinea*. G2. Winter.

9. D. atroviolaceum Rolfe. Illustration: Botanical Magazine, 7371 (1894); American Orchid Society Bulletin **38**: 587, 589 (1969); Millar, Orchids of Papua New Guinea, 40 (1978); Bechtel et al., Manual of cultivated orchid species, 193 (1981).
Stems 15–30 cm, thickened upwards. Leaves 2–3, 7–15 × 2.5–5 cm, ovate-oblong. Raceme 10–20 cm, erect, bearing 3–8 fragrant flowers. Sepals *c.* 2.8 cm, triangular, petals spathulate, slightly longer than sepals, all cream or pale greenish white, purple-spotted outside. Lip pale green, the lateral lobes crimson-veined, margins violet, the central lobe ovate, purple-spotted outside. *New Guinea*. G2. Summer–spring.

*__D. rhodostictum__ Mueller & Kränzlin. Illustration: Botanical Magazine, 7900 (1903). Similar but flowers unspotted, the lip with green veins and a few purple spots on the margins of the lateral lobes. *New Guinea*. G2. Spring–summer.

*__D. johnsoniae__ Mueller. Illustration: Journal of the Royal Horticultural Society **87**: f. 23 (1962); Botanical Magazine n.s., 560 (1969); Millar, Orchids of Papua New Guinea, 35 (1978). Also similar but with more slender stems; lip with longer, ovate central lobe. *New Guinea. Solomon Islands*. G2. Autumn–winter.

*__D. engae__ Reeve. Illustration: The Orchadian **6**: 123–4 (1979). Also broadly similar, but flowers sweetly scented, sepals without spots. *New Guinea*. G2. Summer–spring.

10. D. woodsii Cribb. Illustration: Botanical Magazine n.s., 727 (1977) – as D. fantasticum; The Orchadian **6**: 283 (1981).
Stems to 20 (rarely to 40) cm, slender at the base, thickened upwards. Leaves 2 at the stem apex, *c.* 18 × 2–3.5 cm, elliptic to broadly lanceolate, gradually acuminate. Racemes to 10 cm, hanging, with up to 10 fragrant flowers. Sepals and petals 1–1.2 cm, whitish, bluntly triangular, the petals truncate. Mentum saccate. Lip white flushed with mauve, the lateral lobes flushed with green, the central lobe obtuse and deeply notched. *New Guinea*. G2. Autumn–winter.

*__D. aberrans__ Schlechter. Similar, but smaller in all its parts, the flowers white without any other colour. *New Guinea*. G2. Autumn–spring.

GROUP D (Section *Callista* (Loureiro) Schlechter).
Leaves without sheaths; lip 3-lobed, densely papillose-hairy on the upper surface.
Literature: Schelpe, E.A., Dendrobium section Callista, *Orchid Digest* **45**: 205–10 (1981).

11. D. jenkinsii Lindley. Illustration: Edwards's Botanical Register **20**: t. 1695 (1835); Botanical Magazine, 3643 (1838); Cogniaux & Goossens, Dictionnaire Iconographique des Orchidées, Dendrobium, t. 33 (1904).
Rhizome creeping. Pseudobulbs aggregated, oblong-ovoid, angled, each bearing 1 leaf. Leaves 5–8 × *c.* 2.5 cm, oblong. Racemes from the upper axils of the pseudobulb, arching, with 5–15 flowers which are *c.* 3 cm in diameter. Sepals and petals *c.* 1.5 cm, petals broader than the sepals, at first pale, later deep yellow. Lip longer than the petals, broadly circular, concave, yellow which is deeper at the base, covered in short, white hairs. *Himalaya, Burma, SW China, Laos, Vietnam & Kampuchea*. G2. Spring.

12. D. senile Parish & Reichenbach. Illustration: Botanical Magazine, 5520 (1865); Seidenfaden & Smitinand, Orchids of Thailand **2**(2): t. 7 (1960).
Stem to 10 cm, club-shaped, bearing several leaves, covered, like the leaves, with long, white hairs. Leaves 5–8 cm, lanceolate. Racemes with 1–2 flowers on a short peduncle. Sepals and petals *c.* 2.5 cm, the petals broader than the sepals, all yellow. Lip ovate, as long as the petals, yellow, its surface densely downy. *Burma, Thailand & Laos*. G2. Spring–summer.

13. D. chrysotoxum Lindley. Illustration: Botanical Magazine, 5053 (1858); Cogniaux & Goossens, Dictionnaire Iconographique des Orchidées, Dendrobium, t. 11, 11A (1898); Bechtel et al., Manual of cultivated orchid species, 195 (1981).
Pseudobulbs many-angled, bearing 2–5 leaves. Leaves 3–10 cm, oblong or linear-oblong. Racemes with 8–15 flowers, arching or hanging. Sepals and petals 2–2.2 cm, the petals a little broader than the sepals, all yellow. Lip as long as the petals, circular, the margin conspicuously fringed, orange-yellow, sometimes with a brown blotch at the base. *Himalaya, Burma, SW China (Yunnan), Laos & Thailand*. G2. Winter–spring.

Plants with the lip brown-blotched are sometimes called var. **suavissimum** (Reichenbach) Veitch (*D. suavissimum* Reichenbach).

14. D. densiflorum Lindley. Illustration: Botanical Magazine, 3418 (1835); Cogniaux & Goossens, Dictionnaire Iconographique des Orchidées, Dendrobium, t. 14 (1898); Bechtel et al., Manual of cultivated orchid species, 195 (1981).
Pseudobulbs 25–40 cm, 4-angled, bearing 3–5 leaves. Leaves 7.5–15 × 2–2.5 cm, lanceolate or elliptic-lanceolate. Racemes with many flowers, the scape arching over,

the raceme completely pendent. Sepals and petals 2.5–2.8 cm, the petals somewhat broader than the sepals, all pale yellow. Lip as long as the petals, deep orange-yellow, margins somewhat finely toothed. *Himalaya, Burma, Vietnam*. G2. Spring.

Variants with white sepals and petals sometimes occur and have been cultivated; they are known as var. **schroederi** Anon. (lip orange-yellow) and var. **galliceanum** Linden (lip with white margin).

15. **D. thyrsiflorum** Reichenbach. Illustration: The Garden **30**: 544 (1886); Cogniaux & Goossens, Dictionnaire Iconographique des Orchidées, Dendrobium, t. 18 (1899); Seidenfaden & Smitinand, Orchids of Thailand 2(2): t. 6 (1960).
Very similar to *D. densiflorum* but pseudobulbs not angled, or with many slight angles, sepals and petals white, the lip deep yellow. *Burma, Thailand*. G2. Spring–summer.

Often combined with *D. densiflorum*. The distinctions between the two are slight, and made more obscure by the existence of the varieties of *D. densiflorum* mentioned above. The precise distributions of both species are uncertain.

16. **D. farmeri** Paxton. Figure 15(4), p. 208. Illustration: Annals of the Royal Botanic Garden Calcutta **8**: t. 80 (1898); Cogniaux & Goossens, Dictionnaire Iconographique des Orchidées, Dendrobium, t. 30 (1903); Bechtel et al., Manual of cultivated orchid species, 196 (1981).
Pseudobulbs 20–30 cm, 4-angled, bearing 3–5 leaves. Leaves 8–18 × 3.5–5 cm, ovate-lanceolate. Raceme hanging, with arching scape, and with many flowers. Sepals and petals 2–2.5 cm, petals slightly broader than sepals, all white, often with pink tips. Lip as long as the petals, very densely papillose-hairy above, yellow with whitish margins and a pink apex. *Himalaya, Burma, Thailand & Malaysia*. G2. Spring.

GROUP E (Section *Eugenanthe* Schlechter). Leaves with sheaths; racemes usually from leafless stems; lip not or very obscurely 3-lobed, its sides free from the column foot, often curved upwards around the column, hairy (usually densely so) above.

17. **D. seidenfadenii** Senghas & Bockemühl (*D. arachnites* Reichenbach not Thouars). Illustration: Gardeners' Chronicle **20**: 7 (1896); Botanisk Tidsskrift **65**: 335 (1970); Die Orchidee **29**: cii (1978).
Stem to 10 cm, swollen; flowering stems leafless. Leaves 4–6.5 cm, narrowly lanceolate. Racemes with 2–3 flowers which are not resupinate. Flowers to 6 cm in diameter. Sepals and petals similar, 2–3 cm, orange-red. Lip as long as the petals, lanceolate, acute, orange marked with purple lines, and with ridges near the base only. *Burma*. G2. Spring–summer.

***D. unicum** Seidenfaden. Illustration: Botanical Magazine n.s., 616 (1972); Bechtel et al., Manual of cultivated orchid species, 193 (1981). Very similar, but the stems are longer (to 25 cm) and the lip is oblong-elliptic, brownish orange, veined with brown, and has 3 ridges extending for most of its length. *Laos, Thailand*. G2.

Much material in cultivation as *D. arachnites* could well be *D. unicum*.

18. **D. moschatum** Swartz (*D. calceolaria* Carey; *D. cupreum* Herbert). Illustration: Botanical Magazine, 3837 (1840); Annals of the Royal Botanic Garden Calcutta **8**: t. 84 (1898); Seidenfaden & Smitinand, Orchids of Thailand **2**(2): t. 8 (1960); Bechtel et al., Manual of cultivated orchid species, 200 (1981).
Stems to 2 m, fleshy. Leaves 10–15 cm, lanceolate, acuminate. Racemes hanging, on leafless stems, with 10–15 flowers, axis elongate. Flowers 6–8 cm in diameter. Sepals and petals to 4 cm, pale orange, petals *c*. 2 cm broad, a little broader than the sepals. Lip shorter than the petals, concave, its margins erect, pale orange with 2 reddish brown blotches at the base. *Himalaya, Laos, Burma & Thailand*. G2. Spring–summer.

19. **D. hookerianum** Lindley. Illustration: Botanical Magazine, 6013 (1873); Annals of the Royal Botanic Garden Calcutta **8**: t. 83 (1898).
Stems up to 2.5 m, fleshy, pendent. Leaves 5–15 × 2.5–3 cm, lanceolate or oblong-lanceolate, acute. Racemes hanging, on leafy stems, with 5 or more flowers and an elongate axis. Flowers 7–10 cm in diameter. Sepals and petals to 4 cm, yellow, the petals finely toothed, broader than the sepals. Lip circular, yellow with 2 brown spots near the base, the margins fringed with the fringe *c*. 4 mm deep. *N India*. G2. Autumn.

20. **D. fimbriatum** Hooker. Figure 15(9), p. 208. Illustration: Cogniaux & Goossens, Dictionnaire Iconographique des Orchidées, Dendrobium, t. 9, 9A (1898); Annals of the Royal Botanic Garden Calcutta **8**: t. 82 (1898); Seidenfaden & Smitinand, Orchids of Thailand **2**(2); t. 8 (1960); Bechtel et al., Manual of cultivated orchid species, 196, 197 (1981).
Stems to 1.5 m, erect, fleshy. Leaves 8–15 × 2–3 cm, oblong-lanceolate. Racemes usually borne on leafless stems, with 5 or more flowers and an elongate axis. Flowers to 5.5 cm in diameter. Sepals and petals 2–3.5 cm, yellow-orange, the petals broader than the sepals. Lip orange, deeper orange or with 1 or 2 brown spots near the base, fringed, the fringe at most 2 mm deep. *Himalaya, Burma, Laos, Thailand, Vietnam & Malaysia*. G2. Spring.

Variable. Var. **oculatum** Hooker, with 1 or 2 brown spots on the lip, and var. **gibsoni** (Lindley) Gagnepain (*D. gibsoni* Lindley) with yellow, slightly smaller flowers, are both cultivated.

21. **D. brymerianum** Reichenbach. Figure 15(5), p. 208. Illustration: Botanical Magazine, 6383 (1878); Lindenia **4**: t. 183 (1888); Bechtel et al., Manual of cultivated orchid species, 195 (1981).
Stems to 50 cm, fleshy. Leaves 10–15 × 1.2–2 cm, lanceolate, acute. Racemes with 1–2 flowers, axis very short, borne almost terminally on leafless stems. Flowers 5–6 cm in diameter. Sepals and petals similar in size, 2.5–3.5 cm, yellow. Lip longer than the petals, mostly consisting of a fringe of branched threads, the central solid part small, all pale yellow. *Burma*. G2. Winter–spring.

Var. **histrionicum** Reichenbach (*D. histrionicum* (Reichenbach) Schlechter) with shorter stems, smaller flowers and the lip fringe less conspicuous, is also perhaps in cultivation.

22. **D. chrysanthum** Lindley. Illustration: Edwards's Botanical Register **15**: t. 1299 (1830); Cogniaux & Goossens, Dictionnaire Iconographique des Orchidées, Dendrobium, t. 2 (1897); Annals of the Royal Botanic Garden Calcutta **8**: t. 77 (1898); Bechtel et al., Manual of cultivated orchid species, 195 (1981).
Stems to 1.5 m, fleshy. Leaves 10–17 × 1–2.5 cm, lanceolate. Racemes with 2–3 flowers on a very short axis, borne on leafless stems. Sepals and petals 2–3 × 1.6–2 cm, petals slightly broader than sepals, all yellow, the sepals with reddish, warty ridges on their backs. Lip circular, yellow with 2 reddish brown blotches, shorter than the petals, the margin fringed, more or less hairless and shiny beneath. *Himalaya, Burma, Thailand*. G2. Autumn.

23. D. ochreatum Lindley (*D. cambridgeanum* Paxton). Illustration: Botanical Magazine, 4450 (1849); Cogniaux & Goossens, Dictionnaire Iconographique des Orchidées, Dendrobium, t. 16 (1899).
Very similar to *D. chrysanthum* but stems shorter, the petals little broader than the sepals and the lip densely hairy beneath. *E Himalaya, Thailand*. G2. Spring.

24. D. luteolum Bateman. Illustration: Botanical Magazine, 5441 (1864); Bateman, Second century of orchidaceous plants, t. 185 (1867).
Stems to 40 cm, fleshy. Leaves 6–10 × *c.* 3 cm, oblong-ovate or oblong-lanceolate, obtuse. Racemes with 2–3 flowers on a very short axis, borne on the leafy stems. Flowers 5–6 cm in diameter. Sepals and petals 2.5–2.8 cm, the petals a little broader than the sepals, all pale yellow. Lip oblong-ovate, almost as long as the petals, yellow with fine red lines towards the base. Mentum rather conspicuous. *Burma*. G2. Winter–spring.

25. D. heterocarpum Lindley (*D. aureum* Lindley). Figure 15(7), p. 208. Illustration: Botanical Magazine, 4708 (1853); Annals of the Royal Botanic Garden Calcutta **8**: t. 74 (1898); Cogniaux & Goossens, Dictionnaire Iconographique des Orchidées, Dendrobium, t. 10 (1898); Bechtel et al., Manual of cultivated orchid species, 198 (1981).
Stem to 40 cm, erect, fleshy. Leaves 10–20 × *c.* 2.5 cm, lanceolate or oblong-lanceolate, acute. Racemes with 2–3 flowers on a very short axis, borne on the leafless stems. Flowers 5–7 cm in diameter. Sepals and petals 2.5–3.5 cm, petals broader than sepals, all pale yellow. Lip oblong-ovate, yellow with red-orange lines and veins. *India, Sri Lanka, Burma, Thailand, Malaysia, Indonesia & Philippines*. G2. Spring–summer.

26. D. anosmum Lindley (*D. macrophyllum* Lindley not Richard; *D. superbum* Reichenbach). Illustration: Cogniaux & Goossens, Dictionnaire Iconographique des Orchidées, Dendrobium, t. 20 (1899); Bechtel et al., Manual of cultivated orchid species, 193 (1981).
Stems to 1.2 m, fleshy, hanging. Leaves 12–18 × *c.* 3 cm, oblong-lanceolate or ovate-oblong. Racemes with usually 2 flowers, axis short or absent, borne on leafless stems. Flowers to 10 cm in diameter. Sepals and petals 5–6 cm, petals much broader than sepals, all pinkish to purple. Lip broadly ovate, mauve with 2 deep purple blotches, fringed. *Philippines, Borneo, Moluccas*. G2. Winter–spring.

27. D. loddigesii Rolfe (*D. pulchellum* Loddiges not Lindley). Illustration: Botanical Magazine, 5037 (1858); Bechtel et al., Manual of cultivated orchid species, 199 (1981).
Stems 10–15 cm, thick. Leaves 4–7 × 1.25–2 cm, oblong-lanceolate. Racemes with 1–2 flowers, stalkless on leafless stems. Flowers *c.* 4 cm in diameter. Sepals and petals 2–2.5 cm, petals broader than sepals, all rose red. Lip as long as petals, circular, yellow with a purplish, conspicuously fringed margin. *SW China*. G2. Winter–spring.

28. D. moniliforme (Linnaeus) Swartz (*D. monile* (Thunberg) Kränzlin). Illustration: Edwards's Botanical Register **16**: t. 1314 (1830); Botanical Magazine, 4153 (1845); Bechtel et al., Manual of cultivated orchid species, 200 (1981).
Stem to 30 cm, with the internodes thickened upwards, somewhat beaded. Leaves to 7 cm, lanceolate, acute. Racemes with 2–3 flowers on a short axis, borne on leafy stems. Flowers *c.* 4 cm in diameter. Sepals and petals 2–2.5 cm, the petals somewhat broader than the sepals, all white or white flushed with pink. Lip shorter than the petals, elliptic, acute, white with red spots at the base and a red apex. *Japan, Korea; ?China*. G1. Spring.

29. D. findleyanum Parish & Reichenbach. Figure 15(6), p. 208. Illustration: Botanical Magazine, 6438 (1879); The Garden **49**: 496 (1896); Bechtel et al., Manual of cultivated orchid species, 197 (1981).
Stem to 30 cm, internodes thickened upwards, the stem appearing beaded. Leaves 7–8 × 1.3–3 cm, lanceolate. Racemes with 2 flowers on a short axis, usually borne on the leafless stems. Flowers 5–8 cm in diameter. Sepals and petals 2–2.5 cm, pale rose pink or mauve, rarely white suffused with these tones, the petals broader than the sepals. Lip ovate to almost circular, the centre deep yellow, the margin pale yellow or white, sometimes suffused reddish. *Burma, Thailand*. G2. Winter–spring.

30. D. lituiflorum Lindley. Illustration: Botanical Magazine, 6050 (1873); Bechtel et al., Manual of cultivated orchid species, 199 (1981).
Stem to 60 cm, fleshy. Leaves 7.5–10 × 1.5–2 cm, lanceolate to linear-lanceolate, acute. Racemes with 1–3 flowers on a short axis, borne on leafless stems. Flowers 5–7 cm, in diameter. Sepals and petals 2.5–3.8 cm, petals broader than sepals, all uniformly purple. Lip as long as petals, trumpet-shaped, its basal part more or less tubular, directed downwards, the apical part bell-shaped, directed forwards, purple margined with white, the inner part of the white margin sometimes yellowish. *NE India (Manipur), Burma*. G2. Spring.

31. D. transparens Lindley. Illustration: Botanical Magazine, 4663 (1852); Cogniaux & Goossens, Dictionnaire Iconographique des Orchidées, Dendrobium, t. 27 (1901); Bechtel et al., Manual of cultivated orchid species, 202 (1981).
Stems to 45 cm, rather thin but fleshy. Leaves 8–10 × *c.* 1.2 cm, lanceolate or linear-lanceolate. Racemes with 2–3 flowers and a very short axis, borne on leafless stems. Sepals and petals 1.8–2.5 cm, petals a little broader than the sepals, white or reddish. Lip ovate, shorter than petals, white with 2 red spots towards the base. *Himalaya*. G2. Spring–summer.

32. D. parishii Reichenbach. Illustration: Botanical Magazine, 5488 (1865); Bechtel et al., Manual of cultivated orchid species, 200 (1981).
Stems to 30 cm, hanging or prostrate, fleshy. Leaves to 10 cm, lanceolate to oblong-lanceolate. Racemes usually with 2 flowers on a short axis, borne on leafless stems. Flowers *c.* 5 cm in diameter. Sepals and petals 2.2–3.5 cm, the petals broader than the sepals, all uniformly lilac-purple. Lip shorter than petals, obovate, mauve-purple marked with 2 large dark red-purple spots towards the base. *Burma, SW China, Thailand, Laos & Kampuchea*. G2. Spring–summer.

33. D. aphyllum (Roxburgh) Fischer (*Limodorum aphyllum* Roxburgh; *D. pierardii* Roxburgh). Illustration: Botanical Magazine, 2584 (1825); Annals of the Royal Botanic Garden Calcutta **8**: t. 72 (1898); Cogniaux & Goossens, Dictionnaire Iconographique des Orchidées, Dendrobium, t. 26 (1901); Bechtel et al., Manual of cultivated orchid species, 193 (1981).
Stem to 1 m, slender. Leaves 6–12 × 2–3 cm, lanceolate to linear-lanceolate, acute. Racemes with 2–3 flowers on a very short axis, borne on leafless stems. Flowers 4.5–5.5 cm in diameter. Sepals and petals 2.5–3 cm, petals almost 2 times broader than the sepals, all pale pinkish lilac. Lip broadly ovate, as long as the petals, pale yellow,

with a line of hairs down the middle. *Himalaya, SW China, Burma, Thailand, Vietnam, Laos, Kampuchea & Malaysia.* G2. Spring.

34. D. primulinum Lindley. Illustration: Lindenia **15**: t. 686 (1900); Annals of the Royal Botanic Garden Calcutta **9**: t. 98 (1906); Bechtel et al., Manual of cultivated orchid species, 201 (1981); Die Orchidee **32**: Orchideenbewertung, 9 (1982).
Similar to *D. aphyllum* but leaves unequally 2-lobed at apex, the petals little broader than the sepals and the upper surface of the lip velvety all over. *Himalaya, SW China, Burma, Thailand, Vietnam, Laos & Kampuchea.* G2. Spring.

35. D. pulchellum Lindley (*D. dalhousieanum* Wallich). Illustration: Paxton's Magazine of Botany **11**: t. 145 (1844); Edwards's Botanical Register **32**: t. 10 (1846); Cogniaux & Goossens, Dictionnaire Iconographique des Orchidées, Dendrobium, t. 7 (1897).
Stems to 1.2 m, thickened. Leaf-sheaths red-striped. Leaves 10–15 × *c.* 3 cm, linear-oblong, 2-lobed at apex. Racemes with 5–12 flowers on an elongate axis, borne on leafless stems. Flowers *c.* 7–8 cm in diameter. Petals *c.* 3.5 cm broad, almost twice as broad as the sepals, white. Lip shorter than the petals, slightly concave, white to pale yellow at the base, with 2 red-purple blotches. *Himalaya, Burma, Vietnam.* G2. Spring.

36. D. nobile Lindley. Figure 15(8), p. 208. Illustration: Cogniaux & Goossens, Dictionnaire Iconographique des Orchidées, Dendrobium, t. 1A, B, C (1896–97); Annals of the Royal Botanic Garden Calcutta **8**; t. 71 (1898); Bechtel et al., Manual of cultivated orchid species, 200 (1981).
Stems to 1 m, fleshy. Leaves 7–10 × 1–1.5 cm, lanceolate to ovate-lanceolate. Racemes with 1–2 or rarely 3 flowers on a short axis, borne on leafless stems. Flowers 7–10 cm in diameter. Sepals and petals 3–5 cm, the petals broader than the sepals, all white, white tipped with purple or entirely mauve. Lip almost circular, white or yellowish white, variously marked with pink or mauve. *Himalaya, China, Taiwan.* G2. Spring–summer.

The precise coloration of the flowers is very variable, and several named varieties, based on this feature, have been in cultivation.

37. D. devonianum Paxton. Illustration: Botanical Magazine, 4429 (1849); Cogniaux & Goossens, Dictionnaire Iconographique des Orchidées, Dendrobium, t. 23 (1900); Seidenfaden & Smitinand, Orchids of Thailand **2**(2): t. 9 (1960).
Stems to 1 m, thin but fleshy, hanging. Leaves 7–10 × *c.* 1.5 cm, lanceolate. Racemes usually with 2 flowers on a short axis, borne on leafless stems. Flowers *c.* 5 cm in diameter. Sepals and petals 2–2.3 cm, petals broader than the sepals, white tipped with pinkish purple. Lip circular, as long as the petals, white with 2 large yellow blotches towards the base, the apex pinkish purple, margin conspicuously fringed. *Himalaya, SW China, Burma, Thailand & Vietnam.* G2. Summer.

38. D. falconeri Hooker. Illustration: Botanical Magazine, 4944 (1856); Bateman, Second century of orchidaceous plants, t. 137 (1867).
Stems to 1 m, thin, hanging, the internodes thickened upwards, the stems appearing beaded. Leaves grass-like, 5–8 cm × 3–4 mm. Flowers usually borne singly on the leafless stems, occasionally 2–3 together on a short axis, 5.5–7 cm in diameter. Sepals and petals 3–4.5 cm, the petals broader than the sepals, white tipped with purple. Lip ovate, almost as long as the petals, white, the apex purple and the base with a large purple blotch surrounded by an orange-yellow zone. *NE India, Burma.* G2. Spring–summer.

39. D. pendulum Roxburgh (*D. crassinode* Benson & Reichenbach). Illustration: Cogniaux & Goossens, Dictionnaire Iconographique des Orchidées, Dendrobium, t. 19, 34 (1899); Bechtel et al., Manual of cultivated orchid species, 200 (1981).
Stems to 30 cm, internodes thickened upwards, the stems appearing beaded. leaves 10–12 × 1.5–2 cm, lanceolate. Racemes with 2–3 flowers on a short axis, borne on the leafless stems. Flowers 5–8 cm in diameter. Sepals and petals 3–4 cm × 9–12 mm, the petals broader than the sepals, all white with red tips. Lip ovate to almost circular, the centre deep yellow, the margin white with a reddish apex. *Burma.* G2.

***D. wardianum** Warner. Illustration: Cogniaux & Goossens, Dictionnaire Iconographique des Orchidées, Dendrobium, t. 5, 5A (1897). Similar, perhaps the same species, but with less swollen internodes and larger flowers. *E. Himalaya, Burma.* G2.

40. D. crepidatum Lindley. Illustration: Botanical Magazine, 4993, 5011 (1857); Cogniaux & Goossens, Dictionnaire Iconographique des Orchidées, Dendrobium, t. 40 (1906); Bechtel et al., Manual of cultivated orchid species, 195 (1981).
Stems to 45 cm, fleshy. Leaves 5–10 × 1–1.25 cm, linear-lanceolate, acute. Racemes with 1–3 flowers on a very short axis, borne on leafless stems. Flowers 2.5–4.5 cm in diameter. Sepals and petals 1–2 cm, the petals a little broader than the sepals, white with pink tips. Lip about as long as the petals, ovate, yellow at the base, then white, the margin pink, sparsely hairy above, hairless beneath. *Himalaya, Burma, Thailand & Laos.* G2. Spring.

41. D. gratiosissimum Reichenbach. Illustration: Bechtel et al., Manual of cultivated orchid species, 197 (1981).
Stems to 1 m, fleshy, hanging. Leaves 7–12 × 1–1.5 cm, lanceolate, sheaths reddish. Racemes with 1–3 flowers on a very short axis, borne on leafless stems. Flowers 4–5 cm in diameter. Sepals and petals 1.8–3.5 cm, petals broader than sepals, all white tipped with pink. Lip broadly ovate, white, the base orange or yellow, the apex pink, densely velvety above and beneath. *Burma, Thailand.* G2. Spring.

42. D. amoenum Lindley. Illustration: Botanical Magazine, 6199 (1875); Annals of the Royal Botanic Garden Calcutta **8**: t. 69 (1898).
Stem to 60 cm, slender. Leaves to 10 cm, lanceolate or linear-lanceolate, acute. Racemes with 2–3 flowers on a very short axis, borne on leafless stems. Flowers 3–4 cm in diameter. Sepals and petals 1.7–2.5 cm, the petals finely toothed, sometimes irregularly so, broader than the sepals, all white tipped with purple. Lip as long as the petals, white with a yellow base and purple apex, densely velvety above, sparsely hairy beneath. *Himalaya.* G2. Spring.

Sometimes considered to be the same as *D. aphyllum* (p. 214).

43. D. aduncum Wallich. Illustration: Botanical Magazine, 6784 (1884); Annals of the Royal Botanic Garden Calcutta **8**: t. 67 (1898).
Stems hanging or ascending, red-spotted. Leaves to 8 cm, lanceolate or linear-lanceolate, somewhat red-lined beneath. Racemes with 1–2 flowers on a short axis, borne on leafless stems. Flowers

2.5–3 cm in diameter. Sepals and petals 1.6–1.8 cm, petals scarcely broader than sepals, all pale pinkish mauve. Lip with the free part very small, mostly pouch-like with an oblong projection at the base, whitish, much shorter than the mentum. *India (Sikkim), Bhutan, Burma*. G2. Summer–autumn.

GROUP F (Section *Pedilonum* (Blume) Lindley).
Leaves with sheaths, evenly spread along the stems; racemes with many flowers; lip mostly simple, its margins attached to the column foot for some distance, flat, blunt; ovary not angled or winged.

44. D. amethystoglossum Reichenbach. Illustration: Botanical Magazine, 5968 (1872); Davis & Steiner, Philippine orchids, 118 (1952).
Stems 60–90 cm. Leaves 5–10 cm. Racemes to 10 cm, drooping, borne on leafless stems and bearing 15–many flowers. Upper sepal *c.* 2 cm × 8 mm, lateral sepals to 3 cm, petals *c.* 2 cm, all white. Lip linear-spathulate, the basal part white with a backwardly pointing, fleshy, triangular ridge, the apical part amethyst-purple. *Philippines*. G2. Winter.

***D. victoriae-reginae** Loher. Illustration: Cogniaux & Goossens, Dictonnaire Iconographique des Orchidées, Dendrobium, t. 21 (1899); Orchid Review **20**: f. 6 (1912); Botanical Magazine, 9071 (1925). Similar but with fewer flowers in the raceme, the petals and sepals white in the lower part, violet-blue in the upper, the lip orange-yellow at the base, violet-blue above. *Philippines*. G2. Spring–summer.

45. D. secundum Lindley. Figure 15(10), p. 208. Illustration: Edwards's Botanical Register **15**: t. 1291 (1829); Botanical Magazine, 4352 (1848); Cogniaux & Goossens, Dictionnaire Iconographique des orchidées, Dendrobium, t. 35 (1905); Bechtel et al., Manual of cultivated orchid species, 201 (1981).
Stems to 1 m. Leaves *c.* 10 × 3 cm, broadly oblong, mostly in the upper two-thirds of the stem. Racemes *c.* 12–15 cm, from the uppermost nodes, 1-sided. Flowers 1.5–1.8 cm in length, upper sepal *c.* 7 × 4 mm, petals narrow, all pink or reddish or rarely white. Lip simple, pink or reddish. *Burma, Laos, Vietnam, Kampuchea, Malaysia, Indonesia & Philippines*. G2. Winter.

46. D. bullenianum Reichenbach (*D. topaziacum* Ames). Illustration: Bechtel et al., Manual of cultivated orchid species, 194 (1981).
Stems to 60 cm, longitudinally grooved. Leaves 7–14 × 1.5–2.8 cm, oblong. Racemes short, borne on leafless stems, not 1-sided. Flowers *c.* 1.7 cm in length. Upper sepal lanceolate, *c.* 6 × 3 mm, lateral sepals obliquely triangular, *c.* 1.5 cm, petals *c.* 6 × 2 mm, oblong-spathulate, bluntly pointed, all orange. Lip *c.* 1.5 cm, narrowly elliptic or spathulate. *Philippines*. G2.

GROUP G (Section *Calyptrochilus* Schlechter; Section *Glomerata* Kränzlin).
Leaves with sheaths, uniformly spread along the fleshy stems; racemes short with few flowers; lip simple, its margins attached to the column foot for some distance, its apical part upcurved and fringed.

47. D. lawesii Mueller. Figure 15(11), p. 208. Illustration: Botanical Magazine n.s., 524 (1969); Millar, Orchids of Papua New Guinea, 26 (1978); Bechtel et al., Manual of cultivated orchid species, 198 (1981).
Stems to 50 cm, hanging or erect, leafy for most of their length. Leaves 6–8 × *c.* 1.5 cm, clearly arranged in 2 ranks, lanceolate, gradually pointed. Racemes short with 1–5 (rarely to 10) flowers, occurring in the upper parts of leafless (occasionally leafy) stems. Upper sepal and petals *c.* 7 mm, lateral sepals *c.* 2.5 cm (from tip of sepal to tip of spur-like mentum), all scarlet or purplish. Lip mauve, boat-like, apex upcurved, its margin fringed. *New Guinea*. G2. Winter.

48. D. subclausum Rolfe. Illustration: Feddes Repertorium Beihefte **21**: t. 174, f. 651 & t. 175, f. 653 (1912); Millar, Orchids of Papua New Guinea, 26 (1978); Van Royen, Orchids of the high mountains of New Guinea, 344, 346, 353 (1980).
Habit very variable, stems mostly erect, 50–100 cm, often branched, finely warted on the leaf-sheaths. Leaves numerous, 2.5–7 cm × 2.5–10 mm, clearly arranged in 2 ranks and always twisted. Racemes very short, mostly on leafless stems, with up to 5 (rarely more) flowers. Flowers similar to those of *D. lawesii* but bright orange with pale orange or yellow tips. *New Guinea*. G2. Winter–spring.

D. dichroma Schlechter, *D. flammula* Schlechter, *D. nutriferum* J.J. Smith and *D. phlox* Schlechter, all from New Guinea, may well be the same species, but sufficient information is not yet available.

***D. wentianum** J.J. Smith. Illustration: Van Royen, Orchids of the high mountains of New Guinea, 338, 339, 351, 361 (1980). Similar, but stems hanging, leaves to 2 cm, more or less ovate, often strongly reflexed, flowers much brighter orange. *New Guinea*. G1.

GROUP H (Section *Cuthbertsonia* Schlechter; Section *Leiotheca* Kränzlin in part).
Small tufted plants with crystalline-papillose leaves, sheaths and flower-stalks; racemes each with 1 non-resupinate flower; lip simple, obtuse, boat-shaped, its margins attached to the column foot for some distance.

49. D. cuthbertsonii Mueller (*D. sophronites* Schlechter). Figure 15(12), p. 208. Illustration: Journal of the Royal Horticultural Society **88**: f. 46 (1963); Orchid Review **76**: 248 (1968); The Orchadian **6**: 37, 38 (1978); Millar, Orchids of Papua New Guinea, 14, 15 (1978).
Plant compact, stems *c.* 1–2 cm (rarely to 5) × 4–7 mm, clustered, club-shaped or cylindric and narrowing abruptly upwards. Leaves almost apical, 2–5, linear-lanceolate to almost ovate, dark green and finely crystalline-papillose above, often purplish beneath. Flowers 2.5–4 cm long, 1.3–3.5 cm in diameter, terminal, borne singly on leafless stems, the colour extremely variable, commonly scarlet but frequently purplish, mauve, orange or white, sometimes bicoloured. Lip boat-shaped and strongly concave, with darker marginal markings. *New Guinea*. G1. Summer–spring.

GROUP I (Section *Oxyglossum* Schlechter).
Small tufted plants; leaves few, borne towards the upper parts of the stems, with sheaths; racemes terminal, with 2–4 flowers; lip tightly attached to the column and its foot, simple, sharply pointed; ovary winged or angled.
Literature: Reeve, T.M. & Woods, P.J.B., A preliminary key to the species of Dendrobium section Oxyglossum, *The Orchadian* **6**: 195–208 (1980).

50. D. vexillarius J.J. Smith. Figure 15(13), p. 208. Illustration: Millar, Orchids of Papua New Guinea, 35, 37 (1978), under various specific names; Van Royen, Orchids of the high mountains of New Guinea, 376–93 (1980); Australian Orchid Review **46**: 106 (1981).
Stems 1–25 cm × 3–15 mm, clustered, cylindric, narrowing upwards. Leaves 2–4, 2–10 cm (rarely to 16) × 3–18 mm, oblong to oblong-lanceolate, blunt or shortly pointed, green above, often purplish

beneath. Racemes stalkless from tips of leafy or leafless branches. Sepals and petals 2.5–5 × *c.* 3 cm, spreading, flat, variable in colour, purple, orange or yellow. Lip purplish, blackish or green, the sharp tip bright orange, reflexed. Ovary triangular in section. *New Guinea.* G1. Summer–spring.

A very variable plant both in habit and flower colour.

***D. subacaule** Lindley (*D. oreocharis* Schlechter; *D. tricostatum* Schlechter). Illustration: Orchid Digest **41**: 13 (1977); Northen, Miniature orchids, C-26 (1980). Similar but much smaller, stems and leaves to 1.5 cm, flowers bright orange, 9–18 mm. *New Guinea, Moluccas, Solomon Islands.* G1. Summer–spring.

51. D. hellwigianum Kränzlin (*D. cyananthum* Williams; *D. raphiotes* Schlechter).
Stems 1–8 cm × 2–6 mm, clustered, cylindric. Leaves 3–4, to 15 cm × 2.5 mm, round in section. Racemes with 2–4 flowers, stalkless on leafless stems. Flowers pinkish to purple or bluish, lip-tip acute, bright orange, not reflexed. Ovary 5-winged (the upper 3 wings so close as to give a triangular appearance). *New Guinea.* G1. Summer–spring.

***D. violaceum** Kränzlin. Illustration: Northen, Miniature orchids, C-21 (1980). Similar, but the leaves not round in section, linear, to 25 cm × 3–6 mm, the flowers 3–4.5 cm, purplish, the lip-apex orange. *New Guinea.* G1. Summer–spring.

GROUP J (Section *Stachyobium* Schlechter) Leaves with sheaths; racemes more or less terminal on the leafy stems; lip 3-lobed, fringed, its margin not attached to the column foot.

52. D. delacourii Guillaumin (*D. ciliatum* Hooker not Persoon). Figure 15(14), p. 208. Illustration: Botanical Magazine, 5430 (1864); Seidenfaden & Smitinand, Orchids of Thailand 2(2): 231 (1960); Bechtel et al., Manual of cultivated orchid species, 195 (1981).
Stems to 45 cm, though often shorter, erect. Leaves 10–13 × 2–3 cm, oblong-elliptic, acute. Racemes erect, with several flowers. Sepals and petals 1–1.2 cm, the petals somewhat narrower than the sepals, greenish yellow or yellow. Lip obovate, 3-lobed, as long as the petals, the central lobe fringed with spathulate divisions, all greenish yellow with red veins. *Burma, Thailand, Laos, Vietnam & Kampuchea.* G2. Winter.

GROUP K (Section *Phalaenanthe* Schlechter). Leaves with sheaths; racemes borne on leafless stems; petals much broader than sepals; mentum double, consisting of 2 swellings or humps at right angles to each other.

53. D. bigibbum Lindley. Illustration; Botanical Magazine, 4898 (1856); Bateman, Second century of orchidaceous plants, t. 169 (1867); Lindenia **7**: t. 317 (1891); Bechtel et al., Manual of cultivated orchid species, 194 (1981).
Stem to 50 cm, fleshy, spindle-shaped or cylindric. Leaves 5–12 × 2–3 cm, oblong-lanceolate or lanceolate. Raceme erect with many flowers which are 3–7 cm in diameter. Sepals and petals 2–3.5 cm, petals broader than sepals, broadly rounded, occasionally with a small point, all usually reddish purple, rarely white. Lip shorter than the petals, 3-lobed, with a blunt or notched central lobe which bears many small calluses in the middle. Mentum bluntly conical. *Australia (N Queensland).* G2.

Variable in flower colour and size. *D. superbiens* Reichenbach is the name applied to plants which are almost certainly natural hybrids between *D. bigibbum* and another Australian species, *D. discolor* (p. 218). These plants have been in cultivation (see Lindenia **7**: t. 294, 1891; Cogniaux & Goossens, Dictionnaire Iconographique des Orchidées, Dendrobium, t. 15, 1898) and differ from *D. bigibbum* in having the second hump of the mentum small and the petals relatively narrower (see Blake, S.T., Proceedings of the Royal Society of Queensland **74**: 29–44 1964).

54. D. phalaenopsis Fitzgerald (*D. schroederanum* Gentil). Figure 15(15), p. 208. Illustration: Cogniaux & Goossens, Dictionnaire Iconographique des Orchidées, Dendrobium, t. 4 (1897); Botanical Magazine, 6817 (1885); Lindenia **6**: t. 280 (1890); Bechtel et al., Manual of cultivated orchid species, 200 (1981).
Similar to *D. bigibbum* but more robust, with stems to 60 cm; leaves to 15 cm. Flowers 6–10 cm in diameter. Sepals and petals 4–4.5 cm, petals much broader than sepals, acuminate, usually all red. Lip with a long, acute, central lobe which bears a few, small calluses, usually a deeper red than the petals and sepals. *Indonesia (Timor Laut).* G2. Winter–spring.

Like *D. bigibbum*, this species is very variable in flower colour (shades of red, purple and white), and numerous varieties (cultivars) have been described on the basis of this variation. The origin of the species has long been obscure, but S.T. Blake (reference under *D. bigibbum*) has shown conclusively that it is endemic to Timor Laut and does not occur as a native in Australia or New Guinea.

GROUP L (Section *Ceratobium* Schlechter). Leaves with sheaths; racemes from leafy or leafless stems; sepals and particularly petals often twisted, the petals usually very erect; lip with an entire central lobe.
Literature: Ossian, C.R., A review of the Antelope Dendrobiums (section Ceratobium), *American Orchid Society Bulletin* **50**: 1213 . . . 1455 (1981); **51**: 23 . . . 362 (1982).

55. D. helix Cribb. Illustration: Orchid Review **88**: 144 (1980); The Orchadian **6**: 175 (1980); American Orchid Society Bulletin **51**: 255, 258 (1981).
Stems to 1 m, cane-like. Leaves to 16 × 6.5 cm, elliptic or ovate-elliptic, obtuse and minutely 2-lobed at the apex. Racemes erect with 5 or more flowers. Sepals 2.6–3 cm × 8–12 mm, the upper spirally recurved; petals 3–4.5 cm, erect, spirally twisted 3 or 4 times; all yellow with red-brown blotches. Lip 3-lobed, shorter than the sepals, the central lobe recurved, base mauve, the rest yellow with red-brown spots. *New Guinea (New Britain).* G2.

56. D. tangerinum Cribb. Illustration: Orchid Review **88**: 145 (1980); The Orchadian **6**: 176 (1980); American Orchid Society Bulletin **50**: 1340, 1343 (1981).
Stems to 50 cm, erect, cane-like, clustered. Leaves to 9 × 2.5 cm, oblong-elliptic. Racemes spreading to erect, with *c.* 15 flowers. Upper sepal 2.3–2.5 cm × 6–7 mm, recurved, margins wavy; lateral sepals 1.6–2.3 × 1–1.2 cm; petals 3–3.4 cm × 4–5 mm, erect, spirally twisted 2 or 3 times; all orange-yellow, with dark red venation. Lip 3-lobed, shorter than the sepals, orange-yellow with 3 lilac ridges and a white base. *New Guinea.* G2.

57. D. minax Reichenbach. Illustration: Reichenbach, Xenia Orchidacearum, t. 145 (1868); Das Pflanzenreich **45**: 143 (1910).
Stems to 30 cm. Leaves to 10 × 2 cm, oblong, unequally 2-lobed at the apex. Racemes with 2–3 flowers. Sepals 2.5–2.8 cm, petals to 5 cm, not twisted or with 1 slight twist, all greenish red, intensely red-streaked. Lip 3-lobed, the lateral lobes broadly triangular, the central transversely oblong, broader than long,

with a small point, greenish red with purple lines. *Indonesia (Amboina, Moluccas, Sulawesi).* G2. Summer.

58. D. stratiotes Reichenbach. Illustration: Illustration Horticole **34**: t. 602 (1886); Bechtel et al., Manual of cultivated orchid species, 202 (1981); American Orchid Society Bulletin **50**: 1337, 1343 (1981).
Stems to 60 cm or more. Leaves 8–12 × *c.* 2 cm, oblong, obtuse. Racemes with up to 5 flowers. Upper sepal, *c.* 3.5 cm, laterals *c.* 4.5 cm, petals to 6 cm, all white suffused with green. Lip with diamond-shaped lateral lobes and a central lobe which is broadly ovate or elliptic, longer than broad, acute, white with a greenish yellow base and red veins and spots. *Indonesia (Sunda Islands).* G2. Summer.

59. D. taurinum Lindley. Illustration: Edwards's Botanical Register **29**: t. 28 (1843); Lindenia **13**: t. 621 (1897); American Orchid Society Bulletin **51**: 253, 258 (1982).
Stems to 1.5 m, erect, spindle-shaped. Leaves 10–15 × *c.* 6 cm, oblong or elliptic, rounded or slightly 2-lobed at the apex. Racemes erect with many flowers. Upper sepal *c.* 3 cm, laterals *c.* 4 cm, reflexed; petals 3–4 cm, reflexed; sepals white suffused with green, petals brownish red to pink, slightly twisted. Lip oblong-elliptic, pale brownish red to pink, about as long as the petals, with large lateral lobes and a small, somewhat wavy central lobe. *Philippines.* G2. Autumn–winter.

60. D. crispilinguum Cribb. Illustration: Orchid Review **88**: 146, 147 (1980); The Orchadian **6**: 177 (1980); American Orchid Society Bulletin **51**: 32 (1982).
Stems to 1 m or more, cane-like. Leaves to 12 × 1.5–3.5 cm, lanceolate or ovate-lanceolate, acute. Racemes with up to 20 flowers. Upper sepal recurved and somewhat twisted, 2.6–3 cm; lateral sepals reflexed, *c.* 2.6 cm; petals 2.5–3.3 cm, somewhat twisted; all cream or yellowish. Lip shorter than the petals with oblong lateral lobes and an ovate, very crisped central lobe, white suffused and veined red-purple. *New Guinea.* G2.

61. D. discolor Lindley (*D. undulatum* R. Brown not Persoon). Figure 15(16), p. 208. Illustration: Edwards's Botanical Register **27**: t. 52 (1841); Dockrill, Australian indigenous orchids, 469 (1969); Bechtel et al., Manual of cultivated orchid species, 196 (1981); American Orchid Society Bulletin **51**: 256–9 (1982).
Stems to 5 m, cane-like. Leaves 5–15 × 2–5 cm, ovate or elliptic, obtuse or slightly 2-lobed. Racemes erect with numerous flowers. Sepals and petals 2–5 cm, very wavy, somewhat twisted and reflexed, yellowish brown. Lip shorter than the petals, the central lobe ovate, bent downwards, not wavy, yellowish with 3 white or violet ridges. *Australia (Queensland), New Guinea.* G1. Winter–spring.

Variable in flower colour in the wild. See note under *D. bigibbum* (p. 217).

62. D. mirbelianum Gaudich (*D. wilkianum* Rupp). Illustration: Dockrill, Australian indigenous orchids, 475 (1969); American Orchid Society Bulletin **51**: 32, 147, 152 (1982).
Stems to 1 m, erect. Leaves 8–15 × 1.5–4 cm, ovate, usually suffused with red or purple. Racemes with 4–12 flowers. Sepals 1.8–3 cm, petals slightly longer, not or little twisted, all pale to dark brown. Lip shorter than the petals, 3-lobed, the central lobe oblong-ovate, acute, yellowish green with deep red veins and 4 ridges. *Australia (Queensland); ?New Guinea.* G2.

GROUP M (Section *Oxygenianthe* Schlechter; Section *Nigro-hirsutae* Kränzlin). Leaves with sheaths which are conspicuously brown- or black-hairy; racemes borne on leafy stems; lip usually clearly 3-lobed, the lateral lobes usually arching upwards to the column, usually not hairy.

63. D. formosum (*D. infundibulum* Reichenbach not Lindley). Figure 15(17), p. 208. Illustration: Cogniaux & Goossens, Dictionnaire Iconographique des Orchidées, Dendrobium, t. 8 (1897); Seidenfaden & Smitinand, Orchids of Thailand **2**(2): t. 11 (1960); Orchid Digest **41**: 5 (1977); Bechtel et al., Manual of cultivated orchid species, 197 (1981).
Stems to 45 cm, thick. Leaves 10–12 × *c.* 3.5 cm, ovate-oblong, unequally 2-lobed at the apex. Racemes with 2 or more flowers which are 9–11 cm in diameter. Sepals and petals 4–5 cm, petals much broader than the sepals, all white. Lip longer than petals, scarcely 3-lobed, obovate, wavy, notched, white with yellow markings. Mentum conical, acute. *Himalaya, Burma, Thailand.* G2. Winter–spring.

Variable in the size of the flowers and the precise coloration of the lip.

64. D. infundibulum Lindley. Illustration: Botanical Magazine, 5446 (1864); Cogniaux & Goossens, Dictionnaire Iconographique des Orchidées, Dendrobium, t. 6 (1897); Orchid Digest **41**: 5 (1977); Bechtel et al., Manual of cultivated orchid species, 198 (1981).
Similar to *D. formosum* but with flowers to 8 cm in diameter, sepals and petals at most 4 cm, lip distinctly 3-lobed, irregularly fringed. *Burma, Thailand.* G2. Spring–summer.

Variable; var. **jamesianum** (Reichenbach) Veitch (*D. jamesianum* Reichenbach), which has the lip papillose and marked with orange-yellow, is often grown.

65. D. dearei Reichenbach. Illustration: Cogniaux & Goossens, Dictionnaire Iconographique des Orchidées, Dendrobium, t. 36 (1905); Bechtel et al., Manual of cultivated orchid species, 195 (1981).
Stems to 1 m. Leaves to 6 cm, ovate-oblong, unequally 2-lobed at the apex, hairless. Raceme with 3–6 flowers. Sepals and petals 2.5–3.5 cm, petals broader than sepals, all white. Lip 3-lobed, about as long as petals, the central lobe somewhat deflexed and wavy, white, green at the base. Mentum conical, acute, *c.* 1.5 cm. Ovary 3-keeled. *Philippines.* G2. Spring–summer.

66. D. sanderae Rolfe. Illustration: Botanical Magazine, 8351 (1910).
Similar to *D. dearei* but the lip streaked with purple at the base and the sepals and petals to 4.5 cm. *Philippines.* G2. Autumn–winter.

Apparently sometimes confused with *D. sanderianum* Rolfe, a totally different plant not often seen in cultivation, and with the artificial hybrid *D.* × *ainsworthii* Moore (*D. heterocarpum* × *nobile)*, which is sometimes found in older horticultural literature as '*D.* × *sanderae*' (see Hunt & Summerhayes, Kew Bulletin **20**: 55, 1966).

67. D. draconis Reichenbach (*D. eburneum* Reichenbach). Figure 15(18), p. 208. Illustration: Bateman, Second century of orchidaceous plants, t. 166 (1867): Botanical Magazine, 5459 (1864); Orchid Digest **41**: 5 (1977).
Stems to 45 cm, fleshy. Leaves 6–8 × 1–1.5 cm, lanceolate, apex unequally 2-lobed, with brown hairs on both surfaces. Sepals and petals 3–4 cm, the petals broader than the sepals, all creamy white. Lip longer than the petals, with 2 triangular lateral lobes borne at the base, the central lobe oblong-obovate, acute, somewhat wavy, entire, white with red and yellow lines towards the base. Mentum spur-like, 2–2.5 cm, narrow. *Burma,*

Thailand, Laos, Kampuchea & Vietnam. G2. Spring–summer.

Cleistogamous variants of this species are known (see Pradhan, Orchid Digest **41**: 6, 1977).

68. D. williamsonii Day & Reichenbach. Illustration: Annals of the Royal Botanic Garden Calcutta **5**: t. 9 (1895); Botanical Magazine, 7974 (1904); Bechtel et al., Manual of cultivated orchid species, 203 (1981).
Stems to 25 cm. Leaves 5–10 × 1.5–2.5 cm, linear-oblong, with brown hairs on both surfaces. Racemes with 2 flowers. Sepals and petals 3–3.5 cm, equally broad, white inside, pale yellow outside. Lip longer than the petals with 2 rounded lateral lobes borne about half way along its length, the central lobe conspicuously fringed, creamy white to yellow with a large red-brown spot at the base. Mentum spur-like, *c*. 2.5 cm. *NE India*. G1. Spring.

69. D. bellatulum Rolfe. Illustration: Botanical Magazine, 7985 (1904); Orchid Digest **41**: 5 (1977).
Stems very short, usually less than 5 cm, clustered. Leaves lanceolate or narrowly ovate, apex unequally 2-lobed, with dark brown hairs on both surfaces. Racemes with 1–3 flowers. Sepals and petals 2–2.5 cm, equally broad, all white. Lip ovate, about as long as petals, with a notched, broadly oblong, deflexed central lobe and a thick, papillose ridge down most of its length, red at the base, the rest yellow. Mentum broadly conical, blunt. *SW China, Vietnam, Kampuchea, Laos & Thailand*. G1. Spring.

GROUP N (Section *Rhopalanthe* Schlechter). Stems thickened only at the extreme base, tapering abruptly above this.

70. D. crumenatum Swartz. Figure 15(19), p. 208. Illustration: Botanical Magazine, 4013 (1843); Holttum, Orchids of Malaya, f. 3 (1953).
Stems to 1 m, the lower 4–5 internodes swollen, the rest thin. Leaves fleshy, obtuse or obscurely 2-lobed. Raceme leafless, with many fragrant flowers. Sepals and petals 2.5–3.5 cm, white or white suffused with pink. Lip as long as the petals, 3-lobed, the central lobe wavy and with a prominent crested callus, white with a yellow blotch. *Himalaya to Indonesia*. G2. Can flower at any time.

Flowering in this sparingly cultivated species follows a sudden, though often short, drop in temperature. The flowers last for only 1 day.

GROUP O (Section *Aporum* Schlechter). Stems and leaves strongly laterally flattened, thick, leaves regularly alternating in 2 rows, jointed at the base; flowers produced singly from a cluster of chaffy bracts.

71. D. terminale Parish & Reichenbach. Figure 15(20), p. 208. Illustration: Annals of the Royal Botanic Garden Calcutta **8**: t. 55 (1898).
Stems 10–15 cm. Leaves to 2 × 0.5 cm. Flowers produced subterminally. Upper sepal and petals *c*. 5 × 3 mm, lateral sepals *c*. 13 × 6 mm. Lip simple, notched at the apex, *c*. 10 × 5 mm. *Burma*. G2.

74. EPIGENEIUM Gagnepain

J. Cullen

Epiphytes with long rhizomes bearing distant simple pseudobulbs which are 4-angled, at least when dry, and bear 2 leaves. Leaves leathery, folded when young often unequally 2-lobed at the apex. Racemes erect with 1–many flowers, terminal on the pseudobulbs. Sepals and petals similar, the lateral sepals attached to the short column foot, forming a small mentum. Lip 3-lobed, attached to the base of the column foot and somewhat movable on it. Column short. Pollinia 4, parallel.

A genus of 35 species from Asia, formerly included in *Dendrobium* Swartz *Sarcopodium* Lindley not Schlechtendahl or *Katherinea* Hawkes, of which 3 are occasionally grown. Cultivation in general as for the Indian and Burmese species of *Dendrobium* (p. 207), but these plants are better grown on rafts to accommodate their long rhizomes.
Literature: Summerhayes, V.S., Notes on Asiatic orchids II, *Kew Bulletin* 1957: 259–68; Seidenfaden, G., Orchid genera in Thailand IX, *Dansk Botanisk Arkiv* **34**: 68–82 (1980).

1a. Racemes with 10–15 or more flowers; sepals and petals white with green or yellow tips and maroon bases **1. lyonii**
b. Racemes with 1–2 flowers; sepals and petals brown or greenish white suffused with purple and/or spotted with brown 2
2a. Flowers 8–9 cm in diameter; central lobe of lip rhomboid, tapered towards its base **2. amplum***
b. Flowers 3.5–5 cm in diameter; central lobe of lip transversely oblong, not tapered towards its base **3. rotundatum**

1. E. lyonii (Ames) Summerhayes (*Dendrobium lyonii* Ames; *Sarcopodium lyonii* (Ames) Rolfe; *S. acuminatum* var. *lyonii* (Ames) Kränzlin; *Katherinea acuminata* var. *lyonii* (Ames) Hawkes). Illustration: Das Pflanzenreich **45**: 330 (1910); Davis & Steiner, Philippine orchids, 110 (1952).
Pseudobulbs 4–4.5 cm. Leaves 10–15 × 2–4 cm, oblong or elliptic. Racemes with 10–15 (or more) flowers. Sepals and petals to 4 cm, triangular, white with green or yellow tips and maroon bases. Lip to 3 cm, central lobe triangular, acute, whitish, bases of lateral lobes dark red. *Philippines*. G2.

2. E. amplum (Lindley) Summerhayes (*Dendrobium amplum* Lindley; *Sarcopodium amplum* (Lindley) Lindley; *Katherinea ampla* (Lindley) Hawkes). Illustration: Annals of the Royal Botanic Garden Calcutta **8**: t. 89 (1898); Orchid Review **78**: 117 (1970).
Pseudobulbs to 5 cm, ovoid or spindle-shaped. Leaves 10–15 × *c*. 5 cm, oblong or oblong-lanceolate, acute. Raceme with 1–2 flowers which are 8–9 cm in diameter. Sepals lanceolate, petals linear-lanceolate, all 3–4 cm, greenish suffused brownish purple and/or with brown spots. Lip to 3.5 cm, central lobe rhomboid, tapered to its base, brown, the lateral lobes coloured like the petals. *Nepal, N India*. G2.

***E. coelogyne** (Reichenbach) Summerhayes. Illustration: Bechtel et al., Manual of cultivated orchid species, 211 (1981). Similar, but with larger flowers which are more distinctly spotted and marked; lip *c*. 4.5 cm, dark purple. *Burma*. G2.

Not considered as a distinct species by Seidenfaden who treats it as a synonym of *E. amplum*, but well known in cultivation.

3. E. rotundatum (Lindley) Summerhayes (*Sarcopodium rotundatum* Lindley; *Dendrobium rotundatum* (Lindley) Hooker; *Katherinea rotundata* (Lindley) Hawkes). Illustration: Annals of the Royal Botanic Garden Calcutta **8**: t. 87 (1898).
Pseudobulbs 3–4 cm, ovoid. Leaves to 12 × 2.5 cm, lanceolate or oblong. Sepals and petals *c*. 2.5 cm, brown. Lip oblong *c*. 2.5 cm, the central lobe transversely oblong, not tapered to the base, yellowish and thickened in the middle. *N India, SW China*. G2.

75. BULBOPHYLLUM Thouars

J.C.M. Alexander & P.J.B. Woods

Epiphytic. Pseudobulbs simple, distant or clustered, 1- or 2-leaved. Leaves stalked or stalkless, folded when young, leathery. Flowering stems basal, erect or pendent (in our species), with distant, spreading, sheath-like scales. Flowers 1–many, in spikes, racemes or umbels. Sepals usually almost equal. Lateral sepals joined to column foot, forming a short mentum. Petals usually shorter than sepals, sometimes much reduced. Lip thick, not obviously lobed, sometimes fringed, loosely hinged to column foot. Column short, often with 2 or more apical arms; column foot long and narrow. Pollinia 4.

A genus of perhaps 1000 species from Asia, Australasia, Africa and tropical America. In this work *Cirrhopetalum* is treated as a separate genus.

Being widely distributed, the species of *Bulbophyllum* show a range of cultural requirements. They can be grown in epiphytic compost in pots or on slabs of tree fern or bark. The deciduous species with hard pseudobulbs, mostly native to temperate regions, need a pronounced resting period once growth has ceased. The tropical species require a less marked rest and must not dry out completely. Propagation is by division.
Literature: Seidenfaden. G., Orchid genera in Thailand VIII: Bulbophyllum, *Dansk Botanisk Arkiv* **33**(3), (1979).

Pseudobulbs. 2-leaved: **1**. Cylindric: **3**; spherical and flattened: **2**; ovoid: **4**. Clustered: **2,4**.
Leaves. 10 cm or more: **1,3,4**. 4.5 cm wide or more: **3,4**.
Flowering stem. Pendent: **2**. Thickened or flattened: **1**.
Flowers. Solitary: **4**; more than 10: **1,2**. In umbels: **3,4**.
Sepals. More than 1.5 cm: **3,4**; less than 1.5 cm: **1,2**. Longer than petals: **1–4**; almost equal to petals: **1,4**. Drawn out into long points: **3**. Upper sepal thickened: **1**.
Lip. Ovate to triangular: **3,4**. Fringed towards base: **1**; with apical thread-like hairs: **2**.

1a. Pseudobulbs 2-leaved; flowering stem flat or swollen above **1. falcatum***
b. Pseudobulbs 1-leaved; flowering stem not flat or swollen 2
2a. Flowers 10 or more per raceme; upper sepal 1.5 cm or less **2. barbigerum***
b. Flowers 1–7 per raceme; upper sepal 2 cm or more 3
3a. Pseudobulbs 7 cm or more, cylindric to narrowly fusiform **3. ericssonii**
b. Pseudobulbs 5 cm or less, ovoid or narrowly ovoid **4. lobbii***

1. B. falcatum (Lindley) Reichenbach (*Megaclinium falcatum* Lindley). Illustration: Edwards's Botanical Register **12**: t. 989 (1826); Stewart & Campbell, Orchids of tropical Africa, 61 (1970); Bechtel et al., Manual of cultivated orchid species, 174 (1981).
Pseudobulbs 2–6 cm, ovoid or narrowly ovoid, 4-angled, distant, 2-leaved. Leaves to 16 cm × 6–23 mm, oblong to oblanceolate. Flowering stem to 35 cm, erect, forming a flattened green or reddish blade to 1.4 cm wide; flowers numerous in 2 ranks, 5–20 mm apart, *c.* 5 × 10 mm, green, reddish or purple. Upper sepal to 7 mm, linear; tip wider and thicker; lateral sepals shorter than upper sepal, 2.5–3.5 mm wide, sickle-shaped. Petals 1.7–3 mm, sickle-shaped, linear to oblong, obtuse and often thickened and yellow. Lip *c.* 2 mm, narrowly triangular, purplish. *W Africa to Uganda*. G2. Spring.

***B. leucorrhachis** (Rolfe) Schlechter (*Megaclinium leucorrhachis* Rolfe). Illustration: Botanical Magazine, 7811 (1901). Leaves to 18 × 3 cm. Flowering stem to 18 cm, swollen to *c.* 1.3 cm thick above; bracts triangular, reflexed; flowers *c.* 8 mm, yellow; lip small, tongue-shaped, fringed towards base. *W & C Africa*. G2. Spring.

2. B. barbigerum Lindley. Illustration: Botanical Magazine, 5288 (1861); Schlechter, Die Orchideen, edn 2, 320 (1927); Orchid Review **76**: 318 (1968); Bechtel et al., Manual of cultivated orchid species, 173 (1981).
Pseudobulbs to 3 × 1.5–2.5 cm, spherical, somewhat flattened, clustered, 1-leaved. Leaves to 10 × 2.8 cm, broadly oblong, obtuse. Flowering stem 10–20 cm, erect with *c.* 4 narrowly ovate-acuminate, sheath-like scales *c.* 9 mm long; floral bracts similar; flowers 8–20 mm long, reddish. Sepals 7–13 mm, greenish, narrowly lanceolate; petals minute. Lip *c.* 1 cm, yellowish, linear, with short marginal hairs; apex with a tuft of long, purplish, thread-like, apically swollen hairs; base hinged so that lip and hairs move in the slightest draught. *W & C Africa*. G2. Summer.

***B. careyanum** (Hooker) Sprengel. Illustration: Botanical Magazine, 5316 (1862): Dansk Botanisk Arkiv **33**: 146, 147 (1979); Pradhan, Indian orchids, t. 130 (1979); Bechtel et al., Manual of cultivated orchid species, 173 (1981). Pseudobulbs distant. Flowering stem drooping; flowers many, crowded, yellow. Lateral sepals longer than upper sepal. Lip purple, downy, without long thread-like hairs. *E Himalaya, Burma & Thailand*. G2. Winter.

3. B. ericssonii Kränzlin. Illustration: Botanical Magazine, 8088 (1906).
Pseudobulbs *c.* 9 × 1.5 cm, oblong-cylindric, distant, 1-leaved. Leaf to 20 × 9 cm, broadly elliptic to ovate; apex shortly pointed; base rounded or shortly drawn out. Flowering stem 18–28 cm, erect, with 4 or 5 sheath-like scales *c.* 2 cm long; flowers 4–7, in umbels. Sepals to 11 × 1.5 cm, free, ovate, tapered to a long drawn out, often spirally curved point, greenish with reddish spots. Petals to 4 × 1 cm, narrowly triangular, somewhat sickle-shaped, pale yellow, tapered to a long drawn out, often twisted, reddish point. Lip to 1.5 cm, triangular, strongly recurved, pale yellow with reddish apical markings. *Indonesia (Moluccas)*. G2. Winter.

4. B. lobbii Lindley. Illustration: Botanical Magazine, 4532 (1850); Schlechter, Die Orchideen, edn 2, 325 (1927); Shuttleworth et al., orchids, 88 (1970); Bechtel et al., Manual of cultivated orchid species, 174 (1981).
Pseudobulbs 3–5 × 2–3 cm, broadly ovoid *c.* 8 cm apart, 1-leaved. Leaf to 35 × 9 cm, elliptic to broadly elliptic. Flowering stem 10 cm or more, 1-flowered; flowers to 7.5 cm wide, pale to brownish yellow, flushed with red or brown. Upper sepal to 5 × 1.7 cm, broadly lanceolate, acuminate; veins crimson-dotted; lateral sepals a little shorter and broader, downwardly curved, with purplish veins, apical half brownish. Petals to 3.5 × 1 cm, lanceolate, horizontal, with *c.* 9 purplish veins. Lip *c.* 1.3 × 1 cm, narrowly ovate, curved, basal margins upcurved, purplish with a bright orange circular callus. *NE India, through SE Asia to the Philippines*. G2. Flowers intermittently.

***B. leopardinum** (Wallich) Lindley. Illustration: Botanical Magazinc, 9631 (1941); Dansk Botanisk Arkiv **33**: 29 (1979); Bechtel et al., Manual of cultivated orchid species, 174 (1981). Pseudobulbs clustered. Flowering stem 1–2 cm, with 1–3 flowers; flowers 2–3 cm wide, pale yellowish green, tinged or spotted with pink; petals about two-thirds of the sepal length; lip dark crimson, paler and spotted towards margins. *E Himalaya to Burma & N Thailand*. G2. Spring– summer.

76. CIRRHOPETALUM Lindley
J.C.M. Alexander & P.J.B. Woods
Epiphytic. Pseudobulbs simple, distant or

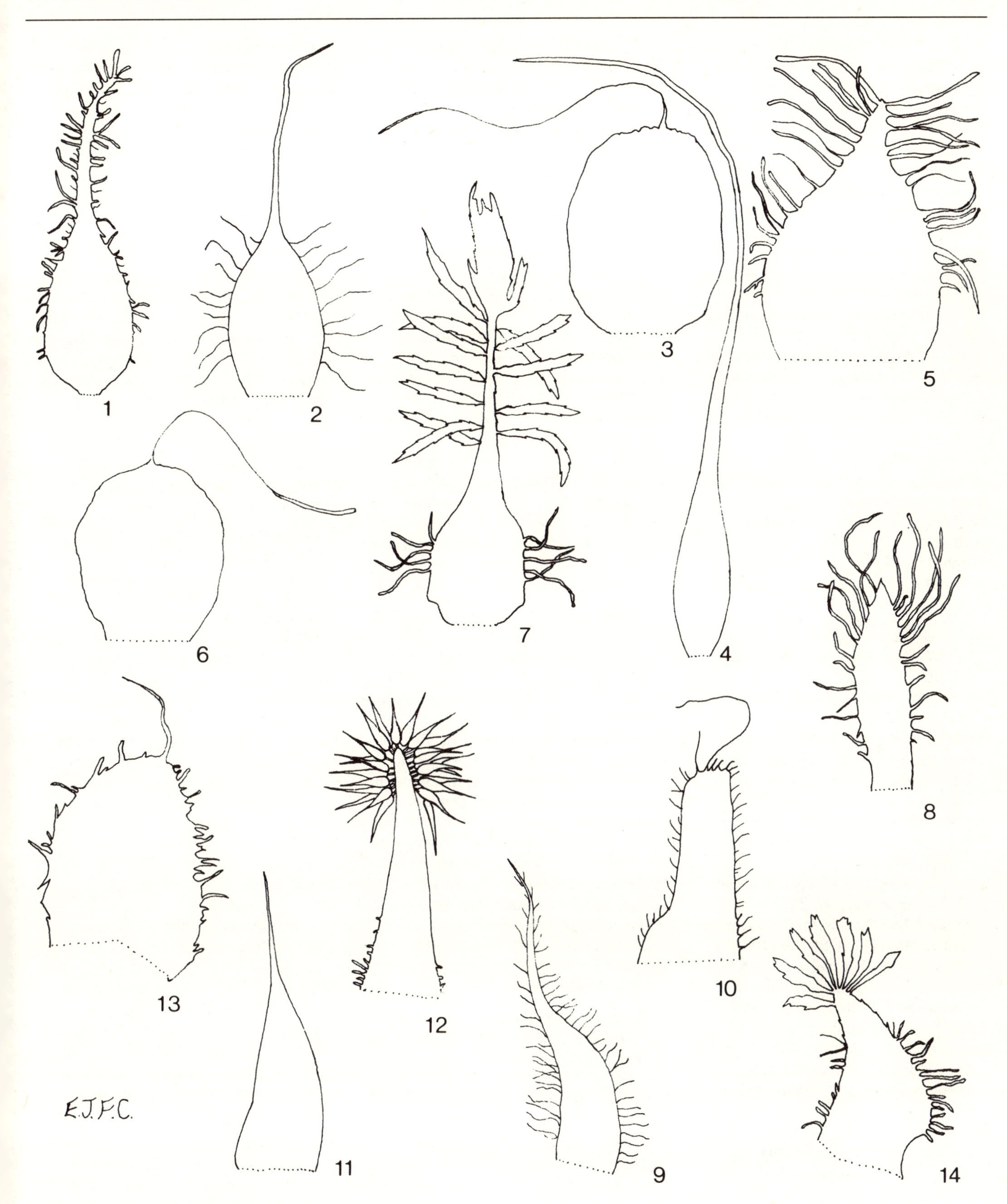

Figure 16. Upper sepals and petals of *Cirrhopetalum* species. 1–7, upper sepals; 8–14, petals. 1 & 8, *C. putidum*. 2 & 9, *C. gracillimum*. 3 & 10, *C. umbellatum*. 4 & 11, *C. medusae*. 5 & 12, *C. ornatissimum*. 6 & 13, *C. picturatum*. 7 & 14, *wendlandianum*. (After Seidenfaden, *Dansk Botanisk Arkiv* **33**(3); not to scale.)

rarely clustered, 1-leaved (in our species). Leaves shortly stalked or stalkless, folded when young. Flowering stems erect or arching, basal, with a few evenly spaced scales. Flowers 1–many in umbels, often forming a regular arc or circle. Lateral sepals much longer than upper sepal, usually lying side by side, often loosely united. Sepals and petals often with marginal hairs or other appendages. Lip small, arched, narrow and fleshy, loosely hinged to column foot. Column short, with a long, narrow foot. Pollinia 4.

A genus of about 40 species from SE Asia and India with a few representatives in Africa and the Pacific islands. Several wild species are morphologically intermediate between *Cirrhopetalum* and *Bulbophyllum* and the 2 genera are often united under the latter name. However, the characteristic *Cirrhopetalum* flower shape and arrangement is evident in most of the cultivated species. Cultivation conditions are similar to those for *Bulbophyllum* (p. 219).

Literature: Seidenfaden, G., Orchid genera in Thailand VIII: Bulbophyllum, *Dansk Botanisk Arkiv* **33**(3), (1979).

Pseudobulbs. 3 cm or more and angled: **1,2,5,7**. Spherical in outline, 4-lobed: **4,5**. Brownish: **1,7**. Clustered: **7**.
Leaves. 7 cm or less: **2,4–6**. 4 cm wide or more: **1,2,5**.
Flowering stem. More than 30 cm: **2,3,7**. Pendent: **3,4**. Arched above: **7**. With large scales: **1,4**. With true leaf: **4**. Slender: **3**.
Flowers. Solitary: **5**; more than 10: **1,6,7**; many: 1.5 cm or less: **2,3,5–7**; 15 cm or more: **1,4,5**.
Bracts. Reaching base of sepals: **1,3**; not reaching base of sepals: **2–7**.
Upper sepal. Less than 1 cm: **2,3,6,7**; more than 2 cm; **1**. Finely tapered: **1,3–5**; abruptly tapered or with an apical hair or bristle: **2,6,7**. Almost circular: **2**; spathulate: **7**. With marginal hairs: **3–7**; with flat appendages: **4**.
Lateral sepals. 5 cm or less: **2–7**; 10 cm or more: **1,3–5**. Finely tapered: **1,3–5**; oblong or linear: **2,6,7**. Warty near base: **5**.
Petals. Finely tapered: **1,3**; oblong or linear: **2,5**. Acute: **5,7**. With apical hair or bristle: **2,6**. With marginal hairs: **2–7**; with concentration or tuft of hairs at tip: **5**; with flat appendages: **4**; with irregular margin: **3**.
Lip. With fleshy auricles: **2,4**. 2-ridged: **5**.
Column. With 2 long projections: **1,2**.

1a. Margin of upper sepal entire (apical bristle may be present) 2
b. Margin of upper sepal irregular, toothed or ciliate 3
2a. Petal margins entire, not hairy **1. medusae**
b. Petal with irregular or hairy margins **2. umbellatum***
3a. Lateral sepals with drawn out, tapered or thread-like tips 4
b. Lateral sepals not drawn out, if long then parallel-sided 6
4a. Petal margins evenly fringed or hairy **3. gracillimum***
b. Petals with small, flat, marginal appendages, or with hairs concentrated into an apical tuft, or denser near apex 5
5a. Petals with 15 or fewer small, flat marginal appendages **4. wendlandianum**
b. Petals with an apical tuft of many hairs **5. ornatissimum***
6a. Lateral sepals about 10 times longer than broad **6. makoyanum**
b. Lateral sepals less than 6 times longer than broad **7. auratum***

1. C. medusae Lindley. Figure 16(4 & 11), p. 221. Illustration: Botanical Magazine, 4977 (1857); Shuttleworth et al., Orchids, 88 (1970); Skelsey, Orchids, 101 (1979).
Pseudobulbs 2–5 × 1–2 cm, ovoid, angled, brownish, *c.* 3 cm apart. Leaf 10–20 × 2–4.5 cm, oblong to elliptic, rounded, sometimes notched. Flowering stem 10–20 cm, erect, with large sheathing scales, bearing many small flowers in a dense umbel; bracts *c.* 2 cm; flowers 7.5–17 cm long, white to pale yellow with pink spots. Upper sepal 2.5–3 cm, linear, finely tapered; lateral sepals to 14 cm, narrow and drawn out, acute. Petals *c.* 5 mm, triangular, tapered, acute. Lip 1–2 mm. Column with 2 long apical projections. *Malaysia, Indonesia, Borneo; ?Philippines.* G2. Winter.

2. C. umbellatum (Forster) Hooker & Arnott (*Bulbophyllum longiflorum* Thouars not Ridley; *C. thouarsii* Lindley; not *B. umbellatum* Lindley). Figure 16(3 & 10), p. 221. Illustration: Botanical Magazine, 4237 (1846); Stewart & Campbell, Orchids of tropical Africa, 69 (1970); Millar, Orchids of Papua New Guinea, 50 (1978).
Pseudobulbs 2.5–4.5 × 1–2 cm, conical to ovoid, angled, 2–4 cm apart. Leaves 8–18 × 3–4 cm, elliptic to oblong, rounded, notched. Flowering stem to 40 cm, erect or arched, with 3–7 flowers; bracts not reaching base of sepals; flowers 4–5 cm long, pale pinkish yellow. Upper sepal 7–10 mm, broadly ovate to circular, concave, purple-tipped, with a long apical bristle; lateral sepals 2–4 cm × 4–7 mm, oblong. Petals 4–10 mm, narrowly triangular, purple-tipped, with an apical bristle; margins hairy. Lip *c.* 7 mm, arched, purple. Column with 2 long apical projections. *E Africa, Indian Ocean Islands, SE Asia, N Australia & Pacific islands.* G2. Summer.

If *Cirrhopetalum* is regarded as part of *Bulbophyllum*, this species must be called *B. longiflorum* Thouars and not *B. umbellatum*.

***C. picturatum** Loddiges. Figure 16(6 & 13), p. 221. Illustration: Botanical Magazine, 6802 (1885); Orchid Review **76**: 319 (1968); American Orchid Society Bulletin **42**(12): back cover (1973). Petals broadly triangular. Lip with distinct fleshy auricles. Projections on column short, broad, 2-toothed. *NE India, Burma, Thailand & Vietnam.* G2. Spring–summer.

3. C. gracillimum Rolfe (*Bulbophyllum psittacoides* (Ridley) Smith). Figure 16(2 & 9), p. 221. Illustration: Dansk Botanisk Arkiv **29**: 36 (1973); Bechtel et al., Manual of cultivated orchid species, 183 (1981).
Pseudobulbs 1–2 cm × 5 mm, ovoid, *c.* 2 cm apart. Leaf to 14 × 2 cm, oblong, rounded. Flowering stem to 35 cm, erect, very slender, bearing 8–10 flowers; bracts very small; flowers 2–2.5 cm long, purplish red. Upper sepal to 1 cm, ovate, concave, hairy on margin, finely tapered; lateral sepals variable, 2 cm or more, finely tapered; petals 5–10 mm, narrowly triangular, finely tapered, hairy on margin. Lip 1.5–2.5 mm, pinkish purple with a paler tip. *Malay peninsula, Indonesia, New Guinea, Solomon & Fiji Islands.* G2. Summer–autumn.

***C. longissimum** Ridley. Illustration: Botanical Magazine, 8366 (1911); Australian Orchid Review **33**: 138 (1968); Bechtel et al., Manual of cultivated orchid species, 184 (1981). Pseudobulbs 3–7 cm apart. Flowering stem arched to pendent, bearing 3–7 flowers; bracts 1–2 cm. Lateral sepals 20–30 cm, finely tapered, pale pink with darker stripes. *Malay peninsula.* G2. Winter.

4. C. wendlandianum Kränzlin (*C. collettii* Hemsley). Figure 16(7 & 14), p. 221. Illustration: Botanical Magazine, 7198 (1891); Sheehan & Sheehan, Orchid genera illustrated, 55 (1979); Northen, Miniature orchids, t. 16 (1980).
Pseudobulbs *c.* 2 × 2 cm, spherical but shallowly 4-lobed, 1.5–3 cm apart. Leaf

4–12 × 2–3 cm, elliptic, obtuse, sometimes notched. Flowering stem 10–15 cm, arched to pendent, bearing *c.* 5 flowers; base sheathed, 1-leaved, later developing into a pseudobulb. Flowers 12–15 cm long; bracts short. Upper sepal 1–2 cm, triangular, finely tapered, yellow with several small, flat, marginal appendages near the tip; margins hairy in lower half. Lateral sepals 8–12 cm, gradually tapered from the base, pink with darker red veins. Petals to 1 cm, triangular; colour, marginal hairs and appendages as in upper sepal. Lip *c.* 5 mm with swollen, bristly, pinkish purple auricles. *Burma & Thailand.* G2. Spring–summer.

5. C. ornatissimum Reichenbach. Figure 16(5 & 12), p. 221. Illustration: Botanical Magazine, 7229 (1892); Dansk Botanisk Arkiv **29**: 28 (1973); Bechtel et al., Manual of cultivated orchid species, 184 (1981). Pseudobulbs 2–5 × 1–2 cm, ovoid, angled, 2–5 cm apart. Leaf 7–16 × 3–6 cm, oblong to elliptic, obtuse or rounded. Flowering stem 10–20 cm, erect, bearing 3–5 flowers; flowers 4–9 cm long; bracts short. Upper sepal 1–1.5 cm, ovate, tapered, hairy on margins, yellowish with purple veins; lateral sepals 5–8 cm, linear-lanceolate, tapered, pinkish yellow with purple veins and spots. Petals 1–1.5 cm, lanceolate to narrowly triangular, with a dense apical tuft of bulbous-based hairs, yellow with purple veins. Lip to 1 cm, 2-ridged, rounded, bristly, purple. *E Himalaya & NE India.* G2. Autumn–winter.

*__C. putidum__ Teijsmann & Binnendijk (*C. fascinator* Rolfe). Figure 16(1 & 8), p. 221. Illustration: Botanical Magazine, 8199 (1908); Kupper & Linsenmaier, Orchids, 55 (1961); American Orchid Society Bulletin **43**: 886, 888 (1974). Pseudobulbs almost spherical, obscurely 4-lobed, smooth. Leaves *c.* 5 × 1.5–3 cm, elliptic. Flowering stem *c.* 10 cm, bearing 1–3 flowers; flowers to 20 cm long, green and purple. Lateral sepals to 18 cm, finely tapered, warty near base. Lip tapered, acute. *E Himalaya, NE India, Laos, Malaysia, Indonesia & Philippines.* G2. Winter.

6. C. makoyanum Reichenbach. Illustration: Botanical Magazine, 7259 (1892); Shuttleworth et al., Orchids, 88 (1970); Graf, Tropica, 715 (1978). Pseudobulbs 1–2 cm × 7–10 mm, broadly ovoid, 1–4 cm apart. Leaf 5–10 × 1.5–3 cm, oblong to elliptic, rounded, notched. Flowering stem 10–25 cm, erect, bearing 10–15 flowers; flowers 3–4 cm long, mostly pale yellow with red spots. Upper sepal *c.* 5 mm, triangular, acuminate, fringed with yellow hairs; lateral sepals *c.* 3.5 cm, linear, parallel-sided. Petals similar to upper sepal but narrower. Lip *c.* 2.5 mm, triangular. *Singapore, Borneo & Philippines.* G2. Winter.

7. C. auratum Lindley (*Bulbophyllum campanulatum* Rolfe). Illustration: Botanical Magazine, 8281 (1909); Dansk Botanisk Arkiv **29**: 98 (1973).
Pseudobulbs 2–3 × 1–1.5 cm, fusiform to ovoid, angled, greenish brown, 1–2 cm apart. Leaf 10–15 × 1.5–2.5 cm, oblong, obtuse, notched. Flowering stem 10–15 cm, erect, arched above, bearing 8–12 flowers; flowers 2–2.5 cm long, mostly pale yellow with reddish veins. Upper sepal 6–7 mm, ovate-acuminate, with an apical hair, margins hairy. Lateral sepals 1.5–2 cm, oblong, sickle-shaped. Petals *c.* 5 mm, triangular, margins hairy. Lip triangular, acute. *Malaysia, Sumatra & Borneo.* G2. Winter.

*__C. mastersianum__ Rolfe. Illustration: Botanical Magazine, 8531 (1913); Dansk Botanisk Arkiv **29**: 84 (1973); Bechtel et al., Manual of cultivated orchid species, 184 (1981). Pseudobulbs ovoid, clustered. Flowering stem 12–30 cm, bearing 6–8 flowers; flowers 4–5 cm long, yellow to orange. Lateral sepals 3–4 × *c.* 1 cm. *Indonesia (Moluccas) & Borneo.* G2. Winter.

77. MALAXIS Swartz

P.J.B. Woods

Perennial herbs, growing in the soil, on rocks, or as epiphytes. Stems fleshy, creeping, the tip leafy and erect, or crowded, or cylindric, or forming pseudobulbs. Leaves few to many, usually deciduous on older stems, broad, often coloured, usually pleated, often unequal at the sheathing base. Inflorescence terminal, erect; scape usually angled or shallowly winged, bracts deflexed. Flowers numerous, small, lip above the column (flowers non-resupinate). Sepals broader than petals, spreading. Petals linear, usually reflexed. Lip flat, entire, usually rounded or ovate, often with a hollow nectary at the base which is auriculate and clasps the column; apical margins often toothed or lobed. Column short, winged. Pollinia 4, in 2 pairs, waxy.

A genus of over 200 species, widely distributed in temperate and tropical regions. Cultivation as for tropical, terrestrial species, with substitution of good leaf-mould (beech or oak) for crushed bark or peat.

1a. Inflorescence dense, the flowers crowded, their stalks less than 8 mm **2. discolor***
b. Inflorescence loose, flowers widely spaced, their stalks 8 mm or more **1. metallica**

1. M. metallica (Reichenbach) Kuntze. Illustration: Botanical Magazine, 6668 (1883); La Belgique Horticole **34**: plate opposite p. 281 (1884).
Stems indistinct, cylindric, leaves up to 6, to 7.5 × 3.8 cm, elliptic to ovate, acute, dark metallic red-purple, with 3–5 veins, margins wavy. Flowers *c.* 1.5 cm in diameter, purplish; stalk and ovary paler, lip pale, apical margin irregularly toothed, auricles triangular. *Borneo.* G2. Late spring.

2. M. discolor (Lindley) Kuntze. Illustration: Botanical Magazine, 5403 (1863); Jayaweera, Flora of Ceylon **2**: 41 (1981). Stems clustered, to *c.* 8 × 1 cm, oblong, leafless after flowering. Leaves 3–6, to 12 × 4 cm, ovate, acuminate, rich purple, pleated, main veins 5–7, margins wavy, sometimes green. Flowers *c.* 3 mm in diameter, at first yellow, changing to orange, then purplish. Lip crescent- or kidney-shaped, margins not toothed, auricles shallow. *Sri Lanka.* G2. Summer.

*__M. calophylla__ (Reichenbach) Kuntze (*Microstylis scottii* Ridley). Illustration: Botanical Magazine, 7268 (1892); Annals of the Royal Botanic Garden Calcutta **8**: t. 20 (1898). Leaves bronze-green, margins paler, with spots, the spots sometimes distributed to the middle of the blade; flowers pink and cream, lip cleft to form 2 teeth, auricles long, lanceolate. *Himalaya to Malaysia and Borneo.* G2. Autumn.

78. LIPARIS Richard

J. Lamond & P.J.B. Woods

Perennial herbs, terrestrial, growing on rocks or epiphytic. Rhizome short. Stems fleshy, thicker at base, sometimes pseudobulbous. Leaves 1–7, membranous or leathery, jointed at the base or not, rolled or folded when young. Inflorescence racemose with few to many flowers. Sepals and petals oblong, reflexed, the petals usually narrower and a little shorter than the sepals. Lip broadly oblong to rounded, sharply bent backwards at or about the middle. Column long and slender, slightly winged. Pollinia 4, in 2 pairs, waxy.

A genus of over 200 species, distributed mostly in the tropics of the Old World. A few species also occur in temperate zones of the northern hemisphere. Many have little to recommend them as garden plants as their flowers are dull greenish or yellowish, but a few, not widespread in cultivation, are particularly attractive. *L. nervosa* should be grown in a compost recommended for terrestrial species whereas *L. viridiflora* should be grown in an epiphytic compost. *L. lilifolia* and *L. loeselii* are probably best grown under alpine-house or cold-frame conditions in a terrestrial compost containing one-third live Sphagnum moss. Literature: Seidenfaden, G., Orchid genera in Thailand IV, *Dansk Botanisk Arkiv* **31**: 1–105 (1976).

1a. Base of leaf unjointed, sheathing the stem; leaves membranous, broadly lanceolate to elliptic or ovate **1. nervosa***
b. Base of leaf jointed, not sheathing the stem; leaves leathery, oblong, lanceolate or oblanceolate **2. viridiflora**

1. L. nervosa (Thunberg) Lindley (*L. elata* Lindley). Illustration: Edwards's Botanical Register **14**: t. 1175 (1828); Maekawa, The wild orchids of Japan in colour, 325 (1971); Luer, Native orchids of Florida, 171 (1972); Dansk Botanisk Arkiv **31**: 33 (1976).
Pseudobulbs 3–7 × *c.* 3 cm, conical. Leaves 2–7, to 25 cm × 12 mm, elliptic to ovate or broadly lanceolate, pleated, acute or acuminate. Inflorescence 15–50 cm, loosely flowered, scape with 5 wings or grooves. Flowers 10–30, purplish. Sepals and petals to 8 × 2–5 mm, petals narrower than sepals. Lip to 8 × 7 mm, broadly oblong to obovate, apex notched, base with 2 short protruberances. *Tropics of America, Africa and Asia*. G2. Summer.

***L. lilifolia** (Linnaeus) Lindley. Illustration: American Orchid Society Bulletin **43**: 199 (1974), **47**: 399 (1978); Luer, Native orchids of the United States and Canada, 309 (1975). Plant to 25 cm. Leaves 2, deciduous. Lip purplish, *c.* 10 × 8 mm. *Eastern N America*. H1–2. Spring–summer.

***L. loeselii** (Linnaeus) Richard. Illustration: Ross-Craig, Drawings of British plants, part 28: t. 52 (1971); Danesch & Danesch, Orchideen Europas Mitteleuropa, edn 3, 207, 211 (1972); Mossberg & Nilsson, Orchids of northern Europe, 63 (1979). Plant to 25 cm. Leaves 2, deciduous. Lip yellowish green, *c.* 5 × 3 mm. *N & C Europe, NE USA*. H1. Spring–summer.

2. L. viridiflora Lindley (*L. longipes* Lindley). Illustration: Annals of the Royal Botanic Garden Calcutta **8**: t. 37 (1898); Dansk Botanisk Arkiv **31**: 85 (1976); Lin, Native orchids of Taiwan **2**: 241 (1977). Pseudobulbs 3–9 × 1–1.5 cm, cylindric, swollen towards the base. Leaves 2, to 28 × 2.5–3.5 cm, oblanceolate, apex shortly pointed. Inflorescence to 27 cm, densely flowered; scape flattened or narrowly winged. Flowers numerous, greenish, lip yellowish orange. Sepals and petals 2–3 × *c.* 1 mm, petals narrower than sepals. Lip to 3 × 2 mm, oblong, apex obtuse. *India, Sri Lanka & SE Asia*. G2. Winter.

79. CALYPSO Salisbury
V.A. Matthews
Terrestrial, 5–20 cm. Rootstock a corm, producing a solitary leaf which dies down in winter. Leaf 3–10 cm, broadly ovate, pleated, rolled when young. Scape bearing 1 terminal flower. Sepals and petals 1.5–2 cm, lanceolate, spreading and ascending, often twisted, purplish pink. Lip 1.5–2.5 cm, inflated and saccate, with 2 small horns at the base, whitish to pale pink, marked with purple and with white or yellow hairs towards the base. Pollinia waxy, flat, in 2 pairs fixed to a detachable viscidium.

A genus with only 1 species, uncommon in cultivation. It succeeds best in a damp, semi-shaded position with plenty of leaf-mould.

1. C. bulbosa (Linnaeus) Oakes. Illustration: Luer, Native orchids of the United States and Canada, t. 95, 96 (1975); Williams et al., Field guide to the orchids of Britain and Europe, 134 (1978).
N temperate regions. H2. Late spring–summer.

80. APLECTRUM (Nuttall) Torrey
V.A. Matthews
Terrestrial, to 55 cm. Rootstock a corm, producing a solitary leaf which dies down in winter. Leaf 10–20 cm, broadly elliptic, pleated, with silvery veins, rolled when young. Scape with 6–15 flowers. Sepals and petals 1–1.4 cm, oblanceolate, brown or yellowish brown, tinged with purple or brown. Lip 1–1.2 cm, 3-lobed, obovate, whitish with purple marks. Pollinia waxy, in 2 pairs, fixed to a detachable viscidium.

A genus with only 1 species which is rather difficult in cultivation. It grows best in a shady damp situation.

1. A. hyemale (Willdenow) Nuttall (*A. spicatum* (Walter) Britton et al.). Illustration: Luer, Native orchids of the United States and Canada, t. 88 (1975).
NE USA, SE Canada. H2. Late spring–early summer.

81. WARREA Lindley
V.A. Matthews
Terrestrial. Pseudobulbs ovoid. Leaves few, in 2 ranks, usually lanceolate, pleated, rolled when young. Raceme loose, with few to several flowers, produced from the base of an immature pseudobulb. Sepals and petals spreading or curved inwards, the lateral sepals joined to the column foot to form a mentum. Petals similar to the upper sepal but smaller. Lip simple, usually with erect sides, disc bearing central ridges. Column stout, somewhat swollen above. Pollinia 4 in 2 pairs on a short, triangular stipe.

A genus of 3 species from C and northern S America. They grow best in a compost like that used for *Phaius* (p. 176). Propagation by division.

1a. Sepals and petals white or pale yellow **1. warreana**
b. Sepals and petals reddish brown **2. costaricensis**

1. W. warreana (Lindley) Schweinfurth (*W. tricolor* Lindley; *W. bidentata* Lindley). Illustration: Lasser, Flora de Venezuela **15**(4): 31 (1970); Pabst & Dungs, Orchidaceae Brasilienses **2**: 230 (1977). Plant to 1 m. Pseudobulbs 4–12 cm with 3–5 flowers. Leaves 30–70 cm, oblong-lanceolate. Raceme erect, usually longer than the leaves, with 6–15 flowers. Flowers somewhat hanging, white or pale yellow. Sepals and petals nearly circular, upper sepal to 3.5 cm. Lip 2–3.5 cm, white or reddish, marked with yellow and reddish purple, wavy at the edges and often with an apical notch. Disc with 3 ridges. *Northwest S America*. G2. Summer.

2. W. costaricensis Schlechter. Plant to 75 cm. Leaves to 70 cm, lanceolate. Raceme erect, about as long as leaves. Flowers reddish brown. Sepals and petals oblong-ovate, *c.* 3 cm. Lip to 3 cm, nearly circular, entire, paler than sepals and petals, with reddish brown markings, disc narrow. *Costa Rica, Panama*. G2.

82. XYLOBIUM Lindley
V.A. Matthews
Epiphytic or rarely terrestrial. Pseudobulbs ovoid or conical, more rarely slender and cylindric. Leaves 1–3 per pseudobulb, stalked, lanceolate to elliptic, pleated, with 3–6 conspicuous veins beneath, rolled

when young. Raceme produced from the base of a mature pseudobulb. Bracts linear, usually exceeding the ovary. Sepals and petals spreading, the lateral sepals broader than the upper and joined to the column foot to form a mentum. Petals similar to the upper sepal but smaller. Lip curved upwards at the base, usually 3-lobed at the apex, the lateral lobes often standing more or less erect, with a central, usually elongate callus. Column cylindric. Pollinia 4 in 2 pairs, the pollinia of each pair unequal, borne on a short, triangular stipe.

A genus of 33 species from C and tropical S America and the West Indies. They should be grown in an epiphytic compost, and should be watered from the time that new leaves appear until November, then only occasionally. They should be freely ventilated during the summer.

Pseudobulbs. Ovoid, conical or shortly cylindric, to 10 cm: **2–6,8–10**; cylindric and elongate, 15–27 cm: **1,7**.
Sepals and petals. Spotted or with other markings: **2,3,6,7**; lacking spots or other obvious markings: **1,4–6,8–10**.
Lip. With warts or papillae: **1,2,4–7**; without warts or papillae: **3,8–10**.

1a. Pseudobulbs with 1 leaf 2
b. Pseudobulbs with 2–3 leaves 6
2a. Pseudobulbs 15–20 cm, cylindric, elongate **1. pallidiflorum**
b. Pseudobulbs to 10 cm, ovoid to conical or shortly cylindric 3
3a. Flowers with spots 4
b. Flowers without spots 5
4a. Leaves with 3 conspicuous veins beneath; raceme hanging **3. colleyi**
b. Leaves with 5–6 conspicuous veins beneath; raceme more or less upright **2. leontoglossum**
5a. Raceme hanging **4. palmifolium**
b. Raceme upright **5. bractescens**
6a. Upper sepal 1–1.5 cm 7
b. Upper sepal 2–2.5 cm 9
7a. Racemes with 6–9 flowers **10. powellii**
b. Racemes with 10–22 flowers 8
8a. Lip creamy brown with pale brown veins **9. hyacinthinum**
b. Lip white, sometimes with pinkish stripes **8. foveatum**
9a. Pseudobulbs 12–27 cm, cylindric, elongate **7. elongatum**
b. Pseudobulbs to 9 cm, ovoid to shortly cylindric 10
10a. Leaves with 3 conspicuous veins beneath **3. colleyi**
b. Leaves with 5 conspicuous veins beneath **6. variegatum**

1. X. pallidiflorum (Hooker) Nicholson. Illustration: Dunsterville & Garay, Venezuelan orchids illustrated **1**: 439 (1959); American Orchid Society Bulletin **43**: 208 (1974).
Epiphytic or rarely terrestrial. Pseudobulbs 15–20 cm with 1 leaf. Leaves to 40 cm. Raceme upright with 3–9 flowers. Flowers white, pale green or yellowish green. Upper sepal *c.* 1.5 cm. Lip *c.* 5 cm, whitish grading to yellowish orange at the base, 3-lobed, the central lobe covered with white papillae or warts. *Northwest S America, West Indies.* G2.

2. X. leontoglossum (Reichenbach) Rolfe. Illustration: Botanical Magazine, 7085 (1889); American Orchid Society Bulletin **43**: 212 (1974); Dunsterville & Garay, Venezuelan orchids illustrated **6**: 443 (1976); Bechtel et al., Manual of cultivated orchid species, 280 (1981).
Epiphytic. Pseudobulbs 3.5–10 cm, ovoid to conical, with 1 leaf. Leaves to 35 cm, with 5–6 veins beneath. Flowers pale to bright yellow with reddish brown or pink spots. Upper sepal 1.8–2 cm. Lip 3-lobed, central lobe usually acute, densely covered with brown warts. *Northwest S America.* G2.

3. X. colleyi (Lindley) Rolfe (*X. brachystachyum* Kränzlin). Illustration: Dunsterville & Garay, Venezuelan orchids illustrated **3**: 331 (1965); Pabst & Dungs, Orchidaceae Brasilienses **2**: 22, 279 (1977).
Epiphytic. Pseudobulbs to 4 cm, ovoid, with 1 or occasionally 2 leaves. Leaves to 50 cm, with 3 veins beneath. Raceme hanging, with 3–6 flowers. Flowers brown or reddish brown, spotted with purple and smelling of cucumber. Upper sepal 2–2.5 cm. Lip *c.* 2 cm, more or less oblong, smooth, dark purple to almost black, shining. *Venezuela, Guyana, Brazil & Trinidad.* G2.

4. X. palmifolium (Swartz) Fawcett (*X. decolor* (Lindley) Nicholson). Illustration: Botanical Magazine, 3981 (1843); Fawcett & Rendle, Flora of Jamaica **1**: t. 23 (1910).
Epiphytic. Pseudobulbs to 7.5 cm, narrowly ovoid, with 1 leaf. Leaves to 42 cm. Raceme somewhat hanging, loose, with few to many flowers. Flowers white or yellowish, scented (often unpleasantly so). Upper sepal *c.* 2 cm. Lip *c.* 1.5 cm, 3-lobed with small lateral lobes, the central lobe rounded or notched, warty, with a callus with 4–5 thickened ridges. *Cuba, Dominican Republic, Trinidad & Jamaica.* G2.

5. X. bractescens (Lindley) Kränzlin. Pseudobulbs *c.* 3 cm, conical, with 1 leaf. Leaves to 27 cm, 3-veined beneath. Raceme flexuous and arching, loose with 5–7 flowers. Flowers dull yellow. Upper sepal *c.* 2 cm. Lip 3-lobed, reddish brown, recurved at the obtuse apex, bearing lines of warts, with the callus apically 3-lobed. *Ecuador, Peru, Brazil.* G2.

6. X. variegatum (Ruiz & Pavon) Garay & Dunsterville (*Maxillaria squalens* Lindley: *X. squalens* (Lindley) Lindley; *X. scabrilingue* Schlechter). Illustration: Botanical Magazine, 2955 (1829); Dunsterville & Garay, Venezuelan orchids illustrated **2**: 343 (1961); Bechtel et al., Manual of cultivated orchid species, 280 (1981).
Epiphytic or terrestrial. Pseudobulbs to 9 cm, ovoid to shortly cylindric, with 2–3 leaves. Leaves to 70 cm, 5-veined beneath. Raceme with 7–many flowers. Flowers white, pale yellow, pale pink or brownish, often marked or tinged with purple. Upper sepal *c.* 2.5 cm. Lip 1–2 cm, 3-lobed, the central lobe with rows of brown or purple papillae. *South C America and northwest S America.* G2.

7. X. elongatum (Lindley) Hemsley. Illustration: American Orchid Society Bulletin **43**: 208 (1974).
Epiphytic or terrestrial. Pseudobulbs 15–27 cm, cylindric, elongate, with 2 leaves. Leaves to 40 cm, with 3–5 veins beneath. Raceme arching or drooping, with 5–20 flowers. Flowers whitish, yellowish or pinkish with red, brown or violet markings. Upper sepal 2–2.5 cm. Lip *c.* 2 cm, usually purplish brown, 3-lobed, central lobe covered with dense papillae. *C America.* G2.

8. X. foveatum (Lindley) Nicholson (*X. stachyobiorum* (Reichenbach) Hemsley). Illustration: American Orchid Society Bulletin **43**: 209 (1974); Dunsterville & Garay, Venezuelan orchids illustrated **6**: 441 (1976).
Epiphytic. Pseudobulbs to 10 cm, more or less ovoid, with 2–3 leaves. Leaves to 40 cm, with 3 veins beneath. Raceme upright to arching, with 20–24 densely packed flowers. Flowers yellowish or sometimes white, fragrant. Upper sepal 1–1.5 cm, lateral sepals each with a dorsal keel projecting at the apex. Lip to 2 cm, 3-lobed, white with pinkish stripes in the basal part. *C America, northern S America, Jamaica.* G2.

9. X. hyacinthinum (Reichenbach) Schlechter. Illustration: Dunsterville & Garay, Venezuelan orchids illustrated **1**:

437 (1959); Lasser, Flora de Venezuela **15**(4): 333 (1970).
Epiphytic. Pseudobulbs to 6 cm, more or less ovoid, with 2–3 leaves. Leaves 25–50 cm. Raceme upright with 10–20 flowers. Flowers greenish cream, smelling of hyacinth. Upper sepal *c.* 1.3 cm, lateral sepals each with a dorsal keel projecting at the apex. Lip *c.* 1 cm, creamy brown with pale brown veins. *Venezuela.* G2.

10. X. powellii Schlechter. Illustration: Hawkes, Encyclopaedia of cultivated orchids, t. XIIB (1965).
Epiphytic. Pseudobulbs 4–5 cm, shortly cylindric, with 2 leaves. Leaves to 20 cm. Racemes upright with 6–9 flowers. Flowers yellow or reddish brown, sometimes tinged with pale green, faintly scented. Upper sepal 1–1.5 cm. Lip *c.* 1.5 cm, 3-lobed, central lobe with 3 ridges, reddish brown. *Costa Rica, Panama.* G2.

83. BIFRENARIA Lindley
V.A. Matthews
Epiphytic. Pseudobulbs usually ovoid, often 4-angled. Leaves 1 per pseudobulb, stalked, rather leathery, linear-lanceolate to oblong, rolled when young. Racemes produced from the bases of the pseudobulbs. Sepals spreading, the laterals fused to the column foot, forming a short mentum. Petals similar to sepals or smaller, spreading or somewhat incurved. Lip joined to column foot, usually 3-lobed with the lateral lobes standing more or less erect, and with a central, usually elongate callus. Column winged above. Pollinia 4 in 2 pairs, each pair on a short, separate stipe attached to a broad, oblong, folded viscidium.

A genus of 10 species from tropical S America, nearly all of which are in cultivation. It is closely related to *Maxillaria*, which differs in producing only 1 flower per inflorescence, and to *Lycaste*, which has larger, more persistent leaves. Cultivation as for *Lycaste* (p. 227).

Flower diameter (when opened out flat). 2–3 cm: **1–4**; 5–9 cm: **5–9**.
Sepals and petals. Basic colour orange or bright yellow: **1–3,5**; basic colour red or purple: **7,9**; basic colour white, pale yellow or green: **4–6,8**. Spotted: **1**.
Lip. Length 1.5 cm: **1,2,4**; length 1.8–3.2 cm; **3,8,9**; length 3.5–5 cm: **5–7**. Entire: **4,9**. Hairy: **5–7**.

1a. Flowers 2–3 cm in diameter when opened out flat — 2
b. Flowers 5–9 cm in diameter when opened out flat — 5
2a. Sepals and petals greyish cream shaded with pink — **4. racemosa**
b. Sepals and petals orange or bright yellow, spotted or not — 3
3a. Sepals and petals spotted — **1. aurantiaca**
b. Sepals and petals not spotted — 4
4a. Lip with a purple patch; sepals and petals ovate, *c.* 10 mm wide — **2. vitellina**
b. Lip lacking a purple patch; sepals and petals lanceolate, 5–7 mm wide — **3. aureo-fulva**
5a. Sepals and petals basically red or purple — 6
b. Sepals and petals basically white, greenish or pale yellowish — 7
6a. Flowers to 9 cm in diameter when opened out flat; lip purple, 3-lobed — **7. tyrianthina**
b. Flowers to 5 cm in diameter when opened out flat; lip whitish or pink, entire — **9. atropurpurea**
7a. Lip hairy — 8
b. Lip not hairy — **8. tetragona**
8a. Sepals and petals basically green or yellowish green — **6. inodora**
b. Sepals and petals basically whitish or pale yellowish — **5. harrisoniae**

1. B. aurantiaca Lindley (*Rudolfiella aurantiaca* (Lindley) Schlechter). Illustration: Botanical Magazine, 3597 (1837); Pabst & Dungs, Orchidaceae Brasilienses **2**: 225, 281 (1977); Bechtel et al., Manual of cultivated orchid species, 271 (1981).
Pseudobulbs more or less ovoid, 3–6 cm, somewhat 4-angled and often flattened, bright or yellowish green, often red-spotted. Leaves 10–25 cm, often red-spotted beneath. Racemes drooping or erect, with 4–18 flowers. Flowers *c.* 2.5 cm in diameter, orange or deep yellow with brown or purplish red spots. Lip 1–1.5 cm, 3-lobed, the central lobe broadly triangular, often notched at apex, callus bright yellow. *Colombia, Venezuela, Guyana, Brazil & Trinidad.* G2.

This species is sometimes treated as a member of *Rudolfiella* Hoehne, a genus of about 9 species, no other members of which are in cultivation.

2. B. vitellina (Lindley) Lindley (*Stenocoryne vitellina* (Lindley) Kränzlin). Illustration: Edwards's Botanical Register **25**: t. 12 (1839); Pabst & Dungs, Orchidaceae Brasilienses **2**: 225, 281 (1977).
Pseudobulbs ovoid, to 4 cm, bluntly angular. Leaves to 30 cm. Racemes drooping or erect, with 5–8 flowers. Flowers orange-yellow, *c.* 2.5 cm in diameter, sepals and petals ovate, *c.* 10 mm wide. Lip 1.1–1.5 cm, 3-lobed, central lobe with a purple patch. *Brazil.* G2. Summer.

3. B. aureo-fulva (Hooker) Lindley (*Stenocoryne aureo-fulva* (Hooker) Kränzlin). Illustration: Pabst & Dungs, Orchidaceae Brasilienses **2**: 225 (1977).
Pseudobulbs ovoid, often wrinkled. Leaves to 15 cm. Racemes with 5–7 flowers. Flowers orange, 2–3 cm in diameter, sepals and petals lanceolate, 5–7 mm wide. Lip *c.* 2 cm, 3-lobed, central lobe lanceolate. *Brazil.* G2. Autumn.

B. aureo-fulva and *B. vitellina* are sometimes included in *Stenocoryne* Lindley, a genus differing in the detailed structure of the pollinia.

4. B. racemosa (Hooker) Lindley. Illustration: Loddiges' Botanical Cabinet **14**: t. 1318 (1828); Pabst & Dungs, Orchidaceae Brasilienses **2**: 225, 281 (1977).
Pseudobulbs ovoid. Racemes erect or drooping, with 4–10 flowers. Flowers greyish cream, shaded with pink, 2–3 cm in diameter. Lip 1.1–1.5 cm, entire, whitish, speckled with pink. *Brazil.* G2.

5. B. harrisoniae (Hooker) Reichenbach. Illustration: Pabst & Dungs, Orchidaceae Brasilienses **2**: 224, 280 (1977); Sheehan & Sheehan, Orchid genera illustrated, 45 (1979); Bechtel et al., Manual of cultivated orchid species, 170 (1981).
Pseudobulbs broadly ovoid, 5–9 cm, weakly or strongly 4-angled. Leaves to 30 cm. Racemes usually erect with 1–3 flowers. Flowers fragrant, 6–7 cm in diameter. Sepals and petals whitish or pale yellowish, often tinged with red. Lip 4–5 cm, 3-lobed, purple to dark red, usually with darker veins, central lobe with a wavy margin, notched at apex, hairy; callus yellow, hairy. *Brazil.* G2. Spring–summer.

Several varieties have been described, differing in flower size and colour.

6. B. inodora Lindley (*B. furstenbergiana* Schlechter). Illustration: American Orchid Society Bulletin **41**: 981 (1972); Pabst & Dungs, Orchidaceae Brasilienses **2**: 280 (1977).
Differs from *B. harrisoniae* in that the pseudobulbs usually have a dark band at the apex, the flowers are a little larger, fragrant or not, and the sepals and petals are green or yellowish green, sometimes tinged with purple. The lip may be white, yellow or pale pink, often suffused with a

darker colour, and the callus is hairless. *Brazil.* G2. Early spring–summer.

7. B. tyrianthina (Loudon) Reichenbach. Illustration: Botanical Magazine, 7461 (1896); Pabst & Dungs, Orchidaceae Brasilienses **2**: 224, 280 (1977).
Differs from *B. harrisoniae* in being more robust, with flowers to 9 cm in diameter, purple. The lip is purple, whitish towards the base and densely hairy. *Brazil.* G2. Spring–summer.

8. B. tetragona (Lindley) Schlechter. Illustration: Edwards's Botanical Register **17**: t. 1428 (1831); Pabst & Dungs, Orchidaceae Brasilienses **2**: 224, 280 (1977).
Pseudobulbs to 9 cm, ovoid, 4-angled. Leaves to 45 cm. Racemes with 3–5 flowers. Flowers fragrant, *c.* 5 cm in diameter. Sepals and petals basically greenish yellow, streaked with brownish purple. Lip *c.* 3 cm, about as wide as long, 3-lobed, whitish, marked with purple towards the base, callus large and conspicuous. *Brazil.* G2. Early summer.

9. B. atropurpurea (Loddiges) Lindley. Illustration: Loddiges' Botanical Cabinet **19**: t. 1877 (1832); Pabst & Dungs, Orchidaceae Brasilienses **2**: 223, 280 (1977); Bechtel et al., Manual of cultivated orchid species, 170 (1981).
Pseudobulbs to 8 cm, ovoid, 4-angled. Leaves to 25 cm. Raceme usually borne horizontally, with 3–5 flowers. Flowers fragrant, *c.* 5 cm in diameter. Sepals and petals dark red flushed with yellow around the middle. Lip 1.8–2.2 cm, more or less entire, whitish or pink. *Brazil.* G2. Spring–summer.

84. LYCASTE Lindley

V.A. Matthews

Epiphytic or terrestrial. Pseudobulbs ovoid or ellipsoid, often laterally compressed, the base enclosed in papery or fibrous sheaths, of which the upper 1 or 2 are often leaf-bearing. Leaves usually 1–3 per pseudobulb, stalked, lanceolate to oblong-elliptic, pleated, rolled when young. Flowers solitary from the bases of the pseudobulbs, scapes shorter than the leaves, with several papery bracts. Lateral sepals broader than the upper, joined to the column foot to form a short mentum. Petals more or less equal to the upper sepal or shorter. Lip 3-lobed, the base continuous with or jointed to the column foot; lateral lobes erect; disc with a thickened callus. Pollinia 4 in 2 pairs, on a linear stipe.

A genus of 45 species from C and tropical S America, of which fewer than half are in cultivation. They require an epiphytic compost and should be grown in a cool to intermediate house with a definite resting period in winter. Watering may begin again when new growth emerges. Spraying should be delayed until after the new leaves have fully developed, to avoid unsightly black blotches.
Literature: Fowlie, J.A., *The genus Lycaste* (1970).

Pseudobulbs. With spines at the apex: **1–5,8,12,16,18**; without spines, or spines vestigial (less than 3 mm): **6,7,9–11,13–15,17,19.**
Lip. Length 2.5–3 cm: **1,2,5,8,9,11,12,19**; length 3–4 cm: **4,6,10**; length 4–5.5 cm: **3,7,13–17**. Basic colour green: **18,19**; basic colour yellow or orange: **1–4,10,15, 17**; basic colour cream or white: **5–12,14**; basic colour reddish, brownish or purple: **10,13,16,17**. Surface of central lobe hairless: **3,4,7–10,12–19**; surface of central lobe hairy: **1,2,5,6,11.** Callus hairless: **1,3,4,7–12,14–19**; callus hairy: **2,5,13.**

1a. Pseudobulbs armed with apical spines more than 3 mm — 2
b. Pseudobulbs unarmed or with vestigial spines less than 3 mm — 10
2a. Lip green — **18. locusta**
b. Lip not green — 3
3a. Central lobe of lip hairless — 4
b. Central lobe of lip hairy, sometimes sparsely so — 7
4a. Petals *c.* 8 cm; lip reddish brown, callus notched at the apex — **16. longipetala**
b. Petals 2–5 cm; lip white to yellow, variously marked, callus not notched at the apex — 5
5a. Lip 4–5 cm, deep yellow with red spots — **3. deppei**
b. Lip 3 cm or less, white, spotted or not — 6
6a. Central lobe of lip narrowly ovate — **12. xytriophora**
b. Central lobe of lip more or less circular — **8. brevispatha**
7a. Apex of callus a flap-like extension protruding over the base of the central lobe of the lip — **1. aromatica**
b. Apex of callus not as above — 8
8a. Central lobe of lip lanceolate — **5. crinita**
b. Central lobe of lip ovate to more or less circular — 9
9a. Callus hairy; lateral lobes of lip larger than the central lobe — **2. cruenta**
b. Callus not hairy; lateral lobes of lip smaller than the central lobe — **4. macrobulbon**
10a. Flowers to 10 cm in diameter — 11
b. Flowers 12–18 cm in diameter — 18
11a. Callus cream, yellow, pale orange, greenish or brownish — 12
b. Callus orange-red or reddish purple, if yellow then spotted — 17
12a. Lip 4–5.5 cm — 13
b. Lip 2.5–3.5 cm — 14
13a. Lip yellowish orange with the apex of the central lobe inrolled; upper sepal 5–6 cm — **15. barringtoniae**
b. Lip creamy white or pale yellow with the apex of the central lobe not inrolled; upper sepal 8–10 cm — **14. ciliata**
14a. Plant hanging; lip green — **19. dyeriana**
b. Plant upright; lip white or yellowish, spotted or not — 15
15a. Central lobe of lip narrowly oblong, densely hairy; upper sepal 6–6.5 cm — **6. lasioglossa**
b. Central lobe of lip ovate, minutely downy or hairless; upper sepal 3–3.5 cm — 16
16a. Lip whitish, heavily spotted with light pink, central lobe hairless — **9. tricolor**
b. Lip creamy white flushed with brownish yellow, central lobe minutely downy — **11. leucantha**
17a. Pseudobulbs unarmed; callus with broadened, truncate apex — **17. denningiana**
b. Pseudobulbs with vestigial spines; callus with a bluntly acute apex — **10. macrophylla**
18a. Lateral sepals *c.* 6.5 cm; sinuses between lateral and central lobes of lip without a projection — **13. skinneri**
b. Lateral sepals 9–9.5 cm; sinuses between lateral and central lobes of lip with a projection — **7. schilleriana**

1. L. aromatica (Hooker) Lindley. Figure 17(1), p. 228. Illustration: Botanical Magazine, 9231 (1931); Fowlie, The genus Lycaste, 14 (1970).
Pseudobulbs 7–10 cm, strongly ribbed, apex with spines. Leaves 30–40 cm, developing after flowering. Scapes 11–15 cm. Flowers 4–6 cm in diameter, smelling of cinnamon. Sepals yellowish green, the upper 2.5–3 cm. Petals orange-yellow, more or less equal to sepals. Lip 2.5–3 cm, orange-yellow, spotted with dark orange, lateral lobes sickle-shaped, central lobe recurved with a finely toothed margin. Callus very shortly hairy with a flap-like extension protruding over the base of the

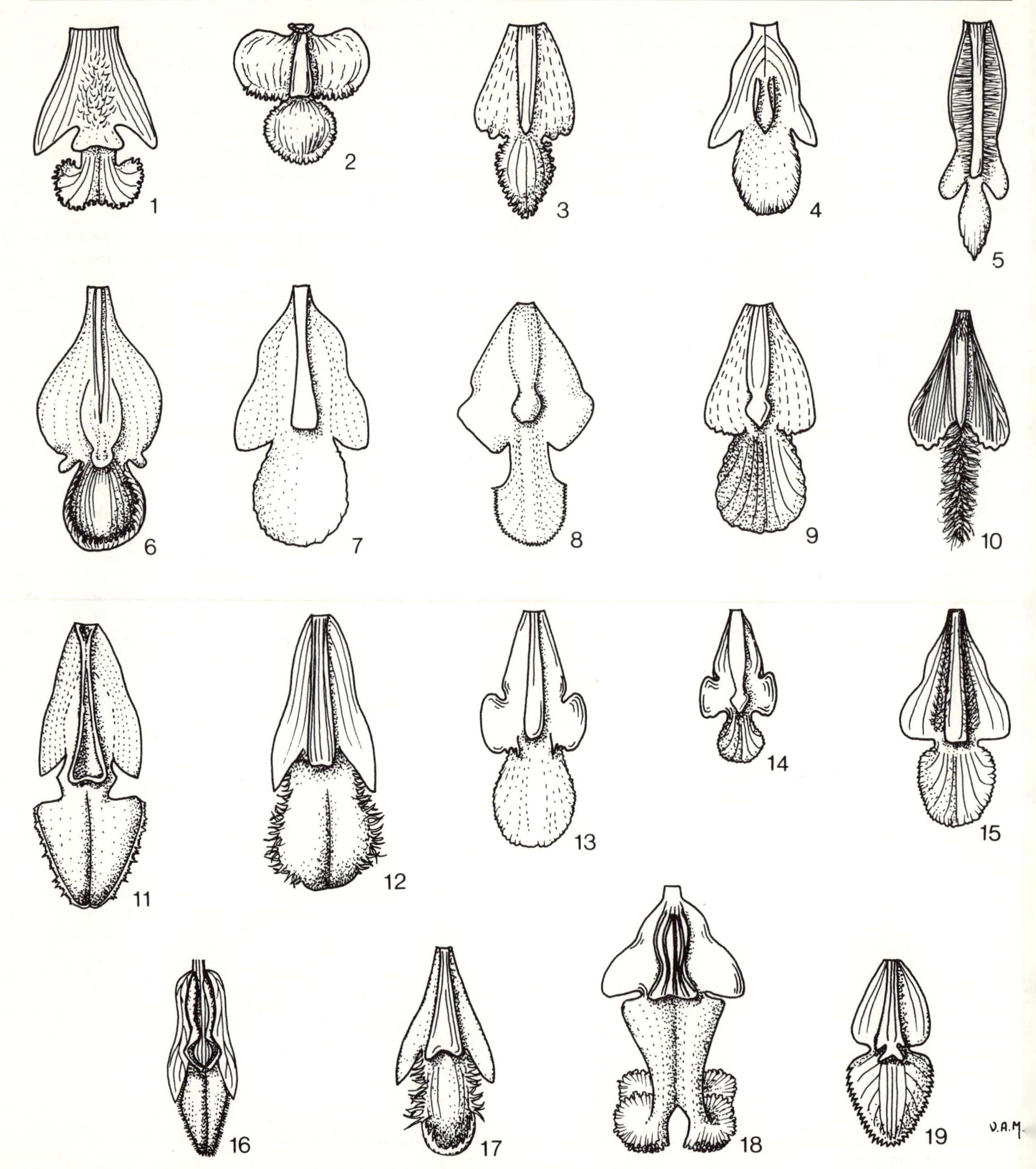

Figure 17. Lips of *Lycaste* species. 1, *L. aromatica*. 2, *L. cruenta*. 3, *L. deppei*. 4, *L. macrobulbon*. 5, *L. crinita*. 6, *L. schilleriana*. 7, *L. brevispatha*. 8, *L. tricolor*. 9, *L. macrophylla*. 10, *L. lasioglossa*. 11, *L. longipetala*. 12, *L. ciliata*. 13, *L. leucantha*. 14, *L. xytriophora*. 15, *L. skinneri*. 16, *L. locusta*. 17, *L. barringtoniae*. 18, *L. denningiana* (not opened out flat). 19, *L. dyeriana*. (Compiled from various sources; not to scale.)

central lobe. Column *c.* 2 cm, hairy below the stigma. *C America*. G2. Spring–summer.

2. L. cruenta (Lindley) Lindley. Figure 17(2), p. 228. Illustration: Edwards's Botanical Register **28**: t. 13 (1842); Fowlie, The genus Lycaste, 26 (1970); Orchid Digest **42**: 209 (1978); Bechtel et al., Manual of cultivated orchid species, 223 (1981).
Pseudobulbs to 10 cm, apex with spines. Leaves 35–45 cm, deciduous at time of flowering. Scapes *c.* 15 cm. Flowers 5–7 cm in diameter, smelling faintly of cinnamon. Sepals greenish yellow, the upper 4–4.5 cm. Petals yellowish orange with red spots towards the base, 3.5–3.8 cm, recurved at apices. Lip *c.* 2.5 cm, orange with red elongate spots towards the base and with a red triangular patch at the base, lateral lobes larger than central lobe, central lobe more or less circular, recurved, margin minutely scalloped, upper surface shortly hairy. Callus ridged, hairy, red at the base. Column 1.5 cm, red, hairy below stigma. *Northern C America*. G2. Spring–summer.

3. L. deppei (Loddiges) Lindley. Figure 17(3), p. 228. Illustration: Loddiges' Botanical Cabinet **17**: t. 1612 (1830); Fowlie, The genus Lycaste, 28 (1970); Bechtel et al., Manual of cultivated orchid species, 223 (1981).
Pseudobulbs to 8 cm, apex with spines. Leaves 30–50 cm. Scapes 12–20 cm. Flowers to 10 cm in diameter. Sepals green with reddish brown spots, the upper 4–7 cm. Petals creamy white with tiny reddish brown spots, shorter than sepals, recurved at apices. Lip 4–5 cm, deep yellow with red spots and pale red stripes towards the base, central lobe recurved, margin minutely scalloped. Callus cream to yellow with pale red markings, hairless. Column *c.* 2.5 cm, whitish spotted with red, hairy. *C America*. G2. Spring–summer.

4. L. macrobulbon (Hooker) Lindley. Figure 17(4), p. 228. Illustration: Botanical Magazine, 4228 (1846); Fowlie, The genus Lycaste, 32 (1970); Orchid Digest **42**: 209 (1978).
Pseudobulbs 5–8 cm, apex with spines. Leaves to 40 cm, developing after flowering. Scapes to 25 cm. Flowers 6–7 cm in diameter, fragrant. Sepals yellowish green, the upper 3.7–4 cm, gently recurved. Petals yellow with orange spots, 3–3.3 cm, recurved at apices. Lip 3–3.3 cm, deep yellow spotted with reddish brown, especially on the central lobe which is recurved with a minutely toothed margin and hairy at the base. Callus with raised margins. Column *c.* 1.8 cm, pale yellow, spotted with red at base. *Colombia, Venezuela*. G2. Spring–summer.

5. L. crinita Lindley. Figure 17(5), p. 228. Illustration: Fowlie, The genus Lycaste, 24 (1970); Orchid Digest **42**: 207 (1978).
Pseudobulbs with apical spines. Leaves to 35 cm. Scapes 8–11 cm. Sepals light green, the upper 2.5–2.8 cm, somewhat recurved. Petals pale yellow, *c.* 2 cm, recurved at apices. Lip *c.* 2.5 cm, pale yellow spotted with orange-red, central lobe lanceolate, hairy. Callus densely hairy on margins. Column *c.* 1.5 cm, hairy at the base. *Mexico*. G2. Summer.

6. L. lasioglossa Reichenbach. Figure 17(10), p. 228. Illustration: Botanical Magazine, 6251 (1876); Fowlie, The genus Lycaste, 30 (1970); Bechtel et al., Manual of cultivated orchid species, 223 (1981).
Pseudobulbs 5–10 cm, slightly ribbed, without apical spines. Leaves 40–50 cm, present at flowering. Scape to 25 cm. Flowers *c.* 10 cm in diameter. Sepals reddish brown, suffused with green outside and at apices, the upper 6–6.5 cm, the laterals with upswept apices. Petals pale yellow, 4–4.3 cm, recurved at apices. Lip 3–3.5 cm, pale yellow, suffused and spotted with red, central lobe recurved, narrowly oblong, densely hairy. Callus cream, tongue-shaped, grooved. Column 2.5–3 cm, hairy above the middle. *Guatemala*. G2. Winter–spring.

7. L. schilleriana Reichenbach (*L. hennisiana* Kränzlin; *L. longisepala* Schweinfurth). Figure 17(6), p. 228. Illustration: Fieldiana **30**(1): 647 (1960); Fowlie, The genus Lycaste, 43 (1970).
Pseudobulbs 7–10 cm, slightly ribbed, without apical spines. Leaves to 40 cm. Scapes 15–20 cm. Flowers 16–18 cm in diameter. Sepals pale green suffused with brown towards the base, upper sepal 9–10 cm, gently recurved. Petals 4–5 cm, white suffused with pale pink, recurved at apices. Lip 4.8–5.2 cm, white, suffused and spotted with pink, central lobe recurved, its margins finely scalloped, sinuses between lateral and central lobes with a tiny projection. Callus yellowish, striped pinkish at the base. Column *c.* 1.8 cm, shortly hairy. *Peru, Colombia & Surinam*. G2. Spring–summer.

8. L. brevispatha (Klotzsch) Lindley (*L. candida* Reichenbach). Figure 17(7), p. 228. Illustration: Reichenbach, Beiträge zu einer Orchideenkunde Central-Amerika's, t. 5 (1866); Fowlie, The genus Lycaste, 36 (1970); Bechtel et al., Manual of cultivated orchid species, 222 (1981).
Pseudobulbs *c.* 6–7 cm, ribbed, apex with spines. Leaves to 50 cm. Scapes 6–10 cm. Flowers 4–5 cm in diameter, fragrant. Sepals brownish green, sometimes spotted with pink, recurved at apices, the upper *c.* 2.5 cm. Petals equal to sepals, paler, sometimes spotted or suffused with pink, apices recurved. Lip 2.5–3 cm, whitish with pink spots, concave towards the base, central lobe with wavy margins. Callus tongue-shaped with 2 ridges, yellowish grading to pink at the base. Column white with pink spots. *Southern C America*. G2. Winter–spring.

9. L. tricolor (Klotzsch) Reichenbach. Figure 17(8), p. 228. Illustration: Fieldiana **26**(1): 555 (1952); Fowlie, The genus Lycaste, 39 (1970).
Pseudobulbs 5–9 cm, unarmed but with sharp apical ridges. Leaves 20–40 cm, deciduous at flowering. Scapes 8–12 cm. Flowers to 5.5 cm in diameter, never opening fully. Sepals greenish brown, sometimes with pink spots, recurved, the upper 3–4.5 cm. Petals shorter and paler than sepals, with pink spots. Lip 2.5–3 cm, whitish, heavily spotted with pink, central lobe with the apex strongly reflexed, margin finely wavy. Callus cream. Column 1.5–2 cm, sparsely hairy in front. *Southern C America*. G2. Spring–summer.

10. L. macrophylla (Poeppig & Endlicher) Lindley (*L. dowiana* Endres & Reichenbach). Figure 17(9), p. 228. Illustration; Fowlie, The genus Lycaste, 49, 53, 54 (1970); Pabst & Dungs, Orchidaceae Brasilienses **2**: 282 (1977).
Pseudobulbs 8–12 cm, ribbed, apex with vestigial spines *c.* 0.3 mm. Leaves to 65 cm. Scapes 8–50 cm. Flowers *c.* 9 cm in diameter, faintly scented or not. Sepals brownish green to dark reddish, recurved at apices, the upper 3–5 cm. Petals cream, suffused or not with yellow or pink, sometimes with pink spots, recurved at apices, shorter than sepals. Lip 3–4 cm, cream or yellowish, often spotted with pink to varying degrees, occasionally completely reddish. Callus tongue-shaped, grooved, yellow with red spots or pale reddish purple. Column *c.* 3 cm, cream, often with 2 purple spots at the base, hairy. *Southern C America & northwest S America*. G2. Late summer.

Related to *L. leucantha* and *L. xytriophora*. Several subspecies are recognised, some of which may be in cultivation.

11. L. leucantha (Klotzsch) Lindley. Figure 17(13), p. 228. Illustration: Paxton's flower garden **2**: 37–8 (1851–2); Fowlie, The genus Lycaste, 46 (1970); Bechtel et al., Manual of cultivated orchid species, 223 (1981).
Pseudobulbs to 6 cm, without apical spines. Leaves 40–50 cm. Scape 15–22 cm. Flowers *c*. 8 cm in diameter. Sepals pale green, sometimes brownish green, recurved at apices, the upper *c*. 4 cm. Petals creamy white, recurved towards apices 3.7–4 cm. Lip 2.7–3 cm, pale brownish yellow, central lobe ovate with a minutely wavy margin, minutely downy. Callus brownish yellow with a shallow groove. Column hairless. *Southern C America*. G2. Winter.

12. L. xytriophora Linden & Reichenbach. Figure 17(14), p. 228. Illustration: Reichenbach, Xenia Orchidacea **3**: t. 241 (1892); Fowlie, The genus Lycaste, 66 (1970); Bechtel et al., Manual of cultivated orchid species, 224 (1981).
Pseudobulbs 7–8 cm, apex with spines. Leaves 40–50 cm. Scapes 4–13 cm. Flowers 7–8 cm in diameter. Sepals pale green suffused with brown, the upper 3–4 cm, apices strongly reflexed. Petals cream suffused with green, shorter than sepals, recurved towards apices. Lip 2.5–3 cm, creamy white suffused with yellow and with mauve inside, central lobe narrowly ovate, sharply deflexed with minutely wavy margins. Callus linear, expanded at apex into an ovate projection, deep yellow grading to red at the base. Column 2.5–2.8 cm, white, hairy below stigma. *Panama, Ecuador*. G2. Summer.

13. L. skinneri (Lindley) Lindley (*L. virginalis* (Scheidweiler) Linden). Figure 17(15), p. 228. Illustration; Kupper & Linsenmaier, Orchids, 83 (1961); Fowlie, The genus Lycaste, 61 (1970); Bechtel et al., Manual of cultivated orchid species, 223 (1981).
Pseudobulbs to 10 cm, slightly ridged, without apical spines. Leaves 50–60 cm. Scapes 15–30 cm. Flowers 12–15 cm in diameter. Sepals cream, variously shaded with pink to lavender, recurved at apices, the upper 7–8 cm. Petals usually darker than sepals, reddish purple, recurved at apices, 4.5–7 cm. Lip 4–5.5 cm, pinkish, unmarked or mottled with purple, central lobe ovate, reflexed. Callus reddish, hairy. Column 3–4 cm, white, spotted with red towards the base, hairy in front. *C. America*. G2. Winter–spring.

Probably the most commonly grown species. About 15 varieties, distinguished mainly by flower colour and size, have been in cultivation at various times, and some may still be available. *L. skinneri* has often been used in hybridisation.

14. L. ciliata (Ruiz & Pavon) Reichenbach (*L. fimbriata* (Poeppig & Endlicher) Cogniaux). Figure 17(12), p. 228. Illustration: Orchid Digest **38**: 144 (1974); Botanical Magazine n.s., 767 (1978).
Pseudobulbs to 12 cm, ridged, without apical spines. Leaves 60–80 cm, present at flowering, then falling. Scapes 8–24 cm. Flowers 9–10 cm in diameter, slightly fragrant, appearing not to open fully. Sepals creamy white suffused with pale green, the upper 8–10 cm. Petals usually paler than sepals, 4.5–5 cm. Lip 4.5–5.5 cm, whitish or pale yellow with a slight green suffusion, central lobe oblong, recurved, lateral margins fringed. Callus with 5–7 ribs, yellowish to pale orange, apex rounded or notched and projecting over the base of the central lobe. Column white, hairy in front. *Northern S America*. G2. Late winter–summer.

15. L. barringtoniae (Smith) Lindley. Figure 17(17), p. 228. Illustration: Fawcett & Rendle, Flora of Jamaica **1**: t. 23 (1910); Fowlie, The genus Lycaste, 74 (1970).
Pseudobulbs 6–8 cm, ribbed, without apical spines. Leaves 35–50 cm, falling at or just before flowering. Scapes 7–8 cm. Flowers 5–7 cm, fragrant, not opening fully. Sepals pale green suffused with brown, the upper 5–6 cm. Petals similar but *c*. 4.5 cm. Lip 4–4.5 cm, yellow-orange, sometimes with red markings, central lobe ovate, fringed, apex inrolled to form a shallow pocket. Callus yellow, with 5 ridges, the notched apex projecting forwards over the base of the central lobe. Column 3–5 cm, white. *Cuba, Jamaica*. G2. Spring–summer.

16. L. longipetala (Ruiz & Pavon) Garay (*L. gigantea* Lindley). Figure 17(11), p. 228. Illustration: Botanical Magazine, 5616 (1866); Dunsterville & Garay, Venezuelan orchids illustrated **1**: 203 (1959); Fowlie, The genus Lycaste, 83 (1970); Bechtel et al., Manual of cultivated orchid species, 223 (1981).
Pseudobulbs 7–12 cm with shallow furrows and apical spines. Leaves to 60 cm, present at flowering. Scapes 50–60 cm. Flowers faintly fragrant, appearing to be only half open. Sepals pale green suffused with brown, the upper 9.5–11 cm. Petals similar in colour, concave, *c*. 8 cm. Lip *c*. 6 cm, reddish brown, central lobe with a lighter orange border, recurved, with a deep central groove, apex folded upwards, lateral margins irregularly minutely toothed. Callus greenish brown with 2 ridges which merge into a notched plate. Column 2.5–3.5 cm, whitish, often orange at base, hairy below. *Northwest S America*. G2. Spring.

17. L. denningiana Reichenbach (*L. cinnabarina* Rolfe). Figure 17(18), p. 228. Illustration: Lindenia **9**: t. 394 (1893); Dunsterville & Garay, Venezuelan orchids illustrated **3**: 167 (1965); Fowlie, The genus Lycaste, 56 (1970).
Pseudobulbs 7–10 cm, furrowed, without apical spines. Leaves 45–70 cm. Scapes to 16 cm. Flowers *c*. 10 cm in diameter. Sepals and petals cream, greenish or brownish yellow, the upper sepal 6–8 cm. Lip *c*. 4 cm, deep orange-red, central lobe ovate to circular, apex folded under and back, margin minutely toothed. Callus orange-red, with 5 ridges and a broadened truncate apex. Column 2–2.5 cm, white. *Northwest S America*. G2.

Fowlie suggests, on the basis of field evidence, that this species may be a hybrid between *L. longipetala* and *L. ciliata*.

18. L. locusta Reichenbach. Figure 17(16), p. 228. Illustration: Botanical Magazine, 8020 (1905); Fowlie, The genus Lycaste, 76 (1970).
Pseudobulbs 6–10 cm, ribbed, apex with spines. Leaves 50–85 cm. Scapes to 30 cm. Flowers *c*. 8 cm in diameter. Sepals and petals pale greyish green, flushed or veined with darker green, sepals *c*. 5 cm, petals *c*. 4 cm with recurved apices. Lip 3.7–4 cm, dark green, lateral lobes lighter, central lobe oblong-ovate, recurved with a hairy border, apex notched. Callus dark green, spathulate with 2 ridges. Column white, hairy in front. *Peru*. G2. Spring–summer.

19. L. dyeriana Rolfe. Figure 17(19), p. 228. Illustration: Botanical Magazine, 8103 (1906).
Plant hanging, Pseudobulbs 4–7.5 cm, angular, without apical spines. Leaves 18–30 cm. Scapes 7–13 cm, hanging. Flowers 4–5 cm in diameter, fragrant, not opening fully. Sepals pale green, the upper *c*. 5 cm. Petals darker green. Lip *c*. 3 cm, green, central lobes ovate-elliptic, reflexed, margin fringed. Callus green, grooved, 2-lobed on each side of the notched apex. Column *c*. 1.7 cm. *Peru*. G2. Summer–autumn.

Fowlie considers that this species has a greater affinity with *Bifrenaria* (p. 226) than with *Lycaste*, a genus in which it is certainly anomalous. However, as the controversy is unresolved, it is retained in *Lycaste* here.

85. ANGULOA Ruiz & Pavon

V.A. Matthews

Terrestrial or occasionally epiphytic. Pseudobulbs oblong, becoming ridged with age. Leaves usually 2–3 per pseudobulb, stalked, pleated, oblong, rolled when young. Scapes often several from the base of the pseudobulb, 15–30 cm, clothed in leafy bracts, bearing 1 flower. Flowers fragrant. Sepals and petals fleshy, concave and incurving, almost concealing the lip. Lip 3-lobed, hinged to the column foot so that it rocks, the lateral lobes erect. Callus linear-lanceolate with 2 apical lobes. Pollinia 4 on a linear stipe which has a slightly expanded viscidium.

A genus of 10 species from northern S America. Cultivation as for *Lycaste* (p. 227).

Flower colour. Yellow: **3**; yellow flushed red or green: **1,2,4**; whitish or pink: **5,6**.
Central lobe of lip. Surface hairless: **2,4–6**; surface hairy: **1,3**.

1a. Flowers whitish or pink — 2
b. Flowers yellow, often flushed red or green — 3
2a. Central lobe of lip wider than callus — **5. uniflora**
b. Central lobe of lip narrower than callus — **6. virginalis**
3a. Central lobe of lip with 2 tooth-like lobules, hairy — 4
b. Central lobe of lip without lobules, hairless — 5
4a. Flowers clear yellow — **3. clowesii**
b. Flowers yellow flushed reddish and with red spots inside or occasionally entirely red inside — **1. ruckeri**
5a. Central lobe of lip oblong, fringed at the apex, not or scarcely recurved — **2. brevilabris**
b. Central lobe of lip more or less 3-lobed, not fringed, recurved — **4. cliftonii**

1. A. ruckeri Lindley. Illustration: Edwards's Botanical Register **32**: t. 41 (1846); Kupper & Linsenmaier, Orchids, 81 (1961).
Flowers dull yellowish flushed with red and with red spots or rarely entirely red inside. Lateral lobes of the lip more or less oblong, central lobe hairy with 2 tooth-like lobules. *Colombia*. G2. Spring–summer.

The albino variant (var. **albiflora** Anon.) with pure white flowers may be in cultivation. Var. **sanguinea** Lindley (Illustration: Botanical Magazine, 5384, 1863) has flowers entirely red inside.

2. A. brevilabris Rolfe. Illustration: Botanical Magazine, 9381 (1935); Orchid Digest **42**: 140 (1978).
Similar to *A. ruckeri* but with smaller flowers flushed with dull reddish green. The lateral lobes of the lip are very broad and the central lobe is oblong, hairless though fringed at the apex, and lacking lobules. *Colombia*. G2. Summer.

3. A. clowesii (*A. clowesii* var. *ruckeri* (Lindley) Foldats). Illustration: Edwards's Botanical Register **30**: t. 63 (1844); Orchid Digest **40**: 132 (1976).
Flowers yellow. Lip whitish to orange-yellow, lateral lobes more or less oblong, central lobe hairy, ovate, reflexed, with 2 tooth-like lobules. *Colombia, Venezuela*. G2. Spring–summer.

Very similar to *A. ruckeri*; in the opinion of some authors the 2 should be treated as varieties of the 1 species.

4. A. cliftonii Rolfe. Illustration: Botanical Magazine, 8700 (1917); Orchid Digest **40**: 132 (1976); Bechtel et al., Manual of cultivated orchid species, 167 (1981).
Flowers yellow splashed and marked with reddish purple. Lip pouched at the base, lateral lobes rounded, central lobe with a basal claw, more or less 3-lobed, hairless. *Colombia*. G2. Spring–summer.

5. A. uniflora Ruiz & Pavon. Illustration: Edwards's Botanical Register **30**: t. 60 (1844); Orchid Digest **40**: 132 (1976); Bechtel et al., Manual of cultivated orchid species, 168 (1981).
Flowers whitish, sometimes flushed with green, spotted and tinged inside with pink. Lip yellowish, spotted with pink, lateral lobes rounded and inrolled, central lobe lanceolate, reflexed, wider than the callus. *Colombia, Venezuela, Peru*. G2. Spring–summer.

Several varieties, based on variation in flower colour, have been described.

6. A. virginalis Schlechter. Illustration: Botanical Magazine, 4807 (1854); Orchid Digest **40**: 132 (1976); Sheehan & Sheehan, Orchid genera illustrated, 37 (1979).
Very similar to *A. uniflora*, differing only in having the central lobe of the lip narrower than the callus. *Colombia*. G2. Summer.

Often considered to be the same as *A. uniflora*. Pink-flowered variants have been found in the wild.

86. PABSTIA Garay

V.A. Matthews

Epiphytic. Pseudobulbs ovoid to oblong. Leaves 2–4 per pseudobulb, lanceolate, pleated, rolled when young. Scapes arising from the bases of the pseudobulbs, clothed in leafy bracts, with 1–4 flowers. Sepals and petals more or less equal or petals smaller. Lip 3-lobed. Column without a foot, hairy in front. Pollinia 4 on a short, linear or oblong stipe.

A genus of 5 species from Brazil usually found under the name *Colax* Lindley. Cultivation as for *Zygopetalum* (p. 232), *Maxillaria* (p. 236) and *Lycaste* (p. 227).

1a. Sepals white — **1. jugosa**
b. Sepals green — 2
2a. Sepals spotted — **2. placanthera**
b. Sepals unspotted — **3. viridis**

1. P. jugosa (Lindley) Garay (*Colax jugosus* (Lindley) Lindley). Illustration: Botanical Magazine, 5661 (1867); Lindenia **8**: t. 372 (1892); Orchid Digest **42**: 144 (1978); Bechtel et al., Manual of cultivated orchid species, 251 (1981).
Pseudobulbs with 2 leaves. Leaves 15–25 cm. Flowers 2–4 in each raceme, 5–7.5 cm in diameter, fragrant. Sepals white, ovate-oblong. Petals white spotted with brownish or reddish purple. Lip *c.* 2.5 cm, lateral lobes rounded and striped with bluish purple, central lobe rounded, striped and marked with bluish purple, velvety. Callus with ridges. *Brazil*. G2. Winter–spring.

The following 2 varieties have been in cultivation, and may still be available:

Var. **rufina** (Reichenbach) Garay (*Colax jugosus* var. *rufinus* Reichenbach). Flowers yellowish green spotted with dark purple and with brown on the lip.

Var. **viridis** (Godefroy-Leboeuf) Garay (*Colax jugosus* var. *punctatus* Reichenbach; *Colax puytdii* Linden & André). Illustration: Illustration Horticole **27**: t. 369 (1880). Flowers greenish yellow spotted with reddish purple.

2. P. placanthera (Hooker) Garay (*Colax placanthera* (Hooker) Lindley). Illustration: Botanical Magazine, 3173 (1832).
Pseudobulbs with 2–3 leaves. Flowers usually solitary. Petals smaller than sepals, yellowish green with brownish purple spots. Lateral lobes of lip pale green streaked with brown, central lobe with a whitish, purple-flushed central area, very shortly hairy. Callus smooth. *Brazil*. G2. Summer.

3. P. viridis (Lindley) Garay (*Colax viridis* (Lindley) Lindley). Illustration: Pabst & Dungs, Orchidaceae Brasilienses **2**: 226, 282 (1977).
Leaves 12–18 cm. Flowers 1–2 per scape. Sepals and petals directed forwards, green, petals with brownish spots which merge towards the centre. Lip with a pale purple, rhomboid central lobe. *Brazil.* G2. Summer.

87. ZYGOPETALUM Hooker
V.A. Matthews
Epiphytic or terrestrial. Pseudobulbs ovoid, sheathed in basal leaves, bearing 2–5 leaves. Leaves lanceolate, pleated, with 5–9 conspicuous veins beneath, rolled when young. Racemes produced from the bases of the pseudobulbs. Sepals and petals spreading, lateral sepals attached to the column foot to form a short mentum. Petals similar to sepals or sometimes smaller. Lip 3-lobed, lateral lobes very small, erect or spreading, with a prominent callus, often ribbed. Pollinia 4 in 2 pairs, stipe variable in length.

A genus of 20 species from C and tropical S America, of which 5 are found in cultivation. They require water throughout the year, though the amount should be reduced in winter. Propagation is usually by division of large plants. All are very prone to attack by thrips, scale and red spider mite.

Petals. 2–2.5 cm: **4,5**; 3–4 cm: **1–3**.
Lip. Hairy: **1–3**; hairless: **1,4,5**. Without spots, stripes or obvious veining: **5**.
Column. Yellow or green: **1,3,4**; white: **2**; purple: **5**.

1a. Lip neither spotted nor striped; callus shaped like a horse's hoof, more or less erect **5. maxillare**
b. Lip spotted or striped; callus not as above 2
2a. Lip hairy, sometimes minutely so 3
b. Lip hairless 5
3a. Petals shorter than upper sepal; column hairless **1. mackayi**
b. Petals equal to upper sepal; column hairy in front 4
4a. Central lobe of lip more or less ovate **2. crinitum**
b. Central lobe of lip kidney-shaped **3. intermedium**
5a. Leaves with 7–9 veins beneath; flowers *c.* 8 cm in diameter; lip 4–4.5 cm **1. mackayi**
b. Leaves with 5 veins beneath; flowers 5–6 cm in diameter; lip 2.5–3 cm **4. brachypetalum**

1. Z. mackayi Hooker. Illustration: Martius, Flora Brasiliensis 3(5): t. 104 (1902); Pabst & Dungs, Orchidaceae Brasilienses **2**: 227, 283 (1977); Gartenpraxis 1980: 272.
Pseudobulbs 4–7 cm with 2–3 leaves. Leaves 30–50 cm, with 7–9 veins beneath. Flowering stems to 75 cm, racemes usually with 5–7 flowers. Flowers fragrant, *c.* 8 cm in diameter. Sepals and petals greenish yellow with purplish or brownish blotches, sepals 4–4.5 cm, petals 3.5–4 cm. Lip 4–4.5 cm, central lobe ovate or fan-shaped, white with interrupted purple radiating lines, hairless or with tiny hairs, margins wavy, apex notched. Callus horseshoe-shaped, ridged, toothed along the front margin. Column *c.* 2 cm, hairless, yellowish green with reddish brown spots. *S & SC Brazil.* G2. Autumn–winter.

2. Z. crinitum Loddiges. Illustration: Botanical Magazine, 3402 (1835); Orchid Digest **36**: 128 (1972); Pabst & Dungs, Orchidaceae Brasilienses **2**: 226, 282 (1977); Bechtel et al., Manual of cultivated orchid species, 280 (1981).
Pseudobulbs 4–7 cm with 3–5 leaves. Leaves 25–40 cm with 7–9 veins beneath. Flowering stems 30–45 cm, raceme with 4–7 flowers. Flowers fragrant, 7–8 cm in diameter. Sepals and petals *c.* 4 cm, yellowish green with purplish brown blotches or bars. Lip 3–4 cm, central lobe broadly obovate, velvety, white or cream with densely hairy bluish purple or reddish veins. Callus yellow, more or less 2-lobed. Column 1.6–2 cm, hairy in front, white with purple stripes. *S Brazil.* G2. Winter.

Several varieties have been described, based mainly on colour variations of the veins of the lip.

3. Z. intermedium Lindley. Illustration: Pabst & Dungs, Orchidaceae Brasilienses **2**: 283 (1977); Bechtel et al., Manual of cultivated orchid species, 280 (1981).
Pseudobulbs 4–8 cm with 3–5 leaves. Leaves 25–50 cm with 7–9 veins beneath. Flowering stems 30–75 cm, raceme with 4–7 flowers. Flowers fragrant, 7–8 cm in diameter. Sepals and petals 3–4 cm, green or yellowish green with reddish brown or purplish markings. Lip 3–4 cm with a short claw, central lobe kidney-shaped, whitish spotted with purple and with purple veins, downy, margin wavy, apex notched. Callus horseshoe-shaped, ridged. Column 1.3–1.5 cm, velvety in front, pale green with purple stripes. *S Brazil, Peru, Bolivia.* G2. Autumn–spring.

4. Z. brachypetalum Lindley. Illustration: Orchid Digest **36**: 128 (1972); Pabst & Dungs, Orchidaceae Brasilienses **2**: 226 (1977).
Pseudobulbs 4–6 cm, somewhat compressed, with 2–3 leaves. Leaves 30–50 cm, with 5 veins beneath. Flowering stems 30–80 cm, raceme with 5–12 flowers. Flowers fragrant, 5–6 cm in diameter. Sepals 2.5–3 cm, petals 2–2.5 cm, all green with brown blotches. Lip 2.5–3 cm, central lobe more or less obovate, white with bluish violet veins and spots, hairless, apex notched. Callus white with purple stripes. Column 1–1.2 cm, yellowish with purple spots and stripes. *S Brazil.* G2. Autumn–winter.

5. Z. maxillare Loddiges. Illustration: Botanical Magazine, 3686 (1839); Orchid Digest **36**: 128 (1972); Pabst & Dungs, Orchidaceae Brasilienses **2**: 227, 283 (1977).
Pseudobulbs 4–8 cm with 2–3 leaves, borne on creeping, branched rhizomes. Leaves 20–40 cm with 5–7 veins beneath. Flowering stems 15–30 cm, racemes with 5–8 flowers, sometimes drooping. Flowers *c.* 5 cm in diameter. Sepals *c.* 2.5 cm, petals 2–2.3 cm, all green with dark brown blotches or bars. Lip with the central lobe more or less circular, bluish purple, hairless. Callus dark purple, more or less erect and shaped like a horse's hoof, ribbed. Column *c.* 1 cm, hairless, bluish purple. *S Brazil, Paraguay, N Argentina.* G2. Can flower at any time of year.

In the wild *Z. maxillare* grows only on tree ferns.

88. ACACALLIS Lindley
V.A. Matthews
Climbing epiphyte with a creeping rhizome bearing pseudobulbs at intervals of 2–5 cm. Pseudobulbs ovoid, 2–5 cm, with 1 or 2 leaves. Leaves 10–20 cm, narrowly elliptic, shortly stalked, rolled when young. Racemes with 3–7 flowers, longer than leaves. Flowers 2.5–5 cm in diameter. Sepals and petals 3.5–4 cm, oval, bluish mauve shading to white outside, flushed with pink inside. Lip 2.5–3 cm, clawed, kidney-shaped with a wavy margin, yellowish brown with a reddish or purplish centre and paler veins. Callus yellowish or pinkish, divided above into finger-like projections. Column 3-angled, with 2 wings and a long slender foot. Pollinia 4, in 2 pairs.

A genus of only 1 species. Best grown in moderate shade in a basket using an epiphytic compost which is kept moist.

1. A. cyanea Lindley (*Aganisia cyanea* (Lindley) Reichenbach). Illustration: Botanical Magazine, 8678 (1916); Dunsterville & Garay, Venezuelan orchids illustrated **2**: 33 (1961); Bechtel et al., Manual of cultivated orchid species, 161 (1981).
Brazil, Venezuela & Colombia. G2.

89. PROMENAEA Lindley
V.A. Matthews
Epiphytic. Pseudobulbs more or less ovoid, compressed. Leaves 1–3 per pseudobulb, pleated, often rather glaucous, folded when young. Scapes produced from the bases of the pseudobulbs. Sepals and petals spreading. Lip 3-lobed with usually narrow, erect lateral lobes. Column with a short foot to which the lip is jointed. Disc of lip with a transverse callus. Pollinia 4, sessile on a broad, rounded viscidium.

A Brazilian genus containing 15 species. Cultivation as for *Zygopetalum* (p. 232). The pans or baskets are best hung near the glass.

1a. Sepals and petals spotted or marked **4. stapelioides***
b. Sepals and petals unspotted 2
2a. Sepals and petals yellow 3
b. Sepals and petals pale green or pale brown **3. microptera**
3a. Lip spotted with red towards the base **1. xanthina**
b. Lip whitish spotted or barred with reddish purple **2. rollissonii**

1. P. xanthina (Lindley) Lindley (*P. citrina* Don). Illustration: Botanical Magazine n.s., 697 (1975); Pabst & Dungs, Orchidaceae Brasilienses **2**: 214 (1977).
Pseudobulbs to 2 cm. Leaves to 7 cm. Scape usually with 1 flower, to 10 cm. Flowers *c.* 3.5 cm in diameter, fragrant. Sepals and petals bright yellow, petals slightly shorter than sepals. Lip bright yellow spotted with red on the lateral lobes and sometimes on the base of the central lobe which is deflexed and more or less circular. Callus 3-lobed, the central lobe with a fleshy prominence and a toothed apex. *Brazil*. G2. Summer.

P. × crawshayana Anon. is a hybrid between *P. xanthina* and *P. stapelioides*. It has pale yellow flowers spotted with reddish brown, especially so on the petals and lip. The central lobe of the lip is ovate and with a small point at the apex. The callus has a tubercled crest.

2. P. rollissonii (Lindley) Lindley. Illustration: Edwards's Botanical Register **24**: t. 40 (1838); Pabst & Dungs, Orchidaceae Brasilienses **2**: 276 (1977).
Pseudobulbs to 2.5 cm. Leaves 7.5–10 cm. Scape usually with 1 flower, arching or somewhat hanging. Flowers 3–5 cm in diameter. Sepals and petals pale yellow, petals shorter and broader. Lip whitish, spotted or barred with reddish purple, markings sometimes confined to the lateral lobes, central lobe slightly broader than long, and with a small point. *Brazil*. G2. Summer.

3. P. microptera Reichenbach. Illustration: Botanical Magazine, 8631 (1915); Pabst & Dungs, Orchidaceae Brasilienses **2**: 276 (1977).
Pseudobulbs 1.5–2.5 cm. Leaves 5–9 cm. Scapes with 1–3 flowers, 5–7 cm, arching. Flowers *c.* 5 cm in diameter. Sepals and petals pale green to pale brown. Lip white or yellowish white, central lobe elliptic-oblong with an acute, recurved apex, marked with purple. Callus spotted with purple and with a tubercled crest. *Brazil*. G2. Summer.

4. P. stapelioides (Link & Otto) Lindley. Illustration; Edwards's Botanical Register **25**: t. 17 (1839); Botanical Magazine, 3877 (1841); Pabst & Dungs, Orchidaceae Brasilienses **2**: 214, 276 (1977).
Pseudobulbs to 2.5 cm. Leaves 7.5–10 cm. Scapes often with 2 flowers, usually longer than the leaves, hanging. Flowers 3–4 cm in diameter. Sepals and petals greenish yellow outside, deep green inside, barred and spotted with dark purple. Lip dark purple, central lobe ovate to circular, with a small point, somewhat concave, paler on the margin. Callus with a 2-lobed crest and a tooth pointing to the base of the lip. *Brazil*. G2. Summer–autumn.

***P. lentiginosa** (Lindley) Lindley. The spotting of the flowers is reddish purple and the spots are more distinct, less coalesced into bars. The lip is similar in colour to the petals and the transverse crest of the callus has a central prominence which is 3-toothed. *Brazil*. G2. Summer–autumn.

90. CHONDRORHYNCHA Lindley
V.A. Matthews
Epiphytic. Pseudobulbs absent. Leaves stalked, jointed to the sheaths, oblong-lanceolate to strap-shaped, folded when young. Scape with 1 flower. Sepals almost equal, elliptic to lanceolate, joined to the column foot to form an acute mentum. Petals obovate-elliptic. Lip not 3-lobed. Callus in centre of disc, with 3 teeth. Pollinia 4.

A genus of about 7 species growing in C and northwestern S America, formerly included in *Warscewiczella* Reichenbach. Cultivation as for *Zygopetalum* (p. 232).

1a. Petal margins with fine teeth **1. flaveola**
b. Petal margins with a long fringe **2. chestertonii**

1. C. flaveola (Linden & Reichenbach) Garay (*C. fimbriata* (Linden & Reichenbach) Reichenbach). Illustration: Saunders' Refugium Botanicum **2**: t. 107 (1872); Dunsterville & Garay, Venezuelan orchids illustrated **2**: 53 (1961); American Orchid Society Bulletin **44**: 1077 (1975).
Leaves to 45 cm, oblong, acute. Scapes erect or arching. Sepals *c.* 2.5 cm, elliptic, long-pointed, sometimes with wavy margins, pale yellow. Petals *c.* 3 cm, margins finely toothed, pale yellow. Lip 3–3.5 cm, fan-shaped, lobed, pale yellow, marked and spotted with brown or red at the base, the margin wavy and fringed. Callus hairless, surrounded by a lobed wall and numerous small calluses. *Colombia, Venezuela, Ecuador & Peru*. G2. Spring–autumn.

2. C. chestertonii Reichenbach. Illustration: Lindenia **9**: t. 405 (1893); Kupper & Linsenmaier, Orchids, 87 (1961).
Very similar to *C. flaveola*, but the sepals and petals are narrower and the petals have a long fringe on the margins. The front of the callus has 3 teeth. *Colombia*, G2. Summer.

91. COCHLEANTHES Rafinesque
V.A. Matthews
Epiphytic. Pseudobulbs absent. Stems short, covered by overlapping leaf-sheaths. Leaves in 2 ranks, borne in a fan, membranous, jointed to basal sheaths, folded when young. Scape shorter than leaves, with 1 flower. Sepals and petals usually spreading, the lateral sepals attached to the column foot. Lip hinged to column foot, 3-lobed or not, often concave. Callus fleshy, with ridges radiating from the base. Column cylindric, with a short foot. Pollinia 4, on a broad viscidium.

A genus of 14 species occuring in tropical America. Cultivation as for *Zygopetalum* (p. 232).

1a. Lip notched at the apex **2. amazonica**
b. Lip with an irregularly toothed or scalloped margin, or if entire then not notched at the apex 2
2a. Callus yellow with ridges of unequal length **1. discolor**

b. Callus white or violet with purple veins and with ridges of more or less equal length 3
3a. Lip to 2.2 cm, with an entire margin; callus oblong, with 5 ridges **3. wailesiana**
b. Lip 2.5–3 cm, with an irregularly toothed or scalloped margin; callus semicircular with more than 11 ridges **4. flabelliformis**

1. C. discolor (Lindley) Schultes & Garay (*Warscewiczella discolor* (Lindley) Reichenbach; *Zygopetalum discolor* (Lindley) Reichenbach). Illustration: Botanical Magazine, 4830 (1855); Orchid Digest **33**: 225 (1969); Sheehan & Sheehan, Orchid genera illustrated, 65 (1979); Bechtel et al., Manual of cultivated orchid species, 185 (1981).
Leaves to 30 cm, oblanceolate to strap-shaped. Scape to 15 cm. Flowers 5–7.5 cm in diameter. Sepals and petals 3–3.5 cm, white, sepals greenish or yellowish towards apex, petals often tinged with lilac towards the apex. Lip more or less 3-lobed, 2.5–4 cm, deep purple, often with a whitish margin, central lobe rounded, entire. Callus yellow, with ridges of unequal length. *Cuba, Honduras, Costa Rica, Panama & Venezuela.* G2. Spring–autumn.

2. C. amazonica (Reichenbach & Warscewicz) Schultes & Garay (*Warscewiczella amazonica* Reichenbach & Warscewicz; *Zygopetalum amazonicum* (Reichenbach & Warscewicz) Reichenbach). Illustration: Lindenia **8**: t. 337 (1892).
Leaves to 30 cm. Scape to 11 cm. Sepals and petals 2.5–3.5 cm, white, reflexed. Lip obscurely 3-lobed, *c.* 5 cm, white, veined with purple or reddish purple in the lower part, central lobe rounded, notched at the apex. Callus with a truncate apex with 3–7 teeth. *Peru, Brazil.* G2.

3. C. wailesiana (Lindley) Schultes & Garay (*Warscewiczella wailesiana* (Lindley) Morren). Illustration: Cogniaux & Goossens, Dictionnaire Iconographique des Orchidées, Warscewiczella, t. 2 (1897); Orchid Digest **33**: 225 (1969); Pabst & Dungs, Orchidaceae Brasilienses **2**: 232, 287 (1977).
Leaves 15–20 cm. Flowers *c.* 5 cm in diameter, scented. Sepals and petals white or cream. Lip to 2.2 cm, roundish, white, stained and striped with violet towards the centre, margin entire. Callus oblong, white striped with purple, with 5 ridges. *Brazil.* G2.

4. C. flabelliformis (Swartz) Schultes & Garay (*Warscewiczella flabelliformis* (Swartz) Cogniaux; *W. cochlearis* (Lindley) Reichenbach). Illustration: Dunsterville & Garay, Venezuelan orchids illustrated **2**: 77, 241 (1961); Pabst & Dungs, Orchidaceae Brasilienses **2**: 232, 286 (1977); Bechtel et al., Manual of cultivated orchid species, 92 (1981).
Leaves to 30 cm, narrowly oblong to oblanceolate. Scape 5–15 cm. Flowers 5–6 cm in diameter, fragrant. Sepals and petals 2.5–3.5 cm, white or greenish white. Lip 2.5–3 cm, wider than long, broadly rounded, white marked with purple and with purple veins, margin irregularly toothed or scalloped. Callus white or violet with purple veins, semicircular with more than 11 ridges. *West Indies, Venezuela, Brazil.* G2. Spring–autumn.

92. PESCATORIA Reichenbach
V.A. Matthews
Epiphytic, without pseudobulbs. Leaves lanceolate, pleated, arranged in 2 ranks, folded when young. Scapes arising from the axils of basal bracts, each with 1 flower. Sepals and petals fleshy, rather concave, lateral sepals joined at the base and inserted on the column foot. Lip 3-lobed, contracted at the base into a claw which merges with the column foot; central lobe convex. Callus semicircular, ribbed. Column with a short basal foot. Pollinia 4 in 2 pairs on a short, linear stipe.

A genus of about 17 species in southern C America and western tropical S America. The name is often misspelled 'Pescatorea'. Cultivation as for *Zygopetalum* (p. 232).
Literature: Fowlie, J.A., A key and annotated check list to the genus Pescatorea, *Orchid Digest* **32**: 86–91 (1968).

Sepals and petals. Striped: **5**.
Lip. Bearing warts and bristles: **5**; bearing warts: **1**: without warts or bristles: **2–4**.
Callus. Yellow marked with brown: **1,2**; reddish brown: **5**; purple or reddish purple: **3**; white, purple in front: **4**.

1a. Sepals and petals striped; lip bearing bristles **5. lehmannii**
b. Sepals and petals without stripes; lip smooth or with warts 2
2a. Lip purple 3
b. Lip white or yellow 4
3a. Callus purple or reddish purple **3. dayana**
b. Callus white, purplish in front **4. wallisii**
4a. Callus purple or reddish purple **3. dayana**
b. Callus yellow 5
5a. Flowers 6–8 cm in diameter; lip bearing warts **1. cerina**
b. Flowers 5–6 cm in diameter; lip without warts **2. lamellosa**

1. P. cerina (Lindley) Reichenbach (*Zygopetalum cerinum* (Lindley) Reichenbach). Illustration: Botanical Magazine, 5598 (1866); Flore des Serres **17**: t. 1815 (1868); Orchid Digest **32**: 87 (1968); Bechtel et al., Manual of cultivated orchid species, 258 (1981).
Leaves 20–60 cm. Scape 4–10 cm. Flowers 6–8 cm in diameter. Sepals white or cream, the laterals each with a greenish yellow blotch near the base. Petals similar to sepals. Lip 2–3 cm, cream to yellow, central lobe oval, warty on the upper surface. Callus semicircular, ribbed, yellowish orange striped with reddish brown. Column white, often with a purple or reddish blotch near the base. *Costa Rica, Nicaragua, Panama.* G2.

2. P. lamellosa Reichenbach. Illustration: Botanical Magazine, 6240 (1876); Die Orchidee **33**(4): 142 (1982).
Differs from *P. cerina* in having slightly smaller flowers (5–6 cm in diameter), and a pale yellow lip with the central lobe more or less round and without warts. The sepals and petals are white to yellowish green and the large callus is yellowish orange striped with brown and sometimes with dark purple. *Colombia.* G2. Summer.

3. P. dayana Reichenbach.
Leaves 20–40 cm. Scape to 6 cm. Flowers 7–8 cm in diameter. Sepals and petals white with green tips. Lip white flushed with purple, with purple lines extending from the base of the purple or reddish purple callus. Column yellow with a reddish band near the base. *Panama, Colombia.* G2. Autumn–winter.

Three varieties are cultivated; some authors recognise them as subspecies, but it appears that they are no more than horticulturally desirable colour forms

Var. **candidula** Reichenbach. Illustration: Orchid Digest **32**: 87 (1968). Sepals and petals completely white.

Var. **rhodacra** Reichenbach. Illustration: Botanical Magazine, 6214 (1876); Orchid Digest **32**: 87 (1968). Sepals and petals with reddish tips.

Var. **splendens** Reichenbach. Sepals and petals with purple tips, the lip deep purple.

4. P. wallisii Reichenbach. Illustration; Flore des Serres **18**: t. 1828 (1869); Brooklyn Botanic Garden Record **23**(2), Handbook on orchids, 40–1 (1967); Orchid

Digest 32: 87 (1968).
Flowers 7–8 cm in diameter. Sepals and petals white tipped with bluish violet. Lip deep violet, central lobe with a white margin and notched apex. Callus white, ribbed, purplish in front. Column white, pale violet below, hairy in front, *Ecuador*. G2.

5. **P. lehmannii** Reichenbach. Illustration: Cogniaux & Goossens, Dictionnaire Iconographique des Orchidées, Pescatorea, t. 2 (1898); Orchid Digest 32: 90 (1968); Die Orchidee 33(4): 140 (1982).
Leaves 30–45 cm. Scape to 15 cm. Flowers 6–9 cm in diameter. Sepals and petals white, yellowish towards the tips, striped longitudinally with reddish purple. Lip purple, central lobe oblong, apex notched, covered with warts and bristle-like whitish or purplish papillae. Callus reddish brown, ribbed. Column purple. *Colombia, Ecuador*. G2. Spring–autumn.

93. BOLLEA Reichenbach

V.A. Matthews

Epiphytic, without pseudobulbs. Leaves several, lanceolate, in 2 ranks, with prominent veins beneath, folded when young. Inflorescence lateral. Scapes shorter than the leaves, each with 1 flower. Sepals and petals free, spreading, somewhat concave, the lateral sepals joined to the base of column which is produced into a short foot. Lip with a narrow claw, movably attached to the column base to form a short mentum, the limb broad, entire. Callus large, ribbed, partly concealing the column. Column short and wide with a short foot, deeply hollowed in front, hooded. Pollinia 4, waxy.

A genus of about 6 species from western tropical S America. Cultivation as for *Zygopetalum* (p. 232).

1a. Lip violet **1. coelestis**
b. Lip yellow **2. lalindei***

1. **B. coelestis** (Reichenbach) Reichenbach (*Zygopetalum coeleste* Reichenbach). Illustration: Botanical Magazine, 6458 (1879); Ospina & Dressler, Orquideas de las Americas, t. 101 (1974).
Leaves 30–60 cm. Flowers 7–10 cm in diameter, fragrant. Sepals and petals pale blue to deep violet blue, darker above the middle, yellowish at the tips, margin wavy. Lip heart-shaped at the base, with recurved tip and margin, dark violet, yellowish towards the base. Callus semicircular, pale brownish or yellow, with many ribs. *Colombia*. G2. Summer.

2. **B. lalindei** (Reichenbach) Reichenbach (*Zygopetalum lalindei* Reichenbach). Illustration: Botanical Magazine, 6331 (1877).
Flowers 6–8 cm in diameter, fragrant. Sepals and petals pale pink or purple, darker above the middle, upper sepal yellowish green at the tip, lateral sepals with yellowish lower margin; petals often with white margins. Lip ovate, narrowly heart-shaped at base, with recurved tip and margin, bright yellow. Callus semicircular, yellow, with many ribs. *Colombia*. G2. Summer–autumn.

***B. patinii** Reichenbach. Flowers larger (*c.* 10 cm in diameter), petals without white margins and with a flesh-coloured callus. *Colombia*. G2. Summer.

94. HUNTLEYA Lindley

V.A. Matthews

Epiphytic, without pseudobulbs. Leaves narrowly lanceolate, arranged in a fan, folded when young. Scapes arising from the leaf axils, shorter than the leaves, each with 1 flower. Sepals and petals more or less similar, spreading, glossy, the lateral sepals attached to the short column foot. Lip 3-lobed with small lateral lobes, attached to the column foot by a long, narrow claw, central lobe curving downwards. Callus semicircular, densely fringed. Column with broad, apical, lateral wings. Pollinia 4 in 2 pairs in which the members are unequal, on a short, reversed-triangular stipe.

A genus from C and northern S America. The number of species is doubtful, some authors maintaining that there are about 10, while others recognise only 4. Cultivation as for *Zygopetalum* (p. 232).

1. **H. meleagris** Lindley (*H. burtii* (Endres & Reichenbach) Pfitzer). Illustration: Dunsterville & Garay, Venezuelan orchids illustrated 3: 147 (1965); Hawkes, Encyclopaedia of cultivated orchids, 249 (1965); Bechtel et al., Manual of cultivated orchid species, 217 (1981).
Leaves 25–40 cm. Flowers 7.5–10 cm in diameter, fragrant. Sepals and petals lanceolate to ovate, pointed, whitish at the base shading into yellow, marked and flushed with reddish brown, usually with a median yellow cross-bar, and often appearing chequered. Lip 2.5–3.5 cm, white at the base, shading upwards to reddish brown, with a small point. Callus white or yellowish. Column 1.5–2 cm, creamy green, often striped with reddish brown, with a semicircular, down-pointing wing on each side and a hood covering the anther cap. *C America, Northwestern S America, Trinidad*. G2. Spring–autumn.

95. ERIOPSIS Lindley

V.A. Matthews

Epiphytic or terrestrial. Pseudobulbs elongate-cylindric or ovoid-oblong. Leaves 2–4 per pseudobulb, pleated, strongly veined, rolled when young. Raceme produced from the base of a pseudobulb, with many flowers. Sepals and petals more or less equal, sepals free or the laterals joined to the column foot to form a short mentum. Lip 3-lobed, lateral lobes erect, larger than the central lobe. Callus composed of 2 or more fleshy plates. Column curved, slightly swollen above. Pollinia 2 or 4, stipe very short, almost non-existent.

A genus of 6 species from southern C America and tropical S America. They should be grown in an epiphytic compost, with frequent watering when in growth. Light shading should be provided when young growth is present, but should be removed when the growth is mature; propagation by division.

1a. Callus of 2 erect fleshy plates with 2 separate lumps in front of each plate **1. biloba**
b. Callus of 2 erect, plate-like horns only **2. sceptrum**

1. **E. biloba** Lindley (*E. rutidobulbon* Hooker; *E. schomburgkii* (Reichenbach) Reichenbach; *E. wercklei* Schlechter). Illustration: Botanical Magazine n.s., 611 (1972); Pabst & Dungs, Orchidaceae Brasilienses 2: 214, 276 (1977).
Epiphytic or terrestrial. Pseudobulbs ranging from elongate-cylindric (to 45 × 3 cm) to ovoid-oblong (to 14 × 8 cm), surface smooth or finely wrinkled, greenish, brown or dark purplish brown, with 2–4 leaves at the apex. Leaves 20–50 cm, narrowly lanceolate to narrowly ovate. Scape 30–110 cm, erect to arching, raceme with 20–35 flowers. Flowers 2.5–5 cm in diameter. Sepals and petals 1.2–2.5 cm, yellow to orange-yellow or brownish yellow, reddish purple or brownish on margins or flushed with red to varying degrees. Lip 1.2–2.5 cm, lateral lobes yellowish with reddish purple spots, central lobe white with reddish purple spots, notched at the apex. Callus of 2 erect fleshy plates with 2 separate fleshy lumps in front of each. Column *c.* 1 cm, greenish yellow. *Tropical America from Costa Rica to Peru & Bolivia*. G2. Summer–autumn.

Extremely variable in flower colour, lip shape and colour of pseudobulbs.

2. E. sceptrum Reichenbach & Warscewicz (*E. sprucei* Reichenbach; *E. helenae* Kränzlin). Illustration: Botanical Magazine, 8462 (1912); Pabst & Dungs, Orchidaceae Brasilienses **2**: 214, 277 (1977).
Epiphytic. Similar to *E. biloba* but generally smaller with flowers *c.* 3.5 cm in diameter, sepals and petals 1.3–1.6 cm, and lip *c.* 1 cm. The callus consists of 2 erect plate-like horns which are cream in colour. *Venezuela, Peru, Brazil.* G2. Summer.

96. MENDONCELLA Hawkes
V.A. Matthews
Epiphytic or terrestrial. Leaves 2–3 per pseudobulb, pleated, rolled when young. Raceme with few flowers, arising laterally from the base of the new growth. Sepals and petals spreading, lateral sepals more or less swollen at the base and joined to the column foot. Lip 3-lobed. Callus formed from a series of longitudinal ribs or plates. Column with distinct wings. Pollinia 4 in 2 pairs on a short, fleshy stipe.

A genus of 9 species, most of which have at some time been included in *Zygopetalum* or *Batemannia* and which will be found in much of the literature under the incorrect name *Galeottia* Richard. They are native to C and tropical S America, and are best grown in small pans, using an epiphytic compost with very free drainage to allow liberal watering during the growing season.

1a. Flower *c.* 5 cm in diameter; central lobe of lip with entire margin; wings of column entire **3. burkei**
b. Flower 7–8 cm in diameter; central lobe of lip fringed; wings of column toothed 2
2a. Central lobe of lip hairy or warty between the veins; callus violet-streaked **1. fimbriata**
b. Central lobe of lip hairless or sparsely hairy; callus orange **2. grandiflora**

1. M. fimbriata (Linden & Reichenbach) Garay. Illustration: Gardeners' Chronicle 1856: 660; American Orchid Society Bulletin **48**: 1223 (1979), **50**: 37 (1981).
Epiphytic. Pseudobulbs 5–6.5 cm, ovoid, with 2 leaves. Leaves 15–30 cm. Raceme with 2–5 flowers. Flowers *c.* 8 cm in diameter. Sepals and petals greenish or brownish yellow striped with reddish brown, acuminate. Lip white, striped longitudinally with violet, lateral lobes erect, fringed, central lobe ovate, recurved, fringed, acuminate, hairy or warty between the veins. Callus violet-streaked, formed of a number of plates arranged like a fan. Column whitish yellow, wings toothed. *Colombia, Venezuela.* G2.

2. M. grandiflora (Richard) Hawkes (*Galeottia grandiflora* Richard). Illustration: Botanical Magazine, 5567 (1866); Annals of the Missouri Botanical Garden **36**: 82 (1949).
Differs from *M. fimbriata* in having flowers 7–8 cm in diameter, the central lobe of the lip hairless or sparsely hairy and less sharply pointed at the apex, and an orange callus. *C America, Colombia.* G2. Spring–summer.

3. M. burkei (Riechenbach) Garay (*Zygopetalum burkei* Reichenbach). Illustration: Dunsterville & Garay, Venezuelan orchids illustrated **3**: 333 (1965); American Orchid Society Bulletin **48**: 1223 (1979).
Terrestrial. Pseudobulbs 5–10 cm, narrowly oblong, with 2–3 leaves. Leaves 20–45 cm. Raceme with 5–6 flowers. Flowers *c.* 5 cm in diameter. Sepals and petals light green or light brown thickly striped or spotted with brown. Lip white, lateral lobes with entire margins, central lobe with recurved sides, margin entire. Callus reddish purple, ribbed. Column yellow or greenish, streaked in front with purple, wings entire. *Guyana, Venezuela.* G2.

97. SCUTICARIA Lindley
V.A. Matthews
Epiphytic. Pseudobulbs small, somewhat stem-like, borne on a stout, branching rhizome. Leaves narrowly cylindric with a longitudinal groove, continuous with the pseudobulbs each of which bears 1 leaf. Raceme with 1–several flowers, arising laterally from a pseudobulb. Sepals and petals spreading, similar, the lateral sepals inserted on the column foot to form a mentum. Lip 3-lobed, joined to the column foot, lateral lobes erect, central lobc concave. Pollinia 2.

A genus of 5 species from tropical S America. They are best grown in epiphytic compost on rafts or blocks of tree fern suspended near the glass. They require ample moisture when growing, but should be kept dry when growth is completed, but not so much as to allow the foliage to shrivel.

1a. Leaves 60–150 cm; raceme with 1–3 flowers; central lobe of lip hairless **1. steelei**
b. Leaves 30–40 cm; raceme with 1 flower; central lobe of lip hairy **2. hadwenii***

1. S. steelei (Hooker) Lindley. Illustration: Pabst & Dungs, Orchidaceae Brasilienses **2**: 240, 296 (1977); American Orchid Society Bulletin **49**: 31 (1980).
Leaves hanging, 60–150 cm. Raceme usually hanging, with 1–3 flowers. Flowers 6–10 cm in diameter, fragrant. Sepals and petals yellow, blotched with reddish brown. Lip 3–4 cm, whitish to pale yellow marked with brownish pink especially on the lateral lobes, central lobe 2-lobed at apex, hairless. Callus yellowish orange with 3–5 apical teeth, hairy. Column whitish spotted with dark pink. *Brazil, Guyana, Colombia & Venezuela.* G2. Flowers throughout the year.

2. S. hadwenii (Lindley) Hooker. Illustration: Botanical Magazine, 4629 (1852); Pabst & Dungs, Orchidaceae Brasilienses **2**: 238 (1977).
Leaves hanging or sometimes more or less erect, 30–45 cm. Raceme with 1 flower, hanging or arching. Flowers 6–10 cm in diameter, fragrant. Sepals and petals yellowish green blotched with reddish brown. Lip pale yellow irregularly spotted with pinkish red, central lobe notched at apex, hairy. Callus with 3 apical teeth. Column whitish yellow flushed and spotted with red. *Brazil, Guyana.* G2. Spring–autumn.

***S. strictifolia** Hoehne. Illustration: Orchid Digest **41**: 235 (1977). Very similar to *S. hadwenii*, differing only in its erect foliage. The two may be the same, as plants as *S. hadwenii* growing in bright light may have the ability to produce erect leaves. *Brazil.* G2.

98. MAXILLARIA Ruiz & Pavon
V.A. Mathews
Epiphytic or terrestrial, with long or short rhizomes clothed in often papery sheaths and generally bearing pseudobulbs which are usually laterally compressed and bear 1–2 leaves. Leaves thin to leathery, with a prominent central vein, folded when young. Scapes produced in the axils of sheaths on the rhizome, solitary or in bundles, each bearing clasping, usually overlapping bracts and 1 flower. Sepals free, the laterals usually oblique and joined to the column foot to form a mentum. Petals similar but smaller, usually joined to the upper sepal. Lip hinged to or continuous with the column foot, often erect and parallel to the column, entire or 3-lobed, usually shorter than the sepals.

Column erect, somewhat curved, without wings. Pollinia 4 in 2 pairs, often 1 large and 1 small in each pair.

A genus of about 250 species from tropical and subtropical America. They should be grown in epiphytic compost; those with long rhizomes are best accommodated on rafts or blocks of tree fern. All should be kept moist throughout the year and should be shaded from bright sunlight. Maxillarias are prone to black spot on the foliage; this can be controlled by adequate ventilation and heating.

Pseudobulbs. Absent: **12.** Bearing 1 leaf: **1–11,13–17,20–24**; bearing 2 leaves: **2,3,16,18–21.** Spaced at intervals along the rhizome: **2,6,8,10,11,16–18,20,21, 24**; clustered: **1–7,9,13–16,19–23.**
Leaves. Linear: **10,11,17,18,23**; relatively broader: **1–9,12–16,19–22,24.**
Sepals. Linear: **1–6**; relatively broader: **7–24.** Length to 3 cm: **8,10–12,16–24**; length 3–8 cm: **1,3–7,9,13–16,18, 19,21**; length 10–15 cm; **2.**
Sepals and petals. Striped: **13.**
Central lobe of lip. Basic colour white, cream, yellow or orange: **2–21,23,24**; basic colour red to maroon: **1,10,16, 17,21–23.** Spotted: **5,7,10,12,14,17, 21**; not spotted: **1–4,6–9,11,13–24.** Entire:**1,9,10,15–17**; 3-lobed: **2–9,11–14,17–24.**

1a. Sepals linear to narrowly strap-shaped, 3–15 cm — 2
b. Sepals ovate to oblong or broadly lanceolate, if narrower, then less than 3 cm — 8
2a. Sepals maroon, yellow at the base, apex maroon-orange — **1. nigrescens**
b. Sepals white, yellow or greenish, sometimes tinged with purple or brown, or red at the base — 3
3a. Sepals 10–15 cm, twisted — **2. fractiflexa**
b. Sepals 3–7.5 cm, rarely twisted — 4
4a. Sepals white on inner surface — **7. venusta**
b. Sepals pale yellow or greenish on inner surface, sometimes tinged with brown, or red at the base — 5
5a. Central lobe of lip white, recurved — **3. setigera**
b. Central lobe of lip deep cream to deep yellow, sometimes with purplish spots, not or scarcely recurved — 6
6a. Central lobe of lip obtuse at the apex, margin wavy, white — **4. luteo-alba**
b. Central lobe of lip acute or mucronate, margin neither wavy nor white — 7
7a. Lip with some purplish spots; pseudobulbs 2.5–5 cm — **5. lepidota**
b. Lip lacking spots; pseudobulbs 5–8 cm — **6. ochroleuca**
8a. Sepals entirely white or cream — 9
b. Sepals not white or cream, or if so then marked with another colour — 12
9a. Flowers *c.* 2 cm in diameter; sepals 1.8–2.5 cm — 10
b. Flowers 9–15 cm in diameter; sepals 3.5–7 cm — 11
10a. Pseudobulbs borne at intervals along the rhizome; lip *c.* 1.2 cm with a downy central lobe — **8. alba**
b. Pseudobulbs clustered; lip 3.5–5 mm, central lobe hairless — **23. densa**
11a. Sepals lanceolate — **7. venusta**
b. Sepals ovate-oblong to triangular-ovate — **9. grandiflora**
12a. Leaves linear, grass-like — 13
b. Leaves lanceolate or broader — 17
13a. Pseudobulbs with 2 leaves — **18. chrysantha**
b. Pseudobulbs with 1 leaf — 14
14a. Lip marked with a number of reddish brown or purplish spots — **10. tenuifolia**
b. Lip unspotted — 15
15a. Pseudobulbs clustered; lip less than 6 mm — **23. densa**
b. Pseudobulbs borne at intervals along rhizome; lip to 1.3 cm — 16
16a. Apex of central lobe of lip yellowish — **17. variabilis***
b. Apex of central lobe of lip white — **11. sanguinea**
17a. Lateral sepals directed upwards, towards upper sepal — 18
b. Lateral sepals directed downwards, away from upper sepal, or spreading horizontally — 23
18a. Plant hanging, with leaves growing in a fan; pseudobulbs absent; sepals and petals pale greenish yellow — **12. valenzuelana**
b. Plant erect or creeping; pseudobulbs present; sepals and petals not greenish yellow, or if so, then striped — 19
19a. Sepals and petals bright pink to orange-red or red — 20
b. Sepals and petals not as above — 21
20a. Pseudobulbs 1–2 cm, borne at intervals along rhizome — **24. sophronitis**
b. Pseudobulbs 2.5–4 cm, clustered — **22. coccinea**
21a. Sepals and petals white, with purple blotches at the base — **15. sanderiana**
b. Sepals and petals white, greenish yellow or purple, lacking purple blotches at the base — 22
22a. Sepals with longitudinal stripes — **13. striata**
b. Sepals with spots towards apex — **14. fucata**
23a. Lip entire — 24
b. Lip 3-lobed — 25
24a. Scapes to 4 cm — **17. variabilis***
b. Scapes 5–25 cm — **16. elatior***
25a. Pseudobulbs with 2 leaves — 26
b. Pseudobulbs with 1 leaf — 29
26a. Central lobe of lip acute — 27
b. Central lobe of lip obtuse or rounded — 28
27a. Sepals and petals spotted — **21. picta***
b. Sepals and petals not spotted — **20. marginata**
28a. Flowers 2–3 cm in diameter — **19. porphyrostele**
b. Flowers 4–4.5 cm in diameter — **18. chrysantha**
29a. Callus hairless — 30
b. Callus downy — **21. picta***
30a. Central lobe of lip bright pink, red, yellow or orange-yellow — 31
b. Central lobe of lip pale yellow with a dark purple margin — **20. marginata**
31a. Pseudobulbs borne at intervals along rhizome; scapes 1–2.5 cm — 32
b. Pseudobulbs clustered; scapes 4–6 cm — **22. coccinea**
32a. Pseudobulbs 1–2 cm; leaves 1–8 cm; central lobe of lip warty — **24. sophronitis**
b. Pseudobulbs 1.5–6 cm; leaves 5–25 cm; central lobe of lip smooth — **17. variabilis***

1. M. nigrescens Lindley. Illustration: Cogniaux & Goossens, Dictionnaire Iconographique des Orchidées, Maxillaria, t. 3 (1899); Dunsterville & Garay, Venezuelan orchids illustrated 3: 189 (1965); American Orchid Society Bulletin **43**: 963 (1974).
Epiphytic. Pseudobulbs to 9 cm, clustered, ovoid, compressed, each with 1 leaf. Leaves strap-shaped, 20–35 cm. Scapes usually erect, to 12 cm. Flowers *c.* 10 cm in diameter. Sepals narrowly strap-shaped, 4.5–6 cm, maroon, yellow at the base and maroon-orange at the apex. Petals similar, but only 3.8–5.5 cm and usually darker than sepals. Lip entire, oblong, obtuse, 1.7–2 cm, deep maroon, flushed with brownish purple. *Colombia, Venezuela.* G2. Summer–autumn.

2. M. fractiflexa Reichenbach. Illustration: Gardeners' Chronicle **31**: 359 (1902). Leaves strap-shaped, to 50 cm. Scapes erect. Flowers *c.* 15 cm in diameter. Sepals and petals linear, 10–15 cm, curved and twisted, yellowish tinged with brown and often purplish at the base and apex. Lip 3-lobed, white, part of the rounded lateral lobes and the central lobe red. Callus white. *Ecuador*. G2.

3. M. **setigera** Lindley. Illustration: Botanical Magazine, 4434 (1849); Dunsterville & Garay, Venezuelan orchids illustrated **3**: 192 (1965); American Orchid Society Bulletin **43**: 963 (1974).
Epiphytic. Pseudobulbs 3–6 cm, clustered, spherical, somewhat compressed, each with 1–2 leaves. Leaves narrow, 12–30 cm. Scapes 7–30 cm. Flowers fragrant, 10 cm or more across. Sepals narrowly strap-shaped, to 6.5 cm, white at base, greenish yellow towards apex. Petals similar, incurved, to 5 cm, with recurved margins. Lip 3-lobed, to 2.5 cm, lateral lobes white striped with purple, central lobe oblong, reflexed, white, margin toothed. Callus yellow, hairy. *Colombia, Venezuela, Guyana & Brazil*. G2. Autumn.

4. M. **luteo-alba** Lindley. Illustration: Cogniaux & Goossens, Dictionnaire Iconographique des Orchidées, Maxillaria, t. 1 (1899); American Orchid Society Bulletin **43**: 962 (1974); Bechtel et al., Manual of cultivated orchid species, 229 (1981).
Epiphytic. Pseudobulbs 3–5 cm, clustered, oblong-ovoid, compressed, each with 1 leaf. Leaves linear-lanceolate, 25–45 cm. Scapes erect, 9–15 cm. Flowers very fragrant, 10 cm or more across. Sepals narrowly strap-shaped, 4–7.5 cm, whitish, or yellow at the base, becoming brownish orange towards the apex. Petals shorter, similar in colour although sometimes with purplish streaks. Lip 3-lobed, to 2.5 cm, downy inside, cream or yellow, lateral lobes purple-streaked, central lobe ovate, obtuse with a white wavy margin. Callus with 2 lateral ridges, downy towards base. *Panama, Colombia, Venezuela & Ecuador*. G2. Summer.

5. M. **lepidota** Lindley. Illustration: Dunsterville & Garay, Venezuelan orchids illustrated **6**: 253 (1976); Bechtel et al., Manual of cultivated orchid species, 229 (1981).
Epiphytic. Pseudobulbs 2.5–5 cm, clustered, oblong, each with 1 leaf. Leaves linear-lanceolate, 15–35 cm. Scapes to 15 cm, often with the flowers hanging. Sepals linear, 4–6 cm, yellow with red at the base and often brown at the tip. Petals similar to sepals but shorter. Lip 3-lobed, *c*. 2 cm, deep cream, spotted with dark purple or maroon towards the apex, reflexed, mealy on the upper surface. Callus grooved, hairy. *Colombia, Ecuador & Venezuela*. G2.

6. M. **ochroleuca** Lindley. Illustration; Pabst & Dungs, Orchidaceae Brasilienses **2**: 235, 291 (1977).
Epiphytic. Pseudobulbs 5–8 cm, compressed, each with 1 leaf. Leaves 25–40 cm. Scapes erect, 5–10 cm, many borne together in a bundle. Flowers fragrant. Sepals 3–3.5 cm, linear, slenderly pointed, pale yellow. Petals similar to sepals but shorter. Lip 3-lobed, pale yellow, central lobe oblong, obtuse, downy, deep yellow at apex. *Brazil*. G2.

7. M. **venusta** Linden & Reichenbach. Illustration: Botanical Magazine, 5296 (1862); Cogniaux & Goossens, Dictionnaire Iconographique des Orchidées, Maxillaria, t. 6 (1902).
Epiphytic. Pseudobulbs 4–7.5 cm, clustered, ovoid, much compressed, each with 1 leaf. Leaves strap-shaped, 30–38 cm. Scapes 10–25 cm. Flowers very fragrant, 12–15 cm in diameter, hanging. Sepals lanceolate, to 6.5 cm, white, usually with recurved margins. Petals similar, to 4.5 cm. Lip 3-lobed, to 2 cm, lateral lobes oblong, pale yellow with red margins, central lobe ovate, obtuse, reflexed, pale yellow with red margin and sometimes with red spots, mealy. Callus oblong, downy. *Colombia, Venezuela*. G2. Summer–autumn.

8. M. **alba** (Hooker) Lindley. Illustration: Hoehne, Flora Brasilica **12**(7): t. 91 (1953); Dunsterville & Garay, Venezuelan orchids illustrated **2**: 197 (1961).
Epiphytic. Pseudobulbs 2.5–5 cm, borne at short intervals along the rhizome, ovoid to narrowly ovoid, strongly compressed, each with 1 leaf. Leaves strap-shaped, 25–40 cm. Scapes 2.5–4 cm. Flowers slightly fragrant, *c*. 2 cm across. Sepals broadly lanceolate, 1.8–2.5 cm, creamy white. Petals similar, 1.6–1.8 cm. Lip 3-lobed towards the apex, hinged to the column foot, *c*. 1.2 cm, pale to bright yellow with an ovate or elliptic central lobe which is downy. Callus strap-shaped, downy, *Most of C America & northern S America*. G2. Spring.

9. M. **grandiflora** (Humboldt, Bonpland & Kunth) Lindley. Illustration: Dunsterville & Garay, Venezuelan orchids illustrated **5**: 185 (1966); Die Orchidee **23**: 92 (1972).
Epiphytic. Pseudobulbs 3.5–7 cm, clustered, ovoid, compressed, each with 1 leaf. Leaves strap-shaped, 25–50 cm. Scapes erect but bent at apex, 10–25 cm. Flowers slightly scented, *c*. 10 cm across, drooping. Sepals ovate-oblong to triangular-ovate, 3.5–5.5 cm, white, sometimes with margins rolled under; lateral sepals with reflexed tips. Petals shorter and narrower than sepals, white, sometimes with pink streaks on the basal half, tips reflexed. Lip entire or very slightly 3-lobed, 2.5–3 cm, lateral lobes (if present) deep pink or purple, streaked paler, central lobe white or yellowish sometimes with a purple margin, wavy, reflexed, mealy. Callus mealy. *Northwestern S America*. G2. Spring–summer.

10. M. **tenuifolia** Lindley. Illustration: Shuttleworth et al., Orchids, 110 (1970); Die Orchidee **23**: 92 (1972); Bechtel et al., Manual of cultivated orchid species, 230 (1981).
Terrestrial or epiphytic. Pseudobulbs 2–6 cm, borne at intervals of 2–4 cm along the rhizome, ovoid, somewhat compressed, each with 1 leaf. Leaves linear, 12–50 cm. Scapes 2–6 cm. Flowers coconut-scented, 4–5 cm in diameter. Sepals narrowly elliptic, 1.7–2.8 cm, dark red, heavily mottled with red or yellow. Petals similar, 1.5–2.4 cm, directed forwards. Lip 1.5–2.5 cm, entire, oblong, reflexed, dark red at base, yellow or whitish towards apex, spotted with purple or reddish brown. Callus downy, dark purple. *Mexico to Costa Rica*. G2. Summer–autumn.

11. M. **sanguinea** Rolfe. Illustration: Hartmann, Introduction to the cultivation of orchids, 89 (1971).
Pseudobulbs to 3 cm, borne at intervals along the rhizome, ovoid, compressed, each with 1 leaf. Leaves linear, to 40 cm. Scapes to 3 cm. Sepals elliptic, to 2.5 cm, dark red or brownish, mottled with red or yellow. Petals similar, to 2 cm. Lip scarcely 3-lobed, to 1.3 cm, elliptic, dark red, pale yellow at base, white at apex. Callus dark red. *Costa Rica; ?Colombia*. G2. Summer.

12. M. **valenzuelana** (Richard) Nash. Illustration: Shuttleworth et al., Orchids, 111 (1970); Bechtel et al., Manual of cultivated orchid species, 231 (1981).
Hanging epiphyte without pseudobulbs. Leaves growing in a fan, 8–20 cm, lanceolate. Scapes to 5 cm, solitary or in bundles in the leaf axils. Sepals ovate, 1–1.4 cm, pale greenish yellow. Petals similar, 8–10 mm. Lip scarcely 3-lobed, *c*. 1 cm, pale brownish yellow with purple spots. Callus in 3 parts, glandular-downy.

C America, northern S America, West Indies. G2. Summer.

13. **M. striata** Rolfe. Illustration: Cogniaux & Goossens, Dictionnaire Iconographique des Orchidées, Maxillaria, t. 4 (1899); Lindenia **9**: t. 398 (1883–84).
Pseudobulbs 4.5–8 cm, clustered, ovoid, compressed, each with 1 leaf. Leaves broadly elliptic, to 25 cm. Scapes more or less erect or arching, to 30 cm. Flowers *c.* 12 cm in diameter. Sepals greenish yellow with reddish purple stripes, 4.5–7 cm, upper sepal oblong-lanceolate, lateral sepals triangular-lanceolate, forming a mentum with the column foot. Petals similar in colour, shorter, lanceolate, often with wavy margins. Lip erect, recurved, slightly 3-lobed towards apex, lateral lobes more or less obovate, white striped with reddish purple, central lobe ovate-lanceolate, white. Callus linear-oblong with many grooves. *Peru, Ecuador.* G2. Summer–autumn.

14. **M. fucata** Reichenbach. Illustration: Botanical Magazine, 9376 (1934); Shuttleworth et al., Orchids, 109 (1970).
Pseudobulbs 3.5–6 cm, clustered, narrowly ovoid, compressed, each with 1 leaf. Leaves narrowly oblong, 30–40 cm. Scapes erect, usually numerous, to 22 cm. Flowers *c.* 6 cm across, fragrant. Sepals 3–3.5 cm, ovate, reflexed at tips, white at the base, purplish in the middle and yellow at the apex, with reddish brown spots towards apex. Petals *c.* 2.5 cm, lanceolate-ovate, similar in colour to sepals but sometimes without spots. Lip 3-lobed, reflexed, lateral lobes yellowish, striped and edged with brown, central lobe ovate-oblong, notched at apex, yellowish, striped, spotted or suffused with brownish red. Callus yellow, strap-shaped, grooved and hairless. *?Ecuador.* G2. Late summer–autumn.

The name is often given erroneously as *M. fuscata* or *M. furcata*.

15. **M. sanderiana** Reichenbach. Illustration: Kupper & Linsenmaier, Orchids, t. 42 (1961); Bechtel et al., Manual of cultivated orchid species, 230 (1981).
Terrestrial or epiphytic. Pseudobulbs 2.5–5 cm, clustered, ovoid, compressed, each with 1 leaf. Leaves broadly oblong, 15–40 cm. Scapes ascending or horizontal, to 25 cm. Flowers 10–15 cm across, fragrant. Sepals white, blotched at base with reddish purple, upper sepal oblong, to 7.5 cm, lateral sepals triangular-ovate, slightly longer than upper sepal, forming a mentum with the column foot. Petals shorter than upper sepal, lanceolate-triangular, reflexed at tips, similar in colour to sepals. Lip entire, erect, recurved, elliptic with a crisped margin, white or yellow, streaked and stained with red or purple. Callus tongue-shaped, yellow, hairless. *Ecuador, Peru.* G2. Summer–early autumn.

16. **M. elatior** Reichenbach. Illustration: Botanical Magazine, 9206 (1930).
Epiphytic. Pseudobulbs 4.5–9 cm, ovoid, compressed, scattered along the rhizome or often solitary at the end of the rhizome, each with 1–2 leaves. Leaves 12–40 cm, lanceolate. Scapes erect, 5–7.5 cm. Sepals 2.2–3.2 cm, oblong-elliptic, reddish yellow, often spotted or striped with red, upper sepal concave, lateral sepals spreading or recurved. Petals 1.7–2.5 cm, dull red. Lip entire, to 2 cm, ovate-oblong, recurved at apex, dark red. Callus *c.* 1 cm, blackish, hairless. *Mexico, Guatemala, Honduras & Costa Rica.* G2. Spring–summer.

***M. elegantula** Rolfe. Pseudobulbs each with 1 leaf and scapes to 25 cm. The sepals and petals are white, yellow toward the tips, flushed and spotted with brownish purple or deep blue, margins revolute. Lip to 1.7 cm, elliptic, apex recurved and minutely toothed and crisped, usually purplish red on margin. *Ecuador, Peru.* G2.

17. **M. variabilis** Lindley. Illustration: Die Orchidee **23**: 92 (1972).
Terrestrial or epiphytic. Pseudobulbs 1.5–6 cm, narrowly ellipsoid to ovoid, borne at 2–3 cm intervals along the rhizome, each with 1 leaf. Leaves 5–25 cm, very narrowly strap-shaped. Scapes 2–2.5 cm. Flowers 2–2.5 cm in diameter. Sepals 6–20 mm, ovate, white, greenish yellow, orange, deep red or purplish, often marked with red. Petals 5–14 mm, oblong-lanceolate, similar in colour to sepals. Lip entire or slightly 3-lobed near apex, up to 1.2 cm, hinged to column foot, erect, usually deep red and often with yellow blotch at apex. Callus strap-shaped, deep red, hairless. *C America, West Indies, Guyana.* G2.

***M. curtipes** Hooker (*M. houtteana* Reichenbach). Illustration: Cogniaux & Goossens, Dictionnaire Iconographique des Orchidées, Maxillaria, t. 2 (1899). Scapes 2–4 cm. Sepals and petals yellowish, overlaid with deep red inside and somewhat blotched towards the base, margins orange. Lip entire, oblong to broadly elliptic, reflexed, deep yellow, spotted and striped with reddish purple. Callus warty. *Mexico, Guatemela, El Salvador & Costa Rica.* G2. Spring.

18. **M. chrysantha** Rodriguez. Illustration: Botanical Magazine, 8979 (1923); Hoehne, Flora Brasilica **12**(7): t. 127 (1953).
Epiphytic. Pseudobulbs 1.5–2.5 cm, ovoid-conical, slightly compressed, produced at intervals of 2.5–4 cm along the rhizome, each with 2 leaves. Leaves 20–35 cm, linear. Scapes 7–10 cm. Flowers 4–4.5 cm in diameter. Sepals 2.8–3 cm, oblong, yellow with reddish or mauve margins. Petals 2.2–2.4 cm, similar in colour to sepals. Lip 3-lobed, recurved at apex, more or less circular, yellowish with violet blotches on the lateral lobes, margin of central lobe wavy. Callus strap-shaped, downy. *Surinam, Brazil.* G2.

19. **M. porphyrostele** Reichenbach. Illustration: Botanical Magazine, 6477 (1880); Cogniaux & Goossens, Dictionnaire Iconographique des Orchidées, Maxillaria, t. 10 (1904); Pabst & Dungs, Orchidaceae Brasilienses **2**: 234, 290 (1977).
Pseudobulbs 2.5–3 cm, clustered, ovoid, compressed, each with 2 leaves. Leaves 12–20 cm, lanceolate. Scapes to 7.5 cm. Flowers *c.* 3 cm across, slightly fragrant. Sepals oblong, incurving, pale yellow. Petals shorter, incurving, pale yellow with purple stripes at the base. Lip 3-lobed, lateral lobes yellow striped with purple, central lobe oblong-circular, yellow. Callus yellow marked with purple. *Brazil.* G2. Winter–spring.

20. **M. marginata** (Lindley) Fenzl. Illustration: Hoehne, Flora Brasilica **12**(7): t. 141 (1953); Pabst & Dungs, Orchidaceae Brasilienses **2**: 290 (1977).
Pseudobulbs 4–6 cm, each with 1–2 leaves. Leaves 12–30 cm, lanceolate. Scapes 7.5–15 cm, erect or arching. Sepals 2–2.5 cm, narrowly oblong, yellow with a dark red margin and a dorsal red line. Petals 1.7–2.2 cm, similar in colour to sepals. Lip 3-lobed, lateral lobes pale yellow streaked with reddish purple, central lobe oblong, reflexed, acute, pale yellow with a scalloped purple margin. Callus strap-shaped, hairless. *Colombia, Ecuador, Brazil.* G2. Summer–autumn.

21. **M. picta** Hooker. Illustration: Botanical Magazine, 3154 (1832); Bechtel et al., Manual of cultivated orchid species, 230 (1981).
Epiphytic. Pseudobulbs 4–7 cm, ovoid, somewhat compressed, clustered or borne at intervals along the rhizome, each with 1–2 leaves. Leaves to 30 cm, narrowly

oblong. Scapes 12–16 cm. Flowers fragrant. Sepals *c.* 3 cm, oblong, incurved, whitish to yellow with purple or dark brown spots usually on the dorsal surface. Petals 2.5–2.8 cm, similar to sepals but with a red basal streak, very incurved. Lip 3-lobed, lateral lobes white with purple marks, central lobe oblong, reflexed, acute, whitish with purple spots. Callus oblong, downy. *Colombia, Brazil.* G2. Spring–summer.

***M. cucullata** Lindley (*M. meleagris* Lindley; *M. praestans* Reichenbach). Illustration: Bechtel et al., Manual of cultivated orchid species, 229 (1981). Sepals 2.5–5 cm, usually deep red, occasionally yellow, brownish, pinkish or almost black, variously striped and spotted with dark red. Petals shorter, similar in colour, with the tips meeting above the column. Lip 3-lobed, hinged to column foot, yellow or whitish with reddish brown or purple spots. or occasionally wholly dark red, central lobe finely warty. *Mexico to Panama.* G2. Summer–autumn.

Very variable in the size of the plant and the flowers and in the width of the sepals and petals.

***M. rufescens** Lindley. Illustration: Dunsterville & Garay, Venezuelan orchids illustrated **2**: 219 (1961); American Orchid Society Bulletin **43**: 966 (1974). Scapes 1–5 cm. Sepals 1–2.5 cm, greenish white, pinkish or yellow, usually flushed with brownish pink or maroon. Petals very slightly shorter, similar in colour, sometimes with deep red splashes on veins. Lip 3-lobed, hinged to column foot, lateral lobes pointed, somewhat sickle-shaped, central lobe oblong, slightly reflexed, usually pale or deep yellow, or reddish, spotted with maroon or brownish-red. Callus brownish yellow or deep red, warty and velvety. *Tropical America.* G2. Autumn.

An extremely variable species.

22. M. coccinea (Jacquin) Hodge (*Ornithidium coccineum* (Jacquin) R. Brown). Illustration: Loddiges' Botanical Cabinet **4**: t. 301 (1819); Hooker, Exotic Flora **1**: t. 38 (1823).
Epiphytic. Pseudobulbs 2.5–4 cm, ovoid, more or less compressed, clustered, each with 1 leaf. Leaves 10–35 cm, lanceolate. Scapes 4–6 cm, borne in bundles. Flowers bright pink to red. Sepals 1.1–1.2 cm, concave, ovate to narrowly ovate. Petals 7–8 mm, similar. Lip 3-lobed, 6–8 mm, the central lobe curved downwards, narrowly ovate, apex acute. Callus fleshy, transverse. *West Indies, Colombia, Venezuela.* G2. Summer.

23. M. densa Lindley (*Ornithidium densum* (Lindley) Reichenbach). Illustration: Hamer, Orquideas de El Salvador **2**: 111 (1974); Bechtel et al., Manual of cultivated orchid species, 229 (1981).
Differs from *M. coccinea* in its very variable flower colour which ranges from greenish white to yellow, purplish, maroon or brownish red. The pseudobulbs are often larger, the leaves linear and the flowers slightly smaller with a lip 3.5–5 mm, whose central lobe is more or less circular to broadly ovate, obtuse at the apex and with a small point. Callus plate-like, obtuse. *C America from Mexico to Honduras.* G2.

24. M. sophronitis (Reichenbach) Garay (*Ornithidium sophronitis* Reichenbach). Illustration: Dunsterville & Garay, Venezuelan orchids illustrated **1**: 237 (1959).
Epiphytic. Pseudobulbs 1–2 cm, borne at 2.5–6 cm intervals along the creeping rhizome, slightly compressed, each with 1 leaf. Leaves 1–8 cm, oblong, thick, apex often asymmetric. Flowers borne on scapes *c.* 1 cm. Sepals and petals ovate, bright orange-red; sepals 9–13 mm, petals smaller. Lip 3-lobed, *c.* 7 mm, yellow to orange-yellow, central lobe oblong, the upper surface warty and margin minutely toothed. Callus smooth, flat. *Venezuela.* G2.

99. MORMOLYCA Fenzl

V.A. Matthews

Epiphytic with creeping rhizomes. Pseudobulbs with 1–4 leaves, subtended by several bracts. Leaves folded when young. Flowers solitary, arising from an axil of a pseudobulb bract. Sepals spreading, more or less equal, free. Petals similar but smaller. Lip simple or 3-lobed. Column without a foot. Pollinia 4 in 2 pairs, waxy.

A genus of 5 species native to C and tropical S America. Its name is often misspelled 'Mormolyce'. Cultivation as for *Maxillaria* (p. 236).

1. M. ringens (Lindley) Schlechter. Illustration: Fieldiana **26**: 591 (1952); Canadian Journal of Botany **37**: 487 (1959); Bechtel et al., Manual of cultivated orchid species, 234 (1981).
Pseudobulbs 2–5 cm, clustered or arising at intervals of 1–2 cm along the rhizome, rounded to ovoid, compressed, each bearing 1 leaf. Leaves 10–35 cm. Sepals 1.5–2 cm, yellowish green to light lavender, striped longitudinally with violet or dark red. Petals similar in colour. Lip *c.* 2 cm, pale purple to dark red, almost erect, jointed with the base of the column, central lobe broadly rounded to oblong-elliptic, curved downwards, hairless or minutely warty towards the margin. Callus triangular, apex entire or 3-toothed. *C America from Mexico to Costa Rica.* G2. Summer–autumn.

The flowers have a characteristic shape resulting from the upper sepal and petals being directed upwards and the lateral sepals and lip downwards.

100. TRIGONIDIUM Lindley

V.A. Matthews

Epiphytic, with creeping rhizomes. Pseudobulbs with 1–5 leaves, ribbed, subtended by several fibrous bracts. Leaves strap-shaped, leathery. Flowers usually solitary, arising from the axils of the pseudobulb bracts. Sepals forming a basal tube, recurved in the upper half. Petals much smaller than sepals. Lip 3-lobed, shorter than petals, almost erect, lateral lobes erect and parallel to column, central lobe fleshy. Callus usually strap-shaped. Pollinia 4 on a short, laterally crescent-shaped stipe.

A genus of 14 species from C and tropical S America. Cultivation as for *Maxillaria* (p. 236).

1. T. egertonianum Lindley (*T. seemanni* Reichenbach). Illustration: Hawkes, Encyclopaedia of cultivated orchids, 475 (1965); Dunsterville & Garay, Venezuelan orchids illustrated **3**: 321 (1965).
Pseudobulbs 4–9 cm, more or less spherical to ellipsoid, compressed, each with 2 leaves. Leaves 20–60 cm. Scapes 15–35 cm. Sepals and petals yellowish green or pinkish with reddish brown or purple veins, upper sepal 2.5–4.5 cm, lateral sepals 2.5–4 cm, petals 1–2 cm with a purple or brown thickening just below the apex. Lip 5–10 mm, green or yellowish brown striped with dark purple or brownish red, central lobe reflexed, warty. *C America, Colombia, Venezuela.* G2. Spring.

101. DICHAEA Lindley

V.A. Matthews

Epiphytes without pseudobulbs. Leaves in 2 ranks, folded when young, the sheaths concealing the stem. Flowers solitary from the leaf axils. Sepals and petals free, similar, though petals usually smaller. Lip joined to the base of the column, clawed, usually 3-lobed, the lateral lobes often recurved, usually without a callus. Pollinia 4, ovoid.

A genus of 40 species occurring in

tropical America and the West Indies. Only 4 species are generally cultivated. They are best grown on a block of tree fern or bark. Cultivation otherwise as for *Angraecum* (p. 254).

1a. Leaf-blade not jointed to the sheath, leaves persistent; lip with the narrow lateral lobes pointing backwards; capsule spiny — 2
b. Leaf-blade jointed to the sheath, leaves deciduous; lip with backwardly pointing lateral lobes; capsule not spiny — 3
2a. Sepals orange, lanceolate, warty on the outer surface — **1. muricata**
b. Sepals pale yellow to pinkish, ovate, not warty — **2. pendula**
3a. Lip 4.5–6 mm, with a callus; leaves green — **3. graminoides**
b. Lip 6–7 mm, without a callus; leaves glaucous — **4. glauca**

1. D. muricata (Swartz) Lindley. Illustration: Schultes, Native orchids of Trinidad and Tobago, 259 (1960); Lasser, Flora de Venezuela **15**(5): 461 (1970).
Stems hanging, 30–50 cm. Leaves 1.5–2 cm, ovate-oblong, acuminate, persistent, not jointed to the sheaths. Sepals orange, lanceolate, 7–10 mm; petals purplish-spotted, 5–8 mm; all warty on the outer surfaces. Lip 5–8 mm, pale purple with 2 recurved, narrow, lateral lobes. Capsule 1–2 cm, densely spiny. *Mexico to Brazil, West Indies*. G2. Winter.

2. D. pendula (Aublet) Cogniaux. Illustration; Lasser, Flora de Venezuela **15**(5): 466 (1970).
Stems hanging. Leaves 1–2.5 cm, elliptic-oblong to lanceolate, each with a small point, persistent, not jointed to the sheaths. Sepals 8–9 mm, pale yellow to pinkish, ovate. Petals *c.* 8 mm, pinkish-buff, sometimes with bluish purple marks near the tips, obovate. Lip 5–9 mm, white or pale purple with bluish purple blotches and bands, with 2 recurved, narrow side lobes. Capsule 1–1.5 cm, spiny. *Southern C America to northern S America, West Indies*. G2. Summer–autumn.

3. D. graminoides (Swartz) Lindley.
Stem erect, 10–30 cm, strongly flattened. Leaves 3–4 cm, linear, jointed to the sheaths, eventually falling. Sepals and petals 5–8 mm, whitish, ovate to lanceolate. Lip 4.5–6 mm, whitish, heart-shaped at the base, shallowly 3-lobed, central lobe almost circular, with a small point. Callus present, linear. Capsule 1.5–2 cm, not spiny. *C and tropical S America, West Indies*. G2.

4. D. glauca (Swartz) Lindley.
Stems ascending or hanging, flattened. Leaves 3.5–7 cm, linear-oblong, glaucous, jointed to the sheaths, eventually falling. Flowers fragrant. Sepals and petals 7–8 mm, generally white but sometimes tinged and spotted with lilac and yellow. Lip 6–7 mm, whitish with a dark red spot at the base, claw broad, lateral lobes very short or absent, central lobe broadly ovate to kidney-shaped. Capsule 1–2 cm, not spiny. *Northern C America, Cuba, Hispaniola & Jamaica*. G2. Summer.

102. ORNITHOCEPHALUS Hooker
V.A. Matthews

Epiphytic, without pseudobulbs. Leaves fleshy, leathery, in fan-shaped clusters, densely overlapping and concealing the stem, jointed to sheaths, the part below the joint often thickened. Racemes erect or hanging, with several–many flowers, arising from the leaf axils. Flowers not resupinate. Sepals usually equal, free. Petals equal to or larger than sepals. Lip joined to the base of the column, simple or 3-lobed; callus present. Rostellum long-beaked. Anther with a long terminal appendage. Pollinia 4 on a slender stipe.

A genus of about 50 species from the American tropics. They are best grown in small shallow pans suspended near the glass, using an epiphytic compost. They require light but must be shaded from the direct rays of the sun. Moisture must be provided throughout the year, particularly during the summer.

1a. Leaves 7.5–15 cm; racemes 22–25 cm — **3. grandiflorus**
b. Leaves 2.5–7.5 cm; racemes to 8 cm — 2
2a. Sepals 2–2.5 mm; lip 4–5 mm, central lobe obtuse — **1. iridifolius**
b. Sepals 3–4 mm; lip 5–6 mm, central lobe acute — **2. gladiatus**

1. O. iridifolius Reichenbach.
Leaves 2.5–7.5 cm, linear-lanceolate. Racemes 4–8 cm, loose, with many flowers. Flowers white. Sepals 2–2.5 mm, almost circular to broadly elliptic, with a keel projecting beyond the tip. Petals 3–3.5 mm, fan-shaped. Lip 4–5 mm, deeply 3-lobed, lateral lobes almost circular, central lobe triangular-ovate. *Mexico, Guatemala*. G2. Summer.

2. O. gladiatus Hooker. Illustration: Hooker, Exotic flora **2**: t. 127 (1824); Howard, Flora of the Lesser Antilles, 214 (1974).
Leaves 2.5–4 cm, lanceolate. Raceme to 5 cm, loose, with few flowers. Flowers greenish yellow to greenish white. Sepals 3–4 mm with a keel projecting beyond the tip. Petals 3–4 mm. Lip 5–6 mm, 3-lobed, central lobe linear-oblong with recurved margins, acute. *C. America, northwest S America, West Indies*. G2. Spring–summer.

3. O. grandiflorus Lindley (*Dipteranthus grandiflorus* (Lindley) Pabst). Illustration: Pabst & Dungs, Orchidaceae Brasilienses **2**: 263, 322 (1977).
Leaves 7.5–15 cm, narrowly oblong. Racemes 22–25 cm with many flowers. Sepals concave, white with a greenish mark at the base, petals similar but a little larger. Lip 3-lobed, white, greenish or yellowish at the base, concave, margins finely crisped. *Brazil*. G2. Summer.

103. SARCOCHILUS R. Brown
V.A. Matthews

Epiphytic. Pseudobulbs absent. Stems covered with the remains of old sheathing leaf-bases. Leaves in 2 ranks, folded when young. Racemes lateral with 3–25 fragrant flowers. Sepals and petals similar, lateral sepals joined to the column foot. Lip hinged to column foot, shallowly pouched, 3-lobed, central lobe much smaller than lateral lobes, with a forward-pointing spur. Surface of lip with a large callus, and some calluses present on the inside walls of the spur. Pollinia 4, in 2 pairs.

A genus of about 12 species, mainly native to Australia. Though good drainage is essential to their culture, they should never be allowed to dry out. Treatment varies somewhat, depending on the origin of the species. Some are best grown in pots in an epiphytic compost, others on rafts or tree fern, close to the glass.

1a. Sepals and petals white, pink or purple, without spots at the base; stems 5–12 cm, unbranched — **1. falcatus***
b. Sepals and petals white spotted with red at the base; stems to 1 m, usually branched — **2. hartmanni***

1. S. falcatus R. Brown. Illustration: Nicholls, Orchids of Australia, t. 452, 453 (1969); Dockrill, Australian indigenous orchids, t. 25 (1969).
Stems 5–7.5 cm. Leaves narrowly oblong, sickle-shaped, 5–14 cm, channelled. Racemes hanging, with 3–12 flowers. Flowers 2–3.5 cm in diameter. Sepals and petals oblong, 8–16 mm, white, sometimes with a purplish midrib on the back. Lip

3.5–6 mm, lateral lobes ovate-oblong, flushed with orange and with red stripes, central lobe broad, about a quarter of the length of the lateral lobes. Spur fleshy, sometimes marked with purple. Callus at opening of spur 2-lobed, spotted, some calluses on inside wall of spur. *Australia.* G2. Autumn–winter.

*__S. ceciliae__ Mueller. Illustration: Nicholls, Orchids of Australia, t. 456, 457 (1969). Stems to 12 cm, leaves often brown-spotted, racemes erect with 6–12 flowers which are 4–6 mm in diameter, facing upwards, pale pink, mauve or purple, lip densely hairy, callus at opening of spur yellow, pronounced calluses present on inside walls of the spur. *Australia.* G2.

2. S. hartmannii Mueller. Illustration: Nicholls, Orchids of Australia, t. 455 (1969); American Orchid Society Bulletin **47**: 1122 (1978); Bechtel et al., Manual of cultivated orchid species, 271 (1981).
Stems to 1 m, usually branched. Leaves linear-oblong, 5–15 cm, channelled. Racemes arching or erect with 5–25 flowers. Flowers 1.5–3.5 cm in diameter. Sepals and petals white spotted with red at the base (rarely completely white or red). Lateral lobes of lip spotted and streaked with red, central lobe yellowish, spur white flushed with yellow. Callus at opening of spur 2-lobed, small calluses on inside walls of spur. *Australia.* G2.

*__S. fitzgeraldii__ Mueller. Illustration: Botanical Magazine n.s., 201 (1953); Nicholls, Orchids of Australia, t. 454 (1969): Bechtel et al., Manual of cultivated orchid species, 271 (1981). Leaves sickle-shaped, racemes hanging, with up to 16 flowers, scape reddish, flowers 2.5–3.5 cm in diameter, lip *c.* 5 mm, callus at opening of spur yellowish with a smaller one behind. *Australia.* G2. Winter–spring.

104. CHILOSCHISTA Lindley
V.A. Matthews

Differs from *Sarcochilus* in that the plants are leafless when flowering, the petals are larger than the sepals and the 4 pollinia are in 2 pairs.

A genus of 3 species found in tropical Asia. Cultivation as for *Sarcochilus* (p. 241).

1. C. lunifera (Reichenbach) J.J. Smith (*Sarcochilus luniferus* (Reichenbach) Hooker). Illustration: Botanical Magazine, 7044 (1899).
Racemes drooping, with many flowers, scapes purple-spotted, shortly hairy. Flowers 1–1.5 cm in diameter. Sepals and petals yellow spotted with red or purple. Lip white, lateral lobes erect, central lobe minute, recurved, with a forward-pointing spur. Disc warty and hairy with 2 thick ridges. Anther cap with 3 spurs. *Sikkim, Nepal & Burma.* G2.

105. PTEROCERAS Hasskarl
V.A. Matthews

Differs from *Sarcochilus* in having the lip movably attached to the column foot, the spur of the lip lacking calluses on its inside walls, and the 4 pollinia in 2 pairs.

Thirty species found in NE India, SE Asia and Malaysia. Cultivation as for *Sarcochilus* (p. 241).

1. P. pallidum (Blume) Holttum (*Sarcochilus pallidus* (Blume) Reichenbach; *S. unguiculatus* Lindley). Illustration: Lindenia **16**: t. 756 (1901).
Leaves oblong, fleshy, up to 20 cm. Raceme usually hanging, axis somewhat thickened. Flowers to 5 cm across, fragrant. Sepals and petals white to pale yellowish. Lip *c.* 6 mm, lateral lobes white streaked with red or purple, central lobe and spur yellowish, spotted with red or purple. *Malay peninsula, Java, Sumatra, Borneo & Philippines.* G2.

106. AERIDES Loureiro
V.A. Matthews

Epiphytic. Pseudobulbs absent. Stem producing long aerial roots which either hang or attach themselves to a support. Leaves usually strap-shaped, thick, in 2 ranks, their bases sheathing the stem, folded when young, occasionally terete. Flowers usually fragrant, in hanging, many-flowered racemes arising from the leaf axils. Sepals and petals spreading, the 2 lateral sepals attached to the column foot. Lip simple or 3-lobed, continuous with the column foot, prolonged at the base into a hollow spur. Column short, usually with a long foot. Pollinia 2 unequally divided, on a single slender stipe.

A genus of some 40 species from tropical Asia. Cultivation as for *Vanda* (p. 249).
Literature: Seidenfaden, G., Contributions to the orchid flora of Thailand V, *Botanisk Tidsskrift* **68**: 68–80 (1973); Wood, J. & Kennedy, G.C., Some showy members of the genus Aerides, *Orchid Digest* **41**: 205–8 (1977).

Leaves. Terete: **1,2**.
Sepals and petals. Basic colour olive green: **6**; white, pale green or yellowish: **1–5**; pinkish or purplish: **4,6,7**. With defined spots at the tips: **3–5**. With wavy margins: **2**.
Lip. Simple: **4,6,7**.
Margin of lip, or if lip 3-lobed, of central lobe. Entire: **1,3,4,7**; toothed or fringed: **2,3,5,6**.

1a. Leaves terete 2
b. Leaves flat 3
2a. Leaves 7.5–10 cm; margins of sepals and petals not wavy; lateral lobes of lip with violet markings **1. cylindrica**
b. Leaves 10–20 cm; margins of sepals and petals wavy; lateral lobes of lip lacking violet markings **2. vandara**
3a. Basic colour of sepals and petals white, pale greenish or yellowish 4
b. Basic colour of sepals and petals olive green, pink or pinkish purple 6
4a. Spur curved upwards towards the rest of the flower **3. odorata***
b. Spur pointing forwards 5
5a. Lip simple with an entire margin **4. multiflora**
b. Lip 3-lobed, the central lobes with a toothed or fringed margin **5. falcata***
6a. Lip simple and fringed or central lobe of lip with a toothed margin **6. crispa***
b. Lip simple and entire or central lobe of lip entire though sometimes notched at apex 7
7a. Flowers 2–3 cm in diameter; lip simple, rounded or slightly notched at the apex **4. multiflora**
b. Flowers 3–5 cm in diameter; lip 3-lobed, or if simple then the apex pointed **7. crassifolia***

1. A. cylindrica Lindley (*Papilionanthe subulata* (Koenig) Garay). Illustration: Orchid Digest **41**: 204 (1977).
Leaves terete, 7.5–10 cm. Raceme with 1 or 2 flowers. Flowers *c.* 2.5 cm in diameter. Sepals and petals white, sometimes flushed with pale pink, ovate-oblong. Lateral lobes of lip white with violet lines, central lobe white or yellowish with violet spots and 3 broad yellow ridges. Spur funnel-shaped, pinkish. *S India, Sri Lanka.* G2.

2. A. vandara Reichenbach (*A. cylindrica* Hooker not Lindley; *Papilionanthe vandarum* (Reichenbach) Garay). Illustration: Botanical Magazine, 4982 (1857); Orchid Digest **41**: 204 (1977); Bechtel et al., Manual of cultivated orchid species, 164 (1981).
Leaves terete, 10–20 cm. Raceme with 1–3 flowers. Flowers *c.* 5 cm in diameter. Sepals and petals white, somewhat reflexed, with wavy margins, petals twisted. Lip white flushed with yellow in the middle, central lobe clawed, expanding into 2 fringed lobes.

Spur curved, 2–2.5 cm. *NE India, Burma.* G2. Spring.

3. A. odorata Loureiro (*A. suavissima* Lindley; *A. cornuta* Roxburgh; *A. ballantiana* Reichenbach). Illustration: Botanical Magazine, 4139 (1845); Botanisk Tidsskrift **68**: 72 (1973); Sheehan & Sheehan, Orchid genera illustrated, 31 (1979).
Stems to 50 cm, in old plants to 150 cm. Leaves 20–25 cm. Racemes dense, with many flowers. Flowers 2–3 cm in diameter. Sepals and petals creamy white tipped with pink or purple. Lateral lobes of lip white or occasionally yellowish, flushed and spotted with pink or purple, especially at the base, central lobe ovate to narrowly oblong, white with a broad central purplish band, margin entire or toothed. Spur curved upwards towards the flower, tipped with green or yellow. *SE Asia, from the Himalaya to Indonesia.* G2. Summer–autumn.

*A. **quinquevulnera** Lindley. Illustration: Orchid Digest **43**: 126 (1979). Similar, but sepals and petals greenish at the base, central lobe of the lip deep pinkish purple with a revolute margin which is always toothed. *Philippines.* G2. Summer–autumn.

Often regarded as a variety of *A. odorata.*

*A. **lawrenceae** Reichenbach. Illustration: Lindenia **9**: t. 401 (1893); Bechtel et al., Manual of cultivated orchid species, 164 (1981). Also similar, but sepals and petals greenish white becoming yellowish, lateral lobes of lip white with toothed margins, central lobe purple, or purple-tipped with lines of purple spots, margin toothed. *Philippines.* G2.

Possibly also merely a variety of *A. odorata.*

4. A. multiflora Roxburgh (*A. affine* Lindley). Illustration: Botanical Magazine, 4049 (1844); Cogniaux & Goossens, Dictionnaire Iconographique des Orchidées, Aerides, t. 2 (1898); Gartenpraxis 1980(11): 512.
Leaves 18–25 cm. Racemes dense with many flowers, arching or hanging. Flowers 2–3 cm in diameter. Sepals and petals white, spotted, flushed and tipped with purplish pink, or occasionally completely pink. Lip pale purplish pink with a deeper central area, simple, oval, with the sides recurved and the apex rounded or slightly notched. Spur pointing forwards. *N India, Bhutan, SE Asia.* G2. Summer–autumn.

5. A. falcata Lindley (*A. larpentae* Reichenbach). Illustration: Botanisk Tidsskrift **68**: 74 (1973); Orchid Digest **41**: 204 (1977).
Stems to 1.8 m. Leaves 15–35 cm. Raceme with many flowers, hanging. Sepals and petals white tipped and sometimes dotted with pink. Lateral lobes of lip white or mauve, sickle-shaped, central lobe pink or purple with a toothed margin. Spur short, greenish, pointing forwards. *NE India, SE Asia.* G2. Spring–summer.

*A. **houlletiana** Reichenbach (*A. falcata* var. *houlletiana* (Reichenbach) Veitch). Illustration: Cogniaux & Goossens, Dictionnaire Iconographique des Orchidées, Aerides, t. 3 (1899); Orchid Digest **41**: 206 (1977). Sepals and petals yellow-buff tipped with reddish purple or brown, upper sepal and petals with finely toothed margins, lateral lobes of lip yellowish, marked with pinkish purple, central lobe pinkish purple, toothed and notched at the apex. *E Thailand, Vietnam, Laos & Kampuchea.* G2. Spring.

6. A. crispa Lindley (*A. brookei* Lindley). Illustration: Orchid Digest **43**: 126 (1979).
Stems to 1.5 m, often purplish. Leaves 15–25 cm. Raceme with many flowers, arching or hanging. Flowers *c.* 5 cm in diameter. Sepals and petals white flushed with pinkish purple. Lateral lobes of lip white striped with pinkish purple, central lobe broadly ovate, deep pinkish purple, margin toothed. Spur compressed, slightly curved. *Himalaya, India, Burma.* G2. Summer.

*A. **flabellata** Downie. Illustration: Botanisk Tidsskrift **68**: 77 (1973); Orchid Digest **41**: 206 (1977). Stems and leaves much shorter, raceme loose, sepals and petals olive green marked with reddish brown, lip simple, clawed, white blotched with red or purple, margin fringed, spur curved upwards. *Burma, Laos, Thailand.* G2. Summer.

A distinct species which may be more closely related to *Vanda.*

7. A. crassifolia Parish & Reichenbach (*A. expansa* Reichenbach). Illustration: Cogniaux & Goossens, Dictionnaire Iconographique des Orchidées, Aerides, t. 1 (1897); Botanisk Tidsskrift **68**: 76 (1973); Orchid Digest **41**: 204 (1977).
Leaves to 20 cm. Racemes with 10–50 flowers. Flowers 3–5 cm in diameter, fragrant. Sepals and petals pinkish purple, deeper towards the apex and white towards the base. Lateral lobes of lip crescent-shaped, pinkish purple, central lobe broadly ovate, often notched at the apex, darker in colour than sepals and petals. Spur with a greenish tip. *Burma, Laos, Thailand.* G2. Spring–summer.

*A. **fieldingii** Williams. Illustration: La Belgique Horticole **26**: facing p. 283 (1876); Botanisk Tidsskrift **68**: 78 (1973); Orchid Digest **41**: 206 (1977); Bechtel et al., Manual of cultivated orchid species, 163 (1981). Racemes branched, lip simple with a pointed tip, spur white. *N India, N Thailand, Laos.* G2. Spring–summer.

*A **jarckiana** Schlechter. Illustration: Botanical Magazine, 9274 (1932); Orchid Digest **43**: 126 (1979); Bechtel et al., Manual of cultivated orchid species, 163 (1981). Lateral lobes of lip very small, rounded, united nearly to their apex with the central lobe which is sharply curved upwards. *Philippines.* G2. Spring.

107. SEDIREA Garay & Sweet

V.A. Matthews

Epiphytic. Pseudobulbs absent. Stem short, *c.* 10 cm. Leaves in 2 ranks, linear-oblong, thick. Raceme axillary, loose, drooping, to 15 cm with few flowers. Flowers *c.* 3 cm in diameter, fragrant. Sepals and petals white, cream or greenish, lateral sepals barred with purplish brown. Lip hinged to the column foot, 3-lobed, lateral lobes triangular, central lobe obovate, recurved, scalloped, white blotched with reddish purple. Spur funnel-shaped, obtuse, pointing forwards. Column *c.* 8 mm, curved, with a short foot. Pollinia 2, ovoid, each with a groove, on a narrowly triangular stipe.

Only 1 species, growing in Japan and Korea. Cultivation as for *Vanda* (p. 249).

1. S. japonica (Linden & Reichenbach) Garay & Sweet (*Aerides japonica* Linden & Reichenbach). Illustration: Orchid Digest **41**: 207 (1977).
Japan, Korea. G1. Summer.

108. RHYNCHOSTYLIS Blume

V.A. Matthews

Epiphytic. Stems short, thick. Leaves fleshy, strap-shaped or linear, in 2 ranks, 2-lobed at the apex. Racemes lateral, erect or drooping, many-flowered. Sepals and petals spreading, petals narrower than sepals. Lip entire, spurred. Column short, dilated at base with a short foot; rostellum beaked. Pollinia 2.

A genus of about 15 species native to tropical Asia often found under the name *Anota* Schlechter. Cultivation as for *Gastrochilus* (p. 252).

1a. Apex of lip 3-lobed — 2
b. Apex of lip entire or notched — 3
2a. Lip pinkish purple, white towards the middle and base, and with 2 ridges

near the base **3. gigantea**
b. Lip completely purple, with 5 ridges near the base **4. violacea**
3a. Racemes erect; sepals and petals tipped with blue **1. coelestis**
b. Racemes drooping; sepals and petals with purple or pinkish markings **2. retusa**

1. R. coelestis (Reichenbach) Veitch. Illustration: Lindenia **7**: t. 300 (1891); Brooklyn Botanic Garden Record **23**(2), Handbook on orchids, 40–1 (1967).
Racemes erect. Flowers *c.* 2 cm in diameter, fragrant. Sepals and petals white tipped with blue. Lip obovate-oblong, blue or violet-blue, white at the base. Spur slightly curved, compressed. *Thailand*. G2. Summer.

2. R. retusa (Linnaeus) Blume. Illustration: Botanical Magazine, 4108 (1844); Orchid Digest **39**: 8 (1975); Bechtel et al., Manual of cultivated orchid species, 269 (1981).
Racemes drooping. Flowers *c.* 2 cm in diameter, fragrant. Sepals and petals white or pinkish, marked with pinkish or purple, lateral sepals broader than upper sepal and petals. Lip variable in shape, usually lanceolate, tip rounded or sometimes notched, mauve. Spur truncate-conical, compressed. *From India and Sri Lanka to the Philippines*. G2. Summer.

'Alba' has completely white flowers and 'Holfordiana' has the sepals and petals spotted with crimson and a crimson lip.

3. R. gigantea (Lindley) Ridley (*R. densiflora* (Lindley) Williams; *Anota densiflora* (Lindley) Schlechter). Illustration: Orchid Digest **43**: 124 (1979); Bechtel et al., Manual of cultivated orchid species, 269 (1981).
Racemes drooping. Flowers 2.5–3 cm in diameter, fragrant. Sepals and petals white spotted with pinkish purple or pale purple. Lip oblong, 3-lobed at apex, pale purple or pinkish purple, white towards the middle and base and with 2 small ridges near the base. Spur very short, obtuse, compressed. *Burma, Thailand, Laos*. G2. Autumn–winter.

Var. **harrisoniana** (Hooker) Holttum has pure white flowers and longer racemes; var. *petotiana* Anon. is probably the same.

4. R. violacea (Lindley) Reichenbach (*Anota violacea* (Lindley) Schlechter; *Saccolabium violaceum* Lindley). Illustration: Edwards's Botanical Register **33**: t. 30 (1847).
Racemes arching or drooping. Flowers 2–2.5 cm in diameter, fragrant. Sepals and petals white spotted with pale violet. Lip oblong, 3-lobed at apex, purple, with 5 ridges near the base. Spur greenish, compressed. *Philippines*. G2. Winter–spring.

109. DORITIS Lindley
V.A. Matthews
Differs from *Phalaenopsis* in that the claw of the lip bears 2 linear appendages and there are 4 pollinia.

A genus of 2 species occurring in tropical Asia. Cultivation as for *Phalaenopsis*.
Literature: Holttum, R.E., Cultivated species of the orchid genus Doritis Lindl., *Kew Bulletin* **19**: 207–12 (1965).

1. D. pulcherrima Lindley (*Phalaenopsis pulcherrima* (Lindley) J.J. Smith; *P. esmeralda* Reichenbach; *P. buyssoniana* Reichenbach). Illustration: Skelsey, Orchids, 112 (1978); Sheehan & Sheehan, Orchid genera illustrated, 84 (1979).
Epiphytic. Leaves to 20 cm, in 2 ranks, oblong, bright green, often purplish beneath. Flowers 2–2.5 cm wide, in erect racemes which are 30–75 cm long with 3–20 flowers. Sepals and petals 1–1.2 cm, pale pink to dark purple, usually somewhat reflexed. Lip 3-lobed with a long claw bearing a linear appendage on each side, appendages pale pink sometimes flushed with yellow, or with purple spots. Lateral lobes erect, rounded, purple or striped with orange-red; central lobe oblong, acute, with a central ridge, purple or reddish. *Burma, Kampuchea, Thailand, Malay peninsula*. G2. Summer–autumn.

The flowers are variable in size and colour and are produced over a period of 2–3 months. Plants which are larger in all their parts and with the sepals and petals not reflexed are usually sold under the name *D. pulcherrima* var. *buyssoniana*, and are tetraploid. A white-flowered variant, 'Alba', is commercially available.

110. PHALAENOPSIS Blume
V.A. Matthews
Epiphytic. Stem short, pseudobulbs absent. Leaves in 2 ranks, usually persistent, fleshy, rarely with petioles. Raceme or panicle lateral, with 1–many flowers and with a straight or zig-zag axis. Sepals almost equal, free, usually spreading; petals similar or larger. Lip relatively small, often more brightly coloured than the sepals and petals, 3-lobed, continuous with the column foot or inserted at a right angle, lateral lobes erect with cushion-like swellings in the middle, central lobe fleshy, usually firmly attached to the lateral lobes. Between the lateral lobes and the base of the central lobe are various calluses and other ornamented processes. Column often constricted in the middle, usually wingless, produced into a short foot at the base. Pollinia 2, more or less spherical.

A genus of 40–55 species occurring in the Himalaya, SE Asia and N Australia. They are best grown in baskets in an epiphytic compost. They lack pseudobulbs and therefore they should not be allowed to dry out and the atmosphere should always be kept moist. Shading is essential in spring and early summer to prevent leaf scorch. Flowering can be encouraged by a drop in temperature.
Literature: Sweet, H.R., *The genus Phalaenopsis* (1980).

Leaves. Mottled or spotted above: **4,5,9,10**.
Inflorescence. Erect or arching: **1–3,6,7, 9–14,16–19**; hanging: **1–5,7,8,10,11, 15,16,18**. With axis winged or flattened: **7,12**.
Sepals and petals. Unmarked or with dots at the base only: **1–6,9,12,15**; barred, blotched or spotted: **7,8,10,11,13,15–19**; longitudinally striped towards the base or apex: **10,14**.
Lip. Less than 2 cm: **6–12,14–16,18,19**; more than 2 cm: **1–5,11–14,16,17**. Central lobe hairless: **1–13,19**; with some hairs: **8,11,14–19**. Warty: **8**. Mobile: **6**. Apex with 2 appendages: **1–5**; lacking 2 appendages: **6–19**. Callus with some backwardly pointing warts or appendages: **11,13**. Callus with 4 bristle-like appendages originating at the base: **6**.

1a. Petals wider than sepals; apex of lip with 2 tendril-like or horn-like appendages 2
b. Petals as wide as or narrower than sepals, if wider then with lilac stripes at the base; apex of lip not as above 8
2a. Leaves purplish beneath, especially when young 3
b. Leaves not purplish beneath 7
3a. Scape purplish 4
b. Scape not purplish 5
4a. Mature leaves green above; petals white **2. aphrodite***
b. Mature leaves spotted above; petals white or pink **4. schilleriana***
5a. Petals pink **4. schilleriana***
b. Petals white 6
6a. Leaves spotted above; central lobe of lip white or yellowish **5. stuartiana**
b. Leaves green above; central lobe of lip deep pinkish purple **2. aphrodite***
7a. Scape purplish; callus horseshoe-shaped **3. sanderiana**

b. Scape green; callus more or less square **1. amabilis**
8a. Claw of lip fused to column foot; central lobe of lip mobile; 4 bristle-like appendages present above the callus **6. parishii**
b. Claw of lip not fused to column foot; central lobe of lip not mobile; bristle-like appendages absent 9
9a. Central lobe of lip anchor-shaped 10
b. Central lobe of lip not anchor-shaped 11
10a. Scape winged above; central lobe of lip without warts **7. cornu-cervi***
b. Scape not winged; central lobe of lip warty **8. mannii**
11a. Callus at junction of lobes of lip peltate **9. equestris***
b. Callus at junction of lobes of lip not peltate 12
12a. Central lobe of lip concave or flat with a high, fleshy central ridge or 5 parallel ridges, if ridge absent then lateral lobes rounded **10. fuscata***
b. Central lobe of lip convex with a thin central ridge, or ridge absent 13
13a. Petals broadly elliptic, length 2 times width or less **11. amboinensis***
b. Petals narrowly obovate, oblanceolate or elliptic, length more than 2 times width 14
14a. Central lobe of lip hairless 15
b. Central lobe of lip with at least some hairs towards the apex 17
15a. Callus with elongate, backwardly pointing warts **13. fasciata**
b. Appendages of callus all pointing forwards 16
16a. Sepals and petals without spots or transverse stripes; upper sepal 2–3.5 cm; lip 1.8–2.8 cm **12. violacea**
b. Sepals and petals with spots or transverse stripes; upper sepal 1.2–1.8 cm; lip 1–1.4 cm **19. maculata***
17a. Upper sepal 1.4–2 cm 18
b. Upper sepal 2–4 cm 19
18a. Sepals and petals white with green tips, sometimes with fine, transverse, reddish purple lines **15. fimbriata**
b. Sepals and petals white, cream or yellow with obvious, brown or brownish red bars and blotches **18. mariae***
19a. Sepals and petals with spots and circles **17. hieroglyphica**
b. Sepals and petals with transverse bars and/or blotches, or with longitudinal stripes 20
20a. Callus between lateral lobes of lip with warts **16. lueddemanniana**
b. Callus between lateral lobes of lip lacking warts 21
21a. Central lobe of lip with entire margin **14. sumatrana***
b. Central lobe of lip with margin toothed towards the apex **18. mariae***

1. P. amabilis (Linnaeus) Blume. Illustration: Kupper & Linsenmaier, Orchids, 111 (1961); Orchid Digest **38**: 192 (1974); Bechtel et al., Manual of cultivated orchid species, 259, 362 (1981); Orchid Digest **46**: 29 (1982).
Leaves elliptic to obovate, 15–50 cm, green above and beneath. Inflorescence simple or branched, often 1 m or more, loose with few to many flowers. Flowers fragrant, 6–10 cm in diameter. Sepals 3–4 cm, elliptic-ovate, white, often pinkish on the back. Petals to 4.5 cm, more or less circular, similar in colour to the sepals. Lip white, to 2.3 cm, lateral lobes oblanceolate with yellow margins and basal red markings, central lobe somewhat cross-shaped with triangular side-arms and 2 terminal, tendril-like appendages with yellow tips. Callus more or less square, yellow, spotted with red. *E Indies, Australia (NE Queensland)*. G2. Autumn–early spring.

There are 3 varieties in cultivation:

Var. **aurea** Rolfe. Most of the lateral lobes, the margin of the central lobe, and the terminal appendages are yellow. *Borneo*.

Var. **moluccana** Schlechter. The central lobe of the lip is linear-oblong with small lateral lobes. *Indonesia (Moluccas, Sulawesi)*.

Var. **papuana** Schlechter. The central lobe of the lip is narrowly triangular. *New Guinea, Australia (NE Queensland)*.

2. P. aphrodite Reichenbach (*P. amabilis* var. *formosa* Shimadzu; *P. amabilis* var. *dayana* Warner & Williams). Illustration: Orchid Digest **38**: 192 (1974); American Orchid Society Bulletin **46**: 217 (1977); Bechtel et al., Manual of cultivated orchid species, 362 (1981).
Leaves broadly elliptic, 20–38 cm, green above, purplish beneath especially when young. Inflorescence simple or branched, arching or hanging, longer than the leaves, purplish. Flowers 7 cm or more in diameter. Sepals to 4 cm, elliptic, white, sometimes with tiny pink dots. Petals similar to the sepals but more or less circular. Lip white, *c.* 3 cm, lateral lobes ovate with yellow margins and dark pink dots and streaks towards the base, central lobe triangular with a dark pink flush at the base and yellow on the side-arms, apex with 2 tendril-like appendages. Callus with notched apex, each side with tooth-like projections, deep yellow with deep pink dots. *Philippines, Taiwan*. G2. Flowering all the year.

According to Wallbrunn (American Orchid Society Bulletin **40**: 228, 1971) it is possible that *P. amabilis, P. aphrodite* and *P. sanderiana* are all one species.

***P. × intermedia** Lindley. Illustration; Die Orchidee **24**: 144 (1973); Rittershausen & Rittershausen, Orchids in colour, 111 (1979); Bechtel et al., Manual of cultivated orchid species, 260, 362 (1981). Leaves elliptic, to 30 cm, green above, brownish purple beneath. Inflorescence to 60 cm, simple or branched, with many flowers. Sepals 2.5–4 cm, ovate-elliptic, white with pink suffusion. Petals 2–3.5 cm, broadly elliptic, white with pink dots at the base. Lip pinkish purple, lateral lobes ovate, purplish spotted with deep pink, central lobe obovate, tapered towards the tip, deep pinkish purple, apex notched or with 2 horn-like or tendril-like appendages. Callus more or less square, yellow spotted with red. *Philippines, Borneo*. G2. Spring.

A hybrid between *P. aphrodite* and *P. equestris*.

3. P. sanderiana Reichenbach. Illustration: Orchid Digest **38**: 221 (1974); Rittershausen & Rittershausen, Orchids in colour, 121 (1979); Sweet, The genus Phalaenopsis, 41 (1980); Bechtel et al., Manual of cultivated orchid species, 362 (1981).
Leaves 15–35 cm, elliptic, dark green above, often marked with silvery grey beneath. Inflorescence to 80 cm but usually less, purplish, simple or branched. Flowers 6–7.5 cm in diameter. Sepals to 3.5 cm, ovate-elliptic, pink, sometimes with white mottling or completely white. Petals similar to sepals but broader. Lip *c.* 3 cm, lateral lobes ovate, white spotted with pink, central lobe triangular with narrowly triangular side-arms, white streaked with brown or purple and sometimes yellow, apex with 2 tendril-like appendages. Callus horseshoe-shaped, white or yellow with brown, red or purple spots. *Philippines*. G2. Spring.

Very variable in flower colour and in the markings on the lip. On this basis several varieties have been recognised but it is generally accepted that they have no botanical standing.

4. P. schilleriana Reichenbach. Illustration: Flore des Serres **5**: t. 1559 (1862–65); Orchid Digest **38**: 221 (1974); Hunt &

Grierson, Country Life book of orchids, 61 (1978); Bechtel et al., Manual of cultivated orchid species, 261, 362 (1981).
Leaves to 45 cm, elliptic, dark green spotted with silvery grey above, purple beneath. Inflorescence to 90 cm and often bearing over 200 flowers, branched, hanging or arched, purplish. Flowers 5–7.5 cm in diameter. Sepals ovate-elliptic, to 3.5 cm, pink fading at margins to white, lateral sepals with deep pink spots towards the base. Petals more or less circular, to 4 cm, similar in colour to the sepals. Lip 2–2.3 cm, white to deep pink, lateral lobes elliptic, often broadly so, with reddish brown spots, central lobe ovate to rounded with 2 curved, horn-like appendages of varying size at the apex. Callus more or less square, yellow. *Philippines*. G2. Winter–spring, sometimes also late summer.

*P. × **leucorrhoda** Reichenbach. Illustration: Cogniaux & Goossens, Dictionnaire Iconographique des Orchidées, Phalaenopsis hybrides, t. 1 (1902); Rittershausen & Rittershausen, Orchids in colour, 113 (1979). Leaves green above, spotted with silvery grey, purplish beneath. Inflorescence usually hanging, to 70 cm, with many flowers, purplish. Sepals 2.2–3.5 cm, elliptic to ovate. Petals 2.5–4 cm, kidney-shaped to almost circular. Sepals and petals white, often flushed with pink, or deep pink with white margins, lateral sepals sometimes with purple dots at the base. Lip 2.2–3 cm, lateral lobes spathulate, white or yellow with reddish purple spots, central lobe white with yellow and purple markings, sometimes entirely purple, variable in shape, tapering towards the apex which has 2 anchor-shaped or tendril-like appendages. Callus irregularly toothed, deep yellow, occasionally paler, densely spotted with dark red. *Philippines*. G2.

A hybrid between *P. aphrodite* and *P. schilleriana*.

5. P. **stuartiana** Reichenbach (*P. schilleriana* var. *alba* Roebelen; *P. schilleriana* var. *vestalis* Reichenbach). Illustration; Orchid Digest **38**: 221 (1974); Der Palmengarten 1979(4): 187 (1979); Bechtel et al., Manual of cultivated orchid species, 261, 362 (1981).
Leaves elliptic-oblong, 15–35 cm, green blotched with grey above, purplish beneath. Inflorescence to 80 cm, with many flowers, branched, hanging. Flowers 5–6 cm in diameter, slightly fragrant. Sepals to 3.5 cm, upper sepal white, lateral sepals white with yellow on basal half, and with brownish red dots. Petals to 3.3 cm, somewhat square to circular, white, often with purplish brown dots at the base. Lip *c.* 2.5 cm, lateral lobes obovate, white at apex, yellow at base with reddish brown spots and with 2 white, horn-like appendages. Callus almost square with an apical projection on each side, orange, spotted. *Philippines*. G2. Winter.

In var. **punctatissima** Reichenbach the dots on the flowers are pale purple.

6. P. **parishii** Reichenbach. Illustration: Botanical Magazine, 5815 (1870).
Leaves to 12 cm, elliptic to obovate. Inflorescence erect or arching, to 15 cm, with few or several flowers. Flowers opening simultaneously. Sepals white, the upper 6–8 mm, elliptic to almost circular, the laterals obovate to almost circular, 7–10 mm. Petals white, elliptic to obovate, 6–7 mm. Lip 10–15 mm, the short claw fused to the column foot, lateral lobes very small, triangular, yellow or whitish spotted with purple or brown, central lobe mobile, purple, triangular. Callus semicircular, fringed at margin. At the junction of the lobes there is a plate-like projection ending in 4 bristle-like appendages. Column white with purple spots. *E Himalaya, India*. G2.

Var. **lobbii** Reichenbach (*P. lobbii* (Reichenbach) Sweet). Illustration: Orchid Digest **37**: 168 (1973); American Orchid Society Bulletin **50**: 33–4 (1981). Lip white with a pair of vertical brown stripes on each side, and the callus margin entire or only slightly toothed.

7. P. **cornu-cervi** (Breda) Blume & Reichenbach. Illustration: Botanical Magazine, 5570 (1866); Rittershausen & Rittershausen, Orchids in colour, 118 (1979); American Orchid Society Bulletin **50**: 32 (1981); Bechtel et al., Manual of cultivated orchid species, 362 (1981).
Leaves to 25 cm, oblong. Inflorescence simple or branched, 10–40 cm, axis flattened. Flowers 3–5 cm in diameter. Sepals 1.8–2.3 cm, elliptic, keeled on the back, upper sepal with recurved margins, yellow or greenish with reddish brown blotches, lateral sepals often with blotches only on the upper half. Petals 7–18 mm, lanceolate, similar in colour to upper sepal. Lip *c.* 8 mm, lateral lobes almost square, with a callus below the truncate apex, whitish with reddish brown stripes, lateral lobes running into base of central lobe to form a flat, semicircular swelling, in front of which the central lobe is constricted and then expanded into a whitish, anchor-shaped portion with recurved lobes. Callus complex with 3 appendages in series, the first upcurved, yellow, the second forked, white, and the third lanceolate, purple; the second and third project over the central lobe. *SE Asia*. G2. Summer.

*P. **pantherina** Reichenbach. Illustration: American Orchid Society Bulletin **38**: 510 (1969); Sweet, The genus Phalaenopsis, 61 (1980). Differs mainly in the larger (*c.* 1.5 cm), differently shaped lip. Below the swelling at the base of the central and lateral lobes, the central lobe has an isthmus which expands into a somewhat anchor-shaped portion with lobes at right angles. The callus appendages are complex. *Borneo*.

8. P. **mannii** Reichenbach (*P. boxallii* Reichenbach). Illustration: Orchid Digest **40**: 207 (1976); Bechtel et al., Manual of cultivated orchid species, 260, 362 (1981); American Orchid Society Bulletin **50**: 30–2 (1981).
Leaves to 35 cm, oblong. Inflorescence usually simple, hanging, with many flowers. Flowers *c.* 5 cm in diameter, opening in succession. Sepals obovate-lanceolate, yellow or green with brownish blotches, keeled on the back towards the apex, upper sepal 2–2.4 cm, margins rolled under, lateral sepals 2.2–2.5 cm. Petals lanceolate, margins rolled under, 1.7–2 cm, similar in colour to sepals. Lip white and purple, 9–11 mm, lateral lobes almost square with a fleshy callus, base of lateral lobes running into the central lobe to form a semicircular swelling from which the variable central lobe expands from a basal isthmus. Central lobe anchor-shaped, the lateral lobes toothed, upper surface warty and with an apical, often hairy callus. The basal callus appendages are complex. *Himalaya, Vietnam*. G2. Summer.

9. P. **equestris** (Schauer) Reichenbach (*P. rosea* Lindley). Illustration: Flore des Serres **6**: t. 1645 (1865–67); Rittershausen & Rittershausen, Orchids in colour, 118 (1979); Die Orchidee **31**: 225 (1980); Bechtel et al., Manual of cultivated orchid species, 259, 362 (1981).
Leaves to 20 cm, oblong. Inflorescence simple or branched, almost erect to arching, purplish, with many flowers. Flowers 2.5–4 cm in diameter. Sepals oblong-elliptic, white to pink, the upper 1–1.7 cm, the laterals 1–1.5 cm. Petals elliptic, 8–15 mm, similar in colour to sepals. Lip deep pink or purplish, 1–1.4 cm, lateral lobes oblong, often with dark

streaks, central lobe ovate with apex pointed, more or less concave in the middle. Callus peltate, 6–8-sided, yellow with red spots. *Philippines, Taiwan*. G2. Spring–autumn.

*P. **lindenii** Loher. Illustration: Orchid Digest **37**: 108 (1973); Bechtel et al., Manual of cultivated orchid species, 260 (1981); Orchid Digest **41**: 6 (1982). Leaves green mottled silvery white. Inflorescence simple, usually green. Lip 1.2–1.4 cm, lateral lobes obovate to strap-shaped, white with tiny orange or reddish dots at the base and 3 purplish lines above, central lobe more or less circular, with a small point, slightly concave in the middle, purplish pink towards the apex with 5–7 darker lines. *Philippines*. G2.

Perhaps a hybrid between *P. equestris* and *P. schilleriana*.

10. P. fuscata Reichenbach. Illustration: Sweet, The genus Phalaenopsis, 64 (1980). Leaves to 30 cm, oblong. Inflorescence to 30 cm, simple or branched, with few flowers. Sepals and petals yellowish green blotched with brown, margins rolled under, sepals 1.4–1.8 cm, petals 1.2–1.5 cm. Lip yellowish with brown markings, 1.1–1.4 cm, lateral lobes almost square, twisted so that they touch, central lobe ovate to elliptic with a central ridge. Callus at junction of lobes with 2 forked appendages. *Malay Peninsula*. G2.

*P. **kunstleri** Hooker. Illustration: Botanical Magazine, 7885 (1903); Orchid Digest **40**: 209 (1976). Similar but with larger flowers *c*. 5 cm in diameter. Sepals and petals yellow at base and apex, brown in the central part, sepals 1.8–2 cm, petals 1.5 cm. Lip yellow, 1–1.2 cm, lateral lobes almost square, central lobe almost circular with 1 or 2 brown stripes on each side of the central ridge. Callus at junction of lobes with 2 forked appendages. *Malaysia*. G2.

Considered to be the same as *R. fuscata* by some authorities.

*P. **cochlearis** Holttum. Illustration: Orchid Digest **40**: 209 (1976). Inflorescence to 50 cm, branched. Sepals and petals white to pale yellow with 2 pale brown bars at the base, upper sepal 1.5–1.8 cm, lateral sepals 1.4–1.5 cm, petals 1.2–1.4 cm. Lip white or yellow, 9–12 mm, lateral lobes narrowly oblong, notched at apex with reddish streaks at the base, central lobe circular, concave, sometimes notched at apex, with reddish brown stripes and 5 shallow ridges. Callus with a pair of 2-lobed appendages. *Sarawak*. G2.

*P. **celebensis** Sweet. Illustration: Sweet, The genus Phalaenopsis, 66 (1980). Leaves mottled with silvery white, sepals and petals white with lilac stripes at the base, sepals to 1.4 cm, petals to 9 mm, slightly wider than sepals. Lateral lobes of lip deep yellow, rounded, with a central fleshy ridge, central lobe up to 9 mm, concave, elliptic-ovate, with a small point at apex. Callus at junction of lobes more or less triangular. *Indonesia (Sulawesi)*. G2.

11. P. amboinensis J.J. Smith. Illustration: Orchid Digest **36**: 88 (1972). Leaves to 25 cm, elliptic to obovate. Inflorescence arching to 45 cm, with few flowers; flowers opening in succession. Sepals and petals white or yellowish with greenish tips and brownish red bars, elliptic to ovate, sepals 2–3 cm, petals shorter. Lip white or yellowish, 2–2.2 cm, lateral lobes oblong, each with an orange spot and a callus, central lobe ovate, the central ridge with a toothed edge. Callus with a pair of 2-lobed appendages. *Indonesia (Amboina)*. G2. Summer.

*P. **gigantea** J.J. Smith. Illustration: Orchid Digest **36**: 38 (1972). Bechtel et al., Manual of cultivated orchid species, 259, 362 (1981); Orchid Digest **41**: 29 (1982). Leaves to 90 cm, oblong-ovate. Flowers *c*. 5 cm in diameter, slightly fragrant, opening simultaneously, numerous, in hanging, dense inflorescences. Sepals and petals greenish or yellow, whitish at the base, blotched with reddish brown or purple. Lip *c*. 1.5 cm, lateral lobes triangular, curved, orange towards the apex, central lobe ovate with some tiny teeth on the lateral margins and an ovoid callus at the apex, striped and blotched with red or purple. Callus between lateral lobes with a pair of 2-toothed, deep yellow appendages. *Borneo*. G2.

*P. **javanica** J.J. Smith (*P. latisepala* Reichenbach). Illustration: Orchid Digest **43**: 57 (1979). Leaves to 22 cm, elliptic. Flowers *c*. 3 cm in diameter. Sepals and petals broadly elliptic, white to yellow with brownish purple spots arranged in longitudinal lines. Lip distinctly clawed, 1–1.8 cm, lateral lobes linear-oblong, grooved in the middle, yellowish, apex with 2 teeth, central lobe elliptic, very fleshy, apex with a few hairs, purple, whitish towards the base. Callus between lobes consisting of a lanceolate, backwardly pointing appendage and a 2-toothed forwardly pointing appendage. *Java*. G2.

12. P. violacea Witte. Illustration: Orchid Digest **36**: 12 (1972); Die Orchidee **30**: 122–3 (1979); Bechtel et al., Manual of cultivated orchid species, 362 (1981). Leaves 20–25 cm, elliptic to obovate. Inflorescence more or less erect or arching, simple, with a few, distant flowers; axis flattened. Flowers 5–7.5 cm in diameter. Sepals and petals bright purple with pale greenish tips. Sepals elliptic, 2–3.5 cm, keeled on the back towards the apex, petals elliptic, 2–3 cm. Lip reddish purple, 1.8–2.8 cm, lateral lobes linear-oblong, marked with yellow, central lobe ovate, abruptly pointed with a central ridge running into an apical callus. Callus between lateral lobes with warts and complex appendages, yellow. *Malay peninsula, Borneo, Sumatra*. G2. Summer.

In var. **alba** Teijsmann & Binnedijk (Illustration: American Orchid Society Bulletin **34**: 206, 1965) the sepals and petals are white with greenish tips.

13. P. fasciata Reichenbach. Illustration: Orchid Digest **40**: 207 (1976); Bechtel et al., Manual of cultivated orchid species, 259 (1981). Leaves 14–20 cm, elliptic. Inflorescence more or less erect or arching, branched, longer than leaves. Flowers *c*. 4 cm in diameter. Sepals and petals yellow to yellowish green with transverse red-brown bars. Sepals ovate-elliptic, 2–3 cm, petals ovate to ovate-elliptic, 2.2–2.6 cm. Lip 2–2.7 cm, lateral lobes strap-shaped, erect, yellow, central lobe pale purple, oblong-ovate, convex with a central ridge running into an apical callus. Callus between lateral lobes orange, with elongate, backwardly pointing warts and a forwardly pointing forked appendage. *Phillippines*. G2.

In cultivation many of the plants offered under the name *P. lueddemanniana* var. *ochracea* are *P. fasciata*.

14. P. sumatrana Korthals & Reichenbach (*P. zebrina* Witte; *P. zebrina* Teijsmann & Binnedijk). Illustration: American Orchid Society Bulletin **37**: 1093 (1968); Orchid Digest **36**: 25 (1972); Bechtel et al., Manual of cultivated orchid species, 362 (1981). Leaves 15–30 cm, oblong to obovate. Inflorescence erect or slightly arching, usually simple. Flowers *c*. 5 cm in diameter. Sepals and petals oblong-lanceolate, whitish to pale yellow with transverse, reddish brown or purplish bars. Sepals 2–4 cm, petals 2.2–3.5 cm. Lip to 2.5 cm, lateral lobes linear-oblong, apex 2-toothed with a third tooth on the inside, cream with yellow or brown margin and often orange spots, central lobe oblong-elliptic, very thick especially towards the hairy apex, convex with a central ridge, white with reddish or purple stripes on each side of the ridge. Callus between the lateral lobes a complex series of forked plates. *Thailand, Malay peninsula, Java, Sumatra & Borneo*. G2.

***P. corningiana** Reichenbach. Illustration: Orchid Digest **40**: 105 (1976). Similar but sepals and petals greenish yellow towards the apex with longitudinal brown or purplish stripes merging into blotches towards the base, upper sepal obovate to oblanceolate, 3–4 cm, lateral sepals ovate, 2.5–3 cm, petals lanceolate 2.5–3.5 cm. Lip *c.* 2 cm, lateral lobes strap-shaped, whitish with a central orange-yellow callus, central lobe deep purplish pink, narrowly elliptic, convex with a central ridge running into a hairy apical callus. Callus between lateral lobes yellow, forked with a central groove. Below the callus is an appendage with finger-like projections. *Malaysia, Sarawak*. G2.

Possibly the same as *P. sumatrana*.

15. P. fimbriata J.J. Smith. Illustration: Sweet, The genus Phalaenopsis, 97 (1980); Bechtel et al., Manual of cultivated orchid species, 259, 362 (1981).
Leaves 14–23 cm, oblong-elliptic. Inflorescence to 30 cm, usually hanging, loose with many flowers. Flowers opening simultaneously. Sepals and petals white, shaded green towards the tips and sometimes with fine transverse reddish purple lines, sepals 1.5–2 cm, petals 1.5–1.8 cm, all elliptic. Lip 1.4–2 cm, lateral lobes oblong, white flushed with reddish purple, convex with a central toothed ridge running into an apical callus with white hairs, lateral margins of central lobe expanded and toothed towards the apex. Callus between the lateral lobes with 3 overlapping plates, each with 2 teeth. *Java, Sumatra*. G2.

16. P. lueddemanniana Reichenbach. Illustration: Orchid Digest **40**: 207 (1976); Bechtel et al., Manual of cultivated orchid species, 362 (1981).
Leaves 20–30 cm, oblong-elliptic. Inflorescence erect or hanging, simple or branched, longer than the leaves. Flowers 5–6 cm in diameter. Sepals and petals white with transverse brownish purple bars, sepals oblong-elliptic, 2–3 cm, petals slightly smaller. Lip pink to purple, 1.8–2.2 cm, lateral lobes oblong, central lobe oblong to ovate with a central ridge running into an apical callus with white hairs. Callus between lateral lobes warty and with a forked appendage. *Philippines*. G2. Summer.

Very variable in the size and colour of the flowers. Var. **delicata** Reichenbach (Illustration: Flore des Serres **6**: t. 1636, 1865–67) has white sepals and petals with narrow brown bars on the apical part and purple bars on the basal part; the lip is purple. Var. **ochracea** Reichenbach (Illustration: Orchid Digest **40**: 207, 1976) has sepals and petals which are yellowish with yellowish brown bars and a basal pale purple flush; this is often confused with *P. fasciata* (p. 247) and many plants cultivated under this name are, in fact, *P. fasciata*. According to Wallbrunn (American Orchid Society Bulletin **40**: 225, 1971), it may be better to consider *P. fasciata, P. hieroglyphica* and perhaps *P. pallens* as subspecies of a variable *P. lueddemanniana*.

17. P. hieroglyphica (Reichenbach) Sweet (*P. lueddemanniana* var. *hieroglyphica* Reichenbach). Illustration: Orchid Digest **40**: 207 (1976); Bechtel et al., Manual of cultivated orchid species, 260, 362 (1981).
Leaves to 30 cm, strap-shaped. Inflorescence more or less erect or arching, simple or branched, with many flowers. Sepals and petals ovate-elliptic, white, greenish or sometimes yellowish with brownish spots and circles. Sepals keeled on the back towards the apex, the upper 2.3–3.8 cm, the laterals 2.4–4.1 cm, petals 2.2–3.3 cm. Lip white or yellowish, 2–2.5 cm, lateral lobes oblong, apex notched, central lobe wedge-shaped with a central ridge running into an apical hairy callus. Callus between lateral lobes with elongate, forwardly pointing warts and 2 forked appendages. *Philippines*. G2. Summer.

18. P. mariae Warner & Williams. Illustration: Botanical Magazine, 6964 (1887); Orchid Digest **35**: 308 (1971); Bechtel et al., Manual of cultivated orchid species, 260 (1981).
Leaves 15–30 cm, strap-shaped. Inflorescence hanging, simple or branched. Flowers 2.5–5 cm in diameter. Sepals and petals oblong-elliptic, white or cream with transverse brownish red bars and blotches, sometimes with purplish spots at the base. Sepals 1.6–2.2 cm, petals 1.5–1.7 cm. Lip pale mauve or purple, lateral lobes broadly strap-shaped, toothed at the apex, central lobe ovate, the margins expanded and toothed towards apex, with a central ridge at the base and a hairy callus at the apex. Callus between lateral lobes with a series of 2-lobed appendages. *Philippines*. G2. Summer.

***P. pallens** (Lindley) Reichenbach (*P. foerstermanii* Reichenbach). Illustration: Orchid Digest **40**: 207 (1976); Sweet, The genus Phalaenopsis, 109 (1980); Bechtel et al., Manual of cultivated orchid species, 261, 362 (1981). Leaves elliptic to obovate, 12–18 cm. Inflorescence erect or arching, simple, with 1 or a few flowers. Flowers *c.* 5 cm in diameter. Sepals and petals oblong-elliptic, yellowish green with transverse brown bars, lines and spots. Sepals with a keel on the back, the upper 1.2–2.3 cm, the laterals 1.5–2.2 cm, petals 1.1–2 cm. Lip 1.3–1.7 cm, lateral lobes oblong, yellowish, central lobe white, narrowly ovate, margins usually expanded and toothed towards the apex, with a central ridge at the base and hairy callus towards the apex. Callus between lateral lobes with a pair of forked appendages. *Philippines*.

Var. **denticulata** (Reichenbach) Sweet has 2 or 3 reddish purple lines on either side of the central ridge of the central lobe of the lip.

19. P. maculata Reichenbach. Illustration: Orchid Digest **35**: 308 (1971), **41**: 29 (1982).
Leaves 15–20 cm, strap-shaped. Inflorescence more or less erect to arching, usually simple, with few flowers. Flowers less than 2.5 cm in diameter. Sepals and petals oblong-lanceolate, cream to greenish white with a few purplish brown or red-brown blotches. Upper sepal 1.3–1.8 cm, lateral sepals 1.3–1.5 cm, petals *c.* 1.2 cm. Lip *c.* 1 cm, lateral lobes diamond-shaped, the apical part with a horseshoe-shaped callus, central lobe red to deep purple, elliptic with faint longitudinal grooves, very thick and fleshy. Callus between side lobes 2-toothed; at the base of central lobe is a shallow, 2-toothed process. *Malay peninsula, Sarawak*. G2.

***P. modesta** J.J. Smith (*P. psilantha* Schlechter). Illustration; Orchid Digest **43**: 212 (1979). Leaves obovate, 11–15 cm. Inflorescence arched, usually simple, with few flowers. Sepals and petals elliptic, white or pale pink, with fine transverse purple stripes towards the base, sepals 1.2–1.5 cm, petals 1.1–1.5 cm. Lip 1.2–1.4 cm, lateral lobes oblong, narrowing towards the notched apex, yellow callus at middle, central lobe oblong, white, purple towards the apex, convex, with a basal ridge and an apical, usually hairless callus. Callus between the lateral lobes with a pair of forked appendages. *Borneo, Sulawesi*. G2.

111. PARAPHALAENOPSIS Hawkes
V.A. Matthews
Differs from *Phalaenopsis* in having terete leaves and often producing more than 1 raceme from the same point on the stem.

A genus of 3 species from W. Borneo (Kalimantan), all of which are in cultivation. Though the genus has been crossed with several other (e.g. *Vanda, Renanthera*), attempts to hybridise it with *Phalaenopsis* have so far failed. Cultivation as for *Phalaenopsis* (p. 244).
Literature: Sweet, H.R., *The genus Phalaenopsis*, 118–23 (1980).

1a. Sepals and petals greenish yellow to yellowish brown **1. denevei**
b. Sepals and petals white or cream, often flushed with pink or violet **2. serpentilingua***

1. P. denevei (J.J. Smith) Hawkes (*Phalaenopsis denevei* J.J. Smith). Illustration: Orchid Digest **37**: 12–13 (1973).
Leaves erect or hanging, to 70 cm. Raceme usually upright with 3–15 flowers, to 15 cm. Flowers *c.* 5 cm in diameter, fragrant. Sepals and petals pale greenish yellow to yellowish brown, with paler, wavy margins, *c.* 2.5 cm. Lateral lobes of lip triangular, yellow at base, purple above, with white margin, central lobe narrowly oblong, somewhat expanded and notched at apex, whitish, reddish or purplish towards apex. Callus more or less square with a slightly toothed margin, yellow with red lines. *W Borneo.* G2. Spring–summer.

2. P. serpentilingua (J.J. Smith) Hawkes (*Phalaenopsis serpentilingua* J.J. Smith). Illustration: Orchid Digest **37**: 12 (1973). Leaves as for *P. denevei.* Raceme usually upright, to 35 cm, with 7–many flowers. Flowers 3–4 cm in diameter, fragrant. Sepals and petals white or cream, sometimes tinged with violet. Lateral lobes of lip pale orange with reddish markings, central lobe narrowly oblong with 2 divergent apical lobes like a snake's tongue, white or pale yellow with transverse yellow and purple marks. Callus with a toothed margin, yellow with red marks. *W Borneo.* G2.

***P. laycockii** (Henderson) Hawkes (*Phalaenopsis laycockii* Henderson). Illustration: Orchid Digest **37**: 12–13 (1973). Raceme with 9–15 flowers which are 5–8 cm in diameter, sepals and petals 3.5–4.5 cm, whitish flushed with pink or pale purple, lateral lobes of lip white flushed with dark purple or red, central lobe narrowly oblong, margins reflexed with 2 short, divergent apical lobes, callus margins untoothed. *W Borneo.* G2.

112. VANDA R. Brown
J. Cullen
Monopodial epiphytes with long or short stems, sometimes scrambling. Leaves in 2 ranks, often fleshy, flat or terete, often lobed or toothed at the apex, folded when young. Racemes axillary with few to many flowers. Sepals and petals similar or the lateral sepals larger than the rest, all more than 1.5 cm, usually spreading. Lip often complex, 3-lobed, the lateral lobes large or small, the central lobe often itself notched or lobed, bearing calluses of various forms, generally with 2 humped ridges at the entrance to the spur. Spur usually present, conical or more rarely cylindric. Column short and thick with or without a foot. Pollinia 2, usually borne on a short, broad stipe and with a large viscidium.

A genus of 50 or more species from the Himalaya and SE Asia. Its classification is confused, as are its relations with other genera. Species No. 1 is often referred to *Euanthe* Schlechter (recognised by its flat flowers with unequal sepals and the particular form of the spurless lip); No. 2 and the species under it have been separated off into the genus *Papilionanthe* (a genus proposed by Schlechter in 1915, but hardly taken up until recently – see Garay, Botanical Museum Leaflets, Harvard University **23**: 369–72, 1974 – recognised by the terete leaves and the extension of the short column into a conspicuous foot); No. 3 is referred by Garay to the genus *Holcoglossum* Schlechter (recognised by its narrow, cylindric spur, grooved leaves, footless, winged column and linear stipe). However, for the present purpose it seems sensible to maintain a wide circumscription of *Vanda*, citing synonyms as appropriate.

The number of species in cultivation is uncertain; 16 are included here, though only 8 are treated fully. A large number of hybrids is also available. Plants should be given as much light as possible, with only light shading in very sunny weather. They require an epiphytic compost and can be successfully grown in pots or on lengths of tree fern or bark. Propagation is by means of pieces of stem cut back to a well-developed aerial root.

Leaves. Terete: **2**; V-shaped in section: **3,4**; flat: **1,5–8**.
Sepals and petals. Spreading: **1–7**; curved forwards: **8**.
Lip. Central lobe with 2 divergent lobes at apex: **7,8**; with a downwardly projecting horn at apex: **8**. Shorter than sepals: **1,5**; as long as or longer than sepals: **2–4, 6–8**.
Spur. Absent: **1,8**; cylindric, *c.* 2.5 cm: **3**; conical and much shorter: **2,4–8**.

1a. Leaves terete or very narrowly V-shaped in section 2
b. Leaves flat 4
2a. Leaves terete **2. teres***
b. Leaves narrowly V-shaped in section with a distinct groove above, at least near the base 3
3a. Spur cylindric, acute, *c.* 2.5 cm; lateral sepals sickle-shaped, longer than the upper **3. kimballiana**
b. Spur much shorter, blunt, conical; lateral sepals similar to the upper **4. amesiana**
4a. Sepals and petals curved forwards; racemes with 2–5 flowers; plants small **8. cristata***
b. Sepals and petals spreading; racemes with 6 or more flowers; plants often large 5
5a. Flower flat with broad, overlapping sepals and petals; base of lip hollowed out, but spur absent **1. sanderiana**
b. Flower not as above; lip with a distinct, if short, spur 6
6a. Central lobe of lip expanded into a kidney-shaped apical part from an oblong base **7. denisoniana***
b. Central lobe of lip oblong, sometimes waisted but then the apical part broadly rounded, not kidney-shaped 7
7a. Flowers 7–10 cm in diameter; lip considerably smaller than the sepals **5. caerulea**
b. Flowers 4–5 cm in diameter; lip about as long as the sepals **6. tricolor***

1. V. sanderiana Reichenbach (*Euanthe sanderiana* (Reichenbach) Schlechter). Illustration: Botanical Magazine, 6983 (1888); Lindenia **12**: t. 547 (1896); Cogniaux & Goossens, Dictionnaire Iconographique des Orchidées, Vanda, t. 12 (1900); Orchid Digest **43**: 84 (1979).
Stem to 60 cm. Leaves to 45 × 5 cm, flat, oblong-linear, sharply truncate and 3-toothed at the apex. Raceme to 30 cm with 5–10 flowers. Flowers 9–10 cm in diameter, flat, sepals and petals overlapping, broadly elliptic; upper sepal to 5.5 cm, violet-blue like the petals, which are *c.* 4 cm; lateral sepals to 6 cm, yellow or greenish with red-brown marbling. Lip shorter than the petals, with a lower, hemispherical, saccate part bearing 2

semicircular lateral lobes, the central lobe broadly kidney-shaped with 3 low ridges. *Philippines*. G2. Autumn–winter.

2. **V. teres** (Roxburgh) Lindley (*Papilionanthe teres* (Roxburgh) Schlechter). Illustration: Botanical Magazine, 4114 (1844); Orchid Digest **43**: 84 (1979); Bechtel et al., Manual of cultivated orchid species, 279 (1981).
Stem to 2 m. Leaves cylindric, terete, upright, rather blunt. Racemes to 30 cm with 3–5 flowers. Flowers 8–10 cm in diameter. Sepals and petals elliptic, lateral sepals white, petals and upper sepal violet or pink. Lip as long as sepals, lateral lobes large, arching over the column, reddish striped with yellow at the base, central lobe obovate, deeply 2-lobed, violet-pink to yellow with red spots towards the base. Spur short, conical. *E Himalaya to Thailand & Laos*. G2. Spring.

*V. **hookeriana** Reichenbach (*Papilionanthe hookeriana* (Reichenbach) Schlechter). Illustration: The Garden **23**: 10 (1883); Illustration Horticole **30**: t. 484 (1883). Similar, but with flowers to 5 cm in diameter, sepals and petals white, flushed and spotted with red, lip with spreading lateral lobes and a fan-shaped, white and red-spotted, shallowly 3-lobed central lobe. *Malaysia, Sumatra, Borneo*. G2. Autumn.

3. **V. kimballiana** Reichenbach (*Holcoglossum kimballianum* (Reichenbach) Garay). Illustration: Lindenia **5**: t. 204 (1889); Botanical Magazine, 7112 (1890); Cogniaux & Goossens, Dictionnaire Iconographique des Orchidées, Vanda, t. 7 (1898).
Stem elongate. Leaves to 20 cm, fleshy, narrowly V-shaped in section, acute. Sepals and petals *c*. 2.5 cm, oblong-elliptic, white, wavy, the lateral sepals longer, sickle-shaped. Lip about as long as the sepals with small, acuminate, yellow, red-spotted lateral lobes and a large, nearly circular, toothed and notched, rose-red central lobe. Spur *c*. 2.5 cm, cylindric, curving downwards, acute. *Burma*. G2. Autumn.

4. **V. amesiana** Reichenbach. Illustration: Botanical Magazine, 7139 (1890); Cogniaux & Goossens, Dictionnaire Iconographique des Orchidées, Vanda, t. 1 (1897); Orchid Digest **43**: 44 (1979).
Stem short. Leaves to 20 cm, fleshy, narrowly V-shaped in section, acute. Raceme to 50 cm, with 15–30 flowers, axis red-spotted. Sepals and petals *c*. 2 cm, broadly oblong, white. Lip slightly longer than sepals, pink with white margin and stripes, lateral lobes erect, short, central lobe broadly ovate, wavy. Spur conical, very short and blunt. *Kampuchea, Burma, SW China*. G1. Winter.

5. **V. caerulea** Lindley. Illustration: Schlechter, Die Orchideen, edn 2, t. 13 (1927); Orchid Digest **43**: 44, 49 (1979); Bechtel et al., Manual of cultivated orchid species, 278 (1981).
Stem short, densely leafy. Leaves to 25 × 2.5 cm, oblong, blunt, flat. Racemes to 45 cm with 7–15 flowers. Flowers 7–10 cm in diameter. Sepals and petals oblong-obovate, clear blue, often with darker tessellation, petals often with twisted claws. Lip much shorter than the sepals, dark violet-blue with small whitish lateral lobes and an oblong, convex, 2–3-ridged central lobe. Spur conical, short, blunt. *E Himalaya, Burma, Thailand*. G2. Winter.

6. **V. tricolor** Lindley (*V. suavis* Reichenbach). Illustration: Botanical Magazine, 4432 (1894); Bechtel et al., Manual of cultivated orchid species, 279 (1981).
Stems to 1 m. Leaves to 45 × 5 cm, oblong. Racemes erect with 8–10 flowers. Flowers *c*. 5 cm in diameter, fragrant. Sepals and petals broadly ovate with broad, claw-like bases, wavy, yellow spotted with dark brown, their backs white. Lip as long as the sepals, lateral lobes small, erect, white, central lobe broadly oblong, waisted at the middle, the basal part broader than the apical, violet-red with purple stripes. Spur short, broad, laterally compressed. *Java*. G2. Winter.

*V. **lamellata** Lindley (*V. boxallii* Reichenbach). Illustration: Cogniaux & Goossens, Dictionnaire Iconographique des Orchidées, Vanda, t. 9 (1898). Similar but with smaller, non-fragrant flowers, sepals and petals *c*. 2 cm, whitish (flushed red in 'Boxallii'), lip shorter than the sepals, the central lobe violet, oblong, notched, with 2–3 conspicuous ridges. *Philippines*. G2. Winter.

*V. **tesselata** (Roxburgh) G. Don (*Epidendrum tesselatum* Roxburgh; *V. roxburghii* R. Brown). Illustration: Botanical Magazine, 2245 (1821); Die Orchidee **22**: 13 (1971); Orchid Digest **43**: 47 (1979); Bechtel et al., Manual of cultivated orchid species, 279 (1981). Also generally similar but with wavy sepals and petals which are tessellated yellow and brown, the central lobe of the lip oblong, slightly waisted, dark violet. *Tropical Asia from SE India to Malaysia*. G2. Winter.

7. **V. denisoniana** Benson & Reichenbach. Illustration: Botanical Magazine, 5811 (1869); Cogniaux & Goossens, Dictionnaire Iconographique des Orchidées, Vanda, t. 8 (1898); Seidenfaden & Smitinand, Orchids of Thailand **4**(1): t. 27 (1963).
Stem short. Leaves to 30 × 2 cm, linear, acutely 2-lobed at the apex. Racemes to 15 cm, with about 6 flowers, arching. Flowers *c*. 5 cm in diameter. Sepals and petals elliptic with broad claws, greenish to cream-white, the lateral sepals somewhat longer than the rest. Lip a little longer than the sepals, white with a yellow patch at the base, lateral lobes small, central lobe oblong at the base, bearing 4–5 ridges, expanding towards the apex to a kidney-shaped, notched portion. Spur oblong, blunt, laterally compressed. *Burma, Thailand*. G2. Spring.

*V. **caerulescens** Griffith. Illustration: Botanical Magazine, 5834 (1870); Orchid Digest **43**: 44 (1979). Similar but with a longer raceme having many flowers, sepals and petals pale violet, the lip darker violet and the spur conical, curved downwards. *Burma, Thailand*. G2. Spring.

*V. **bensonii** Bateman. Illustration: Botanical Magazine, 5611 (1866); Cogniaux & Goossens, Dictionnaire Iconographique des Orchidées, Vanda, t. 2 (1898). Also similar but sepals and petals olive green with brown spots, the lip violet-pink, white at the base, the spur conical. *Burma, Thailand*. G2. Autumn.

*V. **insignis** Blume. Illustration: Botanical Magazine, 5759 (1869); Lindenia **8**: t. 355 (1892); Cogniaux & Goossens, Dictionnaire Iconographique des Orchidées, Vanda, t. 3 (1897). Also broadly similar but with larger flowers in various shades of violet, yellow or white, and with the central lobe of the lip concave and wavy. *Indonesia (Moluccas)*. G2. Autumn.

8. **V. cristata** (Wallich) Lindley. Illustration: Botanical Magazine, 4304 (1847); Gartenflora **20**: t. 680 (1871); Orchid Digest **43**: 44 (1979); Bechtel et al., Manual of cultivated orchid species, 278 (1981).
Stems short, densely leafy. Leaves to 12 × 4 cm, linear, thick and fleshy, flat, 2-lobed at the apex. Racemes short with 2–5 flowers. Sepals and petals to 2.5 cm, curving forwards, oblong, greenish yellow. Lip with short, erect lateral lobes and with an oblong central lobe which is 2-lobed at the apex with the lobes diverging, bearing a downwardly pointing horn-like projection between the lobes, the upper surface with 5

tuberculate ridges, all bright yellow to green with brown stripes. Spur short, blunt, conical. *E Himalaya*. G1. Winter.

***V. alpina** (Lindley) Lindley. Illustration: Annals of the Royal Botanic Garden Calcutta **8**: t. 289 (1898); Hara, Photo-album of plants of E Himalaya, t. 120 (1968). Similar but with hanging flowers, sepals and petals 1.3–1.8 cm, the lip with an oblong, truncate central lobe without a horn-like projection, the base saccate but without a spur. *Himalaya*. G1. Spring.

***V. pumila** Hooker. Illustration: Annals of the Royal Botanic Garden Calcutta **5**: t. 68 (1895) & **8**: t. 288 (1898); Botanical Magazine, 7968 (1904). Also similar but sepals and petals *c*. 3 cm, greenish or yellowish white and the lip with an ovate, concave, red-striped central lobe. *Himalaya*. G1. Summer.

113. VANDOPSIS Pfitzer

J. Cullen

Large monopodial epiphytes. Leaves mainly at the base of the stem, in 2 ranks, large, flat or V-shaped in section, folded when young. Flowers in axillary racemes. Sepals and petals similar, fleshy, widely spreading. Lip 3-lobed, attached to the column by the base and by the small lateral lobes, which have a flap of fleshy tissue joining them above the ridge of the central lobe; central lobe shorter than the petals, often laterally flattened, saccate at the base and bearing a conspicuous ridge which is interrupted near the base. Column short with a projection at its base but without a foot. Pollinia 4 in 2 pairs, borne on a short, broad stipe with reflexed edges, the viscidium broad.

A genus of about 10 species from SE Asia. Cultivation as for *Vanda* (p. 249).

1a. Central lobe of lip as broad as long, not laterally flattened; leaves to 25 cm **3. parishii**
b. Central lobe of lip longer than broad, laterally flattened; leaves 30 cm or more 2
2a. Leaves recurved, flat in section; raceme 25–35 cm **1. gigantea**
b. Leaves straight, V-shaped in section; raceme 1–2 m **2. lissochiloides**

1. V. gigantea (Lindley) Pfitzer (*Vanda gigantea* Lindley; *Stauropsis gigantea* (Lindley) Bentham & Hooker). Illustration: Botanical Magazine, 5189 (1860); Bateman, Second century of orchidaceous plants, t. 142 (1865); Cogniaux & Goossens, Dictionnaire Iconographique des Orchidées, Stauropsis, t. 2 (1898); Orchid Digest **43**: 87 (1979).
Stems thick, fleshy. Leaves to 35 × 6 cm, fleshy, recurved, linear-oblong, flat, slightly unequally 2-lobed at the apex. Raceme 25–35 cm with up to 15 flowers. Sepals and petals 2.5–3 cm, oblong-obovate, yellow with red-brown blotches. Lip longer than broad, mostly yellow, lateral lobes suffused with purple. *Burma, Thailand, Malaysia*. G2. Spring–summer.

2. V. lissochiloides (Gaudichaud) Pfitzer (*Fieldia lissochiloides* Gaudichaud; *Vanda batemanii* Lindley; *Vanda lissochiloides* (Gaudichaud) Lindley; *Stauropsis lissochiloides* (Gaudichaud) Pfitzer). Illustration: Cogniaux & Goossens, Dictionnaire Iconographique des Orchidées, Stauropsis, t. 1 (1898); Orchid Digest **43**: 87 (1979); Bechtel et al., Manual of cultivated orchid species, 279 (1981).
Very large plants. Leaves 30–50 cm, straight, V-shaped in section. Racemes 1–2 m with many flowers. Sepals and petals to 3.5 cm, bright purple or yellowish on the back, yellow with purple spots on the front. Lip longer than broad, short, yellow. *Philippines, Moluccas; ?Thailand*. G2. Summer.

3. V. parishii (Reichenbach) Schlechter (*Vanda parishii* Reichenbach). Illustration: Cogniaux & Goossens, Dictionnaire Iconographique des Orchidées, Vanda, t. 11, 11A (1898); Bechtel et al., Manual of cultivated orchid species, 279 (1981).
Stems short. Leaves to 25 × 6 cm, oblong-elliptic, unequally 2-lobed at the apex. Raceme to 35 cm with 5–6 flowers. Sepals and petals to 2.5 cm, obovate, greenish yellow spotted with brown (rose-pink, whitish at the base in var. **marriottiana** Reichenbach). Lip as broad as long, yellowish with red spots, and a conspicuous, vertical ridge. *Burma, Thailand*. G2. Summer.

Both var. **parishii** and var. **marriottiana** have been in cultivation; the latter is apparently more frequent. Garay (Botanical Museum Leaflets, Harvard University **23**: 374, 1974) suggests that this species should be transferred from *Vandopsis* to the genus *Hygrochilus*.

114. ESMERALDA Reichenbach

J. Cullen

Monopodial epiphytes with elongate, scrambling stems. Leaves in 2 ranks, unequally 2-lobed at their apices, folded when young, close, mainly at the bases of the stems. Flowers in axillary racemes. Sepals and petals similar, spreading. Lip freely movable on its attachment to the column, about as long as the petals, 3-lobed with small, erect, oblong lateral lobes and a larger, flat, ovate, auricled central lobe which has a hollow sac concealed within the tissue at the base (perceptible as a hump on the back of the lip), the surface with 3–5 longitudinal ridges. Column large, fleshy, without a foot. Pollinia 4 in 2 pairs, stipe rapidly broadening to a broad, disc-like viscidium.

A genus of 2 species from the Himalaya and SE Asia. Cultivation as for *Vanda* (p. 249).
Literature: Tan, K., Taxonomy of Arachnis, Armodorum, Esmeralda and Dimorphorchis, *Selbyana* **1**: 1–15 (1975), 365–73 (1976).

1. E. cathcartii (Lindley) Reichenbach (*Vanda cathcartii* Lindley; *Arachnanthe cathcartii* (Lindley) Bentham & Hooker; *Arachnis cathcartii* (Lindley) J.J. Smith). Illustration: Cogniaux & Goossens, Dictionnaire Iconographique des Orchidées, Arachnanthe, t. 2 (1898); Annals of the Royal Botanic Garden Calcutta **8**: t. 278 (1898); Orchid Digest **43**: 89, 92 (1979).
Stem to 2 m, thick, hanging. Leaves to 15 × 3–4 cm. Racemes with 3–5 distant flowers. Sepals and petals to 4.5 cm, ovate to broadly elliptic, rounded at their apices, yellow with dense, red-brown, transverse stripes. Lip white with red stripes, the margin yellow, irregularly toothed and wavy. *E Himalaya*. G2. Spring–summer.

115. DIMORPHORCHIS Rolfe

J. Cullen

Large monopodial epiphytes with elongate stems. Leaves densely set in 2 ranks, folded when young. Flowers in long, loose, axillary racemes with densely woolly axis and flower-stalks, dimorphic, the first 1–4 flowers of each raceme orange-yellow with fine red spots and shorter and broader, more rounded sepals and petals than the rest, which are larger and have wavy sepals and petals which are greenish with large, confluent, red-brown spots; in each type of flower the sepals and petals are similar, spreading and woolly outside. Lip much shorter than the petals, hollowed out at the base and with a thin, high ridge in the centre which ends in a fine, erect point. Column short, without a foot. Pollinia 4 in 2 pairs, borne almost horizontally, the stipe short and broad, broadening a little to the large, disc-like viscidium.

A genus from Borneo with 1 or 2 species. The 1 cultivated species is not common in cultivation, despite Holttum's remark (Orchids of Malay, 619, 1953): 'A large

plant with a dozen inflorescences ranks with *Grammatophyllum* in full flower as one of the most remarkable objects in the Orchid world'. Cultivation as for *Vanda* (p. 249).
Literature: Tan, K., Taxonomy of Arachnis, Armodorum, Esmeralda and Dimorphorchis, *Selbyana* **1**: 1–15 (1975), 365–73 (1976); Soon, P.S., Notes on Sabah orchids, part I, *Orchid Digest* **44**: 193–5 (1980).

1. **D. lowii** (Lindley) Rolfe (*Vanda lowii* Lindley: *Renanthera lowii* (Lindley) Reichenbach; *Arachnanthe lowii* (Lindley) Bentham & Hooker; *Vandopsis lowii* (Lindley) Schlechter; *Arachnis lowii* (Lindley) Reichenbach; *Renanthera rohaniana* Reichenbach). Illustration: Cogniaux & Goossens, Dictionnaire Iconographique des Orchidées, Arachnanthe, t. 4 (1898); Orchid Digest **44**: 194 (1980).
Stems to 2 m. Leaves to 70 × 5–6 cm, oblong, obtuse. Racemes to 2.5 m. Sepals and petals fleshy, 3.5–5 cm. Lip acute, whitish, violet within. *Borneo*. G2. Autumn.
Both var. **lowii** and var. **rohaniana** (Reichenbach) Tan, which has shorter racemes and slightly smaller, paler flowers, have been in cultivation.

116. RENANTHERA Loureiro
J. Cullen
Monopodial epiphytes with stout, branched stems. Leaves spirally arranged or in 2 ranks, well spaced, unequally 2-lobed at their apices, folded when young. Flowers in long, horizontal, axillary panicles. Upper sepal and petals similar, spreading upwards, the lateral sepals longer and broader, spreading downwards, often close together, all more than 1 cm. Lip much smaller than the petals, 3-lobed, with a conical spur and 2 calluses at the junctions of the central and lateral lobes. Column short, without a foot. Pollinia 4, borne on a short, linear-oblong, sometimes slightly waisted stipe, and with an oblong viscidium
A genus of 10 or more species from E Asia. Cultivation as for *Vanda* (p. 249).

1a. Upper sepal 1.3–1.5 cm **4. pulchella**
b. Upper sepal 2 cm or more 2
2a. Central lobe of the lip with 5 yellow calluses at the base in addition to the 2 at the junctions of the lobes **3. imschootiana**
b. Central lobe of the lip with no additional calluses 3
3a. Lateral sepals red with deep crimson blotches; leaves 20–25 cm **2. storiei**
b. Lateral sepals without blotches; leaves 6–12 cm **1. coccinea**

1. **R. coccinea** Loureiro. Illustration: Edwards's Botanical Register **14**: t. 1131 (1828); Botanical Magazine, 2997, 2998 (1830).
Stems scrambling, to 2 m. Leaves 6–12 × 3–3.5 cm, unequally 2-lobed at the apex. Panicle to 70 cm. Upper sepal and petals oblong, obtuse, 2.5–2.8 cm, reddish spotted with scarlet. Lateral sepals oblong-spathulate, obtuse, somewhat wavy, 3.5–4.5 cm, clear scarlet. Lip with low, yellow, red-striped lateral lobes and a scarlet central lobe which is yellowish towards the base. *Thailand, Laos, Vietnam, S China, possibly also in the Philippines, Java & Burma*. G2. Spring, autumn.

2. **R. storiei** (Storie) Reichenbach. Illustration: Botanical Magazine, 7537 (1897); Bechtel et al., Manual of cultivated orchid species, 268 (1981).
Very similar to *R. coccinea*, but stems shorter, leaves 20–25 × 3.5–4 cm, lateral sepals scarlet with crimson blotches. *Philippines*. G2. Summer.

3. **R. imschootiana** Rolfe. Illustration: Botanical Magazine, 7711 (1900); Bechtel et al., Manual of cultivated orchid species, 268 (1981).
Stem erect, scarcely scrambling. Leaves to 7 × 2 cm, oblong, shortly 2-lobed at their apices. Panicle to 45 cm. Upper sepal 2–2.2 cm, petals 3.7–4 cm, yellow spotted with red, lateral sepals clawed, wavy, elliptic, scarlet, or yellow flushed with red. Lip scarlet with 5 yellow calluses at the base of the central lobe. *NE India, Burma, Laos*. G2. Summer.

4. **R. pulchella** Rolfe.
Stem to 20 cm. Leaves 6.5–8 × 1.5–2 cm, in 2 ranks, narrowly oblong, unequally lobed at their apices. Panicle with few branches. Upper sepal 1.3–1.5 cm, lanceolate, obtuse; lateral sepals *c.* 1.7 cm, clawed, oblong-elliptic, obtuse, all yellow. Petals yellow below, red above. Lip with red lateral lobes, the central lobe yellow with 4 distinct teeth at the base. *Burma, perhaps also in Thailand*. G2. Summer.

117. ASCOCENTRUM Schlechter
J. Cullen
Monopodial epiphytes with short leafy stems. Leaves linear, unequally 2-lobed and toothed at the apex, folded when young. Flowers in axillary racemes. Sepals and petals similar, spreading. Lip small, consisting mainly of a vertical, cylindric, blunt pouch bearing at its apex 2 small, erect lateral lobes and a linear-oblong, acute central lobe which is much smaller than the petals. Column short, without a foot. Pollinia 2, spherical, borne on a short linear stipe and with a small, oblong viscidium.
A genus of about 5 species from E Asia (Himalaya to Borneo). Cultivation as for *Gastrochilus*.

1a. Sepals and petals pink to violet-purple; leaves 1.7–2 cm broad **1. ampullaceum**
b. Sepals and petals orange or orange-yellow; leaves rarely exceeding 1 cm in breadth **2. miniatum**

1. **A. ampullaceum** (Lindley) Schlechter (*Saccolabium ampullaceum* Lindley). Illustration: American Orchid Society Bulletin **41**: 820 (1972); Orchid Digest **42**: 180 (1979).
Leaves stiff, 8–12 × 1.7–2 cm. Raceme with many flowers, erect, shortly stalked. Sepals and petals 1–1.2 cm, oblong, blunt, pink to violet-purple. Lip with pink to violet-purple pouch and central lobe (which is horizontal), the lateral lobes whitish. *NE India to Thailand*. G2. Spring.

2. **A. miniatum** (Lindley) Schlechter (*Saccolabium miniatum* Lindley). Illustration: Holttum, Orchids of Malaya, 730 (1953); Orchid Digest **42**: 180 (1979); Bechtel et al., Manual of cultivated orchid species, 169 (1981).
Leaves *c.* 10–20 × 1 cm, fleshy. Raceme shortly stalked. Sepals and petals *c.* 6 mm, orange or orange-yellow. Lip similar in colour, the central lobe bent downwards. *E Himalaya to Borneo & Java*. G2. Spring.

118. GASTROCHILUS D. Don
V.A. Matthews & J. Cullen
Monopodial epiphytes with short stems. Leaves close together at the stem bases, in 2 ranks, fleshy, folded when young. Flowers fleshy, in axillary racemes which are sometimes paired at each node. Sepals and petals similar, spreading. Lip consisting mainly of a vertical pouch with 2 lateral lobes joined to the column and a forwardly pointing, often hairy or fringed central lobe which is kidney- or crescent-shaped. Column short, without a foot. Anther ovate, projecting forwards on the column. Pollinia 2 on a linear stipe and with a small, oblong viscidium.
A genus of 15–20 species from the Himalaya and E Asia, including Japan. Only 3 species are commonly grown. Cultivation as for *Aerides* (p. 242) or *Vanda* (p. 249), but vegetative propagation is difficult.

Literature: Herklots, G.A.C., Nepalese and Indian orchids: Gastrochilus, *Orchid Review* **82**: 354–8 (1974).

1a. Central lobe of the lip with long white hairs on the upper surface **1. bellinus**
b. Central lobe of the lip without such hairs, though it may be papillose or fringed 2
2a. Leaves acute but not notched at the apex; stem 10–30 cm **3. acutifolius**
b. Leaves acute and 2-toothed at the apex; stem 1–3 cm **2. dasypogon**

1. G. bellinus (Reichenbach) Kunze. Illustration: Die Orchidee **23**: 8 (1972); Bechtel et al., Manual of cultivated orchid species, 150 (1981).
Stem usually less than 5 cm, stout. Leaves 15–30 cm, narrowly oblanceolate, curved, deeply notched at the apex. Flowers fragrant, in a cluster-like raceme of 4–7. Sepals and petals 9–14 mm, slightly incurved, greenish or pale yellow with brown or purple spots. Lip *c.* 1.3 cm, central lobe broadly elliptic with a toothed margin, white with purple or red spots, yellow towards the base, upper surface with long white hairs. *Burma, Thailand*. G2. Winter–spring.

2. G. dasypogon (Lindley) Kunze. Illustration: Sheehan & Sheehan, Orchid genera illustrated, 95 (1979); Bechtel et al., Manual of cultivated orchid species, 150 (1981).
Stem 1–3 cm. Leaves 10–15 cm, ovate-oblong, acute, notched at the apex. Raceme almost umbellate. Sepals and petals to 8 mm, obovate, pale yellow or greenish, sometimes spotted with purple. Lip 5–7 mm, central lobe kidney-shaped with a fringed margin, white or yellow with red or purple spots, not hairy. *Himalaya? Thailand & Sumatra*. G2. Autumn.

3. G. acutifolius (Lindley) Kunze (*Saccolabium acutifolium* Lindley). Illustration: Annals of the Royal Botanic Garden Calcutta **8**: t. 302 (1898); Orchid Review **82**: 358 (1974); Die Orchidee **27**: 127 (1976).
Stems 10–30 cm. Leaves to 12 × 2 cm, oblong, acute, not notched at the apex. Raceme with 10–20 flowers, compressed, umbel-like. Sepals and petals *c.* 1 cm, oblong, blunt, greenish yellow, usually with brown-purple spots. Lip white, the central lobe broadly triangular, papillose but not hairy on the upper surface, fringed. *Nepal, NE India*. G2. Winter.

119. SCHOENORCHIS Blume
V.A. Matthews
Differs from *Gastrochilus* in having leaves strap-shaped to terete and a simple or branched raceme with many small flowers; sepals and petals 2–5 mm; lip 3-lobed, spurred but not consisting mainly of a vertical pouch; pollinia 4, united in pairs attached to a long slender viscidium. Rostellum linear, bifid, very conspicuous.

Ten species ranging from the Himalaya to New Guinea. Cultivation as for *Vanda* (p. 249).

1. S. juncifolia Blume.
Stem to 30 cm, hanging, branched. Leaves to 16 cm, terete, often flushed with purple. Raceme to 10 cm, simple, hanging, dense, with many flowers. Sepals and petals oblong, shortly acuminate, bluish violet. Lip longer than sepals, with a more or less curved spur, the central lobe *c.* 2 mm, pale violet or white, recurved. *Sumatra, Java*. G2. Summer.

120. CLEISOSTOMA Blume
J. Cullen
Monopodial epiphytes with elongate stems. Leaves borne all around the stem, terete or flat, folded when young. Flowers numerous in axillary racemes or rarely panicles. Sepals and petals similar or the petals slightly smaller, all spreading, lateral sepals and petals not inserted on the column foot. Lip consisting mainly of a short, broad, blunt, vertical pouch, the blade 3-lobed with small lateral lobes and a larger, usually arrowhead-shaped central lobe. The entrance to the pouch is largely blocked by a complex callus borne on the upper surface of its interior, and there is usually a longitudinal cross-wall from the lower part of its interior as well. Column short with a short foot. Pollinia 4, united in 2 round masses with the stipe linear and the viscidium disc-like or the stipe broadening rapidly from the apex to a saddle-shaped viscidium.

A genus of about 100 species from Asia, occurring from Nepal to New Guinea. The flowers are more interesting than attractive and few are grown. The genus is commonly found under the name *Sarcanthus* Lindley. Cultivation as for *Vanda* (p. 249).
Literature: Garay, L., On the systematics of the monopodial orchids I, *Botanical Museum Leaflets, Harvard University* **23**: 168–76 (1972); Seidenfaden, G., Orchid genera in Thailand II: Cleisostoma, *Dansk Botanisk Arkiv* **29**(3): 1–80 (1975).

1a. Leaves terete **1. appendiculatum***
b. Leaves flat **2. rostratum***

1. C. appendiculatum (Lindley) Jackson (*Sarcanthus appendiculatus* of various authors). Illustration: Annals of the Royal Botanic Garden Calcutta **5**: t. 76 (1895); Hooker's Icones Plantarum **22**: t. 2136 (1893).
Stem to 40 cm, hanging. Leaves 7–10 cm, terete, spirally arranged. Raceme with 10–15 flowers, stalked, hanging, longer than the leaves. Flowers 1.5–2 cm. Sepals and petals oblong, brownish yellow with red-violet veins. Lip with small, erect lateral lobes and a broadly triangular central lobe, rose-pink. Pouch cylindric, bright brownish yellow with purple veins. Stipe linear, viscidium disc-like. *E Himalaya, Burma*. G2. Summer.

***C. filiforme** (Lindley) Garay (*Sarcanthus filiformis* Lindley). Illustration: Bechtel et al., Manual of cultivated orchid species, 184 (1981). Similar, but with longer stems with leaves 20–25 cm borne 1-sidedly, the flowers *c.* 1.1 cm. *E Himalaya to Thailand*. G2. Summer–autumn.

***C. simondii** (Gagnepain) Seidenfaden (*Sarcanthus teretifolius* (Lindley) Lindley). Also similar, but with thick, short, often recurved leaves and a stipe which is narrow at the apex but rapidly broadens to a saddle-shaped viscidium. *Himalaya, Laos, Vietnam, Thailand & S China*. G2. Summer.

2. C. rostratum (Lindley) Garay (*Sarcanthus rostratus* Lindley). Illustration: Lindley, Collectanea Botanica, t. 39 (1821); Edwards's Botanical Register **12**: t. 981 (1826); Dansk Botanisk Arkiv **29**(3): 29 (1975).
Stem to 25 cm, hanging. Leaves to 10 × 1–1.5 cm, flat. Raceme dense, with many flowers, usually shorter than the subtending leaves. Flowers to 1.3 cm. Sepals and petals yellowish green with purplish veins. Lip whitish with short lateral lobes and an acute, upwardly hooked central lobe which is violet-pink. Pouch violet-pink. *S China, Vietnam, Laos & Thailand*. G2. Summer–autumn.

***C. racemiferum** (Lindley) Garay (*Sarcanthus racemifer* (Lindley) Reichenbach; *S. pallidus* Lindley). Illustration: Bechtel et al., Manual of cultivated orchid species, 184 (1981). Similar, but leaves unequally 2-lobed at the apex, flowers in erect panicles which exceed the subtending leaves. *Himalaya to Thailand*. G2. Spring–summer.

121. ANGRAECUM Bory

J. Cullen

Monopodial epiphytes with long or short stems. Leaves in 2 ranks, variable, overlapping and folded at least at the base, or distant and flat or terete, sometimes unequally 2-lobed at their apices, folded when young. Flowers variably resupinate or not resupinate, solitary in the leaf axils, or in axillary racemes. Sepals and petals similar, spreading. Lip stalkless, entire, with a flat or concave spreading blade extending backwards into a conspicuous spur. Column fleshy, short, without a foot, with 2 swellings, between which is the short, inconspicuous, bifid, tooth-like rostellum, beneath the anther. Pollinia 2, spherical, attached either to a short common stipe, or to individual stipes, viscidium single or double.

A genus of over 200 species centred mainly in tropical Africa but extending to Madagascar and Sri Lanka. Garay (reference below) divides it into 18 sections, but as only 6 or 7 species are found in cultivation, these are not described here. Species Nos. 1–3 belong to section *Arachnangraecum* Schlechter; Nos. 4 & 5 to section *Angraecum*; and No. 6 to section *Dolabrifolia* (Pfitzer) Garay.

Cultivation as for *Phalaenopsis* (p. 244), but the smaller species should be suspended near the glass and the thick-leaved species should be given full air and light while avoiding a dry atmosphere. In winter the plants can be allowed to dry out to some extent, but not for long periods.

Literature: Garay, L., Systematics of the genus Angraecum Bory, *Kew Bulletin* **28**: 495–516 (1973).

Leaves. Terete: **3**; overlapping and folded at their bases only: **4,5**; overlapping and folded for most of their length: **6**.

Flowers. Stalkless and single in the leaf axils: **6**; single and stalked, or in racemes in the leaf axils: **1–5**.

Lip. Narrowly ovate, longer than broad: **5**; broadly ovate, broader than long: **1–4**; helmet-like: **6**.

Spur. Less than 1 cm: **6**; between 4 and 15 cm: **1–4**; between 20 and 30 cm: **5**. Sharply bent near the middle: **1,2**.

1a. Leaves overlapping and folded, at least at their bases — 2
b. Leaves flat or terete, not overlapping — 4
2a. Flowers stalkless; sepals and petals less than 1 cm — **6. distichum**
b. Flowers stalked; sepals and petals more than 1 cm — 3
3a. Lip longer than broad, narrowly ovate; spur 20–30 cm — **5. sesquipedale**
b. Lip broader than long, broadly ovate; spur 6–8 cm — **4. eburneum**
4a. Leaves terete — **3. scottianum**
b. Leaves flat — 5
5a. Spur 12–13 cm, tapering, curving forwards — **2. infundibulare**
b. Spur *c.* 4.5 cm, sharply bent at the middle — **1. eichlerianum**

1. A. eichlerianum Kränzlin. Illustration; Botanical Magazine, 7813 (1902); Bechtel et al., Manual of cultivated orchid species, 166 (1981).

Stems to 1 m, hanging, somewhat compressed. Leaves 8–12 × 4.5–5 cm, oblong-elliptic, flat, not overlapping. Flowers solitary or rarely in 2-flowered racemes in the leaf axils, stalk elongate, variably resupinate. Sepals and petals to 4.5 cm, spreading, oblong-lanceolate, acute, yellowish green. Lip broadly ovate, concave, broader than long, with a short point, white shading to greenish yellow at the base, with an elongate ridge. Spur *c.* 4.5 cm, funnel-shaped, sharply bent at the middle, obtuse. Pollinia with a common stipe. *W tropical Africa, from Nigeria to Angola*. G2. Summer.

2. A. infundibulare Lindley (*Mystacidium infundibulare* (Lindley) Rolfe). Illustration: Botanical Magazine, 8153 (1907); Stewart & Campbell, Orchids of tropical Africa, t. 9 (1970); Bechtel et al., Manual of cultivated orchid species, 166 (1981).

Very similar to *A. eichlerianum*, but leaves broader, the lip very broad and concave, the spur 12–13 cm, very wide-mouthed and funnel-shaped, yellowish, curved forwards in the lower half. *Tropical Africa from Zaire to Ethiopia*. G2. Winter.

3. A. scottianum Reichenbach. Illustration: Botanical Magazine, 6273 (1883); Cogniaux & Goossens, Dictionnaire Iconographique des Orchidées, Angraecum, t. 7 (1898); Stewart & Campbell, Orchids of tropical Africa, t. 12 (1970); Bechtel et al., Manual of cultivated orchid species, 167 (1981).

Stems *c.* 30 cm, thin, hanging. Leaves 8–10 cm, terete, not overlapping at the base. Flowers single in the leaf axils, on long stalks, variably resupinate. Sepals and petals 2.5–3 cm, spreading, oblong-linear, acute, greenish white. Lip broader than long, broadly ovate or transversely oblong, with a small point, white. Spur 8–10 cm, narrow, tapering, yellow. Pollinia with a common stipe. *Comoro Islands*. G2. Summer.

4. A. eburneum Bory. Illustration: Botanical Magazine, 5170 (1854); Lindenia **5**: t. 236 (1889); Cogniaux & Goossens, Dictionnaire Iconographique des Orchidées, Angraecum, t. 13 (1898); Die Orchidee **30**: cxxi (1979); Bechtel et al., Manual of cultivated orchid species, 166 (1981).

Stem to 1 m, erect. Leaves 40–50 cm, dense, their bases overlapping and folded, the free part oblong-lanceolate, unequally 2-lobed at the apex. Flowers not resupinate, 8–15 in axillary, 1-sided racemes, stalked, fragrant. Sepals and petals 2.5–3 cm, linear-oblong, acute, spreading, green or whitish green. Lip ovate, broader than long, concave, apiculate, white, often greenish at the base. Spur 6–8 cm, narrow, tapering, white with a green apex. Pollinia with a common stipe. *E tropical Africa, islands of the Indian Ocean*. G2. Autumn–winter.

A variable species. Subsp. **eburneum**, subsp. **superbum** (Thouars) Perrier de la Bathie (*A. superbum* Thouars), which has somewhat larger flowers, and subsp. **giryamae** (Rendle) Senghas & Cribb (*A. giryamae* Rendle), which has a smaller lip and a funnel-shaped spur which is 4–5 cm, are grown. This last subspecies occurs in Kenya and Tanzania.

5. A. sesquipedale Thouars. Illustration: Botanical Magazine, 5113 (1859); Cogniaux & Goossens, Dictionnaire Iconographique des Orchidées, Angraecum, t. 4 (1898); Stewart & Campbell, Orchids of tropical Africa, t. 13 (1970); Bechtel et al., Manual of cultivated orchid species, 167 (1981).

Stems to 1 m, thick, erect. Leaves dense, their bases overlapping and folded, the free part to 30 × 4–5 cm, oblong, unequally 2-lobed at the apex. Flowers usually 2–4 in axillary racemes, rarely single in the leaf axils, stalked, fragrant, resupinate. Sepals and petals to 6 cm, creamy white, spreading, lanceolate, acute, the petals somewhat shorter than the sepals. Lip narrowly ovate, longer than broad, tapering, its sides somewhat toothed, creamy white. Spur 20–30 cm, abruptly tapering from its junction with the lip. Pollinia with separate stipes. *Madagascar*. G2. Winter.

This is the orchid which led Darwin to predict the existence of a moth with a proboscis 20–30 cm long – a species unknown at the time of his prediction but since discovered in Madagascar.

6. A. distichum Lindley. Illustration: Edwards's Botanical Register **21**: t. 1781

(1835); Botanical Magazine, 4145 (1845); Die Orchidee **21**: 246 (1970); Bechtel et al., Manual of cultivated orchid species, 166 (1981).
Small plants, stems to 12 cm. Leaves 8–12 cm, fleshy, folded and overlapping for most of their length, the free part short, obtuse. Flowers solitary in the upper leaf axils, the ovary stalkless, not resupinate. Sepals and petals *c.* 5 mm, white, spreading, the petals slightly shorter and narrower than the sepals. Lip helmet-like, 3-lobed at the apex, the central lobe recurved. Spur *c.* 8 mm, cylindric. *Tropical Africa from Guinea to Uganda and Angola*. G2. Autumn.

122. AERANGIS Reichenbach
J. Cullen
Monopodial epiphytes with short, compressed stems. Leaves leathery, unequally 2-lobed at the apex, folded when young. Racemes with 7–many flowers, hanging from the leaf axils, the axis often conspicuously zig-zag. Bracts small, brownish. Flowers usually white, fragrant. Sepals and petals similar in shape and size (rarely the upper sepal smaller), spreading or somewhat reflexed. Lip smaller than the petals but often broader, projecting backwards into a short or long, curved or twisted and contorted spur, which has a distinct, narrow mouth. Column narrow, without a foot, rostellum large, usually beaked, usually extended at right angles to the column. Pollinia 2, each with a separate stipe or (in all species included here) attached to a single, linear stipe, viscidium oblong, sometimes slightly 2-lobed.

A genus of 35 species from tropical Africa and Madagascar. Cultivation as for *Angraecum* (p. 254).
Literature: Stewart, J.A., A revision of the African species of Aerangis, *Kew Bulletin* **34**: 239–319 (1979).

Habit. Plant very small (leaves 3–5 cm, sepals and petals 5–6 mm): **6.**
Sepals. Upper shorter and blunter than laterals: **5.** 2 cm or more: **1,3,4.**
Spur. Less than 1 cm: **6**; 2–4 cm; **4,5**; 4–10 cm: **2**; more than 10 cm: **1,3.** Contorted and twisted: **3.**

1a. Leaves 3–5 cm, spur 5.5–6 mm, not or only slightly longer than the ovary **6. hyaloides**
b. Leaves larger; spur much longer than the ovary 2
2a. Spur 2–3 cm; upper sepal 5–7 mm, blunt, shorter than the acute lateral sepals (which are 8–10 mm) **5. citrata***
b. Spur 4–27 cm; all sepals similar in size, 1.5–3 cm 3
3a. Spur 15–27 cm, twisted and contorted, the spurs of the individual flowers often twisted around each other **3. kotschyana**
b. Spur 4–17 cm, curved but not twisted or contorted 4
4a. Sepals and petals white tipped with pink; spur brownish; rostellum touching the lower side of the stigmatic cavity which thus appears 2-lobed **4. biloba**
b. Sepals, petals and spur white; rostellum projecting outwards above the stigmatic cavity 5
5a. Raceme to 60 cm or more; leaves 20–25 × 5 cm **1. ellisii***
b. Raceme much shorter; leaves 6–12 × 1.5–3 cm **2. modesta***

1. A. ellisii (Reichenbach) Schlechter (*Angraecum ellisii* Reichenbach). Illustration: Floral Magazine n.s., t. 191 (1875); Lindenia **2**: t. 92 (1886).
Stem short. Leaves 20–25 × 5 cm, oblong-obovate, dark green, unequally 2-lobed at the apex. Racemes 40–65 cm with 18–25 flowers. Sepals and petals 2.5–3 cm, white, spreading or reflexed, oblong-elliptic, acute. Lip similar to the petals but broader and with its margins reflexed. Spur 15–17 cm, curved, tapering. *Madagascar*. G2. Summer–autumn.

Introduced, apparently from Madagascar, in the nineteenth century, but not found there since.

***A. articulata** (Reichenbach) Schlechter (*Angraecum articulatum* Reichenbach). Similar, but with a markedly zig-zag axis to its raceme. *Madagascar*. G2.

Both of these species are somewhat dubious. Perrier de la Bathie (Flore de Madagascar **49**(2): 108, 116, 1939) suggests that both are related to *A. stylosa* (see under 2 below), *A. ellisii* being perhaps a luxuriant variant of it and *A. articulata* a hybrid of it with some other unknown species.

2. A. modesta (Hooker) Schlechter (*Angraecum modestum* Hooker). Illustration: Botanical Magazine, 6693 (1883); Cogniaux & Goossens, Dictionnaire Iconographique des Orchidées, Angraecum, t. 2 (1898).
Stem to 15 cm. Leaves 6–12 × 1.5–3 cm, ovate-oblong, slightly unequally 2-lobed at the apex. Raceme to 30 cm at most, with 6–15 flowers which are very distant. Sepals and petals *c.* 1.2 cm, lanceolate, more or less obtuse, the sepals with 5 nerves, white. Lip broader than the petals, acute. Spur 4–7 cm, curved, tapering. Column with a few short hairs on the upper side. *Madagascar & Comoro Islands*. G2. Spring.

***A. stylosa** (Rolfe) Schlechter (*Angraecum stylosum* Rolfe). Similar but with petals and sepals 1.5–2 cm, the sepals with 7 nerves and the column hairless. *Madagascar & Comoro Islands*. G2. Spring.

3. A. kotschyana (Reichenbach) Schlechter (*Angraecum kotschyanum* Reichenbach; *Angraecum kotschyi* Reichenbach; *Aerangis kotschyi* (Reichenbach) Reichenbach). Illustration: Botanical Magazine, 7442 (1895); Williamson, Orchids of southern central Africa, f. 108 & t. 175 (1977); Ball, Southern African epiphytic orchids, 33 (1978); Kew Bulletin **34**: 255 (1979).
Stem very short. Leaves 10–15 × 4.5–5 cm, oblong-elliptic to oblong-obovate, blunt or slightly 2-lobed at the apex, sometimes spotted with purple above. Raceme to 40 cm with 7–10 (rarely more) flowers. Sepals and petals elliptic-lanceolate, acute, to 2.5 cm, white tinged with pink towards their apices. Lip reversed-heart-shaped, waisted above. Spur 15–27 cm, twisted and contorted, those of the individual flowers often twisted together, slightly broadened towards the brownish apex. *Tropical Africa, from Guinea and Sudan to Zaire and South Africa (Transvaal)*. G2. Autumn.

4. A. biloba (Lindley) Schlechter (*Angraecum bilobum* Lindley). Illustration: Edwards's Botanical Register **27**: t. 35 (1841); Kew Bulletin **34**: 282 (1979).
Stem very short. Leaves to 15 cm, oblong-obovate, deeply and unequally 2-lobed at the apex. Raceme to 25 cm with 7–12 flowers. Sepals and petals *c.* 2 cm, lanceolate or narrowly elliptic, acute, white tipped with pink. Lip similar but broader. Spur *c.* 4 cm, brownish, curved. Anther conspicuously crested. Rostellum extending over the stigmatic cavity, touching its lower side, the cavity thus appearing 2-lobed. *Tropical W Africa, from Senegal to Cameroun*. G2. Summer.

5. A. citrata (Thouars) Schlechter (*Angraecum citratum* Thouars). Illustration: Botanical Magazine, 5624 (1867): Lindenia **5**: t. 238 (1889); Stewart & Campbell, Orchids of tropical Africa, t. 2 (1970).
Stem very short. Leaves 7–15 × 2–3.5 cm, oblong-obovate, very unequally 2-lobed at the apex. Raceme 10–30 cm with 15–60 flowers. Upper sepal 5–7 mm, obtuse,

shorter than the lateral sepals which are clawed, acute and 8–10 mm. Petals slightly shorter than the lateral sepals. All white with a faint yellow tinge. Lip clawed, obovate. Spur 2.5–3 cm, curved, tapering. *Madagascar*. G2. Winter–spring.

*A. **luteo-alba** (Kränzlin) Schlechter var. **rhodosticta** (Kränzlin) Stewart (*Angraecum rhodostictum* Kränzlin; *Aerangis rhodosticta* (Kränzlin) Schlechter). Illustration: Hunt & Grierson, Orchidaceae, 110–11 (1973); Kew Bulletin **34**: 311 (1979). Will key out to *A. citrata* on its short spur, but sepals equal, 1–1.5 cm, column red. *Tropical Africa from Cameroun to Tanzania*. G2.

6. A. **hyaloides** (Reichenbach) Schlechter (*Angraecum hyaloides* Reichenbach). Illustration: Reichenbach, Xenia Orchidacearum **3**: t. 238 (1890).
Plant very small. Leaves 3–5 cm, oblong-obovate, shortly 2-lobed at the apex. Raceme 3–4 cm with 3–4 flowers. Sepals and petals 5–6 mm, whitish, translucent. Lip somewhat broader than petals. Spur 5.5–6 mm, not or scarcely exceeding the ovary, curved. *Madagascar*. G2. Spring.

123. NEOFINETIA Hu

J. Cullen

Very similar to *Aerangis*, but lip distinctly 3-lobed.

1. N. **falcata** (Thunberg) Hu (*Angraecum falcatum* (Thunberg) Lindley; *Angraecopsis falcata* (Thunberg) Schlechter). Illustration: Kitamura et al., Alpine plants of Japan **3**: f. 114 (1978).
Stem very short. Leaves fleshy, to 7 cm. Raceme erect, shorter than the leaves, with 3–7 flowers. Flowers white, *c*. 3 cm in diameter. Lip with short lateral lobes and an oblong central lobe. Spur *c*. 4 cm. *Japan*. G1. Summer.

The only species.

124. CYRTORCHIS Schlechter

J. Cullen

Monopodial epiphytes, stems short or elongate. Leaves in 2 ranks, thick, fleshy and leathery, unequally 2-lobed at their apices, folded when young. Racemes arching or hanging, arising from the axils of the older leaves (or leaf-bases), with several flowers and conspicuous, brownish bracts. Flowers fragrant, especially at night, white fading to dull orange if not pollinated. Sepals and petals similar in size and shape, usually recurved. Lip similar to the petals but prolonged backwards as a conspicuous, tapering spur. Column short, without a foot. Rostellum elongate, conspicuous, beak-like. Pollinia 2, each with a separate stipe, attached to an oblong or linear viscidium.

A taxonomically troublesome genus of 15 species from Africa, of which 2 are occasionally found in cultivation. Cultivation as for *Angraecum* (p. 254).
Literature: Summerhayes, V.S., African orchids XXVII, *Kew Bulletin* **14**: 143–56 (1960).

1a. Stipe considerably broadened upwards; viscidium oblong with a hardened upper part and a translucent lower part **1. arcuata**
b. Stipe scarcely broadened upwards; viscidium linear, entirely translucent **2. monteiroae**

1. C. **arcuata** (Lindley) Schlechter (*Angraecum arcuatum* Lindley; *Listrostachys arcuata* (Lindley) Reichenbach). Illustration: Stewart & Campbell, Orchids of tropical Africa, t. 25 (1970); American Orchid Society Bulletin **49**: 1230 (1980); Bechtel et al., Manual of cultivated orchid species, 192 (1981).
Stem 30–50 cm, thick. Leaves 7–15 × 1.5–2.5 cm, oblong. Racemes with 10–20 densely packed flowers. Sepals and petals 2–4 cm, lanceolate, acute, the petals sometimes a little shorter than the sepals. Spur 3–6 cm, curved, tapering. *S and tropical Africa northwards to Sierra Leone and Kenya*. G2. Spring–summer.

A very variable species, divided by Summerhayes into 5 subspecies; it is not known which of these are in cultivation.

2. C. **monteiroae** (Reichenbach) Schlechter (*Listrostachys monteiroae* Reichenbach). Illustration: Botanical Magazine, 8026 (1905).
Stem to 60 cm, leafy. Leaves 15–17 × 4.5–5.5 cm, oblong. Raceme with 10–15 rather distant flowers. Sepals and petals 1.5–2 cm, lanceolate. Spur 4–5 cm, curved, brownish yellow towards the apex. *Tropical Africa, from Sierra Leone to Angola and eastwards to Uganda*. G2. Spring.

125. DIAPHANANTHE Schlechter

J. Cullen

As for *Cyrtorchis* but sepals and petals erect, lip considerably larger than them and bearing a conspicuous tooth-like callus just in front of the narrow mouth of the spur.

1a. Leaves in a close tuft: lip fringed **1. pellucida**
b. Leaves distributed along the stem; lip not fringed **2. bidens**

1. D. **pellucida** (Lindley) Schlechter (*Angraecum pellucidum* Lindley; *Listrostachys pellucida* (Lindley) Reichenbach). Illustration: Cogniaux & Goossens, Dictionnaire Iconographique des Orchidées, Listrostachys, t. 1 (1906); Bechtel et al., Manual of cultivated orchid species, 204 (1981).
Stem compressed, leaves in a close tuft, each unequally 2-lobed at the apex. Flowers white or brownish pink. Lip triangular, conspicuously fringed. *Tropical Africa from Sierra Leone to Uganda*. G2. Winter.

2. D. **bidens** (Reichenbach) Schlechter (*Listrostachys bidens* Reichenbach). Illustration: Botanical Magazine, 8014 (1905); Bechtel et al., Manual of cultivated orchid species, 204 (1981).
Stems to 40 cm, leafy along their length. Leaves very conspicuously unequally 2-lobed. Flowers brownish pink. Lip triangular, notched at the apex, with a small point within the notch. *Guinea*. G2. Summer.

126. EURYCHONE Schlechter

J. Cullen

Monopodial epiphytes with very short stems. Leaves narrowly elliptic to oblanceolate, unequally 2-lobed at their apices, folded when young. Racemes lateral, with 5–10 flowers, arching, shorter than the leaves. Sepals and petals similar, thin, spreading. Lip about as long as the sepals but much broader, funnel-shaped, very obscurely 3-lobed, extended backwards as a wide-mouthed spur which is thickened and recurved at its apex. Column without a foot. Rostellum beaked, conspicuous. Pollinia 2, almost spherical, borne on a narrow, linear stipe, viscidium large, oblong or oval.

A genus of 2 species from tropical Africa. Cultivation as for *Angraecum* (p. 254).

1. E. **rothschildiana** (O'Brien) Schlechter (*Angraecum rothschildianum* O'Brien). Illustration: Gardeners' Chronicle **34**: 131 (1903); Stewart & Campbell, Orchids of tropical Africa, t. 29 (1970).
Stem to 7 cm. Leaves 7–14 × 2–2.5 cm. flowers fragrant. Sepals and petals 2–2.5 cm, each white with a pale green band. Lip to 2.5 cm wide, white with the centre bright green shading to brown or purple at the base, its sides curving upwards towards the column. Spur to 2.5 cm. *Tropical Africa, from Sierra Leone to Uganda*. G2. Summer.

127. POLYSTACHYA de Jussieu

J. Cullen

Epiphytic. Pseudobulbs variable, simple, ovoid, bearing 2 or more leaves. Leaves lanceolate, pleated, dark green, folded when young. Raceme or panicle with many flowers, its axis often finely downy, terminal. Flowers not resupinate (in ours). Sepals and petals similar in size and shape, arching forwards. Lip 3-lobed, not or scarcely exceeding the petals, often hairy or papillose. Column with a distinct foot, forming a conspicuous mentum. Pollinia 4, in 2 superposed pairs, almost spherical, borne on a short stipe with a circular viscidium.

A genus of about 200 species from the tropics, centred in Africa. Only 1 species is commonly available. Cultivation as for *Eulophia*.

1. P. affinis Lindley (*P. bracteosa* Lindley). Illustration: Botanical Magazine, 4161 (1845); Bechtel et al., Manual of cultivated orchid species, 265 (1981).
Pseudobulbs 3–5 cm, compressed, circular in outline, each bearing 2 leaves. Leaves to 20 cm, thin, lanceolate, acute. Flowers in a raceme (rarely slightly branched), axis finely downy. Sepals and petals *c.* 1 cm, yellow with brown stripes. *W Africa*. G2. Summer.

A few other species may occur from time to time in specialist collections.

128. GALEANDRA Lindley

J. Cullen

Epiphytic. Pseudobulbs ovoid or stem-like, compound, bearing 2–7 leaves. Leaves rolled when young. Raceme terminal, open, with few to many flowers. Sepals and petals similar, spreading or reflexed. Lip somewhat cup-shaped, rolled around and partially concealing the column, with a short, downwardly directed or straight spur. Column without a foot. Pollinia 2, laterally ovoid, stipe more or less triangular, short and broad.

A genus of about 8 species from C & S America. They may be grown in epiphytic or terrestrial compost in pots. Cooler conditions are required during the resting season. Plants are very susceptible to attack by thrips and red spider mite. Propagation by division.

Literature: Rolfe, R.H., Galeandras, *Gardeners' Chronicle* **12**(2): 430–1 (1892); Teuscher, H., Die Gattung Galeandra, *Die Orchidee* **26**: 1–5 (1975).

1a. Pseudobulbs ovoid: flowers *c.* 5 cm in diameter **1. batemanii***
b. Pseudobulbs elongate, stem-like; flowers *c.* 10 cm in diameter **2. devoniana**

1. G. batemanii Rolfe (*G. baueri* misapplied). Illustration: Edwards's Botanical Register **26**: t. 49 (1840).
Pseudobulbs ovoid, each with 5–7 leaves. Leaves 15–20 cm, linear-lanceolate, acute. Racemes arching, with 12–many flowers which are *c.* 5 cm in diameter. Sepals and petals reflexed, yellow or greenish, flushed with brown. Lip violet-red margined with white, deeply notched at the apex. Spur straight. *Mexico, Guatemala*. G2. Summer.

Genuine *G. baueri* Lindley, from French Guiana, is probably not in cultivation: it has fewer flowers and a yellow, purple-streaked or purple-flushed lip.

***G. lacustris** Barbosa-Rodriguez. Illustration: Orchid Review **83**: 165 (1975); Die Orchidee **26**: 3 (1975); Bechtel et al., Manual of cultivated orchid species, 214 (1981). Similar but with sepals and petals pale green suffused with purple, and the lip white with purple spots. *N & W Brazil, Peru*.

***G. nivalis** Masters. Also similar but with olive green sepals and petals and the lip white with a central violet blotch. *N & W Brazil*.

2. G. devoniana Schomburgk. Illustration: Botanical Magazine, 4610 (1851); Dunsterville & Garay, Venezuelan orchids illustrated **2**: 157 (1961); Die Orchidee **26**: 2 (1975); Bechtel et al., Manual of cultivated orchid species, 214 (1981).
Pseudobulbs cylindric, stem-like, each with 2–5 leaves. Leaves to 20 × 1 cm, linear. Raceme arching, with few flowers. Flowers *c.* 10 cm in diameter. Sepals and petals linear, brown with greenish margins. Lip white, lined with violet, with 2 low ridges on the surface. Spur lightly curved. *Tropical S America, from Venezuela to N Brazil*. G2. Summer.

129. EULOPHIA R. Brown

J. Cullen

Large, terrestrial plants. Pseudobulbs compound, sometimes stem-like or swollen. Leaves 2–several on each pseudobulb, pleated and with prominent veins, rolled when young. Racemes lateral with 5–many flowers. Sepals and petals similar to each other or the petals larger than the sepals, all spreading or the sepals sometimes reflexed. Lip pivoting freely on the base of the column, 3-lobed, shortly spurred or saccate at the base. Column without a foot. Pollinia 2, ovoid or laterally ovoid, stipe short and broad, viscidium almost circular.

A genus of about 300 species, mostly from tropical Africa. Several of those cultivated have at one time or other been placed in the genus *Lissochilus* R. Brown. Cultivation can be difficult, as most species are found in seasonally damp places. Abundant moisture is required during the growing season and drier conditions with maximum light while the plants are resting. Propagation by seed or by division.

1a. Petals much larger than the sepals, broadly obovate to almost circular 2
b. Petals similar in size to the sepals, relatively much narrower **1. guineensis***
2a. Petals yellowish or greenish yellow; flowers *c.* 3 cm in diameter; bracts lanceolate **4. streptopetala**
b. Petals white or white flushed with pink, or purple; flowers 4 cm or more in diameter; bracts ovate 3
3a. Sepals green; petals lilac-purple with darker veins; lateral lobes of the lip purple **2. gigantea**
b. Sepals purple; petals white flushed with pink; lateral lobes of the lip green striped with purple **3. horsfallii**

1. E. guineensis Lindley. Illustration: Botanical Magazine, 2467 (1824).
Pseudobulbs to 5 cm, each with 2–3 leaves. Leaves 30–45 cm, narrowly elliptic, narrowed to the base, acute. Racemes with 5–15 flowers. Flowers to 6 cm in diameter. Sepals and petals similar in size, narrow, acuminate, either whitish pink or greenish or brownish purple. Lip with the central lobe whitish pink to purple, streaked and spotted with darker purple. *Tropical Africa, from Gambia to Angola and Uganda*. G2. Autumn.

***E. alta** (Linnaeus) Fawcett & Rendle (*E. longifolia* Humboldt, Bonpland & Kunth: *E. woodfordii* (Lindley) Rolfe). Illustration: Hoehne, Flora Brasilica **12**(6): t. 1 (1942); Hawkes, Encyclopaedia of cultivated orchids, 208 (1965); Dunsterville & Garay, Venezuelan orchids illustrated **1**: 150 (1959); Ospina & Dressler, Orquideas de las Americas, f. 141 (1974). Leaves 100–120 cm, raceme with many flowers which are 3–5.5 cm in diameter, greenish or brown, the central lobe of the lip purple. *Tropical Africa from Ghana and Sudan to Angola and Zimbabwe, tropical America from Mexico to Argentina*. G2. Autumn.

2. E. gigantea (Welwitsch) R. Brown (*Lissochilus giganteus* Welwitsch).

Illustration: Gardeners' Chronicle 3: 616–7 (1888).
To 5 m or more. Leaves to 1.2 m × 10 cm, bright green with prominent yellow midrib. Raceme with 20–100 flowers, bracts ovate, wrapped around the bases of the flower-stalks. Flowers 6–7 cm in diameter. Sepals green. Petals larger, lilac-purple with darker veins. Lip mostly purple throughout. *W tropical Africa from Congo to Angola*. G2. Spring–summer.

3. E. horsfallii (Bateman) Summerhayes (*Lissochilus horsfallii* Bateman; *L. porphyroglossa* Reichenbach; *E. porphyroglossa* (Reichenbach) Bolus). Illustration: Botanical Magazine, 5486 (1865).
Very like *E. gigantea* but to 2 m, raceme with 60–100 flowers which are 4–6 cm in diameter, sepals purple, petals white or white flushed with pink, lateral lobes of the lip green striped with purple. *Tropical Africa from Sierra Leone to Mozambique*. G2. Winter.

Sometimes found under the name *Lissochilus roseus* Lindley – a name of doubtful application.

4. E. streptopetala Lindley (*Lissochilus krebsii* Reichenbach). Illustration: Botanical Magazine, 2931 (1829), 5861 (1870); Bolus, Orchids of S Africa 3: t. 9–10 (1913); Flowering plants of South Africa 21: t. 820 (1941); Gibson, Wild flowers of Natal (coastal region), t. 25 (1975).
Stem to 1.5 m. Leaves to 50 cm × 7 mm, lanceolate, bright green, with a prominent midrib. Raceme with up to 50 flowers, bracts lanceolate, not wrapped around the bases of the flower-stalks. Flowers *c.* 3 cm in diameter. Sepals green striped with purple. Petals yellow or greenish yellow. Lip yellow, the lateral lobes purplish or brownish. *South Africa, Zimbabwe*. G2. Winter.

130. OECEOCLADES Lindley

J. Cullen

Plants very like *Eulophia* but pseudobulbs simple, each bearing 1–3 leathery leaves which are folded when young. Pollinia 2, ovoid or pear-shaped, stipe very short or absent, viscidium semicircular or oblong.

A genus of about 30 species from the tropics, mostly in Africa and Madagascar, formerly known as *Eulophidium* Pfitzer. Cultivation as for *Eulophia* (p. 257).
Literature: Summerhayes, V.S., The genus Eulophidium Pfitzer, *Bulletin du Jardin Botanique de l'Etat, Bruxelles* 27: 391–403 (1957); Garay, L.A. & Taylor, P., The genus Oeceoclades Lindl., *Botanical Museum Leaflets, Harvard University* 24: 249–74 (1976).

1a. Each pseudobulb with 1 non-pleated, irregularly variegated leaf **1. maculata**
b. Each pseudobulb with 2–3 pleated, green leaves **2. saundersiana**

1. O. maculata (Lindley) Lindley (*Eulophia maculata* (Lindley) Reichenbach; *Eulophidium maculatum* (Lindley) Pfitzer; *Eulophidium ledienii* N.E. Brown). Illustration: Gartenflora 37: t. 1288 (1888); Ospina & Dressler, Orquideas de las Americas, t. 141 (1974).
Pseudobulbs each bearing 1 leaf. Leaves 15–25 × 2.5–4 cm, not pleated, fleshy, leathery, irregularly variegated. Racemes 7–15 cm with several flowers. Sepals and petals 8–10 mm, pale brownish green. Lip 3-lobed with the central lobe itself 2-lobed, brownish green with 2 reddish or purple spots on the sides. *Tropical America, tropical Africa*. G2. Summer–autumn.

2. O. saundersiana (Reichenbach) Garay & Taylor (*Eulophidium saundersianum* (Reichenbach) Summerhayes). Illustration: Reichenbach, Xenia Orchidacearum 2: t. 173 (1873).
Pseudobulbs each with 2–3 leaves, the leaves deciduous some distance above the pseudobulb, the persistent portion splitting and forming a stiff, fibrous crown. Leaves 10–18 × 5–7.5 cm, pleated, green. Raceme 10–15 cm with many flowers. Sepals and petals 1–1.2 cm, green, the petals marked with black lines. Lip 4-lobed, green with a few black lines. *Tropical Africa from Cameroun to Uganda*. G2. Autumn?

131. EULOPHIELLA Rolfe

J. Cullen

Rhizomatous epiphytes. Pseudobulbs large, compund, each bearing 3–6 pleated, prominently veined leaves, rolled when young. Raceme basal or lateral with 7–25 flowers with prominent bracts. Flowers large, almost circular in outline. Sepals and petals all similar or the petals a little smaller, lateral sepals attached to the column foot. Lip 3-lobed, not spurred, attached directly to the column foot, without an obvious claw. Column with a conspicuous foot. Pollinia 2, ovoid or ellipsoid, stipe very short and broad, oval or conical.

A genus of 4 species from Madagascar, where they occur in seasonally damp places. Cultivation as for *Eulophia* (p. 257).
Literature: Bosser, J. & Morat, P., Contribution a l'étude des Orchidaceae de Madagascar IX: Les Genres Grammangis et Eulophiella, *Adansonia* 9: 299–309 (1969); Kennedy, G.C., The genus Eulophiella, *Orchid Digest* 36: 120–2 (1972).

1a. Leaves 100 × 8–10 cm or more; flower-stalk 4.5–5 cm; flowers 8–10 cm in diameter **1. roempleriana**
b. Leaves 60–80 × 3.5–5 cm; flower-stalk 2–3 cm; flowers 4–5 cm in diameter **2. elizabethae**

1. E. roempleriana (Reichenbach) Schlechter (*Grammatophyllum roemplerianum* Reichenbach; *E. peetersiana* Kränzlin; *E. hamelinii* Rolfe). Illustration: Botanical Magazine, 7612, 7613 (1898); Gardeners' Chronicle 23: 200 (1898); Orchid Digest 36: 120 (1972).
Pseudobulbs *c.* 20 cm, spindle-shaped, without fibrous leaf remains. Leaves 100 × 8–10 cm or more. Axis of raceme green, bracts brown. Flowers 8–10 cm in diameter. Sepals pink tinged purplish and with purple tips. Petals pink. Lip with large, pink lateral lobes which curve over the white column, the central lobe purple or white margined with purple and with 3–4 orange-yellow ridges. *Madagascar*. G2. Spring–summer.

2. E. elizabethae Linden & Rolfe. Illustration: Botanical Magazine, 7387 (1894), n.s., 656 (1973); Orchid Digest 36: 120 (1972).
Pseudobulbs *c.* 15 cm, spindle-shaped, covered with fibrous leaf remains. Leaves 60–80 × 3.5–5 cm. Axis of raceme and bracts red. Flowers 4–5 cm in diameter. Sepals red lined with white outside, white within. Lip with small lateral lobes, white, red at the base and with a large yellow blotch on the central lobe. *Madagascar*. G2. Spring.

132. CYRTOPODIUM R. Brown

J. Cullen

Terrestrial or epiphytic. Pseudobulbs often stem-like, compound, bearing several leaves which are rolled when young and pleated, with conspicuous veins. Raceme or panicle with many flowers, lateral. Sepals and petals similar or the petals somewhat broader and shorter, the lateral sepals attached to the column foot. Lip 3-lobed, conspicuously clawed, the claw and lateral lobes spreading, the central lobe directed downwards. Column with a small foot. Pollinia 2, ovoid or ellipsoid, stipe triangular, short and broad.

A genus of about 30 species from C & S America. Cultivation as for *Eulophia* (p. 257).

1a. Sepals and petals equal, unspotted; lip yellow **1. andersonii**
b. Sepals narrower than petals, all with red-brown lines or spots; lateral lobes of lip dark red **2. punctatum***

1. C. andersonii (Lambert) R. Brown (*Cymbidium andersonii* Lambert). Illustration: Botanical Magazine, 1800 (1816); Bechtel et al., Manual of cultivated orchid species, 192 (1981).
Pseudobulbs stem-like to 1.5 m. Leaves to 50 cm, linear and tapering. Panicle with many flowers and ovate, acute, inconspicuous bracts. Flowers *c.* 5 cm in diameter. Sepals and petals spreading, more or less equal, yellow faintly flushed with green. Lip entirely yellow or orange-yellow, the central lobe spoon-shaped, wavy. *West Indies, Venezuela.* G2. Spring–summer.

2. C. punctatum (Linnaeus) Lindley (*Epidendrum punctatum* Linnaeus). Illustration; Hoehne, Iconografia de orchidaceas do Brasil, t. 153 (1949); Ospina & Dressler, Orquideas de las Americas, f. 139 (1974); Sheehan & Sheehan, Orchid genera illustrated, 77 (1979); Bechtel et al., Manual of cultivated orchid species, 192 (1981).
Like *C. andersonii* but with conspicuous bracts, sepals greenish yellow with brown stripes or spots, narrower than the yellow, red-spotted petals, and lip with dark red lateral lobes, the central lobe edged with red. *Tropical America from Costa Rica to C Brazil.* G2. Spring–summer.

***C. virescens** Reichenbach & Warming. Illustration: Botanical Magazine, 7396 (1895); Hoehne, Iconografia de orchidaceas do Brasil, t. 166 (1949). Pseudobulbs to 10 cm, each with 3–4 leaves, the flowers *c.* 3 cm in diameter, borne in a raceme, and the lip fleshy. *S Brazil and Paraguay.* G2. Spring–summer.

133. ANSELLIA Lindley
J. Cullen
Large epiphytes. Pseudobulbs compound, elongate and stem-like, bearing several leaves. Leaves folded when young. Raceme or panicle with 20–many flowers, terminal. Sepals and petals similar or the petals somewhat broader than the sepals, all spreading, the lateral sepals attached to the column foot, forming a small mentum. Lip without a spur, 3-lobed, the lateral lobes wrapped around the column, the central lobe with 2–3 ridges. Pollinia 2, each 2-lobed, appearing as 4, ovoid, stipe very short and broad, narrowly triangular.

A genus of 1 species from the southern half of Africa. Plants require careful potting, using terrestrial compost, and establishment, as they suffer from root disturbance. They should not be allowed to dry out completely during the resting season.
Literature: Summerhayes, V.S., African orchids IX, *Kew Bulletin for 1937*: 461–3 (1938).

1. A. africana Lindley (*A. gigantea* Reichenbach var. *nilotica* (N.E. Brown) Summerhayes). Illustration: Botanical Magazine, 4965, excluding f. 3 (1854); Sheehan & Sheehan, Orchid genera illustrated, 39 (1979); Bechtel et al., Manual of cultivated orchid species, 168 (1981).
Pseudobulbs to 80 cm. Leaves 4–7, 15–30 × 3–4 cm, linear to narrowly elliptic, prominently veined. Panicle to 75 cm with 20–many flowers. Flowers 2–2.5 cm in diameter. Petals broader than sepals or as broad as them, all greenish to yellow, usually with brownish or purplish spots. Lip 3-lobed, the lateral lobes greenish striped with brown, the central lobe yellow, oblong to circular with 2 bright yellow ridges. *Tropical and southern Africa.* G2. Winter.

134. CYMBIDIUM Swartz
J. Cullen
Epiphytic or terrestrial. Pseudobulbs obvious or inconspicuous, compound, with 2–several leaves. Leaves leathery, conspicuously veined, folded when young. Raceme lateral with 2–many flowers. Sepals and petals usually similar, spreading or curving forwards. Lip free or somewhat united to the base of the column, scarcely to obviously 3-lobed. Column narrow, without a foot. Pollinia 2, laterally ovoid, stipe very short and broad, triangular.

A genus of about 50 species from Asia. Most of the cultivated Cymbidiums are hybrids, many of them complex. Species Nos. 9–10 are sometimes separated into the genus *Cyperorchis* Blume.

They should be grown in pots in an epiphytic compost and in cool conditions with a marked drop in temperature at night, which initiates flower-bud formation. They should be shaded from hot sunshine in the summer and in general should be well ventilated. Plants should be repotted after flowering. Propagation by division and commercially by meristem culture.
Literature: Hunt, P.F., Notes on Asiatic orchids V, *Kew Bulletin* **24**: 75–99 (1970); Pradhan, G.M., The Cyperorchis species of northern India, *Orchid Digest* **40**: 115–17 (1976).

Flowers. 8 or more in each inflorescence: **1–5,10**; inflorescence with up to 7 flowers: **6–9**.
Sepals and petals. Directed forwards, not widely spreading: **9,10**; widely spreading, flower up to 5 cm in diameter: **1–3**; widely spreading, flower more than 5 cm in diameter: **4–8**. White or cream, sometimes flushed with pink: **5,6,8,9**; yellow or greenish, rarely brown, often with dark lines or blotched: **1–4,7,10**.
Raceme. Arching or hanging: **1,2,10**; erect or spreading: **3–9**.

1a. Flowers 8 or more in each well-developed inflorescence 2
b. Flowers 1–7 in each well-developed inflorescence 7
2a. Flowers to 5 cm in diameter 3
b. Flowers more than 5 cm in diameter 6
3a. Sepals and petals directed forwards, the flower not opening widely **10. elegans**
b. Sepals and petals spreading, the flower opening widely 4
4a. Racemes hanging or arching 5
b. Racemes erect or spreading **3. ensifolium***
5a. Leaves distinctly stalked; lip diamond-shaped, scarcely 3-lobed, usually pink to purple **1. devonianum**
b. Leaves not stalked; lip clearly 3-lobed, yellowish or reddish **2. simulans***
6a. Sepals and petals white or white flushed with pink **5. insigne**
b. Sepals and petals green or yellowish, sometimes marked with red, purple or brown lines and spots. **4. giganteum***
7a. Sepals and petals olive green to yellow, faintly spotted and lined with red or purple; leaves to 15 cm **7. tigrinum**
b. Sepals and petals white or ivory white sometimes flushed with pink; leaves 30–70 cm 8
8a. Sepals and petals erect, forwardly directed, the flower not opening widely **9. mastersii**
b. Sepals and petals spreading, the flower widely open 9
9a. Column bright crimson; lateral lobes of the lip large, more or less folding over the column **8. erythrostylum**
b. Column not crimson; lateral lobes of

the lip not extending over the column
6. eburneum

1. C. devonianum Paxton. Illustration: Botanical Magazine, 9327 (1933); Hawkes, Encyclopaedia of cultivated orchids, 192 (1965); Orchid Digest **41**: 8 (1977); Bechtel et al., Manual of cultivated orchid species, 190 (1981).
Pseudobulbs with 2–5 leaves. Leaves 30–40 × *c.* 3 cm, distinctly stalked, oblanceolate. Racemes equalling or exceeding the leaves, arching, with many flowers. Flowers 2.5–4 cm in diameter. Sepals and petals yellowish green or brown with pale purple streaks, spreading. Lip diamond-shaped, scarcely 3-lobed, variable in colour, usually pink to purplish, often with darker spots. *NE India.* G1. Spring–summer.

2. C. simulans Rolfe (*C. aloifolium* misapplied). Illustration: Orchid Digest **42**: 126 (1978); Bechtel et al., Manual of cultivated orchid species, 189 (1981).
Pseudobulbs scarcely obvious. Leaves 30–60 × 1.5–3 cm, not stalked, apex unequally notched. Racemes arching with many flowers which are 3.5–4.5 cm in diameter. Sepals and petals brownish yellow, each with a purple median stripe, spreading. Lip distinctly 3-lobed, yellowish or reddish with red lines and with 2 curved and interrupted ridges, finely downy. *From E Himalaya to S China and Malaysia.* G1. Spring.

Often found under the name *C. aloifolium* (Linnaeus) Swartz (see Rolfe, Orchid Review **25**: 175, 1917).

***C. pendulum** Swartz. Leaves larger (60–80 × 1.8–2.5 cm), acute at the apex, sepals and petals dark brown and lip with straight, continuous ridges. *India, Burma.* G1. Spring.

3. C. ensifolium (Linnaeus) Swartz. Illustration: Botanical Magazine, 1751 (1815).
Pseudobulbs scarcely obvious. Leaves 60–100 × 2–4 cm, linear to sword-like, acute. Raceme erect or spreading, with 8–12 flowers, exceeding the leaves. Flowers to 4.5 cm in diameter, fragrant. Sepals and petals greenish yellow with reddish brown lines, spreading. Lip clearly 3-lobed, white or yellowish with irregular, red-brown spots. *India, China, Japan and probably elsewhere in SE Asia.* G1. Summer.

***C. floribundum** Lindley (*C. pumilum* Rolfe). Smaller, with leaves 30–55 × 1.5–2 cm, the raceme shorter than the leaves, the sepals and petals sometimes brown and the lip white to pink, dotted with red and tipped with yellow. *W China.* G1. Summer.

4. C. giganteum Wallich. Illustration: Botanical Magazine, 4844 (1855); Bechtel et al., Manual of cultivated orchid species, 190 (1981).
Pseudobulbs ovoid, several-leaved. Leaves linear, tapering to the acute apex, 60–75 × 2–4 cm. Raceme erect or spreading, with 8–15 flowers which are 8–10 cm in diameter. Sepals and petals green or yellowish green with red or brown stripes. Lip yellow, spotted with reddish brown, hairy. *Himalaya, W China.* G1. Winter.

This species forms part of a complex of variants distributed in the Himalaya, W China and the northern part of SE Asia. Though various species have been described, they are possibly no more than selections from the general variation. The most important are:

***C. grandiflorum** Griffith (*C. hookerianum* Reichenbach). Illustration: Botanical Magazine, 5574 (1866). Flowers 9–13 cm in diameter (occasionally as much as 16 cm?), sepals and petals olive green, unstriped, lip yellow blotched with purple. *Himalaya, W China.* G1. Winter.

***C. lowianum** (Reichenbach) Reichenbach. Illustration: Orchid Digest **42**: 124 (1978); Bechtel et al., Manual of cultivated orchid species, 191 (1981). Flowers 9–11.5 cm in diameter, sepals and petals greenish yellow with faint red-brown stripes, lip variable, yellow, white or orange, usually with a V-shaped scarlet mark. *Burma.* G1. Winter.

***C. tracyanum** Rolfe. Illustration: Botanical Magazine n.s., 56 (1949); Bechtel et al., Manual of cultivated orchid species, 191 (1981). Flowers fragrant, 10–15 cm in diameter, sepals and petals yellowish suffused and lined with red-brown, lip cream to yellow lined with purple or brown. *Burma, Thailand, Vietnam.* G1. Winter.

Sometimes considered to be a hybrid between *C. giganteum* and *grandiflorum*.

***C. wilsonii** (Cooke) Rolfe. Illustration: Botanical Magazine n.s., 704 (1976). Flowers *c.* 9 cm in diameter, sepals and petals green lined with reddish brown, lip yellowish white spotted with red-brown on the ridges and near the apex, lateral lobes lined with red-brown. *W China.*

Taylor & Woods consider this to be a cultivar of *C. giganteum* (in the notes to the illustration cited).

***C. longifolium** D. Don. A somewhat more distinct species with leaves to 1.5 cm broad, flowers fragrant, 6–8 cm in diameter, sepals and petals pale green with faint brownish red lines, lip white or pinkish yellow with a central crimson line and crimson blotches. *Himalaya, W China.* G1. Spring–summer.

5. C. insigne Rolfe (*C. sanderi* invalid). Illustration: Botanical Magazine, 8312 (1910).
Pseudobulbs spherical. Leaves 50–100 × 1.5–2 cm, linear, tapering to the acute apex. Raceme erect, exceeding the leaves, with 10–15 flowers. Flowers 7.5–10 cm in diameter. Sepals and petals white or white flushed with pink, their bases red-spotted. Lip white, flushed, lined and spotted with pink and with 2 yellow ridges. *Vietnam, Laos.* G1. Spring–summer.

6. C. eburneum Lindley. Illustration: Botanical Magazine, 5126 (1859); Orchid Digest **42**: 126 (1978); Bechtel et al., Manual of cultivated orchid species, 191 (1981).
Pseudobulbs scarcely evident. Leaves 45–60 × *c.* 2 cm, linear, tapering to the acute apex. Raceme erect with 1–2 flowers. Flowers usually 7.5–10 cm in diameter, fragrant. Sepals and petals white, spreading. Lip usually white with yellow ridges or a yellow centre, occasionally spotted with red or purple, the lateral lobes extending up to the column but not overlapping it, downy. *Himalaya, Burma, Vietnam.* G1. Spring.

Variable in the colour of the lip (see Pradhan, Orchid Digest **40**: 69–71, 1976).

7. C. tigrinum Hooker. Illustration: Botanical Magazine, 5457 (1864); Orchid Digest **42**: 129 (1978); Bechtel et al., Manual of cultivated orchid species, 191 (1981).
Pseudobulbs almost spherical, with 3–4 leaves. Leaves to 15 × 2.5 cm, broadly linear, acute. Racemes erect, exceeding the leaves, with 3–6 flowers. Flowers usually 5–7 cm in diameter (rarely larger). Sepals and petals olive green to yellow, faintly spotted and lined with red or purple, spreading. Lip mostly white, lined and spotted with purple, the margins often mostly purple. *Burma, Thailand.* G1. Summer–autumn.

8. C. erythrostylum Rolfe. Illustration: Botanical Magazine, 8131 (1907); Orchid Digest **42**: 129 (1978).
Pseudobulbs scarcely obvious. Leaves 30–45 × 1–1.5 cm, linear, tapering to the

acute apex. Raceme erect, loose, with 4–6 flowers. Flowers 7.5–11 cm in diameter. Sepals and petals white. Lip white shaded with yellow and with red lines, the lateral lobes large, curved over the column and more or less overlapping. Column bright crimson. *Vietnam*. G1. Winter.

9. C. mastersii Lindley (*Cyperorchis mastersii* (Lindley) Bentham). Illustration: Edwards's Botanical Register **31**: t. 50 (1845); Orchid Digest **41**: 116 (1976) & **42**: 127 (1978); Bechtel et al., Manual of cultivated orchid species, 191 (1981).
Pseudobulbs scarcely obvious. Leaves 45–70 × 1.5 cm, linear, tapering to the acute apex. Racemes erect with 4–6 (rarely 7) flowers. Flowers fragrant, *c.* 6 cm long, the sepals and petals forwardly directed, white or ivory white, sometimes flushed with pink. Lip white, sometimes spotted with pink, with a yellow centre, downy. *E Himalaya*. G1. Winter.

10. C. elegans (Blume) Lindley (*Cyperorchis elegans* Blume). Illustration: Botanical Magazine, 7007 (1888); Orchid Digest **42**: 127 (1978).
Pseudobulbs scarcely obvious. Leaves 40–70 × 1–2.5 cm, numerous, linear, tapering to the acute apex. Raceme arching, with many flowers. Flowers 4–5 cm long, the sepals and petals forwardly directed, little spreading, pale to tawny yellow. Lip yellow marked with 2 orange lines, sometimes spotted with red, downy and sparsely ciliate. *Himalaya, W China*. G1. Winter.

135. GRAMMANGIS Reichenbach

J. Cullen

Like *Cymbidium* but the lateral sepals fused at the base and attached to the column foot forming a small but distinct mentum. Petals much smaller than the sepals and differently coloured.

1. G. ellisii Reichenbach. Illustration: Botanical Magazine, 5179 (1860); Bechtel et al., Manual of cultivated orchid species, 215 (1981).
Sepals *c.* 4 × 2 cm, yellow barred and blotched with brown. Petals white with pink tips. Lip white with reddish lines, central lobe small. *Madagascar*. G2. Summer.

136. GRAMMATOPHYLLUM Blume

J. Cullen

Large epiphytes. Pseudobulbs compound, distinct, or elongate and stem-like, with 2–several leaves. Leaves rolled when young. Raceme large, lateral, with many flowers; lower flowers sometimes sterile and abortive. Sepals and petals similar, spreading. Lip small, clawed, 3-lobed, the lateral lobes large, extending up to the column, the central lobe smaller, bearing several ridges. Column without a foot. Pollinia 2, laterally ovoid to almost spherical, stipe short and broad, divided into 2 branches above.

A genus of about 10 species from SE Asia. They are strong-growing plants requiring epiphytic compost and careful watering. *G. scriptum* benefits from cooler conditions during the resting period. Propagation by division in early spring.

1a. Pseudobulbs to 14 cm, distinct, with a few leaves at the apex: flowers to 8 cm in diameter. **1. scriptum**
b. Pseudobulbs elongate, stem-like, bearing several leaves along their length; flowers to 10 cm in diameter **2. speciosum**

1. G. scriptum Blume (*G. fenzlianum* Reichenbach; *G. measuresianum* Weather; *G. multiflorum* Lindley; *G. rumphianum* Miquel). Illustration: Botanical Magazine, 7507 (1896); Sheehan & Sheehan, Orchid genera illustrated, 101 (1979).
Pseudobulbs to 14 cm, ridged, bearing 2–3 leaves at the apex. Leaves very narrowly elliptic, 30–35 × *c.* 8 cm. Raceme very long, with many flowers which are all fertile, and up to 8 cm in diameter. Sepals and petals pale green or greenish yellow, suffused or spotted with brown or purplish brown. Lip small, hairy. *Borneo, Philippines, Indonesia (Moluccas)*. G2. Summer.

Variable as regards the colour of the lip, which varies from white to yellow and is striped with brown or purple.

2. G. speciosum Blume. Illustration: Botanical Magazine, 5157 (1860).
Pseudobulbs elongate, stem-like, leafy along their length. Leaves 45–60 × 3 cm, broadly linear, acute. Raceme 2 m or more with many flowers, the lower of which are abortive and sterile with 2 sepals, 2 petals, no lip and an abortive column. Fertile flowers *c.* 10 cm in diameter, sepals and petals pale greenish yellow with dull orange-brown spots. Lip small, hairy, yellow striped with red-brown. *Malaysia, Indonesia, Philippines*. G2. Autumn–winter.

137. MORMODES Lindley

J. Cullen

Epiphytic. Pseudobulbs conspicuous, compound, broadly spindle-shaped, each bearing 2–several leaves. Leaves strongly veined, rolled when young. Raceme lateral, from the central nodes of the pseudobulb, erect or arching, with inconspicuous bracts. Flowers variously coloured, often fragrant. Sepals and petals more or less equal, the upper 3 often converging to form a hood over the column. Lip clawed, not spurred, entire or 3-lobed, often with the margins rolled under. Column cylindric, always twisted to 1 side of the flower which is thus asymmetric. Pollinia 2, ovoid, the stipe wineglass-shaped, longer than the pollinia and broader than them at the apex, the viscidium almost circular.

A genus of about 50 species from C & S America which is taxonomically very confused, owing to the wide and little-understood variation in flower colour, shape and size. A few species are grown as curiosities. Cultivation as for *Catasetum* (p. 262).

Literature: Pabst, G.F.J., An illustrated key to the species of the genus Mormodes, *Selbyana* **2**: 149–55 (1978).

1a. Lip clearly 3-lobed (lateral lobes may be small) **1. maculata***
b. Lip entire 2
2a. Lip hairy above; flower usually uniformly dark purple **2. hookeri**
b. Lip hairless; flower usually of some other colour or combination of colours 3
3a. Lip flat or with the sides curved upwards; pseudobulbs 8–10 cm **3. aromatica**
b. Lip with the sides distinctly rolled under; pseudobulbs to 30 cm 4
4a. Lip ovate, tapering to the apex **4. colossus**
b. Lip broader than long to circular, rounded to notched (though with a small point) at the apex **5. buccinator***

1. M. maculata (Kotzsch) Williams (*Cyclosia maculata* Klotzsch; *M. pardinum* Bateman). Illustration: Botanical Magazine, 3879 (1842), 3900 (1843); Bechtel et al., Manual of cultivated orchid species, 234 (1981).
Pseudobulbs 8–10 cm with 4–5 leaves. Flowering stem to 40 cm, raceme with many flowers. Flowers *c.* 6 cm in diameter, yellow or pale yellow, with or without red-brown spots. Sepals and petals lanceolate, acute. Lip *c.* 5 cm with the margins rolled under, 3-lobed from the middle, the lateral lobes conspicuous. *Mexico*. G2. Autumn.

***M. lineatum** Lindley. Illustration: Die Orchidee **22**: 45 (1971). Similar but with slightly smaller flowers (*c.* 5 cm in diameter) and the lip *c.* 3 cm, with very small, incurved, tooth-like lateral lobes. *Guatemala.* G2. Autumn.

*****M. luxatum** Lindley. Also similar but larger, with flowers *c.* 7 cm in diameter, the 2 upper sepals and upper petal converging to form a hood over the column, flowers sometimes white. *Mexico.* G2. Autumn.

2. M. hookeri Lemaire. Illustration: Botanical Magazine, 4577 (1851). Pseudobulbs oblong-ovoid. Flowering stem *c.* 30 cm, raceme with 4–5 distant flowers. Sepals and petals reflexed, dark purple, their sides themselves reflexed. Lip pointing forwards, entire, its sides rolled under, hairy on the upper surface, purple but paler than sepals and petals. *Panama.* G2. Winter.

3. M. aromatica Lindley. Illustration: Edwards's Botanical Register **29**: t. 56 (1843); Hamer, Las Orquideas de El Salvador **2**: 124–5 & t. 14 (1974). Pseudobulbs 8–10 cm with 4–5 leaves. Flowering stem weak, to 15 cm, with 5–8 rather distant flowers. Flowers *c.* 3 cm in diameter, yellowish deeply flushed with violet-brown and with darker spots. Sepals and petals similar, elliptic, acute. Lip entire, flat or with the sides curved upwards, with a triangular apex. *Mexico, El Salvador.* G2. Autumn.

4. M. colossus Reichenbach. Illustration: Botanical Magazine, 5840 (1870); Bechtel et al., Manual of cultivated orchid species, 233 (1981).
Pseudobulbs to 30 cm. Flowering stem to 70 cm, with 6–10 distant flowers which are up to 15 cm in diameter. Sepals and petals lanceolate, acuminate, reflexed, bright yellow, red towards the base. Lip ovate, acuminate, entire, the sides rolled under, bright yellow marked with red at the base. *Costa Rica.* G2. Spring.

5. M. buccinator Lindley (*M. lentiginosa* Hooker). Illustration: Botanical Magazine, 4455 (1849), 8041 (1905); Couret, Orquideas Venezolanas, 39–42 (1977). Pseudobulbs to 30 cm. Flowering stems to 25 cm with 7–10 flowers. Flowers extremely variable in colour, bright yellow to dark red-brown, variably spotted and lined. Sepals and petals narrowly elliptic, acute, spreading or reflexed at first, later curving forwards. Lip 2–2.5 cm wide, entire, broader than long, its sides very markedly rolled under, forming a trumpet shape. *Venezuela.* G2. Spring.

*****M. igneum** Lindley & Paxton. Similarly variable in colour, but the lip 3–3.5 cm wide. *C America and northern S America.*

138. CATASETUM Richard

J. Cullen

Epiphytes with large, compound pseudobulbs bearing several leaves. Leaves large, pleated, rolled when young. Racemes arising from the bases of the pseudobulbs, erect or arching, with few to many flowers. Flowers usually not resupinate, monomorphic and bisexual, or, more usually, dimorphic and unisexual, very rarely trimorphic (male, female and bisexual). Sepals and petals similar, free, spreading or incurved. Lip variable in size, saccate or not. Column large, without a foot, usually acute in male and bisexual flowers, with (in all species treated below except No. 1) 2 processes (antennae) arising from near the apex and directed backwards onto the lip in the male flowers. Pollinia 2, ovoid or ellipsoid, the stipe elongate, linear or strap-shaped, its sides usually folded upwards, viscidium circular or broadly oblong.

The descriptions given here (with the exception of that for No. 1) refer to the male flowers; the female flowers are less well known and their identification is difficult. Plants bearing female flowers were at one time referred to the genus *Monachanthus* Lindley. The production of male and female flowers appears to be controlled by light and nutrition (see Dodson, Annals of the Missouri Botanical Garden **49**: 35–56, 1962). The antennae borne by the male flowers are sensitive, and touching of them by an insect causes the pollinia to be forcibly ejected. For more details on pollination mechanisms in the genus see Hills, Williams & Dodson, Biotropica **4**: 61–9 (1972). The genus is divided into 3 genera, *Clowesia, Dressleria* and *Catasetum*, by Dobson, Selbyana **1**: 130–8 (1975). Under this arrangement, species No. 1 covered here belongs to *Clowesia*, the rest to *Catasetum*.

Catasetum in the broad sense is a genus of about 70 species from C & S America; about a dozen of these are grown for the sake of their bizarre male flowers.

The plants require sharp drainage and full light, combined with a humid atmosphere. They should be rested when the pseudobulbs are ripened. Propagation by seed or by division.

Literature: Mansfeld, R., Die Gattung Catasetum, *Feddes Repertorium* **30**: 257–75 (1932), **31**: 99–125 (1933).

Flowers. Always bisexual: **1**; usually unisexual: **2–12**.

Lip. (fig. 18, p. 263). Flat, not hollowed or saccate: **6,8**: hollowed or saccate but with a flat blade: **1,4,5,7,9–12**; entirely hollowed, with no flat blade: **2,3**. With conspicuous fringed or ciliate margins: **1,5,7–11**; with more or less entire margins: **2–4,6,12**.

Antennae. Asymmetric: **2–5**; symmetric: **6–12**.

1a. Flowers all bisexual; column without antennae **1. russellianum***
b. Flowers mostly unisexual; column in the male flowers with 2 antennae 2
2a. Lip not at all saccate or hollowed 3
b. Lip distinctly saccate, the position of the sac and the degree of saccation variable 4
3a. Lip broadly obovate with 3 teeth at the apex, the laterals narrowly triangular, incurved, the central shorter, straight; margins of the rest of the lip entire (figure 18(6), p. 263) **6. cernuum**
b. Lip triangular with rounded corners, the apex entire, the margins lacerate-ciliate (figure 18(8), p. 263) **8. trulla**
4a. Antennae not symmetric 5
b. Antennae symmetric 8
5a. Lip entirely helmet-shaped, hollowed out completely, without a flat blade 6
b. Lip variably saccate, but with a flat blade 7
6a. Basal angles of the lip projecting inwards, touching each other or almost so, toothed (figure 18(2), p. 263) **2. integerrimum**
b. Basal angles of the lip not projecting inwards, entire (figure 18(3), p. 263) **3. macrocarpum**
7a. Lip semicircular, truncate at the base, not lobed (figure 18(4), p. 263) **4. pileatum**
b. Lip oblong, 3-lobed, the central lobe narrow, not truncate at the base (figure 18(5), p. 263) **5. saccatum**
8a. Lip broadly obovate, fan-shaped, bearing a large humped callus in front of the mouth of the sac (figure 18(7), p. 263) **7. fimbriatum**
b. Lip variously shaped, not as above, either without a callus or with the callus behind the mouth of the sac 9
9a. Lip small, deeper than long, containing a callus composed of hair-like teeth which almost entirely fill the sac (figure 18(9), p. 263) **10. microglossum**

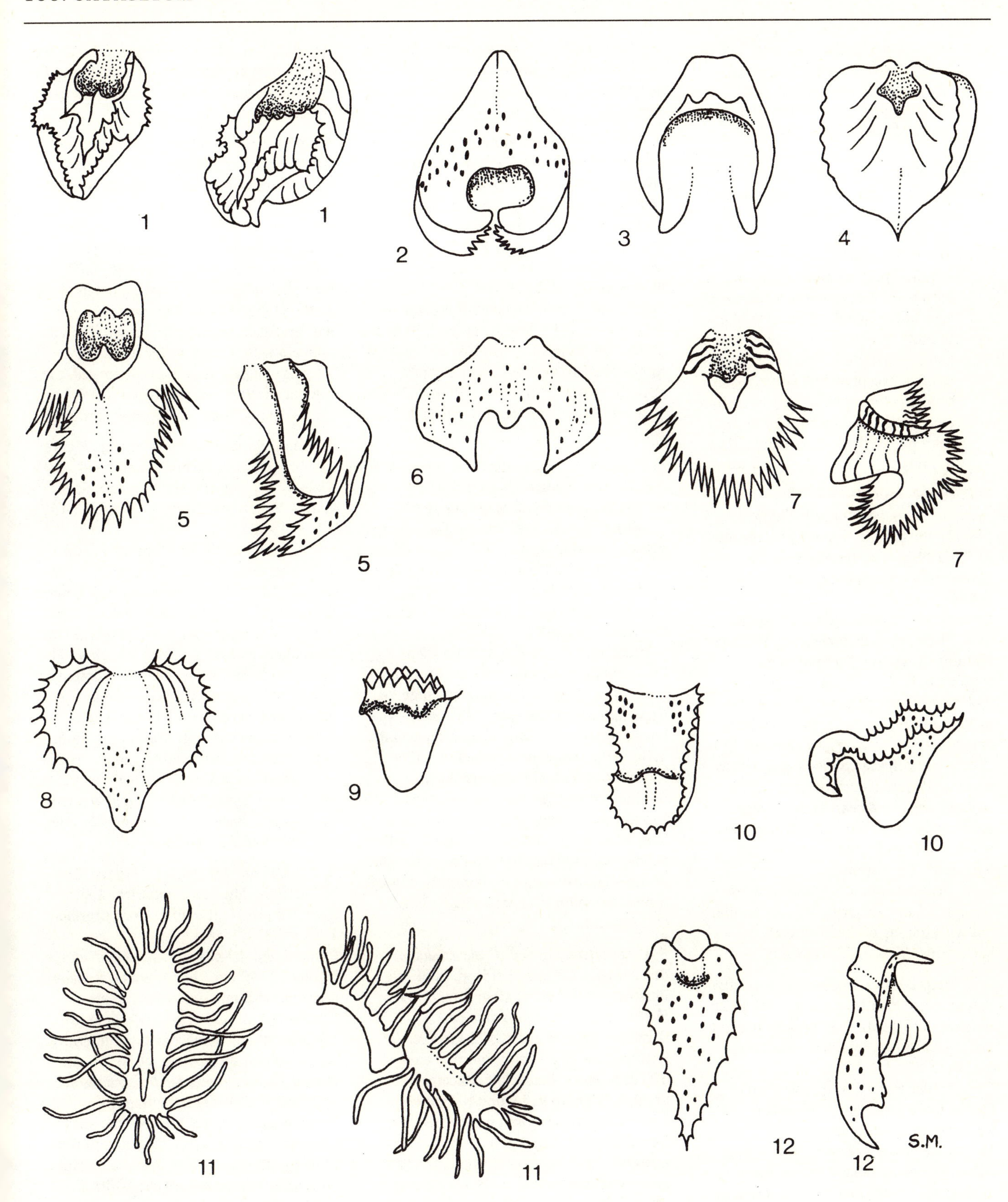

Figure 18. Lips of *Catasetum* species. 1, *C. russellianum*. 2, *C. integerrimum*. 3, *C. macrocarpum*. 4, *C. pileatum*. 5, *C. saccatum*. 6, *C. cernuum*. 7, *C. fimbriatum*. 8, *C. trulla*. 9, *C. microglossum*. 10, *C. atratum*. 11, *C. barbatum* (from 2 aspects). 12, *C. callosum*. (Compiled from various sources; not to scale.)

b. Lip various, longer than deep, callus not as above, or absent 10
10a. Lip narrowly ovate, its margin distantly and shortly toothed, green with red spots (figure 18(12), p. 263) **12. callosum**
b. Lip oblong, its margin, at least in part, deeply, closely and coarsely toothed, not green with red spots 11
11a. Lip margin almost entirely divided into fleshy, blunt, hair-like teeth, its terminal portion narrowly triangular with a few teeth, with a small saccate area in the middle (figure 18(11), p. 263) **11. barbatum**
b. Lip margin coarsely and sharply toothed except for the entire, semicircular terminal part; saccation extending from the base for more than half of the lip (figure 18(10), p. 263) **9. atratum**

1. C. russellianum Hooker (*Clowesia russelliana* (Hooker) Dodson. Figure 18(1), p. 263). Illustration: Botanical Magazine, 3777 (1840); Dunsterville & Garay, Venezuelan orchids illustrated **4**: 41–2 (1966).
Pseudobulbs ovoid-ellipsoid. Leaves to 36 × 11 cm, broadly lanceolate. Raceme arching with many fragrant, non-resupinate flowers. Sepals and petals 3–4 cm, somewhat incurved, ovate-oblong to elliptic, white with green stripes. Lip oblong, deeply saccate for about one-third of its length, the sac projecting forwards under the ovate blade; margins lacerate-toothed, the blade bearing an elongate, double ridge with toothed margins; the whole white with green lines. *C America, Venezuela*. G2. Autumn.

*__C. thylaciochilum__ Lemaire (*Clowesia thylaciochilum* (Lemaire) Dodson). Similar, but with the sac of the lip curved backwards. *Mexico*. G2. Autumn.

2. C. integerrimum Hooker (*C. maculatum* Batcman not Knuth). Figure 18(2), p. 263. Illustration: Edwards's Botanical Register **26**: t. 42 (1840); Botanical Magazine, 3823 (1840).
Pseudobulbs to 15 cm, spindle-shaped. Leaves to 40 cm, elliptic-lanceolate. Racemes erect, to 35 cm, with 6–12 non-resupinate flowers. Sepals 4.5–4.8 cm, elliptic or oblong-elliptic, acute, green outside, green suffused or spotted with red inside. Petals similar but slightly smaller, Lip helmet-shaped, wholly hollowed, the basal angles projecting inwards and touching each other or almost so, toothed, the whole greenish yellow or yellow, sometimes with red spots outside, greenish with brown spots inside. Antennae asymmetric. *Mexico, Guatemala, Honduras & Venezuela*. G2. Autumn.

3. C. macrocarpum Knuth (*C. tridentatum* Hooker). Figure 18(3), p. 263. Illustration: Edwards's Botanical Register **10**: t. 840 (1824); Botanical Magazine, 2559 (1825), 3329 (1834); Hoehne, Iconografia de Orchidaceas do Brasil, t. 170 (1949); Dunsterville & Garay, Venezuelan orchids illustrated **2**: 60–1 (1961).
Similar to *C. integerrimum* but raceme to 45 cm with 4–10 flowers, lip with its basal angles not projecting inwards, and not toothed except for the 3-toothed apex, entirely yellow within and without. *Tropical S America from Trinidad to Brazil*. G2. Autumn–winter.

4. C. pileatum Reichenbach (*C. bungerothii* N.E. Brown). Figure 18(4), p. 263. Illustration: Botanical Magazine, 6998 (1888); Cogniaux & Goossens, Dictionnaire Iconographique des Orchidées, Catasetum, t. 1, 1A (1897); Hoehne, Iconografia de Orchidaceas do Brasil, t. 163 (1949); Dunsterville & Garay, Venezuelan orchids illustrated **1**: 70–1 (1959).
Pseudobulbs 15–25 cm, spindle-shaped. Leaves large, lanceolate. Raceme to 30 cm, arching, with 4–10 resupinate(?) flowers. Upper sepal and petals to 5 cm, erect, lateral sepals spreading, oblong-lanceolate, acuminate, white, pale yellow or white spotted with red. Lip semicircular in outline, truncate at the base, unlobed, saccate with a small sac whose mouth is just below the lip attachment, white, pale yellow or rarely red, the interior of the sac orange-red. Antennae asymmetric. *Tropical S America, south to Ecuador and N Brazil*. G2. Autumn.

5. C. saccatum Lindley (*C. christyanum* Reichenbach). Figure 18(5), p. 263. Illustration: Botanical Magazine, 8007 (1905); Hoehne, Iconografia de Orchidaceas do Brasil, t. 172 (1949).
Pseudobulbs to 20 cm, fusiform. Leaves to 30 cm or more, lanceolate. Raceme arching, with up to 18 non-resupinate flowers. Upper sepal and petals close, erect, lateral sepals spreading, all 5–5.5 cm, narrowly lanceolate, acuminate, reddish green with darker red spots. Lip oblong, 3-lobed, its margins rolled downwards, lacerate-ciliate, the central lobe oblong, its basal part saccate, all reddish brown with a white callus in front of the mouth of the sac. Antennae asymmetric. *Tropical S America from E Peru to Guyana*. G2. Winter.

The most commonly cultivated variant is var. *christyanum* (Reichenbach) Mansfeld, which is described above.

6. C. cernuum (Lindley) Reichenbach (*C. trifidum* Hooker). Figure 18(6), p. 263. Pseudobulbs large. Leaves elliptic. Raceme to 30 cm, arching, bearing 10–15 non-resupinate flowers which are themselves downwardly directed. Sepals and petals incurved, up to 3 cm, the petals usually a little shorter than the sepals, all oblong, acute, green with dark brown spots. Lip flat, greenish spotted with brown, broadly obovate with 3 teeth at the apex, the laterals elongate-triangular, incurved, the central shorter, straight. Antennae symmetric. *S Brazil*. G2. Spring–summer.

7. C. fimbriatum (Morren) Lindley. Figure 18(7), p. 263. Illustration: Botanical Magazine, 7158 (1891); Dunsterville & Garay, Venezuelan orchids illustrated **4**: 42–3 (1966).
Pseudobulbs to 20 cm, ellipsoid. Leaves 20 cm or more, elliptic. Raceme to 40 cm, arching, with 7–15 non-resupinate flowers. Upper sepal and petals close, erect, lateral sepals pointing downwards, all to 3.5 cm, lanceolate, acuminate, with margins rolled downwards, yellow or greenish, spotted or striped with purple. Lip broadly obovate, fan-shaped, saccate near the base with a large humped callus in front of the mouth of the sac, margins lacerate-ciliate, yellow or greenish spotted with purple towards the base. Antennae symmetric. *Tropical and subtropical S America*. G2. Spring–summer.

A very variable species.

8. C. trulla Lindley. Figure 18(8), p. 263. Illustration: Edwards's Botanical Register **27**: t. 34 (1841); Hoehne, Flora Brasilica **12**(6): t. 63 (1942).
Pseudobulbs 12–20 cm, variable in shape. Leaves 35–40 cm, oblong-linear. Raceme arching, with many non-resupinate flowers. Sepals and petals spreading, the upper sepal and petals erect, the lateral sepals outwardly directed, all to 2.5 cm, oblong-elliptic, green. Lip flat, triangular with rounded corners, the margin lacerate-ciliate except towards the apex, the whole whitish green tinged with red-brown in the centre, the apical part blunt, brown. Antennae symmetric. *S Brazil*. G2. Autumn.

9. C. atratum Lindley. Figure 18(10), p. 263. Illustration: Edwards's Botanical Register **26**: t. 63 (1838); Botanical Magazine, 5202 (1860).

Pseudobulbs 10–13 cm, spindle-shaped. Leaves to 30 cm, narrowly elliptic or lanceolate. Raceme to 30 cm, arching, bearing 12–15 resupinate(?) flowers. Sepals and petals to 2.5 cm, all more or less directed downwards, oblong, acute, green, densely marked with small brown transverse stripes. Lip oblong, saccate for two-thirds of its length from the base, margins lacerate-ciliate or lacerate-toothed, yellowish green with some brown spots and a white apical part. Antennae symmetric. *S & E Brazil.* G2. Spring.

10. C. microglossum Rolfe. Figure 18(9), p. 263. Illustration: Botanical Magazine, 8514 (1913).
Pseudobulbs to 10 cm, spindle-shaped. Leaves to 30 cm, oblong-elliptic to oblanceolate. Raceme to 65 cm, arching, with many non-resupinate flowers. Upper sepal and petals erect, lateral sepals directed downwards, all to 2.5 cm, oblong-lanceolate, acute, purplish. Lip small, consisting mostly of an obconic sac which is deeper than long, with lacerate margins and containing a callus which terminates in numerous teeth or hair-like processes, yellow. Antennae symmetric. *Borders of Peru and Colombia.* G2.

11. C. barbatum (Lindley) Lindley (*Myanthum barbatum* Lindley; *M. spinosum* Hooker). Figure 18(11), p. 263. Illustration; Botanical Magazine, 3802 (1841); Dunsterville & Garay, Venezuelan orchids illustrated **3**: 54–5 (1965).
Pseudobulbs 10–12 cm, narrowly conical to spindle-shaped. Leaves lanceolate, wavy. Raceme to 45 cm, arching or erect(?), bearing numerous non-resupinate flowers. Upper sepal and petals erect, lateral sepals spreading, all to 3 cm, narrowly oblong, acute, green with small, dark purple, transverse bars. Lip oblong, white or pink, with a small sac at about the middle, the margin deeply divided into fleshy, blunt, hair-like teeth, with a narrowly triangular apical part and a callus divided usuallly into 3 fleshy teeth situated behind the mouth of the sac. Antennae symmetric. *Northern S America.* G2. Spring.

The variant most commonly found in cultivation is var. **spinosum** (Hooker) Rolfe, with the callus as described above.

12. C. callosum Lindley. Figure 18(12), p. 263. Illustration: Botanical Magazine, 6648 (1882).
Pseudobulbs to 8 cm, spindle-shaped. Leaves 15–30 cm, oblanceolate, greyish. Raceme to 35 cm, arching or erect, with 10–15 resupinate(?) or non-resupinate flowers. Upper sepal and petals erect, lateral sepals directed downwards, all 4–4.5 cm, narrowly lanceolate, acute, brownish. Lip narrowly ovate, green with red spots, its margin finely and distantly toothed, with a small sac near the base whose mouth is sited just in front of a humped callus. Antennae symmetric. *Venezuela, Guyana, Peru.* G2. Winter.

139. CYCNOCHES Lindley

J. Cullen

Epiphytic. Pseudobulbs cylindric, tapering slightly towards the apex, compound. Leaves pleated, rolled when young. Racemes with several flowers, lateral from the axils of the upper leaves. Flowers not resupinate, usually unisexual. Sepals and petals similar, smaller in the female flowers than in the male. Lip fleshy and undivided (all female flowers and male flowers of No. 1) or variously divided or conspicuously appendaged (male flowers of Nos. 2 and 3), with a basal callus. Column short and fleshy in the female flowers, elongate, broadened above and arched in male flowers. Pollinia 2, ovoid, stipe narrowly ovoid, longer and narrower than the pollinia, viscidium oblong, 2-lobed.

A genus of 7 species from C & S America, whose classification is complicated and based almost entirely on the male flowers, which are more frequently seen than the female. Gregg (Selbyana **2**: 217, 1978) reports that 'plants growing in full sunlight with adequate moisture and embedded in a suitable substrate are robust and generally produce female flowers. Less robust plants often growing in shade or lacking adequate nutrition tend to produce male flowers'. Cultivation as for *Catasetum* (p. 262).
Literature: Allen, P.H., The swan orchids: a revision of the genus Cycnoches, *Orchid Journal* **1**: 173–84, 225–30, 349–54, 397–403 (1952).

1a. Male inflorescence with up to 9 flowers: lip of male flowers simple and entire **1. ventricosum***
b. Male inflorescence with many flowers: lip of male flowers lobed or with marginal teeth or appendages 2
2a. Lip of male flower linear or lanceolate, flat or wavy with a forwardly pointing process on the claw and 2 linear, rounded teeth on each side **3. pentadactylon**
b. Lip of male flower clawed, ovate to circular, the margin drawn out into several processes which are club-shaped or forked at their apices **2. egertonianum***

1. C. ventricosum Bateman.
Male flowers 9–15 cm in diameter. Sepals and petals green or yellowish green, spreading. Lip 6–6.5 cm, simple, fleshy, stalkless or with an elongate claw which may be narrowly winged, white marked with dark green at the base. Column 2.8–3.2 cm. *Tropical America, from S Mexico to French Guiana.* G2. Summer.

A variable species, divided by Allen into several varieties; the most commonly grown is var. **chlorochilon** (Klotzsch) Allen (*C. chlorochilon* Klotzsch) which has large flowers with a stalkless lip (see Hoehne, Flora Brasilica **12**(6): t. 101, 1942; Dunsterville & Garay, Venezuelan orchids illustrated **2**: 81–2, 1961; Couret, Orquideas Venezolanas, 36, 1977; Sheehan & Sheehan, Orchid genera illustrated, 71, 1979; Bechtel et al., Manual of cultivated orchid species, 189, 1981).

***C. loddigesii** Lindley. Illustration: Hoehne, Flora Brasilica **12**(6): t. 100 (1942); Couret, Orquideas Venezolanas, 37 (1977). Similar, but with larger flowers with greenish brown sepals and petals and the lip with a conspicuously winged claw. *Venezuela & NE Brazil.* G2. Summer.

2. C. egertonianum Bateman. Illustration: Hoehne, Flora Brasilica **12**(6): t. 109 (1942); Bechtel et al., Manual of cultivated orchid species, 189 (1981).
Flowers 5–6 cm in diameter, Sepals and petals greenish or greenish brown, spotted with brown or purple, sometimes yellow or almost white (var. **aureum**), more rarely pink. Lip of male flowers stalked, green or white, the apical part concave, ovate or circular, the margin with several elongate, club-shaped teeth which may be forked at their apices. Column *c.* 20 mm. *C & S America.* G2. Autumn.

Variable in flower colour; both var. **egertonianum**, with greenish or greenish brown, darker spotted perianth, and var. **aureum** (Lindley) Allen (*C. aureum* Lindley), with yellow to almost white perianth, are grown.

***C. maculatum** Lindley. Similar but with the apical part of the lip drawn out into a strap-shaped lobe, and the marginal teeth not forked at their apices. *Venezuela.*

3. C. pentadactylon Lindley. Illustration: Flora Brasilica **12**(6): t. 104 (1942); Bechtel et al., Manual of cultivated orchid species, 189 (1981).

Flowers 7–10 cm in diameter. sepals and petals green blotched with dark brown. Lip fleshy and rigid, linear-lanceolate, flat or wavy, clawed and with a forwardly pointing process on the upper surface of the claw, the blade narrow, with 2 large, linear, rounded teeth on each side. *E & C Brazil*. G2. Spring.

140. LOCKHARTIA Hooker

J. Cullen

Epiphytic, without pseudobulbs. Stems erect, covered with 2-ranked, strongly keeled leaves which consist mainly of overlapping sheaths, the free blades small, folded when young. Flowers borne laterally, either singly or in open, few-flowered racemes, usually conspicuously bracteate with stem-clasping bracts. Sepals and petals similar, 7–15 mm, spreading or somewhat reflexed, yellow. Lip simple or more frequently complex with 2 lateral lobes and a central lobe which is often further divided and warty at its base. Column spreading, winged towards the apex, without a foot. Pollinia 2, attached directly to a small, circular viscidium.

Lockhartia is a genus of about 20 species from C & S America. Only 1 species is readily available in Europe, though others are grown in the USA (see Teuscher, H., The genus Lockhartia, American Orchid Society Bulletin **43**: 399–405, 1974). Cultivation as for *Oncidium* (p. 282).

1. **L. oerstedtii** Reichenbach (*L. verrucosa* Reichenbach; *L. robusta* Schlechter). Illustration: Saunders' Refugium Botanicum **2**: t. 76 (1869); Bechtel et al., Manual of cultivated orchid species, 222 (1981).
Stems 15–30 cm. Leaves rounded or somewhat acute at the apex. Petals and sepals golden yellow. Lip with blunt, downwardly deflected, yellow and often red-spotted lateral lobes and a yellow central lobe which is often red-spotted at the base, oblong or reversed-heart-shaped, deeply notched at the apex and with 2 lateral, somewhat reflexed, lobe-like prolongations near its junction with the lateral lobes. *C America*. G2. Summer.

*__L. lunifera__ Reichenbach. Similar but with smaller flowers (sepals *c.* 7 mm), the petals slightly smaller than the sepals and the lateral lobes of the lip sickle-shaped. *N & E Brazil*.

141. ACINETA Lindley

J. Cullen

Epiphytic with distinct, simple, ovoid or spindle-shaped, grooved pseudobulbs, each bearing 2–4 leaves. Leaves large, pleated, rolled when young. Flowers not resupinate, fleshy, bell-shaped, in lateral, hanging racemes which arise from the bases of the pseudobulbs. Sepals and petals similar in size or the sepals somewhat larger, incurved. Lip clawed, 3-lobed, the lateral lobes erect and borne near the middle of the claw, reaching up to the column, the central lobe relatively small, with a prominent callus between the lateral lobes. Column finely downy on the back, with a short foot to which the lateral sepals are attached. Pollinia 2, ellipsoid, stipe broadly linear, almost as long as the pollinia, the viscidium with a truncate or rounded basal part.

A genus of 10–15 species from C & S America. They should be grown in suspended baskets or rafts so that the hanging racemes can develop properly. An epiphytic compost should be used and abundant moisture provided when in growth. A lower temperature is required during the resting season. Propagation is by seed or division.

Literature: Schlechter, R., Die Gattung Acineta, *Orchis* **11**: 21–47 (1917).

1a. Sepals and petals red, or pale brown and red-spotted; flowering in spring **3. superba**
b. Sepals and petals yellow, with or without red spots; flowering in autumn 2
2a. Flowers 4–5 cm in diameter; pseudobulbs with 2 leaves; racemes with 12–15 flowers **1. barkeri**
b. Flowers *c.* 6 cm in diameter; pseudobulbs with 3–4 leaves; racemes usually with more than 15 flowers **2. chrysantha**

1. **A. barkeri** Lindley. Illustration: Botanical Magazine, 4203 (1846).
Pseudobulbs *c.* 10 cm, with 2 leaves, ovoid. Leaves to 50 × 10 cm. Raceme with 12–15 flowers which are 4–5 cm in diameter. Sepals and petals ovate, golden yellow. Lip yellow with a purple-red spot or spots near the base. *S Mexico*. G2. Autumn.

2. **A. chrysantha** (Morren) Lindley & Paxton (*A. densa* Lindley & Paxton). Illustration: Botanical Magazine, 7143 (1890); Bechtel et al., Manual of cultivated orchid species, 161 (1981).
Like *A. barkeri* but larger; pseudobulbs with 3–4 leaves, racemes usually with more than 15 flowers, petals red-spotted, lip with very prominent callus. *Tropical America from Mexico to Colombia*. G2. Autumn.

3. **A. superba** (Humboldt, Bonpland & Kunth) Reichenbach (*A. humboldtii* Lindley). Illustration: Dunsterville & Garay, Venezuelan orchids illustrated **1**: 40–1 (1959); Bechtel et al., Manual of cultivated orchid species, 161 (1981).
Pseudobulbs with 3 leaves. Racemes loose, with 6–12 flowers which are *c.* 6 cm in diameter. Sepals and petals red, or pale brown and red-spotted. Lip yellow, reddish towards the base, and with dark purple spots. *Tropical S America from Venezuela to Ecuador*. G2. Spring.

The variant with brownish, red-spotted flowers is referred to as var. **fulva** (Hooker) Schlechter (see Botanical Magazine, 4156, 1845).

142. PERISTERIA Hooker

J. Cullen

Epiphytes with simple, ovoid or spindle-shaped pseudobulbs which are grooved and bear 3–5 leaves. Leaves stalked, often large, pleated, rolled when young. Racemes arising from the bases of the pseudobulbs, erect or hanging, with few to many flowers. Flowers spherical to bell-shaped, not widely open, often fragrant, resupinate or not. Sepals and petals similar or the petals somewhat smaller, fleshy, ovate to circular, incurved. Lip complex, joined to the base of the column by a basal part which bears 2 lateral lobes which stand upright on either side of the column, and an erect or outwardly curving anterior part which is distinguished from the basal part by a waist which is sharply bent or folded; the anterior part is freely movable on the rest. Column thick and fleshy, sometimes with a small wing on either side, the rostellum prominently beaked. Pollinia 2, ellipsoid, attached independently to the triangular viscidium, stipe absent.

A genus of about 10 species from C & S America. All can be grown like *Acineta*, but *P. elata*, which has erect racemes, is suitable for pot-culture, and will tolerate a terrestrial compost.

1a. Raceme erect, stiff, with 12–25 flowers **1. elata**
b. Raceme hanging, with 4–13 flowers 2
2a. Column with a narrow, downwardly pointing wing on either side; basal part of the lip with a crested surface **3. pendula**
b. Column without wings; basal part of lip smooth **2. cerina**

1. P. elata Hooker. Illustration: Botanical Magazine, 3116 (1832); Dunsterville & Garay, Venezuelan orchids illustrated **1**: 238–9 (1959); Ospina & Dressler, Orquideas de las Americas, t. 159 (1974); Sheehan & Sheehan, Orchid genera illustrated, 131 (1979).
Pseudobulbs to 13 × 10 cm. Leaves to 95 cm, narrowly elliptic, acute. Raceme to 1.5 m, held stiffly erect, with 12–25 flowers. Sepals and petals waxy, white, flower 4.5–5 cm in diameter. Lip with anterior part entire, curving outwards. Column unwinged. *Tropical America from Costa Rica to Venezuela and Colombia*. G2. Summer.

Known as the dove orchid because of a resemblance between the column and the lateral lobes of the lip and a dove with outstretched wings.

2. P. cerina Lindley. Illustration: Edwards's Botanical Register **23**: t. 1953 (1837); Hoehne, Flora Brasilica **12**(6): t. 119 (1942).
Pseudobulbs to 11 × 5 cm. Leaves to 45 cm, narrowly elliptic, acute. Racemes hanging, 10–15 cm, with 7–13 flowers which are *c*. 3 cm in diameter. Sepals and petals yellow. Lip with the basal part smooth, the anterior part erect, somewhat hooded, the margins orange-yellow and crisped. Column unwinged. *Trinidad, N Brazil*. G2. Summer.

3. P. pendula Hooker. Illustration: Botanical Magazine, 3479 (1836).
Vegetatively similar to *P. cerina*. Racemes hanging, with 4–7 flowers which are 4.5–5 cm in diameter. Sepals and petals white or pale yellow, spotted with red. Column with a narrow, downwardly directed wing from the apex on either side. *Tropical S America from Trinidad to N Brazil and N Peru*. G2. Spring.

143. LACAENA Lindley

J. Cullen

Epiphytes with simple, ovoid or spindle-shaped, grooved pseudobulbs, each bearing 2 leaves. Leaves stalked, elliptic, pleated, rolled when young. Racemes hanging, arising from the bases of the pseudobulbs, with 8–30 flowers. Flowers openly bell-shaped, not resupinate. Sepals and petals similar, incurved. Lip attached to the column foot, complex in shape, 3-lobed, with a callus between the short, erect, basal, lateral lobes, the central lobe hairless, clawed, ovate or oblong, its anterior part not freely movable on the rest. Pollinia 2, ellipsoid, stipe linear, longer than the pollinia, somewhat expanded above, the viscidium with an acute base.

A genus of 2 species from C America; both are grown in conditions like those for *Acineta* (p. 266).
Literature: Jenny, R., Die Gattung Lacaena Lindley, *Die Orchidee* **30**: 55–61 (1979).

1a. Flowers 4.5–5 cm in diameter, lip with a hairy callus and ovate central lobe **1. bicolor**
b. Flowers 2.5–3 cm in diameter, lip with a hairless callus and oblong central lobe **2. spectabilis**

1. L. bicolor Lindley. Illustration: Botanical Magazine n.s., 330 (1959); Ospina & Dressler, Orquideas de las Americas, f. 155 (1974).
Pseudobulbs 6–8 cm, ovate. Leaves 20–25 cm, elliptic. Racemes with 10–30 flowers which are fragrant and 4.5–5 cm in diameter. Sepals and petals greenish white, greenish yellow or pinkish, scurfy outside. Lip white with a hairy, purple callus and purple spots on the shortly stalked, ovate central lobe. *Mexico, Guatemala, Honduras*. G2. Spring–summer.

2. L. spectabilis (Klotzsch) Reichenbach. Illustration: Botanical Magazine, 6516 (1880).
Like *L. bicolor* but racemes with 8–12 flowers which are 2.5–3 cm in diameter, petals and sepals pinkish white, lip with hairless callus and an oblong, spade-like, conspicuously stalked, purple central lobe. *Mexico*. G1. Spring.

144. LUEDDEMANNIA Reichenbach

J. Cullen

Like *Lacaena* but pseudobulbs with 3–5 leaves, the sepals and petals somewhat more spreading and the lip simpler, 3-lobed, with curving, obliquely triangular lateral lobes and a narrowly triangular, downy central lobe, bearing a humped, downy callus.

Cultivation as for *Acineta* (p. 266).

1. L. pescatorei (Lindley) Linden & Reichenbach. Illustration: Botanical Magazine, 7123 (1890).
Sepals broader than petals, brown and scaly-hairy outside, yellow striped with red inside, petals yellow, lip yellow with red spots. *Colombia*. G2. Summer.

145. HOULLETIA Brongniart

J. Cullen

Usually epiphytic with simple pseudobulbs to 6 cm, each bearing 1 leaf. Leaves stalked, elliptic, acute, pleated, rolled when young. Raceme erect or arching, arising from the base of the pseudobulb, with 4–10 non-resupinate flowers which are often fragrant, axis hairless. Sepals and petals widely spreading, similar or the petals slightly smaller. Lip complex, attached to the column foot, divided by a narrow waist into a clawed basal part which is narrowly oblong and bears on each side a narrow, pointed lateral lobe which curves backwards and upwards, then forwards; anterior part stalkless or shortly stalked, oblong or ovate, arrowhead-shaped at the base with pointed lobes. Pollinia 2, irregularly ellipsoid, stipe linear, longer than pollinia, slightly expanded above, viscidium elongate, acute. Rostellum conspicuously beaked.

A genus of about 10 species from C & S America. Cultivation as for *Acineta* (p. 266) except that Houlletias are best grown in pots, as the large leaves make suspension difficult.

1a. Sepals and petals oblong, of uniform colour, dark red or dark brown, not spotted or blotched **3. odoratissima**
b. Sepals and petals ovate to elliptic, with conspicuous brown blotches on a yellow ground 2
2a. Raceme arching; lip whitish with violet spots; petals each with a large tooth on the inner margin **1. lansbergii**
b. Raceme erect; lip yellow with red or brown spots; petals not toothed **2. brocklehurstiana***

1. H. lansbergii Linden & Reichenbach. Illustration: Botanical Magazine, 7362 (1894); Fieldiana **26**: 525 (1953).
Pseudobulbs *c*. 6 cm, ovate. Leaves to 40 cm, elliptic. Raceme arching, with 5–10 flowers, each *c*. 8.5 cm in diameter. Sepals and petals ovate to elliptic, yellow with conspicuous brown spots, each petal with a single tooth on the upper margin. Lip whitish with violet spots. *Costa Rica*. G2. Autumn.

2. H. brocklehurstiana Lindley. Illustration: Botanical Magazine, 4072 (1844); Hoehne, Flora Brasilica **12**(6): t. 136 (1942).
Vegetatively similar to *H. lansbergii*. Raceme erect with 5–10 flowers, each *c*. 7 cm in diameter. Sepals and petals ovate to elliptic, yellow with brown spots, petals untoothed. Basal part of the lip yellow with red spots, the anterior part purple. *E Brazil*. G2. Autumn–winter.

***H. picta** Linden & Reichenbach. Illustration: Botanical Magazine, 6305 (1877). Similar, but the bases of the sepals and petals entirely brown and the anterior part of the lip yellow with very dark brown spots. *Colombia.*

***H. wallisii** Linden & Reichenbach. Illustration: Ospina & Dressler, Orquideas de las Americas, f. 154 (1974). Also similar but with the lateral sepals united and the petals rather narrow. *Colombia.*

3. H. odoratissima Lindley & Paxton. Illustration: Lindenia 7: t. 324 (1891). Vegetatively similar to *H. lansbergii*. Raceme erect with 6–9 flowers each *c.* 7 cm in diameter. Sepals and petals oblong, dark brown or dark red, petals entire. Lip mostly white. *Colombia.* G2. Autumn.

A variant with red sepals and petals has been called var. **antioquiensis** Anon.

146. POLYCYCNIS Reichenbach

J. Cullen

Like *Houlletia* but plants not so large, the axis of the raceme downy, lip with the basal part somewhat clawed and bearing upcurving, oblong-triangular lateral lobes, the central lobe heart-shaped. Column slender, arching, abruptly widened at the apex.

Cultivation as for *Acineta* (p. 266).

1. P. barbata Reichenbach. Illustration: Botanical Magazine, 4479 (1849). Raceme-axis and flower-stalks purple, downy. Sepals and petals 2.5–2.7 cm, yellowish flushed with red and with brown-purple spots, lip similar in colour, prominently but sparsely hairy. *Costa Rica, Colombia.* G2. Winter.

147. PAPHINIA Lindley

J. Cullen

Epiphytes with rather small, simple pseudobulbs, each usually bearing 2 leaves. Leaves pleated, rolled when young. Racemes hanging, arising from the bases of the pseudobulbs, each with 1–5 flowers which are resupinate and 8–14 cm in diameter. Sepals and petals widely spreading, the petals somewhat smaller than the sepals. Lip complex, 3-lobed, smaller than the petals, the claw and basal part ridged and crested, bearing the erect, forwardly directed lateral lobes; the central lobe movable on the rest, diamond-shaped, bearing a tuft of club-shaped, hair-like processes on the margin towards and at the apex. Column narrow with small wings or teeth towards the apex. Pollinia 2, ellipsoid, their surfaces roughened, stipe linear-oblong with the sides curved upwards near the top, viscidium very small. Rostellum conspicuously beaked.

A genus of about 7 species from C & S America. They require warm, humid conditions when growing and infrequent watering when resting. They are best grown in epiphytic compost in small pans, and given some shade during the summer. Literature: Jenny, R., Die Gattung Paphinia Lindley, *Die Orchidee* **29**: 207–15 (1978).

1a. Flowers to 8 cm in diameter; lip red with white crests, ridges and hair-like processes **1. cristata**
b. Flowers 12–14 cm in diameter; lip dark purple, white towards the base, hair-like processes white **2. grandiflora**

1. P. cristata Lindley. Illustration: Fieldiana **26**: 527 (1952); Hoehne, Flora Brasilica **12**(7): t. 3 (1953).
Pseudobulbs *c.* 4 cm, somewhat compressed. Leaves 10–15 cm, lanceolate, acute. Flowers to 8 cm in diameter. Sepals and petals translucent white with brown or red stripes or spots. Lip red with white ridges, crests and hair-like processes. *Tropical S America, from Colombia and Trinidad to N Brazil.* G2. Autumn.

2. P. grandiflora Rodriguez. Illustration: Hoehne, Flora Brasilica **12**(7): t. 1 (1953). Pseudobulbs to 5 cm. Leaves 20–25 cm, lanceolate to elliptic. Flowers 12–14 cm in diameter. Sepals and petals white, striped and spotted with red-purple, especially towards the apex. Lip dark purple, white towards the base and with white hair-like processes. *NW Brazil.* G2. Autumn.

The flowers are said to produce a strong, rather unpleasant smell.

148. STANHOPEA Hooker

J. Cullen

Epiphytic with small, simple, 1-leaved pseudobulbs. Leaves large, leathery, elliptic, stalked, pleated, rolled when young. Racemes arising from the bases of the pseudobulbs, hanging, with 2–10 flowers which have prominent, coloured bracts. Flowers usually fragrant, not resupinate. Sepals and petals similar, or petals sometimes a little smaller, all spreading or reflexed, the lateral sepals joined at the base. Lip complex, very fleshy, usually clearly divided into 3 parts, the basal part hollowed or saccate, the middle usually bearing a horn-like lateral lobe on either side, the interior part simple or 3-toothed or 3-lobed at the apex, jointed to the median part; more rarely the whole lip forming an irregular pouch. Column arched, slender, often winged towards the apex. Pollinia 2, ellipsoid, stipe linear, longer than pollinia, somewhat expanded above, its sides turned upwards, viscidium elongate.

A genus of from 25 to 50 species whose classification is chaotic. Most of the species are poorly known, variable and difficult to distinguish. About 10 species are reputedly in cultivation; they should be grown in the same manner as *Acineta* (p. 266).
Literature: Dodson, C.H. & Frymire, G.F., Preliminary studies in the genus Stanhopea, *Annals of the Missouri Botanical Garden* **48**: 137–72 (1961); Dodson, C.H., Clarification of some nomenclature in Stanhopea, *Selbyana* **1**: 46–55 (1975).

1a. Lip not divided into 3 parts, forming an irregularly margined pouch **1. ecornuta**
b. Lip clearly divided into 3 parts, not as above 2
2a. Lateral lobes (horns) borne on the basal part of lip **2. grandiflora**
b. Horns borne on the middle part of lip 3
3a. Horns less than 1 cm **3. lewisae**
b. Horns 2–3 cm 4
4a. Basal part of lip globular or saccate, its hollow centre linked to the middle part by a short groove; anterior part usually 3-toothed or 3-lobed **4. insignis***
b. Basal part of the lip oblong or boat-shaped, hollowed only near its base, the hollow linked to the middle part by an elongate groove; anterior part usually entire **5. oculata***

1. S. ecornuta Lemaire. Illustration: Botanical Magazine, 4885 (1855); Fieldiana **26**: 530 (1952); Annals of the Missouri Botanical Garden **48**: 145 (1961); Die Orchidee **27**: 173 (1976).
Leaves 35–40 cm. Racemes with 1–3 flowers. Sepals and petals creamy white, the petals usually spotted with purple towards the base, sepals 4.5–6.5 cm. Lip simple, mostly yellow, forming an irregularly margined pouch, without obvious lateral lobes. *C America, from Guatemala to Costa Rica.* G2. Summer.

2. S. grandiflora (Loddiges) Lindley not Reichenbach (*S. eburnea* Lindley). Illustration: Botanical Magazine, 3359 (1834); Hoehne, Flora Brasilica **12**(6): t. 117 (1942); Annals of the Missouri Botanical Garden **48**: 153 (1961); Couret, Orquideas Venezolanas, 28 (1977).
Leaves to 30 cm. Raceme with 2–4 flowers,

stalks and ovaries with sparse, short, blackish hairs. Sepals *c.* 8 cm, white. Petals white, sometimes spotted with violet around their margins, violet at the base. Lip clearly 3-partite, the basal part purple-spotted, hollowed and bearing small, horn-like lateral lobes, the middle part solid, oblong, whitish spotted and streaked with purple, the anterior part ovate, acute, white. *Tropical S America from Trinidad to N Brazil.* G2. Summer.

3. S. lewisae Ames & Correll. Illustration: Fieldiana **26**: 533 (1952); Annals of the Missouri Botanical Garden **48**: 155 (1961). Leaves 40–50 cm. Racemes with 3–5 flowers. Sepals and petals creamy white with reddish or purple spots, sepals 5–5.7 cm. Lip with a short, hollow, deep yellow, purple-spotted basal part, the middle part bearing horns which are less than 1 cm long on each side, the anterior part broadly ovate, obtuse, white densely spotted with red. *Guatemala.* G2. Summer.

A relatively recently described species whose distribution in cultivation is uncertain. It is possibly a naturally occurring hybrid between *S. ecornuta* and *S. jenishiana* (see Oestereich, Orchidata **7**: 209–10, 1969).

4. S. insignis Hooker. Illustration: Botanical Magazine, 2948, 2949 (1829); Annals of the Missouri Botanical Garden **48**: 158 (1961).
Leaves 20–30 cm. Racemes with 2–5 flowers. Sepals and petals yellowish white spotted with red or purple, the sepals *c.* 7 cm. Basal part of the lip spherical, saccate, purple, linked to the yellow, red-spotted middle part by a very short groove; horns 2–3 cm, borne on the middle part; anterior part broadly ovate, weakly 3-lobed at the apex, yellowish white spotted with red. Column broadly winged. *SE Brazil.* G2. Autumn

S. insignis is the best known of a complex of species which are very difficult to distinguish among themselves. Three other species of this complex are reputedly in cultivation:

***S. hernandezii** (Knuth) Schlechter (*S. devoniensis* Lindley). Sepals 5.5–6.5 cm, yellow or orange-yellow spotted with red; lip white spotted with purple, the anterior part broadly ovate with 3 equal, blunt teeth; column unwinged or very narrowly winged. *Guatemala.* G2. Autumn.

Easily recognised, looking like a smaller version of *S. tigrina.*

***S. saccata** Bateman. Illustration: Fieldiana **26**: 535 (1952); Annals of the Missouri Botanical Garden **48**: 161 (1961). Sepals 5–5.5 cm, greenish white to cream spotted with red or brown; lip orange-yellow spotted with red or brown, the anterior part 3-lobed at the apex, the central lobe shorter than the laterals; column slightly winged. *Guatemala, El Salvador.* G2. Autumn.

***S. tigrina** Lindley. Illustration: Botanical Magazine, 4197 (1845); Annals of the Missouri Botanical Garden **48**: 159 (1961); Die Orchidee **31**: clxv (1980); Bechtel et al., Manual of cultivated orchid species, 275 (1981). Sepals *c.* 9 cm, yellowish white with confluent red-purple spots; lip yellow with violet spots, the anterior part broadly ovate with 3 equal, acute teeth; column very broadly winged. *S Mexico.* G2. Autumn.

Very distinct, with the largest flowers in the genus.

5. S. oculata (Loddiges) Lindley. Illustration: Botanical Magazine, 5300 (1862); Hoehne, Flora Brasilica **12**(6): t. 115 (1942); Sheehan & Sheehan, Orchid genera illustrated, 161 (1979); Bechtel et al., Manual of cultivated orchid species, 275 (1981).
Leaves 30–45 cm. Racemes with 4–10 flowers. Sepals 5.5–7 cm, sepals and petals variable in colour, white, yellow or orange, usually red-spotted. Basal part of the lip oblong when viewed from the side, hollowed only near its base, where the sides are notched, linked to the middle part by an elongate groove; middle part bearing horns which are 2–3 cm long; anterior part broadly ovate, acute or obtuse, not usually 3-lobed. Column winged. *Tropical America from S Mexico to Venezuela and N Peru.* G2. Summer–autumn.

As with *S. insignis*, this is the best known of a complex of species, of which 2 more (and perhaps others) are in cultivation:

***S. jenishiana** Reichenbach (*S. bucephalus* misapplied; *S. graveolens* misapplied; *S. grandiflora* Reichenbach, not (Loddiges) Lindley). Illustration: Botanical Magazine, 5278 (1861), 8517 (1913); Annals of the Missouri Botanical Garden **48**: 165 (1961). Sepals and petals orange-yellow, sparsely purple-spotted, sepals *c.* 7 cm; basal part of lip boat-shaped when viewed from the side, its sides not notched; column winged. *Ecuador to C Peru.* G2. Summer.

***S. wardii** Loddiges. Illustration: Botanical Magazine, 5289 (1862); Fieldiana **26**: 537 (1952); Annals of the Missouri Botanical Garden **48**: 167 (1961); Couret, Orquideas Venezolanas, 29–31 (1977). Sepals and petals orange-yellow with red-purple spots near the base, sepals *c.* 7 cm; basal part of the lip oblong when viewed from the side, its sides not notched; column very broadly winged. *Tropical America, from S Mexico to Venezuela & Peru.* G2. Summer.

149. GONGORA Ruiz & Pavon

J. Cullen

Epiphytic with simple, ovoid, grooved pseudobulbs each usually with 2 leaves. Leaves elliptic to narrowly elliptic, stalked or not, pleated, rolled when young. Racemes hanging, arising from the bases of the pseudobulbs, with many flowers. Flowers often fragrant, resupinate. Sepals spreading, the upper attached to the column for about half its length or more. Petals smaller than the sepals, attached to the side of the column for part of their length. Lip borne on the column foot, complex, fleshy, strongly keeled, divided into 2 parts, the basal part often bearing narrow lateral lobes (horns), the anterior part forwardly directed, usually acuminate. Anther at the apex of the column. Pollinia 2, elongate-ovoid, stipe linear, longer than the pollinia, viscidium narrow. Rostellum prominently beaked.

A genus of about 25 species from Tropical America. A few of them are cultivated in the same manner as *Acineta* (p. 266), for the sake of their bizarre flowers.

1a. Lip without horns, its apex forming an upcurved hook; apex of petals truncate, the angles obtuse or prolonged into curved points **1. galeata**
b. Lip with distinct horns on the basal part, apex acute or acuminate, not upcurved; petals acute 2
2a. Petals very small, less than half the length of the column **2. truncata**
b. Petals more than half the length of the column 3
3a. Sepals yellow with red spots or mostly red; basal part of the lip with 2 spreading lobes at the base as well as 2 horns at its apex **3. quinquenervis**
b. Sepals mostly purple, made up of confluent purple spots on a white or greenish ground; basal part of lip with 2 horns at its apex but without lobes at the base **4. bufonia**

1. G. galeata (Lindley) Reichenbach (*Maxillaria galeata* Lindley; *Acropera loddigesii* Lindley). Illustration: Botanical Magazine, 3653 (1834); Bechtel et al.,

Manual of cultivated orchid species, 215 (1981).
Pseudobulbs *c.* 6 cm, ovoid. Leaves 25–35 cm. Sepals to 2.2 cm, ovate, concave, brownish yellow. Petals brown, joined to the column only at the base, truncate at the apex, the angles obtuse or prolonged into curved points. Lip yellow spotted with red, hornless, the apex prolonged into an upwardly curving, hook-like point. *Mexico*. G2. Summer.

2. G. truncata Lindley. Illustration: Edwards's Botanical Register **31**: t. 56 (1845); Addisonia **2**: t. 46 (1917).
Pseudobulbs 6 cm or more. Leaves to 30 cm. Sepals to 2.3 cm, ovate, pale yellow spotted with purple-brown. Petals very small, less than half the length of the column, acute. Lip yellow, unspotted, the basal part with 2 lobes at its base and 2 upcurving horns at its apex, the anterior part curved, abruptly acute. *S Mexico*. G2. Summer.

3. G. quinquenervis Ruiz & Pavon (*G. maculata* Lindley). Illustration: Botanical Magazine, 3687 (1835); Lindenia **5**: t. 208 (1889); Die Orchidee **30**: cxxiii–cxxiv (1979).
Pseudobulbs 5–12 cm. Leaves 20–60 cm. Sepals 1.5–2.4 cm, yellow with red spots or red, occasionally yellow with whitish spots, lanceolate, acuminate. Petals purplish, acute, attached to the column for about half their length. Lip mostly yellow with 2 spreading lobes near its base and 2 upcurving horns at the apex of the basal part, the anterior part very strongly keeled, acuminate. *C & S America*. G2. Summer.

4. G. bufonia Lindley. Illustration: Edwards's Botanical Register **26**: t. 2 (1841); Hoehne, Flora Brasilica **12**(6): t. 129 (1942).
Pseudobulbs 6 cm or more. Leaves to 30 cm, pale whitish green. Sepals *c.* 2.3 cm, mostly purple, the colour made up of confluent purple blotches on a white or greenish ground. Petals similar in colour to the sepals, attached to the column for about half their length, acute. Lip purple, the basal part with 2 upcurving horns at its apex, the anterior part keeled, acuminate. *S & E Brazil*. G2. Spring–summer.

150. CIRRHAEA Lindley

J. Cullen

Like *Gongora* but pseudobulbs each with 1 leaf, upper sepal and petals free from the back of the column, the lip 3-lobed, the lateral lobes directed backwards, the central lobe narrowly triangular, directed forwards, the whole with a narrow arrowhead shape. Anther on the back of the column.

Cultivation as for *Acineta* (p. 266).

1. C. dependens Reichenbach (*C. tristis* Lindley). Illustration: Hoehne, Iconografia de orchidaceas do Brasil, t. 191 (1949).
Sepals 2–2.5 cm, greenish with purple spots or flushed with purple, petals smaller, similar in colour. Lip violet. *Brazil*. G2. Summer.

151. CORYANTHES Hooker

J. Cullen

Epiphytic, with small, simple, grooved, ovoid or almost spindle-shaped pseudobulbs, each bearing 1–2 leaves. Leaves lanceolate, pleated, rolled when young. Racemes hanging, arising from the bases of the pseudobulbs, with 2–6 flowers. Flowers fragrant. Upper sepal reflexed, smaller than the 2 lateral sepals which are spreading with their apices folded over backwards. Petals erect, smaller than the sepals, twisted or with wavy margins. Lip extremely complex, divided into 3 parts. The basal part, which consists of a stalk borne at an acute angle to the ovary, terminating in a hollow, cup-like structure whose axis is at right angles to the stalk, is confluent with the column foot and bears 2 white, peg-like, nectaries near its base. The middle part, which is narrow and keeled, arises from the margin of the cup. The anterior part is bucket-like, hollowed out and with 3–5 lobes towards the apex, which lies close to the apex of the column. Column slender, its apex reflexed at right angles to the rest. Pollinia 2, ellipsoid, folded backwards along the surface of the curved stipe but arching away from it, viscidium narrowly triangular, almost as long as the stipe.

A genus of about 12 species from C America and the northern part of S America with striking, if bizarre, flowers which are difficult to describe; study of the illustrations cited is helpful in understanding the structure of the lip and column and their spatial relations. Nectar drips from the glands on the basal part of the lip and fills up the bucket-like anterior part, from where it overflows via the apical lobes, which form a spout.

Cultivation as for *Acineta* (p. 266); the plants should not be allowed to dry out completely during the resting period.
Literature: Kennedy, G., Some members of the genus Coryanthes, *Orchid Digest* **42**: 31–7 (1978).

1a. The sides of the middle part of the lip corrugated and ridged; lateral sepals 12–13 cm **1. macrantha**
b. The sides of the middle part of the lip smooth; lateral sepals to 9 cm 2
2a. Basal and middle parts of the lip finely downy outside; anterior part of the lip yellowish brown inside and out **2. speciosa**
b. Basal and middle parts of the lip not downy outside; anterior part of the lip yellow marbled with purple, at least within **3. maculata**

1. C. macrantha (Hooker) Hooker. Illustration: Botanical Magazine, 7692 (1900); Dunsterville & Garay, Venezuelan orchids illustrated **3**: 63 (1961); Couret, Orquideas Venezolanas, 35 (1977); Orchid Digest **42**: 35 (1978).
Pseudobulbs to 12 cm, narrow. Leaves to 30 cm. Raceme usually with 2 flowers. Lateral sepals 12–13 cm, yellow with many elongate red spots. Petals with wavy margins, reddish yellow, red-spotted at the base. Basal part of lip with a purplish stalk *c.* 2.5 cm, the cup orange-red; middle part yellow with red spots or entirely red, its sides ridged and corrugated; anterior part yellow, densely red-spotted. *Tropical S America, from E Peru to Trinidad*. G2. Spring.

2. C. speciosa Hooker. Illustration: Hoehne, Flora Brasilica **12**(6): t. 121, 122 (1942); Fieldiana **26**: 543 (1953); Couret, Orquideas Venezolanas, 34 (1977); Orchid Digest **42**: 35 (1978).
Pseudobulbs 7–15 cm. Leaves 35–55 cm. Raceme with 2–5 flowers. Sepals yellow, the laterals 6–8 cm. Petals yellow, somewhat twisted and with wavy margins. Basal part of the lip with a brownish stalk 1–1.2 cm, and a brownish cup, downy outside; middle part brown, its sides smooth, downy; anterior part pale yellowish brown. *Tropical America, from Guatemala to Peru & E Brazil*. G2. Spring?

3. C. maculata Hooker. Illustration: Botanical Magazine, 3102 (1831), 3747 (1845); Orchid Digest **42**: 32 (1978).
Pseudobulbs to 12 cm. Leaves to 30 cm. Raceme with 3–6 flowers. Sepals *c.* 9 cm, yellowish. Petals yellowish, margins wavy. Basal part of the lip with a yellowish or purplish stalk *c.* 1 cm, the cup yellow, hairless outside; middle part yellow or purplish, its sides smooth, hairless; anterior part yellow outside, yellow marbled with purple or rarely entirely purple within. *Guyana*. G2. Spring–summer.

152. TRICHOCENTRUM Poeppig & Endlicher
J. Cullen
Epiphytic. Pseudobulbs small, simple, to 1 cm, each with 1 leaf. Leaves thick, narrowly elliptic, acute, folded when young. Racemes spreading or arching, arising from the bases of the pseudobulbs, with 1–2 flowers which are 2.5 cm or more in diameter. Sepals and petals similar in size and colour, spreading. Lip spurred, with a narrow basal part which often bears a number of ridges and a spreading, usually broad, notched blade. Column short, without a foot, conspicuously winged, the wings variable in shape, sometimes coarsely fringed. Anther downy. Pollinia 2, ovoid, stipe strap-shaped, longer and broader than the pollinia, viscidium small, circular.

A genus of about 20 species from C & S America. Cultivation as for *Paphinia* (p. 268), but good drainage is essential.

1. T. albo-coccineum Linden (*T. albo-purpureum* Reichenbach; *T. alboviolaceum* Schlechter). Illustration: Botanical Magazine, 5688 (1868); Lindenia **16**: t. 748 (1901); Hoehne, Flora Brasilica **12**(7): t. 165 (1953).
Leaves to 8 cm. Flowers to 5 cm in diameter. Sepals and petals oblong-obovate, olive green outside, brown fading to yellow at the tips inside. Lip notched at the apex, mostly white with 2 reddish violet patches at the sides near the base and with 4 yellow ridges. Spur about a quarter the length of the lip. Wings on column erect, acuminate. *E Peru & Brazil.* G1–2. Summer–autumn.

***T. orthoplectron** Reichenbach. Similar but with leaves 7–14 cm and the lip without ridges but with 5 violet lines. *Brazil.* G1–2.

***T. pfavii** Reichenbach. Illustration: Gardeners' Chronicle **17**: 117 (1882); Illustration Horticole **33**: t. 587 (1886). Flowers slightly smaller with white, brown-spotted sepals and petals, the lip white with a red spot near the base, the spur very short, the wings of the column oblong, truncate. *Costa Rica.* G1–2. Autumn.

***T. tigrinum** Linden & Reichenbach. Illustration: Illustration Horticole **24**: t. 282 (1877); Botanical Magazine, 7380 (1894); Bechtel et al., Manual of cultivated orchid species, 277 (1981). Flowers *c.* 6 cm in diameter, sepals and petals yellow spotted with brown; the lip white with lateral violet spots, 3 yellow ridges and 2 small yellow subulate teeth on the margins near the ridges; the spur very short and the wings of the column coarsely fringed. *C America.* G1–2. Spring.

153. IONOPSIS Humboldt, Bonpland & Kunth
J. Cullen
Small rhizomatous epiphytes. Pseudobulbs simple, to 1.5 cm, each with 1 leaf; some leaves borne directly on the rhizomes. Leaves oblong-linear to narrowly elliptic, fleshy or not, folded when young. Flowers in panicles arising from the apices of the pseudobulbs. Upper sepal free, the lateral sepals united at their bases and forming a spur under the flower. Petals similar to the upper sepal. Lip with a narrow claw which usually bears 2 ridges and a reversed-heart-shaped blade. Column short, unwinged, without a foot. Pollinia 2, ellipsoid, stipe linear, slightly broadened above, longer than the pollinia, viscidium small, oblong, acute, forming an acute angle with the stipe.

A genus of 8 species widely distributed in tropical and subtropical America. Plants can be grown successfully on cork bark or blocks of tree fern hung near the glass in a warm house. Frequent spraying is necessary during the summer. Propagation by division.

1. I. utricularioides (Swartz) Lindley (*I. tenera* (Steudel) Lindley). Illustration: Ames, Orchidaceae **1**: t. 5 (1905); Fieldiana **26**: 598 (1953); Dunsterville & Garay, Venezuelan orchids illustrated **1**: 174–5 (1959); Bechtel et al., Manual of cultivated orchid species, 217 (1981).
Pseudobulbs cylindric. Leaves flat, not very fleshy, narrowly elliptic. Sepals and petals to 8 mm, pinkish white. Lip pinkish white with the claw intense purple, the blade deeply notched. *Tropical and subtropical America from USA (Florida) to Peru and Paraguay.* G2. Spring & autumn.

I. paniculata Humboldt, Bonpland & Kunth, from Brazil, is sometimes treated as a separate species; it is larger than *I. utricularioides*, with oblong, fleshy leaves, sepals and petals to 12 mm, and the blade of the lip less deeply notched.

154. RODRIGUEZIA Ruiz & Pavon
J. Cullen
Epiphytes with short or long rhizomes. Pseudobulbs distant or close, simple, each with 1–3 leaves. Leaves oblong, folded when young. Racemes with few to many flowers, arching or somewhat hanging, arising from the apices of the pseudobulbs. Upper sepal and petals similar, incurved, the lateral sepals fused at their bases, and forming a short spur. Lip directed forwards with a distinct claw and a notched or 2-lobed blade, bearing 2 ridges on the claw and a single appendage projecting back into the sepal spur. Column short with 2 large or small wings at the apex, without a foot. Pollinia 2, broadly ovoid, stipe very narrow, longer than pollinia, viscidium small, oblong, borne at an acute angle to the stipe.

A genus of about 30 species from C & S America, concentrated in Brazil. Cultivation as for *Paphinia* (p. 268), but *R. decora* should be grown on a raft or on a piece of bark. All are subject to attack by red spider mite and thrips.

1a. Rhizome long, pseudobulbs separated; lip with claw longer than blade and bearing 2 elongate, erect, toothed ridges; column wings elongate, ciliate at their ends **1. decora**
b. Rhizome short, pseudobulbs close or touching; lip with claw shorter than blade, the ridges represented by 2 swellings; column wings minute, not ciliate 2
2a. Raceme 1-sided; flowers usually pink, occasionally almost orange or purple **2. secunda**
b. Raceme not 1-sided; flowers white with a yellow blotch on the lip **3. venusta**

1. R. decora (Lindley) Reichenbach. Illustration: Hoehne, Iconografia de Orchidaceas do Brasil, t. 239 (1949); Bechtel et al., Manual of cultivated orchid species, 270 (1981).
Rhizome long, pseudobulbs *c.* 10 cm apart, each compressed and with 1 leaf. Leaves 6–9 cm, oblong. Raceme with 6–10 flowers, arching, loose. Flowers *c.* 5 cm in diameter. Sepals and petals oblong, acute, yellow or white with red-purple spots. Lip much longer than sepals, the claw longer than the blade and bearing 2 toothed erect white- and red-spotted or rarely entirely red ridges; blade white, broader than long, deeply notched. Column wings elongate, ciliate at their ends. *S & C Brazil.* G2. Autumn.

2. R. secunda Humboldt, Bonpland & Kunth. Illustration: Dunsterville & Garay, Venezuelan orchids illustrated **1**: 376–7 (1959); Couret, Orquideas Venezolanas, 58–9 (1977); Sheehan & Sheehan, Orchid genera illustrated, 153 (1979); Bechtel et al., Manual of cultivated orchid species, 270 (1981).

Rhizomes short, pseudobulbs close or touching, each with 2–3 leaves. Leaves to 15 cm, oblong. Raceme arching, 1-sided, with many closely packed flowers. Sepals and petals ovate, pink, rarely orange or purple. Lip scarcely longer than the sepals, pink, with a short claw gradually broadening into the oblong blade which is slightly notched; ridges on the claw in the form of 2 swellings. Wings of the column minute, not ciliate. *Tropical America, from Panama to Ecuador, Trinidad & French Guiana*. G2. Spring–summer.

3. **R. venusta** (Lindley) Reichenbach (*R. fragrans* Lindley). Illustration: Hoehne, Iconografia de Orchidaceas do Brasil, t. 240 (1949).
Similar to *R. decora* but with very fragrant flowers with narrower, acute, white sepals and petals, the lip white with a yellow centre, its blade almost circular, all borne in a raceme which is not 1-sided. *E Brazil*. G2. Autumn.

155. COMPARETTIA Poeppig & Endlicher

J. Cullen

Rhizomatous epiphytes. Pseudobulbs small, simple, close or touching, each with 1 leaf. Leaves narrowly elliptic, folded when young. Racemes arising from the bases of the pseudobulbs, arching or somewhat hanging, loose, with 6–10 flowers. Upper sepal and petals similar or the petals slightly broader, erect; lateral sepals united at the base and prolonged into a backwardly pointing spur. Lip much larger than the petals, with a short claw and broad blade, and with 2 appendages projecting back into the sepal spur. Column unwinged and without a foot. Pollinia 2, ovoid, stipe longer than pollinia, broadened above, viscidium small, circular or heart-shaped.

A genus of about 8 species from tropical America. Cultivation as for *Rodriguezia* (p. 271).

1a. Sepals and petals more than 1.5 cm; spur 3–5 cm **3. macroplectrum**
b. Sepals and petals to 1.5 cm; spur 1–2 cm 2
2a. Sepals and petals rose-pink to violet **2. falcata**
b. Sepals and petals bright shining scarlet throughout **1. coccinea**

1. **C. coccinea** Lindley. Illustration: Edwards's Botanical Register **24**: t. 68 (1838); Hoehne, Iconografia de Orchidaceas do Brasil, t. 238 (1949); Bechtel et al., Manual of cultivated orchid species, 188 (1981).
Pseudobulbs 2–2.5 cm. Leaves to 10 cm, oblong, acute. Flowers bright shining scarlet. Sepals and petals *c*. 1.3 cm. Lip yellow outside. Spur to 1.8 cm. *S & C Brazil*. G2. Summer.

2. **C. falcata** Poeppig & Endlicher (*C. rosea* Lindley). Illustration: Botanical Magazine, 4980 (1857); Fieldiana **26**: 603 (1953); Dunsterville & Garay, Venezuelan orchids illustrated **1**: 86–7 (1959); Bechtel et al., Manual of cultivated orchid species, 188 (1981).
Vegetatively very like *C. coccinea*. Sepals and petals *c*. 1.3 cm, rose-pink to violet. Lip deeper violet than the petals inside, pink or violet outside. Spur to 2 cm. *Tropical America, from S Mexico & the West Indies to Bolivia*. G2. Winter.

3. **C. macroplectrum** Reichenbach & Triana. Illustration: Botanical Magazine, 6679 (1883); Lindenia **14**: t. 664 (1899); Sheehan & Sheehan, Orchid genera illustrated, 69 (1979); Bechtel et al., Manual of cultivated orchid species, 188 (1981).
Vegetatively like *C. coccinea*. Flowers clear violet with purple spots. Sepals and petals 1.5–2 cm. Lip violet outside. Spur 3–5 cm, very conspicuous. *Tropical S America from Ecuador to N Brazil*. G2. Autumn.

156. TRICHOPILIA Lindley

J. Cullen

Epiphytic. Pseudobulbs simple, small or large, each with 1 leaf. Leaves more or less elliptic or elliptic–oblong, acute, folded when young. Racemes arising from the bases of the pseudobulbs, arching or somewhat pendent with 1–5 flowers which are often fragrant. Sepals and petals similar, spreading, narrow, sometimes twisted. Lip rolled around the column at its apex, forming a tube, obscurely 3-lobed, margins wavy. Column elongate, without a foot, with 3 fringed lobes at the apex. Pollinia 2, ovoid to almost spherical, stipe linear, slightly broadened above, longer than pollinia, viscidium small, forming an acute angle with the stipe.

A genus from C and northern S America of an uncertain number (15–30?) of doubtfully distinguished species. Various species, formerly included in the genus, have now been placed elsewhere (see Garay, Orquideologia 7: 191–6, 1972). Five species of the reduced genus are found in cultivation.

They are grown in epiphytic compost which must be well drained in pots or tied on to sections of bark. Only occasional watering is required in the winter. Propagation by division.

1a. Petals and sepals conspicuously twisted **4. tortilis**
b. Petals and sepals not twisted 2
2a. Pseudobulbs almost circular in outline, 3–4 cm; sepals and petals 6–7 cm **1. suavis**
b. Pseudobulbs elongate in outline, 5–13 cm; sepals and petals 3–6 cm 3
3a. Lip red within **5. marginata**
b. Lip white with a yellow blotch and/or red-brown spots within 4
4a. Sepals and petals 5.5–6 cm; lip white with a yellow blotch within **2. fragrans**
b. Sepals and petals to 5 cm; lip white with a faint yellow blotch and red-brown spots within **3. galeottiana**

1. **T. suavis** Lindley & Paxton. Illustration: Botanical Magazine, 4654 (1852); Lindenia **9**: t. 423 (1893); Sheehan & Sheehan, Orchid genera illustrated, 165 (1979); Bechtel et al., Manual of cultivated orchid species, 277 (1981).
Pseudobulbs 3–4 cm, almost circular in outline, compressed. Leaves to 20 cm, elliptic, broadened abruptly to almost cordate at the base. Raceme with 2–3 flowers. Sepals and petals 6–7 cm, white, sometimes faintly yellowish or greenish, rarely red-spotted. Lip white with a yellow blotch inside and red spots on the central lobe, or rarely (var. **alba** Anon.) without spots. *Costa Rica, Panama & Colombia*. G2. Spring.

2. **T. fragrans** (Lindley) Reichenbach. Illustration: Cogniaux & Goossens, Dictionnaire Iconographique des Orchidées, Trichopilia, t. 3 (1900); Dunsterville & Garay, Venezuelan orchids illustrated **3**: 316–17 (1965); Bechtel et al., Manual of cultivated orchid species, 277 (1981).
Pseudobulbs 10–13 cm, elongate in outline, compressed. Leaves to 17 cm, elliptic, tapered to the base. Racemes with 2–5 flowers. Sepals and petals 5.5–6 cm, white or greenish white. Lip white with a yellow blotch inside. *Tropical America from the West Indies to Bolivia*. G1–2. Winter.

3. **T. galeottiana** A. Richard (*T. picta* Lemaire). Illustration: Illustration Horticole **6**: t. 225 (1859); Cogniaux & Goossens, Dictionnaire Iconographique des Orchidées, Trichopilia, t. 5 (1900).
Like *T. fragrans*, but racemes with 2–3 flowers, sepals and petals up to 5 cm,

greenish, sometimes with brown lines, lip white outside, yellowish or whitish with red-brown spots inside. *Mexico*. G2. Summer.

4. T. tortilis Lindley. Illustration: Botanical Magazine, 3739 (1839); Cogniaux & Goossens, Dictionnaire Iconographique des Orchidées, Trichopilia, t. 6 (1900); Bechtel et al., Manual of cultivated orchid species, 278 (1981).
Pseudobulbs 4–8 cm, elongate in outline, compressed. Leaves 10–15 cm, elliptic, tapered to the base. Raceme with 1–2 flowers. Sepals and petals up to 6 cm, conspicuously twisted, green, striped and spotted with red-brown. Lip white with red-brown spots inside. *C America from S Mexico to Honduras*. G2. Winter.

An albino variant, with pure white flowers, has been in cultivation.

5. T. marginata Henfrey (*T. coccinea* Lindley; *T. crispa* Lindley). Illustration: Paxton's Flower Garden **2**: t. 54 (1851–52); Botanical Magazine, 4857 (1855); Cogniaux & Goossens, Dictionnaire Iconographique des Orchidées, Trichopilia, t. 1, 2, 2A (1900).
Pseudobulbs 5–7 cm, elongate in outline, compressed. Leaves to 17 cm, elliptic, narrowed to the base. Raceme with 2–3 flowers. Flowers rather variable in colour. Sepals and petals to 6 cm, white or greenish white, variously tinged, spotted or lined with red. Lip white outside, mostly red within, the margin sometimes white. *C. America, Colombia*. G2. Spring.

157. COCHLIODA Lindley

J. Cullen

Epiphytic, with short rhizomes. Pseudobulbs simple, compressed, each with 1 or 2 leaves. Leaves linear to oblong, usually blunt, folded when young. Flowers in erect, arching or hanging racemes or rarely panicles, arising laterally. Sepals and petals spreading. Lip with an erect claw which is fused to the column base and a spreading and usually 3-lobed blade bearing 2 or 4 swellings near the base, the lateral lobes rounded and often reflexed, the central lobe pointing forwards, often laterally expanded from a narrow base, notched or 2-lobed. Column erect, curved, without a foot, with 2–3 lobes at the apex; stigmatic surface divided into 2. Pollinia 2, broadly ovoid, stipe linear, somewhat longer than the pollinia, viscidium oblong, almost as long as the stipe and forming an acute angle with it.

A genus of about 6 species from S America. Cultivation as for *Odontoglossum* (p. 274).

1a. Lateral sepals united; lip without obvious lateral lobes, tapering to its tip **4. sanguinea**
b. Lateral sepals free; lip with obvious lateral lobes, the central either oblong or enlarged laterally from a narrow base 2
2a. Bract equalling the ovary and its stalk; flower white or cream **2. densiflora**
b. Bract much shorter than the ovary and its stalk; flowers red to rose-purple 3
3a. Central lobe of lip oblong, not expanded laterally from a narrow base; column 3-lobed at apex **3. rosea**
b. Central lobe of lip expanded laterally from a narrow base; column finely toothed at apex **1. vulcanica***

1. C. vulcanica (Reichenbach) Veitch. Illustration: Botanical Magazine, 6001 (1872); Cogniaux & Goossens, Dictionnaire Iconographique des Orchidées, Cochlioda, t. 1 (1898).
Pseudobulbs 2.5–6 cm, clustered, compressed, each with 2 leaves. Leaves 7.5–15 × 1.5–3.5 cm, elliptic-oblong to oblong-linear. Raceme with 6–many flowers, almost erect. Bract shorter than the ovary. Flowers *c.* 4 cm in diameter, rosy purple. Sepals and petals *c.* 2.2 cm, lanceolate-oblong, acute. Lip fused to the lower half of the column, 3-lobed, the central lobe with a narrow base and laterally expanded, 2-lobed, apical part. Column red, finely toothed at the apex. *Ecuador, Peru*. G1. Autumn–winter.

C. noezliana Rolfe. Flowers scarlet with yellow blotches on the lip, the column 3-lobed at the apex. *Peru, Bolivia*.

2. C. densiflora Lindley.
Pseudobulbs to 5 cm, compressed, each with 1 leaf. Leaves to 14 × 1.8 cm, oblong. Racemes with *c.* 9 flowers, almost erect. Bract as long as the ovary and its stalk. Flowers *c.* 3 cm in diameter, white or cream. Sepals and petals *c.* 1.5 cm, the lateral sepals narrower than the rest. Lip fused to the lower half of the column, 3-lobed, the central lobe with a narrow base and laterally expanded, 2-lobed apical part. Column with 2 auricles at the apex. *Peru*. G1.

3. C. rosea (Lindley) Bentham. Illustration: Cogniaux & Goossens, Dictionnaire Iconographique des Orchidées, Cochlioda, t. 3 (1898); Bechtel et al., Manual of cultivated orchid species, 185 (1981).
Pseudobulbs 3–5 cm, compressed, clustered, each with 1–2 leaves. Leaves 5.5–20 × 1–2.5 cm, elliptic-oblong to linear. Raceme with up to 20 flowers, erect or arched, rarely branched above. Bract much shorter than the ovary. Flowers 2–3.5 cm in diameter, rose-red. Sepals and petals *c.* 1.1 cm, narrowly elliptic to lanceolate. Lip fused to the lower third of the column, 3-lobed, the central lobe oblong. Column 3-lobed at the apex. *Ecuador, Peru*. G1. Spring.

The name *C. stricta* Cogniaux has been applied to plants resembling *C. rosea*, but thought to originate in Colombia. It is not certain what these plants are, or whether any is in cultivation now.

4. C. sanguinea (Reichenbach) Bentham.
Pseudobulbs 3–5 cm, ovoid, compressed, each with 2 leaves. Leaves to 25 × 1.5 cm, linear. Raceme with many flowers, almost erect, often branched above. Bract much shorter than the ovary. Flowers *c.* 3 cm in diameter, red, the apex of the column and the base of the lip white. Sepals and petals *c.* 1.5 cm, the lateral sepals united for about half their length. Lip small with scarcely perceptible lateral lobes and a tapering central lobe, fused to the lower half of the column. Column weakly 3-lobed at apex. *Ecuador, Peru*. G2. Autumn.

Sometimes placed in the genus *Symphyglossum*, where its correct name is *S. sanguinea* (Reichenbach) Schlechter.

158. GOMESA R. Brown

J. Cullen

Epiphytic, with short rhizomes. Pseudobulbs simple, clustered, oblong, compressed, each with 2 leaves. Leaves oblong or very narrowly elliptic, folded when young. Racemes arising from the bases of the pseudobulbs, arching, many-flowered. Sepals and petals spreading, greenish yellow, the upper sepal and the petals erect, the lateral sepals fused at the bases to some extent, spreading downwards. Lip greenish yellow, shorter than the sepals, with an erect basal part and a downwardly deflected, ovate or oblong-ovate blade which bears 2 large, finely toothed ridges from the base to the middle, which reach up to and slightly clasp the column. Column white, reddish around the rostellum, erect, without a foot. Pollinia 2, ovoid, stipe oblong, shorter than pollinia, viscidium small, making an acute or right angle with the stipe.

A genus of about 8 species from S America, mostly from Brazil. They are grown in a similar manner to *Odontoglossum* but require more heat in winter.

1a. Lateral sepals united only at the extreme base; blade of lip oblong-ovate **1. crispa**
b. Lateral sepals united for more than half their length; blade of lip ovate **2. planifolia**

1. G. crispa (Lindley) Klotzsch & Reichenbach. Illustration: Pabst & Dungs, Orchidaceae Brasilienses 2: f. 1914 (1977); Sheehan & Sheehan, Orchid genera illustrated, 97 (1979); Bechtel et al., Manual of cultivated orchid species, 215 (1981).
Pseudobulbs 6–10 cm. Raceme many-flowered. Flowers *c.* 2 cm in diameter. Sepals and petals 9–10 mm, wavy, oblong, rounded at the apex, the lateral sepals united only at the base. Lip oblong-ovate. *Brazil.* G1. Spring.

2. G. planifolia (Lindley) Klotzsch & Reichenbach (*G. recurva* Loddiges, not R. Brown). Illustration: Loddiges' Botanical Cabinet 7: t. 660 (1822); Botanical Magazine, 3504 (1836); Pabst & Dungs, Orchidaceae Brasilienses 2: f. 1922 (1977).
Very like *G. crispa*, but sepals and petals scarcely wavy, rather broad, the lateral sepals united for about two-thirds of their length, the blade of the lip ovate, rounded. *Brazil, Paraguay, Argentina.* G1. Spring–summer.

The flowers of this species are said to be fragrant.

159. HELCIA Lindley
J. Cullen
A genus related to *Trichopilia* but keying near *Gomesa*, differing from the latter in its 1-leaved pseudobulbs, large flowers *c.* 7 cm in diameter, and the 3-lobed lip, whose small lateral lobes reach up to the column and whose central lobe is much larger, conspicuously wavy and without ridges extending to the column.

Cultivation as for *Odontoglossum*.

1. H. sanguinolenta Lindley (*Trichopilia sanguinolenta* (Lindley) Reichenbach). Illustration: Botanical Magazine, 7281 (1893); Bechtel et al., Manual of cultivated orchid species, 216 (1981).
Sepals and petals yellow with large brown spots, lip white with violet or purple spots. *Ecuador.* G2. Winter.

160. ODONTOGLOSSUM Humboldt, Bonpland & Kunth
J. Cullen
Rhizomatous epiphytes. Pseudobulbs simple, compressed, each bearing 1–3 leaves. Leaves often produced on the rhizome as well as on the pseudobulbs, variable in shape, folded when young. Flowers showy, in racemes or panicles, sometimes fragrant, bracts often conspicuous. Sepals spreading, free or the lateral sepals united for a short distance at the base. Petals similar to, or shorter than the sepals. Lip complex, usually shorter than the sepals, with a short claw ascending parallel to the column (sometimes united to it) with or without small lateral lobes and with complex keels and swellings and a variably shaped blade which is bent backwards near its base. Column long and slender, not thickened at the base, without a foot, often with auricles near the apex. Pollinia 2, ellipsoid, stipe linear or linear-oblong, somewhat longer than pollinia, viscidium almost as long as the stipe and forming an acute angle with it, narrowly oblong or triangular.

A large genus of about 200 species from C & S America, in need of further study. A very large number of species has been available in cultivation, but only about 20 are widely available today. Many hybrids have been raised in the past, and many species are variable in flower colour, which has formed the basis for the recognition of numerous cultivars in some species. Species of *Odontoglossum* have been hybridised with species of several other genera.

Plants should be grown in an epiphytic compost. They require humid conditions combined with good ventilation, and may require shading during the hottest months. Watering should be reduced in winter. Propagation is by division, which is sometimes difficult, or by seed.
Literature: Escobar, R., El Genero Odontoglossum, *Orquideologia* **11**: 21–57, 119–60, 257–302 (1976); Halbinger, F., Odontoglossum y Generos Afines en Mexico y Centroamerica, *Orquidea* **8**: 155–282 (1982).

Rhizome. Long, pseudobulbs distant: **7,18**; short, pseudobulbs clustered: **1–6,8–17, 19–20.**
Pseudobulbs. Usually bearing 1 leaf: **1–5, 7,8**; usually bearing 2–3 leaves: **6, 9–20.**
Inflorescence. Panicle: **2,9–11,19**; raceme: **1,3–8,11–18,20.**
Lip. (Figure 19, p. 275.) Blade waisted: **4,7–13,20**; blade not waisted: **1–3,5,6, 14–20.** Blade longer than broad: **1–13, 20**; blade broader than long: **14–19.** Blade distinctly notched at the apex: **7,15,18.**
Column. Without auricles: **1,3,5,6,10,18,19**; with flap-like, entire auricles: **4,13,14,16,17**; with flap-like, toothed or finely toothed auricles: **7,9,11,12,15,20**; with hair-like auricles: **2**; with auricles divided into a number of tooth-like processes: **8.**

1a. Pseudobulbs mostly bearing 1 leaf 2
b. Pseudobulbs mostly bearing 2–3 leaves 8
2a. Column with auricles near the apex 3
b. Column without auricles 6
3a. Sepals and petals conspicuously undulate-crisped; rhizome extensive, pseudobulbs distant **7. brevifolium**
b. Sepals and petals not undulate-crisped: rhizome short, pseudobulbs clustered 4
4a. Lip narrowing from just above its base (Figure 19(2), p. 275); column auricles formed by 2 hair-like processes **2. cirrhosum**
b. Lip broadening from just above the base; column auricles various, not as above 5
5a. Blade of lip toothed to lacerate, usually waisted in the middle, the base bearing a keel formed from several tooth-like processes (Figure 19(8), p. 275) **8. hallii**
b. Blade of lip entire, not waisted, the base bearing 2 entire keels (Figure 19(4), p. 275) **4. cervantesii**
6a. Blade of lip narrowly triangular to heart-shaped, acuminate (Figure 19(1), p. 275); raceme with more than 4 flowers **1. cordatum**
b. Blade broadly ovate or circular, rounded to the apex; raceme with 1–4 flowers 7
7a. Blade of lip irregularly toothed or lacerate (Figure 19(3), p. 275) **3. stellatum**
b. Blade entire though wavy (Figure 19(5), p. 275) **5. rossii**
8a. Column without auricles 9
b. Column with auricles 12
9a. Blade of lip distinctly waisted just below its middle (Figure 19(10), p. 275) **10. reichenheimii**
b. Blade various, not waisted as above 10
10a. Blade of lip longer than broad,

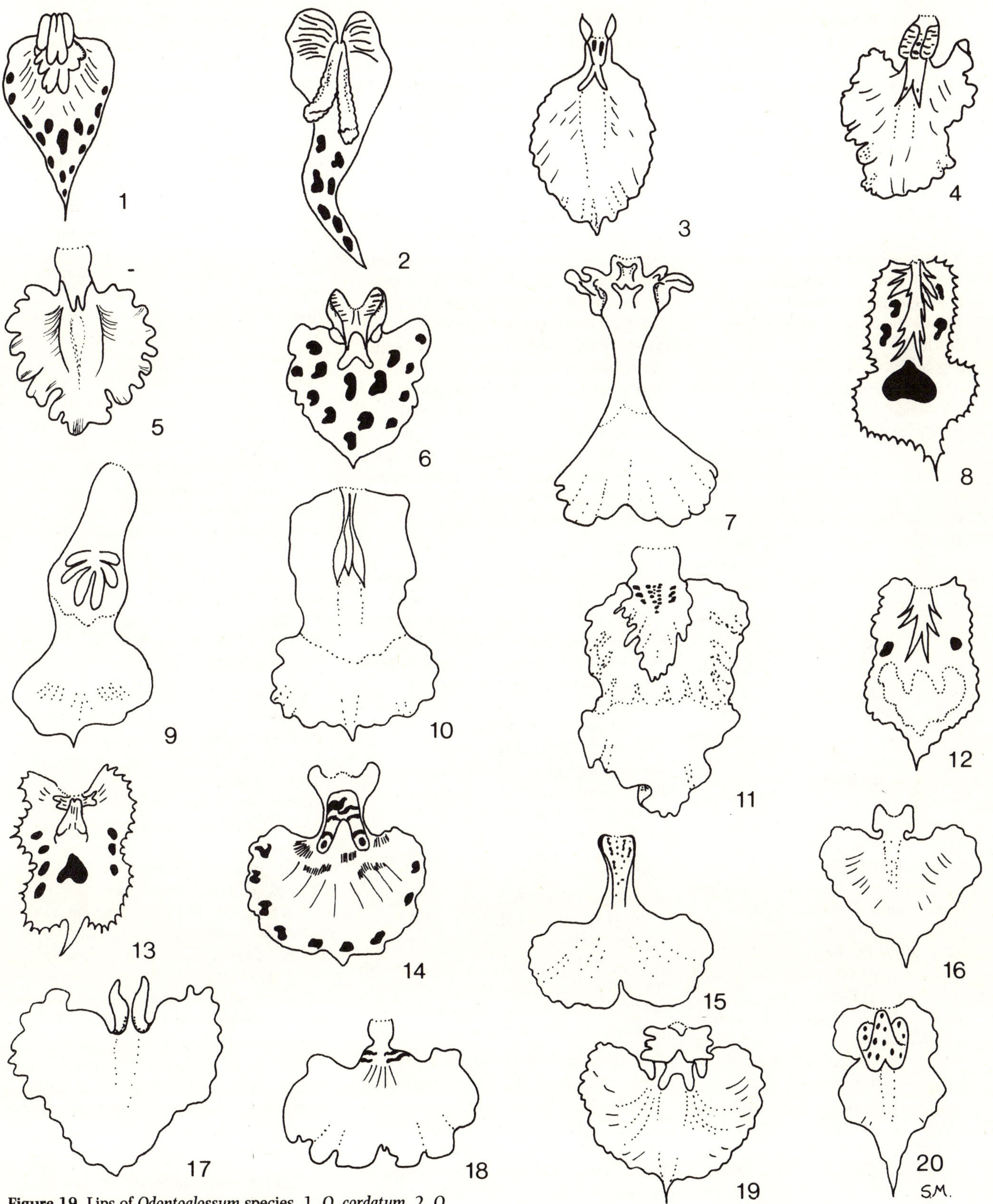

Figure 19. Lips of *Odontoglossum* species. 1, *O. cordatum*. 2, *O. cirrhosum*. 3, *O. stellatum*. 4, *O. cervantesii*. 5, *O. rossii*. 6, *O. nebulosum*. 7, *O. brevifolium*. 8, *O. hallii*. 9, *O. laeve*. 10, *O. reichenheimii*. 11, *O. harryanum*. 12, *O. spectatissimum*. 13, *O. crispum*. 14, *O. grande*. 15, *O. pendulum*. 16, *O. bictoniense*. 17, *O. uroskinneri*. 18, *O. londesboroughianum*. 19, *O. cariniferum*. 20, *O. pulchellum*. (Compiled from various sources; not to scale.)

tapering to the apex, not notched (Figure 19(6), p. 275) **6. nebulosum**
- b. Blade broader than long, notched at the apex 11
- 11a. Blade of lip shining yellow barred with brown at the base, sometimes with additional brown spots on the surface **18. londesboroughianum**
- b. Blade mostly white, purple at the base **19. cariniferum**
- 12a. Blade of the lip broader than long, usually notched at the apex 13
- b. Blade longer than broad or as broad as long, tapering to the apex, not notched 14
- 13a. Sepals *c.* 2.5 cm, petals similar to sepals, white or pink, sometimes with a few reddish spots; inflorescence usually with more than 8 flowers **15. pendulum**
- b. Sepals 5–8.5 cm, petals shorter and broader, all yellow, the sepals barred with brown, the petals brown in the lower half; raceme with 4–8 flowers **14. grande**
- 14a. Blade of the lip triangular to heart-shaped, tapering gradually to the apex 15
- b. Blade of the lip oblong, waisted or not, tapering abruptly to the apex 16
- 15a. Blade of the lip 2.8 × 2.8 cm or more, pink, irregularly veined with white; petals ovate-elliptic **17. uroskinneri**
- b. Blade up to 2 × 2.5 cm, violet, white or pink; petals elliptic-lanceolate to lanceolate **16. bictoniense**
- 16a. Flowers to 3 cm in diameter, pure white except for the yellow, red-spotted keels on the lip; leaves oblong-linear **20. pulchellum**
- b. Combination of characters not as above 17
- 17a. Blade of the lip distinctly waisted 18
- b. Blade oblong, not waisted 19
- 18a. Sepals and petals to 3.5 cm, yellow or green with brown bars or red spots; panicle many-flowered **9. laeve***
- b. Sepals and petals at least 4 cm, dark brown with yellow veining; raceme with 4–12 flowers **11. harryanum**
- 19a. Sepals and petals white or pink, spotted with brown, elliptic, overlapping, the petals finely toothed **13. crispum**
- b. Sepals and petals yellow with brown blotches, oblong-lanceolate or oblanceolate, not overlapping, petals wavy, not finely toothed **12. spectatissimum**

1. O. cordatum Lindley. Figure 19(1), p. 275. Illustration: Lindenia **9**: t. 430 (1893); Hamer, Orquideas de El Salvador **2**: 153 & t. 17 (1974); Williams et al., Orchids for everyone, 139 (1980); Bechtel et al., Manual of cultivated orchid species, 236 (1981).
Rhizome short, pseudobulbs clustered, 4.5–7.5 cm, each with 1 leaf. Leaves 9–30 × 3–4.5 cm, narrowly elliptic, acute. Raceme erect with few to many flowers. Sepals 3.5–5 cm × 6–12 mm, lanceolate, acuminate, petals 2.5–4 cm × 7–11 mm, ovate-lanceolate to elliptic-lanceolate, acuminate; all greenish, whitish or yellowish blotched with brown. Lip with a short claw with complex yellow keels which extend on to the blade; blade narrowly triangular to heart-shaped, acuminate, finely scalloped, white at the base with pink around the keels, the apex yellow spotted with brown. Column very shortly downy, without auricles. *C. America, Venezuela.* G1. Summer.

2. O. cirrhosum Lindley. Figure 19(2), p. 275. Illustration: Botanical Magazine, 6317 (1877); Gartenflora **41**: t. 1383 (1892); Cogniaux & Goossens, Dictionnaire Iconographique des Orchidées, Odontoglossum, t. 9 (1897); Bechtel et al., Manual of cultivated orchid species, 236 (1981).
Rhizome short, pseudobulbs clustered, 5–8 cm, each with 1 leaf. Leaves 10–30 × 2.5–3 cm, oblong-linear. Panicle arching, with many flowers. Sepals *c.* 4 cm × 7 mm, lanceolate, acuminate, terminating in a recurved point, petals broader and shorter; all white or cream with red-brown or red-purple blotches. Lip with a short claw bearing 2 small lateral lobes, yellow lined with red and with 2 forwardly projecting keels; blade white spotted reddish, narrowly heart-shaped to triangular, long-acuminate. Column bearing 2 hair-like auricles near the apex. *Ecuador, Peru.* G1. Spring.

3. O. stellatum Lindley. Figure 19(3), p. 275. Illustration: Hamer, Orquideas de El Salvador **2**: 159 & t. 18 (1974).
Rhizome short, pseudobulbs clustered, 2–6 cm, each with 1 leaf. Leaves 6.5–15 cm × 8–25 mm, ovate to elliptic or oblanceolate. Raceme erect with 1–2 flowers. Sepals and petals similar, 2–2.8 cm × 3–5 mm, yellowish barred with brown, sometimes entirely brown, lanceolate, acute. Lip with a narrow claw bearing keels which extend onto the base of the blade as a short bifid plate; blade white or pink tinged with mauve, triangular-ovate, obtuse, margin toothed to lacerate. Column papillose, without auricles. *Mexico, Guatemala, El Salvador.* G1. Winter–spring.

4. O. cervantesii Llave & Lexarza. Figure 19(4), p. 275. Illustration: Cogniaux & Goossens, Dictionnaire Iconographique des Orchidées, Odontoglossum, t. 16 (1897); Journal of the Orchid Society of Great Britain **20**: 9 (1971); Williams et al., Orchids for everyone, 138 (1980); Bechtel et al., Manual of cultivated orchid species, 236 (1981).
Rhizome short, pseudobulbs clustered, 2–6.5 cm, each with 1 leaf. Leaves 4–30 × 1–3.5 cm, variable, ovate-lanceolate to elliptic-oblong. Raceme with 1–6 flowers, erect or arching, loose. Flowers fragrant. Sepals lanceolate, petals elliptic, otherwise similar, 1.8–3 cm × 6–20 mm, white to pink, marked in the lower third with concentrically arranged red spots and bars. Lip with a thick claw, white or pink, purple striped at the sides and with a yellow, papillose callus extending on to the blade as 2 points; blade heart-shaped, irregularly toothed, with a few spots. Column papillose with 2 circular auricles. *S Mexico, Guatemala.* G1. Winter–spring.

5. O. rossii Lindley. Figure 19(5), p. 275. Illustration: Edwards's Botanical Register **25**: t. 48 (1839); Cogniaux & Goossens, Dictionnaire Iconographique des Orchidées, Odontoglossum, t. 6, 6A (1897).
Rhizome short, pseudobulbs loosely clumped, 3–6 cm, each with 1 leaf. Leaves 5–20 × 1.5–4 cm, elliptic or elliptic-lanceolate. Raceme with 2–4 flowers, erect or arched. Sepals 2.5–4.5 cm × 5–11 mm, oblong-elliptic to oblong-lanceolate, acute or acuminate, petals 2.5–3.8 cm × 8–19 mm, broadly elliptic to oblong-elliptic, obtuse to acute; all yellow, white or pink, the sepals and the bases of the petals spotted with brown. Lip with a narrow claw bearing a yellow, brown-spotted callus which is notched at the apex with divergent tips; blade circular or nearly so to ovate, white or pink, wavy. Column without auricles. *C America southwards to Nicaragua.* G1. Spring.

6. O. nebulosum Lindley (*O. apterum* misapplied). Figure 19(6), p. 275. Illustration: Lindenia **8**: t. 350 (1892).
Rhizome short, pseudobulbs clustered, to 10 cm, each with 2 leaves. Leaves to 20 × 6 cm, elliptic or narrowly elliptic. Raceme with 5–7 flowers, loose, arching.

Sepals and petals *c.* 3 cm, ovate, sepals acute, petals broader and more rounded at the apex, all white with red or greenish spots in the lower half. Lip shortly clawed, the claw bearing 2 yellow keels which project as divergent teeth over the base of the triangular (longer than broad), acute, slightly wavy, white, red- or yellow-spotted blade. Column without auricles. *Mexico.* G1. Summer.

7. O. brevifolium Lindley (*O. chiriquense* Reichenbach; *O. coronarium* Lindley; *Otoglossum brevifolium* (Lindley) Garay & Dunsterville). Figure 19(7), p. 275. Illustration: Cogniaux & Goossens, Dictionnaire Iconographique des Orchidées, Odontoglossum, t. 18 (1897); Botanical Magazine, 7687 (1899), 8725 (1917).
Rhizome creeping, pseudobulbs distant, 4–11 cm, each with 1 leaf. Leaves 10–30 × 6–9 cm, ovate to elliptic. Raceme with few to many flowers, erect or arching. Sepals and petals similar, 1.6–3 × 1.6–2 cm, obovate or obovate-oblong, obtuse, brown edged with yellow or yellow blotched with brown, margins conspicuously undulate-crisped. Lip with an indistinct claw bearing small lateral lobes and irregular keels, the blade oblong, usually waisted, whitish or yellow with transverse brown bars near the base. Column with conspicuous, variably developed auricles which are finely toothed. *Tropical America from Panama to Peru.* G1. Flowering irregular.

8. O. hallii Lindley. Figure 19(8), p. 275. Illustration: Botanical Magazine, 6237 (1876); Lindenia **4**: t. 158 (1888); Bechtel et al., Manual of cultivated orchid species, 237 (1981).
Rhizome short, pseudobulbs 5–10 cm, each with usually 1 leaf. Leaves 13.5–30 × 2–4.5 cm, elliptic-oblong. Raceme with many flowers, loose, arching. Sepals 3.5–6 cm, ovate-lanceolate, acuminate, petals shorter and broader; all yellow with large purple-brown blotches. Lip with a short claw bearing a large crest of several yellow, tooth-like keels which extend on to the blade; blade white, variously spotted with red or purple, oblong, usually waisted, toothed to lacerate. Column with auricles divided into a number of tooth-like processes. *Colombia, Ecuador, Peru.* G1. Spring.

9. O. laeve Lindley. Figure 19(9), p. 275. Illustration: Botanical Magazine, 6265 (1876); Bechtel et al., Manual of cultivated orchid species, 237 (1981).
Rhizomes short, pseudobulbs clustered, 5–12 cm, each with 2 leaves. Leaves 15–45 × 2.5–5.5 cm, oblong to narrowly elliptic, rounded to almost acute. Panicle with many flowers, arching. Sepals and petals 2.5–3.5 cm × 6–9 mm, linear-elliptic to linear–oblanceolate, yellowish blotched and barred with reddish brown. Lip with a short, thickened claw bearing 2–5 inconspicuous keels which run on to the blade, blade oblong, waisted, the basal part pale purple, the rest white, apex with a short point. Column without a ledge below the stigmatic surface and with finely toothed auricles. *S Mexico, Guatemala.* G1. Summer.

***O. confusum** Garay (*O. schroederianum* Reichenbach; *Miltonia schroederiana* (Reichenbach) Veitch). Illustration: Cogniaux & Goossens, Dictionnaire Iconographique des Orchidées, Miltonia, t. 3 (1897). Similar, but with green, red-spotted sepals and petals, the lip red in the lower half and the column with a transverse ledge below the stigmatic surface. *Costa Rica.* G1. Winter.

10. O. reichenheimii Linden & Reichenbach. Figure 19(10), p. 275. Illustration: Bechtel et al., Manual of cultivated orchid species, 239 (1981).
Very similar to *O. laeve* but the lip as in figure 19(10) and the column entirely without auricles. *Mexico.* G1. Summer.

For further information on the complex involving *O. laeve, O. reichenheimii* and *O. confusum* see Garay, Die Orchidee **6**: 213–18 (1962), and Garay & Stacy, Bradea **1**(40): 393–424 (1974) in which all these species are considered to form part of the genus *Oncidium*.

11. O. harryanum Reichenbach. Figure 19(11), p. 275. Illustration: Lindenia **3**: t. 142 (1887); Cogniaux & Goossens, Dictionnaire Iconographique des Orchidées, Odontoglossum, t. 11 (1897); Williams et al., Orchids for everyone, 140 (1980).
Rhizome short, pseudobulbs clustered, 6–8 cm, each bearing 2 leaves. Leaves to 45 × 2–4 cm, oblong to oblong-elliptic, acute to obtuse. Raceme with 4–12 flowers. Sepals 4.5 × 1.3–2.5 cm, elliptic-oblong to elliptic, acute, petals somewhat smaller; all brown to dark brown, mottled darker, with yellow veins and with purple lines on a white ground at the bases of the petals. Lip with a short claw with a prominent yellow callus which is lacerate towards the front, blade oblong, waisted below the middle, yellowish at the base, the sides and the centre with purple lines on a white or yellow ground, the apical part white. Column with small, finely toothed auricles. *Colombia, Peru.* G1. Summer–autumn.

12. O. spectatissimum Lindley (*O. triumphans* Reichenbach). Figure 19(12), p. 275. Illustration: Cogniaux & Goossens, Dictionnaire Iconographique des Orchidées, Odontoglossum, t. 8, 8A, 8B (1897); Dunsterville & Garay, Venezuelan orchids illustrated **5**: 218–19 (1972).
Rhizome short, pseudobulbs clustered, 4–8 cm, each with 2 leaves. Leaves to 40 cm, linear-elliptic, acute. Raceme with 10–20 flowers, erect. Sepals and petals similar, to 4 cm, acute, margins undulate-crisped, yellow with numerous brown blotches. Lip with a short claw bearing a fan-shaped crest which is divided into processes of which the inner 2 are the longest, yellow; blade oblong, acute, white or rarely yellow, the basal part pink or reddish, the apex brown, lacerate. Column with finely toothed auricles. *Colombia, Venezuela.* G1. Spring.

13. O. crispum Lindley (*O. alexandrae* Bateman). Figure 19(13), p. 275. Illustration: Cogniaux & Goossens, Dictionnaire Iconographique des Orchidées, Odontoglossum, t. 1 (1897); Schlechter, Die Orchideen, edn 2. t. 8 (1927); Williams et al., Orchids for everyone, 31 (1980); Bechtel et al., Manual of cultivated orchid species, 236 (1981).
Rhizome short, pseudobulbs clustered, 4–8 cm, each with 2 leaves. Leaves to 40 cm, linear-elliptic, acute. Raceme with 8–20 flowers, arched. Sepals and petals 3–4 cm, elliptic, acute, overlapping, margins wavy, petals toothed, white or pink spotted with brown. Lip with a short claw bearing a fan-shaped, laciniate, yellow keel with 2 longer projections in the middle, the blade oblong, wavy, finely toothed, acute, white, pink or pale purple, with brown spots. Column with 2 rounded, entire auricles. *Colombia.* G1. Winter.

This species is very variable and many selected variants (some of them perhaps accidental hybrids) were grown in the past. The species has also been used widely in hybridisation.

14. O. grande Lindley (*Rossioglossum grande* (Lindley) Garay & Kennedy). Figure 19(14), p. 275. Illustration: Schlechter, Die Orchideen, edn 2, t. 9 (1927); Fieldiana **26**: 615 (1953); Die Orchidee **31**: clxviii (1980); Williams et al., Orchids for everyone, 140 (1980).
Rhizome short, pseudobulbs 4–10 cm, each

with 2 (rarely 1) leaves. Leaves 10–40 × 3–6.5 cm, elliptic to lanceolate, leathery, acute. Raceme with 4–8 flowers, erect. Sepals 5.5–8.5 × 1–2 cm, lanceolate, acuminate and recurved at the apex, yellow, heavily barred with brown, petals oblanceolate to oblong, rather blunt at the apex, brown in the lower half, yellow above, margins wavy. Lip more or less without a claw, with 2 small lateral lobes at the base, bearing a callus with variable horns and projections; blade broader than long, tapered to the base, not or scarcely notched at the apex, white with reddish, brownish or yellowish spots near the base and on the margins. Column very finely downy with 2 oblong or rounded auricles. *Mexico, Guatemala*. G1. Winter.

Rather variable in flower colour. This species and 5 others compose the genus *Rossioglossum* Garay & Kennedy (Orchid Digest **40**: 139, 1976) – see Die Orchidee **31**: clxvi–clxx (1980); the other species are found only in specialist collections.

15. O. pendulum (Llave & Lexarza) Bateman (*O. citrosmum* Lindley). Figure 19(15), p. 275. Illustration: Cogniaux & Goossens, Dictionnaire Iconographique des Orchidées, Odontoglossum, t. 19 (1897); Williams et al., Orchids for everyone, 139 (1980); Bechtel et al., Manual of cultivated orchid species, 238 (1981).
Rhizome short, pseudobulbs clustered, 5–8 cm, each with 2 leaves. Leaves to 30 × 7 cm, oblong-elliptic, obtuse. Raceme with many lemon-scented flowers, dense, arching. Sepals and petals similar, *c.* 2.5 cm, ovate, shortly clawed, obtuse, white or pink, sometimes with a few reddish spots. Lip with an elongate claw bearing 2 yellow, red-spotted lateral lobes, without distinct keels; blade broader than long, heart-shaped at the base, notched at the apex, purple or pink. Column with 2 toothed auricles. *Mexico*. G1. Spring and autumn.

16. O. bictoniense (Bateman) Lindley. Figure 19(16), p. 275. Illustration: Edwards's Botanical Register **26**: t. 66 (1840); Hamer, Orquideas de El Salvador **2**: 151 & t. 18 (1974); Williams et al., Orchids for everyone, 138 (1980); Bechtel et al., Manual of cultivated orchid species, 235 (1981).
Rhizome short, pseudobulbs clustered, to 10 cm, each with 2–3 leaves. Leaves 10–45 × 1.5–5.5 cm, elliptic-oblong to linear. Raceme with many flowers, rarely branched, erect. Sepals and petals similar, 1.8–2.7 cm × 5–8 mm, elliptic-lanceolate to lanceolate, acute, pale yellowish green banded or spotted with brown. Lip with a short claw bearing a finely downy callus which clasps the column at the base and forms convex lobes at the apex, the blade 2 × 2.5 cm, heart-shaped, tapering to the apex, violet, white or pink, margin crisped. Column with 2 flap-like auricles. *Mexico, Guatemala, El Salvador*. G1. Winter–spring.

17. O. uroskinneri Lindley. Figure 19(17), p. 275. Illustration: Cogniaux & Goossens, Dictionnaire Iconographique des Orchidées, Odontoglossum, t. 3, 3A (1897); Bechtel et al., Manual of cultivated orchid species, 239 (1981).
Similar to *O. bictoniense* but leaves lanceolate, sepals and petals 2.5–3 cm × 7–13 mm, dark red to greenish with brown bars and spots, blade of the lip 2.8 cm or more, pink, veined or spotted with white. *Guatemala, Honduras*. G1. Summer–autumn.

18. O. londesboroughianum Reichenbach. Figure 19(18), p. 275. Illustration: Illustration Horticole **30**: t. 497 (1883).
Rhizome creeping, pseudobulbs distant, 6–8 cm, each bearing 2 (rarely 1) leaves. Leaves to 45 cm, lanceolate. Raceme with many flowers, occasionally branched, arching. Sepals and petals similar, 1.5–2 cm, ovate, obtuse, yellow with concentrically arranged red-brown spots towards the base. Lip with a short, keeled claw and a blade which is broader than long, heart-shaped at the base, notched at the apex, usually shining yellow, rarely spotted with brown, barred with brown at the base. Column without auricles. *Mexico*. G1. Winter.

19. O. cariniferum Reichenbach. Figure 19(19), p. 275. Illustration: Williams et al., Orchids for everyone, 138 (1980); Bechtel et al., Manual of cultivated orchid species, 235 (1981).
Rhizome short, pseudobulbs close, 8–10 cm, each with 2 leaves. Leaves to 30 × 3–3.5 cm, oblong-elliptic, acute. Panicle with many flowers, arching. Sepals and petals similar, *c.* 3.5 cm, lanceolate, acute, purplish brown. Lip with a short claw bearing 2 small, oblong, purple lateral lobes and 2 divergent, purple keels; blade broader than long, heart-shaped at the base, notched and with a small point within the notch at the apex, white. Column without auricles. *Tropical America, from Costa Rica to Venezuela*. G1. Spring.

20. O. pulchellum Lindley. Figure 19(20), p. 275. Illustration: Botanical Magazine, 4104 (1844); Cogniaux & Goossens, Dictionnaire Iconographique des Orchidées, Odontoglossum, t. 18 (1897); Hamer, Orquideas de El Salvador **2**: 157 & t. 17 (1974); Bechtel et al., Manual of cultivated orchid species, 238 (1981).
Rhizome short, pseudobulbs clustered, to 10 cm, each with 2 leaves. Leaves 10–35 × 8–15 mm, oblong-linear, acute. Raceme with 6–10 flowers, loose, arching. Flowers very fragrant. Sepals 1–2 cm × 6–13 mm, obovate to elliptic, the laterals fused at their bases for *c.* 5 mm, white. Petals 1.3–2 cm × 7–15 mm, obovate, white. Lip complex, its claw ascending, the rest sharply curved outwards and then under, claw with a large, yellow, red-spotted callus extending as 2 blunt projections over the blade; blade waisted, acute, white, margins crisped. Column with 3 coarsely toothed, pinkish auricles. *Mexico, Guatemala, El Salvador*. G1. Winter.

161. ASPASIA Lindley

J. Cullen

Rhizomatous epiphytes. Pseudobulbs simple, elliptic to oblong, somewhat compressed, often furrowed, bearing 1–2 leaves. Leaves lanceolate to linear-lanceolate, acuminate, folded when young. Raceme lateral with 1–10 flowers. Sepals spreading, pointed, the upper united to the base of the column. Petals usually similar to the sepals but broader, both united to the base of the column. Lip united to the column by its lateral margins for about half the length of the column or more, then bent at right angles into a broad, 3-lobed blade, the lateral lobes small, the base bearing 2–4 keels. Column cylindric, short, without a foot. Pollinia 2, ellipsoid, stipe slightly longer than pollinia, narrowly club-shaped, viscidium oblong, slightly shorter than the stipe and forming an acute angle with it.

A genus of 5 species widely distributed in C and northern S America. Cultivation as for *Odontoglossum* (p. 274).
Literature: Williams, N.H., The taxonomy of Aspasia, *Brittonia* **26**: 333–46 (1974).

1a. Free part of column, above the junction with the lip, much shorter than the basal part; apex of lip entire, pointed, not curved upwards **3. lunata**
b. Free part of column as long as or longer than the basal part; apex of lip notched, often curved upwards 2
2a. Lip 1.7–2.1 cm wide, its apex curving

upwards; anther cap not beaked **1. epidendroides**
b. Lip 1.3–1.5 cm wide, its apex not curving upwards; anther cap prominently beaked **2. variegata**

1. A. epidendroides Lindley. Illustration: Botanical Magazine, 3962 (1842); Fieldiana **26**: 625 (1953); Bechtel et al., Manual of cultivated orchid species, 169 (1981).
Pseudobulbs 5–12 cm, flattened on one side, convex on the other, bearing 2 leaves. Leaves 13–30 cm. Sepals and petals 1.5–2.5 cm, elliptic-obovate, yellow or yellow-green with transverse brown bars. Lip 1.7–2.1 cm wide, hairless at the base, white with purple markings towards the apex, yellow towards the base where there is a callus of 2–4 ridges, the apex notched, curving upwards. Lower half of the column united to the lip. Anther cap not beaked. *C America, Colombia*. G2. Spring.

2. A. variegata Lindley. Illustration: Edwards's Botanical Register **22**: t. 1907 (1836); Botanical Magazine, 3679 (1838); Pabst & Dungs, Orchidaceae Brasilienses **2**: t. 2082 (1977).
Similar to *A. epidendroides* but pseudobulbs and leaves a little smaller, lip 1.3–1.5 cm wide, downy at the base, its apex not curving upwards, and anther cap prominently beaked. *Tropical S America from Colombia to C Brazil*. G2. Spring.

3. A. lunata Lindley. Illustration: Lindenia **14**: t. 669 (1899); Pabst & Dungs, Orchidaceae Brasilienses **2**: t. 2080 (1977).
Pseudobulbs compressed. Leaves 14–25 cm, thin. Sepals 2.4–4.5 cm, linear-lanceolate, yellow-brown with dark red spots. Petals shorter and broader than the sepals. Lip 2–4 × 1.7–4 cm, oblong, conspicuously waisted, the apex entire, with a small point, the margin toothed, white on the margin and on the apical part, violet in the centre. Column united to the lip for most of its length, the free part very short. Anther cap not beaked. *C Brazil*. G2. Spring.

162. ADA Lindley
J. Cullen

Epiphytic, with short rhizomes. Pseudobulbs simple, ovoid, tapered, somewhat compressed, each with 1–3 leaves. Leaves lanceolate to narrowly elliptic, acute, folded when young. Racemes erect, arising from the bases of the pseudobulbs, with 8–13 shortly stalked, rather erect flowers. Sepals clawed, very narrowly elliptic, acuminate, orange; petals similar but shorter, sometimes with a dark red line or spots inside; all incurved, spreading only at their tips. Lip shorter than the petals, very narrowly elliptic, acuminate, with 1–2 elongate, raised ridges, the lip spreading away from the short column. Column without a foot. Pollinia 2, ovoid, stipe slightly longer than the pollinia, expanded above, viscidium shorter than the stipe and forming an acute angle with it.

A genus of 2 very similar species (often considered as 1), both of which are cultivated. Conditions as for *Odontoglossum* (p. 274).

1. A. aurantiaca Lindley. Illustration: Botanical Magazine, 5435 (1864); Cogniaux & Goossens, Dictionnaire Iconographique des Orchidées, Ada, t. 1 (1897); Bechtel et al., Manual of cultivated orchid species, 161 (1981).
Pseudobulbs 6–10 cm. Leaves 15–25 cm. Sepals 3–3.5 cm, petals 2–2.5 cm. Lip orange, bearing 2 elongate, irregularly toothed ridges. *Tropical S America, from Venezuela to Ecuador*. G1. Winter.

***A. lehmannii** Rolfe. Very similar (and perhaps not distinct), but with a white lip bearing an orange ridge. *Colombia*. G1. Autumn.

163. BRASSIA Lindley
J. Cullen

Epiphytic, with stout, creeping rhizomes which give rise to leaves as well as pseudobulbs. Pseudobulbs large, simple, ovoid, compressed, each bearing 1–3 leaves. Leaves leathery, variable in shape, folded when young. Racemes arising from the bases of the pseudobulbs, arching, with 6–many flowers. Sepals spreading, linear-lanceolate, acuminate, the laterals often longer than the upper, all 6 or more times longer than broad. Petals similar to the sepals but shorter. Lip shorter than the sepals, stalkless at the base of the column, directed away from it at an obtuse angle, simple, bearing a keel which is grooved or composed of 2 ridges. Column without a foot. Pollinia 2, ovoid, attached independently to the small, oblong viscidium, stipe absent.

A genus of about 50 species from tropical and subtropical America, notable for its spider-like flowers with long, narrow petals and sepals. Six species, all belonging to section *Brassia* are available in Europe and are dealt with below; several other species, belonging to the section *Glumacea* Lindley are grown in the USA (see Teuscher, Memoires du Jardin Botanique de Montreal **55**: 15–21, 1962, reprinted in American Orchid Society Bulletin **42**: 1089–94, 1973); these have recently been placed in the genus *Ada*.

Plants should be grown in hanging baskets or on rafts or pieces of bark, in epiphytic compost. The atmosphere should be humid, though watering can be reduced during the winter. Propagation by division or by seed.

Literature: Kooser, R.G. & Kennedy, G.C., The genus Brassia, *Orchid Digest* **43**: 164–72 (1979).

1a. Blade of lip with green warts or excrescences **5. verrucosa**
b. Blade without warts or excrescences 2
2a. Blade of lip conspicuously broadened towards the apex so that its maximum breadth is at least twice the breadth at the base 3
b. Blade not or little broadened, its broadest part little broader than the base 4
3a. Blade of lip white or yellowish with purple spots; petals yellowish green with purple-brown spots towards the base **1. maculata**
b. Blade yellow with red-brown spots; petals yellow in the upper half, brown in the lower **2. gireoudiana**
4a. Ridges of the keel of the lip each ending in a separated tooth **4. caudata**
b. Ridges of the keel not ending in separated teeth **3. lanceana***

1. B. maculata R. Brown. Illustration: Botanical Magazine, 1691 (1814); Hamer, Orquideas de El Salvador 1: 81 & t. 5 (1974); Orchid Digest **43**: 167 (1979).
Pseudobulbs 6–8 cm, each with 2 leaves. Leaves to 30 cm, oblong-linear. Raceme with 10–15 flowers. Sepals 6–7.5 cm, greenish yellow with a few brown-purple spots near the base; petals 4–5.5 cm, similar in colour to the sepals. Lip white or yellowish, purple-spotted, parallel-sided near the base then widely expanded towards the rounded or obtuse, though mucronate apex. Ridges of the lip yellow, grooved, slightly downy. *C America, West Indies*. G2. Autumn and spring.

2. B. gireoudiana Reichenbach & Warscewicz. Illustration: Reichenbach, Xenia Orchidacearum **1**: t. 32 (1855); Orchid Digest **43**: 166 (1979); Bechtel et al., Manual of cultivated orchid species, 172 (1981).

Similar to *B. maculata* but sepals 10–13 cm, bright yellow with brown spots, petals *c.* 5 cm, bright yellow in the upper half, brown in the lower, lip yellow with red-brown spots and deep yellow, hairless ridges. *Costa Rica, Panama*. G2. Summer.

3. B. lanceana Lindley. Illustration: Botanical Magazine, 3577 (1837); Dunsterville & Garay, Venezuelan orchids illustrated **4**: 38–9 (1966); Orchid Digest **43**: 166 (1979).
Pseudobulbs 10–12 cm, each with 2 leaves. Leaves to 30 cm, lanceolate. Racemes loose, with 7–12 flowers. Sepals *c.* 6 cm, yellow with brown spots towards the base, petals *c.* 3 cm, similar in colour to the sepals. Lip oblong, acuminate, pale yellow-green with a few brown spots near the base. Ridges yellow, downy. *Tropical S America from Colombia to E Brazil*. G2. Autumn.

***B. lawrenceana** Lindley. Illustration: Edwards's Botanical Register **27**: t. 18 (1841); Dunsterville & Garay, Venezuelan orchids illustrated **2**: 48–9 (1961); Orchid Digest **43**: 166 (1979). Similar but with sepals *c.* 7.5 cm, yellow with red-brown spots, petals *c.* 4 cm, lip clear yellow or green, unspotted, sometimes a little widened towards the base. *Tropical S America, southwards to N Brazil and NE Peru*. G2. Summer.

4. B. caudata (Linnaeus) Lindley. Illustration: Fieldiana **26**: 625 (1953); Dunsterville & Garay, Venezuelan orchids illustrated **4**: 36–7 (1966); Luer, Native orchids of Florida, t. 76, f. 5, 6 (1972); Sheehan & Sheehan, Orchid genera illustrated, 51 (1979).
Pseudobulbs 6–15 cm, each with 2 leaves. Leaves to 35 cm, oblong-elliptic, lanceolate or oblong-lanceolate. Raceme loose, with 7–12 flowers. Sepals orange-yellow, spotted with red-brown, the upper 3.5–7.5 cm, the laterals 7.5–18 cm. Petals 1.5–3.5 cm, similar in colour to the sepals. Lip oblong, broadening slightly towards the apex, yellow or greenish with a few brown blotches towards the base. Ridges 2, hairless, each ending in a distinct, separated tooth. *Tropical and subtropical America from USA (Florida) to Bolivia and N Brazil*. G2. Autumn & spring.

5. B. verrucosa Lindley (*B. brachiata* Lindley). Illustration: Edwards's Botanical Register **33**: t. 29 (1847); Cogniaux & Goossens, Dictionnaire Iconographique des Orchidées, Brassia, t. 1 (1897); Fieldiana **26**: 629 (1953); Orchid Digest **43**: 169 (1979); Bechtel et al., Manual of cultivated orchid species, 173 (1981).
Pseudobulbs 6–10 cm, clustered, each with 2 leaves. Leaves to 45 cm, oblong-elliptic to oblanceolate. Raceme with 6–15 flowers, loose. Sepals yellowish green with dark brown spots towards the base, the upper 8–15 cm, the laterals 10–20 cm. Petals 4–8.5 cm, similar in colour to the sepals. Lip white, red-spotted, bearing green warts or excrescences, oblong at the base, widened towards the apex. Ridges ending in attached teeth. *Tropical America from S Mexico to Venezuela*. G2. Spring–summer.

164. MILTONIA Lindley
J. Cullen
Epiphytic with rhizomes. Pseudobulbs clustered or distant, grey-green or yellow-green, simple, usually compressed, each with 1–2 leaves. Leaves oblong, oblong-linear, or oblong-lanceolate, usually acute, folded when young. Racemes arising from the bases of the pseudobulbs, usually erect, with 1–several flowers; bracts sometimes conspicuous. Flowers often rather flat. Sepals and petals similar. Lip spreading away from the column base, usually simple, usually larger than the sepals, with 2 or more longitudinal ridges (or rarely a patch of hairs) at the base. Column short, with a narrow wing on each side at the base, without a foot, with auricles at the apex. Pollinia 2, ovoid to ellipsoid, stipe variable, shorter or longer than pollinia, viscidium shorter than the stipe and forming an acute angle with it.

A genus of about 20 species from C & S America, of which 13 are generally available. Species Nos. 1–5 are sometimes separated off as the genus *Miltoniopsis* Godefroy. Much hybridisation has been done within the genus, and many hybrids are cultivated.

Cultivation as for *Odontoglossum* (p. 274) but the plants require a higher winter temperature.

Rhizome. Long, pseudobulbs distant: **1–5**; short, pseudobulbs clustered: **6–13**.
Pseudobulbs. Greyish green, surrounded by the bases of a number of leaves: **1–5**; yellow-green, not surrounded by leaf-bases: **6–13**.
Raceme. With 4 or more flowers: **1–3,8–13**; with fewer than 4 flowers: **4–7**. Axis flattened: **7**.
Sepals. White, pink, red or purple: **1–6,8**; green to yellow, often spotted with brown: **7–13**.
Lip. Longer than sepals: **1–8**; shorter than sepals: **9,13**; about as long as sepals: **10–12**. Clearly waisted: **3–5,9–11**; with oblong claw and abruptly widened blade: **7,12**; widening gradually from the base and more or less flat: **1,2,6,8**; widening gradually from the base, the sides curving up to the column: **13**. With 2 backwardly projecting auricles: **1,2**.

1a. Pseudobulbs and leaves greyish green, the pseudobulbs surrounded by the bases of a number of leaves or leaf-sheaths; pseudobulbs usually with 1 leaf, usually distant — 2
b. Pseudobulbs and leaves yellowish green, the pseudobulbs not surrounded by leaf-bases; pseudobulbs usually with 2 leaves and clustered — 6
2a. Lip with 2 backwardly pointed projections at the base, the blade not waisted — 3
b. Lip without 2 backwardly pointed projections at the base, the blade waisted — 4
3a. Sepals and petals white, acuminate, the petals usually with a deep purple blotch at the base; leaves striped with dark green lines beneath — **2. roezlii**
b. Sepals and petals entirely pink, rounded to a blunt point; leaves without dark green lines beneath — **1. vexillaria**
4a. Sepals and petals oblong or oblong-obovate, strongly undulate-crisped, purple with white tips — **5. warsewiczii**
b. Sepals and petals ovate to elliptic, not or scarcely undulate-crisped, mainly white — 5
5a. Flowers *c.* 7 cm in diameter; lip with 2 small reddish or violet spots near the base, the rest white — **3. endresii**
b. Flowers to 5.5 cm in diameter; lip mostly violet, the margins white — **4. phalaenopsis**
6a. Sepals and petals of a uniform colour, white, yellow, pink or purple, not spotted or blotched — 7
b. Sepals and petals blotched or spotted with brown on a yellow or green ground — 10
7a. Bract much shorter than the ovary and flower-stalk together; base of lip with 5–9 ridges — **8. regnellii**
b. Bract very large and conspicuous, as long as the ovary and flower-stalk together, sheathing; base of the lip with 3 ridges or a patch of hairs — 8
8a. Flowers 6–12 in a raceme; lip, sepals

and petals yellow, lip with a patch of hairs at the base **9. flavescens**
b. Flowers solitary (rarely 2); lip, sepals and petals white, greenish yellow, purple or reddish, lip with 3 ridges 9
9a. Axis of raceme flattened; lip white with purple lines, oblong at the base, abruptly widened towards the apex **7. anceps**
b. Axis of raceme terete; lip purple or reddish, widening gradually from the base **6. spectabilis**
10a. Sides of the lip curving upwards to meet the column **13. candida**
b. Lip flat, or if somewhat concave, the sides not meeting the column 11
11a. Blade of the lip narrowly oblong at the base, abruptly widened to an apical part which is broader than long **12. cuneata**
b. Blade oblong, waisted towards the apex, the part beyond the waist longer than broad 12
12a. Sepals and petals little spreading, directed forwards; lip purple or pink with darker spots **10. russelliana**
b. Sepals and petals widely spreading; base of the lip purple with a median yellow stripe, apical part white **11. clowesii**

1. M. vexillaria (Bentham) Nicholson (*Odontoglossum vexillarium* Bentham; *Miltoniopsis vexillaria* (Bentham) Godefroy-Lebeuf. Illustration: Cogniaux & Goossens, Dictionnaire Iconographique des Orchidées, Miltonia, t. 1 (1900); Schlechter, Die Orchideen, edn 2, t. 10 (1927); Bechtel et al., Manual of cultivated orchid species, 233 (1981).
Pseudobulbs 5–6 cm, grey-green, distant, compressed, each with 1 leaf, subtended by 3–4 leaf-bases on each side. Leaves 15–25 cm, grey-green, linear-oblong. Raceme with 4–6 flowers, loose. Sepals and petals *c.* 4 cm, oblong-elliptic, rounded to a blunt point, pink. Lip much larger than the sepals, almost circular, with 2 backwardly directed projections and 3 ridges at the base, deeply notched at the apex, very variable in colour. *Colombia, N Ecuador*. G2. Spring–summer.

A very attractive species with the lip varying from pink to white, usually with red lines and a yellow blotch towards the base. Many cultivars exist and are widely grown.

2. M. roezlii (Reichenbach) Nicholson (*Odontoglossum roezlii* Reichenbach; *Miltoniopsis roezlii* (Reichenbach) Godefroy-Lebeuf). Illustration: Lindenia **2**: t. 78 (1886); Cogniaux & Goossens, Dictionnaire Iconographique des Orchidées, Miltonia, t. 6 (1900); Sheehan & Sheehan, Orchid genera illustrated, 121 (1979); Bechtel et al., Manual of cultivated orchid species, 233 (1981).
Similar to *M. vexillaria* but leaves with dark green longitudinal lines beneath, sepals and petals acuminate, white, the petals with a purple blotch at the base, lip with 5 ridges at the base. *Colombia*. G2. Winter.

3. M. endresii (Reichenbach) Nicholson (*M. superba* Schlechter). Illustration: Cogniaux & Goossens, Dictionnaire Iconographique des Orchidées, Miltonia, t. 10 (1900).
Pseudobulbs 5–6 cm, grey-green, subtended by 3–4 leaf-bases on each side, each 1-leaved, compressed, usually distant on the rhizome. Leaves 15–25 cm, grey-green, linear-oblong. Raceme with 4–6 flowers, loose. Flowers *c.* 7 cm in diameter. Sepals and petals 3–3.5 cm, ovate-elliptic, obtuse, white with red or purple stripes at the base. Lip larger than the sepals, broadly obovate, slight waisted near the base, deeply notched and with a small point in the sinus at the apex, white with 2 small red or purple blotches at the base, and with 3 short papillose ridges. *Costa Rica*. G2. Winter.

4. M. phalaenopsis (Linden & Reichenbach) Nicholson (*Odontoglossum phalaenopsis* Linden & Reichenbach; *Miltoniopsis phalaenopsis* (Linden & Reichenbach) Garay & Dunsterville). Illustration: Lindenia **7**: t. 334 (1891): Cogniaux & Goossens, Dictionnaire Iconographique des Orchidées, Miltonia, t. 11 (1900); Bechtel et al., Manual of cultivated orchid species, 233 (1981).
Pseudobulbs 2.5–4 cm, ovoid, grey-green, usually distant on the rhizome, subtended by 3–4 leaf-bases on each side, each with 1 leaf. Leaves 15–22 cm, green, linear. Raceme with 2–4 flowers, loose. Flowers to 5.5 cm in diameter. Sepals and petals ovate-elliptic, white. Lip larger than the sepals, broadly oblong, waisted towards the base, notched and with a small point within the sinus at the apex, white, mostly overlaid by red-purple in the form of streaks or lines which join together, covering most of the surface; callus with 3 short teeth at the base. *Colombia*. G2. Autumn.

5. M. warsewiczii Reichenbach (*Odontoglossum weltonii* invalid; *Oncidium fuscatum* Reichenbach; *M. vexillaria* var. *warsewiczii* (Reichenbach) Schlechter; *Miltoniopsis warsewiczii* (Reichenbach) Garay & Dunsterville). Illustration: Botanical Magazine, 5843 (1870); Lindenia **8**: t. 384 (1892); Cogniaux & Goossens, Dictionnaire Iconographique des Orchidées, Miltonia, t. 8 (1900); Bechtel et al., Manual of cultivated orchid species, 253 (1981).
Similar to *M. phalaenopsis* but sepals and petals oblong or oblong-obovate, conspicuously undulate-crisped, mostly purple, the tips white, lip mostly purple, the margins white, usually with a yellow blotch or variegation in the centre. *Peru, Ecuador, Colombia*. G2. Spring.

6. M. spectabilis Lindley. Illustration: Edwards's Botanical Register **23**: t. 1992 (1837); Cogniaux & Goossens, Dictionnaire Iconographique des Orchidées, Miltonia, t. 2, 2A (1900); Bechtel et al., Manual of cultivated orchid species, 232 (1981).
Pseudobulbs to 7 cm, clustered, ovoid, compressed, each with 2 leaves. Leaves to 15 cm, narrowly linear-oblong, green. Raceme with 1 or 2 flowers and a terete axis; bracts exceeding the flower-stalk and ovary together, sheathing. Sepals and petals to 4.5 cm, spreading, oblong, obtuse, uniformly white, red or purple. Lip broadly obovate, widening gradually from the base, purple or reddish, with 3 yellow ridges at the base. *Venezuela, E Brazil*. G2. Summer.

7. M. anceps Lindley (*Odontoglossum anceps* (Lindley) Klotzsch). Illustration: Botanical Magazine, 5572 (1866); Pabst & Dungs, Orchidaceae Brasilienses **2**: f. 2050 (1977).
Pseudobulbs 4.5–5.5 cm, each with 2 leaves, yellow-green, oblong, strongly compressed, clustered. Raceme with 1 flower, a flattened axis and the bract sheathing, exceeding the ovary and flower-stalk together. Sepals and petals 3.5–3.7 cm, spreading, oblong-lanceolate, obtuse, uniformly greenish yellow. Lip oblong, slightly waisted, exceeding the sepals, white with purple lines, and with 3 ridges which are yellow, each with a single red stripe at the base. *SE Brazil*. G2. Spring.

8. M. regnellii Reichenbach (*Oncidium regnellii* (Reichenbach) Reichenbach). Illustration; Botanical Magazine, 5436 (1864); Cogniaux & Goossens, Dictionnaire Iconographique des Orchidées, Miltonia, t. 7, 7A, 7B (1900); Hoehne, Iconografia de Orchidaceas do Brasil, t. 267 (1949); Bechtel et al., Manual of cultivated orchid species, 232 (1981).
Pseudobulbs 5–8 cm, each with 2 leaves, yellow-green, clustered, ovoid. Leaves to 30 cm, linear-lanceolate, acute. Raceme with 4–7 flowers, erect, loose. Sepals

and petals *c.* 3.5 cm, ovate-oblong, acute, white or yellow. Lip broadly obovate, larger than the sepals, obscurely 3-lobed, blunt at the apex, white or violet with white margins, with 7–9 low ridges at the base. *S Brazil.* G2. Summer.

9. M. flavescens (Lindley) Lindley (*Cyrtochilum flavescens* Lindley). Illustration: Gartenflora **39**: t. 1328 (1890); Hoehne, Iconografia de Orchidaceas do Brasil, t. 263 (1949); Pabst & Dungs, Orchidaceae Brasilienses **2**: f. 2051 (1977).
Pseudobulbs 6–8 cm, each with 2 leaves, clustered, yellow-green, oblong. Leaves to 30 cm, linear-oblong, acute. Raceme with 6–12 flowers, erect, loose, bracts conspicuous, sheathing, longer than the ovary and flower-stalk together. Sepals and petals *c.* 3.5 cm, yellow, linear-oblong or linear-lanceolate, acute. Lip oblong, waisted, shorter than the sepals, pale yellow with reddish lines. *S & E Brazil, Paraguay, N Argentina.* G2. Summer–autumn.

10. M. russelliana Lindley. Illustration: Edwards's Botanical Register **22**: t. 1830 (1836); Pabst & Dungs, Orchidaceae Brasilienses **2**: f. 2058 (1977).
Pseudobulbs clustered, yellow-green. Leaves linear-oblong. Racemes several-flowered, erect, bracts inconspicuous. Sepals and petals 2.5–3 cm, oblong, acute, not spreading but directed forwards, green or greenish yellow heavily marbled or spotted with brown. Lip about as long as sepals, oblong, somewhat waisted towards the apex, purple or pink with darker spots, with many ridges at the base. *S Brazil.* G2. Winter–spring.

11. M. clowesii Lindley (*Odontoglossum clowesii* (Lindley) Lindley; *Brassia clowesii* (Lindley) Lindley). Illustration: Botanical Magazine, 4109 (1844); Cogniaux & Goossens, Dictionnaire Iconographique des Orchidées, Miltonia, t. 4 (1900); Bechtel et al., Manual of cultivated orchid species, 231 (1981).
Pseudobulbs 5–8 cm, clustered, yellow-green, elongate-ovoid, each with 2 leaves. Leaves to 30 cm, linear-lanceolate, acute. Raceme with 3–7 flowers, erect, loose, bracts inconspicuous. Sepals and petals 3–3.5 cm, spreading, oblong, acute, greenish yellow with brown spots. Lip about as long as sepals, waisted, the anterior part ovate, acute, violet in the lower part, white towards the apex, with several short ridges at the base. *E Brazil.* G2. Autumn.

12. M. cuneata Lindley (*M. speciosa* Klotzsch). Illustration: Edwards's Botanical Register **31**: t. 8 (1845); Cogniaux & Goossens, Dictionnaire Iconographique des Orchidées, Miltonia, t. 9 (1900); Pabst & Dungs, Orchidaceae Brasilienses **2**: f. 2055 (1977); Bechtel et al., Manual of cultivated orchid species, 232 (1981).
Pseudobulbs 5–8 cm, oblong-ovoid, yellow-green, clustered, each with 2 leaves. Leaves to 30 cm, oblong-lanceolate. Raceme with 5–8 flowers, erect. Sepals and petals 3–3.5 cm, oblong, acute, the margins somewhat wavy, yellow spotted with brown. Lip about as long as the sepals, narrowly oblong at the base, abruptly widened to an obovate, heart-shaped apical part which is broader than long, slightly notched at the apex, white, with 2 ridges at the base. *E Brazil.* G2. Spring.

13. M. candida Lindley (*Oncidium candidum* (Lindley) Reichenbach). Illustration: Hoehne, Iconografia de Orchidaceas do Brasil, t. 262 (1949); Pabst & Dungs, Orchidaceae Brasilienses **2**: f. 2053 (1977); Bechtel et al., Manual of cultivated orchid species, 231 (1981).
Pseudobulbs 5–8 cm, yellow-green, clustered, oblong-ovoid, each with 2 leaves. Leaves to 30 cm, linear-lanceolate, acute. Raceme with 3–7 flowers, erect, loose. Sepals and petals 4–4.5 cm, oblong, acute, greenish yellow with many brown spots. Lip shorter than the sepals, white or rarely pink, sometimes purplish in the centre, its sides rolled upwards to meet the column, somewhat wavy, with 3 ridges at the base. *E Brazil.* G2. Autumn.

165. ONCIDIUM Swartz

J. Cullen

Rhizomatous epiphytes. Pseudobulbs usually present, simple, usually bearing 1–3 leaves and subtended by a number of leaves or sheathing leaf-bases. Leaves variable, fleshy, leathery, sometimes terete, folded when young. Flowers usually numerous in racemes or panicles. Sepals and petals usually spreading, usually clawed, similar in size or the petals larger, the lateral sepals free or variably united at the base. Lip entire or 3-lobed, often waisted, very variable in shape and size, firmly united with the base of the column from which it diverges, with a complex tuberculate or ridged callus. Column with a fleshy plate below the stigma but without a foot, and with distinct, variably shaped auricles or wings on either side. Rostellum short or beaked. Pollinia 2, ovoid to spherical, stipe linear, longer than pollinia, viscidium small, making an acute angle with the stipe.

A large genus (*c.* 450 species?) from subtropical and tropical America, of which over 100 species have been in cultivation. It is sometimes difficult to distinguish from both *Miltonia* and *Odontoglossum*, and it has been suggested that it should be broken up into a number of genera. Garay & Stacy (reference below), however, prefer to maintain it as a single genus. They divide the genus into many sections, some of which are interpolated into the present account with very short descriptions.

The species are very variable as to their requirements in cultivation. They are usually grown in epiphytic compost in containers or on rafts. Some have a definite resting period while others do not, and treatment should vary accordingly. Propagation is by seed, division or the removal of small bulbs from mature plants.
Literature: Kränzlin, F., *Das Pflanzenreich* **80**: 25–290 (1922); Garay, L.A., A reappraisal of the genus Oncidium Sw., *Taxon* **19**: 443–67 (1970); Garay, L.A. & Stacy, J.E., Synopsis of the genus Oncidium, *Bradea* **1**(40): 393–424 (1974).

1a. Sepals, petals and lip fleshy (section *Cyrtochilum*) 2
b. Sepals fleshy or not; petals and lip not fleshy 4
2a. Lip oblong, tapering slightly to the apex (figure 20(3), p. 283) **3. superbiens**
b. Lip broadly triangular at the base, abruptly narrowed to an acuminate apex 3
3a. Sepals and petals more than 4 cm, petals yellow **1. macranthum**
b. Sepals and petals less than 2 cm, petals reddish brown **2. microchilum**
4a. Pseudobulbs absent or rudimentary and less than 4 cm, rarely present and large, when the plant is very large (panicle up to 2 m) 5
b. Pseudobulbs present, usually more than 4 cm, always large and conspicuous for the size of the plant 13
5a. Sheaths enveloping the pseudobulbs all leaf-bearing (section *Onusta*) **4. onustum**
b. Sheaths enveloping the pseudobulbs mostly without leaf-blades 6
6a. Leaves terete (section *Cebolleta*) **5. cebolleta**
b. Leaves flat 7
7a. Leaves leathery; callus of lip hairless

Figure 20. Lips of *Oncidium* species. 1, *O. macranthum*. 2, *O. microchilum*. 3, *O superbiens*. 4, *O. onustum*. 5, *O. cebolleta*. 6, *O. altissimum*. 7, *O. splendidum*. 8, *O. nanum*. 9, *O cavendishianum*. 10, *O. bicallosum*. 11, *O. harrisonianum*. 12, *O. divaricatum*. 13, *O. papilio*. 14, *O. kramerianum*. 15, *O. pubes*. 16, *O. maculatum*. 17, *O. hastatum*. 18, *O. ornithorrhynchum*. 19, *O. cheirophorum*.

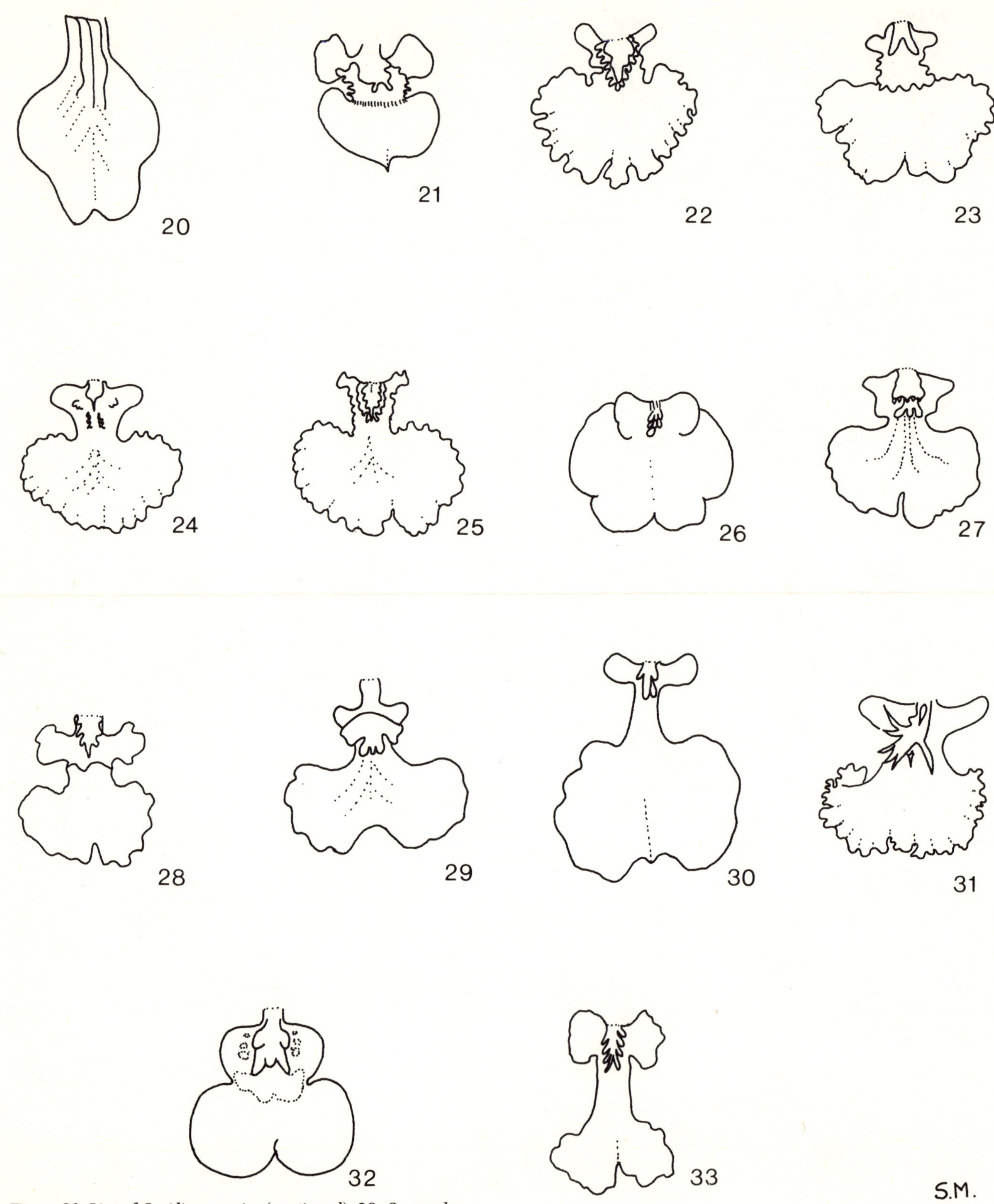

Figure 21. Lips of *Oncidium* species (continued). 20, *O. concolor*. 21, *O. longipes*. 22, *O. crispum*. 23, *O. forbesii*. 24, *O. gardneri*. 25, *O. marshallianum*. 26, *O. varicosum*. 27, *O. flexuosum*. 28, *O. excavatum*. 29, *O. ampliatum*. 30, *O. tigrinum*. 31, *O. leucochilum*. 32, *O. sphacelatum*. 33, *O. wentworthianum*. (Compiled from various sources; not to scale.)

with many variable tubercles (section *Plurituberculata*) 8
b. Leaves fleshy; callus of lip downy, swollen (section *Pulvinata*) 12
8a. Pseudobulbs absent or at most 1.5 cm 9
b. Pseudobulbs present and larger 11
9a. Sepals and petals unspotted, greenish yellow or yellow, bordered with yellow or crimson; flowers in racemes **10. bicallosum**
b. Sepals and petals yellow spotted with reddish brown or purple-brown; flowers in panicles 10
10a. Petals to 9 mm; lip 11 mm at its widest **8. nanum**
b. Petals more than 1.2 cm; lip more than 1.5 cm at its widest **9. cavendishianum**
11a. Panicle to 2 m or more with many branches and many flowers; sepals and petals to 2 cm **6. altissimum**
b. Panicle much smaller with few branches (rarely unbranched) and with 9–20 flowers; sepals and petals 2.5–3 cm **7. splendidum**
12a. Flowers to 1.8 cm in diameter; leaves grey-green, 7.5–15 cm **11. harrisonianum**
b. Flowers more than 2 cm in diameter; leaves green, 20–30 cm **12. divaricatum***
13a. Upper sepal and petals erect, antenna-like (section *Glanduligera*) 14
b. Upper sepal and petals not as above 15
14a. Axis of raceme flattened above **13. papilio**
b. Axis of raceme not flattened above **14. kramerianum**
15a. Column downy with auricles in the form of forwardly projecting fleshy arms on the sides of the stigma (section *Waluewa*) **15. pubes**
b. Column hairless with membranous auricles 16
16a. Sepals, petals and lip spread in 1 plane, the lip little more conspicuous than the rest (section *Stellata*) 17
b. Sepals, petals and lip not spread in 1 plane, the lip much more conspicuous than the rest 18
17a. Lip white with red lines at the base, yellow towards the apex, lateral lobes triangular, pointing forwards, (figure 20(16), p. 283) **16. maculatum**
b. Lip mostly pink, lateral lobes oblong or almost square, directed outwards (figure 20(17), p. 283) **17. hastatum**
18a. Rostellum elongate, proboscis-like (section *Rostrata*) 19
b. Rostellum short, not projecting 20
19a. Lip, sepals and petals mostly pink except for the yellow ridges on the lip **18. ornithorrhynchum**
b. Lip, sepals and petals yellow except for the white ridges on the lip **19. cheirophorum**
20a. Callus of the lip with 2–4 parallel tubercles or ridges (section *Concoloria*) **20. concolor**
b. Callus of lip with 3, 5 or 7 serial or scattered tubercles, rarely with accessory tubercles or papillae as well 21
21a. Lateral sepals longer than lip (section *Barbata*) **21. longipes**
b. Lateral sepals shorter than the lip 22
22a. Lateral sepals united at the base to varying extents 23
b. Lateral sepals entirely free from each other 28
23a. Petals large, much more conspicuous than sepals (section *Crispa*) 24
b. Petals and sepals similar in size (section *Synsepala*) 27
24a. Wings of column sharply and conspicuously toothed **22. crispum**
b. Wings of column entire 25
25a. Sepals, petals and lip chestnut brown with yellow marbling on the margins; flowers usually in a raceme **23. forbesii**
b. At least the lip mostly yellow, sometimes with red-brown spots at the base or on the margins; flowers in panicles 26
26a. Lip 3 × 4.5 cm; lateral sepals united for half their length **24. gardneri**
b. Lip *c.* 4 × 5 cm; lateral sepals united for a third of their length **25. marshallianum**
27a. Blade of lip 2.5–4 × 3.5–5.5 cm with 3 notches towards the apex, producing 4 lobes (figure 21(26), p. 284) **26. varicosum**
b. Blade of lip 1.5–2 × 1.5–1.8 cm with 1 deep notch at the apex **27. flexuosum**
28a. Petals wider and more conspicuous than the sepals (section *Excavata*) **28. excavatum**
b. Petals and sepals similar in size and colour 29
29a. Bracts small, scale-like, inconspicuous (section *Oblongata*) 30
b. Bracts large and conspicuous (section *Planifolia*) 32
30a. Lip white; sepals and petals to 1.7 cm **31. leucochilum**
b. Lip yellow; sepals and petals 2–7.5 cm 31
31a. Pseudobulbs with 1 leaf; sepals and petals 6–7.5 cm; auricles of the column toothed **29. ampliatum**
b. Pseudobulbs with 2 leaves; sepals and petals 2–3 cm; auricles of the column entire **30. tigrinum**
32a. Leaves 35–60 cm; lip wider across the central lobe than across the lateral lobes (figure 21(32), p. 284) **32. sphacelatum**
b. Leaves 20–30 cm; lip wider across the lateral lobes than across the central lobe (figure 21(33), p. 284) **33. wentworthianum**

Section **Cyrtochilum** (Humboldt, Bonpland & Kunth) Lindley.
Flowers conspicuous, sepals, petals and lip all fleshy; column with fleshy auricles.

1. O. macranthum Lindley (*O. hastiferum* Reichenbach & Warscewicz). Figure 20(1), p. 283. Illustration: Botanical Magazine, 5743 (1868); The Garden **24**: t. 416 (1883); Cogniaux & Goossens, Dictionnaire Iconographique des Orchidées, Oncidium, t. 14 (1898); Bechtel et al., Manual of cultivated orchid species, 246 (1981).
Pseudobulbs 7–15 cm, oblong-conical, each with 2 leaves. Leaves 25–50 × 2.5–5 cm, oblanceolate to oblong, acute. Panicle to 3 m, each branch with 2–5 flowers. Sepals 2.5–3.5 cm, clawed, the laterals somewhat longer than the upper, dull yellow-brown. Petals circular to ovate, very shortly clawed, crisped-undulate, a little shorter than the sepals, golden yellow. Lip smaller than the petals, broadly triangular, abruptly narrowed to the recurved, acuminate apex, red or brownish purple towards the base, yellowish with red lines towards the apex, with a large white callus with 6–7 erect teeth. *Colombia, Ecuador, Peru.* G1. Summer.

2. O. microchilum Lindley. Figure 20(2), p. 283. Illustration: Fieldiana **26**: 655 (1953); Bechtel et al., Manual of cultivated orchid species, 246 (1981).
Pseudobulbs 2.5–3.5 cm, spherical, compressed, each with 1 leaf. Leaves 12–25 × 3–6.5 cm, fleshy, elliptic, margins irregularly scalloped. Panicle to 1.5 m with many flowers. Sepals 1.2–1.4 cm, shortly clawed, circular to elliptic, reddish brown, the upper very concave. Petals almost clawless, oblong–elliptic, otherwise similar to sepals. Lip complex, very broadly triangular, much shorter than the petals, white spotted with red; callus dark red, covering most of the centre, with 5 or more blunt tubercles. Column with 2 slender,

sickel-shaped wings. *Mexico, Guatemala*. G2. Summer.

3. O. superbiens Riechenbach. Figure 20(3), p. 283. Illustration: Botanical Magazine, 5980 (1872); Cogniaux & Goossens, Dictionnaire Iconographique des Orchidées, Oncidium, t. 31 (1904); Bechtel et al., Manual of cultivated orchid species, 249 (1981).
Pseudobulbs 6–10 cm, clustered, each usually with 2 leaves. Leaves to 70 cm, linear-oblanceolate, acute. Panicle to 6 m, twining, with many flowers. Sepals clawed, ovate to almost circular, the upper 1.6–2 cm, the laterals somewhat longer, all greenish brown or brown, sometimes with yellow margins. Petals smaller than the sepals, oblong-ovate to almost circular, shortly clawed, margins crisped-undulate, yellow or white with reddish or brown bars in the lower half. Lip shorter than the petals, oblong, the apex recurved, purple, with a complex, toothed callus. Column with sickle-shaped wings. *Peru, Colombia, Ecuador & Venezuela*. G1. Winter.

Section **Onusta** Garay & Stacy.
Pseudobulbs well developed, bearing leaves, or leafless with a terminal appendage; sheaths surrounding the pseudobulbs all leaf-bearing; leaves leathery or fleshy, finely toothed towards the apex.

4. O. onustum Lindley. Figure 20(4), p. 283. Illustration: Lindenia **11**: t. 498 (1895).
Pseudobulbs 2–4 cm, subtended by several leaves with blades, green mottled with brown, bearing 1 or 2 leaves or a terminal appendage. Leaves 5–12.5 × 1–1.8 cm, oblong or linear-oblong. Raceme (rarely branched) to 40 cm, often 1-sided, with many flowers. Sepals 7–8 mm, ovate or ovate-elliptic, yellow, the upper hooded. Petals *c*. 2 times as large as the sepals, almost circular, wavy, yellow. Lip larger than the petals, 3-lobed, the lateral lobes small, the central lobe almost circular, bearing a 3-lobed callus. Column with large, crescent-shaped auricles. *Panama, Colombia, Ecuador & Peru*. G2. Spring.

Section **Cebolleta** Lindley.
Pseudobulbs small, each bearing 1 leaf; leaves terete in section.

5. O. cebolleta (Jacquin) Swartz. Figure 20(5), p. 283. Illustration: Botanical Magazine, 3568 (1837); Dunsterville & Garay, Venezuelan orchids illustrated **2**: 246–7 (1961); Pabst & Dungs, Orchidaceae Brasilienses **2**: t. 1945 (1977); Bechtel et al., Manual of cultivated orchid species, 242 (1981).
Pseudobulbs 1.5–2 cm, conical to almost spherical, each with 1 leaf. Leaves 7–40 cm, fleshy, terete, slightly grooved, erect, often tinged or spotted with purple. Panicle up to 1.2 m with many flowers, its stalk spotted with purple. Sepals 6–10 mm, obovate, greenish yellow with red-brown spots. Petals similar, margins wavy. Lip 3-lobed, with lateral lobes rather large, oblong or obovate, the central lobe larger than the sepals, shortly clawed, kidney-shaped, deeply notched at the apex, yellow, callus consisting of a sharp projecting ridge surrounded by tubercles. Column with conspicuous, oblong, sometimes 2-lobed auricles. *Tropical America, from Mexico and the West Indies to N Argentina*. G2. Spring.

Section **Plurituberculata** Lindley.
Pseudobulbs absent or very small for the size of the plant; leaves leathery; callus of lip with many tubercles.

6. O. altissimum (Jacquin) Swartz. Figure 20(6), p. 283. Illustration: Edwards's Botanical Register **19**: t. 1651 (1834).
Pseudobulbs 5–7 cm, ovoid, compressed, each with 2 leaves. Leaves 25–30 cm, oblong or oblong-lanceolate, acute. Panicle to 2 m or more, with many flowers. Sepals and petals *c*. 2 cm, narrowly lanceolate or oblanceolate, somewhat wavy, greenish yellow with red or red-brown spots. Lip 3-lobed with small, oblong lateral lobes and a clawed, kidney-shaped central lobe which is slightly notched at the apex and bears a complex callus with about 10 tubercles, yellowish with a reddish band across the narrowest part. Column with small, triangular, entire auricles. *West Indies*. G2. Spring–summer.

For a discussion of the use of the name *O. altissimum* see Garay & Stacy, Bradea 1(40): 395–6 (1974).

7. O. splendidum Duchartre. Figure 20(7), p. 283. Illustration: Cogniaux & Goossens, Dictionnaire Iconographique des Orchidées, Oncidium, t. 7 (1897); Bechtel et al., Manual of cultivated orchid species, 248 (1981).
Pseudobulbs 4–5 cm, clustered, almost spherical, each bearing 1 leaf. Leaves to 100 × 3.5 cm, oblong-elliptic, usually purple-tinged. Panicle to 1.5 m with many flowers. Sepals *c*. 2.5 cm, shortly clawed, linear-elliptic to oblanceolate, wavy, yellow-green with broad, red-brown bands. Petals similar to sepals but up to 3 cm. Lip much larger than the sepals, 3-lobed, the lateral lobes very small, the central up to 4 cm wide, almost stalkless, kidney-shaped, deeply notched, somewhat wavy, yellow, with a callus composed of 3 conspicuous, bright yellow ridges. Column with small wings. *Guatemala*. G2. Winter–spring.

8. O. nanum Lindley. Figure 20(8), p. 283. Illustration: Dunsterville & Garay, Venezuelan orchids illustrated **1**: 272–3 (1959); Pabst & Dungs, Orchidaceae Brasilienses **2**: t. 1948 (1977); Bechtel et al., Manual of cultivated orchid species, 246 (1981).
Pseudobulbs to 7 mm, cylindric, each with 1 leaf. Leaves 7–24 × 2–7 cm, elliptic or oblong, acute. Panicle to 25 cm with many flowers. Sepals 7–9 mm, obovate, yellow spotted with orange, purple or brown. Petals similar but a little smaller. Lip slightly longer than the sepals, 3-lobed, the lateral lobes small, the central lobe transversely oblong to kidney-shaped, notched at the apex and with a complex, tuberculate callus at the base. Column with oblique, slightly toothed auricles. *Tropical S America from Colombia to Peru and C Brazil*. G2. Spring.

9. O. cavendishianum Bateman. Figure 20(9), p. 283. Illustration: Cogniaux & Goossens, Dictionnaire Iconographique des Orchidées, Oncidium, t.11 (1898); Bechtel et al., Manual of cultivated orchid species, 242 (1981).
Pseudobulbs absent or very small. Leaves 15–45 × 5–13 cm, elliptic to broadly lanceolate, acute or subobtuse. Panicle 60–150 cm, with many flowers. Flowers very fragrant. Sepals 1.2–1.7 cm, obovate, obtuse, margins undulate-crisped, yellow spotted with red. Petals similar but clawed. Lip larger than the sepals, deeply 3-lobed, the lateral lobes conspicuous, obovate, the central lobe large, transversely oblong to kidney-shaped, notched at the apex, yellow with red spots at the base; callus consisting of 5 tubercles. Column with sickle-shaped, red-spotted auricles. *S Mexico, Guatemala, Honduras*. G2. Spring.

10. O. bicallosum Lindley. Figure 20(10), p. 283. Illustration: Edwards's Botanical Register **29**: t. 12 (1843); Botanical Magazine, 4148 (1845); Bechtel et al., Manual of cultivated orchid species, 241 (1981).
Pseudobulbs very small or absent. Leaves 15–35 × 4–8.5 cm, oblong-elliptic, obtuse. Raceme (rarely branched) 20–65 cm with many flowers. Flowers fragrant. Sepals 1.5–2 cm, obovate to spathulate, obtuse,

greenish yellow to deep yellow, rarely margined with red. Petals similar but with crisped margins. Lip larger than the sepals, deeply 3-lobed, the lateral lobes rather small, obovate, the central lobe transversely oblong to kidney-shaped, yellow, with a callus of 2 swellings and a variable number of tubercles, which are white with red spots. Column wings fleshy, sickle-shaped. *S Mexico, Guatemala, El Salvador*. G2. Autumn.

The following species, also belonging to this section, are sometimes grown:

O. carthagenense (Jacquin) Swartz. Illustration: Bechtel et al., Manual of cultivated orchid species, 242 (1981). Leaves spotted with red, flowers white or pale yellow, densely spotted with red. *USA (Florida), West Indies, C America and northern S America*. G2. Summer.

O. pumilum Lindley. Leaves 5–10 cm, panicle little longer than the leaves, flowers yellow, to 7 mm in diameter. *S Brazil, Paraguay*. G1. Spring.

O. stramineum Bateman. Flowers whitish yellow, lip white, spotted with red. *Mexico*. G2. Spring.

Section **Pulvinata** Lindley.
Pseudobulbs small, subtended by sheaths without blades; callus of the lip downy.

11. O. harrisonianum Lindley. Figure 20(11), p. 283. Illustration: Edwards's Botanical Register **19**: t. 1569 (1833); Pabst & Dungs, Orchidaceae Brasilienses **2**: t. 1955 (1977); Bechtel et al., Manual of cultivated orchid species, 244 (1981).
Pseudobulbs to 2.5 cm, almost spherical, each bearing 1 leaf. Leaves 7.5–15 cm, greyish green. Panicle with 15–20 flowers, little branched, erect, loose, exceeding the leaves. Sepals and petals 6–9 mm, yellow with red longitudinal lines. Lip longer than the sepals, 3-lobed, the lateral lobes small, the central lobe clawed, transversely oblong or kidney-shaped, notched at the apex, yellow with a callus formed of 5 concave lobes. Wings of column sickle-shaped. *E Brazil*. G2. Autumn–winter.

12. O. divaricatum Lindley. Figure 20(12), p. 283. Illustration: Pabst & Dungs, Orchidaceae Brasilienses **2**: t. 1951 (1977); Bechtel et al., Manual of cultivated orchid species, 243 (1981).
Pseudobulbs 2.5–4 cm, spherical, each bearing 1 leaf. Leaves 20–30 cm, oblong, obtuse. Panicle to 1.2 m, very branched, with many flowers. Sepals and petals 1–1.3 cm, clawed, obovate, brownish with yellow apices. Lip larger than the sepals, 3-lobed, with large lateral lobes and a rather small central lobe which is notched at the apex, all pale yellow, reddish at the base, with a callus consisting of 4 swellings. Wings of the column sickle-shaped, deflexed. *E Brazil*. G2. Autumn.

***O. pulvinatum** Lindley. Illustration: Cogniaux & Goossens, Dictionnaire Iconographique des Orchidées, Oncidium, t. 10 (1897). A much larger plant with the lateral lobes of the lip relatively smaller and undulate-crisped. *S & E Brazil*.

Section **Glanduligera** Lindley.
Pseudobulbs conspicuous; upper sepal and petals linear, erect, antenna-like, lateral sepals curving downwards, much broader than the upper.

13. O. papilio Lindley. Figure 20(13), p. 283. Illustration: Edwards's Botanical Register **11**: t. 90 (1825); Botanical Magazine, 3733 (1840); Cogniaux & Goossens, Dictionnaire Iconographique des Orchidées, Oncidium, t. 3 (1896); Dunsterville & Garay, Venezuelan orchids illustrated **2**: 254–5 (1961); Bechtel et al., Manual of cultivated orchid species, 247 (1981).
Pseudobulbs 3–5 cm, clustered, spherical, each bearing 1 leaf. Leaves 12–25 × 4–7 cm, ovate to elliptic, acute, green mottled with brown. Raceme or panicle to 1.2 m, erect, few-flowered, with only 1 or 2 flowers open at any one time, its axis conspicuously flattened towards the apex. Upper sepal and petals 8–13 cm, erect, linear, slightly expanded towards the apex, acute, with slightly wavy margins, purplish brown slightly mottled with greenish yellow. Lateral sepals oblong-lanceolate, crisped-undulate, brown with yellow transverse bands. Lip shorter than the lateral sepals, 3-lobed, the lateral lobes rather small, the central lobe clawed, almost circular to kidney-shaped, notched at the apex, mottled yellow and brown with a yellow centre, and with an obscurely 3-lobed callus. Wings of column oblong, lacerate, ending at the top in capitate, fleshy teeth. *Tropical S America from Trinidad to Peru*. G2. Flowering throughout the year.

14. O. kramerianum Reichenbach. Figure 20(14), p. 283. Illustration: Lindenia **6**: t. 246 (1890); Cogniaux & Goossens, Dictionnaire Iconographique des Orchidées, Oncidium, t. 26 (1900); Bechtel et al., Manual of cultivated orchid species, 245 (1981).
Very similar to *O. papilio*, but smaller, the axis of the panicle not flattened above, the central lobe of the lip rather shortly clawed. *Tropical America from Costa Rica to Ecuador*. G2. Autumn.

Section **Waluewa** (Regel) Schlechter.
Pseudobulbs conspicuous; lateral sepals united for most of their length; column downy with curved, forwardly projecting fleshy arms.

15. O. pubes Lindley. Figure 20(15), p. 283. Illustration: Edwards's Botanical Register **12**: t. 1007 (1826); Pabst & Dungs, Orchidaceae Brasilienses **2**: t. 1967 (1977); Bechtel et al., Manual of cultivated orchid species, 247 (1981).
Pseudobulbs 5–7 cm, cylindric, narrowed above, each bearing 1 leaf. Leaves 12–16 cm, lanceolate to elliptic, acute. Panicle 30–50 cm, erect, with many flowers. Sepals and petals 1–1.3 cm, obovate-oblong, brown with yellow transverse bars, the lateral sepals united for the whole of their length. Lip slightly shorter than the sepals, 3-lobed with small, spreading lateral lobes and a large, almost circular central lobe, yellow spotted with red-brown and bearing a large, tuberculate, downy callus. *Brazil*. G2. Summer.

Section **Stellata** Kränzlin.
Pseudobulbs conspicuous; sepals and petals and lip in 1 plane, the lip not very different from the sepals and petals, giving the flower the appearance of a 6-pointed star.

16. O. maculatum (Lindley) Lindley (*Cyrtochilum maculatum* Lindley). Figure 20(16), p. 283. Illustration: Cogniaux & Goossens, Dictionnaire Iconographique des Orchidées, Oncidium, t. 19 (1898).
Pseudobulbs 7–10 cm, ovoid, compressed, bearing 2 leaves. Leaves 18–25 × 2.5–5 cm, linear to oblong or oblong-elliptic, obtuse or acute. Raceme or panicle erect, up to 1 m. Flowers fragrant. Sepals and petals 1.5–3 cm, oblong-elliptic to oblong-lanceolate, pale yellow or yellow-green blotched with brown. Lip oblong, 3-lobed at about the middle, the lateral lobes directed forwards, triangular, the central lobe oblong, notched or rounded at the apex, white with red stripes at the base, yellow at the apex, with a callus of 4 united ridges which end in upright teeth. Wings of column small and inconspicuous. *Mexico, Guatemala, Honduras*. G2. Summer.

17. O. hastatum (Bateman) Lindley (*Odontoglossum hastatum* Bateman). Figure 20(17), p. 283. Illustration: Bechtel et al., Manual of cultivated orchid species, 244 (1981).

Similar to *O. maculatum* but the lip mostly pink with oblong or almost square, outwardly directed lateral lobes, the central lobe ovate, acute. *S Mexico*. G2.

Section **Rostrata** Rolfe.
Pseudobulbs conspicuous; rostellum elongate, proboscis-like.

18. O. ornithorrhynchum Humboldt, Bonpland & Kunth. Figure 20(18), p. 283. Illustration: Edwards's Botanical Register **26**: t. 10 (1840); Botanical Magazine, 3912 (1841); Cogniaux & Goossens, Dictionnaire Iconographique des Orchidées, Oncidium, t. 22 (1899); Bechtel et al., Manual of cultivated orchid species, 247 (1981).
Pseudobulbs 2.5–9 cm, ovoid or ellipsoid, each bearing 2–3 leaves. Leaves 10–40 × 1–3 cm, linear-lanceolate to linear-elliptic, acute. Panicle up to 50 cm, erect. Sepals and petals 7–11 mm, shortly clawed, oblong–elliptic to oblanceolate-elliptic, pink or pinkish purple, obtuse. Lip 3-lobed, longer than the sepals, the lateral lobes small, rolled downwards and clasping the bases of the lateral sepals, the central lobe clawed, obovate, deeply notched at the apex, pink or pinkish purple with a small, reddish brown-spotted callus. Wings of the column broadly triangular from a narrow base, finely toothed. *C America, from S Mexico to Costa Rica*. G2. Autumn.

19. O. cheirophorum Reichenbach. Figure 20(19), p. 283. Illustration: Botanical Magazine, 6278 (1877); Lindenia **3**: t. 126 (1887); Bechtel et al., Manual of cultivated orchid species, 242 (1981).
Pseudobulbs 3–4 cm, ovoid, each with 1 leaf. Leaves to 15 cm, linear-oblong, acute. Panicle to 25 cm, erect or arching, with many fragrant flowers. Sepals and petals 6–13 mm, ovate, rounded at the apex, yellow. Lip 3-lobed, longer than the sepals, the lateral lobes spreading, rounded, the central lobe stalkless, oblong to almost circular, notched at the apex, yellow with a conspicuous white callus with 5 teeth. Wings of the column obovate, entire. *Costa Rica, Panama, Colombia*. G2. Winter.

Section **Concoloria** Kränzlin.
Pseudobulbs conspicuous; lateral sepals united; lip entire and bearing a callus made up of 2 or 4 ridges.

20. O. concolor Hooker. Figure 21(20), p. 284. Illustration: Botanical Magazine, 3752 (1840); Cogniaux & Goossens, Dictionnaire Iconographique des Orchidées, Oncidium, t. 32 (1904); Pabst & Dungs, Orchidaceae Brasilienses **2**: t. 1989 (1977); Bechtel et al., Manual of cultivated orchid species, 242 (1981).
Pseudobulbs to 5 cm, ovoid, grooved, each bearing 2 leaves. Leaves to 15 cm, linear-lanceolate. Raceme with 6–12 flowers, arching. Sepals and petals 2.5–3 cm, oblong-elliptic, acute, yellow, the lateral sepals united for about half their length. Lip larger than the sepals, shortly clawed, diamond-shaped, slightly notched at the apex, yellow, bearing 2 red or orange ridges. Wings of the column oblong, entire, directed forwards. *SE Brazil, N Argentina*. G2. Spring.

Section **Barbata** Pfitzer.
Pseudobulbs conspicuous; lateral sepals variously united, longer than the lip; callus of lip compound, composed of an uneven number of plates or tubercles.

21. O. longipes Lindley. Figure 21(21), p. 284. Illustration: Botanical Magazine, 5193 (1860); Bechtel et al., Manual of cultivated orchid species, 246 (1981).
Pseudobulbs 2–3 cm, elongate-ovoid, each with 2 leaves. Leaves 10–15 cm, oblong, acute. Raceme with 2–6 flowers, loose, scarcely exceeding the leaves. Sepals and petals 1.4–1.8 cm, oblong, rounded and finally pointed at the apex, yellow spotted with red-brown below or dark purple-brown throughout. Lip 3-lobed, the lateral lobes broadly obovate, attached to the broadly kidney-shaped central lobe by a broad, toothed claw, yellow with a central brown spot, bearing a large, yellow-spotted ridge with 3 teeth at its apex, the central tooth larger and blunter than the other 2. Wings of the column small, toothed. *SE Brazil*. G2. Spring.

Section **Crispa** Pfitzer
Pseudobulbs conspicuous; sepals and petals clawed, crisped, the lateral sepals variously united, the petals more conspicuous than the sepals; lip with a callus formed from an uneven number of tubercles.

22. O. crispum Loddiges. Figure 21(22), p. 284. Illustration: Botanical Magazine, 3499 (1836); Edwards's Botanical Register **23**: t. 1920 (1837); Cogniaux & Goossens, Dictionnaire Iconographique des Orchidées, Oncidium, t. 6, 6A (1897); Bechtel et al., Manual of cultivated orchid species, 242 (1981).
Pseudobulbs to 10 cm, ovoid, grooved, each bearing 2 leaves. Leaves to 20 cm, narrowly lanceolate acute. Panicle with many flowers, loose, arching, axis bluish green. Sepals 2.5–3 cm, oblong, crisped, brown with yellow spots towards the base. Petals similar but broader. Lip a little longer than the sepals, obscurely 3-lobed with very small lateral lobes and a large, kidney-shaped, strongly crisped central lobe which is brown with a yellow centre, bearing a prominent callus with an uneven number of teeth. Wings of the column oblong, finely toothed. *E Brazil*. G2. Autumn–spring.

Autumn-flowering material with rather small flowers may prove to be *O. enderianum* Masters (see Fowlie, Orchid Digest **43**: 190–5, 1979).

23. O. forbesii Hooker. Figure 21(23), p. 284. Illustration: Botanical Magazine, 3705 (1839); Cogniaux & Goossens, Dictionnaire Iconographique des Orchidées, Oncidium, t. 1 (1896); Pabst & Dungs, Orchidaceae Brasilienses **2**: t. 2024 (1977); Bechtel et al., Manual of cultivated orchid species, 244 (1981).
Very similar to *O. crispum* but the inflorescence unbranched, the sepals and petals chestnut brown with yellow marbling around the margins and the column with entire, red-spotted wings. *E. Brazil*. G2. Autumn.

24. O. gardneri Lindley. Figure 21(24), p. 284. Illustration: Pabst & Dungs, Orchidaceae Brasilienses **2**: t. 2013 (1977); Bechtel et al., Manual of cultivated orchid species, 244 (1981).
Very similar to *O. crispum* but sepals and petals oblanceolate, brown with yellow stripes on the margin, the lip with very small lateral lobes, the central lobe *c.* 3 × 4.5 cm, mostly yellow with a zone of red-brown spots around the margin, brown at the base, column with entire wings. *E. Brazil*. G2. Summer.

25. O. marshallianum Reichenbach. Figure 21(25), p. 284. Illustration: Botanical Magazine, 5725 (1868); Cogniaux & Goossens, Dictionnaire Iconographique des Orchidées, Oncidium, t. 8 (1897); Pabst & Dungs, Orchidaceae Brasilienses **2**: t. 2008 (1977).
Pseudobulbs 7–10 cm, ovoid, each with 2 leaves. Leaves 20–30 cm, oblong-lanceolate, acute. Panicle to 1.5 m with many flowers, erect. Sepals 2.3–2.5 cm oblong-obovate, greenish yellow with brown spots or transverse stripes. Petals to 3 cm, notched at the apex, golden yellow with dark brown spots in the centre. Lip 3-lobed with small, yellow and red-spotted lateral lobes and a large, broadly kidney-shaped, notched, yellow

central lobe bearing a large, many-toothed callus. Wings of column oblong, entire. *E Brazil.* G2. Spring–summer.

Section **Synsepala** Pfitzer.
Pseudobulbs conspicuous; sepals and petals similar, inconspicuous, the lateral sepals united for part of their length; lip with a callus made up of an uneven number of tubercles.

26. O. varicosum Lindley. Figure 21(26), p. 284. Illustration: Cogniaux & Goossens, Dictionnaire Iconographique des Orchidées, Oncidium,t. 18, 18A (1898); Schlechter, Die Orchideen, edn 2, t. 11 (1927); Pabst & Dungs, Orchidaceae Brasilienses **2**: t. 2029 (1977).
Pseudobulbs 8–10 cm, elongate-ovoid, grooved, each bearing 2 leaves. Leaves 15–25 cm, oblong, acute. Panicle 80–150 cm, with many flowers, loosely branched. Flowers very variable in size. Sepals and petals 5–7 mm, oblong, lateral sepals united for about half their length, yellowish green spotted with brown. Lip 3-lobed, much larger than the sepals, the lateral lobes small, rounded, the central lobe 2.5–4 × 3.5–5.5 cm, 3-notched towards the apex to produce 4 small lobes, golden yellow, bearing a callus with several teeth. Column with entire wings. *E & C Brazil.* G2. Winter.

27. O. flexuosum Loddiges Figure 21(27), p. 284. Illustration: Botanical Magazine, 2203 (1821); Reichenbach, Flora Exotica **2**: t. 94 (1834); Pabst & Dungs, Orchidaceae Brasilienses **2**: t. 2024 (1977); Bechtel et al., Manual of cultivated orchid species, 243 (1981).
Pseudobulbs 3–4 cm, compressed, rather distant, each bearing 1–2 leaves. Leaves 10–20 cm, oblong, somewhat acute. Panicle to 80 cm, with many flowers. Sepals and petals oblong, acute, yellow with red-brown spots at the base, the lateral sepals united at the base. Lip 3-lobed, much longer than the sepals, the lateral lobes small, curved erect, the central lobe 1.5–2 × 1.5–1.8 cm, clawed, kidney-shaped, notched at the apex, golden yellow, bearing a many-toothed callus. Wings of the column small, entire. *SE Brazil, Paraguay & Argentina.* G2. Autumn–winter.

Section **Excavata** Kränzlin.
Pseudobulbs conspicuous; sepals and petals clawed, the lateral sepals free, the petals broader and more conspicuous than the sepals; callus of lip with an uneven number of tubercles.

28. O. excavatum Lindley. Figure 21(28), p. 284. Illustration: Botanical Magazine, 5293 (1862); Cogniaux & Goossens, Dictionnaire Iconographique des Orchidées, Oncidium, t. 20 (1899); Bechtel et al., Manual of cultivated orchid species, 243 (1981).
Pseudobulbs 7–12 cm, oblong-ovoid, somewhat compressed, each with 1–2 leaves. Leaves 30–50 × 2.5–4 cm, linear-oblong, acute. Panicle to 1.5 m, erect,with many flowers. Sepals 1.2–1.6 cm, oblanceolate-oblong, rounded and abruptly acute, the laterals narrower than the upper, all yellow spotted with red towards the base. Petals similar but slightly longer and much broader. Lip longer than the sepals, 3-lobed, the lateral lobes small, the central lobe 1.2–2.2 × 2 cm, clawed, yellow, red towards the base with a complex toothed and grooved callus. Wings of the column oblong, slightly notched. *Ecuador, Peru.* G2. Spring.

Section **Oblongata** Kränzlin.
Pseudobulbs conspicuous; lateral sepals free; bracts small and inconspicuous, scale-like.

29. O. ampliatum Lindley. Figure 21(29), p. 284. Illustration: Edwards's Botanical Register **20**: t. 1699 (1835); Fieldiana **30**: 859 (1961); Dunsterville & Garay, Venezuelan orchids illustrated **2**: 242–3 (1961); Bechtel et al., Manual of cultivated orchid species, 240 (1981).
Pseudobulbs 3–7.5 cm, spherical or pear-shaped, compressed, each bearing 1 leaf. Leaves 8–50 × 2.5–10 cm, variable in shape, obtuse or broadly rounded. Panicle to 80 cm with many flowers. Upper sepal 7–9 mm, obovate, hooded, the lateral sepals longer and narrower, all yellow with small brown spots. Petals distinctly clawed, yellow with small brown spots. Lip longer than the sepals, 3-lobed, the lateral lobes small, the central shortly clawed, kidney-shaped, notched at the apex, yellow, with a small callus which has 2 broad lateral lobes and 3 smaller lobes in front. Wings of the column toothed. *Tropical America from Guatemala to Peru.* G2. Spring.

30. O. tigrinum Llave & Lexarza. Figure 21(30), p. 284. Illustration: Illustration Horticole **22**: t. 221 (1875); Cogniaux & Goossens, Dictionnaire Iconographique des Orchidées, Oncidium, t. 4, 4A, 4B (1897); Bechtel et al., Manual of cultivated orchid species, 249 (1981).
Pseudobulbs 7–9 cm, spherical, with 2 leaves. Leaves 20–30 cm, narrowly oblong, acute, thinly leathery. Panicle to 80 cm, little branched, with many fragrant flowers. Sepals and petals similar, 2.5–3.5 cm, spreading, oblong to obovate, yellow with heavy brown spotting and marbling. Lip much larger than the sepals, 3-lobed, the lateral lobes small, the central lobe clawed, spherical to kidney-shaped, golden yellow, with a large, 3-lobed callus. Wings of the column oblong, entire. *Mexico.* G2. Winter.

31. O. leucochilum Bateman. Figure 21(31), p. 284. Illustration: Cogniaux & Goossens, Dictionnaire Iconographique des Orchidées, Oncidium, t. 24 (1899).
Pseudobulbs 5–13 cm, ovoid, compressed, each with 1–2 leaves. Leaves 10–60 × 1.5–4.5 cm, oblong, obtuse to acute. Panicle to 3 m. Sepals and petals 1.3–2.3 cm, similar, oblong-elliptic to oblanceolate, bright green or greenish white blotched with red. Lip larger than the sepals, 3-lobed, the lateral lobes small, oblong to ovate, the central clawed, transversely oblong to kidney-shaped, notched at the apex, white, sometimes with red lines, with a purplish callus which ends in 5–9 slender teeth, some of them upcurved. Wings of the column oblong, irregularly scalloped. *Mexico, Guatemala, Honduras.* G2. Winter–spring.

The following species, also belonging to this section, is grown occasionally.
O. reflexum Lindley. Sepals and petals conspicuously reflexed, violet-brown tipped with yellow. *Mexico.* G2. Autumn.

Section **Planifolia** Bentham & Hooker.
Pseudobulbs conspicuous; bracts conspicuous, as long as or little shorter than the stalked ovaries; lip with a callus composed of an uneven number of tubercles.

32. O. sphacelatum Lindley. Figure 21(32), p. 284. Illustration: Edwards's Botanical Register **28**: t. 30 (1842); Dunsterville & Garay, Venezuelan orchids illustrated **4**: 196–7 (1966); Bechtel et al., Manual of cultivated orchid species, 248 (1981).
Pseudobulbs to 20 cm, ovoid-ellipsoid, tapering, each with 2 leaves. Leaves to 100 × 3.5 cm, linear-oblong or linear-lanceolate, acute. Panicle to 1.5 m, with many flowers, erect. Sepals 10–20 × 3–6.5 mm, clawed, elliptic to elliptic-obovate, yellow with reddish brown blotches. Petals similar but not clawed. Lip about as long as the sepals, 3-lobed, wider across the central lobe than across the lateral lobes, central lobe large, semicircular, notched at the apex and often

at the sides as well, yellow, with a downy callus of 5–7 tubercles, some of which are divergent. Column with narrow, brown-bordered, slightly scalloped wings. *C America, Venezuela*. G2. Spring.

33. O. wentworthianum Lindley. Figure 21(33), p. 284. Illustration: Paxton's Flower Garden, f. 127 (1851–52); Die Orchidee **31**(5): clxxxv–clxxxvi (1980); Bechtel et al., Manual of cultivated orchid species, 249 (1981).
Pseudobulbs 7–10 cm, ovoid-ellipsoid, compressed, often mottled with brown, each bearing 2 leaves. Leaves 13–35 × 1.5–3 cm, linear-oblong to lanceolate. Panicle up to 1 m, with many flowers, arching. Sepals and petals 1.4–2.2 cm, spreading or somewhat reflexed, elliptic or elliptic-obovate, deep yellow irregularly blotched with brown. Lip 3-lobed, about as long as the sepals, wider across the lateral lobes than across the central lobe, the lateral lobes small, curved forwards, scalloped on their margins, the central lobe reversed-heart-shaped, deeply notched at the apex, yellow, with a fleshy reddish brown callus consisting of 3 teeth flanked by ridges. Wings of column triangular, finely scalloped, bordered with red-brown spots. *Mexico, Guatemala*. G2. Summer.

The following species, also belonging to this section, is occasionally grown:
O. incurvum Baker. Illustration: Edwards's Botanical Register **31**: t. 64 (1845); Cogniaux & Goossens, Dictionnaire Iconographique des Orchidées, Oncidium, t. 29 (1900); Bechtel et al., Manual of cultivated orchid species, 245 (1981). Sepals, petals and lip white with red spots. *Mexico*. Autumn.

166. SIGMATOSTALIX Reichenbach
J. Cullen
Small, rhizomatous epiphytes. Pseudobulbs simple, clustered or distant, each bearing 1–2 leaves. Leaves very narrow, grass-like, folded when young. Raceme (rarely branched) erect, arising from the bases of the pseudobulbs. Sepals and petals similar, widely spreading to reflexed. Lip conspicuously clawed, usually simple, with a callus near the base. Column very slender, arched, widened above, without a foot. Pollinia 2, ovoid, stipe oblong, expanded above with its sides upturned, longer than the pollinia and the small, almost circular viscidium.

A genus of about 12 species from C & S America. Cultivation as for *Odontoglossum* (p. 274).
Literature: Kränzlin, F., *Das Pflanzenreich* **80**: 301–12 (1922).

1. S. graminea (Poeppig & Endlicher) Reichenbach. Illustration: Reichenbach, Xenia Orchidacearum **1**: t. 8 (1854).
Pseudobulbs 7–12 mm, clustered, oblong-cylindric to cylindric-ellipsoid. Leaves 3–5 cm × 2 mm, narrowly linear, acute or obtuse, obliquely notched at the apex. Raceme erect or arching. Sepals 2–2.5 mm, oblong or elliptic-lanceolate, acute. Petals slightly broader than sepals, both yellow with purple spots. Lip shortly and broadly clawed, blade ovate to almost square, notched, with rounded angles at the base. *Peru*. G2.

GLOSSARY

achene. A small, dry, indehiscent, 1-seeded fruit, in which the wall is of membranous consistency and free from the seed.

acuminate. With a long, slender point (figure 22(24), p. 292).

adpressed. Closely applied to a leaf or stem and lying parallel to its surface but attached only at the base.

adventitious. (1) Of roots: arising from a stem or leaf, not from the primary root derived from the radicle of the seedling. (2) Of buds: arising somewhere other than in the axil of a leaf.

aggregate fruit. A collection of small fruits, each derived from a single free carpel, closely associated on a common receptacle, but not united. *Ranunculus* and *Rubus* provide familiar examples.

alternate. Arising singly, 1 at each node; not in opposite pairs or whorled (figure 22(2), p. 292).

anatropous. Describes an ovule which turns through 180° in the course of development, so that the micropyle is near the base of the funicle (figure 25(2), p. 295).

androgynophore. A stalk which raises the stamens and carpels some distance above the petals in some flowers.

annual. A plant which completes its life-cycle from seed to seed in less than a year.

anther. The uppermost part of a stamen, containing the pollen (figure 24(3 & 4), p. 294).

apical. Describes the attachment of an ovule to the apex of a 1-celled ovary (figure 25(8), p. 295).

apiculate. With a small point.

apomictic. Reproducing by asexual means, though often through the agency of seeds, which are produced without the usual sexual fusion.

arachnoid. Describes hairs which are soft, long and entangled, suggestive of cobwebs.

aril. An outgrowth from the region of the hilum, which partly or wholly envelops the seed; it is usually fleshy.

ascending. Prostrate for a short distance at the base, but then curving upwards so that the remainder is more or less erect; sometimes used less precisely to mean pointing obliquely upwards.

attenuate. Drawn out to a fine point.

auricle. A lobe, normally 1 of a pair, at the base of the blade of a leaf, bract or sepal (figure 2(1), p. 32).

awn. A slender but stiff bristle on a glume, sepal or fruit (figure 2(6), p. 32).

axil. The upper angle between a leaf-base or leaf-stalk and the stem that bears it (figure 22(1), p. 292).

axile. A form of placentation in which the cavity of the ovary is divided by septa into 2 or more cells, the placentas being situated on the central axis (figure 25(10), p. 295).

axillary. Situated in or arising from an axil (figure 22(1), p. 292).

back-cross. A cross between a hybrid and a plant similar to one of its parents.

basal. (1) Of leaves: arising from the stem at or very close to its base. (2) Of placentation: describes the attachment of an ovule to the base of a 1-celled ovary (figure 25(6 & 7), p. 295).

basifixed. Attached to its stalk or supporting organ by its base, not by its back (figure 24(3), p. 294).

berry. A fleshy fruit containing 1 or more seeds embedded in pulp, as in the genera *Berberis*, *Ribes* and *Phoenix*. Many fruits (such as those of *Ilex*) which look like berries and are usually so called in popular speech, are in fact, *drupes*.

biennial. A plant which completes its life-cycle from seed to seed in a period of more than 1 year but less than 2.

bifid. Forked; divided into 2 lobes or points at the tip.

bilaterally symmetric. Capable of division into similar halves along 1 plane and 1 only (figure 24(9), p. 294).

bipinnate. (1) Of a leaf: with the blade divided pinnately into separate leaflets which are themselves pinnately divided (figure 22(18), p. 292). (2) Of an inflorescence: see p. 10 (Bromeliaceae).

bract. A leaf-like or chaffy organ with a flower in its axil or forming part of an inflorescence, differing from a foliage leaf in size, shape, consistency or colour (figure 23(2), p. 293).

bracteole. A small, bract-like organ which occurs on the flower-stalk, above the bract, in some plants.

bulb. A seasonally dormant underground bud, usually fairly large, consisting of a number of fleshy leaves or leaf-bases.

bulbil. A small bulb, especially one borne in a leaf axil or in an inflorescence.

calyx. The sepals; the outer whorl of a perianth (figure 24(1), p. 294).

campylotropous. Describes an ovule which becomes curved during development and lies with its long axis at right angles to the *funicle* (figure 25(4), p. 295).

capitate. Compact and approximately spherical, head-like.

capitulum. An inflorescence consisting of small flowers (florets), usually numerous, closely grouped together so as to form a 'head', and often provided with a common involucre of bracts.

capsule. A dry, dehiscent fruit derived from 2 or more united carpels and usually containing numerous seeds.

carpel. One of the units (sometimes interpreted as modified leaves) situated in the centre of a flower and together constituting the gynaecium or female part of the flower. If more than 1, they may be free or united. They contain ovules and bear a stigma (figure 24(1 & 2), p. 294).

caruncle. A soft, usually oil-rich appendage attached to a seed near the hilum.

catkin. A spike or spike-like panicle, often pendent, consisting of numerous unisexual flowers without perianths, each subtended by a fairly conspicuous bract.

chromosome. One of the microscopic, thread-like or rod-like bodies consisting of nucleic acid and containing the genes, which become visible in a cell nucleus shortly before cell division.

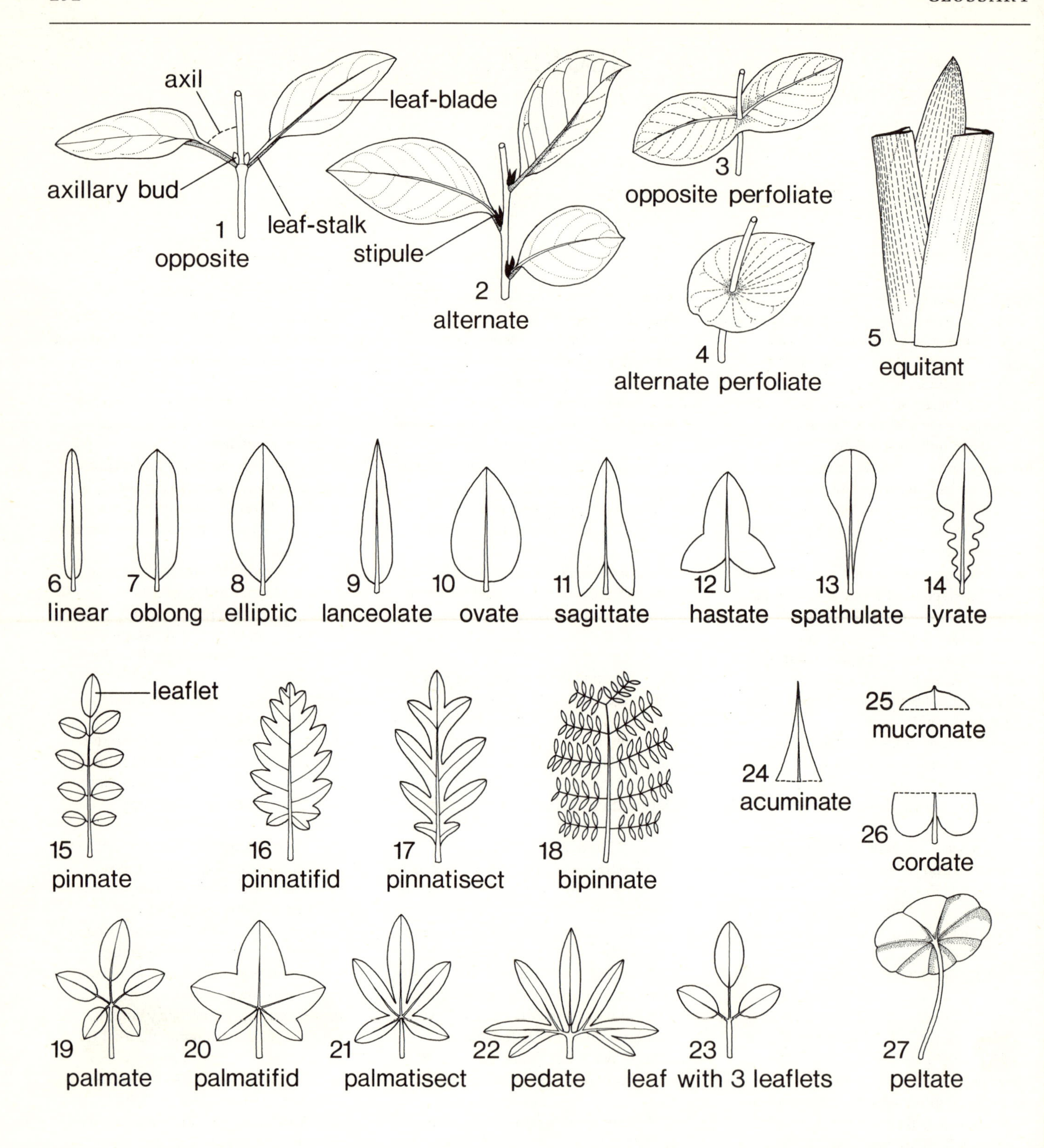

Figure 22. Leaves. 1–5, Leaf insertion types. 6–14, Leaf-blade outlines. 15–23, Leaf dissection types. 24–26, Leaf apex and base shapes. 27, Attachment of leaf-stalk to leaf-blade.

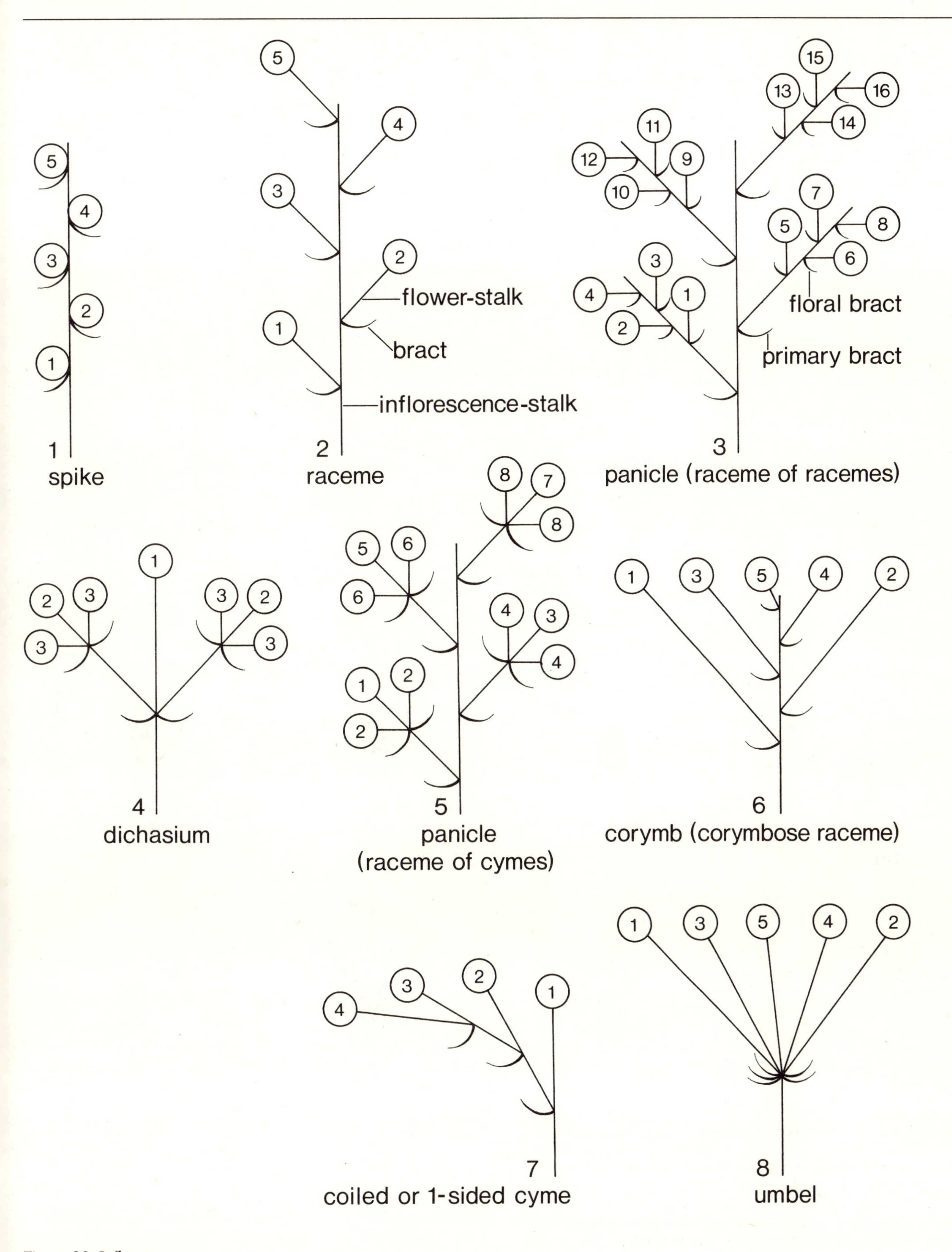

Figure 23. Inflorescences.

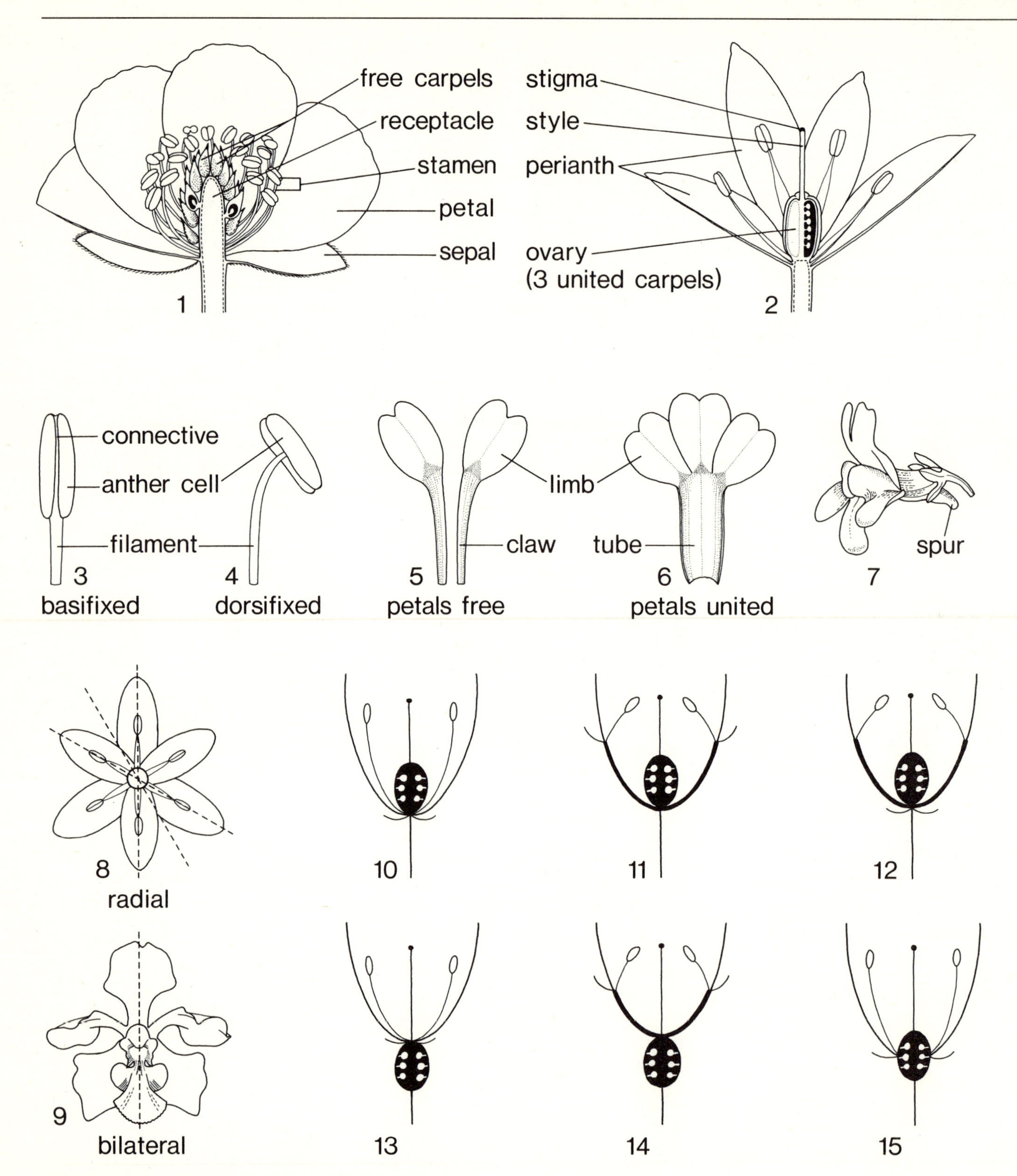

Figure 24. 1,2, Two flowers illustrating floral parts. 3,4, Two stamens showing alternative types of anther attachment. 5–7, Some terms relating to petals. 8,9, Floral symmetry, planes of symmetry shown by broken lines. 10–15, Position of ovary. 10–12, Superior ovaries. 13,14, Inferior ovaries. 15, Half-inferior ovary. 11, Perigynous zone bearing sepals, petals and stamens. 12, Perigynous zone bearing petals and stamens. 14, Epigynous zone bearing sepals, petals and stamens.

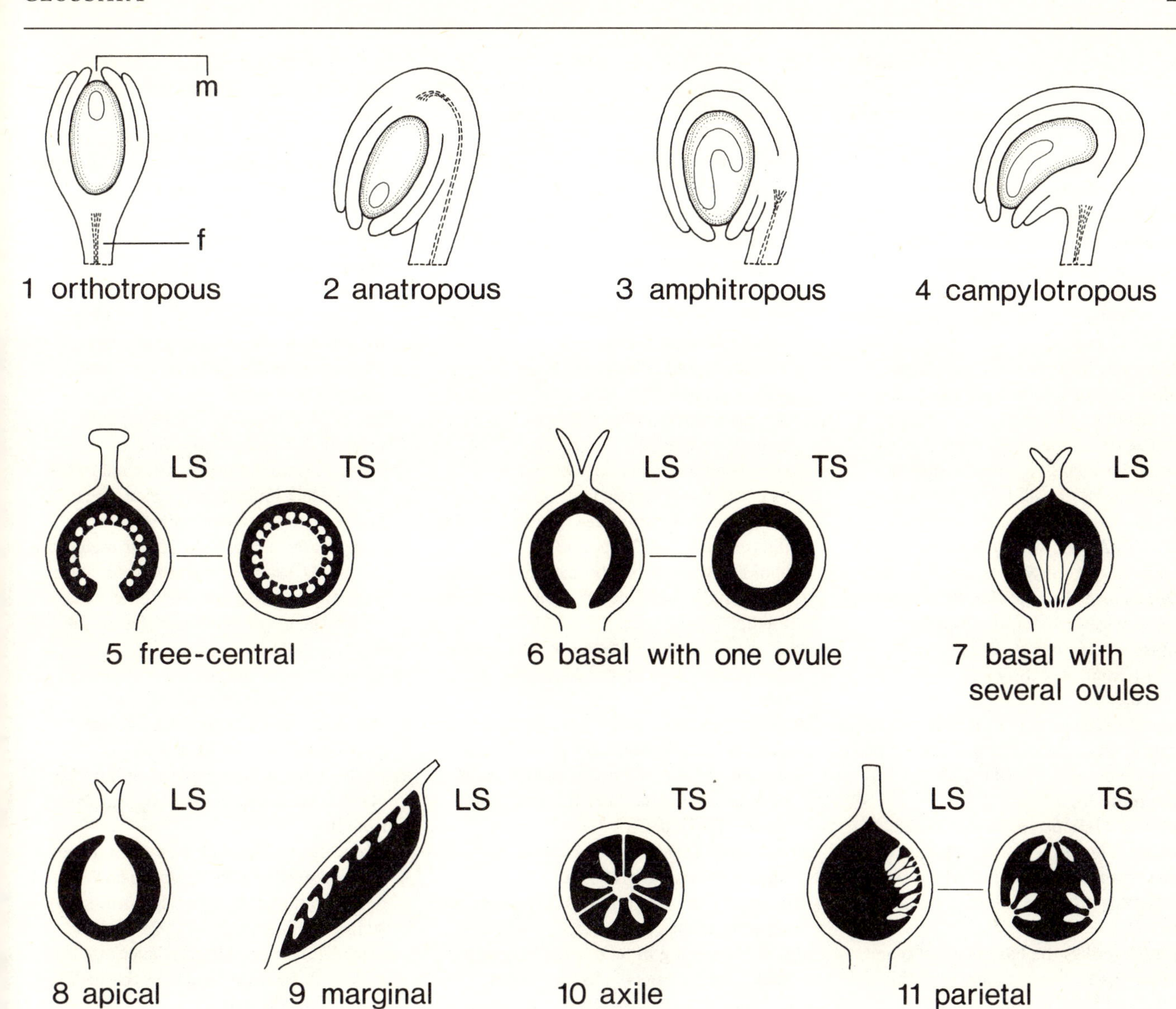

Figure 25. Ovules and placentation. 1–4, Ovule forms (f, funicle; m, micropyle). 5–11, Placentation types (LS, longitudinal section; TS, transverse section).

ciliate. Fringed on the margin with usually fine hairs.
circinate. Coiled at the tip, so as to resemble a crozier.
cladode. A branch which takes on the functions of a leaf (the leaves being usually vestigial). It may be flattened, as in *Ruscus*, or needle-like, as in many species of *Asparagus*.
claw. The narrow base of a petal or sepal, which widens above into the limb (figure 24(5), p. 294).
cleistogamous. Describes a flower with reduced corolla, which does not open but sets seed by self-pollination.
clone. The sum-total of the plants derived from the vegetative reproduction of a single individual, all having the same genetic constitution.
column. (1) A solid structure in the centre of a flower, consisting of the style and stigmas united to the stamen or stamens. It is characteristic of the Orchidaceae (figure 12(16–20), p. 142). (2) see p. 54 (Gramineae: *Stipa*).
compound. (1) Of a leaf: divided into separate leaflets. (2) Of an inflorescence: bearing secondary inflorescences in place of single flowers. (3) Of a fruit: derived from more than 1 flower.
compressed. Flattened from side to side.
connective. The tissue which separates the 2 lobes of an anther, and to which the filament is attached (figure 24(3), p. 294).
cordate. Describes the base of a leaf-blade which has 2 rounded lobes on either side of the central sinus (figure 22(26), p. 292).
corm. An underground, thickened stem-base, often surrounded by papery leaf-bases, and superficially resembling a bulb.
corolla. The petals; the inner whorl of the perianth (figure 24(1), p. 294).
corymb. A broad, flat-topped inflorescence. In the strict sense the term indicates a raceme in which the lowest flowers have stalks long enough to bring them to the level of the upper ones (figure 23(6), p. 293), but the term *corymbose* is often used to indicate a flat-topped, cymose inflorescence.
costapalmate. See p. 65 (Palmae).
cotyledon. One of the leaves preformed in the seed.
crisped. (1) Of hairs: strongly curved, so that the tip lies near the point of attachment. (2) Of leaves, leaflets or petals: finely and complexly wavy.
crownshaft. See p. 65 (Palmae).
culm. See p. 31 (Gramineae).
cultivar. A variant of horticultural interest or value, maintained in cultivation, but not conveniently assignable to an infraspecific category in botanical classification. A cultivar may arise in cultivation, or may be brought in from the wild. Its distinguishing name can be Latin in form, but is more usually in a modern language, and is written enclosed in single inverted commas, e.g. 'Alba', 'Madam Lemoine', 'Frühlingsgold', 'Beauty of Bath'.
cupule. A group of bracts, united at least at the base, surrounding the base of a fruit or a group of fruits.
cyme. An inflorescence in which the terminal flower opens first, other flowers being borne on branches arising below it (figure 23(4, 5 & 7), p. 293).
decumbent. More or less horizontal for most of its length, but erect or semi-erect near the tip.
decurrent. Continued down the stem below the point of attachment as a ridge or ridges.
dehiscent. Splitting, when ripe, along one or more predetermined lines of weakness.
dichasium. A form of cyme in which each node bears 2 equal lateral branches (figure 23(4), p. 293).
dichotomous. Dividing into 2 equal branches; regularly forked.
dioecious. With male and female flowers on separate plants.
diploid. Possessing in its normal vegetative cells 2 similar sets of chromosomes.
disc. (1) A variously contoured, ring-shaped or circular area (sometimes lobed) within a flower, from which nectar is secreted. (2) The flat central part of an orchid lip.
dissected. Deeply divided into lobes or segments.
dorsifixed. Attached to its stalk or supporting organ by its back, usually near the middle (figure 24(4), p. 294).
double. Of flowers: with petals much more numerous than in the normal wild type, often replacing the stamens.
drupe. An indehiscent fruit in which the outer part of the wall is soft and usually fleshy, but the inner part stony. A drupe may be 1-seeded as in *Prunus* or *Juglans*, or may contain several seeds, as in *Ilex*. In the latter case each seed is enclosed in a separate stony endocarp and constitutes a *pyrene*.
drupelet. A miniature drupe forming part of an aggregate fruit.
ellipsoid. As *elliptic* but applied to a solid body.
elliptic. About twice as long as broad, tapering equally both to the tip and the base (figure 22(8), p. 292).
embryo. The part of a seed from which the new plant develops; it is distinct from the endosperm and seed-coat.
endocarp. The inner, often stony layer of a fruit wall in those fruits in which the wall is distinctly 3-layered.
endosperm. A food-storage tissue found in many seeds, but not in all, distinct from the embryo and serving to nourish it and the young seedling during germination and establishment.
entire. With a smooth, uninterrupted margin; not lobed or toothed.
epicalyx. A group of bracts attached to the flower-stalk immediately below the calyx and sometimes partly united with it.
epigynous. Describing a flower, or preferably the petals, sepals and stamens (or perianth and stamens) of a flower in which the ovary is inferior (figure 24(13 & 14), p. 294).
epigynous zone. A rim or cup of tissue on which the sepals, petals and stamens are borne in some flowers with inferior ovaries (figure 24(14), p. 294).
epiphyte. A plant which grows on another plant but does not derive any nutriment from it.
equitant. Used of leaves folded so that they are V-shaped in section at the base, the bases overlapping regularly, as in many Iridaceae (figure 22(5), p. 292).
exocarp. The outer, skin-like layer of a fruit wall in those fruits in which the wall is distinctly 3-layered.
filament. The stalk of a stamen, bearing the anther at its tip (figure 24(3 & 4), p. 294).
filius. Used with authority names to distinguish between father and son when both have given names to species, e.g. Linnaeus (C. Linnaeus, 1707–1778), Linnaeus filius (C. Linnaeus, 1741–1783, son of the former).
floret. A small flower, aggregated with others into a compact inflorescence.
follicle. A dry dehiscent fruit derived from a single free carpel, and with a single line of dehiscence.
free. Not united to any other organ except by its basal attachment.
free-central. A form of placentation in which the ovules are attached to the central axis of a 1-celled ovary (figure 25(5), p. 295).
fruit. The structure into which the gynaecium is transformed during the ripening of the seeds; a *compound fruit* is

derived from the gynaecia of more than 1 flower. The term 'fruit' is often extended to include structures which are derived in part from the receptacle (*Fragaria*), epigynous zone (*Malus*) or inflorescence-stalk (*Ficus*) as well as from the gynaecium.

funicle. The stalk of an ovule (figure 25(1), p. 295).

fusiform. Spindle-shaped; cylindric, but tapered gradually at both ends.

glandular. Of a hair: bearing at the tip a usually spherical knob of secretory tissue.

glaucous. Green strongly tinged with bluish grey.

glume. A small bract in the inflorescence of a grass (Gramineae) or sedge (Cyperaceae); also used in a narrower sense to denote the 2 small bracts at the base of a grass-spikelet (figure 2, p. 32).

graft-hybrid. A plant which, as a consequence of grafting, contains a mixture of tissues from 2 different species. Normally the tissues of 1 species are enclosed in a 'skin' of tissue from the other species.

gynaecium. The female organ (carpels) of a single flower, considered collectively, whether they are free or united.

half-inferior. Of an ovary: with its lower part inferior and its upper part superior (figure 24(15), p. 294).

hastate. With 2 acute, divergent lobes at the base, as in a mediaeval halberd (figure 22(12), p. 292).

hastula. See p. 65 (Palmae).

herb. A plant in which the stems do not become woody, or, if somewhat woody at the base, do not persist from year to year.

herbaceous. Of a plant: possessing the qualities of a herb as defined above.

hilum. The scar-like mark on a seed indicating the point at which it was attached to the funicle.

hybrid. A plant produced by the crossing of parents belonging to 2 different named groups (e.g. genera, species, subspecies, etc.). An F_1 hybrid is the primary product of such a cross. An F_2 hybrid is a plant arising from a cross between 2 F_1 hybrids (or from the self-pollination of an F_1 hybrid).

hydathode. A water-secreting gland immersed in the tissue of a leaf near its margin.

hypocotyl. That part of the stem of a seedling which lies between the top of the radical and the insertion of the cotyledon(s).

hypogynous. Describing a flower, or, preferably the petals, sepals and stamens (or perianth and stamens) of a flower in which the ovary is superior and the petals, sepals, and stamens (or perianth and stamens) arise as individual whorls on the receptacle (figure 24(12), p. 274).

incised. With deep, narrow spaces between the teeth or lobes.

included. Not projecting beyond the organs which enclose it.

indefinite. More than 12 and possibly variable in number.

indehiscent. Without preformed lines of splitting; opening, if at all, irregularly by decay.

induplicate. See p. 65 (Palmae).

inferior. Of an ovary: borne beneath the sepals, petals and stamens (or perianth and stamens) so that these appear to arise from its top (figure 24(13 & 14), p. 294).

inflorescence. A number of flowers which are sufficiently closely grouped together to form a structural unit (figure 23, p. 293).

infraspecific. Denotes any taxonomic unit below species level, such as subspecies, variety and form. To be distinguished from *subspecific*, which means relating to subspecies only.

internode. The part of a stem between 2 successive nodes.

interprimary vein. See p. 75 (Araceae).

involucre. A compact cluster of bracts at or near the base of a flower or inflorescence; sometimes reduced to a hair or ring of hairs (Gramineae, p. 31).

keel. A narrow ridge, suggestive of the keel of a boat, developed along the midrib (or rarely other veins) of a leaf, sepal or glume.

laciniate. With the margin deeply and irregularly divided into narrow and unequal teeth.

lanceolate. 3–4 times as long as wide and tapering more gradually towards the tip (figure 22(9), p. 292).

layer. To propagate by pegging down on the ground a branch from near the base of a shrub or tree, so as to induce the formation of adventitious roots.

leaflet. One of the leaf-like components of a compound leaf (figure 22(15), p. 292).

lemma. The lower and usually stouter of the 2 horny or membranous bracts which enclose the floret of a grass (Gramineae) (figure 2, p. 32).

lenticel. A small, slightly raised interruption of the surface of the bark or of the outer, corky layer of a fruit, through which air can penetrate to the inner tissues.

ligule. (1) A small membranous flap (more rarely a line of hairs) at the base of a leaf-blade (figure 2, p. 32). (2) A whitish area, superficially somewhat similar to the ligule of a grass, at the base of the upper surface of the leaf-blade in members of the Cyperaceae (figure 1, p. 11).

limb. A broadened part, furthest from the base, of a petal, corolla or similar organ, which has a relatively narrow basal part – the *claw* or tube (figure 24(5 & 6), p. 294).

linear. Parallel-sided and many times longer than broad (figure 22(6), p. 292).

lip. (1) That petal in an orchid flower which differs from the other 2; it usually occupies the lowest position in the flower and serves as an alighting place for insects (figure 11, p. 141). (2) A staminode or petal, suggestive of the lip of an orchid, in the flower of some other monocotyledonous families (e.g. Zingiberaceae).

lodicale. Small, scale-like structures in the floret of most grasses (Gramineae) which occur between the lemma and the reproductive organs. By their swelling they cause the lemma and palea to diverge at flowering time (figure 2(8), p. 32).

marginal. Placentation found in a single, free carpel, in which the ovules are attached to suture, like peas in a pod (figure 25(9), p. 295).

mentum. See p. 144 (Orchidaceae).

mericarp. A carpel, usually 1-seeded, released by the break-up at maturity of a fruit formed from 2 or more joined carpels.

mesocarp. The central, often fleshy layer of a fruit wall in those fruits in which the wall is distinctly 3-layered.

micropyle. A pore in the seed-coat (figure 25(1), p. 195).

monoecious. With separate male and female flowers on the same plant (in flowering plants male flowers may contain non-functional carpels and *vice versa*).

monopodial. A type of growth pattern in which the terminal bud continues growth from year to year (in this volume used only with reference to Orchidaceae – see p. 143).

mucronate. With a short narrow point at the apex (figure 22(25), p. 292).

nectary. A nectar-secreting gland.

neuter. Without either functional male or female parts.

node. The point at which 1 or more leaves or floral members are attached to an axis.

nut. A 1-seeded indehiscent fruit with a woody or bony wall.
nutlet. A small nut, usually a component of an aggregate fruit.
obconical. Shaped like a cone, but attached at the narrow end.
oblanceolate. As *lanceolate* but attached at the more gradually tapered end.
oblong. With more or less parallel sides and about 2–5 times as long as broad (figure 22(7), p. 292).
obovate. As *ovate* but attached at the narrower end.
obovoid. As *ovoid* but attached at the narrower end.
obsolete. Rudimentary; scarcely visible; reduced to insignificance.
opposite. Denotes 2 leaves, branches or flowers attached on opposite sides of the axis at the same node.
orthotropous. Describes an ovule which stands erect and straight (figure 25(1), p. 295).
ovary. The lower part of a carpel, containing the ovule(s) (i.e. excluding style and stigma); the lower, ovule-containing part of a gynaecium in which the carpels are united (figure 24(2), p. 294).
ovate. With approximately the outline of a hen's egg (though not necessarily blunt-tipped) and attached at the broader end (figure 22(10), p. 292).
ovoid. As *ovate* but applied to a solid body.
ovule. The small egg-shaped body from which a seed develops after pollination (figure 25, p. 295).
palea. The upper, and usually smaller and thinner, of the 2 bracts enclosing the floret of grass (Gramineae) (figure 2, p. 32).
palmate. Describes a compound leaf composed of more than 3 leaflets, all arising from the same point, as in the leaf of *Aesculus* (figure 22(19), p. 292).
palmatifid. Lobed in a palmate manner, with the incisions pointing to the place of attachment, but not reaching much more than half way to it (figure 22(20), p. 292).
palmatisect. Deeply lobed in a palmate manner, with the incisions almost reaching the base (figure 22(21), p. 292).
panicle. A compound raceme, or any freely branched inflorescence of similar appearance (figure 23(3 & 5), p. 293).
papillose. Covered with small blunt protuberances (papillae).
parietal. A form of placentation in which the placentas are borne on the inner surface of the walls of a 1-celled ovary (figure 25(11), p. 295).
pedate. With a terminal lobe or leaflet, and on either side of it, an axis curving outwards and backwards, bearing lobes or leaflets on the outer side of the curve (figure 22(22), p. 292).
peltate. Describes a leaf or other flat structure with stalk attached other than at the margin (figure 22(27), p. 292).
perennial. Persisting for more than 2 years.
perfoliate. Describes a pair of stalkless, opposite leaves of which the bases are united, or a single leaf in which the auricles are united, so that the stem appears to pass through the leaf or leaves (figure 22(3 & 4), p. 292).
perianth. The calyx and corolla considered collectively, used especially when there is no clear differentiation between calyx and corolla; used also to denote a calyx or corolla when the other is absent (figure 24(2), p. 294).
perigynous. Describing a flower, or, preferably the petals, sepals and stamens (or perianth and stamens) of a flower in which the ovary is superior and the petals, sepals and stamens (or perianth and stamens) are borne on the margins of a rim or cup which itself is borne on the receptacle below the ovary (it often appears as though the sepals, petals and stamens (or perianth and stamens) are united at their bases) (figure 24(11 & 12), p. 294).
perigynous zone. The rim or cup of tissue on which the sepals, petals and stamens (or perianth and stamens) are borne in a perigynous flower (figure 24(11&12), p. 294).
petal. A member of the inner perianth whorl (corolla) used mainly when this is clearly differentiated from the calyx. The petals usually function in display and often provide an alighting place for pollinators (figure 24(1), p. 294).
phyllode. A leaf-stalk taking on the function and, to a variable extent, the form of a leaf-blade.
pinnate. Denotes a compound leaf or frond in which distinct leaflets are arranged on either side of the axis (figure 22(15), p. 292). If these leaflets are themselves of a similar compound structure the leaf or frond is termed *bipinnate.*
pinnatifid. Lobed in a pinnate manner, with the incisions reaching not much more than half way to the axis (figure 22(16), p. 292).
pinnatisect. Deeply lobed in a pinnate manner, with the incisions almost reaching the axis (figure 22(17), p. 292).
pistillode. See p. 65 (Palmae).
placenta. A part of the ovary, often in the form of a cushion or ridge, to which ovules are attached.
placentation. The manner of arrangement of the placentas (figure 25(5–11), p. 295).
pollen-sac. One of the cavities in an anther in which the pollen is formed; in flowering plants each anther normally contains 4 pollen-sacs, 2 on either side of the connective and separated by a partition which shrivels at maturity.
pollinium. Regularly shaped masses of pollen formed by a large number of pollen grains cohering (Orchidaceae: figure 12, p. 142).
polyploid. Possessing in its normal vegetative cells more than 2 sets of chromosomes.
protandrous. With anthers beginning to shed their pollen before the stigmas in the same flower are receptive.
protogynous. With stigmas becoming receptive before the anthers in the same flower shed their pollen.
pseudobulb. A swollen, above-ground internode, green (at least when young), characteristic of epiphytic Orchidaceae (figure 8, p. 138).
pulvinus. A swollen region at the base of a leaflet, leaf-blade or leaf-stalk.
pyrene. A small nut-like body enclosing a seed, 1 or more of which, surrounded by fleshy tissue, make up the fruit of, for example, *Ilex.*
raceme. An inflorescence consisting of stalked flowers arranged on a single axis, the lower opening first (figure 23(2), p. 293).
rachilla. See p. 31 (Gramineae).
radially symmetric. Capable of division into 2 similar halves along 2 or more planes of symmetry (figure 24(8), p. 294).
radicle. The root preformed in the seed and normally the first visible root of a seedling.
raphe. A perceptible ridge or stripe on some seeds.
receptacle. The tip of an axis to which the floral parts or (when present) perigynous zone are attached (figure 24(1), p. 294).
reduplicate. See p. 65 (Palmae).
reflexed. Bent sharply backwards from the base.
resupinate. See p. 144 (Orchidaceae).
rhizome. A horizontal stem, situated underground or on the surface, serving the purpose of food storage or vegetative reproduction or both; roots or stems arise from some or all of its nodes.

rootstock. The compact mass of tissue from which arise the new shoots of a herbaceous perennial. It usually consists mainly of stem tissue, but is more compact than is generally understood by *rhizome*.
rostellum. See p. 143 (Orchidaceae).
ruminate. See p. 65 (Palmae).
runner. A slender, above-ground stolon with very long internodes.
sagittate. With a backwardly directed basal lobe on either side, like an arrowhead (figure 22(11), p. 292).
samara. A dry indehiscent fruit, part of which is extended as a wing, as in *Acer* or *Ulmus*.
saprophytic. Dependent for its nutrition on soluble organic compounds in the soil. Saprophytic plants do not photosynthesise and lack chlorophyll; some plants, however, are *partially saprophytic* and combine the 2 modes of nutrition.
scale-leaf. A reduced leaf, usually not photosynthetic.
scape. A leafless flower-stalk or inflorescence-stalk arising from ground level.
scarious. Dry and papery, often translucent.
schizocarp. A fruit which, at maturity, splits into its constituent mericarps.
scion. A branch cut from 1 plant to be grafted on the rooted stock of another.
seed. A reproductive body adapted for dispersal, developed from an *ovule* and consisting of a protective covering (the seed-coat), an embryo and, usually, a food reserve.
semi-parasite. A plant which obtains only part of its nourishment by parasitism.
sepal. A member of the outer perianth whorl (calyx) when 2 whorls are clearly differentiated as calyx and corolla, or when comparison with related plants shows that the corolla is absent. The sepals most often function in protection and support of other floral parts (figure 24(1), p. 294).
septum. An internal partition.
sheath. The part of a leaf or leaf-stalk which surrounds the stem, being either tubular or with free but overlapping edges.
shrub. A woody plant with several stems or branches from near the base, and of smaller stature than a tree.
simple. Not divided into separate parts.
sinus. The gap or indentation between 2 lobes or teeth.
spadix. A spike with numerous florets borne on a usually fleshy axis which may project beyond the topmost floret, often wholly or partly enclosed by a large bract (*spathe*) (figure 4, p. 77).
spathe. A large bract at the base of a flower or inflorescence and wholly or partly enclosing it. The term is used in some families to denote collectively 2 or 3 such bracts (figure 4, p. 77).
spathulate. With a narrow basal part, which towards the apex is gradually expanded into a broad blade (figure 22(13), p. 292).
spike. (1) An inflorescence or subdivision of an inflorescence consisting of stalkless flowers arranged on a single axis (figure 2, p. 32 & figure 23(1), p. 293).
spikelet. A small spike forming one of the units of a complex inflorescence (figure 2, p. 32).
spur. An appendage or prolongation, more or less conical or cylindric, often at the base of an organ. The spur of a corolla or single petal or sepal is usually hollow and often contains nectar (figure 24(7), p. 294).
stamen. The male organ, producing pollen, generally consisting of an anther borne on a filament (figure 24(1), p. 294).
staminode. An infertile stamen, often reduced or rudimentary or with a changed function.
stellate. Star-like, used particularly of branched hairs.
stigma. The part of a style to which pollen adheres, normally differing in texture from the rest of the style (figure 24(2), p. 294).
stipe. (1) See p. 143 (Orchidaceae). (2) See p. 75 (Araceae).
stipule. An appendage, usually 1 of a pair, found beside the base of the leaf-stalk in many flowering plants, sometimes falling early, leaving scars. In some cases the 2 stipules are united; in others they are partly united to the leaf-stalk.
stock. A rooted plant, often with the upper part removed, on to which a *scion* may be grafted.
stolon. A far-creeping, more or less slender, above-ground or underground stem giving rise to a new plant at its tip and sometimes at intermediate nodes.
style. The usually slender upper part of a carpel or gynaecium, bearing the *stigma* (figure 24(2), p. 294).
subtend(ed). Used of any structure (e.g. a flower) which occurs in the axil of another organ (e.g. a bract); in this case the bract subtends the flower. A diagram in figure 22(1), p. 292, shows a leaf subtending a bud.
subulate. Narrowly cylindric, and somewhat tapered to the tip.
sucker. An erect shoot originating from a bud on a root or a rhizome sometimes at some distance from the stem of the parent plant.
superior. Of an ovary: borne at the morphological apex of the flower so that the petals, sepals and stamens (or perianth and stamens) arise on the receptacle below the ovary (figure 24(10–12), p. 294).
suture. A line marking an apparent junction of neighbouring organs.
sympodial. A type of growth pattern in which the terminal bud ceases growth (often by the production of an inflorescence), further growth being carried on by lateral buds (in this volume used only with reference to Orchidaceae – see p. 143).
tendril. A thread-like structure which, by its coiling growth, can attach a shoot to something else for support.
terete. Approximately circular in cross-section; not necessarily perfectly cylindric, but without grooves or ridges.
throat. The part of a calyx or corolla transitional between tube and limb.
triploid. Possessing in its normal vegetative cells 3 similar sets of chromosomes. Most triploid plants are highly sterile.
truncate. As though with the tip cut off.
tuber. A swollen underground stem or root used for food storage.
tubercle. A small, blunt, wart-like protuberance.
tunic. The dead covering of a bulb or corm.
umbel. An inflorescence in which the flower-stalks arise together from the top of an inflorescence-stalk; this is a *simple umbel* (figure 23(8), p. 293). In a *compound umbel* the several stalks arising from the top of the inflorescence-stalk terminate not in flowers, but in secondary umbels.
undivided. Without major divisions or incisions, though not necessarily entire.
urceolate. Shaped like a pitcher or urn, hollow and contracted at or just below the mouth.
vascular bundle. A strand of conducting tissue, usually surrounded by softer tissue.
vein. A vascular strand, usually in leaves or floral parts and visible externally.
venation. The pattern formed by the veins.
versatile. Of an anther: flexibly attached to the filament by its approximate mid-point so that a rocking motion is possible.

viscidium. See p. 143 (Orchidaceae).

viviparous. Bearing young plants, bulbils or leafy buds which can take root when they detach; these can occur anywhere on the plant, and may be interspersed with, or wholly replace, flowers in the inflorescence.

whorl. A group of more than 2 leaves or floral organs inserted at the same node.

wing. A thin, flat extension of a fruit, seed, sepal or other organ.

INDEX

Synonyms, and names mentioned only in observations, are printed in *italic* type.

Acacallis Lindley, 232
cyanea Lindley, 233
Acanthorrhiza warscewiczii Wendland, 67
Acanthostachys Klotzsch, 19
strobilacea (Schultes) Klotzsch, 19
Aceras Brown, 165
anthropophorum (Linnaeus) Aiton, 165
Acineta Lindley, 266
barkeri Lindley, 266
chrysantha (Morren) Lindley, 266
densa Lindley & Paxton, 266
humboldtii Lindley, 266
superba (Humboldt, Bonpland & Kunth) Reichenbach, 266
var. fulva (Hooker) Schlechter, 266
Acorus Linnaeus, 85
calamus Linnaeus, 85
gramineus Solander, 85
Acropera loddigesii Lindley, 269
Actinophloeus macarthurii (Wendland) Beccari, 73
Ada Lindley, 279
aurantiaca Lindley, 279
lehmannii Rolfe, 279
Aechmea Ruiz & Pavon, 20
amazonica Ule, 21
benrathii Mez, 22
blumenavii Reitz, 22
bromeliiflora (Rudge) Baker, 21
burchelii Baker, 19
caerulea Morren, 21
calyculata (Morren) Baker, 22
caudata Lindmann, 22
var. variegata Foster, 22
chantinii (Carrière) Baker, 21
coelestis (Koch) Morren, 22
var. albo-marginata Foster, 22
distichantha Lemaire, 21
excavata Baker, 21
fasciata (Lindley) Baker, 21
fulgens Brongniart, 21
var. discolor (Morren) Brongniart, 21
gamosepala Wittmack, 23
gracilis Lindmann, 22
legrelliana (Baker) Baker, 22
lindenii (Morren) Baker, 22
var. makoyana Mez, 22
lueddemanniana (Koch) Mez, 21
marmorata (Lemaire) Mez, 23
miniata (Beer) Baker, 21
var. discolor (Baker) Baker, 21
nudicaulis (Linnaeus) Grisebach, 22
var. aureo-rosea (Antoine) Smith, 22
ortgiesii Baker, 22
platyphylla Hassler, 21
purpureo-rosea (Hooker) Wawra, 21
recurvata (Klotzsch) Smith, 22
var. benrathii (Mez) Reitz, 22
var. ortgiesii (Baker) Reitz, 22
rhodocyanea invalid, 21
suaveolens Knowles & Westcott, 21
tinctoria (Martius) Mez, 21
Aegilops Linnaeus, 44
ovata Linnaeus, 44
Aerangis Reichenbach, 255
articulata (Reichenbach) Schlechter, 255
biloba (Lindley) Schlechter, 255
citrata (Thouars) Schlechter, 255
ellisii (Reichenbach) Schlechter, 255
hyaloides (Reichenbach) Schlechter, 256
kotschyana (Reichenbach) Schlechter, 255
kotschyi (Reichenbach) Reichenbach, 255
luteo-alba (Kränzlin) Schlechter var. rhodosticta (Kränzlin) Stewart, 256
modesta (Hooker) Schlechter, 255
rhodosticta (Kränzlin) Schlechter, 256
stylosa (Rolfe) Schlechter, 255
Aerides Loureiro, 242
affine Lindley, 243
ballantiana Reichenbach, 243
brookei Lindley, 243
cornuta Roxburgh, 243
crassifolia Parrish & Reichenbach, 243
crispa Lindley, 243
cylindrica Hooker, 242
cylindrica Lindley, 242
expansa Reichenbach, 243
falcata Lindley, 243
var. *houlletiana* (Reichenbach) Veitch, 243
fieldingii Williams, 243
flabellata Downie, 243
houlletiana Reichenbach, 243
japonica Linden & Reichenbach, 243
jarckiana Schlechter, 243
larpentae Reichenbach, 243
lawrenceae Reichenbach, 243
multiflora Roxburgh, 243
odorata Loureiro, 243
quinquevulnera Lindley, 243
suavissima Lindley, 243
vandara Reichenbach, 242
Afromomum Schumann, 127
melegueta Schumann, 128
Aganisia cyanea (Lindley) Reichenbach, 233
Aglaonema Schott, 94
acutispathum Brown, 95
brevispathum (Engler) Engler, 95
forma brevispathum, 95
forma hospitum, 95
commutatum Schott, 95
costatum Brown, 95
forma concolor Nicolson, 95
forma costatum, 95
forma immaculatum (Ridley) Nicolson, 95
forma virescens (Engler) Nicolson, 95
crispum (Pitcher & Manda) Nicolson, 95
modestum Engler, 95
nitidum (Jack) Kunth, 96
pictum (Roxburgh) Kunth, 95
roebelinii Pitcher & Manda, 95
Agrostis Linnaeus, 50
nebulosa Boissier & Reuter, 50
spica-venti Linnaeus, 51
Aira Linnaeus, 46
capillaris Host, 46
caryophyllea Linnaeus, 46
elegans Gaudin, 46
elegantissima Schur, 46
Allagoptera Nees, 75
arenaria (Gomes) Kuntze, 75
Alocasia Necker, 101
× *chantrieri* André, 102
cucullata (Loureiro) Don, 101
cuprea (Koch & Bouché) Koch, 101
indica misapplied, 101
var. *metallica* misapplied, 102
jenningsii Veitch, 100
korthalsii Schott, 103
lindenii Rodigas, 90
longiloba Miquel, 102
lowii Hooker, 102
var. *grandis*, 103
var. *veitchii* (Henderson) Engler, 102
macrorrhiza (Linnaeus) Don, 101
var. *rubra* (Hasskarl) Furtado, 102
marshallii Bull, 100
micholitziana Sander, 103
odora (Loddiges) Spach, 102
plumbea (Koch) Van Houtte, 102
portei Schott, 102
putzeysii Brown, 103
sanderiana Bull, 102
× *sedenii* Veitch, 102
thibautiana Masters, 103
veitchii (Lindley) Schott, 102
watsoniana Masters, 102
zebrina Koch & Veitch, 102
Alopecurus Linnaeus, 49
pratensis Linnaeus, 49
Alpinia Roxburgh, 128
calcarata Roscoe, 129
cernua Sims, 129

galanga (Linnaeus) Willdenow, 129
malaccensis (Burmann) Roscoe, 129
mutica Roxburgh, 129
nutans misapplied, 129
officinarum Hance, 129
purpurata (Veillard) Schumann, 129
rafflesiana Baker, 128
sanderae Sander, 128
speciosa (Wendland) Schumann, 129
tricolor Sander, 128
vittata Bull, 128
zerumbet (Persoon) Burtt & Smith, 129
zingiberina Hooker, 129
Ambrosina ciliata Roxburgh, 111
Amomum Roxburgh, 127
cardamomum misapplied, 127
compactum Maton, 127
kepulaga Sprague & Burkill, 127
Amorphophallus Blume, 89
bulbifer (Curtis) Blume, 90
rivieri Durieu, 89
Anacamptis Richard, 165
pyramidalis Richard, 165
Ananas Miller, 25
bracteatus (Lindley) Schultes, 25
var. tricolor (Bertoni) Smith, 25
comosus (Linnaeus) Merrill, 25
var. variegatus (Lowe) Moldenke, 25
sativus Schultes, 25
var. *bracteatus* (Lindley) Mez, 25
Ananassa bracteata Lindley, 25
Andropogon Linnaeus, 39
argenteus de Candolle, 39
gerardii Vitman, 40
ischaemum Linnaeus, 39
laguroides de Candolle, 39
saccharoides Swartz, 39
Angraecopsis falcata (Thunberg) Schlechter, 256
Angraecum Bory, 254
arcuatum Lindley, 256
articulatum Reichenbach, 255
bilobum Lindley, 255
citratum Thouars, 255
distichum Lindley, 254
eburneum Bory, 254
subsp. giryamae (Rendle) Senghas & Cribb, 254
subsp. superbum (Thouars) Perrier de la Bathie, 254
eichleranum Kränzlin, 254
ellisii Reichenbach, 255
falcatum (Thunberg) Lindley, 256
giryamae Rendle, 254
infundibulare Lindley, 254
hyaloides Reichenbach, 256
kotschyanum Reichenbach, 255
kotschyi Reichenbach, 255
modestum Hooker, 255
pellucidum Lindley, 256
rhodostictum Kränzlin, 256
rothschildianum O'Brien, 256
scottianum Reichenbach, 254
sesquipedale Thouars, 254
stylosum Rolfe, 255
superbum Thouars, 254
Anguloa Ruiz & Pavon, 231
brevilabris Rolfe, 231
cliftonii Rolfe, 231
clowesii Lindley, 231
var. *ruckeri* (Lindley) Foldats, 231
ruckeri Lindley, 231
var. albiflora Anon., 231
var. sanguinea Lindley, 231
uniflora Ruiz & Pavon, 231
virginalis Schlechter, 231
Anoectochilus Blume, 162
regalis Blume, 162
roxburghii (Wallich) Lindley, 162
setaceus (Blume) Lindley, 162
Anota densiflora (Lindley) Schlechter, 244
violacea (Lindley) Schlechter, 244
Ansellia Lindley, 259
africana Lindley, 259
gigantea Reichenbach var. *nilotica* (Brown) Summerhayes, 259
Anthoxanthum Linnaeus, 48
gracile Bivona, 48
Anthurium Schott, 80
acaule (Jacquin) Schott, 85
acaule misapplied, 85
aemulum Schott, 82
amnicola Dressler, 85
andraeanum André, 83
bakeri Hooker, 85
× carneum Regel, 84
× chelsiense Brown, 84
clarinervium Matuda, 83
clavigerum Poeppig & Endlicher, 82
corrugatum Sodiro, 82
crassinervium (Jacquin) Schott, 84
crassinervium misapplied, 84
crenatum Kunth, 85
crystallinum Linden & André, 83
× *cultorum* Birdsey, 84
digitatum (Jacquin) Don, 82
elegans Engler, 82
ellipticum Koch & Bouché, 84
enneaphyllum (Vellozo) Stellfeld, 82
× ferriense Masters & Moore, 84
fissum Regel, 82
forgetii Brown, 83
fortunatum Bunting, 82
fraternum misapplied, 84
fraternum Schott, 84
holtonianum Schott, 82
hookeri Kunth, 84
× *hortulanum* Birdsey, 85
huegelii Schott, 84
huixtlense Matuda, 84
kalkbreyeri Brown, 82
lilacinum Dressler, 85
lindenianum Koch & Augustin, 84
magnificum Linden, 83
nymphaeifolium Koch & Bouché, 84
ornatum Schott, 84
palmatum (Linnaeus) Don, 82
panduratum Martius, 82
papilionense invalid, 82
papillosum Markgraf, 82
pedatum Humboldt, Bonpland & Kunth, 82
pentaphyllum (Aublet) Don, 82
var. bombacifolium (Schott) Madison, 82
var. *digitatum* (Jacquin) Don, 82
var. pentaphyllum, 82
pictamayo invalid, 82
podophyllum, (Chamisso & Schlechtendahl) Kunth, 82
polyschistum Schultes & Idrobo, 82
putumayo invalid, 82
radicans Koch, 83
regale Linden, 83
× roseum Closon, 84
× *rothschildianum* Bergman & Veitch, 85
rugosum Schott, 84
scandens (Aublet) Engler, 85
scherzerianum Schott, 85
schlechtendahlii Kunth, 84
tetragonum (Hooker) Schott, 84
undatum Kunth, 82
veitchii Masters, 83
warocqueanum Moore, 83
watermaliense Bailey, 83
Anubias Schott, 94
afzelii Schott, 94
lanceolata misapplied, 94
Apera Adanson, 51
spica-venti (Linnaeus) Palisot de Beauvois, 51
Aplectrum (Nuttall) Torrey, 224
hyemale (Willdenow) Nutall, 224
spicatum (Walter) Britton et al., 224
Araceae, 75
Arachnanthe cathcartii (Lindley) Bentham & Hooker, 251
lowii (Lindley) Bentham & Hooker, 252
Arachnis cathcartii (Lindley) Smith, 251
lowii (Lindley) Reichenbach, 252
Archontophoenix Wendland & Drude, 72
alexandrae (Mueller) Wendland & Drude, 72
var. beatricae (Mueller) Bailey, 72
cunninghamiana (Wendland) Drude, 72
Areca lutescens misapplied, 72
Arecaceae, 65
Arecastrum (Drude) Beccari, 74
romanzoffianum (Chamisso) Beccari, 75
var. australe (Martius) Beccari, 75
Aregelia ampullacea (Morren) Mez, 17
burchellii (Baker) Mez, 19
carolinae (Beer) Mez, 17
concentrica (Vellozo) Mez, 18
marechalii Mez, 17
marmorata (Baker) Mez, 18
micrope (Mez) Mez, 19
morreniana (Antoine) Mez, 17
pineliana (Lemaire) Mez, 17
princeps Baker, 17
spectabilis (Moore) Mez, 18
tristis (Beer) Mez, 17
Arenga Labillardière, 71
engleri Beccari, 71
Arisaema Martius, 107
atrorubens (Aiton) Blume, 108
candidissimum Smith, 108
concinnum Schott, 109
consanguineum Schott, 109
dracontium (Linnaeus) Schott, 109
griffithii Schott, 108
japonicum Blume, 108
praecox Koch, 108
ringens (Thunberg) Schott, 108
sasenzoo misapplied, 108
serratum (Thunberg) Schott, 108

sikokianum Franchet & Savatier, 108
speciosum (Wallich) Martius, 108
triphyllum (Linnaeus) Schott, 108
Arisarum Targioni-Tozzetti, 107
proboscideum (Linnaeus) Savi, 107
simorrhinum Durieu, 107
vulgare Targioni-Tozzetti, 107
Arpophyllum Llave & Lexarza, 179
giganteum Lindley, 179
spicatum Llave & Lexarza, 179
Arrhenatherum Palisot de Beauvois, 47
avenaceum Palisot de Beauvois, 47
elatius (Linnaeus) Presl, 47
Arthrostylidium longifolium (Fournier) Camus, 59
Arum Linnaeus, 104
creticum Boissier & Heldreich, 105
dioscoridis Sibthorp & Smith, 105
dracunculus Linnaeus, 106
guttatum Wallich, 106
hygrophilum Boissier, 105
italicum Miller, 106
maculatum Linnaeus, 105
orientale Bieberstein, 105
ovatum Linnaeus, 109
palaestinum Boissier, 105
pictum Linnaeus, 105
venosum Aiton, 106
xanthorrhizon Jacquin, 100
Arundina Blume, 178
bambusifolium Lindley, 178
chinensis Blume, 178
graminifolia (Don) Hochreutiner, 178
revoluta Hooker, 178
speciosa Blume, 178
Arundinaria Michaux, 59
amabilis McClure, 59
anceps Mitford, 60
angustifolia (Mitford) Houzeau de la Haie, 61
aristata Gamble, 60
auricoma Mitford, 62
borealis (Hackel) Makino, 64
chino (Franchet & Savatier) Makino, 61
var. *argenteo-striata* (Regel) Makino, 61
disticha (Mitford) Pfitzer, 61
falcata Nees, 62
falconeri (Munro) Rivière, 60
fastuosa (Mitford) Makino, 58
fortunei invalid, 62
var. *aurea* invalid, 62
gauntletii invalid, 61
gracilis invalid, 62
graminea (Bean) Makino, 61
hindsii Munro, 61
hookeriana Munro, 62
humilis Mitford, 61
japonica Steudel, 64
jaunsarensis Gamble, 60
laydeckeri Bean, 61
longifolia Fournier, 59
maling Gamble, 60
marmorea (Mitford) Makino, 62
murielae Gamble, 60
narihira Makino, 58
narihita Makino, 58
niitikayamensis misapplied, 60
nitida Mitford, 60
nobilis Mitford, 60
palmata (Burbidge) Bean, 63
pantlingii misapplied, 60
pumila Mitford, 61
pygmaea (Miquel) Mitford, 61
quadrangularis (Fenzi) Makino, 63
racemosa misapplied, 60
ragamowskii (Nicolson) Pfitzer, 63
ramosa (Makino) Makino, 64
simonii (Carrière) Rivière, 62
var. *albo-striata* Bean, 62
var. *heterophylla* (Makino) Nakai, 62
var. *variegata* Hooker, 62
spathacea (Franchet) McClintock, 60
spathiflora Trinius, 60
tessellata (Nees) Munro, 60
vagans Gamble, 64
vaginatus Hackel, 61
variegata (Miquel) Makino, 62
var. *viridis* Makino, 62
veitchii (Carrière) Brown, 63
viridi-striata (André) Makino, 62
Arundo Linnaeus, 35
conspicua Forster, 36
donax Linnaeus, 35
richardii Endlicher, 35
Ascocentrum Schlechter, 252
ampullaceum (Lindley) Schlechter, 252
miniatum (Lindley) Schlechter, 252
Aspasia Lindley, 278
epidendroides Lindley, 279
lunata Lindley, 279
variegata Lindley, 279
Ataenidia conferta (Bentham) Milne-Redhead, 136
Avena Linnaeus, 46
fragilis Linnaeus, 48
ludoviciana Durieu, 47
sativa Linnaeus, 47
sterilis Linnaeus, 47
Avenula planiculmis (Schrader) Sauer & Chmelitschek, 47

Bambusa Schrader, 59
angulata invalid, 36
angustifolia Mitford, 61
argentea invalid, 62
castillonis Carrière, 57
disticha Mitford, 61
erecta invalid, 61
fastuosa Mitford, 58
glaucescens (Willdenow) Holttum, 59
gracilis misapplied, 61
longifolia (Fournier) McClure, 59
marmorea Mitford, 62
metake Miquel, 64
multiplex (Loureiro) Steudel, 59
nana Roxburgh, 59
palmata Burbidge, 63
quadrangularis Fenzi, 63
ragamowskii Nicholson, 63
ramosa Makino, 64
ruscifolia Munro, 58
tessellata Munro, 63
viridi-glaucescens Carrière, 58
Barkeria Knowles & Westcott, 191
elegans Knowles & Westcott, 191
lindleyana Anon., 191
skinneri (Lindley) Richard & Galeotti, 191
spectabilis Lindley, 191
Barlia Parlatore, 165
robertiana (Loiseleur) Greuter, 165
Belosynapsis kewensis Hasskarl, 28
Biarum Schott, 106
eximium (Schott & Kotschy) Engler, 107
tenuifolium (Linnaeus) Schott, 107
var. abbreviatum (Schott) Engler, 107
var. tenuifolium, 107
Bifrenaria Lindley, 226
atropurpurea (Loddiges) Lindley, 227
aurantiaca Lindley, 226
aureo-fulva (Hooker) Lindley, 226
furstenbergiana Schlechter, 226
harrisoniae (Hooker) Reichenbach, 226
inodora Lindley, 226
racemosa (Hooker) Lindley, 226
tetragona (Lindley) Schlechter, 227
tyrianthina (Louden) Reichenbach, 227
vitellina (Lindley) Lindley, 226
Billbergia Thunberg, 23
caespitosa Lindmann, 24
chantinii Carrière, 21
chlorosticta Saunders, 24
decora Poeppig & Endlicher, 24
distachya (Vellozo) Mez, 24
fasciata Lindley, 21
iridifolia (Nees & Martius) Lindley, 24
var. concolor Smith, 23
leptopoda Smith, 24
liboniana De Jonghe, 24
lietzei Morren, 24
marmorata Lemaire, 23
nutans Regel, 24
var. schimperiana (Baker) Mez, 24
purpureo-rosea Hooker, 21
pyramidalis (Sims) Lindley, 24
var. concolor Smith, 24
pyramidata Beer, 22
rhodocyanea Lemaire, 21
saundersiae Dombrain, 24
thyrsoidea Schultes, 24
zebrina (Herbert) Lindley, 24
Blephariglottis ciliaris (Linnaeus) Linnaeus, 170
Bletia Ruiz & Pavon, 175
catenulata Ruiz & Pavon, 175
graminifolia Don, 178
purpurea (Lamarck) De Candolle, 175
sanguinea Poeppig & Endlicher, 175
shepherdii Hooker, 175
sherrattiana Lindley, 175
verecunda (Salisbury) Brown, 175
watsonii Hooker, 175
Bletilla Reichenbach, 172
hyacinthina (Smith) Pfitzer, 172
sinensis (Rolfe) Schlechter, 172
striata (Thunberg) Reichenbach, 172
Boesenbergia Kuntze, 124
ornata (Brown) Smith, 124
pandurata (Roxburgh) Schlechter, 124
rotunda (Linnaeus) Mansfeld, 124
vittata (Brown) Loesener, 124
Bollea Reichenbach, 235
coelestis (Reichenbach) Reichenbach, 235
lalindei (Reichenbach) Reichenbach, 235
patinii Reichenbach, 235

Bonatea Willdenow, 169
speciosa (Linnaeus) Willdenow, 169
Bothriochilus Lemaire, 172
bellus Lemaire, 172
macrostachyus (Lindley) Williams, 172
Bothriochloa Kuntze, 39
ischaemum (Linnaeus) Keng, 39
saccharoides (Swartz) Rydberg, 39
Bouteloua Lagasca, 37
curtipendula (Michaux) Torrey, 37
gracilis (Kunth) Steudel, 37
oligostachya Torrey, 37
Brachychilum Petersen, 125
horsfieldii (Brown) Petersen, 125
Brachypodium Palisot de Beauvois, 43
sylvaticum (Hudson) Palisot de Beauvois, 44
Brahea Martius, 68
armata Watson, 68
calcarea Liebmann, 68
dulcis (Humboldt, Bonpland & Kunth) Martius, 68
edulis Wendland & Watson, 69
Brassavola Brown, 197
acaulis Lindley & Paxton, 198
cordata Lindley, 198
cucullata (Linnaeus) Brown, 198
cuspidata Hooker, 198
digbyana Lindley, 198
flagellaris Rodrigues, 198
fragrans Lemaire, 198
glauca Lindley, 198
lineata Hooker, 198
nodosa (Linnaeus) Lindley, 198
perrinii Lindley, 198
subulifolia Lindley, 198
tuberculata Hooker, 198
Brassia Lindley, 279
brachiata Lindley, 280
caudata (Linnaeus) Lindley, 280
clowesii (Lindley) Lindley, 282
gireoudiana Reichenbach & Warscewicz, 279
lanceana Lindley, 280
lawrenceana Lindley, 280
maculata Brown, 279
verrucosa Lindley, 280
Briza Linnaeus, 52
maxima Linnaeus, 52
media Linneaus, 52
minor Linnaeus, 52
Bromelia bicolor Ruiz & Pavon, 20
carolinae Beer, 17
comosa Linnaeus, 25
iridifolia Nees & Martius, 24
nudicaulis Linnaeus, 22
pyramidalis Sims, 24
tristis Beer, 17
zebrina Herbert, 24
Bromeliaceae, 10
Bromus Linnaeus, 45
briziformis Fischer & Meyer, 45
danthoniae Trinius, 45
japonicus Thunberg, 46
lanceolatus Roth, 45
madritensis Linnaeus, 45
squarrosus Linnaeus, 46
Broughtonia Brown, 190
coccinea Lindley, 190
domingensis (Lindley) Rolfe, 190
lilacina Henfrey, 190
sanguinea (Swartz) Brown, 190
Bulbophyllum Thouars, 219
barbigerum Lindley, 220
campanulatum Rolfe, 223
careyanum (Hooker) Sprengel, 220
ericssonii Kränzlin, 220
falcatum (Lindley) Reichenbach, 220
leopardinum (Wallich) Lindley, 220
leucorhachis (Rolfe) Schlechter, 220
lobbii Lindley, 220
longiflorum Thouars, 222
psittacoides (Ridley) Smith, 222
Burbidgea Hooker, 128
nitida misapplied, 128
schizocheila Hackett, 128
Butia (Beccari) Beccari, 74
capitata (Martius) Beccari, 74
eriospatha (Drude) Beccari, 74
yatay (Martius) Beccari, 74

Caladium Ventenat, 99
argyrites Lemaire, 99
bicolor (Aiton) Ventenat, 99
× *hortulanum* Birdsey, 99
humboldtii Schott, 99
lilliputiense Rodigas, 99
lindenii (André) Madison, 99
marmoratum Mathieu, 99
picturatum Koch, 99
poecile Schott, 99
puberulum Engler, 100
schomburgkii Schott, 99
Calamagrostis Adanson, 50
canescens (Weber) Roth, 50
epigejos (Linnaeus) Roth, 51
lanceolata Roth, 50
Calamus Linnaeus, 71
ciliaris Blume, 71
Calanthe Brown, 173
amamiana Fukuyama, 175
bicolor Lindley, 175
brevicornu Lindley, 175
cardioglossa Schlechter, 174
discolor Lindley, 175
var. flava Yatabe, 175
furcata Lindley, 174
gracilis (Lindley) Lindley, 175
var. venusta (Schlechter) Maekawa, 175
herbacea Lindley, 174
izu-insularis (Satomi) Ohwi, 174
lurida Decaisne, 175
masuca (Don) Lindley, 173
reflexa Maximowicz, 175
regnieri Reichenbach, 173
rosea (Lindley) Bentham & Hooker, 173
rubens Ridley, 173
sieboldii Decaisne, 175
striata (Banks) Brown, 175
sylvatica (Thouars) Lindley, 174
tricarinata Lindley, 175
triplicata (Willemet) Ames, 174
veratrifolia (Willdenow) Brown, 174
venusta Schlechter, 175
vestita Lindley, 173
Calathea Meyer, 129
albicans Schumann, 133
altissima (Poeppig & Endlicher) Koernicke, 135
bachemiana Morren, 131
crocata Morren & Jorisenne, 131
cylindrica (Roscoe) Schumann, 131
fasciata (Linden) Koernicke, 135
grandifolia Lindley, 131
insignis invalid, 133
lancifolia Boom, 133
leopardina (Bull) Regel, 133
lietzii Morren, 133
lindeniana Wallis, 133
makoyana Morren, 133
medio-picta (Morren) Regel, 133
micans (Mathieu) Koernicke, 133
ornata (Linden) Koernicke, 133
pavonii Koernicke, 133
picturata (Linden) Koch & Linden, 133
princeps (Linden) Regel, 135
roseo-picta (Linden) Regel, 135
rotundifolia (Koch) Koernicke, 135
var. fasciata (Linden) Petersen, 135
rufibarba Fenzl, 135
sanderiana invalid, 133
tubispatha Hooker, 133
undulata Linden & André, 135
vandenheckei (Lemaire) Regel, 133
variegata (Koch) Koernicke, 135
veitchiana Hooker, 135
warscewiczii (Mathieu) Koernicke, 135
wiotiana Morren, 135
wiotii (Morren) Regel, 135
zebrina (Sims) Lindley, 135
var. *binotii* Bailey, 135
Calla Linnaeus, 89
aethiopica Linnaeus, 97
palustris Linnaeus, 89
Callisia Linnaeus, 28
elegans Moore, 29
fragrans (Lindley) Woodson, 29
multiflora (Martens & Galeotti) Standley, 29
navicularis (Ortgies) Hunt, 29
repens Linnaeus, 29
warscewicziana (Kunth & Bouché) Hunt, 29
Callopsis Engler, 104
volkensii Engler, 104
Calypso Salisbury, 224
bulbosa (Linnaeus) Oaks, 224
Canistrum Morren, 20
lindenii (Regel) Mez, 20
var. roseum (Morren) Smith, 20
roseum Morren, 20
Canna Linnaeus, 129
flaccida Salisbury, 129
× *generalis* Bailey, 129
glauca Linnaeus, 129
indica Linnaeus, 129
iridiflora Ruiz & Pavon, 129
orchioides Bailey, 129
Cannaceae, 129
Caraguata cardinalis (André) André, 16
latifolia Beer, 16
zahnii Hooker, 16
Carex Linnaeus, 116
acutiformis Ehrhart, 117
baccans Nees, 116

baldensis Linnaeus, 116
buchananii Berggren, 117
comans Berggren, 117
elata Allioni, 116
firma Host, 116
flagellaris Colenso, 117
foliosissima misapplied, 117
fraseri Andrews, 116
hudsonii Bennett, 116
japonica misapplied, 117
morrowii Boott, 117
pendula Hudson, 116
pseudocyperus Linnaeus, 117
riparia Curtis, 117
stricta Goodenough, 116
vilmorinii Mottet, 117
Caryota Linnaeus, 71
mitis Loureiro, 71
urens Linnaeus, 71
Catasetum Richard, 262
atratum Lindley, 264
barbatum (Lindley) Lindley, 265
var. spinosum (Hooker) Rolfe, 265
bungerothii Brown, 264
callosum Lindley, 265
cernuum (Lindley) Reichenbach, 264
christyanum Reichenbach, 264
fimbriatum (Morren) Lindley, 264
integerrimum Hooker, 264
macrocarpum Kunth, 264
maculatum Bateman, 264
microglossum Rolfe, 265
pileatum Reichenbach, 264
russellianum Hooker, 264
saccatum Lindley, 264
var. christyanum (Reichenbach) Mansfeld, 264
thylaciochilum Lemaire, 264
tridentatum Hooker, 264
trifidum Hooker, 264
trulla Lindley, 264
Catimbium de Jussieu, 128
Catopsis penduliflora Wright, 13
Cattleya Lindley, 191
aclandiae Lindley, 192
amethystoglossa Linden & Reichenbach, 192
aurantiaca (Lindley) Don, 192
bicolor Lindley, 193
bowringiana O'Brien, 192
bulbosa Lindley, 193
citrina (Llave & Lexarza) Lindley, 190
dormaniana Reichenbach, 192
elatior Lindley, 193
elongata Rodriguez, 193
forbesii Lindley, 192
granulosa Lindley, 193
guttata Lindley, 193
var. *leopoldii* (Lemaire) Rolfe, 193
harrisoniae Bateman, 192
harrisoniana Lindley, 192
intermedia Graham, 192
labiata Lindley, 193
var. dowiana (Bateman & Reichenbach) Veitch, 194
var. gaskelliana (Reichenbach) Veitch, 194
var. labiata, 193
var. mendellii (Backhouse) Reichenbach, 194
var. mossiae (Lindley) Hooker, 194
var. percivalliana Reichenbach, 194
var. schroederae (Reichenbach) Anon., 194
var. trianae Linden & Reichenbach, 194
var. warscewiczii (Reichenbach) Reichenbach, 194
lawrenciana Reichenbach, 193
leopoldii Lemaire, 193
loddigesii Lindley, 192
var. *harrisoniana* Rolfe, 192
luteola Lindley, 193
maxima Lindley, 194
nobilior Reichenbach, 193
pauper (Vellozo) Stellfeld, 192
rex O'Brien, 194
schilleriana Reichenbach, 193
skinneri Bateman, 192
superba Lindley, 193
tigrina Richard, 193
velutina Reichenbach, 192
violacea (Humboldt, Bonpland & Kunth) Rolfe, 193
walkeriana Gardner, 193
Cautleya Hooker, 126
gracilis (Smith) Dandy, 126
lutea (Royle) Hooker, 126
robusta misapplied, 127
spicata (Smith) Baker), 126
Cenchrus Linnaeus, 43
ciliaris Linnaeus, 43
Cephalanthera Richard, 159
damasonium (Miller) Druce, 159
longifolia (Linnaeus) Fritzsch, 160
rubra (Linnaeus) Richard, 160
Chamaedorea Willdenow, 71
elegans Martius, 71
Chamaerhops Linnaeus, 67
excelsa misapplied, 66
excelsa Thunberg, 67
fortunei Hooker, 66
humilis Linnaeus, 67
hystrix Pursh, 66
martiana Wallich, 66
ritchiana Griffith, 69
Chasmanthium Link, 34
latifolium (Michaux) Yates, 35
Chiloschista Lindley, 242
lunifera (Reichenbach) Smith, 242
Chimonobambusa Makino, 62
falcata (Nees) Nakai, 62
hookeriana (Munro) Nakai, 62
marmorea (Mitford) Makino, 62
quadrangularis (Fenzi) Makino, 63
Chionochloa Zotov, 36
conspicua (Forster) Zotov, 36
Chloris Swartz, 36
elegans Kunth, 37
virgata Swartz, 37
Chondrorhyncha Lindley, 233
chestertonii Reichenbach, 233
fimbriata (Linden & Reichenbach) Reichenbach, 233
flaveola (Linden & Reichenbach) Garay, 233
Chrysalidocarpus Wendland, 72
lutescens Wendland, 72
Chrysopogon Trinius, 39
fulvus (Sprengel) Chiovenda, 39
Chusquea Kunth, 64
andina Philippi, 64
breviglumis Philippi, 64
culeou Desvaux, 64
var. tenuis McClintock, 64
cumingii Nees, 64
quila misapplied, 64
Chysis Lindley, 177
aurea Lindley, 177
var. bractescens (Lindley) Allen, 177
bractescens Lindley, 177
laevis Lindley, 177
limminghii Linden & Reichenbach, 177
Cirrhaea Lindley, 270
dependens Reichenbach, 270
tristis Lindley, 270
Cirrhopetalum Lindley, 220
auratum Lindley, 223
collettii Hemsley, 222
fascinator Rolfe, 223
gracillimum Rolfe, 222
longissimum Ridley, 222
makoyanum Reichenbach, 223
mastersianum Rolfe, 223
medusae Lindley, 222
ornatissimum Reichenbach, 223
picturatum Loddiges, 222
putidum Teijsmann & Binnendijk, 223
thouarsii Lindley, 222
umbellatum (Forster) Hooker & Arnott, 222
wendlandianum Kraenzlin, 222
Cleisostoma Blume, 253
&appendiculatum (Lindley) Jackson, 253
filiforme (Lindley) Garay, 253
racemiferum (Lindley) Garay, 253
rostratum (Lindley) Garay, 253
simondii (Gagnepain) Seindenfaden, 253
Clowesia russelliana (Hooker) Dodson, 264
thylaciochilum (Lemaire) Dodson, 264
Cochleanthes Rafinesque, 233
amazonica (Reichenbach & Warscewicz) Schultes & Garay, 234
discolor (Lindley) Schultes & Garay, 234
flabelliformis (Swartz) Schultes & Garay, 234
wailesiana (Lindley) Schultes & Garay, 234
Cochlioda Lindley, 273
densiflora Lindley, 273
noezliana Rolfe, 273
rosea (Lindley) Bentham, 273
sanguinea (Reichenbach) Bentham, 273
stricta Cogniaux, 273
vulcanica (Reichenbach) Veitch, 273
Cochliostema Lemaire, 27
jacobianum Koch & Linden, 27
odoratissimum Lemaire, 27
Cocos Linnaeus, 73
australis misapplied, 74
capitata Martius, 74
eriospatha Drude, 74
nucifera Linnaeus, 74
plumosa Hooker, 75
romanzoffianum Chamisso, 75
weddelliana Wendland, 75
yatay Martius, 74
Coelia bella (Lemaire) Reichenbach, 172
macrostachya Lindley, 172
Coeloglossum Hartmann, 169

viride (Linnaeus) Hartmann, 169
Coelogyne Lindley, 179
asperata Lindley, 182
barbata Griffith, 181
cinnamomea Teijsmann & Binnendijk, 182
corrugata Wright, 182
corymbosa Lindley, 182
cristata Lindley, 182
dayana Reichenbach, 181
elata misapplied, 180
fimbriata Lindley, 180
flaccida Lindley, 181
fuliginosa Lindley, 180
fuscescens Lindley, 182
var. brunnea (Lindley) Lindley, 182
graminifolia Parish & Reichenbach, 182
holochila Hunt & Summerhayes, 180
huettneriana misapplied, 182
lactea Reichenbach, 182
lawrenceana Rolfe, 180
massangeana Reichenbach, 181
miniata (Blume) Lindley, 180
nervosa Richard, 182
nitida (Wallich) Lindley, 182
ochracea Lindley, 182
ovalis Lindley, 180
pandurata Lindley, 181
parishii Hooker, 181
pogonioides Rolfe, 183
punctulata Lindley, 182
speciosa (Blume) Lindley, 180
stricta (Don) Schlechter, 181
stricta misapplied, 180
swaniana Rolfe, 181
tomentosa Lindley, 181
trinervis Lindley, 182
veitchii Rolfe, 181
viscosa Reichenbach, 182
wallichiana Lindley, 183
xyrekes, 180
Coix Linnaeus, 37
lacryma-jobi Linnaeus, 37
Colax jugosus Lindley, 231
var. *punctatus* Reichenbach, 231
var. *rufinus* Reichenbach, 231
placanthera (Hooker) Lindley, 231
puytdii Linden & André, 231
viridis (Lindley) Lindley, 232
Collinia elegans (Martius) Liebmann, 71
Colocasia Schott, 100
affinis Schott, 100
antiquorum Schott, 100
esculenta (Linnaeus) Schott, 100
Commelina Linnaeus, 26
alpestris Standley & Steyermark, 26
benghalensis Linnaeus, 26
coelestis Willdenow, 26
var. bourgeaui Clarke, 26
communis Linnaeus, 26
dianthifolia Delile, 26
diffusa Burmann, 26
elegans Kunth, 26
elliptica Kunth, 26
erecta Linnaeus, 26
nudiflora misapplied, 26
tuberosa Linnaeus, 26
Commelinaceae, 25
Commelinantia anomala (Torrey) Tharp, 26
pringlei (Watson) Tharp, 26
Comparettia Poeppig & Endlicher, 272
coccinea Lindley, 272
falcata Poeppig & Englicher, 272
macroplectrum Reichenbach & Triana, 272
rosea Lindley, 272
Copernicia Martius, 69
alba Morong, 69
prunifera (Miller) Moore, 69
Cortaderia Stapf, 35
conspicua (Forster) Stapf, 36
jubata (Lemoine) Stapf, 35
richardii (Endlicher) Zotov, 35
selloana (Schultes & Schultes) Ascherson & Graebner, 35
Coryanthes Hooker, 270
macrantha (Hooker) Hooker, 270
maculata Hooker, 270
speciosa Hooker, 270
Corybas Salisbury, 160
dilatatus (Rupp & Nicholls) Rupp, 160
Costaceae, 120
Costus Linnaeus, 120
afer Ker Gawler, 122
cuspidatus (Nees & Martius) Maas, 122
igneus Brown, 122
lucanusianus Braun & Schumann, 122
malortieanus Wendland, 122
speciosus (Koenig) Smith, 122
Cryosophila Blume, 67
warscewiczii (Wendland) Bartlett, 67
Cryptanthus Otto & Dietrich, 18
acaulis (Lindley) Beer, 18
var. argenteus Beer, 18
var. ruber Beer, 18
bivittatus (Hooker) Regel, 18
emergens Lindman, 19
zonatus (Visiani) Beer, 18
forma fuscus (Visiani) Mez, 18
forma viridis Mez, 18
× *Cryptbergia meadii* Wilson & Wilson, 24
Cryptochilus Wallich, 200
lutea Lindley, 201
sanguinea Wallich, 200
Cryptocoryne Wydler, 110
affinis Brown, 112
aponogetifolia Merrill, 112
axelrodii Rataj, 111
balansae Gagnepain, 112
becketii Trimen, 111
bertelihansenii Rataj, 112
blassii De Wit, 111
caudata Brown, 111
ciliata (Roxburgh) Schott, 111
var. ciliata, 111
var. latifolia Rataj, 111
cordata Griffith, 111
crispatula Engler, 112
drymorrhiza Zippelius, 111
elata Griffith, 111
evae Rataj, 111
griffithii misapplied, 111
griffithii Schott, 111
haerteliana Milkuhn, 112
hejnyi Rataj, 111
johorensis Ridley, 111
kerrii Gagnepain, 111
koenigii Schott, 109
kwangsiensis Li, 112
lancifolia Schott, 109
lingua Engler, 112
longicauda Engler, 111
longispatha Merrill, 112
lucens De Wit, 111
lutea Alston, 111
minima Ridley, 111
nevillii misapplied, 111
ovata (Linnaeus) Schott, 109
parva De Witt, 111
petchii Alston, 111
pontederiifolia Schott, 111
purpurea Ridley, 111
retrospiralis (Roxburgh) Engler, 112
siamensis Gagnepain, 111
var. *schneideri* Schöpfel, 111
sinensis Merrill, 112
spathulata Engler, 112
stonei Rataj, 111
sulphurea De Wit, 111
thwaitesii Schott, 112
tonkinensis Gagnepain, 112
undulata Wendt, 111
usteriana Engler, 112
walkeri Schott, 111
wendtii De Wit, 111
willisii Baum, 111
× willisii Reitz, 111
yunnanensis Li, 112
Cryptophoranthus Rodrigues, 207
atropurpureus (Lindley) Rolfe, 207
dayanus Rolfe, 207
maculatus Rolfe, 207
Ctenanthe Eichler, 136
compressa (Dietrich) Eichler, 136
kummeriana (Morren) Eichler, 136
lubbersiana (Morren) Eichler, 136
oppenheimiana (Morren) Schumann, 136
setosa (Roscoe) Eichler, 136
Curcuma Linnaeus, 127
albiflora Thwaites, 127
amada Roxburgh, 127
augustifolia Roxburgh, 127
aromatica Salisbury, 127
domestica Valeton, 127
heyneana Valeton, 127
mangga Valeton & Van Zijp, 127
longa Linnaeus, 127
petiolata Roxburgh, 127
purpurascens Blume, 127
roscoeana Wallich, 127
xanthorrhiza Roxburgh, 127
zedoaria (Christmann) Roscoe, 127
Cyanotis Don, 28
barbata Don, 28
cristata Schultes, 28
fasciculata Schultes, 28
foecunda Hasskarl, 28
hirsuta Fischer & Meyer, 28
kewensis (Hasskarl) Clark, 28
moluccana (Roxburgh) Merrill, 28
nodiflora (Lamarck) Kunth, 28
somaliensis Clarke, 28
speciosa (Linnaeus) Hasskarl, 28

veldhoutiana invalid, 31
Cyclosia maculata Klotzsch, 261
Cycnoches Lindley, 265
aureum Lindley, 265
chlorochilon Klotzsch, 265
egertonianum Bateman, 265
var. aureum (Lindley) Allen, 265
var. egertonianum, 265
loddigesii Lindley, 265
maculatum Lindley, 265
pentadactylon Lindley, 265
ventricosum Bateman, 265
var. chlorochilon (Klotzsch) Allen, 265
Cymbidium Swartz, 259
aloifolium (Linnaeus) Swartz, 260
andersonii Lambert, 259
devonianum Paxton, 260
eburneum Lindley, 260
elegans (Blume) Lindley, 261
ensifolium (Linnaeus) Swartz, 260
erythrostylum Rolfe, 260
floribundum Lindley, 260
giganteum Wallich, 260
grandiflorum Griffith, 260
hookerianum Reichenbach, 260
insigne Rolfe, 260
longifolium Don, 260
lowianum (Reichenbach) Reichenbach, 260
mastersii Lindley, 260
pendulum Swartz, 261
pumilum Rolfe, 260
sanderi invalid, 260
simulans Rolfe, 260
tigrinum Hooker, 260
tracyanum Rolfe, 260
wilsonii (Cooke) Rolfe, 260
Cymbopogon Sprengel, 40
nardus (Linnaeus) Rendle, 40
Cymophyllus fraseri (Andrews) Mackenzie, 116
Cynosurus Linnaeus, 51
echinatus Linnaeus, 51
Cyperaceae, 114
Cyperorchis elegans Blume, 261
mastersii (Lindley) Bentham, 261
Cyperus Linnaeus, 115
albostriatus Schrader, 116
alternifolius misapplied, 115
diffusus misapplied, 116
elegans misapplied, 116
eragrostis Lamarck, 115
esculentus Linnaeus, 115
var. sativus Boekeler, 115
flabelliformis Rottboell, 115
involucratus Rottboell, 115
isocladus Kunth, 115
longus Linnaeus, 115
maximus Anon., 115
papyrus Linnaeus, 115
vegetus Willdenow, 115
Cypripedium Linnaeus, 149
acaule Aiton, 151
album Aiton, 151
arietinum Brown, 151
calceolus Linnaeus, 150
var. calceolus, 150
var. parviflorum (Salisbury) Fernald, 150
var. pubescens (Willdenow) Correll, 150
californicum Gray, 151
canadensis Michaux, 151
candidum Willdenow, 150
cordigerum Don, 150
debile Reichenbach, 151
ebracteatum Rolfe, 151
elegans Reichenbach, 151
fasciculatum Watson, 151
fasciolatum Franchet, 151
flavum Hunt & Summerhayes, 151
formosanum Hayata, 151
franchetii Rolfe, 150
guttatum Swartz, 152
himalaicum Hemsley, 151
irapeanum Llave & Lexarza, 151
japonicum Thunberg, 151
knightae Nelson, 151
laevigatum Bateman, 155
luteum Aiton, 150
luteum Franchet, 151
macranthon Swartz, 150
margaritaceum Franchet, 151
microsaccus Kränzlin, 150
montanum Lindley, 150
parviflorum Salisbury, 150
plectrochilum Franchet, 151
pubescens Willdenow, 150
reginae Walter, 151
speciosum Rolfe, 150
spectabile Reichenbach var. *dayanum* Lindley, 156
spectabile Salisbury, 151
tibeticum Hemsley, 151
veganum Cockerell & Barker, 150
×ventricosum Swartz, 151
wilsonii Rolfe, 151
yatabeanum Makino, 152
Cyrtochilum flavescens Lindley, 282
maculatum Lindley, 287
Cyrtopodium Brown, 258
andersonii (Lambert) Brown, 259
punctatum (Linnaeus) Lindley, 259
virescens Reichenbach & Warming, 259
Cyrtorchis Schlechter, 256
arcuata (Lindley) Schlechter, 256
monteiroae (Reichenbach) Schlechter, 256
Cyrtostachys Blume, 73
lakka Beccari, 73
renda Blume, 73

Dactylis Linnaeus, 51
glomerata Linnaeus, 51
Dactylorrhiza Nevski, 167
elata (Poiret) Soó, 167
foliosa (Solander) Soó, 167
fuchsii (Druce) Soó, 168
incarnata (Linnaeus) Soó, 168
latifolia (Linnaeus) Soó, 168
maculata (Linnaeus) Soó, 168
subsp. *ericetorum* (Linden) Hunt & Summerhayes, 168
subsp. *fuchsii* (Druce) Hylander, 168
majalis (Reichenbach) Hunt & Summerhayes, 168
sambucina (Linnaeus) Soó, 168
Danthonia cunninghamii Hooker, 36
Dendrobium Swartz, 207
aberrans Schlechter, 212
aduncum Wallich, 215
×*ainsworthii* Moore, 218
amethystoglossum Reichenbach, 216
amoenum Lindley, 215
amplum Lindley, 219
anosmum Lindley, 214
aphyllum (Roxburgh) Fischer, 214
arachnites Reichenbach, 213
atroviolaceum Rolfe, 212
aureum Lindley, 214
bellatulum Rolfe, 219
bigibbum Lindley, 217
brymerianum Reichenbach, 213
var. *histrionicum* Reichenbach, 213
bullenianum Reichenbach, 216
calceolaria Carey, 213
cambridgeanum Paxton, 214
chrysanthum Lindley, 213
chrysotoxum Lindley, 212
var. suavissimum (Reichenbach) Veitch, 212
ciliatum Hooker, 217
crassinode Benson & Reichenbach, 215
crepidatum Lindley, 215
crispilinguum Cribb, 218
crumenatum Swartz, 219
cupreum Herbert, 213
cuthbertsonii Mueller, 216
cyananthum Williams, 217
dalhousieanum Wallich, 215
dearei Reichenbach, 218
delacourii Guillaumin, 217
densiflorum Wallich, 212
var. galliceanum Linden, 213
var. schroederi Anon., 213
devonianum Paxton, 215
dichroma Schlechter, 216
discolor Lindley, 218
draconis Reichenbach, 218
eburneum Reichenbach, 218
engae Reeve, 212
falconeri Hooker, 215
falcorostrum Fitzgerald, 211
farmeri Paxton, 213
fimbriatum Hooker, 213
var. gibsoni (Lindley) Gagnepain, 213
var. oculatum Hooker, 213
findleyanum Parish & Reichenbach, 214
flammula Schlechter, 216
forbesii Ridley, 212
formosum Roxburgh, 218
gibsoni Lindley, 213
gratiosissimum Reichenbach, 215
helix Cribb, 217
hellwigianum Kränzlin, 217
heterocarpum Lindley, 214
histrionicum (Reichenbach) Schlechter, 213
hookerianum Lindley, 213
infundibulum Lindley, 218
var. jamesianum (Reichenbach) Veitch, 218
jamesianum Reichenbach, 218
jenkinsii Lindley, 212
johnsoniae Mueller, 212
kingianum Bidwill, 211
lawesii Mueller, 216
lituiflorum Lindley, 214

loddigesii Rolfe, 214
luteolum Bateman, 214
lyonii Ames, 219
macrophyllum Lindley, 214
macrophyllum Richard, 212
minax Reichenbach, 217
mirbelianum Gaudich, 218
monile (Thunberg) Kränzlin, 214
moniliforme (Linnaeus) Swartz, 214
moschatum Swartz, 213
nobile Lindley, 215
nutriferum Smith, 216
ochreatum Lindley, 214
oreocharis Schlechter, 217
parishii Reichenbach, 214
pendulum Roxburgh, 215
phalaenopsis Fitzgerald, 217
phlox Schlechter, 216
pierardii Roxburgh, 214
primulinum Lindley, 215
pugioniforme Cunningham, 211
pulchellum Lindley, 215
pulchellum Loddiges, 214
raphiotes Schlechter, 217
rhodostictum Mueller & Kränzlin, 212
rotundatum (Lindley) Hooker, 219
ruppianum Hawkes, 211
sanderae Rolfe, 218
sanderianum Rolfe, 218
schroederianum Gentil, 217
section *Aporum* Schlechter, 219
section *Callista* (Loureiro) Schlechter, 212
section *Calyptrochilus* Schlechter, 216
section *Ceratobium* Schlechter, 217
section *Cuthbertsonia* Schlechter, 216
section *Dendrocoryne* (Lindley) Schlechter, 211
section *Eugenanthe* Schlechter, 213
section *Glomerata* Kränzlin, 216
section *Latouria* (Blume) Schlechter, 211
section *Leiotheca* Kränzlin, 216
section *Nigro-hirsutae* Kränzlin, 218
section *Oxygenianthe* Schlechter, 218
section *Oxyglossum* Schlechter, 216
section *Pedilonum* Blume, 216
section *Phalaenanthe* Schlechter, 217
section *Rhizobium* Schlechter, 211
section *Rhopalanthe* Schlechter, 219
section *Stachyobium* Schlechter, 217
secundum Lindley, 216
seidenfadenii Senghas & Bockemühl, 213
senile Parish & Reichenbach, 212
sophronites Schlechter, 216
speciosum Smith, 211
var. *fusiforme* Bailey, 211
spectabile (Blume) Miquel, 212
stratiotes Reichenbach, 218
suavissimum Reichenbach, 212
subacaule Lindley, 217
subclausum Rolfe, 216
subgenus *Dendrocoryne* Lindley, 211
superbiens Reichenbach, 217
superbum Reichenbach, 214
tangerinum Cribb, 217
taurinum Lindley, 218
terminale Parish & Reichenbach, 219
tetragonum Cunningham, 211
thyrsiflorum Reichenbach, 213
topaziacum Ames, 216
transparens Lindley, 214
tricostatum Schlechter, 217
undulatum Brown, 218
unicum Seidenfaden, 213
vexillarius Smith, 216
victoriae-reginae Loher, 216
violaceum Kränzlin, 217
wardianum Warner, 215
wentianum Smith, 216
wilkianum Rupp, 218
williamsonii Day & Reichenbach, 219
woodsii Cribb, 212
Dendrocalamus Nees, 59
asper (Schultes) Heyne, 59
Dendrochilum Blume, 184
cobbianum Reichenbach, 184
filiforme Lindley, 184
glumaceum Lindley, 184
latifolium Lindley, 184
Deschampsia Palisot de Beauvois, 46
cespitosa (Linnaeus) Palisot de Beauvois, 46
flexuosa (Linnaeus) Trinius, 46
Diacrium bidentatum (Lindley) Hemsley, 189
Diaphananthe Schlechter, 256
bidens (Reichenbach) Schlechter, 256
pellucida (Lindley) Schlechter, 256
Dichaea Lindley, 240
glauca (Swartz) Lindley, 241
graminoides (Swartz) Lindley, 241
pendula (Aublet) Cogniaux, 241
muricata (Swartz) Lindley, 241
Dichorisandra Mikan, 27
hexandra (Aublet) Standley, 27
musaica var. *undata* (Koch & Linden) Miller, 27
reginae Linden & Rodigas, 27
thyrsiflora Mikan, 27
Dieffenbachia Schott, 96
amoena misapplied, 97
×bausei Masters & Moore, 97
bowmanii Carrière, 97
humilis Poeppig, 96
leopoldii Bull, 96
maculata (Loddiges) Don, 96
oerstedii Schott, 96
perfecta invalid, 97
picta Schott, 96
reginae invalid, 97
sanguine (Jacquin) Schott, 96
×splendens Bull, 96
weirii Berkeley, 97
Digitaria Haller, 40
sanguinalis (Linnaeus) Scopoli, 40
Dimorphorchis Rolfe, 251
lowii (Lindley) Rolfe, 252
var. lowii, 252
var. rohaniana (Reichenbach) Tan, 252
Dinema polybulbon (Swartz) Lindley, 190
Diplothemium maritimum Martius, 75
Dipteranthus grandiflorus (Lindley) Pabst, 241
Disa Bergius, 170
grandiflora Linnaeus, 171
uniflora Bergius, 171
Doritis Lindley, 244
pulcherrima Lindley, 244
var. *buyssoniana* Anon., 244
Dracula Luer, 204
bella (Reichenbach) Luer, 204
chimaera (Reichenbach) Luer, 204
erythrochaete (Reichenbach) Luer, 204
radiosa (Reichenbach) Luer, 204
vampira (Luer) Luer, 204
Dracunculus Schott, 106
canariensis Kunth, 106
crinitus Schott, 106
muscivorus (Linnaeus) Parlatore, 106
vulgaris Schott, 106
Dyckia Schultes, 13
brevifolia Baker, 14
frigida Hooker, 14
remotiflora Otto & Dietrich, 14

Echinaria Desfontaines, 51
capitata (Linnaeus) Desfontaines, 51
Echinochloa Palisot de Beauvois, 41
polystachya (Humboldt, Bonpland & Kunth) Hitchcock, 41
Ehrharta Thunberg, 34
erecta Lamarck, 34
Elaeis Jacquin, 75
guineensis Jacquin, 75
Eleocharis Brown, 115
acicularis (Linnaeus) Roemer & Schultes, 115
Elettaria Maton, 128
cardamomum (Linnaeus) Maton, 128
Elymus Linnaeus, 44
canadensis Linnaeus, 44
virginicus Linnaeus, 44
Encholirion saundersii Carrière, 15
Encyclia Hooker, 186
adenocaula (Llave & Lexarza) Schlechter, 189
advena Dressler, 190
alata (Bateman) Schlechter, 189
aromatica (Bateman) Schlechter, 189
baculus (Reichenbach) Dressler & Pollard, 188
boothiana (Lindley) Dressler, 189
bractescens (Lindley) Hoehn, 189
brassavolae (Reichenbach) Dressler, 189
citrina (Llave & Lexarza) Dressler, 190
cochleata (Linnaeus) Lemee, 188
cordigera (Humboldt, Bonpland & Kunth) Dressler, 190
fragrans (Swartz) Lemee, 188
glumacea (Lindley) Pabst, 188
hanburii (Lindley) Schlechter, 190
mariae (Ames) Hoehne, 190
michuacana (Llave & Lexarza) Schlechter, 190
oncidioides misapplied, 190
oncidioides (Lindley) Schlechter, 190
patens Hooker, 189
polybulbon (Swartz) Dressler, 190
prismatocarpa (Reichenbach) Dressler, 189
radiata (Lindley) Dressler, 188
selligera (Lindley) Schlechter, 190
tampensis (Lindley) Small, 189
varicosa (Lindley) Schlechter, 190
vespa (Vellozo) Dressler, 188
vitellina (Lindley) Dressler, 189
var. majus Veitch, 189
Ensete Horaninow, 117
edule Horaninow, 117

superbum (Roxburgh) Cheesman, 118
ventricosum (Welwitsch) Cheesman, 117
Epidendrum Linnaeus, 185
arachnoglossum Reichenbach, 186
atropurpureum misapplied, 190
capartianum Lindley, 190
ciliare Linnaeus, 186
citrinum (Llave & Lexarza) Reichenbach, 190
cnemidophorum Lindley, 186
costaricense Reichenbach, 186
diffusum Swartz, 186
elegans (Knowles & Westcott) Reichenbach, 191
eximium Williams, 186
falcatum Lindley, 186
floribundum Humboldt, Bonpland & Kunth, 186
ibaguense Humboldt, Bonpland & Kunth, 186
latilabrum Lindley, 186
loefgrenii Cogniaux, 186
macrochilum Hooker, 190
medusae (Reichenbach) Veitch, 186
myrianthum Lindley, 186
nemorale Lindley, 189
nocturnum Jacquin, 186
× o'brienianum Rolfe, 186
odoratissimum Lindley, 189
oerstedtii Reichenbach, 186
osmanthum Rodriguez, 190
paniculatum Ruiz & Pavon, 186
parkinsonianum Hooker, 186
pentotis Reichenbach, 188
punctatum Lindley, 259
radicans Ruiz & Pavon, 186
skinneri Lindley, 191
stamfordianum Bateman, 185
tessellatum Roxburgh, 250
variegatum Hooker, 188
verrucosum Lindley, 189
verrucosum Swartz var. *myrianthum* (Lindley) Ames & Correll, 186
virgatum Lindley, 190
xanthinum Lindley, 186
Epigeneium Gagnepain, 219
amplum (Lindley) Summerhayes, 219
coelogyne (Reichenbach) Summerhayes, 219
lyonii (Ames) Summerhayes, 219
rotundatum (Lindley) Summerhayes, 219
Epipactis Zinn, 158
atropurpurea Rafinesque, 159
atrorubens (Hoffman) Besser, 159
gigantea Hooker, 159
helleborine (Linnaeus) Crantz, 159
latifolia (Linnaeus) Aiton, 159
palustris (Miller) Crantz, 159
phyllanthes Smith, 159
royleana Lindley, 159
Epipremnum Schott, 86
aureum (Lindley & André) Bunting, 86
pinnatum (Linnaeus) Engler, 86
Eragrostis Wolf, 37
curvula (Schrader) Nees, 37
japonica (Thunberg) Trinius, 37
mexicana (Hornemann) Link, 37
trichodes (Nuttall) Wood, 37
Eria Lindley, 200
bractescens Lindley, 200
convallarioides Lindley, 200
coronaria (Lindley) Reichenbach, 200
floribunda Lindley, 200
fragrans Reichenbach, 200
javanica (Swartz) Blume, 200
spicata (Don) Handel-Mazzetti, 200
stellata Lindley, 200
Erianthus ravennae (Linnaeus) Palisot de Beauvois, 39
Eriochloa Kunth, 40
villosa (Thunberg) Kunth, 40
Eriopsis Lindley, 235
biloba Lindley, 235
helenae Kränzlin, 236
rutidobulbon Hooker, 235
sceptrum Reichenbach & Warscewicz, 236
schomburgkii (Reichenbach) Reichenbach, 235
sprucei Reichenbach, 236
wercklei Schlechter, 235
Erythea armata (Watson) Watson, 68
edulis (Wendland & Watson) Watson, 69
Esmeralda Reichenbach, 251
cathcartii (Lindley) Reichenbach, 251
Euanthe sanderiana (Reichenbach) Schlechter, 249
Eulophia Brown, 257
alta (Linnaeus) Rendle & Fawcett, 257
gigantea (Welwitsch) Brown, 257
guineensis Lindley, 257
horsfallii (Bateman) Summerhayes, 258
longifolia Humboldt, Bonpland & Kunth, 257
maculata (Lindley) Reichenbach, 258
porphyroglossa (Reichenbach) Bolus, 258
streptopetala Lindley, 258
woodfordii (Lindley) Rolfe, 257
Eulophidium ledienii Brown, 258
maculatum (Lindley) Pfitzer, 258
saundersianum (Reichenbach) Summerhayes, 258
Eulophiella Rolfe, 258
elizabethae Linden & Rolfe, 258
hamelinii Rolfe, 258
peetersiana Kränzlin, 258
roempleriana (Reichenbach) Schlechter, 258
Eurychone Schlechter, 256
rothschildiana (O'Brien) Schlechter, 256

Fargesia spathacea Franchet, 60
Fascicularia Mez, 19
bicolor (Ruiz & Pavon) Mez, 20
pitcairniifolia (Verlot) Mez, 20
Festuca Linnaeus, 52
eskia de Candolle, 53
glacialis (Hackel) Richter, 53
glauca Lamarck, 53
punctoria Sibthorp & Smith, 53
valesiaca Gaudin, 53
Fieldia lissochiloides Gaudichaud, 251
Fosterella Smith, 13
penduliflora (Wright) Smith, 13

Galeandra Lindley, 257
batemanii Rolfe, 257
baueri Lindley, 257
baueri misapplied, 257
devoniana Schomburgk, 257
lacustris Rodriguez, 257
nivalis Masters, 257
Galeottia Richard, 236
grandiflora Richard, 236
Gastrochilus Don, 252
acutifolius (Lindley) Kunze, 253
bellinus (Reichenbach) Kunze, 253
dasypogon (Lindley) Kunze, 253
humblotii (Reichenbach) Schlechter, 176
ornatus (Brown) Valeton, 124
panduratus (Roxburgh) Ridley, 124
tuberculosa (Thouars) Schlechter, 176
vittatus (Brown) Valeton, 124
Gaudinia Palisot de Beauvois, 47
fragilis (Linnaeus) Palisot de Beauvois, 48
Geogenanthus Ule, 27
poeppigii (Miquel) Faden, 27
undatus (Koch & Linden) Mildbraed & Strauss, 27
Gibasis Rafinesque, 28
pellucida (Martens & Galeotti) Hunt, 28
schiedeana (Kunth) Hunt, 28
Globba Linnaeus, 122
atrosanguinea Teijsmann & Binnendijk, 122
bulbifera Roxburgh, 122
marantina Linnaeus, 122
schomburgkii Hooker, 122
winitii Wright, 122
Glyceria Brown, 53
aquatica (Linnaeus) Wahl, 53
maxima (Hartmann) Holmboe, 53
Gomesa Brown, 273
crispa (Lindley) Klotzsch & Reichenbach, 274
planifolia (Lindley) Klotzsch & Reichenbach, 274
recurva Loddiges, 274
Gongora Ruiz & Pavon, 269
bufonia Lindley, 270
galeata (Lindley) Reichenbach, 269
maculata Lindley, 270
quinquenervis Ruiz & Pavon, 270
truncata Lindley, 270
Goodyera Brown, 161
colorata (Blume) Blume, 161
decipiens (Hooker) Hubbard, 161
hispida Lindley, 161
menziesii Lindley, 161
oblongifolia Rafinesque, 161
repens (Linnaeus) Brown, 161
var. ophioides Fernald, 161
Gramineae, 31
Grammangis Reichenbach, 261
ellisii Reichenbach, 261
Grammatophyllum Blume, 261
fenzlianum Reichenbach, 261
measuresianum Weathers, 261
multiflorum Lindley, 261
roemplerianum Reichenbach, 258
rumphianum Miquel, 261
scriptum Blume, 261
speciosum Blume, 261
Guillania purpurata Veillard, 128
Guzmania Ruiz & Pavon, 16
cardinalis (André) Mez, 16
lingulata (Linnaeus) Mez, 16
var. cardinalis (André) Mez, 16
var. minor (Mez) Smith, 16
var. splendens (Planchon) Mez, 16

minor Mez, 16
monostachia (Linnaeus) Mez, 16
musaica (Linden & André) Mez, 16
peacockii (Morren) Mez, 16
zahnii (Hooker) Mez, 16
Gymnadenia Brown, 168
conopsea (Linnaeus) Brown, 168
odoratissima (Linnaeus) Brown, 168

Habenaria Willdenow, 169
bonatea Reichenbach, 169
carnea Brown, 170
ciliaris Linnaeus, 170
militaris Reichenbach, 170
psycodes (Linnaeus) Sprengel, 170
radiata Thunberg, 170
rhodocheila Hance, 170
Hakonechloa Makino, 35
macra (Munro) Makino, 36
Hartwegia purpurea Lindley, 199
Hechtia pitcairniifolia Verlot, 20
Hedychium Koenig, 124
aurantiacum Roscoe, 125
chrysoleucum Hooker, 125
coccineum Smith, 125
var. angustifolium Roxburgh, 125
var. carneum Roscoe, 125
coronarium Koenig, 125
var. *chrysoleucum* (Hooker) Baker, 125
var. *flavescens* (Roscoe) Baker, 125
var. maximum (Roscoe) Baker, 125
densiflorum Wallich, 125
flavescens Roscoe, 125
flavum misapplied, 125
forrestii Diels, 125
gardnerianum Ker Gawler, 125
greenei Smith, 125
speciosum Wallich, 125
spicatum Smith, 125
Hedyscepe Wendland & Drude, 72
canterburyana (Moore & Mueller) Wendland & Drude, 72
Helcia Lindley, 274
sanguinolenta Lindley, 274
Helicodiceros muscivorus (Linnaeus) Engler, 106
Heliconia Linnaeus, 120
Helictotrichen Besser, 47
sempervirens (Villars) Pilger, 47
Hexisea Lindley, 185
bidentata Lindley, 185
Hierochloe Brown, 48
odorata (Linnaeus) Palisot de Beauvois, 48
redolens (Vahl) Roemer & Schultes, 48
Himantoglossum Koch, 165
hircinum (Linnaeus) Sprengel, 165
longibracteatum (Bernardi) Schlechter, 165
Hohenbergia strobilacea Schultes, 19
Holcoglossum kimballianum (Reichenbach) Garay, 250
Holcus Linnaeus, 46
mollis Linnaeus, 46
Hologyne miniata (Blume) Pfitzer, 180
Holoschoenus vulgaris Link, 114
×Homalocasia miamiensis, 90
Homalomena Schott, 90
lindenii (Rodigas) Lindley, 90
wallisii Regel, 90
Hoplophytum calyculatum Morren, 22
coeleste Koch, 22
lindenii Morren, 22
Hordeum Linnaeus, 45
hystrix Roth, 45
jubatum Linnaeus, 45
Hormidium baculus (Reichenbach) Brieger, 188
boothianum (Lindley) Brieger, 189
brassavolae (Reichenbach) Brieger, 189
cochleatum (Linnaeus) Brieger, 188
fragrans (Swartz) Brieger, 188
glumaceum (Lindley) Brieger, 188
mariae (Ames) Brieger, 190
prismatocarpum (Reichenbach) Brieger, 189
radiatum (Lindley) Brieger, 188
variegatum (Hooker) Brieger, 188
Houlletia Brongniart, 267
brocklehurstiana Lindley, 267
lansbergii Linden & Reichenbach, 267
odoratissima Lindley & Paxton, 268
var. antioquiensis Anon., 268
picta Linden & Reichenbach, 268
wallisii Linden & Reichenbach, 268
Howea Beccari, 73
belmoreana (Moore & Mueller) Beccari, 73
forsteriana (Moore & Mueller) Beccari, 73
Huntleya Lindley, 235
burtii (Endres & Reichenbach) Pfitzer, 235
meleagris Lindley, 235
Hyophorbe indica Gaertner, 72
Hyparrhenia Fournier, 40
cymbaria (Linnaeus) Stapf, 40
Hystrix Moench, 44
patula Moench, 44

Imperata Cirillo, 38
cylindrica (Linnaeus) Räuschel, 38
Indocalamus Nakai, 63
tessellatus (Munro) King, 63
Ionopsis Humboldt, Bonpland & Kunth, 271
paniculata Humboldt, Bonpland & Kunth, 271
tenera (Steudel) Lindley, 271
utricularioides (Swartz) Lindley, 271
Ischnosiphon bambusaceus (Poeppig & Endlicher) Koernicke, 137
Isochilus Brown, 179
linearis (Jacquin) Brown, 179
Isolepis cernua (Vahl) Roemer & Schultes, 115

Jubaea Humboldt, Bonpland & Kunth, 74
chilensis (Molina) Baillon, 74
spectabilis Humboldt, Bonpland & Kunth, 74
Juncaceae, 10
Juncus Linnaeus, 10
effusus Linnaeus, 10

Kaempferia Linnaeus, 123
elegans (Wallich) Baker, 124
galanga Linnaeus, 124
gilbertii Bull, 124
ornata Brown, 124
ovalifolia misapplied, 123
ovalifolia Roscoe, 123
pandurata Roxburgh, 124
parishii Hooker, 123
pulchra Ridley, 123
roscoeana Wallich, 124
rotunda Linnaeus, 123
vittata Brown, 124
Karatas chlorosticta Baker, 17
fulgens (Lemaire) Antoine, 19
innocentii (Lemaire) Antoine, 19
marmorata Baker, 18
meyendorfii Antoine, 17
Katherinea acuminata var. *lyonii* (Ames) Hawkes, 219
ampla (Lindley) Hawkes, 219
rotundata (Lindley) Hawkes, 219
Kentia belmoreana Moore & Mueller, 73
canterburyana Moore & Mueller, 73
forsteriana Moore & Mueller, 73
Koeleria Persoon, 48
cristata (Linnaeus) Persoon, 48
gracilis Persoon, 48
macrantha (Ledebour) Schultes, 48
phleioides (Villars) Persoon, 48

Lacaena Lindley, 267
bicolor Lindley, 267
spectabilis (Klotzsch) Reichenbach, 267
Laelia Lindley, 174
acuminata Lindley, 196
albida Lindley, 196
anceps Lindley, 196
autumnalis Lindley, 196
var. furfuracea (Lindley) Rolfe, 196
boothiana Reichenbach, 196
cinnabarina Lindley, 195
crispa (Lindley) Reichenbach, 195
crispata (Thunberg) Garay, 195
crispilabia Richard, 196
digbyana (Lindley) Bentham, 196
flava Lindley, 195
furfuracea Lindley, 196
glauca (Lindley) Bentham, 198
gouldiana Reichenbach, 196
grandiflora Lindley, 196
grandis Lindley & Paxton, 195
var. *tenebrosa* Gower, 195
harpophylla Reichenbach, 195
jongheana Reichenbach, 196
lobata (Lindley) Veitch, 195
longipes Reichenbach, 196
lundii Reichenbach & Warming, 196
majalis Lindley, 196
peduncularis Lindley, 196
perrinii Bateman, 195
pumila (Hooker) Reichenbach, 196
purpurata Lindley, 195
regnellii Rodrigues, 196
rubescens Lindley, 196
rupestris Lindley, 195
speciosa (Humboldt, Bonpland & Kunth) Schlechter, 196
tenebrosa Rolfe, 195
xanthina Lindley, 195
Laeliopsis Lindley, 190
domingensis Lindley, 190
Lagenandra Dalzell, 109
insignis Trimen, 109
koenigii (Schott) Thwaites, 109
lancifolia (Schott) Thwaites, 109
ovata (Linnaeus) Thwaites, 109
ovata misapplied, 109

praetermissa De Wit, 109
thwaitesii Engler, 109
Lagurus Linnaeus, 51
ovatus Linnaeus, 51
Lamarckia Moench, 51
aurea (Linnaeus) Moench, 51
Lampra volcanica Bentham, 28
Lamprococcus miniatus Beer, 21
Languas Small, 128
Leleba multiplex (Loureiro) Nakai, 59
Lemna, 112
Lemnaceae, 112
Leptotes Lindley, 199
bicolor Lindley, 199
unicolor Rodriguez, 199
Licuala Thunberg, 68
grandis Wendland, 68
spinosa Thunberg, 68
Limodorum aphyllum Roxburgh, 214
Lindmania penduliflora (Wright) Stapf, 13
Liparis Richard, 223
elata Lindley, 224
lilifolia Lindley, 224
loeselii Richard, 224
longipes Lindley, 224
nervosa (Thunberg) Lindley, 224
viridiflora Lindley, 224
Lissochilus giganteus Welwitsch, 257
horsfallii Bateman, 258
krebsii, Reichenbach, 258
porphyroglossa Reichenbach, 258
roseus Lindley, 258
Listera Brown, 160
cordata (Linnaeus) Brown, 160
ovata (Linnaeus) Brown, 160
Listrostachys arcuata (Lindley) Reichenbach, 256
bidens Reichenbach, 256
monteiroae Reichenbach, 256
pellucida (Lindley) Reichenbach, 256
Livistona Brown, 67
australis (Brown) Martius, 68
chinensis (Jacquin) Martius, 68
decipiens Beccari, 68
rotundifolia (Lamarck) Martius, 67
Lockhartia Hooker, 266
lunifera Reichenbach, 266
oerstedtii Reichenbach, 266
robusta Schlechter, 266
verrucosa Reichenbach, 266
Lophochloa phleoides (Villars) Reichenbach, 48
Lueddemannia Reichenbach, 267
pescatorei (Lindley) Linden & Reichenbach, 267
Luzula de Candolle, 10
luzuloides (Lamarck) Dandy, 10
nivea (Linnaeus) de Candolle, 10
sylvatica (Hudson) Gaudin, 10
Lycaste Lindley, 227
aromatica (Hooker) Lindley, 227
barringtoniae (Smith) Lindley, 230
brevispatha (Klotzsch) Lindley, 229
candida Reichenbach, 229
ciliata (Ruiz & Pavon) Reichenbach, 230
cinnabarina Rolfe, 230
crinita Lindley, 229
cruenta (Lindley) Lindley, 229
denningiana Reichenbach, 230
deppei (Loddiges) Lindley, 229
dowiana Endres & Reichenbach, 229
dyeriana Rolfe, 230
fimbriata (Poeppig & Endlicher) Cogniaux, 230
gigantea Lindley, 230
hennisiana Kränzlin, 229
lasioglossa Reichenbach, 229
leucantha (Klotzsch) Lindley, 230
locusta Reichenbach, 230
longipetala (Ruiz & Pavon) Garay, 230
longisepala Schweinfurth, 229
macrobulbon (Hooker) Lindley, 229
macrophylla (Poeppig & Endlicher) Lindley, 229
schilleriana Reichenbach, 229
skinneri (Lindley) Lindley, 230
tricolor (Klotzsch) Reichenbach, 229
virginalis (Scheidweiler) Linden, 230
xytriophora Linden & Reichenbach, 230
Lysichiton Schott, 88
americanus Hultén & St John, 89
camschatcensis (Linnaeus) Schott, 89
camschatcensis misapplied, 89

Macodes Lindley, 161
petola (Blume) Lindley, 162
Macrochordium recurvatum Klotzsch, 22
Malaxis Swartz, 223
calophylla (Reichenbach) Kuntze, 223
discolor (Lindley) Kuntze, 223
metallica (Reichenbach) Kuntze, 223
Maranta Linnaeus, 136
arundinacea Linnaeus, 136
bachemiana invalid, 136
bicolor Ker Gawler, 136
kegeliana invalid, 131
leuconeura Morren, 136
var. kerchoviana Morren, 137
var. massangeana Morren, 137
makoyana Morren, 133
Marantaceae, 129
Marcgravia paradoxa misapplied, 86
Masdevallia Ruiz & Pavon, 201
abbreviata Reichenbach, 203
amabilis Reichenbach, 203
astuta Reichenbach, 204
backhousiana Reichenbach, 204
barleana Reichenbach, 204
bella Reichenbach, 204
biflora Regel, 203
caloptera Reichenbach, 203
chimaera Reichenbach, 204
civilis Reichenbach & Warscewicz, 203
coccinea Lindley, 203
coriacea Lindley, 202
corniculata Reichenbach, 203
echidna Reichenbach, 204
ephippium Reichenbach, 202
erinacea Reichenbach, 202
erythrochaete Reichenbach, 204
estradae Reichenbach, 202
fenestrata Hooker, 207
harryana Reichenbach, 203
horrida Garay, 202
ignea Reichenbach, 203
infracta Lindley, 202
leontoglossa misapplied, 202
leontoglossa Reichenbach, 203
lindenii André, 203
macrura Reichenbach, 202
maculata Klotzsch & Karsten, 202
melanopus misapplied, 203
militaris Reichenbach, 203
muscosa Reichenbach, 204
peristeria Reichenbach, 202
polysticta Reichenbach, 203
racemosa Lindley, 203
radiosa Reichenbach, 204
rolfeana Kraenzlin, 202
rosea Lindley, 203
schlimii Lindley, 203
triaristella Reichenbach, 205
tovarensis Reichenbach, 202
vampira Luer, 204
veitchiana Reichenbach, 203
velifera Reichenbach, 202
Massangea hieroglyphica Carrière, 15
Maxillaria Ruiz & Pavon, 236
alba (Hooker) Lindley, 238
chrysantha Rodriguez, 239
coccinea (Jacquin) Hodge, 240
cucullata Lindley, 240
curtipes Hooker, 239
densa Lindley, 240
elatior Reichenbach, 239
elegantula Rolfe, 239
fractiflexa Reichenbach, 237
fucata Reichenbach, 239
galeata Lindley, 269
grandiflora Humboldt, Bonpland & Kunth, 238
houtteana Reichenbach, 239
lepidota Lindley, 238
luteo-alba Lindley, 238
marginata (Lindley) Fenzl, 239
meleagris Lindley, 240
nigrescens Lindley, 237
ochroleuca Lindley, 238
picta Hooker, 239
porphyrostele Reichenbach, 239
praestans Reichenbach, 240
rufescens Lindley, 240
sanderiana Reichenbach, 239
sanguinea Rolfe, 238
setigera Lindley, 238
sophronitis (Reichenbach) Garay, 240
squalens Lindley, 225
striata Rolfe, 239
tenuifolia Lindley, 238
valenzuelana (Richard) Nash, 238
variabilis Lindley, 239
venusta Linden & Reichenbach, 238
Megaclinium falcatum Lindley, 220
leucorrhachis Rolfe, 220
Meiracyllium Reichenbach, 201
gemma Reichenbach, 201
trinasutum Reichenbach, 201
wendlandii Reichenbach, 201
Melica Linnaeus, 53
altissima Linnaeus, 53
ciliata Linnaeus, 53
nutans Linnaeus, 54
uniflora Retzius, 53
Mendoncella Hawkes, 236

barkeri (Reichenbach) Garay, 236
fimbriata (Linden & Reichenbach) Garay, 236
grandiflora (Richard) Hawkes, 236
Mibora Adanson, 50
minima (Linnaeus) Desvaux, 50
Microcoelum Burret & Potztal, 75
martianum (Drude) Burret, 75
weddellianum (Wendland) Moore, 75
Microstylis scottii Ridley, 223
Milium Linnaeus, 49
effusum Linnaeus, 49
Miltonia Lindley, 280
anceps Lindley, 281
candida Lindley, 282
clowesii Lindley, 282
cuneata Lindley, 282
endressii (Reichenbach) Nicholson, 281
flavescens (Lindley) Lindley, 282
phalaenopsis (Linden & Reichenbach) Nicholson, 281
regnelli Reichenbach, 281
roezlii (Reichenbach) Nicholson, 281
russelliana Lindley, 282
schroederiana (Reichenbach) Veitch, 277
speciosa Klotzsch, 282
spectabilis Lindley, 281
superba Schlechter, 281
vexillaria (Bentham) Nicholson, 281
var. warscewiczii (Reichenbach) Schlechter, 281
warscewiczii Reichenbach, 281
Miltoniopsis phalaenopsis (Linden & Reichenbach) Garay & Dunsterville, 281
roezlii (Reichenbach) Godefroy, 281
vexillaria (Bentham) Godefroy, 281
warscewiczii (Reichenbach) Garay & Dunsterville, 281
Miscanthus Andersson, 38
sacchariflorum (Maximowicz) Hackel, 38
sinensis Andersson, 39
Molinia Schrank, 36
caerulea (Linnaeus) Moench, 36
Monstera Adanson, 87
adansonii Schott var. laniata (Scott) Madison, 87
deliciosa Liebmann, 87
latevaginata misapplied, 86
pertusa (Linnaeus) De Vries, 87
Mormodes Lindley, 261
aromatica Lindley, 262
buccinator Lindley, 262
colossus Reichenbach, 262
hookeri Lemaire, 262
igneum Lindley & Paxton, 262
lentiginosa Hooker, 262
lineatum Lindley, 262
luxatum Lindley, 262
maculata (Klotzsch) Williams, 261
pardinum Bateman, 261
Mormolyca Fenzl, 240
ringens (Lindley) Schlechter, 240
Muhlenbergia Schreber, 36
mexicana (Linnaeus) Trinius, 36
Musa Linnaeus, 118
acuminata Colla, 118
arnoldiana De Wildeman, 117
basjoo Siebold, 118
cavendishii Paxton, 118
coccinea Andrews, 118
ensete Gmelin, 117
japonica Thibaut & Keteleer, 118
mannii Hooker, 119
ornata Roxburgh, 118
× paradisiaca Linnaeus, 118
rosacea Jacquin, 118
rosacea misapplied, 118
rosea misapplied, 118
sanguinea Hooker, 119
× sapientium Linnaeus, 118
superba Roxburgh, 118
uranoscopus Loureiro, 118
velutina Wendland & Drude, 118
ventricosa Welwitsch, 117
violacea Baker, 118
zebrina Planchon, 118
Musaceae, 117
Myanthum barbatum Lindley, 265
spinosum Hooker, 265
Myrobroma fragrans Salisbury, 172
Mystacidium infundibulare (Lindley) Rolfe, 254

Nageliella Williams, 199
purpurea (Lindley) Williams, 199
Nanodes medusae Reichenbach, 186
Nanorrhops Wendland, 69
ritchiana (Griffith) Aitchison, 69
Nastus tessellatus Nees, 60
Neanthe bella invalid, 71
Neofinetia Hu, 256
falcata (Thunberg) Hu, 256
Neoregelia Smith, 16
ampullacea (Morren) Smith, 17
carolinae (Beer) Smith, 17
forma tricolor (Foster) Smith, 17
chlorosticta (Baker) Smith, 17
concentrica (Vellozo) Smith, 18
fosteriana Smith, 17
marmorata (Baker) Smith, 18
morreniana (Antoine) Smith, 17
pineliana (Lemaire) Smith, 17
forma phyllanthidea (Baker) Smith, 17
princeps (Baker) Smith, 17
sarmentosa (Regel) Smith, 17
var. *chlorosticta* (Baker) Smith, 17
spectabilis (Moore) Smith, 18
tristis (Beer) Smith, 17
Nephthytis Schott, 90
afzelii Schott, 90
triphylla Bailey, 104
Nicolaia Horaninow, 128
elatior (Jack) Horaninow, 128
imperialis Horaninow, 128
speciosa (Blume) Horaninow, 128
Nidularium Lemaire, 18
acanthocrater Morren, 18
ampullaceum Morren, 17
burchellii (Baker) Mez, 19
carolinae Baker, 17
cyaneum Linden & André, 17
fulgens Lemaire, 19
innocentii Lemaire, 19
var. innocentii, 19
var. lineatum (Mez) Smith, 19
var. paxianum (Mez) Smith, 19
var. striatum Wittmack, 19
var. wittmackianum (Harms) Smith, 19
laurentii Regel, 18
lindenii Regel, 20
microps Mez, 19
pinelianum Lemaire, 17
seidelii Smith & Reitz, 19
spectabile Moore, 18
Nigritella Richard, 168
nigra (Linnaeus) Reichenbach, 168
Nipponocalamus simonii (Carrière) Nakai, 62

Octomeria Brown, 207
crassifolia Lindley, 207
gracilis Lindley, 207
graminifolia Brown, 207
grandiflora Lindley, 207
juncifolia Rodriguez, 207
Odontoglossum Humboldt, Bonpland & Kunth, 274
alexandrae Bateman, 277
anceps (Lindley) Klotzsch, 281
apterum misapplied, 276
bictoniense (Bateman) Lindley, 278
brevifolium Lindley, 277
cariniferum Reichenbach, 278
cervantesii Llave & Lexarza, 276
chiriquense Reichenbach, 277
cirrhosum Lindley, 276
citrosmum Lindley, 278
clowesii (Lindley) Lindley, 282
confusum Garay, 277
cordatum Lindley, 276
coronarium Lindley, 277
crispum Lindley, 277
grande Lindley, 277
hallii Lindley, 277
harryanum Reichenbach, 277
hastatum Bateman, 287
laeve Lindley, 277
londesboroughianum Reichenbach, 278
nebulosum Lindley, 276
pendulum Llave & Lexarza, 278
phalaenopsis Linden & Reichenbach, 281
pulchellum Lindley, 278
reichenheimii Linden & Reichenbach, 277
roezlii Reichenbach, 281
rossii Lindley, 276
schroederianum Reichenbach, 277
spectatissimum Lindley, 277
stellatum Lindley, 276
triumphans Reichenbach, 277
uroskinneri Lindley, 278
vexillarium Bentham, 281
weltonii invalid, 281
Oeceoclades Lindley, 258
maculata (Lindley) Lindley, 258
saundersiana (Reichenbach) Garay & Taylor, 258
Oncidium Swartz, 282
altissimum (Jacquin) Swartz, 286
ampliatum Lindley, 289
bicallosum Lindley, 286
candidum (Lindley) Reichenbach, 282
carthagenense (Jacquin) Swartz, 287
cavendishianum Bateman, 286
cebolleta (Jacquin) Swartz, 286

cheirophorum Reichenbach, 288
concolor Hooker, 288
crispum Loddiges, 288
divaricatum Lindley, 287
enderianum Masters, 288
excavatum Lindley, 289
flexuosum Loddiges, 289
forbesii Hooker, 288
fuscatum Reichenbach, 281
gardneri Lindley, 288
harrisonianum Lindley, 287
hastatum (Bateman) Lindley, 287
hastiferum Reichenbach & Warscewicz, 285
incurvum Baker, 290
kramerianum Reichenbach, 287
leucochilum Bateman, 289
longipes Lindley, 288
macranthum Lindley, 285
maculatum (Lindley) Lindley, 287
marshallianum Reichenbach, 288
microchilum Lindley, 285
nanum Lindley, 286
onustum Lindley, 286
ornithorrhynchum Humboldt, Bonpland & Kunth, 288
papilio Lindley, 287
pulvinatum Lindley, 287
pumilum Lindley, 287
pubes Lindley, 287
reflexum Lindley, 289
regnellii (Reichenbach) Reichenbach, 281
section Barbata Pfitzer, 288
section Cebolleta Lindley, 286
section Concoloria Kränzlin, 288
section Crispa Pfitzer, 288
section Cyrtochilum (Humboldt, Bonpland & Kunth) Lindley, 285
section Excavata Kränzlin, 289
section Glanduligera Lindley, 287
section Oblongata Kränzlin, 289
section Onusta Garay & Stacy, 286
section Planifolia Bentham & Hooker, 289
section Pluritubercalata Lindley, 286
section Pulvinata Lindley, 287
section Rostratae Rolfe, 288
section Stellata Kränzlin, 287
section Synsepala Pfitzer, 289
section Waluewa (Regel) Schlechter, 287
sphacelatum Lindley, 289
splendidum Duchartre, 286
stramineum Bateman, 287
superbiens Reichenbach, 286
tigrinum Llave & Lexarza, 289
varicosum Lindley, 289
wentworthianum Lindley, 290
Ophrys Linnaeus, 162
apifera Hudson, 163
aranifera Hudson, 163
fuciflora (Schmidt) Moench, 163
fusca Link, 163
subsp. iricolor (Desfontaines) Swartz, 163
holoserica (Burmann) Greuter, 163
insectifera Linnaeus, 162
iricolor Desfontaines, 163
litigiosa Camus, 163
lutea (Gouan) Cavanilles, 163
mammosa Desfontaines, 163
musciflora Hudson, 162
oestrifera Bieberstein, 164
subsp. *heldreichii* (Schlechter) Soó, 164
scolopax Cavanilles, 164
subsp. cornuta (Steven) Camus, 164
subsp. heldreichii (Schlechter) Nelson, 164
subsp. orientalis (Renz) Nelson, 164
speculum Link, 163
sphegodes Miller, 163
subsp. litigiosa (Camus) Becherer, 163
subsp. mammosa (Desfontaines) Soó, 163
tenthredinifera Willdenow, 163
umbilicata Desfontaines, 164
Orchidaceae, 137
Orchis Linnaeus, 165
anatolica Boissier, 166
coriophora Linnaeus, 166
elodes Grisebach, 168
ericetorum (Linden) Marshall, 168
italica Poiret, 167
joo-iokiana (Makino) Maekawa, 166
maderensis Summerhayes, 167
mascula Linnaeus, 166
militaris Linnaeus, 167
morio Linnaeus, 166
papilionacea Linnaeus, 166
provincialis Balbis, 166
purpurea Hudson, 167
quadripunctata Tenore, 166
ustulata Linnaeus, 167
Orontium Linnaeus, 89
aquaticum Linnaeus, 89
Ornithidium coccineum (Jacquin) Brown, 240
densum (Lindley) Reichenbach, 240
sophronitis Reichenbach, 240
Ornithocephalus Hooker, 241
gladiatus Hooker, 241
iridifolius Reichenbach, 241
grandiflorus Lindley, 241
Otatea (McClure & Smith) Calderon & Soderstrom, 58
acuminata (Munro) Calderon & Soderstom, 59
aztecorum (McClure & Smith) Calderon & Soderstrom, 59
Otoglossum brevifolium (Lindley) Garay & Dunsterville, 277

Pabstia Garay, 231
jugosa (Lindley) Garay, 231
var. rufina (Reichenbach) Garay, 231
var. viridus (Godefroy) Garay, 231
placanthera (Hooker) Garay, 231
viridis (Lindley) Garay, 232
Palmae, 65
Pandanaceae, 112
Pandanus Parkinson, 113
pygmaeus Thouars, 113
sanderi Masters, 113
utilis Bory, 113
veitchii Masters & Moore, 113
Panicum Linnaeus, 41
capillare Linnaeus, 41
clandestinum Linnaeus, 41
miliaceum Linnaeus, 41
obtusum Humboldt, Bonpland & Kunth, 41
virgatum Linnaeus, 41
Paphinia Lindley, 268
cristata Lindley, 268
grandiflora Rodriguez, 268
Paphiopedilum Pfitzer, 153
acmodontum Wood, 157
× *ang-thong* Fowlie, 154
appletonianum (Gower) Rolfe, 156
argus (Reichenbach) Stein, 158
barbatum (Lindley) Pfitzer, 156
bellatulum (Reichenbach) Stein, 154
boxalli (Reichenbach) Pfitzer, 155
bullenianum (Reichenbach) Pfitzer, 156
callosum (Reichenbach) Pfitzer, 156
var. sanderae Anon., 156
chamberlainianum (Sander) Stein, 158
charlesworthii (Rolfe) Pfitzer, 155
ciliolare (Reichenbach) Pfitzer, 158
concolor (Bateman) Pfitzer, 154
curtisii (Reichenbach) Stein, 158
dayanum (Lindley) Pfitzer, 156
delenatii Guillaumin, 154
druryi (Beddome) Stein, 155
elliotianum misapplied, 154
exul (Ridley) Kerchove, 155
fairrieanum (Lindley) Pfitzer, 156
glanduliferum misapplied, 155
glaucophyllum Smith, 158
godefroyae (Godefroy) Stein, 154
haynaldianum (Reichenbach) Stein, 158
hirsutissimum (Lindley) Stein, 155
hookerae (Reichenbach) Stein, 157
var. *volonteanum* (Masters) Hallier, 157
insigne (Wallich) Pfitzer, 155
var. sanderae Reichenbach, 156
var. sanderianum Sander, 156
javanicum (Lindley & Paxton) Pfitzer, 157
var. *virens* (Lindley & Paxton) Pfitzer, 157
lawrenceanum (Reichenbach) Pfitzer, 156
var. hyeanum (Reichenbach) Pfitzer, 156
× leeanum (Reichenbach) Kerchove, 155
leucochilum (Rolfe) Fowlie, 154
linii Schoser, 156
lowii (Lindley) Stein, 158
mastersianum (Reichenbach) Pfitzer, 157
× maudiae Anon., 156
niveum (Reichenbach) Stein, 154
parishii (Reichenbach) Stein, 158
philippinese (Reichenbach) Stein, 155
praestans (Reichenbach) Pfitzer, 155
purpurascens Fowlie, 157
purpuratum (Lindley) Stein, 157
rothschildianum (Reichenbach) Stein, 154
spicerianum (Reichenbach) Stein, 155
stonei (Hooker) Stein, 155
sukhakulii Schoser & Senghas, 158
superbiens (Reichenbach) Stein, 158
subsp. ciliolare (Reichenbach) Wood, 158
tonsum (Reichenbach) Stein, 157
venustum (Sims) Pfitzer, 157
Victoria-mariae invalid, 158
victoriae-reginae (Sander) Wood, 158
subsp. chamberlainianum (Sander) Wood, 158
subsp. glaucophyllum (Smith) Wood, 158
villosum (Lindley) Stein, 155
violascens Schlechter, 156
virens misapplied, 157
virens (Reichenbach) Pfitzer, 157

volonteanum (Sander) Pfitzer, 157
wardii Summerhayes, 158
wolterianum (Kraenzlin) Pfitzer, 157
Papilionanthe hookeriana (Reichenbach) Schlechter, 250
subulata (Koenig) Garay, 242
teres (Roxburgh) Schlechter, 250
vandarum (Reichenbach) Garay, 242
Paraphalaenopsis Hawkes, 249
denevei (Smith) Hawkes, 249
laycockii (Henderson) Hawkes, 249
sepentilingua (Smith) Hawkes, 249
Paspalum Linnaeus, 40
ceresia (Kuntze) Chase, 40
elegans Roemer & Schultes, 40
membranaceum Lamarck, 40
Pecteilis Rafinesque, 169
radiata (Thunberg) Rafinesque, 170
Peltandra Rafinesque, 98
sagittifolia (Michaux) Morong, 98
virginica (Linnaeus) Kunth, 98
Pennisetum Richard, 43
alopecuroides (Linnaeus) Sprengel, 43
cenchroides Richard, 43
ciliare (Linnaeus) Link, 43
macrourum Trinius, 43
orientale Richard, 43
setaceum (Forsskahl) Chiovenda, 43
villosum Fresenius, 43
Peristeria Hooker, 266
cerina Lindley, 267
elata Hooker, 267
pendula Hooker, 267
Pescatoria Reichenbach, 234
cerina (Lindley) Reichenbach, 234
dayana Reichenbach, 234
var. candidula Reichenbach, 234
var. rhodacra Reichenbach, 234
var. splendens Reichenbach, 234
lehmannii Reichenbach, 235
lamellosa Reichenbach, 234
wallisii Reichenbach, 234
Phaenosperma Bentham, 54
globosa Bentham, 55
Phaeomeria Lindley, 128
magnifica (Roscoe) Schumann, 128
speciosa (Blume) Merrill, 128
Phaius Loureiro, 176
albus Wallich, 178
bicolor Lindley, 176
blumei Lindley, 176
flavus (Blume) Lindley, 176
grandifolius Loureiro, 176
humblotii Reichenbach, 176
maculatus Lindley, 176
mishmensis (Lindley) Reichenbach, 176
tankervilleae (Banks) Blume, 176
tuberculosus (Thouars) Blume, 176
wallichii Hooker, 176
Phalaenopsis Blume, 244
amabilis (Linnaeus) Blume, 245
var. aurea Rolfe, 245
var. *dayana* Warner & Williams, 245
var. *formosa* Shimidzu, 245
var. moluccana Schlechter, 245
var. papuana Schlechter, 245
amboinensis Smith, 247
aphrodite Reichenbach, 245
boxalli Reichenbach, 246
buyssoniana Reichenbach, 244
celebensis Sweet, 247
cochlearis Holttum, 247
corningiana Reichenbach, 248
cornu-cervi (Breda) Blume & Reichenbach, 246
denevei Smith, 249
equestris (Schauer) Reichenbach, 246
esmeralda Reichenbach, 244
fasciata Reichenbach, 247
fimbriata Smith, 248
foerstermanii Reichenbach, 248
fuscata Reichenbach, 247
gigantea Smith, 247
hieroglyphica (Reichenbach) Sweet, 248
× intermedia Lindley, 245
javanica Smith, 247
kunstleri Hooker, 247
latisepala Reichenbach, 247
laycockii Henderson, 249
× leucorrhoda Reichenbach, 246
lindenii Loher, 247
lobbii (Reichenbach) Sweet, 246
lueddemanniana Reichenbach, 248
var. delicata Reichenbach, 248
var. *hieroglyphica* Richard, 248
var. ochracea Reichenbach, 248
maculata Reichenbach, 248
mannii Reichenbach, 246
mariae Warner & Williams, 248
modesta Smith, 248
pallens (Lindley) Reichenbach, 248
var. denticulata (Reichenbach) Sweet, 248
pantherina Reichenbach, 246
parishii Reichenbach, 246
var. lobbii Reichenbach, 246
psilantha Schlechter, 248
pulcherrima (Lindley) Smith, 244
rosea Lindley, 246
sanderiana Reichenbach, 245
schilleriana Reichenbach, 245
var. *alba* Roebelen, 246
var. *vestalis* Reichenbach, 246
serpentilingua Smith, 249
stuartiana Reichenbach, 246
var. punctatissima Reichenbach, 246
sumatrana Korthals & Reichenbach, 247
violacea Witte, 247
var. alba Teijsmann & Binnendijk, 247
zebrina Teijsmann & Binnendijk, 247
zebrina Witte, 247
Phalaris Linnaeus, 49
arundinacea Linnaeus, 49
canariensis Linnaeus, 49
minor Retzius, 49
paradoxa Linnaeus, 49
Philodendron Schott, 91
andreanum Devansaye, 93
angustisectum Engler, 92
asperatum Koch, 94
auritum invalid, 103
bipennifolium Schott, 93
bipinnatifidum Endlicher, 92
cannifolium Kunth, 93
cordatum misapplied, 93
cordatum (Vellozo) Kunth, 94
× corsinianum Senoner, 92
devansayeanum Linden, 94
domesticum Bunting, 94
elegans Krause, 92
erubescens Koch & Augustin, 93
gloriosum André, 94
hastatum misapplied, 94
ilsemannii Sander, 94
imbe Endlicher, 94
lacerum (Jacquin) Schott, 92
laciniatum (Vellozo) Engler, 92
mamei André, 94
martianum Engler, 93
melanochrysum Linden & André, 93
micans Koch, 93
ornatum Schott, 94
oxycardium Schott, 93
panduraeforme misapplied, 93
pedatum (Hooker) Kunth, 92
pertusum Kunth & Bouché, 87
sagittifolium misapplied, 94
scandens Koch & Sello, 93
subsp. oxycardium (Schott) Bunting, 93
subsp. scandens, 93: forma scandens, 93; forma micans (Koch) Bunting, 93
selloum Koch, 92
sodiroi invalid, 94
squamiferum Poeppig, 93
trifoliatum invalid, 103
verrucosum Schott, 93
wendlandii Schott, 93
Phleum Linnaeus, 50
pratense Linnaeus, 50
Phoenix Linnaeus, 70
canariensis Chabaud, 70
dactylifera Linnaeus, 70
hanceana Naudin, 70
humilis Royle, 70
loureirii Kunth, 70
reclinata Jacquin, 70
roebelenii O'Brien, 70
spinosa Schumacher, 70
sylvestris (Linnaeus) Roxburgh, 70
theophrastii Greuter, 70
Pholidota Hooker, 184
articulata Lindley, 184
chinensis Lindley, 185
imbricata Lindley, 185
pallida Lindley, 185
ventricosa (Blume) Reichenbach, 185
Phragmipedium (Pfitzer) Rolfe, 152
boissierianum (Reichenbach) Rolfe, 153
caricinum (Lindley) Rolfe, 152
caudatum (Lindley) Rolfe, 152
lindenii (Lindley) Dressler & Williams, 152
lindleyanum (Lindley) Rolfe, 152
longifolium (Warscewicz & Reichenbach) Rolfe, 152
sargentianum (Rolfe) Rolfe, 152
schlimii (Linden & Reichenbach) Rolfe, 152
warscewiczianum (Reichenbach) Schlechter, 152
Phragmites Adanson, 35
australis (Cavanilles) Steudel, 35
communis Trinius, 35
vulgaris (Lamarck) Crépin, 35

Phrynium confertum (Bentham) Schumann, 136
cylindricum Roscoe, 131
Phyllostachys Siebold & Zuccarini, 56
aurea (Carrière) Rivière, 57
aureosulcata McClure, 57
bambusoides Siebold & Zuccarini, 57
boryana Mitford, 58
castillonis (Carrière) Mitford, 57
congesta Rendle, 57
edulis invalid, 57
edulis misapplied, 58
fastuosa (Mitford) Nicholson, 58
flexuosa (Carrière) Rivière, 57
formosana Hayata, 57
henonis Mitford, 58
heterocycla (Carrière) Mitford, 57
forma nabeshimana (Muroi) Muroi, 57
forma pubescens (Lehaie) McClintock, 57
kumasasa (Zollinger) Munro, 58
mazelii Rivière, 57
mitis misapplied, 57, 58
nevinii misapplied, 57
nidularia Munro, 58
nigra (Loddiges) Munro, 57
var. henonis (Mitford) Stapf, 58
var. *punctata* Bean, 58
puberula Makino, 58
pubescens Lehaie, 57
quiloi (Carrière) Rivière, 57
reticulata misapplied, 57
ruscifolia (Munro) Satow, 57
viridi-glaucescens (Carrière) Rivière, 58
viridis (Young) McClure, 58
Physosiphon Lindley, 201
guatemalensis Rolfe, 201
loddigesii Lindley, 201
tubatus (Loddiges) Reichenbach, 201
Pinellia Tenore, 109
pedatisecta Schott, 109
ternata (Thunberg) Breitenbach, 109
Piptatherum Palisot de Beauvois, 54
miliaceum (Linnaeus) Cosson, 54
paradoxum (Linnaeus) Palisot de Beauvois, 54
Pironneava luddemanniana Koch, 21
Pistia Linnaeus, 112
stratiotes Linnaeus, 112
Pitcairnia L'Heritièr, 13
andreana Linden, 13
flavescens Baker, 13
heterophylla (Lindley) Beer, 13
var. albiflora Standley & Smith, 13
lepidota Regel, 13
xanthocalyx Martius, 13
Platanthera Richard, 169
bifolia (Linnaeus) Richard, 169
chlorantha (Custer) Reichenbach, 169
ciliaris (Linnaeus) Lindley, 170
psycodes (Linnaeus) Lindley, 170
Platyclinis cobbiana (Reichenbach) Hemsley, 184
filiformis (Lindley) Bentham, 184
glumacea (Lindley) Bentham, 184
latifolia (Lindley) Hemsley, 184
Pleioblastus Nakai, 61
angustifolius (Mitford) Nakai & Okamura, 61
argenteo-striatus (Regel) Nakai, 61
chino (Franchet & Savatier) Makino, 61
forma angustifolius (Mitford) Muroi, 61
forma humilis (Makino) Suzuki, 61
forma vaginatus (Hackel) Muroi & Okamura, 61
var. *laydekeri* (Bean) Nakai, 61
var. *vaginatus* (Hackel) Suzuki, 61
distichus (Mitford) Muroi & Okamura, 61
fortunei invalid, 62
var. *aurea* invalid, 62
gracilis (Makino) Nakai, 61
gramineus (Bean) Nakai, 61
hindsii (Munro) Nakai, 61
humilis (Mitford) Nakai, 61
forma humilis, 61
kongosanensis Makino, 62, 64
maximowiczii (Rivière) Nakai, 61
pumilus (Mitford) Nakai, 61
pygmaeus (Miquel) Nakai, 61
var. distichus (Mitford) Nakai, 61
shibuyanus Nakai, 62
simonii (Carrière) Nakai, 62
forma *variegata* (Hooker) Muroi, 62
var. *heterophyllus* (Makino) Nakai, 62
vaginatus (Hackel) Nakai, 61
variegatus (Miquel) Makino, 62
forma glaber (Makino) McClintock, 62
var. viridis (Makino) McClintock, 62
viridi-striatus (André) Makino, 62
forma *vagans* (Gamble) Muroi, 64
Pleione Don, 182
albiflora Cribb & Tang, 184
bulbocodioides (Franchet) Rolfe, 183
× confusa Cribb & Tang, 184
delavayi (Rolfe) Rolfe, 183
formosana Hayata, 183
forrestii misapplied, 184
forrestii Schlechter, 184
henryi (Rolfe) Schlechter, 183
hookeriana (Lindley) Williams, 183
humilis (Smith) Don, 183
lagenaria Lindley, 183
limprichtii Schlechter, 183
maculata (Lindley) Lindley, 183
praecox (Smith) Don, 183
pricei Rolfe, 183
reichenbachiana (Moore & Veitch) Williams, 183
yunnanensis (Rolfe) Rolfe, 183
Pleurothallis Brown, 205
atropurpurea Lindley, 207
gelida Lindley, 206
ghiesbrechtiana Richards & Galeotti, 206
grandis Rolfe, 206
grobyi Lindley, 205
immersa Linden & Reichenbach, 205
macrophylla Humboldt, Bonpland & Kunth, 205
ornata Reichenbach, 206
pectinata Lindley, 205
quadrifida (Llave & Lexarza) Lindley, 206
roezlii Reichenbach, 205
saurocephala Loddiges, 206
schiedei Reichenbach, 206
tribuloides Lindley, 205
Poa Linnaeus, 52
aquatica Linnaeus, 53
chaixii Villars, 52
glauca Vahl, 52
Poaceae, 31
Polycynis Reichenbach, 268
barbata Reichenbach, 268
Polypogon Desfontaines, 50
monspeliensis (Linnaeus) Desfontaines, 50
Polystacha de Jussieu, 257
affinis Lindley, 257
bracteosa Lindley, 257
Ponerorchis joo-iokiana (Makino) Maekawa, 166
Porroglossum Schlechter, 204
echidnum (Reichenbach) Garay, 204
muscosum (Reichenbach) Schlechter, 205
Pothos aureus Linden & André, 86
celatocaulis Brown, 86
Promenaea Lindley, 233
citrina Don, 233
× crawshayana Anon., 233
lentiginosa (Lindley) Lindley, 233
microptera Reichenbach, 233
rollissonii (Lindley) Lindley, 233
stapelioides (Link & Otto) Lindley, 233
xanthina (Lindley) Lindley, 233
Pseudosasa Nakai, 64
japonica (Steudel) Makino, 64
Pteroceras Hasskarl, 242
pallidum (Blume) Holttum, 242
Pterostylis Brown, 171
banksii Hooker, 171
curta Brown, 171
nutans Brown, 171
pedunculata Brown, 171
Ptychosperma Labillardière, 73
alexandrae Mueller, 72
cunninghamiana Wendland, 72
elegans (Brown) Blume, 73
macarthurii (Wendland) Nicholson, 73
Puya Molina, 12
alpestris (Poeppig) Gay, 13
chilensis Molina, 13
coerulea Lindley, 12
var. violacea (Brongniart) Smith, 12
gigas André, 12
spathacea (Grisebach) Mez, 12
venusta Philippi, 12
violacea (Brongniart) Mez, 12
whytei Hooker, 13
Pyrrheimia fuscata (Loddiges) Hasskarl, 28

Quesnelia Gaudichaud, 23
liboniana (De Jonghe) Mez, 23
marmorata (Lemaire) Read, 23
seideliana Smith & Reitz, 23

Ravenala Adanson, 119
madagascariensis Sonnerat, 119
Ravenalopsis Nakai, 120
Renanthera Loureiro, 252
coccinea Loureiro, 252
imschootiana Rolfe, 252
lowii (Lindley) Reichenbach, 252
pulchella Rolfe, 252
rohaniana Reichenbach, 252
storiei (Storie) Reichenbach, 252
Renealmia monostachia Linnaeus, 46
usneoides Linnaeus, 14
Restrepia Humboldt, Bonpland & Kunth, 206

antennifera Humboldt, Bonpland & Kunth, 206
elegans Karsten, 206
guttulata Lindley, 206
maculata Lindley, 206
Rhaphidophora Hasskarl, 86
aurea (Linden & André) Birdsey, 86
celatocaulis (Brown) Knoll, 86
decursiva (Wallich) Schott, 86
Rhaphidophyllum Wendland & Drude, 66
hystrix (Pursh) Wendland & Drude, 66
Rhapis Linnaeus, 67
excelsa (Thunberg) Henry, 67
flabelliformis Aiton, 67
humilis Blume, 67
Rhektophyllum Brown, 90
mirabile Brown, 90
Rhodospatha Poeppig, 87
picta Nicholson, 87
Rhodostachys bicolor Ruiz & Pavon, 20
pitcairniifolia (Verlot) Baker, 20
Rhoeo spathacea (Swartz) Stearn, 30
Rhopalostylis Wendland & Drude, 72
baueri Wendland & Drude, 73
cheesemanii Beccari, 73
sapida Wendland & Drude, 72
Rhynchelytrum Nees, 42
repens (Willdenow) Hubbard, 42
roseum (Nees) Stapf & Hubbard, 42
Rhyncholaelia digbyana (Lindley) Schlechter, 198
glauca (Lindley) Schlechter, 198
Rhynchostylis Blume, 243
coelestis (Reichenbach) Veitch, 244
densiflora (Lindley) Williams, 244
gigantea (Lindley) Ridley, 244
var. harrisoniana (Hooker) Holttum, 244
var. *petotiana* Anon., 244
retusa (Linnaeus) Blume, 244
violacea (Lindley) Reichenbach, 244
Richardia africana Kunth, 97
elliottiana Watson, 98
pentlandii Watson, 98
Rodriguezia Ruiz & Pavon, 271
decora (Lindley) Reichenbach, 271
fragrans Lindley, 272
secunda Humboldt, Bonpland & Kunth, 271
venusta (Lindley) Reichenbach, 272
Roscoea Smith, 125
alpina Royle, 126
auriculata Schumann, 126
capitata var. *scillifolia* misapplied, 126
cautleoides Gagnepain, 126
chamael Gagnepain, 126
humeana Balfour & Smith, 126
intermedia Gagnepain, 126
longifolia Baker, 126
procera Wallich, 126
purpurea misapplied, 126
purpurea Smith, 126
var. *procera* (Wallich) Baker, 126
scillifolia (Gagnepain) Cowley, 126
sikkimensis Gentil, 126
sinopurpurea Stapf, 126
yunnanensis Loesener, 126
Rossioglossum grande (Lindley) Garay & Kennedy, 277
Rostraria Trinius, 48
cristata (Linnaeus) Tzvelev, 48
Rudolfiella aurantiaca (Lindley) Schlechter, 226

Sabal Adanson, 69
adansonii Guersent, 70
blackburniana Glazebrook, 69
minor (Jacquin) Persoon, 70
palmetto (Walter) Loddiges, 70
Saccharum Linnaeus, 39
ravennae (Linnaeus) Murray, 39
Saccolabium acutifolium Lindley, 253
ampullaceum Lindley, 252
miniatum Lindley, 252
violaceum Lindley, 244
Sarcanthus appendiculatus Anon., 253
filiformis Lindley, 253
pallidus Lindley, 253
racemifer (Lindley) Reichenbach, 253
rostratus Lindley, 253
teretifolius (Lindley) Lindley, 253
Sarcochilus Brown, 241
ceciliae Mueller, 242
falcatus Brown, 241
fitzgeraldii Mueller, 242
hartmannii Mueller, 242
luniferus (Reichenbach) Hooker, 242
pallidus (Blume) Reichenbach, 242
unguiculatus Lindley, 242
Sarcopodium acuminatum var. *lyonii* (Ames) Kränzlin, 219
amplum (Lindley) Lindley, 219
lyonii (Ames) Rolfe, 219
rotundatum Lindley, 219
Sasa Makino & Shibata, 63
albo-marginata (Miquel) Makino, 63
forma *nana* Makino, 63
borealis (Hackel) Makino, 64
cernua misapplied, 63
disticha (Mitford) Camus, 61
humilis (Mitford) Camus, 61
japonica (Steudel) Makino, 64
kurilensis (Ruprecht) Makino & Shibata, 63
palmata (Burbidge) Camus, 63
forma nebulosa (Makino) Suzuki, 63
pygmaea (Mitford) Camus, 61
ramosa (Makino) Makino & Shibata, 64
senanensis (Franchet & Savatier) Rehder, 63
senanensis misapplied, 63
tessellata (Munro) Makino & Shibata, 63
tsuboiana Makino, 63
vagans misapplied, 64
variegata (Miquel) Camus, 62
veitchii (Carrière) Rehder, 63
Sasaella Makino, 64
ramosa (Makino) Makino, 64
Sasamorpha Nakai, 64
borealis (Hackel) Makino, 64
purpurascens (Hackel) Nakai, 64
var. *borealis* misapplied, 64
tessellata (Munro) Nakai, 63
Satyrium Swartz, 171
nepalensis Don, 171
var. ciliata (Lindley) King & Pantling, 171
Sauromatum Schott, 106
guttatum (Wallich) Schott, 106
nubicum Schott, 106
venosum (Aiton) Kunth, 106
Scaphyglottis Poeppig & Endlicher, 199
amethystina (Reichenbach) Schlechter, 199
Schismatoglottis Zollinger & Moritzi, 91
concinna Schott, 91
neo-guineensis André, 91
picta Schott, 91
Schizocasia portei (Schott) Engler, 102
regnieri Linden & Rodigas, 102
sanderiana (Bull) Engler, 102
Schoenoplectus lacustris (Linnaeus) Palla, 114
Schoenorchis Blume, 253
juncifolia Blume, 253
Schomburgkia Lindley, 196
crispa Lindley, 197
crispa misapplied, 197
gloriosa Reichenbach, 197
lyonsii Lindley, 197
superbiens (Lindley) Rolfe, 197
tibicinis Lindley, 197
undulata Lindley, 197
Scindapsus Schott, 86
aureus (Linden & André) Engler, 86
pictus Hasskarl var. argyraeus Engler, 86
Scirpus Linnaeus, 114
cernuus Vahl, 115
gracilis Koch, 115
holoschoenus Linnaeus, 114
lacustris Linnaeus, 114
subsp. lacustris, 115
subsp. tabernaemontani (Gmelin) Palla, 115
maritimus Linnaeus, 114
savii Sebastiana & Mauri, 115
tabernaemontani Gmelin, 114
Scuticaria Lindley, 236
hadwenii (Lindley) Hooker, 236
steelei (Hooker) Lindley, 236
strictifolia Hoehne, 236
Seaforthia elegans Brown, 73
elegans misapplied, 72
Sedirea Garay & Sweet, 243
japonica (Linden & Reichenbach) Garay & Sweet, 243
Semiarundinaria Nakai, 58
fastuosa (Mitford) Makino, 58
var. yashadake (Makino) Makino, 58
yashadake Makino, 58
Serapias Linnaeus, 164
cordigera Linnaeus, 164
gregaria Godfrey, 165
longipetala Tenore, 164
neglecta de Notaris, 164
olbia Verguin, 165
orientalis invalid, 164
pseudocordigera Moricand, 164
vomeracea (Burman) Briquet, 164
subsp. orientalis Greuter, 164
Serenoa Hooker, 68
repens (Bartram) Small, 68
serrulata (Michaux) Nicholson, 68
Setaria Palisot de Beauvois, 42
glauca misapplied, 42
italica (Linnaeus) Palisot de Beauvois, 42
pallidefusca (Schumacher) Stapf & Hubbard, 42
plicatilis (Hochstetter) Hackel, 42
pumila (Poiret) Roemer & Schultes, 42

viridis (Linnaeus) Palisot de Beauvois, 42
Setcreasea purpurea Boom, 31
striata invalid, 29
Shibataea Nakai, 58
kumasasa (Zollinger) Nakai, 58
Siderasis Rafinesque, 27
fuscata (Loddiges) Moore, 28
Sigmatostalix Reichenbach, 290
graminea (Poeppig & Endlicher) Reichenbach, 290
Sinarundinaria Nakai, 60
murielae (Gamble) Nakai, 60
nitida (Mitford) Nakai, 60
Sitanion Rafinesque, 44
hystrix (Nuttall) Smith, 44
Sobralia Ruiz & Pavon, 178
cattleya Reichenbach, 178
decora Bateman, 178
leucoxantha Reichenbach, 178
macrantha Lindley, 178
sessilis Lindley, 178
xantholeuca Reichenbach, 178
Sophronitella Schlechter, 199
violacea (Lindley) Schlechter, 199
Sophronites Lindley, 198
cernua Lindley, 198
coccinea (Lindley) Reichenbach, 199
grandiflora Lindley, 199
violacea Lindley, 199
Sorghastrum Nash, 39
nutans (Linnaeus) Nash, 39
Sparganium Linnaeus, 113
emersum Rehmann, 113
erectum Linnaeus, 113
minimum Wallroth, 113
ramosum Hudson, 113
simplex Hudson, 113
Spartina Schreber, 36
pectinata Link, 36
Spathicarpa Hooker, 104
sagittifolia Schott, 104
Spathiphyllum Schott, 87
candidum (Bull) Brown, 88
candidum misapplied, 88
cannifolium (Dryander) Schott, 88
clevelandii invalid, 88
cochlearispathum (Liebmann) Engler, 88
commutatum Schott, 88
floribundum (Linden & André) Brown, 88
friedrichsthalii misapplied, 88
kochii misapplied, 88
patinii (Hogg) Brown, 88
phryniifolium Schott, 88
wallisii Regel, 88
Spathoglottis Blume, 176
aurea Lindley, 177
fortunei Lindley, 177
kimballiana Hooker, 177
plicata Blume, 177
vieillardii Reichenbach, 177
Spiranthes Richard, 160
aestivalis (Poiret) Richard, 161
australis Lindley, 161
autumnalis Richard, 160
cernua Richard, 161
gracilis (Bigelow) Beck, 161
lacera (Rafinesque) Rafinesque, 161
sinensis (Persoon) Ames, 161
spiralis (Linnaeus) Chevallier, 160
Spirodela, 112
Spodiopogon Trinius, 38
sibiricus Trinius, 38
Strachyphrynium jagorianum (Koch) Schumann, 136
Stanhopea Hooker, 268
bucephalus misapplied, 269
devoniensis Lindley, 269
eburnea Lindley, 268
ecornuta Lemaire, 268
grandiflora (Loddiges) Lindley, 268
grandiflora Reichenbach, 269
graveolens misapplied, 269
harnandezii (Kunth) Schlechter, 269
insignis Hooker, 269
jenishiana Reichenbach, 269
lewisae Ames & Correll, 269
oculata (Loddiges) Lindley, 269
saccata Bateman, 269
tigrina Lindley, 269
wardii Loddiges, 269
Stauropsis gigantea (Lindley) Bentham & Hooker, 251
lissochiloides (Gaudichaud) Richard, 251
Stelis tubatus Loddiges, 201
Stenocoryne aureo-fulva (Hooker) Kränzlin, 226
vitellina (Lindley) Kränzlin, 226
Stenoglottis Lindley, 170
fimbriata Lindley, 170
longifolia Hooker, 170
Stenospermation Schott, 86
popayanense Schott, 87
wallisii Masters, 87
Stipa Linnaeus, 54
calamagrostis (Linnaeus) Wahlenberg, 54
gigantea Link, 54
pennata Linnaeus, 54
splendens Trinius, 54
tirsa Steven, 54
Strelitzia Aiton, 119
alba (Linnaeus) Skeels, 119
augusta Wright, 119
caudata Dyer, 119
juncea Link, 119
nicolai Regel & Koernicke, 119
parvifolia Aiton, 119
reginae Aiton, 119
var. juncea (Link) Ker Gawler, 119
var. parvifolia (Aiton) Anon., 119
Strelitziaceae, 119
Stromanthe Sonder, 137
porteana Grisebach, 137
var. *oppenheimiana* invalid, 136
sanguinea Sonder, 137
Syagrus weddelliana (Wendland) Beccari, 75
Symphyoglossum sanguinea (Reichenbach) Schlechter, 273
Symplocarpus Salisbury, 89
foetidus (Linnaeus) Nuttall, 89
Synandrospadix Engler, 104
vermetoxicus (Grisebach) Engler, 104
Syngonium Schott, 103
albolineatum invalid, 104
angustatum Schott, 104
auritum (Linnaeus) Schott, 103
gracile (Miquel) Schott, 104
oerstedianum Schott, 104
podophyllum Schott, 104
var. *oerstedianum* (Schott) Engler, 104
riedelianum Schott, 104
vellozianum Schott, 104

Taniatherum Nevski, 44
crinitum (Schreber) Nevski, 45
Tapeinochilus Miquel, 122
ananassae Hasskarl, 122
pungens Teijsmann & Binnendijk, 122
Tetramicra bicolor (Lindley) Bentham, 199
Thalia Linnaeus, 137
dealbata Fraser, 137
geniculata Linnaeus, 137
Thamnocalamus aristatus (Gamble) Camus, 60
falcatus (Nees) Camus, 62
falconeri Munro, 60
spathaceus (Franchet) Soderstrom, 60
spathiflorus (Trinius) Munro, 60
tessellatus (Nees) Soderstrom & Ellis, 60
Thunia Reichenbach, 177
alba Reichenbach, 178
bensoniae Hooker, 178
bracteata (Roxburgh) Schlechter, 176
marshalliana Reichenbach, 178
Tillandsia Linnaeus, 14
araujei Mez, 14
bromeliifolia Rudge, 21
concentrica Vellozo, 18
cyanea Koch, 15
distachia Vellozo, 24
filiformis invalid, 14
lindeniana Regel, 15
lindenii Morren, 15
lindenii Regel, 15
lingulata Linnaeus, 16
morreniana Regel, 15
musaica Linden & André, 16
splendens Brongniart, 15
stricta Solander, 14
tenuifolia Linnaeus, 14
tricolor Schlechtendal & Chamisso, 15
usneoides (Linnaeus) Linnaeus, 14
Tinantia Scheidweiler, 26
anomala (Torrey) Clarke, 26
erecta (Jacquin) Schlechtendal, 27
fugax Jacquin, 27
pringlei (Watson) Rohweder, 26
Trachycarpus Wendland, 66
caespitosus Roster, 66
dracocephalus Ching & Hsu, 66
excelsa invalid, 66
excelsus (Thunberg) Wendland, 67
fortunei (Hooker) Wendland, 66
var. surculosa Henry, 66
khasyanus (Griffith) Wendland, 66
martianus (Wallich) Wendland, 66
nanus Beccari, 66
takil Beccari, 66
wagnerianus Roster, 66
Tradescantia Linnaeus, 30
albiflora Kunth, 30
× *andersoniana* Ludwig & Rohweder, 31
blossfeldiana Mildbraed, 30
canaliculata Rafinesque, 31

cerinthoides Kunth, 30
crassula Link & Otto, 30
fluminensis Vellozo, 30
navicularis Ortgies, 29
ohiensis Rafinesque, 31
pallida (Rose) Hunt, 31
pexata Moore, 31
sillamontana Matuda, 31
spathacea Swartz, 30
subaspera Ker Gawler, 31
velutina misapplied, 31
virginiana Linnaeus, 31
zebrina Bosse, 30
Triaristella (Reichenbach) Luer, 205
reichenbachii Brieger, 205
Trichocentrum Poeppig & Endlicher, 271
albo-coccineum Linden, 271
albo-purpureum Reichenbach, 271
alboviolaceum Schlechter, 271
orthoplectron Reichenbach, 271
pfavii Reichenbach, 271
tigrinum Linden & Reichenbach, 271
Tricholaena Schrader, 42
teneriffae (Linnaeus) Link, 42
Trichopilia Lindley, 272
coccinea Lindley, 273
crispa Lindley, 273
fragrans (Lindley) Reichenbach, 272
galeottiana Richard, 272
marginata Henfrey, 273
picta Lemaire, 272
sanguinolenta (Lindley) Reichenbach, 274
suavis Lindley & Paxton, 272
var. alba, 272
tortilis Lindley, 273
Trichosma suavis Lindley, 200
Trigonidium Lindley, 240
egertonianum Lindley, 240
seemannii Reichenbach, 240
Tripogandra Rafinesque, 29
multiflora (Swartz) Rafinesque, 29
Tripsacum Linnaeus, 38
dactyloides (Linnaeus) Linnaeus, 38
Typha Linnaeus, 114
augustifolia Linnaeus, 114
latifolia Linnaeus, 114
minima Funck, 114
shuttleworthii Koch & Sonder, 114
Typhaceae, 113

Uniola Linnaeus, 34
latifolia Michaux, 35
paniculata Linnaeus, 34

Vanda Brown, 249
alpina (Lindley) Lindley, 251
amesiana Reichenbach, 250
batemanii Lindley, 251
bensonii Bateman, 250
boxallii Reichenbach, 250
caerulea Lindley, 250
caerulescens Griffith, 250
cathcartii Lindley, 251
cristata (Wallich) Lindley, 250
denisoniana Benson & Reichenbach, 250
gigantea Lindley, 251
hookeriana Reichenbach, 250
insignis Blume, 250
kimballiana Reichenbach, 250
lamellata Lindley, 250
lissochiloides (Gaudichaud) Lindley, 251
lowii Lindley, 252
parishii Reichenbach, 251
pumila Hooker, 251
roxburghii Brown, 250
sanderiana Reichenbach, 249
suavis Reichenbach, 250
teres (Roxburgh) Lindley, 250
tessellata (Roxburgh) Don, 250
tricolor Lindley, 250
Vandopsis Pfitzer, 251
gigantea (Lindley) Pfitzer, 251
lissochiloides (Gaudichaud) Pfitzer, 251
lowii (Lindley) Schlechter, 252
parishii (Reichenbach) Schlechter, 251
var. marriottiana, 251
var. parishii, 251
Vanilla Miller, 172
aphylla Blume, 172
fragrans Ames, 172
planifolia Andrews, 172
pompona Schiede, 172
Vriesea Lindley, 15
fenestralis Linden & André, 16
gigantea Gaudichaud, 15
hieroglyphica (Carrière) Morren, 15
longibracteata (Baker) Mez, 16
psittacina (Hooker) Lindley, 16
saundersii (Carrière) Mez, 15
splendens (Brongniart) Lemaire, 15
var. formosa Witte, 16
tessellata (Linden) Morren, 15
xiphostachys Hooker, 15

Warrea Lindley, 224
bidentata Lindley, 224
costaricensis Schlechter, 224
tricolor Lindley, 224
warreana (Lindley) Schweinfurth, 224
Warscewiczella amazonica Reichenbach & Warscewicz, 234
cochlearis (Lindley) Reichenbach, 234
discolor (Lindley) Reichenbach, 234
flabelliformis (Swartz) Cogniaux, 234
wailesiana (Lindley) Morren, 234
Washingtonia Wendland, 69
filifera (Linden) Wendland, 69
robusta Wendland, 69
Weldenia Schultes, 28
candida Schultes, 28
Wolffia, 112
Wolfiella, 112

Xanthosoma Schott, 99
atrovirens Koch, 100
barilletii Carrière, 100
brasiliense (Desfontaines) Engler, 100
hastifolium Koch, 100
helleborifolium (Jacquin) Schott, 100
hoffmannii Schott, 100
holtonianum Schott, 100
jacquinii Schott, 99
lindenii (André) Engler, 99
pilosum Koch & Augustin, 100
sagittifolium (Linnaeus) Schott, 100
undipes (Koch) Schott, 99
violaceum Schott, 100
Xylobium Lindley, 224
brachystachyum Kränzlin, 225
bractescens (Lindley) Kränzlin, 225
colleyi (Lindley) Rolfe, 225
decolor (Lindley) Nicholson, 225
elongatum (Lindley) Hemsley, 225
foveatum (Lindley) Nicholson, 225
hyacinthinum (Reichenbach) Schlechter, 225
leontoglossum (Reichenbach) Rolfe, 225
pallidiflorum (Hooker) Nicholson, 225
palmifolium (Swartz) Fawcett, 225
powellii Schlechter, 226
scabrilingue Schlechter, 225
squalens (Lindley) Lindley, 225
stachyobiorum (Reichenbach) Hemsley, 225
variegatum (Ruiz & Pavon) Garay & Dunsterville, 225

Yadakea japonica (Steudel) Makino, 64
Yushania anceps (Mitford) Lin, 60
aztecorum McClure & Smith, 59

Zantedeschia Sprengel, 97
aethiopica (Linnaeus) Sprengel, 97
albo-maculata (Hooker) Baillon, 98
subsp. albo-maculata, 98
subsp. macrocarpa (Engler) Letty, 98
subsp. valida Letty, 98
elliottiana (Watson) Engler, 98
jucunda Letty, 98
oculata (Lindley) Engler, 98
pentlandii (Watson) Wittmack, 98
rehmannii Engler, 98
Zea Linnaeus, 38
mays Linnaeus, 38
Zebrina pendula Schnizlein, 30
Zingiber Boehmer, 123
cassumunar Roxburgh, 123
mioga (Thunberg) Roscoe, 123
officinale Roscoe, 123
purpureum Roscoe, 123
spectabile Griffith, 123
zerumbet (Linnaeus) Smith, 123
Zingiberaceae, 120
Zygopetalum Hooker, 232
amazonicum (Reichenbach & Warscewicz) Reichenbach, 234
brachypetalum Lindley, 232
burkei Reichenbach, 236
coeleste Reichenbach, 235
cerinum (Lindley) Reichenbach, 234
crinitum Loddiges, 232
discolor (Lindley) Reichenbach, 234
intermedium Lindley, 232
lalindei Reichenbach, 235
mackayi Hooker, 232
maxillare Loddiges, 232